18th European Frequency and Time Forum

(EFTF 2004)

5 – 7 April 2004

Organised by The IEE Satellite Systems and Applications Professional Network.

University of Surrey, Guildford, UK

Author Disclaimer

Copyright and Copying

Conditions of Acceptance of Advertisements

The Institution reserves the right to refuse to insert any advertisements (even if ordered and paid for) and/or to make alterations necessary to maintain its standards.

It is not guaranteed that any advertisement will be placed in any specified position or on any specified page unless agreement has been entered into and the agreed surcharge paid.

Every effort will be made to avoid errors but no responsibility will be accepted for any mistakes that may arise in the course of publication of any advertisement. These mistakes may include non-insertion, insertions other than those ordered and errors and omissions within the advertisement.

Notice to cancel any advertisement must be received in writing ten days prior to its next scheduled appearance, otherwise a charge will be made.

No responsibility will be accepted for repetitive errors unless the advertiser's correction has been accepted in respect of that error.

No responsibility will be accepted for loss or damage alleged to arise from errors within advertisement copy, non-appearance of an advertisement or delay in forwarding box number replies.

Advertisers are required to ensure that the content of advertisements conforms with all legislation currently in force affecting such matters. They shall further indemnify the publisher in respect of any claims, costs and expenses that may arise from anything contained within the advertisement and published on their behalf by the Institution of Electrical Engineers.

The placing of an order or contract for insertion of an advertisement in any newspaper or journal published by the Institution of Electrical Engineers whether in writing or by verbal or telephone instructions will be deemed an acceptance of each and all the above conditions.

Published by the Institution of Electrical Engineers, London

ISBN 0 86341 384 6 ISSN 0537-9989

IC 499

Printed in Cambridge University Press, UK

Becoming a member of the IEE

To take of advantage of the discounted Member rates and the wide range of membership benefits, products and services offered by the IEE, why not join the largest engineering body in Europe? The IEE is an innovative international organisation for electronics, electrical, manufacturing and IT professionals, with specifically tailored products, services and qualifications to meet the needs of today's technology industry.

If you would like to find out more about becoming a member of the IEE, view membership subscription fees or use the on-line membership application facility, please visit: **www.iee.org/membership**

You can also apply for Affiliate membership (for those who are not qualified electrical engineers, or who do not work in engineering, but have an interest in IEE activities).

There is a short waiting period from applying for membership to having your membership confirmed. However, you can still register for the event at member rate by posting or faxing in the attached registration form, quoting your *five-digit membership application reference number in the "Membership No:" line. (* You will be given this five-digit number on completing the on-line membership application and payment section)

You can review the status of your membership application by visiting the IEE Membership web page and scrolling down to the "trouble shooter" link at the bottom of the page. This also gives the answers to many general membership questions.

For further information, please contact +44 (0) 1438 765 675 to request an information pack.

EFTF 2004 is organised by

EUROPEAN FREQUENCY AND TIME FORUM EXECUTIVE COMMITTEE

Claude Audoin
LHA / CNRS
France
audoin_c@club-internet.fr

Raymond Besson
Laboratoire de Chronométrie
Electronique Piézoélectricité
France
rbesson@ens2m.fr

Marcel Ecabert
FSRM
Switzerland
ecabert@fsrm.ch

Lute Maleki
Jet Propulsion Laboratory
USA
lute.maleki@jpl.nasa.gov

Fritz Riehle
Physikalisch-Technische
Bundesanstalt
Germany
fritz.riehle@ptb.de

Bernard Schlueter
FSRM
Germany
schlueter@fsrm.ch

Pierre Thomann
Observatoire Cantonal de Neuchâtel
Switzerland
Pierre.Thomann@ne.ch

Michael Underhill
University of Surrey
UK
m.underhill@eim.surrey.ac.uk

EFTF 2004 LOCAL ORGANISING COMMITTEE

Michael Underhill (Chairman)
University of Surrey
UK
m.underhill@eim.surrey.ac.uk

Jeremy Everard
University of York
UK
jkae@ohm.york.ac.uk

Clive Green
Quartzlock (UK) Ltd
UK
cgreen@quartzlock.com

David Salt
C-MAC MicroTechnology
UK
DavidSalt@cmac.com

Peter Whibberley
National Physical Laboratory
UK
peter.whibberley@npl.co.uk

TECHNICAL PROGRAMME COMMITTEE (SCIENTIFIC COMMITTEE)

Jeremy Everard (Chairman)
University of York
UK
jkae@ohm.york.ac.uk

Jean-Pierre Aubrey
Oscilloquartz SA
Switzerland
aubry@oscilloquartz.com

Claude Audoin
LHA / CNRS
France
audoin_c@club-internet.fr

Andreas Bauch
Physikalisch-Technische
Bundesanstalt
Germany
andreas.bauch@ptb.de

Elio Bava
Politecnico di Milano
Italy
bava@elet.polimi.it

Laurent-Guy Bernier
METAS
Switzerland
laurent-guy.bernier@metas.ch

Roger Bourquin
Laboratoire de Chronométrie
Electronique Piézoélectricité
France
rbourquin@ens2m.fr

Emmanuel Bigler
ENSMM-LCEP
France
bigler@ens2m.fr

Jean-Simon Boulanger
National Research Council of Canada
Canada
jean-simon.boulanger@nrc.ca

Jörg Hahn
European Space Agency
The Netherlands
joerg.hahn@esa.int

Daniel Hauden
LPMO – CNRS
France
dhauden@lpmo.edu

Kurt Hilty
METAS
Switzerland
kurt.hilty@metas.ch

Carl Hruska
York University
Canada
carl.hruska@mathstat.yorku.ca

Kalevi Kalliomäki
University of Oulu
Finland
kalevi.kalliomaki@ees2.oulu.fi

Masama Kihara
NTT Network Innovation Laboratories
Japan
kihara@exa.onlab.ntt.co.jp

Dieter Kirchner
Technische Universität Graz
Austria
kirchner@inw.tu-graz.ac.at

Peter Krempl
AVL List GmbH
Austria
peter.krempl@avl.com

Maleki Lute
Jet Propulsion Laboratory
USA
lute.maleki@jpl.nasa.gov

Michel Brunet
Centre National d'Etudes Spatiales
France
michel.brunet@cnes.fr

Giovanni Busca
KYTIME Sàrl
Switzerland
gs.busca@bluewin.ch

André Clairon
BNM - SYRTE
France
andre.clairon@obspm.fr

Gerrit De Jong
NMi Van Swinden Laboratory
The Netherlands
gdejong@nmi.nl

Andrea De Marchi
Politecnico di Torino
Italy
demarchi@polito.it

François Deyzak
ONERA
France
deyzac@onera.fr

Noël Dimarcq
BNM - SYRTE
France
noel.dimarcq@obspm.fr

Christopher Ekstrom
US Naval Observatory
USA
ekstrom@atom.usno.navy.mil

Pekka Eskelinen
Helsinki University of Technology
Finland
pekka.eskelinen@hut.fi

Demetrios Matsakis
US Naval Observatroy
USA
dnm@orion.usno.navy.mil

Bernd Neubig
Germany
BNeubig@axtal.com and
BNeubig@t-online.de

Rodriguez Juan Palacio
Real Instituto y Observatorio de la Armada
Spain
jpalacio@roa.es

Christophe Salomon
Ecole Nationale Supérieure
France
Christophe.Salomon@lkb.ens.fr

Giorgio Santerelli
BNM – SYRTE
France
giorgio.santarelli@obspm.fr

Jörn Stenger
Physikalische-Technische Bundesanstalt
Germany
joern.stenger@ptb.de

Patrizia Tavella
Istituto Elettrotecnico Nacionale Galileo Ferraris
Italy
tavella@ien.it

Pierre Thomann
Observatoire Cantonal de Neuchâtel
Switzerland
Pierre.Thomann@ne.ch

Dmitry Tsarapkin
MPEI(TU)
Russia
tsarap@srv-vmss.mpei.ac.ru and
ts6161@gagarinclub.ru

Steve Feltham
European Space Agency
The Netherlands
stephen.feltham@esa.int

Michael Garvey
Symmetricom TRC
USA
rmgarvey@ieee.org

Patrick Gill
National Physical Laboratory
UK
patrick.gill@npl.co.uk

Vincent Giordano
CNRS – LPMO
France
giordano@lpmo.edu

Aldo Godone
IEN Galileo Ferraris
Italy
godone@ien.it

Michel Granveaud
BNM – SYRTE
France
michel.granveaud@obspm.fr

Yuri Gulyaev
Russian Academy of Sciences
Russia
gulyaev@cplire.ru

Pierre Uhrich
BNM – SYRTE
France
pierre.uhrich@obspm.fr

Adolph Ulianov
Kvarz Institute of Electronic
Measurements
Russia
kvarz@kvarz.ru

Michael Underhill
University of Surrey
UK
m.underhill@eim.surrey.ac.uk

Rainer Unverdross
Unverdross Technik
Germany
unverdross@unverdross.de

Jacques Vanier
Universite de Montreal
Canada
jac.vanier@sympatico.ca

Walter Zingg
Rolex SA
Switzerland
walter.zingg@rolex.com

The permanent EFTF secretariat is held by FSRM.

Foreword by the Chairman of the Local (UK) Organising Committee Professor Mike Underhill

As Chairman (of the LOC) it was my great pleasure to welcome the 18th EFTF to Guildford at the University of Surrey (UniS) between the 5th and 7th of April 2004 this year. As a member of the University part time since 1966 and full time since 1992, it was a great relief to find that the conference of 200 attendants (from more than 20 countries) and its venue in Guildford this year was considered to have been a great success! This note is an opportunity to give my personal thanks to all those who contributed so much to make it so.

Thanks first to my colleagues on the Executive Committee for selecting the United Kingdom for this year's EFTF and for providing the encouragement, help and support that a LOC Chairman needs. The membership of the EFTF Executive Committee is given overleaf together with the 'mission statement' of the EFTF and a brief history of the EFTF with a list of its previous conferences. From this you will see that Guildford 2004 is the second time that the EFTF has been in the UK; the first was in Brighton in 1996.

Next I would like to thank the University of Surrey for making available its excellent conference facilities. The only pity is that such a large conference as ours had to be 'out of term (semester) time'; the presence of our undergraduate students would have shown just how successful and in demand that UniS is. I have also included below a brief history of the University and also of our own on-campus 'spin-out' company Surrey Satellite Technology Limited (SSTL). Thanks also go to SSTL for the visits to the satellite control centre and manufacturing facilities. These I know came as considerable surprise to some in respect of just how much had been achieved by 'the only non-government independent satellite launching company in the world!'

The visit to the National Physical Laboratory (NPL) on the day following the conference was much enjoyed by us all. The NPL is responsible for 'UTC' and this made it a particularly appropriate place for a 'Frequency and Time Community' to visit. Seeing (a direct descendent of) 'Newton's apple tree' was also a high spot.

The local organising committee also are to be thanked for all their work and particularly for making the Exhibition a success, particularly since our industry in general is still regarded as having a tough time (at least in some respects). The 'British' theme for the food and wine at the Banquet also seemed to be well received, and hopefully countered at least some of the prevalent perceptions in this area! The LOC is listed overleaf.

As always thanks are due to the 'Scientific Committee'. Together with the authors, this very international committee is responsible for the high quality of the oral and poster papers at the EFTF, and this year was no exception. We were pleased to have 64 oral papers and a very creditable set of 76 poster papers. Our particular thanks to this year's SC Chairman, Jeremy Everard. The Scientific committee is listed overleaf.

Finally I am sure we would all like to thank the IEE in the shape of Simon Blows and Nadine Richards. Their tireless behind-the-scenes efforts and helpful and cheerful presence at the conference contributed greatly to a very successful and enjoyable EFTF 2004.

Mike Underhill.

The European Frequency and Time Forum

Purpose

The European Frequency and Time Forum (EFTF) is an international conference and exhibition, providing information on recent advances and trends of scientific research and industrial development in the fields of Frequency and Time.
The fact that time can be measured so precisely, far better than any other physical parameter, represents a technological asset of great importance.
Many space programmes and terrestrial applications require some form of stable timing for successful operation. Typical examples are digital network synchronisation, spread spectrum communication, range and range-rate determination, navigation and position location, very long baseline interferometry, scientific research (gravity and relativity experiments), time transfer by satellites, Earth observation and geodesy.

EFTF Meetings since 1987

Year	Location
1987 to 1991	alternately in Besançon (France) and Neuchâtel (Switzerland)
1992	Noordwijk, The Netherlands
1993	Neuchâtel, Switzerland
1994	Munich, Germany
1995	Besançon, France
1996	Brighton, United Kingdom
1997	Neuchâtel, Switzerland
1998	Warszawa, Poland
1999	Besançon, France (with IEEE-FCS)
2000	Torino, Italy
2001	Neuchâtel, Switzerland
2002	St-Petersburg, Russia
2003	Tampa, USA (with IEEE-FCS)
2004	Guildford, United Kingdom

The University of Surrey

The University of Surrey was established on 9 September 1966 with the grant of its Royal Charter but its roots go back to the late 19th century. The forerunner of the University, the Battersea Polytechnic Institute (founded 1891 in South London), first admitted students in 1894. In 1956 it was designated a 'college of advanced technology' and renamed Battersea College of Technology.

At the beginning of the sixties the College had outgrown its main building in Battersea Park road and in 1962 it had already decided to move to Guildford. In 1963 'Robbins Report' proposed that most of the colleges of advanced technology should become universities awarding their own degrees. The greenfield site for the University-designate was acquired from Guildford Cathedral, Guildford Borough Council and the Onslow Village Trust in 1965, and the move from Battersea was completed in1970.

Since 1966 the University in Guildford has not only expanded in numbers (now there are some 12,000 students and 1300 academic staff), but it has also 'federated' with the Roehampton Institute London, which in 2000 became The University of Surrey Roehampton. In addition, The University of Surrey has links with St Mary's College – a College of the University of Surrey, Wimbledon School of Art and Farnborough College of Technology, where it has validated courses and accredits degrees. It also validates courses at North East Surrey College of Technology (Nescot), Guildford School of Acting Conservative, Guildford College of Further & Higher Education, King Edward VII Hospital Department of Staff Development the Nuclear Department at HMS Sultan, St John's Seminary, Southern Theological & training scheme (SEETS), the Pre-Retirement Association and SHL (UK) Ltd. There are now over 90,000 graduates of the University working in all parts of the world. (The University consistently comes top in the UK Universities league table of graduate employment!)

Recent major developments on the University campus include the Austin Pearce Building (1997) the Duke of Kent building (1998), Millennium House, University Court (2000), the Daphne Jackson building which houses the Advanced Technology Institute (2002) and the Surrey Management School (2003). A new residence for postgraduate students (International House) was opened in the autumn of 2003.

One of the most significant site developments by the University has been the Surrey Research Park which currently accommodates over 100 companies employing 2,500 staff engaged in research and development activities, many which relate closely to the Universities own 'Schools'.

On site, but now (since 1995) an independent company in its own right, Surrey Satellite Technology Ltd provides 'the only non-government independent satellite launching facilities in the world'. It is a 'one-stop shop' capable of making a satellite to order, and arranging a launch, typically with a maximum programme time of 18 months. SSTL also transfers know-how by training operators and educating its customers in space technology if required. Since 1981 when the first 'micro-satellite' was launched a total of 23 satellites have been launched (an average rate of .one a year) with a success rate of over 90%. SSTL has done much to enhance the global reputation of the University of Surrey.

CONTENTS

The IEE is not, as a body, responsible for the opinions expressed by individual authors or speakers

Sensors

Time Scale and Time Transfer

3A - Optical Frequency Standards & Measurements I

3B - Time & Frequency Remote Comparison I

Atomic Frequency Standards and Clocks

5B - Time & Frequency Dissemination & Distribution

Authors List

Peltonen, J
Peng, H-M
Pereira Dos Santos, F
Perfiliev, A A
Petit, G
Pettiti, V
Picard, F
Piester, D
Pietsch, U
Pilshchicov, S A
Pisarevsky, Y V
Plimmer, M D
Polewka, T
Polzikova, N I
Pottie, P E
Powers, E
Prevedelli, M
Prokhorova, I G
Pugla, S

Quessada, A

Radecki, K
Raevski, A O
Rashmi,
Ray, J
Reina, F
Reiter, C
Ressler, H
Richard, G
Richard, J Y
Riebe, M
Riehle, F
Riley, W J
Robertson, I D
Robinson, H
Rochat, P
Rokos, H
Roos, C
Rosenbusch, P
Rotova, O
Rovera, G D
Rubiola, E
Rudenko, S A
Ruoskanen, J

Saalaoui, J
Saccoccio, M
Safonova, E
Salomon, C
Salzenstein, P
Santarelli, G
Satyanarayana , S
Schafer, W
Schauperl, R
Schildknecht, T
Schmidt-Kaler, F
Schnatz, H
Schwarz, K
Schwedes, C
Schwindt, P
Segal, D
Senior, K
Serkola, A
Sesia, I
Seydel, E
Shah, V
Sharma, R K
Shelkovnikov, A
Shmaliy, Y
Shu-Liao, C
Shvetsov, A
Silver, J D
Sin-Peng, H M
Sirmain, C
Slavov, D
Sodagar, A M
Sojdr, L
Sokolov, Y L
Sood, K N
Sorrentino, F
Spassov, L
Stacey, P W
Stehlin, X
Stein, S R
Steplewski, W
Sterr, U
Sthal, F
Stoehr, H
Svelto, C
Swanson, T B
Szymaniec, K

Taitchenatchev, A
Takamoto, M
Tanner, C E

Tasset, T N
Tavella, P
Taziev, R
Thieme, B
Thomann, P
Tobar, M E
Tremine, S
Trenque, N
Tsarapkin, D P
Tseng, W H
Tu, K Y

Uhrich, P
Underhill, M J

Valat, D
Velcheva, R
Veretin, V S
Vergov, I
Vernier, D
Vernotte, F
Vian, C
Viswanathan, A N
Vitta, M
Von Zanthier, J
Vorontsov, V

Wallerand, J P
Wallnofer, W
Walther, H
Wan, Y
Wang, G
Wang, J
Wang, L S
Wasiak, D
Webster, S A
Weiss, K
Weiss, M
Weyers, S
Wilpers, G
Wolf, P
Wynauds, R

Yakimov, A
Yamni, K
Yolkin, G A
Yu Nevsky, A
Yudin, V

Zanello, R
Zanon, T
Zhai, Z
Zhang, S
Zhang, V
Zhang, W
Zhang, Y
Zhgoon, S
Zhong, Z
Zondy, J J
Zoschg, D
Zucca, C

THEORETICAL AND EXPERIMENTAL INVESTIGATION ON BEAM RESONATORS OF GALLIUM ORTHOPHOSPHATE GAPO4

L. DELMAS, F. STHAL, E. BIGLER AND R. BOURQUIN,
FEMTO-ST LCEP UMR CNRS 6174
ENSMM, 26 Chemin de l'Epitaphe, 25030 Besançon cedex – FRANCE
e-mail contact: fsthal@ens2m.fr

Abstract

We report progress in theoretical and experimental investigations on vibrating beam resonators made with Gallium Orthophosphate (GaPO4) crystals.
Three kinds of vibrating modes are investigated: length extensional, flexural and torsional. Analytical models to simulate vibrating beam resonators with a rectangular cross-section have been developed and validated by comparison with Quartz data on extensional, flexural and torsional GaPO4 resonators.
Temperature-compensated cuts have been found theoretically for all three kinds of beam resonators in GaPO4. Experimental devices have been fabricated by ultrasound machining on Z-cut plates. Experimental results on vibrating mode identification and comparison with finite element analysis are presented.

Keywords: GaPO4, beam, resonator

I. Introduction

The vibrating modes of beam resonators can be used in many kinds of physical sensors like: accelerometers, pressure or temperature sensors, viscosimeter...
The quartz-homeotypic piezoelectric crystal which is GaPO4 has an interesting potential for this kind of physical sensors applied at high temperature. Due to the lack of α–β phase transition most material properties are stable up to 970°C.
This paper reminds the analytical studies of the three vibrating modes of beam resonators: length extensional, flexural and torsional modes. Temperature-compensated cuts have been found theoretically for all three kinds vibrating modes of beam resonators in GaPO4. Experimental devices have been made by ultrasound machining in Z-cut plates.
A comparison of vibrating mode identification between experimental results and finite element analysis are also presented.

II. Analytical models

Fig. 1 defines the geometry of the beam resonators with rectangular cross-section.

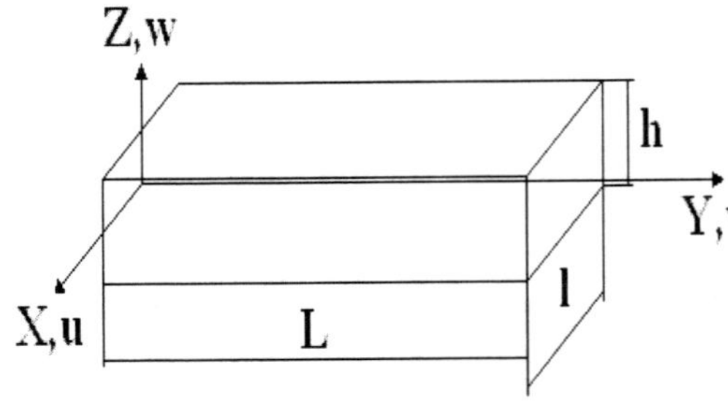

Fig. 1: Definition of the geometry of the beam resonators.

u,v,w : Mechanical displacements along X, Y, Z axis, respectively.
L, l, h: Dimensions of the beam, length, width and thickness, respectively
S_{ij},C_{ij} : Elastic constants [1]
c: Acoustic wave velocity
ρ: Density of the beam
I: Inertia of the beam
A: Cross-sectional area of the beam

Fig. 2 represents the rotation angles as they are defined in IEEE Std 176-1987.

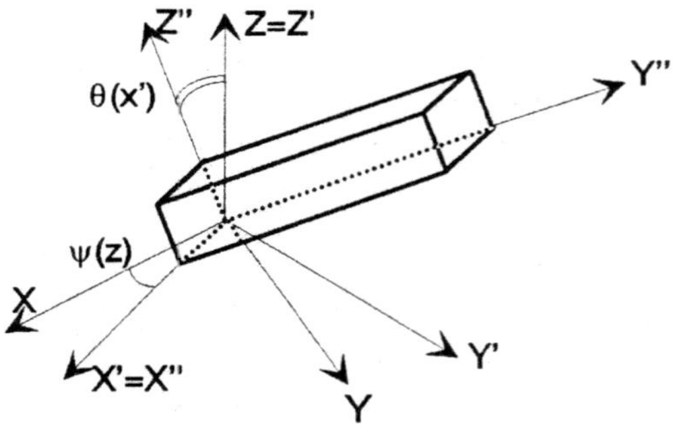

Fig.2 : Orientation in double rotation.

The resonant frequencies of length extensional, flexural and torsional modes are determined from the equations of motion (1), (2), (3), respectively. However, the Analytical Model (AN) is built without taking into account the piezoelectric effect and mass loading as a first approach.

$$(1)\ c^2 \frac{\partial^2 v}{\partial t^2} - \frac{d^2 v}{dy^2} = 0 \ \text{ with } c^2 = \frac{1}{s_{22}\rho}$$

(2) $\frac{1}{s_{22}} I \frac{\partial^4 w}{\partial y^4} - \rho A \frac{\partial^2 w}{\partial t^2} = 0$ where s_{22}: Compliance constant for Y axis

(3) $\rho I_y \frac{\partial^2 \alpha}{\partial t^2} - C_t \frac{\partial^2 \alpha}{\partial y^2} = 0$ with C_t: torsional constant and α rotation angle versus Y axis

The resonant frequencies depend on the boundary conditions. The formulae for the three vibrating beam modes are reported in Table 1 for two boundary conditions [2][3].

	Fixed-Free	***Fixed-Fixed***
Length extensional	$f_n = \frac{2n-1}{4L} c$	$f_n = \frac{n}{2L} c$
Flexural $f = \frac{\lambda^2}{2 \pi L^2} \sqrt{\frac{I}{S_{22} \rho A}}$	$1+\cos(\lambda) cosh(\lambda) = 0$	$1-\cos(\lambda) cosh(\lambda) = 0$
Torsional	$f = \frac{n}{4 y_o} \sqrt{\frac{C_t}{\rho I_y}}$	$f = \frac{n}{2 y_o} \sqrt{\frac{C_t}{\rho I_y}}$

Table 1: Resonant frequencies in function of the boundary conditions for the three vibrating modes.

With the formulae of the resonant frequencies and using the approximate method which consists in varying elastic constants, beam dimensions and mass density versus temperature, we obtain an analytical expression of the first Temperature Coefficients of Frequency (TCF) (Table 2) for the different vibrating modes of beam resonators.

	TCF1
Length extensional	$\frac{1}{2}\left(\frac{\dot{x}}{x} + \frac{\dot{z}}{z} - \frac{\dot{S}_{22}}{S_{22}} - \frac{\dot{y}}{y}\right)$
Flexural	$\frac{1}{2}\left(\frac{\dot{x}}{x} + 3\frac{\dot{z}}{z} - 3\frac{\dot{y}}{y} - \frac{\dot{S}_{22}}{S_{22}}\right)$
Torsional	$\frac{1}{2}[-\frac{\dot{y}}{y} + \frac{\dot{C}_{66}}{C_{66}} + 3\frac{\dot{z}}{z} - \frac{h}{1-h}\frac{h}{h}$.. ..$-2\frac{x^2}{x^2\ z^2}\frac{\dot{x}}{x} - 2\frac{z^2}{x^2\ z^2}\frac{\dot{z}}{z}]$

Table 2: Expression of the first TCF for the three vibrating beam modes.

These expressions of the TCF allow to find the temperature-compensated cuts. The temperature-compensated cuts of quartz are found again [4-5] for all of the three vibrating beam modes and the temperature-compensated cuts are found for the GaPO4. (Fig. 3a, 3b and 3c).

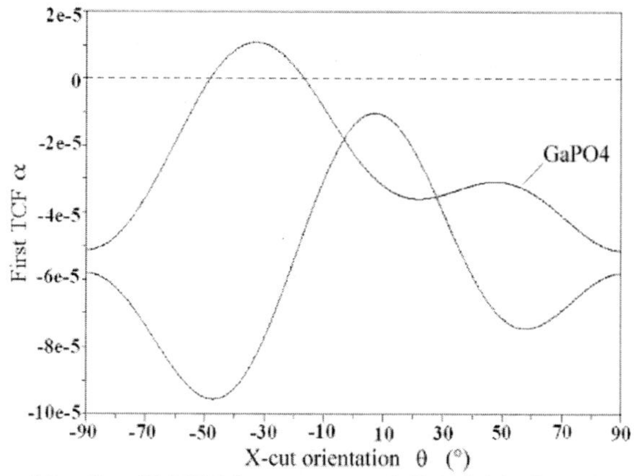

Fig. 3a: 1st TCF for quartz and GaPO4 in length extensional mode.

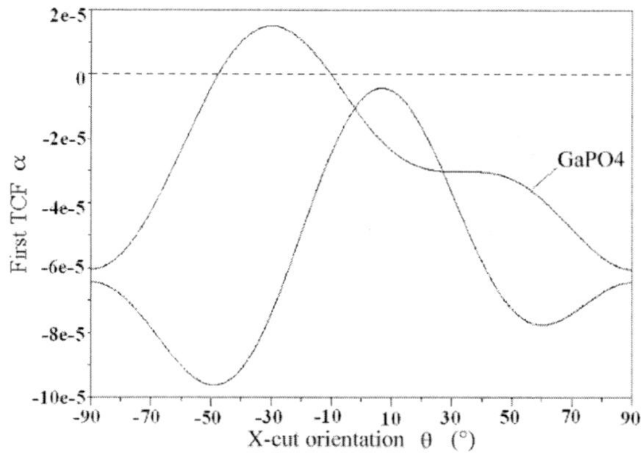

Fig. 3b: 1st TCF for quartz and GaPO4 in flexural mode.

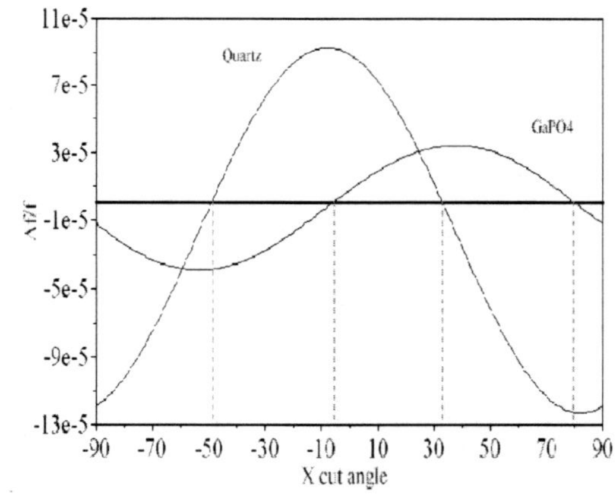

Fig. 3c: 1st TCF for quartz and GaPO4 in torsional mode.

Table 3 reports the temperature-compensated cuts for quartz and GaPO4 beam resonators.

	Quartz	***GaPO4***
Length extensional	θ = 5° (minimisation)	θ=-10.2° or θ=-48°
Flexural	θ =5° (minimisation)	θ=-16.7° or θ=-48°
Torsional	θ=-34° (ϕ=60°) or θ=32° (ϕ=0°)	θ=-7°

Table 3: Orientation angle of temperature-compensated cuts for quartz and GaPO4.

III. Finite element analysis

A finite element analysis is made to compare and to check if the hypothesis which consists in neglecting the piezoelectric effect is right or not. In Table 4 is reported the different modes and their resonant frequencies for a beam resonator with rectangular cross-section.

Fig. 4 shows the Finite Element Model (FEM) of the beam resonators with piezoelectric effect but no mass loading. In this Finite Element Model, we use a reference element which allows the simulation of piezoelectricity effect.

Fig.4: Resonator beams by FEM.

Modes (Hz)	*Resonant frequencies for Free -Free boundary conditions by FEM*	*Resonant frequencies for Fixed -Free boundary conditions by FEM*
1st Z - Flexural	21 055	3 314.8
1st X - Flexural	43 108	7 027
2nd Z - Flexural	60 486	21 203
1st Torsional	71 768	36 500
2nd X -Flexural	113 742	42 007
3rd Z – Flexural	126 297	62 189
1st Length extensional	145 841	72 978
2nd Torsional	147 705	110 537

Table 4: Resonant frequencies obtained by Finite Element Model for a beam resonator which dimensions are 13.5 x 2 x 1 mm (L; l; h).

IV. Experimental devices

Experimental devices have been made by ultrasound machining in Z-cut plates. The electrodes are deposed by sputtering method and the design of the electrodes allows to obtain beam resonators vibrating in length extensional modes (Fig. 5). The boundary conditions for the experimental devices are free-free. The electrodes are designed to generate an electrical field along X-axis in order to have length extensional modes.

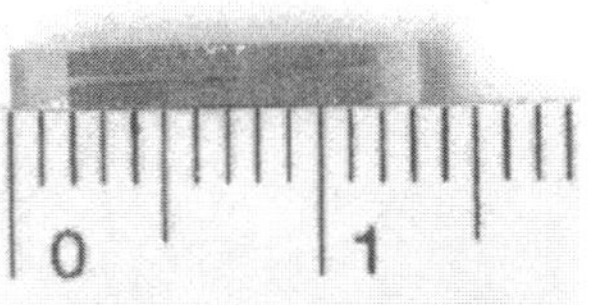

Fig. 5: Z-cut of beam resonator vibrating in length extensional mode (13.5 x 2 x 1 mm).

Fig. 6a and Fig. 6b show experimental device which allows to measure the resonant frequencies.

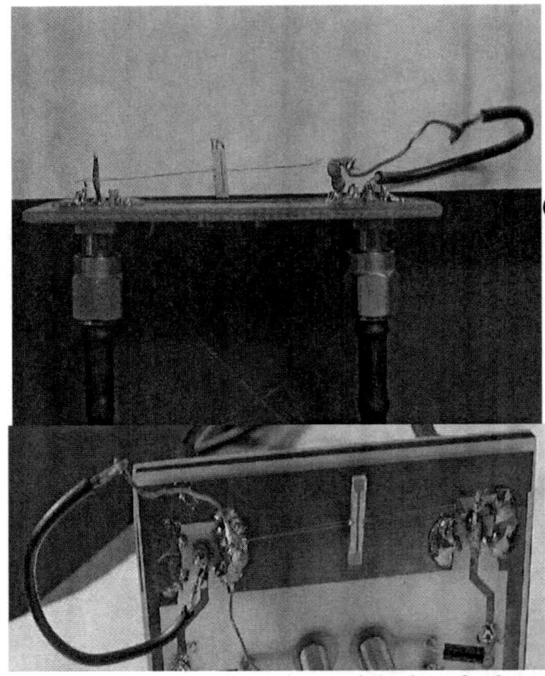

(a)

(b)

Fig. 6a & Fig. 6b: Experimental devices for length extensional mode.

Fig. 7a and Fig. 7b represent the amplitude of the first resonant frequency of the beam resonator vibrating in length extensional mode. The features of the beam resonator in Fig. 7a are 13.5 mm of length, 2 mm of width and 1 mm of thickness. For Fig.7b the length of the beam is 7.6 mm, the width is 2 mm and the thickness is 1 mm.

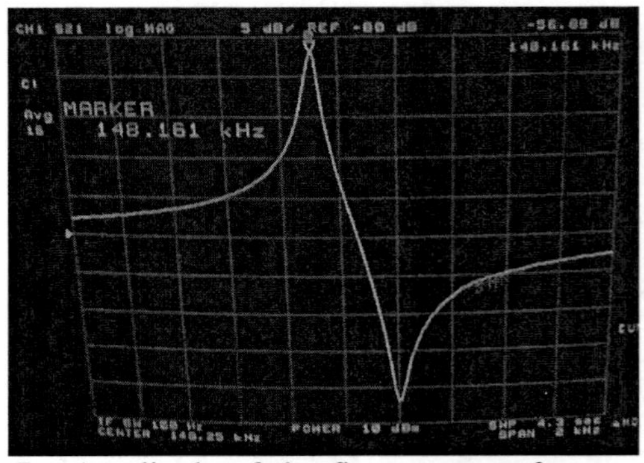

Fig. 7a: Amplitude of the first resonant frequency for a beam resonator in length extensional mode (L = 13.5 mm).

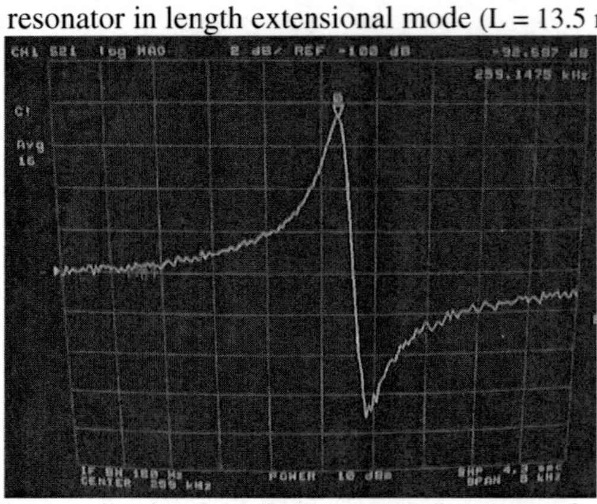

Fig. 7b: Amplitude of the first resonant frequency for a beam resonator in length extensional mode (L = 7.6 mm).

Table 5 summerizes the results of the three analyses for a free-free beam resonators in the first length extensional mode.

Resonant frequency of the first mode (Hz)	*Analytical model without piezoelectric effect and no mass loading*	*FEM with piezoelectric effect and no mass loading*	*Measure*
Length extensional L = 13.368 mm	147 835	147 546	148 161

Table 5: Results for beam resonator in length extensional mode in free-free boundary conditions.

Table 6 reports the error of analytical model versus FEM and experimental measurement.

%Error	*AN/FEM*	*AN/Measure*	*FEM/Measure*
L = 13.368 mm	0.2	0.3	0.4

Table 6: Percentage of error between models and measures.

Then, we notice the hypothesis which consists in neglecting the piezoelectricity effect as first approach is right.

V. Conclusion

The experimental measures and the results of the two different kinds of model give the same results. Moreover, we notice that the hypothesis which consists in neglecting the piezoelectric effect is right.

The experimental results are only for length extensional modes, but some other experimental measures in flexural modes are made on tuning fork devices. The conclusion on length extensional modes allows to think that the results for the other vibrating modes will follow the model.

Some measures to check the temperature-compensated cuts are also in progress.

Acknowledgements

This work is supported by the European Community under contract G5RD-CT-2002-00709.

References

[1] C. Reiter, H Thanner, W. Wallnöfer, P.M. Worsh and P.W. Krempl, « Elastic Constants and temperature-compensated orientation of GaPO4 », Proc. 15th EFTF, Neuchâtel 2001, pp.50-54.

[2] L.Delmas, F. Sthal, E. Bigler, B. Dulmet, R. Bourquin, « Temperature-compensated cuts for vibrating beam resonators of gallium orthophosphate GaPO4 », joint meeting 17th EFTF & 2003 IEEE FCS, pp.663 - 667 Tampa 2003.

[3] L. Delmas, F. Sthal, E.Bigler, B. Dulmet, R. Bourquin, « Piezoelectric vibrating beam resonators revisited: flexural and torsional modes of gallium orthophosphate GaPO4 », 5th WCU, PARIS 2003.

[4] J.P. Buchanan, « Handbook of piezoelectric crystal for radio and equipment designers », WADC Technical report 54-248, december 1954, pp. 24-31.

[5] W.G. Cady, « Piezoelectricity », McGraw Hill, Book Company, Inc, New York, 1946.

PARAMETRICALY EXITED BAW COMPOZITE RESONATOR STRUCTURES

G.D.MANSFELD, N.I.POLZIKOVA, A.O.RAEVSKII, I.G.PROKHOROVA
Institute of Radioengineering and Electronics, Russian Academy of Sciences
Mokhovaya 11, Build.7, 125009, Moscow, Russia
polz@mail.cplire.ru

The general approach for calculation and analyzing frequency and field impedance dependencies of BAW composite resonator structures with parametrically exited layers is developed. The BAW resonator consisting of two piezoelectric layers was considered. The first dielectric layer was the electric input transducer of the structure and was connected with the source of signal frequency ω. The second one was a piezoelectric parametric layer placed in alternating uniform electric field of frequency Ω. Different mechanisms of non-linearity in parametric piezoelectric layers are analyzed. First or them is due to nonlinear piezoeffect, the second one is due to interaction of acoustic and electron density oscillations in piezoelectric semiconductor. The acoustic wave propagation in parametric layers is accompanied by the electric field wave and electron density wave of frequency ω. In the external electric field along with acoustic oscillations of main frequency ω the oscillations of combined frequency ω-nΩ are excited. Parametric resonance make take place when the frequency of the applied voltage Ω equals $2\omega_l/n$, where ω_l is the frequency of one of the composite resonator modes, l is an integer.

The calculations of an electrical input impedance of the structures involved were based on the solution of the system of equations: wave equation for mechanical displacement, Poisson equation for electric displacement and hydrodynamic equation for electron density in semiconductor. The system was added with mechanical and electric boundary conditions. The analysis of the results obtained confirms that at some amplitude of the voltage (threshold voltage) parametric amplification of acoustic waves may take place. In this case the sign of the real part of the input electric impedance changes. The analytical expressions for threshold voltages were obtained. The change in applied voltage amplitude governs the rearrangement of the resonator spectrum.

Numerical calculations of the structures composed of ZnO as a transducer and $LiNbO_3$ or GaAs as parametric medium confirmed the possibility of parametric amplification and wave generation. The frequency range near 400 MHz was considered The calculated threshold values were 77 kV/m and 11kV/m for structures with $LiNbO_3$ and GaAs respectively. Small deviations of pumping frequency from double frequency of main resonance change the threshold values, the frequency of the amplified peak and its amplitude. Frequency interval of parametric instability region grows up with the increase in the pumping field.

Parametrically exited composite resonator structures may serve as a source of electromagnetic oscillations.

The work was partly supported by Russian Foundation for Basic Research, Grant No 02-02-17150.

MODEL FOR A QUARTZ-CRYSTAL TUNING FORK WITH COUPLING BETWEEN THE ARMS

H. Hantz, F. Sthal, R. Bourquin
FEMTO-ST, LCEP department, ENSMM, 26 chemin de l'Épitaphe
25000 BESANÇON, FRANCE

Keywords: Tuning fork, coupling, model.

Abstract

We developed a method for studying a tuning fork with non-perfect clamp and coupling between the arms. The model consists of torsion springs linking each beam with the base and a torsion spring linking the beams together. This model allows to consider the case of two different beams to study imperfections. Full determination of motional parameters is presented.

1 Introduction

Most of the actual models [1, 2] are dealing with half the tuning fork and don't take attention of the effect of a beam on the other. Moreover, they are mainly devoted to calculate frequencies and don't reflect accurately the shape of the beams.
For convenience in this paper, the tuning fork is assumed to be in a Z' quartz cut with the beams along the Y' axis of the quartz (singly rotated $Z + \theta$ cut in IEEE Std 176-1987).
The ratio length to thickness of the beam is supposed large, so that the rotating inertia and the shear stress can be neglected.

2 Mechanical model

The model consists of two beams linked themselves at their clamps by torsion spring R_{12} and to the base by R_1 and R_2 (fig. 1). Each beam is considered as ideal and described by usual equation (Bernouilli's model) but we consider that there is rotation of the section of a beam at its clamp. The driving of the beam consists of a torque applied on the section at the end of the electrodes.

2.1 Beam equation with damping and bulk driving

To take into account the energy loss in the material, a damping factor η is introduced in the expression of the stress T_2. When a driving electric field E_1 is applied, the total stress is:

$$T_2 = Y\varepsilon_2 + \eta\frac{\partial\varepsilon_2}{\partial t} - \frac{d_{12}}{s_{22}^E}E_1 \quad (1)$$

Where $Y = 1/s_{22}^E$ is the Young modulus and ε_2 the strain.
Then the bending moment $M(x_2, t)$ is integrated:

$$M(x_2, t) = \int_{-b/2}^{b/2}\int_{-h/2}^{h/2} T_2 x_i \, dx_3 \, dx_1 \quad (2)$$

The index i (1 or 3) denotes the direction of vibration. b, h are the widths along x_1 and x_3.

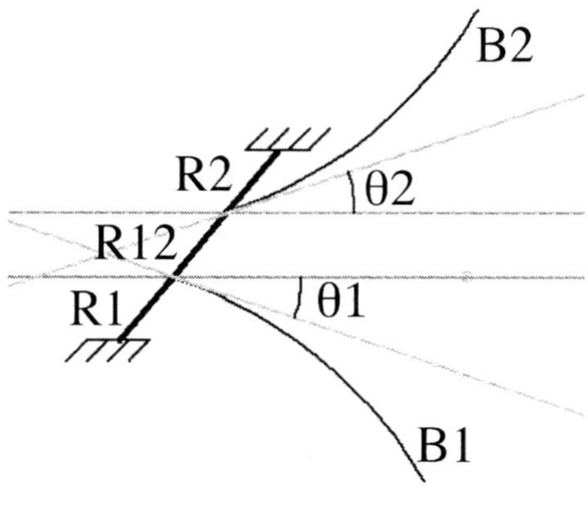

Figure 1: Model of tuning fork

with $\varepsilon_2 = \dfrac{-x_i}{R_c}$ and $\dfrac{1}{R_c} = \dfrac{\partial^2 u_i}{\partial x_2^2}$

This leads to the result:

$$M(x_2, t) = -IY\frac{\partial^2 u_i}{\partial x_2^2} - I\eta\frac{\partial^3 u_i}{\partial x_2^2 \partial t} - \frac{d_{12}}{s_{22}^E}F(x_2, t) \quad (3)$$

Where $F(x_2, t) = \iint E_1 x_i \, dx_1 \, dx_3$ and I is the moment of inertia.

Then we consider the beam equilibrium equation:

$$\rho A\frac{\partial^2 u_i}{\partial t^2}dx_2 = \frac{\partial^2 M(x_2, t)}{\partial x_2^2}dx_2 \quad (4)$$

ρ is the mass per volume unit and A the area of the cross section.
The calculous of $\partial^2 M(x_2 t)/\partial x_2^2$ gives us:

$$\frac{\partial^2 M(x_2, t)}{\partial x_2^2} = -IY\frac{\partial^4 u_i}{\partial x_2^4} - I\eta\frac{\partial^5 u_i}{\partial x_2^4 \partial t} - \underbrace{\frac{d_{12}}{s_{22}^E}\frac{\partial^2 F(x_2, t)}{\partial x_2^2}}_{\approx 0 \text{ if } R_c \text{ great}} \quad (5)$$

The last term in (5) disappears because the dependence of the electric field with respect to x_2 is negligible, excepted at the end of electrodes.
We obtain:

$$\rho A\frac{\partial^2 u_i}{\partial t^2} = -IY\frac{\partial^4 u_i}{\partial x_2^4} - I\eta\frac{\partial^5 u_i}{\partial x_2^4 \partial t} \quad (6)$$

Let say $u_i(x_2,t)=\delta(x_2)e^{\jmath(\omega t)}$, the equation becomes:

$$\frac{\partial^4\delta}{\partial x_2^4}\left(1+\frac{\jmath\omega\eta}{Y}\right)=\lambda_0^4\delta \quad \text{with } \lambda_0^4=\frac{\rho A\omega^2}{IY} \tag{7}$$

With $\lambda^4=\lambda_0^4/\left(1+\frac{\jmath\omega\eta}{Y}\right)$, we finaly have $\frac{\partial^4\delta}{\partial x_2^4}=\lambda^4\delta$ admitting solutions of the type:

$$\delta(x_2)=a\cos\lambda x_2+b\sin\lambda x_2+c\cosh\lambda x_2+d\sinh\lambda x_2 \tag{8}$$

Assuming the same time dependence ($e^{\jmath\omega t}$) for E_1, the moment (3) becomes:

$$M(x_2)=-IY\frac{\partial^2\delta(x_2)}{\partial x_2^2}-\jmath I\eta\omega\frac{\partial^2\delta(x_2)}{\partial x_2^2}-\frac{d_{12}}{s_{22}^E}F(x_2) \tag{9}$$

At the discontinuities of the electrodes, the driving effect of the electric field appears as a localized external torque M_d:

$$M_d=\frac{d_{12}}{s_{22}^E}\left[F(x_2)\right] \tag{10}$$

where $[F(x_2)]$ is the discontinuity of $F(x_2)$.

2.2 Single beam equation and boundary conditions

We use the beam equation (8) for the shape of the beam and the relevant boundary conditions (11) to represent the rotation of the base, the torque due to the springs and the driving moment. If the length L of the beam is fully electroded, we have:

$$\begin{aligned}
\delta(0) &= 0\\
\frac{\partial\delta}{\partial x}(0) &= \theta\\
I(Y+\jmath\omega\eta)\frac{\partial\delta^2}{\partial x^2}(0) &= R\theta+M_d\\
I(Y+\jmath\omega\eta)\frac{\partial\delta^2}{\partial x^2}(L) &= M_d\\
I(Y+\jmath\omega\eta)\frac{\partial\delta^3}{\partial x^3}(L) &= 0
\end{aligned} \tag{11}$$

R is the stiffness of the torsion spring in the model.
As we can see, a fifth unknown θ, the rotation of the clamping edge of the beam, is introduced along a fifth boundary condition. Substituting the expression (8) in the conditions (11) gives us the system to solve:

$$\{S\}*\{c\}=\{d\} \tag{12}$$

Where $\{S\}$:

$$\left\{\begin{array}{ccccc}
1 & 0 & 1 & 0 & 0\\
0 & \lambda & 0 & \lambda & -1\\
-\lambda^2 & 0 & \lambda^2 & 0 & -R/\alpha\\
-\cos\lambda L & -\sin\lambda L & \cosh\lambda L & \sinh\lambda L & 0\\
\sin\lambda L & -\cos\lambda L & \sinh\lambda L & \cosh\lambda L & 0
\end{array}\right\} \tag{13}$$

with $\alpha=I(Y+\jmath\omega\eta)$

$$\text{and}\{c\}=\left\{\begin{array}{c}a\\b\\c\\d\\\theta\end{array}\right\}\ \{d\}=\left\{\begin{array}{c}0\\0\\M/I(Y+\jmath\omega\eta)\\M/I(Y+\jmath\omega\eta)\\0\end{array}\right\} \tag{14}$$

2.3 Tuning fork

The reaction torque at the clamp of the beams is given by the equation (15) for beam B1 and equation (16) for beam B2.

$$M_1=R_1\theta_1+R_{12}(\theta_1-\theta_2) \tag{15}$$

$$M_2=R_2\theta_2-R_{12}(\theta_1-\theta_2) \tag{16}$$

In this case, using (15), (16), (8) and the boundary conditions (11) on each beam, the matrix $\{S\}$ and the vectors $\{c\}$ and $\{d\}$ are:

$$\{c\}:\left\{\begin{array}{c}a_1\\b_1\\c_1\\d_1\\\theta_1\\a_2\\b_2\\c_2\\d_2\\\theta_2\end{array}\right\}\ \{d\}:\left\{\begin{array}{c}0\\0\\M_1/I_1(Y+\jmath\omega\eta)\\M_1/I_1(Y+\jmath\omega\eta)\\0\\0\\0\\M_2/I_2(Y+\jmath\omega\eta)\\M_2/I_2(Y+\jmath\omega\eta)\\0\end{array}\right\} \tag{17}$$

$$\{S\}:\left\{\begin{array}{cc}S1 & S2\\S3 & S4\end{array}\right\} \tag{18}$$

With $\{S2\}$ and $\{S3\}$:

$$\left\{\begin{array}{ccccc}
0&0&0&0&0\\
0&0&0&0&0\\
0&0&0&0&0\\
0&0&0&0&0\\
0&0&0&0&\beta
\end{array}\right\}$$

Where $\beta=R_{12}/I_i(Y+\jmath\omega\eta)$ and I_i is I_1 in $\{S2\}$ and I_2 in $\{S3\}$.

$\{S1\}$:

$$\left\{\begin{array}{ccccc}
1 & 0 & 1 & 0 & 0\\
0 & \lambda_1 & 0 & \lambda_1 & -1\\
-\lambda_1^2 & 0 & \lambda_1^2 & 0 & \alpha_1\\
-\cos\lambda_1 L_1 & -\sin\lambda_1 L_1 & \cosh\lambda_1 L_1 & \sinh\lambda_1 L_1 & 0\\
\sin\lambda_1 L_1 & -\cos\lambda_1 L_1 & \sinh\lambda_1 L_1 & \cosh\lambda_1 L_1 & 0
\end{array}\right\}$$

$\{S4\}$:

$$\left\{\begin{array}{ccccc}
1 & 0 & 1 & 0 & 0\\
0 & \lambda_2 & 0 & \lambda_2 & -1\\
-\lambda_2^2 & 0 & \lambda_2^2 & 0 & \alpha_2\\
-\cos\lambda_2 L_2 & -\sin\lambda_2 L_2 & \cosh\lambda_2 L_2 & \sinh\lambda_2 L_2 & 0\\
\sin\lambda_2 L_2 & -\cos\lambda_2 L_2 & \sinh\lambda_2 L_2 & \cosh\lambda_2 L_2 & 0
\end{array}\right\}$$

With $\alpha_i=(-R_i-R_{12})/I_i(Y+\jmath\omega\eta)$

2.4 Stiffness calculation of torsion springs

To calculate the frequencies of the tuning fork, we need to estimate R_1, R_2 and R_{12}.
These three values can be obtained with a single static finite elements analysis (fig. 2) in the case of a symmetrical system. In this case, we assume $R_1=R_2$. Then a force is applied at the

free end of B1. In this case, M_1 is known and $M_2 = 0$. From (16) we easily obtain $R_{12} = R_2\theta_2/(\theta_1 - \theta_2)$ and from (15) we have $R_1 = R_2 = M_1/(\theta_1 + \theta_2)$.

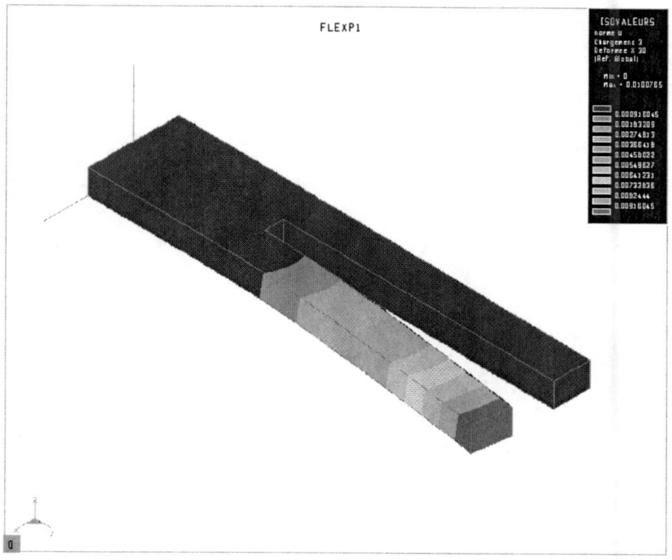

Figure 2: Static Finite Element Analysis

θ_1 and θ_2 are calculated with a mean plan determined by the four corners of the clamping section.

2.5 Mode shapes and frequencies

To obtain the eigenfrequencies, the free response of the system is calculated. We just have to solve the homogenous system $det\{S\} = 0$.

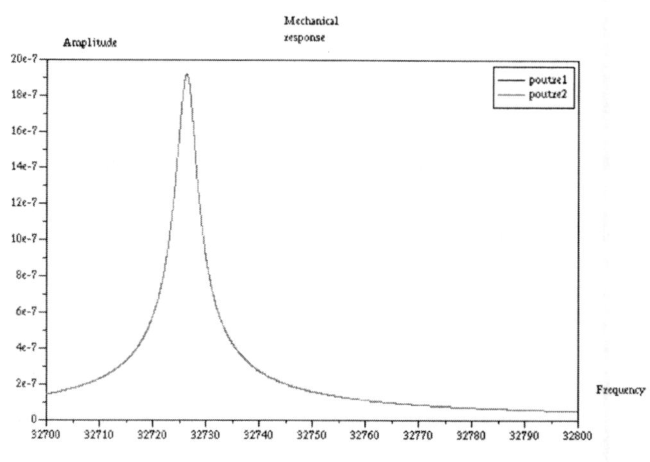

Figure 3: Frequency response of a tuning fork

To obtain the forced response (fig. 3), (12) is solved in a frequency range near the eigenfrequency ω_0 of the device. The frequency is set in ω. Thus we obtain the coefficients (complex values) of the beam equation (8) from which we can compute the amplitude and phase of each beam for the given frequency. By sweeping the frequency, phase and amplitude responses of each beam can be plotted. From the amplitude, the quality factor Q is calculated and can be adjusted by changing η. This part of the work is mainly numerical due to the size of the system to solve.

3 Motional parameters and driving moment calculation

From this point, the only thing needed for the calculus of the driving moment (used for the amplitude of vibration) and of the motional parameters is the electric potential V_e in the cross section of the beam. This potential can be obtained by several methods (analytical or numerical) [3, 4].

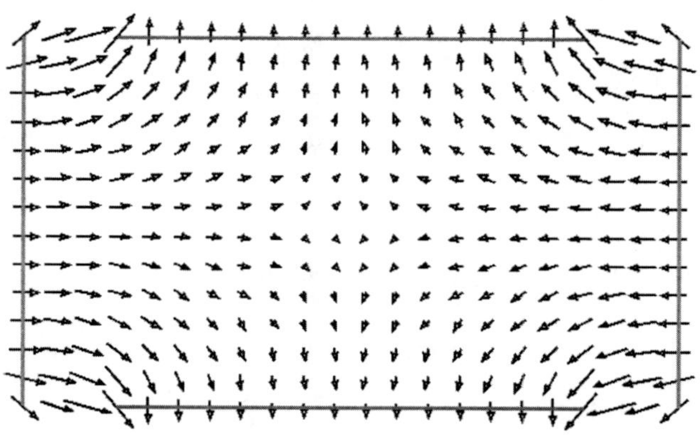

Figure 4: Electric field calculated by finite differencies

3.1 Driving moment

As seen in the beam equation section, the driving moment M_d is:

$$M_d = \frac{d_{12}}{s_{22}^E} \iint E_1 x_i \, dx_1 \, dx_3 \tag{19}$$

3.2 Motional parameters

We restrict our attention to the case of devices made of materials having a small electromechanical coupling factor.
The motional parameters are obtained from the mode shape of the beams and simplified Lewis formulas. For the static capacitance, the formula is:

$$C_S = \frac{L_e}{V_e^2} \int_{-h/2}^{h/2} \int_{-b/2}^{b/2} \left(\epsilon_{11} \left(\frac{\partial V_e}{\partial x} \right)^2 + \epsilon_{33} \left(\frac{\partial V_e}{\partial z} \right)^2 \right) dx \, dz \tag{20}$$

where L_e is the length of electrodes.

For the dynamic capacitance, we use:

$$C_1 = \frac{e_{11}^2}{\omega_0^2 V_e^2 h b \rho} \frac{\left(\int_{L_e} \frac{\partial^2 u_i(y)}{\partial y^2} dy \right)^2 \left(\iint x_i \frac{\partial V_e}{\partial x} dx \, dz \right)^2}{\int_0^L u_i(y)^2 \, dy} \tag{21}$$

The dynamic resistance and inductance can be calculated from C_1:

$$R_1 = \frac{1}{C_1 Q \omega_1} \tag{22}$$

$$L_1 = \frac{1}{C_1 \omega_1^2} \tag{23}$$

4 Results for an existing device

4.1 Tested device

The device actually used to test the model is a watch tuning-fork working at 32 764 Hz. The mode of vibration is in the plane. The geometry and size of the resonator are depicted in (fig. 5).

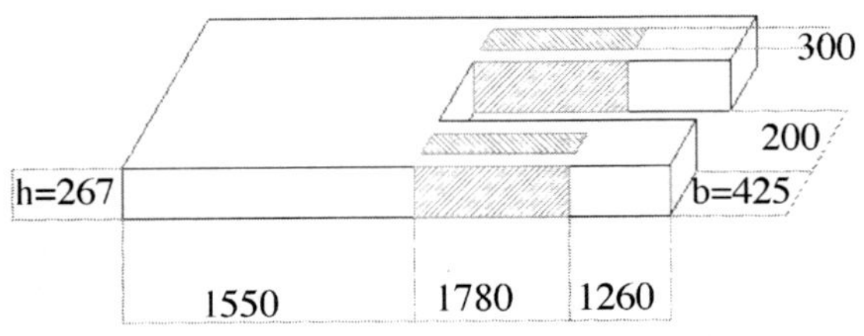

Figure 5: Dimensions of the device (in μm)

All the measurements have been made with the device in the air.

4.2 Simulation and results

For the simulation, different methods are used and compared together. The first is a dynamic finite element analysis for eigen frequencies, the second is a "perfect beam" simulation with damping, the third is a cantilever beam with half the width added to the length and the fourth is our model.
From the static finite element analysis, we have found the springs constants $R_1 = 0.287$ and $R_{12} = 0.079$.
The results are shown in the table (1) and compared to the experimental values.

	Meas.	Model.	Perfect	Canti.	FEM
Freq. (Hz)	32724	32720 to 32794	42510	37145	33400
C_0 (pF)	0.7	0.92	0.92		
C_1 (fF)	0.45	1.17	3.96		
R_1 (MΩ)	2.27	0.88	0.2		
L_1 (kH)	52.5	20.2	3.5		

Table 1: Results for a 32 kHz tuning fork

4.3 Comments

The column model corresponds to our model. It may be noticed that the calculated frequency is strongly depending on the value of spring constant R_1, the indicated frequency range corresponds to a variation of R_1 of 1%.
The data in the third column (Perfect) are obtained by taking $R_1 = R_2 = \inf$. The significant differences of values of motional parameters in the two columns are due to the modification of the mode shape. The diminution of the curvature in the region near the clamping of the beam induces an increase of motional inductance and resistance.
The cantilever model shown here is an empirical model for the frequency of tuning fork. It is based on the frequency formula for a beam, in which we use the length of the beam increased with half its width to obtain the frequency. Its purpose is not to give the shape of the beam, and therefore we prefer not using it for calculation of motional parameters.
Calculating the motional parameters in the case of the finite element analysis would have need the extraction of the displacements of points along the central fiber of the beam to fit them to the beam equation. We didn't do that work due to time limitation.

4.4 Effect of dissymetry

Based on the previously tested device, we introduced a $1\mu m$ variation of the length of one beam. We can easily observe the influence of the coupling (R_{12}) between the arms. With no coupling (fig. 6), the device exhibits the comportment of two distinct beams. With a small coupling, we have a small influence of a beam on the other and a frequency drift of the two resonant frequencies (fig. 7). With the calculated coupling, the two beams vibrate at the same frequency (fig. 8).

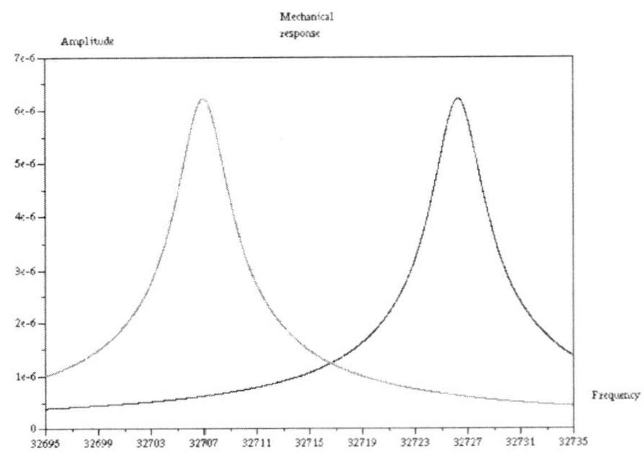

Figure 6: Assymetrical tuning fork without coupling

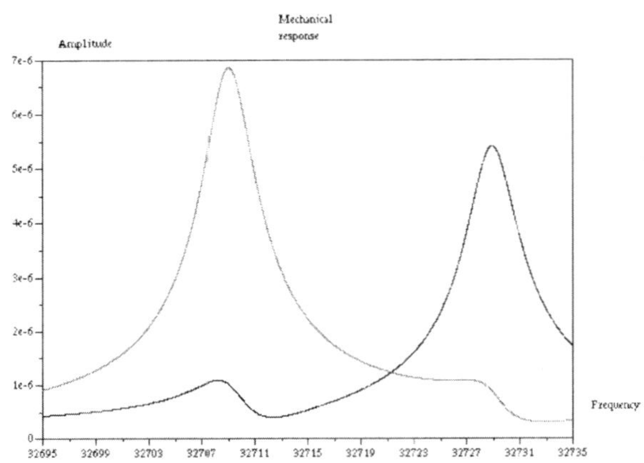

Figure 7: Assymetrical tuning fork with small coupling

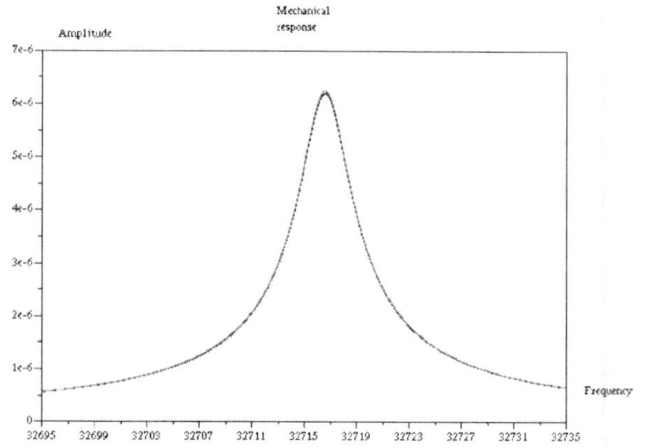

Figure 8: Assymetrical tuning fork with normal coupling

5 Conclusion

This model gives a significant correction with respect to the classical model that supposes perfect clamping of beams. It can be easily used to study the effect of dissimetry of the beams in tuning fork resonators. It can be also easily extended in the cases of structures having more than two vibrating beams coupled together [5].

Acknowledgement

This work was supported by the french ministry of defence (DGA).

References

[1] H. Itoh, T. Yuara. "An analysis of frequency of a quartz crystal tuning fork by Sezawa's Theory", *Proc. 52nd IEEE Ann. Freq. Cont. Symp.*, pp. 921–925, (1998).

[2] J. R. Kusters, C. A. Adams, H. E. Karier, P. W. Ward. "Analytical and experimental investigations of 32 kHz quartz tuning forks", *Proc. 30th IEEE Ann. Freq. Cont. Symp.*, pp. 175–183, (1976).

[3] J. Söderkvist. "Electric equivalent circuit for flexural vibrations in piezoelectric materials", *IEEE Trans. on UFFC*, **Vol. 37 N°6**, pp. 577–586, (nov. 1998).

[4] J. Hermann. "Determination of the electromechanical coupling factor of quartz bars vibrating in flexure or length extension", *Proc. 29th IEEE Ann. Freq. Cont. Symp.*, pp. 26–34, (1975).

[5] O. Bel, R. Bourquin, A. Jeanroy. "Two axes quartz angular rate sensor", *Proc. 56th IEEE Ann. Freq. Cont. Symp.*, pp. 214–219, (2002).

DRIVE LEVEL DEPENDENCE IN QUARTZ CRYSTAL RESONATORS AT LOW DRIVE LEVELS: A REVIEW

R. Brendel [a], M. Addouche [b], P. Salzenstein [a], E. Rubiola [c], and Y. S. Shmaliy [d]

[a] FEMTO–ST – Laboratoire de Physique et Métrologie des Oscillateurs
32, avenue de l'Observatoire – F25044 Besançon cedex – France
brendel@lpmo.edu – Fax +33 (0)3 81 95 39 98

[b] Laboratoire d'Astrophysique de l'Observatoire de Besançon
41, avenue de l'Observatoire – F25030 Besançon cedex – France

[c] ESSTIN – LPMI Université H. Poincaré
2, avenue Jean Lamour – F54519 Vandœuvre lès Nancy – France

[d] FIMEE Guanajuato University
p.o. Box 215-A, Prol. Tampico S/N 36790, Salamanca Gto – Mexico

Keywords: Quartz, Resonator, Drive Level Sensitivity

Abstract

Usually, starting of oscillation in a quartz crystal oscillator requires a resonator's input power in the range of – 20 dBm, but under storage a phenomenon known as Drive Level Dependency (DLD) or Drive Level Sensitivity (DLS) may appear that prevents the starting of oscillation. Several studies performed in the past have shown that at low drive level some quartz resonators may exhibit a large increase of their series resistance preventing the starting of oscillation. This work reviews the studies and results obtained for nearly fifty years on very low drive level sensitivity of quartz. The various mechanisms and models based on the hypothesis of moving particles and surface defects in the resonator inducing resistance increase and its relation with noise mechanism are reviewed as well. Also, the paper describes several experimental set-ups, and measurement procedures used to obtain very low drive level motional parameters. This work is a contribution to understand the problem of starting quartz after a long storage period. Some preliminary results of the series resistance measured at very low drive level are also presented.

1 Introduction

For about fifty years, engineers and manufacturers have been faced with the particular behaviour of certain oscillators that do not start despite the fact that the resonator's parameters meet the specifications at normal drive level [3]. It was found that the reason why this happens comes from a large increase of the resonator's series resistance at low drive level that cannot be compensated by the negative resistance of the feedback amplifier [12]. The number of names this phenomenon has been given shows that it is very common: authors often refer to "Starting Resistance" [3], "Hard Starting Characteristics", "Sleeping Sickness" [4], "Current Dependency of Crystal Resistance" [21], "Low Level Drive Sensitivity" [8], "Second Level of Drive" [15]. Although the name "Drive Level Dependency" [29] is widely used in this context, it is often used to mean the anisochronism effect occurring at high drive level, to avoid misunderstanding with this non-linear phenomenon that has a quite different origin, it seems wiser to use a different expression such as "Drive Level Sensitivity" (DLS) [10] used here[1]. Very early, DLS has been attributed to surface defects coming from microscopic scraps of various origin often associated with a sticky surface coating [3] or surface scratches [8,21]. The increase of the resonator series resistance is one of the most obvious effects of these surface defects, but other parasitic effects should have the same origin such as intermodulation in monolithic filters [8,14], ageing [4,6], frequency jumps [10], or noise [2,10,30]. On the other hand, the extreme sensitivity of the resonator characteristics to a tiny surface modification has been turned to account by using it as a sensor for the characterization of bonding forces between a surface and gold particles [9], stainless steel [31], or polystyrene spheres [16].

2 DLS behaviour and characteristics

Many experiments have been reported during several decades, some of the most demonstrative will be summarized here. For example, a monotonic decrease of the series resistance as the drive level increase (Fig. 1) [21], sometime, series resistance starts low, increases up to a maximum then decreases (Fig. 2) [21], most of the time, the phenomenon is not reversible, and exhibits an hysteresis with unpredictable threshold as shown in Fig. 3 [21]. Note that the resistance change is accompanied with a change of the resonance frequency either negative or positive as in this figure. Also, asymmetrical behaviour as shown in Fig. 4 has been reported [8].

[1] A more rigorous and unambiguous definition of the effect would require to call it "Low drive levels DLD" or "DLD at low drive levels". Here the term "Drive Level Sensitivity" or DLS should be considered only as a short cut.

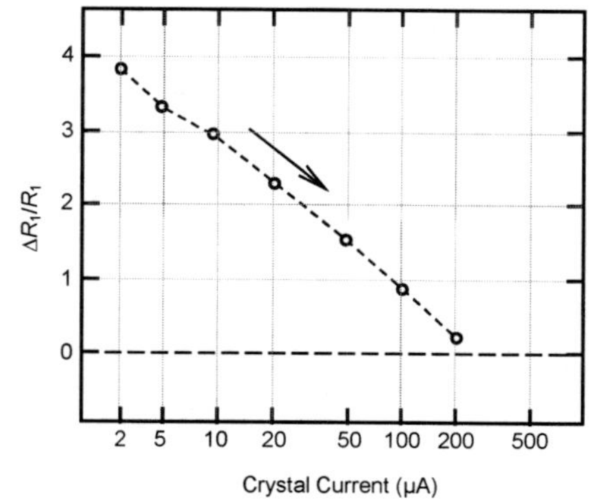

Figure 1: Monotonic decrease [21].

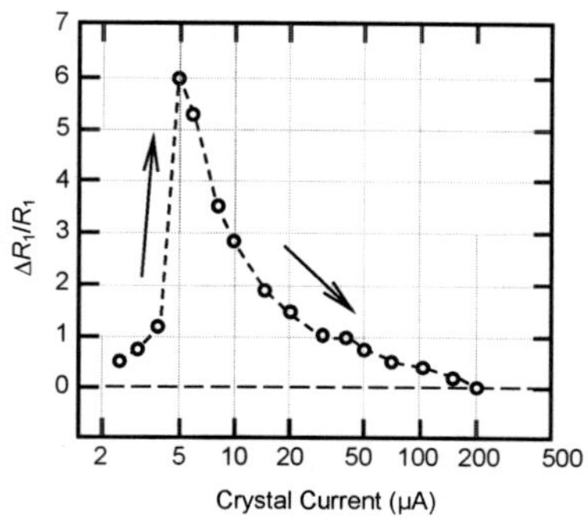

Figure 2: Maximum resistance for a particular drive level [21].

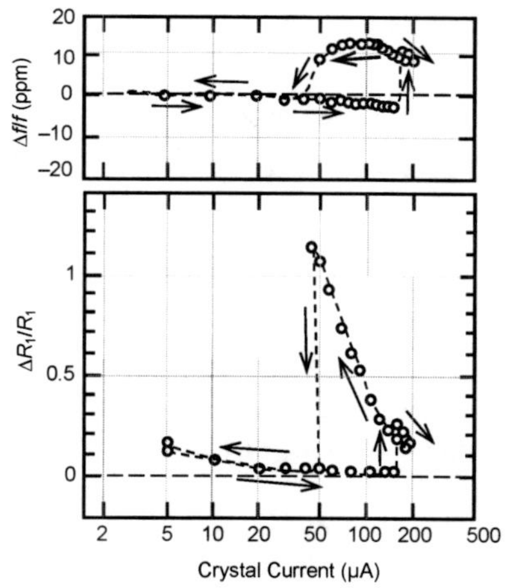

Figure 3: Hysteresis of the resonance frequency and series resistance [21].

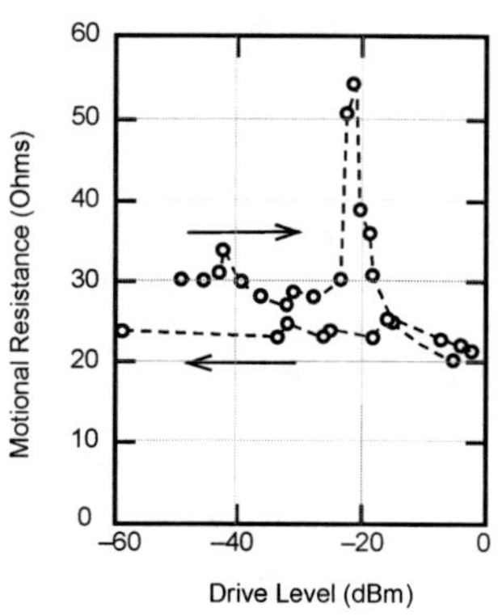

Figure 4: Peak and hysteresis of the series resistance [8].

A large number of experiments carried out for decades have led to the following observations:

- Increase of the series resistance is always associated with a positive or negative frequency shift.
- The DLS "signature" strongly depends on the temperature [21].
- The DLS behaviour can be modified or suppressed permanently or temporarily by overdriving the resonator [3,21], by polishing, etching or cleaning the crystal, this latter process being often considered as the most efficient [7,15].
- The most frustrating aspect of this phenomenon is its lack of reproducibility, hence crystals seemingly identical may be drive sensitive or not, and DLS of crystals apparently cured may reappear after a long time of inactivity.

3 Origin of the DLS

A lot of works and efforts have been put into understanding the origin of the phenomenon, and very early the attention has been focused on the surface imperfections as a possible cause of the DLS. Among the most often reported surface defects involved one can cite:

- Particles of metal or quartz or abrasive,
- Thin coat of resin or oil,
- Surface scratches,
- Flaking of quartz surface or metal electrode,
- Poorly adhesive electrodes or blisters,
- Surface stress.

Various experiments have proved the relationship of cause and effect between the surface pollution and DLS. For example, talc blown in the vicinity of an unsealed quartz resonator may induce DLS [5]. Another interesting and dramatic demonstration of the correlation between surface contamination and DLS has been reported a few years ago [6]: small squares in Figure 6 is a record of the relative resistance vs. drive level of a 100 MHz 5th overtone AT-cut crystal that doesn't exhibit a noticeable DLS. The same resonator has been opened and the surface has been sprinkled with alumina particles as shown in Fig. 5, after the resonator has been resealed, it presents an important DLS (small circles in Fig. 6). Once the resonator has been reopened, cleaned and resealed, it approximately recovered its original state (small triangles in Fig. 6).

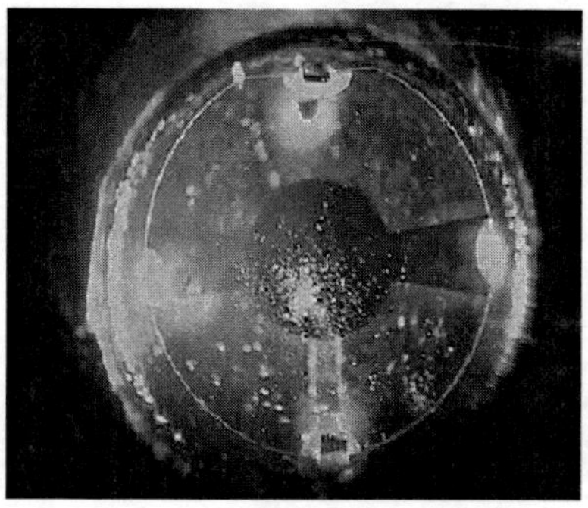

Figure 5: Particles of alumina on the surface of a resonator [6].

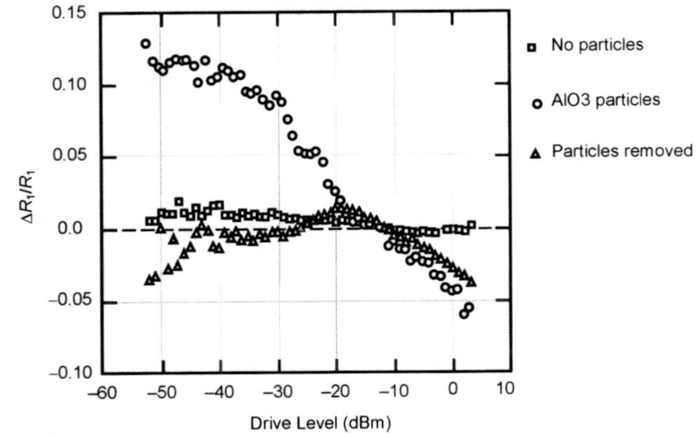

Figure 6: Effect of a surface contamination [6].

4 Mechanism of the DLS

Many attempts have been made to explain the mechanism responsible for the DLS. It should be noted first that a particle by oneself, that is without any bound with the surface, cannot induce the observed phenomena. On the other hand, if the particle were tightly bound to the surface, it would act as a mass loading and should induce a negative frequency shift which is often refuted by experiments. So, observed phenomena can be explained only if the particle is bound to the surface by an elastic force that can be due to a thin sticky coating of oil or resin, or any other attractive force such as Van der Waals, electrostatic, capillarity forces for example [31] (Fig. 7). In this case, as the surface moves back and forth under the shear motion, the bounded particle acts as a small oscillating system (Fig. 7c).

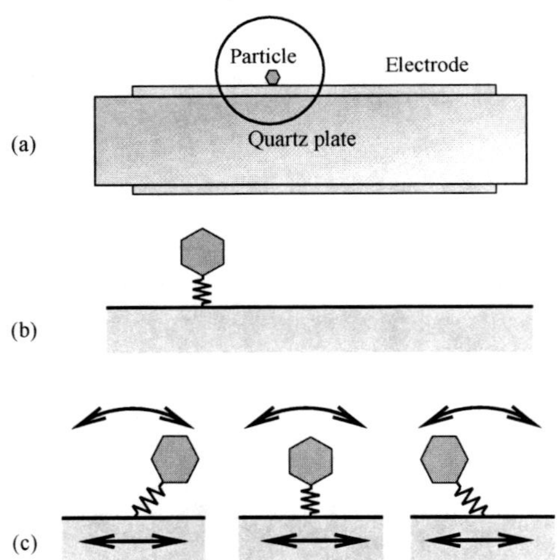

Figure 7: Particle bound to the surface by an elastic force.

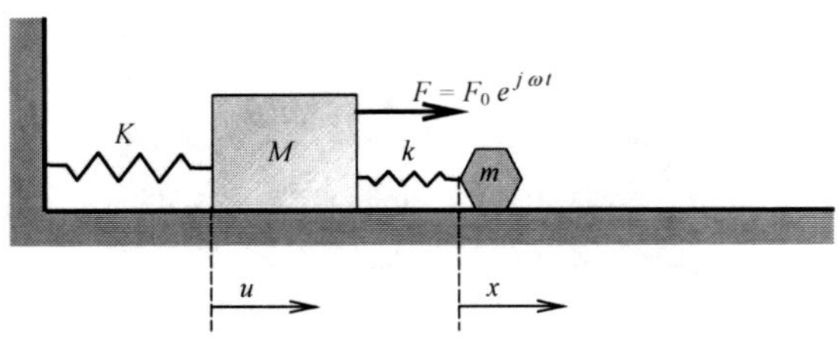

Figure 8: Simple mechanical model [9].

4.1 Simple mechanical model

A simple model of coupled oscillators has been proposed on the basis of a mechanical analogy represented in Fig. 8 [9]. The large spring–mass system at left mimics the resonator motion, while the small one at right represents the particle motion. The shear motion of the surface acts as an external driving force F. Using Newton's law with the system parameters defined here, the motion of the two masses can be obtain under the form of a differential system [Equation (1)].

$$\begin{cases} M\ddot{u} + Ku + k(u - x) = F_0 e^{j\omega t} \\ m\ddot{x} + k(x - u) = 0 \end{cases} \tag{1}$$

Searching a harmonic solution of the system, the resonant frequency of the coupled system [Equation (2)] can be expressed as a function of the resonant frequencies of the isolated spring–mass systems and of the stiffness ratio α [Equation (3)].

$$\omega_c^2 = \frac{\omega_p^2 + (1+\alpha)\omega_r^2}{2}\left[1 \pm \sqrt{1 - \frac{4\omega_p^2\omega_r^2}{\omega_p^2 + (1+\alpha)\omega_r^2}}\right] \tag{2}$$

$$\omega_p^2 = \frac{k}{m} \qquad \omega_r^2 = \frac{K}{M} \qquad \alpha = \frac{k}{K} \tag{3}$$

By plotting the coupled system resonant frequency as a function of the stiffness ratio, it turns out in Fig. 9 that for a weak particle binding the resonant frequency should increase while for a strong coupling, the frequency should decrease as for the mass loading effect.

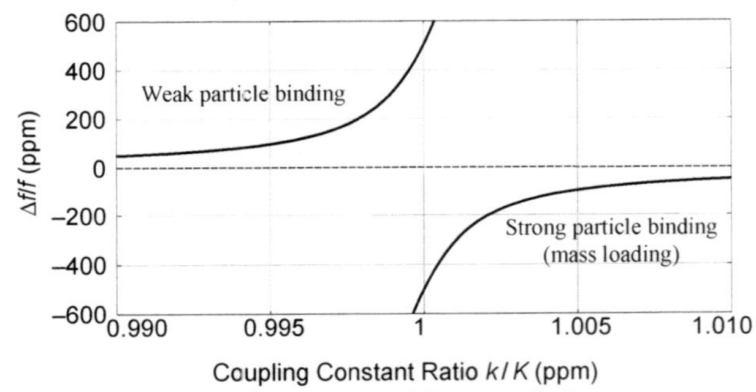

Figure 9: Resonant frequency of the coupled system.

4.2 Surface trapped particle model

Although the simple coupling system we just presented does explain that the resonant frequency can either decrease or increase, it doesn't give information on the damping mechanism. Even if linear damping terms were introduced in the resonator and particle spring–mass equivalent systems, they would induce a linear equivalent damping term independent of the drive level. Thus, a drive level sensitive damping term must invoke some non-linear mechanism such as proposed by Dworsky [8] assuming that the particles trapped in some surface imperfections should experience inelastic collisions with scratch walls as illustrated in Fig. 10 for example, thus inducing the required non-linear damping term.

Figure 11 shows the mechanical equivalent coupled oscillator that looks like the previous one with damping terms added to the spring–mass system.

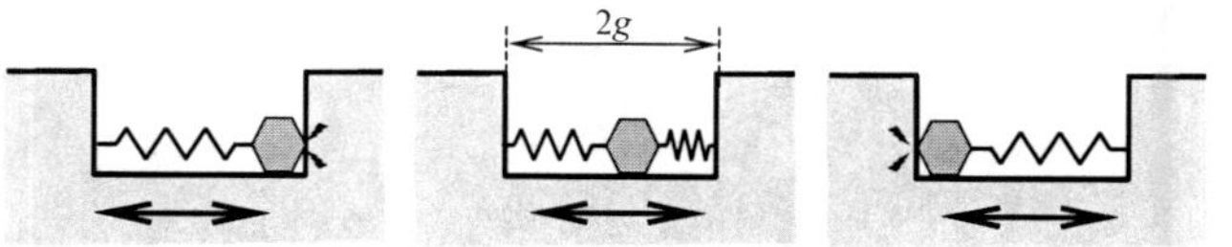

Figure 10: Surface trapped particle [8].

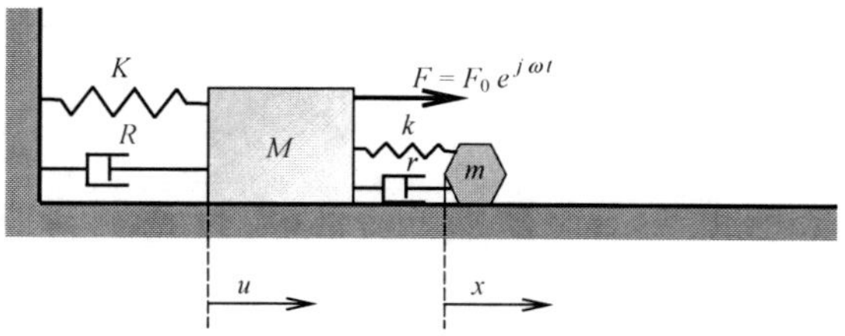

Figure 11: Mechanical equivalent of a trapped particle [8].

Motion Equations (4) are derived from the Newton's law where the particle damping factor has been given the form of Equation (5) suggested by some statistical and physical considerations.

$$\begin{cases} M\ddot{u} + R\dot{u} + Ku + k(u - x) + r(\dot{u} - \dot{x}) = F_0 e^{j\omega t} \\ m\ddot{x} + k(x - u) + r(\dot{x} - \dot{u}) = 0 \end{cases} \tag{4}$$

$$r = r_0 \left[1 + \left(\frac{x - u}{g} \right)^n \right] \tag{5}$$

The mechanical coupled system previously presented has an electrical equivalent circuit represented in Fig. 12 where the non-linear damping term of the particle is modelled by a non-linear resistor the resistance of which depends on the driving current [Equation (6)].

$$r = r_0 \left[1 + \left(\frac{i(r)}{g} \right)^n \right] \tag{6}$$

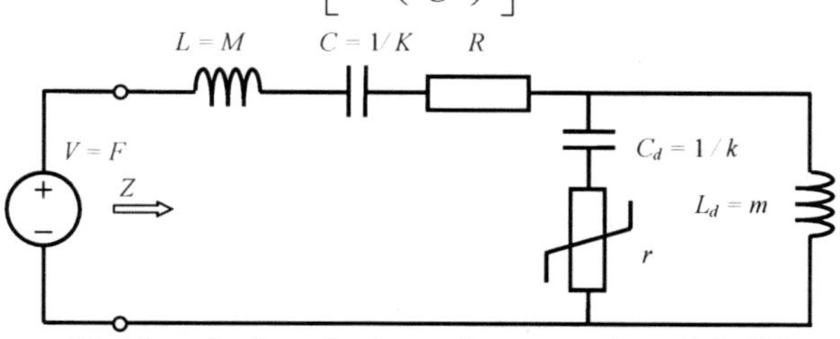

Figure 12: Electrical equivalent of a trapped particle [8].

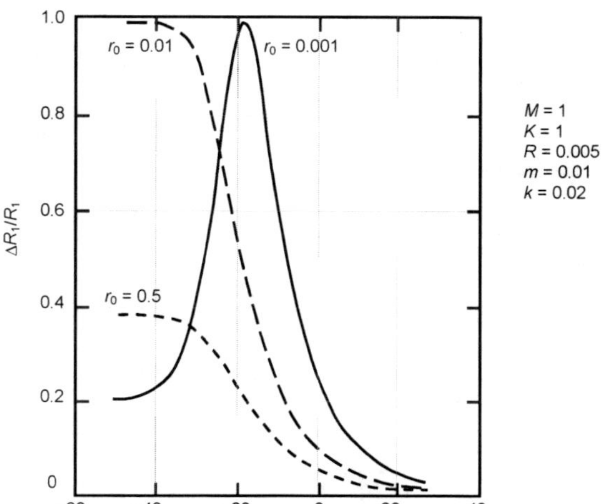

Figure 13: Modelling of a drive level sensitive resistance [8].

The equivalent resistance of the coupled system can be obtained from the real part of the equivalent impedance Z calculated by using numerical method. According to the value given to the linear particle damping term r_0, Fig. 13 shows that the model satisfactorily depicts various behaviours observed experimentally except the hysteresis phenomenon.

4.3 Physical model

Up to now, the models used give only qualitative explanations on the observed phenomenon. By using a physical distributed model for the resonator [10] it is possible to express its electrical admittance under the form given by Equation (7) where the first term of the right hand side represents the series–branch admittance. The parameter K_q depends on the geometry, on the physical constants, on the orientation, and on the particular vibration mode of the crystal plate. The second term is the static or parallel capacitance of the resonator. In this model, the crystal losses are introduced under the form of an imaginary part in the expression of the series resonant frequency [Equation (8)].

$$Y = \frac{j\omega K_q}{\omega_s^2 - \omega^2} + j\omega C_0 \tag{7}$$

$$\omega_s = \omega_s^r + j\omega_s^i \tag{8}$$

The influence of a particle located on the resonator can be calculated assuming it acts as a small spring–mass–dashpot system inducing an equivalent surface force F as shown in Fig. 14.

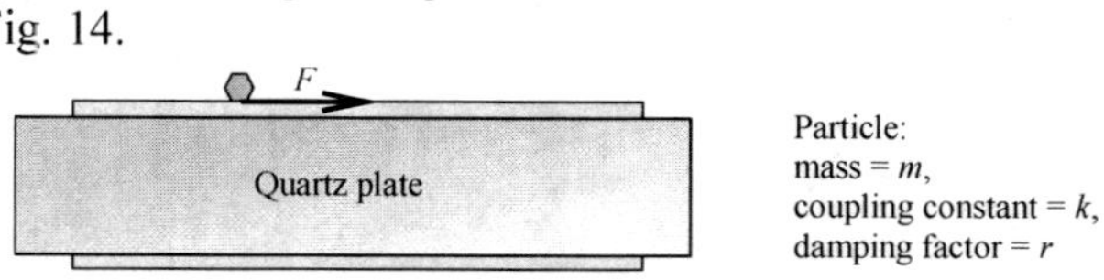

Figure 14: Action of a lumped spring–mass–dashpot particle on a distributed resonator model [10].

In this case, it can be shown that the particle introduces a perturbation term in the denominator of the series–branch admittance [Equation (9)], that has the form given by Equation (10). The perturbation term can be expressed as a function of the particle resonant frequency, and the mass ratio of the particle to the mass of the actively vibrating quartz region [Equation (11)].

$$Y = \frac{j\omega K_q}{\omega_s^2 - \omega^2 - \frac{2F}{mu}} + j\omega C_0 \tag{9}$$

$$\frac{2F}{mu} = 2\alpha \frac{\omega^2(\omega_p^2 + j\omega \frac{r}{m})}{\omega_p^2 - \omega^2 + j\omega \frac{r}{m}} \tag{10}$$

$$\alpha = \frac{m}{M} \qquad \omega_p^2 = \frac{k}{m} \tag{11}$$

Figure 15 shows the relative series resistance change of the resonator as a function of the particle damping to mass ratio for two values of the particle to resonator mass ratio. Note that a mass ratio of only 1 ppm would induce a 100 % increase of the series resistance and this value would dramatically increase if the particle resonant frequency is close from the resonator one.

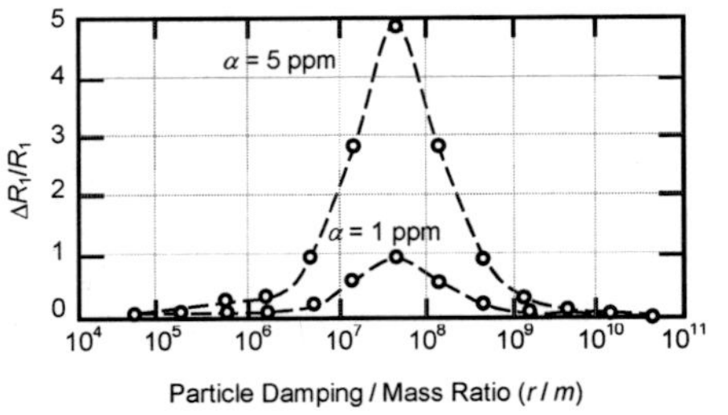

Figure 15: Physical modelling of a drive level sensitive resistance [10].

Besides of its influence on the pure shear mode equivalent admittance previously described, a localized irregularity can also be responsible for resonant mode distortions due to small parasitic flexural motions that induce additional vibration energy losses [11]. Furthermore, the strong dependence of the parasitic effect with the particular location of the irregularity should explain the hysteresis or other irreproducible behaviour of the DLS as well as the high drive level curing mechanism [10].

5 Experimental methods and results

From the experimental point of view, several methods can be used to measure DLS. In the earliest experiments, the crystal impedance meter was the most common method used [3]. Then, an easier way was to use an oscillator with variable gain as schematically depicted in Fig. 16. At low gain, the negative resistance of the amplifier cannot compensate for the resonator loss so that the oscillator doesn't start. By progressively increasing the amplifier gain up to reach the oscillation level it is possible to know if the crystal is drive level sensitive or not. Nevertheless, this fast and inexpensive method doesn't give dependency with drive level and is mainly used for screening of crystals to specified acceptance limits. [17,29].

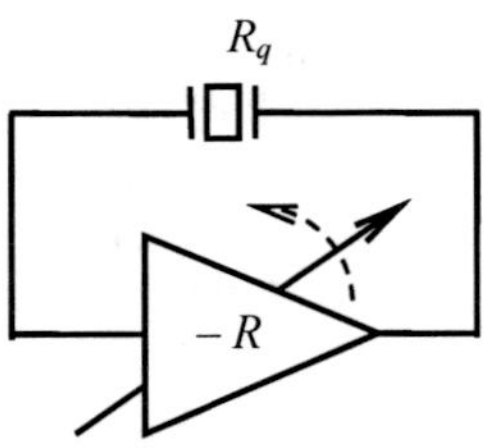

Figure 14: Active method for DLS measurement [29].

More accurate measurement of DLS can be achieved by using passive transmission methods such as the popular IEC-444 Pi-network (Fig. 17) widely used to measure the resonator motional parameters [13,20,32].

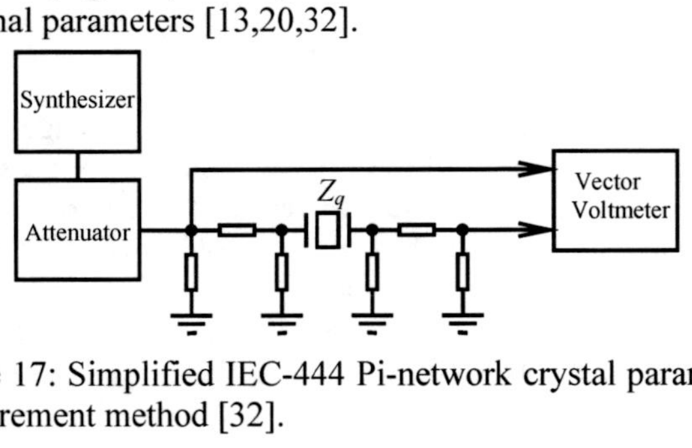

Figure 17: Simplified IEC-444 Pi-network crystal parameters measurement method [32].

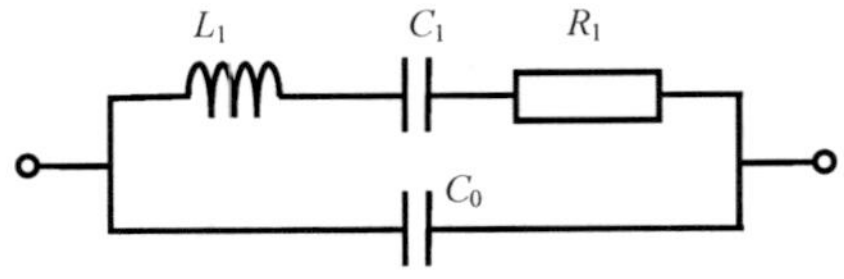

Figure 18: Crystal equivalent circuit.

The parameters of the resonator equivalent circuit (Fig. 18) can be calculated from the amplitude and phase transfer function of the measurement network the crystal is inserted in [28]. For high frequency resonators, those parameters can be obtained from the scattering parameters measurement by using automatic network analysers [1,23,24,25].
With the development of smart instrumentation and computer aided measurement techniques it is now very convenient to use network/impedance analysers that enable the user to vary several parameters such as the input signal magnitude and to calculate the equivalent circuit parameters from the impedance magnitude and phase records [18,19]. In the present case, the experiments have been performed by using an Agilent 4395A Network/Spectrum/Impedance Analyser and the 43961A Impedance Kit [22,33]. To improve the accuracy of the measurements, the resonator under test has to be isolated from the environmental perturbations mainly due to temperature fluctuations. To this end, the impedance test kit has been complemented with a precision oven keeping the crystal at its turnover point (Fig. 19, 20).

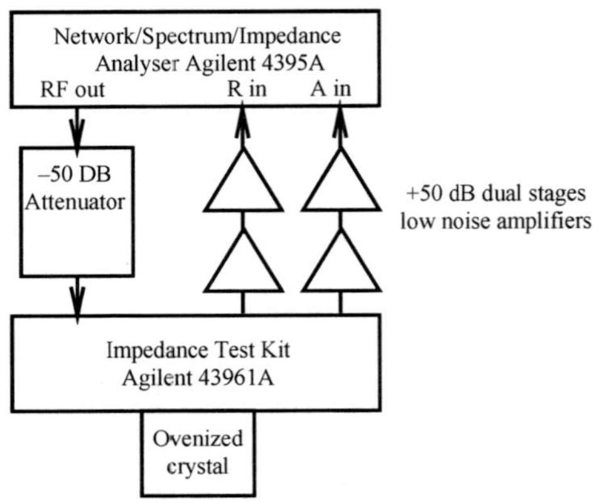

Figure 19: Low drive level crystal parameters measurement.

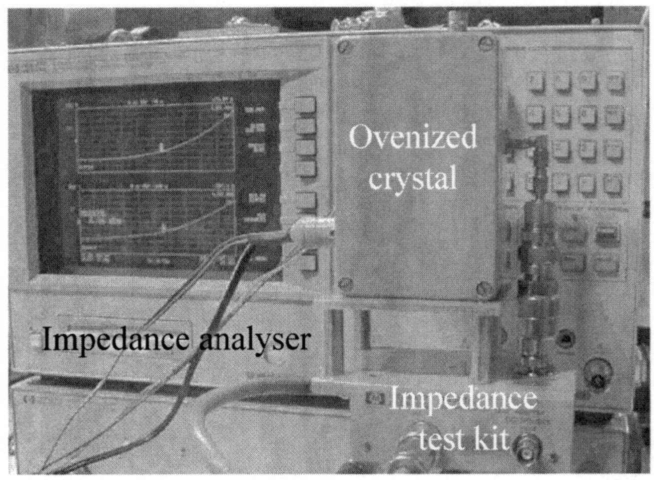

Figure 20: Experimental set-up.

Besides, so as to lower the drive level range initially set between – 50 to + 15 dBm, the experimental set-up has been modified as shown in Fig. 19: A wide band 50 dB attenuator is inserted in the RF input signal and two dual-stage low noise amplifiers are inserted in the measuring ports of the test kit allowing measurements from approximately – 100 dBm to - 35 dBm. It should be noted that a careful attention has to be paid to the calibration procedure to guarantee the measurement accuracy.

Figure 21 represents two typical records of the impedance the crystal parameters are calculated from. Of course, as the drive level decreases, the records become more and more noisy and several measurements are required to increase by averaging the signal to noise ratio. Figure 22 presents the drive level sensitivity of a 10 MHz SC-cut 3rd overtone quartz resonator. Nevertheless, the experimental set-up is not yet accurate enough to obtain satisfactory measurements of the crystal reactive parameters L_1 and C_1, and this plot is only a preliminary result that is still to be confirmed by additional experiments.

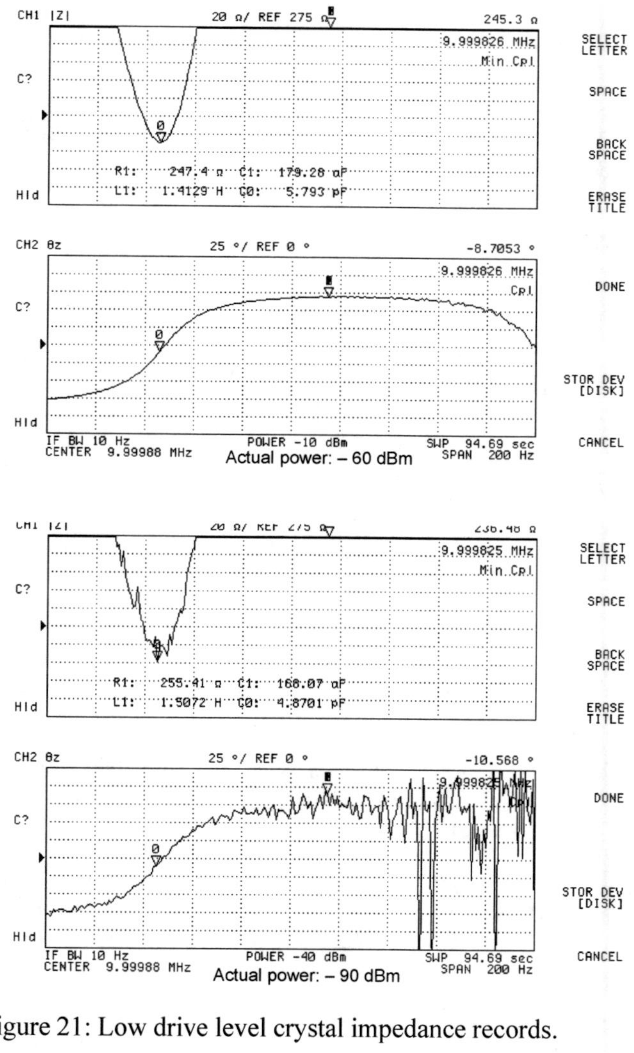

Figure 21: Low drive level crystal impedance records.

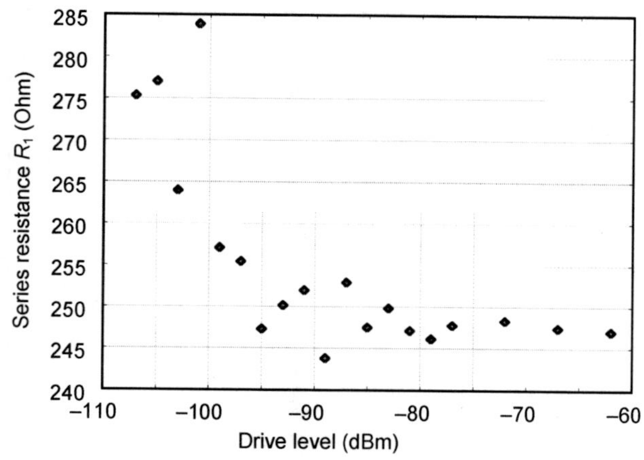

Figure 22: Low drive level sensitivity of a 10 MHz SC-cut 3rd overtone quartz resonator.

6 Noise and DLS

The main reason why these experiments are currently carried out is to check for a possible correlation between drive level sensitivity and noise of the resonators that should have the same origin. One of the possible mechanisms relating these two phenomena has been suggested in the past [30]. It assumed that some contaminant species are randomly trapped to and released from N possible surface sites at rates r_0 and r_1 respectively, each trapped particle causing an average relative frequency shift $\Delta f/f$ (Fig. 23).

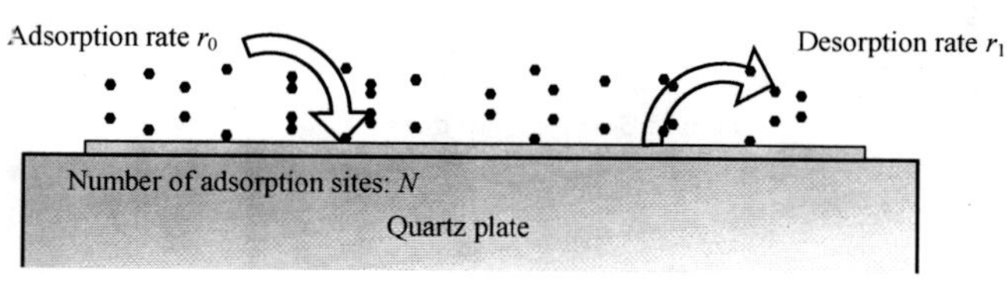

Figure 23: Noise induced by a contaminant species [30].

In this case it has been shown that the spectral density of phase fluctuations has the form given by Equation (12) represented in Fig. 24.

$$S_\phi(f) = \frac{8 r_0 r_1}{N(r_0 + r_1)} \left(\frac{\Delta f}{f}\right)^2 \frac{1}{(r_0 + r_1)^2 + (2\pi f)^2} \qquad (12)$$

Experimental verification of the expected correlation between drive level sensitivity and resonator noise has also been investigated [2,6]. The result presented in Fig. 25 seems demonstrate that there is a positive correlation between DLS measured by the slope of the resistance change and the noise measured by the residual phase noise at 100 Hz from the carrier [2].

Nevertheless, these promising results have not been confirmed by another set of experiments performed on a large sample of resonators as shown in Fig. 26 [6].

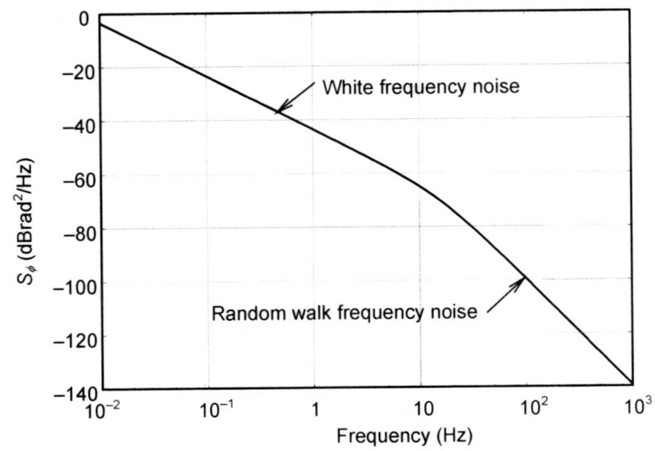

Figure 24: Spectral density of phase fluctuations for a 525 MHz resonator [30].

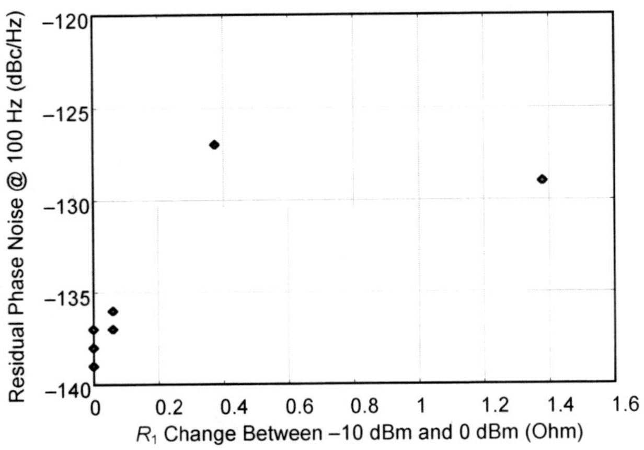

Figure 25: Noise vs. DLS of a 100 MHz AT-cut quartz [2].

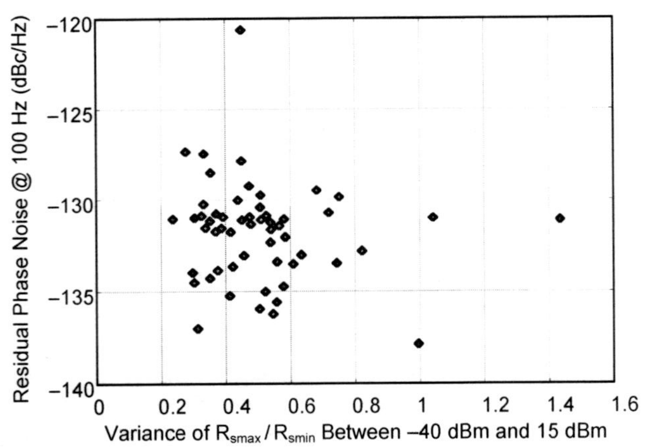

Figure 26: Noise vs. DLS of a 100 MHz AT-cut quartz [6].

Thus, the question of a possible correlation between noise and surface defects that could be disclosed by measuring DLS is still open. A theoretical noise model based on a non-linear series resistance R_1 (Fig. 18) whose value is a function or the crystal current as shown in Equation (13) has been introduced in the past [27] and is still under consideration [26].

$$R_1 = R_{10}\left(1 + \frac{\gamma}{I^n}\right) \tag{13}$$

In Equation (13):
R_{10} is the normal drive level resistance,
γ is the DLS coefficient,
I is the resonator drive current amplitude,
n is an integer number ($1 \le n \le 4$).

7 Conclusion

In summary, the low level drive sensitivity of quartz resonators is mainly expressed by a change of the series resonant frequency and series resistance that may exhibit various and often irreproducible behaviours. The correlation between DLS and surface defects is now well established and several models have been developed to describe the physical mechanism. Nevertheless, the question of the correlation between DLS and resonator noise is still open. Many other questions deserves careful attention such as:

- What happens at very low drive level, near the thermal noise floor?
- How behave the other motional parameters at these low drive level?
- Are the mechanisms responsible for the DLS the same as for the noise?
- How the assumed correlation can be demonstrated?

The work in progress will try to give an answer to these questions.

References

[1] J. P. Aubry, E. Gerard, S. Lechopier. "S. Y. Parameters Method for Accurate Measurements of Bulk Wave Crystal Resonators at Frequencies up to 2 GHz", *Proc. 37th AFCS*, pp. 306–316, (1983).

[2] P. Bates. "Drive Level Dependence Versus Residual Phase Noise of Fifth Overtone AT Cut Quartz Crystals", *Proc. 54th IEEE IFCS*, pp. 233–237, (2000).

[3] M. Bernstein. "Increased Crystal Unit Resistance at Oscillator Noise Level", *Proc. 21st AFCS*, pp. 244–258, (1967).

[4] V. E. Bottom. "Introduction to Quartz Crystal Unit Design", *Van Nostrand Reinhold Co.*, pp. 208–219, (1982).

[5] V. E. Bottom. "The Behavior of Quartz Resonators at Low Drive Levels", *Proc. 5th Crystal Conference*, pp. 245–249, (1983).

[6] E. E. Carlson, T. E. Wickard. "A Study on the Measured Correlation of Drive Level Dependency and Phase Noise of Quartz Crystal Resonators", *Proc. 55th IEEE IFCS*, pp. 338–344, (2001).

[7] R. C. Chittick, D. M. Hyland. "A Preliminary Investigation into a Novel Approach for the Removal of Drive Level Dependence in Quartz Crystal Resonators", *Proc. 11th EFTF*, pp. 173–177, (1997).

[8] L. Dworsky, R. Kinsman. “A Simple Model for Quartz Resonator Low Drive Level Sensitivity and Monolithic Filter Intermodulation” *IEEE Trans. on UFFC*, **41** (2), pp. 261–268, (1994).

[9] G. L. Dybwad. “A sensitive New Method for the Determination of Adhesive Bonding Between a Particle and a Substrate”, *Journal of Applied Physics*, **58** (7), pp. 2789–2790, (1985).

[10] E. P. EerNisse. “An Analysis of Drive Level Sensitivity in Thickness Shear Quartz Resonators”, *Proc. 50th IEEE IFCS*, pp. 346–356, (1996).

[11] E. P. EerNisse, E. Benes, M. Schmid. “The Role of Localized Rotational Imbalance in Drive Level Dependence Phenomena”, *Proc. 56th IEEE IFCS*, pp. 2–7, (2002).

[12] E. A. Gerber. “VHF Crystal Grinding”, *Electronics*, pp. 161–163, (March, 1954).

[13] E. Hafner. “The Piezoelectric Crystal Unit – Definitions and Methods of Measurement”, *Proc. of the IEEE*, **57** (2), pp. 179–203, (1969).

[14] W. H. Horton, R. C. Smythe. “Experimental Investigations of Intermodulation in Monolithic Crystal Filters”, *Proc. 27th AFCS*, pp. 243–245, (1973).

[15] J. E. Knowles. “On the Origin of ‘Second Level of Drive’ Effect in Quartz Oscillators”, *Proc. 29th AFCS*, pp. 230–236, (1975).

[16] R. M. Lec, J. Sorial. “The Study of Colloidal Systems with TSM Piezoelectric Sensors”, *Proc. 55th IEEE IFCS*, pp. 838–845, (2001).

[17] L. Lindberg, G. Portnoff. “Test Oscillator for Study of Drive Level Dependence of Quartz Crystal”, *Proc. 48th IEEE IFCS*, pp. 273–280, (1994).

[18] H. Matthey. “Measurement of Crystal Parameters Using a Network Analyzer in Direct Impedance Mode”, *Proc. 7th EFTF*, pp. 177–180, (1993).

[19] P. E. Morley, R. J. T. Marshall, R. J. Williamson. “Crystal Parameter Measurement Using a Hewlett Packard HP4195A Network/Spectrum Analyzer”, *Proc. 3rd EFTF*, pp. 133–138, (1989).

[20] B. Neubig. “Measurement of Quartz Crystal Units up to 500 MHz and Above by the Use of a Pi Network with Error Correction”, *Proc. 11th Quartz Devices Conference*, pp. 60–68, (1989).

[21] S. Nonaka, T. Tuuki, K. Hara. “The Current Dependency of Crystal Unit Resistance”, *Proc. 25th AFCS*, pp. 139–147, (1971).

[22] K. Okada, T. Sekino. “The Impedance Measurement Handbook. A Guide to Measurement Technology and Techniques”, *App. Note No.* 5950-3000E, available from http://we.home.agilent.com, (2003).

[23] R. C. Peach, A. J. Dyer, A. Byrne, S. P. Doherty. “New Method for the Measurement of Quartz Crystal Resonator Parameters”, *Proc. 36th AFCS*, pp. 297–301, (1982).

[24] R. C. Peach, S. E. Morris. “A System for Precision Parameter Measurements on Quartz Crystal Resonators and Bipoles”, *Proc. 39th AFCS*, pp. 527–534, (1985).

[25] H. S. Pustarfi, W. L. Smith. “An Automatic Crystal Measurement System”, *Proc. 27th AFCS*, pp. 63–72, (1973).

[26] Y. S. Shmaliy, R. Brendel. “An Asymptotical Noise Model of a Crystal Oscillator with a Non-Linear Resonator”, in these proceedings.

[27] S. A. Trushliakov, Y. S. Shmaliy. “Influence of Equivalent Losses of Quartz Crystal Resonator on its Excitation in Oscillator”, *Radiotechnika*, **44** (7), pp. 41–44, (1989, in Russian).

[28] R. J. Williamson. “An Improved Method for Measuring Quartz Crystal Parameters”, *IEEE Trans. on UFFC*, **34** (6), pp. 681–689, (1987).

[29] J. S. Yerna. “Resistance – Measurement of Quartz Crystal at Very Low Drive Levels”, *Proc. 38th AFCS*, pp. 499–506, (1984).

[30] Y. K. Yong, J. R. Vig. “Resonator Surface Contamination – A Cause of Frequency Fluctuations?” *IEEE Trans. on UFFC*, **36** (4), pp. 452–458, (1989).

[31] Q. Zhang, R. Lec, K. Pourrezaei. “The Study of an Interaction of Solid Particles with Various Surfaces Using TSM Sensors”, *Proc. 57th IEEE IFCS*, pp. 1001–1006, (2003).

[32] “Basic Method for Measurement of Resonance Frequency and Equivalent Series Resistance of Quartz Crystal Units by Zero Phase Technique in a Pi_Network”, IEC Publication 444, (1973).

[33] “HP 4395A Network/Spectrum/Impedance Analyzer Operation Manual”, *Hewlett Packard Part No.* 04395-90000, (1997).

THE INFLUENCE OF LATTICE DISTORTION ON THE TEMPERATURE FREQUENCY DEPENDENCY

EBERHARD SEYDEL

Vita-Chemie GmbH, Düppelstrasse 13, D-14163 Berlin, Germany
Tel: +49/30/8027077 Fax: +49/30/80496955 E-mail: es.quartz@t-online.de

Keywords: lattice distortion, activity dips, X-ray measurement

Abstract

How the experience shows with a stable production flow and an exact equivalent geometric shape of the blanks, the temperature frequency behaviour of the finished crystals can be different. It is known that the first and the third order coefficient of the temperature frequency curve vary for different manufacturers. This effect looks like a variation of the optimal cut angle for the same specification. The position of the activity dips over temperature changes also. Therefore it is necessary to give the blank quality more attention.
The effects of different cut angles are well known and can be controlled by common X-ray measurements.

The present work correlates the lattice deformations with the changes in the behaviour of the activity dips. The measurement of the lattice constants and the rocking curves for different blank producers (and raw material producers) corresponds with the temperature frequency behaviour and the theoretical calculations.

1 Introduction

In addition to the publication on the EFTF 2000 in Turin [1], now the influences of lattice deformations on the vibration behaviour of crystals were calculated. The calculation program is extended with three different models for lattice distortions. Initialized was the idea by the known effects of different zero cut angles and activity dip results for mechanically equal blanks. Blanks with the same design but from different blank sources, means growing facilities, show such effects. First X-ray tests of such crystals have shown significant differences. Models for the lattice deformations were calculated to determine the influence of these effects on the electrical parameters of quartz crystals.
Until now a precise X-ray test of the involved parameters is not practicable. Q-value (Infrared) and etch channel density are not suited for continuous tests. The as-grown surface is a sufficient tool for the control of some parameters, but not applicable at the finished blank. A hard-line control in the growing area and blank manufacturing is the only way for a stable production of critical specifications.

2 X-ray Measurements

X-ray determination of the lattice constants for different material sources is shown in **table 1**. Significant differences in activity dips were measured between sample No 4 and sample No 6. For the measurement the reflexes 026, 413 and 50-1 were used (fig. 1). These reflexes guarantee a sufficient accuracy of the calculated values. This method and also the used standard ω-scan, takes a lot of time and is not applicable for production requirements.

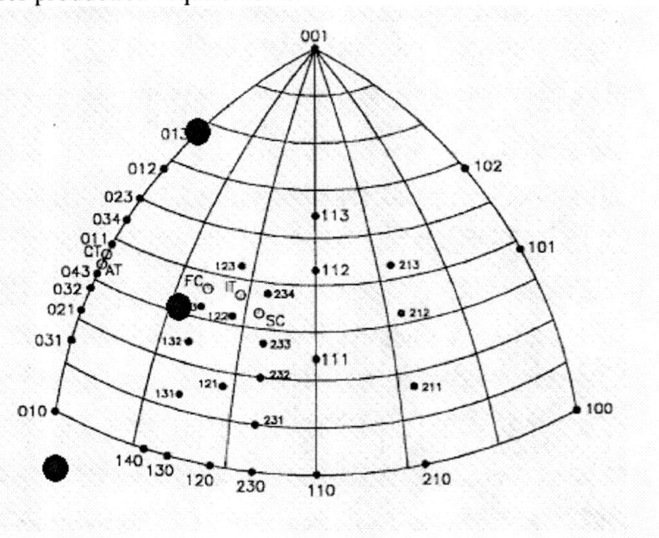

Figure 1: location of the used reflexes in the stereographic projection

The Q-value of all samples was specified better than 2.2 Mio with the IR-method (IEC 758).
The differences in the lattice constants are not significant enough to use them for fast determination and sorting processes. The lower cell volume is connected with better material; it means there exist less lattice distortions in the integrated volume.
More important and significant was the measurement of the peak form. Differences in the ω-scan from normal peak forms to multiple peaks (fig.2a-c) can be seen. The deformations appear on the 026 peak or on all three used peaks.

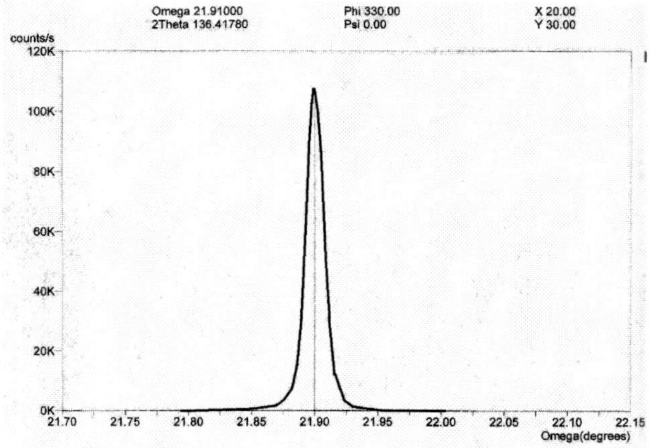

Figure 2a

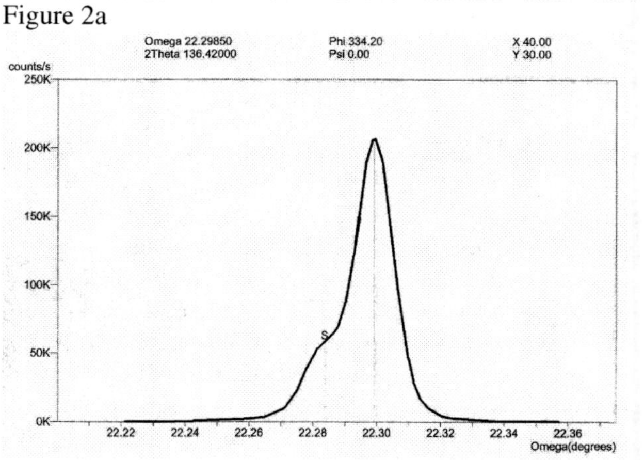

Figure 2b

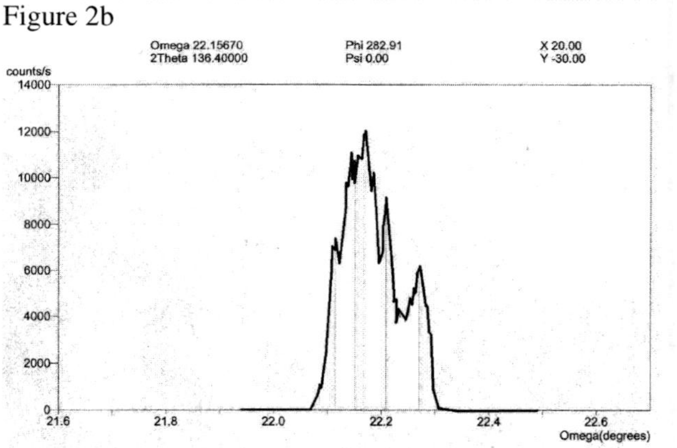

Figure 2c

Such deformed curve form can be seen only, if a large parallel beam is used and the full blank area is integrated.

The lattice distortions can be caused by the growing process or by the blank manufacturing process. To determine the real origin of the effect blanks from the same blank manufacturer but with different materials were compared. All blanks should show the same peakform if the problem would be caused in the blank manufacturing. Otherwise a problem in the growing process would be announced. Two tests were performed and the result showed clearly that the rough material causes the effect.

The measurement of the cut angle at different points of the blanks confirmed the existence of lattice differences in blanks from different production lots which was selected by yield in activity dips. But all these tests are not applicable for production requirements.

There are different possibilities to generate those types of X-ray peaks.

First: The seed material can induce a high number of deformations. This can partly be seen in the number of etch channels and will produce a reflection like in figure 2c. In the market material exists from very low numbers of etch channels up to thousands per square centimetre. The infrared Q-value will be influenced not much; it is an integrated value over the volume and the volume of the etch channel is less 1% of the total volume.

Second: There are unlinearities during the growing process; means power breaks or irregularities by other parameters inducing an inconstant growing speed over the complete autoclave or only in parts of it. This can be seen as wider peaks or double peaks like in fig1b. The effect can occur in any material and is not detectable with etch channels or with Q-value measurement.

The material properties should be constant over the autoclave run under very well controlled conditions. But all the control procedures in the growing facilities and rejects of grown material are very cost intensive.

At this point starts the problem for the blank customer. A precise test for those dislocations is not practicable on each blank.

3 Calculations

The FED-program, described in [1], was used with several modifications to calculate lattice disturbances.

To model those effects in the theoretical calculations, three simplified models were made and compared with the normal undisturbed calculation.

First: Linear change of the crystal parameters (growing with constant deposition rate, which means faster growing in the end of the process)

Second: Change in form of a step in the crystal parameters (power break etc. in the growing)

Third: Change in the crystal parameters in special points (etch channel)

One of the most interesting points is the influence of these lattice changes on the activity dips. Calculations were done for the three models of a fundamental mode crystal in the range around 17MHz. A design for maximum coupling at 25 C was searched to use it for the calculations (fig 3a). For the same design, all calculations were done. Rectangles mark the influenced area in all graphs. The small gray rectangel corresponds to the undisturbed case for comparison.

Figure 3b shows the effect of a continuous change of the lattice parameters, e.g. induced by a constant deposition rate. The coupling point with the unwanted mode is moved in temperature. The area of influence on the main mode is mainly constant.

Abrupt changes in the parameters, e.g. induced by power breaks during the growing process (fig.3c), give another picture. Here is the temperature of the coupling also changed, like the linear case, but the amplitude reacts much more. The area of influence is twice.

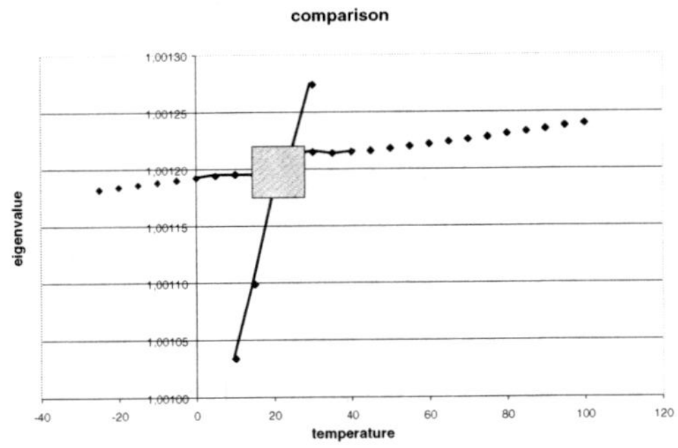

Figure 3a

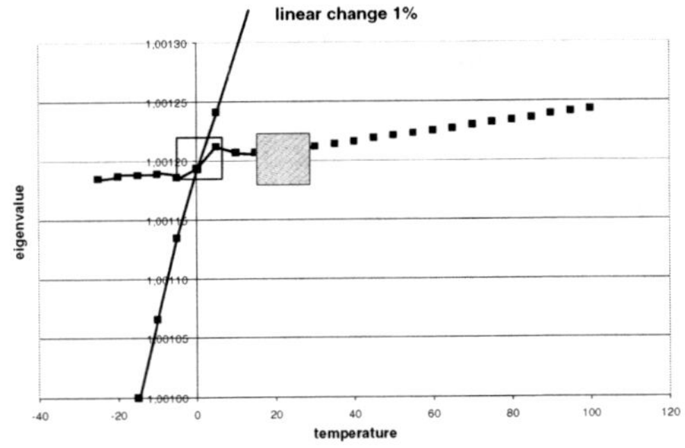

Figure 3b

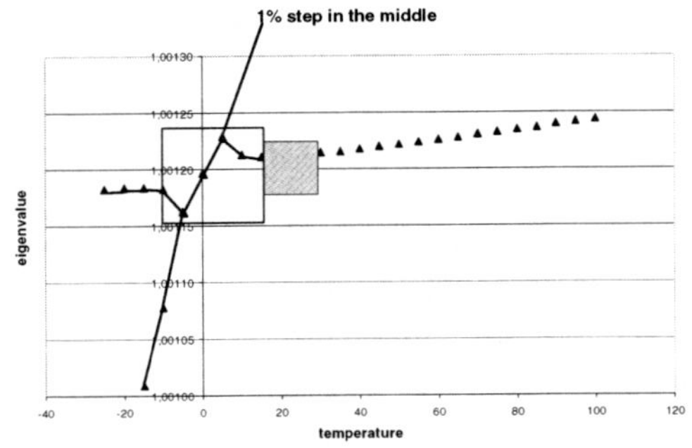

Figure 3c

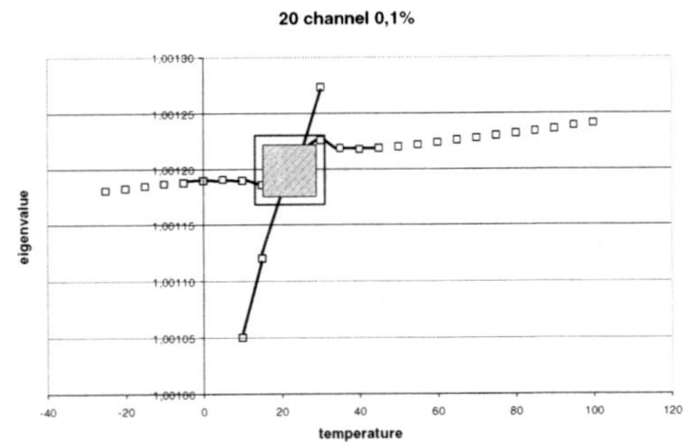

Figure 3d

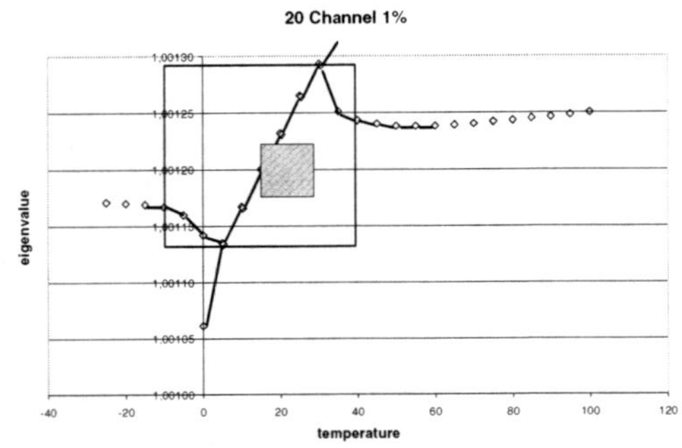

Figure 3e

Etch channels with low disturbance change the curve form slightly (fig 3d).
But normally are the parameters much more changed inside the etch channels. This disturbs the temperature frequency curve (fig.3e) in a higher degree, like edge effects, up to deformations.
A change in the size of the influenced area or a change in temperature of the coupling point can cause, that the crystal runs out of specification. This effect can work in negative direction, as well. Bad material produces less activity dips, but not a better phase noise.
Unpleasant is in fact the instability. It is not possible to produce all the time material with the same distortions. The only way to avoid the problems is a stable design for excellent material.

4 Perspective

Calculations and first measurements show a significant influence of differences in the growing process on the temperature frequency behaviour, especially activity dips, on the ageing process and phase noise, of course. This is of high interest for TCXO- and OCXO-producers. In the next time the work will be concentrated on a fast and secure technique to indicate relevant lattice distortions in the blank. Different methods will be tested and combined. Important part will be an electrical measurement of activity dips independently of the holder system and the use of vibration topography, to eliminate other effects.

5 Resume

It will be necessary to use better controlled blank material for stable yield in the TCXO-Crystal production and a better yield in the ageing and phase noise. One method to reach this target is a very intensive control of all processes down to the growing. Another method will be the development of a new procedure for a usable incoming control to detect those lattice distortions.

Literature

[1] E. Seydel, The Influence of the Surface Structure on the Vibration Amplitude and the Equivalent Electrical Data of Quartz Crystals – a Theoretical Overview, Proc. 14th Europ. Frequ. Time Forum, pp.103-107, 2000.

[2] E.Seydel, A.Landmann, The influence of surface structure on amplitude deviation and electrical parameters of quartz crystals, Proc. 9th Europ. Frequ. Time Forum, pp.450-453, 1995.

[3] E.Seydel, L.Weitzel, High Frequency Fundamental mode resonators-equivalent electrical data dependence on electrode properties and blank surface, Proc. 10th Europ. Frequ. Time Forum, 1996.

[4] Christian Rivas Mujica, Estudio de los parámetros utilisados en el proceso de electrodifusión de de cristales sintéticos de cuarzo, Universidad Simón Bolívar, 1999

Sample No	a	c	Cell volume	c/a	Producer	Remarks
1	4,9130	5,4051	112,988	1,10014	1	
9	4,9134	5,4051	113,006	1,10006	1	
10	4,9137	5,4050	113,017	1,09999	1	
12	4,9131	5,4048	112,987	1,10008	2	
2	4,9131	5,4052	112,993	1,10016	2	
11	4,9132	5,4049	112,993	1,10007	2	
13	4,9134	5,4050	113,003	1,10006	2	
3	4,9134	5,4049	113,004	1,10003	3	
4	4,9131	5,4052	112,991	1,10017	3a	double peak in 026
5	4,9132	5,4052	112,997	1,10014	4	
6	4,9130	5,4050	112,984	1,10015	5	
7	4,9132	5,4051	112,998	1,10011	6	Multiple peaks in all directions
8	4,9131	5,4049	112,989	1,10010	6a	Normal peak

Table 1 (TK activity dips differences between sample 4 and sample 6)

MICROFABRICATED ATOMIC FREQUENCY REFERENCES

J. Kitching[*], S. Knappe[*], L. Liew[†], J. Moreland[†], H. G. Robinson[*], P. Schwindt[*‡], V. Shah[*‡] and L. Hollberg[*]

*Time and Frequency Division, National Institute of Standards and Technology, Boulder, CO 80305
[†]Electromagnetics Division, National Institute of Standards and Technology, Boulder, CO 80305
[‡] The University of Colorado, Boulder, CO 80309

Keywords: Atomic, wafer bonding, clock, compact, microfabrication.

Abstract

We describe a design for a microfabricated atomic frequency reference with a volume of several cubic millimetres and a power dissipation in the range of tens of milliwatts. It is anticipated that this frequency reference will be capable of achieving a fractional frequency instability below 10^{-11} at integration times of hours.

1 Introduction

Atomic frequency references are being used in an increasing number of real-world applications. Much of this growth is a result of the exceptional long-term frequency stability routinely achieved by atomic standards, combined with improved miniaturization and power management. The most recent generation of compact atomic frequency standards [1,2] were developed primarily for the synchronization of wireless communication networks. These devices, with a volume of roughly 100 cm^3, achieve long-term fractional frequency instabilities below 1×10^{-11} and consume several watts of power while operating. While certainly useful for many applications in addition to wireless network synchronization, these frequency references cannot be applied in hand-held, portable units because their large power dissipation is not compatible with battery power. Examples of such portable applications are receivers for global navigation satellite systems (GNSS) and wireless communication devices.

Since many devices rely on high data transfer rates and information portability, precision timing in battery-operated devices is highly desirable. One example is jam-resistant global positioning system (GPS) receivers for the military. Because of the extremely low power of the signal broadcast by GPS satellites, receivers are highly susceptible to intentional jamming and unintentional interference from other RF sources transmitting in the same frequency band. Because of the larger bandwidth over which the military P(Y) signal is transmitted, it is considerably less susceptible to jamming than the civilian C/A signal. However, since the P(Y) code repeats only every seven days, a P(Y) receiver needs a better local clock than in a C/A receiver in order to narrow down its search window and reduce the time required to find the code match [3]. Existing P(Y) receivers usually have to first lock onto the civilian C/A signal, which eliminates the anti-jam advantage of the larger bandwidth P(Y) signal. If the receiver's local clock were capable of determining the time to within 1 ms over several days, it would be possible for a receiver to lock onto the P(Y) signal directly without first acquiring the C/A signal. Thus, a significant advantage in resistance to jamming would be achieved.

The frequency-reference physics package we are developing [4] is the first atom-based reference to present significant potential for battery operation. In addition, its small size and amenability to wafer-level fabrication and assembly make it appealing for commercialization and integration into other devices.

2 Microfabricated Vapour Cells

The heart of the atomic clock physics package is a vapour cell containing a combination of alkali atoms and a buffer gas to reduce the wall-induced decoherence of the hyperfine oscillation. Vapour cells have traditionally been fabricated using conventional glass-blowing techniques, which have two important drawbacks. Firstly, it is difficult to make small cells because of the increasing importance of surface tension in shaping the melted glass at small sizes. Secondly, the cells must be made one by one, leading to substantial fabrication cost and difficulty integrating the cells with other clock components.

We have developed a method of cell fabrication [5] based on techniques usually applied to microelectromechanical systems (MEMS). The cells are formed by sandwiching an etched Si wafer between two transparent glass wafers, as shown in Figure 1. A Si wafer, typically a few hundred micrometers thick, is lithographically patterned and etched by use of wet-chemical (KOH) or deep-reactive-ion etching. An example of the wet-chemical etching process is outlined in Figure 2. By use of one of these processes, holes are etched through the wafer with a square cross-section of sides roughly 0.6 mm. However, the highly scalable nature of the etching process would allow holes as small as a few tens of microns to be created simply by changing the etch mask.

Once the holes are created in the Si wafer, glass is attached to one side using the technique of anodic (or field-assisted) bonding [6]. Developed by Wallis and Pomerantz in 1969, this process can be used to bond flat wafers of borosilicate glass to a variety of materials including other glasses, metals

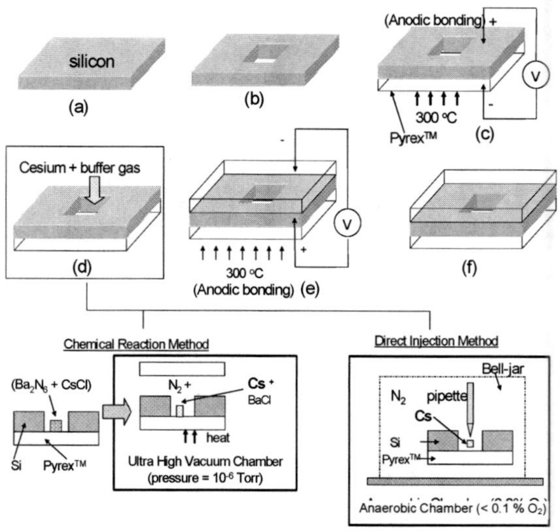

Figure 1 Schematic showing the steps in wafer-level fabrication of an alkali cell.

and Si. The bonding process is carried out by placing the two clean wafers in contact in a dust-free environment. The sample is then heated to approximately 300 °C and a few hundred volts of potential difference is applied across the wafer pair (Figure 1c). Because of the high temperature, impurity ions in the glass (such as K^+ and Na^+) begin to drift in the electric field, leaving behind electrons. The electrons form a space-charge field, which attracts one material to the other and creates a strong bond between them. The bond is fully hermetic and ideal for confining alkali atoms inside the cell, as far as we have been able to ascertain.

Once the initial piece of glass is bonded onto one side of the Si wafer, Cs (or Rb) is then deposited into the cell (Figure 1d and subpanels). This is carried out with one of two methods. The first method involves the chemical reaction of BaN_6 and CsCl in a high-vacuum environment. These two materials are both soluble in water and are deposited into the cell preform in solution. The water is then evaporated and the preform, with remaining chemicals in solid form, is placed into a high-vacuum chamber. The chamber is evacuated and backfilled with a buffer gas at an appropriate pressure. The sample is then heated to 150 °C, at which point the chemicals react and create Cs, BaCl and N_2. A second glass piece is then pushed up against the top of the sample, and the cell is heated further with an electric field applied to seal the Cs and buffer gas inside the cell. The residual N_2 gas produced by the reaction presumably is pumped or diffuses away before the cell is sealed since the final buffer-gas pressure in the cell roughly matches the pressure in the chamber during bonding.

The second technique of cell filling involves the use of an anaerobic chamber, essentially an airtight glove box with the water and oxygen reacted away. The cell preform is placed in the anaerobic chamber and Cs is added by breaking open a Cs ampoule inside the chamber and injecting some of the liquid metal using a micropipette. The cell preform is then placed inside a bell jar (inside the anaerobic chamber) that is

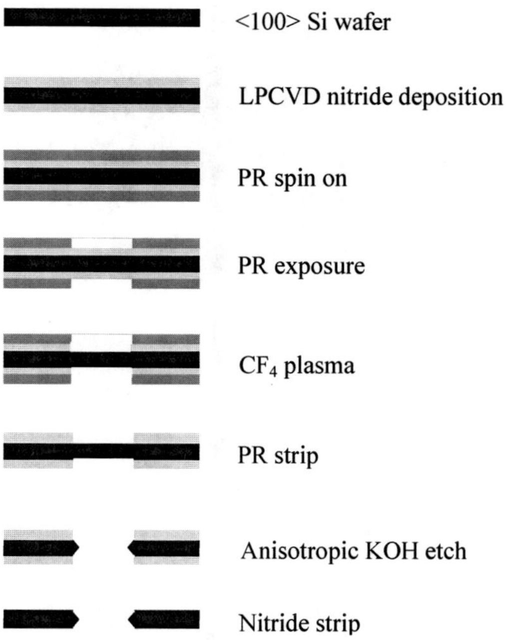

Figure 2 Steps to carry out anisotropic chemical etching of Si for cell fabrication. Silicon Nitride is deposited on a Si wafer using liquid-phase chemical vapour deposition (LPCVD). Photoresist (PR) is spun onto the nitride and exposed to ultra-violate (UV) light through a mask. The patterned PR and nitride is plasma-etched using CF_4 and the PR is stripped off leaving the Si exposed. An anisotripic KOH etch is then used to etch holes in the exposed Si. Finally the nitride is removed.

backfilled with the desired buffer gas. The second window is then attached, again using anodic bonding.

After the final bonding step, the cells can be diced into individual components. A cell fabricated using the first method described above is shown in Figure 3. It should be clear that the process outlined in could be easily implemented at the wafer level. Lithographic patterning, etching and bonding of entire wafers are routinely done in the MEMS field and cell filling could be carried out either with an automated Cs dispenser (anaerobic chamber technique) or simultaneous deposition of chemical solution (chemical reaction technique).

3 Physics Package Design

The development of wafer-level processing of planar cells allows for marked change in the design of frequency-reference physics packages. For the first time it becomes possible to assemble physics packages in an integrated, vertically-stacked structure. This type of structure allows not only for extremely small size but also for a corresponding reduction in power dissipation. In addition, this assembly method has the potential to drastically reduce the cost of manufacturing physics packages.

(a)

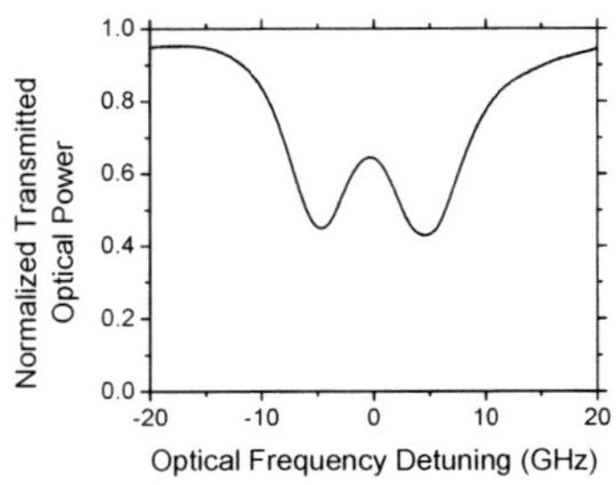

(b)

Figure 3 (a) Photograph of a micromachined Cs vapour cell fabricated by anodic bonding. (b) Optical absorption resonance indicating the presence of Cs in the cell, along with approximately 250 Torr of buffer gas.

The frequency references we are developing are based on the microwave transition between the F=3, m_F=0 and F=4, m_F=0 hyperfine sublevels of the $6S_{1/2}$ ground state of ^{133}Cs. A coherence between these two levels can be generated through the phenomenon of coherent population trapping (CPT) [7,8,9] in a Λ-system (see Figure 4a). The CPT resonance is excited using light from a diode laser modulated through the injection current at one-half the Cs hyperfine splitting of 9.192 GHz [10] (see Figure 4b). The two first-order sidebands on the optical spectrum therefore create a Λ-system with the atoms on the D2 optical transition at 852 nm. When the laser modulation frequency is scanned near the first subharmonic of the hyperfine splitting, a resonance is observed by monitoring the total transmitted power through the cell with a Si PIN photodiode. This signal is used to determine when the local oscillator (LO) is on-resonance with the atoms.

A schematic of one possible design of a fully integrated physics package is shown in Figure 5. A die containing a vertical-cavity surface-emitting laser (VCSEL) is bonded onto a substrate patterned with gold (Figure 5, layer a). The VCSEL is used because of its low power requirements (typically < 5 mW for most devices), high modulation efficiency and availability of single-mode devices at the 852 nm D2 transition in Cs. The light emitted by the VCSEL is conditioned by an optics assembly (Figure 5, layers b-f) attached to the baseplate. This optics assembly attenuates and collimates the light and change the light polarization from linear to circular. The cell assembly (Figure 5, layers g-k) is

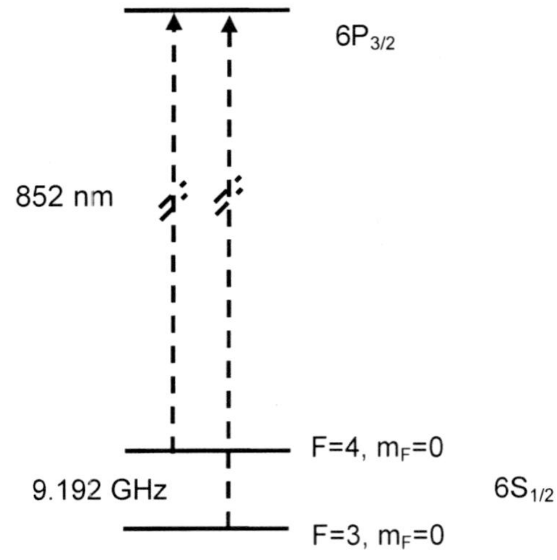

(a)

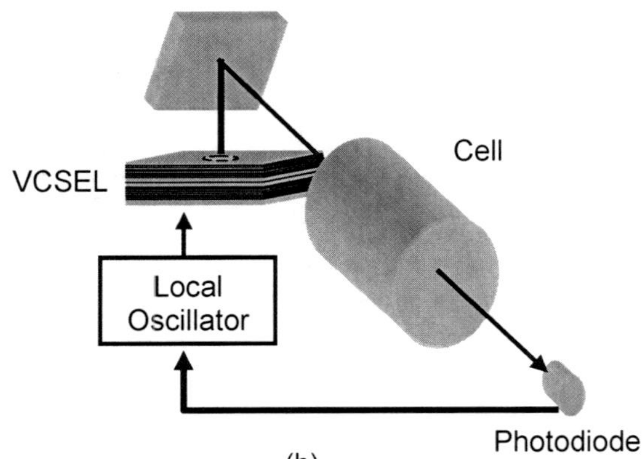

(b)

Figure 4 (a) Part of the Cs atom level spectrum showing the states relevant for CPT excitation. (b) Schematic of the experimental implementation based on a modulated diode laser.

placed on top of the optics assembly. Because of the small optical path length in the cell, a temperature of about 80 °C is required to provide an optimal signal. One way to accomplish this is to attach integrated heaters and temperature sensors to the cell structure. Finally a photodiode assembly (Figure 5, layers l-m) is mounted onto the top of the structure to detect the light power transmitted through the cell. An example of how physics packages might be assembled at the wafer level is shown in Figure 6.

Power dissipation is a critical aspect of any design of a portable atomic frequency reference. A typical AA battery yields about 2000 mW-hours of power and therefore a few tens of milliwatts would be a reasonable goal for the power dissipation of a portable, battery-operated atomic clock. This is a challenging target since the cell must be held at a temperature several tens of Celsius degrees above ambient.

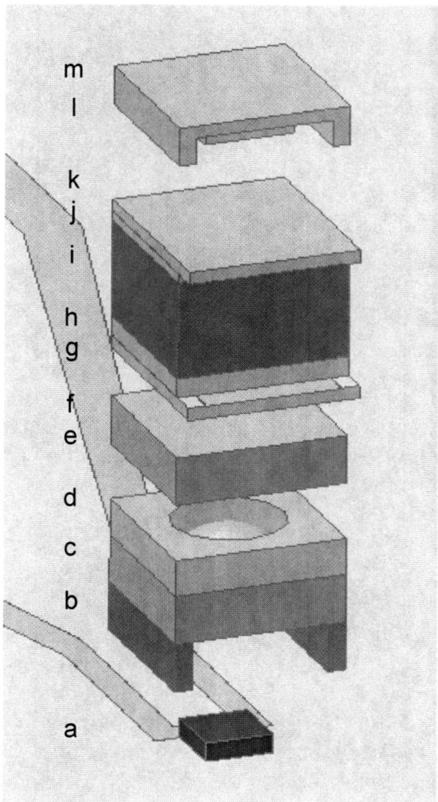

Figure 5: Schematic of one possible design of a microfabricated atomic clock physics package. Layer a is the laser, layers b-f are the optics assembly, layers g-k is the cell assembly and layers l-m are the photodiode assembly.

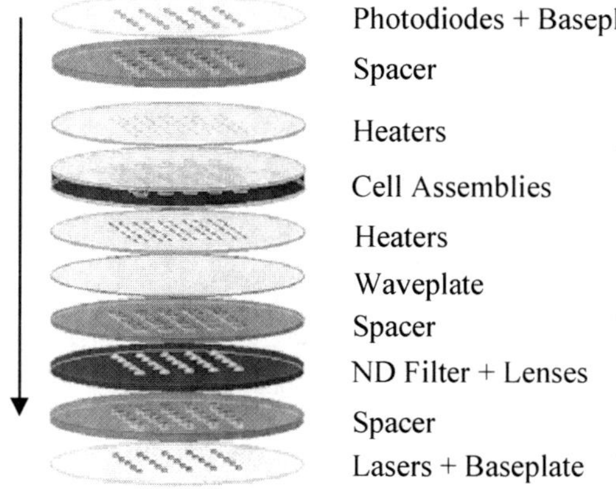

Figure 6 Wafer-level assembly of microfabricated frequency-reference physics packages.

The power dissipation is in fact the primary consideration in determining how big the structure can be; larger cells radiate and conduct more power for a given temperature difference between the cell and the surroundings.

Despite the drawback of high cell temperature with regard to power dissipation, one rather fortuitous circumstance resulting from the small cell size is that the cell is operated above the range of temperatures typically specified for commercial devices. The cell must be actively temperature stabilized in order achieve good long-term frequency stability, and the high cell temperature obviates the need for a cooling mechanism, which is typically far less efficient than heating. As a result the small size of the cell is compatible with low-power temperature stabilization.

For the design in Figure 5, in which the cell is heated independently from the baseplate, the major heat loss channels are conduction through the cell support structure and electrical connections, conduction and convection through the air surrounding the structure, and radiation. Conduction through the air can be largely eliminated by packaging the structure in a vacuum enclosure. The power dissipated by radiation is given by

$$\dot{Q}_{rad} = \alpha\sigma\left(T_1^4 - T_0^4\right)A, \qquad (1)$$

where α is the surface emissivity, σ is the Stefan-Boltzmann constant, T_1 is the cell temperature, T_0 is the ambient temperature and A is the surface area of the device. If the interior of the vacuum enclosure were coated with a material such as gold, which has a radiative emissivity of about 0.02, the radiated power would be approximately 0.4 mW for T_1-T_0 = 100 K.

Conduction through the support structure is perhaps the most important source of power dissipation. In the design shown in Figure 5, the cell is held away from the baseplate by two thin supports of rectangular cross section, A_s and height L. The power conducted through a support is given by

$$\dot{Q}_{cond} = I\left(T_1 - T_0\right)\frac{A_s}{L}, \qquad (2)$$

where I is the thermal conductivity of the material. Polymer photoresist materials such as SU-8 have low thermal conductivity (about 0.2 W/(m·K)) and are also micromachinable. For supports 1.5 mm long, 0.1 mm wide and 0.5 mm high, the power dissipated to maintain temperature difference of 100 K is 12 mW. Conduction through the electrical connections can be minimized by making them thin and long. A gold trace 2 μm high, 50 μm wide and 2 mm long would dissipate only about 2 mW to support a temperature difference of 100 K between its ends.

Since the laser wavelength depends on the temperature of the device, the laser temperature is typically actively stabilized. This requires power to heat the laser but because of the small size of the laser die, this power is substantially smaller than that required to heat the cell. The contact area of the laser on the baseplate is about 0.1 mm^2. If the laser were mounted on a thermally insulating polymer substrate, roughly 6 mW would be required to heat the laser to 100 K above ambient.

Nonthermal sources of power dissipation within the physics package include the laser operation (< 5 mW), RF modulation (typically hundreds of microwatts) and detector bias (very small). Overall, therefore, it appears that a power dissipation of the order of tens of milliwatts is possible with this compact clock design. A summary of the physics package power budget is shown in Table 1.

Table 1 Summary of power budget of physics package.

Source	Power (mW)
Cell heating ($\Delta T = 100$ K)	20
Laser heating ($\Delta T = 100$ K)	6
Laser DC	4
Laser RF	0.1
Total	**30.1**

4 Anticipated Short-Term Frequency Instability

The short-term instability of vapour-cell atomic frequency references is determined by a number of factors. Perhaps the most important is the resonance linewidth, which depends on both the frequency of collisions of the alkali atoms with the walls of the cell and also the pressure of the buffer gas used to prevent frequent wall collisions. Theoretical estimates based on diffusion in a buffer gas and complete depolarization on wall collisions indicate that a linewidth of near 1 kHz should be possible in a cell with dimensions of about 1 mm [11]. Experimental measurements confirm these predictions [12].

A second important factor is the resonance contrast, which we define as the ratio of the change in power due to the CPT resonance to the total absorbed power. For excitation on the D1 line of Rb, contrast values above 10 % have been observed [13]. Finally, the noise on the measured signal is determined fundamentally by the photon shot noise. For one microampere of detected photocurrent, the signal-to-noise ratio should be approximately 1×10^5, assuming a contrast of 10 % and an absorption of 50 %. This leads to a fundamental short-term fractional frequency instability of roughly 1×10^{-12} at one second of integration. Real devices are expected to fall short of this mark due to technical noise and additional linewidth-broadening mechanisms such as power broadening and spin-exchange broadening, but this analysis nevertheless gives an indication of what might be possible in a millimeter-scale device. Based on these numbers, a long-term instability below 1×10^{-11} should be easily achievable.

5 Integration With Other Components

In addition to the physics package, two other components are required to enable a fully functional atomic frequency reference: a local oscillator and a control system. The local oscillator provides the initial (unstable) RF signal that is locked to the atomic resonance. Leading candidates to realize a compact, low-power local oscillator are thin-film bulk-acoustic wave resonators [14] and SiC-based nanoresonators [15]. Both of these technologies can be used to achieve high Q-factors (> 1000) at gigahertz frequencies and can be excited with low circulating power levels.

The control electronics carry out the servo systems required to keep the system locked and stable. In the large-scale CPT frequency references currently operating in our laboratory, four servo systems are required. Two of these are temperature servos that stabilize the laser and cell temperatures. The remaining two are lock-in-based servos that stabilize the laser frequency onto the optical transition and the LO frequency onto the microwave transition. We anticipate that a microprocessor-based digital servo system would be appropriate for the frequency reference control. An alternative would be an application-specific integrated circuit (ASIC) in which an analogue circuit was implemented. The power required to run the local oscillator and control circuitry has not been evaluated with a high degree of certainty but levels in the range of tens of milliwatts are not out of the realm of possibility.

6 Conclusions

We have described here a fundamentally new design for compact atomic frequency references based on MEMS microfabrication techniques. Advantages of this technique include small size, low power dissipation, low-cost mass-production through wafer-level processing and a high degree of scalability. These features may enable atomic frequency references to be integrated into portable, battery-operated devices used for global positioning and wireless data communications.

Acknowledgements

This work is supported by NIST and the Defence Advanced Research Projects Agency (DARPA). This work is a contribution of NIST, an agency of the US government, and is not subject to copyright.

References

[1] P. J. Chantry et al., "Miniature laser-pumped cesium cell atomic clock oscillator," *Proc. 1996 IEEE Int. Freq. Cont. Symp*, 1002-1010 (1996).

[2] J. Ho, I. Pascaru, C. Stone and T. McClelland, "New Rubidium frequency standard designs for telecommunications applications," *Proc. 1998 IEEE Int. Freq. Cont. Symp*, 80-83 (1998).

[3] H. Fruehoff, "Fast "direct-P(Y)" GPS signal acquisition using a special portable clock," *Proc. 33rd Ann. Precise Time and Time Interval (PTTI) Meeting*, pp. 359-369 (2001).

[4] S. Knappe, L. Liew, V. Shah, P. Schwindt, J. Moreland, L. Hollberg and J. Kitching, "A micromachined atomic clock," to be published.

[5] L. Liew, S. Knappe, J. Moreland, H. G. Robinson, L. Hollberg and J. Kitching, "Micromachinined alkali atom vapor cells," *Appl. Phys. Lett.* **84**, 2694-2696 (2004).

[6] G. Wallis and D. I. Pomerantz, "Field assisted glass-metal sealing," *J. Appl. Phys.* **40**, 3946-3949 (1969).

[7] W. E. Bell, A. L. Bloom, "Optically driven spin precession," *Phys. Rev. Lett.* **6**, 280-281 (1961).

[8] G. Alzetta, A. Gozzini, L. Moi, G. Orriols, "An experimental method for the observation of R. F. transitions and laser beat resonance in oriented Na vapor," *Il Nuovo Cim.* **36B**, 5-20 (1976).

[9] E. Arimondo, "Coherent population trapping in laser spectroscopy," *Progress in Optics XXXV,* E. Wolf, ed., 257-354 (1996).

[10] J. Kitching, S. Knappe, N. Vukicevic, L. Hollberg, R. Wynands, and W. Weideman, "A microwave frequency reference based on VCSEL-driven dark line resonances in Cs vapor," *IEEE Trans. Instrum. Meas.* **49**, 1313-1317 (2000).

[11] J. Kitching, S. Knappe and L. Hollberg, "Miniature vapor-cell atomic frequency references,"*Appl. Phys. Lett.* **81**, 553-555 (2002).

[12] S. Knappe, L. Hollberg and J. Kitching, "Dark-line atomic resonances in sub-millimeter structures, " *Opt. Lett.* **29**, 388-390 (2004).

[13] M. Stahler et al., "Coherent Population Trapping Resonance in a Thermal ^{85}Rb vapor: D1 versus D2 line excitation," *Opt. Lett.,* **27**, 1472-1474 (2002).

[14] K. M. Lakin, J. S. Wang, "Acoustic bulk wave composite resonators," *Appl. Phys. Lett.* **38**, 125-127 (1981).

[15] X. M. H. Huang, C. Zorman, M. Mehregany, M. L. Roukes, "Nanodevice motion at microwave frequencies," *Nature* **421**, 496 (2003).

Observation of Ramsey fringes with optical CPT pulses

T. Zanon, S. Guérandel, E. de Clercq, D. Holleville, N. Dimarcq and A. Clairon
BNM-SYRTE – CNRS UMR8630 – Observatoire de Paris
61 avenue de l'Observatoire, 75014 Paris, France
Email: thomas.zanon@obspm.fr

***Abstract*— The linewidth of CPT atomic resonances observed in a vapor cell are usually limited by optical saturation and collisions due to buffer gas. We have applied here the Ramsey method with optical CPT pulses generated by a phase-locked device. Fringes are then produced giving narrow atomic signals only dependent on interrogation time between the two pulses. The possibility to obtain narrow fringe widths below 100 Hz with short or long pulses is shown experimentally and analysed using the classical Schrödinger state representation.**

key words: Coherent Population Trapping, Lamb-Dicke regime, optical coherence, Ramsey pulses

INTRODUCTION

Coherent Population Trapping (CPT) or resonant Raman interaction is a well known effect commonly used to produce narrow atomics clock resonances below 100 Hz in a three level system with a buffer gas [7], [9]. The method of interrogation is usually done with continuous excitation of the two ground states defining the clock transition and an excited state coupled by two coherent laser fields (Fig. 1). The atom-laser interaction in this 3-level system is well described by the optical Bloch equations. The CPT linewidth (FWHM) coming from the stationary solution of this system is at resonance [10] :

$$\Delta\nu_{1/2} = \frac{2\gamma_c}{\sqrt{1+S}}\sqrt{(1+3S+\frac{\Gamma}{\gamma_c}S^*)(1+\frac{\Gamma}{\gamma_c}S^*)} \qquad (1)$$

where Ω_1, Ω_2 are the Rabi frequencies ($\propto$ laser field amplitudes), γ_c is the hyperfine coherence relaxation term due to buffer gas, $\Gamma = 2\pi \times 500$ MHz is the relaxation term due to spontaneous emission broadened by collisions with buffer gas. We define two saturation parameters

$$S = \frac{\Omega_1^2\Omega_2^2}{\Omega_1^2+\Omega_2^2}\frac{1}{\Gamma^2/4}, \qquad S^* = \frac{\Omega_1^2/2+\Omega_2^2/2}{\Gamma^2/4}$$

driving the transitions $|1\rangle \mapsto |3\rangle$ and $|2\rangle \mapsto |3\rangle$. S is the Einstein saturation parameter coming from the classical rate equations.

Interference between amplitudes of transition in the two optical pumping channels cancel both fluorescence and absorption when the resonance condition between all three states is fulfilled (Fig. 1). The possibility to use continuous CPT beams to observe narrow resonances below 500 Hz used for metrology was already demonstrated in several groups [8], [9]. With usual intensity lower than 100 nW/cm^2, CPT signals are mainly broadened by the finite relaxation rate γ_c of the hyperfine coherence due to the buffer gas but with poor signal to noise ratio. An alternative method to get round of the saturation limitation given by equation (1) is to generate optical CPT pulses to observe Ramsey fringes only limited by the interrogation time T [2], [3].

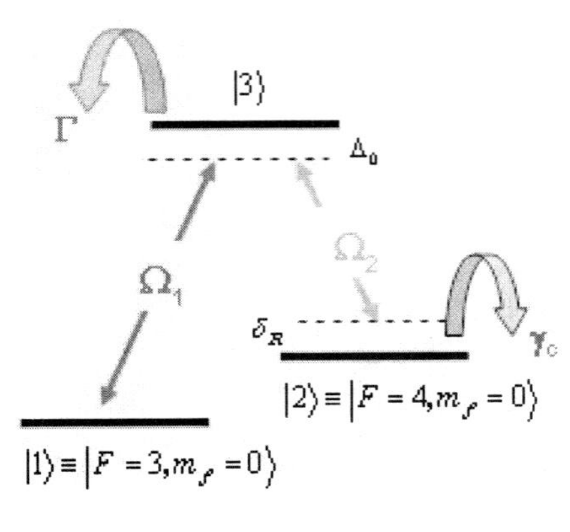

Fig. 1. Three level system under two coherent laser fields excitation for CPT. Δ_0 is the common optical detuning and δ_R is the Raman detuning.

In the first part of the paper, we present the phase-locked device required to insure coherent laser beams for the CPT phenomenon and the experimental setup used to generate the optical pulses. In the second part, we analyse the experimental fringes with the assumption that the pulses are sufficiently short to use the Schrödinger states representation. At the end,

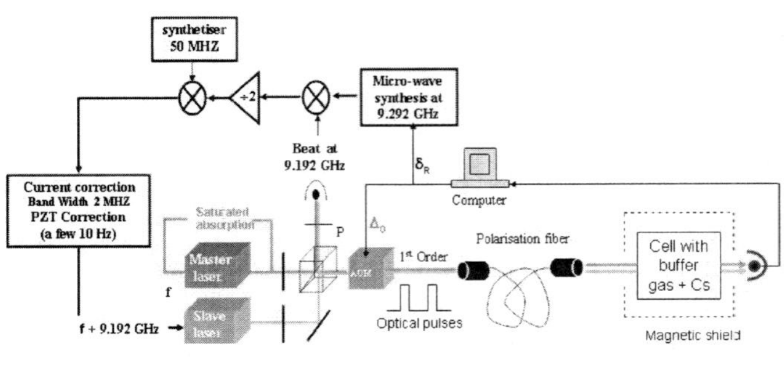

Fig. 2. Hot vapor cell CPT experiment. The AOM for the double pass not shown in the figure is including into the saturated absorption.

we will show that optical Bloch equations are a powerful tool to analyse all experimental situations taking into account the spontaneous emission driving atoms in a more realistic closed

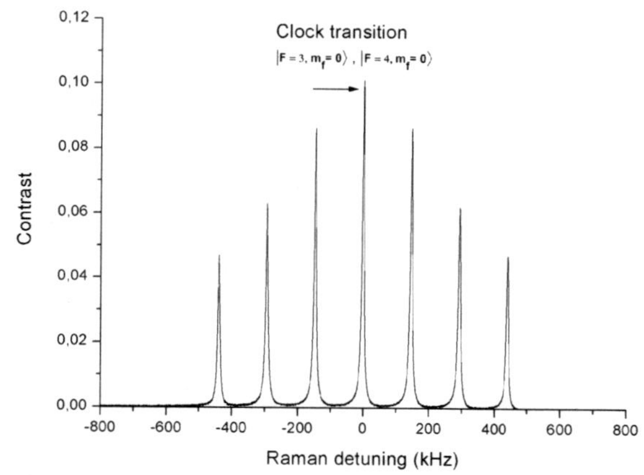

Fig. 3. Experimental CPT transitions observed in a 20 μT static magnetic field using orthogonal laser polarisations.

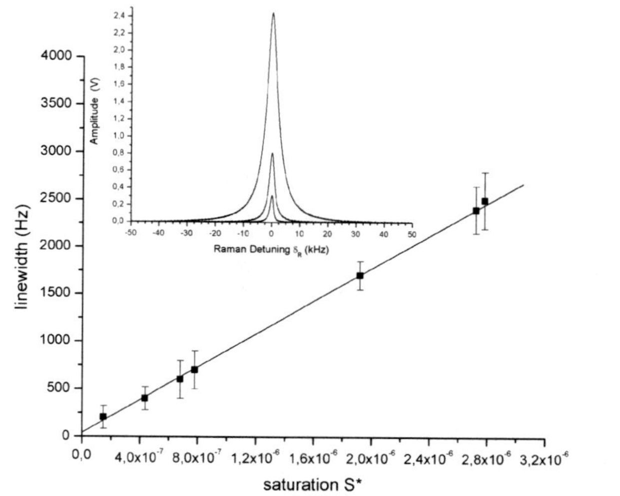

Fig. 4. CPT resonance linewidth versus saturation parameter. Insert gives the amplitudes of the clock transition for different Rabi frequencies.

system.

II Experiment

II.1 Phase-locked lasers

In order to realise CPT phenomenon, we generate two coherent laser beams with a frequency offset given by the cesium clock frequency at 9.192 GHz. Fig. 2 shows the experimental setup. Two extended cavity laser diodes (55 mm long) using Littrow configuration are spectrally narrowed below 100 kHz. Tuned to 894 nm for D1 transitions in cesium, they provide about 10 mW optical power. Phase-locked lasers are ideal for producing only two optical frequencies without other sidebands that may generate distortion and frequencies shifts on the CPT resonance. The frequency reference is the sum of a constant 9.292 GHz locked to a H-maser and the frequency of a digital synthesizer (DDS) 50 MHz. We control the small Raman detuning δ_R by changing another DDS frequency driving the reference chain.

The two beams orthogonally polarised are splitted by a polarising beam splitter cube to realise the reference beat note at 9.192 GHz for the phase-lock loop and the copropagating beams for CPT. The master laser is frequency shifted to the $6S_{1/2}, F = 4 \rightarrow 6P_{1/2}, F' = 3$ through a 80 MHz acousto-optic modulator (AOM). This lock scheme avoids the modulation of the master laser frequency which could be seen by the atoms.

II.2 Experimental setup and results

A longitudinal static magnetic field is maintained at about 20 μT in a glass cell (50 mm long, 25 mm diameter) containing the atomic cesium vapor. Two magnetic shields surround the coil in order to protect CPT transitions from any stray magnetic field fluctuations. The cesium cell is filled with N_2 buffer gas at 23 Torr which cancels the first order Doppler broadening (Lamb-Dicke regime) and increases atomic transit time in the light beams. By choosing orthogonal laser polarisations, we enhance the contrast of the seven CPT transitions and reduce the sensitivity to the residual bad polarisations. For coherent pulses generation, a second 80 MHz AOM computer driven simultaneously switches the laser beams. Optical fields sent through the gas cell are then detected.

First, we have done a continuous interrogation to detect the seven CPT transitions between Zeeman sublevels (Fig. 3) where the clock transition is defined by $|1\rangle \equiv |F = 3, m_f = 0\rangle, |2\rangle \equiv |F = 4, m_f = 0\rangle$. We have studied the linewidth $\triangle\nu_{1/2}$ and amplitude of this transition versus laser intensity sent through the cell. The linear dependance well fitted by equation (1) clearly shows the effect of the saturation (Fig. 4). By extrapolation at zero intensity, the residual linewidth due to the buffer gas is estimated at $\Delta\nu_{1/2} = 2\gamma_c \approx 80$ Hz$\pm$40 Hz.

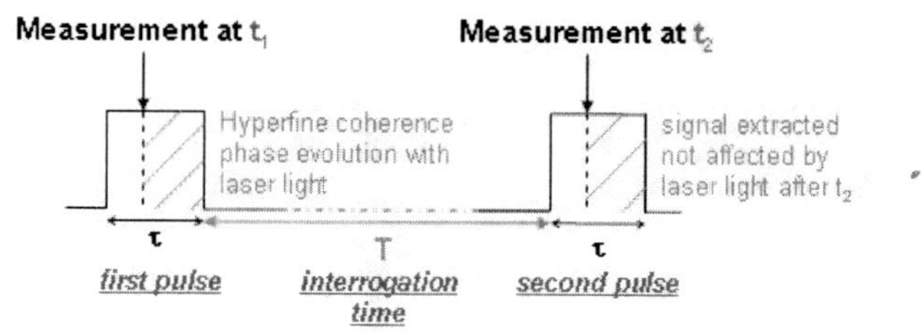

Fig. 5. Extraction of the clock signal by measurement of amplitudes $V(t_i)$ on each pulse.

We can also observe optical coherences related to the dynamical evolution of the hyperfine coherence (i.e. Raman coherence) in the fundamental hyperfine structure of Cs with optical pulses for different interrogation times starting from $T = 1$ ms to $T = 10$ ms. We build the atomic signal on the first or the second pulse by measuring the amplitude of the transmitted light through the cell for different Raman detunings δ_R. We suppress the background signal by subtracting the amplitudes $V(t_i)$ of the fringes at two different times t_1, t_2 on each pulse keeping only the modulation of the transmitted light (Fig. 5).

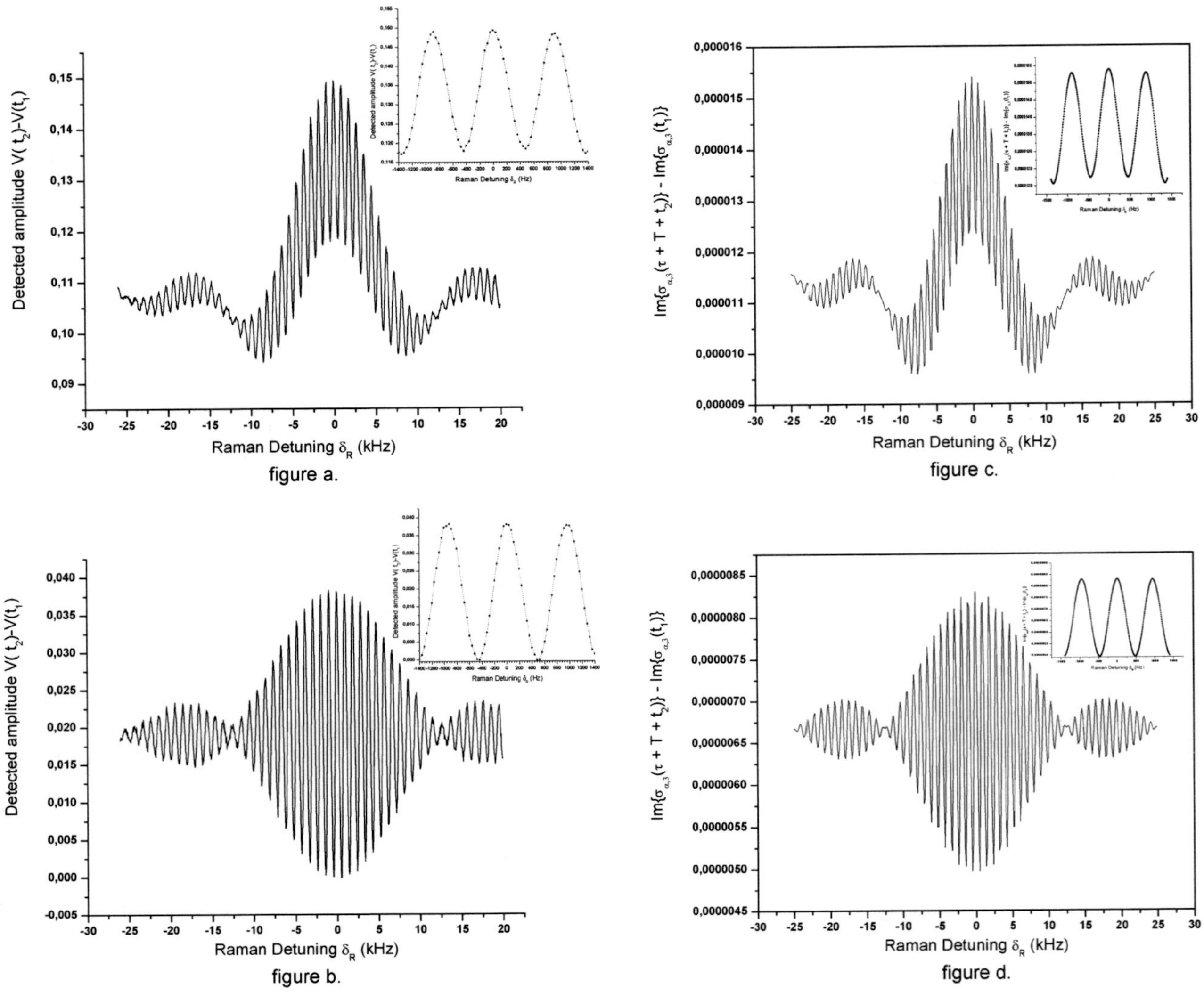

Fig. 6. Experimental (6.a and 6.b) and theoretical (6.c and 6.d) Raman-Ramsey fringes for $\tau = 80\ \mu$s and $\Omega_1 \approx \Omega_2 = 2\pi \times 200$ kHz. In the first horizontal line, clock signal is extracted at $t_1 = 12\ \mu$s and $t_2 = 80\ \mu$s. In the second line the signal is extracted at $t_1 = t_2 = 12\ \mu$s. Inserts give the central fringe (each curve is an average over 8 scans).

We generate optical CPT pulses with the common AOM (fig. 2). Experimentally a typical interrogation sequence is made using $\Omega_1 \sim \Omega_2 \approx 2\pi \times 200$ kHz with $\tau = 80\ \mu$s pulse width and $T = 1$ ms dead time. The result is given in figure 6.a and 6.b. We now have for the linewidth of the central fringe $\Delta\nu_{1/2} = \frac{1}{2T} = 500$ Hz from the separated oscillating fields Ramsey method. Such a method for signal construction allows us to follow the evolution of fringes with time. Oscillations are present during each laser pulse, it means that if we change the reading time t_1 for $V(t_1)$ on first pulse and t_2 for $V(t_2)$ on the second pulse, we also change the amplitude of oscillations and modify the lineshape of fringes.

We then apply a sequence with $\tau = 5$ ms, $T = 1$ ms, and $\Omega_1 \sim \Omega_2 \approx 2\pi \times 400$ kHz. We only change the pulse duration while keeping the same reading times t_1 and t_2 from the signals $V(t_i)$. As we can see, the linewidth of Raman-Ramsey fringes generated are no longer limited by the saturation effect which is about 2.5 kHz in the continuous case. Line shapes of fringes are modified from a round shape (Fig. 6.b) to a sharp one (Fig. 7.a). Small asymmetries on the fringes are now under investigation.

Two regimes are clearly put in evidence. For short pulses and weak intensities, the dynamical evolution is only driven by the fields which induce coupling. For long pulses and strong intensities, the effect of the stationary solution on the line shape of fringes has to be taken into account.

III Schrödinger representation and interpretation of fringes

To describe the physical situation acting on CPT transient, we use the simple Schrödinger point of view. Starting with the usual three level system under two coherent fields (fig. 1), we study the dynamical effect using a time-dependent wavefunc-

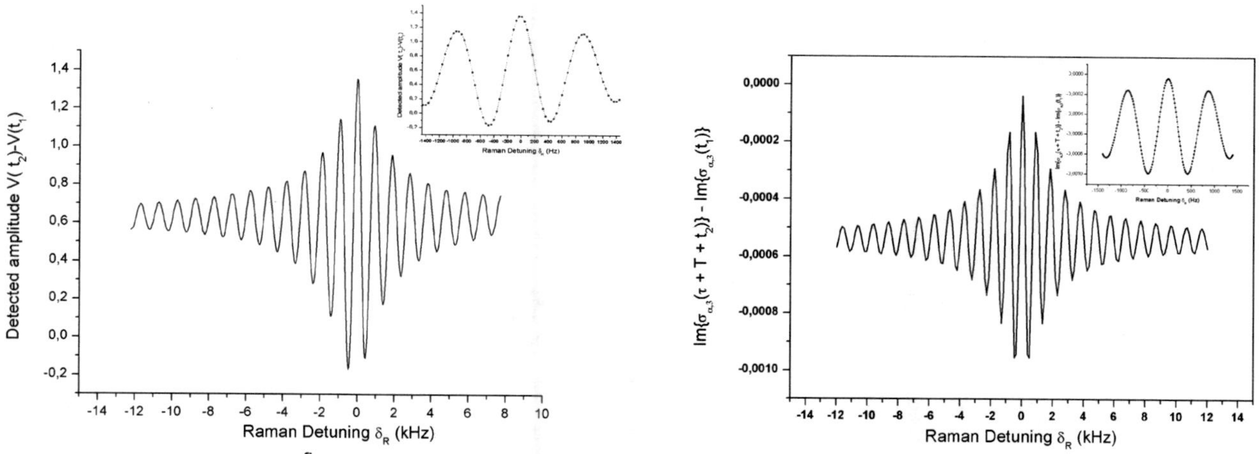

figure a. figure b.

Fig. 7. (a) Experimental and (b) theoretical Raman-Ramsey fringe patterns. Calculations are made for $\tau = 5$ ms, $t_1 = t_2 = 12$ μs, and $\Omega_1 \approx \Omega_2 = 2\pi \times 400$ kHz. Notice small asymmetries on experimental fringes, averaged over 8 scans.

tion :

$$\psi(t) = \sum_{n=1}^{3} C_n(t) \exp[-i\omega_n t]|n\rangle \tag{2}$$

The evolution of the wavefunction is given by :

$$i\hbar\frac{d}{dt}|\psi(t)\rangle = H|\psi(t)\rangle \tag{3}$$

We take into account the effect of the spontaneous emission by adding an imaginary term $i\frac{\Gamma}{2}$ (i.e. an opened system) and $i\gamma_c$ for Raman coherence, the hamiltonian H is now written in the Rotating Wave Approximation :

$$H = \begin{pmatrix} 0 & 0 & \Omega_1 \\ 0 & \delta_R - i\gamma_c & \Omega_2 \\ \Omega_1 & \Omega_2 & \Delta_0 - i\frac{\Gamma}{2} \end{pmatrix}$$

The amplitudes for transition probabilities are then [5], [6]:

$$\begin{aligned} i\dot{c}_1(t) &= \Omega_1 c_3(t) \\ i\dot{c}_2(t) &= \Omega_2 c_3(t) + (\delta_R - i\gamma_c)c_2(t) \\ i\dot{c}_3(t) &= \Omega_1 c_1(t) + \Omega_2 c_2(t) + (\Delta_0 - i\tfrac{\Gamma}{2})c_3(t) \end{aligned}$$

In our case $\Omega_{1,2} \ll \Gamma$ ($\approx 2\pi \times 500$ MHz), the excited level $c_3(t)$ reaches its asymptotic value which follows adiabatically the evolution of $c_1(t)$ and $c_2(t)$. It means that the amplitude for the excited level is clearly a sum of two channels which could interfere. We then find the state evolution under laser fields [5] taking $c_1(0) = 1$ and $c_2(0) = 0$:

$$c_1(t) = \exp -(\gamma_c + \Gamma s)\tfrac{t}{2}\ \exp i(-\delta_R + \Delta_0 s)\tfrac{t}{2} \times \left(\cos\frac{\omega t}{2} + \frac{i}{\omega}\left[(\Omega_1^2 - \Omega_2^2)\frac{\Delta_0 + i\Gamma/2}{\Delta_0^2 + \Gamma^2/4} + \delta_R - i\gamma_c\right]\sin\frac{\omega t}{2}\right)$$

$$c_2(t) = \exp -(\gamma_c + \Gamma s)\tfrac{t}{2}\ \exp i(-\delta_R + \Delta_0 s)\tfrac{t}{2} \times \frac{i}{\omega}\left[2\Omega_1\Omega_2\frac{\Delta_0 + i\Gamma/2}{\Delta_0^2 + \Gamma^2/4}\right]\sin\frac{\omega t}{2}$$

$$c_3(t) = -\frac{\Omega_1 c_1(t) + \Omega_2 c_2(t)}{\Delta_0 - i\Gamma/2}$$

with the saturation parameter :

$$s = \frac{\Omega_1^2/2 + \Omega_2^2/2}{\Delta_0^2 + \Gamma^2/4}$$

and the complex angular frequency :

$$\omega^2 = \left[(\Omega_1^2 - \Omega_2^2)\frac{\Delta_0 + i\Gamma/2}{\Delta_0^2 + \Gamma^2/4} + \delta_R - i\gamma_c\right]^2 + \left[2\Omega_1\Omega_2\frac{\Delta_0 + i\Gamma/2}{\Delta_0^2 + \Gamma^2/4}\right]^2$$

Experimentally, we detect transmissions of laser $\alpha = 1, 2$ which are related to the imaginary part of the optical coherence $\sigma_{\alpha,3}(t) = c_\alpha(t)c_3^*(t)$. Our theoretical clocksignal of the Raman-Ramsey fringes given by measuring the imaginary part of the optical coherence at t_1 on the first pulse and at t_2 on the second pulse is calculated as:

$$\begin{aligned} P_{detected} &= Im\{\sigma_{\alpha,3}(\tau + T + t_2)\} - Im\{\sigma_{\alpha,3}(t_1)\} \\ &= V(t_2)_{experimental} - V(t_1)_{experimental} \end{aligned}$$

The original detection method based on the optical coherence driven by laser fields allows us to follow this evolution by measuring at any time the amplitude of transition on each pulse without destroying the interferences.

The linewidth of the central fringe clearly scales as the inverse of the dead time T :

$$\Delta\nu_{1/2} = \frac{1}{2T} \tag{4}$$

We can compare experimental fringes for short pulses given in Fig. 6.a and 6.b to the theoretical calculation in Fig. 6.c and 6.d. The qualitative agreement between the four curves justifies the choice of this representation for short pulses and weak fields.

Schrödinger representation is valid if the pulse width τ is short compared to the time needed to reach the steady state solution. This time is given by t_{sat} as the inverse of the saturation :

$$t_{sat} \approx \frac{1}{\Gamma s} \tag{5}$$

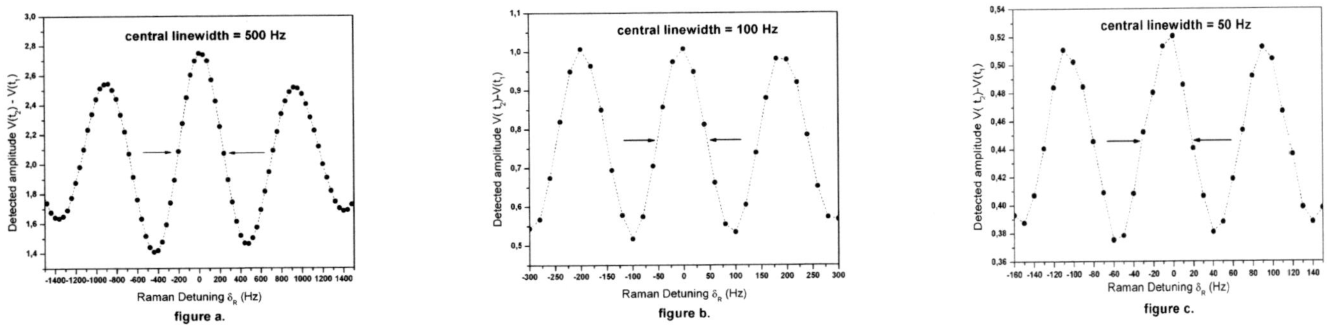

Fig. 8. Experimental Raman-Ramsey fringes for $\tau = 5$ ms and $\Omega_1 \approx \Omega_2 = 2\pi \times 400$ kHz with three different times $T = 1$, 5 and 10 ms from left to right. Notice the loss of fringe contrast by increasing the dead time T while keeping the same Rabi frequencies (all curves are averaged over 8 scans).

In any case, the Schrödinger representation and a simple Hamiltonian describing field-induced coupling will be the core for future calculations (i.e. general density matrix formalism).

From the expressions of $c_{1,2}(t)$ on Fig 6, the saturation still plays an important role because the relaxation of the hyperfine coherence between the two ground states at any time t is proportional to $\exp -(\gamma_c + \Gamma s)\frac{t}{2}$. Because of the relatively strong broadening of the excited state to about 500 MHz by the presence of a buffer gas, we can reduce the loss of signal when increasing the dead time T by sending much more power in the cell.

However the exact dynamical evolution of the Raman Ramsey fringes should take into account the relaxation terms which bring back the atoms down to the ground states. The density matrix formalism is then the natural choice for a more realistic three-level closed system which can be written as [1], [4] :

$$\frac{d}{dt}\sigma_{\alpha,\beta} = \frac{i}{\hbar}[\sigma_{\alpha,\beta}H^{\dagger} - H\sigma_{\alpha,\beta}] + \Re\sigma_{33} \tag{6}$$

and the matrix source describing relaxation from upper $|3\rangle$ level to lower $|1\rangle$ and $|2\rangle$ levels :

$$\Re = \begin{pmatrix} \Gamma_{31} & 0 & 0 \\ 0 & \Gamma_{32} & 0 \\ 0 & 0 & 0 \end{pmatrix}$$

We only give here the numerical results in Fig 7.b once time again in agreement with the experimental results given in Fig 7.a. The small difference between the two curves could be induced by a residual optical detuning not clearly understood at this time. Finally we succeeded in measuring central linewidths from 500 Hz to 50 Hz (Fig. 8 a, b, c) in the case of long pulses and high intensities corresponding to 1, 5 and 10 ms pulse separation time.

Conclusion

Linewidth of continuous CPT resonances is limited by saturation effect even for small laser intensities as low as 100 nW/cm^2. However such a low intensity induces a poor signal to noise ratio. We show that optically coherent pulses create narrow fringes with linewidths determined only by $\frac{1}{2T}$ where T is the interrogation time between the two pulses during free induction decay of the hyperfine coherence (Ramsey method). The results are in very good qualitative agreement with the calculations coming from a simple view of a Schrödinger representation. This model gives more physical insight of the relaxation terms which evolve in a realistic three level system (i.e. optical Bloch equations including all feeding terms). Finally, we also show an experimental record of promising 50 Hz width Ramsey fringe.

Acknowledgment

We address special thanks to Electronic group at BNM-SYRTE for help and advice on the experiments. T. Zanon is grateful to MENRT for providing research studentship.

References

[1] G. Orriols, *Nonabsorption resonances by nonlinear coherent effects in a three level system*, Il Nuovo Cimento, B 53 vol 1, p 1 (1979).

[2] J.E. Thomas, S. Ezekiel, C.C. Leiby, Jr.R.H. Picard, C.R. Willis, *Ultrahigh-resolution spectroscopy and frequency standards in the microwave ad far-infrared region using optical lasers*, Opt. Letters, vol 6, No 6, p 298 (1981).

[3] J.E. Thomas, P.R.Hemmer, S. Ezekiel, C.C. Leiby, Jr.R.H. Picard, C.R. Willis, *Observation of Ramsey fringes using a stimulated Raman transition in a sodium atomic beam*, Phys. Rev. Lett, vol 48, No 13, page 867 (1982).

[4] P.R. Hemmer, G.P. Ontai, S. Ezekiel, *Precision studies of stimulated-resonance Raman interactions in an atomic beam*, J. Opt. Soc. Am. B, vol 3, No 2, p 219 (1986).

[5] K. Moler, D.S. Weiss, M. Kasevich, S. Chu, *Theoretical analysis of velocity-selective Raman transitions*, Phys. Rev. A, vol 45, No 1, page 342 (1992).

[6] M.S. Shariar, P.R. Hemmer, D.P. Katz, A. Lee, M.G. Prentiss, *Dark-state based three-element vector model for the stimulated Raman interaction*, Phys. Rev. A, vol 55, No 3, p 2272 (1997).

[7] S. Brandt, A. Nagel, R. Wynands, D. Meschede, *Buffer gas induced linewidth reduction of coherent dark resonances to below 50 Hz*, Phys. Rev. A, vol 56, R1063-R1066 (1997).

[8] J. Vanier, A. Godone, F. Levi, *Coherent population trapping : Dark lines and coherent microwave emission*, Phys. Rev. A, vol 58, No 3, p 2345 (1998).

[9] T. Lindvall, M. Merimaa, I. Tittonen, E. Ikonen, *All-optical atomic clock based on dark states of ^{85}Rb*, J. Opt. Soc. Am. B, vol 20, No 2, p 273 (2003).

[10] T. Zanon, S. Guérandel, E. de Clercq, D. Holleville, N. Dimarcq, A. Clairon, Proc. Conf. Laser and Opt. Quant., J. Phys. IV France (2004).

EVALUATION OF NPL CSF1 FREQUENCY STANDARD

K. Szymaniec, W. Chalupczak, D. Henderson

National Physical Laboratory,
Queens Road, Teddington, TW11 0LW, UK

Keywords: atomic frequency standard, laser cooling, caesium fountain.

Abstract

In this contribution a new caesium fountain frequency standard at the National Physical laboratory (NPL-CsF1) is described. Procedures for evaluation of the systematic frequency shifts are presented. The NPL-CsF1 has a short-term stability of $1.4\times10^{-13}\ \tau^{-1/2}$, predominantly due to the local oscillator phase noise. The accuracy, of 1 part in 10^{15} is limited by the uncertainty of the frequency shift due to collisions between cold atoms.

1 Introduction

A new generation of primary frequency standards (PFS), based on atomic fountains, has become operational, following the development of techniques in laser cooling and manipulation of atoms [1-4]. In atomic fountains, laser cooled atoms are launched upward, rising about a metre, resulting in an increased interaction time, on the order of a second. Consequently, the line Q-factor of 10^{10}, higher than in conventional beam clocks, has led to substantial improvements in the performance of PFSs. A short-term stability of $10^{-13}\ \tau^{-1/2}$ and an accuracy of 1 in 10^{15} are typical for these devices, with scope for improvement of another order of magnitude. To date, four National Measurement Institutes have been contributing to the evaluations of the TAI time interval using a fountain PFS. The contribution of these fountain PFSs is leading to the international time scale, TAI being a more accurate representation of Terrestrial Time (whose step interval is the SI second). In addition, several further NMIs have been developing atomic fountains.

At the National Physical Laboratory, a new caesium fountain standard (NPL-CsF1) is now operational [5]. In the last year an evaluation of the stability and the systematic frequency shifts has been conducted on NPL-CsF1. At the same time, NPL-CsF1 has been linked to the NPL time scale (UTC(NPL)) and has been used for direct measurements of optical frequency standards that are under development.

NPL-CsF1 is described in Section 2, including an analysis of the detection noise and the short-term stability. The identified systematic effects are listed in Section 3, together with estimates of their uncertainties. Results of frequency measurements and comparisons are presented in Section 4, followed by conclusions in Section 5.

2 Description of the apparatus

In NPL-CsF1, atoms are collected and cooled in a magneto-optical trap (MOT). They are launched and cooled further as a moving molasses. During their ballistic flight the atoms twice pass through a microwave cavity (Ramsey cavity). The two microwave interactions, together with the intervening period (Ramsey time) constitute a Ramsey interaction. The atomic state after the Ramsey interaction is detected by means of laser induced fluorescence. The detection of the distribution of the atom population between the hyperfine components of the ground state is highly efficient, as each atom scatters 10^4 photons during the process.

The NPL-CsF1 apparatus can be divided into the following parts: the vacuum vessel with magnetically shielded flight tube, the laser system, the source of the interrogating microwave field and the electronics for data acquisition and control of the experiment.

2.1 Vacuum vessel

There are three functional sections of the vacuum system (see Fig.1): the trapping and cooling region in the centre; the flight tube above, containing the Ramsey cavity; and the detection chamber below.

The cooling chamber is constructed from an A316 stainless-steel 6-way DN63 cross, modified with an additional 4 arms in the horizontal plane, fitted with DN40 flanges. A total of 8 view-ports are mounted on the horizontal arms, providing access for cooling lasers, for MOT diagnostics and for the microwaves used for the selection of the $m_F=0$ population. Two additional view-ports provide access along the vertical direction. The (0,0,1) cooling geometry is used, with two horizontal pairs of counter-propagating beams and one vertical pair. The central cross is connected to a reservoir containing a 5g ampoule of caesium. A Cs vapour at a pressure of 10^{-6} Pa is maintained in the trapping and cooling region. The transition pieces connecting the flight tube and the detection chamber are lined with graphite tubes serving as getter pumps to minimize the Cs pressure outside the cooling chamber. A set of anti-Helmholtz coils is mounted between the vertical flanges of the 10-way cross, to produce a field gradient of 1.0 mT/cm for the MOT. An additional three pairs of coils are used to reduce the effect of the Earth's magnetic field to less than 0.2 μT.

The flight tube, above the cooling region, is made from aluminium and has a diameter of 20 cm. A doubly wound, 72 cm long solenoid is placed immediately outside the flight tube to produce a highly uniform magnetic field (C-field), required to define the quantisation axis and to separate the magnetic components of the Cs ground state hyperfine transition. A microwave cavity, made of OFHC copper, is mounted 18 cm above the lower end of the flight tube. In order to isolate it thermally from the vacuum vessel, the cavity is held on a ceramic plate, which is supported by four aluminium rods. The flight tube is screened from undesired magnetic fields; three layers of magnetic shielding surround the flight tube, each a mu-metal cylinder with 2 mm thick walls.

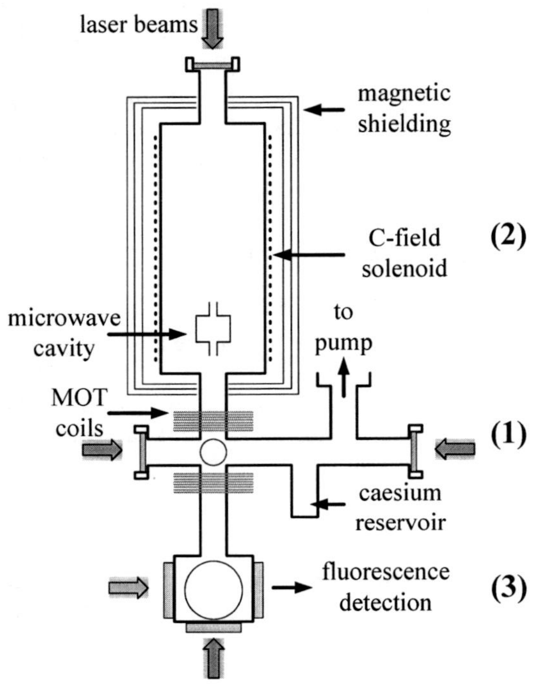

Fig 1: Schematic layout of the Cs-F1 vacuum vessel.

The detection chamber is a DN63 cube with five view-ports, providing access for detection laser beams and for fluorescence collection optics.

The vacuum vessel is pumped by two 50 l/s ion pumps maintaining the background pressure (outside the MOT region) below 10^{-7} Pa.

2.2 Lasers and optical system

Lasers operating at 852 nm (Cs D2 line) are used to capture, cool, manipulate and detect the caesium atoms. One extended cavity diode laser (ECDL) is frequency stabilised to the F=4→F'=5 transition using a saturated absorption signal from a room temperature Cs cell and the FM locking technique. A small part of its output is amplified in four slave diodes by injection locking, to provide cooling beams, while the remainder is used for detection. A second ECDL is stabilised, in the same way as the first, to the F=3→F'=4 transition. This laser provides repumping light for both cooling and detection. The detuning and intensity of the cooling light is controlled by acousto-optic modulators in a double pass configuration. In addition, mechanical shutters are used to eliminate residual scattering of resonant light during the Ramsey time. Polarisation-maintaining optical fibres transfer the cooling and repumping light to the fountain from a separate optical table. At the output of the fibres the cooling beams are collimated to an $e^{-1/2}$ diameter of 1.5 cm with a peak intensity of 5 mW/cm^2. One of the vertical beams is truncated to 1 cm diameter by the aperture in the Ramsey cavity.

Detection is carried out by collecting fluorescence from the cycling transition, F=4→F'=5. A polarising fibre transmits the detection beam directly from the ECDL laser to the detection chamber. There, it is collimated and split into two beams, each shaped into a rectangular cross-section of 0.8×2.0 cm. The beams are circularly polarised and retro-reflected. The lower part (approximately 1/3) of the upper beam is blocked at the retro-mirror so that the resulting travelling wave removes, by radiation pressure, the F=4 atoms fluorescing in the upper beam. The repumper light is combined with the lower detection beam; the F=3 atoms are pumped to the F=4 state and scatter photons on the cycling transition F=4→F'=5. Approximately 3% of the light scattered in the two detection zones is imaged onto two photodiodes, one dedicated to each detection channel.

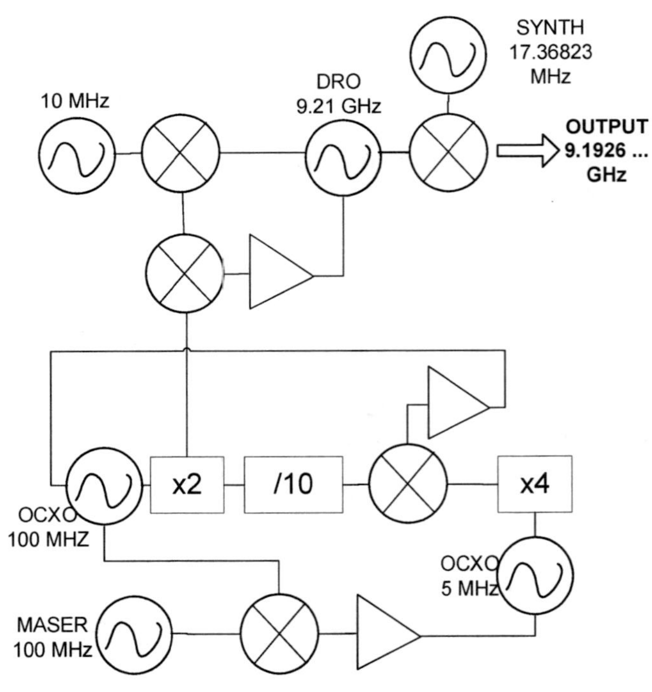

Fig. 2: Diagram of the 9.2 GHz synthesis.

2.3 Microwave interrogation

Figure 2 illustrates the local oscillator (LO) synthesis for Ramsey interaction. The microwave output is produced by mixing a dielectric resonant oscillator (DRO) with an HP3325B synthesizer. The DRO is phase-locked to a hydrogen maser. There are two intermediate oscillators, each phase-locked with individual time constants chosen to optimise the phase noise at all offsets from the 9.192 GHz carrier. The frequency of the maser is continuously measured against UTC(NPL).

A sampling-downconverter is used to phase-lock the DRO to the doubled output of a 100 MHz oven-controlled crystal (OCXO), in an offset loop, using a second HP3325B synthesizer. The 100 MHz OCXO is locked to the 5 MHz OCXO, using a digital divider. The 5 MHz OCXO is locked to the hydrogen maser by a phase measurement of the (locked) 100 MHz OCXO. Both HP3325B synthesizers were phase locked to the hydrogen maser via the 5 MHz OCXO.

By replacing the 10 MHz synthesizer with 7.368 MHz, and omitting the final mixer, the synthesis scheme is easily adapted to produce an offset lock where the DRO runs directly on the caesium resonance. This arrangement provides higher power and was used to explore the susceptibility of the fountain to stray "leakage" fields at the caesium resonance frequency.

2.4 Fountain operation and data acquisition

The NPL-CsF1 operates in a sequence, with the following steps: collection of atoms in the MOT; launching and cooling in optical molasses; state selection; Ramsey interaction; and detection. A typical cycle lasts 1.5 s.

Around 10^8 atoms is loaded into the MOT in 550 ms. Then, before launching, a 50 ms delay allows the eddy currents induced in the magnetic shields to decay. During the first 10 ms of the delay, the cooling lasers are off, to avoid any displacement of the atomic cloud by uncontrolled magneto-optical forces. As the atoms are still relatively hot (40 μK) the cloud expands, in the dark, to approximately 1 cm diameter. The reduction of the atomic number density improves the efficiency of the cooling in the molasses phase and lowers the collisional frequency shift during the Ramsey time. To launch the atoms, the descending and ascending vertical beams are detuned by –4.8 MHz and +4.8 MHz respectively. To optimise the launch efficiency, the horizontal beams are turned off for 0.3 ms. Once the horizontal beams are restored, the intensities of all the cooling beams are slowly reduced and their detuning increased from 2 Γ to 8 Γ. Within 1.4 ms a temperature of 2 μK is reached and then all the cooling light is extinguished.

At this stage the atomic population is approximately evenly distributed among all the magnetic sublevels of the F=4 state, with about 12% of the population in the m_F=0 state. These atoms are transferred, with 95% efficiency, to the (F=3, m_F=0) state by a 5 ms pulse of a travelling microwave. Atoms remaining in the F=4 are pushed by the radiation pressure from a 0.5 ms pulse of a vertical beam, tuned to the resonance. The selected (F=3, m_F=0) atoms continue their flight into the magnetically shielded flight tube. To avoid mixing of the magnetic sublevels near the shields' end-caps, where the magnetic field is highly inhomogeneous, additional loops of current are employed. The apogee of the ballistic flight is 31 cm above the Ramsey cavity centre, giving more than 0.5 s between interactions with the microwave field and hence Ramsey fringes narrower than 1 Hz.

After completion of the Ramsey interaction some 10^6 atoms reach the detection region. As described earlier, atomic populations of the F=4 and F=3 states are detected by the upper and lower detection beams, respectively. The normalised fraction of the F=4 atoms, $N_4/(N_4+N_3)$, is used as the measure of the probability of transition, F=3→F=4, during the Ramsey interaction.

The central Ramsey fringe is used to lock the frequency, ν, of the LO to the Cs resonance. Every fountain cycle the LO frequency is toggled between ν+Δν/2 and ν-Δν/2 (where Δν is the full width at half maximum of the fringe). The transition probability for F=3→F=4 is measured and the difference between the LO frequency and the Cs frequency is calculated every two cycles. This frequency correction is then fed back to the LO.

The feedback is provided by digital control of the synthesiser whose nominal frequency is 17.36823 MHz. The mean difference of that synthesiser from 17.36823 MHz is used to calculate the fractional frequency difference of the maser from the SI frequency, taking account of the known offsets in the caesium resonance. The values for the synthesis offset, during a measurement session, are recorded for later analysis.

2.5 Short-term stability

The frequency measurement data are subject to noise due to the detection and the microwave generation processes. Fig. 3 shows the Allan deviation of the NPL-CsF1 frequency with a hydrogen maser (Datum MHM 2010) used as a reference. For predominantly white noise processes, the Allan deviation $\sigma_y(\tau)$ can be estimated from the measured signal-to-noise ratio (*S/N*) according to:

$$\sigma_y(\tau) = \frac{2}{\pi}\frac{1}{Q}\left(\frac{S}{N}\right)^{-1}\sqrt{\frac{T_c}{\tau}} \tag{1}$$

where T_C is the duration of a single fountain cycle and τ is the averaging time. For the fountain operating at the top of the Ramsey fringe (where the LO noise contribution is negligible), the *S/N* exceeds 1000. The *S/N* is, however, reduced to 600 if measured on the slope of the fringe, due to the contribution from the phase noise of the LO. The resulting value of $\sigma_y(\tau) = 1.4\times10^{-13}\ \tau^{-½}$, is in very good agreement with that obtained experimentally.

The Allan deviation departs from the $\tau^{-1/2}$ dependence for averaging times longer than 10^4 seconds (fig. 3). This is due to hydrogen maser instabilities and drift. These instabilities can be a limitation for some comparative measurements of systematic effects, when long averaging times are required to achieve sufficient statistical resolution. In such cases the fountain is operated in a different sequence, where one (and only one) parameter is alternated between two values every two cycles (alternating mode). Two separate subsets of measured frequencies are collected and one subset is compared (referenced) to the other. Figure 4 shows the Allan deviation of such a measurement, where the fountain was operating, respectively, at high and low density. The poorer value of the short-term stability is expected as, for each data subset, only half as many data points is acquired in a given time compared with the non-alternating mode.

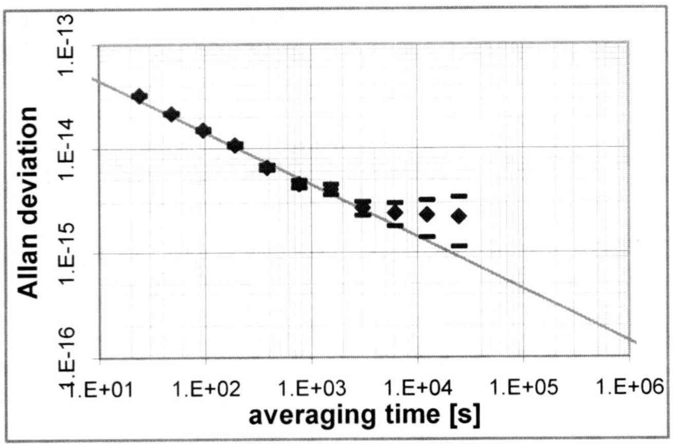

Fig. 3: Allan deviation of the frequency difference NPL-CsF1–HM3 between the fountain and maser HM3. For averaging times over 10^4 s the maser instabilities dominate.

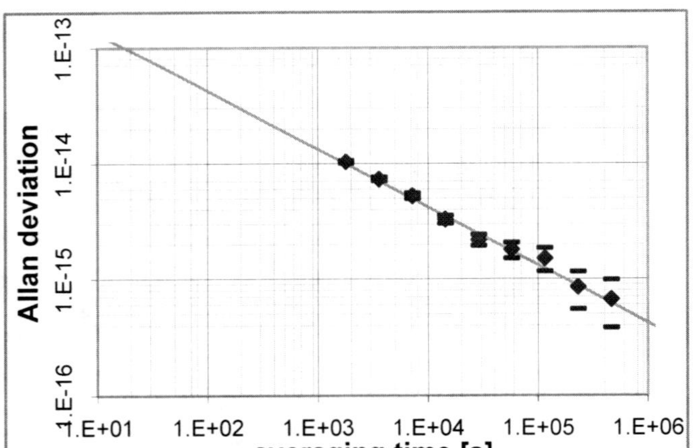

Fig. 4: Allan deviation of the frequency difference between two data subsets recorded during operation in the alternating mode. The short-term stability is poorer ($4\times10^{-13}\ \tau^{-1/2}$), but longer term stability is not limited by maser drift.

3 Systematic frequency shifts and their uncertainties

The frequency of the Cs ground state hyperfine transition is offset from that of the unperturbed transition by several systematic effects. These offsets are evaluated, together with their uncertainties. Some make a negligible contribution to the total offset; however, their uncertainty contributes to the total uncertainty. A list of systematic effects, identified for NPL-CsF1, is given below.

3.1 Second order Zeeman effect

The frequency of the "clock" transition ($F=3, m_F=0 \rightarrow F=4, m_F=0$), although free from the linear Zeeman shift, must be corrected for the quadratic shift resulting from the applied static magnetic field in the flight tube. This correction is calculated using measurements of the frequency shift of the field-sensitive transition ($F=3, m_F=1 \rightarrow F=4, m_F=1$), denoted by $f_{1-1}^{(1)}$. In general, frequencies of hyperfine transitions in the presence of an external static magnetic field are given by the Breit-Rabi formula [6]:

$$\nu_{(3,m_F \to 4,m_F)} = \nu_0 \left[1 + \frac{m_F}{4} x + \left(1 - \left(\frac{m_F}{4} \right)^2 \right) \frac{x^2}{2} + \ldots \right] \qquad (2)$$

where the parameter x is given by:

$$x = (g_j - g_i)\mu_B B / (h\nu_0) \qquad (3)$$

and B is the magnetic field induction. The relation between the linear frequency shift $f_{1-1}^{(1)}$ and the required quadratic shift of the clock transition can be derived from (2):

$$f_{0-0}^{(2)} = 8 \cdot \left(f_{1-1}^{(1)} \right)^2 \qquad (4)$$

The shifts are fractional frequency shifts, relative to ν_0, and the quadratic contribution to the shift of the (3,1→4,1) transition is neglected.

Using the Ramsey interaction, a time-averaged value of the linear Zeeman shift is obtained and $\left\langle f_{1-1}^{(1)} \right\rangle^2$ is used to calculate $f_{0-0}^{(2)}$ from the formula (4). However, a bias is introduced to $f_{0-0}^{(2)}$, if the atoms experience an inhomogeneous magnetic field. A measure of the inhomogeneity is given by the variance:

$$\left\langle \left(f_{1-1}^{(1)} \right)^2 \right\rangle - \left\langle f_{1-1}^{(1)} \right\rangle^2 = \sigma_f^2 \qquad (5)$$

and the bias is equal to $8 \times \sigma_f^2$.

It is possible to map the magnetic field by launching the atomic cloud to various heights and measuring the frequency $\left\langle f_{1-1}^{(1)} \right\rangle$ (see fig. 5). The values of $f_{1-1}^{(1)}$ for each point above the cavity can then be calculated, whereupon $\left\langle \left(f_{1-1}^{(1)} \right)^2 \right\rangle$ can be calculated from the atoms' trajectory. In NPL-CsF1 the bias introduced by inhomogeneity in the magnetic field is at the 10^{-19} level and is neglected.

For non-monokinetic atomic samples, the contrast of the Ramsey pattern is reduced; interference occurs due to groups of atoms with different velocities having different Ramsey interaction times. The interference causes the contrast to be reduced towards the tail of the Rabi envelope. For fringes measured on the (3,1→4,1) transition, this effect, together with a small variation of B over the ballistic trajectory, results in a displacement of the central fringe from the centre of the Rabi pedestal. Therefore, the field mapping procedure of launching the atomic cloud to various heights is used to identify the central Ramsey fringe.

The uncertainty of the frequency correction due to the second order Zeeman effect is dominated by the temporal instability of B. The temporal peak-to-peak variation of B is less than 30 pT over several weeks of fountain operation, leading to an uncertainty of less than 5×10^{-17}.

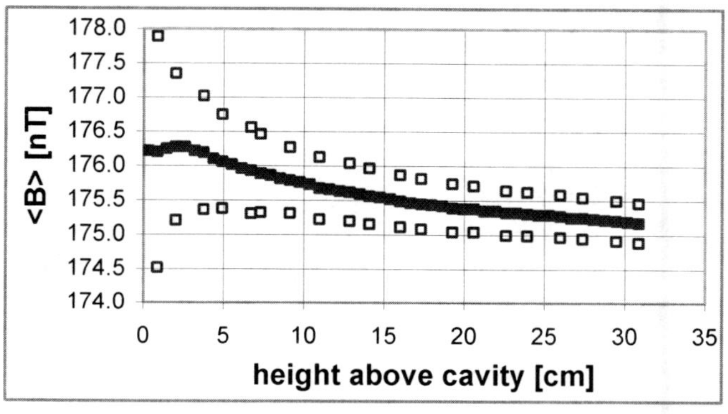

Fig. 5: The time integral of the magnetic field over the path, for atoms launched to different heights $<B>$ (full squares). The magnetic field values are calculated from the positions of the central Ramsey fringe for the (3,1→4,1) transition. To avoid misinterpretation of the fringes, the values of $<B>$ calculated for fringes adjacent to the central one are also shown (open squares).

3.2 Frequency shift due to collisions

Collisions between cold caesium atoms give rise to a shift of the frequency of the "clock" transition [7]. Theoretical calculations show that the shift displays a resonant dependence on the temperature of colliding atoms for samples cooled to about 100 nK [8]. In NPL-CsF1, however, where the temperature of the launched atoms is 1-2 μK, the approximation is made that the collisional frequency shift is proportional only to the atomic number density in the cloud. An extrapolation of the frequency, as directly measured, to the zero-density value is problematic, for two reasons. First, there is only a limited range over which the atomic density can be varied without a significant degradation of the S/N. (A density reduction by factor of 2 does not significantly change the stability, as the short-term of NPL-CsF1is dominated by the LO.) Second, the atomic density is not controlled directly; the detected fluorescence is controlled and hence the number of atoms N_{at}, reaching the detection zone (to the extent that the detection process is linear). A technique to link, unambiguously, changes in density with changes in N_{at} has recently been demonstrated [9]. In the current configuration of NPL-CsF1 it is not possible to implement that technique. However, it is possible to demonstrate the linearity between N_{at} and the frequency shift. In fig. 6 fractional frequency differences between NPL-CsF1 and one of our masers (HM3) are plotted for several values of N_{at}. The number of detected atoms was varied by changing only the power or detuning of the selecting pulse. Other parameters, such as the MOT loading time or the intensity of the cooling beams remained, unchanged. Given the statistical uncertainty of the data collected for each value of N_{at}, it is possible to calculate the uncertainty of the intercept for $N_{at}=0$. From this uncertainty it is estimated that, by applying a linear dependence between N_{at} and the collisional frequency shift, the extrapolation procedure may lead to a systematic bias not greater than 8×10^{-16}.

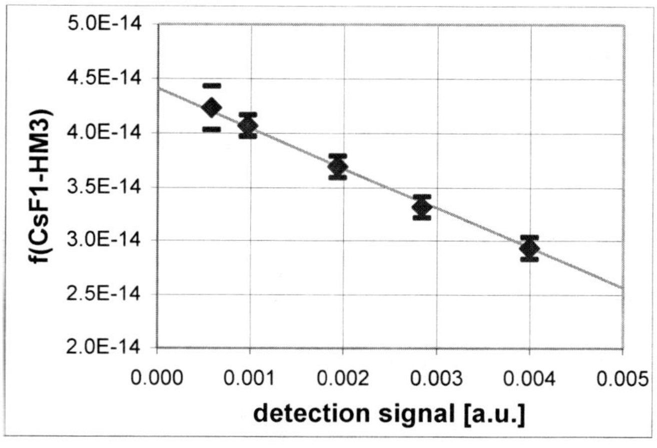

Fig. 6: The fractional frequency difference NPL-CsF1–HM3 as a function of the detected atom number N_{at}. The maser frequency drift over the measurement time was removed. The straight line is a least square fit to the experimental data.

It should be noted that no use is made of the linear fit from fig. 6 as a transfer function between the detection signal and the correction for the spin-exchange shift; the utility of the linear fit is solely to validate our extrapolation procedure. During a measurement campaign the fountain is operated continuously in the alternating mode, correcting for the collisional shift (see Section 4). The LO corrections together with N_{at} are recorded for the two sets of data corresponding to the high and low atom number densities. Typically, the fountain data is averaged in 30-minute blocks and the extrapolated frequency is calculated from the averaged ratio between signals, for the high and low density respectively (fig. 7). (The 30-minute averaging time is chosen in order that the noise from the H-maser reference is white frequency noise, before the flicker-floor is reached, and makes a negligible contribution.) The accuracy of the measurement is not affected by the long-term stability of either the atom number or the ratio. Nevertheless, the N_{at} detection signal is stable within 10% over a 30-day measurement campaign.

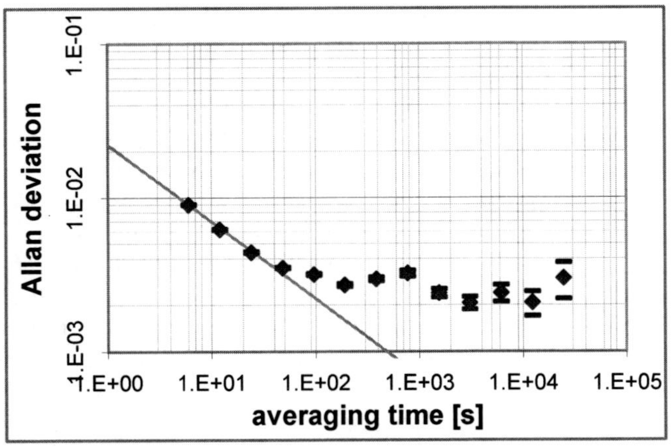

Fig. 7: The stability of the atom number ratio. For averaging times over 100 seconds it is better than 0.5%. The correction for collisional shift is calculated every 1800 seconds.

The NPL-CsF1 operates at a background pressure below 10^{-7} Pa. At this level the frequency shift caused by collisions with molecules of the background gas [10] is two orders of magnitude smaller than that due to the collisions with cold caesium atoms and being negligible, is omitted from the uncertainty budget.

3.3 Black-body radiation

Assuming that a 'black-body' radiator at temperature T surrounds atoms in the flight tube, the frequency of the standard is shifted [11] according to:

$$\Delta\nu = \gamma\left(\frac{T}{300}\right)^4\left[1+\eta\left(\frac{T}{300}\right)^2\right] \tag{4}$$

In order to evaluate the shift due to black-body radiation, the temperature of the vacuum vessel is monitored, and the published values of the coefficients γ and η [12] are used. NPL-CsF1 has four platinum resistors placed on the flight tube, two at the bottom and two at the top. The temperature measurement is traceable to the temperature scale at NPL. The peak-to-peak variation in the temperature of the vacuum vessel is less than 0.1 K. However, the lower part of the flight tube is warmer, by nearly 2 K, than the upper part due to the heat dissipated by the MOT coils. Taking this value as an estimate of the temperature uncertainty, the uncertainty of the shift due to black-body radiation is calculated to be 4×10^{-16}.

3.4 Microwave leakage

The frequency of NPL-CsF1 is shifted by the residual microwave field experienced by the atoms during the Ramsey time; the shift is linearly dependent on the microwave power. By changing the microwave field amplitude in the Ramsey cavity from $\pi/2$ to $11\pi/2$ (a power increase of a factor of 121), the extent of the shift is quantified (see fig.8). (The other shifts that depend on microwave power can be shown to be less than 10^{-16}. See Section 3.5.)

The difference in the shift for $\pi/2$ and $11\pi/2$ is found to be within ±30% of the average value over a 30 day period. This range is used to calculate the uncertainty due to microwave leakage. The value of the frequency shift, extrapolated to the $\pi/2$ level, is below 8×10^{-16} with an uncertainty of less than 3×10^{-16}.

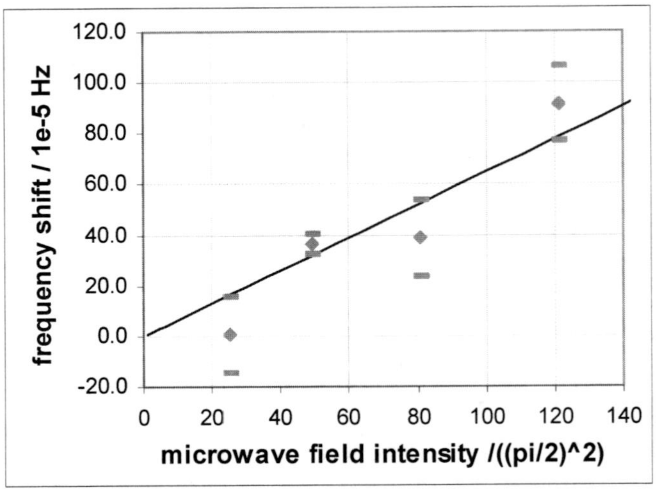

Fig. 8: Frequency difference NPL-CsF1-HM3 measured for microwave intensities corresponding to 5, 7, 9, and $11\pi/2$ pulses experienced by atoms in the Ramsey cavity.

3.5 Adjacent transitions, Rabi pulling, Ramsey pulling

The shift due to Rabi pulling is estimated following the analysis in [13]. After selecting the $m_F=0$ sublevel using the microwave selection pulse, the frequency applied to the Ramsey cavity is scanned over the transition frequencies of the other magnetic sub-levels. The populations of the $m_F\neq0$ sublevels are found to be less than 0.2% of the population of the $m_F=0$ sublevel. The Rabi pulling is calculated to be less than 10^{-20} and is omitted from the uncertainty budget.

Ramsey pulling is estimated using the approach developed in [14]. The worst case is considered; the $m_F=1$ and -1 atoms accumulate in the Ramsey interferometer asymmetric phases of π and $-\pi$, respectively. For the 1230 Hz separation of the $(0\rightarrow0)$ and $(1\rightarrow1)$ transition in NPL-CsF1 and the corresponding reduction of the $(1\rightarrow1)$ excitation to 4×10^{-4}, a negligible frequency shift of less than 10^{-16} is calculated.

3.6 Cavity pulling

The resonance frequency of the Ramsey cavity is offset from the atomic resonance by 800 kHz (±100 kHz). The offset is measured periodically. The cavity loaded Q is 12,500. The factional frequency shift is calculated using the theory presented in [15] to be 1×10^{-16}. The contribution to the cavity pulling due to the maser action of atoms in the cavity (radiation damping) is negligible for the number of atoms used [16].

3.7 Spectral impurities

The synthesis described in Section 2.3 contains spectral impurities (spurs) on the caesium resonance frequency; these are due to coupling of the phase locking loops to the ac supply (offset at harmonics of 50 Hz) and also to the 9.21 GHz DRO frequency and alternate sideband, from the output of the final mixer (offset at 17.368 and 34.736 MHz).

The spurs with offsets at 50 Hz and harmonics are more than 80 dB below the carrier and appear to be approximately the same on both sides of the carrier (double-sideband). The theory presented in [17] predicts a negligible shift in fractional frequency of less than 10^{-18} due to such a spur, in the worst case of it being a single-sideband spur.

The spur at 17.386 MHz offset has 20 dB higher power than the caesium resonance frequency. The theory [17] predicts a fractional frequency shift of several parts in 10^{15}. However, the relative amplitude of the field due to this spur is strongly reduced by the microwave cavity, resulting in a negligible shift in fractional frequency of a few parts in 10^{18}.

3.8 Distributed phase shift

The variation of the phase of the microwave field, in the horizontal plane, across the aperture in the Ramsey cavity results in a first order Doppler shift An estimate of the frequency shift is calculated by taking the worst-case values of the phase shift for a similar cavity [18] and accounting for the effect on the shift caused by the horizontal expansion of the atomic cloud. The vertical size of the atomic cloud is calculated from the effect produced by switching off the microwaves at a variable delay, relative to the launch. It is assumed that the horizontal and vertical size of the cloud is the same. The fractional frequency shift is estimated to be 3×10^{-16}.

3.9 AC Stark effect due to resonant light

During the Ramsey time, mechanical shutters, placed at the input of the optical fibres, block the laser beams used for cooling and detection. In order to estimate the AC Stark shift due to stray light coupled into the fibres, the fountain is operated alternately using two values of the laser detuning during the Ramsey time, 0.5 Γ and 10 Γ ($\Gamma = 2\pi\times5.3$ MHz). The measured frequency difference is consistent with zero within a statistical uncertainty of 2×10^{-15}. Thus, it is estimated that the AC Stark effect due to resonant light does not change the fractional frequency by more than 1×10^{-16}.

3.10 Gravity

The fountain frequency is shifted by the gravity potential. The gravity potential is calculated from the height relative to the surface of constant gravity potential, the geoid (orthometric height) [19].

The orthometric height is obtained by two methods; from survey data relative to the Ordinance Datum Newlyn (ODM) (whose position is known relative to the geoid) and from extended GPS position data using a C/A code receiver. The GPS receiver provides height relative to the WGS84 ellipsoid. The model of the EGM96 geopotential, published by the NGA/NASA [20], was used to obtain the height difference between WGS84 and the geoid. The GPS-based measurement of the orthometic height agrees with the ODM-based height, which has an uncertainty of approximately 1 meter in the laboratory where NPL-CsF1 is housed. The uncertainty in the shift due to the gravity potential is estimated to be 1×10^{-16}

Effect	Bias ($\times10^{-15}$)	Uncertainty ($\times10^{-15}$)
2nd order Zeeman	142.7	<0.1
AC Stark (BBR)	-17.1	0.4
Collisions	-7.7	0.8
μ-w leakage	0.8	0.3
Cavity pulling	-	0.1
μ-w spectrum	-	<0.1
Cavity phase shift	-	0.3
Rabi, Ramsey pull.	-	0.1
AC Stark (lasers)	-	0.1
Gravity	1.6	0.1
Total (1σ)		1.0

Table 1: A summary of systematic effects considered in this evaluation together with their fractional frequency uncertainties.

4 Frequency measurements

The frequency of NPL-CsF1 is referenced to a hydrogen maser HM3, which in turn is linked to the maser HM1. HM1 is the master clock representing UTC(NPL) and is reported to BIPM by a two-way satellite time and frequency link (TWSTFT). The two masers are located 0.8 km apart and a 10

MHz link is maintained by an underground cable. The round-trip phase delay is continuously monitored, and demonstrates that the uncertainty in the fractional frequency, in relating the frequency of NPL-CsF1 to HM1, is negligible over a 30-day period. It is necessary to operate NPL-CsF1 in 30-day campaigns in order to reduce the satellite link uncertainty to less than 10^{-15} for one evaluation period. Two such campaigns have been completed (MJD 53054-53084 and 53089-53119).

During a campaign the fountain is run in the alternating mode between high and low atom number. The correction for the collisional frequency shift is calculated using a continuous set of 30 minute averages. Additional noise due to the correction procedure and the link between HM1 and HM3 results in a short-term stability of the NPL-CsF1–HM1 frequency difference of 6×10^{-13} in 1s, which means that a statistical uncertainty of 1×10^{-15} is reached after 4 days of averaging. Although the short-term stability is degraded by the continuous correction of the collisional shift, there is no reliance on long-term stability of the atom number or on any previously measured coefficient between the atom number and the collisional frequency shift.

Other than the collisional shift, the corrections are sufficiently stable over the campaign period for one value to be applied. However, the validity of that value is, in some cases, checked during a campaign. (All the corrections are checked at the start and end of a campaign. A detailed map of the magnetic field is made at these times.) The temperature of the vacuum vessel is recorded at one minute intervals. The microwave leakage level and the value of $\langle B \rangle$ for a 31 cm launch is checked every two days.

During the five-day period MJD 52995-53000, the NPL-CsF1 fountain was running simultaneously with the primary standard of the PTB (PTB CSF1). The two standards were compared using TWSTFT with the link dominating the uncertainty of the comparison. The fountains were found to agree within the uncertainty in the comparison.

5 Conclusions

A report on the construction and operation of the NPL-CsF1 fountain primary frequency standard has been presented. The NPL-CsF1 now realises the SI second with an uncertainty in the fractional frequency of 1×10^{-15}. The standard is contributing to the representation of TT, TAI and the reference to the SI second for optical frequency measurements at NPL.

It is planned to use NPL-CsF1 to improve the stability of UTC(NPL).

Acknowledgements

The authors thank Y. Ovchinnikov for assistance and helpful discussions and P. Whibberley for careful reading of the manuscript. This work was funded by the Time Programme of the UK National Measurement System.

References

[1] A. Clairon, G. Santarelli, S. Ghezali, P. Laurent, S. Lea, M. Bahoura, E. Simon, S. Weyers, K. Szymaniec, *Proc. 5th Symp. on Frequency Standards and Metrology* (World Scientific 1996) pp 49-59.

[2] S. Weyers, U. Huebner, R. Schroeder, Ch. Tamm, A. Bauch. *Metrologia,* **38**, pp. 343-352, (2001).

[3] S.R. Jefferts, J. Shirley, T.E. Parker, T.P. Heavner, D.M. Meekhof, C. Nelson, F. Levi, G. Constanzo, A. De Marchi, R drullinger, L. Hollberg, W.D. Lee, F.L. Walls. *Metrologia,* **39**, pp. 321-336, (2002).

[4] F. Levi, L. Lorini, D. Calonico, A. Godone. *IEEE Trans. Instr. Meas.,* **52**, pp. 267-271, (2003).

[5] K. Szymaniec, W. Chalupczak, D. Henderson. in *Proc.17th EFTF-FCS,* Tampa, Florida (2003).

[6] N.F. Ramsey. *Molecular Beams,* Oxford, (1956).

[7] S. Ghezali, P. Laurent, S.N. Lea, A. Clairon. *Europhys. Lett.,* **36**, pp. 25-30, (1996).

[8] P.J. Leo, P.S. Julienne, F.H. Mies, C.J. Williams. *Phys. Rev. Lett.,* **86**, pp. 3743-3746, (2001).

[9] F. Pereira, Dos Santos, H. Marion, S. Bize, Y Sortais, A. Clairon. *Phys. Rev. Lett.,* **89**, 233004, (2002).

[10] M. Arditi, T.R. Carver. *Phys. Rev.* **124**, pp. 800-809, (1961).

[11] W.D. Itano, L. Lewis, D. Wineland. *Phys. Rev. A,* **45**, pp. 1233-1235, (1982).

[12] E. Simon, P. Laurent, A. Clairon. *Phys. Rev. A,* **57**, pp. 436-439, (1998).

[13] J. Vanier, C. Audoin. *The Quantum Physics of Atomic Frequency Standards,* Adam Hilger, 1989, p. 835.

[14] L.S. Cutler, C.A. Flory, R.P. Giffard, A. De Marchi. *J. Appl. Phys.* Pp. 2780-2792, (1991).

[15] J. Vanier, C. Audoin. *The Quantum Physics of Atomic Frequency Standards,* Adam Hilger, 1989, p. 830.

[16] C. Fertig, K. Gibble. *Phys. Rev. Lett.,* **85**, pp. 1622-1625, (2000).

[17] J. Vanier, C. Audoin. *The Quantum Physics of Atomic Frequency Standards,* Adam Hilger, 1989, p. 818.

[18] A. Khursheed, G. Vecchi, A. De Marchi. *IEEE Trans. Ultrason. Ferroel. Freq. Contr,* **43**, pp. 201-210, (1996).

[19] N.K. Pavlis, M.A. Weiss. *Metrologia* **40**, pp. 66-73 (2003).

[20] http://earth-info.nga.mil/GandG/wgsegm/egm96.html

MEASUREMENT OF THE BLACKBODY SHIFT IN THE IEN CS FOUNTAIN

F. Levi , D. Calonico, L. Lorini, S. Micalizio, A. Godone and H. Kalakhety*

Istituto Elettrotecnico Nazionale "Galileo Ferraris", Strada delle Cacce 91, 10135 Torino, Italy

Contact e-mail: levi@ien.it

Keywords: Cs Fountain, Blackbody Radiation Shift.

Abstract

This paper reports a review of the Blackbody Radiation (BBR) shift theory for the Cs atom hyperfine clock transition, together with a measurement of the Blackbody Radiation shift in an atomic cesium fountain.
In the first part a derivation of the BBR theory from the AC Stark shift theory is reported, together with a numerical evaluation of BBR shift parameters for the Cs clock transition.
The second part presents a measurements of the BBR shift for the Cs clock transition in the IEN Cs fountain (IEN-CsF1). The measurement is performed operating the fountain at different temperatures. Results are then discussed and compared with those obtained with different techniques.

1 Introduction

The accuracy level of cesium fountain frequency standards is presently quoted around $1\cdot10^{-15}$ [1, 2, 3]. Four main effects contributes to the fountain uncertainty: the density shift caused by cold Cs atomic collisions, the Zeeman shift induced by the quantization magnetic field, the AC Stark shift induced by the Blackbody Radiation (BBR) of the environment (the AC Zeeman shift is three order of magnitude lower, hence the clock frequency is not corrected for at the present accuracy level), and the red shift due to the Earth's gravitational potential.
The formula currently used to evaluate the BBR shift in the atomic frequency standards is [4]:

$$\frac{\Delta\nu}{\nu_0}=\beta\left(\frac{T}{T_0}\right)^4\left[1+\varepsilon\left(\frac{T}{T_0}\right)^2\right] \qquad (1)$$

We point out that the formula (1) can be directly derived by the AC Stark shift theory, considering the electric field associated to the BBR. However, the T^4 dependent term in (1) is due to the mean-squared value of the electric field (static approximation) and then the value of the parameter β can also be derived using DC Stark effect experimental results.
In the third paragraph of the paper we present an experimental evaluation of the BBR shift in IEN-CsF1 fountain. To obtain a direct measurement of the BBR shift we operate the IEN-CsF1 fountain at two different temperatures a direct measurement of the BBR shift is obtained. The β value is directly estimated measuring the fountain frequency difference versus a H maser used as a stable reference.

2 BBR shift theoretical evaluation

2.1 AC Stark shift theory

The formula (1) can be directly derived from the AC stark shift theory applied to the Cs clock transition.
The energy shift ΔW_α induced by a time dependent electric field on the state $|\alpha\rangle$ is [5]:

$$\Delta W_\alpha=\sum_k\frac{|d_{\alpha k}|^2}{h}\int_0^\infty\frac{\nu_\alpha-\nu_k}{(\nu_\alpha-\nu_k)^2-\nu^2}\mathcal{E}_\nu^2(\nu)d\nu \qquad (2)$$

where $d_{\alpha k}$ is the dipole moment of the atomic transition connecting the level α with each level k, ν_k and ν_α are the energy of the levels α and k in frequency units and $\mathcal{E}_\nu(\nu)$ is the electric field per frequency unit. The BBR shift can be calculated considering the AC Stark energy shifts ΔW_α and ΔW_β on the hyperfine levels ($|\alpha\rangle=|F=3,m_F=0\rangle,|\beta\rangle=|F=4,m_F=0\rangle$) involved in the Cs clock transition, caused by the Blackbody induced electric field $\mathcal{E}_\nu(\nu)$. In this case $\mathcal{E}_\nu^2(\nu)$ can be expressed in function of the spectral density of a Planck radiator $E_\nu(\nu,T)$.

$$\mathcal{E}_\nu^2(\nu)=\frac{1}{\varepsilon_0}E_\nu(\nu,T)=\frac{8\pi h\nu^3}{\varepsilon_0 c^3}\frac{1}{e^{h\nu/k_BT}-1} \qquad (3)$$

At T = 300 K the maximum of $E_\nu(\nu,T)$ holds for ν = 17.6 THz (Wien law); in the Cs case the minimum value for $\nu_\alpha-\nu_k=-\nu_{\alpha k}$ holds for the D_1 transition (355 THz). With

* Mr. Kalakhety is granted by ICTP TRIL Programme.

these conditions the integrand of the expression (2) can be series expanded and the integral can be solved as a series of $(T/T_0)^n$.

The first term in the expansion leads to a T^4 dependence and to the β expression [4]:

$$\beta = -\frac{4\sigma T_0^4}{\nu_0 h^2 \varepsilon_0 c}\left\{\sum_k \frac{|d_{\beta k}|^2}{\nu_{\alpha k}} - \sum_k \frac{|d_{\alpha k}|^2}{\nu_{\beta k}}\right\} \quad (4)$$

The second order term of the expansion leads to a T^6 dependence and then to the ε expression [4].

$$\varepsilon = \frac{40\pi^2}{21}\left(\frac{kT_0}{h}\right)^2 \left\{\frac{\sum_k \frac{|d_{\beta k}|^2}{\nu_{\beta k}^3} - \sum_k \frac{|d_{\alpha k}|^2}{\nu_{\alpha k}^3}}{\sum_k \frac{|d_{\beta k}|^2}{\nu_{\beta k}} - \sum_k \frac{|d_{\alpha k}|^2}{\nu_{\alpha k}}}\right\} \quad (5)$$

The value of β and ε for the Cs clock transition can be numerically evaluated using the dipole moment matrix elements ($d_{\alpha k}$) and frequency transitions values ($\nu_{\alpha k}$).

2.2 Connection with DC Stark shift theory

The β coefficient can be also related to DC Stark shift measurements. In fact the BBR shift can be approximated by the DC Stark shift induced by the mean-squared value of time dependent electric field $E(t)$ related to BBR:

$$\Delta\nu = kE^2 \approx k\langle E^2(t)\rangle \quad (6)$$

This "static" approximation corresponds to the T^4 approximation for the AC Stark shift theory. In fact, the mean-squared value of $E(t)$ can be expressed by:

$$\langle E^2(t)\rangle = \frac{4\sigma T_0^4}{\varepsilon_0 c}\left(\frac{T}{T_0}\right)^4 \quad (7)$$

where, at 300 K, is [4]:

$$\langle E^2(t)\rangle \approx (832.2\ \text{V/m})^2 \quad (8)$$

Comparing the equations (1) and (7) the following relation for β is obtained:

$$\beta = k\frac{4\sigma T_0^4}{\nu_0 \varepsilon_0 c} \quad (9)$$

In virtue of (9), a DC Stark shift experiment providing the value of k for the Cs hyperfine clock transition $|F=3, m_F=0\rangle \Rightarrow |F=4, m_F=0\rangle$ can be used also to estimate β.

In fact the k value can be expressed in function of the difference of the static polarizability (α_0) of the two hyperfine levels involved in the Cs clock transition

$$k = -\frac{1}{2h}\{\alpha_0(|\beta\rangle) - \alpha_0(|\alpha\rangle)\} = -\frac{8\alpha_{10}}{7h} \quad (10)$$

where α_{10} is the scalar differential polarizability introduced in [6].

2.3 Hyperfine mixing on Cs ground state and excited levels

As it was pointed out by Feitchner [8], the hyperfine interaction mixes the states $|nL_JF\rangle$ with states with higher n and same J and F. This means that n cannot be considered a good quantum number any more. The contribution of the hyperfine mixed states to the absolute polarizability α_0 is negligible but on the contrary is relevant to the calculation of the differential polarizability α_{10} and then to the evaluation of BBR shift parameters β and ε.

The mixed states can be expressed with the following notation: [8]:

$$\begin{aligned} |6\tilde{S}_{1/2}, F\rangle &= |6S_{1/2}, F\rangle + \sum_{n=7}^{\infty} a_{n1/2F}|nS_{1/2}, F\rangle \\ |6\tilde{P}_{J'}, F\rangle &= |6S_{J'}, F\rangle + \sum_{n=7}^{\infty} b_{nJ'F}|nS_{1/2}, F\rangle \\ |7\tilde{P}_{J}, F\rangle &= |7S_{J'}, F\rangle + \sum_{\substack{n=6\\ n\neq 7}}^{\infty} c_{nJ'F}|nS_{1/2}, F\rangle \end{aligned} \quad (11)$$

The $a_{n1/2'F}$, $b_{nJ'F}$, $c_{nJ'F}$ coefficients depends on the frequency transitions connecting, respectively, the S and P states with different quantum number n and on the s-wavefunction calculated in the coordinates origin $\psi_{nS}(0)$.

The Cs model used for BBR shift calculation took into account the mixed states reported in (11).

2.4 Calculation of β and ε parameters

Following the expressions (4) and (5), for the β and ε calculation we limited the contribution in the summation to the BBR AC stark induced shift due to the excited states ($|6\tilde{P}_{1/2}\rangle, |6\tilde{P}_{3/2}\rangle, |7\tilde{P}_{1/2}\rangle, |7\tilde{P}_{3/2}\rangle, |8P_{1/2}\rangle, |8P_{3/2}\rangle$). In the case of the levels $|6\tilde{P}_{1/2}\rangle, |6\tilde{P}_{3/2}\rangle$ (corresponding to Cs D_1 and D_2 transitions) was calculated the different contribution of each hyperfine sublevel.

The isotropic nature of BBR was explicitly taken into account by averaging over the whole solid angle.

To numerically evaluate β, we used up to 20 dipole matrix element values, expanded in term of $3j$ and $6j$ symbols and reduced matrix elements, and 18 transition frequency values, coming from the most recent theoretical and experimental works [8, 9, 10, 11].

As a result of this calculation, the evaluated value of β is [12]:

$$\beta = (1.49 \pm 0.07)\cdot 10^{-14} \quad (12)$$

The uncertainty was evaluated propagating the uncertainty of the input values; the main contribution to the uncertainty is due to the contributions of hyperfine mixing coefficients.
From the expression (5) the ε value has been calculated as [12]:

$$\varepsilon = 1.4\cdot 10^{-2} \quad (13)$$

2.5 Discussion of the results

The theoretical approach leading to the values reported in (12) and (13) is basically the same developed by Feitchner [8] and adopted by Itano [4] to provide the well known values, $\beta = 1.69\cdot 10^{-14}$ and $\varepsilon = 1.4\cdot 10^{-2}$.
The discrepancy in the β value estimation arises mainly by the dipole matrix elements used as input values for the theory. In fact the data available 30 years ago, and used in [8], are considerable over-estimated with respect to most recent values used for the present calculation. As the older input values reported in [8] come without uncertainty a more comprehensive comparison is not possible.
Using the same input values used for the calculation of the result reported in (12) and (13), is possible to calculate the cesium ground state scalar polarizability, provided that the core polarizability (α_0^c) contribution is taken into account [13]:

$$\alpha_0(|\alpha\rangle) = \frac{2}{h}\sum_k \frac{|d_{\alpha k}|^2}{\nu_{\alpha k}} + \alpha_0^c \quad (14)$$

Also in this case the result here reported is in agreement with the most recent theoretical and experimental works (see Table 1).

α_0 (10^{-39} Cm/V)	Ref.
6.587 ± 0.031	[14]
6.596 ± 0.013	[13]
6.600 ± 0.016	This work
6.611 ± 0.009	[9]

Table 1. Comparison between different α_0 values reported in literature and the result of our work

3 BBR shift measurement

In this section we report a BBR shift measurement using the IEN Cs fountain primary standard. A detailed description of the standard is reported in [15].
This experiment relies on differential frequency measurements: the fountain was operated alternatively at two different temperatures T_1 and T_2 and the β value was calculated from the difference of the two measured frequencies with respect to the H maser, taken as reference.
The key points of the experiment are a careful estimation of the H maser stability and frequency drift and an accurate evaluation of the fountain biases when operating in a differential mode.
The estimation of the frequency change due to the different BBR shift when the fountain is operated at two different temperature is provided by a least square model which also estimates and rejects the effect of the maser drift. The β value is then calculated from these measurements and from the effective BBR temperature of the fountain, taking into account the T^6 contribution (using the ε value (13) from the theory).
For the uncertainty evaluation, particular care was then applied to the evaluation of possible temperature dependent shifts, which can result in systematic biases.

3.1 Fountain cavity tuning at different temperatures

The IEN-CsF1 Ramsey cavity is thermally tuned to the Cs resonance frequency; the high thermal coupling between the fountain structure sections do not allow the change of the temperature in the atoms interaction zone without changing the cavity tuning. Hence during the BBR shift tests the cavity is operated out of resonance.
The Ramsey cavity has a tuning sensitivity of 150 kHz/°C and is tuned to the Cs frequency at T = 343 K; a 15 K change of the cavity operation temperature requires a 20 dB increase of microwave power to couple the optimum power level for Ramsey spectroscopy ($\pi/2$ pulse). The operation at a high microwave level could arise in several related frequency shifts (e.g. due to microwave leakage, spectrum spurious, excitation of unwanted modes, etc.). Differential measurements, performed operating the fountain (with the cavity tuned at T = 343 K) alternatively with a $\pi/2$ pulse and a $5\pi/2$ pulse showed that no shift at the 10^{-15} level was introduced by the high microwave power level.
However, a further cancellation of possible residual power related shifts was achieved operating the fountain cavity at two different temperature T_1 and T_2, equally spaced from the temperature which tunes the cavity to the Cs resonance frequency. T_1 and T_2 values during the experiments were respectively about 328 K and 358 K.

3.2 BBR temperature measurements

The fountain inner structure is a very good blackbody radiator. It can be considered as an enclosure with two small

holes at the ends; the top end glass window is extremely dark at the blackbody spectrum peak wavelength.
The fountain structure temperature is evaluated using the measurements of four type T thermocouples; a servo system allows to set and to keep constant the structure temperature at the 0.05 K level.

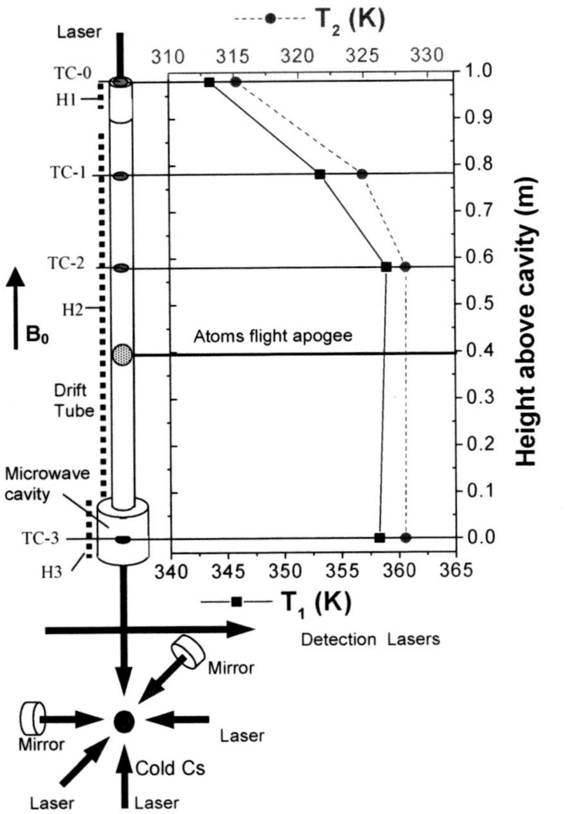

Figure 1. Schematic structure of the fountain, showing the thermocouples measurements points and the typical temperature distribution when operated at T_1 and T_2.

Calculation of the effective blackbody radiation temperature is performed interpolating the temperature profile along the structure using the thermocouples measurements (see Figure 1): the average radiation temperature at each structure point is calculated integrating the temperature profile, obtained interpolating the measured values, over the solid angle (this accounts for the structure holes contribution). The averaged temperature values are then used to calculate the time averaged temperature seen by the atoms during their ballistic flight.
The effect of the limited emissivity (values within 0.5 and 0.8 are reported for the oxidized copper) of the inner surface of the fountain structure is difficult to evaluate, as the level of oxidization and roughness of the surface is not well known.
The limited emissivity can modify the effective contribution to the BBR of the upper part of the fountain structure, which has not a homogenous temperature. The multiple scattering of photons by the inner surface can result both in a larger contribution of the colder part and in a more homogenous BBR temperature. To account for that the BBR temperature was then evaluated with two different models which can be considered as limiting cases. In the first model a modified temperature profile (figure 1) was used setting the upper window position close to TC-1 height; in the second model a homogeneous temperature profile is used with a temperature value taken as the average of the TC-2 and TC-3 (figure 1) readings.
The value of β was calculated using the profile reported in figure 1 and the results coming from the limiting cases were used for the uncertainty evaluation.

3.3 Frequency measurements

The H maser used as a reference for the β measurements is stable at medium and long term. After a drift removal, the fountain versus H maser frequency stability shows an Allan deviation as low as $3\cdot10^{-16}$ at 1 week, probably limited by the fountain stability itself. Taking advantage from this performance, a fountain run can be as long as few weeks.
Each fountain run is composed by at least four subset of frequency measurements, alternated at T_1 and T_2; each subset is at least five days long. The data analysis process allows to evaluate and reject the maser drift: the data are processed with a 3 parameters least squares model.
The model uses the following equation sets

$$\begin{cases} At_i + B = y_i \\ At_j + B' = y_j \end{cases} \tag{15}$$

where y_i and y_j are the frequency value of the measurements respectively at T_1 and T_2, t is the epoch of the measurement datum, A estimates the frequency drift of the maser and B-B' the difference of the fountain frequency when operated at T_1 and T_2. The least square model provide also an estimation of the uncertainty related to the fit parameter A, B and B'.

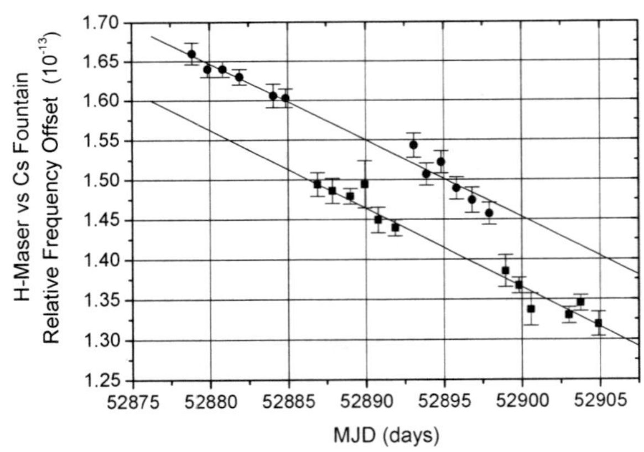

Figure 2. Relative frequency measurements of fountain vs. H maser during a BBR shift measurement run. Each point, plotted versus the measurement epoch, averages 1 day of measurement . Data represented with circles are taken at

T_1 = 328 K, data represented with squares are taken at T_2 = 358 K.

This maser drift evaluation was also compared with the evaluation versus the UTC(NIST) and UTC(NPL) remote time scales obtained with the Two Way Satellite Time and Frequency Transfer (TWSTFT). These evaluations are in agreement with each other at the 10% level.

3.4 Temperature related shifts

The fountain frequency shifts that are temperature dependent had been carefully analyzed, as they can bias the β measurement.
Despite of the careful thermal insulation of the fountain structure, operation at two different temperatures can result in different C-field values, due mainly to the thermal dilation of the C-Field solenoid. We mapped (see [15] for the map technique details) the C-field at the two temperature conditions, estimating a difference in the quadratic Zeeman frequency shift less than 10^{-16}.
Changing the fountain operating temperature could affect the copper surface outgasing (which is mainly composed by molecular hydrogen); the difference of the frequency shift due to the background gas collision in the two temperature conditions was calculated as low as $1\cdot10^{-16}$, using collisional theories and the spin exchange cross sections reported in the literature [16].
The fountain structure is heated with an electric resistor fed with audio-frequency power. The magnetic field produced by the heater is highly shielded by the skin effect induced in the copper tube thickness; however the quadratic Zeeman shift caused by the residual magnetic field penetrating inside the tube depends on the audio frequency power applied and then is different at T_1 and T_2 temperatures. This residual effect was evaluated less then $3\cdot10^{-16}$, as results by differential measurements performed with the heater pulsed at 50% duty cycle on or off during the Ramsey time.
High microwave power feeding of the cavity can excite modes other than the TE_{011}, leading to a frequency pulling shift. A numerical analysis was carried on to calculate the gain and resonance transition of unwanted modes. Results shows that the main pulling contribution is due to the cavity mode TM_{111}, lying about 100 MHz away from the Cs frequency resonance. The pulling effect introduced by the TM_{111} mode is driven mainly through the excitation of $\Delta F = 1$, $\Delta m_F = \pm 1$ transitions.
The shift induced by this pulling is evaluated to be less than $1\cdot10^{-16}$.

3.5 Other shifts and Type B uncertainty budget

Atomic density instability during fountain runs can lead to additional uncertainty in differential frequency evaluation. The uncertainty contribution to β evaluation of atomic density instability was evaluated as $4\cdot10^{-16}$.
Table 2 summarizes the Type B uncertainty contributions on the β values.

Effect	Uncertainty on β
Radiation temperature	$4\cdot10^{-16}$
Atomic density fluctuations	$4\cdot10^{-16}$
AC and DC Zeeman effect	$4\cdot10^{-16}$
Microwave leakage	$<1\cdot10^{-16}$
Cavity pulling	$<1\cdot10^{-16}$
Background gas	$\leq1\cdot10^{-16}$
Total	$7\cdot10^{-16}$

Table 2. Type B uncertainty contribution to the β estimation

The reported uncertainty due to the radiation temperature includes also the additional contribution of the emissivity effect analysis.

3.6 Experimental Results

The β final result comes from the weighted average of the values in three different runs. The total uncertainty is evaluated as the quadratic average of the Type A (coming from the least square fit) and Type B ones.

	$\beta(\cdot10^{-14})$	$u_A\ (\cdot10^{-14})$	$u_B\ (\cdot10^{-14})$
Run 1	-1.50	0.19	0.07
Run 2	-1.47	0.17	
Run 3	-1.40	0.10	
Weighted Average	-1.43	0.08	0.07

Table 3. Values of β and related uncertainty obtained in three different fountain runs

The final value is:

$$\beta = (-1.43 \pm 0.11)\cdot10^{-14} \qquad (16)$$

The fountain frequency was also measured at various temperatures to look for possible systematic effects, the results are reported in Figure 3 together with a T^4 function fit. Since in this experiment the differential technique on the microwave power can not be applied, the measurement accuracy is reduced. The fit however leads to a value of $\beta = (-1.5 \pm 0.2)\cdot10^{-14}$, in agreement with the more accurate differential measurement reported.

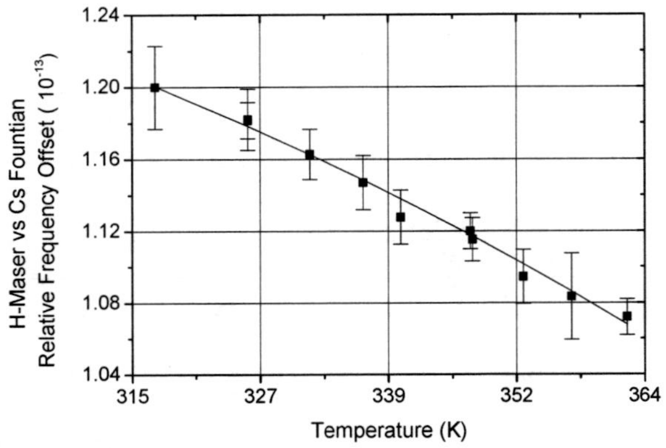

Figure 3. Measurements of the fountain frequency at various temperatures (H-Maser drift removed). Solid line: data fit using a T^4 function

3.7 Results discussion

The experimental value obtained with the direct BBR shift measurements on the IEN Cs fountain (16) agrees with the theoretical estimation reported in (12) and with a previous direct measurement [17]. However its discrepancy with high accuracy evaluations of β coming from DC Stark shift experiments [7, 18, 19] is not negligible (see Figure 4).
At the moment there is no theoretical evidence to believe in a breakdown of the equivalence of AC and DC Stark approach to the BBR shift β parameter evaluation.
However, in the DC Stark effect measurements [7, 18, 19], which are in agreement among them, a pure quadratic dependence of the shift to the electric field was assumed up to 10^6 V/m values. In these case the hyperpolarizability term can be no more negligible, as it was suggested by recent calculations [20].

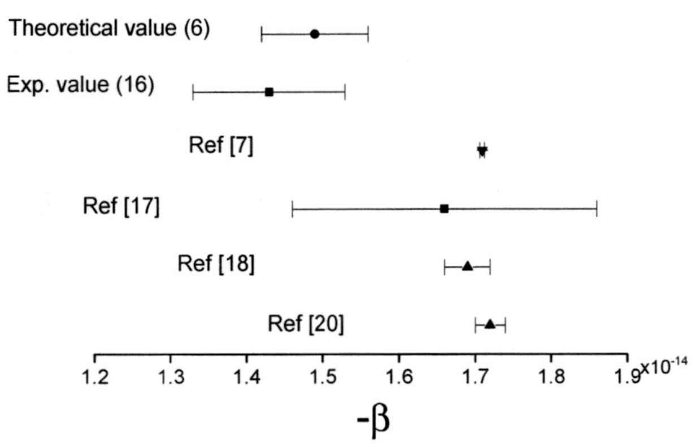

Figure 4. Values of β obtained with the theory re-evaluation and the experimental measurement reported in this work, compared with values reported in literature

4 Conclusions

The BBR shift parameter β was measured with a $1.1\cdot10^{-15}$ accuracy for the first time operating a Cs atomic fountain at different temperatures.
The experimental value agrees with a theoretical results coming from a re-evalution of assessed theories using the most recent experimental values as input values.
However there is a $3\cdot10^{-15}$ discrepancy between this value with respect to highly accurate values coming from DC Stark shift measurements. This discrepancy has an impact on Cs fountain primary frequency standards at the $2\text{-}3\cdot10^{-15}$ level, depending on the operating temperatures of the fountains themselves.
Further theoretical and experimental work is then required to push the accuracy level of the Cs fountain primary frequency standards down to the 10^{-16} area. In particular theoretical studies on hyperpolarizability are of fundamental importance as well as direct measurements of the BBR shift on Cs atomic fountains using larger temperature.

Acknowledgements

The authors wish to thank P. Tavella for useful suggestions about the least square model used for data analysis.

References

[1] A. Clairon, S. Ghezali, G. Santarelli, P. Laurent, S. Lea, M. Bahoura, E. Simon, S. Weyers, K. Szymaniec, Proc. of 5th Symp. on Freq. Stand. and Metr., Woods Hole MA, (1995). pp. 49-59

[2] S. R Jefferts, D. M. Meekhof, J. Shirley, T. E. Parker, C. Nelson, F. Levi, G. Costanzo, A. De Marchi, R. Drullinger, L. Hollberg, W. D. Lee, F. L. Walls, Metrologia **39**, 321 (2002).

[3] S. Weyers, U. Huebner, R. Schroeder, C. Tamm, A. Bauch, Metrologia **38**, 343 (2001)

[4] W. M. Itano, L. L. Lewis and D. J. Wineland, Phys. Rev. A **25**, 1233 (1982).

[5] C. H. Townes, A. L. Schwalow, *Microwave spectroscopy*, Dover publication (1975)

[6] P.G.H. Sandards, Proc. Phys. Soc **92**, 857 (1967); J.R.P. Angel, P.G.H. Sandards Proc. Roy. Soc A **305**, 125 (1968)

[7] E. Simon, P. Laurent and A. Clairon, Phys. Rev. A **57**, 436 (1998).

[8] J. D. Feichtner, M. E. Hoover and M. Mizushima, Phys. Rev. **137**, A702 (1965)

[9] J. M Amini, H. Gould, Phys. Rev. Lett. **91** 153001 (2003)

[10] M. S. Safronova, W. R. Johnson, A. Derevianko, Phys. Rev. A **60**, 4476 (1999).

[11] S. A. Blundell, W. R. Johnson, J. Sapirstein, Phys. Rev. A **43**, 3407 (1991).

[12] S. Micalizio, D. Calonico, F. Levi, L. Lorini, A. Godone, to be published in Phys. Rev. A **69** (2004).

[13] A. Derevianko, W. R. Johnson, M. S. Safronova, J.F. Babb, Phys. Rev. Lett **65**, 3589 (1999).

[14] A. Derevianko, S.G. Porsev, Phys. Rev. A **65**, 53403 (2002).

[15] F. Levi, L. Lorini, D. Calonico, A. Godone, IEEE Trans. on Instr. Meas. **52**, 2, (April 2003).

[16] J. Vanier and C. Audoin, *The Quantum Physics of Atomic Frequency Standards,* Adam Hilger, Bristol-Philadelphia, (1989).

[17] A. Bauch and R. Schroder, Phys. Rev. Lett. **78**, 622 (1997).

[18] J. R. Mowat, Phys. Rev. A **5**, 1059 (1972).

[19] R. D. Haun and J. R. Zacharias, Phys. Rev. **107**, 107(1957).

[20] V. Pal'chikov, Y. Domnin, and A. Novoselov, J. Opt. B: Quantum Semiclass. Opt. 5, S131 (2003)

FIRST OBSERVATION OF FESHBACH RESONANCES AT VERY LOW MAGNETIC FIELD IN A ^{133}CS FOUNTAIN.

H. Marion, S. Bize, L. Cacciapuoti,* D. Chambon, F. Pereira dos Santos, G. Santarelli, P. Wolf, and A. Clairon
BNM-SYRTE, Observatoire de Paris, 61 Avenue de l'Observatoire, 75014 Paris, France.

A. Luiten and M. Tobar
The University of Western Australia, School of Physics, 35 Stirling Hwy, Crawley, Western Australia.

S. Kokkelmans† and C. Salomon
Laboratoire Kastler Brossel, ENS, 24 rue Lhomond, 75005 Paris, France.

Keywords - **Atomic fountain, cryogenic oscillator, frequency measurement, cold collisions, adiabatic passage, Feshbach resonance.**

Abstract - **For a long time, one of the main limitations of cesium atomic fountains has been the cold collision frequency shift [2, 7, 10, 11]. By using a method based on a transfer of population by adiabatic passage (AP) [12, 16, 18] allowing to prepare cold atomic samples with a well defined ratio of atomic density as well as atom number we have a better measurement of this effect. By improving the method we found out some unexpected properties of cold collisions, confirming that ^{133}Cs is a rich atom to study cold collisions. Those results lead to fine comparisons with cold collision theory and may constraint some parameters of cesium.**

INTRODUCTION.

Frequency standards are used in many industrial (Global Positioning System, navigation) and scientific pursues (fundamental physic tests [5, 15]). Primary standards are today based on cesium and are contributing to the realization of the second. The most accurate are using cold atoms to allow long interaction time. In cold atomic gases collisions play an important role by shifting the clock frequency. Thus, it is crucial to control and to study accurately cold collisions for clock uncertainty budget. Nevertheless, there is a tradeoff between the clock stability and its accuracy. In this paper we will describe improvement of the a technic based on fast AP that allows to reconcile those two goals. For high density atomic sample, such as in Bose-Einstein condensates (BEC) [1], manifestations of collision such as molecular Feshbach resonances [8, 23] are observed. We found out, in our fountain, such phenomenon at very low magnetic field and density. Theoretical tools will be remind and a numerical simulation of some of our data will be presented.

THE DOUBLE FOUNTAIN: RECENT IMPROVEMENTS AND PERFORMANCES.

Atom loading improvement.

At Paris observatory, there is a set of three atomic fountains, using cesium. In particular, our one is a double fountain [4] able to work alternatively with rubidium and cesium. In the last year, on the cesium part, we improved the loading rate by supplementing a transverse collimation of the atomic beam in addition to the present chirp cooling (see FIG. 1). It consists in a 2D optical molasses with two laser beams, recycling the light in a "zigzag" way [19, 21]. This allows us to cool 10^9 atoms in 200 ms in an optical molasses. The 2D molasses increase the loading rate by a factor of ~ 10. Regular parameters of the double fountain are a magnetic field in the interrogation zone of 2 mG, a loading duration of ~ 200 ms that allows to load 10^9 atoms within a fountain cycle of ~ 1.2 s. The launch velocity is ~ 4 m.s^{-1} that imply a launch height of $\sim$ 0.8 m.

Frequency stability.

A sapphire cryogenic oscillator (SCO) [13] from the University of Western Australia is now used as a filter for one of our H-maser. The SCO is an experiment by itself [24], but is very reliable and can be used routinely to drive the fountain. The SCO is weakly phase-locked on the H-maser with a time constant of about an hour. Therefore, the resulting signal reflects the good short term stability of the SCO while in long time scales it reproduce the characteristics of the H-maser. The SCO delivers a signal of about 12 GHz.

At BNM-SYRTE most of the microwave synthesizer use a 100 MHz signal to generate the 9.2 GHz necessary to interrogate the atoms. Therefore, a down-converter has been developed to transform the SCO signal into a 100 MHz one. An evaluation has shown that the fountain was limited by the phase noise of the microwave setup. Henceforth, a new microwave synthesizer [6] able to down-convert directly the 12 GHz to 9.2 GHz has been built in order to minimize the phase noise. It achieves a frequency stability of $\sim 3\ 10^{-15}$ at 1s. In

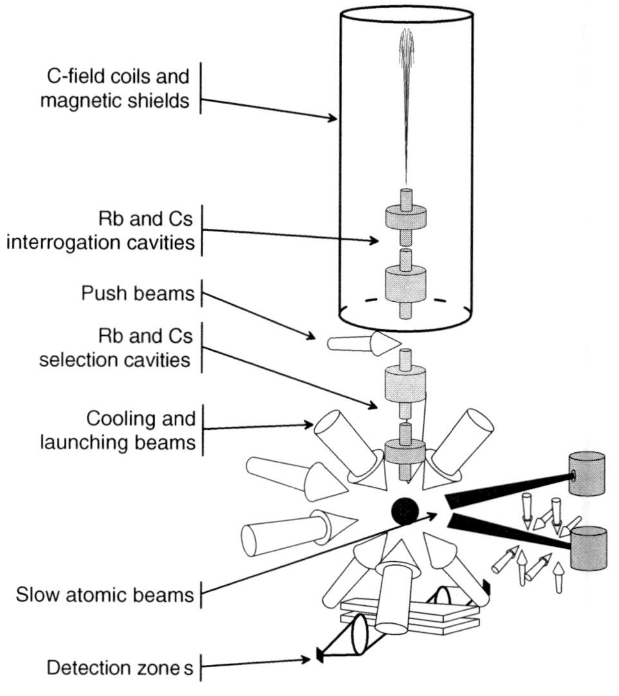

FIG. 1: Schematic of the double fountain both base on ^{87}Rb and ^{133}Cs, so that there is two pairs of cavity finely tuned on hyperfine atomic frequency of corresponding atoms and two ovens. On the cesium part, a transversal collimation of the atomic beam has been improved in addition of the present chirp cooling. This consists in a 2D recycling light, pair of laser beam, used in a "zigzag" way. This allows to load around 10^9 atoms in 200 ms in an optical molasses.

order to take benefits of its excellent short term stability ($\sim 5\ 10^{-16}$ at 1s) a large number of trapped atoms is required [20]. With $\sim 10^7$ detected atoms we were able to reach 1.6 10^{-14} at 1 s, the best frequency stability for a primary standard to date. With this performances a resolution of 10^{-16} is achievable in 6 hours.

FIG. 2 shows the Allan deviation of the fountain driven by the SCO. This is a differential measurement [22], this means that a full atom number configuration (■) is alternated with a half atom number configuration (•) each 60 fountain cycles. The (■) represents the fractional frequency deviation of the corrections applied to the interrogating microwave generated with the SCO signal to stay at resonance with the atoms. This is performed with the maximum atom number (N_{at}) which allows the 1.6 10^{-14} at 1s stability. On short term (10 s to 100 s) the allan variance is dominated by the Cs fountain noise. After 200 s, the drift of the SCO is observed. The Allan deviation of the frequency difference between low ($N_{at}/2$) and high(N_{at}) density (▲) averages as white frequency noise for all times and reaches $\sim 2\ 10^{-16}$ for a 20000 s measurement. This leads to a good statistical evaluation of the mean difference in very reasonable time. The SCO drift is efficiently rejected and atom number dependent effects (i.e. cavity pulling [3] and collisional shift) are measured with a statistical uncertainty of 10^{-16} in

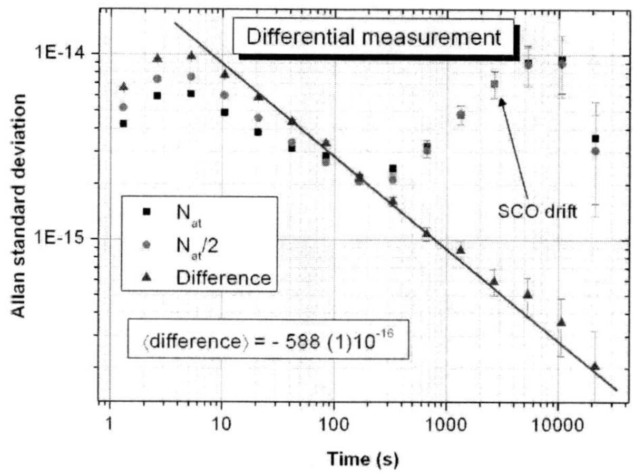

FIG. 2: Shows the Allan deviation of the fountain driven by the SCO. This is a differential measurement of the collisional shift, this means that a full atom number configuration (■) is alternated with a half atom number configuration (•) each 60 fountain cycles. The (■) represents the fractional frequency deviation of the corrections applied to the interrogating microwave generated with the SCO signal to stay at resonance with the atoms. This is performed with the maximum atom number which allows the 1.6 10^{-14} at 1 s stability. The Allan deviation of the difference (▲) averages well for all times and has also a good stability. This leads to a good statistical evaluation of the mean difference in a very reasonable time.

one day. This behavior indicates that determination of the collisional frequency shift is only limited by the frequency stability of the fountain.

ACCURATE CONTROL OF COLD COLLISION FREQUENCY SHIFT.

In previous work [18], we demonstrated the possibility to control and evaluate the cold collision shift at the percent level by using the adiabatic passage method. Cold collisions represent the main systematic effect that shifts the frequency and restricts the accuracy in cesium fountains. At that time, our collisional shift for full density, was around 10^{-14}. Therefore, the corresponding uncertainty was near 10^{-16}.

Adiabatic passage performances.

To achieve a stability around $\sim 2\ 10^{-14}$ at 1 s it is necessary to load a high number of atoms. For example the collisional shift corresponding to N_{at} of the (■) curve implies a frequency shift of about $\sim 1.2\ 10^{-13}$. Thus, an AP at the percent level is no longer enough to stay at the 10^{-16} level of uncertainty on this effect. To perform AP the amplitude of the microwave into the selection cavity has to be modulated and its frequency has to be swept. It is critical that the synthesizer used to chirp the selection microwave stops the frequency sweep at resonance for the half AP. Our model (Mar-

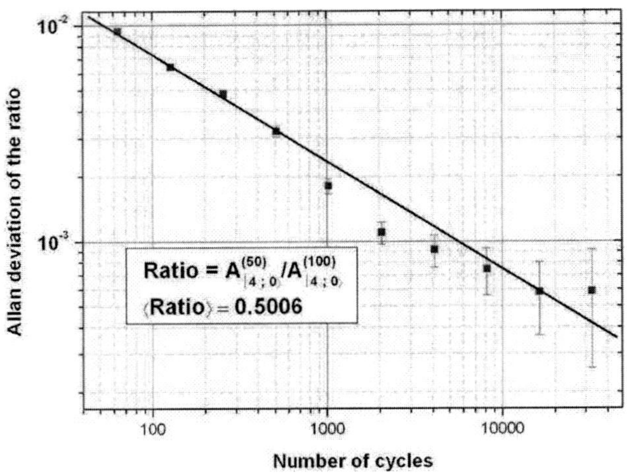

FIG. 3: Allan deviation of the ratio of atom numbers in the $|4;0\rangle$ obtained by alternating each 60 fountain cycle a full atom number configuration over half atom number configuration on the $|4;0\rangle$ state as a function of cycle number. The Allan deviation of the ratio averages as white noise and reaches $\sim 10^{-3}$ resolution after 1 hour. It leads to evaluate AP at the 10^{-3} level.

coni 2030) was not accurate enough and failed to be reproducible. It has been replaced by a DDS more easily frequency controllable. According to [18], one observable for AP evaluation is the ratio of atom number at full and half density. Each measurement is daily analyzed using this criterion to verify that the AP method is properly working.

FIG. 3 shows the Allan deviation of the ratio of atom numbers in the $|4;0\rangle$ obtained by alternating each 60 fountain cycle a full atom number configuration over half atom number configuration as a function of cycle number. The Allan deviation of the ratio averages as white noise and reaches $\sim 10^{-3}$ resolution after 1 hour. It leads to evaluate AP at the 10^{-3} level such as:

$$\frac{A^{(50)}_{|4;0\rangle}}{A^{(100)}_{|4;0\rangle}} = \frac{\tilde{n}^{(50)}_{|4;0\rangle}}{\tilde{n}^{(100)}_{|4;0\rangle}} = \frac{1}{2} \times (1 + 1\ 10^{-3}) \tag{1}$$

Where $\tilde{n}^{(100)}_{|4;0\rangle}$ ($\tilde{n}^{(50)}_{|4;0\rangle}$) is the density of the $|4;0\rangle$ atoms for full atom number (half atom number) and $A^{(100)}_{|4;0\rangle}$ ($A^{(50)}_{|4;0\rangle}$) is the detected atom number on the $|4;0\rangle$ for full atom number (half atom number).

The deviation measured at the 10^{-3} level could be induced by a non linearity in the detection. This will be investigated in a future work.

Similarly, we also check routinely the same ratio on the $F = 3$ level. Here, we find a larger deviation:

$$\frac{A^{(50)}_{|3;0\rangle}}{A^{(100)}_{|3;0\rangle}} = \frac{\tilde{n}^{(50)}_{|3;0\rangle}}{\tilde{n}^{(100)}_{|3;0\rangle}} = \frac{1}{2} \times (1 + 3\ 10^{-3}) \tag{2}$$

The departure of 3 10^{-3} from the 1/2 ratio is due to residual populations in the $|3;m_F\rangle$ states. A fraction of the atoms is coming from the push beam (see FIG. 1, i.e. a laser beam that finalize to the selection process) which is de-pumping atoms from the $|4;m_F\rangle$ states to the $|3;m_F\rangle$ states. The remanning fraction seem to be launched in the $F = 3$ state. Attempts to remove these atoms with extra repumper beams where partially unsuccessful. At this point, there is nothing much to do to prevent it but to look at the contribution of those spurious atoms on the clock shift.

Towards an adiabatic passage at the 10^{-3} level.

An accurate control of the collisional shift is possible only if the contribution of the 0.3 % atoms populating states different from the clock states is correctly evaluated. To measure this, we intentionally populated the atomic cloud using AP with $|3;m_F \neq 0\rangle$ states in addition of the $|3;0\rangle$ clock state. An additional microwave synthesizer tuned on a $|4;m_F \neq 0\rangle \rightarrow |3;m_F \neq 0\rangle$ transition is combined in the selection cavity with the regular selection synthesizer. A magnetic field is applied during the selection by AP in order to prevent parasitic excitations between zeeman sub-levels. To select, for example $m_F = 1$ state, one has to detune from resonance the extra synthesizer of the corresponding first order Zeeman shift value.

The measurement method alternate three different fountain configurations. The first one is a full AP selection of the $|3;0\rangle$ state, the second is a half AP selection of the $|3;0\rangle$ state. The first configurations give the collisional shift due to the clock states. The third one is a full AP selection of the $|3;0\rangle$ state plus full AP selection of one of the $|3;m_F \neq 0\rangle$ states. The last configuration combined with the two others lead to a measurement the collisional shift due to the extra population. The respective frequency shifts are:

$$\Delta\nu_1 = \tilde{n}_{00}K_{00} \tag{3}$$

$$\Delta\nu_2 = \frac{\tilde{n}_{00}}{2}K_{00} \tag{4}$$

$$\Delta\nu_3 = \tilde{n}_{00}K_{00} + \tilde{n}_{m_Fm_F}K_{m_Fm_F} \tag{5}$$

Where $\tilde{n}_{00}$ ($\tilde{n}_{m_Fm_F}$) represents the atom density of the clock states (m_F states), and K_{00} ($K_{m_Fm_F}$) is the cold collisional shift coefficient resulting of the extrapolated to zero density for the clock states (m_F states).

Under regular operating conditions, we evaluate the contribution to the clock shift due to the collisional shift of the extra $|3;m_F \neq 0\rangle$ atoms per detected atom ($\nu_{m_Fm_F}$ in Hz.(detected atoms)$^{-1}$) normalized to the collisional shift of the clock states $|3;0\rangle$ atoms per detected atoms (ν_{00}) as a function of the m_F value. This ratio (R_{0m_F}) can be expressed as:

$$R_{0m_F} = \frac{\nu_{m_Fm_F}}{\nu_{00}} \tag{6}$$

$$= \frac{\Delta\nu_3 - \Delta\nu_1}{2(\Delta\nu_1 - \Delta\nu_2)}\frac{N_{00}}{N_{m_Fm_F}} \tag{7}$$

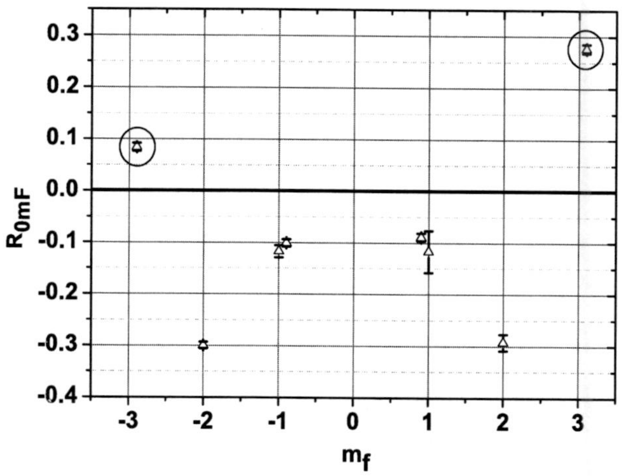

FIG. 4: displays R_{0m_F} as a function of the m_F value for regular clock conditions. The contribution to the cold collision shift of the $|m_F| = 1;\ 2$ are negative and equal for the same absolute values. At most the contribution of spurious $|3; m_F \neq 0\rangle$ is 1/3 of the contribution of clock states. The contribution of the $|m_F| = 3$ is different.

Were N_{00} ($N_{m_F m_F}$) is the detected atoms number for the clock states (m_F state) and can be expressed as $N_{00} = A_{|3;0\rangle} + A_{|4;0\rangle}$. A verification that ν_{00} is field independent has been performed so that it could be used as a reference. It is important to ensure that the density is the same whatever the could is made of. Thus a verification that the spatial distribution of the different Zeeman sub-levels is homogenous into the cloud has been performed. By selecting one m_F at the time and pulsing the Ramsey cavity with different durations, we access at different cloud slice. It shows that the peak density are equal to less than 7 %.

FIG. 4 displays R_{0m_F} as a function of the m_F value for regular conditions ($B \simeq 2\ mG$). The contribution to the cold collision shift of the $|m_F| = 1;\ 2$ are negative and equal for the same absolute values. At most the contribution of spurious $|3; m_F \neq 0\rangle$ is 1/3 of the contribution of clock states. Thus, the collisional shift is controlled at the 10^{-3} level. Surprisingly the contribution of the $|m_F| = 3$ is different, which was unpredicted at such low magnetic field. This observation is a clear indication of the presence of Feshbach resonances.

FESHBACH RESONANCES.

Experimental results.

We thus looked for the ratio (R_{0m_F}), for each absolute value of the m_F as a function of magnetic field. We found out three Feshbach resonances (FIGs. 5, 6 and 7) deeply field dependent. It gives a direct access to the amplitude and the width of each resonance. The field is very well controlled so that there is no way that error bars could mix up the shapes. For some values of the magnetic field the frequency shift due to $|3; m_F\rangle \neq 0$ population can be as large as the standard clock shift. Let's remind that our regular magnetic field is 2 mG. One has to be very cautious by choosing a value of magnetic field in an experiment using cold ^{133}Cs. The $|m_F| = 2$ resonance is obviously a multiple resonance, but the $|m_F| = 1$ and specially $|m_F| = 3$ look to be more simple. Assuming that $|m_F| = 3$ resonance is a simple resonance we try to understand the line shape with a simple model.

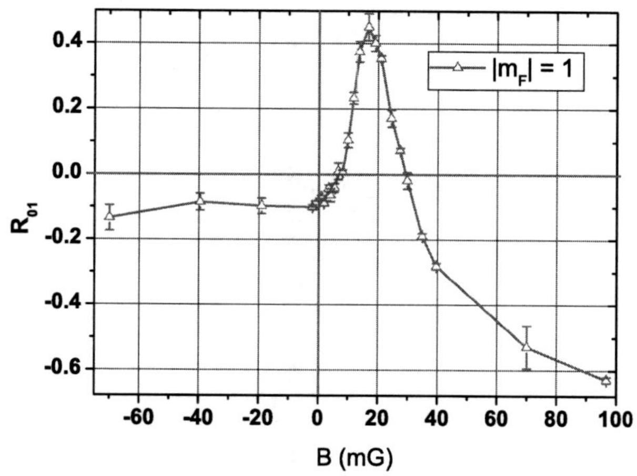

FIG. 5: Ratio R_{01} as a function of the magnetic field. It represents a Feshbach resonance for $|m_F| = 1$.

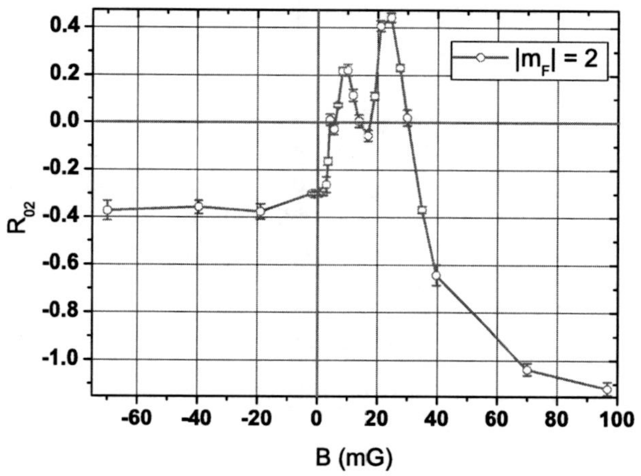

FIG. 6: Ratio R_{02} as a function of the magnetic field. It represents a Feshbach resonance for $|m_F| = 2$. This is obviously a multiple resonance.

Theoretical approach: model for S-matrix.

In order to describe the measured resonances, we need to account for the inelastic collision processes. Feshbach resonances in cold atomic systems are usually considered only in the context of elastic scattering, with one open channel only [14, 17]. In such a situation, only an elastic resonance width Γ_e is involved, and the incoming channel for the atoms is simultaneously the exit channel. Here we consider a situation with more than one open channel. Atoms which exit in

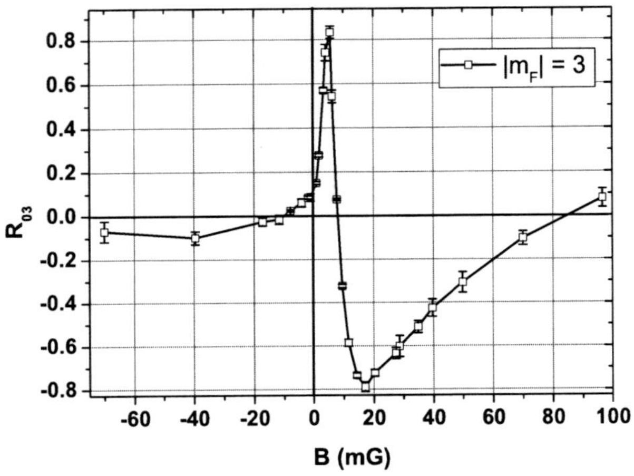

FIG. 7: Ratio R_{03} as a function of the magnetic field. It represents a Feshbach resonance for $|m_F| = 3$. This resonance has a more simple shape. Assuming that it is a simple resonance we gone to simulate the scattering process.

a different channel than the incoming channel will gain additional kinetic energy, and are considered as lost. This decay process can be mediated via the molecular Feshbach state, and gives rise to an additional energy scale, the inelastic resonance width Γ_i [9].

The S-matrix for such an inelastic resonance can be written as

$$S(k) = S^{bg}(k)\left(1 - \frac{i\Gamma_e}{E - \nu + \frac{1}{2}i\Gamma_e + \frac{1}{2}i\Gamma_i}\right) \quad (8)$$

with $E = \hbar^2k^2/m$ the relative kinetic energy between the colliding atoms, $\nu = \Delta\mu(B - B_0)$ the detuning, and $S^{bg}(k) = \exp(-2ika_{bg})$ the background (or direct) part of the S-matrix. Here B_0 is the magnetic field value of resonance, and $\Delta\mu$ the difference between the magnetic moments of the two particles in the incoming channel and the molecular Feshbach state. The background scattering length $a_{bg} = a^r_{bg} + ia^i_{bg}$ contains not only a real part a^r_{bg}, but also an imaginary part a^i_{bg} since there is also a direct contribution to the inelastic decay.

The energy widths are in principle a function of the relative wavenumber k, since they depend on the overlap between the molecular wavefunction corresponding to the Feshbach state, and the wavefunction in the corresponding open channel. In case of the elastic energy width, the energy dependence of the incoming channel gives rise to linear dependence in wavenumber, i.e. $\Gamma_e = c_e k$ with c_e a constant [17]. For the inelastic energy width, however, the energy dependence of the corresponding open channels can be safely neglected over the energy range of interest. Therefore Γ_i can be taken energy-independent.

Monte Carlo model.

The frequency shift of the clock transition due to the $|3; m_F \neq 0\rangle$ state population is given by Eq. (9):

$$\frac{\delta\omega_{\beta\alpha}}{2\pi} = \frac{\hbar\rho_{\gamma\gamma}}{mk} Im\left\{S_{\alpha\gamma}(k)S^\dagger_{\beta\gamma}(k) - 1\right\} \quad (9)$$

where α and β refer to the $|3;0\rangle$ and $|4;0\rangle$ clock states respectively. $\rho_{\gamma\gamma}$ is the atomic density of the $|3; m_F \neq 0\rangle$ state (denoted as γ). To be consistent with our initial intent to analyze the $|m_F| = 3$ data as a single Feshbach resonance, we will assume that a resonance occurs for only one of the two entrance channels ($\alpha\gamma$ and $\beta\gamma$). The other entrance channel contribute only with a constant background part, independent on the magnetic field. This factor can be formally accounted for in the background term of the channel where the resonance occurs. Therefore, we are assuming that $S_{\alpha\gamma} = 1$ (resp. $S_{\beta\gamma} = 1$) while $S_{\beta\gamma}(k)$ (resp. $S_{\alpha\gamma}$(k)) follows Eq. (8).

Eq. (9) clearly shows that the clock shift depends on the wavenumber k. The total clock shift is obtained by suitably averaging Eq. (9) over the duration of the interrogation, taking into account the atomic cloud space and velocity distribution. This is done using a Monte Carlo simulation where the atomic trajectories are randomly sampled according to the space and velocity distributions as measured in the experiment (with CCD imaging and analysis of time of flight signals). The simulation accounts for the distribution of collisional energy, for the time varying (decreasing) atomic density, for the non-uniform response of the final transition probability to a time varying perturbation (sensitivity function), for the truncation of the atomic cloud crossing the microwave resonator.

FIGs. 8 and 9 represent examples of calculated line shape of the Feshbach resonance for various parameters. For all cases, we take $a_{bg} = 0$, $B_0 = 0$ and $\Delta\mu = 2\mu_B$. Also, we assume that the resonance occurs on the $\beta\gamma$ channel. The initial distribution is gaussian with $\sigma = 3.5$ mm. The velocity distribution is $\propto (1 + v^2/v_c^2)^{-b}$ with $v_c = 10$ mm.s^{-1} and $b = 2.1$. The vertical axis represents the fractional frequency shift due to the γ ($|3; m_F\rangle$) state for 10^8 atoms launched in this state. This corresponds to a typical effective atomic density of 2.2 10^7 cm^{-3}. In FIG. 8, $\Gamma_i = 0$. C_e is equal to $0.4 \times E_{rec}/k_{rec}$, $4 \times E_{rec}/k_{rec}$, $40 \times E_{rec}/k_{rec}$ and $400 \times E_{rec}/k_{rec}$ where $E_{rec} = \hbar^2k^2_{rec}/2m$ and $k_{rec} = 2\pi/\lambda$, $\lambda = 852$ nm. For large values of C_e, the shape and the ("natural") width of the resonance is determined by C_e. The resonance has a symmetrical dispersive shape. For small values of C_e, the shape and the width of the resonance is determined by the collision energy distribution. The line shape shows a strong asymmetry. The calculation clearly indicates that the top of the sharp feature corresponds to the magnetic field that meets the resonant condition (vanishing difference between the molecular bound state energy and the dissociation threshold). Similarly,

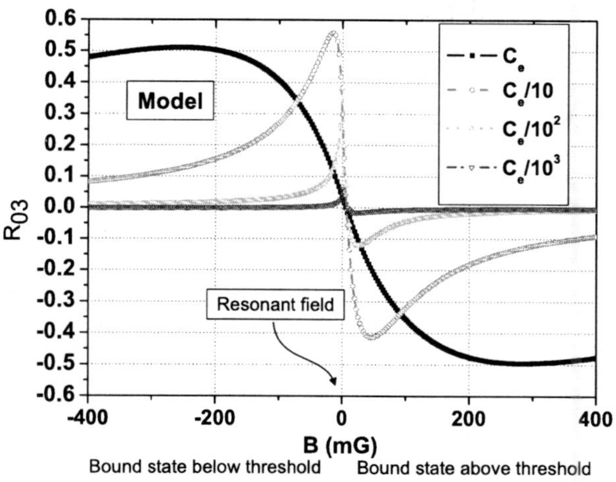

FIG. 8: For $\Gamma_i = 0$. For large values of C_e, the shape and the ("natural") width of the resonance is determined by C_e. The resonance has a symmetrical dispersive shape. For small values of C_e, the shape and the width of the resonance is determined by the collision energy distribution. The line shape shows a strong asymmetry. The calculation clearly indicates that the top of the sharp feature corresponds to the magnetic field that meets the resonant condition.

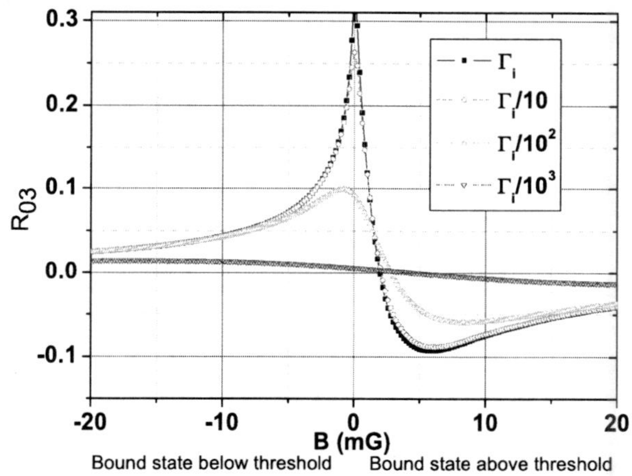

FIG. 9: illustrates the effect of parameter Γ_i. C_e is kept constant. The resonance shape is computed for different Γ_i amplitudes. The comparison with the measured line shape clearly indicates that the observed resonance are not strongly affected by the inelastic processes. The calculation shows that the sign of the clock shift is inverted when the resonance occur on the $\alpha\gamma$ channel.

FIG. 9 illustrates the effect of parameter Γ_i. C_e is kept constant and equal to $0.4 \times E_{rec}/k_{rec}$. The resonance shape is computed for Γ_i equal to $0.04 \times E_{rec}$, $0.4 \times E_{rec}$, $4 \times E_{rec}$ and $40 \times E_{rec}$. The comparison with the measured line shape clearly indicates that the observed resonance are not strongly affected by the inelastic processes. Finally, the calculation (as well as direct consideration of Eq. (8) shows that the sign of the clock shift is inverted when the resonance occur on the $\alpha\gamma$ channel.

The measured $|m_F| = 3$ resonance is fitted (see

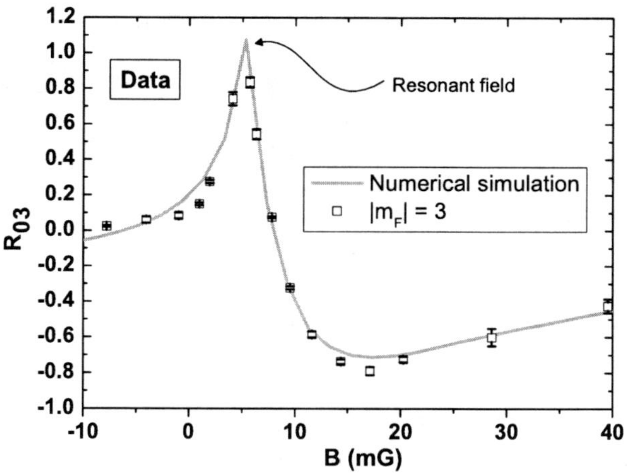

FIG. 10: The $|m_F| = 3$ data are in good agreement with a resonance dominated by the collision energy. The Monte Carlo model fits well the point and indicates the resonant field.

FIG. 10) using the above model leaving B_0, $\Delta\mu$, C_e, Γ_i, a_{bg} as free parameters. The model also includes the possibility of a slow variation of the background scattering length with the magnetic field through the addition of a linear and quadratic term as a function of the magnetic field. We checked that the following analysis is not significantly affected by this additional term. Finally, the model includes a scale factor for the atomic density. Similarly, the inclusion of this scale factor has no influence on the following analysis.

The simple consideration of the sign of the data indicates that the resonance occurs in the $\beta\gamma$ channel involving the upper clock state $|4;0\rangle$ and the $|3;3\rangle$ state. FIG. 10 shows the optimized fit to the data. The qualitative agreement with the model is quite satisfactory. For $\Delta\mu$, C_e and Γ_i, the analysis does not lead to precise value, due to the sensitivity of these parameters to the atomic cloud space and velocity distribution. We find $\Delta\mu \simeq 1.5\mu_B$, $C_e \sim 0.4E_{rec}/k_{rec}$ and $\Gamma_i \lesssim 0.4 \times E_{rec}$. Conversely, B_0 is tightly constrained and well decorrelated from the other parameters. Although an uncertainty on B_0 cannot be easily extracted from our (non-linear) fitting procedure, we can safely state that B_0 is constrained to within 1 mG (to be understood as a 1σ error bar). We find $B_0 = 5 \pm 1$ mG. The corresponding uncertainty on the energy of the molecular bound state causing the resonance is ~ 100 nK.

CONCLUSION.

To sum-up, our primary goal is achieved, the collisional shift is controlled at the 10^{-3} of its value. A calibration of Zeeman sub-states contribution to the clock shift as a function of the field has been performed and teaches us magnetic field values to avoid. Under regular clock conditions an upper limit for their contribution shows an upper limit compatible with AP at the 10^{-3} level. This makes reachable the goal of primary standards at the 10^{-16} uncertainty level. Feshbach resonances have been observed for the first time at very low magnetic field and with a very good resolution. A Monte Carlo simulation has been performed and could fit properly some of experimental data. This constrains some parameters of the theory of collisions.

Acknowledgments: The authors wish to thank K. Williams for fruitful discussions. This work was supported in part by BNM and CNRS. BNM-SYRTE and Laboratoire Kastler-Brossel are Unités Associées au CNRS, UMR 8630 and 8552. S.K. acknowledges supported from the Netherlands Organization for Scientific Research (NWO).

---

* Present address: Dipartimento di Fisica, Universita degli studi di Fireze, via Sansone 50019 Steso Firentino (FI), Italy

† Present address: Eindhoven University of Technology, P.O. Box 513, 5600 MB Eindhoven, The Netherlands

[1] M.H. Anderson *et al.*, Science **269**, 198 (1995).
[2] M. Bijlsma *et al.*, Phys. Rev. A **49**, R4285 (1994).
[3] S. Bize *et al.*, IEEE Trans. on Instr. and Meas. **50**, 503 (2001)
[4] S. Bize *et al.*, in *Proc. of the 6th Symposium on Frequency Standards and Metrology* (World Scientific, Singapore, 2001), p 53.
[5] S. Bize *et al.*, Phys. Rev. Lett. **90**, 150802 (2003).
[6] D. Chambon *et al.*, Proceeding EFTF 2004.
[7] C. Chin *et al.*, Phys. Rev. A **63**, 033401 (2001).
[8] H. Feshbach, Ann. of Phys. (N.Y.) **5**, 357 (1958); **19**, 287 (1962).
[9] H. Feshbach, *Theoretical Nuclear Physics* (John Wiley and Sons, New York, 1992).
[10] S. Ghezali *et al.*, Europhys. Lett. **36**, 25 (1996).
[11] K. Gibble and S. Chu, Phys. Rev. Lett. **70**, 1771 (1993).
[12] M.M.T. Loy, Phys. Rev. Lett. **32**, 814 (1974).
[13] A. Luiten *et al.*, IEEE Trans. on Instr. and Meas. **44**, 132 (1995).
[14] B. Marcelis, E.G.M. van Kempen, B.J. Verhaar, and S.J.J.M.F. Kokkelmans, cond-mat/0402278.
[15] H. Marion *et al.*, Phys. Rev. Lett. **90**, 150801 (2003).
[16] A. Messiah, Quantum Mechanics, **2**,637 (1959).
[17] A.J. Moerdijk, B.J. Verhaar, and A. Axelsson, Phys. Rev. A **51**, 4852 (1995).
[18] F. Pereira Dos Santos *et al.*, Phys. Rev. Lett. **89**, 233004 (2002).
[19] E. Rasel *et al.*, Eur. Phys. J. D **7**, 311-316 (1999).
[20] G. Santarelli *et al.*, Phys. Rev. Lett. **82**, 4619 (1999)
[21] F. Shimizu *et al.*, Chem. Phys. **145**, 327 (1990).
[22] Y. Sortais *et al.*, Phys. Rev. Lett. **85**, 3117 (2000).
[23] E. Tiesinga *et al.*, Phys. Rev. A **47**, 4114 (1993).
[24] P. wolf *et al.*, Phys. Rev. Lett. **90**, 060402 (2003).

AN ANALYSIS OF QUARTZ CRYSTAL RESONATORS MADE FROM A CONVEX PIEZOELECTRIC PLATE OF ARBITRARY SHAPE

S.S. Nedorezov, O.Yu. Shmaliy, [*]B. Dulmet

Ukrainian Engineering-Pedagogical Academy, Universitets'ka Str., 16, Kharkiv, Ukraine, oshmaliy@ic.kharkov.ua
[*] FEMTO-ST, Dept LCEP, ENSMM

Keywords: Crystal resonator, eigenvibrations, frequency spectrum.

Abstract

The paper addresses an analysis of the transversely varying thickness modes of a piezoelectric resonator affected by a convex anisotropic boundary surface. We show that anisotropy of the boundary surface strongly influences the eigenvibrations and capacitances. The relations to calculate the frequency spectrum, relative amplitudes of vibrations and motional capacitances are derived. A numerical simulation is provided for the crystal resonator of SBTC-cut.

1 Introduction

Accurate calculation of the frequency spectrum and equivalent electrical parameters of the piezoelectric resonator is important if its anharmonics are employed in the multimode oscillatory system or as sensors of environment. Yet, such knowledge allows for making steps ahead avoiding problems with the frequency jumps [3]. During several decades, the mathematical theory of eigenvibrations in resonators has undergone great changes forwarded to serve engineering needs. Even so, determination of resonator parameters sometimes is not accurate and new mathematical models are still of great interest.

In this report, an analysis is given of the eigenvibrations and capacitances of a convex piezoelectric plate of an arbitrary shape. It is shown that the frequency spectrum is determined by the curvature of the boundary surface at the point of a maximum. The formula is derived for the eigenfrequencies of the vibrations and capacitances, assuming arbitrary orientation of the direction of major curvature of a surface with respect to the self axes of a piezoelectric plate. It is noticed that anisotropy of the piezoelectric surface affects essentially the resonator parameters that may be used to optimize the resonator performance.

2 Eigenvibrations of a piezoelectric resonator made from a plate of arbitrary shape

Let us examine a piezoelectric resonator (Fig. 1) with a convex plate of an arbitrary shape

$$z = U(x_1, x_2). \tag{1}$$

We suppose the surface to be convex at the point $M_0(x_{10}, x_{20})$ and $U(M_0) = L_0$, where L_0 is a maximum thickness of a plate. The maximum and minimum radius of the surface curvature at the point M_0 is R_1 and R_2, respectively. Yet, we induce the inter direction angle γ of the major surface curvature and self-axes of the piezoelectric plate.

The first terms of the extension of (1) into the Taylor series with respect to a small value $\sim 1/R$ make a principle contribution to the frequency spectrum, so that in the following we suppose that

$$U(x_1,x_2) = U(x_{10},x_{20}) + \frac{1}{2}\sum_{i,j=1}^{2}\frac{\partial^2 U(M_0)}{\partial x_i \partial x_j}(x_i - x_{i0})(x_j - x_{j0}) + \ldots \tag{2}$$

Following [3], the vibration frequency spectrum of a resonator is obtained by a solution of the effective differential equation, which looks like

$$-d_1\frac{\partial^2\Psi}{\partial x_1^2} - d_2\frac{\partial^2\Psi}{\partial x_2^2} + \omega_n^2\left(1 - \frac{1}{L_0}\sum_{i,j=1}^{2}\frac{\partial^2 U(M_0)}{\partial x_i \partial x_j}(x_i - x_{i0})(x_j - x_{j0})\right)\times \Psi(x_1,x_2) = \omega^2\Psi(x_1,x_2), \tag{3}$$

where d_1 and d_2 are eigenvalues of a tensor $d_{\alpha\beta}$ dependent on the whole set of material constants of the piezoelectric crystal ([3]), ω_n is the fundamental frequency of the nth overtone. The high order terms of the extension (3) determine small unequidistant amendments to the spectrum. To calculate such amendments, one needs to use the most common equations (see, for example, [1]).

We now substitute $x_i = \sqrt{d_i}\xi_i + x_{i0}$ $(i = 1,2)$ in the (3) and arrive at the equation:

$$-\frac{\partial^2\Psi}{\partial \xi_1^2} - \frac{\partial^2\Psi}{\partial \xi_2^2} + \omega_n^2\left(1 - \frac{1}{L_0}\sum_{i,j=1}^{2}\frac{\partial^2 U(M_0)}{\partial x_i \partial x_j}\sqrt{d_i}\xi_i\sqrt{d_j}\xi_j\right)\times \Psi(\xi_1,\xi_2) = \omega^2\Psi(\xi_1,\xi_2). \tag{4}$$

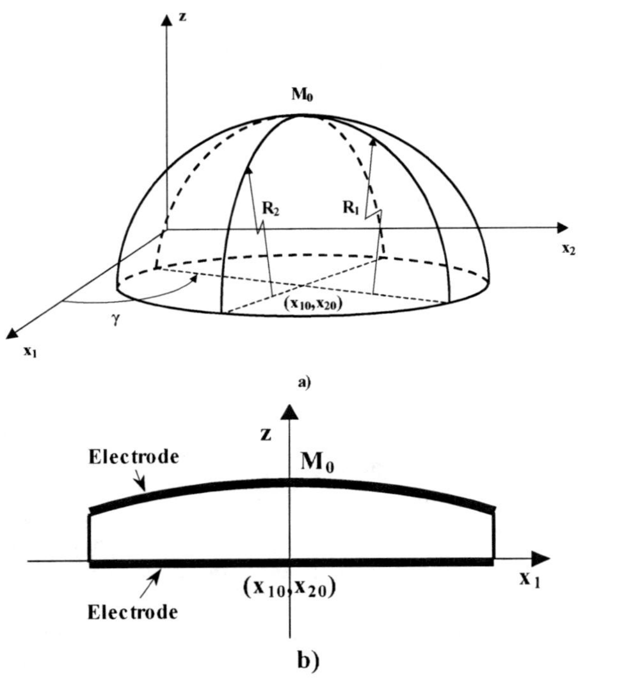

Fig.1. Crystal resonator with a convex piezoelectric plate of an arbitrary shape, $M_0(x_{10},x_{20})$ marks the point of a maximum, R_1 and R_2 are the major and minor radiuses **of curvature of the surface at the point of maximum, γ is** the angle between the direction of major curvature of surface and self-axes of the piezoelectric plate: a) common view, b) cross-section.

We now transform the coordinate system in a manner to exclude the product of $\xi_1\xi_2$ in (4), by introducing

$$\begin{cases}\xi_1 = \xi_1\cos\beta - \xi_2\sin\beta \\ \xi_2 = \xi_1\sin\beta + \xi_2\cos\beta\end{cases}, \qquad (5)$$

where

$$\beta = \frac{1}{2}arctg\left(\frac{2\dfrac{\partial^2 U(M_0)}{\partial x_1\partial x_2}\sqrt{d_1d_2}}{\dfrac{\partial^2 U(M_0)}{\partial x_1^2}d_1 - \dfrac{\partial^2 U(M_0)}{\partial x_2^2}d_2}\right).$$

It is now convenient to express the partial derivatives of the function U via the radiuses R_1 and R_2. In doing so, we get

$$\frac{\partial^2 U(M_0)}{\partial x_1^2} = -\frac{\cos^2\gamma}{R_1} - \frac{\sin^2\gamma}{R_2},$$

$$\frac{\partial^2 U(M_0)}{\partial x_1\partial x_2} = \left(\frac{1}{R_2}-\frac{1}{R_1}\right)\sin\gamma\cos\gamma, \qquad (6)$$

$$\frac{\partial^2 U(M_0)}{\partial x_2^2} = -\frac{\sin^2\gamma}{R_1} - \frac{\cos^2\gamma}{R_2}.$$

As a result of such a manipulation, we obtain the new equation, namely:

$$\frac{\partial^2\Psi}{\partial\xi_1^2}+\frac{\partial^2\Psi}{\partial\xi_2^2}+\left[\omega^2-\omega_n^2-\frac{\omega_n^2 d_1}{L_0R_1}\left(c(\beta)\xi_1^2+c\left(\beta+\frac{\pi}{2}\right)\xi_2^2\right)\right]\times\Psi(\xi_1,\xi_2)=0, \qquad (7)$$

where

$$c(\beta)=\left[\chi(\gamma)\cos^2\beta+\frac{\sqrt{\varepsilon}}{2}(1-\delta)\sin 2\gamma\sin 2\beta+\varepsilon\chi\left(\gamma+\frac{\pi}{2}\right)\sin^2\beta\right], \qquad (8)$$

$$\chi(\gamma)=\cos^2\gamma+\delta\sin^2\gamma \qquad (9)$$

and

$$\beta=\frac{1}{2}arctg\left(\frac{\sqrt{\varepsilon}(1-\delta)\sin 2\gamma}{\chi(\gamma)-\varepsilon\chi\left(\gamma+\frac{\pi}{2}\right)}\right), \qquad (10)$$

$$\varepsilon=\frac{d_2}{d_1},\ \delta=\frac{R_1}{R_2}.$$

We now find frequency spectrum ω_{nkl} of eigenvibrations:

$$\omega_{nkl}^2=\omega_n^2+2\omega_n\sqrt{\frac{d_1}{L_0R_1}}\times\left(\sqrt{c(\beta)}\left(k+\frac{1}{2}\right)+\sqrt{c\left(\beta+\frac{\pi}{2}\right)}\left(l+\frac{1}{2}\right)\right), \qquad (11)$$

where $k,l=1,2,3,\ldots$,

$$b_1=\frac{\omega_n^2 d_1}{L_0R_1}c(\beta),\ b_2=\frac{\omega_n^2 d_1}{L_0R_1}c\left(\beta+\frac{\pi}{2}\right). \qquad (12)$$

We write now correspondent eigenfunctions Ψ_{kl} in a manner:

$$\Psi_{kl}=\left(\frac{\sqrt{b_1b_2}}{\pi^2 d_1d_2}\right)^{1/4}\frac{1}{\sqrt{2^{k+l}k!l!}}e^{-\frac{\zeta_1^2+\zeta_2^2}{2}}H_k(\zeta_1)H_l(\zeta_2), \qquad (13)$$

where

$$\zeta_1=b_1^{1/4}\left(\frac{1}{\sqrt{d_1}}(x_1-x_{10})\cos\beta+\frac{1}{\sqrt{d_2}}(x_2-x_{20})\sin\beta\right),$$

$$\zeta_2=b_2^{1/4}\left(-\frac{1}{\sqrt{d_1}}(x_1-x_{10})\sin\beta+\frac{1}{\sqrt{d_2}}(x_2-x_{20})\cos\beta\right) \qquad (14)$$

and $H_k(\xi)$ are the Hermite polynomials.

3 Eigenvibrations of a piezoelectric resonator made from a plate with an ellipsoid surface

Let us examine the aforementioned crystal resonator (Fig. 2.a) with a convex plate and ellipsoid surface. The system

$\xi_1\ \xi_2\ \xi_3$ is rotated on Euler's angles φ, ψ, θ with respect to x_1x_2z system (Fig. 2.b). The relative radiuses in the direction of ξ_1, ξ_2 and ξ_3 are r_1, r_2 and r_3, respectively.
The equation for surface in the x_1x_2z system is

$$\tilde{U}(x_1, x_2) = -e_0 \times \left(1 + \frac{1}{e_0}\left(e_{11}x_1^2 + e_{22}x_2^2 + e_{12}x_1x_2 + e_1x_1 + e_2x_2\right)\right), \quad (15)$$

where

$$e_0 = -\frac{1}{\sqrt{a_1}}, \qquad e_1 = \frac{a_{21}}{2a_1}, \qquad e_2 = \frac{a_{22}}{2a_1},$$

$$e_{11} = \frac{1}{8\sqrt{a_1^3}}\left(4a_1a_{34} - a_{21}^2\right),$$

$$e_{12} = \frac{1}{8\sqrt{a_1^3}}\left(4a_1a_{33} - 2a_{21}a_{22}\right),$$

$$e_{22} = \frac{1}{8\sqrt{a_1^3}}\left(4a_1a_{35} - a_{22}^2\right).$$

and

$$a_1 = \frac{sin^2\theta\, sin^2\psi}{r_1^2} + \frac{cos^2\theta\, sin^2\psi}{r_2^2} + \frac{cos^2\psi}{r_3^2},$$

$$a_{21} = 2\left(\left[\frac{1}{r_3^2} - \frac{1}{r_2^2}\cos^2\theta - \frac{1}{r_1^2}\sin^2\theta\right]\sin\psi\sin\varphi\cos\psi + \left[\frac{1}{r_1^2} - \frac{1}{r_2^2}\right]\cos\theta\cos\varphi\sin\theta\sin\psi\right),$$

$$a_{22} = 2\left(\left[\frac{1}{r_1^2}\sin^2\theta + \frac{1}{r_2^2}\cos^2\theta - \frac{1}{r_3^2}\right]\cos\psi\cos\varphi\sin\psi - \left[\frac{1}{r_2^2} + \frac{1}{r_1^2}\right]\cos\theta\sin\varphi\sin\theta\sin\psi\right),$$

$$a_{33} = 2\left(\left[-\frac{1}{r_1^2}\cos\psi\sin^2\varphi + \frac{1}{r_1^2}\cos^2\varphi\cos\psi - \left(\frac{1}{r_2^2}\cos^2\varphi + \frac{1}{r_2^2}\sin^2\varphi\right)\cos\psi\right]\sin\theta\cos\theta + \left[-\frac{1}{r_3^2}\sin^2\psi + \frac{1}{r_1^2}\cos^2\theta - \frac{1}{r_2^2}\cos^2\theta\cos^2\psi - \frac{1}{r_1^2}\sin^2\theta\cos^2\psi + \frac{1}{r_2^2}\sin^2\theta\right]\sin\varphi\cos\varphi\right),$$

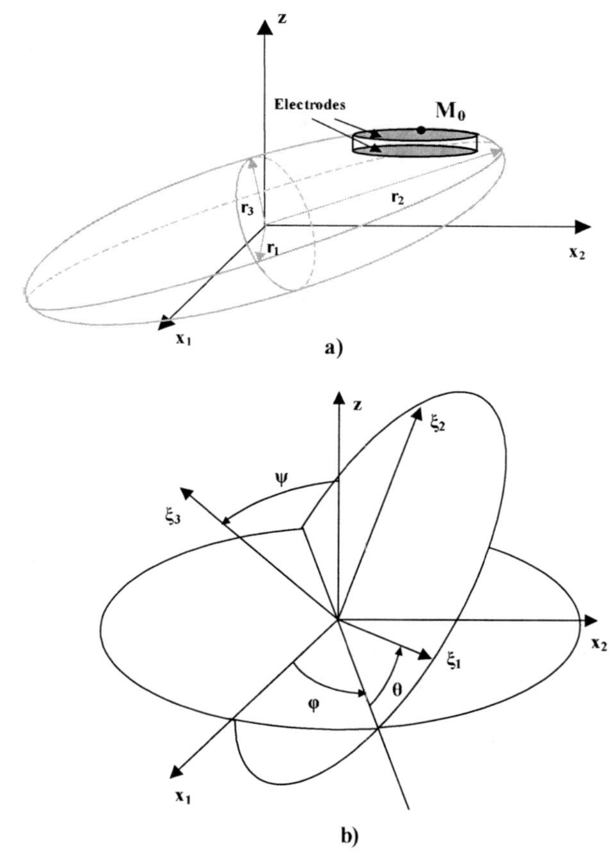

Fig.2. a) Crystal resonator with a convex piezoelectric plate with ellipsoid surface, $M_0(x_{10}, x_{20})$ is the point of maximum; b) Rotated coordinate system.

$$a_{34} = \left(\left[-\frac{2}{r_1^2} + \frac{2}{r_2^2}\right]\cos\theta\cos\varphi\sin\theta\cos\psi\sin\varphi + \left[\frac{1}{r_2^2}\cos^2\theta + \frac{1}{r_1^2}\sin^2\theta\right]\cos^2\psi\sin^2\varphi + \left[\frac{1}{r_1^2}\cos^2\theta + \frac{1}{r_2^2}\sin^2\theta\right]\cos^2\varphi + \frac{1}{r_3^2}\sin^2\psi\sin^2\varphi\right),$$

$$a_{35} = \left(\left[\frac{2}{r_1^2} - \frac{2}{r_2^2}\right]\cos\theta\cos\varphi\sin\theta\cos\psi\sin\varphi + \left[\frac{1}{r_1^2}\sin^2\theta + \frac{1}{r_2^2}\cos^2\theta\right]\cos^2\psi\cos^2\varphi + \left[\frac{1}{r_1^2}\cos^2\theta + \frac{1}{r_2^2}\sin^2\theta\right]\sin^2\varphi + \frac{1}{r_3^2}\sin^2\psi\cos^2\varphi\right),$$

The point of maximum for the $\tilde{U}(x_1, x_2)$ is $M_0(x_{10}, x_{20})$, where

$$x_{10} = \frac{e_2e_{12} - 2e_1e_{22}}{4e_{11}e_{22} - e_{12}^2}, \qquad x_{20} = \frac{e_1e_{12} - 2e_2e_{11}}{4e_{11}e_{22} - e_{12}^2}. \quad (16)$$

We move the coordinate system on the value $L_0 - M_0$ along the axe z and find the equation of the surface in the new coordinate system:

$$U(x_1,x_2) = \bar{U}(x_1,x_2) + L_0 - M_0. \quad (17)$$

The major and minor radiuses or curvature of $U(x_1,x_2)$ can be found through the next formulas:

$$\begin{cases} 1/R_1 = 2\left(e_{11}\cos^2\gamma + e_{12}\cos\gamma\sin\gamma + e_{22}\sin^2\gamma\right) \\ 1/R_2 = 2\left(e_{11}\sin^2\gamma - e_{12}\cos\gamma\sin\gamma + e_{22}\cos^2\gamma\right) \end{cases}, \quad (18)$$

and

$$\gamma = arctg\left(\frac{\cos\theta\,\sin\varphi + \sin\theta\,\cos\psi\,\cos\varphi}{\cos\theta\,\cos\varphi - \sin\theta\,\cos\psi\,\sin\varphi}\right). \quad (19)$$

We now notice that straightforward substituting (18) and (19) into (8) - (14), leads to the frequency spectrum ω_{nkl} and eigenvibrations Ψ_{kl} of a piezoelectric plate with an arbitrary oriented ellipsoidal surface.

3 Calculation of the static and motional capacitances of a piezoelectric resonator

Differential equation of the driven problem may be written [3] in the form

$$\sum_{n=1,3,5}^{\infty}\left(\frac{\partial^2\Psi_1^n}{\partial\xi_1^2} + \frac{\partial^2\Psi_1^n}{\partial\xi_2^2} - \left[\omega_n^2 + \frac{\omega_n^2 d_1}{L_0 R_1}\left(c(\beta)\xi_1^2 + c\left(\beta+\frac{\pi}{2}\right)\xi_2^2\right)\right]\times \Psi_1^n(\xi_1,\xi_2) - \rho\ddot{\Psi}_1^n\right) = \rho\omega^2(-1)^{\frac{n-1}{2}}\frac{e_{26}}{c^{(1)}}\frac{4V}{n^2\pi^2}, \quad (20)$$

where e_{26} is the piezoelectric constant, $c^{(1)}$ is the eigenvalues of the Christoffel's equation referred to the plate axes, V is the driving voltage, ρ is the mass density of the plate.

Using the following form for the solution of the (20)

$$\Psi_1^n = \Psi(\xi_1,\xi_2)\sin\left(\frac{n\pi x_2}{L}\right)e^{i\omega t}, \quad (21)$$

we write the equation of linear forced vibrations of the contoured resonator in the form

$$\frac{\partial^2\Psi}{\partial\xi_1^2} + \frac{\partial^2\Psi}{\partial\xi_2^2} + \left[\omega^2 - \omega_n^2 - \frac{\omega_n^2 d_1}{L_0 R_1}\left(c(\beta)\xi_1^2 + c\left(\beta+\frac{\pi}{2}\right)\xi_2^2\right)\right]\times \Psi(\xi_1,\xi_2) = \rho\omega^2(-1)^{\frac{n-1}{2}}\frac{e_{26}}{c^{(1)}}\frac{4V}{n^2\pi^2}, \quad (22)$$

Steady-state solution of (22) may be written as a sum of eigensolutions (13); thus

$$\Psi = \sum_k\sum_l H^{nkl}\Psi_{nkl}. \quad (23)$$

Substituting from (23) into (22) and employing the homogeneous form of (22) for each eigensolution, we obtain

$$\sum_k\sum_l H^{nkl}\left(\omega^2 - \omega_{nkl}^2\right)\Psi_{nkl} = \omega^2(-1)^{\frac{n-1}{2}}\frac{e_{26}}{c^{(1)}}\frac{4V}{n^2\pi^2}, \quad (24)$$

from which, with the help of the orthogonality of the Ψ_{kl}, we obtain

$$H^{nkl} = \frac{(-1)^{\frac{n-1}{2}} e_{26}4VF_{nkl}}{c^{(1)}n^2\pi^2\left[1-\left(\omega_{nkl}^2/\omega^2\right)\right]}, \quad (25)$$

where

$$F_{nkl} = \int_{D_e}\Psi_{nkl}dx_1dx_3. \quad (26)$$

Thus now we have the series representation of the steady-state solution for the linear forced vibrations of the contoured resonator. In the vicinity of a resonance, say the NKLth mode, one term in the sums in (24) with (21) dominates and other are negligible. Following [3] in the vicinity of said resonance the steady-state solution may be written in the form

$$\Psi_1 = H^{NKL}\sin\frac{N\pi x_2}{L}\Psi_{NKL}e^{i\omega t} - \frac{e_{26}Vx_2}{c^{(1)}L}e^{i\omega t},$$

$$\varphi = \frac{Vx_2}{L}e^{i\omega t} + \frac{e_{26}}{\varepsilon_{22}}H^{NKL}\Psi_{NKL}\times \left(\sin\frac{N\pi x_2}{L} - (-1)^{\frac{N-1}{2}}\frac{2x_2}{L}\right)e^{i\omega t}, \quad (27)$$

where φ is the electric potential.

The admittance Y_{NKL} of this contoured resonator operating in the NKLth mode is obtained by first substituting from (28) into

$$D_2 = e_{26}\Psi_{1,2} - \varepsilon_{22}\varphi_{,2}, \quad (29)$$

and then substituting from (26) into

$$I = -\int_{D_e}\dot{D}_2dx_1dx_3, \quad (30)$$

with the result

$$Y_{NKL} = \frac{e_{26}^2 + c^{(1)}\varepsilon_{22}}{c^{(1)}}i\omega\int_{D_e}\frac{1}{L}dx_1dx_3 + \frac{e_{26}^2 4F_{NKL}G_{NKL}}{\left(\omega_{NKL}^2/\omega^2 - 1\right)c^{(1)}N^2\pi^2}i\omega, \quad (31)$$

where

$$G_{NKL} = \int_{D_e}\frac{2}{L}\Psi_{NKL}dx_1dx_3. \quad (32)$$

The static and motional capacitances defined, respectively, by

$$C_0 = \frac{e_{26}^2 + c^{(1)}\varepsilon_{22}}{c^{(1)}}\int_{D_e}\frac{1}{L}dx_1dx_3,$$

$$C_{NKL} = \frac{e_{26}^2 4F_{NKL}G_{NKL}}{\left(\omega_{NKL}^2/\omega^2 - 1\right)c^{(1)}N^2\pi^2}. \quad (33)$$

4 Numerical simulation

In this section we use (13) and (33) to calculate amplitude distribution and motional capacitances of a doubly-rotated piezoelectric plate of a SBTC-cut crystal resonator, namely oriented by $yxbl/+16.3°/-34.5°$.

Fig. 3 illustrates amplitude distribution for the anharmonic with the index $nkl = 302$ of the B-mode vibration of the resonator [2] with the spherical radius of curvature $R = 260$ mm, plate thickness at the centre $L_0 = 1.0679$ mm and gold electrodes of thickness 800 A.

Fig. 4 illustrates amplitude distribution for the anharmonic with the index $nkl = 302$ of the B-mode vibration of the resonator with the ellipsoid surface (Fig. 2) with radiuses $r_1 = 130$ mm, $r_2 = 260$ mm, $r_3 = 260$ mm, plate thickness at the centre $L_0 = 1.0679$ mm and gold electrodes of thickness 800 A.

Fig. 5 illustrates amplitude distribution for the anharmonic with the index $nkl = 302$ of the B-mode vibration of the resonator with the ellipsoid surface (Fig. 2), rotated on the angle $\varphi = 30°$, with radiuses $r_1 = 130$ mm, $r_2 = 260$ mm, $r_3 = 260$ mm, plate thickness at the centre $L_0 = 1.0679$ mm and gold electrodes of thickness 800 A.

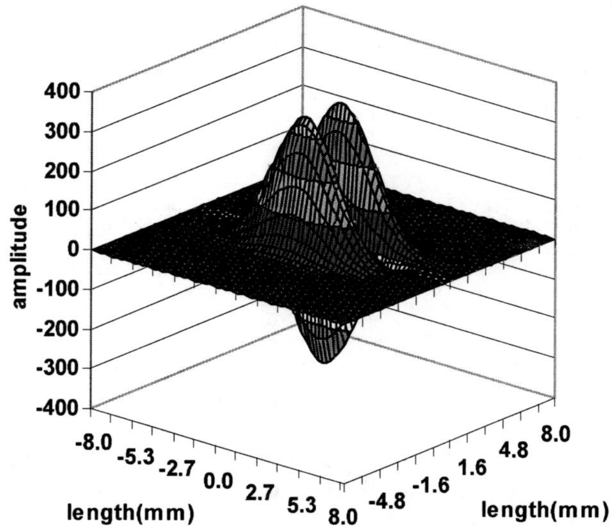

Fig.4. Distribution of amplitude of the B-mode of vibrations for (3,0,2) overtone for a ellipsoidal plate.

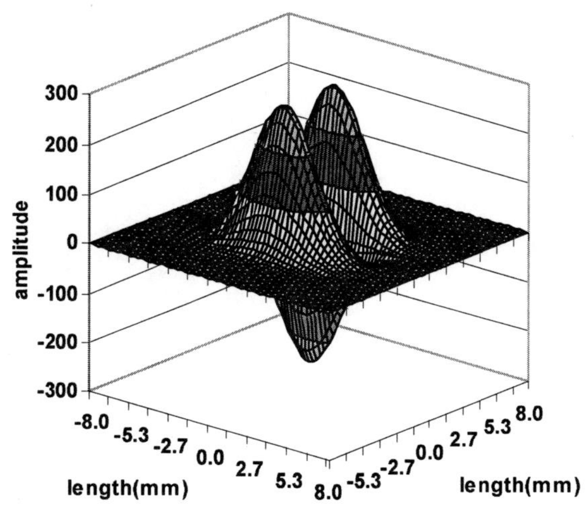

Fig.3. Distribution of amplitude of the B-mode of vibrations for (3,0,2) overtone for a spherical plate.

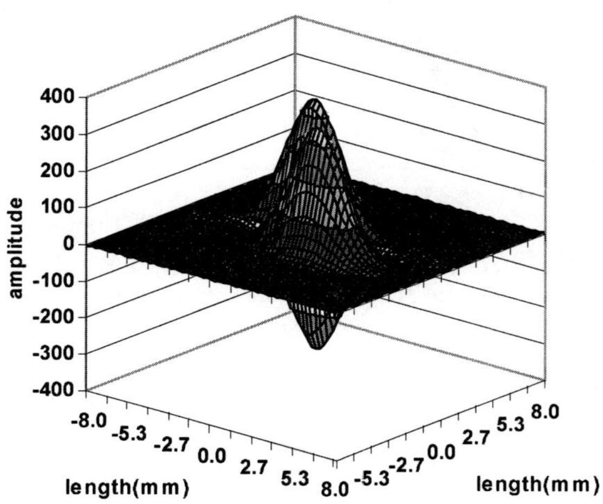

Fig.5. Distribution of amplitude of the B-mode of vibrations for (3,0,2) overtone for a rotated ellipsoidal plate.

Note that the changing the shape of the surface of piezoelectric plate from spherical to ellipsoidal influences on the amplitude distribution by increasing the height and decreasing the thickness of peeks on Fig. 4.

Fig. 6, 7 and 8 show the influence of the shape of the piezoelectric surface on the motional capacitances of the various harmonic and anharmonic overtones of the B-modes of vibrations for the case of spherical, ellipsoidal and rotated ellipsoidal plates, respectively. Corresponding resonator parameters were taken from previous cases of Fig. 3, 4 and 5, respectively.

Conclusions

In this report, a new mathematical model of piezoelectric resonator is presented for a piezoelectric plate of an arbitrary shape. The axes of major curvature of a surface are arbitrarily oriented with respect to the self axes of a piezoelectric plate. In the particular case of coincidence of these axes such a problem was successfully solved in [2]. Considered model allows for accounting the additional parameters of a manufactured plate: angle γ and the radius ratio $\delta = \frac{R_1}{R_2}$.

We intnd using this model to optimize performance of the

piezoelectric resonator with the fundamental vibration mode and the anharmonic sensors of environment.

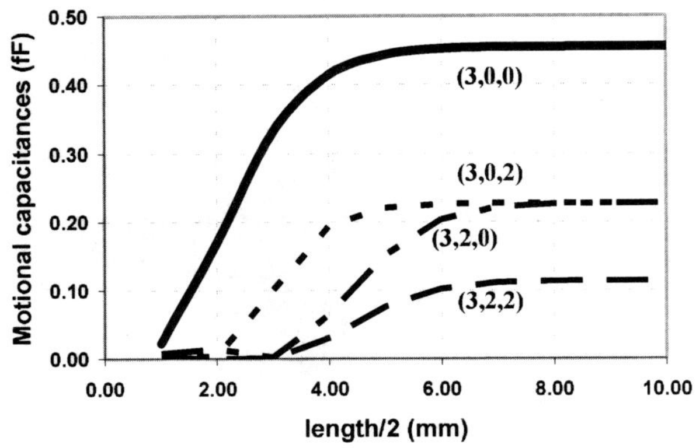

Fig.6. Motional capacitances versus length/2 of square electrodes for the B-mode of vibrations for a spherical plate.

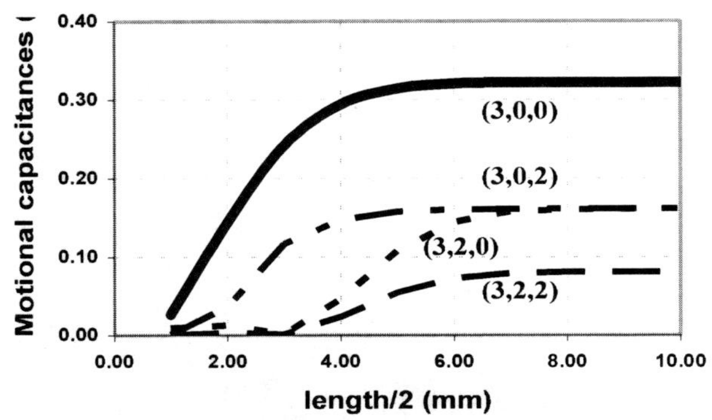

Fig.7. Motional capacitances versus length/2 of square electrodes for the B-mode of vibrations for a ellipsoidal plate.

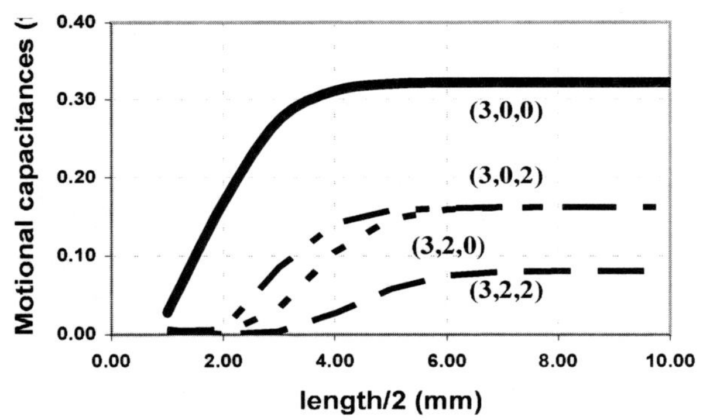

Fig.8. Motional capacitances versus length/2 of square electrodes for the B-mode of vibrations for a rotated ellipsoidal plate.

References

[1] S. S. Nedorezov and E. A. Ganenko, "Frequency spectrum of the piezoelectric resonators", *Proc. Int. Symp. on AFCSG,* St. Petersburg, Russia, pp. 89—92, (1998).

[2] B. K. Sinha, "Doubly rotated contoured quartz resonators", *IEEE Trans. Ultrason., Ferroelect., Feq. Contr.*, **48 (5)**, pp. 1162–1180, (2001).

[3] D. S. Stevens and H. F. Tiersten, "An analysis of doubly-rotated quartz resonators utilizing essentially thickness modes with transverse variation", *J. Acoust. Soc. Am,* **79 (6)**, pp. 1811–1826, (1986).

ACOUSTO-OPTIC METHODS FOR INVESTIGATION OF ELASTIC PROPERTIES OF CRYSTALS FROM THE LANGASITE FAMILY

VICTOR KLUDZIN, SERGEI KULAKOV, VLADIMIR KULAKOV
St.Petersburg State University of Aerospace Instrumentation
B. Morskaya Str, 67, St. Petersburg, Russia, 190000
weconf@home.ru

A great interest recently shown to crystals from the langasite family is a result of the combination of their unique properties and promising feature for those materials used in resonators, filters, and sensors in many fields of technology. Crystals from this family have very versatile compositions and types, they are produced at many different production facilities, and therefore, they require fast methods for the estimation and measurements of their elastic characteristics and properties. An acousto-optic method, which uses the phenomenon of a monochromatic optical wave diffraction by a periodical structure generated by a traveling acoustic wave, may be one of the most promising characterization methods. It requires minimum number of crystal samples and orientations, at the same time it yields complete information on elastic properties.

Specific conditions for visualization of acoustic waves and distributions in the crystals are discussed and formulated. Depending on the measurement plane, this method may show either the Schlieren images visualizing acoustic field distributions in the crystal, or the Schaeffer-Bergmann patterns visualizing the angular distributions of inverse velocities for two or three acoustic velocities simultaneously. That is why only one properly oriented sample can give all independent (6) elastic modulii of a crystal with the trigonal symmetry.

The photoelastic properties of the crystals from the langasite family can be measured by comparing them to the properties of a known material (fused quartz, for example).

Possible applications of the crystals from the langasite family require that the coefficients of elastic nonlinearity determining attenuation of acoustic waves in selected directions should be known. The acousto-optic method for the measurement of elastic nonlinearity parameters is most convenient, complete and visual, because it implements spectral analysis most naturally.

The measurement results for the elastic and photoelastic properties of the LGS and LGT crystals obtained by the acousto-optic method are presented. These results are compared with previously known data.

The research reported herein has been sponsored in part by the United States Army through its European Research Office.

STUDY OF PHYSICAL CAUSES OF INTERMODULATION DISTORTIONS IN HFF CRYSTAL RESONATORS

I. Abramzon, S. Pilshchicov, O.Rotova, S.Rudenko

"Magic Crystal Ltd", 18 K. Marx av., bld. 29, Omsk 644046, Russia
E-mail: mxl@omsknet.ru

Keywords: Quartz filter, HFF crystal, intermodulation, non-linearity

Abstract

The paper is devoted to experimental study of non-linearity of HFF crystal units and its influence on the quartz filter intermodulations. On statistical treatment of the measurement data the regression coefficients linking the crystal parameters non-linearity with the intermodulation level have been defined. The origin of the crystal non-linearity was explored through standard fabrication process stages. It was deduced that a main source of the non-linearity most probably arises during the electrodes deposition as the film imperfections creating some flicking capacitance and resistance depending of crystal vibration amplitude. Some ways of reduction of the crystal non-linearity are proposed and discussed in the paper.

1 Introduction

The intermodulation distortions in quartz filters being a substantial factor of degradation of sensitivity and selectivity of FM receivers remain in a focus of scientist and engineer efforts for many last years. As a main source of the phenomena non-linearity properties of the quartz resonators is a major subject of the investigations. In a number of works devoted to the crystals non-linearity the frequency and motional resistance drive level dependence (DLD) are considered as only evidence of the non-linearity, however without strong correlation with the intermodulations level [1, 2].

Physical causes of the frequency DLD and anomalies in motional resistance vs. dissipated power behavior were thoroughly explored in numerous works. Starting from quartz material non-linearity and the structure imperfections as main causes of the phenomena in the first works some authors came later to idea of influence on the main vibration mode of some surface masses capable of movement under the crystal vibrations [3]. Based on this idea physical model allowed satisfactory explanation of some non-linear phenomena in the crystal resonators was developed. At the same time apparent couple of the non-linearity effects with fabrication process of the resonator has not been revealed.

The goal of the present work was an attempt to develop a model linking the filter intermodulations with non-linearity of the crystal resonator parameters basing on statistical analysis of the experimental data. Upon this method of the non-linearity evaluation we attempted to deduce main fabrication process factors producing the non-linear effects as well as to develop a physical model explaining observed phenomena. On the base of obtained knowledge of the phenomena some ways of "improvement" of linearity of HFF crystals were proposed and evaluated.

2 Phenomenon model of intermodulations in the quartz filter

Since the intermodulation distortions of a quartz filter result from passing of two different frequency signals through the crystal unit the non-linearity of its parameters should define level of the intermodulation spectral products. Schematic drawing of simplest quartz filter is depicted in Fig.1. In general case the equivalent circuit of the crystal resonator consists of non-linear motional resistance R1, motional capacitance C1, motional inductance L1 and shunt capacitance C_0 which values depend on amplitude of passing through the filter signals. The crystal's resonance frequency f_0 fully defined by above parameters has however own dependence on the signal amplitude known as DLD.

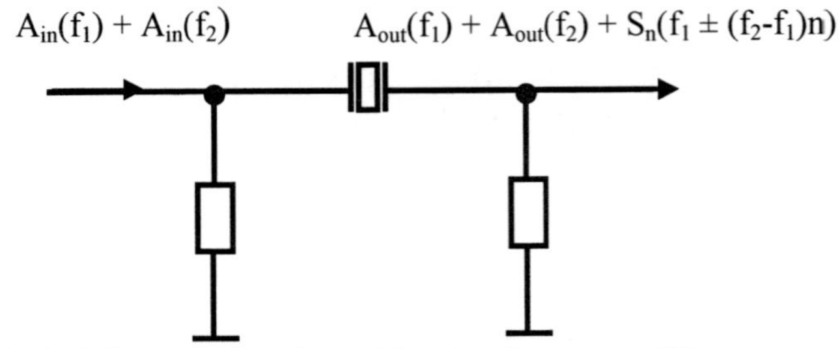

Fig.1. Schematic drawing of the simplest quartz filter.

In dependence on order of the crystal non-linearity the output spectra of two harmonic signals passing through the filter contains different set of spurious modes. In case of the third-order non-linearity a two-harmonic signal $A^{in}(f_1)$, $A^{in}(f_2)$ produces besides the same signals $A^{out}(f_1)$, $A^{out}(f_2)$ with their odd harmonics a set of intermodulation products with frequency $f_1 \pm (f_2 - f_1)n$ $(n = 1,2...\infty)$ and amplitude S_n, depending on the input signals and the non-linearity value of the crystal parameters.

Relationships between the crystal non-linearity and the intermodulation products level can be defined on solving the differential equations for passing the signals through the non-linear resonance circuit. Another approach can be based on statistical analysis of experimental data obtained with measurements of intermodulation spectra of a number of crystal units with different values of their non-linearity.

The third-order nonlinearity of the crystal parameters means its quadratic dependence on the driving current and can be determined via the non-linearity coefficients calculated as: $Nf = R1(\Delta f/f)/P$, $Nr = \Delta R1/P$, $Nc = R1(\Delta C1/C)/P$, $N_L = R1(\Delta L1/L)P$, $Ns = R1(\Delta Cs/Cs)P$, where Δf, $\Delta R1$, $\Delta C1$, $\Delta L1$, ΔCs are deviation of the parameters caused by change of the driving power from 0 to P; R1, C1, L1, f1, Cs – equivalent parameters measured at low power.

One should note that correct estimation of the non-linearity coefficients requires monotonic and smooth shape of the parameter versus drive level behavior as well as correspondence of the measurement power P to amplitude of signals passing though the quartz filter. Obviously for in-band and out-of band signals brining essentially different values of the current through the crystal unit appropriate values of the measurement power should be used resulting in different non-linearity coefficients.

Using the defined coefficients relative level of the intermodulation product S_n/A^{out} can be linked with the crystal non-linearity through a set of regression equations:

$$(S_n/A^{out})_i = K_r N_{ri} + K_C N_{Ci} + N_{Li} + K_f N_{fi} + K_S N_{Si}, \quad (1)$$

where $(S_n/A^{out})_i$- intermodulation n-product level measured with i–crystal unit having non-linearity coefficients N_{fi}, N_{ri}, N_{Ci}, N_{Li}, N_{Si}; K_r, K_C, K_L, K_f, K_S – coefficients of influence of the parameter non-linearity on the intermodulation product level.

Applying the correlation and regression analysis to the experimental data obtained with sufficient number of the units one can determine the influence of all the non-linearities on the filter intermodulation level.

As it follows from above expressions the influence coefficients do not depend on the crystal non-linearity but are stuck with the filter configuration and the circuit parameters.

On the base of considered method we studied influence on the filter intermodulation of the HFF crystals operating at 65-160 MHz frequencies and fabricated with the standard process. The carried out researches didn't reveal any nonlinearity of the shunt capacitance. The DLD of the crystal resonance frequency appeared to be conventional for the kind of units and didn't indicated distinct correlation with the intermodulation level for the most cases but seldom ones where sharp anomalies of the DLD were observed.

At the same time the motional parameters of the crystals revealed noticeable changes under the driving power essentially varying between the units and correlating with the filter intermodulation level. Typical dependences of the crystal parameters are depicted in fig.2.

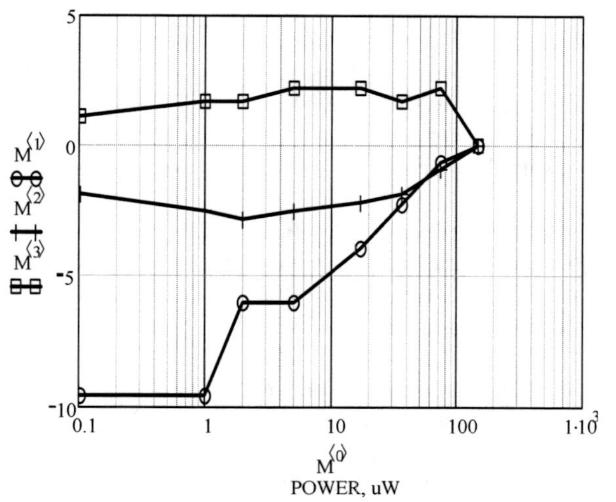

a)

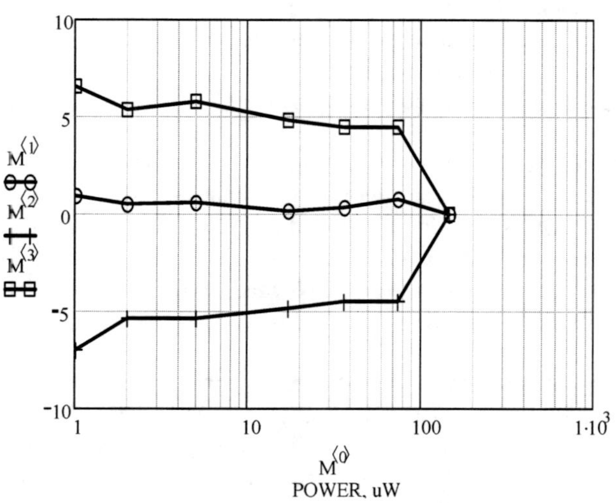

b)

Fig. 2. Typical DLD of equivalent parameters of HFF resonators N_r (corresponds to $M^{\langle 1 \rangle}$), N_L (corresponds to $M^{\langle 2 \rangle}$), N_C (corresponds to $M^{\langle 3 \rangle}$): a) unit #1 with "high power" non-linearity; b) unit #2 with "low power" non-linearity

As one can see from the curves unit #1 indicates greater non-linearity at low power level producing out-of band filter intermodulation, while unit #2 is more sensitive to high power variation that should be a cause of in-band intermodulation. Another corollary following from the researches is that the capacitance and inductance non-linearity are of almost equaled values and opposite sign, i.e. $N_C = - N_L$. This fact along with above conclusions allows study of the crystal non-linearity basing only on two coefficients N_r and N_C (or N_r and N_L).

3 Experimental study of the non-linearity origins through the fabrication process

As it follows from previous analysis the non-linearity of the crystals vary significantly between the units that is obvious result of influence on the parameters of the fabrication factors. To evaluate influence of the factors we've traced the non-linearity through the standard fabrication process.

As possible contributors into the crystal's non-linearity the main process stages such as chemical etching the inverted mesa-structure, base-plating the electrodes and final frequency adjustment were investigated.

To define possible impact of the etched surfaces into the non-linearity we measured the non-linearity coefficients N_r, N_C for two groups of HFF crystals after base-plating procedure. The first group consisted of the units coated with Cr-Ag electrodes just after the chemical etching. Second one contained only "linear" units selected from the first group and recoated with new electrodes after chemical removing first ones. Statistical treatment of measured non-linearity coefficients for the average value and the variance doesn't reveal essential difference of the group that is evidence of negligible contribution to the non-linearity of the etched surface conditions and priority of the base-plating factor.

Basing on the observed results a set of experiments with different electrodes geometry, material kind and the deposition methods were realized to link value of the influence with parameters of base-plating process. Table 1 represents average values N_r, N_C measured at ranged from 10 to 150 uW power in dependence on the electrodes features.

As one can see from the data best linearity is achieved with the Al and Ni-electrodes at some degradation of the parameter with raise of the film thickness. Taking into account essentially worse adhesion properties of Ag film we can presume that its greater influence on linearity is may be by detach from the crystal surface of some film parts mismatching the adhesion layer. The detached film parts forced to movement by the crystal plate vibrations

Material of the electrodes	*Mass loading %*	*Deposition method*	*Average values of non-linearity coefficients*	
			N_r, Ohm/W	*N_C, Ohm/W*
Cr- Ag	1.5	Thermal evaporation	$4.47*10^{-3}$	$3.24*10^{-3}$
Cr- Ag	3.0	Same	$6.84*10^{-3}$	$4.73*10^{-3}$
Cr- Ni	3.0	Magnetron	$4.72*10^{-3}$	$1.92*10^{-3}$
Al	0.7	Thermal evaporation	$2.49*10^{-3}$	$1.44*10^{-3}$
Al	1.5	Same	$4.7*10^{-3}$	$2.17*10^{-3}$

Table 1: Average values N_r, N_C in dependence on electrodes features.

create flicking capacitance of the crystal-electrode junction modulating the motional reactance of the crystal unit. The presumption was confirmed by experiment with deposition on the crystal units of Ag-electrodes without adhesive layer. Measurement of the unit linearity indicated cardinal degradation of the parameter.

Dependence of the non-linearity on the electrode sizes displayed inverse proportion at the high driving power that can be obviously explained by decrease of the driving current density followed by lower vibration amplitude. At the same time we didn't noticed any dependence on the electrode sizes at low levels of the driving power.

To explore influence on the crystal non-linearity of the final adjustment procedure we tested a few groups of the units base-plated with Al and Cr-Ag electrodes and final plated with Al, Au or Ag materials. On analysis of the experimental data we've come to a conclusion of small impact of the procedure on the non-linearity for all the materials that probably can be explained by rather small mass loading value of about 0.1% of the crystal frequency.

4 Physical model of HFF crystal non-linearity and the process improvement

On the base of the experimental results one can deduce some features regarding nature of the HFF crystal non-linearity as well as its origin in the fabrication process. Obtained data on the crystal motional reactance behavior (changes under the driving power, independence on ambient temperature and dependence on the electrode adhesion properties) allowed presumption of the non-linearity model based on influence of poor adhesive parts of the electrode film on the crystal parameters. Forced to vibration by the crystal resonance mode the electrodes free parts create the flicking capacitance in the crystal-film junction modulating the motional reactance of the crystal. Besides the vibration of the electrode parts causes fluctuation of the plate mode energy dissipation in the film resulting in flicking resistance included into the crystal circuit. Electrical model of the non-linearity is depicted in fig. 4.

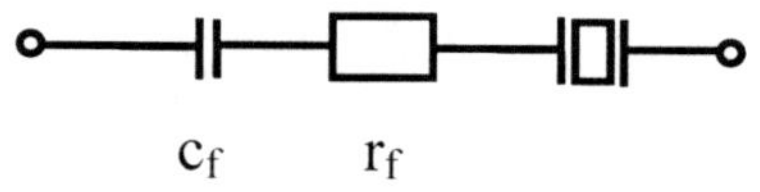

Fig. 4. Electrical model of the crystal unit non-linearity: Cf –the flicking capacitance of the crystal-electrode junction; Rf – the flicking resistance.

Average value of the flicking capacitance decreases with raise of the vibration amplitude that results in increase of the motional inductance, resistance and resonance frequency. The raise of the parameters vs. driving power was observed during described above experiments (fig. 2).

The electrodes deposition procedure is proved to be a main source for the poor adhesive film parts that explains the fact that the non-linearity depends on the film material and the deposition method. The Cr-Ag structure appeared to be worse option for the high-linear unit process comparing with aluminum or nickel materials.

Obviously removing the free fractions off the electrodes surface should improve the crystal linearity. That can be done by "cleaning" the electrode surfaces with ion-plazma etching procedure. Results of application of this method are depicted in fig. 5. As one can see noticeable improvement of the linearity appears at evaporation depth of about 30% of the electrode mass.

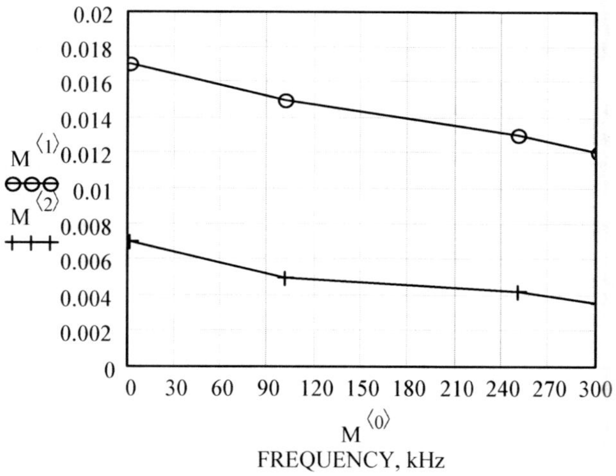

Fig. 5. Dependence of non-linearity coefficients N_r (corresponds to $M^{\langle 1 \rangle}$), N_C (corresponds to $M^{\langle 2 \rangle}$) on the electrode mass removing at initial mass-loading of 1 MHz.

Conclusion

1 The intermodulation distortions of quartz filters can be estimated on measurement of DLD of equivalent parameters of the crystal units R1, C1 and Ll. Meantime the unit's frequency DLD indicates only essential anomalies of the linearity.

2 Physical model of the non-linearity based on idea of flicking capacitance of the crystal surface-electrode junction created by poor adhesive electrode parts and flicking under the crystal vibrations was proposed and confirmed during the research.

3 An effective way of reduction of the non-linearity is a choice of high adhesive materials for the electrodes deposition and the final plating as well as plasma cleaning the electrodes surface from the poor adhesive fractions.

References

[1] E.Hafner, "Some phenomena in VHF crystal units", *Proc. 10th AFCS*, pp. 182-189, (1956).

[2] L.Dworsky, G.Kinsman, "A Simple Model for Quartz Crystal Resonator Low Level Drive Sensetivity and Monolotic Filter Intermodulation", *Transactions on UFFC*, **vol. 41**, pp.261-269, (1994).

[3] W.C. Horton, R.C.Smyth, "Experimental investigation of intermodulation in the quartz filters", *Proc. 27s AFCS*, pp. 180-186, (1972).

THE EFFECT OF ELECTRODES ON THE SURFACE ACOUSTIC WAVE PROPAGATION IN ELASTIC SOLIDS

Ji Wang*, Jingbo Lin*, Yongping Wan†, Zheng Zhong†

*Department of Engineering Mechanics and Materials Science, School of Engineering, Ningbo University, 1211 Zhengning Xilu Road, Ningbo, Zhejiang 315211, China
E-mail: **{wangji, linjingbo}@ nbu.edu.cn**
†Key Laboratory of Mechanics of MOE, School of Aeronautics, Astronautics, and Mechanics, Tongji University, 1239 Siping Road, Shanghai 200092, China

Keywords: surface, acoustic, waves, electrode, velocity

Abstract

Electrodes on the surface of an elastic substrate have important effects on the surface acoustic wave (SAW) propagation, and related changes like wave velocity and mode shapes due to the presence of electrodes should be considered in the analysis so the information available to practical applications can be accurate and adequate. Because the essential characteristics of surface acoustic wave propagation is obtained from semi-infinite solids, practical applications related to device analysis and design are derived with assumptions and approximations through simplifying the electrodes and reduction of numbers of variables.

In an attempt to improve the accuracy and complexity of SAW analysis, a simple electrode as a thin metal layer over an infinite isotropic elastic solid is considered for a demonstrative study. With precise solutions from three-dimensional equations of elasticity for surface acoustic wave propagation, we can extend the results to discontinuous periodic electrodes, or the interdigital transducers (IDTs), for better estimation of the effect on wave velocity and mode shapes. In a manner similar to our previous work on bulk acoustic wave resonators, simple but effective concepts like the thickness and mass ratios are introduced for a direct estimation and easy solution of wave properties. The method will be further extended to anisotropic materials that are widely used in surface acoustic wave devices as substrates for better and practical analytical approaches closer to demands of design process. Eventually, we want to further apply these methods to the two-dimensional theory we are working on so the analysis of surface acoustic wave propagation, both analytical and numerical, can be improved for the device design.

1 Introduction

Surface acoustic wave resonator design requires a detailed analysis on essential properties of waves like the phase velocity, which can only be obtained for semi-infinite elastic solids [1-4]. The solutions from three-dimensional elasticity and piezoelectricity equations have been widely used for the study of materials for SAW devices [5], and the solutions are also part of the evolving theory for SAW analysis. Current analytical methods for SAW resonators are very limited in accuracy and functions because there are no analytical methods can analyze the surface acoustic waves propagating in finite elastic solids, so we have turned to finite element method and boundary element method for numerical solutions often with further simplifications [5-8]. However, even the powerful numerical methods cannot be enough in the SAW analysis because the size of the problems usually exceeds the capacity of computers we can find in a typical industrial and research environment. As an alternative, many analyses are done with one period of the complicated structures, and these results offer some best approximations for engineering applications.

These facts and reality show the difficulties in the SAW analysis, and it reminds us that SAW problems are far from well studied, as we compare it with bulk acoustic wave problems we are familiar with. By a careful examination of surface acoustic waves in finite solids, the similarity with bulk acoustic waves should offer some hints on a possible theory for the two-dimensional analysis. A careful comparison with plate theories [9-11] invites us to work out a similar method, and this has been realized in our recent studies [12-14]. With the arrival of the two-dimensional theory, we need to turn to some important complications in a typical SAW resonator model for simplification and possible solutions. Naturally, the electrodes on the top of the SAW resonator substrate, although discontinuous in their usual layout, are the primary problem we need to work on.

In this paper, we started with a thin metal layer on top of semi-infinite isotropic elastic solids to represent the simplest electrode configuration. We want to use the results of this study for real electrodes that are in the form of IDTs for practical consideration. We leave applications of the results in the two-dimensional theory in this study to future work.

2 Surface acoustic waves in a semi-infinite isotropic solid with a thin metal layer on top

In practical SAW resonators, electrodes are generally the discrete IDTs, and the substrate is usually a finite piezoelectric solid. As a result, the analysis is actually very complicated and challenging from engineering point of view, and the problem has been a focus of research for decades in frequency device industry. For surface acoustic waves propagating in a semi-infinite solid, the simplest electrode layout is a thin metal layer on the top of solids, as shown in Fig. 1. In this study, the substrate and electrode are considered isotropic.

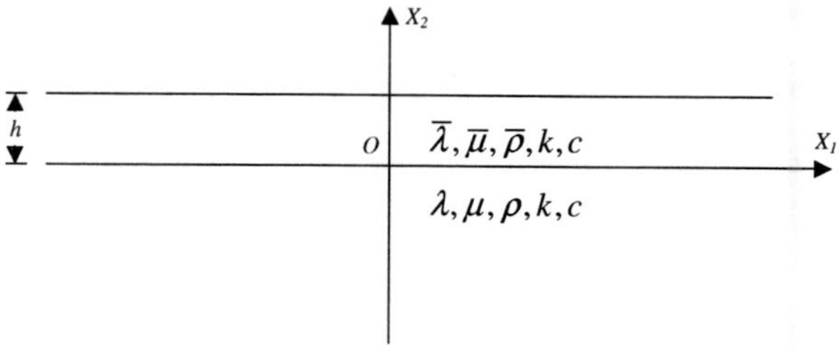

Fig. 1 Semi-infinite solid with a thin top layer

For surface acoustic waves propagate in semi-infinite solids along x_1 direction, we assume the displacements are

$$u_1 = Ae^{k\beta x_2}e^{i(kx_1-ct)},\quad u_2 = Be^{k\beta x_2}e^{i(kx_1-ct)}, \tag{1}$$

and in top thin metal layer we assume the displacements have two terms as

$$\bar{u}_1 = \left(\bar{A}_1e^{-k\bar{\beta}x_2} + \bar{A}_2e^{k\bar{\beta}x_2}\right)e^{i(kx_1-ct)},\quad \bar{u}_2 = \left(\bar{B}_1e^{-k\bar{\beta}x_2} + \bar{B}_2e^{k\bar{\beta}x_2}\right)e^{i(kx_1-ct)}, \tag{2}$$

where $A, B, \bar{A}_i, \bar{B}_i, k, \beta, \bar{\beta}, c, t$ are amplitudes of the substrate, amplitudes of the thin layer, wave number, wave vector of the bottom layer, wave vector of the top thin layer, phase velocity, and time, respectively.

With displacements specified in (1), we can obtain strains in semi-infinite solids in abbreviated notations as

$$S_1 = ikAe^{k\beta x_2}e^{i(kx_1-ct)},\quad S_2 = k\beta Be^{k\beta x_2}e^{i(kx_1-ct)},\quad S_6 = k(\beta A + iB)e^{k\beta x_2}e^{i(kx_1-ct)}. \tag{3}$$

Similarly for the strains in the top layer, from (2) we have

$$\bar{S}_1 = ik\left(\bar{A}_1e^{-k\bar{\beta}x_2} + \bar{A}_2e^{k\bar{\beta}x_2}\right)e^{i(kx_1-ct)},$$
$$\bar{S}_2 = k\bar{\beta}\left(\bar{B}_1e^{-k\bar{\beta}x_2} + \bar{B}_2e^{k\bar{\beta}x_2}\right)e^{i(kx_1-ct)},$$
$$\bar{S}_6 = k\left[\bar{\beta}\left(\bar{A}_1e^{-k\bar{\beta}x_2} + \bar{A}_2e^{k\bar{\beta}x_2}\right) + i\left(\bar{B}_1e^{-k\bar{\beta}x_2} + \bar{B}_2e^{k\bar{\beta}x_2}\right)\right]e^{i(kx_1-ct)}. \tag{4}$$

With the strains in (3) and (4), we can define the stresses in substrate layer as

$$T_1 = k[(\lambda + 2\mu)iA + \lambda\beta B]e^{k\beta x_2}e^{i(kx_1-ct)},$$
$$T_2 = k[\lambda iA + (\lambda + 2\mu)\beta B]e^{k\beta x_2}e^{i(kx_1-ct)}, \tag{5}$$
$$T_6 = k\mu(\beta A + iB)e^{k\beta x_2}e^{i(kx_1-ct)},$$

and in the electrode layer

$$\bar{T}_1 = k\left[(\bar{\lambda} + 2\bar{\mu})i\left(\bar{A}_1e^{-k\bar{\beta}x_2} + \bar{A}_2e^{k\bar{\beta}x_2}\right) + \bar{\lambda}\bar{\beta}\left(\bar{B}_1e^{-k\bar{\beta}x_2} + \bar{B}_2e^{k\bar{\beta}x_2}\right)\right]e^{ik(x_1-ct)},$$
$$\bar{T}_2 = k\left[\bar{\lambda}i\left(\bar{A}_1e^{-k\bar{\beta}x_2} + \bar{A}_2e^{k\bar{\beta}x_2}\right) + (\bar{\lambda} + 2\bar{\mu})\bar{\beta}\left(\bar{B}_1e^{-k\bar{\beta}x_2} + \bar{B}_2e^{k\bar{\beta}x_2}\right)\right]e^{ik(x_1-ct)},$$
$$\bar{T}_6 = k\bar{\mu}\left[\bar{\beta}\left(\bar{A}_1e^{-k\bar{\beta}x_2} + \bar{A}_2e^{k\bar{\beta}x_2}\right) + i\left(\bar{B}_1e^{-k\bar{\beta}x_2} + \bar{B}_2e^{k\bar{\beta}x_2}\right)\right]e^{ik(x_1-ct)}. \tag{6}$$

The equations of motion for elastic solids with given displacement fields are

$$T_{1,1} + T_{6,2} = \rho\ddot{u}_1,\quad T_{6,1} + T_{2,2} = \rho\ddot{u}_2, \tag{7}$$

and substitution of (5) and (6) into (7) for electrode layer we have

$$\left[(\bar{\lambda} + 2\bar{\mu}) - \bar{\mu}\bar{\beta}^2 - \bar{\rho}c^2\right]\bar{A}_1 + \bar{B}_1(\bar{\lambda} + \bar{\mu})i\bar{\beta} = 0,$$
$$(\bar{\lambda} + \bar{\mu})i\bar{\beta}\bar{A}_1 + \left[\bar{\mu} - (\bar{\lambda} + 2\bar{\mu})\bar{\beta}^2 - \bar{\rho}c^2\right]\bar{B}_1 = 0, \tag{8}$$

and

$$\left[(\bar{\lambda} + 2\bar{\mu}) - \bar{\mu}\bar{\beta}^2 - \bar{\rho}c^2\right]\bar{A}_2 - (\bar{\lambda} + \bar{\mu})i\bar{\beta}\bar{B}_2 = 0,$$
$$(\bar{\lambda} + \bar{\mu})i\bar{\beta}\bar{A}_2 - \left[\bar{\mu} - (\bar{\lambda} + 2\bar{\mu})\bar{\beta}^2 - \bar{\rho}c^2\right]\bar{B}_2 = 0. \tag{9}$$

For the substrate, we have the familiar surface acoustic wave equations as

$$\left[(\lambda + 2\mu) - \mu\beta^2 - \rho c^2\right]A - (\lambda + \mu)i\beta B = 0,$$
$$(\lambda + \mu)i\beta A - \left[\mu - (\lambda + 2\mu)\beta^2 - \rho c^2\right]B = 0. \tag{10}$$

Equations (8) and (9) are actually the surface acoustic waves we are familiar with, and the solutions are also the same except the difference resulted from the negative argument of displacements in (1). Without further elaboration, we have solutions from (8) and (9) as

$$\bar{\beta}_1^2 = 1 - \frac{c^2}{\bar{c}_L^2},\ \bar{\beta}_2^2 = 1 - \frac{c^2}{\bar{c}_T^2},$$
$$\left(\frac{\bar{B}_1}{\bar{A}_1}\right)_1 = i\bar{\beta}_1,\ \left(\frac{\bar{B}_1}{\bar{A}_1}\right)_2 = \frac{i}{\bar{\beta}_2}, \tag{11}$$
$$\left(\frac{\bar{B}_2}{\bar{A}_2}\right)_1 = -i\bar{\beta}_1,\ \left(\frac{\bar{B}_2}{\bar{A}_2}\right)_2 = -\frac{i}{\bar{\beta}_2},$$

where

$$\bar{c}_L^2 = \frac{\bar{\lambda} + 2\bar{\mu}}{\bar{\rho}},\ \bar{c}_T^2 = \frac{\bar{\mu}}{\bar{\rho}}, \tag{12}$$

are the longitudinal and transverse wave velocity of the top layer. As a result, the displacements given in (2) can be rewritten as

$$\bar{u}_1 = \left(\bar{A}_{11}e^{-x_2k\bar{\beta}_1} + \bar{A}_{12}e^{-x_2k\bar{\beta}_2} + \bar{A}_{21}e^{x_2k\bar{\beta}_1} + \bar{A}_{22}e^{x_2k\bar{\beta}_2}\right)e^{ik(x_1-ct)},$$
$$\bar{u}_2 = \left(i\bar{\beta}_1\bar{A}_{11}e^{-x_2k\bar{\beta}_1} + \frac{i}{\bar{\beta}_2}\bar{A}_{12}e^{-x_2k\bar{\beta}_2} - i\bar{\beta}_1\bar{A}_{21}e^{x_2k\bar{\beta}_1} - \frac{i}{\bar{\beta}_2}\bar{A}_{22}e^{x_2k\bar{\beta}_2}\right)e^{ik(x_1-ct)}, \tag{13}$$

where the displacement amplitudes $\bar{A}_{ij}(i, j = 1,2)$ are to be determined.

For the semi-infinite elastic solid as the substrate, we have the familiar surface acoustic wave solutions as

$$u_1 = \left(A_1e^{-x_2k\beta_1} + A_2e^{-x_2k\beta_2}\right)e^{ik(x_1-ct)},$$
$$u_2 = \left(i\beta_1A_1e^{-x_2k\beta_1} + \frac{i}{\beta_2}A_2e^{-x_2k\beta_2}\right)e^{ik(x_1-ct)}, \tag{14}$$

with decaying parameters, which are unique in surface acoustic waves to guarantee the exponential decaying of the displacements in the thickness direction and relate to material properties like the density and elastic constants,

$$\beta_1^2 = 1 - \frac{c^2}{c_L^2}, \beta_2^2 = 1 - \frac{c^2}{c_T^2},$$
$$\left(\frac{B}{A}\right)_1 = -i\beta_1, \left(\frac{B}{A}\right)_2 = -\frac{i}{\beta_2}, \tag{15}$$

and velocities

$$c_L^2 = \frac{\lambda + 2\mu}{\rho}, c_T^2 = \frac{\mu}{\rho}. \tag{16}$$

Referring to Fig.1, continuity conditions at the interface are

$$u_1 = \bar{u}_1, u_2 = \bar{u}_2, T_2 = \bar{T}_2, T_6 = \bar{T}_6 \text{ at } x_2 = 0, \tag{17}$$

and traction-free boundary conditions on surface of the electrode are

$$\bar{T}_2 = 0, \bar{T}_6 = 0 \text{ at } x_2 = -h. \tag{18}$$

With known displacements and stresses in (5), (6), (13), and (14), applying boundary conditions in (17) and (18), we have the boundary condition equations as

$$\left[\lambda - (\lambda + 2\mu)\beta_1^2\right]A_1 - 2\mu A_2 - \left[\bar{\lambda} - (\bar{\lambda} + 2\bar{\mu})\bar{\beta}_1^2\right]\bar{A}_{11} + 2\bar{\mu}\bar{A}_{12} -$$
$$\left[\bar{\lambda} - (\bar{\lambda} + 2\bar{\mu})\bar{\beta}_1^2\right]\bar{A}_{21} + 2\bar{\mu}\bar{A}_{22} = 0,$$

$$\mu\left[2\beta_1 A_1 + \left(\beta_2 + \frac{1}{\beta_2}\right)A_2\right] -$$

$$\bar{\mu}\left[2\bar{\beta}_1\bar{A}_{11} + \left(\bar{\beta}_2 + \frac{1}{\bar{\beta}_2}\right)\bar{A}_{12} - 2\bar{\beta}_1\bar{A}_{21} - \left(\bar{\beta}_2 + \frac{1}{\bar{\beta}_2}\right)\bar{A}_{22}\right] = 0,$$

$$\left(\beta_1 A_1 + \frac{A_2}{\beta_2}\right) - \bar{\beta}_1\bar{A}_{11} - \frac{\bar{A}_{12}}{\bar{\beta}_2} + \bar{\beta}_1\bar{A}_{21} + \frac{\bar{A}_{22}}{\bar{\beta}_2} = 0,$$

$$A_1 + A_2 - \bar{A}_{11} - \bar{A}_{12} - \bar{A}_{21} - \bar{A}_{22} = 0,$$

$$\left[\bar{\lambda} - (\bar{\lambda} + 2\bar{\mu})\bar{\beta}_1^2\right]\bar{A}_{11}e^{hk\bar{\beta}_1} - 2\bar{\mu}\bar{A}_{12}e^{hk\bar{\beta}_2} +$$
$$\left[\bar{\lambda} - (\bar{\lambda} + 2\bar{\mu})\bar{\beta}_1^2\right]\bar{A}_{21}e^{-hk\bar{\beta}_1} - 2\bar{\mu}\bar{A}_{22}e^{-hk\bar{\beta}_2} = 0,$$

$$2\bar{\beta}_1\bar{A}_{11}e^{hk\bar{\beta}_1} + \left(\bar{\beta}_2 + \frac{1}{\bar{\beta}_2}\right)\bar{A}_{12}e^{hk\bar{\beta}_2} - 2\bar{\beta}_1\bar{A}_{21}e^{-hk\bar{\beta}_1} -$$

$$\left(\bar{\beta}_2 + \frac{1}{\bar{\beta}_2}\right)\bar{A}_{22}e^{-hk\bar{\beta}_2} = 0. \tag{19}$$

It is clear that the surface acoustic waves existing in the substrate can be solved from (19) with material properties like the elastic constants and densities of both materials. Of course, the thickness of the metal layer is also an important parameter. It is natural for us to relate the surface acoustic waves in this study to well-known Love wave [1-4], which is the transverse, or SH, surface waves in a similarly layered media. To have Love wave in the top layer, there is a requirement that the transverse wave velocity of the elastic substrate, c_T, should be larger than that of the electrode, $\bar{c}_T$. Indeed, for the surface acoustic waves in this study to exist in the substrate, similar requirement on the wave velocity of each material is also needed in our case. One important feature from this study is that the surface acoustic waves here is also dispersive, just like the surface acoustic waves in typically layered structures. As a result, to have surface acoustic waves in the substrate of such a layered structure, the transverse wave velocity of the electrode layer material should be larger.

To simplify (19), we define the thickness to wavelength ratio and utilize the relationship between wavenumber and wavelength as follows

$$H = \frac{h}{\lambda_l}, k\lambda_l = 2\pi. \tag{20}$$

Through solving (19), we can evaluate the effect of the top layer on the surface acoustic waves in the layered structure.

3 Numerical results

To demonstrate the application of equations we present in the previous section, we calculated the effect of a layered structure with following material parameters,

$c_T = 3400\,\text{m/s}, c_L = 5880, \bar{c}_T = 5300, \bar{c}_L = 9600.$

The relation between the surface acoustic wave velocity in terms of the transverse wave velocity in the substrate, c_T, and the electrode layer thickness is given in Fig. 2.

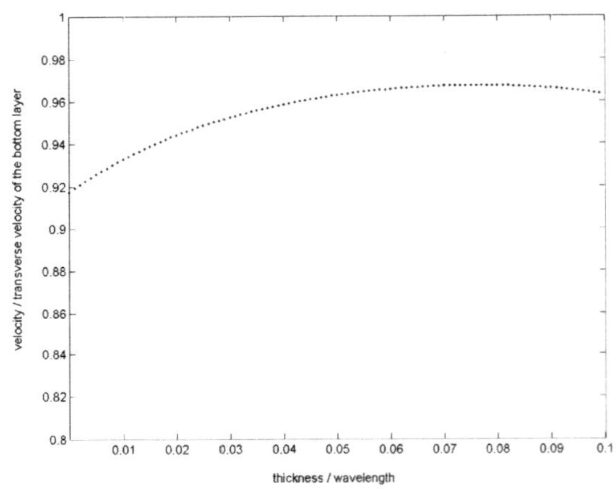

Fig. 2 Surface acoustic wave velocity versus electrode thickness.

It is clear from Fig. 2 that the surface acoustic wave velocity in the structure will increase as thickness of the top layer increases, but there will be a turnover point. It is easy to understand that the gradual dominance of the surface acoustic waves in one media will change the wave velocity, which is being bounded between the transverse wave velocities of the two different materials.

4 Discussions

We analyzed the surface acoustic waves in a layered structure for the consideration of the effect of a thin metal electrode. Although the devices we are familiar with have discontinuous interdigital transducers (IDTs) that should be considered as discrete, the results here nonetheless will be used for the studies of surface acoustic waves from a different approach. This is part of our new effort in SAW analysis represented by the two-dimensional theory for surface acoustic waves in finite solids, and the combination of our work on the piezoelectric considerations and electrode effects will make the two-dimensional theory more practical for engineering problems.

Acknowledgements

This research is supported by *Qian* River Fellow Fund of Zhejiang Province established by Zhejiang Provincial Government and Ningbo University and administered by Ningbo University.
This research is also supported by the National Natural Science Foundation of China (Project No. 10125209).

References

1. J. D. Achenbach: *Wave Propagation in Elastic Solids*, North-Holland/American Elsevier, 1987.
2. K. F. Graff: *Wave Motion in Elastic Solids*, Dover, 1991.
3. B. A. Auld: *Acoustic Fields and Waves in Solids* (I & II), Krieger Publishing Company, Malabar, Florida, 1990.
4. D. Royer and E. Dieulesaint: *Elastic Waves in Solids* (I & II), Springer, 2000.
5. K.-Y. Hashimoto: *Surface Acoustic Wave Devices in Telecommunications: Modelling and Simulation*, Springer, Berlin, 2000.
6. Yook-Kong Yong: "Analysis of periodic structures for BAW and SAW resonators", *Proceedings of 2001 IEEE Ultrasonics Symposium*, Oct 7-10, 2001, Atlanta, GA, USA, 781-790.
7. S. Yoon, J.-D. Yu, S. Kanna, M. Oshio, and M. Tanaka: "Finite element analysis of the substrate thickness on traveling leaky surface acoustic waves", *Proceedings of 2003 IEEE International Ultrasonics Symposium*, 1696-1699.
8. R.D. Mindlin: "High frequency vibrations of piezoelectric crystal plates", *Intl. J. Solids Struct.*, **8**:891-906, 1972.
9. P.C.Y. Lee, S. Syngellakis, and J. P. Hou: "A two-dimensional theory for high-frequency vibrations of piezoelectric crystal plates with or without electrodes", *J. Appl. Phys.*, **61**(4):1249-1262, 1987.
10. R. C. Peach: "A normal mode expansion for the piezoelectric plates and certain of its applications", *IEEE Trans. UFFC*, **35**(5):593-611, 1988.
11. J. Wang and K.-Y. Hashimoto: "A two-dimensional theory for the analysis of surface acoustic waves in anisotropic elastic solids", *Proceedings of 2003 IEEE International Ultrasonics Symposium*, 637-640, 2003.
12. Ji Wang and Jingbo Lin, "A two-dimensional theory for surface acoustic wave analysis in finite piezoelectric solids", *J. Intelligent Material Structures and Systems*, submitted.
13. Ji Wang and Jingbo Lin, "An approximate determination of surface acoustic wave velocity in anisotropic piezoelectric solids and its applications in two-dimensional analysis", *Acata Mechnica Sinica*, submitted.
14. Vicent Laude, Alexandre Reinhardt, Mikael Wilm, Abdelkrim Khelif, and Sylvain Ballandras, "Fast FEM/BEM simulation of SAW devices via asymptotic waveform evaluation", *IEEE Trans UFFC*, **51**(3):359-363, 2004.

CHEMICAL ANALYSIS OF LGS AND LGT CRYSTALS BY ICP-AES

M. ASSOUD[1,2], J.J. BOY[2,*], K. YAMNI[1], A. ALBIZANE[1]

[1]Laboratoire des Matériaux Catalyse et Environnement, FST Mohammedia BP 146 - 20650 Mohammedia Maroc
[2]LCEP, ENSMM & FEMTO-ST - 26 Chemin de l'Epitaphe- 25000 Besançon France
*Email : jjboy@ens2m.fr

Keywords: Langasite family, Alkali fusion, ICP-AES, IR spectrum, Czochralski method

Abstract

In this paper we report the method used to dissolve the Langasite and Langatate crystals. Then the characterization of chemical impurities like Al, Fe, Na, Li... incorporated in the crystal during the pooling or coming from strating oxides (La_2O_3, Ga_2O_3, Ta_2O_3 and SiO_2) is discussed. Finally, the comparison with the quartz crystal material has been done as well in the IR range as in term of chemical impurities.

1 Introduction

Crystals of the Langasite family (LGS: $La_3Ga_5SiO_{14}$, LGT: $La_3Ga_{5.5}Ta_{0.5}O_{14}$ or LGN: $La_3Ga_{5.5}Nb_{0.5}O_{14}$) belong to the same class (32) of the quartz crystal. It has attracted a lot of attentions because of the combination of their properties such as high piezoelectric coupling, temperature compensation and low acoustic losses... It appears to be new candidates to achieve a large pass bandwidths for filters or to improve frequency stability for oscillators [1]. Another very important property is that the compound melts congruently so that large single crystals of langasite can be produced by the conventional Czochralski melt pulling technique. This makes commercial scale production feasible. By growing the crystals from the melt, the possibility of a certain numbers of impurities introduced in the crystals is not to exclude, although they used a highly pure La_2O_3, Ga_2O_3 and SiO_2. LGS is isostructural to $Ca_3Ga_2Ge_4O_{14}$. It is crystallized in the space group P321, and its lattice is composed of 4 kinds of sublattices represented as $A_3BC_3D_2O_{14}$. "La" atoms occupied the decahedral site (A), "Ga" atoms the octahedral (B) and tetrahedral sites (C), and "Ga" and "Si" atoms are in the smallest tetrahedral one (D) [2,3]. In LGT, La^{3+} occupies the A sites, Ga^{3+} fully occupies the C and D sites, and Ta^{5+} the B sites [4,5]. Finally, no kind of helicoidal structure observed in quartz exists in LGT.

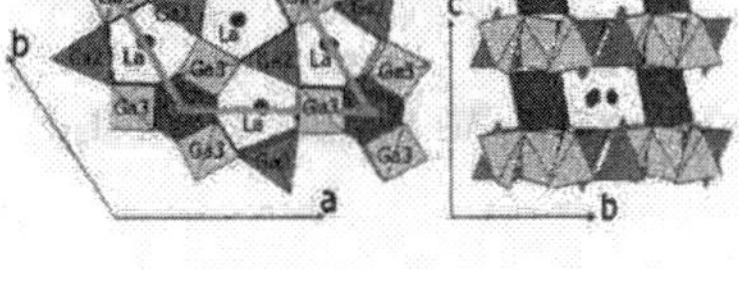

Figure 1: (a,b) and (b,c) projections

As for quartz crystal, the quality of the single crystal is usually determined by Infra-Red spectrometry and ICP-AES spectroscopy. This last technique is usually used to characterize the contents of impurities incorporated in the lattice when a complete dissolution of the crystal is achieved.

Unfortunately, a perfect crystal does not exist and contains impurities or defects depending on the growth conditions and on the quality of the starting materials. Its physical properties are also strongly depending on the nature of the defects.

So, in this paper we report the method of sample's preparation and the characterization of certain impurities in LGS and LGT crystals.

2 Sample preparation

For ICP spectroscopy, the sample is always introduced as a liquid in the instrument. To achieve this state, an acid digestion is often used.

The samples can be dissolved by direct acid attack or by fusion technique. In this last method, the sample is weighed into a metal crucible, mixed with a suitable flux and fused over a hot flame. The resulting melt is leached in either water or acid.

Before each dissolution, we need to protect working atmosphere, two principal causes of uncertainty having to take into account due to the very low contents observed in the analyzed crystals:

1) it is possible to involve significant fluctuations on the results when the contents are close to the limits of the apparatus (between tens of ppb and a few ppm), especially for alkaline elements and sometimes trivalent ions. To overcome this problem, it is necessary to perform several different measurements for each element.

2) even if a weak accidental contamination is not detectable, it consequently can drastically modify results. In order to minimize these uncertainties, the blank and calibration solutions were prepared under the same conditions as the sample solution. Only under these conditions, the level of pollution will be the same for all of them.

Step one: scouring

The solid sample of LGx, placed in a polyethylene vessel, is initially cleaned successively by:

- acetone (to remove the adhesives used for sawing operations)
- 65 % nitric acid solution during 20 minutes.

Step two: washing

This washing is carried out in a Teflon beaker by a 12 molar chlorydric acid solution. The rinsing is realized by ultra pure water which has a resistivity of 18 MΩ cm.

Step three: melt of the mixture

After grinding the sample, the powder is weighed and mixed with Lithium metaborate and lithium tetraborate. The ratio of the flux added in the mixture must be correlated to the mass of the sample in order to realize the full dissolution of the weighed mass of crystal. According to the mass of the Lithium metaborate and lithium tetraborate, 0.6g and 0.4g respectively are used for 0.2g of LGS. Then the sample is placed in a platinum crucible and heated during 45 min. at 1100°C. The same procedure is used for LGT with 2:3 ratio.

Step four: chemical attack

The chemical attack is directly realized in the crucible where is synthesized the borax bead by using nitric acid.

Last step: preparation of the sample solution

The matrix and the foreign elements are dissolved in an 8% nitric acid, up to 200 ml.

3 ICP-AES method

Usually, the element contents are measured by ICP-AES techniques (Inductively Coupled Plasma - Atomic Emission Spectrometry) in order to detect very low concentrations as in synthetic crystals (about few ppm or a dozen ppb for alkali ions). ICP-AES is a photometric method using plasma at very high temperature (between 5,000 and 10,000 K) so that each element is excited and emits characteristic photons. The concentrations are determined by comparison with a standard solution, prepared under the same conditions.

As for quartz crystal, we want to determine the existence of trace elements in the lattice of LGx crystals such as Al, Fe, Na, or K. Table I summarizes some results obtained from different varieties of crystal samples.

Real impurities concentrations depend first on starting materials and materials of the crucible as Rhodium for example which can be etched out of the Pt/Rh crucible and incorporated in the crystal to give a deep reddish brown color [3, 6]. We note that generally the levels of impurities are lower than in the case of quartz surely due to the difference of the growth method. It is known that the synthetic quartz crystal is obtained by the hydrothermal method for which it is used solvent and aqueous solutions in contrary to the Czochralski one.

The oldest sample (LGS1 from FOMOS) was grown more than ten years ago, explaining its very high content of foreign elements in contrary to the LGS2 sample, nevertheless reddish. We have to note that this color can not be attributed to the existence of Rhodium because it did not appear in the ICP analysis.

In the case of LGT (grown by Chai et al), we see that the contents of impurities is lower than in the other cases.

Content (in mg/l)	Al	Fe	Na	K
LGS1	66	35	30	60
LGS2	4.5	5	5	5
LGT	1.6	4	20	8
Quartz	4.64	8.75	2.7	1.43

Table I: defects contents for different samples

4 Energy dispersive X-ray analysis

Moreover to tentatively confirm the presence of some impurities, we used the energy dispersive X-ray analysis (EDS). Known to evaluate the stoichiometry of such a material, the accuracy of this analysis is usually not sufficient to detect impurities present as traces. Nevertheless, we have clearly confirmed the presence of the most important classes of defects related to the Al^{3+} ions (Figure 2).

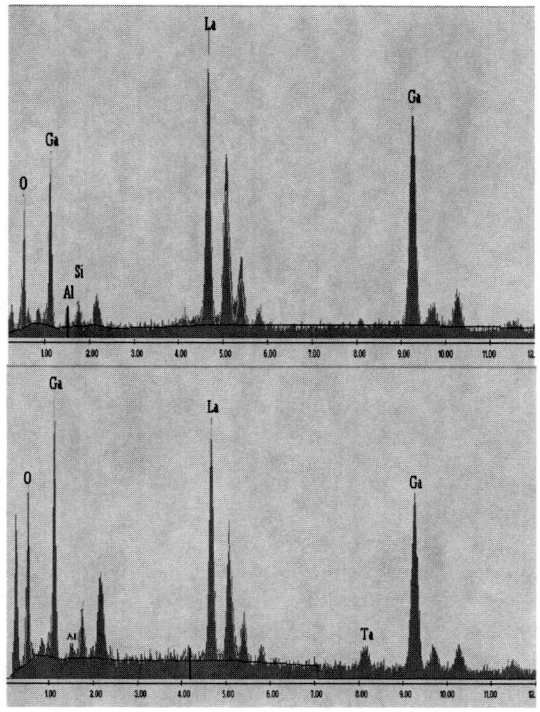

Figure 2: EDAX of: (a) LGS, (b) LGT

Indeed, the existence of cavities is shown especially in the vicinity of the D site (mixed occupation between Ga and Si) where we can find alkali ions or protons, inserted within the lattice in order to balance out the electrical neutrality. As in the case of the quartz crystal, trivalent Aluminum (Al^{3+}) ion is the principal substitutional impurities which replaces tetravalent Silicon ion during the crystal growth. Though the ionic radius of Al is almost equal to this of Ga and that Al and Ga belong to the same group III, we think that the impurities of Al can take the place of Si. This was confirmed by EDAX analysis, which shows that the Si level is lower than the Si level of the stoichiometric composition, at least in the case of LGS. And, in the case of LGT, we can note that the atomic theoretical and experimental percentages are closely the same to prove that the stoichiometric composition of this crystal will be reached.

5 Fourier transformation infrared analysis

Almost of the defect centers can be correlated either to absorption bands in the middle IR range or to some final properties of the crystal resonator. So, the evaluation of each impurity content appears as a fundamental necessity to correlate various properties of the material.

The IR spectra of Y-cut polished samples of LGS and LGT crystals are made without polarization and compared with this of a same orientated sample of quartz ; the results are shown in Figure.3. For LGS, we see a strong absorption at $3405cm^{-1}$, indicating the existence of the hydroxyl group. We note that the spectra of the two crystals LGS and LGT do not contain sharp absorption bands seen in the infra-red spectrum of quartz and corresponding to the presence of OH groups. For the $5437cm^{-1}$ peak, we note that it vanishes when the polarization is parallel to the Z-axis.

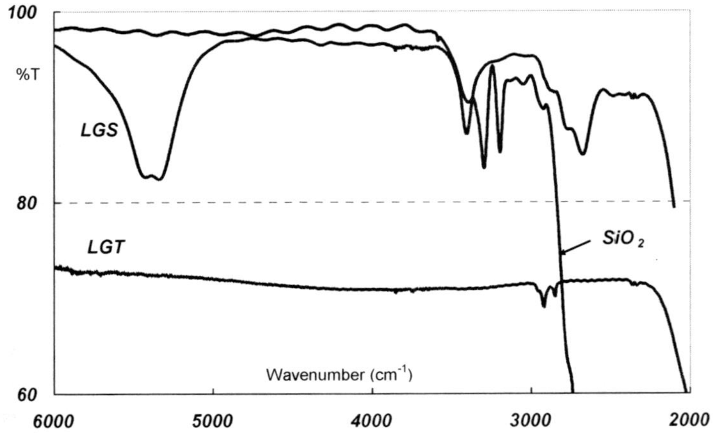

Figure 3: IR spectra of LGS, LGT and quartz samples

6 Other results

Before using the alkaline fusion method, we have tried to dissolve the crystals in different acids as for the mineralization of the quartz crystal. We have observed that a white insoluble part remains after each acid attack, even after long contact with acids like HNO_3, HCl, H_3PO_4 or H_2SO_4.
The different experiments using acid attacks are summarized in table II.

Mixture	Ratio	Observation
HF	-	White precipitate
HF / HCl / HNO_3	1 / 1 /1	id
HF / HCl / H_2O	1 / 1 /1	id
HF / HCl / H_2O	1 / 3 /10	id
HCl / HNO_3	1 / 1	id
H_3PO_4 / H_2O	2 / 1	id
HCl / HNO_3	1/3 / 2/3	id
HCl / HNO_3+H_2O_2	1/3 / 2/3 + 10ml	id

Table -II-Digestion by acid solutions

7 Conclusion

In conclusion, we have to note that Langasite and Langatate crystals are completely dissolved by alkaline fusion. The principal advantages of this method are simplicity and time saving compared to other dissolution methods. Moreover, the flux with Lithium offers also the advantage of being used as buffer to decrease certain effects of the matrix.
The chemical composition of impurities included in the lattice of these crystals was determined by ICP-AES. High level of certain impurities contents has been confirmed by EDS method.

Acknowledgements

The authors wish to acknowledge Professor P. Zecchini for his critical reading of the manuscript and thank Professor Y. Pizarevski for our fruitful discussions.

References

[1] H. Takeda, K. Shimamura, T. Kohno: T. Fukuda, "Growth and characterization of La3Nb0.5Ga5.5O14 single crystals", J. of crystal growth ***169*** **(1996)** 503-508.
[2] J. Bohm, R. B. Heimann, M. Hengst, R. Roewer, J. Schindler: "Czochralski growth and characterization of piezoelectric single crystals with langasite structure: LGS, LGN and LGT", J. Crystal Growth ***204*** **(1999)** 128-136.
[3] B. Chai, J.L. Lefaucheur: "Growth and evaluation of large size LGS, LGN and LGT single crystals", **(1998)** IEEE International Frequency Control Symposium.
[4] I.H. Jung, J. M. Ko, K. B. Shim, T.Fukuda, K. H. Auh: "Influence of starting melt composition on La3(TaxGa1-x)Ga5O14 crystals grown by Czochralski method", J. of crystal growth ***220*** **(2000)** 275-280.
[5] H. Takeda, K. Sugiyama, K. Inaba, K. Shimmamura, T. Fukeda, Jpn. J. Appl. Phys. ***36*** **(1997)** 919.
[6] L. Gheorghe, S. Georgescu, I. Mateescu, L. Dumitrache, E. Borca, C. Ranea: " Synthesis, growth and X-ray diffraction studies of langasite crystals ", **(2002)** IEEE International Frequency Control Symposium and PDA Exhibition

LOW-LOSS NSPUDT SAW FILTERS WITH INCLINED REFLECTORS ON QUARTZ

S. A. Doberstein*, A. Shevtsov, S. Zhgoon

*ONIIP, Russia, 644009, Omsk-9, 231 Maslennikov str., e-mail: info@oniip.ru, fax: 7-381-2-51-49-87

Keywords: NSPUDT SAW filters, inclined reflectors

Abstract

This paper presents the low-loss natural single phase unidirectional transducer (NSPUDT) SAW filters, which use SAW propagation direction reversal without the reversal of NSPUDT directivity. In this design the input and output NSPUDTs are placed in parallel acoustic tracks. SAW transmission between these tracks and propagation direction reversal are provided by a double reflection from the inclined reflectors. An asymmetric Quartz crystal orientation was chosen where conventional IDTs with the $\lambda/4$ electrodes have SAW unidirectional radiation (λ is the SAW wavelength on the central frequency of the filter). The 56 MHz and 172 MHz experimental filters with the LC-matching showed an insertion loss of 8-10 dB and low ripple within the passband.

1 Introduction

For development of the low-loss NSPUDT SAW filters the reversing a directivity of one of the NSPUDT is commonly used [1-4]. This complicates the manufacture technology and reduces design flexibility. Another way of obtaining the low-loss proposes the reversal of NSPUDT directivity. In this case inclined NSPUDTs are used and SAW transmission between them is provided by a single reflection of SAW from a polished edge of a piezoelectric substrate [5]. However this solution can be used only when the IDTs are placed symmetrically at specific angles equal to ±25° in relation to the X-axis on ST-Quartz. These features limit the design flexibility and application of NSPUDT SAW filters. This paper presents another design concept of low-loss NSPUDT SAW filters which uses SAW propagation direction reversal without the reversal of NSPUDT directivity. The design uses a simple single-level technology and increases the design flexibility of low-loss NSPUDT SAW filters.

2 Operation principle and design concept

A filter design similar to the widely known structure with a double reflection [6] but an asymmetric ST-Quartz crystal orientation with SAW propagation +25° relative to the X-axis is used as a piezoelectric substrate. A SAW unidirectional radiation and detection is possible by a conventional IDT with an electrode width of $\lambda/4$ along this direction [1]. In the suggested design the input and output NSPUDTs are placed in parallel acoustic tracks. The inclined reflectors are placed on the side of the primary SAW radiation direction from the NSPUDTs. SAW transmission between the parallel tracks is provided by a double reflection of SAW from inclined reflectors (Fig. 1). As seen in Fig. 1 in this design the low-loss NSPUDT SAW filter is realized by the SAW propagation direction reversal without the reversal of NSPUDT directivity. To obtain the low-loss SAW filter with inclined reflectors it is necessary to provide a 90° SAW reflection in the inclined reflectors. For this purpose we must choose an optimal angle β between a perpendicular to reflector electrode and SAW propagation direction on ST (X+25°) Quartz. A solution of this problem is illustrated in Fig. 2. The inclined reflectors are characterized by an angle β and pitch d. The angle β is

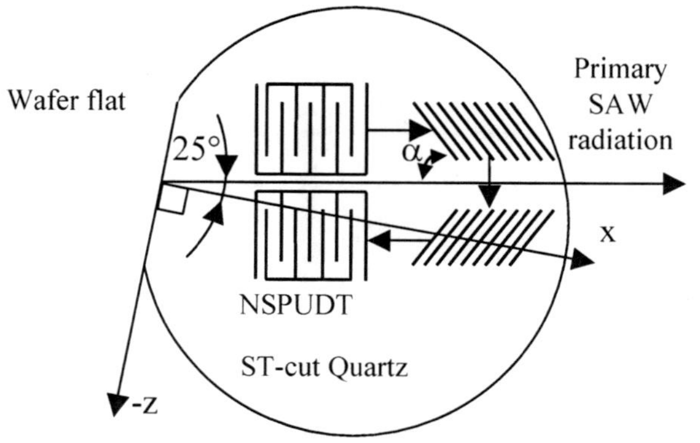

Figure 1: Design concept of the low-loss NSPUDT SAW filter with inclined reflectors

determined from a condition when a phase shift of an incident SAW (on segment L_{inc}) equals to a phase shift of a reflected SAW (on segment L_{ref}):

$$(\omega/V_{inc})L_{inc}=(\omega/V_{ref})L_{ref}, \quad L_{inc}/L_{ref}= V_{inc}/V_{ref},$$

$$\beta = \mathrm{arctg}(L_{inc}/L_{ref})=\mathrm{arctg}(V_{inc}/V_{ref}).$$

Here $\omega = 2\pi f_o$, V_{inc}, V_{ref} is a velocity of the incident and reflected SAW, respectively; f_o is a synchronism frequency of the IDT.

The pitch d is determined from a condition when a phase shift of the incident SAW on a segment L is equal to 2π:

$$(\omega\cdot V_{inc})\cdot L=2\pi, \quad L= V_{inc}/f_o,$$

$$d=L\cdot\cos(\beta)=(V_{inc}/f_o)\cdot\cos(\beta)=\lambda\cdot\cos(\beta).$$

Here $\lambda = V_{inc} / f_o$ is an IDT period.

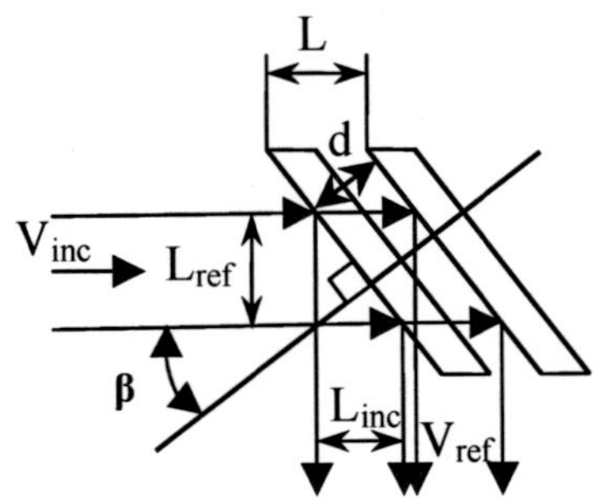

Figure 2: The incident and 90° reflected SAW in the inclined reflectors

The calculated data for the incident and 90° reflected SAW on ST(X+25°) Quartz are presented in Table 1 (here k^2 is electromechanical coupling coefficient).

Parameter	*Incident SAW*	*Reflected SAW*
Euler angles	(0°,132.75°,25°)	(0°,132.75°,115°)
Phase velocity	3249 m/s	3455 m/s
k^2	0.135%	0.03%
β	43.2°	
d	0.73 λ	

Table 1: Calculated data for the incident and 90° reflected SAW on ST(X+25°) Quartz

3 Experimental results

The experimental samples of the filters were developed on two central frequencies f_0: 56 MHz and 172 MHz. The design parameters for these samples are given in Table 2.

Parameter	f_0 *=56 MHz*	f_0 *=172 MHz*
IDT period (λ)	58 μm	18.7 μm
Number of IDT electrodes	402	402
Number of the reflector electrodes	300	300
Aperture of IDT	5800 μm	1870 μm
Aluminum thickness	0.4 μm 1.7 μm	0.4 μm 0.6 μm
Gap between IDTs	2900 μm	936 μm
α	46.8°	46.8°

Table 2: Design parameters of the filter samples

The measured frequency response of the 172 MHz unmatched filter in a 50-Ω system has shown in Fig. 3. An aluminum thickness h was 0.4 μm. The insertion loss for the 56 MHz filter and the 172 MHz filter was 16 and 17 dB respectively.

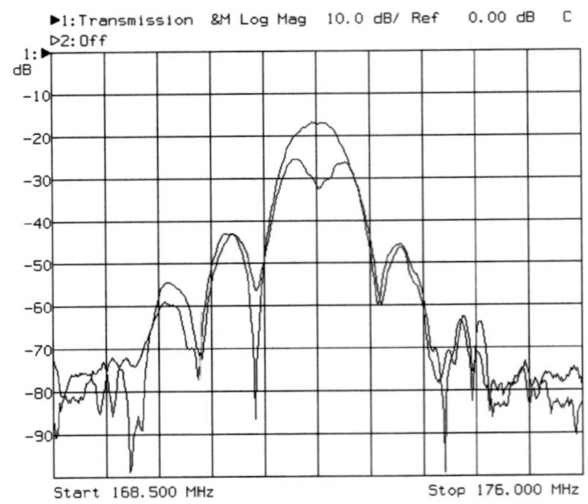

Figure 3: Measured frequency response of the 172 MHz unmatched low-loss NSPUDT filter (h=0.4 μm)

To demonstrate that the NSPUDTs used the asymmetric crystal orientation for achieving the unidirectivity the device was fabricated on ST-Quartz substrate that had been rotated on 180°. In this case the inclined reflectors will be placed on the opposite side of the primary SAW radiation direction from the NSPUDTs (Fig. 1). The frequency response of the filter has a typical dip of 15 dB in a passband (Fig. 3). This characterizes the directivity of the NSPUDT for ST (X+25°) Quartz with an aluminum thickness of 0.4 μm.

To decrease the insertion loss it is necessary to match the filters with a 50-Ω system. Matching was provided by the LC elements (Fig. 4). Fig 5,6 show the measured input/output impedance characteristics of the 172 MHz filter after matching. As will be seen from Fig. 5,6 at the center frequency of 172 MHz the input/output impedances of the filter are close to real value of 50 Ω.

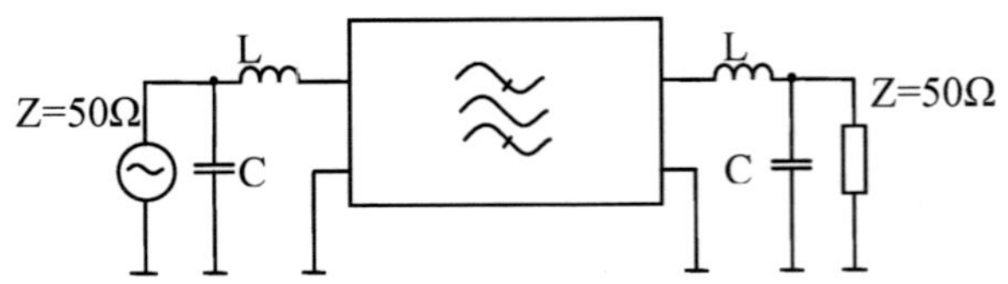

Figure 4: Matching circuit of the filter

Fig. 7 shows the measured frequency response of the filter matched with a 50-Ω system. The insertion loss was reduced to 11.5 for the 56 MHz filter and to 12.5 dB for the 172 MHz filter. The ripple in the passband became very low. The 3-dB bandwidth was 178 kHz for the 56 MHz filter and 550 kHz for the 172 MHz filter. The sidelobe suppression was about 20 dB.

Evidently the directivity of the NSPUDTs and SAW reflection from the inclined reflectors depend on an aluminum thickness. The measured frequency responses of the 55.5 MHz and 172 MHz matched filters with an aluminum thickness of 1.7 μm and 0.6 μm, respectively, have shown in Fig. 8, 9.

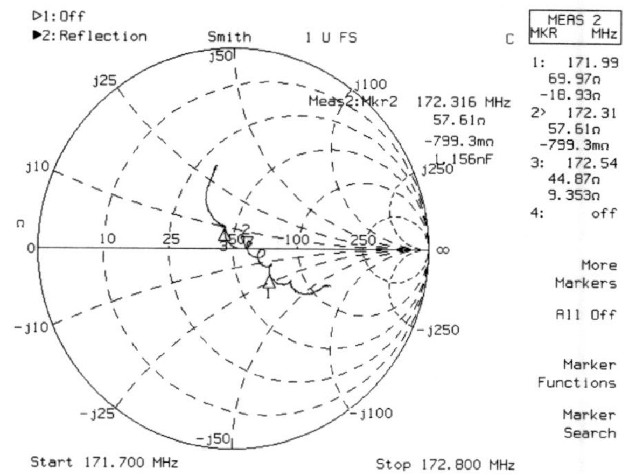

Figure 5: Measured input impedance characteristics of the 172 MHz filter after matching

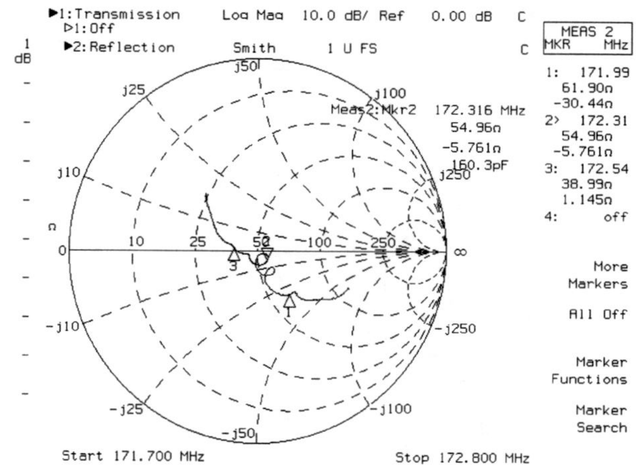

Figure 6: Measured output impedance characteristics of the 172 MHz filter after matching

The insertion loss was reduced to 8.2 for the 55.5 MHz filter and 10.5 dB for the 172 MHz filter. Fig. 9 shows measured frequency response of the 172 MHz filter with an opposite directivity. The typical dip in the passband came to 45 dB, 30 dB more than in the case when the thickness was 0.4 μm (Fig. 3).

As seen in Fig. 8, 9 the sidelobe suppression was about 20 dB. The calculations show that increasing sidelobe suppression is achieved by decreasing the number of the electrodes in one of IDTs. The measured frequency response of the 56 MHz filter (h=0.4 μm) with the number of the electrodes reduced to 160 in an output IDT has shown in Fig. 10. The filter provides the sidelobe suppression over 30 dB. The insertion loss was increased to 12 dB by means of the degradation of the NSPUDT directivity with the reduced number of electrodes.

Figure 8: Measured frequency response of the 55.5 MHz low-loss NSPUDT filter with matching (h=1.7 μm)

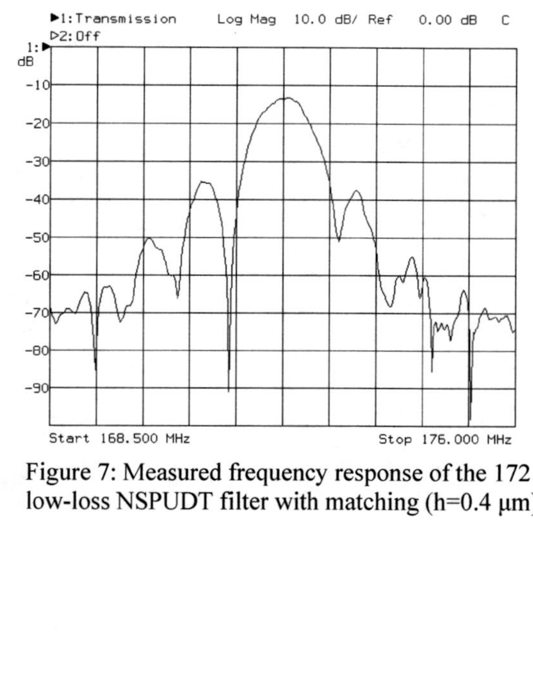

Figure 7: Measured frequency response of the 172 MHz low-loss NSPUDT filter with matching (h=0.4 μm)

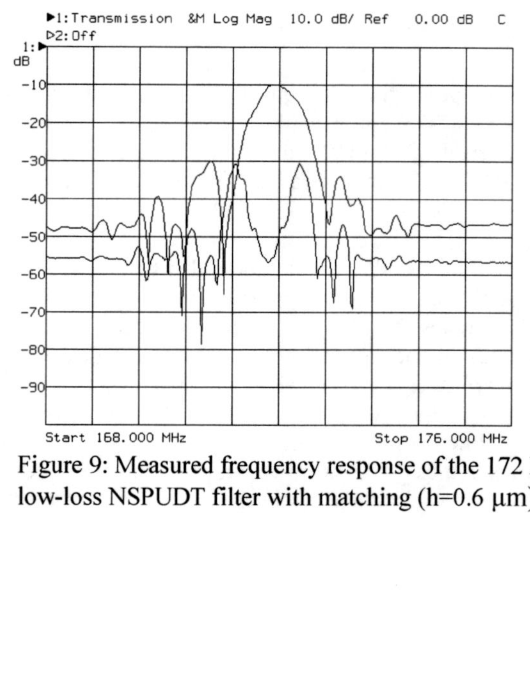

Figure 9: Measured frequency response of the 172 MHz low-loss NSPUDT filter with matching (h=0.6 μm)

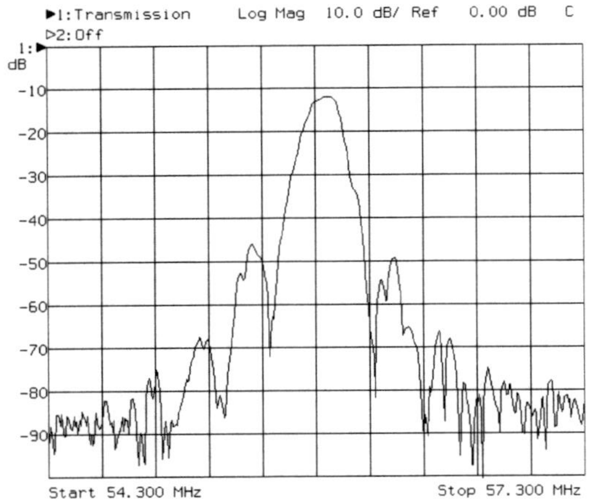

Figure 10: Measured frequency response of the filter with reduced number of the electrodes in one of IDTs

4 Temperature characteristics of the low-loss NSPUDT filter with inclined reflectors

The temperature characteristics were investigated on the 172 MHz filter with an aluminum thickness of 0.4 μm. We measured a fractional change of the center frequency $\Delta f/f_o$ and change of the insertion loss versus temperature. The range of the temperature variation was from -60°C to +60°C (Fig. 11). The measured temperature coefficient of frequency (TCF) was ≈ -8.3 ppm/°C, which is close to the calculated TCF=-11.9 ppm/°C for ST(X+25°) Quartz [2]. The insertion loss was changed to ≈ ±4 dB. This is associated with the temperature dependence of a power flow angle of the NSPUDTs. This effect degrades the reflection of the inclined reflectors. To improve the temperature stability of the NSPUDT filters we can increase an angle Θ of the rotated Y-cut Quartz and simultaneously decrease a chosen angle of SAW propagation direction in relation to the X-axis [2]. It is necessary for NSPUDT operation. This will lead to a re-calculation of the angle β for the inclined reflectors.

5 Conclusion

We have presented a design of the low-loss NSPUDT SAW filter with SAW propagation direction reversal but without the reversal of NSPUDT directivity. The conventional IDTs with the λ/4 electrodes are used in the filter. These input and output IDTs are placed in parallel acoustic tracks on the ST(X+25°) Quartz. Unidirectional SAW transmission between the parallel tracks is provided by a double reflection of SAW from the inclined reflectors. The presented concept of the low-loss NSPUST SAW filter essentially simplifies the filter design and increases the development flexibility of the similar filters. The 56 MHz and 172 MHz experimental samples of the filters had an insertion loss of 8-10 dB and very low ripple with matching. The temperature characteristics were shown. Evidently the presented results are not limiting values for the investigated d... high, si... s not e... f the lc... above ir...

A...

W... and re...

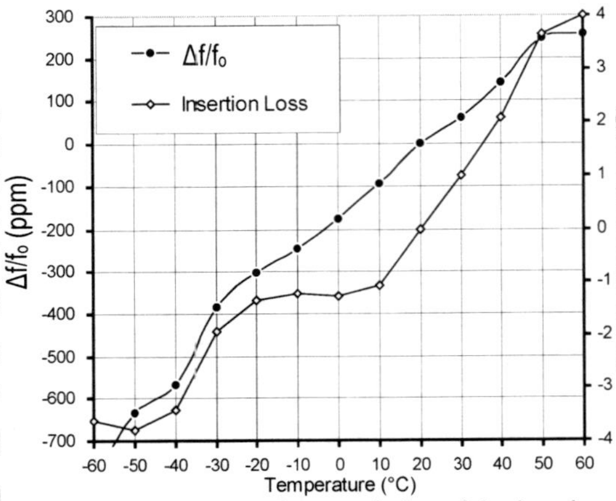

Figure 11: Temperature characteristics of the low-loss NSPUDT SAW filter

R...

[1] ... ional Transducer: A New Low-Loss SAW Transducer", in *Proceedings of the IEEE Ultrasonics Symposium*, pp. 58-63,(1985).

[2] C.S. Lam and D. Gunes, "A Low-Loss SAW Filter Using Two Finger Per Wavelength Electrodes on the NSPUDT Orientation of Quartz", in *Proceedings of the IEEE Ultrasonics Symposium*, pp. 185-188, (1993).

[3] K. Yamanouchi, M. Takeuchi, H. Odagawa and M. Tanaka, "Low-Loss SAW Filter Using Thickness Difference Type of IDT on the NSPUDT Orientation Substrate", in *Proceedings of the IEEE Frequency Control Symposium*, pp. 537-541, (1995).

[4] M. Takeuchi, H. Odagawa, M. Tanaka and K. Yamanouchi, "SAW Transducer Configurations for Reversing the Directivity of NSPUDT Substrates", in *Proceedings of the IEEE Ultrasonics Symposium*, pp. 17-22, (1995).

[5] S. Doberstein, "Low-Loss NSPUDT SAW Filter without Reversing the Directivity of NSPUDT", in *Proceedings of the IEEE Ultrasonics Symposium*, pp. 29-32, (1999).

[6] Herbert Matthews, Surface Wave Filters. Design, Construction, and Use, 2nd ed., John Willey & Sons, New York, ch. 9, (1977).

THE DIFFERENT AGEING BEHAVIOUR OF FUNDAMENTAL AND OVERTONE MODES

EBERHARD SEYDEL

Vita-Chemie GmbH, Düppelstrasse 13, D-14163 Berlin, Germany
Tel: +49/30/8027077 Fax: +49/30/80496955 E-mail: es.quartz@t-online.de

Keywords: ageing, crystal processing, crystal design.

Abstract

The general understanding describes the ageing of quartz crystals as a change of the frequency over time for all vibration modes in the same size and direction. In reality, a different ageing behaviour of fundamental mode and overtone modes can be noticed. This effect depends on the blank design and on the electrode design.
In the oscillator, only one mode can be measured over time and not the different behaviour of individual modes.

The present work shows the influence of varied design parameters on the ageing behaviour. Theoretical calculations are used to determine the mode shape and the direction of the frequency shift for these effects. The calculation correlates with experimental values in ageing and drive level dependence. It can be shown, that electrode design, electrode material and deposition technology has an important influence on the direction of the ageing. Migration effects of the electrode material can explain this behaviour.

The measurement of different modes at the same time can give an indication of the direction for the long time ageing in a short time.

1 Introduction

Different parameters, like mechanical stress, cleanness, diffusion processes etc, influence the ageing of quartz crystals. All of them change the frequency of the vibrator in different ways. Ageing can be described by summation of several exponential functions ($a*e^{-bt}$) with different time constants and amplitude factors. There are processes with very high time constants and it is possible, that the ageing curve looks stable, but after one year, the behaviour changes significantly (fig1). One effect with a high time constant will be described below.

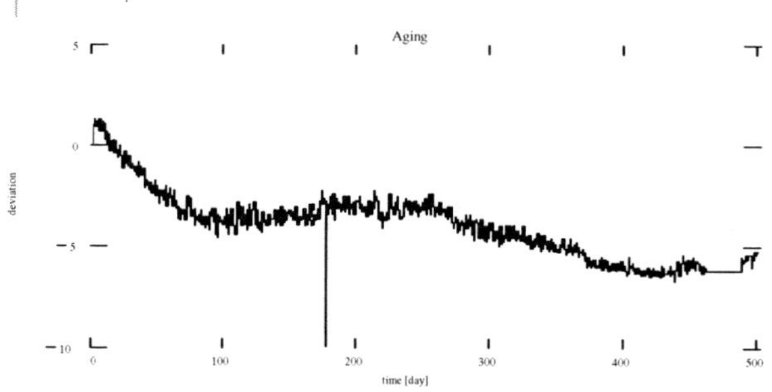

Figure 1

2 Electrode effects

2.1 Electrode design

Crystals from different plating processes were controlled in ageing over longer time. The used crystals were made from the same blank lots and all processes, excluded plating, were the same. Plating process 1 produced a very sharp edge of the electrode and plating process 2 produced a soft thickness profile on the border of the electrode (fig 2).

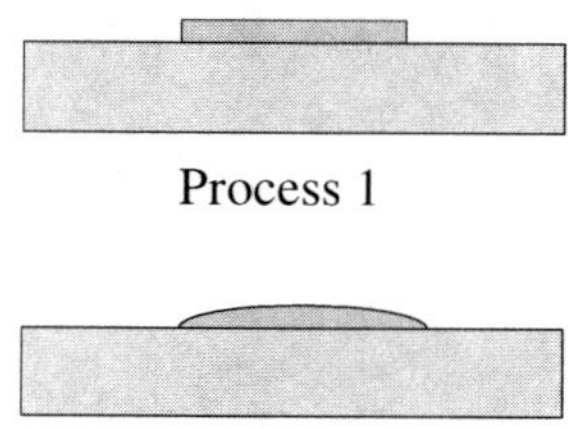

Figure 2

Normally, an ageing measurement is done in an oscillator at nominal frequency for only one mode (active measurement). In contrast, a passive measurement was used to measure the crystal each time at three different frequencies: fundamental mode, third and fifth overtone.
For the two process types, a significant difference was measured. For plating process 1 all three modes have exact the same curve form (fig.3).
Different behaviour could be seen for plating process 2 (fig.4). The ageing for 3rd and 5th overtone runs in the same negative direction whereas the fundamental mode runs in positive direction, nearly linear, means exponential with a high time constant. Fundamental and overtone modes show a totally different behaviour.

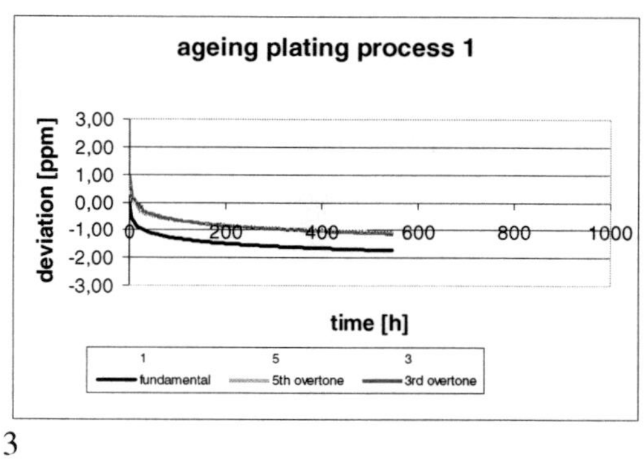

Figure 3

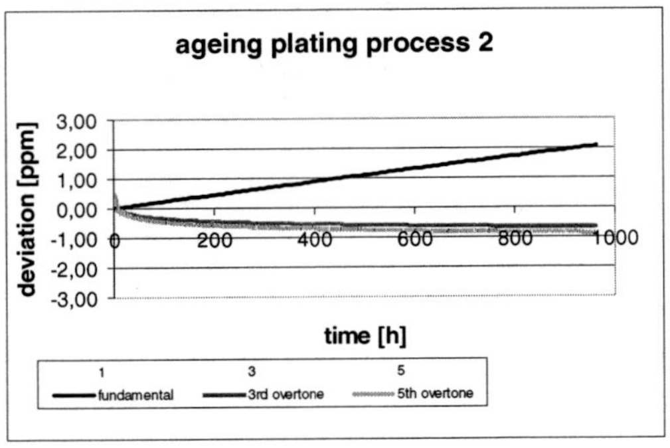

Figure 4

2.2 Electrode and vibration amplitude

To understand this effect, it is necessary to understand the amplitude distribution over the blank for all measured modes. The results are shown in fig 5 to 7. Significant differences can be seen at the marked electrode edges. For the 3rd and 5th overtone, the vibration is concentrated under the electrode with a sufficient energy trapping. The fundamental mode is not trapped. The value of amplitude on the electrode edge is around 50% of the value in the centre.

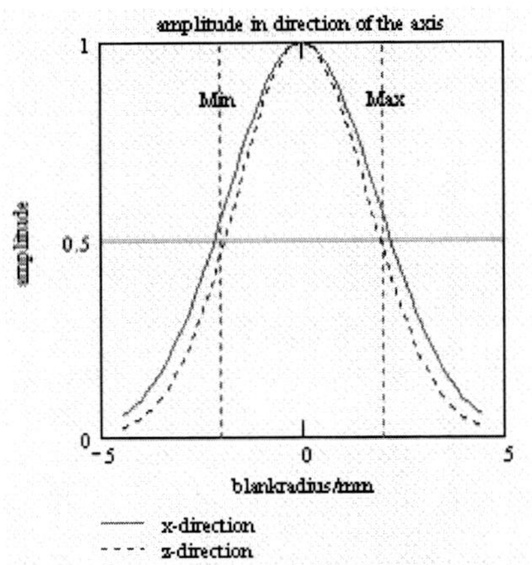

Figure 5 fundamental

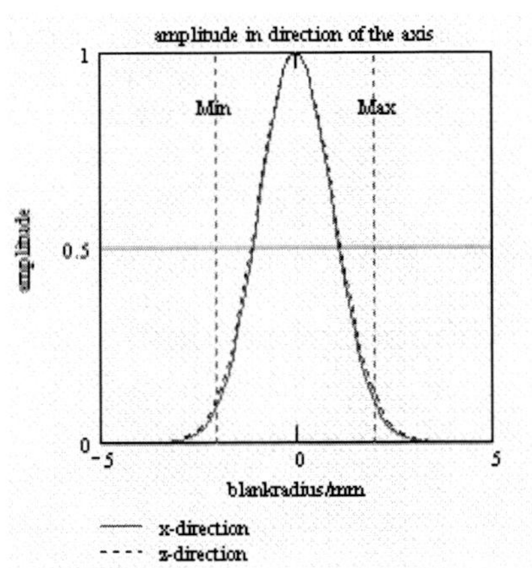

Figure 6 3rd overtone

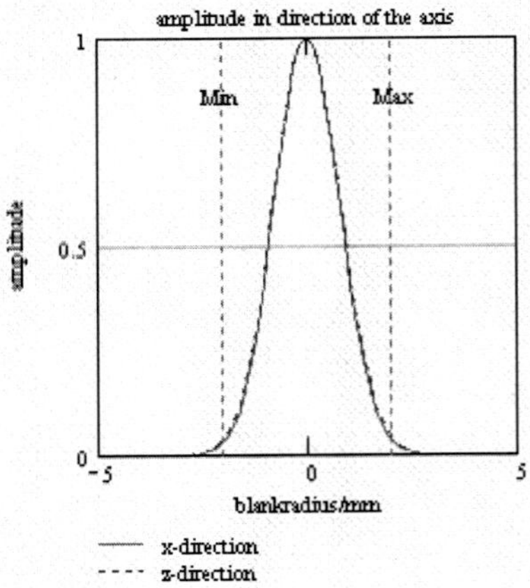

Figure 7 5th overtone

The amplitude of the vibration on the electrode edges induces the different ageing behaviour of fundamental and overtone modes. That's why the mass distribution on the edge of the electrode influences the frequency in the fundamental mode much more than in the overtone modes. Theoretical calculations show the influence of the mass distribution. The direction of the frequency shift for different mass distributions but constant electrode mass was calculated for the fundamental mode. The effect of the roughness is not significant.

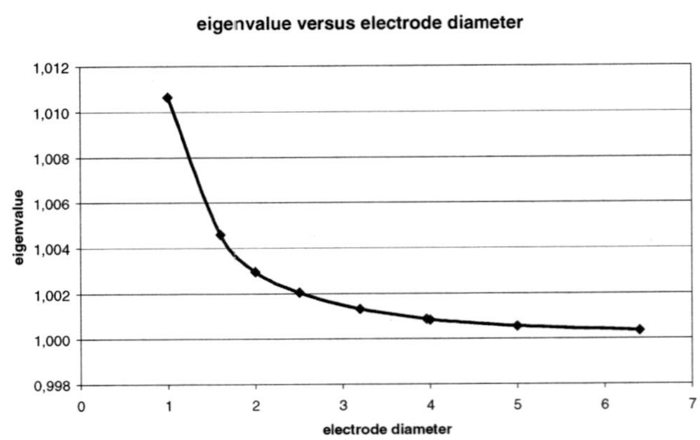

Fig.8 Eigenvalue versus electrode diameter (electrode mass constant)

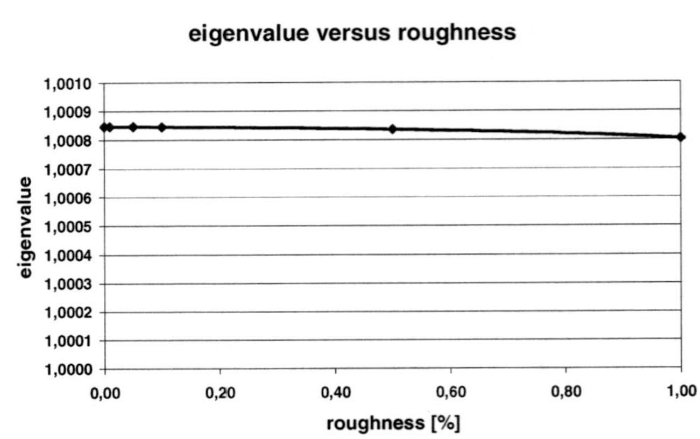

Fig.9 Frequency shift versus roughness (electrode mass constant)[3]

2.2 Migration effect

In the area of very thin electrode film, the material builds islands over time and especially at higher temperatures. Later after a time of migration, balls of gold or cubes of silver grow on the surface. This process depends on temperature and vibration amplitude. The effect is known for gold and in a slightly different form for silver (white films). The electrode material influences the size of the effect. At the end of the process (fig.10) , formed balls or cubes, it is possible that the crystals fall asleep, means a high DLD arises.

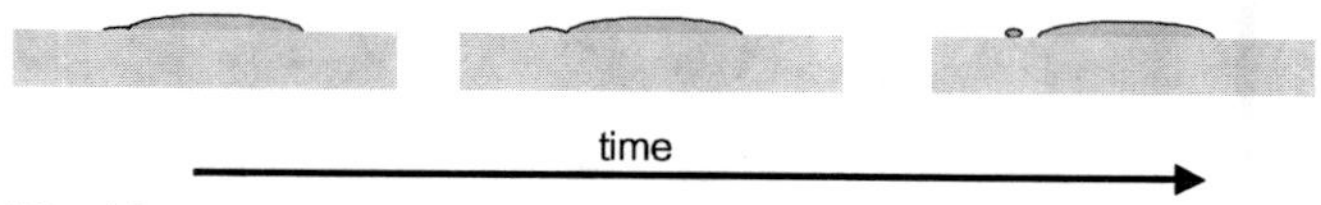

Fig 10

Such very thin electrode areas (less than 200 Å, not more electrical conductive) are caused by the plating process.
This can be:
- Base plating and additional fine plating with the same or larger electrode diameter
- Base plating with wrong electrodes
- One shot plating

At all, the important point is a too large distance between mask and blank. In this case, a scattering takes place and produces a metal dust around the wanted electrode area.
If the process is very well controlled, this is a method to produce a positive part in the ageing.
It is possible to calculate a better long time forecast for the crystal ageing using the different amplitude deviations of the modes (fundamental, overtone, in harmonic) and under consideration of the different electrode influence on the modes.

2.3 Perspective

Other mechanical parameters, like the displacement of the bevel or of the contour, change the amplitude deviation and the different ageing parameters become another weight (amplitude). Figure 11 shows the real amplitude deviation for a crystal with displacement of the contour. The measurement was done with a mechanical damping method. The vibration amplitude on the electrode border in this case is much higher than in the normal case and the stress of the mounting points can influence the vibration. Using the ageing of the inharmonic mode frequencies (fig 12 and 13), differences in the ageing to the normal behaviour will be seen earlier.

3. Resume

The blank design influences not only the electrical data of the final crystal but also the ageing behaviour. It is necessary to find a compromise between the needed electrical data and a good ageing. Useful for the design procedure are tools, which in a first step calculate the amplitude deviation theoretically and in a second step measure the real situation on the final blank. Both methods are presented and can be used for a faster and better ageing design.

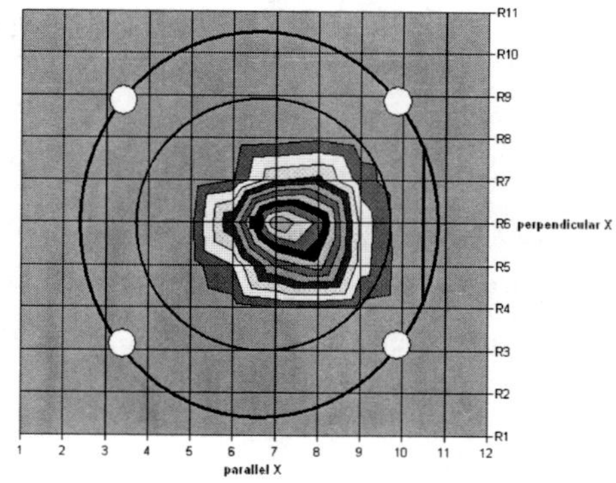

Fig.11

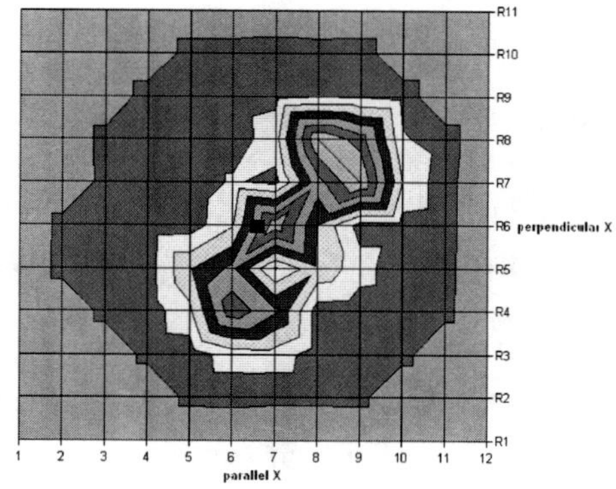

Fig.12

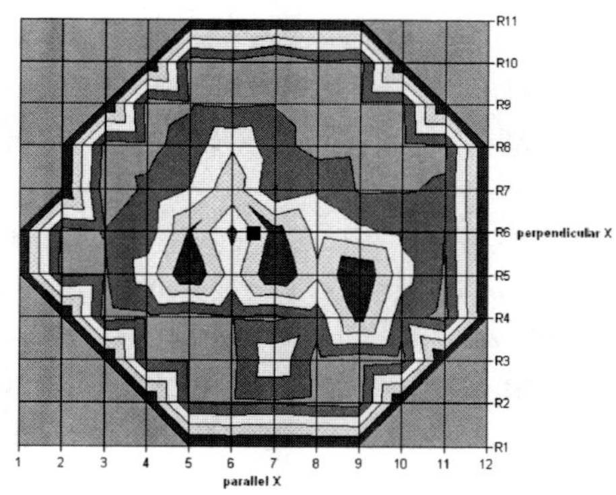

Fig.13

Literature

[1] E. Seydel, The Influence of the Surface Structure on the Vibration Amplitude and the Equivalent Electrical Data of Quartz Crystals – a Theoretical Overview, Proc. 14th Europ. Frequ. Time Forum, pp.103-107, 2000.

[2] E.Seydel, A.Landmann, The influence of surface structure on amplitude deviation and electrical parameters of quartz crystals, Proc. 9th Europ. Frequ. Time Forum, pp.450-453, 1995.

[3] E.Seydel, L.Weitzel, High Frequency Fundamental mode resonators-equivalent electrical data dependence on electrode properties and blank surface, Proc. 10th Europ. Frequ. Time Forum, 1996.

THE INFLUENCE OF THE MATERIAL CONSTANTS ON LGS RESONANT STRUCTURES CHARACTERISTICS

IRINA MATEESCU, LIVIU DUMITRACHE, CRISTINA BRAN, National Institute of Materials Physics, Bucharest-Magurele, PO Box MG-7, Romania
SERBAN GEORGESCU, LUCIAN GHEORGHE, National Institute for Lasers, Plasma and Radiation Physics, Bucharest-Magurele, PO Box MG-11, Romania

Based on Tiersten's analysis [1], an accurate calculation of the coupling coefficient dependence on geometrical dimensions and mass-loading for quartz and langasite monolithic structures was performed [2].

In this paper we have studied the influence of the material constants on the attenuation characteristics of two-pole langasite monolithic filters and on electrical parameters of langasite resonators. This study was demand by the real dispersion of material constant values for langasite crystals grown in various laboratories. The dispersion of the constants is probably done by the in-homogeneity produced during the langasite growth process.

The calculations, based on Tiersten's analysis, in which was retained only thickness-shear and flexure modes of vibration, shown that the influence of elastic constant c_{66}, piezoelectric constant e_{26} and Voight's elastic constant γ_{11} is more important as others material constants on coupling coefficient behavior of langasite monolithic filters.

The effect of these material constants was estimated through the sensitivity coefficients defined as: $(\Delta k/k)/(\Delta c_{66}/c_{66})$, $(\Delta k/k)/(\Delta e_{26}/e_{26})$, $(\Delta k/k)/(\Delta \gamma_{11}/\gamma_{11})$.

The behavior of the coupling coefficient as a function of the individual material parameters for several langasite crystals, point out that the filter bandwidths have a large change (~ 20%) as a result of a small dispersion (1-2%) of the material constants. We have to mention that the maximum change of coupling coefficient and, consequently of bandwidth, for the same dispersion of the material constants, is obtained in the case of wide bandwidths.

In this paper we present too the influence of the material constants on electrical characteristics of langasite resonators. The dependence of the electrical parameters of langasite resonators for an arbitrary variation of material constants was calculated.

In this study, similar with that performed on langasite monolithic filters, we selected from the literature the values of the elastic constants c_{66} and piezoelectric constant e_{26} for some langasite samples grown in several laboratories.

As a conclusion, it is worthwhile to mention that the electrical characteristics of langasite monolithic filters and resonators are sensitive to the material constants that are scattered among the world producers.

[1] H.F.Tiersten, Linear piezoelectric plate vibrations, Plenum Press, New York, 1969
[2] I.Mateescu, G.Pop, C.Ghita, A new theoretical approach for the coupling coefficient in the monolithic structures, Proceedings of the IEEE Int. Frequency Control Symposium, Pasadena, USA,1998, pp.271-277.

PAPER WITHDRAWN

THERMOLUMINESCENCE INVESTIGATIONS IN Ge-DOPED CULTURED QUARTZ CRYSTALS

Harish Bahadur
National Physical Laboratory
Hillside Road, New Delhi-110012, INDIA.
E-mail: hbahadur@mail.nplindia.ernet.in

Keywords: Thermoluminescence, infrared spectra Ge-doped quartz, cultured quartz, natural quartz.

Abstract

Thermoluminescence spectral characteristics of a Ge-doped quartz crystal show the presence of a temperature-glow peak at 200 EC. This peak is not observable in conventionally grown quartz undoped with Ge. Results have been compared and discussed with the near infrared spectral characteristics of natural and conventionally grown quartz crystals.

1. Introduction

Quartz oscillator crystals play an essential role in almost all the frequency control appliances. The impurity-related point defects present in the crystalline material of quartz have a major ramification with regard to the device operation. Among the prominent impurities, Al is the most pervading impurity in quartz. In addition, Ge, Ti, Fe etc are some of the other substitutional impurities which can isomorphously replace silicon in quartz lattice. In all quartz, Al and Ge are always present in varying proportions. While Al $^{3+}$ in quartz acts as an acceptor atom and needs a charge compensation, Ge $^{4+}$ exists as a neutral atom.

Quartz crystals when exposed to ionizing radiation, exhibit both transient and steady state frequency and Q offsets. These offsets are related to radiation induced modifications of impurity-related point defects. Upon exposure to ionizing radiation, the electrons and holes produced move through the crystal until they recombine or are trapped by these impurities present in the quartz material. While the electrons are trapped at Ge-sites the holes are captured by Al ions to form Al-hole centers. At room temperature, the alkali charge compensator M^+ (Li^+ or Na^+) ion diffuses away and stabilizes the Ge-defect site forming $[Ge(A)e^-/M^+]^0$ and $[Ge(C)e^-/M^+]^0$. These centers are also designated as Ge (A)-Li and Ge (C)-Li, Ge (A)-Na and Ge (C)-Na.

Point defects in quartz have been studied extensively using a variety of techniques [1-17]. Thermally stimulated luminescence (TSL) is a sensitive technique used to characterize point defects in wide-band gap materials. The experiment consists of irradiating a sample to produce various defects and then heating a constant rate while monitoring any emitted light due to charge recombination. During heating, as the characteristic temperatures are reached, specific defects become unstable and thus release charge which then recombines with the opposite charge at the other defect sites. The temperature of these "glow peaks", their associated activation energies, and their spectral dependences are the usual information obtained from the TSL experiments.

Despite a large amount of research work already done to date on various impurity-related point defects in quartz crystals, an exact understanding of these defects and their behavior is still lacking. Thermoluminescence (TL), ESR, near infrared, optical and optical absorption, and resonance frequency of quartz oscillators are some of the properties of crystalline quartz that have been studied for a long time and still under study. It is now well realized that the source of TL is quartz is the presence of impurities and their traps. In this communication, we shall report our investigations carried out on a Ge-doped quartz crystal grown in a water rich environment at Bell Laboratories, Murray Hill, NJ. The crystal was grown hydrothermally at Bell Laboratories, Murray Hill, NJ. . The results show that a variety of complex hydroxyl defect centers in addition to the conventional growth defect bands occur in the Ge-doped crystalline quartz . We have conducted thermoluminescence investigations on as-grown and irradiated Ge-doped quartz crystal having Ge and Al concentration ~ 3000 and 100 ppm respectively. Its near infrared absorption investigations and the irradiation effects studied earlier [18-21] have shown that the crystal, in addition to the conventional-growth defects produced bands at 3481 cm^{-1}, 3656 cm^{-1}. Intense irradiation at 77 K in prior 300 K-irradiated sample produces another band absorbing at 3400 cm^{-1} which is a different species of defect center than the one absorbing at this frequency as one of the conventional growth defect bands.

2. Experimental

Crystalline quartz was powdered and tested in the TL-glow curves set up at Oklahoma State University (OSU) in the laboratory of Professor S.W.S. McKeever. Irradiation was done at 300 K again using 1.75 MeV electron beam from a Van de Graaff electrostatic accelerator of the OSU. The irradiation at this temperature would allow the movement of both the alkali charge compensators at Al-sites as well as protons from the growth defect bands.

3. Results and Discussion

Figures 1 and 2 presents a typical glow-curve of a cultured quartz crystal undoped with Ge and the Ge-doped quartz respectively. Comparison of Figures 1 and 2 reveals that the glow-peaks occur at 100BC, and 310BC are typical of cultured quartz while the peak at 200BC occurs in the Ge-doped quartz. Figures 3 and 4 represent the near infrared absorption spectra of a typical cultured quartz and the Ge-doped respectively. The near IR spectra also show that the Ge-doped sample (Figure 4) exhibits a large number of peaks superposed over a broad background absorption due to water molecules. A comparison of Figures 3 and 4 shows that the Ge-doped sample (Figure 4) also exhibits the bands presents in the conventionally grown undoped cultured quartz. AT this stage, we realized that the near IR and TL spectra of natural quartz should also be compared. As a comparison, Figure 5 and 6 show the near infrared absorption spectrum of a typical natural Brazilian quartz and Figure 6 shows the TL spectrum of a natural Brazilian quartz observed by Durrani *et.al.*[22]. It may be clearly seen that the TL spectra of natural quartz shown in Figure 6 shows the glow-peak at 200 BC. We thus expect that the peak at 200 BC shown in Figure 2 is perhaps due to the presence of Ge in quartz crystal. Natural crystals are know to contain a good amount of Ge as an impurity. We have in our earlier studies on natural quartz shown that all the natural quartz contain a peak at 3481 cm^{-1} which is found in Ge-doped quartz. The peaks has been assigned to be related to an hydroxyl defect that traps Li as one of the grown-in impurity. It is known that both natural as well as cultured quartz contain Li as one of the charge compensators of the electron excess defect of aluminum. We have noticed that the Li gets trapped at some of the hydroxyl defects as well. The peak position does not seem to affected by sweeping the unswept sample with either Na or H with in the resolution limit of spectrophotometer. It is logical to believe that sweeping would not affect any Ge-related impurity because Ge in the as-grown quartz would not require any charge compensation as Al^{3+}. Table (I) lists the various impurity-related near infrared bands in typical Brazilian quartz in its unswept, Na-swept and H-swept conditions. For a description of sweeping technique, reader is referred to reference [23].

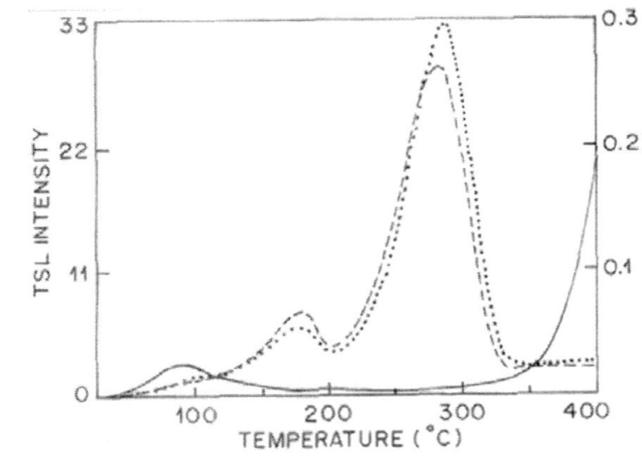

Figure 1. Temperature glow curve for an unswept quartz from Toyo Communications (Ref. 24)

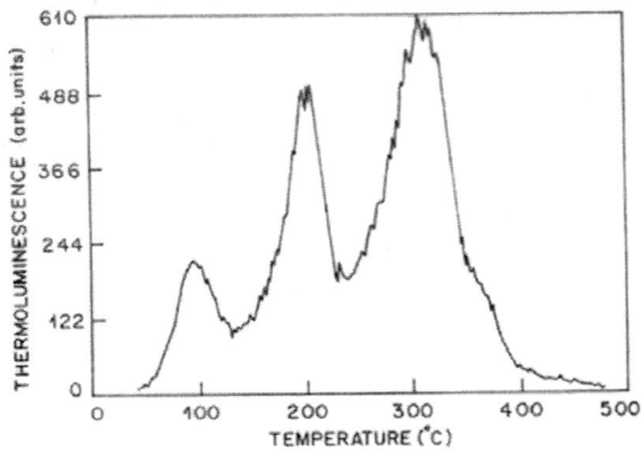

Figure 2. Temperature glow curve of the Ge-doped quartz.

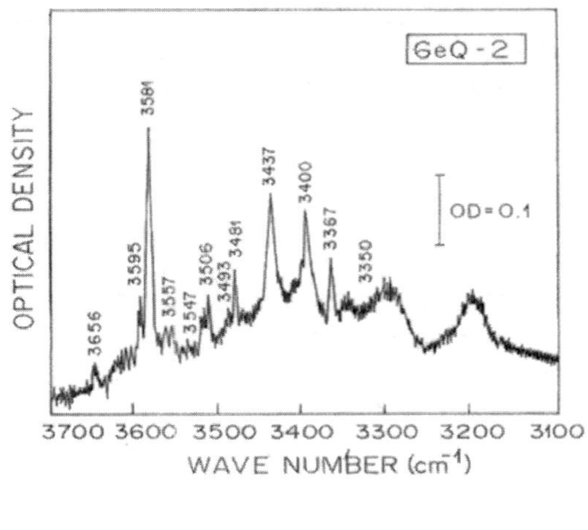

Figure 3. Near infrared spectrum of the Ge-doped quartz.

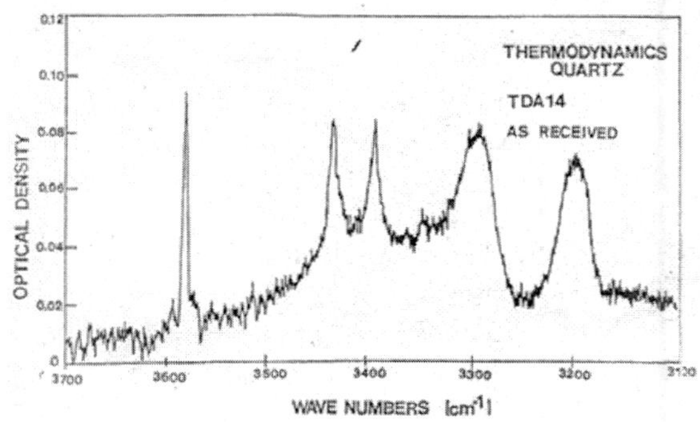

Figure 4. Near infrared spectrum of a conventionally grown cultured quartz undoped with Ge.

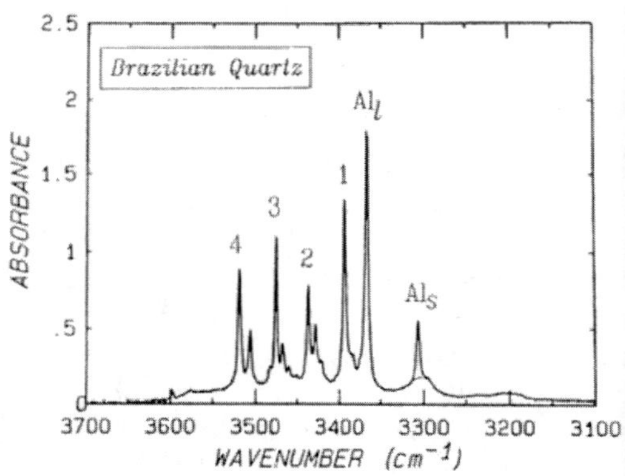

Figure 5. Near infrared spectrum of a natural Brazilian quartz crystal.

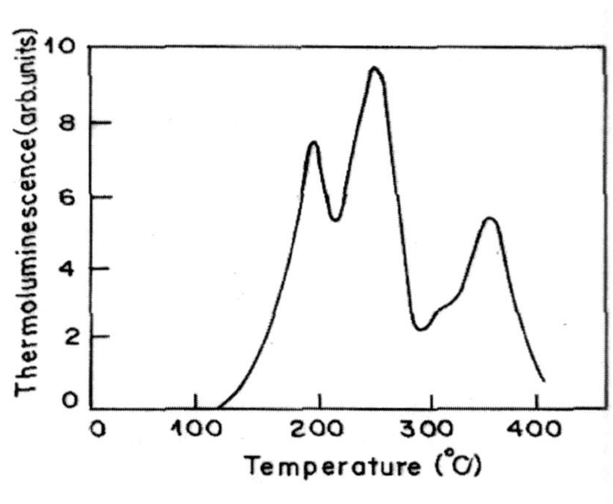

Figure 6. Temperature glow curve a natural Brazilian quartz

TABLE (I)

Positions near infrared bands in a typical natural Brazilian unswept quartz crystal.

Wave number (cm^{-1})	Notation
3240	...
3306	Al_s
3367	Al_l
3383	...
3395	Li-*n*1*a*
3420	Li-*n*2*c*
3430	Li-*n*2*b*
3437	Li-*n*2*a*
3451	Na-*n*2*a*
3460	Li-*n*3*c*
3468	Li-*n*3*b*
3476	Li-*n*3*a*
3480	Li-*n*3*d*
3507	Li-*n*4*b*
3520	Li-*n*4*a*
3575	...
3595	...

4. Summary

Ge-doped cultured quartz crystal in its theremolumineswcence spectrum (TSL) has exhibited a new temperature-glow peak at 200 EC in addition to the peaks at 100EC and 310EC. While the peaks at 100EC and 310EC are well known to occur in conventionally grown cultured quartz, the peak at 200EC is likely to be due to Ge doped in the quartz crystal. A discussion has been presented based on the TSL spectrum of natural quartz and the results have been compared with the near infrared spectral absorption characteristics of natural, conventionally grown cultured and the Ge-doped cultured quartz crystals.

Acknowledgments

The author thanks Dr. Vikram Kumar, Director of the National Physical Laboratory, New Delhi for his encouragement in the work. The work was done in the laboratory of Professor J.J. Martin at the Oklahoma State University (OSU). The author thanks Professor Martin and Dr. S.W. S. McKeever of OSU for his help in the TSL measurements.

References:

1 J.C. King, "Radiation effects in quartz crystals"- Special Session. Proc. Annual Frequency Control Symp. 27 (1973).

2. J.C. King, Guest Editor "Radiation effects in quartz crystals" Rad. Effects 26, (1975). Special Issue.

3. J.C. King, "Hardening quartz resonators to ionizing radiation-a review" Proc. Int. Conf. Evaluation, Space Environment on Materials, Centre Spatial de Toulouse, France June 17-21, (1974).

4. J.C. King and D.R. Koehler, "Radiation effects on resonators" Precision Frequency Control, *Academic, Orlando, FL* 1, 147-159 (1985).*[see cited references]*

5 L.E. Halliburton, J.J. Martin and D.R Koehler, "Properties of piezoelectric materials" Precision Frequency Control, Academic, Orlando, FL 1, 1-45 (1985).*[see cited references]*

6. J.A. Weil, "The aluminum centers in a-quartz" Rad. Effects, **26**, 261-265, (1975).[see cited references]

7. J.A. Weil, "A review of electron spin resonance spectroscopy and its applications to the study of paramagnetic defects in crystalline quartz" *Phys. Chem. Min.* **10,** 149- 165 (1984).

8. S.W.S McKeever, "Thermoluminescence in quartz and silica" *Rad. Prot. Dosimetry* **8**, 81-98 (1984).

9. L.E. Halliburton, "Defect models and radiation damage mechanisms in alpha quartz" *Crys. Lattice Defects and Amor. Mat.* **12**, 163-190 (1985)*[see cited references]*

10. D.B. Fraser, " Impurities and anelasticity in crystalline quartz" *Physical Acoustics*, Academic Press, **5**, 59- 110 (1968).

11. W.A. Sibley, J.J. Martin, M.C. Wintersgill and J.D. Brown ," The effect of radiation on the OH-infrared absorption of quartz crystals" *J Appl. Phys.* **50**, 5449- 5452 (1979).

12 R.N. Brown and A. Kahan , "Optical absorption of irradiated quartz in the near IR, *J. Phys. Chem. Solids*, **36**, 467-476 (1975).

13. J.J. Martin, Ho B. Hwang, H. Bahadur and G.A. Berman, " Room temperature acoustic loss peaks in quartz" *J.Appl. Phys.* **65,**4666 -4671(1989).

14. Harish Bahadur,"Hydroxyl defects and electrodiffusion (sweeping) in natural quartz crystals" *J. Appl. Phys.* **73**, 7790-7797 (1993).

5. Harish Bahadur, "Ion exchange and radiation response of H-related point defects in natural quartz crystals" *J. Mat. Res.* **9**, 1789-1801, (1994).

16. A.Kats,"Hydrogen in alpha quartz" *Philips Res. Rept.***17**, 133- 195 (1962).

17. L.E. Halliburton, M. Markes, J.J. Martin, S.P. Doherty, N. Koumvakalis, W.A. Sibley, A.F. Armington and R.N. Brown, "Radiation effects in synthetic quartz:The role of electrodiffusion and radiation induced mobility of interstitial ions" *I EEE Trans. Nucl. Sci.* **NS-26**, 4851-4856 (1979).*[see cited references*].

18. Harish Bahadur, "Near infrared spectral characteristics of Ge-doped cultured quartz", *J. Appl. Phys.* **75,** 4741-4743 (1994)

19. Harish Bahadur, "Infrared characterization of Ge-doped cultured quartz crystals", Phys. Rev.B, **50**, 717-721(1994).

20. Harish Bahadur, "Hydroxyl defects in Ge-doped quartz: Defect dynamics and radiation effects, Phys. Rev. B, **52**, 7065-7075 (1995).

21. Harish Bahadur, "Irradiation effects on Ge-H, Al-H and other H-related point defects in culturedquartz crystals", Phys. Rev.B **43**, 2085-2095 (1996).

22. S.A. Durrani, P.J. Groom, K.A.R. Khazal and S.W.S. McKeever, J.Phys. D. Appl. Phys, **10**, 1351 (1977).

23. Harish Bahadur," Infrared characterization of natural and cultured quartz:The effect of electrodiffusion and irradiation" *J. Appl. Phys.* **66**, 4973- 4982 (1989).

24. L E Halliburton, J J Martin and W A Sibley, *"Study of defects produced by the growth, post treatment and fabrication of quartz"* Oklahoma state university, **RADC-TR-83-309**, January 1984.

CHARACTERIZATION OF THIN FILMS OF ZnO PREPARED BY SOL-GEL PROCESSES

Harish Bahadur*$, Ram Kishore*, K.N. Sood*, Rashmi*, R.K. Sharma* A. Basu*, D. Haranath*, Harish Chander* and Sudhir Chandra**

*National Physical Laboratory, Hillside Road, New Delhi-110012, INDIA
** Center for Applied Research in Electronics, Indian Institute of Technology, Hauz Khas, New Delhi-110016, INDIA.

$ Corresponding author: E-mail: hbahadur@mail.nplindia.ernet.in

Keywords: Thin films, Zinc oxide, sol-gel, XRD, Ellipsometry, Electron Microscopy

Abstract

Thin films of ZnO have been grown using sol-gel technique. The precursor materials for sol preparation include zinc acetate or zinc nitrate. Characterization was done using optical absorption, ellipsometry, X-ray diffraction, photoluminescence and scanning electron microscopy.

1. Introduction

Zinc oxide, in its thin film form, continues to be a versatile material of significant interest in various technological applications such as piezoelectric transducer [1], transparent conductor [2], surface acoustic wave devices [3,4] and a variety of sensors [5,6]. Consequently, thin films of ZnO are grown by a variety of methods such as vacuum evaporation [7], sputtering [8], MOCVD [2,9], sol-gel [10] chemical solution deposition [11], chemical vapour deposition [12] and spray pyrolysis [13] etc. It is known that ZnO has a wurtzite structure having a number of alternating planes composed of fourfold coordinated O^{2-} and Zn^{2+} ions stacked alternatively along *c*-axis. It can thus be considered as two interpenetrating hexagonal closed-packed lattices of Zn and O. The preparation of ZnO thin films has been the subject of continuous research for a long time because the properties of ZnO films depend upon the method of preparation.

Because of its high piezoelectric coefficient, ZnO has a great potential for applications as vibration sensor. We aim at using thin films of ZnO for MEMS-based acoustic resonator. Characterization and preparation of thin films of piezoelectric materials thus forms an important part of our project of integrating such films into MEMS based acoustic sensing devices. We have prepared polycrystalline thin films of ZnO by using two different routes of sol-gel process. An estimation of the characteristics of the films would not only be cost effective in selection of the film for device fabrication but will also make the device reliable.

2. Film Growth: Sol-gel Technique

The films were grown using sol-gel technique. The growth procedure consisted of first making the surface of the silicon substrate hydrophilic by boiling the wafer in 70% HNO_3 followed by washing with de-ionized water and subsequent drying. This oxidized the surface of Si wafer. The sol was prepared by two different routes. The first route involved dissolving 10 gm of zinc nitrate [Zn $(NO_3)_2$. $6H_2O$] in 100 ml of ethylene glycol monomethyl ether [CH_3O-CH_2-CH_2OH] to form the zinc compound. The other sol was prepared by dissolving 10 gm of zinc acetate [Zn $(CH_3COO)_2$. $2H_2O$] in 100 ml of isopropyl alcohol at 84 °C. This was followed by clarifying the turbid solution by adding a few drops of diethanolamine. The films were prepared by spinning the sol at 4000 rpm. This was followed by drying the coating at 120 °C for 2 minutes. Films were prepared using either of the two sols. Multiple coatings were done to obtain the workable thickness of the film using both the sols. The final heating was done at 450 °C for 1 hour in air.

3. Results and Discussion

The thin films of ZnO grown from both the routes were investigated. The techniques used for characterization involved XRD, optical absorption, ellipsometry, photoluminescence, and scanning electron microscopy with EDS attachment for microanalysis. In the following, we present and discuss our results.

3.1 Optical Absorption

Optical absorption data on the deposited films was measured using Shimadzu UV3101 PC spectrophotometer. For measurement in the UV and optical region, the films were coated onto a substrate of amorphous quartz. The absorption peak was noticed at 372 nm for the ZnO films prepared using zinc nitrate precursor. The peak in the film prepared using zinc acetate route was also centered around this region. This absorption was used to estimate the band gap, which was found to be 3.3 eV [14]. Infrared absorption spectra at this stage have not shown any significant information about the characteristics of the films.

3.2 Ellipsometric Measurements

Ellipsometric measurements were carried out on some films of ZnO deposited on Si substrates to characterize them for refractive index, extinction coefficient and thickness. The measurements were made on a manual research ellipsometer (Rudolph Research, USA) with an angle of 60 ° and the light of wavelength 546.1 nm. The films were found to have refractive indices in the range of 1.92 to 2.03, with an extinction coefficient of 0.58 to 0.68 and thickness 80-90 nm. This thickness was obtained in ten coatings. Thus, the films had refractive indices in agreement with the reported values of 1.9 to 2.1 for ZnO in the visible region.

3.3 X-ray Diffraction

X-ray diffraction experiments were done in grazing incidence geometry on a Bruker -AXS X-ray Diffractometer (Model D8 Advance) using CuKα radiation. Diffraction patterns were recorded at a fixed incidence angle of 1.5 ° and 2θ were scanned from 25 ° to 70 °. Figure 1 shows the XRD pattern of formation for the ZnO film grown on Si substrate using zinc nitrate as the precursor. The diffraction pattern shows the formation of hexagonal ZnO with (100), (002), (101), (102), (110), (103), (200), (112) and (201) planes resolved clearly. However, no preferential orientation was observed. Figure 2 and 3 show the XRD pattern for the two different films grown by using zinc acetate precursor. The diffraction patterns show formation of hexagonal ZnO. The hkl indices have been marked on the diffraction patterns. The film labeled ZnO/Si -B (Figure 3) showed additional unidentified diffraction peaks corresponding to *d* values of 0.3476nm, 0.1679nm and 0.1655nm. The film prepared using nitrate precursor (Figure 1) did not show any preferred orientation, whereas the film prepared using acetate precursor (ZnO/Si-A) has some degree of preferred orientation along 002 direction. It must be mentioned here that the maximum piezoelectric effect occurs along this direction.

3.4 Scanning Electron Microscopy

Microstructural analysis of the sol-gel grown ZnO films on Si substrates were carried out using LEO-0440 SEM equipped with ISIS 300 Oxford microanalysis system (EDS attachment). The specimens were mounted on a 15 mm diameter stub and a thin layer of gold was sputtered onto it. The stubs were then

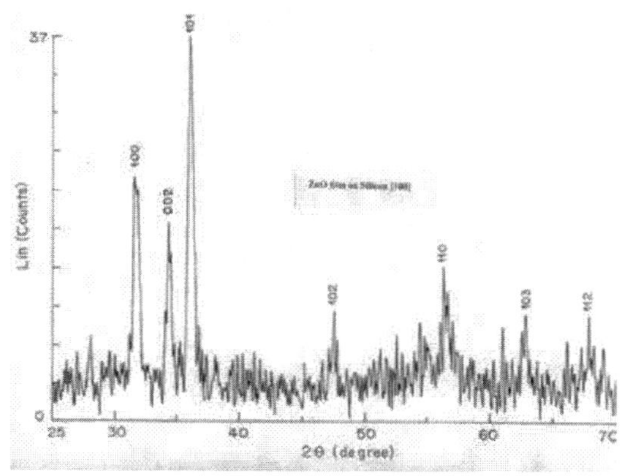

Fig.1 XRD pattern of a ZnO film on Si substrate using zinc nitrate precursor.

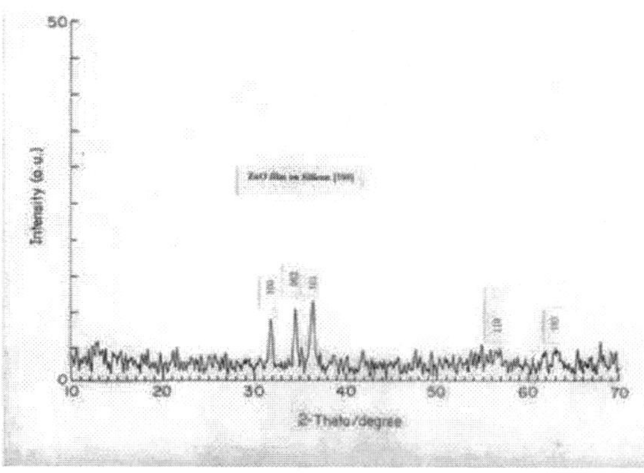

Fig.2 XRD pattern of a ZnO film on Si substrate using zinc acetate precursor

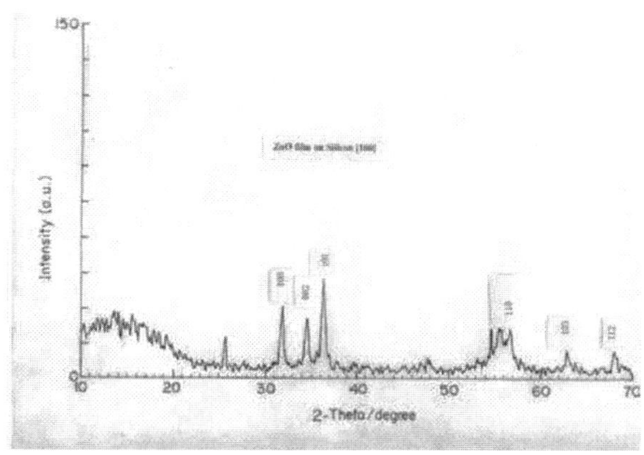

Fig. 3 XRD pattern of another ZnO film labeled ZnO/Si-B on Si substrate using zinc acetate precursor. The diffraction pattern shows some additional unidentified peaks

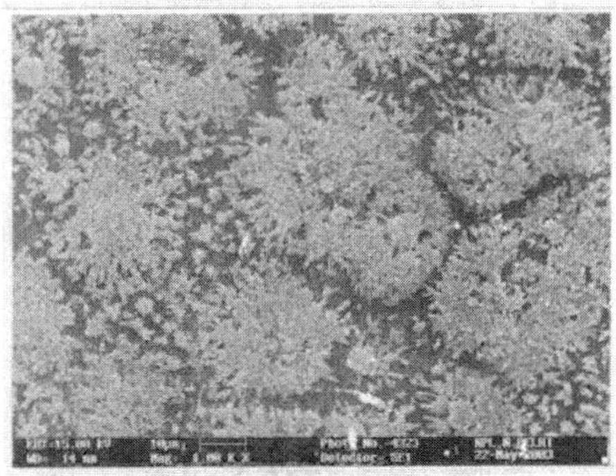

Fig.4 Scanning electron micrograph of ZnO film grown on Si substrate using zinc nitrate precursor exhibiting the dendrites

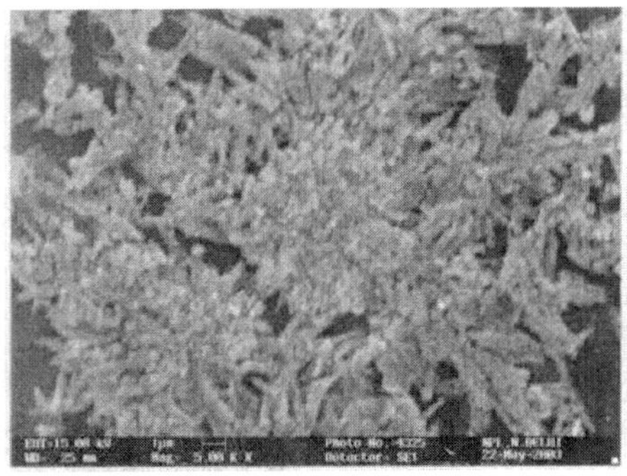

Fig.5 magnified micrograph of one of the dendrite of Fig. 4

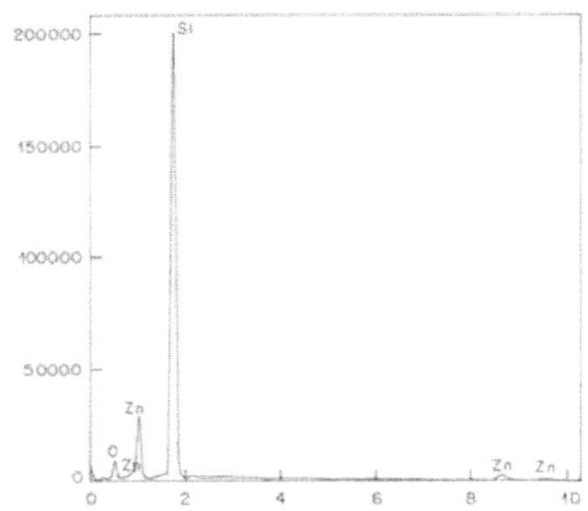

Fig.6 EDS spectrum of the micrograph of Fig. 4

fixed on the viewing stage of the SEM having a *xyz* movement as well as rotation and tilt facility. The specimen was scanned thoroughly at a lower to see the uniformity of the film. The film prepared by zinc nitrate precursor showed a polycrystalline structure having dendritic growth with agglomeration of dendrites in some areas as shown in Figure 4. Figure 5 shows a magnified micrograph of one of the dendrites The EDS analysis (Figure 6) showed the presence of zinc, oxygen and silicon; no other impurity was found within the EDS detection limit. The point analysis in the center of some of the agglomerates was also carried out and it showed higher concentration of zinc in it.

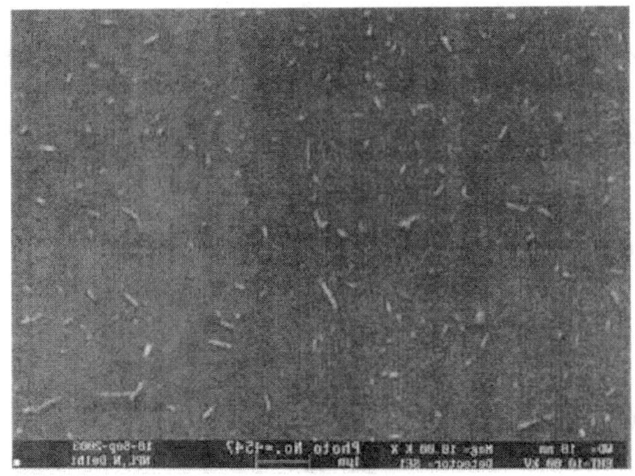

Fig.7 Scanning electron micrograph of a ZnO film grown on Si substrate using zinc acetate precursor showing a much smoother surface with uniform structure of needle shaped particles distributed throughout the scanned area.

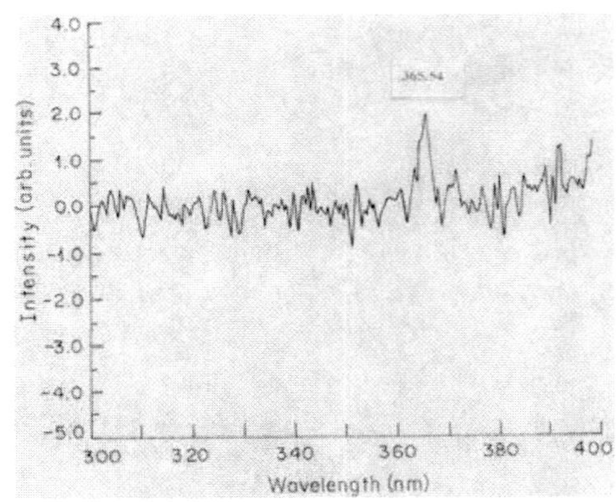

Fig. 8 Excitation spectrum of the ZnO film grown on a quartz substrate using zinc nitrate precursor.

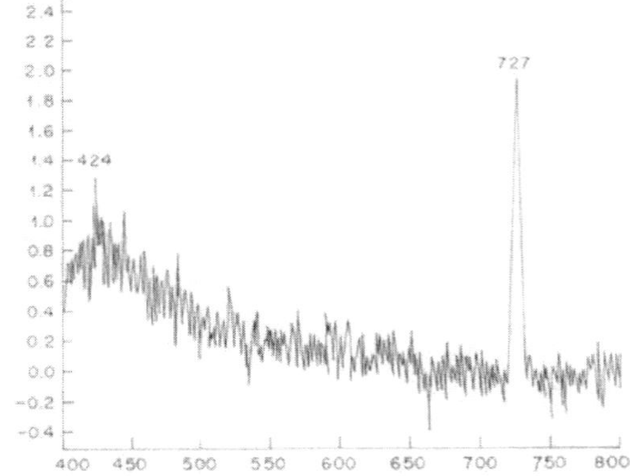

Fig. 9 Emission spectrum of the same ZnO film of which the excitation spectrum is shown in Fig. 8

Figure 7 shows the microstructure of the film grown by using zinc acetate precursor. The films so prepared have shown uniform structure with needle shaped particles distributed throughout the area scanned. The needles have their length between 0.2 to 0.8 μm. The EDS analysis of these films revealed the presence of zinc, silicon, nitrogen and oxygen. The traces of nitrogen could broadly be attributed to the fact that diethanolamine was used in the film preparation. The films thus consisted mainly of ZnO. The micrographs suggest that the films prepared by using zinc nitrate show a rapid and random crystallization compared to that using zinc acetate.

3.5 Photoluminescence Measurements

Photoluminescence measurements were conducted on the ZnO films grown by both the precursors using the Perkin-Elmer LS-55 Luminescence Spectrometer. The excitation of the films was done using the standard source emitting the ultraviolet radiation at 365 nm. Following the excitation, the emission spectra were recorded in the visible range of electromagnetic spectrum. Figure 8 shows the excitation spectrum of the film of ZnO grown by using zinc nitrate precursor.

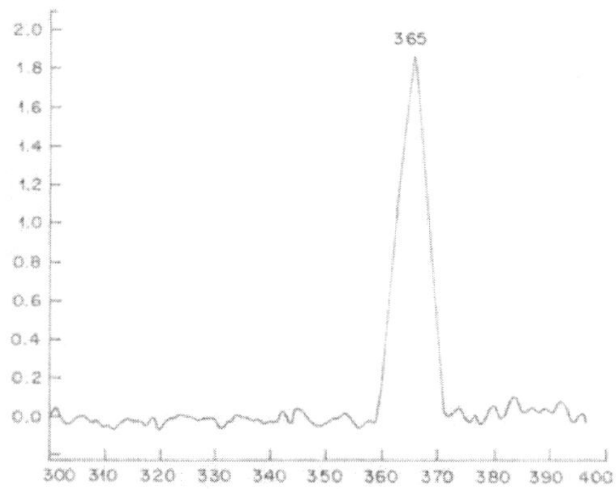

Fig. 10 Excitation spectrum of a ZnO film on a Si substrate using zinc acetate precursor.

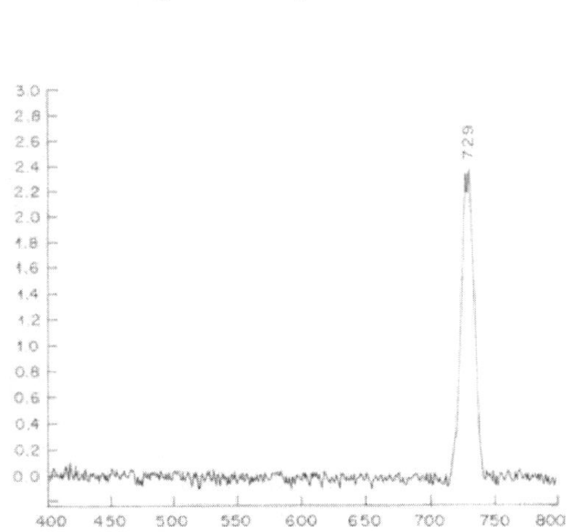

Fig. 11 Emission spectrum of the ZnO film of which the excitation spectrum is shown in Fig. 10

Visual inspection of this sample under UV exposure showed that it was exhibiting light in the red region. The emission spectrum of the same sample is shown in Figure 9. Figure 9 is the plot of the spectrum averaged out for ten times to reduce the background noise. In the emission spectrum of Figure 9, two emission peaks may be clearly noticed ; one at 424.02 nm and the other at 726.63 nm. The excitation spectrum for the film grown by using zinc acetate precursor is shown in Figure 10 and its emission spectrum in Figure 11. In Figure 11, we observe that the lower wavelength emission is missing but the peak at higher wavelength is nearly at the same position. It is positioned at 729.16 nm. The nonappearance of the conventional lower wavelength peak would be perhaps due to the fact that the film was not sufficiently thick to exhibit the excitation and emission signals. The peak at the lower wavelength (~ 425 nm) is known to be due to oxygen vacancies during the growth. These oxygen vacancies trap electrons and thus become paramagnetic and give the characteristic electron spin resonance signal. When the material gets exposed to ultraviolet, these vacancies then become susceptible to trap a hole generated by the UV exposure [15,16]. The recombination of hole with the electron already trapped at the oxygen vacancies emits light, which is generally in the green-blue region. The peak at the 727 nm was initially thought to correspond with the red color emission observed during the exposure of the film to ultraviolet. However, subsequent studies indicate that this is probably the second harmonic of the exciting UV wavelength. Continued work in this direction is expected to reveal the complete picture. However, the nearly equal intensity of 727 nm peak in both types of samples is perhaps an experimental artifact.

4. Summary and Conclusions

The present work shows that ZnO thin films using sol-gel methods can be prepared by using zinc nitrate as well as zinc acetate as the precursors. This was confirmed by several experimental characterization techniques. Optical absorption, ellipsometric measurements have shown that the band gap, refractive index match with the standard values of ZnO. The XRD investigations reveal that the films grown with zinc nitrate although showed the hexagonal structure but did not exhibit any preferred orientation with regard to any axis during their growth. The dendrite features on the micrographs showed their random and rapid crystallization. On the other hand, using zinc acetate as precursor, the films showed a much smoother topographical features and also a preferential orientation with respect to (002) planes. It appears that the films with zinc nitrate precursor grow in an island mode while the films with zinc acetate grow in monolayers. Photoluminescence spectrum has shown that the films had a few oxygen vacancies, which have given a characteristic emission spectrum peak at 425 nm. The present studies indicate that, out of the two precursors used (zinc nitrate and zinc acetate) for sol-gel preparations, the film prepared by using zinc acetate appears to be a better choice for device fabrication, as it tends to grow along the 002 planes and has better surface topographical features. Future work in this direction is necessary and is continuing.

ACKNOWLEDGEMENTS

This work was done under a collaborative program between the National Physical Laboratory (NPL), New Delhi and the Indian Institute of Technology, Delhi (IITD) for promoting research in the areas of mutual interest. The authors thank Dr. Vikram Kumar, Director, NPL and Dr. S.C. Garg, Head, Radio and Atmospheric Science Division of NPL for their encouragement and support in the work.

References:

1. O. Yamazaki, T. Mitsuyu, K. Wasa, "ZnO thin film SAW devices" *IEEE Trans. Sonics and Ultrasonics*, **SU-27**, pp 369-379 (1989).
2. J. Hu and G. Gordon, " Textured aluminum- doped zinc oxide thin films from atmospheric pressure chemical-vapor deposition", *J. Appl. Phys.* **71**, pp.880- 890 (1992).
3. F.S. Hickernell, "Zinc oxide films for acoustoelectric device applications" *IEEE Trans. Sonics and Ultrasonics* **SU-32**, pp 621-629. (1980),
4. V. Gupta and A. Mansingh, "Influence of post deposition annealing on structural and optical properties of sputtered zinc oxide film" *J. Appl. Phys.* **80 (2)** 1063-1073 (1996)
5. S.P.S. Arya and O.N. Srivastava, "Investigation on synthesis, characterization, and electronic behavior of zinc oxide thin films in relation to fabrication of oxygen gas sensor" *Cryst. Res. Technol.* **23** pp.669-675. (1988).
6. Weon-Pil Tai and Jae-Hee Oh, "Humidity sensing behaviors of nanocrystalline Al-doped ZnO films prepared by sol-gel process" *J. Materials Science: Materials in Electronics* **13,** pp.391-394. (2002).
7. Jin Ma, Feng Ji, Hong-lei Ma and Shu-Yung Li, "Preparation and characterization of ZnO films by an evaporating method" *J. Vacuum Sci. Tech.* **A13**, pp.92-94 (1995).
8. T. Minani, H. Nanto and S. Takata, "Optical properties of aluminum-doped zinc oxide thin films prepared by rf magnetron sputtering" Jpn. J. Appl. Phys. 24(8), L605–607 (1985).
9. J.S. Kim, H.A. Marzouk, P.J. Reucroft and C.M. Hamrin, "Characterization of high quality c-axis oriented ZnO thin films grown by metal organic vapor deposition using zinc acetate as source material". *Thin Solid Films*, **217**, pp.133-137 (1992).
10. M. Ohyama, H. Kouzuka, and T. Yoko, "Sol-gel preparations of ZnO films with extremely preferred orientation along (oo2) plane from zinc acetate solution" " *Thin Solid Films*", **306**, pp.78-85 (1997).
11. T. Saeed and Paul O'Brien, "Deposition and Characterization of ZnO films grown by chemical bath deposition" *Thin Solid Films*, **271**, pp.35- 8(1995).
12. L.G. Mar, P.Y. Timbrell and R.N. Lamb, "An XPS study of zinc oxide thin film grown on copper using zinc acetate as a precursor" *ibid*, **223**, pp. 341–347 (1993).
13. M. de la L. Olvera, A. Maldonado, Y. Matsumoto, R. Asomoza, M. Melééndez-Lira, D. R Acosta. "Chromium-doped ZnO thin films deposited by chemical spray: chromium effect"., *J. Vac. Sci. Technol.* **A 19 (5)** pp. 2097-2101 (2001).
14. Harish Bahadur, Ram Kishore, K.N. Sood, Rashmi, D.K. Suri, M.Kar, A.Basu, R.K. Sharma, G. Bose and Sudhir Chandra, "Preparation and characaterization of ZnO thin films on silicon substrate using sol-gel process "*Physics of Semiconductor Devices (IWPSD-2003), Ed. K.N.Bhat and A. Das Gupta, Narosa Publishing House, New Delhi,* **Vol. 12**, pp. 298-300, (2003).
15. K. Vanheusden, W.L. Warren, C.H. Seager, D.R. Tallant, J.A. Voigt and B.E. Gnade, "Mechanisms behind green photoluminescence in ZnO phosphor powders", *J. Appl. Phys.* **79**, pp. 7983-7990 (1996)
16. C.H. Lin, Bi-Shiou Chiou, C.H. Chang and J.D. Lin, "The effect of ZnS content on the luminescence and electrical properties of ZnO:Zn phosphor", *J. Materials Science: Materials in Electronics*, **13**, pp.1-5 (2002).

INFLUENCE OF THE SURFACE ROUGHNESS ON THE SORPTION SENSITIVIT OF UART RESONATORS

V.GEORGIEVA, L.SPASSOV, E.MANOLOV

IINSTITUE OF SOLID STATE PH SICS,
BULGARIAN ACADEM OF SCIENCES,
72 T ARIGRADSKO CHAUSSEE, SOFIA 1784, BULGARIA,
TEL: +359 2 7144 214, E-MAIL: LA AROVA@ISSP.BAS.BG

Keywords uartz resonator gas sensor quartz crystal microbalance

Abstract

The aim of the investigation is to improve resonator s sensitivity by increasing the surface of sorption.

The sensitivity of quartz resonator SnO_2 structure for registration of ammonium concentration in vapour phase is studied. The different effective surface of a quartz resonator wafers with thin SnO_2 deposited on it are used for determining its influence on the sorption ability. For this purpose, quartz substrates with different roughness have been prepared using abrasive materials with different size of the grain from 1 m to 14 m. In this way the average effective area of the electrodes has been changed. The surface s roughness of the quartz substrate and those of the SnO_2 layers on it were estimated by using Scanning Electron Microscopy (SEM). The quality comparison among different surfaces of quartz substrate roughness and SnO_2 as a result of use of various abrasive is carried out.

The investigated structures have been formed on quartz substrates with different roughness on which consequently were deposited Au electrodes by vacuum thermal evaporation with a thickness of 120 nm. The thin 90nm SnO_2 were prepared by RF magnetron sputtering.

The influence of the effective surface of quartz resonator with thin SnO_2 on it of sorption ability has been studied.

So created structures were used for NH_3 monitoring in a range concentration from 10 ppm to 5000 ppm. Correlation has been established between sorption ability of SnO_2 layers as a function of the effective electrode surface.

Observed dependence could be used for improving the sensitivity of resonator - SnO_2 structures used for NH_3 registration and preparing of NH_3 sensor elements.

1 Introduction

It is well known that, quartz crystals may be used as sensor elements in the quartz crystal microbalance (CM). Usually CMs are covered with sensitive material, which interacts with the ambient gas. The sorption process leads to mass loading of the CM and results in a change of the resonator s frequency. This parameter is used for measuring. In the case of AT - cut, the correlation and frequency change is given by Ketzman [1].

The quartz resonator thin layer system, sensitivity, is to a great extend, determined by the kind of the layer and its sorption ability.

Tin oxide is most widely studied for use in gas sensor application at the account of its special surface properties: thermodynamical stability and possibility for enhancement of the sensitivity by introducing impurity [2]. The layers can be obtained by different methods with various properties:

- by pulsed laser deposition (PLD) technique using Sn target under ambient of oxygen gas [3].
- by atomic layers chemical vapour deposition like (ALCVD) from $SnCl_4$ and H_2O at 180 300^0C [4].
- by metal-organic chemical deposition (MOCVD) tetratertiarybutoxytin at 350^0C [5] or tetraethyltin and oxygen at 600 ^0C [6].
- by dual ion beam sputtering (DIBS)[7] and others.

In each of these methods, subsequent annealing and a controlled gas atmosphere is particularly important for the stabilization of the film [7].

The properties of tin oxide thin films and their sensitivity are in great dependence of their chemical compositions, microstructures and porosity. All these properties are determined by the films preparation. The tin oxides prepared by ALCVD [4] are sensitive to CO in air, the others obtained by MOCVD [5] posses response ability to toxic gases such as: H_2S, CH_4, NO_2, H_2, and CO, while the ones fabricated by DIBS [7] are suitable for H_2S detection.

Some of the sensitivity properties of SnO_2 might be changed by surface chemical modification [8,9,10] .

The most important parameters that influence the sensitivity of tin layers are: thickness, grain size, porosity and point defects. The cracks in the tin films also have an important role. Usually, they offer short circulating pathways for sensing gases and this way the transfer between the surface and the bulk of the films can be controlled much easily. The cracking in the films can be controlled by doping additives and by smoothening the surface of the underlying substrate [11].

Sensitivity is also affected by the surface roughness.

The aim of the investigation is studying the influence of the surface roughness of the thin SnO_2 layers on sorption properties of quartz resonators to ammonium.

2 E periments

The AT - cut of quartz is the most commonly used cut for quartz crystal oscillator applications (frequency control applications) [12].

The resonator structures were created on this quartz plate cut with 8mm diameter and 4mm diameter on gold electrodes.

The wafers were divided into three groups and treated with abrasive materials with different size of the grains.

The samples from the first group were finally treated with cerium oxide with the size of the grain 1 m, the second group was treated with abrasive of silicon carbide with the size of the grain 7 m, and the third one with grain size of 14 m.

After mechanical process and the chemical cleaning all samples were undergone to standard technological steps for creation of quartz resonators.

The electrodes were formed from two layers: thin Cr sublayer thickness 100 , deposited by RF sputtering of Cr target - and upper Au layer prepared by resistive evaporation.

Over the quartz piezoelements, thin tin layer was deposited by RF sputtering from sintered SnO_2 target in argon ambient with thickness of 900 . The exact conditions of thin layer deposition are published in [13]. The obtained structures with different roughness were maintained subsequently over tempered solution of NH_3 with various concentrations, from 10ppm to 5000ppm. The resonators frequency changes, which are results of the sorption were measured. A detailed description of the experimental unit is presented in [13].

3 Results and discussion

The roughness of the samples was determined by Scanning Electron Microscopy (SEM) with magnification of 5000.

Fig. 1 SEM image showing the surface morphology of uncovered quartz plate, treated by 1 m grain sized abrasive.

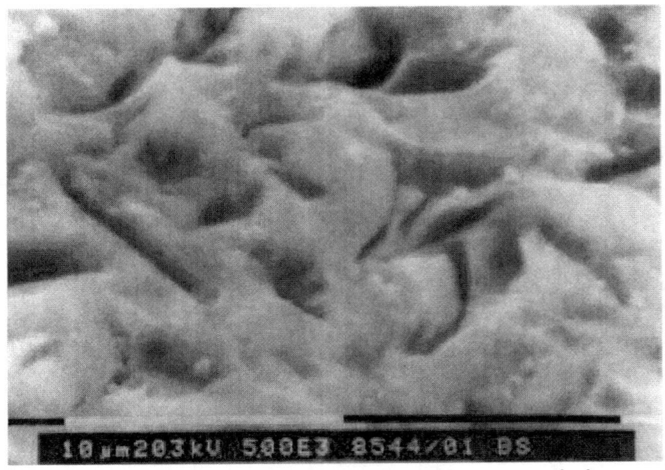

Fig. 2 SEM image showing the surface morphology of uncovered quartz plate, treated by 14 m grain sized abrasive.

Fig. 3 SEM image showing the surface morphology of SnO_2 layer deposited on quartz plate treated by 1 m grain sized abrasive.

Fig. 4 SEM image showing the surface morphology of SnO_2 layer deposited on quartz plate treated by 14 m grain sized abrasive.

This way, quality control of the surfaces is achieved. Fig.1 and fig.2 show SEM images of the surface morphology of quartz plates treated with grains of the abrasive 1 m and 14 m. The difference between the two surfaces is obvious.

After metal electrode formation and thin tin layer deposition the surface is altered a tendency towards smoothening is observed - fig.3 and fig.4. They illustrate the surface image of the same samples presented on fig1 and fig.2, but after thin layer deposition. A comparison between fig.1 and fig.3, and fig2 and 4 shows that, the surface change remains visible. The difference between the samples surfaces treated with an abrasive altering the size of the grain is preserved after all processes (fig.2 and fig.4 respectively). The same tendency is obtained with the samples treated with 3 m and 7 m abrasive grain.

The frequency change dependence of the resonators structures prepared with different roughness vs. ammonium concentration is shown on fig.5.

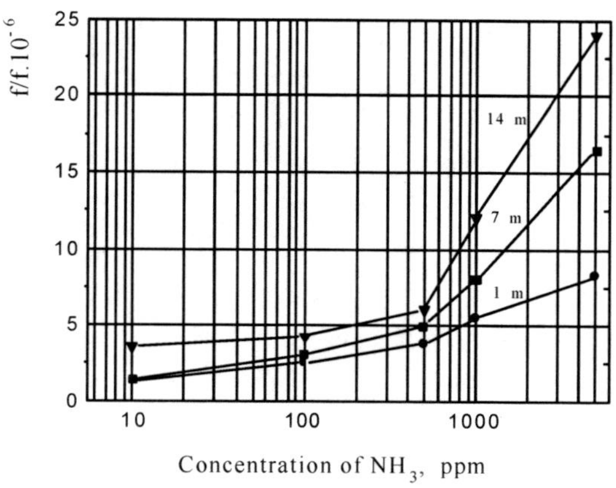

Fig.5 Dependence of quartz resonator s f/f vs. ammonia concentration.

It is obvious that, with increasing NH_3 concentration, the relative frequency change (f/f) for each sample rises in all the investigated range. Resonators formed on a rougher surface obtained much bigger changes. The alteration at the lower concentrations (10ppm) became three times bigger - form f/f 1,31.10 ^{6} to f/f 3,6. 10 ^{6} - referring to the resonators prepared on a smoother surface (grain 1 m), compare to the ones formed on a surface treated by an abrasive with 14 m grain size. The same differences concerning higher ammonium concentration - up to 5000 ppm - were found (form f/f 8.10 ^{6} to f/f 24. 10 ^{6}). It is evident that, increasing the roughness of the surface leads to raising the resonators sensitivity by increasing the effective area for gas molecules. This way, the absorption became bigger.

4 Conclusion

It was determined that the surface roughness may take a significant part by increasing the effective area where the absorption occurs. The control of surface morphology became important aiming to ensure more sensitive structure.

References

[1]. T.Ketzman, "Piezoelectric sensors for use as pollution detectors, meteorology monitors and research instruments", *oc S*, 25, pp 102- 103, (1971).

[2]. U. Schraum, C.E.O. Roesky. "Temperature dependence of an ammonium sensor in humid air based on a cryptophane coated quartz microbalance", *Sen o an c a o* , B.57, N1 3, pp. 233-237, (1999).

[3]. C.K.Kim, M.Choi, I.H. Noh, J.H.Lee, C.Hong, H.B.Chae, G.E.Jang, H.D. Park, "A study on thin film gas sensor based on SnO_2 prepared by pulsed laser deposition method", *Sen o an c a o* , B: Chemical, Vol. 77(1 -2), pp. 463-467, (2001).

[4]. A. Rosental, A. Tarre, A. Gerst, T. Uustare, V. Sammelselg, "Atomic layer chemical vapor deposition of SnO_2 for gas sensing applications", *Sen o an c a o* , B: Chemical, Vol. 77(1-2), pp. 297 - 300, (2001).

[5]. J.R.Brown, P.W.Haycock, L.M.Smith, A.C.Jones, E.W.Williams, "Response behavior of tin oxide thin film gas sensors grown by MOCVD", *Sen o an c a o* , B: Chemical, Vol. 63 (1-2), pp. 109 - 114, (2000).

[6]. Sang Woo Lee, Ping Ping Tsai, Haydn Chen, "Comparison study of SnO_2 thin and thick film gas sensors", *Sen o an c a o* , B: Chemical, Vol. 67 (1-2), pp. 122 - 127, (2000).

[7]. ong Sahm Choe, "New gas sensing mechanism for SnO_2 thin-film gas sensors fabricated by using dual ion beam sputtering", *Sen o an c a o* , B: Chemical, Vol. 77 (1-2), pp. 200 - 208, (2001).

[8]. Kenji Wada, Makoto Egashira, "Hydrogen sensing properties of SnO_2 subjected to surface chemical modification with ethoxysilanes", *Sen o an c a o* , B: Chemical, Vol. 62 (3), pp. 211 - 219, (2000).

[9]. Shiyong hao, Peihai Wei, Shenhao Chen, " Enhancement of trimethylamine sensitivity of MOCVD - SnO_2 thin film gas sensor by thorium", *Sen o an c a o* , B: Chemical, Vol. 62 (2), pp. 117 - 120, (2000).

[10]. Wu uanda, Tong Maosong, He iuli, hang ushu, Dai Guorui, "Thin film sensors of SnO_2 CuO - SnO_2 sandwich structure to H_2S", *Sen o an c a o* , B: Chemical, Vol. 79 (2-3), pp. 187 - 191, (2001).

[11]. henan Tang, Philip C.H.Chan, Rajnish K. Sharma, Guizhen an, I-Ming Hsing, Johnny K.O.Sin, "Investigation and control of microcracks in tin oxide gas sensing thin-films", *Sen o an c a o* , B 79, pp. 39 - 47, (2001).

[12]. Ron F. Schmitt, John W. Allen, John F. Vatelino, Jesse Parks, Chao hang, "Bulk acoustic wave modes in quartz for sensing measuring-induced mechanical and electrical property changes", *Sen o an c a o* , B 76, pp. 95 - 102 (2001).

[13]. V. Lazarova, L. Spassov, V. Georgiev, S. Andreev, E. Manolov, L. Popova, " uartz resonators with SnO_2 thin film as acoustic gas sensors for NH_3", *ac* , Vol. 47, pp. 1423 1425 (1996).

FURTHER RESULTS OF QCM USING FOR HEAVY METALS DETERMINATION IN WATER

KRZYSZTOF WEISS IEEE member* WOJCIECH STEPLEWSKI* JAN DOJLIDO** BARBARA OBMINSKA** -*Tele and Radio Research Institute **Institute of Meteorology and Water Management

At the previous EFTF the initial results of heavy metals determination in water using QCM were presented [1]. The proposed analytical method relies on electrolytic deposition of metals from water solution on quartz resonator electrode. The quartz resonator is located in the manner that its one surface is immersed in solution and the opposite surface is free. The metal deposition process is caused (provoked) by electric potential between quartz crystal electrode and another electrode immersed in solution. The gradual increase of potential results in progressive deposition of metals correctly to electrochemical series. Many technical problems were resolved during the method elaboration. The optimal materials and shape of electrodes were determined. The professional equipment was utilized. As the best method of heavy metals determination in water proved the method of deposition and dissolution metal layer in one process. The investigations were carried out for silver, copper, nickel and lead. The method resolution for individual cations and for ions mixture was determined. The method resolution did appear lower as anticipated previously. For individual cations it was 40÷100 μg/L (depending on cation). For mixtures it was over 100 μg/L. The experiments of sensitivity increase were carried out but without satisfactory results. Also the tests of heavy metals determination in mineral and pipe water were carried out. The results of these investigations are described in the paper.

EXPERIMENTAL ESTIMATION OF FUNCTIONAL CAPABILITIES AND TEMPERATURE ERROR OF MICROWAVE SENSORS OF COMBUSTION PROCESSES.

E.SAFONOVA, V.BOLOZNEV,
Kazan State Technical University, Russia,
Kul-Gali st., 7"a", 15, Kazan-420141, Russia.
E-mail: kate@c-tel.net, ksafonova@yandex.ru.

ABSTRACT

Temperature regime of the sensors, controlling flame electron density N_e in chambers of thermal power engines, has been investigated in this work. By detailed analysis of electrical and thermal processes, suggestions, improving thermal regime of the sensors, reducing temperature error and increasing their life time, were made. Experiment has proved the opportunity of using microwave sensors for advanced flame diagnostics in all air get engines and other thermal power engines (TPE).

1. INTRODUCTION

There is a high interest in studying properties of microwave sensors of autodyne type and its behavior under high temperatures and aggressive medium influence. It is very important to design such construction of the sensors and use correct materials and methods of temperature error compensation, which is a main problem in flame diagnostics in TPE and prevent such effects as vibrations, chemical and thermo erosion, soot deposition on the sensor's fire surface, surface currents etc .

In previous papers [1,2,3,4,5,6] detailed description of sensor's construction and it's principle of operation were done and also analysis of thermal conditions of the microwave active oscillating sensor (AOS), which is meant for process of burning monitoring in TPE chambers, was made. Operation principle of AOS based on technological parameters of burning and TPE regime interconnection with an electron density and flame permittivity. Heat effect of AOS nodes- such as antenna, resonator and active element (Gunn's diode) on monitor signal frequency was established for 3 situations: quasi-static mode of burning with medium temperature T_i [4], temperature fluctuations [4] and fast (start) heating [5]. Requirements for AOS construction, which provides retention of capacity for work and overheating diode prevention, were set.

On the basis of made analysis quantitative estimations were received. They are necessary for realization of methods of useful signal extraction from its totality with temperature disturbance, fluctuation noise decrease.

Was shown that independent control of AOS temperature or its parts, without taking into account received in [1,2,3,4,5,6] dependences, doesn't give complete information about this disturbance. Different types of thermostable AOS constructions were suggested.

2. PURPOSE OF THE WORK

The purpose of the present work is sensor's capacity for work and its new constructions experimental verification; main characteristics estimation in "cold" and heated conditions.

3. EXPERIMENTAL STAGES:

Experimental program includes:

- manufacturing and testing of two types of the sensor's constructions;
- experimental bed modification, providing imitation of combustion chamber conditions;

- spatial sensitivity of AOS estimation, supposedly close by sizes for a near zone of antenna;
- sensor's "cold" sensitivity measurement;
- investigation of a transfer function under operation mode imitation;
- estimation of oscillation frequency temperature dependence;
- verification of the diode's heat protection efficiency.

4. SENSOR'S CONSTRUCTION

Basic construction of the sensor has been described in [1,2]. From electric point of view it consists from antenna with aperture turned into combustion chamber, resonator and waveguide – both are coaxial. In operating position plane of flush-mounted antenna coincides with an inner surface of the chamber wall, which provides undisturbed monitoring.

Resonator is quarter- or half-wave, and main reason, that determines its total length with waveguide, is a diode's heat protection. For the heat transfer decrease from the heated up to few hundreds degrees fire surface, which is spatially coincides with antenna, sensor should be long enough, metallic surfaces – have to be thin, dielectric should have small heat conductivity η and permittivity ε_r. The last one affects on the wave length and, consequently, on the resonator length too. From the all said above, the best from suggested sensor's constructions is the one, which forms by the porous ceramics tube with sputtered (up to skin-layer thickness) outer and inner surfaces. Tube or better to say its thick wall (ratio of diameters is about three), transforms into coaxial resonator with a dielectric filling. So, in such construction, dielectric should fulfill requirements on strength too besides all others.

In our case we've used glassy dielectric made by mixture of quartz dust with liquid glass, this mixture foams under the heating (while making). Its parameters:

$\varepsilon_r \approx (1{,}2 \ldots 1{,}4)$,

$tg\delta > 10^{-3}$,

$\eta \approx 1{,}2 \cdot 10^{-6}\, W/m^2 \cdot deg$,

temperature conductivity $a \approx 10^{-5}\, m^2/sec$.

Because of a low strength of dielectric, walls turned from bronze, their thickness are about 0,1mm.

Two types of the sensors, different one from another by diameters and features of the dielectric forming, presented on the photo 1 (both are shown below on the left), all others are prototypes of the present sensors. Operating frequencies are about 2,65GHz.

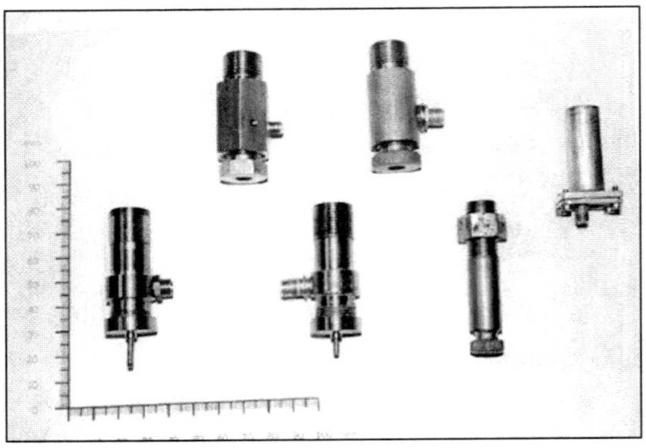

Photo1. Thermo-stable microwave sensors for technological parameters diagnostics in different TPEs.

5. EXPERIMENT

Let's take a look at the experimental part of the work.

Spatial sensitivity of the sensor, which was mounted on a flat or semicylinder screen, has been evaluated in 2 stages:

1) sensor's reaction at probe and angular reflector, which were carrying into near zone of antenna in a different direction and on a different angles (imitation of an opposite chamber wall and constructive elements of the chamber);
2) sensor's reaction at the dielectric samples, including probing zone total filling by dielectric (imitation of an averaged permittivity changes).

Frequency variations under these influences are shown on the Fig.1 (a,b,c,), Fig. 2 of enclosure correspondingly and experimental bed is shown on photo 1.

General view of the curves (Fig.3 - enclosure) is typical for autodynes in the distance measurement mode; maximums stand one from another on a wave length. Chamber sizes of the jet engines made by different companies are shown on abscissa axis. As we can see from the graphs, an effective zone of antenna encloses the cross-section till the flow axis of existing chambers, and not only near wall chamber zone, as we were written in our early articles. By the data received from specialists, working with combustion chambers, we know, that flow structure in a cross-section in symmetric, so, the sensor is able to monitor (more or less) full section of the flow. This fact proves that we have received sufficient sensitivity for our purposes.

Sensor's operating characteristics were investigated on the following testing beds (photos 2, 3, 4) in collaboration with "Jet engines and industrial burners laboratory" (JEIB). Simulated experiment will be described here. It was an attempt of measuring signal and temperature error independent estimation.

Electron density growth was provided by variations of combustion materials chemical composition. Metal powder (aluminium) was injected into the flame. It gave immediate and extremely fast electron density growth while temperature growth was inessential and not so fast. From the other side, sensor's reaction (oscillator's reaction) at the temperature growth under shorted antenna (shielded) antenna was discovered-without electron density variations.

Made estimations are not so precise, because they were made on the testing bed with gas burners and not with a real jet engine during the operating mode, but they give us an ideas about sensor's construction modification. Next step in our work will be an experiment with bimetallic antenna, which will give more precise results.

Photos (3, 4) and fig 1, 2, dedicating to this part of experiment, are showing the operating process imitation by two gas burners with temperature $T_{max} = 600...2500^0 C$.

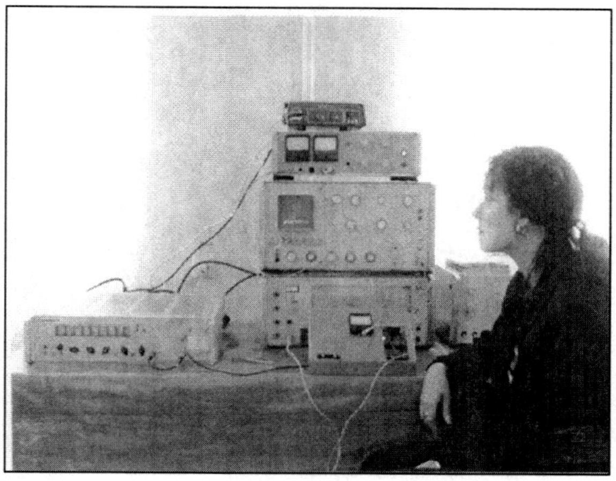

photo2. Testing bed № 1

Photo 3. Testing bed № 2

Photo 4. Testing bed № 3

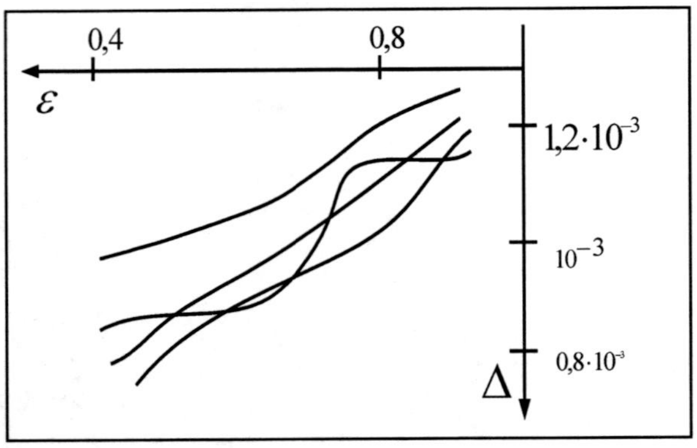

Fig.1 Experimental dependence of measuring signal frequency (sensor №2) under combustion process registration (testing bed № 2)

All the curves were received under the same conditions. Transfer function slop is about $df/d\varepsilon \sim (1{,}5...5) MHz / per \quad unit \cdot$

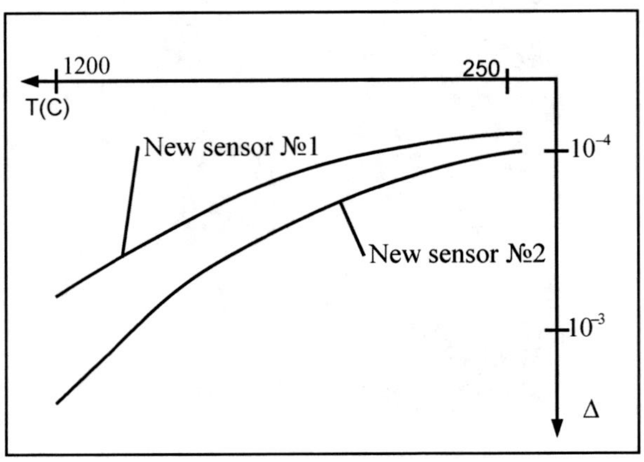

Fig.2 Sensors frequency shift under intensive chamber wall heating (testing bed № 3)

6. CONCLUSIONS

- Constructions of the created sensors provide thermo-protection of the diode (work capacity) under flame temperature up to $2500^\circ C$ [6], longevity of the sensors increased into tens times and more;
- Systematic temperature error reached few percents and could be considered, using data about flame temperature and received during the work analytical dependences [6],
- Random error and it's spectrum are depend from thermo inertia of constructional materials a lot; their choice help to improve signal/noise ratio into most important and informative part of the spectrum of the monitoring signal [6];
- Analysis has shown that direct compensation and essential reduction under processing for the dynamic error seems quite complicated still.

As a main conclusion, we can say, that microwave AOS could be recommended for the wide usage systems of monitoring for rocket and jet engines.

REFERENCES:

1. V.V. Boloznev, A.Sh.Chabdarov, "Permittivity Sensor for an Ionized Gas Flow", Kazan' Aviation Institute.Translated from Pribory i Tekhnika Eksperimenta, № 5 pp.149-152, September-October, 1992.
2. Boloznev V., Safonova E., Microwave oscillator as a sensor of electrophysics flow parameters // Proc. Joint meeting of the 13th European frequency and time forum and 1999 IEEE international frequency control symposium, Besancon, France, 1999.
3. Safonova E.V. Minimization of an autodyne sensor's error, stipulated by the antenna's heating // Proc. Joint meeting of the 14th European frequency and time forum and 2000 IEEE international frequency control symposium EFTF Torino, Italy, 2000.- p.241-245.
4. Boloznev V., Safonova E.V. Determination of the frequency of oscillations in the autodyne's sensor under the regular and casual temperature mode variations // Proc. 2000 IEEE/EIA international frequency control symposium and exhibition, Kansas, USA, 2000.-p.96-99.
5. Safonova E. Autodyne microwave sensor of combustion process: behavior under fast heating // Proc. Joint meeting of the 15th European frequency and time forum and 2001 IEEE international frequency control symposium, Neufchatel, Switzerland, 2001.-p.84-88.
6. Safonova E. Microwave sensors of an electrophysical parameters of flames // Thesis of Joint meeting of the 16th European frequency and time forum and 2002 IEEE international frequency control symposium, St.Peterburg, Russia 2002.-p.109.
7. Кирдяшев К.П. Радиофизические исследования плазмы космических двигательных установок, «Радиотехника», 2003г., №11.
8. Зарко В., Вдовин Д., Перов В. Методические проблемы измерения скорости горения твердых топлив с использованием СВЧ-излучения, «Физика горения и взрыва», 2000, т.36, №1.

ENCLOSURE

1. Estimation of the near–field zone of antenna

Stable generation of the microwave sensor for the Gunn diode AA736A were observed on the frequency f=2,651802HHz under the voltage U=6,7V, current I=0,2A

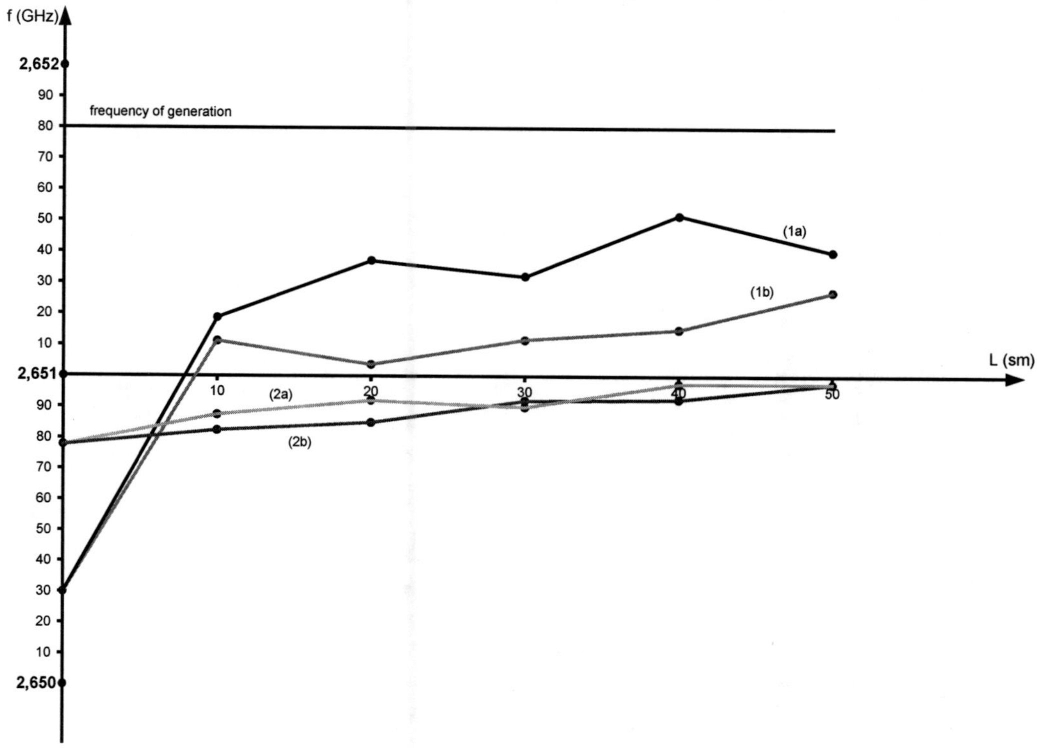

a)

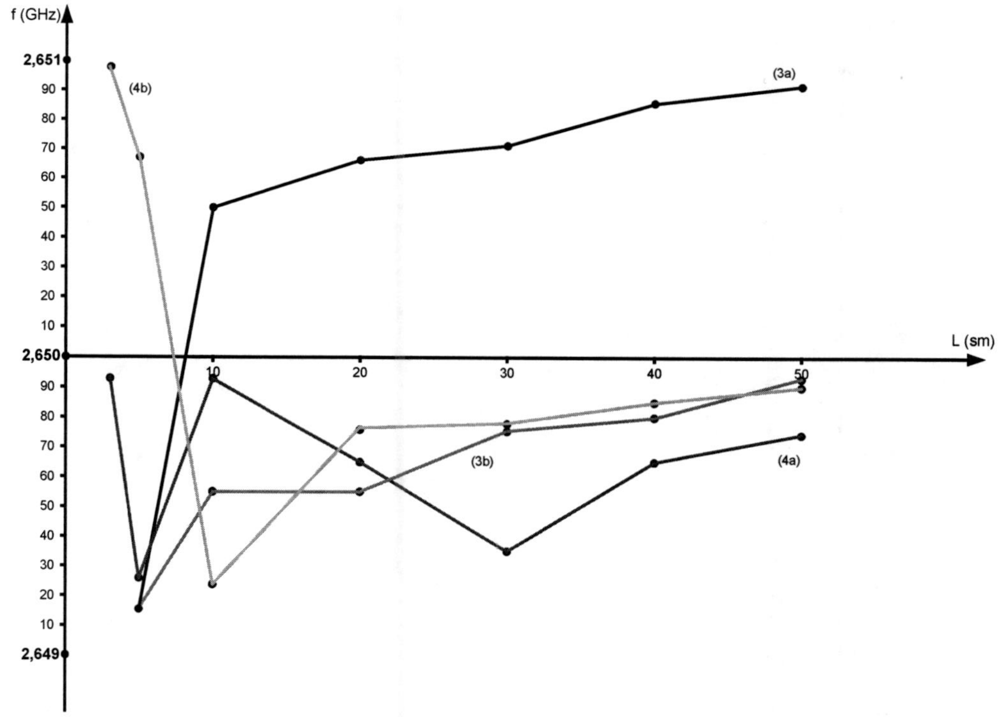

b)

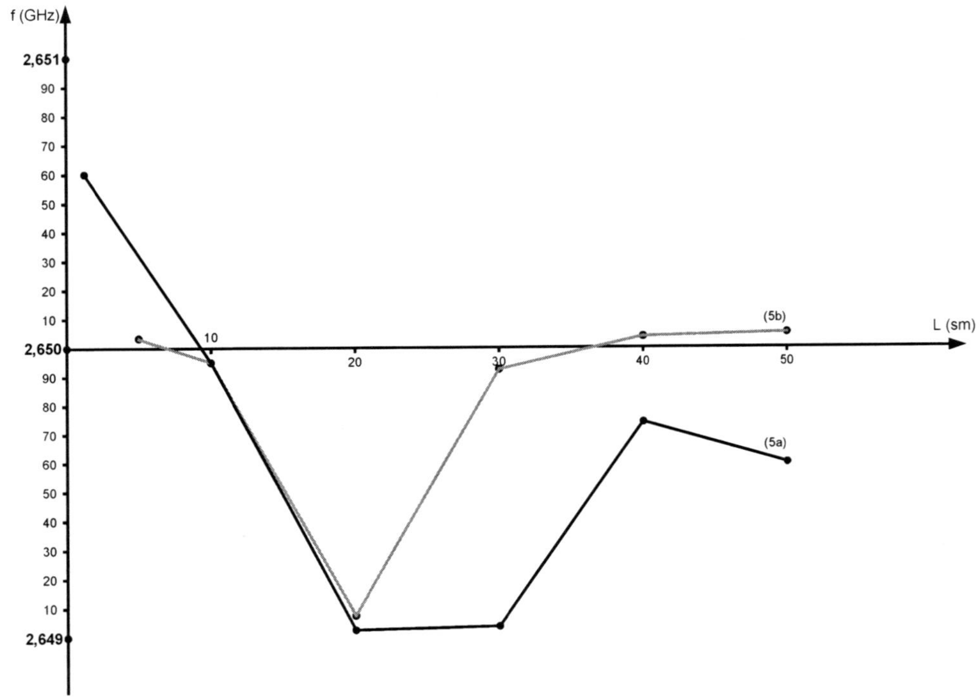

c)

Fig 1. Sensor's reaction at probe and angular reflector, which were carrying into near zone of antenna in a different direction and on a different angles (imitation of an opposite chamber wall and constructive elements of the chamber);

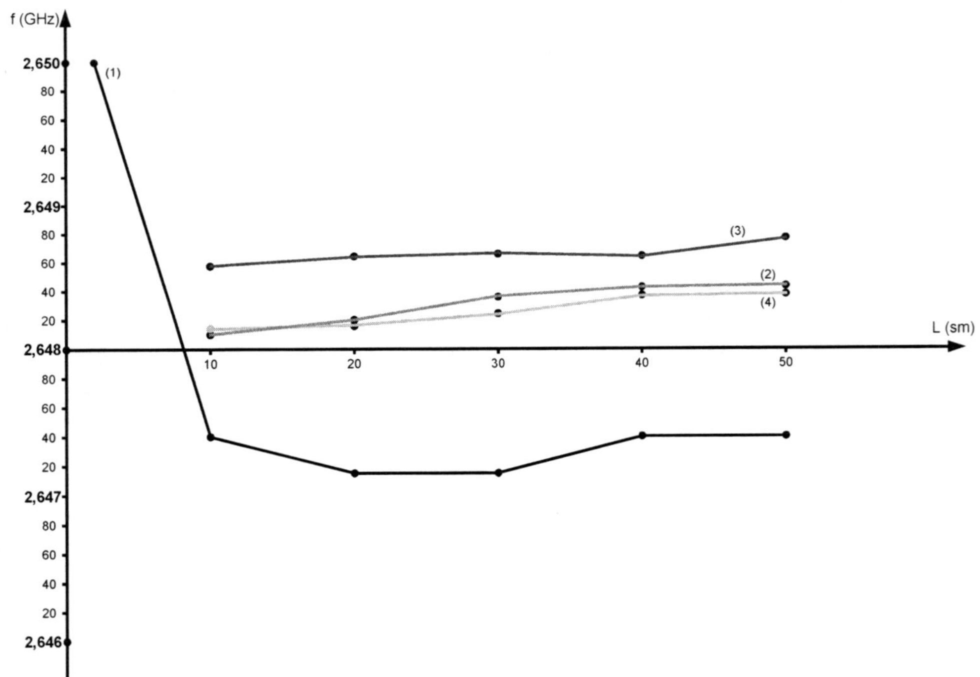

Fig. 2. Sensor's reaction at the dielectric samples, including probing zone total filling by dielectric (imitation of an averaged permittivity changes).

2. Frequency variations of an autodyne from the reflectors position in the near-field zone

Length, sm	Frequency 1	Frequency 2	Frequency 3
In air	2,651863	2,651835	2,651854
40	2,651876	2,651885	2,651716
35	2,651872	2,651796	2,651783
30	2,651862	2,651846	2,651763
25	2,651847	2,651836	2,651785
20	2,651863	2,651867	2,651777
15	2,651776	2,651695	2,651746
10	2,651889	2,651896	2,651800
5	2,651772	2,651785	2,651783

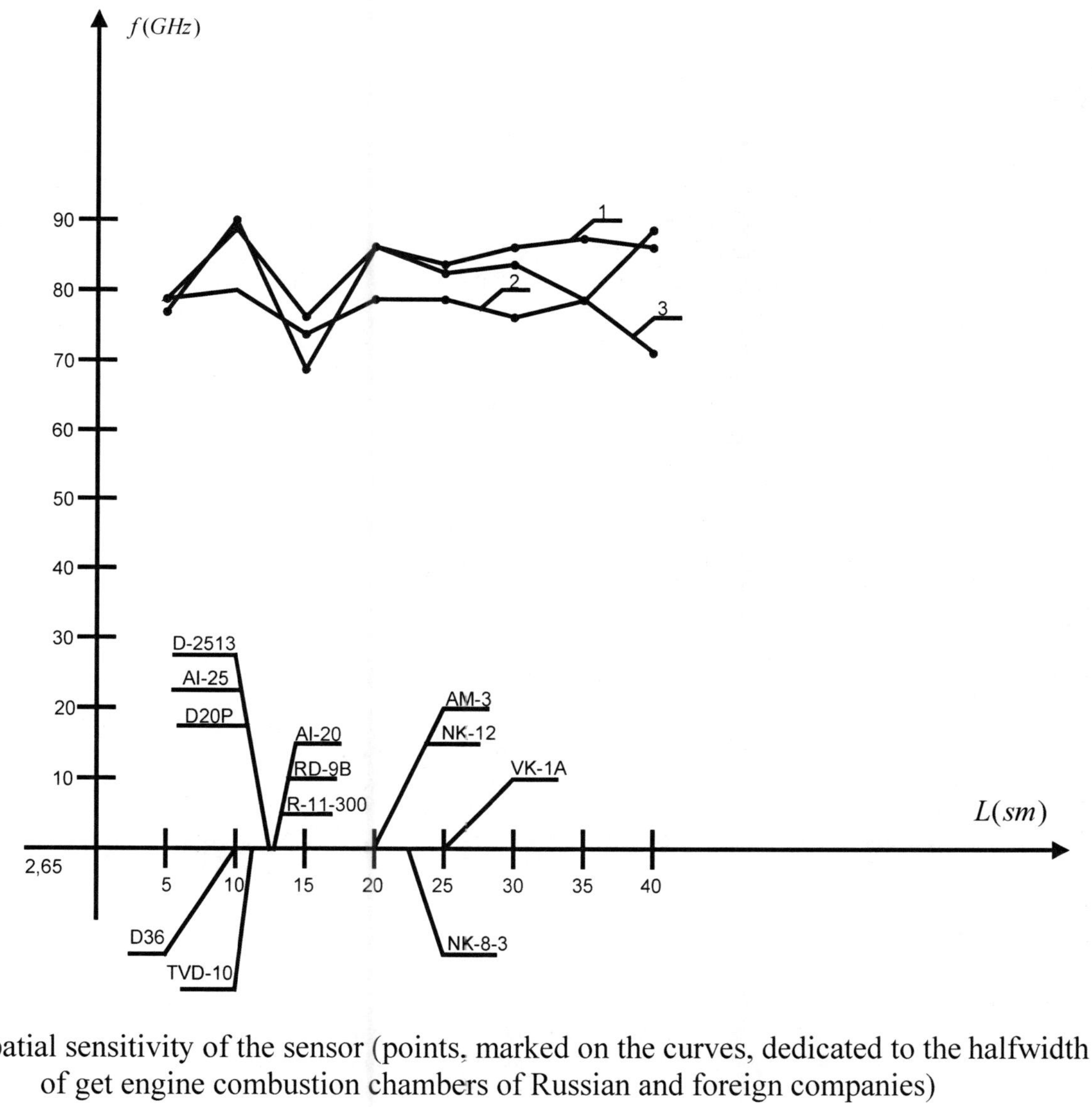

Fig.3. Spatial sensitivity of the sensor (points, marked on the curves, dedicated to the halfwidth of get engine combustion chambers of Russian and foreign companies)

A QUARTZ-CRYSTAL-SENSOR OSCILLATOR APPROPRIATE FOR HIGHLY VISCOUS LIQUID MEASUREMENTS

Ivan D. Avramov *

* Institute of Solid State Physics
72 Tzarigradsko Chaussee Blvd., 1784 Sofia, Bulgaria
e-mail: iavramov@issp.bas.bg
FAX: ++ (3592) 975 36 32

Keywords: crystal sensors, oscillators, liquid measurements.

Abstract

This paper describes a sensor oscillator appropriate for quartz crystal microbalance (QCM) based measurements in highly viscous liquids. The oscillator is stabilized by a recently developed impedance transformation QCM filter network which greatly reduces electrical load by the sustaining amplifier. In addition, it provides 0 phase and a well behaved symmetric resonance at the QCM series resonant frequency f_s over three orders of magnitude variation range of the motional resistance R_m. The sustaining amplifier has a constant pure resistive input and output impedance to eliminate phase shifts due to QCM impedance changes with liquid load. Also it allows precise gain and phase control for accurate adjustment of the oscillator frequency at f_s and extraction of R_m information at the measured liquid load. The oscillator was tested with different water based glycerol solutions in the 40wt% to 100wt% range. The worst-case frequency error was within 5% in this viscosity range.

1 Introduction

Thickness shear resonators are widely used as mass sensitive quartz crystal microbalances (QCM) for accurate characterization of liquid properties in chemical and biological sensors. The mechanical interaction of the QCM with the liquid causes a frequency down shift of the resonator's series resonant frequency f_s and an increase in its motional resistance R_m [1, 2]. Both effects are a function of the density and viscosity of the characterized liquid, therefore, accurate measurements of f_s and R_m under liquid load are necessary for adequate measurement accuracy and resolution. The easiest and most commonly used method for that is to connect the QCM to an oscillator circuit, measure the frequency of oscillation [3] and extract R_m information from the voltage driving a loss proportional gain control circuit [3, 4, 5]. Unfortunately, QCM sensor oscillators have several drawbacks which affect measurement accuracy and limit their use in practical sensor systems:

1. The oscillation frequency f_o generally is not equal to f_s. The reason is that under liquid load, R_m increases from a few hundred Ω to tens of kΩ and becomes comparable with the static capacitance C_o. This causes a substantial phase shift at resonance and moves f_o away from f_s causing a significant measurement error, especially at high R_m values [6]. Sophisticated calibration techniques are necessary to improve measurement accuracy [4]. One way of solving the problem is to compensate C_o eliminating its effect on f_o. This is done either by using transmission-line balanced bridge oscillators [7] or PLL systems [8], both requiring increased circuit complexity.

2. An oscillator is a nonlinear selflimiting device in which the sustaining amplifier is driven into gain compression to provide stable amplitude of oscillation. This makes extraction of R_m information from the loop loss very difficult and inaccurate. In addition, a compressed amplifier generates additional phase shifts that are hard to predict and compensate for. Automatic gain control (AGC) circuits also cause parasitic phase shifts and are difficult to control over a large R_m variation range. Typically such circuits are limited to R_m values not exceeding 1 kΩ and again calibration is necessary for better accuracy [4, 5].

This paper presents a simple quartz crystal liquid sensor oscillator (QCLSO) which attempts to solve the above problems. It operates at or very close to f_s and provides R_m information with reasonable accuracy over a large R_m variation range to allow measurements at highly viscous liquid loads. This is achieved by exerting external operator control over the oscillation frequency and oscillator gain for tuning the QCLSO over a narrow oscillation limit around f_s and operating the sustaining amplifier in its linear region far below saturation for accurate R_m readings. Measurement error is limited by the subjective readings of the operator and do not exceed 5% over a large viscosity variation range.

2 The impedance transformation QCM filter network

The heart of the proposed oscillator is a two-port QCM based impedance transformation network (ITN) which has zero phase and a well behaved symmetric resonance at the QCM series resonance frequency f_s. In this network, the QCM is connected in parallel with an inductor L_o, compensating for the static capacitance C_o, and embedded between two transformers as shown in Fig. 1 a). The role of the transformers is twofold: On one hand, the QCM sees on both sides a high electrical impedance. In this way it is loaded only by the liquid and additional degradation of its Q due to electrical loading by the low impedance of the measurement system, typically 50 or 75Ω, is avoided. On the other hand, compensation of C_o by L_o is enhanced and the series resonant behavior of the QCM is greatly improved even at very high R_m values. Practically, the transformers in Fig. 1 a) can be replaced by a low-pass and a high-pass LC network (Fig. 1 b)), both providing an appropriate impedance transformation ratio. This approach was recently developed in [9] and has the following features:

1. By choosing appropriate values for L1, C1 and L2, C2, a desired impedance transformation ratio is easily achieved. The transformation ratio has an influence on the filter loaded Q and insertion loss for a given input and output load impedance, as well as on the loss sensitivity with regard to small changes of R_m. The filter loss versus R_m data of a 9 MHz QCM in Fig. 2 was obtained for a transformation ratio of 25, meaning that a pure resistive 50Ω input and output load impedance Z_o of the measurement system is transformed into a pure resistive Z_{tr}=1.25kΩ on both sides of the QCM. This transformation ratio has been chosen for maximum loss sensitivity at R_m values exceeding 1kΩ to achieve maximum R_m measurement resolution at highly viscous loads. If another R_m range requires maximum measurement resolution, then this can be achieved by selecting different L1, C1 and L2, C2 values [9].

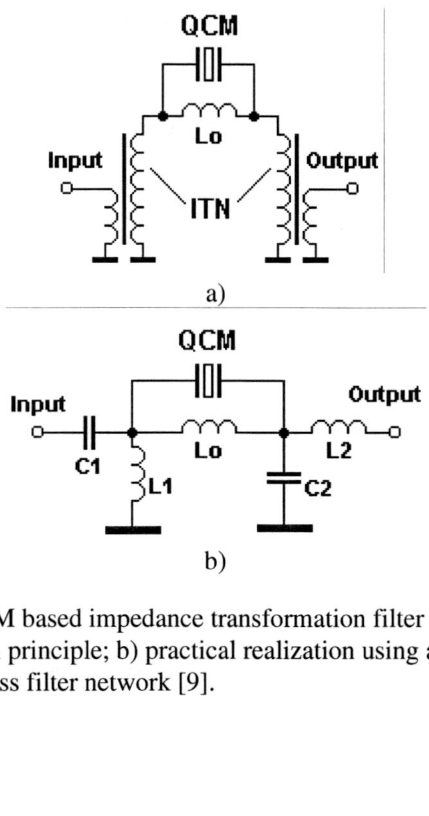

Fig. 1. QCM based impedance transformation filter network: a) operation principle; b) practical realization using a low-pass and high-pass filter network [9].

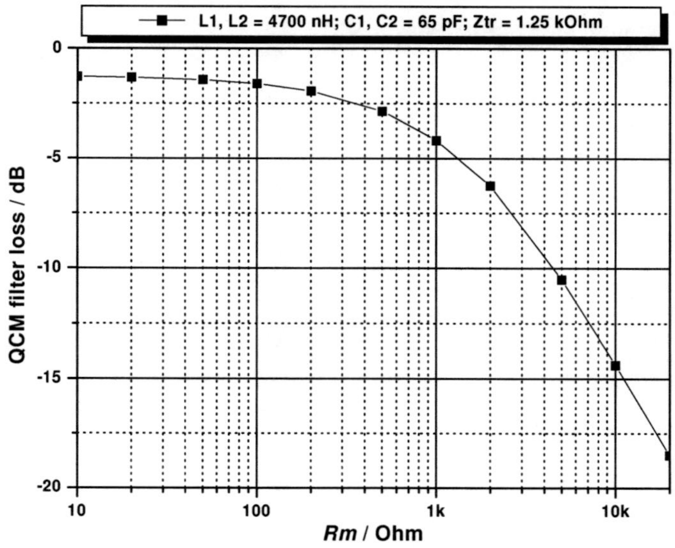

Fig. 2. QCM filter insertion loss versus R_m for a 9 MHz QCM at a transformation ratio of 25 chosen for maximum loss sensitivity at R_m exceeding 1kΩ (highly viscous liquids).

2. The input and output impedance of the QCM filter remains pure resistive over the entire R_m variation range [9]. If the filter is loaded on both sides with a pure resistive load, then no phase shifts, associated with input and output reactances can occur. In an oscillator circuit, this condition is accomplished by loading the filter with a sustaining amplifier featuring a resistive input and output impedance as shown further in this paper.

3. Zero phase of the filter network from Fig. 1 b) is achieved at f_s and appears in the middle of a symmetric resonance regardless of the R_m value. Thus, the circuit maintains 0 phase with sufficiently high precision over the entire R_m variation range [9]. This feature is illustrated by the S21 data plots in Fig. 3 a) and b) indicating the electrical behavior at R_m=38Ω and 20kΩ, respectively. Note that the QCM response at series resonance sits symmetrically in the middle of the notch response formed by L_o, C_o while the out of band high- and low-pass behavior is formed by L1, C1 and L2, C2 as shown in Fig. 3 b).

4. For a given QCM and input and output load impedance, the 0-phase adjustment needs to be performed only once. Then the electrical behavior will be maintained throughout all measurements, provided that the filter component values in Fig. 1 b) remain constant. In that respect, care should be taken to avoid temperature induced shifts of the external inductors in the filter. Practically, this should not be a problem. Since the viscosity of the measured liquids greatly depends on temperature, QCM-based liquid measurements require constant temperature conditions anyway. Those could be applied to the QCM filter assembly too.

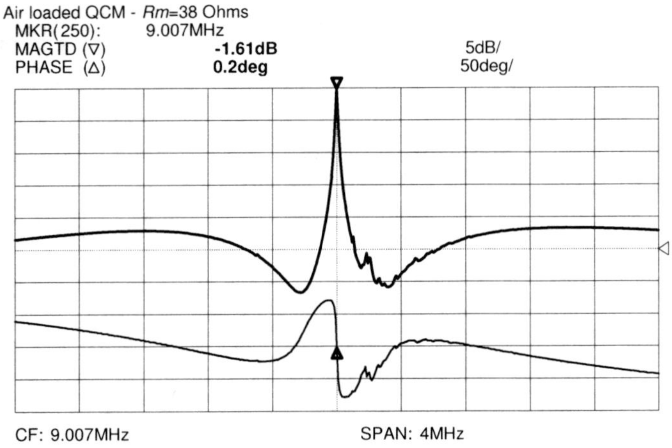

a)

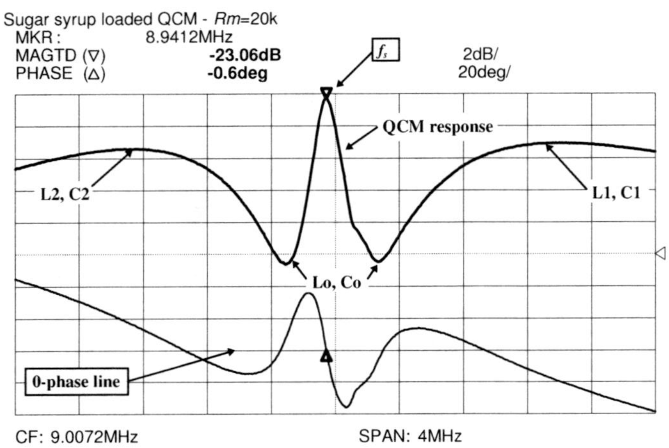

b)

Fig. 3. Magnitude (upper curves) and phase (lower curves) responses of a 9 MHz filter obtained with a 50Ω network analyzer (NWA) in a S21 transmission measurement: a) air loaded QCM (R_m=38Ω, loss is -1.61 dB); b) QCM loaded with highly viscous sugar syrup (R_m=20kΩ, loss is -23.06 dB). Note that, in both cases, the phase at series resonance is 0.

5. Due to circuit simplicity, linear analysis software provides very accurate simulation results which allow easy optimization for a required transformation ratio and precise prediction of the filter loss behavior versus R_m variation (see Fig. 2). These data are important for the design of the gain controlled sustaining amplifier in the proposed oscillator since R_m information is obtained from the value of the amplifier gain, (equal to the filter loss), necessary for oscillation to start at the QCM series resonance. Alternatively, such data can be obtained experimentally by replacing the QCM and L_o from Fig. 1 b) with a variable ohmic resistor [9], substituting R_m at series resonance, and recording the filter loss for a set of resistance values. In both cases, the ohmic resistance of the used inductors should be taken into account to avoid errors in the high R_m range.

3 The quartz crystal liquid sensor oscillator (QCLSO) circuit

The operation principle of the proposed QCLSO is shown by its block diagram in Fig. 4. It is a classical feedback-oscillator loop stabilized with the QCM filter, described in the previous section. The filter loss in the loop is compensated for by a sustaining amplifier (A) incorporating a variable phase shifter (VPS) and a variable gain control (VGC) circuit. The VPS and VGC circuits are operator controlled by means of external voltages to allow adjustment of the oscillator frequency at the QCM series resonant frequency where loop loss is minimum and $f_o = f_s$, and to adjust the gain at that frequency just above the limit, necessary for oscillation to start. On one hand, by operating the QCLSO close to the oscillation limit, the sustaining amplifier is operated far below saturation in its linear region. On the other hand, the gain control voltage, calibrated in R_m data, directly provides the R_m value at the oscillation limit. After subsequent high-impedance buffering (B), the output signal is detected and its magnitude is monitored on the indicator. The buffered oscillation frequency is measured by a frequency counter.

3.1 Measurement procedure

The measurement procedure is fairly simple. It aims at finding and recording the minimum filter loss frequency where $f_o = f_s$. Once this frequency is found, the gain is adjusted at the oscillation limit and the corresponding R_m proportional gain control voltage is recorded. This is performed by altering the VPS and VGC voltages and monitoring oscillation amplitude on the indicator from Fig. 4. The indicator reading versus QCLSO

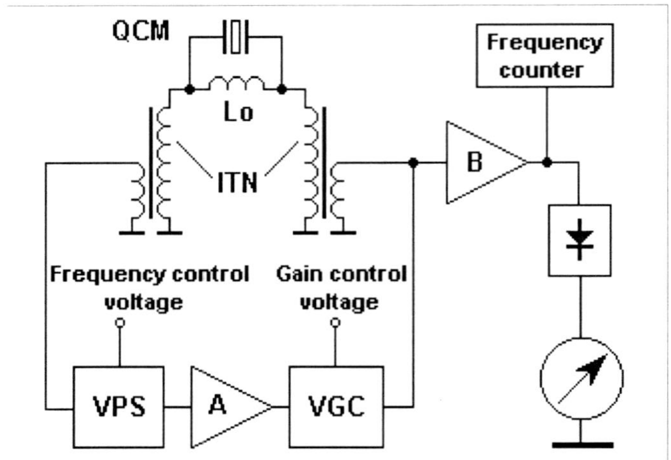

Fig. 4. Block diagram of the QCLSO.

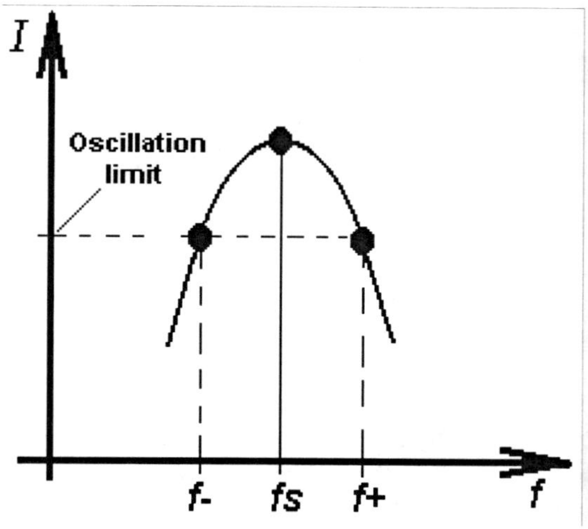

Fig. 5. Indicator (I) reading versus VPS shifted oscillation frequency f.

frequency is indicated in Fig. 5. The measurement starts with setting the gain control voltage to minimum gain (no oscillation, indicator reading is 0). The liquid is applied to one face of the QCM in the liquid cell and gain is increased until oscillation starts. Then the VPS voltage is varied until the indicator reading reaches its maximum. At that point, gain is reduced again until a condition is reached, where, by varying the VPS voltage, oscillation starts at f_-, goes through the maximum at f_s and collapses at f_+. Gain should be reduced in such manner that the two edge frequencies f_- and f_+ are kept as close as possible. After recording f_- and f_+, the f_s value of interest can be calculated as $(f_+ + f_-)/2$. The motional resistance is obtained with sufficient accuracy by tuning the QCLSO at f_s and reducing the gain until oscillation is bound to collapse. In this condition, the gain control voltage is recorded and the R_m value is obtained from the calibration data for this voltage. The data plots in Fig. 6, experimentally obtained for the considered QCLSO, provide

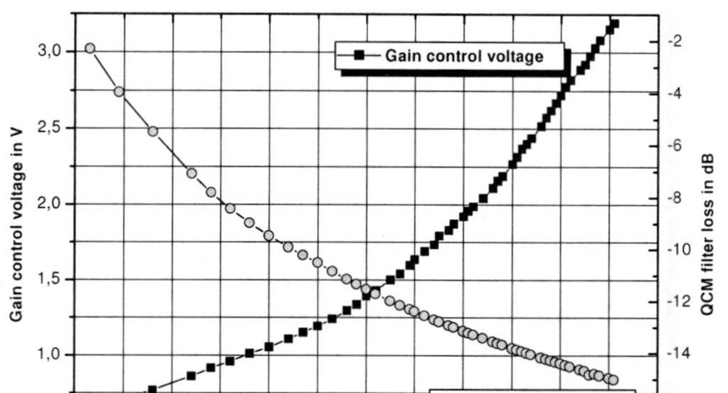

Fig. 6. Experimental relationship between R_m, QCM filter loss at f_s and the sustaining amplifier gain control voltage.

the required relationship between R_m, QCM filter loss at f_s (equal to the amplifier gain at the oscillation limit), and the gain control voltage.

3.2 Sustaining amplifier design

The sustaining amplifier (SA) for the QCLSO described in the previous section should fulfill the following requirements:

A. It should have a pure resistive input and output impedance equal to the load impedances which the QCM filter was designed for, (see Section 2). A 50Ω impedance value is very convenient since the QCM-filter adjustments and active circuit evaluations can be performed by means of a readily available 50Ω NWA.

B. To fulfill the phase condition of oscillation, the SA should have 0 phase too. In addition, a voltage controlled variable phase shift on both sides of the 0-phase point should be provided to allow for the measurement procedure under Section 3.1 and for possible cable length compensation between QCM filter and SA circuit. Also, the gain should remain constant over the entire phase variation range for accurate R_m measurements.

C. The SA should be able to provide a sufficient amount of voltage controlled variable gain to cover the entire loss variation range of the QCM filter (see Fig. 2). The phase should remain constant with gain variation.

D. Finally, the SA should be as linear as possible to avoid parasitic phase shifts, associated with different amplitudes of oscillation around the QCLSO loop.

Extensive studies within this work have shown that dual-gate MOS field effect transistors (DGMOSFETs) in a common-source configuration, ideally meet the above requirements. On one hand, they have extremely high input and output impedances with negligible reactance values up to several tens of MHz to practically cover the entire operation range of QCM based liquid sensors. Loading the gate and the drain of such a transistor with a 50Ω ohmic resistor perfectly meets requirement A. On the other hand, DGMOSFETs have two gates. If the signal is applied to Gate 1, the gain can be controlled over a 20 dB variation range by applying a variable control voltage to Gate 2. And since these devices use a cascode configuration, the isolation between both gates is excellent, meaning that transistor phase is not influenced by the gain control. This fulfills requirement C. Finally, DGMOSFETs are well known for their excellent linearity which satisfies requirement D.

The simplified circuit of the SA used in the QCLSO is shown in Fig. 7. It consists of two DGMOSFETs, FET1 and FET2, each providing 180 deg. of phase for a total loop phase of 0. The

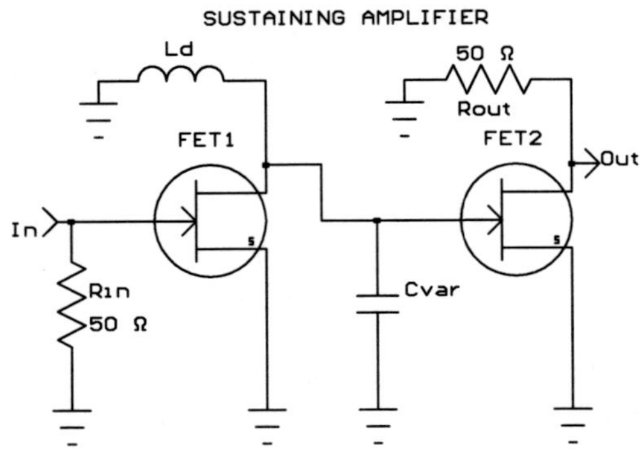

Fig. 7. Simplified SA schematic.

resistors Rin and Rout provide the required 50Ω load to the QCM filter. The VPS is formed by the inductor Ld and the variable capacitance C_{var} which is a varactor. Variable gain is accomplished by a control voltage applied to Gate 2 (not shown here) of one of the transistors, as explained above. The gain and phase responses of the SA have been optimized for the 9 MHz QCLSO using linear circuit simulation software.

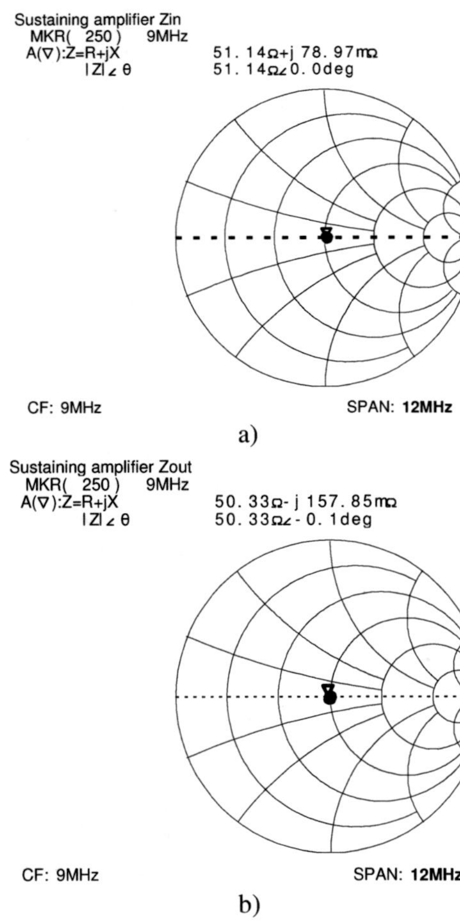

Fig. 8. Input (a) and output (b) impedances of the SA circuit from Fig. 7 over a 12 MHz frequency span around 9 MHz.

The SA input and output impedances, measured with a NWA in the 3 to 15 MHz frequency range are shown in Fig. 8 a) and b), respectively. The black spots in the middle of the Smith charts indicate a pure ohmic 50Ω impedance. The series reactances are in the mΩ range and, therefore, negligible.

The data plots in Fig. 9 illustrate the variable gain control performance of the SA, obtained in a S21 transmission measurement. The gain was varied between the minimum 2 dB value (lower gain curve) and the maximum 21 dB (upper gain curve). No measurable phase change as a result gain variation was observed. Finally, Fig. 10 shows the phase variation performance of the SA. When the phase was varied by 17 degrees on both sides of the 0-degree value at 9 MHz, the gain variation was within a few hundredths of a dB, also negligible.

The measurements in Fig. 8 through 10 were repeated at input levels in the -50 to -20 dBm range within the linear region of SA operation. The -20 dBm level was still about 10 dB lower

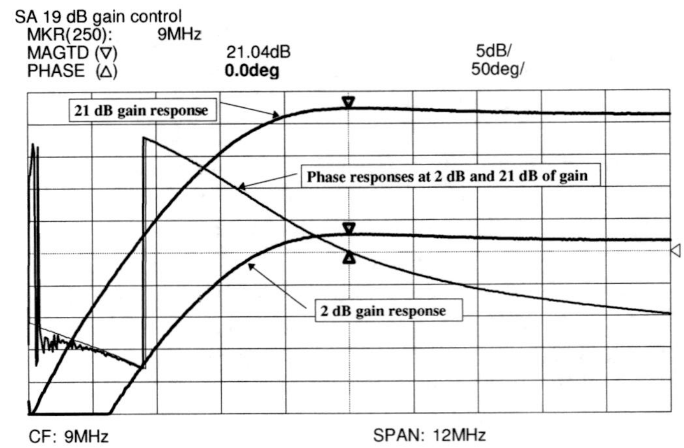

Fig. 9. Variable gain performance of the SA from Fig. 7. Triangular markers are positioned at 9 MHz

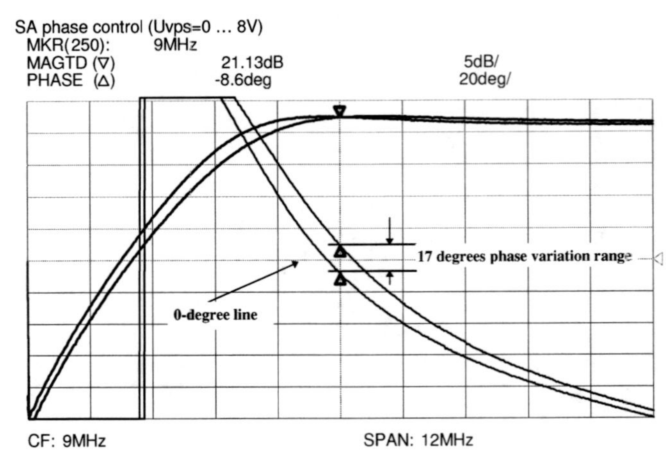

Fig. 10. SA performance at 21 dB of gain at 9 MHz with the phase varied within 17 deg. of variation range around zero.

than the saturation level. No measurable deviation from the above performance was observed.

3.3 QCLSO open loop performance

To study the linear behavior of the 9 MHz QCLSO, an open loop simulation was performed. In Fig. 11, the QCM filter, in which the quartz crystal is represented by its Butterworth-van-Dyke circuit and its equivalent electrical parameters R_m, L_m, C_m and C_o, is cascaded with the SA from Fig. 7. The S21 transmission behavior is calculated for different R_m values to simulate variable liquid loading. The results are shown in Fig. 12 for two different R_m values (38 Ω for air loading and 5kΩ for a highly viscous liquid load). It is evident that the loop loss changes between 14.5 and 1.1 dB and oscillator Q varies as a result of R_m variation. However, the loop phase remains 0 at resonance. This is an indication that the SA has no effect on the loop phase when liquid load is varied in these wide limits and, with sufficient gain, the QCLSO will oscillate at the maximum of the resonant curve where $f_o = f_s$.

The simulation results in Fig. 13 illustrate the phase variation

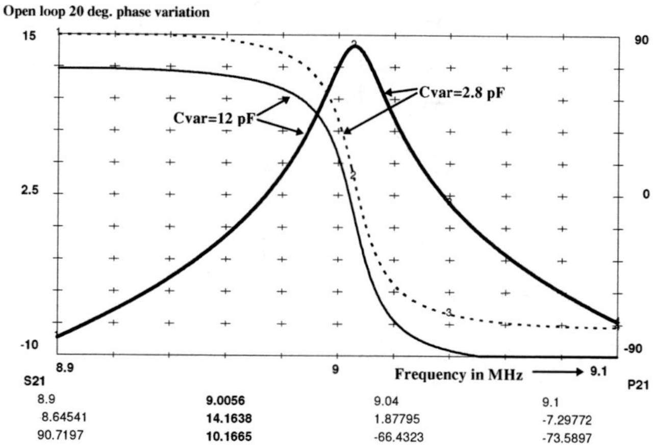

Fig. 13. Simulated QCLSO open-loop S21 magnitude and phase at R_m = 38 Ω and C_{var} from Fig. 11, varied between 2.8 pF (dashed phase line) and 12 pF (solid phase line).

capabilities of the QCLSO for performing the measurement procedure according to Section 3.1. In this simulation, the varactor capacitance C_{var} was varied in the 2.8 to 12 pF range to provide about 20 deg. variable phase shift around the 0-phase point at resonance. The solid line resonance in Fig. 13 represents the open-loop S21-magnitudes for both capacitance values, overlaid on each other. Excellent gain constancy is observed indicating that the R_m measurement accuracy will not be compromised by the phase variation in the loop.

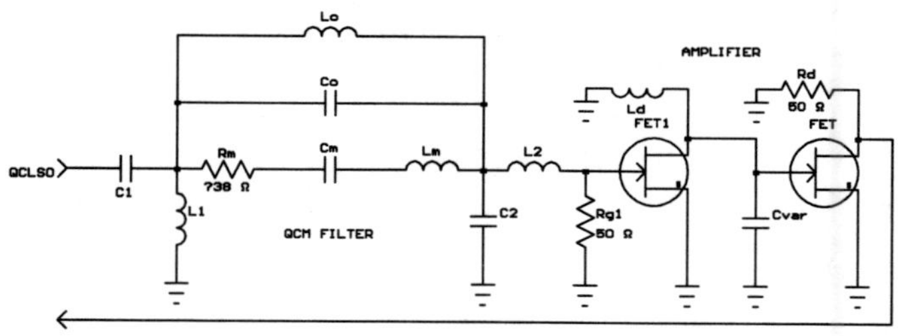

Fig. 11. QCLSO schematic as used in the open loop simulations.

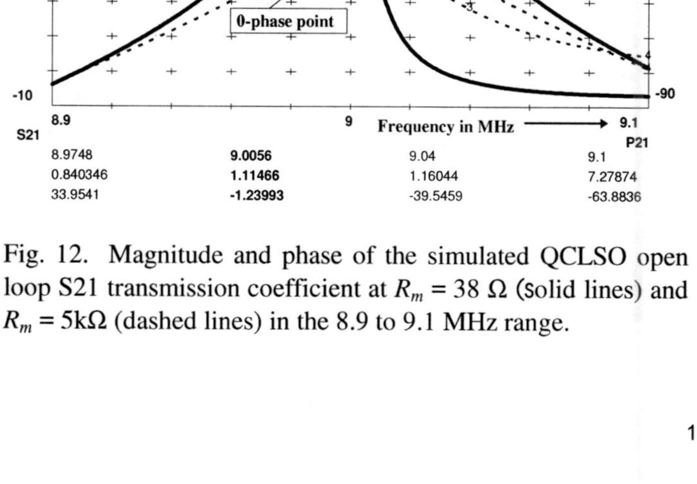

Fig. 12. Magnitude and phase of the simulated QCLSO open loop S21 transmission coefficient at R_m = 38 Ω (solid lines) and R_m = 5kΩ (dashed lines) in the 8.9 to 9.1 MHz range.

4 Experiments

To test the QCLSO performance at highly viscous liquids, water based glycerol solutions with 6 different concentrations (Conc. 1 through 6) were prepared. The lowest and the highest concentrations were about 40wt% and 100wt% glycerol, respectively. The measurements were performed with single-face loading of the 9 MHz QCM at open air and room temperature conditions. Since the viscosity of the test solutions strongly depends on temperature and humidity of the ambient air, and no climatic chamber was available, the exact R_m and f_s values of the QCM at each concentration were first obtained in a NWA admittance measurement and then used as a reference for the evaluation of the QCLSO measurement error. After each NWA measurement, the liquid loaded QCM was disconnected from the NWA and connected to the QCSLSO to perform the measurement according to the measurement procedure described in Section 3.1. Table 1 compares the results from the motional resistance measurements and Table 2 is a comparison of the frequency shifts at each of the 6 liquid loads including air loading. For the sake of simplicity, the results in Table 2 are presented as frequency shifts with regard to the air loaded case, rather than indicating the absolute frequency values. As shown in Table 1, for R_m values exceeding 1kΩ (concentrations larger than 40wt%) the QCLSO measurement accuracy is very good,

QCM Loading	R_m/Ohm (NWA)	R_m/Ohm (QCLSO)
Air	38	100
Conc. 1	793	900
Conc. 2	1307	1300
Conc. 3	1779	1800
Conc. 4	2551	2600
Conc. 5	3012	3100
Conc. 6	5882	6200

Table 1. Comparison of the NWA and QCLSO motional resistance measurement data.

QCM Loading	Δf_s/ppm (NWA)	Δf_s/ppm (QCLSO)	Freq. error (QCLSO)
Air	0	0	-
Conc. 1	448.5	448.5	0%
Conc. 2	753.8	789.2	4.7%
Conc. 3	1055.2	1095.6	3.8%
Conc. 4	1500.3	1548.6	3.2%
Conc. 5	1939.9	1963.7	1.2%
Conc. 6	3260.9	3271.4	0.03%

Table 2. Comparison of the series resonant frequency shifts obtained in the NWA and QCLSO measurements.

although all values have been rounded to a 100Ω accuracy because of the subjective reading error of the operator. For R_m values below 1kΩ, the discrepancy between NWA and QCLSO results increases dramatically since the QCLSO is operated over the flat and insensitive filter loss area in Fig. 2. A similar behavior shows also the frequency error in Table 2. The smallest error, obtained at the highest glycerol concentration, is explained by the fact that the initial QCM filter adjustment was performed for minimum error at the highest viscosity. The error increase at lower concentrations is attributed to QCM-filter phase error. It is caused by a slight displacement of the QCM series resonance (as a result of QCM liquid loading) from the center of the LC-filter's low-Q resonance formed by L1, C1, L2 and C2 from Fig. 1 b) [9]. Still, the worst-case QCLSO frequency error, measured at Conc. 2, does not exceed 5% over this very high viscosity variation range.

Summary and conclusions

This paper has presented a simple operator controlled oscillator circuit for measuring the motional resistance R_m and the series resonant frequency f_s of a liquid loaded QCM over a large viscosity range. By tuning the oscillator frequency at the center of the QCM series resonance, where loop loss is minimum, and by reducing the sustaining amplifier gain to maintain oscillation over a narrow limit on both sides of series resonance, separate readings for f_s and R_m are obtained. Measurement error depends on the subjective operator readings and does not exceed 5% in the worst case. Substantial error reduction is possible by designing the QCM filter in the oscillator for operation over a narrower viscosity range of interest.

Acknowledgements

The author wishes to gratefully acknowledge Dr. Ralf Lucklum from the Otto-von-Guericke University in Magdeburg, Germany for encouraging this work and for the most helpful discussions.

References

[1] K. K. Kanazawa and J. G. Gordon, "Frequency of a quartz microbalance in contact with liquid," *Anal. Chem.*, vol. 57, pp. 1770-1771, (1985).

[2] S. J. Martin, K. 0. Wessendorf, C. T. Gebert, G. C. Frye, R. W. Cernosek, L. Casaus, and M. A. Mitchell, "Measuring liquid properties with smooth- and textured-surface resonators", Proc. *1993 IEEE Int. Freq. Contr. Symp.*, pp. 603-608.

[3] F. Josse, Z. A. Shana, D. E. Radtke and D. T. Haworth, "Analysis of piezoelectric bulk-acoustic-wave resonators as detectors in viscous conductive liquids", *IEEE Trans. Ultrason., Ferroelect., Freq. Contr.*, vol. 37, no. 5, pp. 359-368, Sept. (1990).

[4] R. Borngraeber, J. Schroeder, R. Lucklum and P Hauptmann, "Is an oscillator based measurement adequate in a liquid environment?", *IEEE Trans. Ultrason., Ferroelect., Freq. Contr.*, vol. 49, no. 9 pp.199-206, Sept. (1991).

[5] F. Eichelbaum, R. Borngraeber, R. Lucklum, P Hauptmann and S. Roessler, "Oszillatoren fuer Quartz-Crystal-Microbalance-Sensoren in Fluessigkeiten *Technisches Messen*, 12, pp. 434-444, (1998).

[6] K. O. Wessendorf, "The lever oscillator for use in high resistance resonator applications", Proc. *1993 IEEE Int Freq. Contr. Symp.*, pp. 711-717.

[7] E. Benes, M. Schmid, M. Groeschl, P. Berlinger, H Novotny and K. C. Harms, "Solving the cable problem between crystal sensor and electronics by use of a balanced bridge oscillator circuit", *Proc. 1999 Joint Meeting Eur. Time and Freq. Forum and IEEE Int Freq. Control Symp.*, pp. 1023-1030.

[8] A. Arnau, T. Sogorb and Y. Jimenez, "A continuous motional series resonant frequency monitoring circuit and a new method of determining Butterworth-Van Dyke parameters of a quartz crystal microbalance in fluid media", *Rev. Sci. Instrum.*, Vol. 71, No. 6, pp. 2563 2571, June (2000).

[9] I. D. Avramov, "A 0-phase circuit for QCM-based measurements in highly viscous liquid environments" *IEEE Sensors Journal* (2004), under review.

HOW LONG DOES A CLOCK ERROR REMAIN INSIDE TWO THRESHOLD BARRIERS? AN EVALUATION BY MEANS OF STOCHASTIC PROCESSES

G. Panfilo[†°], P. Tavella[°], C. Zucca[*]

[*] Università degli Studi di Torino, Dip. Matematica, Via Carlo Alberto 10, Torino, Italy, zucca@dm.unito.it.
[°]IEN Galileo Ferraris, Strada delle Cacce 91, Torino, Italy, tavella@ien.it.
[†]Politecnico di Torino, Corso degli Abruzzi 24, Torino, Italy, panfilo@ien.it. Fax +39-011-3919259

Keywords: clock error, SDE, PDE, synchronization.

Abstract

In this paper we study the atomic clock error by using the stochastic differential equations. In particular, we focus on the evaluation of the "survival probability", that is the probability that clock error exceeds the limit of permissible error at a certain time after synchronization. The results are applied to experimental cases where the typical noise parameters of atomic clocks are considered. We also determine the optimal calibration intervals of measuring instruments as GNSS clock making use of the discussed results.

1 Introduction

The behavior of an atomic clock can be studied by modeling its phase and frequency errors by means of stochastic processes. A formal clock model, using the theory of stochastic differential equations, was presented in [6] including two typical noises, the white and "random walk" frequency noises. That approach allowed to determine the exact solution for the clock error that was also written in recursive form useful for simulations and discrete clock model applications.

Here we focus on another problem which is generally referred to as the evaluation of the "survival probability". In metrological language it is related to a set of questions: which is the probability that, at a certain time after synchronization, the clock error does exceed a threshold value? If the clock is on a GNSS satellite, how often should it be re-synchronized by a ground control station to be confident, at a certain percentage level, that its phase error does not exceed a threshold of permissible error? More in general, how long a measuring instrument can maintain its good calibration condition before accumulating an unacceptable error after calibration?

To face with these questions we use suitable mathematical tools. The "survival probability" corresponds to the probability that the stochastic process describing the clock error has not crossed the threshold level corresponding to the permissible error at a certain instant t. The analytical expression of the survival probability cannot be determined in closed form but it is known that it is the solution of a partial differential equation.

Hence, here we present two approaches to the evaluation of the survival probability in the case of two symmetric thresholds:

1) We solve numerically the partial differential equation verified by the survival probability.

2) We use a Monte Carlo method that simulates the clock error evolution in the presence of two thresholds.

The proposed methods are discussed and compared for the two considered clock noises. An application to experimental cases where the typical noise parameters of cesium atomic clocks is considered. An estimation of the interval during which the clock error remains within the threshold limits is determined, enlightening its dependence on the value of the thresholds and on the requested confidence level.

The case of GNSS clocks and to the determination of the optimal calibration intervals in the frame of industrial quality assurance are also addressed.

2 Clock model and survival probability

Typically the clock signal is affected by five types of random noises that are known in metrological literature as: White phase modulation, Flicker phase modulation, White frequency modulation (WFM), that produces a Brownian motion (BM, also called Wiener process) on phase, Flicker frequency modulation, Random walk frequency modulation (RWFM) which produces an integrated Brownian motion (IBM) on the phase. In this paper we use the two terms Brownian motion and Wiener process with the same meaning. The presence of different kinds of random noises and their mixture properties is different from clock to clock. From experimental evidence it appears that on cesium clock the predominant noises on the phase error are the WFM and the RWFM that correspond, in mathematical language, to the Wiener process and to the Integrated Wiener process respectively. In this paper we consider the atomic clock model based on these two main noises. We use the stochastic processes and stochastic differential equations and we study the first passage time of the stochastic process across two fixed constant thresholds. This topic is strictly related to the survival probability between two absorbing thresholds.

We can interpret the survival probability as the probability that the clock error has not yet exceeded the limit of permissible errors at a certain time after synchronization.

2.1 Mathematical model of the clock error

Let $X_1(t)$ be a stochastic process describing the time evolution of the atomic clock phase error. It can be viewed as the solution of a dynamical system, written as a system of stochastic differential equations (SDE) whose driving terms are two independent standard Wiener processes $\{W_1(t),\ t\geq 0\}$ and $\{W_2(t),\ t\geq 0\}$

$$\begin{cases} dX_1(t) = (X_2(t) + \mu_1)dt + \sigma_1 dW_1(t) \\ dX_2(t) = \mu_1 dt + \sigma_2 dW_2(t), \end{cases} \qquad (1)$$

with initial conditions $X_1(0) = x$, $X_2(0) = y$, where μ_1 and μ_2 are constants that in the clock case represent what is generally referred to as the deterministic phenomena of the atomic clocks. In particular μ_1 is related to the constant initial frequency offset, while μ_2 represents what is generally indicated by frequency ageing or drift. The positive constants σ_1 and σ_2 represent the diffusion coefficients of the two noise components and give the intensity of each noise. The relationship between the diffusion coefficients and the more familiar Allan variance $\sigma_y(\tau)$ [1] used in time metrology is known [4]. The stochastic differential equation (1) is constituted by a deterministic part, expressed in terms of the infinitesimal increment dt, and a random part characterized by the noise component $dW(t)=\xi(t)dt$ where $\xi(t)$ is the white noise.

In (1) $W_1(t)$ is the Wiener noise acting on the phase X_1 driven by a white noise on the frequency. Since the Wiener process can be thought of as the integral of a white noise, the phase error X_1 is the integral of the frequency error affected by white noise. $W_2(t)$ represents the Wiener process on the frequency (the so called "Random Walk FM") and contributes with an Integrated Wiener process on the phase. Note that the initial conditions x and y should not be confused with the metrological notation where usually x represents the phase and y represents the frequency. Here we consider x and y as numbers but they can be considered as random variables (cf. Section 3.2). X_1 represents the phase deviation, while the frequency deviation results to be $\dot{X}_1$. The second component X_2 is only a part of the clock frequency deviation, i.e. what is generally called the "random walk" component.

Since equation (1) is a bidimensional strictly linear stochastic differential equation, it is possible to obtain its solution in closed form [6,7] given by

$$\begin{cases} X_1(t) = x + (y+\mu_1)t + \mu_2 \dfrac{t^2}{2} + \sigma_1 W_1(t) + \sigma_2 \displaystyle\int_0^t W_2(s)ds \\ X_2(t) = y + \mu_2 t + \sigma_2 W_2(t) \end{cases} \qquad (2)$$

This solution can be written in iterative form, that results very useful for further processing of the data and for simulations [6].

2.2 Survival probability

Let $X(t)$ be a homogeneous stochastic Markov process starting in x at time $t = 0$, $\{X(t),\ t\geq 0;\ X(0) = x\}$ and let $\tau_{(-n,m)}$ be the first passage time of $X(t)$ through two constant absorbing thresholds $-n$ and m, with $n,m>0$

$$\tau_{(-n,m)} = \inf\left\{ \begin{array}{l} t\geq 0 : X(t)\notin(-n,m);\ X(\theta)\in(-n,m)\ \forall\theta<t; \\ X(0)=x \end{array} \right\}$$

Let $p(t,x)$ be the probability that the process starting in x at time $t = 0$ does not reach any of the thresholds in the time interval $(0,t)$

$$p(t,x) = P\left(\tau_{(-n,m)} \geq t \mid X(0) = x\right)$$

This function is called *survival probability*. The survival probability represents the probability that the clock error has not exceeded the limit of permissible errors at a certain time after synchronization assumed here as time zero. The classical approach to the study of the survival probability requests to consider the "infinitesimal generator" [2] of the process. Indeed, for any generic multidimensional stochastic differential equation

$$dX_t = \mu(X_t)dt + \sigma(X_t)dW_t$$

where $\mu : \Re^k \to \Re^k$, $\sigma : \Re^k \to \Re^{k\times k}$ are measurable functions, $x\in\Re^k$, one can associate a functional named "infinitesimal generator"

$$L_t = \frac{1}{2}\sum_{i,j=1}^{k}\left(\sigma(x)\sigma^T(x)\right)_{i,j}\frac{\partial^2}{\partial x_i \partial x_j} + \sum_{i=1}^{k}\mu_i(x)\frac{\partial}{\partial x_i}\cdot \qquad (4)$$

Here σ is a matrix containing the diffusion coefficients and μ is the drift vector. The survival probability $p(t,x)$ is the solution of $Lp = \frac{\partial p}{\partial t}$ when one introduce suitable initial and boundary conditions. In the case of the clock model (1) we have the bidimensional process; we indicate with $p(t,x,y)$ the survival probability where (x,y) are the initial conditions of X_1 and X_2, respectively. In order to obtain the survival probability $p(t,x,y)$ the PDE to be solved is:

$$\begin{cases} \dfrac{\partial p}{\partial t} = y\dfrac{\partial p}{\partial x} + \sigma_1^2 \dfrac{1}{2}\dfrac{\partial^2 p}{\partial x^2} + \sigma_2^2\dfrac{1}{2}\dfrac{\partial^2 p}{\partial y^2} & t\in[0,T] \\ p(t,-n,y)=0 & (x,y)\in D = \{[-n,m]\times[-\infty,\infty]\} \\ p(t,m,y)=0 & \\ p(t,x,-\infty)=0 & \\ p(t,x,\infty)=0 & \\ p(0,x,y) = \begin{cases} 0 & x=-n,m \quad y\to\pm\infty \\ 1 & (x,y)\in(-n,m)\times(-\infty,\infty) \end{cases} & \end{cases}$$

where D is the domain of possible values for the initial conditions (x,y) and $(-n,m)$ are the thresholds. In order to solve numerically the equation we consider a finite domain for y, that is $y\in(-Y,Y)$. Note that the initial problem involving a SDE has been transformed in a problem in terms of PDE thanks to the introduction of the infinitesimal operator.

3 Numerical solution of the survival probability

The solution of the equation governing the survival probability is not known in closed form, except for the simple case of Brownian motion. Then, here we deal with some numerical methods and we consider the following two different approaches:

1) The numerical solution of the PDE that we perform according to different techniques:
   a) The finite differences method: a method based on the discretization of the domain and on the substitution of the continuous derivatives with discrete approximations [10].
   b) The finite elements method: a method based on the discretization of the domain and on the approximation of the solution with the linear combination of the trial function [11].
2) The Monte Carlo method applied to solution (2): it consists in the simulation of a given number of sample paths of the considered stochastic processes and in the identification of the first passage time $\tau_{(-n,m)}$ of each simulated trajectory across the two constant thresholds $-n$ and m [9].

We apply these three methods to the cases of Brownian motion, integrated Brownian motion [9] and to the bidimensional model (1) composed by both processes BM and IBM. The numerical results obtained are then discussed with the aim to validate their reliability by means of the comparison of different estimates obtained via different techniques.

3.1 Brownian motion

For the Brownian motion (Wiener process) the analytical expression of the survival probability is known in closed form [9]. Here we compare the analytical expression of the survival probability with the numerical solution obtained via the Monte Carlo method and the finite differences method implemented with discretization steps $h_x = 0.1$ and $h_t = 0.01$ on the variable x and the time t respectively. For the Monte Carlo method we simulate $N = 10^5$ sample paths with discretization step $\tau = 0.01$. In Figure 1 we plot the survival probability in the case of standard parameters $\sigma=1$, $\mu=0$ with the choice $n=1$, $m=1$, and $(x,y)=(0,0)$. Note that the results obtained with the three different methods perfectly agree.

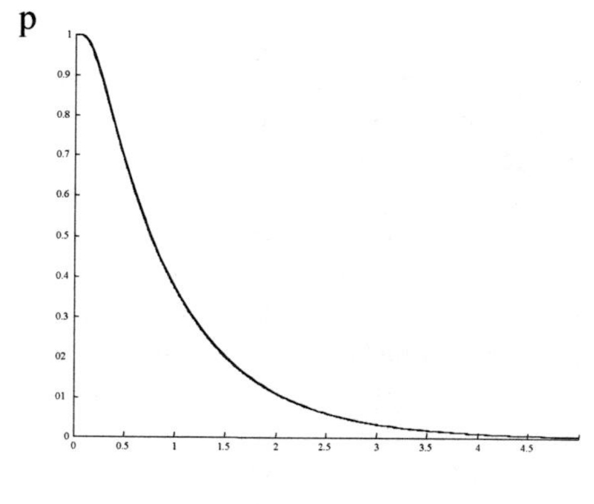

Figure 1. Survival probabilities obtained with the analytical solution, the Monte Carlo method and the finite differences method in case of BM with $\sigma=1$, $\mu=0$, $n=1$, $m=1$, $(x,y)=(0,0)$ (all lines are superimposed).

3.2 Integrated Brownian Motion (IBM)

The IBM is defined by the SDE (1) considering $\sigma_1 = 0$, $\mu_1 = 0$ and $\mu_2 = 0$. For this process the analytical expression of the survival probability is not available, then we look for numerical evaluations. Without loosing generality we apply the numerical methods to the IBM with $\sigma_2=1$. In Figure 2 is plotted the survival probability obtained via the Monte Carlo method and via the finite differences method implemented with discretization steps $h_x = 0.04$, $h_y = 0.5$ and $h_t = 0.05$ or $h_t = 0.01$. We can notice that the three curves agree to a large extent but small discrepancies arise in managing very small values of time t.

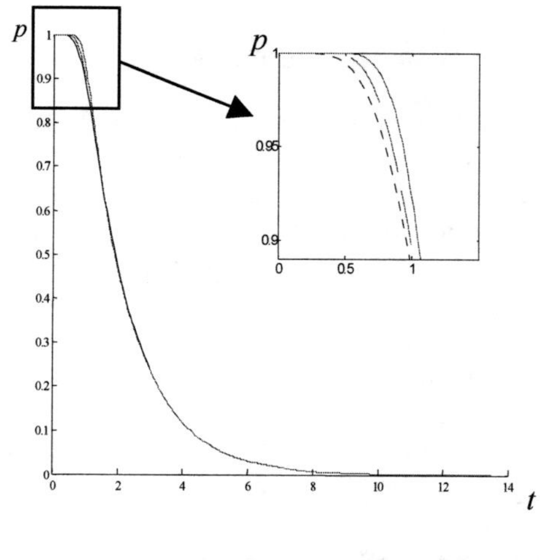

Figure 2. Survival probabilities obtained via the Monte Carlo method (solid line) simulating $N = 10^5$ trajectories with discretization step $\tau = 0.01$ and the finite differences method implemented with two different time steps $h_t = 0.05$ (dotted line) or $h_t = 0.01$, fixing $h_x = 0.04$, $h_y = 0.5$. (dashed line) in case of IBM with the choice $n=1$, $m=1$, and $(x,y)=(0,0)$.

3.3 Bidimensional model

In order to get a numerical approximation of the survival probability of the bidimensional clock process (1) we consider the finites element method (FEM), implemented with discretization steps $h_x = 0.01$, $h_y = 0.02$ and $h_t = 0.003$, together with the Monte Carlo and finite differences methods. This last method was implemented with discretization steps $h_x = 0.2$, $h_y = 0.5$ and $h_t = 0.01$ while the Monte Carlo method simulating $N = 10^5$ trajectories with discretization step $\tau = 0.01$. Without loosing in generality we deal with the parameter choice $\mu_1=\mu_2=0$ and $\sigma_1=\sigma_2=1$. In Figure 3 the survival probability obtained via these three different numerical methods is represented and the three shapes perfectly agree.

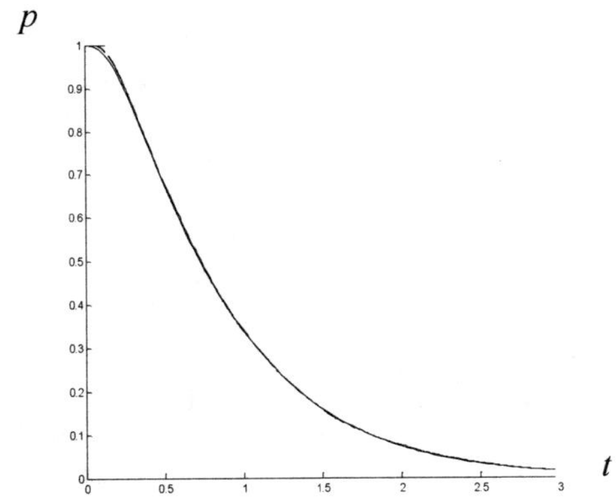

Figure 3: Survival probability for the bidimensional model (1) with the choice n=1, m=1, and (x,y)=(0,0) obtained via the Monte Carlo method, the finite differences method and the finites element method (FEM).

4 Application to atomic clock

In the introduction we pointed out the question: which is the probability that, a certain time after synchronization, the cesium clock error does exceed a threshold value? In order to give an answer to this question we dealt with the survival probability which tell us the allowed time before re-synchronizing the clock, to be sure, at a certain percentage level, that the error has not exceeded two fixed constant boundaries. In fact, if we evaluate the abscissa t corresponding to a survival probability, for example $prob$ = 0.95, we can estimate the time t that we should wait before the process crosses one of the two boundaries with probability 0.95.

The diffusion coefficients σ_i appearing in the equation parameters of the clock model, characterize the atomic clock considered and can be estimated via the Allan variance [4]. For a cesium clock the typical values estimated are:

$$\sigma_1 = 3.05\frac{ns}{\sqrt{days}} \qquad \sigma_2 = 0.305\frac{ns}{\sqrt{(days)^3}} \quad . \tag{5}$$

As in the previous section, here we present separately the results for the Brownian motion, for the integrated Brownian motion and for the bidimensional model combining both these noises.

For the BM the analytical solution of the survival probability with parameters $\sigma_1 = 3.05$, $\sigma_2 = 0$ gives us the time t (in days) that can elapse before the clock error exceeds the boundaries for different percentage levels and for different values of the thresholds. In Table 1 the obtained results are reported.

Thresholds\Prob [ns]	*90% [days]*	*95% [days]*	*99% [days]*
10	2.8	2	1.3
30	25.5	19	12.1
50	69.6	53.2	34.1
100	279	213	136
300	2500	1920	1220
500	6990	5340	3410

Table 1. Values of survival time t (in days) corresponding to different percentage levels of the survival probability and to different choices of the thresholds [nanoseconds] (BM case with $\mu = 0$ and $\sigma = 3.05$)

From Table 1 we can infer that a Cs clock error affected only by white frequency noise ($\sigma_1 = 3.05$) can remain about 279 days within the threshold of ± 100 ns at 90% confidence level, or about 136 days if we want to be confident at 99% that the thresholds have not been yet crossed.

An other type of random noise is the frequency Random Walk which can be modeled by the integrated Brownian Motion (IBM) on the phase error. In Table 2 we report the time t (in days) that can elapse before the clock error exceeds the boundaries for different percentage levels and for different values of the thresholds applied to pure IBM with parameters $\sigma_1=0$ and $\sigma_2 = 0.305$ and obtained by the use of the numerical methods explained in the previous section.

Thresholds\Prob [ns]	*90% [days]*	*95% [days]*	*99% [days]*
10	10.6	9.4	7.8
30	21.9	19.6	16.2
50	30.8	27.5	22.7
100	48.9	43.7	36.1

Table 2: Values of survival time t (in days) corresponding to different percentage levels of the survival probability and to different choices of the thresholds [nanoseconds] for the IBM.

From Table 2 we can deduce that, fixing the maximum error allowed at ± 10 ns, we can wait 9.4 days being sure at 95% that the error has not yet exceed the permissible error.

For the bidimensional model with the both noises with the values of the parameters as in (5) we obtain the Table 3 by using the numerical methods explained in the previous section

Thresholds\Prob [ns]	90% [days]	95% [days]	99% [days]
10	2.9	2.2	1.4
30	16.3	13.6	9.9
50	26.8	23.2	18.3
100	46.3	41	33.5

Table 3: Values of survival time t (in days) corresponding to different percentage levels of the survival probability and to different choices of the thresholds [nanoseconds] for the bidimensional model.

In this case we can notice that for small threshold values the bidimensional model behaves very similarly to the BM model. For larger values of the boundary the bidimensional model behaves mostly as the IBM model.

5 Application to GNSS clock

In this section we consider the application to GNSS clock. In the Galileo GNSS project the clock expected on board are Rubidium clocks . Whit the following specified values for the Allan Variance :

a) White frequency: $\sigma_y^2(\tau)=\left(5\times10^{-12}\tau^{-1/2}\right)^2$

b) Random Walk frequency: $\sigma_y^2(\tau)=\left(1\times10^{-13}\tau^{1/2}\right)^2/day$

c) Flicker frequency: $\sigma_y^2(\tau)=\left(5\times10^{-14}\right)^2$ (6)

Here we consider only the white frequency and the random walk frequency noises that corresponds to our bidimensional model (1). From the Allan variance in (6) we obtain the following values for the diffusion coefficients:

$$\sigma_1=1.5\frac{ns}{\sqrt{days}} \qquad \sigma_2=15\frac{ns}{\sqrt{(days)^3}}$$

As in the previous section we are interested in the time t (in days) that can elapse before the GNSS clock error exceeds a certain threshold for different percentage levels and different values of the thresholds. The results obtained are reported in Table 4.

Thresholds\Prob [ns]	90% [days]	95% [days]	99% [days]
1.5	0.17	0.14	0.1
5	0.47	0.41	0.31
10	0.78	0.69	0.57

Table 4: Values of survival time t (in days) corresponding to different percentage levels of the survival probability and to different choices of the thresholds [nanoseconds] for the GNSS rubidium clock.

For example, we can wait 0.14 days (201.6 minutes) with a probability of 95% that the error has not yet exceeded the thresholds of $\pm$ 1.5 ns.

We have considered the clock model composed by only two noises: white frequency and the random walk frequency. In [5,3] we can find similar results for the same problem where with all the three clock noises are considered.

6 Conclusions

In this paper we describe a model of the atomic clock error using stochastic differential equations trying to answer to the question: which is the probability that, a certain time after synchronization, the clock error does exceed a threshold value? To this scope we evaluated the survival probability that, in metrological language, is the probability that a certain time after synchronization the clock error exceeds a threshold limit. We then applied this study to cesium atomic clocks and to the clock expected on the Galileo system. In particular, we get that for a cesium clock modeled by (1) we can wait about 2 days being sure at 95% that the error has not yet exceed the permissible error of the 10 ns. In case of a GNSS Rubidium clock the same level of a confidence is ensured only for 0.69 days.

Similar studies were applied to the determination of the optimal calibration interval of measuring instruments. The use of a calibrated instruments and the need of periodical re-calibration is in fact very important in metrological industrial laboratories. If the degradation of the calibration condition can be modelled by the accumulated effect of random disturbances, a model given by a Wiener process (or Integrated Wiener Process) can fit quite well [8] and can give a quantitative insight on the necessity of a new calibration.

Acknowledgements

The authors wish to thank Prof. Laura Sacerdote from the Dip. Mathematics, University of Torino and Prof. Lorenzo Galleani of Politecnico of Torino for helpful discussions on the clock model.

References

[1] D.W. Allan, "Time and Frequency (Time –Domain) Characterization, Estimation and Prediction of Precision Clocks and Oscillators", *Proc. IEEE Transactions on Ultrasonic, Ferroelectrics, and Frequency Control*, **34**, N. 6, pp. 647-654 (1987).

[2] P. Baldi, *Equazioni differenziali stocastiche applicazioni.* U.M.I., Bologna, Pitagora Editrice (2000).

[3] G. Busca, Q. Wang. Time prediction accuracy for Space Clock. *Metrologia,* **40**, N. 3, pp. 265-269, (2003).

[4] J. W. Chaffee, "Relating the Allan Variance to the Diffusion Coefficients of a linear Stochastic Differential Equation Model for Precision Oscillators". *IEEE Transactions on Ultrasonic, Ferroelectrics, and Frequency Control*, **34**, N. 6, pp. 655-658 (1987).

[5] J. Delporte, F. Vernotte, T. Tournier, M. Brunet. "Modelisation and extrapolation of time deviation: application to the estimation of the datation stability of a navigation payload". Proc. 15th European frequency and Time Forum (EFTF), pp. 245-249 (2001).

[6] L. Galleani, L. Sacerdote, P. Tavella, C. Zucca. "A mathematical model for the atomic clock error", *Metrologia,* **40**, N. 3, pp. 257-264, (2003).

[7] P. E. Kloeden, E. Platen, *Numerical solution of stochastic differential equation.* Berlin Heidelberg, Springer-Verlag (1992).

[8] E. Nunzi, G. Panfilo, P.Tavella, P. Carbone, D. Petri. Stochastic an Relative Methods for the Determination of Optimal Calibration Intervals. *IEEE Instrumentation and Measurement Technology Conference (IMCT 2004).* 18-20 may 2004 Como, Italy.

[9] G. Panfilo, P. Tavella, C. Zucca. "Stochastic processes for modelling and evaluation atomic clock behaviour", *Advanced Mathematical and Computational Tools in Metrology (AMCTM) VI*, pp. 229-239, (2004).

[10] W. H. Press et al., *Numerical Recipes in Fortran.* 2 ed. Camb. Univ. Press (1995).

[11] A. Quarteroni, A. Valli, *Numerical Approximation of Partial Differential Equations.* Berlin Heidelberg. Springer series in computational mathematics (1997).

LONG TERM STABILITY OF MEDIUM PRICE COMMERCIAL GPS RECEIVERS

J.Mannermaa [2], K.Kalliomäki [1], T.Mansten [1]

[1] MIKES, Electricity Group, Otakaari 7B, FI-02150, Espoo, Finland
[2] Nokia Mobile Phones, P.O.Box 68, FI-33721, Tampere, Finland

Keywords: GPS receiver, long term stability, time difference, phase jump

Abstract

Long-term phase stability of GPS receivers is important in time-keeping applications. MIKES and Nokia Mobile Phones have studied in co-operation the performance of medium price commercial GPS receivers both in the short term and long-term sense. MIKES has run five GPS receivers of two manufacturers for over five years. An automatic measurement system has collected data from those receivers and from environmental conditions every 10 minutes, having a Cs-clock as a master one. By comparing time differences of the receivers, both short term and long-term stability can be evaluated assuming that all receivers do not drift into the same direction. Long-term stability seems to be about 1...5 ns per year but unexpected transient phase jumps occurred in some receivers when they were rebooted. The magnitude of these jumps varies from 10 to 50 ns. The older receivers show a diurnal phase variation of 10...20 ns due to ionospheric effects. According to Modified Allan Variance (MDEV) the long-term stability has been 10^{-13} diurnally and 10^{-15} annually. Thermal effects seem to be rather small though the ambient temperature can vary from –30°C to +30°C.

1 Introduction

The results of the studies of the short-term stability were published in Proceedings of 14th EFTF-conference (year 2000) /2/. As a granted matter the studies were continued with the suitable receivers having a target of the longer era than 5 years for the non-stop measurements.

Most time-keeping laboratories rely just on GPS-time since the availability of the signal from the other system, GLONASS, is not rather high. The GPS system time is properly supervised and controlled but the received signal is sometimes degraded and corrupted due to interferences and disturbances from different sources. However, in the near future, there probable will be a complementary or competitive Global Navigation Satellite System or Galileo, available for civil user, too. There has, already presently, been a serious intention to connect the GPS and Galileo system closely together by inter-timing the system clocks. A non-determined issue is if the timing is carried out by exploiting the navigation data or even by the synchronisation of their system times.

But, currently, the GNSS time is based on the GPS signal and system time and it is widely used as a comparator to monitor or even to supervise the local, nationwide or even global times. If there are several GPS-receivers in a laboratory, preferably from different manufacturers, it is advisable annually, to check the measured time series. In this paper we have compared our GPS-receiver data since January 1997 with a Cs clock as "flywheel".

Measured GPS data were recorded every ten minutes and the diurnal measurements were averaged over 24 hours to resolve phase jumps or irregularities at nanosecond level. Time (phase) differences between all the receivers were computed pair-wisely. The long-term drifts of our receivers were estimated and phase jumps in certain receiver types were detected, by using this method with respective power spectra and with the Modified Allan Variance (MDEV) and Allan Time Variance (TVAR).

2 Measurement system

An automatic measuring system with sixteen analogue and sixteen digital channels records and saves to database the phase differences of all clocks and receivers every 10 min /1/.

3 Measured GPS receivers

In the long-term stability studies were used several receivers as follows:

Navstar (later Navsymm) XR5, type A120-001G2 receivers with serial numbers:

- 58482 (GPS1)
- 64580 (GPS2b)
- 74298 (GPS2), bought during Apr 94 and Apr 96.

Later (after Oct 98), was used Motorola Remote GPS, type of GCPRP12103 receivers with the serial numbers:

- 5700942 (GPS3)
- 5700944 (GPS4)

4 Analysis methods

All measurements were made by the basic sampling rate of 10 minutes. These values were exploited to calculate two-hour averages and the diurnal averages. These averages were used to compute time differences pair-wisely between desired frequency standards as a function of MJD (Mean Julian Day).

Moreover, the Maximum Entropy Method (MEM) was applied to calculate power spectra that were used to estimate e.g. the noise processes in the measured time series.

Finally, the Modified Allan Variance (MDEV) and Allan Time Variance (TVAR) were employed to find out the long-term stability of receivers.

5 Measurements and results

As a main feature in these studies was considered the pair-wisely computed time difference between a pair of receivers. The GPS1 receiver was selected as a reference one. In Fig. 1, there is shown the time difference between the GPS1 and GPS2 receivers.

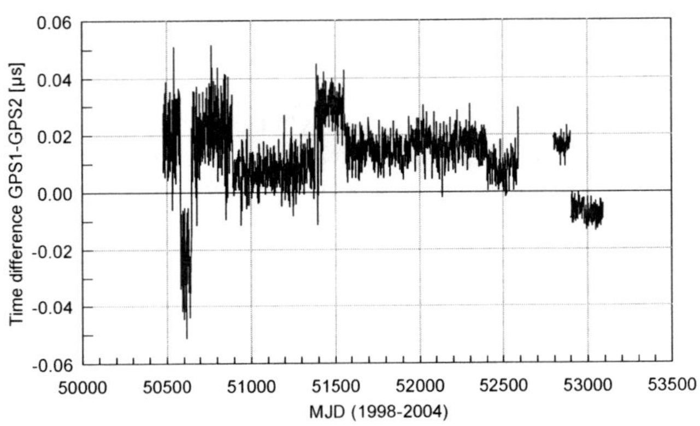

Fig. 1: Time difference between the GPS1 and GPS2 receivers represented as a function of MJD

The behaviour of the time difference between the GPS1 and GPS3 receivers is represented in Fig. 2.

To observe possible seasonal effects in the time difference between the GPS1 and GPS3 receivers, their mutual correlation was searched after. Seasonal influences were described by sinus curve. An ocular comparison can be made based on Fig. 3.

The result of the time differences of the GPS1-GPS4 and GPS3-GPS4 receivers is shown in Fig. 4.

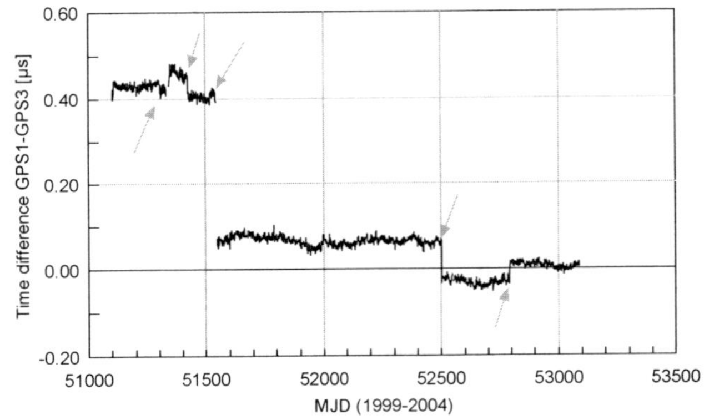

Fig. 2: Raw time difference between the GPS1 and GPS3 receivers represented as a function of MJD. The hardware caused, unexpected phase jumps are pointed out by arrows

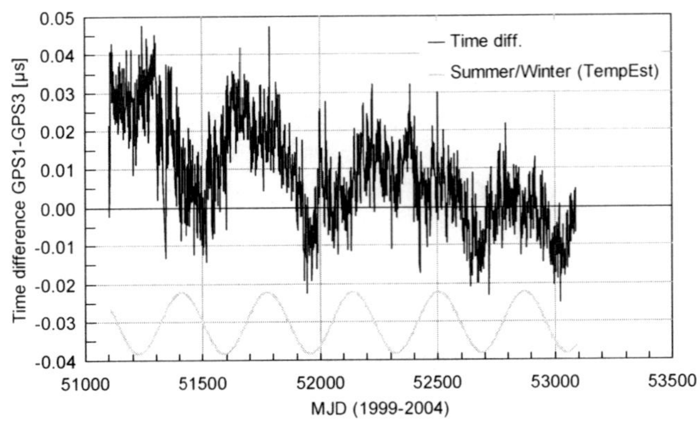

Fig. 3: Corrected time difference between the GPS1 and GPS3 receivers represented as a function of MJD with sinus curve describing the expected seasonal effects

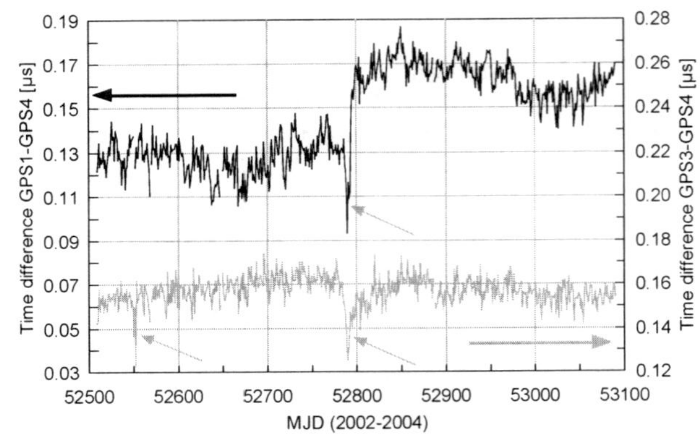

Fig. 4: Time difference between the GPS1 and GPS4 (darker curve) and GPS3 and GPS4 (lighter curve) receivers represented as a function of MJD. Short arrows point out the unexpected jumps and long arrows point the respective y-axes.

To detect the noise processes in the time differences of the different receiver pairs, all power spectra of them were computed by the maximum entropy method. Typical spectra are described in Figs. 5 and 6.

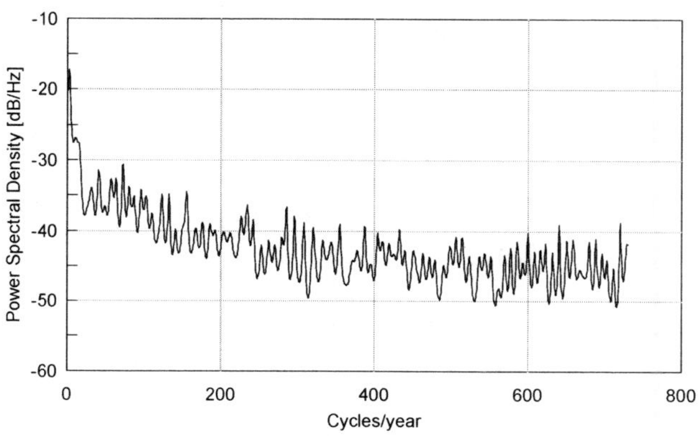

Fig. 5: Power spectrum of Time Difference between the GPS1 and GPS3 receivers calculated by the Maximum Entropy Method

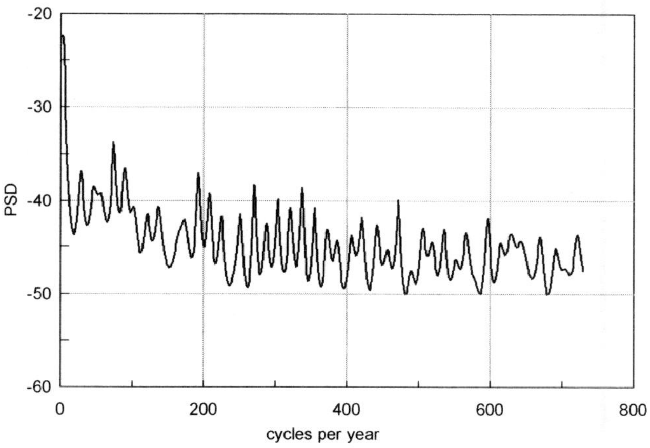

Fig. 6: Power spectrum of Time Difference between the GPS1 and GPS4 receivers calculated by the Maximum Entropy Method

To consider the long-term stability of the receivers, the Allan MDEV and TVAR were computed. The respective results are shown in Fig. 7.

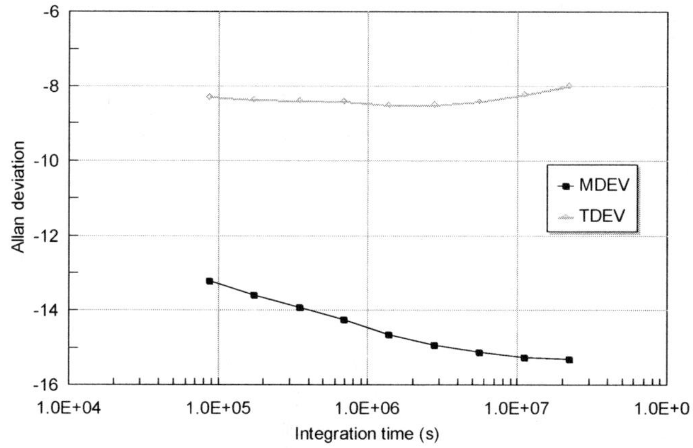

Fig. 7: Allan MDEV and TDEV of the time difference of two receivers (GPS1 and GPS3) as a function of the integration time (see Fig. 3)

6 Conclusions

In the case of the GPS2 receiver, some unknown software problems were met because the tracking was lost, typically once per year. Rebooting of the GPS2 receiver caused usually a phase jump (Fig. 1). At first, a couple of 50 ns phase jumps and later, some 15...25 ns jumps were detected. At MJD 52587, the GPS2 receiver "died" and it was replaced with a similar receiver, denoted by GPS2b. It was observed to have the same features in performance and the detected phase jumps were alike in respect of the original GPS2 receiver, referring also to software problems in the more recent Navstar units.

When operated over five years, the GPS3 receiver has performed 5 phase jumps due to the PPS driver. The amplitude of the "tick" was high enough but the rise time was slow, around 500 ns. This caused the respective changes into delay(s) as a function of the changes in the loading impedance. Because natural explanations to all phase jumps were found, it was concluded that the GPS3 receiver is fit to behave this way under the conditions in question (Fig. 3).

Any perfect correlation between the time difference behaviour and seasonal effects was not found but rather a remarkable one, however (Fig. 3). The worst anomaly exists at around the MJD 52600 where there should be the minimum value but no clear one can be observed in the time difference curve. Otherwise, the correlation seems to be rather high although the decreasing slope of the curve disturbs the visual consideration.

GPS4 appears to be quite similar to GPS3. It performed two phase jumps but recovered automatically by itself within few days (Fig. 4).

By applying a linear regression to the long-term time differences, significant drifts between the GPS1-GPS3 and GPS1-GPS4 curves were found, -15 and -14 ps/day, respectively. The time difference between the GPS3 and

GPS4 receivers had no drift. There was a drift of 4 ps/day between GPS1and GPS2.

Based on the results, the conclusion is that the drift of Navstar receivers is greater than the drift of Motorola receivers.

The oldest receiver, GPS1 (Navstar XR5) has operated continuously for ten years with only one small phase jump at MJD 52739 due to service.

Noise processes were searched after the values of the power spectra, couple of which are represented in Figs. 5 and 6. In the curve in Fig. 5, there can only be found except the noise floor, a 1/f-noise component. The curve in Fig. 6 also provided a $1/f^2$ component.

A typical long-term stability of the receivers is shown in Fig. 7 and this result is consistent with the data available in the literature of the field. By the diurnal integration time, the long-term stability is about 10^{-13} descending onto the lower level than 10^{-15} by the annual integration time.

References

[1] K. Kalliomäki, T. Mansten, A computerized time and frequency standards monitoring system, Proceedings of EFTF4 (Neuchatel, Switzerland, 13 - 15 March, 1990), pp. 177 - 179

[2] J. Mannermaa, K. Kalliomäki, T. Mansten, S. Turunen, Short-term stability of crystal oscillators in commercial GPS receivers, Proceedings of 2000 EFTF Meeting (Turin, Italy, March 2000), pp. 424-428

THE DEVELOPMENT OF A KALMAN FILTER CLOCK ALGORITHM

J A DAVIS AND P W STACEY
National Physical Laboratory, Queens Road, Teddington, Middlesex, TW11 0LW
C A GREENHALL
Jet Propulsion Laboratory, California Institute of Technology
4800 Oak Grove Dr. MS 298-100, Pasadena, CA 91109

NPL is developing a new clock algorithm to combine measurements from its three active hydrogen masers and caesium fountain clock. The algorithm takes advantage of recent developments, in particular the use of covariance reduction techniques [1]. This algorithm differs from previous Kalman filter clock algorithms in that it has been designed to operate near optimally in the presence of flicker frequency modulation (FFM). The aim of the work is to produce a composite clock that over a wide range of averaging times has near optimal stability. This is achieved by modelling the FFM approximately by a linear combination of Markov noise processes. Each Markov process is included in the Kalman filter and contributes an additional component to the state vector. Both the validity of the model and the effectiveness of adding these additional components to the state vector are examined.

To apply a Kalman filter clock algorithm to real clock data, it is first necessary to estimate the noise parameters used in the process covariance matrix. An iterative method had been developed previously [2] to resolve the magnitude of noise parameters from a measurement time series. This method is extended through the use of wavelet variances, and then applied to a "three cornered hat" of hydrogen maser measurements. Several aspects of the implementation of the algorithm are discussed, including the mechanism for adding and removing clocks, and the treatment of missing data.

The performance of the new algorithm is examined when applied to simulated measurements and also to measurements from NPL's three hydrogen masers. The latter is achieved through comparisons against NPL's caesium fountain clock and against external UTC timescales using two-way time transfer measurements.

[1] C A Greenhall, "Forming stable timescales from the Jones-Tryon Kalman filter", Metrologia **40,** 2003, S335-S341.

[2] P M Harris, J A Davis, M G Cox and S L Shemar, "Least-squares analysis of time series data and its application to two-way satellite time and frequency transfer measurements", Metrologia **40**, 2003, S342-S347.

CALIBRATION OF THE IEN-PTB TWSTFT LINK WITH A PORTABLE REFERENCE STATION

F. Cordara*, L. Lorini*, V. Pettiti*, P. Tavella*, D. Piester†, J. Becker†, T. Polewka†, G. De Jong, O. Koudelka††, H. Ressler††, B. Blanzano††, C. Karel††**

*Istituto Elettrotecnico Nazionale (IEN) "G. Ferraris", Strada delle Cacce 91, I-10135 Torino, Italy
†Physikalisch-Technische Bundesanstalt (PTB), Bundesallee 100, D-38116 Braunschweig, Germany
**NMi Van Swinden Laboratorium (VSL), P.O. Box 654, 2600AR Delft, The Netherlands
††Department of communication and wave propagation, Technical University Graz (TUG) and Joanneum Research GmbH, Graz, Austria

Contact e-mail: lorini@ien.it, time@ptb.de

Keywords: TWSTFT, Galileo, Time Link, TAI.

Abstract

During 2003 the IEN-PTB Two-Way Satellite Time and Frequency Transfer (TWSTFT) link was calibrated using a portable reference station. The calibration activity was conducted in the framework of the Galileo System Test Bed Version 1 (GSTB V1), under a contract with Joanneum Research G.m.b.H. (Austria). The calibration constant was determined with an uncertainty lower than 1 ns.
Only few months after the calibration, the satellite provider moved the TWSTFT service to a different satellite; this caused changes non-reciprocal delays of the link (Sagnac and earth stations delays) and the calibration constant had to be re-evaluated. A recalculation of the Sagnac delays is presented together with a re-evaluation of earth stations delays with different measurements techniques.

1 Introduction

In the last decade the Two-Way Satellite Time and Frequency Transfer (TWSTFT) demonstrated to be one of the most powerful tools for time and frequency transfer [1].
When TWSTFT is used for remote time scale comparisons, an absolute calibration of the synchronization link is required. In fact non-reciprocal delays in the transmitting-receiving equipment can amount to few hundreds of nanoseconds.
Different approaches are available for a TWSTFT link calibration: the comparison with a parallel calibrated time link (e.g. a GPS link), the direct measurement of ground station non-reciprocal delays with a satellite simulator, or the differential delay measurement with a portable reference station [2].
The portable reference station approach has demonstrated up to now the best performances (accuracy at the 1-ns-level) for an absolute TWSTFT link calibration [2]. However, this is a very challenging activity, mainly by the logistic point of view.

1.1 The calibration activity

The GSTB V1 is the first experimental phase of the Galileo project, funded by the European Space Agency (ESA). One of the main GSTB V1 products is an experimental time scale (Experimental Galileo System Time, E-GST) with optimized features for navigation and time-dissemination purposes.
The infrastructure devoted to E-GST generation is the experimental precise timing station (E-PTS) which is located at IEN, Italy, and reuses part of the IEN infrastructure, namely the clocks and the time transfer equipment. The E-GST time scale has to be steered to TAI with a quite demanding performance [3], and the calibration of time transfer links connecting IEN with external laboratories (PTB and NPL) involved in GSTB V1 is requested for this purpose.
In this framework, dedicated funding were provided by ESA for high accuracy calibration of remote time transfers with external UTC laboratories participating in the project. The calibration of the TWSTFT link between the IEN and PTB, which is the pivot laboratory in Europe for TAI [4], with a portable station, was decided.
The calibration activity was performed under a contract with Joanneum Research G.m.b.H. and the Institute of Applied System Technology of Technische Universitaet Graz – TUG. The Joanneum Research G.m.b.H. had the possibility to use the portable station identified as TUG02 which was already used in a calibration experiment between European TWSTFT stations. Reported accuracy of that calibration was about 1 ns [2].
The calibration activity was carried out during May-June 2003.

1.2 Link recalibration after the satellite change

In September 2003 the satellite provider moved the TWSTFT service for the European laboratories to a different satellite.

This caused a change of the IEN-PTB non-reciprocal delays of the link, making the calibration value, calculated as the result of the calibration activity with the TUG02 portable station, no more valid. To save this high accuracy calibration, the delay change due to the satellite switch was evaluated with a work in close cooperation between IEN, PTB and VSL

2 The IEN-PTB link calibration

2.1 Principle of operations

The calibration of a TWSTFT link with a portable station relies on two common clock experiments between the portable station and the local station at both sides of the link. A mandatory condition for a correct calibration is that the intrinsic non-reciprocal delays of the portable station equipment stay constant during the calibration trip. An evaluation of possible delay changes of the portable station during the calibration activity is usually performed starting and ending the trip at the same location with a closure measurement. This calibration procedure requires, that the TX and RX satellite spots are overlapping. Otherwise, if the TX and RX spots are separated, a slightly different calibration technique has to be applied [5].
The common clock experiment at each location provides a differential delay between the local and the portable station. Link calibration is calculated using the S = 0 formula defined by the ITU recommendation ITU-R TF.1153 [6], which combines the two calibration constants determined with the same reference station at both link sides and the position dependent delays (e.g. Sagnac delay).

2.2 The TUG02 portable station

The TUG02 portable station is a complete TWSTFT measurement system, developed for calibration purpose, that can be transported hosted in a car trailer. It is composed by a specially modified VSAT 1.8 m antenna (details reported in [2]), a Ku-band transceiver and a local delay monitor device (satellite simulator). The set-up can be assembled in a few hours by two people.
The indoor equipment is made of a SATRE modem (no. 036), a portable PC for measurement automation and data acquisition, and distribution amplifiers for delivering the reference frequencies and 1PPS signals. Cable connections between outdoor and indoor devices are possible up to 50 m length, ensuring various installation possibilities with the same cable configuration.
Phase coherent frequency and 1PPS signals, related to those driving the local TWSTFT station, should be provided to the indoor equipment of the portable station.

2.3 The calibration trip

The calibration trip was organized in three stages with two installations at IEN, including the closure measurement, and one at PTB.
As the journey from Torino to Braunschweig needs more than one day of car travel, the calibration measurements took place on Friday 30th and Saturday 31th May 2003 (MJD 52789 and 52790) in Torino, on Monday 2nd and Tuesday 3rd June (MJD 52792 and 52793) in Braunschweig, and on Friday 6th June (MJD 52796) again in Torino.
The portable station used the MITREX code 3, which was not used in the European session at the moment of the calibration. This allowed the portable station to participate to the regular schedule (Monday, Wednesday, Friday, form 14:00 to 14:30 UTC at the moment of the calibration); however the satellite provider gave the possibility to use the satellite channel also outside the schedule and some additional measurements were taken.

2.4 Installation and operation at IEN

The IEN location offered a quite clear visibility to the satellite Intelsat 706 (IS706), so it was possible to accommodate the outdoor equipment of the travelling station on the green, just in front of the Time and Frequency laboratory. The IEN01 TWSTFT antenna and transceiver are located on the roof of the building hosting the Time and Frequency laboratory; the distance between the two antennas was not more than 10 m.
Using the 50 m cable assembly of TUG02, the indoor equipment of the portable station was installed inside the IEN time and frequency laboratory.

Figure 1. The TUG02 portable station installed at IEN

Both stations (IEN01 and TUG02) were driven by the IEN H maser M1(IEN) frequency and 1PPS signals and internal delay measurements related the modems outputs to both UTC(IEN) and M1(IEN) reference signals. Although the internal measurements allowed to use both M1(IEN) and UTC(IEN) as the common clock reference, the data reduction was performed using M1(IEN), which allowed for a simpler measurement scheme and a better stability.

2.5 Installation and operation at PTB

To achieve a good visibility of IS706, the PTB01 TWSTFT antenna and transceiver are located on a roof top in a distance of about 400 m to the time and frequency laboratory where the modem and automation systems are located. The TUG02 antenna and transceiver were mounted side by side to the PTB outdoor equipment. The indoor set-up was located in the same building (in a room below the antenna).
Because no reference frequency and 1PPS were available at the TUG02 set-up, 1PPS and frequency were supplied by a caesium clock (model HP5071A, ID C9) from the PTB clock ensemble, which was brought to the TUG02 set-up only for the duration of the experiment. Usually, the clock is hosted in the time and frequency laboratory and continuously measured versus UTC(PTB) which was used as the common time scale in the calibration experiment. UTC(PTB) – C9 is shown in the lower part of Figure 2. In the upper part the residuals to a linear fit is represented. The arrows indicate when the clock was moved from the time and frequency laboratory to the TUG01 set-up.

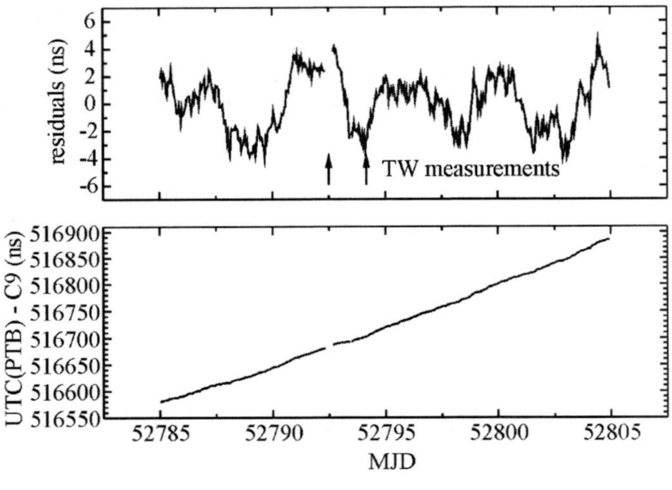

Figure 2. UTC(PTB)-C9 time differences measurements and residuals after a linear fit.

The TUG02 modem output was compared to UTC(PTB) by a linear interpolation of UTC(PTB)-C9 data during the calibration periods.

2.6 Calibration results

The station calibration constant (defined as CALR in [6]) was calculated as the average of the common clock TWSTFT measurements, taking into account the delays of the modem reference with respect to the common clock reference (defined as REFDELAY in [6]).

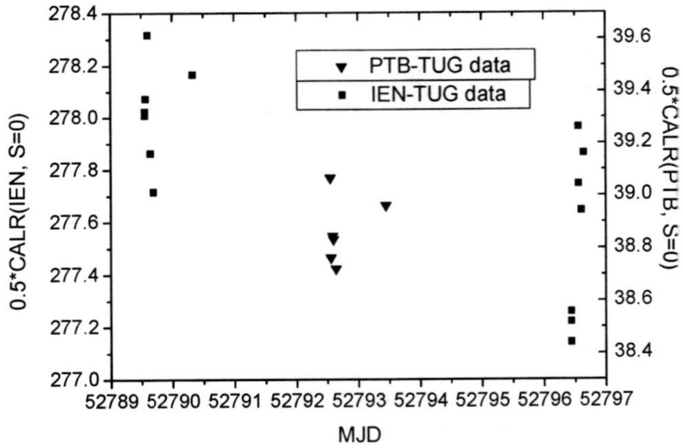

Figure 3. Calibration measurement data used for CALR(IEN) and CALR(PTB) calculation

The CALR(IEN) was obtained as the average of 14 measurements (6 on MJD 52789, 1 on MJD 52790, 6 on MJD 52796); its value is [7]:

$$\text{CALR(IEN)} = -555.6 \text{ ns} \pm 0.7 \text{ ns } (1\sigma). \quad (1)$$

The uncertainty was limited by the REFDELAY(IEN) evaluation. The source of the instability was identified in the high thermal sensitivity of MITREX modem used in the IEN01 set-up. In fact, the TUG02 equipment, installed nearby the IEN01 modem, caused some thermal instabilities [7].
The CALR(PTB) value was calculated as the average of six measurements (5 on MJD 52792, 1 on MJD 52793); its value is [7]:

$$\text{CALR(PTB)} = -77.7 \text{ ns} \pm 0.3 \text{ ns } (1\sigma). \quad (2)$$

The uncertainty was limited by the TWSTFT measurement instability and is the standard deviation of the six recorded measurements at PTB [7].
As the calibration constants come from co-location experiments, to calculate the actual calibrated TW differences for the IEN-PTB link the contribution of Sagnac delays has to be explicitly taken into account. The complete formula which has to be used in this case is reported in [6] and specified in the TW measurement file with the S = 0 (Calibration Switch).

2.7 Comparison with Circular T calibration of the IEN-PTB TWSTFT link

At the beginning of the year 2002 (MJD 52297), with the aim of using the IEN-PTB TWSTFT link for TAI, the BIPM had calculated a calibration constant using the Circular T.
The comparison between the TWSTFT link and another calibrated link provides a CALR value which includes all contributions to the link delay calibration. The formula which has to be used in this case for the calculation of the calibrated TWSTFT differences is different from the S = 0 case, and is specified with S = 1 [6].

The calibration value obtained with Circular T at MJD 52297 was:

CALR(IEN01) = – CALR(PTB01) = – 253 ns ± 5 ns (1σ). (3)

To compare the link calibration value obtained with the portable station, which provides a CALR with S = 0 and the calibration value obtained with Circular T, which provides a CALR with S =1, the following relation has to be used:

CALR(IEN01, S=1) = 0.5·[CALR(IEN01, S=0) – CALR(PTB01, S=0)] + SAGNAC (IEN-PTB). (4)

Using the Sagnac delay calculated for the IEN-PTB link through the IS706 satellite, which was used at the epoch of both calibrations:

SAGNAC(IEN-PTB) = –15.090 ns (5)

the comparison shows a very good agreement between the two methods:

CALR(IEN01, S=1) = – 253 ns ± 5 ns (1σ) with Circ.T
CALR(IEN01, S=1) = – 254.0 ns ± 0.4 ns (1σ) with TUG02.

3 Satellite change

During August 2003 the satellite provider (Intelsat) dismissed the service on the Transatlantic Two Way link with the satellite IS706. After a few weeks, Intelsat moved the TWSTFT service to a different satellite namely the IS903 at position 34° 30' W. During the whole period of non-operation of the transatlantic link, the European link on IS706 was kept operating and the TW community eventually decided to switch both European and Transatlantic link to IS903 on MJD 52798.
At both laboratories (PTB and IEN), the visibility of the new satellite is good and the satellite could easily be pointed. However the switch to the new satellite required the recalculation of the link calibration constant; in fact, the Sagnac delay changed due to the different satellite position and earth station delays could change due to different transmitting (TX) and receiving (RX) frequencies used on the new satellite; moreover at PTB the RX frequency change on the European link required a replacement of the downlink hardware.
The Sagnac delay change can be accurately recalculated (with uncertainties of few picoseconds), using theoretical formulas, meanwhile the earth station delay change has to be evaluated with experimental measurements.

3.1 Sagnac delay recalculation for the satellite IS903

The new Sagnac delay can be calculated using the satellite position and the coordinates of the earth stations. In the table below the Sagnac delay changes are reported for each active TWSTFT link where VSL, PTB and IEN are involved.

LAB	Sagnac Change (ns) VSL-LAB	Sagnac Change (ns) PTB-LAB	Sagnac Change (ns) IEN-LAB
VSL	0.000	-3.736	2.121
PTB	3.736	0.000	5.858
NPL	-3.029	-6.765	-0.908
IEN	-2.121	-5.858	0.000
OCA	-3.294	-7.030	-1.172
ROA	-15.906	-19.642	-13.784
OP	-3.200	-6.936	-1.079
USNO	-16.782	-20.518	-14.660
NIST	3.187	-0.550	5.308

Table 1. Sagnac delay change at the satellite switch for TWSTFT links involving IEN, PTB and VSL

The Sagnac delay change affects the CALR value when it is expressed using the calibration switch S = 1. In this case it becomes

CALR(IEN01, S=1) = – 247.15 ns ± 0.4 ns (1σ) (6)

after the satellite change.

3.2 ESDVAR calculation at the satellite change

Following the prescriptions in [6], the earth station delay variation has to be reported in the TWSTFT measurement file in the ESDVAR column. As the ESDVAR is defined as an incremental value, the delay change due to the satellite switch from IS706 to IS903 will be reported in this work as ΔESDVAR.
At VSL the earth station delay change due to the different TX and RX frequencies between the satellites IS706 and IS903 was evaluated using a satellite simulator (SATSIM) [8]. In this case the SATSIM device is very suitable to evaluate possible delay changes; in fact the knowledge of the SATSIM absolute delay is not required because the ΔESDVAR measurement relies on the stability of the SATSIM delay itself.
The ΔESDVAR(VSL) value evaluated with the SATSIM is:

ΔESDVAR(VSL) = 0.2 ns. (7)

Because neither at IEN nor at PTB a SATSIM device is installed, additional TWSTFT measurements between IEN, PTB, and VSL were recorded just before and after the satellite switch (MJD 52898 at 14:00 UTC).
The differential earth station delay for the IEN-PTB link is:

$$0.5 \cdot [\Delta ESDVAR(IEN) - \Delta ESDVAR(PTB)]. \quad (8)$$

Itcan be evaluated with different methods, but the contribution due to the single station (the ΔESDVAR(IEN) or the ΔESDVAR(PTB)) can be calculated only with respect to VSL.

Triangulation measurement provides in fact two independent equation sets and three ΔESDVAR unknowns, therefore the solution requires the ΔESDVAR(VSL) value evaluated with the SATSIM technique.
Solving the triangulation equations we achieve to:

$$\Delta ESDVAR(IEN) = -0.1 \text{ ns} \pm 1.0 \text{ ns}$$
$$\Delta ESDVAR(PTB) = +64.2 \text{ ns} \pm 1.0 \text{ ns}. \qquad (9)$$

The main limit of this approach is the stability of the reference time scales and the possibility that closing errors can affect the results. Using a conservative approach a uncertainty of 1 ns was therefore assigned to IEN and PTB ESDVAR(IEN) and ESDVAR(PTB) values with this method.
As it is become clear from this calculation, the time step introduced by the change in the TX and RX frequencies at IEN and VSL is well below the uncertainty . On the contrary, at PTB the time step introduced by the RX hardware replacement is notably higher.
The differential earth station delay change 0.5·[ESDVAR(IEN) - ΔESDVAR(PTB)] can be evaluated with an higher accuracy by using high stability H-masers as references or by comparing TWSTFT with high performance GPS-based time transfer tools.

3.3 Differential earth station delay change evaluation using H-masers comparison

Both IEN and PTB drive their TWSTFT modems using the frequency delivered by their H-masers. The link to UTC(IEN) and UTC(PTB) is made with a second measurement and reported as the REFDELAY.
Disregarding the REFDELAY values one can access to the comparison between the two time scales generated by the H-masers, which have a very good short/medium-term stability. Thus, it is possible to use the H-masers as frequency flywheel to evaluate the time step at the satellite switch.

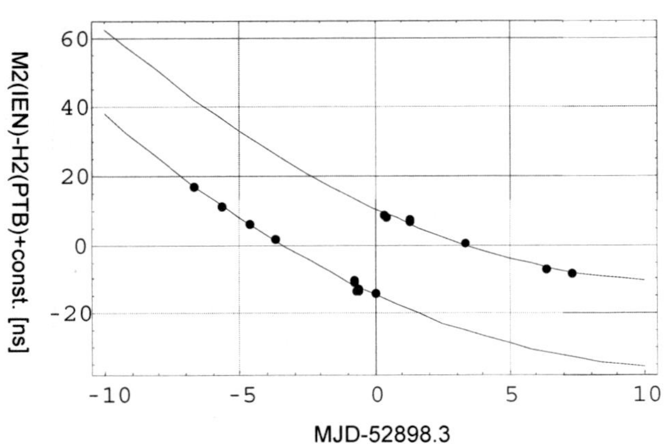

Figure 4. M2(IEN)-H2(PTB) measurements via TWSTFT. The lines represents the fitting functions (10)

Measurement data spanning up to 20 days have been fitted (using a linear least square model) with the following equation set, which also accounts for the masers frequency drift:

$$\begin{cases} At_i^2 + Bt_i + C = x(IS706)_i \\ At_i^2 + Bt_i + C' = x(IS903)_i \end{cases} \qquad (10)$$

where t_i are the epochs of the measurement data, $x(IS706)_i$ and $x(IS903)_i$ are the TWSTFT measurement data with IS706 and IS903, respectively, A and B estimates the frequency drift and offset between the two masers and C'-C estimates the time jump at the satellite change.
The estimated earth stations differential delays is

$$0.5 \cdot [\Delta ESDVAR(IEN) - \Delta ESDVAR(PTB)] = -31.0ns \pm 0.9ns \qquad (11)$$

Differences in fit values obtained with slightly different (longer or shorter) periods are not relevant.
The uncertainty was evaluated as a result of the least square analysis; the resulting uncertainty would have taken advantage by alternating the measurements with the two different satellites; unfortunately it was not possible (PTB needed to change the receiving hardware) and the resulting uncertainty is quite large.

3.4 Differential earth station delay change evaluation using TAIP3

TAIP3 is a pilot experiment, promoted by the BIPM, to test the use of the GPS P3 code measurements for remote clock comparisons using geodetic GPS receivers as an input to TAI [9]. P3 is the ionosphere-free combination of the P1 and P2 codes available on the two frequencies L1 and L2 in use by the GPS system.
We consider the difference between UTC(IEN) and H2(PTB) which were the time scales driving the geodetic receivers at IEN and PTB at that time. One can easily evaluate H2(PTB)-UTC(IEN) also from the TW measurements taking into account only the REFDELAY(IEN), whereas the REFDELAY(PTB) is left out of consideration (for detailed information see [6]). H2(PTB)-UTC(IEN) differences coming from TWSTFT measurements (spanning the period MJD 52890-52906) were compared to corresponding GPS data evaluated with the TAIP3 algorithm.
The TAIP3 algorithm provides one measurement every 16 minutes consisting of data from 4-6 satellites in common view (depicted as full circles in Figure 5). The data were adjusted to the TWSTFT data collected before the satellite change (dashed vertical line) in the following way. The 11 data points around each TWSTFT epoch were fitted using a linear regression model and the difference of the midpoint to the corresponding TWSTFT data (open circles) was determined.

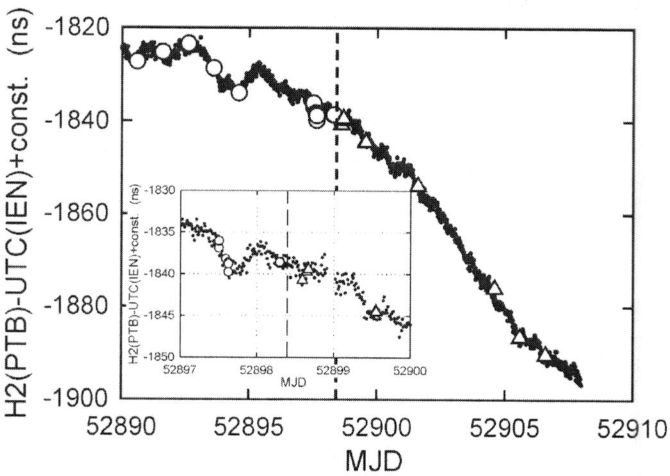

Figure 5. H2(PTB)-UTC(IEN) as evaluated with TWSTFT and GPS TAIP3. The vertical dashed line separates the IS706 and IS906 activity periods

The time step at the satellite change was calculated minimizing the sum of the squared differences between the TWSTFT and the TAIP3 averaged data. The TWSTFT data after the satellite change (triangles) were adjusted to the GPS data in the same way.
The earth station differential delay obtained by this calculation is:

$$0.5 \cdot [\Delta ESDVAR(IEN) - \Delta ESDVAR(PTB)] = -30.6 \pm 0.3\, ns \quad (12)$$

3.5 Differential earth station delay change evaluation using Precise Point Positioning

The Precise Point Positioning (PPP) is a post-processing approach using un-differenced dual frequency pseudo-range and carrier phase observations coming from a single geodetic GPS receiver, together with the high-quality GPS products such as those provided by the International GPS Service (IGS). Taking advantage of the precise satellite clock estimates as well as of the precise satellite coordinates in the IGS products, the PPP is able to allow stand-alone receiver clock offset estimates with sub-nanosecond uncertainty.
RINEX files coming out of the IEN and PTB receivers, together with the IGS final products, were processed with the PPP algorithm, providing the time differences between the GPS time (as provided by the IGS products) and UTC(IEN) and H2(PTB), respectively.
The PPP software version used for this work was implemented by the Geodetic Survey Division (GSD) of Natural Resources Canada (NRCan) and its used was granted to IEN [10].
As the PPP algorithm provides clock differences with respect to the same reference (a GPS based time scale) for both UTC(IEN) and H2(PTB), the time differences UTC(IEN)-H2(PTB) can be easily calculated. These time differences are spaced by 30 seconds, which is the sample rate of the receivers.

Afterwards, the UTC(IEN)-H2(PTB) time differences are hourly averaged, around the TWSTFT measurement epoch, and compared with the data coming from the TWSTFT measurements themselves. For the PPP-TWSTFT comparison a 4 days period across the satellite switch was selected.
Looking at the data reported in Figure 6 the constant C-C', which represents the time step due to the satellite change, was adjusted minimizing the sum of the squared differences between GPS and TWSTFT data.

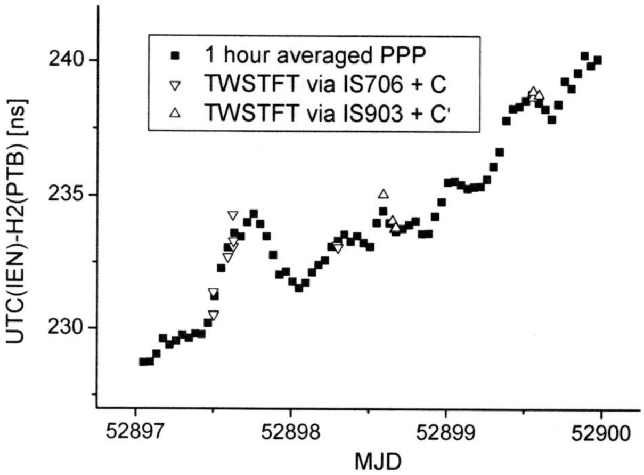

Figure 6. H2(PTB)-UTC(IEN) as evaluated with TWSTFT and PPP.
Squares: PPP data.
Triangles: TWSTFT data, already corrected for the constants C and C'

The estimated earth stations differential delay with this method is

$$0.5 \cdot [\Delta ESDVAR(IEN) - \Delta ESDVAR(PTB)] = -31.1 ns \pm 0.2\, ns \quad (13)$$

3.6 Results comparison

The comparison between 0.5·[ΔESDVAR(IEN)-ΔESDVAR(PTB)] values obtained with three accurate methods, shows that the results (see Figure 7) are in agreement within 500 ps. This implies that the quality of the IEN-PTB calibration with TUG02 is only slightly degraded and the actual uncertainty can be evaluated to be as low as 1 ns after the satellite switch.

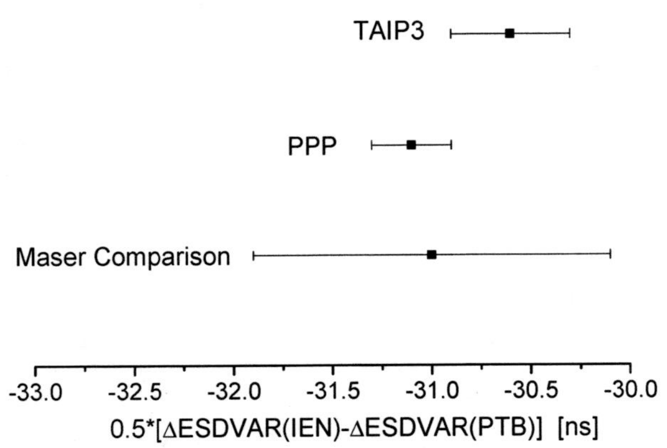

Figure 7. Comparison between the earth station differential value obtained with three different methods

However, as PTB and IEN are involved in TWSTFT links with other European laboratories, for the evaluation of the time step due to the satellite switch involving these links the ΔESDVAR(IEN) and ΔESDVAR(PTB) values are required. Taking advantage from the evaluations reported in the previous sections, since the March 2nd, 2003 (MJD 53066) the following values were adopted to update the ESDVAR values in the ITU files:

$$\Delta ESDVAR(VSL) = +0.2 \text{ ns}$$
$$\Delta ESDVAR(IEN) = -0.1 \text{ ns}$$
$$\Delta ESDVAR(PTB) = +61.1 \text{ ns} \quad (14)$$

where the ΔESDVAR(VSL) comes from the measurements with the SATSIM, the ΔESDVAR(IEN) value from the UTC(IEN)-UTC(VSL) comparison (see equation (9)) and the ΔESDVAR(PTB) value from the comparison with TAIP3 (see equation (12)), combined with the ΔESDVAR(IEN) value.

4 Conclusions

The calibration with the TUG02 portable station provided a calibration of the IEN-PTB link differential delay with an uncertainty of less than 1 ns. After the satellite change, an accurate analysis was required to evaluate the Sagnac and the earth stations delay changes which could have affected the link calibration. Using three different methods the link calibration after the satellite change was re-evaluated with 1 ns uncertainty.

The calibrated TWSTFT IEN-PTB link is currently used for the E-GST generation algorithm, in the frame of the GSTB V1 experiment.

As the IEN-PTB TWSTFT link is also involved in TAI computation, the uncertainty of the link calibration with TUG02 is reported in the BIPM Circular T bulletin since April 2004 [11].

Acknowledgements

The authors acknowledge the financial support of ESA in the framework of GSTB-V1. The authors thank D. Orgiazzi for providing the carrier phase data calculated with the PPP algorithm and the Geodetic Survey Division (GSD) of Natural Resources Canada (NRCan) for providing the PPP software.

References

[1] Kirchner, D., "Two-Way Satellite Time and Frequency Transfer (TWSTFT): "Principle, Implementation and Current Performance", in *Review of Radio Science 1996-1999*, Editor W. Ross Stone, Oxford University Press, Oxford, UK. (1999)

[2] D. Kirchner, H. Ressler, P. Hetzel, A. Soering: "The calibration of three European TWSTFT stations using a portable station and comparison of TWSTFT and GPS common-view measurements", *Proceedings of the 27th Annual Precise Time and Time Interval (PTTI)*. Reston (Virginia), December 1998, pp. 365-376.

[3] F. Cordara, R. Costa, L. Lorini, D. Orgiazzi, V. Pettiti, P. Tavella, G. Graglia, E. Detoma, J. Hahn: "The generation of the experimental Galileo system time in the Galileo System Test Bed V1". *Proceedings of the 2003 IEEE Intern. Frequency Control Symposium and 17th European Frequency and Time Forum*, Tampa (Florida, USA), May 2003, pp. 306-311.

[4] See for example: J. Azoubib and W Lewandowski, "22rd BIPM TWSTFT Report". Available at ftp://ftp2.bipm.fr/pub/tai/publication/twstft22.pdf.

[5] D. Piester, A. Bauch, J. Becker, and T. Polewka "An update of PTB's activity on time and frequency" . To be appeared in the *Proceedings of 35th Annual Precise Time and Time Interval (PTTI), Systems and Applications Meeting*. San Diego (California), December 2003.

[6] Recommendation ITU-R TF.1153-1. "The operational use of two-way satellite time and frequency transfer employing PN time codes", ITU, Radiocommunication Study Group, Geneva, last update 2003.

[7] O. Koudelka, H. Ressler, B. Blanzano, C. Karel, "Two-Way-Satellite-Transfer Calibration Campaign", Joanneum Research (2003). Confidential Report

[8] G. De Jong, R. Van Bemmelen: "Evaluations and Improvements in the calibration of a TWSTFT station using SATSIM". *Proceedings of the 34th Annual Precise Time and Time Interval (PTTI)*. Reston (Virginia), December 2002, pp. 391-404.

[9] P. Defraigne and G. Petit, "Time transfer to TAI using geodetic receivers", *Metrologia*, **40** (2003), pp. 184-188

[10] J Kouba, P. Heroux, "Precise Point Positioning using IGS orbit and clock products", *GPS solutions*, **5** , n. 2 (2001)

[11] BIPM, Circular T n. 195. Available at ftp://ftp2.bipm.fr/pub/tai/publication/cirt.195.

A RE-REVISITED THREE CORNERED HAT METHOD FOR ESTIMATING CLOCK INSTABILITIES

F. Vernotte*, J. Delporte†, M. Brunet†

* Observatoire de Besançon - Université de Franche-Comté/CNRS - FRANCE
E-mail: francois.vernotte@obs-besancon.fr

† Centre National d'Études Spatiales[1] - FRANCE

Keywords: Time stability, three cornered hat, clock correlations, power law spectral density, multivariance method.

Abstract

The three cornered hat is a well known method for estimating the instabilities of each clock in a set of three clocks. The first assumption lies in the hypothesis of total uncorrelation between the clocks [1, 2]. A revisited method has been proposed which no longer assumes the uncorrelation hypothesis, but yields a more realistic approach estimating the covariance matrix by ensuring suitable properties such as positive variances [3]. Finally, we propose a new method which goes further by assuming a power law model for the spectral density of frequency deviation of each clock.

The method will be detailed and different examples of clocks, more or less correlated, will be given. The limitation of this method in terms of correlations will also be studied.

1 Introduction

The intercomparison of a set of 3 clocks yields the frequency instabilities of each clock differences: $y_1(t) - y_2(t)$, $y_2(t) - y_3(t)$ and $y_3(t) - y_1(t)$ (see figure 1). However, the third difference may be easily deduced from the other ones:

$$y_3(t) - y_1(t) = -\left[y_2(t) - y_3(t)\right] - \left[y_1(t) - y_2(t)\right]. \quad (1)$$

Thus, the system is underdetermined: the knowledge of the 2 independent quantities, for instance $y_1(t) - y_2(t)$ and $y_2(t) - y_3(t)$ cannot lead to the determination of the 3 independent unknowns $y_1(t)$, $y_2(t)$ and $y_3(t)$. The goal of this work consists then in using one or several supplementary hypothesis in order to estimate these 3 quantities from the clock intercomparisons, or, at least, to deduce their statistical characteristics such as their Allan variances.

[1] This study has been partially financed by CNES (contract N# 02/7860/00).

The methods described by Gray and Allan [1], followed by Fest, Groslambert and Gagnepain [2] yield an estimation of the Allan variances of the 3 clocks if we assume that the signals of the clocks are completely **uncorrelated** (see figure 2). If we neglect the correlations, the problem is very easy [1] :

$$\begin{cases} \sigma^2_{y_1-y_2}(\tau) = \sigma^2_{y_1}(\tau) + \sigma^2_{y_2}(\tau) \\ \sigma^2_{y_2-y_3}(\tau) = \sigma^2_{y_2}(\tau) + \sigma^2_{y_3}(\tau) \\ \sigma^2_{y_3-y_1}(\tau) = \sigma^2_{y_3}(\tau) + \sigma^2_{y_1}(\tau) \end{cases} \quad (2)$$

therefore :

$$\begin{cases} \sigma^2_{y_1}(\tau) = \frac{1}{2}\left(\sigma^2_{y_1-y_2}(\tau) + \sigma^2_{y_3-y_1}(\tau) - \sigma^2_{y_2-y_3}(\tau)\right) \\ \sigma^2_{y_2}(\tau) = \frac{1}{2}\left(\sigma^2_{y_1-y_2}(\tau) + \sigma^2_{y_2-y_3}(\tau) - \sigma^2_{y_3-y_1}(\tau)\right) \\ \sigma^2_{y_3}(\tau) = \frac{1}{2}\left(\sigma^2_{y_2-y_3}(\tau) + \sigma^2_{y_3-y_1}(\tau) - \sigma^2_{y_1-y_2}(\tau)\right). \end{cases} \quad (3)$$

Fest, Groslambert and Gagnepain suggest a similar method, based on the "Allan covariance" (two-sample cross variance) defined by [2] :

$$\mathrm{Cov}_{y_1,y_2} = \frac{1}{2}\left\langle \left(y_1(t_k+\tau) - y_1(t_k)\right)\left(y_2(t_k+\tau) - y_2(t_k)\right)\right\rangle. \quad (4)$$

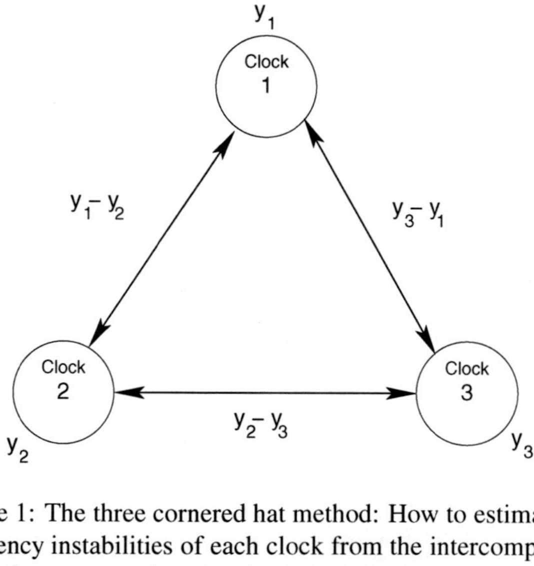

Figure 1: The three cornered hat method: How to estimate the frequency instabilities of each clock from the intercomparison of the frequencies of each pair of clocks?

For example, by applying the Allan covariance to $y_1 - y_2$ and $y_2 - y_3$, we get:

$$\begin{aligned}\mathrm{Cov}_{y_1-y_2,y_2-y_3} &= \frac{1}{2}\langle (y_1(t_k+\tau) - y_2(t_k+\tau) \\ &\quad -y_1(t_k) + y_2(t_k)) \\ &\quad \times (y_2(t_k+\tau) - y_3(t_k+\tau) \\ &\quad -y_2(t_k) + y_3(t_k))\rangle\,. \end{aligned} \tag{5}$$

If the clock signals are uncorrelated, we find:

$$\left\{\begin{aligned}\mathrm{Cov}_{y_3-y_1,y_1-y_2} &= \tfrac{1}{2}\langle(-y_1(t_k+\tau)+y_1(t_k)) \\ &\quad \times (y_1(t_k+\tau) - y_1(t_k))\rangle \\ &= -\sigma^2_{y_1}(\tau) \\ \mathrm{Cov}_{y_1-y_2,y_2-y_3} &= \tfrac{1}{2}\langle(-y_2(t_k+\tau)+y_2(t_k)) \\ &\quad \times (y_2(t_k+\tau) - y_2(t_k))\rangle \\ &= -\sigma^2_{y_2}(\tau) \\ \mathrm{Cov}_{y_2-y_3,y_3-y_1} &= \tfrac{1}{2}\langle(-y_3(t_k+\tau)+y_3(t_k)) \\ &\quad \times (y_3(t_k+\tau) - y_3(t_k))\rangle \\ &= -\sigma^2_{y_3}(\tau). \end{aligned}\right. \tag{6}$$

Unfortunately, it is relatively difficult to justify such an hypothesis, particularly if the clocks are in the same environment. When this hypothesis is not valid, non-physical results may be obtained such as negative Allan variances.

Then, Premoli and Tavella proposed another method based on the minimal correlations between the clocks ensuring the positiveness of the variances [3]. This method assumes that one of the 3 clocks, used as a reference, is both more stable and better known, in term of stability, than the other 2 ones.

On the other hand, the aim of our new method consists in assuming a power law model for the spectral density of frequency deviation of each clock and we will consider that each clock exhibits a maximum of stability over one duration range: one is stable for short term, the second one is stable for middle term and the last one is stable for long term. Of course, this method would better work with clocks exhibiting the same stability.

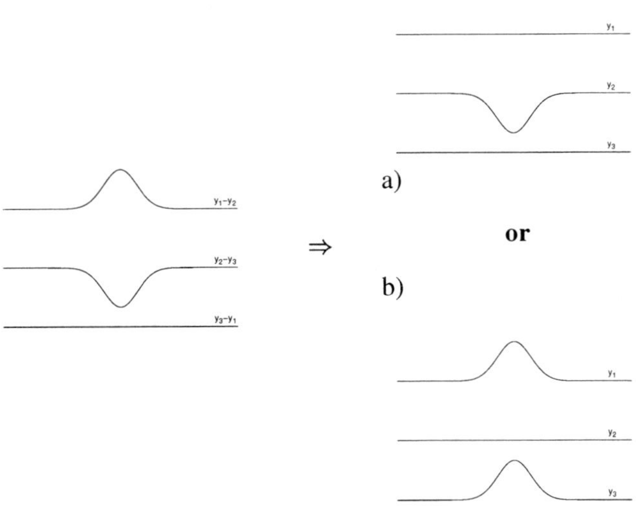

Figure 2: If both $y_1 - y_2$ and $y_2 - y_3$ exhibit a "bump", this feature may be due to y_2 (a), or both y_1 and y_3 (b). If the assumption of uncorrelation is assumed, only the solution (a) must be considered.

2 Method

2.1 Influence of the correlations

Let us imagine 2 clocks partially correlated:

$$\left\{\begin{aligned} y_1 &= y_{10} + \alpha_1 y_{12} \\ y_2 &= y_{20} + \alpha_2 y_{12} \end{aligned}\right. \tag{7}$$

where y_{10} and y_{20} are the independent noises (uncorrelated) of respectively clock 1 and clock 2 and y_{12} the common noise of these clocks.

The variance (whatever this variance is: classical or Allan variance) of each signal may be written as:

$$\left\{\begin{aligned} \sigma^2_{y_1} &= \sigma^2_{y_{10}} + \alpha_1^2\sigma^2_{y_{12}} \\ \sigma^2_{y_2} &= \sigma^2_{y_{20}} + \alpha_2^2\sigma^2_{y_{12}} \end{aligned}\right. \tag{8}$$

The covariance of y_1 and y_2 may also be calculated:

$$\mathrm{Cov}_{y_1,y_2} = \alpha_1\alpha_2\sigma^2_{y_{12}}. \tag{9}$$

Now, let us consider the intercomparison between clock 1 and clock 2:

$$y_1 - y_2 = y_{10} - y_{20} + (\alpha_1 - \alpha_2)y_{12}. \tag{10}$$

Then, its variance is:

$$\sigma^2_{y_1-y_2} = \sigma^2_{y_1} + \sigma^2_{y_2} - 2\mathrm{Cov}_{y_1,y_2} = \sigma^2_{y_{10}} + \sigma^2_{y_{20}} + (\alpha_1-\alpha_2)^2\sigma^2_{y_{12}}. \tag{11}$$

In order to simplify this problem, let us assume that the third clock is totally uncorrelated with the other 2 ones. What will be the consequence of this correlation if we use the method of Gray and Allan? The system (3) becomes:

$$\left\{\begin{aligned} \hat\sigma^2_{y_1}(\tau) &= \sigma^2_{y_1} - \alpha_1\alpha_2\sigma^2_{y_{12}} \\ \hat\sigma^2_{y_2}(\tau) &= \sigma^2_{y_2} - \alpha_1\alpha_2\sigma^2_{y_{12}} \\ \hat\sigma^2_{y_3}(\tau) &= \sigma^2_{y_3} + \alpha_1\alpha_2\sigma^2_{y_{12}}. \end{aligned}\right. \tag{12}$$

Therefore, if α_1 and α_2 have the same sign (positive correlation), clocks 1 and 2 seem to be more stable than they are, whereas clock 3 seems to be less stable. Furthermore, if $\alpha_1 = \alpha_2 = 1$, the common part of clocks 1 and 2 disappears completely, and y_{12} is attributed to clock 3:

$$\left\{\begin{aligned} \hat\sigma^2_{y_1}(\tau) &= \sigma^2_{y_10} = \sigma^2_{y_1} - \sigma^2_{y_{12}} \\ \hat\sigma^2_{y_2}(\tau) &= \sigma^2_{y_20} = \sigma^2_{y_2} - \sigma^2_{y_{12}} \\ \hat\sigma^2_{y_3}(\tau) &= \sigma^2_{y_3} + \sigma^2_{y_{12}}. \end{aligned}\right. \tag{13}$$

On the other hand, if α_1 and α_2 have opposite signs (anti-correlation), clocks 1 and 2 seem to be less stable than they are, whereas clock 3 seems to be more stable. If $\alpha_1 = -\alpha_2 = 1$ (or the opposite), the anti-correlation y_{12} is counted twice for clocks 1 and 2 whereas it is subtracted from clock 3:

$$\left\{\begin{aligned} \hat\sigma^2_{y_1}(\tau) &= \sigma^2_{y_10} + 2\sigma^2_{y_{12}} = \sigma^2_{y_1} + \sigma^2_{y_{12}} \\ \hat\sigma^2_{y_2}(\tau) &= \sigma^2_{y_20} + 2\sigma^2_{y_{12}} = \sigma^2_{y_2} + \sigma^2_{y_{12}} \\ \hat\sigma^2_{y_3}(\tau) &= \sigma^2_{y_3} - \sigma^2_{y_{12}}. \end{aligned}\right. \tag{14}$$

In this case, the estimation of $\sigma^2_{y_3}(\tau)$ may be negative if $\sigma^2_{y_3} < \sigma^2_{y_{12}}$ **: it is a symptom that we will use for detecting the correlations between clocks 1 and 2!**

2.2 Noise level estimation

The originality of this method lies in the use of this hypothesis: we assume that the frequency spectral density of each clock is modeled by power laws:

$$\begin{cases} S_{y_1}(f) = \sum_{\alpha=-2}^{+2} h_{\alpha 1} f^{\alpha} \\ S_{y_2}(f) = \sum_{\alpha=-2}^{+2} h_{\alpha 2} f^{\alpha} \\ S_{y_3}(f) = \sum_{\alpha=-2}^{+2} h_{\alpha 3} f^{\alpha}. \end{cases} \quad (15)$$

On the same hand, the Allan variance of these signals may be written as a linear combination of the responses of the Allan variance for the different noise types, denoted $\Phi_\alpha(\tau)$ (for instance, $\Phi_0(\tau)$ is the response of the Allan variance for the white FM noise, e. g. $1/(2\tau)$):

$$\begin{cases} \sigma^2_{y1}(\tau) = \sum_{\alpha=-2}^{+2} h_{\alpha 1}\Phi_\alpha(\tau) \\ \sigma^2_{y2}(\tau) = \sum_{\alpha=-2}^{+2} h_{\alpha 2}\Phi_\alpha(\tau) \\ \sigma^2_{y3}(\tau) = \sum_{\alpha=-2}^{+2} h_{\alpha 3}\Phi_\alpha(\tau). \end{cases} \quad (16)$$

Practically, we will begin by calculating the Allan variances of the 3 intercomparisons: $\sigma^2_{y_1-y_2}(\tau)$, $\sigma^2_{y_2-y_3}(\tau)$ et $\sigma^2_{y_3-y_1}(\tau)$. Then, we estimate the noise levels $h_{\alpha ij}$ for each independent $\sigma^2_{y_i-y_j}(\tau)$. Finally, it is possible to deduce the individual noise levels $h_{\alpha i}$, for each clock i, thanks to the transposition of the Gray-Allan transform (3) for the noise levels:

$$\hat{h}_{\alpha i} = \frac{1}{2}\left(h_{\alpha ij} + h_{\alpha ki} - h_{\alpha jk}\right) \quad (17)$$

where the hat $\hat{h}_{\alpha i}$ denotes the estimate of the real parameter $h_{\alpha i}$.

The exploitation of the results, i. e. the estimates $h_{\alpha i}$, must rely on the estimation of their uncertainties:

$$\Delta\hat{h}_{\alpha i} = \frac{1}{2}\sqrt{\Delta h^2_{\alpha ij} + \Delta h^2_{\alpha ki} + \Delta h^2_{\alpha jk}} \quad (18)$$

where the uncertainties $\Delta h_{\alpha ij}$ are obtained by the multivariance method [4]. Since the $h^2_{\alpha ij}$ estimates are χ^2 distributed with a high equivalent degree of freedom, they may be considered as Gaussian samples. Therefore, the uncertainties of the individual noise level estimates $\hat{h}_{\alpha i}$ may be assessed by the root of the squared sum of the uncertainties $h_{\alpha ij}$.

Noise level	Clock 1	Clock 2	Clock 3
h_0	300	0	6
h_{-1}	1	0.5	0
h_{-2}	0	$1000{\cdot}10^{-5}$	$50{\cdot}10^{-5}$

Table 1: Noise levels chosen for the simulations (arbitrary units).

2.3 Procedure

We propose then the following procedure:

1. Estimation of the noise levels $h_{\alpha ij}$ of the differences between the clocks and their uncertainties $\Delta h_{\alpha ij}$ thanks to the multivariance method [4].
2. Deduction of the noise levels $h_{\alpha i}$ of the individual clocks.
3. Estimation of the uncertainties thanks to relationship (18).
4. Detection of the negative noise levels:
   - noise level set to 0 (minimal correlation assumption),
   - estimation of the negative correlation,
   - correction of the noise levels of the other clocks.

3 Simulation

We have checked the efficiency of the method by applying it to simulations. We generated 100 realizations of the same noise processes in 3 cases: without correlations between the clocks, with positive correlations between 2 clocks and with negative correlations between the same 2 clocks.

3.1 Uncorrelated clocks

In order to simulate clocks, each of them showing an optimal stability over one duration range (see figure 3), we choose the noise levels reported in table 1, for durations $1 \le \tau \le 4096$.

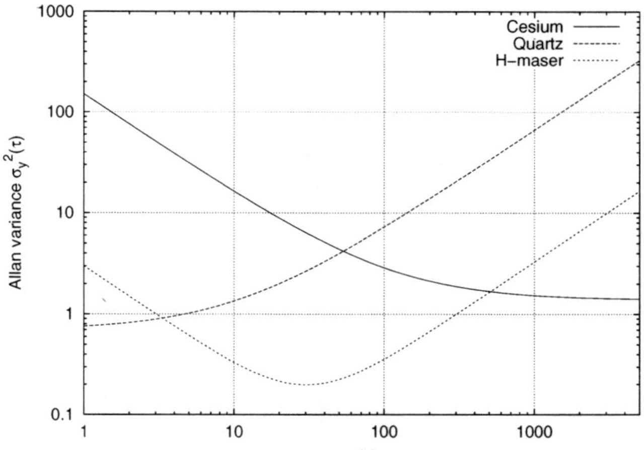

Figure 3: Theoretical Allan variance graphs of the 3 clocks chosen for the simulation.

3.1.1 Example of one realization

Let us analise one of the realizations:

- First, we generate a frequency deviation sequence of 16384 samples for each of the 3 clocks, using the noise levels of table 1.

- For simulating the intercomparisons, we calculate the differences between the clocks for each time values: $y_1(t) - y_2(t)$, $y_2(t) - y_3(t)$ and $y_3(t) - y_1(t)$.

- We apply the multivariance method over the 3 intercomparison sequences. Figure 4 shows the Allan variance curves of these 3 simulated intercomparisons. The results are reported in table 2(a).

- We deduce the individual noise levels of the clocks from the intercomparison noise levels. For instance, in order to estimate the flicker FM noise level of clock 2, $\hat{h}_{-1,2}$, we apply relationship (17):

$$\begin{aligned} \hat{h}_{-1,2} &= \frac{1}{2}\left(h_{-1,1,2} + h_{-1,2,3} - h_{-1,3,1}\right) \\ &= \frac{1}{2}(1.02 + 0.49 - 0.76) \\ &= 0.375. \end{aligned}$$

- We assesses the uncertainties of the individual noise level estimates from the intercomparison noise level uncertainties by applying relationship (18). For instance, the uncertainty of the flicker FM noise level estimate of clock 2 is:

$$\begin{aligned} \Delta\hat{h}_{-1,2} &= \frac{1}{2}\sqrt{\Delta h_{-1,1,2}^2 + \Delta h_{-1,2,3}^2 + \Delta h_{-1,3,1}^2} \\ &= \frac{1}{2}\sqrt{0.31^2 + 0.06^2 + 0.30^2} \\ &= 0.22 \end{aligned}$$

The final results for this realization are reported in table 2(b). Only one noise level is too low to be measured: $h_{-2,3}$ is considered as null[2] whereas $50 \cdot 10^{-5}$ were entered for the simulation. The uncertainty $\Delta\hat{h}_{-2,3}$ is greater than this value: $75 \cdot 10^{-5}$ may be considered as the sensitivity limit of this method.

[2]A measure is considered as null if it is lower than its uncertainty.

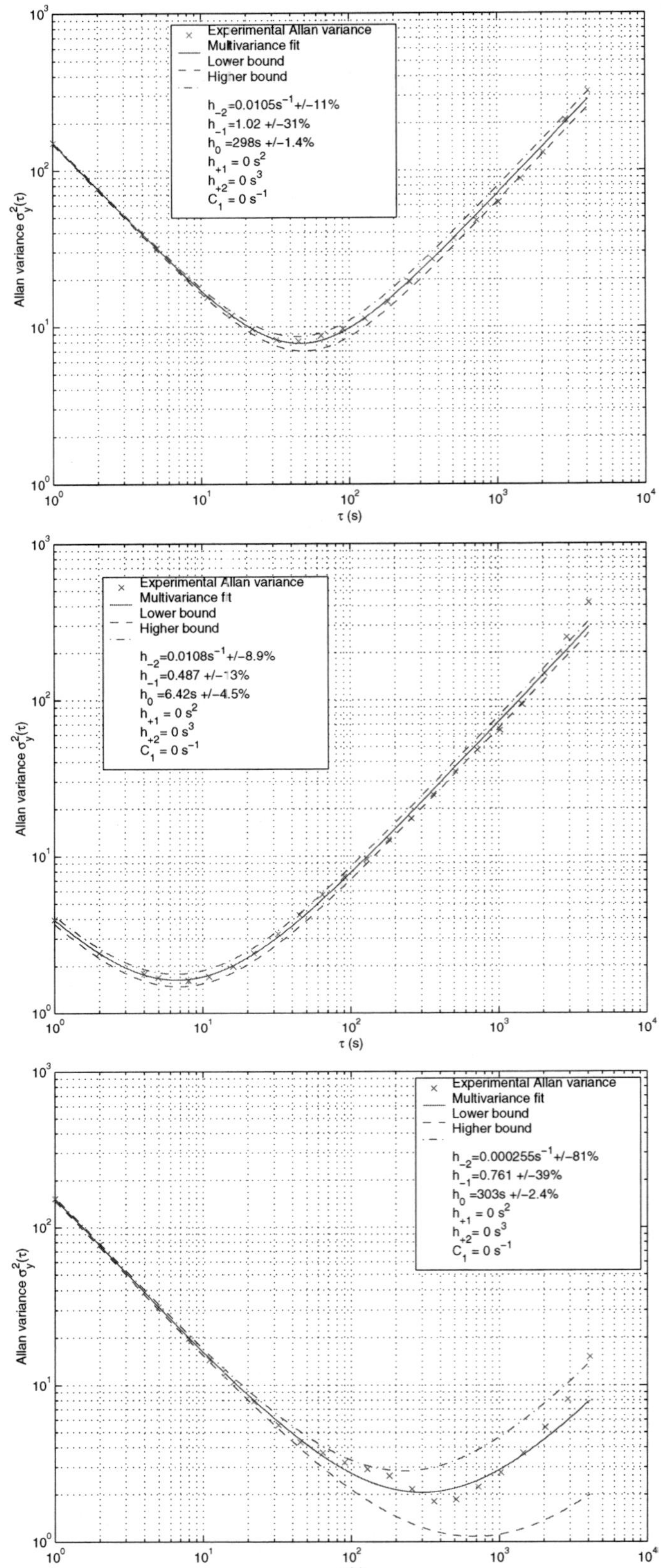

Figure 4: Allan variance graphs for one realization of $y_1 - y_2$ (above), $y_2 - y_3$ (middle), $y_3 - y_1$ (below).

a)

Noise levels	Clocks 1-2	Uncertainty	Clocks 2-3	Uncertainty	Clocks 3-1	Uncertainty
h_0	298	4	6.4	0.3	303	7
h_{-1}	1.0	0,3	0.49	0.06	0.8	0.3
h_{-2}	$1054 \cdot 10^{-5}$	$113 \cdot 10^{-5}$	$1084 \cdot 10^{-5}$	$97 \cdot 10^{-5}$	$25 \cdot 10^{-5}$	$21 \cdot 10^{-5}$

b)

Noise levels	Clock 1	Uncertainty	Clock 2	Uncertainty	Clock 3	Uncertainty
h_0	297	4	0	4	6	4
h_{-1}	0.6	0.2	0.4	0.2	0	0.2
h_{-2}	0	$75 \cdot 10^{-5}$	$1056 \cdot 10^{-5}$	$75 \cdot 10^{-5}$	**0**	$75 \cdot 10^{-5}$

Table 2: (a) Estimation of the noise levels of the clock differences for the realization shown in figure 4. The uncertainties are estimated by the multivariance method. (b) Estimation of the noise levels of the individual clocks obtained by the method of Gray and Allan. The uncertainties are calculated from relationship (18). A noise level lower than its uncertainty is considered as null (for example h_{-2} for clock 2). The bold values point out the undetermined noise levels (such as h_{-2} for clock 3).

3.1.2 Results for 100 realizations

We generated 100 realizations like the one described in section 3.1.1. We performed the same procedure for estimating the individual noise levels for each clock. Table 3 reports the average and the standard deviation of the noise level estimates over all the realizations.

These estimations are compared to a direct estimation of the noise level over the individual frequency deviation sequence of each clock (first item of section 3.1.1), as if the stability of each clock would be measured with an ideal reference clock.

For clock 1, we may notice that the uncertainty of the h_{-2} estimate obtained by the three cornered hat method is much higher than the one obtained by the direct measurement. But, in both cases, this level may be considered as null, in accordance with the level entered for the simulation.

For clock 2, the direct measurement seems to detect a white PM noise whereas there is not such a noise. The uncertainty of this noise level is much higher with the three cornered hat method, but here also, it doesn't have any consequence since this noise level is null.

For clock 3, the uncertainties obtained by the three cornered hat method are from 10 to 20 times higher than the ones obtained by the direct method. Thus, the estimation of the stability of this clock is difficult, because it is much more stable than clock 1 for short term (white FM), much more stable than clock 2 for long term (random walk FM), and much more stable than both clocks 1 and 2 for middle term (flicker FM). As a consequence, the three cornered hat method is not sensitive enough for detecting the random walk FM noise.

Method	h_0	σ_{h_0}	h_{-1}	$\sigma_{h_{-1}}$	h_{-2} $\times 10^{-5}$	$\sigma_{h_{-2}}$
Clock 1	300		1		0	
Dir. meas.	302	4	0.86	0.19	1	3
3 cor. hat	300	4	1.11	0.27	-43	92
Clock 2	0		0.5		1000	
Dir. meas.	**0.25**	0.11	0.52	0.05	910	181
3 cor. hat	-0.48	1.89	0.64	0.28	875	224
Clock 3	6		0		50	
Dir. meas.	5.94	0.10	0.008	0.012	45	10
3 cor. hat	6.67	1.72	-0.108	0.227	**84**	95

Table 3: Comparison of the noise levels obtained by a direct measurement of the individual clocks from an ideal clock reference (Dir. meas.) and by the three cornered hat method from the clock differences (3 cor. hat). Each noise level is the average of the noise levels over 100 realizations, and σ_{h_α} is its standard deviation. The theoretical noise levels entered for the simulation are in the rows of the corresponding clocks, according to table 1. The bold values point out the false measurements (for example $h_0 = 0.25 \pm 0.11$ instead of 0 for clock 2 - dir. meas.), or the undetermined noise levels (such as h_{-2} for clock 3 - 3 cor. hat).

Method	h_0	σ_{h_0}	h_{-1}	$\sigma_{h_{-1}}$	h_{-2} $\times 10^{-5}$	$\sigma_{h_{-2}}$
Clock 1	300		0.7 + *0.3* = 1		0	
Dir. meas.	302	4	0.87	0.20	0.9	2.5
3 cor. hat	300	4	**0.77**	0.27	-14	79
Clock 2	0		0.2 + *0.3* = 0.5		1000	
Dir. meas.	**0.27**	0.09	0.51	0.04	949	123
3 cor. hat	-0.16	1.55	**0.27**	0.23	918	193
Clock 3	6		0		50	
Dir. meas.	5.93	0.11	0.009	0.013	44	11
3 cor. hat	6.35	1.39	**0.258**	0.181	**57**	75

Table 4: Comparison of the noise levels obtained by a direct measurement and by the three cornered hat method. The italic value of the theoretical level h_{-1} for clocks 1 and 2 shows the correlation between these clocks: a flicker FM is common to clock 1 and 2 ($h_{-1} = 0.3$) but two independent flicker FM affect also these 2 clocks ($h_{-1} = 0.7$ for clock 1 and $h_{-1} = 0.2$ for clock 2). The bold values point out the false measurements (for example $h_{-1} = 0.26 \pm 0.18$ instead of 0 for clock 3 - 3 cor. hat), or the undetermined noise levels (such as h_{-2} for clock 3 - 3 cor. hat).

3.2 Positive correlations

The flicker FM noise level is modified for clocks 1 and 2: $h_{-1,1} = 0.7$ and $h_{-1,2} = 0.2$ arbitrary units. Of course, these two noises are independent (uncorrelated). But we add the flicker FM noise ($h_{-1} = 0.3$ arbitrary unit) to both clocks 1 and 2: the new noise levels are then $h_{-1,1} = 0.7 + 0.3 = 1$ and $h_{-1,2} = 0.2 + 0.3 = 0.5$ arbitrary units.

The direct measurement retrieves these noise levels. However, according to relationship (13), only the independent flicker FM are measured by the three cornered hat method for clocks 1 and 2. **The correlation between clock 1 and clock 2 is attributed to clock 3** ($h_{-1,3} = 0.26 \pm 0.18$)**!**

3.3 Negative correlations

Now, the independent flicker FM level are $h_{-1,1} = 0.7$ and $h_{-1,2} = 0.8$ arbitrary units. The common flicker FM noise ($h_{-1} = 0.3$ arbitrary unit) is added to clock 1 and subtracted to clock 2. Therefore, the actually noise levels are then $h_{-1,1} = 0.7 + 0.3 = 1$ and $h_{-1,2} = 0.8 + |-0.3| = 1.1$ arbitrary units.

Once again, the direct measurement gives estimates very close to these theoretical values. In the case of the three cornered hat method, according to relationship (14), the correlation is counted 2 times for clocks 1 and 2: $h_{-1,1} \approx 0.7 + 2 \times 0.3 = 1.3$ and $h_{-1,1} \approx 0.8 + 2 \times 0.3 = 1.4$. The correlation is reported to clock 3 **with a negative sign** ($h_{-1,3} \approx -0.3$)!

Method	h_0	σ_{h_0}	h_{-1}	$\sigma_{h_{-1}}$	h_{-2} ×10^{-5}	$\sigma_{h_{-2}}$
Clock 1	300		$0.7 + \vert +\mathit{0.3} \vert = 1.0$		0	
Dir. meas.	301	4	0.87	0.18	2	3
3 cor. hat	299	4	**1.38**	0.34	-18	97
Clock 2	0		$0.8 + \vert -\mathit{0.3} \vert = 1.1$		1000	
Dir. meas.	**0.59**	0.18	1.10	0.06	939	171
3 cor. hat	0.30	2.04	**1.48**	0.32	905	231
Clock 3	6		0		50	
Dir. meas.	5.95	0.10	0.008	0.012	45	10
3 cor. hat	6.23	1.88	**-0.360**	0.262	**62**	92

Table 5: Comparison of the noise levels obtained by a direct measurement and by the three cornered hat method. The italic value of the theoretical level h_{-1} for clocks 1 and 2 shows the anti-correlation between these clocks: a flicker FM is common to clock 1 and 2 ($h_{-1} = 0.3$) but with an opposite sign. The bold values point out the false measurements, the undetermined noise levels or the non-physical values (such as the negative h_{-1} level for clock 3 - 3 cor. hat).

Therefore, if a negative noise level is found for one clock, it must be interpreted as an anti-correlation of the other 2 clocks, and subtracted from their corresponding noise levels. For example, in this case, $h_{-1,3}$ must be subtracted from both $h_{-1,1}$ and $h_{-1,2}$.

4 Conclusion

Positive correlations between two clocks can not be detected by this method: the corresponding noise level of these clocks are underestimated and all the correlation is considered as a noise of the third clock.

On the other hand, in the case of anti-correlation between two clocks, the corresponding noise levels of these clocks are overestimated: the noise level of the correlation is counted twice. However, this noise level is subtracted from the corresponding noise level of the third clock. Therefore, if the real noise level of the third clock is lower than the noise level of the correlation, a negative noise level appears: this is the typical signature of an anti-correlation! In this case, the real noise levels may be estimated: the negative noise level must be set to 0 (minimization of the correlation), and subtracted from the noise levels of the other 2 clocks.

References

[1] J.E. Gray and D.W. Allan. "A method for estimating the frequency stability of an individual oscillator", In *Proceedings of the 28th Annual Symposium on Frequency Control*, pp. 243–246, May 1974.

[2] D. Fest, J. Groslambert, and J. J. Gagnepain. "Individual characterization of an oscillator by means of cross-correlation or cross-variance method", *IEEE Transactions on Instrumentation and Measurement*, 32(2):447–450, September 1983.

[3] A. Premoli and P. Tavella. "A revisited three-cornered hat method for estimating frequency standard instability", *IEEE Transactions on Instrumentation and Measurement*, 42(1):7–13, February 1993.

[4] F. Vernotte, E. Lantz, J. Groslambert, J.-J. Gagnepain. "Oscillator noise analysis : multivariance measurement. Characterization of frequency stability: Uncertainty due to finite number of measurements", *IEEE Transactions on Instrumentation and Measurement*, 42(2):342–350, Avril 1993.

PROGRESS IN TEMPERATURE STABILIZATION OF GPS ANTENNAS

A. Lisowiec*, L. Nafalski*

*Tele & Radio Research Institute, Warsaw, Poland
alisow@itr.org.pl

Keywords: time transfer, GPS, temperature stabilization.

Abstract

Temperature variations cause signal delay changes in GPS receiver and antenna and can be of the order of 0.1 ns/°C. Temperature stabilization of GPS antenna and receiver can increase the accuracy of time transfer by 2 to 3 ns. A temperature stabilization system for Motorola Timing 2000 GPS antenna has been developed. The temperature regulator and insulating layer fit inside the antenna cover. The gain pattern of the antenna remains unchanged and its temperature is stabilized to ±0.5°C around +59.5°C. Tests with new antenna confirmed the gain in accuracy that could be expected.

1 Introduction

Time transfer by GPS Common View method cancels out most sources of error like SV clock errors, ephemeris errors, tropospheric and ionospheric errors. What is left is equipment noise, multipath and the influence of environmental conditions. The most pronounced is the influence of temperature on signal delay in GPS antenna and receiver. This influence has been known for several years and a few designs of thermally stabilized GPS antennas and receivers appeared on the market in the past few years. Some of these designs are no longer offered – 3S Navigation TSA 100 product. Some, like the temperature stabilization system of Motorola GCNAC1111A antenna can only be offered to laboratories that supply this antenna as it is no longer manufactured.

New type of antenna that is gaining popularity in GPS time transfer systems is Timing 2000. It has similar gain to GCNAC1111A, is equipped with bell like cover to prevent from snow gathering and has N-type signal connector at the base for direct connection to low loss cable. It has been used by some of the laboratories dealing with time transfer (AO Borowiec in Kurnik, Poland, ITR in Warsaw, Poland). As the previous experiments have showed [1], signal delay variations in Motorola designs are on the order of 0.1ns/°C. Diurnal and seasonal signal delay variations due to changes in outdoor temperature can easily exceed 2ns. It is thus well worth to stabilize the antenna electronics temperature.

2 Construction of temperature stabilization system

The construction of Timing 2000 antenna is much better suited to thermal stabilization than previous GCNAC1111A type. The antenna microstrip and dielectric are placed on the top side of a metal plate which purpose is presumably to eliminate multipath while the antenna electronics is placed on the bottom side of it. It is thus sensible to stabilize the temperature of the metal plate. The antenna with temperature stabilization system in various stages of disassembly has been shown in figures 1, 2 and 5. The temperature regulator electronics has been placed on the bottom side of the metal plate, away from antenna electronics as shown in fig. 1. Two medium power transistors heat this plate to 60°C. To prevent the heat loss, the original metal screws connecting the shielding plate to aluminum die cast bottom housing have been replaced by teflon screws. The rigidity of the whole antenna structure has been preserved by filling the extra space with insulating material, shown in fig. 5. As the antenna dielectric is mounted on the upper side of the metal plate, the gain pattern of the antenna remains almost unchanged which has been confirmed by comparing it with the gain pattern of intact antenna using the Motorola VP Oncore receiver and software supplied in evaluation kit.

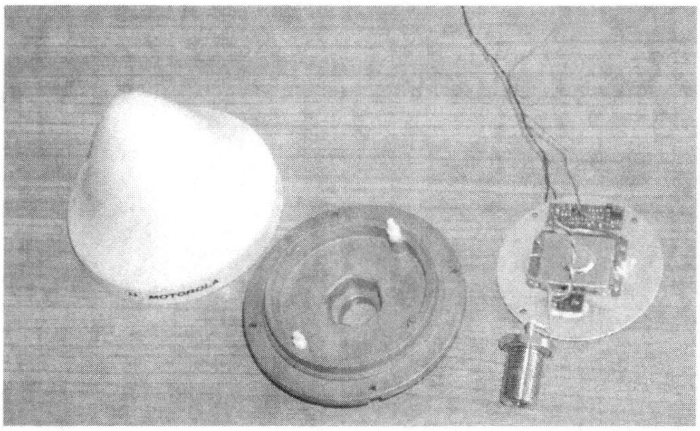

Figure 1. The antenna electronics at the bottom of the shielding plate with temperature stabilization electronics mounted close to it.

The tests of the gain of the antenna revealed the effect that was spotted with older types of Motorola GPS antennas [1], namely that increasing the antenna temperature decreases slightly its gain. This may be one of the reasons why temperature has an influence on the signal delay in the antenna.

The effectiveness of the temperature stabilization, shown in fig. 2 has been measured by placing the antenna in temperature chamber and changing its temperature from –30°C to 55°C.

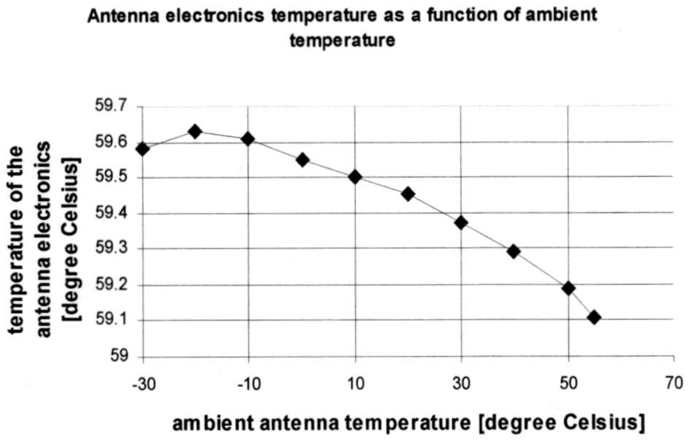

Figure 2. Temperature at the antenna plate as a function of ambient temperature

The temperature of the antenna substrate has been measured by small thermistor mounted close to it, shown in figure 3.

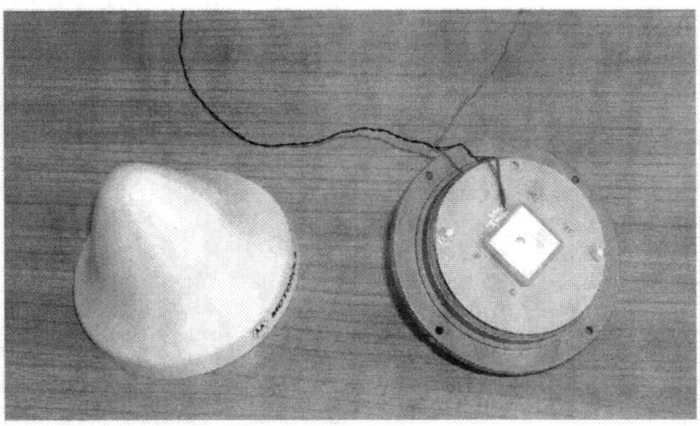

Figure 3. Antenna microstrip structure with thermistor mounted close to it.

In the final product the thermistor is not included in the temperature stabilization system but there is a simple way to control the proper functioning of the circuit. The temperature regulator has a high gain so the heating current is at maximum or minimum value when the antenna temperature has not yet been stabilized. When the antenna temperature has reached its final value, the current falls off rapidly to some value between I_{max} and I_{min}. Temperature stabilization can be monitored by observing the heating current, which has been shown in fig. 4 as a function of external temperature. The current level between $0.02*I_{max}$ and $0.9*I_{max}$ signals a steady state in which antenna temperature has been stabilized. A simple heating current sensing circuit with LED ON/OFF indicator has been presented in [1]. It can be placed at the output of heating current power supply which has not to be well regulated.

The temperature stabilization time after heating current supply power switch on depends on the ambient temperature and at 0°C is 40 minutes.

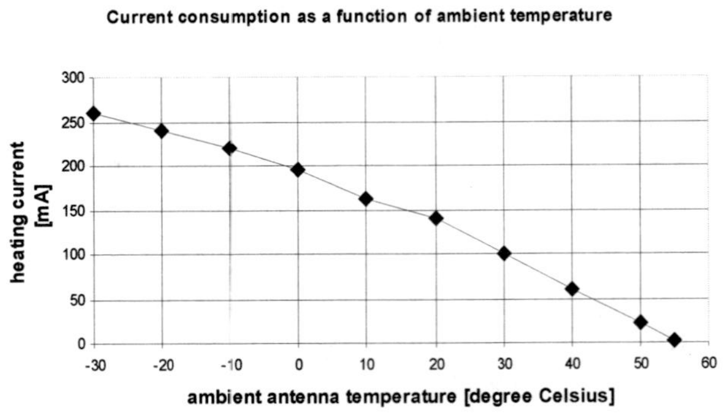

Figure 4. Current consumption of the antenna temperature stabilization electronics as a function of ambient temperature.

The insulation material, shown in fig. 5, provides rigidity to the antenna structure, but its most important characteristic is loss at 1.5 GHz. A good choice is highly-foamed polyethylene because of its low dielectric constant. Care has to be taken to test the available of the polyethylene samples as even the small amount of impurities seriously increases the loss.

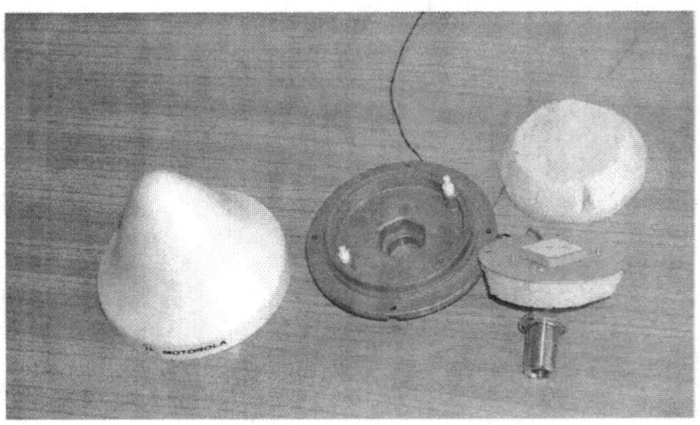

Figure 5. Upper and lower temperature insulation layers.

The thickness of the bottom insulating layer determining its temperature resistance has to chosen carefully. Too thick layer degrades the performance in warm weather condition with high solar irradiation while too thin layer provides not enough isolation at low outdoor temperature days.

3 Measurements of the signal delay variations

To measure the gain in time transfer accuracy that could be expected with temperature stabilization, GPS Common View multi-channel observations have been performed. Two laboratories took part in this experiment. One was Tele & Radio Research Institute Time and Frequency Measurement Laboratory in Warsaw with temperature stabilized Motorola Timing 2000 GPS antenna described above and the other was CBR, TP S. A. (the research laboratory of Polish Telecommunications), also in Warsaw. These two laboratories are placed about 9 km apart. Both are equipped with TTS2 time transfer systems running identical software and are routinely performing time transfer experiments according to the BIPM schedule.

The computed time differences between two laboratories have been shown in fig. 6 with mutual drift of two HP 5071A clocks situated in the laboratories removed. During the experiment the heating of the antenna at Tele & Radio Research Institute has been turned on and off several times. At the time when the experiment was performed the outdoor temperature was relatively constant at 0°C. Turning on and off of the heating current varied the antenna temperature by about 60 degrees. The grey areas in fig. 6 indicate the periods when the heating current was on. Each of this periods lasted about 3 hours.

Although it is difficult from fig. 6 to quantitatively estimate the change of signal delay per 1 °C there is a visible influence of antenna temperature on signal delay.

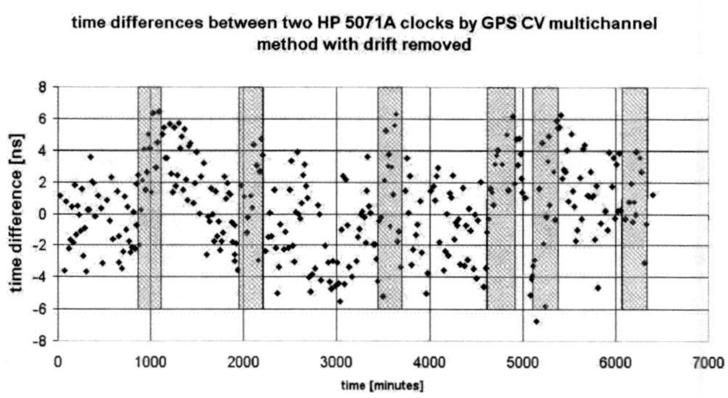

Figure 6. GPS CV time transfer experiment between ITR, Warsaw and CBR TP S.A., Warsaw laboratories. Grey areas show periods when the GPS antenna at ITR has been heated to 60°C.

4 Conclusion

Experiments with temperature stabilized Motorola GPS antenna have clearly showed that temperature has a visible influence on the accuracy of GPS CV time transfer. Diurnal and seasonal temperature variations can degrade the accuracy by as much as 2 to 3 ns. Temperature stabilization of the antenna, that eliminates this effect is technically simple and such antennas can work reliably for many years.

References

[1] A. Lisowiec, A. Smolarski, J. Nawrocki "Improving the accuracy of gps time transfer by thermal stabilization of gps antenna and receiver", Proc. of The 15th EFTF, 2001, Neuchatel.

GPS SMOOTHED P3 CODE FOR TIME TRANSFER

H.M. Peng and C.S. Liao

National Standard Time and Frequency Lab, Telecommunication Laboratories,
12, Lane 551, Min-Tsu Road Sec.5, Yang-Mei, Taoyuan, Taiwan, 326, R.O.C.
E-mail: hmpeng@cht.com.tw, Fax: 886 3 424 5474

Keywords: GPS, smoothed P3 code, time transfer.

Abstract

In this paper, international time transfer using GPS smoothed P3 code is proposed and tested. The traditional GPS common-view technique uses single channel C/A code for computation; therefore, the ionospheric delay can not be eliminated but only be reduced by an ionosphere model. Recently, GPS P3 code is used in time transfer for TAI computation to overcome this significant error; however, this observation still has large variations when the satellites are at low elevation angles. To acquire more stable time comparison results, we adopt some carrier smoothing techniques to reduce this noise. Two smoothing techniques based on different considerations are tested in our paper. The first method uses the concept of averaging; in other word, the mean difference of P3 and L3 measurements is regarded as the cycle ambiguity and L3 is used to substitute for P3 code for the further computation. To ensure the estimate ambiguity is acceptable, the span of data should be as long as possible after detecting and isolating cycle slips and outliers. In the second technique, the concept of filtering is adopted for data smoothing. The weighting average of outputs from two filters, forward filter and backward filter based on forward and backward directions of measurements, is used for time comparison. The IGS precise ephemerides are also used to reduce the satellite orbital errors in order to obtain better time comparison results. For verifying our programs, we processed some GPS data for international time comparison between laboratories and the corresponding P3 code and carrier phase time transfer results are also shown for comparison. The test data show that using smoothed P3 code can reduce the uncertainty of GPS time comparison and the output data interval can be shortened since the noise level caused by the code measurements is lowered.

1 Introduction

The international time and frequency transfer is a key issue for maintaining a universal time scale. Nowadays, two kinds of satellites are used to create comparison links with their corresponding structures. One type of communication link is created by geostationary satellites and two-way technique and the other one is built by navigation satellites, such as GPS, with one-way method. Both of these links have advantages and drawbacks. The former one has better performance; however, the communication fee for continuous operation is also expensive. On the other hand, although there is no additional pay for using signals from navigation satellites, noises and errors in signal transmission between satellites and receivers can not be eliminated under the one-way link structure. Hence, how to reduce the noises and errors generated in the transmission is the only way to improve the performance of one-way time transfer.

In the conventional GPS time transfer technique, the common view concept is adopted to reduce the signal transmission errors. Only one satellite is tracked using C/A code measurements at a pre-specified epoch by GPS receivers and time comparison results between the synchronized 1PPS signals and laboratory references are obtained. Under this circumstance, the errors in signal transmission for nearby laboratories are assumed common and able to be reduced by difference operation between laboratories. The uncertainty of this common view structure is about 3 nanoseconds. Although this strategy can be used to eliminate some common errors, the site-dependent errors and random noise from C/A code are not reduced under this structure. For obtaining more stable results, it is necessary to make some improvements to the conventional technique. In this paper, several modifications are tested for decreasing the uncertainty of time transfer. P code is substituting for the traditional C/A code as the core observables [2]. Since P code is modulated on both L1 and L2 carriers, more accurate ionospheric delay can be obtained. Since the noise level of P code is still larger than that of carrier phase, the carrier smoothing technique are used. Furthermore, the multi-channel structure is adopted, too. Accordingly, for reducing the satellite-dependent errors, IGS precise ephemerides are used.

In this paper, two carrier smoothing techniques are studied. The first one is a kind of averaging method. The code and carrier phase measurements are pre-processed by Melbourne-Wübbena linear combination to remove the unacceptable data. After that, the mean difference between P3 and L3 arcs is regarded as the cycle ambiguity; thus, the L3 carrier phase measurements can be used to substitute for the P3 code for time transfer. On the other hand, the second scheme for obtaining the carrier smoothed code is designed based on a filtering structure. The GPS P3 code and time differences of L3 code are measurements while the clock phase and phase difference are the state variables. The estimate states are calculated using the standard Kalman filtering formulae both in the forward direction and backward direction of data. Then, the weighted averaging of the two estimate state variables are the final results for computing the phase difference between the local reference and satellite clock.

This paper is organized as follows. In the next section, two carrier smoothing techniques are described and compared. Since the noise level of smoothed code is lower, the improvement can be revealed from the common view results. The benefit of using IGS ephemerides will be discussed in section 3. Combining these two modifications, we will show our result is better than the existent GPS code-based time transfer techniques in section 4. Carrier phase time transfer results are also shown for comparison. The conclusion is given in the last section.

2 Description of carrier smoothing techniques

In this paper, two smoothing techniques are studied for different considerations. The first one just uses the P code to resolve the carrier phase cycle ambiguities. This method is a typical procedure for generating carrier smoothed code but only measurements are considered in this model. The other scheme adopts the Kalman filtering structure while the forward filter and backward filter are used together for data smoothing. This method uses a model for local reference; hence, it can be used to predict the clock behaviour. The details of these two methods are described as follows.

The ionosphere-free P code P_3 and carrier phase L_3 are modelled as:

$$P_3 = \frac{1}{f_1^2 - f_2^2}(f_1^2 P_1 - f_2^2 P_2), \qquad (1)$$

$$L_3 = \frac{1}{f_1^2 - f_2^2}(f_1^2 L_1 - f_2^2 L_2), \qquad (2)$$

where the subscripts 1 and 2 mean the associated carriers. In the real time data processing, the common carrier smoothing equation is written as [4]

$$SP_3(k) = \frac{1}{k}P_3(k) + \frac{k-1}{k}(SP_3(k-1) + L_3(k) - L_3(k-1)) \qquad ,(3)$$
$$SP_3(1) = P_3(1)$$

The above Equation (3) is an iterative format of an averaging equation. Since we always post-process our data, the carrier smoothed ionosphere-free code SP_3 can be rewritten as:

$$SP_3(k) = L_3(k) + (\overline{P}_3 - \overline{L}_3), \qquad (4)$$

where the $\overline{P}_3, \overline{L}_3$ are the mean values of the acceptable P_3 and L_3 measurements in an observation arc. Their difference is regarded as the cycle ambiguity; hence, the arc must be as long as possible and without any cycle slip in order to obtain an acceptable ambiguity value. Using this smoothed P3 code, we can compute the time differences between local reference and satellite clocks for time comparison.

The filtering structure is adopted in the second scheme. The system and measurement model is assumed as

$$\mathbf{x}(k+1) = \begin{bmatrix} 1 & 1 \\ 0 & 1 \end{bmatrix}\mathbf{x}(k) + \mathbf{w}(k), \qquad (5)$$

$$\begin{bmatrix} \widetilde{P}_3(k) \\ \widetilde{L}_3(k) - \widetilde{L}_3(k-1) \end{bmatrix} = \begin{bmatrix} 1 & 0 \\ 0 & 1 \end{bmatrix}\mathbf{x}(k) + \mathbf{v}(k), \qquad (6)$$

where $\mathbf{x} = [x_1 \quad x_2]^T$ represents clock phase and phase difference; ~ is used to represent the residues between the measurements and known values, such as the geometric distance, satellite's clock error, and the tropospheric delay, and the process noise $\mathbf{w}(k)$ and measurement noise $\mathbf{v}(k)$ are assumed white and the corresponding covariance matrices $\mathbf{Q}(k)$ and $\mathbf{R}(k)$ are determined by the observation data. Using the formulae of Kalman filter [3] in both forward and backward time sequences, the estimate state variables are obtained. Combining the results from the forward filter and backward filter, we can get the phase differences between local clock and satellite references for time transfer.

The above two methods are used in our programs for obtaining carrier smoothed code measurements. The comparing results are shown in section 4.

3 Testing of IGS Ephemerides

For reducing the satellite-dependent errors, the IGS ephemerides are adopted in our program. The products are generated by IGS with latency about 17 hours and within 5 cm accuracy. This product supplies the satellites coordinates with 15 minutes interval of one day. For generating the satellite coordinates in any time, we use the seventeenth degree polynomial interpolation [4], while the Bernese software [5] is used to generate the reference orbit for comparison. The differences between the reference orbit and our generated orbit of PRN 1 on MJD 53005 are shown in Figure 1.

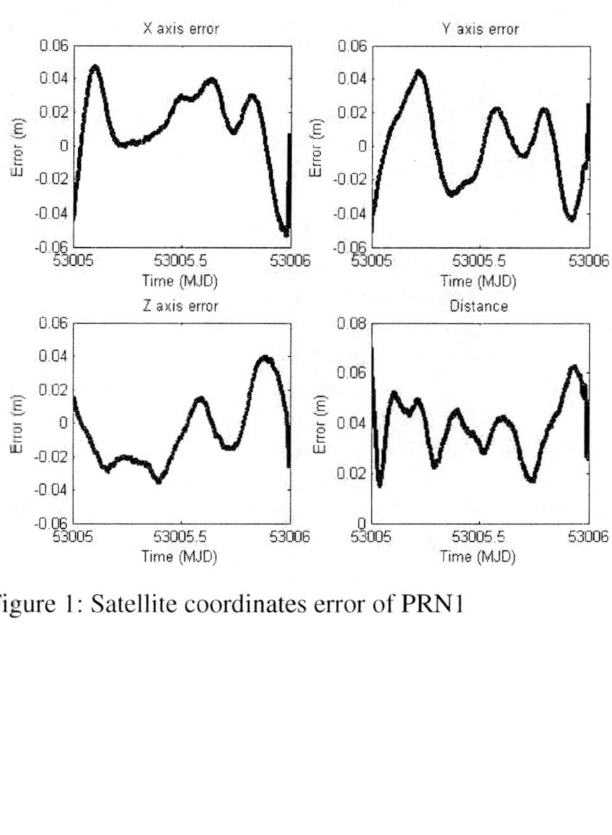

Figure 1: Satellite coordinates error of PRN1

It is noted that daily satellite coordinates are given in the IGS products; hence, the ephemerides of the next day are necessary for computing orbits after 22:00. On the other hand, we do not need the orbit file of the previous day for computing the coordinates before 2:00 based on our test data. This might be caused by the orbit generation procedure of IGS.

4 Time Comparison Tests

The uncertainty of GPS time transfer is direct affected by the corresponding observation. Three kinds of GPS observations are shown in Figure 2. Figure 2(a) shows the time transfer results computed by GPS P3 code. The Allan deviation of these results is 2.12×10^{-9} at $\tau=30\,\text{sec}$. On the other hand, Figure 2(b) and 2(c) show REF - GPS results computed by smoothed P3 code of method 1 and 2 and their corresponding Allan deviations at $\tau=30\,\text{sec}$ are 2.05×10^{-11} and 2.10×10^{-11}, respectively. Figure 2(d) shows the two different smoothed codes in one plot for comparison. Based on these data, the REFGPS values defined in CGGTTS format could be obtained by the linear fit method.

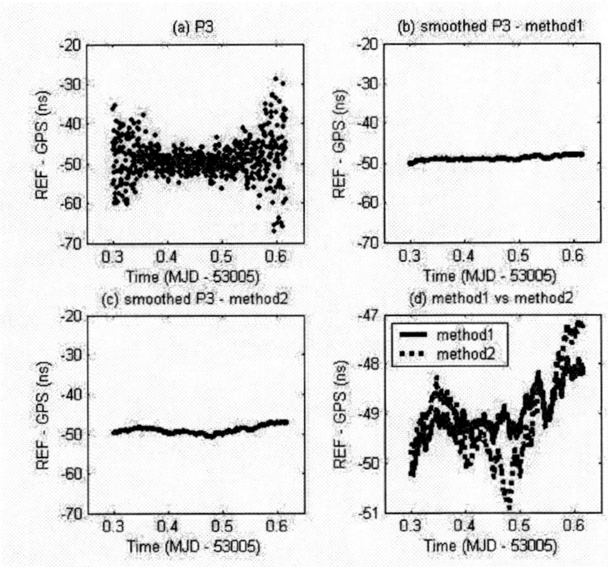

Figure 2: REF – GPS from different observations

Figure 3 shows the REFGPS values defined in CGGTTS format [1]. These values are obtained by computing the middle points of the linear fit straight lines of every 13 minutes data. Three features are shown in the figure. The circle is used to represent the middle point of linear fit and the slope of the fitting line is also shown. The error bar of one sigma is given to demonstrate the noise level of the raw measurements. Based on these data, we can conclude that the linear fit process for computing REFGPS is not necessary for smoothed P3 code. Thus, the data interval of REFGPS can be shortened to display the short term behaviour of the local reference.

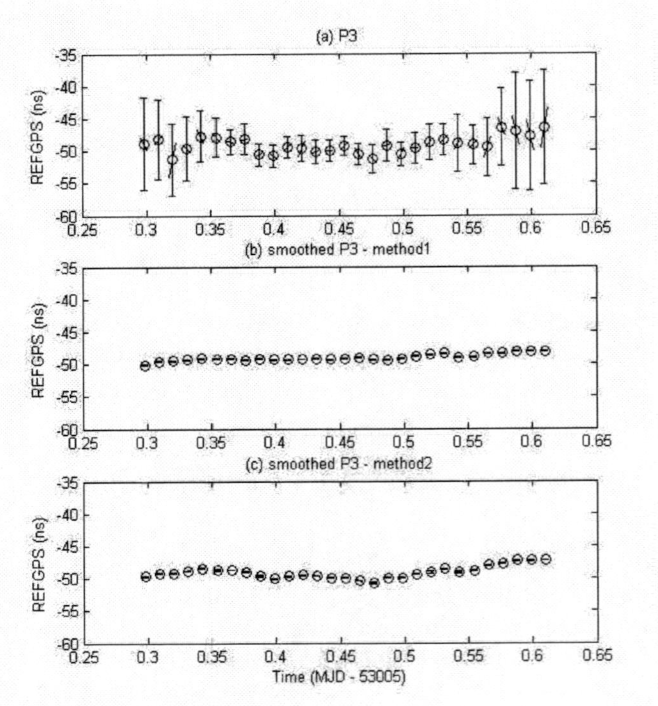

Figure 3: Time comparison results from different codes

To compare the effect caused by the broadcast ephemeris and IGS products, we use the program developed by Dr. Defraigne of ROB. Figure 4(a) is the computed REFGPS values by our program using IGS product and P3 code and Figure 4(b) is the results from ROB program. It is obvious that using IGS products can reduce the satellite-dependent error.

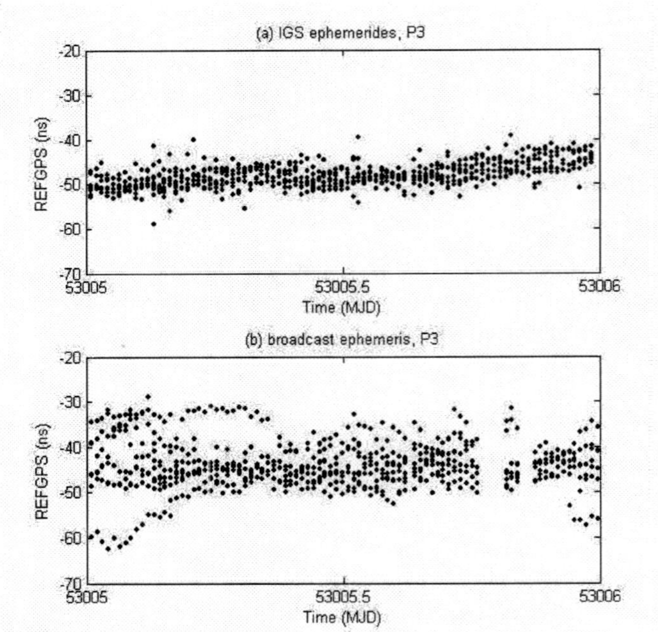

Figure 4: Comparison between using IGS and broadcast ephemerides

Figure 5 shows the phase differences induced by different code measurements. The upper plot is the phase differences resulted by the smoothed P3 code with method 1 and P3 code, and the lower plot shows the results induced by smoothed P3 code of method 1 and method 2. The standard deviations of these two data are 0.46 and 0.21 ns.

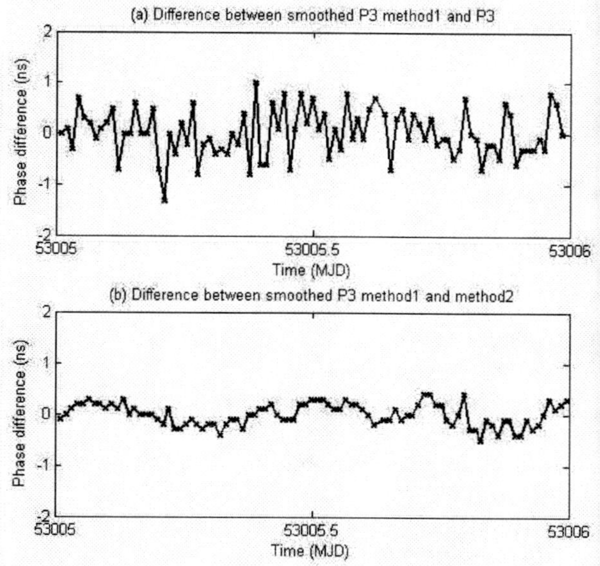

Figure 5: Time comparison between using different codes

For demonstrating the improvements of using smoothed P3 code and IGS ephemerides for time transfer, GPS data from four laboratories are processed and shown in Figures 6-8.

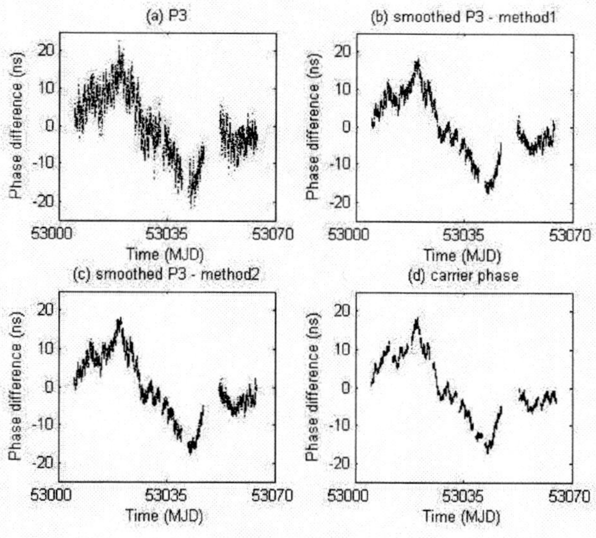

Figure 6: Time comparison between TL and USNO

Four kinds of GPS observables are used for time comparison and shown in figures. The P code time comparison results are generated by the ROB program with broadcast ephemeris, the results from two smoothed P3 code measurements are generated by TL programs, and the time transfer results from GPS carrier phase observables are processed by Bernese GPS software V4.2 [5]. In these figures, the offset are removed for display.

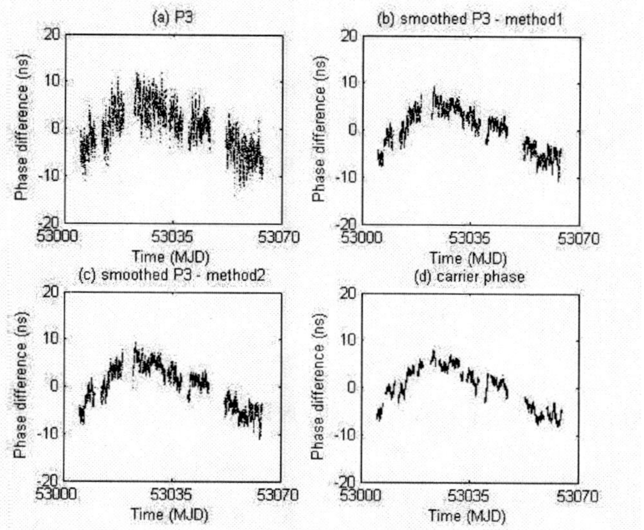

Figure 7: Time comparison between CRL and USNO

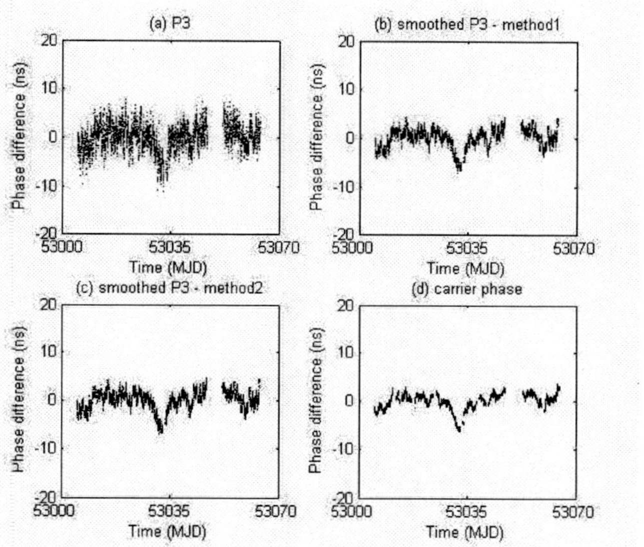

Figure 8: Time comparison between PTB and USNO

Figures 9-11 show the stability analyses of the time comparisons. From the analyses, we can conclude that using smoothed P3 code and IGS ephemerides for international time comparison can reduce the uncertainty from about 2×10^{-12} to 5×10^{-13} at $\tau = 960$ seconds.

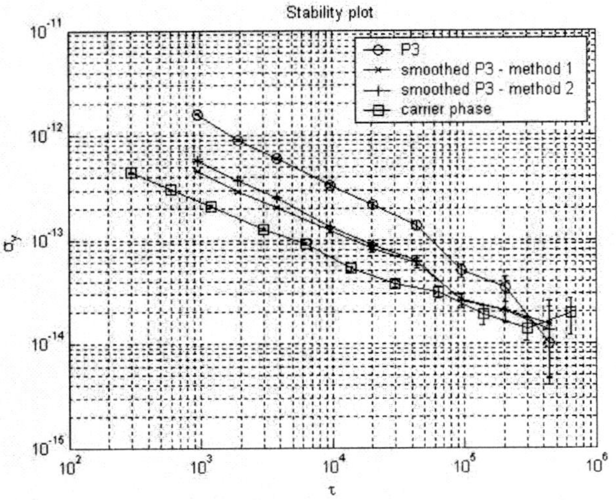

Figure 9: Stability plot of TL-USNO

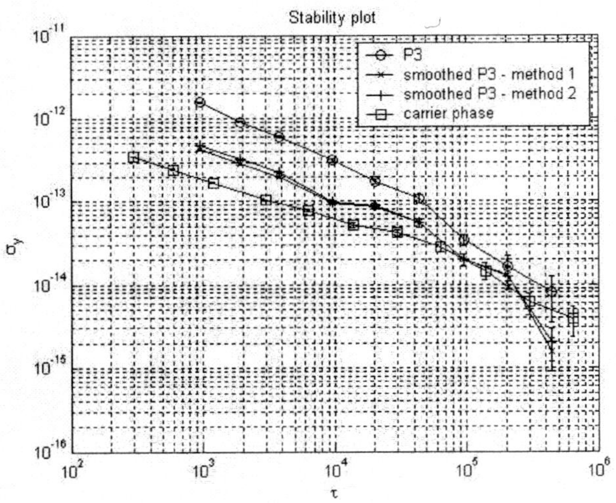

Figure 10: Stability plot of CRL-USNO

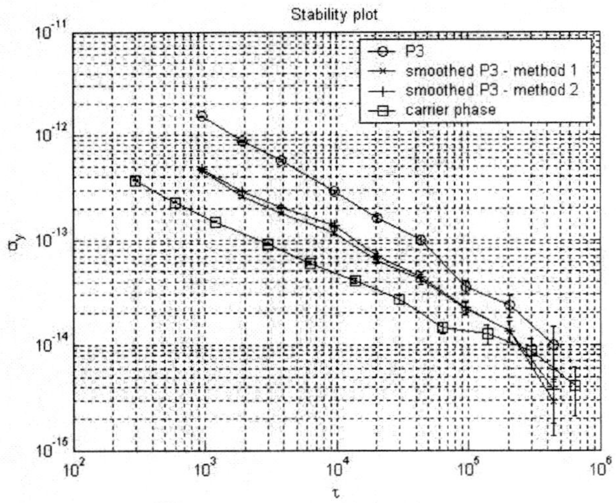

Figure 11: Stability plot of PTB-USNO

5 Conclusion

In this paper, we use GPS smoothed P3 code and IGS ephemerides for time comparison. Two carrier smoothing schemes are tested and the IGS ephemerides are also adopted to reduce the satellite-dependent errors. The first carrier smoothing method uses the P3 code to estimate the cycle ambiguity and L3 carrier phase can then be used for time comparison. The other method processes the P3 code and L3 code by the forward filter and backward filter. The estimate of state variable is then regarded as the smoothed code for the further data processing. Based on our tests, the improvement of using smoothed P3 code measurement and IGS ephemerides is obvious. Furthermore, since the noise on the code measurements can be reduced, the data interval of time comparison can be shortened.

References

[1] D. W. Allan and C. Thomas, "Technique directives for standardization of GPS time receiver software", *Metrologia*, 31, pp. 69-79, 1994.

[2] P. Defraigne, C. Bruyninx, J. Clarke, J. Ray and K. Senior, "Time Transfer to TAI Using Geodetic Receivers", *Proceeding of Symposium EFTF*, pp. 517-521, 2001.

[3] Arthur Gelb, editor, *Applied optimal estimation*, The MIT Press.

[4] B. Hofmann-Wellenhof, H. Lichtenegger, and J. Collins, *Global positioning system: Theory and practice*, fourth, revised edition, Springer Wien New York, 1997.

[5] U. Hugentobler, S. Schaer, and P. Fridez, editors, *Operation manual of Bernese GPS software version 4.2*, 2001.

WAAS, EGNOS, MSAS, and SNAS for Telecom Sync Applications

HUGO FRUEHAUF
FEI-Zyfer Inc.1585 S. Manchester Ave.Anaheim CA, 92802 USA
hxf@fei-zyfer.com

The WAAS, EGNOS, MSAS, and SNAS GPS-Augmentation Systems bridge the gap between the military applications of GPS and the needs of the civil aviation navigation community. These USA, Europe, and Asian systems consist of a network of differential GPS ground stations, using principally the Inmarsat-3 geo-synchronous satellite constellation as the means of relaying the differential corrections to the users. In the process of providing precise navigation signals, GPS time is maintained through the Inmarsat-3 uplink/downlink networks. It is this functionality that provides for robust time retrieval for the time and frequency synchronization needs of the telecommunications community. This paper will discuss:

- The basic makeup and performance of the system
- How time and frequency can be extracted for the telecom terminals
- The added functionality and noise rejection that can be obtained by using an 18" dish antenna
- How the system may offer the first true backup to the GPS-C/A signal for telecom users

NIST AND OP GPS RECEIVER CALIBRATIONS SPANNING TWENTY YEARS: 1983 - 2003*

M.A. Weiss[1], V. Zhang[1], W. Lewandowski[2], P. Uhrich[3], D. Valat[3]

[1] National Institute of Standards and Technology, Boulder CO, USA mweiss@boulder.nist.gov, vzhang@boulder.nist.gov
[2] Bureau International des Poids et Mesures,92312 Sèvres, FRANCE wlewandowski@bipm.org
[3] Bureau National de Métrologie - Systèmes de Référence Temps Espace, Observatoire de Paris 75014 Paris, FRANCE
Pierre.Uhrich@obspm.fr , David.Valat@obspm.fr

Keywords: Calibration, common-view time transfer, Global Positioning System, GPS receiver, International Atomic Time.

Abstract

We present results from two calibrations performed in 2003 between the Global Positioning System (GPS) common-view receivers located at the National Institute of Standards and Technology (NIST) and at the Bureau National de Métrologie – Systèmes de Référence Temps Espace (BNM-SYRTE) of the Observatoire de Paris (OP) in the context of calibrations performed over 20 years, 1983-2003. We also present several years of continuous comparisons between receivers located at each of the two labs: NIST and OP. These results show that the best GPS receivers in use have delay variations with peak differences of under 5 ns over a year. This contributes to defining the current practical limits of GPS common-view time transfer. Since GPS common-view time transfer is still used for the majority of links between laboratories contributing to International Atomic Time (TAI), the noise and uncertainties in common-view affect the short-term performance of TAI, for averaging times from 5 to 30 days.

1. Introduction

One of the goals of this paper is to present the long-term stability of the best common-view GPS receivers. We do this by giving long-term comparisons of such receivers, both among co-located receivers, and among receivers calibrated by means of a travelling receiver.

GPS common-view time transfer is still the dominant mode for establishing the links between laboratories contributing to TAI. The instability of the receiver delays is probably the largest contributor to the instability in TAI in short-term, from 5 days out to about 30 days.

We can see this in Figure 1: the modified Allan deviation (MDEV) [8] of TAI against the NIST scale AT1e. AT1e is a post-processed time scale dominated by five Hydrogen Masers (H-Masers), all in environmental chambers controlling temperature and humidity. The stability of AT1e is below 3 part in 10^{16} at 10 day, about 10^{6} seconds [5]. This is well below the apparent flicker phase modulation (PM) noise dominated deviation in Figure 1, from 5 days out to almost 40 days.

The data in Figure 1 are consistent with a flicker PM noise model for the first three points, because the slope on the log-log plot is –1. The first three points are for averaging times of 5, 10, and 20 days. At 40 days, there seems to be some flattening as the clock noise begins to contribute. Flicker PM at the levels in Figure 1 from 5 to under 40 days is typical of GPS common-view time transfer [2,4,6]. The long term stability between AT1e and TAI in Figure 1 is dominated probably by either random-walk frequency modulation , giving a slope of +½, or linear frequency drift, giving a slope of +1.

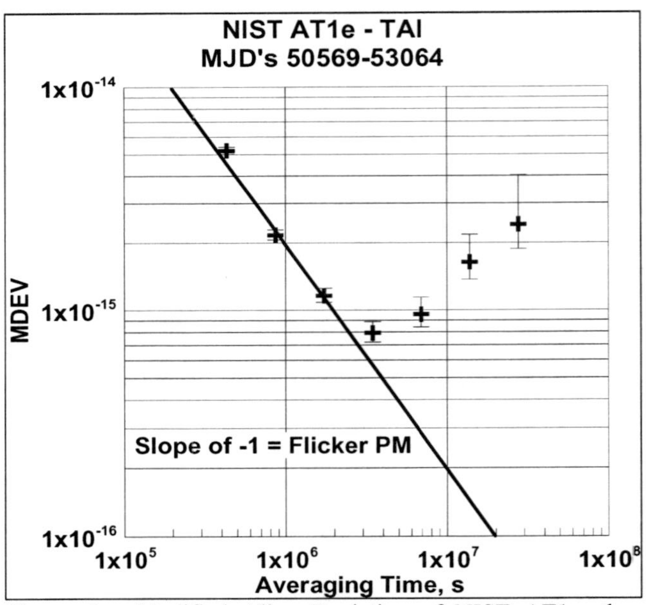

Figure 1: Modified Allan Deviation of NIST AT1e, the NIST post-processed H-Maser scale, against TAI. The data cover almost seven years, from May 1, 1997 to February 29, 2004.

The data in Figure 1 cover almost seven years. Since the clocks contributing to TAI improve continuously, the stability of TAI should improve also with time. Figure 2 gives MDEV of NIST AT1e against TAI for about the last 2 years. The long term stability is markedly improved over Figure 1. The short term, from 5 to 20 d, is somewhat improved but still dominated by flicker PM, hence by transfer noise. The improvement in transfer is probably due to the increased use of Two-Way Satellite Time and Frequency (TWSTFT) [1] among a few of the labs, since common-view GPS time transfer has not improved significantly in the last few years. Unfortunately, TWSTFT is expensive both in equipment costs and staff time. It is unlikely that this method will be used by most links contributing to TAI.

Many receiver systems have temperature dependant delays in the receivers, the antenna cables, and the antennas[3,7]. The impedance matching in the antenna cable can produce slowly changing instabilities of the order of tens of nanoseconds [9]. Hence, it is important to routinely calibrate receiver delays against each other, by doing common-clock experiments with a travelling receiver.

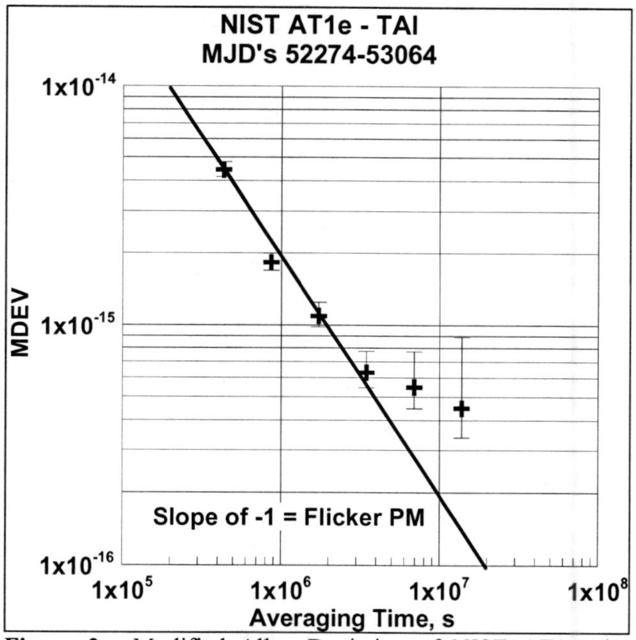

Figure 2: Modified Allan Deviation of NIST AT1e, the NIST post-processed H-Maser scale, against TAI. The data cover over 2 years, from December 31, 2001 to February 29, 2004.

2. OP – NIST Calibrations

One of the longest histories of such calibrations is that between NIST, Boulder, Colorado, USA, and OP, Paris, France. The primary receivers at the two institutions have remained the same over that time: NBS10 at NIST, and NBS51 at OP. Since these receivers have been used to link the respective labs UTC offsets, changes in delays correspond to corrections that need to be added to the difference in the UTC values.

We present the results of these calibrations in both Table 1 and Figure 3, in the form of the correction needing to be added to UTC(NIST) – UTC(OP). We give the corrections from calibrations along with their uncertainties since the original NBS-type receiver was first placed at OP in July, 1983.

We see that the variation in delays is 6.3 ns over 20 years. We also see there may be changes up to 4.4 ns in periods of several months, January to April 1988, though the uncertainties on those calibrations overlap. Also note that the two calibrations of 2003 have a good agreement. There seems to be a consistency in the differential delay over the last few years.

Date	*d*/ns	*u(d)*/ns
July 1983	0.0	2.0
September 1986	0.7	2.0
October 1986	–1.4	2.0
January 1988	–3.8	3.0
April 1988	0.6	3.0
March 1995	–3.7	1.0
May 1996	–0.7	1.5
May 2002	–5.0	3.0
July 2003	–5.6	1.9
December 2003	-5	3

Table 1. Some past calibrations between NIST and OP: *d* are differential time corrections to be added to [*UTC*(NIST)-*UTC*(OP)], and *u(d)* are estimated uncertainties for the periods of comparisons.

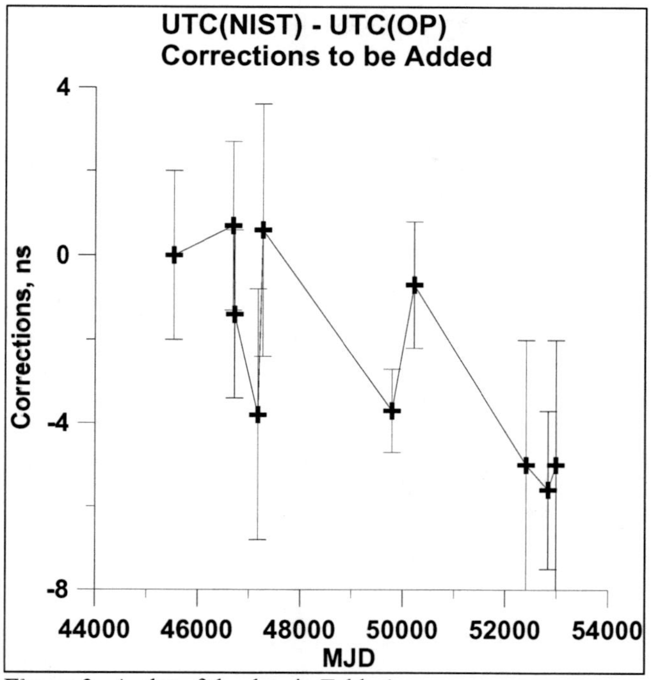

Figure 3: A plot of the data in Table 1.

3. Local Common-Clock Comparisons

There have been many GPS receivers at both NIST and BNM-SYRTE of OP. We show here the deviations between some of them in each lab. Figure 4 gives the variations of several receivers at BNM-SYRTE against the long-term primary receiver, NBS51. Each point in the figure is computed every day from a linear fit of the common-views between the receivers, which follow the BIPM International Tracking Schedule. These are all NBS-type receivers, although produced by various manufacturers.

As indicated in the figure, receiver data have been offset to separate the time series. We see that over many years receiver delays generally changed by only a few nanoseconds, with some exceptions. There was a step in the data of VSL15 against NBS51 that lasted over a year. There were some large deviations at the end of the A400 data against NBS51, which exhibit the end of life of A400. Annual terms of amplitude 5 ns appear in the A263 data against NBS51 and at the end of the TTR01 data against NBS51, of amplitude 2.5 ns. This might partly come from the common NBS51 reference receiver, since the phase is similar. Such an annual variation can be detected only from long-term studies as this one.

Figure 5 shows data for receivers at NIST against NBS10, the primary receiver. We see a significant difference in stability among receivers. Some of the multi-channel receivers (labelled M1, M2, and M3) vary significantly over the year 2003. The NIST M3 receiver, however, appears to be as stable as the NBS-type in long term. The M3 receiver needs to be calibrated, however. Again we see that the best receivers vary under 5 ns over a year. This is consistent with previously published data. Local common-clock data at NIST were previously published spanning the years 1991 – 1997 [10]. That paper reported typical variations of about 5 ns over a year, often with an annual signature. There were, of course, larger deviations at times.

4. Conclusions

The delays of the most stable GPS common-view time transfer receivers vary typically by a few nanoseconds over years, generally by less than 5 ns peak-to-peak. Some changes in the receiver delays, such as annual terms, can be detected only from such long-term studies. Receiver calibrations are essential to maintain accuracy and long-term stability in TAI. However, TAI is currently dominated by common-view time transfer instabilities from 5 to 30 d. We need a significantly more stable time-transfer system than the current GPS common-view system, if we are to compare the clocks contributing to TAI at shorter averaging times. TWSTFT is significantly more stable, but requires more resources as well.

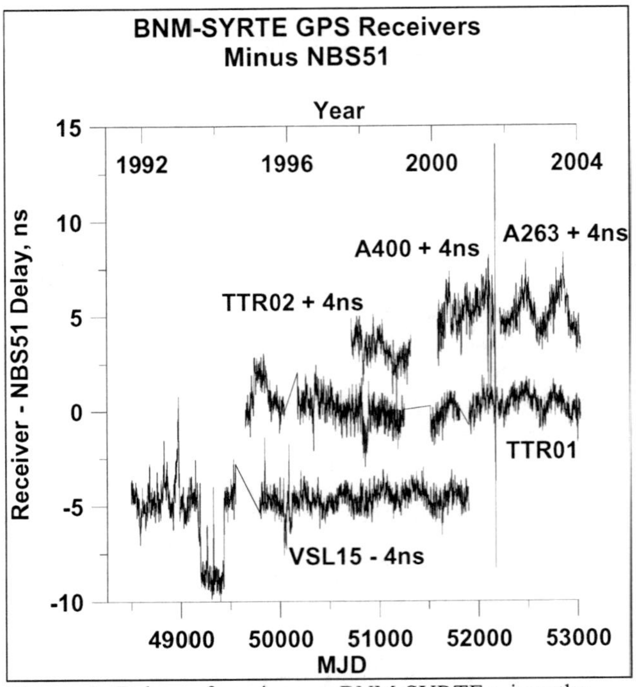

Figure 4: Delays of receivers at BNM-SYRTE minus the delay of NBS51, the primary receiver. The data have offsets added to them as indicated. Note that the vertical spike after MJD 52000 is solely from the data of the A400 receiver against NBS51.

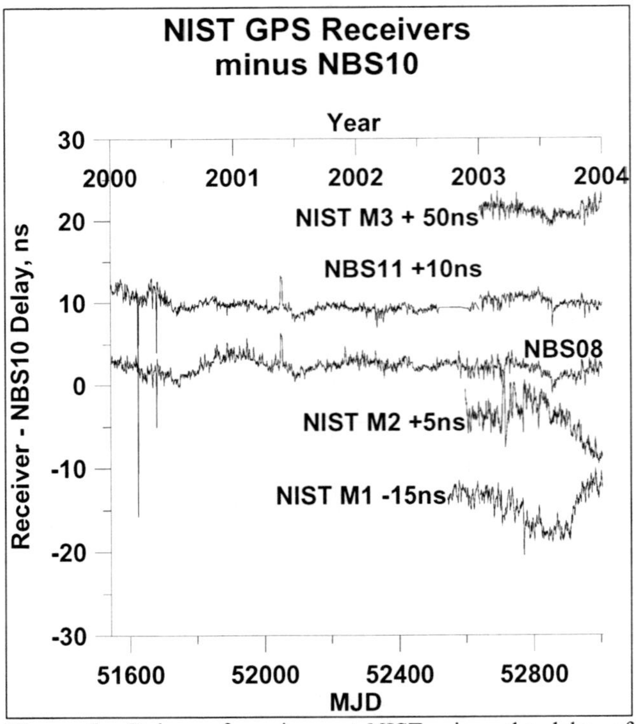

Figure 5: Delays of receivers at NIST minus the delay of NBS10, the primary receiver. NBS08 has no offset added to it, though other receivers have offsets as indicated.

Acknowledgements

We thank Dr. T. Parker and A. Novick for help in reading the paper. We report here data covering decades. We thank also the many people who have contributed to the long term development of GPS timing receivers. At NIST, OP and the BIPM the names include: D. Davis, R. Tourde, G. Fréon, C. Thomas, and D.W. Allan.

References

[1] J. Azoubib, W. Lewandowski, "Uncertainties of time links used for TAI," *Proc. 34th annual Precise Time and Time Interval (PTTI) Applications and Planning Meeting*, 2002, pp. 413-422 (2003).

[2] W. Lewandowski, J. Azoubib, "Time Transfer and TAI," *Proc. 2000 IEEE International Frequency Control Symposium (IFCS)*, pp 586-597 (2000).

[3] W. Lewandowski, R. Tourde, "Sensitivity to external temperature of some GPS time receivers," *Proc. 22nd annual Precise Time and Time Interval (PTTI) Applications and Planning Meeting*, 1990, pp. 307-316 (1991).

[4] C. Thomas, "The accuracy of International Atomic Time TAI," *Proc. 11th European Forum on Time and Frequency(EFTF)*, 1997, pp. 283-289 (1997).

[5] T.E. Parker, "Performance of a Hydrogen Maser Ensemble and Characterization of Frequency Standards," in *Proc. 1999 Joint Mtg. IEEE IFCS and EFTF Conf.*, pp. 173-176 (1999).

[6] T.E. Parker, D.A. Howe, and M.A. Weiss, "Accurate Frequency Comparisons at the 10^{-15} Level," *Proc. 1998 IEEE IFCS.*, pp. 265-272 (1998).

[7] G. Petit, C. Thomas, Z. Jiang, P. Uhrich, and F. Taris, "Use of GPS Ashtech Z12T Receivers for Accurate Time and Frequency Comparisons," *IEEE Trans. on Ultrasonics, Ferroelectrics, and Frequency Control*, **vol. 46**, #4, July 1999, 941-949 (1999).

[8] D.B. Sullivan, D.W. Allan, D.A. Howe, F.L. Walls, ed., *Characterization of Clocks and Oscillators*, NIST Tech Note 1337, U.S. Government Printing Office, Washington D.C., USA, (1990).

[9] M. A. Weiss, "Long Term Effects of Antenna Cables on GPS Timing Receivers," *Proc. 2000 IEEE IFCS*, pp. 637-641 (2000).

[10] M.A. Weiss, V. Zhang, L. Nelson, V. Hanns, M.G.L. Regalado, "Delay Variations in Some Timing Receivers," *Proc. 1997 IEEE IFCS*, pp 304-312 (1997).

OSCILLATOR STABILITY TESTS ON BOARD ARMORED VEHICLES

Heikki Heikkila[1], Pekka Eskelinen[2], Jukka Ruoskanen[3], Jouni Peltonen[3], Arvi Serkola[3], Pekka Hautala[2] and Petri Kuosmanen[2]

[1]Armoured Brigade, FINLAND
[2]Helsinki University of Technology, FINLAND,
[3]Defense Force's Technical Research Centre, FINLAND
Email: heikki.heikkila@mil.fi

Keywords: test arrangement, oscillator test, armoured vehicle, vibration and acceleration

Abstract

In several field tests and experiments, the authors have shown that armoured vehicles are demanding platforms of electronic or microwave oscillators. Recent measuring efforts evaluating these problems are related to the armoured vehicular installation itself and stability issues. The paper shows some tested alternatives, test and recording arrangements and selected examples of measured three dimensional vibration spectra.

1. Introduction

Much of the performance of military electronics is coupled to the characteristics of respective time- or frequency-defining oscillators [1]. However, the environment of mobile armoured vehicle oscillators, either used in communication or computation systems or in radars [2], may well be one of the harshest [3]. Temperature extremes are frequently met. Vehicle speeds exceed 80 km/h also in severe terrain. Only coarse suspension is typically at hand and thus shocks and vibration tend to be a major concern [4]. This is particularly true if we e.g. want to maintain a synchronous fast frequency hopping VHF radio network, where we have to take care of a number of interacting oscillators [5]. A similar situation arises if our mobile radar performs moving target processing down to low Doppler frequencies [6] and we still need to maintain full speed due to tactical reasons. Additional stress of varying vibration spectra is caused by the weapon launch. Already proper protection of electronics against destruction due to brute collisions requires careful design. Unfortunately indeed, signal intelligence can make use of distorted oscillator spectra as a way of identifying units individual weapon platforms through their electronic fingerprints [7]. Despite armoured vehicles are physically large when compared to many electronics packages under consideration, we seldom find enough empty and unobstructed mounting space inside [8] whereby shock absorbing or vibration damping gets complicated.

2. Armoured vehicle as an oscillator platform

The authors have experimented with various armoured platforms in oscillator stability tests. An example of demanding vehicles is the Russian T-72 main battle tank, because of limited space and the severe vibration of platform.

Figure 1: T-72 main battle tank

There is not enough space inside the armored vehicle for mounting electronic systems or test equipment. For additional measurement devices is the turrets upper surface the only suitable mounting place.

Vibrations of the platform were measured in field tests carried out in true operational environment. There were 7 acceleration analyzer mounted on the turret of the vehicle (Fig. 2). The acceleration values were measured in typical operational situations.

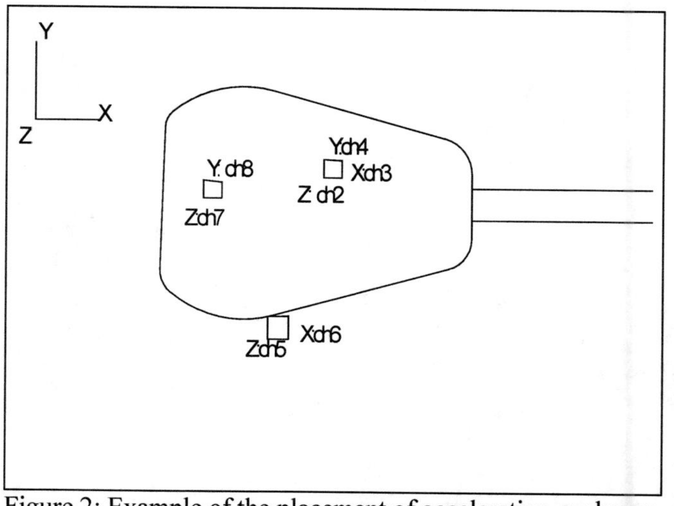

Figure 2: Example of the placement of acceleration analyzers on the turret surface

Measured vertical acceleration peak values exceed 10 m/s^2 with averages of 5-7 m/s^2 and typical time constants of 0.5-2 seconds, of course depending on terrain profile and speed. Momentary frequency offsets have been found to be up to 1×10^{-5} and shock-based drift around 1×10^{-7}/s.

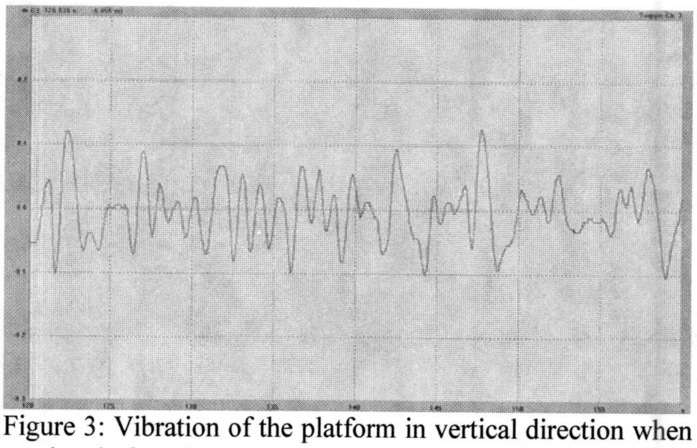
Figure 3: Vibration of the platform in vertical direction when the platform is moving in the operating environment

3. Oscillator stability test arrangements

Laboratory instruments and indoor vehicle test platforms are everyday facilities used in military oscillator work. Computer simulation tools enabling circuit development and mechanical vibration analysis have been available for a couple of decades. More recently we have seen dynamic simulation tools for moving vehicles. Despite these advances, oscillator performance evaluations continuously require extensive and well-prepared experimenting out in the field, in the particular foreseen operating environment. This implies that the frequency-generating device under test is seldom tailored to the available mobile platform and we have to take unprotected hardware out to the range. Often quite provisional arrangements are needed in order to make such experiments [9].

In oscillator stability tests the circumstances were really difficult. Tests were carried out in the desert. Temperature in test field varied between -10 to + 10 °C. During this research, all environmental problems concentrated to moisture, variation in temperature and dust. To remove some of these problems, the computers and the other measuring devices were placed in a container or in warmed tent.

Figure 4: Container and tent for the research

The most demanding part of the arrangements was the electric power system. Electric power was produced by two aggregates. The aggregates were 4 kVA and 8 kVA. One aggregate (4 kVA) provided supply for the measuring devices and the other (8 kVA) for heaters, lightning and etc.

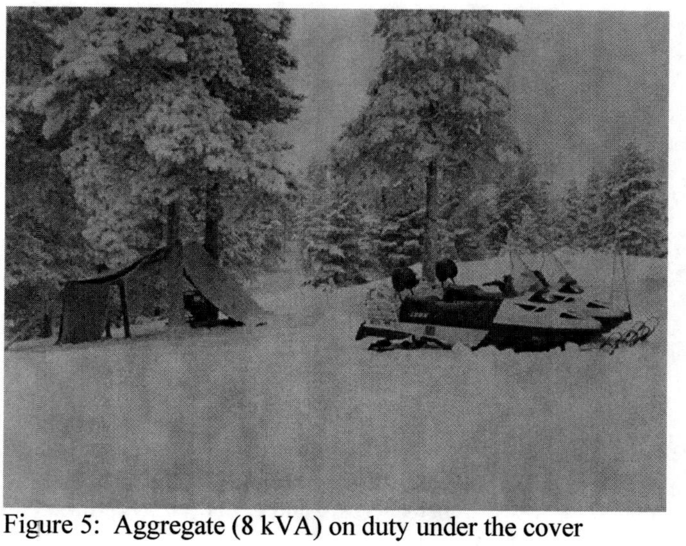
Figure 5: Aggregate (8 kVA) on duty under the cover

Figure 6: Aggregate (4 kVA) on duty on board the vehicle

Figure 8: Transmitter mounting on the main gun from the front side

4. Test transmitter integration to the armoured vehicle

Vibration is a demanding challenge, when a moving armoured vehicle is the oscillator stability tests platform. To eliminate vibration and for an exact alignment, the tank's stabilized main gun remains the only mounting place for the transmitter. By using the vehicle's fire control system, it is possible to point the transmitter beam exactly towards the receiver even when the vehicle is moving. In that way the authors could perform the experiments fluently.

Figure 7: The test transmitter was integrated on the armoured vehicle's main gun.

To integrate the transmitter to the main gun, a special stand had to be designed. The stand keeps the transmitter stable. Battery supply was included in the box of the transmitter.

Figure 9: Transmitter mounting on the main gun from the right side

5. Stability test

Oscillator reactions were tested in five situations. Those situations are most common in operational environment.

- standing vehicle (engine off)
- standing vehicle (engine on)
- forward moving vehicle
- reward moving vehicle
- forward moving vehicle (full speed)

In all cases the test arrangements were identical. The transmitter was integrated on the tank's main gun and the receiver was on a distant tripod. The receiver was aligned towards the moving vehicle as is shown in figure 10.

Figure 10: A view of the oscillator stability test runs

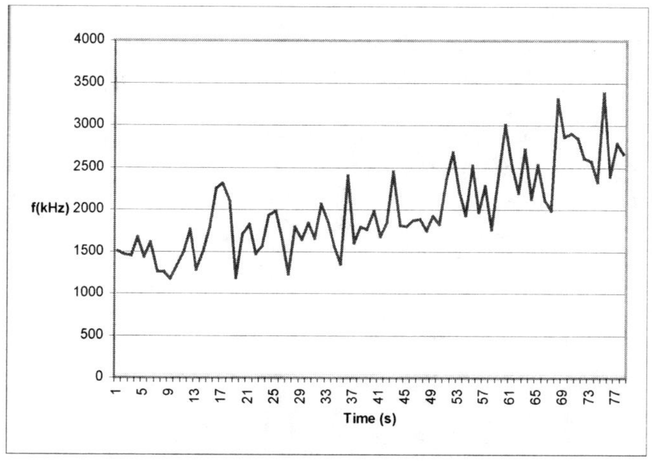

Figure 13: Forward moving vehicle

6. Results

The results of measurements in different operational situations are shown in figures 11 to 15. They describe frequency variations as a function of time.

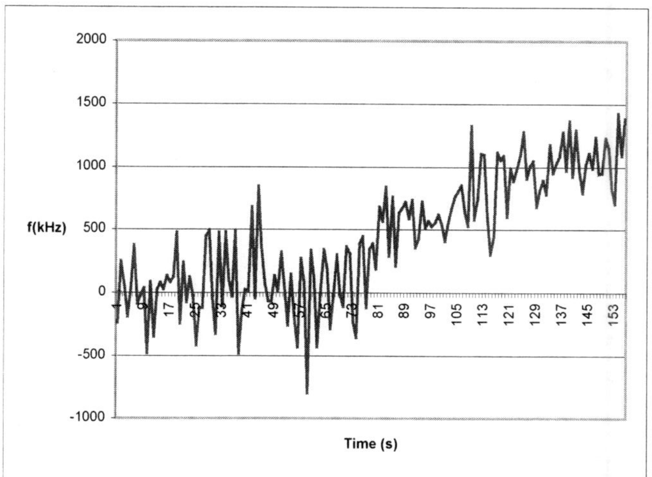

Figure 11: Standing vehicle (engine off)

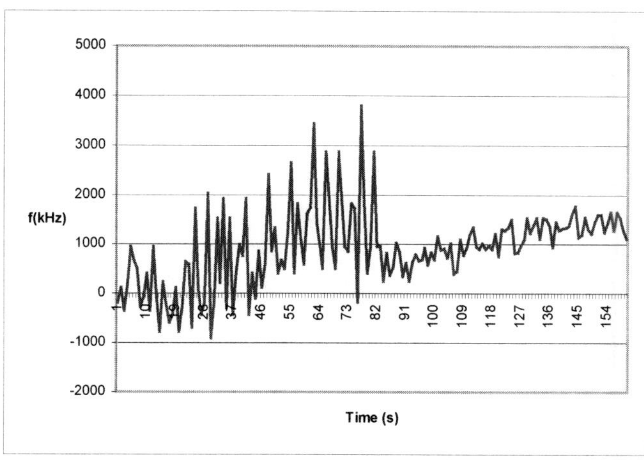

Figure 12: Standing vehicle (engine on)

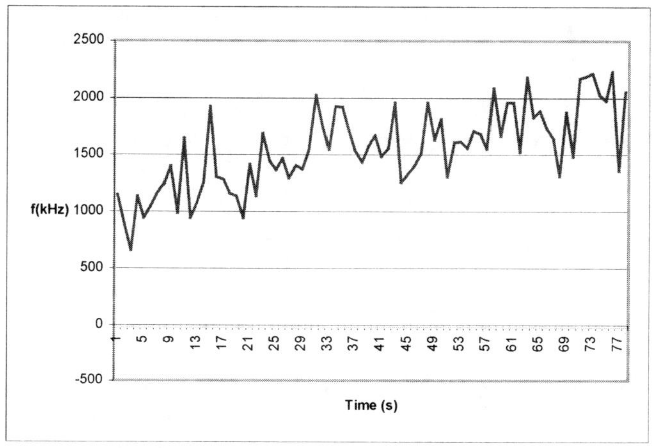

Figure 14: Reward moving vehicle

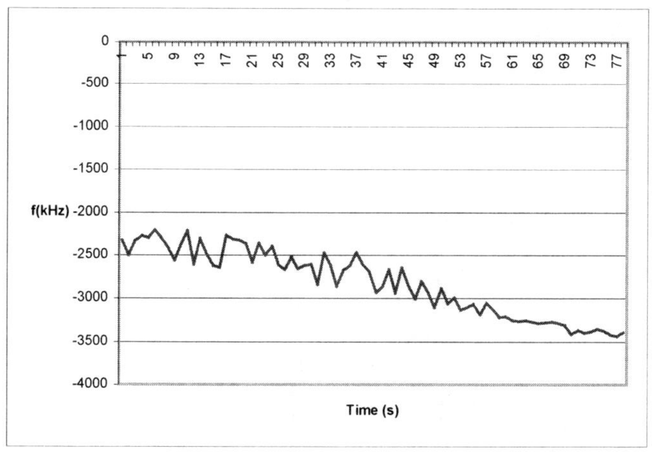

Figure 15: Forward moving vehicle (full speed)

The results are summarized in figure 16. The light curve presents frequency difference in ppm and the darker curve indicates vehicle motion. In chronological order steps are:

1. standing vehicle (engine off)
2. standing vehicle (engine on)
3. forward moving vehicle
4. reward moving vehicle
5. forward moving vehicle (full speed)

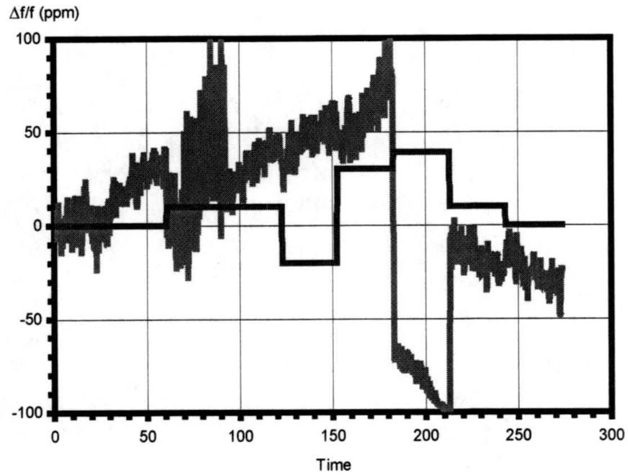

Figure 16: Results of the oscillator stability test

7. Conclusions

The hypothesis, that armoured vehicles are some of the harshest oscillator installation platforms seems to be right. The tests which were made in true operational environments and with an original armoured vehicle show that the most difficult situation for oscillator stability is forward moving with full speed in severe terrain.

References

[1]Eskelinen P. and Eskelinen H., "Shock tests with a phase locked crystal oscillator", Proc. 14th EFTF, March 2000,Turin, pp. 324-328

[2]Lacomme P., Hardange J., Marchais J., and Normant E., Air and Spaceborne Radar Systems. Norwich, SciTech Publishing, 2001

[2]Neubig B, and Briese W., Das Grosse Quarzkochbuch, Feldkirchen, Franzis-Verlag, 1997, p. 116

[3] Heikkila H., Eskelinen P., Ruoskanen J. and Hautala P., "Upgrading Armored Vehicle Sensor Systems", IEEE Aerospace and Electronic Systems Magazine, vol 19, Number 2, 2004, 6 p.

[4]Kihara M., Ono S. and Eskelinen P., Digital Clocks for Synchronization and Communications, Norwood, Artech House, 2002

[5]Chevalier F., Principles of Radar and Sonar Signal Processing, Norwood, Artech House, 2002, p. 272

[6]Vakin S., Shustov L., and R. Dunwell, Fundamentals of Electronic Warfare. Norwood, Artech House, 2001

[7]Ruoskanen J., Eskelinen P., and Heikkila H., "Target Detection Trials with A Millimeter Wave Radar System", IEEE Aerospace and Electronic Systems Magazine, vol 18, Number 11, 2003

[8]Eskelinen H., and Eskelinen P., Microwave Component Mechanics, Norwood, Artech House 2003, pp. 165-167

[9]Eskelinen P., "Experiments with a cesium-locked time and frequency reference transmitter", Proc. 14th EFTF, March 2000, Turin, pp 204-208

TIME EFFECTIVE METHODS OF CALCULATION OF DYNAMIC PARAMETERS OF SYNCHRONIZATION SIGNAL

Andrzej Dobrogowski, Michal Kasznia

Institute of Electronics and Telecommunications
Poznan University of Technology
ul. Piotrowo 3A, 60-965 Poznan, Poland
dobrog@et.put.poznan.pl, mkasznia@et.put.poznan.pl
phone: +48 61 6652 293, fax: +48 61 6652 572

Keywords: timing signal, Allan deviation, time deviation

Abstract

In the paper the time effective methods of calculation of dynamic parameters of synchronization signals are proposed. First the idea of dynamic Allan deviation (DADEV) and dynamic time deviation (DTDEV) is presented. Then the methods of data preprocessing enabling short in time calculation of this parameters are described. The results of DADEV and DTDEV calculation performed for several different data sequences are presented and compared. The best solutions of computation procedure considering the accuracy and time of calculation are suggested.

1 Introduction

Allan deviation ADEV and time deviation TDEV are the parameters, which describe the quality of synchronization signal in the telecommunication network. The parameters allow to assess the variations of time interval provided by the synchronization signal and recognize the type of phase noise affecting the signal. The estimates of the parameters are calculated for a series of observation intervals using the sequence of time error samples measured at the network interface.

A new approach to the evaluation of synchronization signal using Allan deviation was proposed by L. Galleani and P. Tavella at the Joint Meeting of IEEE FCS and EFTF in 2003 [7]. This approach called dynamic Allan deviation (DADEV) consists in the calculation of the Allan deviation values for the segments (slices) of time error sequence with changing initial point of calculation. The results of calculation are plotted in the form of 3D graph and allow to recognize the variations of the phase noise affecting the analyzed signal. Such computation procedure causes the increase in time of calculation.

In the paper the time effective methods of calculation of dynamic Allan deviation DADEV and dynamic time deviation DTDEV in the telecommunication applications are presented. The methods consist in the initial preprocessing of the whole time error sequence used for the calculation. The squares of appropriate differences of time error samples are calculated first. Then the calculations of the parameter for each segment with shifting initial instant are performed using the preprocessed data. The procedure proposed allows to save the time of calculation in opposite to the time consuming independent calculation for each segment using raw time error data. The methods of time error data preprocessing based on averaging and decimation presented by authors of this paper at EFTF in 2000 and 2001 (used for time effective ADEV and TDEV computation) [5, 6] are also used in the proposed procedure of the dynamic deviations calculation. The results of calculation of the dynamic parameters performed with and without data preprocessing for several data sequences are presented and discussed. The relations between the length of time error data sequence, data segments (slices) and maximum observation intervals, important for reliable parameter estimation in the telecommunication applications (fulfilling the telecom standards and recommendations) are also discussed in the paper.

2 Idea of dynamic ADEV and dynamic TDEV

The idea of dynamic parameters is simple. Instead of one curve representing the values of ADEV or TDEV as a function of observation interval τ, the set of the curves plotted in the form of three-dimensional graph as a function of observation interval τ and time t is considered. As result we can recognize the changes of the type of phase noise affecting analyzed timing signal. Changing slope of the parameter's curve indicates the changes of the noise type [7].

In order to obtain this form of graph, the following computation procedure is performed. First, the data sequence (time series of time error measured at some network interface) is divided into equal segments (slices) with the length of T_s. We can consider overlapping data segments (Fig. 1) and non-overlapping data segments (Fig. 2). The arrangement of the segments depends on the number of segments, their length and the length of the whole data sequence. Then the calculation of the parameter's value for required range of observation interval for each data segment is done. The results of calculation are plotted in a form presented in Fig. 3.

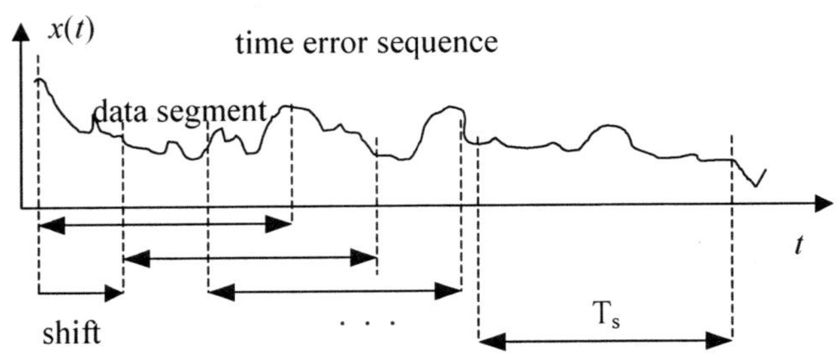

Fig. 1. Overlapping segments of time error sequence

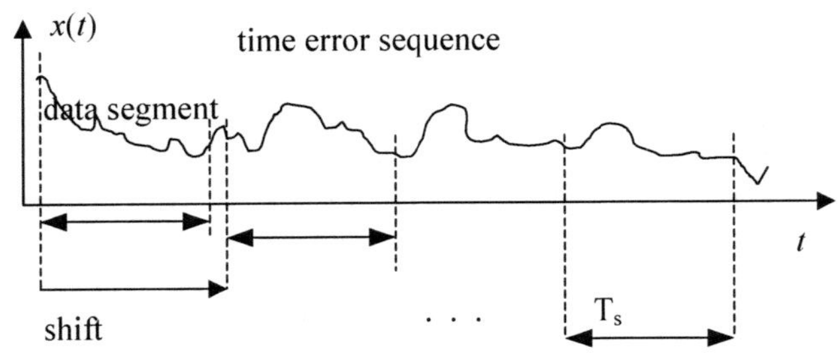

Fig. 2. Non-overlapping segments of time error sequence

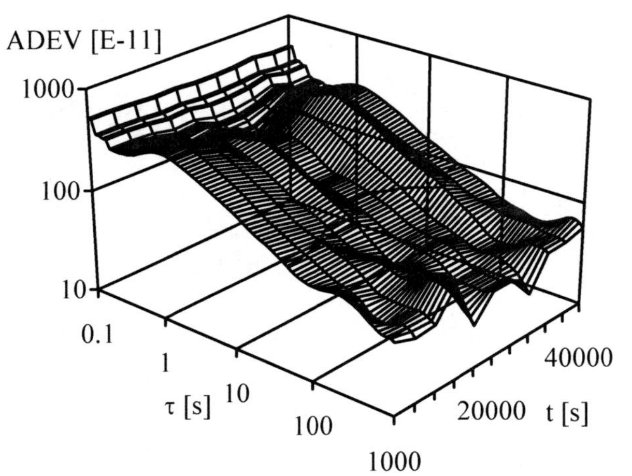

Fig. 3. Example plot of dynamic Allan deviation

Computing the dynamic ADEV and the dynamic TDEV we must take into consideration the relations between the length of time error data sequence T, length of data segments (slices) T_s, and maximum observation interval τ_{max}. Theoretically, the observation time T should be infinite. The telecommunication standards and recommendations require, that the length of data sequence used for ADEV or TDEV calculation must be 12 to 15 times longer than the maximum observation interval τ_{max} [1, 2, 3]. In practice, the value of the parameter's estimate can be computed, if the data sequence length is two times longer than τ_{max} for ADEV and three times longer for TDEV. Unfortunately, in this case we obtain poor quality of the parameter's estimate.

It may happen, that the length of data segment T_s, does not fulfills the telecom recommendations, i.e. T_s is shorter than twelve times τ_{max}. In such situation we have two choices. The first choice is to calculate the parameter's value for each segment for assumed range of observation intervals (including τ_{max}), but realizing rather poor quality of the estimate for longer observation intervals (of course, T_s has to be two or three times longer than τ_{max}). The second choice is to shorten the maximum observation interval τ_{max} to the value of $T_s/12$ (in accordance with recommendations) and to calculate the parameter's value for smaller range of observation intervals.

3 Time effective methods of DADEV and DTDEV calculation

The formulae for the estimators of the Allan deviation ADEV and the time deviation TDEV take the form:

$$A\hat{D}EV(\tau)=\sqrt{\frac{1}{2n^2\tau_0^2(N-2n)}\sum_{i=1}^{N-2n}(x_{i+2n}-2x_{i+n}+x_i)^2} \quad (1)$$

$$T\hat{D}EV(\tau)=\sqrt{\frac{1}{6n^2(N-3n+1)}\sum_{j=1}^{N-3n+1}\left[\sum_{i=j}^{j+n-1}(x_{i+2n}-2x_{i+n}+x_i)\right]^2} \quad (2)$$

where $\{x_i\}$ is a sequence of N samples of time error function $x(t)$ taken with interval τ_0; $\tau=n\tau_0$ is an observation interval.

The amount of time error samples necessary for the reliable estimation (when the recommendations are fulfilled) and the complication of the estimator formulae result in rather very long computing time of the estimates. The problem of time-consuming computation becomes even more serious in the case of the dynamic ADEV and dynamic TDEV, if the calculations are performed independently for each data segment. Therefore we propose some procedures, which make the computation process time effective.

3.1. Initial preprocessing of the whole data sequence – square preprocessing

Before the parameter's value calculation for a current observation interval $\tau=n\tau_0$ (i.e. current n), the new data sequence is created. The item in the sequence created for ADEV computation has a form:

$$a_i=(x_{i+2n}-2x_{i+n}+x_i)^2 \quad (3)$$

For TDEV computation the estimator formula (2) is changed in order to simplify the summing [4, 8] and takes the form:

$$T\hat{D}EV(n\tau_0)=\sqrt{\frac{1}{6}\cdot\frac{1}{N-3n+1}\cdot\frac{1}{n^2}\sum_{j=1}^{N-3n+1}S_j^2(n)} \quad (4)$$

where

$$S_j(n)=S_{j-1}(n)-x_{j-1}+3x_{j+n-1}-3x_{j+2n-1}+x_{j+3n-1}$$

and

$$S_1(n)=\sum_{i=1}^{n}(x_{i+2n}-2x_{i+n}+x_i)$$

Therefore the item of the new sequence is:

$$a_j = S_j^2 \quad (5)$$

The preprocessing is performed for the whole data sequence. Then the calculation of the parameter for the current observation interval for each segment is performed. As result, only the operation of summing (using previously computed squares of appropriate differences) is performed. In this situation the order of computation is changed. The calculation of the parameter's value for the current n for each data segment is performed first. Then the data preprocessing as well as the parameter's value computation for the next observation interval are carried out.

3.2. Independent computation with data reduction

In order to reduce the calculation time for the observation intervals greater than 10 s, the reduction of the data sequence is performed. The new sequence contains decimated data (in the case of ADEV calculation) or averaged data (in the case of TDEV calculation). The new sequence is 10 times shorter then the original one for the observation intervals greater than 10 s and 100 times shorter for the observation intervals greater than 100 s. Notice, that small error (about 5%) may appear as the results of reduction [5, 6]

3.3. Data reduction with square preprocessing

Both methods presented above may be joined together. In the first step the data reduction using decimation (for ADEV) or averaging (for TDEV) is performed. Then the new sequence containing squares of appropriate differences using previously reduced data is created.

4 Experiment

The methods of data preprocessing presented in the Section 3 were examined in the calculation experiment. Two cases of dynamic parameters' computation were considered in the experiment: computation with overlapping and with non-overlapping data segments. In the first case we considered 11 overlapping data segments with the length of 4000 s and the shift of 800 s. The data sequences (short series) have the length of 12000 s in this case. In the second case we considered 11 non-overlapping data segments with the length of 4000 s and the shift of 4000 s. The data sequences (long series) have length of 44000 s in this case. The time error samples were measured with the sampling interval τ_0=1/30 s for each sequence. Both parameters, dynamic ADEV and dynamic TDEV were computed for each case and each data sequence. The parameter's values were computed for 81 observation intervals from τ_{min}=0.1 s till τ_{max}=1000 s for each data segment. The personal computer with Pentium IV 1.4 GHz microprocessor was used for computations.

Several data sequences were used for DADEV and DTDEV computation. First two sequences (denoted as WPM1 and WPM2) represent the samples of white phase noise, that changes its standard deviation for some period in the middle of sequence. The length of inserted subsequence is 2000 s for the short data series (WPM1) and 8000 s for the long data series (WPM2). The WPM1 data series is presented in Fig. 4. The results of DADEV and DTDEV calculation are presented in Fig. 5 and Fig. 6 respectively.

Fig. 4. Short data sequence representing white phase noise with abrupt change of its standard deviation (WPM1)

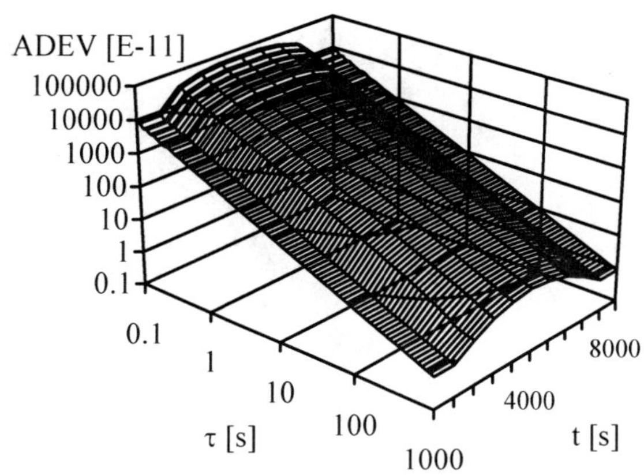

Fig. 5. DADEV for the WPM1 data sequence

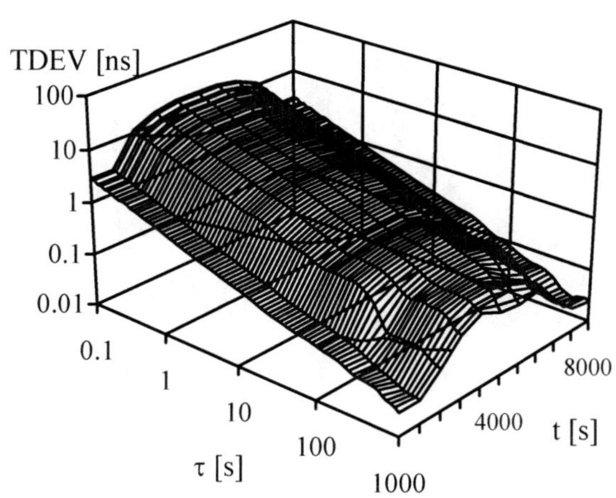

Fig. 6. DTDEV for the WPM1 data sequence

The WPM2 data series is presented in Fig. 7. The results of DADEV and DTDEV calculation are presented in Fig. 8 and Fig. 9 respectively.

Fig. 7. Long data sequence representing white phase noise with abrupt change of its standard deviation (WPM2)

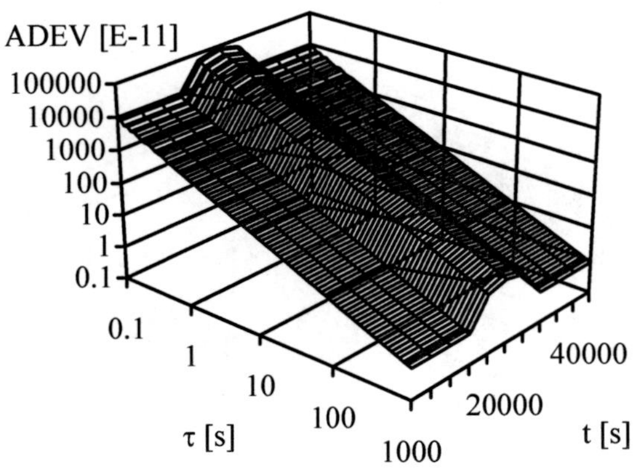

Fig. 8. DADEV for the WPM2 data sequence

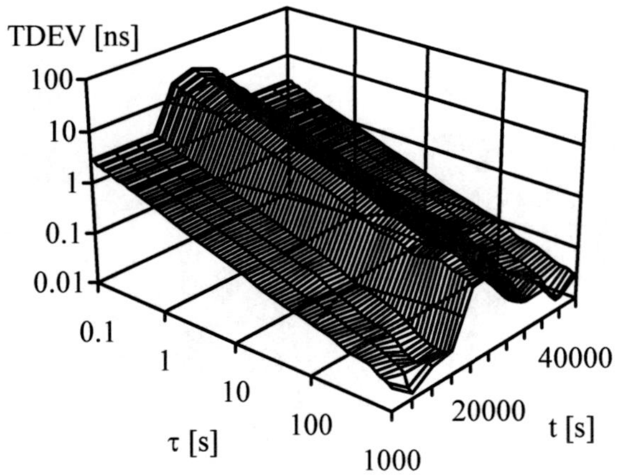

Fig. 9. DTDEV for the WPM2 data sequence

Next pair of sequences (denoted as WPFM1 and WPFM2) represents white phase noise which changes its type to the white frequency noise in the middle of the sequence. The WPFM1 data series is presented in Fig. 10. The results of DADEV and DTDEV calculation are presented in Fig. 11 and Fig. 12 respectively.

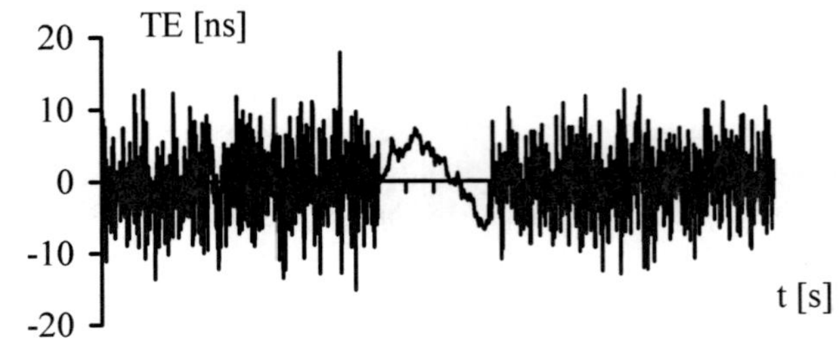

Fig. 10. Short data series representing white phase noise having inserted white frequency noise (WPFM1)

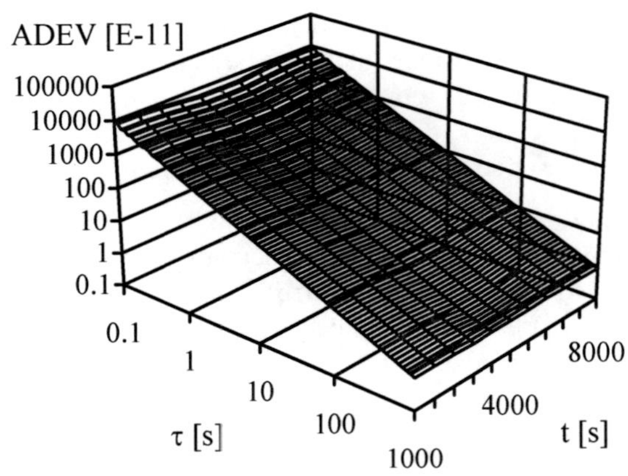

Fig. 11. DADEV for the WPFM1 data sequence

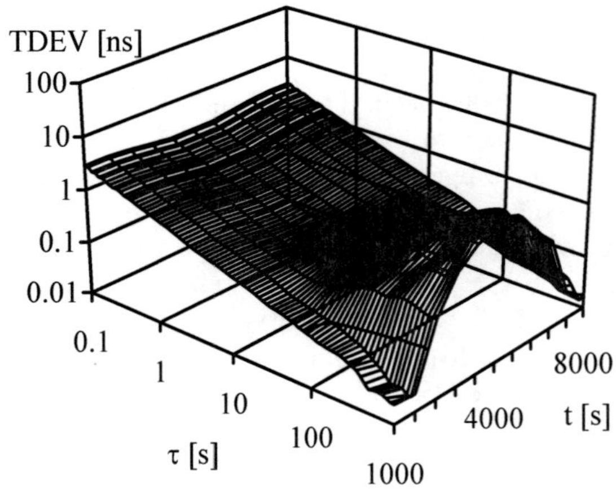

Fig. 12. DTDEV for the WPFM1 data sequence

The WPFM2 data series is presented in Fig. 13. The results of DADEV and DTDEV calculation are presented in Fig. 14 and Fig. 15 respectively.

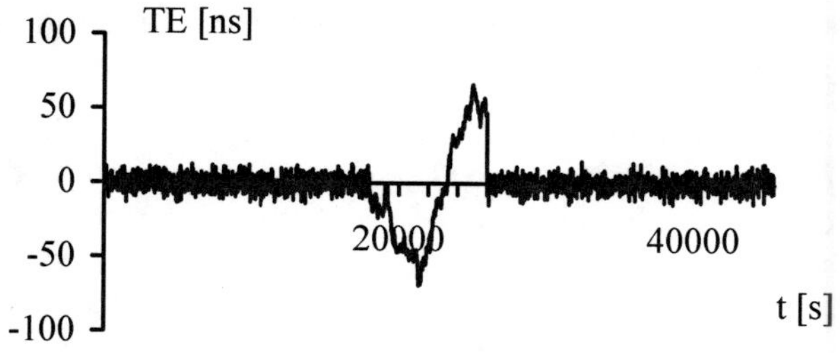

Fig. 13. Long data series representing white phase noise having inserted white frequency noise (WPFM2)

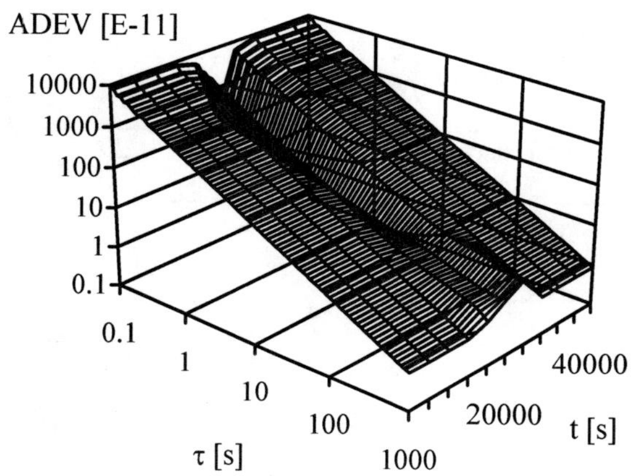

Fig. 14. DADEV for the WPFM2 data series

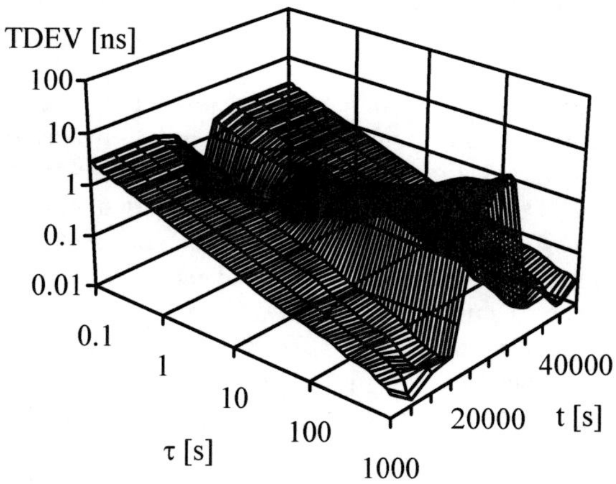

Fig. 15. DTDEV for the WPFM2 data series

Last pair of sequences (denoted as GPS1 and GPS2) comes from the comparisons of an oscillator working as a time server disciplined by GPS signals with another GPS oscillator working as a reference. The GPS1 short data series is presented in Fig. 16. The results of DADEV and DTDEV calculation are presented in Fig. 17 and Fig. 18 respectively.

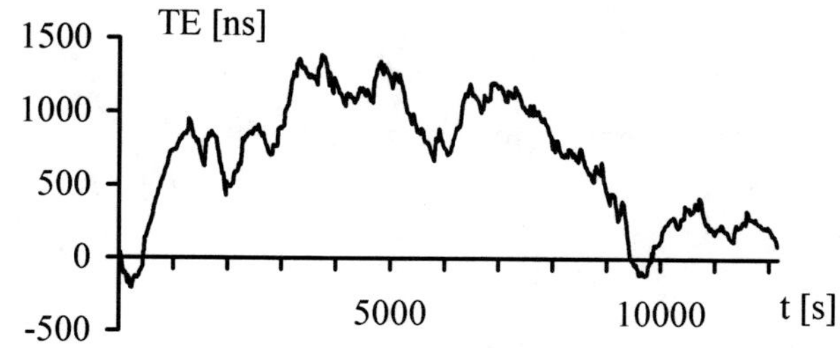

Fig. 16. Short data series from GPS measurement (GPS1)

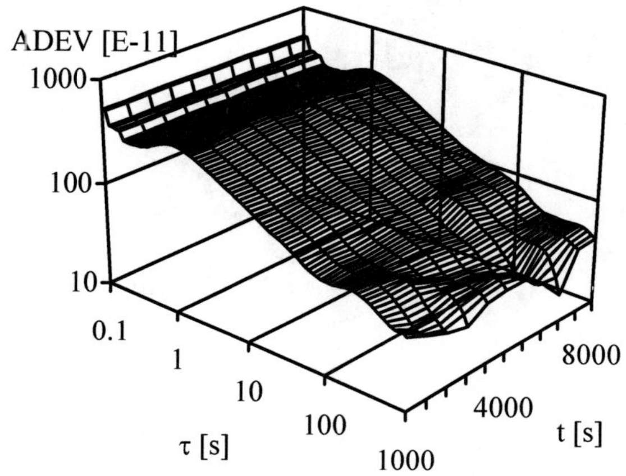

Fig. 17. DADEV for the GPS1 data series

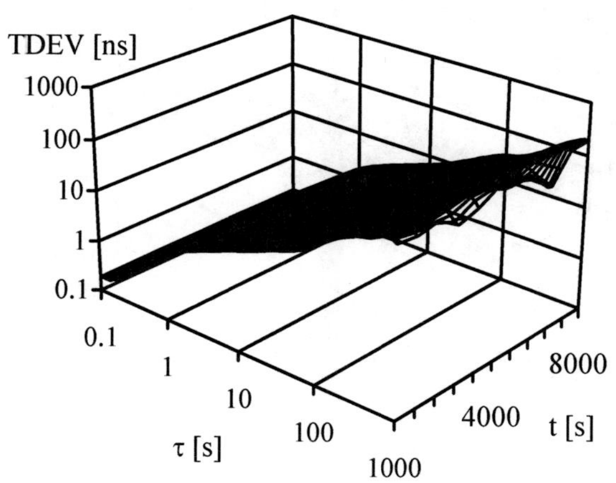

Fig. 18. DTDEV for the GPS1 data series

The GPS2 long data series is presented in Fig. 19. The results of DADEV and DTDEV calculation are presented in Fig. 20 and Fig. 21 respectively.

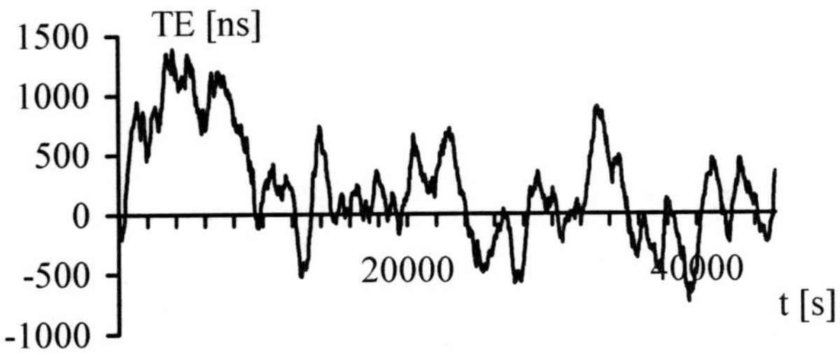

Fig. 19. Long data series from GPS measurement (GPS2)

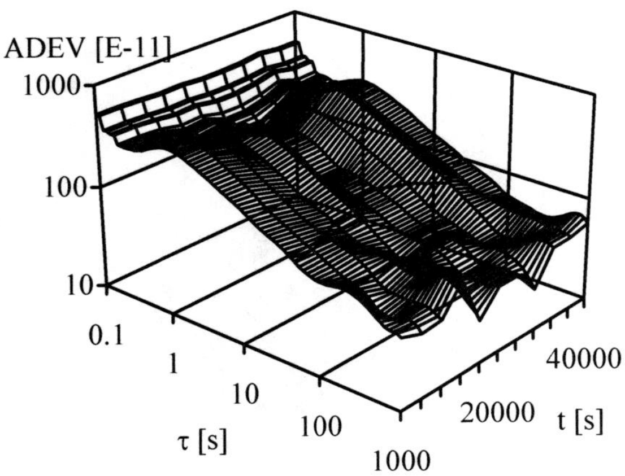

Fig. 20. DADEV for the GPS2 data series

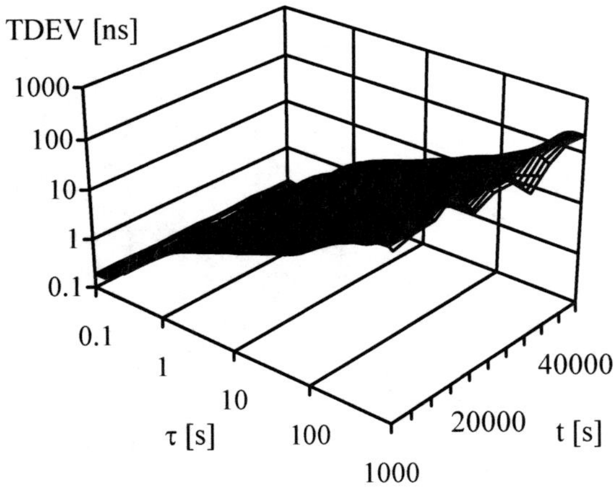

Fig. 21. DTDEV for the GPS2 data series

Changes of the surface on the 3D plots representing dynamic parameters (especially in the case of DTDEV, Fig. 9, 12, and 15) allow to recognize the changes of the phase noise type. The plots of dynamic TDEV computed for non-overlapping data segments are better than DADEV for these goals of analysis of timing signal.

The computations of DADEV and DTDEV for the data sequences presented above were performed using the methods proposed in the Section 3. The time of the parameters' computation is presented in the Table 1 – the case of overlapping segments – and in the Table 2 – the case of non-overlapping segments. For comparison in the Tables the time of computation without data preprocessing and data reduction is included.

method of computation	time of computation [min:sec]	
	DADEV	DTDEV
independent	39:01	52:39
with square preprocessing	20:50	24:10
with square preprocessing and data reduction	11:25	13:30
independent with data reduction	21:46	29:42

Table 1: Time of DADEV and DTDEV computation for overlapping data segments

method of computation	time of computation [h:min:sec]	
	DADEV	DTDEV
independent	38:44	52:05
with square preprocessing	56:01	1:10:07
with square preprocessing and data reduction	30:06	37:50
independent with data reduction	21:53	29:43

Table 2: Time of DADEV and DTDEV computation for non-overlapping data segments

Each method of data preprocessing (square preprocessing, data reduction and both joined together) results in shortening of the calculation time in the case of overlapping data segments. The shortest time was obtained for calculation with squares preprocessing and data reduction (decimation for ADEV and averaging for TDEV). The time of computation in this case was 11 min 25 s for DADEV and 13 min 30 s for TDTEV. It is almost four times shorter then in the case of independent computation. In the case of non-overlapping segments the square preprocessing causes longer calculation time than independent calculation. The data being the results of preprocessing are not used repeatedly in the calculation process, in opposite to the case of overlapping segments. Additional operations in the preprocessing procedure cause longer overall computation time. The shortest time was obtained for the independent computation with data reduction.

5 Conclusions

The dynamic approach to timing signal evaluation using Allan deviation and time deviation allows recognizing the variations of the phase noise affecting the analyzed signal.

The data preprocessing using initial computation of squares of differences enable shorter parameter calculation time only in the case of overlapping data segments. In the second case, for non-overlapping data segments this type of preprocessing is not time effective.

Application of data reduction methods brings good results in both cases – overlapping and non-overlapping data segments. In general we suggest as a right choice:

- calculation with square preprocessing and data reduction in the case of overlapping data segments;
- independent calculation (without square preprocessing) and data reduction in the case of non-overlapping data segments.

References

[1] ANSI T1.101-1999, Synchronization interface standard

[2] ETSI EN 300 462, "Generic requirements for synchronization networks" (1998).

[3] ITU-T Rec. G.810, "Considerations on timing and synchronization issues" (08/96).

[4] S. Bregni, "Synchronization of Digital Telecommunications Networks", J. Wiley & Sons, 2002

[5] A. Dobrogowski, M. Kasznia, "Modification of the raw data set for time efficient ADEV and TDEV assessment", Proceedings of the 14th European Frequency and Time Forum, pp. 338-342, Torino 14-16 March 2000.

[6] A. Dobrogowski, M. Kasznia, "Impact of time error data preprocessing for TDEV assessment", Proceedings of the 15th European Frequency and Time Forum, pp. 463-467, Neuchatel 6-8 March 2001.

[7] L. Galleani, P. Tavella, "The characterization of clock behavior with the dynamic Allan variance", Proceedings of the 2003 IEEE International Frequency Control Symposium and 17th European Frequency and Time Forum, pp. 239-244, Tampa 5-8 May 2003.

[8] M. Kasznia, "Some Approach to Computation of ADEV, TDEV and MTIE", Proceedings of the 11th European Frequency and Time Forum, pp. 544-548, Neuchatel 4-6 March 1997.

HIGHLY-ACCURATE REAL-TIME SYNTONIZATION BY USE OF A SINGLE-FREQUENCY RECEIVER WITH GPS CARRIER PHASE

C.L. Cheng*, F.R. Chang*, L.S. Wang, K.Y. Tu*****

* Dept. of Electrical Engineering, National Taiwan University, Taiwan
** Inst. of Applied Mechanics, National Taiwan University, Taiwan
*** NSTF Lab., TL, Chunghwa Telecom. Co., Ltd., Taiwan
RM207, Dept. of Electrical ENG, National Taiwan University, Taipei 10617, Taiwan, R.O.C.
Tel: +886-2-23635251ext. 207, Fax: +886-2-23638247, Email: jialung-cheng@umail.hinet.net

Keywords: *GPS, Syntonization, Carrier Phase*

Abstract

The purpose of this paper is to develop a low-cost, highly-accurate and real-time syntonization system composed of a single-frequency receiver with GPS carrier phase. For averaging time of one day under the configuration, our experimental results showed that the accuracy of the steered clock could be improved from about two parts in 10^9 to about three parts in 10^{14}, and the stability of the syntonized clock could be improved from about nine parts in 10^{10} to about four parts in 10^{14}. Three improvements and advantages in our methodology were given. First, we developed a forecasting model with recurrent neural networks to correct atmospheric errors in real time. The model was more accurate than the existed ionosphere model and troposphere model for a single-frequency receiver. Furthermore, it was available simply and anticipative day-to-day partially, irrelevant to user locations. Second, the low-cost clock could be automatically steered to obtain the very high frequency accuracy and stability in the short term as well as in the long term. Experimental results showed the stability of per second was about nine parts in 10^{12}. Compared with the commercial GPS Disciplined Oscillator (GPSDO), the short term stability (1s) was improved about ten times. Third, the frequency performance of the syntonization system by use of low-cost GPS engines, inexpensive clocks and less communication effects was as good as the commercial atomic clock. So, the frame of syntonization was sound, reliable and cost-effective.

1 Introduction

The frequency source (FS) plays a key role in many applications, such as telecommunication networks, power systems, navigation systems, automotive systems, instrumentation systems, and Doppler radars, etc. The capability of using GPS carrier rather than C/A code with the commom-view technique to transfer precise time and frequency has been recognized and described by many researchers [1-3]. Because the frequency of the carrier phase is roughly 1000 times higher than that of C/A code , time and frequency dissemination using carrier phase have much greater resolution in principle. To achieve the highest frequency accuracies from the GPS system, the atmospheric propagation errors shall be neutralized and considered within the adjustment process. Nowadays on the one hand, the atmospheric errors may be minimized by a multi-channel receiver with/or common view method, while they are either more expensive, or less reliable than single-frequency receivers. Therefore we introduce a low-cost and less-communication-dependent reliable accurate frequency calibration or syntonization system composed of a single-frequency receiver with GPS carrier phase. On the other hand, single-frequency receivers usually include a correction for the atmosphere delay based on an ionosphere model and a troposphere model which are built into the GPS system. Nevertheless, these models are expected to remove about 40%~75% of the atmospheric effects on average. Since the parameters of these models are estimated in advance and then transmitted to the GPS satellites, it cannot anticipate day-to-day random fluctuations, and cannot be completely accurate. Alternatively, various organizations develop detailed and accurate models of the atmosphere based on GPS observations for single-frequency users to make atmospheric effects as small as possible. However, these models vary in accuracy, input data, computational complexity, and user location, so the trade-off among the different models depends on the individual circumstances of the user. In order to improve the above drawbacks, we develop a low-cost, highly-accurate and real-time syntonization system composed of a single-frequency receiver with GPS carrier phase. The scheme could achieve the traceability of frequency dissemination. The GPS receiver Ashtech G12 was modified in order to estimate the real-time average frequency deviation of the steered Oven-Controlled Crystal Oscillator (OCXO) with respect to the GPS system time by performing the time-difference, all-satellites-in-view average and atmospheric correction. The neural network MPC (Model Predictive Controller) was employed to implement the syntonization system. The steered clock was then syntonized with the GPS system time by way of the D/A converters.

With the above mentioned method, the steered clock (OCXO)

showed that the accuracy could be improved from about two parts in 10^9 to about three parts in 10^{14}, and the stability of the syntonized clock could be improved from about nine parts in 10^{10} to about four parts in 10^{14} for averaging time of one day. Our experiments revealed that the proposed architecture is sound and cost-effective. The potential roles of our system include the FSs for telecommunication networks, calibration laboratories, power systems, navigation systems, instrument calibration, and others.

This paper is organized as follows:

1) Section 2 presents models of GPS carrier phase observations.
2) Section 3 presents the development of recurrent neural networks for the forecasting model with atmosphere delay.
3) Section 4 describes the system architecture.
4) The experimental results are illustrated in Section 5.
5) Finally, the concluding remarks are given in Section 6.

2 The model of GPS carrier phase observables

The typical model of GPS carrier phase observables [4, 5] is referred to equation (1):

$$\Phi_A^j = \rho_A^j + c(dt^j - dT_A) + \lambda N_A^j - d_{ion}^j + d_{trop}^j + \varepsilon_A^j \qquad (1)$$

where

Φ_A^j: Carrier phase measurement of the receiver A from the j-th GPS satellite;

ρ_A^j: True distance between the receiver A and the j-th GPS satellite;

c: Speed of light;

dt^j: Clock bias of the j-th satellite;

dT_A: Clock difference between the GPS time and receiver A clock;

λ: GPS carrier phase wavelength;

N_A^j: Initial phase integer ambiguity;

d_{ion}^j: Ionospheric delay;

d_{trop}^j: Tropospheric delay;

ε_A^j: Unmodeled errors primarily due to multi-path, temperature variation, physical factors, etc.

The unit of the phase observable Φ_A^j in the equation is meter.

To study the frequency syntonization, we would like to first examine the behavior of the oscillator. Hence, the GPS receiver's internal clock will be replaced by an external one. Under this arrangement, the term dT_A in equation (1) represents the time difference between the GPS clock and the external clock A. In our system, the A (user) site performs the time-difference (differences between two epochs) on carrier phase observations to obtain the phase difference with respect to the GPS system time. If the satellite signal is continuously tracked and there is no cycle slip occurring, the cycle ambiguities N_A^j remain a constant. After the time difference, equation (1) is then rewritten as equation (2):

$$\delta\Phi_A^j = \delta\rho_A^j + c\delta dt^j - c\delta dT_A - \delta d_{ion}^j + \delta d_{trop}^j + \delta\varepsilon_A^j \qquad (2)$$

where $\delta(\cdot)$ denotes the operator for differences between two epochs. The δdT_A represents the phase difference with respect to the GPS system time. The unmodeled ionospheric delay δd_{ion}^j and tropospheric delay δd_{trop}^j cannot be eliminated and they are regard as error of the frequency offset.

On the other hand, the B (host) site, stratum 1, installs the same GPS receiver and performs the time difference that is the same as the A (user) site. Additionally, the A (user) site performs the single difference between two receivers at the post processing. The two receivers are denoted by A and B, respectively. The satellite is denoted by j. After the time-difference and the single-difference, the equation (2) is further adapted for equation (3):

$$\Delta\delta\Phi_{AB}^j = \Delta\delta\rho_{AB}^j - c\Delta\delta dT_{AB} + \Delta\delta\varepsilon_{AB}^j \qquad (3)$$

where $\Delta(\cdot)$ represents the operator for differences between receivers with the same satellite. Due to the strong correlation between the un-modeled ionospheric and tropospheric delays of the two receivers in the local area, the terms d_{ion}^j and d_{trop}^j in equation (1) are then eliminated. Since dT_A and dT_B are both referring to the same GPS time, their difference $\Delta\delta dT_{AB}$ in (3) is the phase difference between external clock A and external clock B.

3 Recurrent neural networks

Atmospheric forecasting is an example of a signal processing problem which is challenging due to small sizes, high noise, non-stationarity, and non-linearity. Neural networks have been very successful in a number of signal processing applications and pattern recognition applications. The prediction of future events from noisy time series data is commonly done using various forms of statistical models [6].

3.1 Predicting noisy time series data

For noisy time series prediction, neural networks typically take a delay embedding of previous input which is mapped into a prediction. However, high noise, high non-stationarity time series prediction is fundamentally difficult for the method:

1) The problem of learning is fundamentally ill-posed, i.e. there are infinitely many models which fit the training data well, but few of these generalize well. In order to form a more accurate model, it is desirable to use as large a training set as possible. However, for the case of highly non-stationary data, increasing the size of the training set results in more data with statistics that are less relevant to the task at hand being used in the creation of the model.

2) The high noise and small datasets make the models prone to over-fitting. Random correlations between the inputs and outputs can present great difficulty. The methods typically do not explicitly address the temporal relationship of the inputs – e.g. they do not distinguish between those correlations which occur in temporal order, and those which do not.

In our approach we consider recurrent neural networks (RNNs). Recurrent neural networks employ feedback connections and have the potential to represent certain computational structures in a more parsimonious fashion [7]. RNNs address the temporal relationship of their inputs by maintaining an internal state. RNNs are biased towards learning patterns which occur in temporal order - i.e. they are less prone to learning random correlations which do not occur in temporal order.

The use of a recurrent neural network is important from the viewpoint of the curse of dimensionality because the RNN can take into account greater history of the input. Trying to take into account greater history with a multi-layer perceptron (MLP) by increasing the number of delayed inputs results in an increase in the input dimension. This is problematic, given the small number of data points available. The use of a scaling function for the series conversion is also important from the view point of the efficiency of learning. As will be seen later, the output of the scaling function always fell within a specified range into a single input for the RNNs.

3.2 System II details

A high-level block diagram of system II used is shown in figure 1. The raw time series values of phase error function are $\Phi(k), k = 1,2,...., N$ where $\Phi(k)$, phase error function $\in \Re$ These denote the phase error time series of the receiver external clock. The first difference for the time series of phase error function, $\delta\Phi(k)$, is taken as follows:

$$\delta\Phi(k) = \Phi(k) - \Phi(k-1) \tag{4}$$

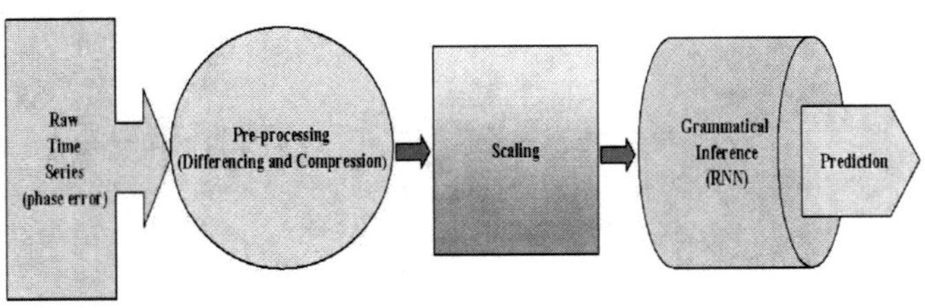

Figure 1: A high-level block diagram of system II used

This produces $\delta\Phi(k), \delta\Phi(k) \in \Re, k = 1,2,..., N-1$. In order to compress the dynamic range of the series and reduce the effect of the outliers, a log transformation of data is used:

$$m(k) = sign\ (\delta\Phi(k))(\log(\ |\delta\Phi(k)| + 1)) \tag{5}$$

resulting in $m(k), k = 1,2,..., N-1, m(k) \in \Re$. As is usual, a delay embedding of this series is then considered [8].

$$\mathbf{M}\ (k; d_1) = (m(k), m(k-1), m(k-2),..., m(k-d_1+1)) \tag{6}$$

where d_1 is the delay embedding dimension and equal to 1 for the experiments reported here. $\mathbf{M}(k;d_1)$ is a state vector. This delay embedding forms the input to the scaling function. Hence, the scaling function input is a vector of the last d_1 values of log transformed differenced time series. The output of the scaling function is always fall within a specified range. Then, the neural network training can be made more efficient on the network inputs and targets.

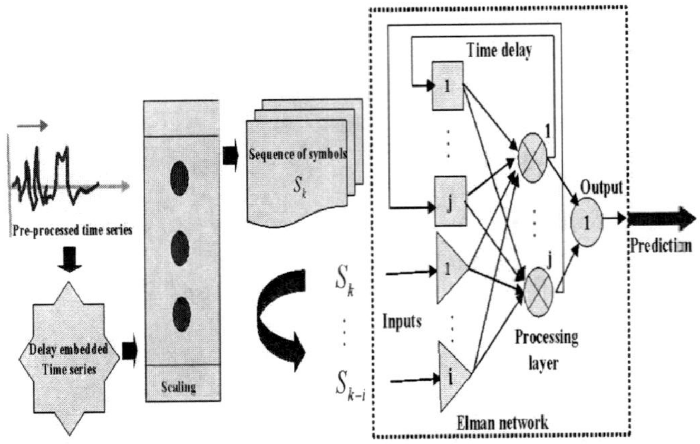

Figure 2: An Elman neural network is trained on sequence of scaling time series (The pre-processed, delay embedded series is converted into specified range using a scaling function).

The scaling function can be represented by the following equation [9]:

$$S(k) = g(\mathbf{M}\ (k, d)) \tag{7}$$

where $S(k) \in [1,2,3,..., n_s]$ and n_s is the number of embedded time series for the scaling function. If the scaling function is used to preprocess the training set data, then each time the trained network is used with new inputs they should be preprocessed with the same parameters of scaling that were computed for the training set.

An Elman recurrent neural network is then used which is trained on the sequence of outputs from the scaling function. The Elman network was chosen because it is suitable for the grammatical inference style problem, and because it has been shown to perform well in comparison to other recurrent architectures. The Elman neural network has feedback from each of the hidden nodes to all of the hidden nodes, as shown in figure 2. For the Elman network:

$$\mathbf{O}(\ k+1) = \mathbf{C}^{T}\mathbf{z}_{\mathbf{k}} + \mathbf{C}_{0} \tag{8}$$

and

$$\mathbf{z}_{\mathbf{k}} = F_{n_h}(\mathbf{A}\mathbf{z}_{\ \mathbf{k}-1} + \mathbf{B}\mathbf{u}_{\ \mathbf{k}} + \mathbf{b}) \tag{9}$$

where $\mathbf{C}$ is a $n_h \times n_0$ vector representing the weights from the hidden (processing) layer to the output nodes, n_h is the number of the hidden nodes, n_0 is the number of the output nodes, $\mathbf{C_0}$ is a constant bias vector, $\mathbf{z_{K,}} \in \Re^{n_h}$, is an $n_h \times 1$ vector, denoting the outputs of the hidden layer neurons. $\mathbf{u_k}$ is a $d_2 \times 1$ vector as follows, where d_2 is embedding dimension used for the recurrent neural network:

$$\mathbf{u_k} = \begin{bmatrix} S(k) \\ S(k-1) \\ \vdots \\ S(k-d_2+1) \end{bmatrix} \tag{10}$$

$\mathbf{A}$ and $\mathbf{B}$ are matrices of appropriate dimensions which represent the feedback weights from the hidden layer to the hidden notes and the weights from the input layer to the hidden layer respectively. F_{n_h} is a $n_h \times 1$ vector containing the sigmoid functions. One benefit of sigmoid function is that it reduces the effect of extreme input values, thus providing some degree of robustness to the network. The $\mathbf{b}$ is a $n_h \times 1$ vector, denoting the bias of each hidden layer neuron. $\mathbf{O}(k)$ is a $n_0 \times 1$ vector containing the outputs of the network. n_0 is 1 throughout this paper. Thus, for the completed system:

$$\mathbf{O}(k+1) = F_1(\delta\Phi(k), \delta\Phi(k-1), ..., \delta\Phi(k-4)) \tag{11}$$

which can be considered in terms of original time series:

$$\mathbf{O}(k+1) = F_2(\Phi(k), \Phi(k-1), ..., \Phi(k-5)) \tag{12}$$

Note that this is not a static mapping due to the dependence on the hidden state of the recurrent neural network. However, the forms of the nonlinear function F_1, F_2 are not apparent from the neural network.

3.3 Atmospheric delay forecasting of RNNs

We propose a new methodology of atmospheric forecasting for a single-frequency GPS receiver. In order to develop a low-cost, highly-accurate and real-time syntonization system composed of a single-frequency receiver with GPS carrier phase, we construct the hybrid system with RNNs for divining in real time. A hybrid-system block diagram with RNNs is shown in figure 3. After the process of initialization and modeling, system I generates the time series with atmospheric delay variation according to carrier phase and the receiver parameters at the moment. The real-time estimation time series are then sent to system II and system III by all satellites in view averaging. System II would be ability to predict the change of atmosphere delay in real time relatively, and the time series of output are as another input of system III. So, the real –time synthesized anticipation of the atmosphere delay is available by way of system II and the best average frequency offset correctional estimation may change significantly over time as a result of the non-stationary of the series through system III. Due to less sensors of atmosphere measurement for a single-frequency receiver, the hybrid forecasting system with RNNs provides the syntonization system with more accurate atmospheric correction compared with the existed method in GPS receivers. Besides, this system reduces the communication and interdependence between host site and user site. Therefore, the sytonization system by performing highly-accurate correction individually is sound, reliable and cost-effective at most of time.

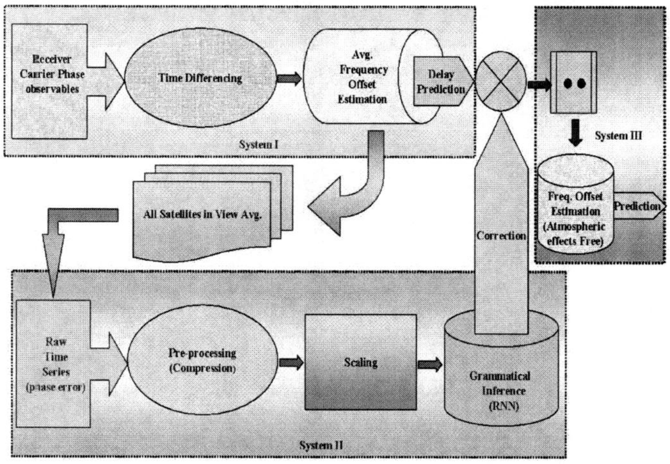

Figure 3: A hybrid-system block diagram of RNNs

4 The architecture of syntonization system

4.1 Low-cost syntinozation system

The functional block diagram of our syntonization system is shown as in Figure 4. This system includes the DDS (Direct Digital Synthesizer), a D/A converter, a low-cost modified G12 GPS receiver, a forecasting model of atmosphere correction, and a notebook PC mainly. In order to estimate the offsets of the steered clock with respect to the GPS system time, the user clock is connected to the modified GPS receiver. Hence, the original internal quartz oscillator in the receiver is replaced. By the help of the frequency synthesizer, i.e. the DDS (Direct Digital Synthesizer) manufactured by NOVATECH, the signal of the external clock can be appropriately converted and supplied to the GPS receiver. The behavior of the clock then comes into view from the GPS observations. The average frequency offset of the syntonized clock with respect to the GPS system time can be estimated by performing the time difference (i.e., (2)) on carrier phase observations. Equation (2) can be further expressed as equation (13):

$$\delta\Phi_A^j - \delta\rho_A^j - c\delta dt^j + \delta d_{ion}^j - \delta d_{trop}^j = -c\delta T_A + \delta\varepsilon_A^j \tag{13}$$

The left-hand side of the equation (13) is time difference data of the known measurement. The coordinates of the GPS antenna are predetermined by IGS (International GPS

Service), and the coordinates and the satellite clock error of the j-th GPS satellites are obtained from the broadcast navigation message.

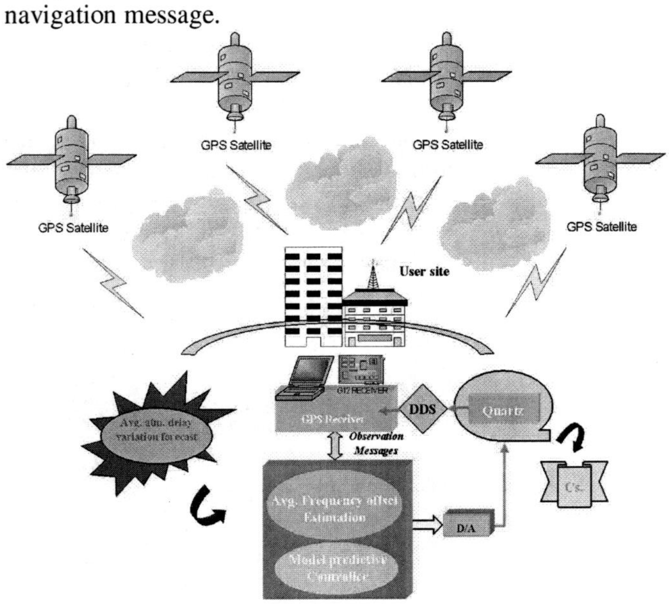

Figure 4: The functional block diagram of our syntonization system

Furthermore, the ionospheric delay (δd^j_{ion}), the tropospheric delay (δd^j_{trop}) would be eliminated from atmospheric delay forecasting model composed of RNNs. Some noise errors affecting the estimation of the average frequency offset may occur in the evaluation of the right-hand side. This term $\delta dT_A(t_i)$ ($= dT_A(t_i) - dT_A(t_{i-1})$, where $t_i = t_{i-1} + \tau$) can be obtained by averaging (13) for all in view GPS observations. As previously mentioned, since $\delta dT_A(t_i)$ is the phase difference between the GPS system time, the associated estimate frequency offset $\hat{y}'_r(t_i)$ is shown in equation (14):

$$\hat{y}'_r(t_i) = \frac{\delta dT_A(t_i)}{\tau} - \frac{\delta\varepsilon^j_A}{c\tau} \tag{14}$$

In general, the fine frequency tuning can be performed on the inexpensive oscillator through voltage control. Because of the environmental effects such as vibration, temperature, pressure, and humidity, the desired frequency output is not always under a constant voltage.

In our system, the frequency offset $\hat{y}_r(t_i)$ is chosen as the input variable of the model predictive controller. An incremental voltage $\Delta v(t_i)$ will be generated to update the voltage for steering the oscillator.

$$v_{i+1} = v_i + \Delta v \tag{15}$$

For the low-cost quartz or rubidium oscillator, the frequency offset may vary over very large scales. The log transformation of data is used to compress the values of the estimation $\hat{y}_r(t_i)$. The Compression function is shown as in equation (16):

$$w(t_i) = sign\,(\hat{y}_\tau(t_i))(a\log(|\hat{y}_\tau(t_i)| + b)) \tag{16}$$

where a and b are the scaling factor and the offset factor, respectively. We decide to adopt model predictive controller (MPC) for steering the external clock, because the MPC can systematically take into account real plant constraints in real time. It is also robust with respect to modeling errors, over and under parameterization, and sensor noise [10].

The neural network predictive controller that is implemented in the highly-accurate real-time sytonization uses a neural network model of a nonlinear plant to predict future plant performance. The controller then calculates the control input that will optimize plant (OCXO) performance over a specified future time horizon. The first step in MPC is to determine the neural network plant model. Next, the plant model is used by the controller to predict future performance. The following section will describe the system identification process.

The first stage of model predictive control is to train a neural network to represent the forward dynamics of the OCXO. The prediction error between the compressed signal of the OCXO output phase error and the compressed signal of the neural network output is used as the neural network training signal. The process is represented by the following figure 5. The neural network plant model uses previous inputs and previous plant outputs to predict future values of the plant output. The network can be trained offline in batch mode, using data collected from the operation of the plant.

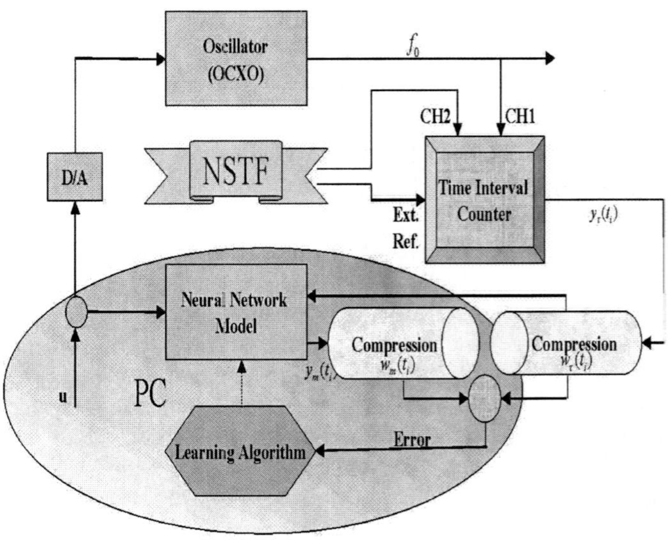

Figure 5: The process of the MPC system identification

The second stage is predictive control. The neural network model predicts the plant response over a specified time horizon. The predictions are used by numerical optimization program to determine the control signal that minimizes the following performance criterion over the specified horizon.

$$J=\sum_{k=N_1}^{N_2}(y_m(t+k)-y_r(t+k))^2+\rho\sum_{k=1}^{N_u}(u'(t+k-1)-u'(t+k-2))^2 \quad (17)$$

where N_1, N_2 and N_u define the horizons over which the tracking error and the control increments are evaluated. The u' variable is the tentative control signal, y_r is desired response and y_m is the network model response. The ρ value determines the contribution that the sum of the squares of the control increments has on the performance index.

The following block diagram illustrates the model predictive control process. The controller consists of the neural network plant model and the optimization block. The optimization block determines the values of u' that minimize J, and then the optimal u is input to the plant.

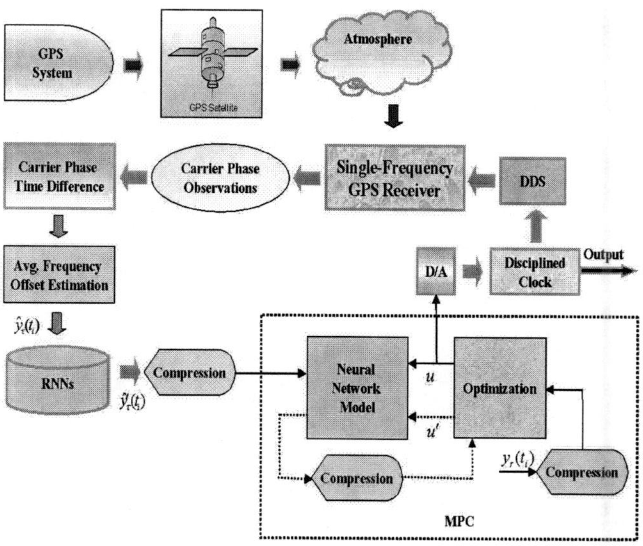

Figure 6: The model predictive control block diagram

4.2 Initialization and modeling

On the other hand, the host (master) station also contains high-performance frequency source with national standard of time and frequency, the direct digital synthesizer (DDS), an modified G12 GPS receiver, and an industrial PC which are the same as the user site owns. The carrier phase data and other GPS observations messages are then passed to the user site through communication network, e.g. the PSTN or the internet. The average frequency offset of the user clock with respect to the master clock can be estimated by performing the time difference (i.e. (2)) and then single difference (i.e. (3)) on carrier phase observations. In the process, the biases and errors from satellites and receivers can be significantly reduced. Equation (3) can be further expressed as follows:

$$\Delta\delta\Phi_{AB}^{j} - \Delta\delta\rho_{AB}^{j} = -c\Delta\delta dT_{AB} + \Delta\delta\varepsilon_{AB}^{j} \quad (18)$$

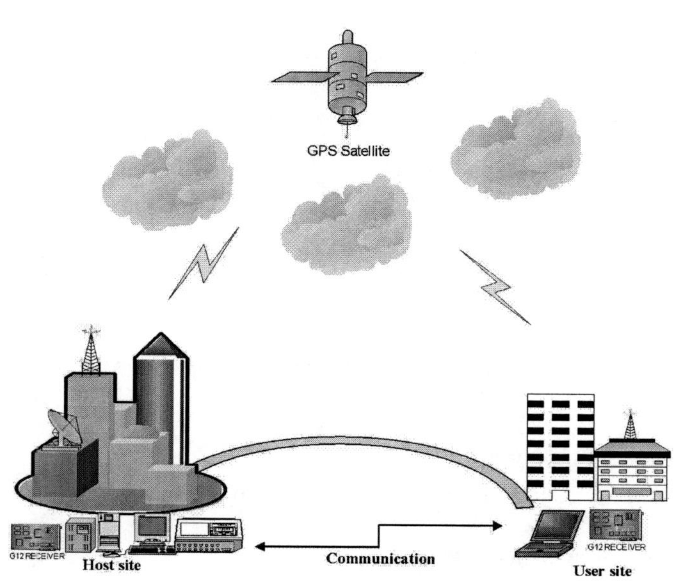

Figure 7: The initialization and modeling architecture of atmosphere

This term $\Delta\delta dT_{AB}(t_i)(=\delta dT_A(t_i)-\delta dT_B(t_i)$, where $t_i=t_{i-1}+\tau$) can be obtained by averaging equation (18) for all-in-view GPS observations. Then, the associated estimate average frequency offset $\hat{y}_r''(t_i)$ is:

$$\hat{y}_r''(t_i) = \frac{\Delta\delta dT_{AB}(t_i)}{\tau} - \frac{\Delta\delta\varepsilon_{AB}^{j}}{c\tau} \quad (19)$$

To attain the low-cost highly-accurate training data of atmospheric variation for common users, we can first obtain the frequency stability and accuracy by performing equation (14) in real-time processing at both sites. Next, we acquire the reference destination of atmosphere modeling by subtracting $\hat{y}_r''(t_i)$ from $\hat{y}_r'(t_i)$ in the post-processing at user site. Since the master station possesses the same degree of time and frequency as GPS system time does. The initialization and modeling architecture of atmosphere is shown as in figure 7. Therefore, our model will not vary in accuracy by user location. The model creation is adapted to individual circumstances of the user site. The correction method of the syntonization system is then more accurate and easily-used than the other models for a single-frequency receiver.

The difficulty with our approach is the reduction in the already small number of training data. The size of the training set controls a noise vs. non-stationary trade-off [11]. If the training set is too small, the noise makes it harder to estimate the appropriate mapping. If the training set is too large, the non-stationary of data will mean more data with statistics that are less relevant for the task at hand will be used for creating the estimator. The entire training/test set window is moved forward 1 day and process is repeated, as depicted in figure 8. For the problem of noisy data, we use early stopping to control over-fitting.

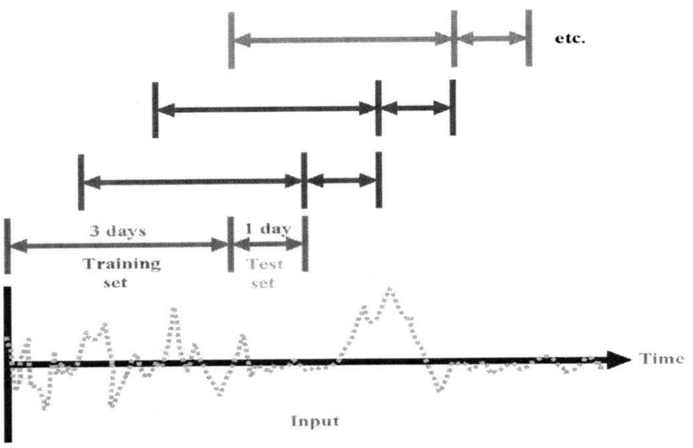

Figure 8: A depiction of the training and test sets used

5 Experimental results

The basic experimental structure for tests is shown as figure 4 in Section 4.1. The Ashtech ™ G-12 single-frequency GPS receivers installed in our system were not designed for time and frequency applications. They have no interface ports for external clocks. In order to use the G-12 receivers to establish the system of frequency syntonization, we replaced the 20.460MHZ internal quartz oscillator of the receiver with the external frequency source and connected into a G-12 receiver through a DDS manufactured by NOVATECH™ , model DDs5m. The oven controlled crystal oscillator manufactured by Datum™, model FTS 1130, was used as syntonized clock. The software, including the model predictive controller and forecasting model of RNNs, communication interfaces between time interval counter (TIC) and PC, data collection, were programmed in C++ language on Professional Windows 2000 and executed on an industrial PC manufactured by NOVATECH™ or a mobile computer manufactured by Acer. The data used for frequency accuracy and stability analysis were measured every one second by a TIC manufactured by SRS™ , model SR620.

The IEEE recommended modified Allan deviation (MDEV) was adopted for the analysis of frequency stability under our experiments. The confidence intervals applied to the frequency stability analysis were $\pm 68\,\%$ in our tests. In addition, we estimated the frequency accuracy by performing the linear-least-square fit on phase difference data over the observable intervals. The linear-least-square fit, in fact, is commonly used in many frequency calibration or syntonization systems, such as the FMAS of the NIST (U.S.).

We examined the performance of the free running OCXO used in our system and a GPSDO manufactured by Trimble ThunderBolt^{TM} , in order to compare the results with the performance of the controlled OCXO under atmospheric correction and the common view (CV) method. These frequency stability analysis are shown in figure 9.

Figure 10 shows the phase difference between the free running OCXO clock and the primary clock.

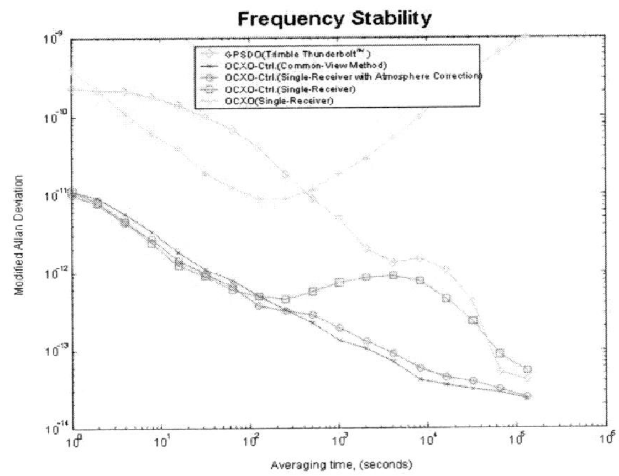

Figure 9: Frequency stability analysis of the free running OCXO, GPSDO(Thunderbolt ^{TM}), OCXO-Ctrl., OCXO-Ctrl.(with the correction of the RNN or CV method)

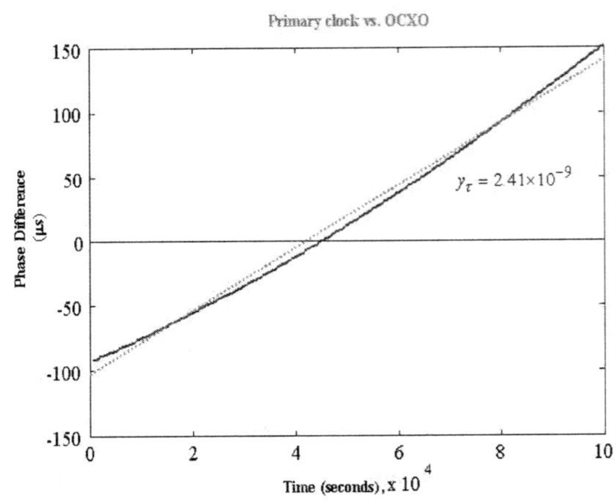

Figure 10: Phase difference between free running OCXO and primary clock with linear-fit line.

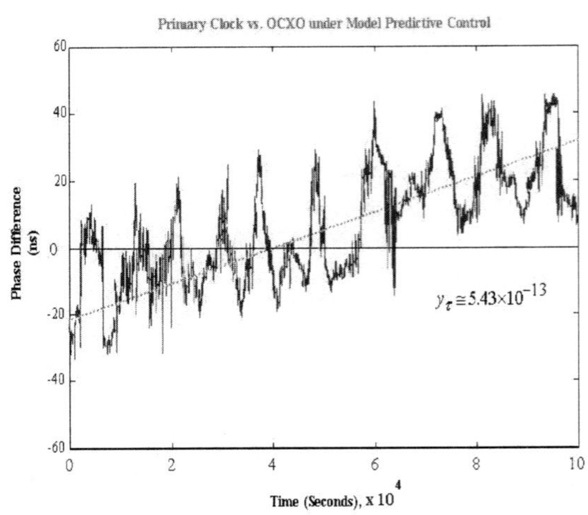

Figure 11: Phase difference between primary clock and OCXO-Ctrl. (MPC) under atmospheric effects with linear-fit line.

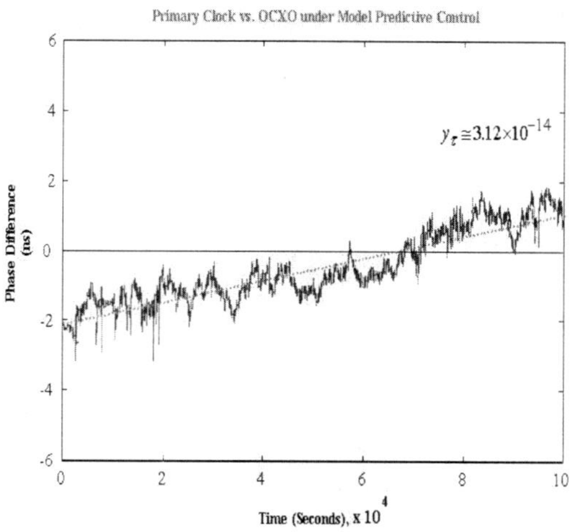

Figure 12: Phase difference between primary clock and OCXO-Ctrl. (MPC) under atmospheric model correction with linear-fit line.

The accuracy of this OCXO is about 2.41×10^{-9} for averaging time of about one day. Figure11 was represented the phase difference between primary clock and the controlled OCXO with atmospheric effects. The accuracy of the OCXO is about 5.43×10^{-13} for averaging time of about one day. Furthermore, the frequency stability of the OCXO is significantly degraded over the averaging time between the 100s to the 60,000s in figure 9. Figure12 was represented the phase difference between primary clock and the controlled OCXO with atmospheric model (RNNs) correction. The accuracy of the OCXO is about 3.12×10^{-14} for averaging time of about one day and the stability of per second was about nine parts in 10^{12}. In addition, we could find the frequency stability of the OCXO is significantly improved over the averaging time between the100s to the 60,000s in figure 9. From above results, we demonstrated our low-cost scheme composed of a single-frequency receiver with GPS carrier phase to achieve the highly-accurate and real-time syntonization system.

6 Conclusions

In this paper, a new low-cost, highly-accurate and real-time GPS carrier phase syntonization system based on a single-frequency receiver is presented. The scheme can achieve the traceability of frequency dissemination. We develop a forecasting model with recurrent neural networks to correct atmospheric errors in real time. The model was more accurate than the existed ionosphere model and troposphere model for a single-frequency receiver used. Furthermore, it was available simply and anticipative day-to-day partially, irrelevant to user locations. In addition, we observed that model predictive control of neural networks is robust and adaptive for our frequency syntonization by GPS carrier phase measurements. The inexpensive steered clock can be automatically syntonized to obtain the very high frequency accuracy and stability in the short term as well as in the long term. So, the frame used in syntonization or calibration for apparatus was sound, reliable and cost-effective for common users.

Acknowledgement

We gratefully acknowledge the NSTF Lab., TL, Chunghwa Telecom. Co., Ltd., Taiwan, R.O.C. for experiment supporting of this paper.

References

[1] G. Petit and C. Thomas, "GPS frequency transfer using carrier-phase measurements," Proc. IEEE Freq. Contr. Symp., pp. 1151-1159, 1996.

[2] K. Larson, and J. Levine, "Time-transfer using GPS carrier phase methods," Procccdings 29th Annual Precise Time and Time Interval (PTTI) Meeting, Long Beach, California, 1997.

[3] C. Bruyninx, P. Defraigne, J. M. Sleewaegen, and P. Paquet, "Frequency transfer using GPS: comparative study of code and carrier phase analysis results," Proc. 30th Precise Time and Time Interval Meeting, 1998.

[4] B. Hofmann-Wellenhof, H. Lichtenegger, and J. Collins, "Global positioning system: theory and practice," Springer-Verlag Wien, New York, USA, 1994.

[5] D. Wells, "Guide to GPS positioning," Canadian GPS Associates, 1996.

[6] C.W.J. Granger and P. Newbold., "Forecasting economic time series," Academic Press, San Diego, second edition, 1986.

[7] J.L. Elman, "Distributed representations, simple recurrent networks and grammatical structure," Machine Learning, 7(2/3):195-226, 1991.

[8] G.U. Yule, "On a method of investigating periodicities in disturbed series with special reference to Wolfer's sunspot numbers," Philosophical Transactions Royal Society London Series A, 226:267-298, 1927.

[9] C. Lee Giles, Steve Lawrence, A.C. Tsoi, "Noisy time series prediction using a recurrent neural network and grammatical inference," Machine Learning, Vol. 44, pp161-183, Jul./Aug., 2001.

[10] Soloway, D. and P.J. Haley, "Neural Generalized Predictive Control," Proceedings of the IEEE International symposium on Intelligent Control, pp. 277-281, 1996.

[11] J. E. Moody, " Economic forecasting: Challenges and neural network solutions," Proceedings of the international symposium on Artificial neural network, Hsinchu, Taiwan, 1995.

THE RECENT ACTIVITIES ON TWO-WAY SATELLITE TIME TRANSFER AT TL

Huang-Tien LIN, Wen-Hung Tseng, Hsin-Ming Peng, Chia-Hsu Liao

National Standard Time & Frequency Lab., Telecommunication Laboratories
12, Lane 551, Min-Tsu Road Sec. 5 Yang-Mei, Taoyuan, Taiwan 326, R.O.C.
Email: linht@cht.com.tw, Fax: +886-3-4245474

Keywords: TWSTFT, TAI, GPS CV.

Abstract

The researches of Two Way Satellite Time and Frequency Transfer (TWSTFT) at TL are very active in the recent years. In Asia Pacific Rim region, TL has joined the Asia-Pacific networks (being organized and sponsored by CRL) since June 2000, and performed the C-band experiments of NML(Australia)-TL and NIST(USA)-TL from September 2002 to April 2003. Moreover, the Ku-band TWSTFT experiments of CRL-TL, CRL-NML and NML-TL links, using PAS-8 satellite, are in progress.

For the intercontinental cases, the major time transfer link between Europe and Asia is the PTB-CRL GPS CV single channel currently. The time-transfer connection between North America and Asia is also not well planned. In order to enhance the time transfer connection between Europe and Asia, TL and VSL (the Netherlands) have cooperated and succeeded in the set up of the first Europe-Asia TWSTFT link, VSL-TL link, and the regular experiments have been performed from March 2003. For the cooperation between North America and Asia, the USNO-TL link is also under planning. The recent activities and the future works on TWSTFT at TL will be illustrated in this paper.

1 Introduction

Two Way Satellite Time and Frequency Transfer (TWSTFT) has currently become one of the major time-transfer techniques [1]. The time- transfer data obtained from several regular experiments in Europe, North America [2] and Asia have been used to calculate the international atomic time (TAI). However, the major time transfer link between Europe and Asia is the PTB-CRL GPS CV single channel currently [3]. The time-transfer connection between North America and Asia is also not well planned.

In order to participate in the TWSTFT networks and contribute to the calculation of TAI, TL has purchased and maintained the related equipment, including four earth stations, one Atlantis modem, three SATRE modems and some RF facilities. The information and application of these facilities are described in Table 1. All the transceivers are located outdoors under the temperature range of 10~35°C all year around. TWSTFT modems and the external counter are located in a temperature-controlled room. Reference signal of the TWSTFT experiments is UTC(TL), which is based on one Cesium clock (HP5071A) been steered by a AOG micro phase stepper.

In Asia-Pacific Rim region, TL has joined the Asia-Pacific TWSTFT network, which is organized and sponsored by CRL, since June 2000. Up to the present, the Ku-band TWSTFT links that have been set up at TL include two CRL-TL links (using JCSAT-1B and PAS_8 satellites, separately) and NML-TL link (using PAS-8 satellite). The CRL-TL JCSAT-1B link, together with other two Asia-pacific (i.e. CRL-NMIJ and CRL-NTSC) links, has been introduced into the computation of TAI since January 2002 [4].

The C-band experiments between NML, NIST and TL had been performed from September 2002 to April 2003. This link was suspended and the investigation about the influence of the variations in ionospheric delay to the C-band time transfer results is in progress.

For the European and North American cooperation, the first Asia-Europe TWSTFT link, that is TL-VSL link, has been performed from March 2003 to the present. The USNO-TL link is also under planning. The TWSTFT links that have been set up at TL are shown in Fig.1. The recent activities, some experiment results and the future works on TWSTFT at TL will be illustrated in this paper.

2 The CRL-TL Link with JCSAT-1B satellite

The first Ku-band CRL/TL link, with JCAT-1B satellite, was set up in June 2000. For the present, a 2.4 m earth station and an Atlantis modem at TL are used for this link, and a half-hour session (including two or three 5-min observations) is performed every Tuesday and Friday. Fig.2 shows the time difference results between UTC(CRL) and UTC(TL) for the MJD period 52650 to 53070. The experiment data is also available at CRL's web-site. The TWSTFT result of this link has been involved into the calculation of TAI since January 2002.

Besides, CRL is developing a multi-channel time transfer modem and several Asian laboratories, including CRL, NMIJ,

KRISS, NML, TL, SPRING, NTSC, have joined this program for the performance tests. TL has one set of the CRL modem and the CRL-TL link with JCSAT-1B satellite is also utilized for the trial experiments.

3 The CRL-TL and NML-TL links with PAS-8 satellite

For extending the TWSTFT network to Australia and USA, CRL have rent two transponder (one for East Asia beam, and another for USA and Australia beam) of PAS-8 satellite. In order to set up the Ku-band TWSTFT link with laboratories in Australia and USA, TL has maintained a 1.8 m earth station pointed to PAS-8. Under the assistance of CRL, the regular CRL-TL and NML-TL PAS-8 links have been performed since March 2003. Atlantis modem and SATRE modem are used for the CRL-TL link and NML-TL link, respectively.

Since three TWSTFT links (CRL-TL, CRL-NML and NML-TL) composed a closed loop and two CRL-TL (JCSAT-1B and PAS-8) links are both in action, it would be interesting to further analyze and compare the data obtained from different links. This is worth to work on in the near future.

Moreover, TL and USNO are also discussing to set up a North America-Asia TWSTFT link trough PAS-8. Just like the case of NML-TL PAS-8 link, it's possible to use one transponder for uplinking from North America and delivering signal to Asia, and use another transponder for uplinking from Asia and delivering signal to North America.

4 The VSL-TL link with PAS-4 satellite

The topic of establishing a Europe-Asia link was first discussed in the 9th TWSTFT meeting of the CCTF WG [5]. Both TL and VSL have been devoted to this task. After several satellites were surveyed, the PAS-4 satellite at 72°E was found to be suitable for the TL-VSL link. The pretest was performed successfully on February 2003. The follow-on tests also showed that the performance of the TL-VSL link was good. Therefore, the regular experiment has been undertaken from March 2003 to the present. A 2.4 m earth station and a SATRE modem at TL are applied for this link.

Basically, the TWSTFT session between TL and VSL was performed twice, every Tuesday and Thursday, per week. The monthly schedule was confirmed by the end of the previous month. Each individual session started at 0700 UTC and lasted for 30 minutes. Both TL and VSL have put the measurement files about this link on their anonymous FTP sites separately:
" ftp.stdtime.gov.tw " and " ftp.nmi.nl ".

The clock difference results obtained by TWSTFT and Circular-T data of the TAI standard dates are calculated and illustrated in Fig. 3. Because the TWSTFT data of TL-VSL link are unevenly spaced by intervals of 2, 5, or 7 days, they are linearly interpolated to give the data for the TAI standard dates with 5-day intervals. The time delay differences of the TWSTFT link have not been calibrated. Fig. 4 shows the difference of the TWSTFT and Circular-T data for the MJD period 52700 to 53070. The RMS value is less than 5.0 ns. Although the elevation from VSL was only 4.9 degrees, the performance of this link is still good.

5 Summary and future works

In recent years, TL has purchased many effective facilities and actively set up several TWSTFT links. In Asia-Pacific region, two CRL-TL links (using JCSAT-1B and PAS-8 satellites) and one NML-TL link (PAS-8) are maintained.

For the intercontinental cases, the VSL-TL link has been set up. Although the elevation from VSL was only 4.9 degrees, the performance of this link is still good. This would encourage other European laboratories to establish the Asian TWSTFT links. The USNO-TL link is also under planning.

For the future work, TL intends to set up the TWSTFT links with some other institutes in North America and Europe. By doing this, the intercontinental TWSTT network will be more extended and complete. TL will also proceed with the investigation of earth station delay measurement, temperature effect and other related topics, so that the instability of earth station delay can be reduced.

Acknowledgments

The authors would like to thank all colleagues of CRL, NML and VSL participating in the TWSTFT experiments for their great help.

References

[1]D. Kirchner, "Two-Way Time Transfer via Communication Satellites," in Proc. of the IEEE, 79, pp.983-990, 1991.

[2] J. Azoubib, D Kirchner, W. Lewandowski, et. al, "Two-Way Satellite Time Transfer Using INTELSAT 706 On a Regular Basis: Status and Data Evaluation," in Proc. of the 30th PTTI, 1998, p.393.

[3] J. Azoubib, W. Lewandowski, "23nd BIPM TWSTFT Report", 13 September 2002.

[4] J. Azoubib, W. Lewandowski, "20th BIPM TWSTFT Report", 20 March 2002.

[5] "REPORT OF THE 9th MEETING OF THE CCTF WORKING GROUP ON TWO-WAY SATELLITE TIME AND FREQUENCY TRANSFER"

Table 1. TWSTFT equipment of TL

Equipment	Description
1.8 m earth station	• Ku-band Antenna • 8W transmitter • Point to PAS-8 • SFJ1-50A IF-band cable • Used for CRL-TL and NML-TL links
2.4 m earth station (A)	• Ku-band Antenna • 8W transmitter • Point to JCSAT-1B • Used for the CRL-TL link
2.4 m earth station (B)	• Ku-band Antenna • 8W transmitter • Point to PAS-4 • Used for the VSL-TL link
4.6 m earth station	• C-band Antenna • 20W transmitter • Point to NSS-513 • Used for the NML-NIST-TL link (suspended for the present)
Modem	• SATRE 63, SATRE 66, SATRE 73 (TimeTech) • ATLANTIS 135 (AOA)
Satellite-simulator	• SATSIM system (TimeTech)
External counter	• HP 5370B

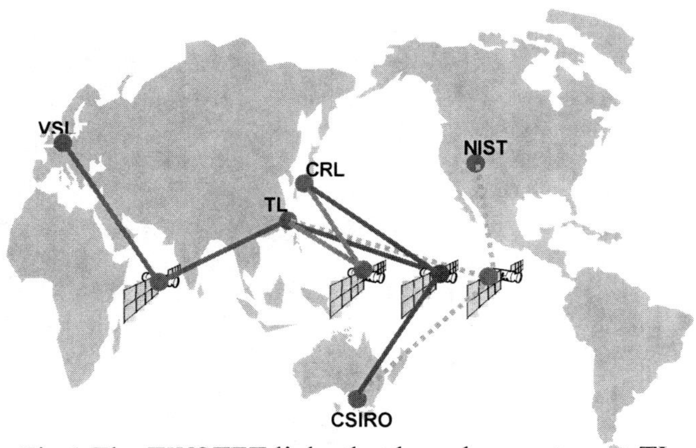

Fig.1 The TWSTFT links that have been set up at TL

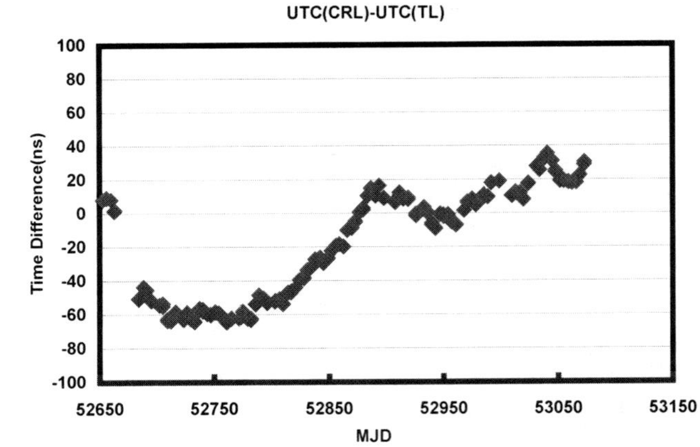

Fig. 2. TWSTFT results of UTC(CRL)-UTC(TL) for the MJD period 52650-53070.

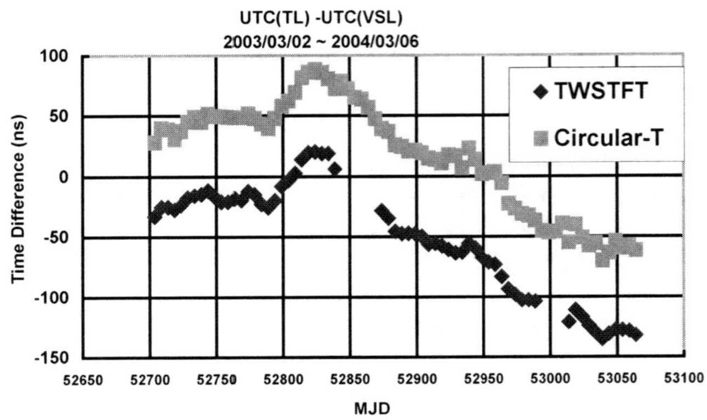

Fig. 3 Results of UTC(TL)-UTC(VSL) by TWSTFT and Circular-T data for the MJD period 52700-53070.

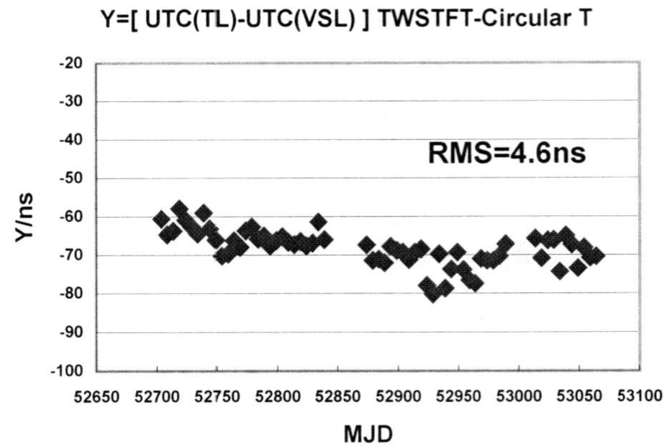

Fig. 4 Differences between TWSTFT results and Circular-T data for VSL-TL link.

PRACTICAL PERFORMANCE OF THE UTC(CH.R) REAL TIME REALIZATION OF UTC(CH) AND PROSPECTS FOR IMPROVEMENT

Laurent-Guy Bernier, Gregor Dudle

METAS - Swiss Federal Office of Metrology and Accreditation
Lindenweg 50, CH-3003 Bern-Wabern, Switzerland
e-mail: laurent-guy.bernier@metas.ch fax: +41-31-3233573

Keywords: timescales, steering, prediction methods

Abstract

UTC(CH.R) is a hardware real-time realization of UTC(CH) generated from the reference cesium clock by means of a micro-phase-stepper. The purpose of the steering algorithm is to keep UTC(CH.R) as close as possible to UTC(CH) by means of a time-frequency control loop which must be stable despite the delay of 1 d between the epoch of the last computed state and the epoch of the daily steering. The paper presents the practical performance of the UTC(CH.R) realization based on measurement data recorded from April 2002 to April 2003. The performance of the UTC(CH.R) steering algorithm is compared against simulation results that apply alternative steering algorithms to the same free running clock data. In conclusion plans to use a hydrogen maser instead of a cesium clock for the generation of a future, improved UTC(CH.R) are discussed.

1 Introduction

The METAS time & frequency metrology laboratory is an exception among NMI laboratories contributing to TAI and UTC in the sense that UTC(CH) is not a physical clock but a computed timescale. All the cesium clocks, including the reference clock (REF), are free running. The inputs to the time scale algorithm are the time differences between the reference clock and the other hardware clocks measured by means of a 5 MHz phase comparator. The output of the time scale algorithm are the daily computed states of each clock versus the time scales UTC(CH) and TA(CH). TA(CH) is a free running time scale while UTC(CH) is steered monthly to UTC. Both time scales are generated using the same time scale algorithm and are based on the same clock data. The advantage of a computed UTC(CH) proceeds from the fact that a time scale is more stable and more reliable than a single physical clock. The obvious disadvantage is that UTC(CH) is not readily available in real-time. Although it is always possible to perform both time and frequency calibration measurements against the REF clock and to refer the calibration results *a posteriori* to UTC(CH) by processing the clock data, it is more convenient to use a hardware clock which tracks UTC(CH) with a known uncertainty and to get instant results. Hence the development of UTC(CH.R) in our laboratory. UTC(CH.R) is a real-time realization of UTC(CH) generated from the REF clock by means of a micro-phase-stepper (MPS) and a daily steering algorithm. The difficulty in realizing the closed-loop steering control of UTC(CH.R) arises from the fact that, at the epoch of the steering, the most recent time scale data available is for epoch UTC 00:00 of the day before. This is illustrated on the timing diagram of Figure 1. The epoch variable is noted n in day units and refers to UTC 00:00 of the current MJD. We define the constant $u = 1$ d to simplify the notation. The steering is performed at hh:00 UTC every day. The epoch of the current daily steering is therefore noted $n + h$ with h = hh/24 d. Every night at UTC 00:00 the local clocks data for the previous day becomes available but not the remote clocks data. This is why the *Autotime* system computes the timescale products only up to UTC 00:00 of the previous day. This leaves a whole day to solve the eventual remote file transfer problems and to collect the data from the remote clocks. However, the consequence is that the delay in the measurement arm of the control loop is $u + h$ i.e. more than one day. This configuration is problematic from the point of view of the stability of the control loop. In order to increase the bandwidth of the steering control loop and to decrease the rms deviation without falling into instability problems, we have opted for a predictive control loop, i.e. the steering error signal is based on the prediction of the steered clock state UTC(CH.R)-UTC(CH) from the epoch of last computation up to the epoch of the daily steering, i.e. a $u+h$ prediction interval. The closed loop predictive

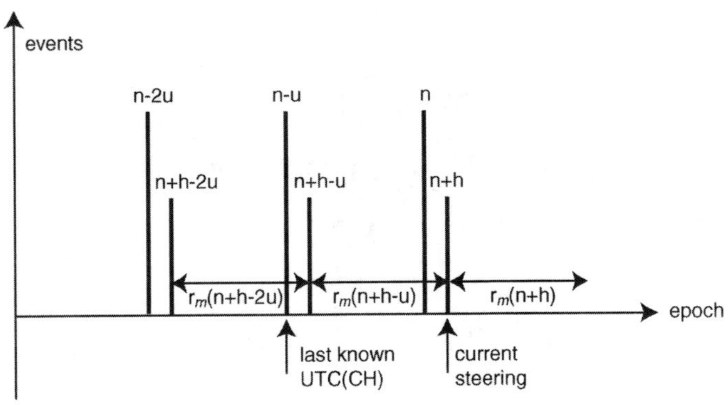

Figure 1: Timing diagram of UTC(CH.R) steering. Constant definitions: $u = 1$ d, $h = \text{hh}/24$ d.

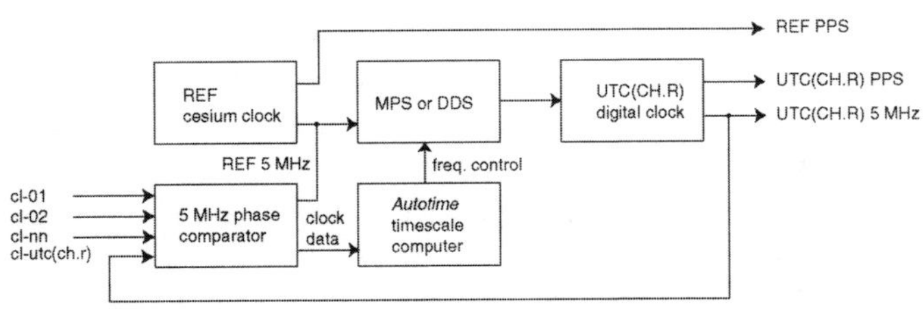

Figure 2: System block diagram of UTC(CH.R) steering.

steering algorithm was reported in [1]. The prediction algorithm itself was reported in [2] . Up to February 2002, UTC(CH.R) was based on a classical, non predictive, control loop reported in [1]. The disadvantage of the non predictive control loop was a very long closed loop time constant, necessary to avoid instability. Sudden changes such as frequency steps of the REF clock or steerings of UTC(CH) used to generate long transients. Our hope was that the new predictive algorithm would yield a stable control loop with a shorter time constant. The new steering algorithm was implemented into our in-house software package *Autotime* which is also used for the timescale generation. The state of the steered UTC(CH.R) clock is measured and recorded in the same way as the free running clocks, i.e. via a 5 MHz phase comparison system as shown of Figure 2. The system has been running continuously during most of 2002 and 2003. We are presently testing a new DDS (Direct Digital Synthesizer) to replace the MPS which is no longer available commercially. In this paper we present the experimental results obtained with UTC(CH.R) as recorded from April 2002 to April 2003. The observed experimental deviation of UTC(CH.R) versus UTC(CH) is discussed and compared with simulation results based on several alternative steering algorithms. Plans to use our new hydrogen maser instead of the REF clock for the generation of UTC(CH.R) are also discussed.

2 UTC(CH.R) Algorithm

We define r_m as the rate correction programmed into the MPS so that the input rate r_i is transformed into the output rate $r_o = r_i + r_m$. The steering algorithm has a feedforward term and a feedback term. The new rate correction $r_m(n+h)$ to be programmed into the MPS starting at the epoch of the current steering is given by

$$r_m(n+h) = -r_f(n-u, N_2) - \Delta r_m(n+h) \tag{1}$$

where the feedforward term $r_f(n-u, N_2)$ is the moving average filtered rate of the free running REF clock versus UTC(CH)

$$r_f(n-u, N_2) = \frac{x_f(n-u) - x_f(n-u-N_2)}{N_2}. \tag{2}$$

where N_2 is the averaging interval over which the frequency offset of the REF clock is estimated. The feedforward term corrects the frequency error of the REF clock versus UTC(CH). The proportional feedback term $\Delta r_m(n+h)$, on the other hand, is a small supplementary rate adjustment of the MPS that corrects the time error of the steered clock UTC(CH.R)

$$\Delta r_m(n+h) = \frac{\hat{x}_s(n-u+N_1)}{N_3}. \tag{3}$$

$\hat{x}_s(n-u+N_1)$ is the predicted time error of UTC(CH.R) projected from the last known epoch $n-u$ up to the epoch $n+h$ of the current steering which implies a prediction interval $N_1 = u+h$. $N_3 = 1$ d is the time constant for the correction of the time error. The prediction is performed using the simple prediction algorithm described in [1],

$$\hat{x}_s(n-u+N_1) = x_s(n-u) + \Delta \tag{4}$$

where

$$\Delta = h \times \Delta r_m(n+h-2u) + u \times \Delta r_m(n+h-u). \tag{5}$$

(4) and (5) can be described as follows. The prediction $\hat{x}_s(n-u+N_1)$ depends on the last known state $x_s(n-u)$ of the steered clock UTC(CH.R) extrapolated by the feedback rate correction $\Delta r_m(n+h-2u)$ that was applied at this epoch. After an interval h the steering of yesterday occurred and then the new feedback rate correction $\Delta r_m(n+h-u)$ was applied for a full day, until the current steering. Figure 3 shows the resulting UTC(CH.R)-UTC(CH) recorded daily from MJD 52381 to MJD 52755. The rms deviation from the mean is 10.5 ns and the mean offset is -9 ns. As discussed in the next section, the simulation of this algorithm predicts a smaller rms deviation and a negligible offset. Therefore it seems that there was a problem in the practical implementation of the algorithm. This will be discussed later on. Nevertheless the rms deviation, although not optimal, is quite good. Moreover, Figure 4 representing the Allan deviation of UTC(CH.R)

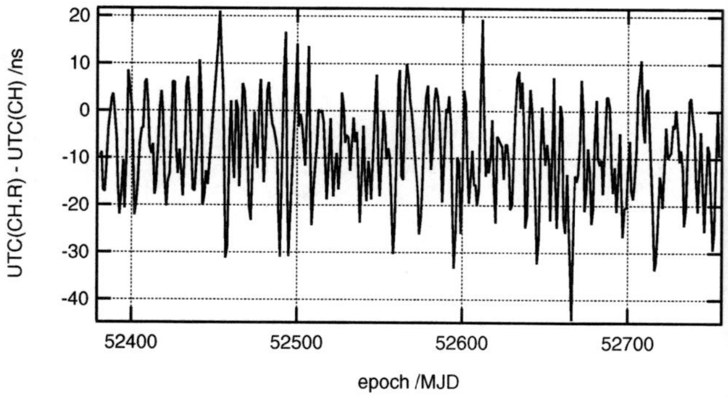

Figure 3: UTC(CH.R)-UTC(CH) from MJD 52381 to MJD 52755. Rms deviation from the mean: 10.5 ns. Offset: -9 ns.

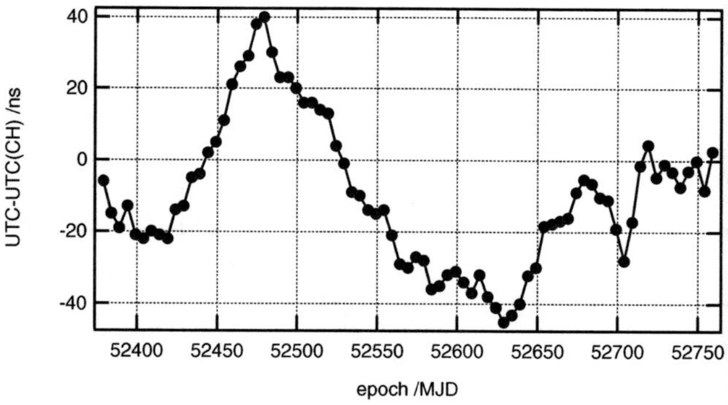

Figure 4: UTC-UTC(CH) from MJD 52379 to MJD 52759 as published by BIPM in Circular T.

shows that the closed-loop time constant is approximately 2 d. This is remarkable for a stable control loop with a delay of more than one day in the measurement branch. Figure 4 shows UTC-UTC(CH), as published by BIPM in Circular T, for the same period. Assuming a 20 ns uncertainty (1 sigma) of UTC(CH.R) vs UTC(CH) which includes both the noise and the bias, and assuming an uncertainty of 20 ns (1 sigma) uncertainty of UTC(CH) vs UTC, i.e. half the peak to peak variation during this period, we can state that the uncertainty of UTC(CH.R) as a real-time realization of UTC via UTC(CH) is approximately 30 ns (1 sigma).

3 UTC(CH.R) Simulations

In order to understand better the behavior of the steering control loop and to optimize the parameters, we have recently performed a simulation of the UTC(CH.R) steering system. The simulation is based on the actual, recorded, experimental free running REF clock data. For the current day, the simulated state $x_s(n)$ of the steered clock UTC(CH.R) is computed as

$$x_s(n) = x_s(n-u) + \Delta_1 + \Delta_2 \qquad (6)$$

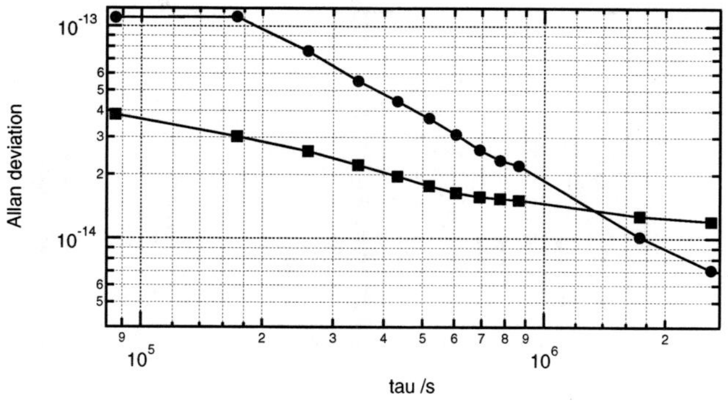

Figure 5: Allan deviation of the steered clock UTC(CH.R) (circles) and of the free running REF clock (squares).

where

$$\Delta_1 = x_f(n) - x_f(n-u) \qquad (7)$$

is the recorded daily increment of the free running REF clock and where

$$\Delta_2 = h \times r_m(n+h-2u) + u \times r_m(n+h-u) \qquad (8)$$

represents the simulated action of the MPS which is split into two periods: before and after the steering that occurred at epoch $n+h-u$. The current rate correction $r_m(n+h)$ is computed using the algorithm variant selected for a given simulation run. Variant A is a feedforward feedback predictive algorithm, in principle identical to the one implemented in *Autotime*. In the simulation (1), (2) and (3) apply. The prediction is given by

$$\hat{x}_s(n-u+N_1) = x_s(n-u) + r_s \times N_1 + c \qquad (9)$$

instead of (4), where

$$r_s = r_f(n-u, N_2) + r_m(n+h-2u). \qquad (10)$$

(9) and (10) express the fact that the prediction starts from epoch $n-u$ which implies that the MPS rate correction applied was $r_m(n+h-2u)$. However, the steering of epoch $n+h-u$ occurred during the prediction interval. Therefore the prediction must be corrected by a term

$$c = u \times [r_m(n+h-u) - r_m(n+h-2u)]. \qquad (11)$$

which represents the effect of the correction rate change over the last 24 h up to the epoch of the current steering.

On the other hand, variant B is a feedback only predictive algorithm with integrator where $r_m(n)$ is given by

$$r_m(n+h) = r_m(n+h-u) - \Delta r_m(n+h) \qquad (12)$$

instead of (1). (12) is the recursive definition of an integrator since $-\Delta r_m(n+u)$ is the very last correction term while $r_m(n+h-u)$ is the sum of all the previous correction terms. Table 1 summarizes the results of the steering control algorithm simulations. In all simulations the steering

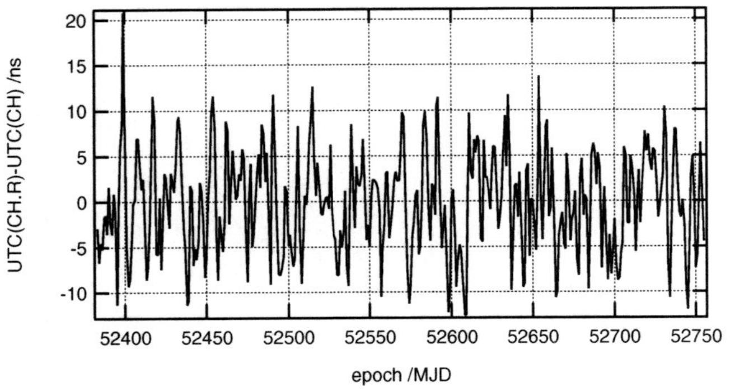

Figure 6: Simulation of the steered clock UTC(CH.R) using the steering control algorithm, variant A, $N_1 = 1.16$ d, $N_2 = 15$ d, $N_3 = 0.8$ d, rms deviation from UTC(CH) = 5.7 ns.

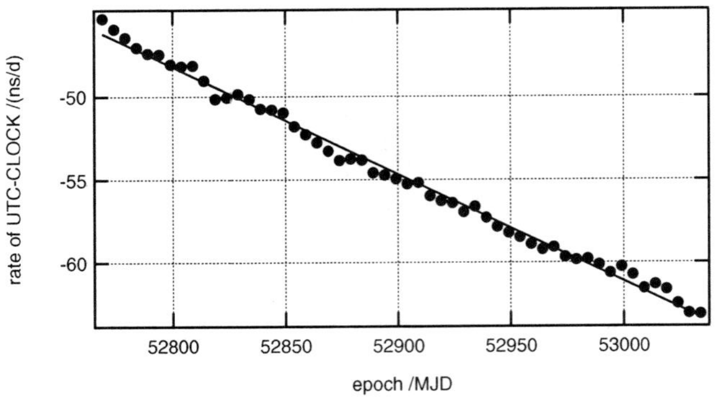

Figure 7: Daily rate of METAS hydrogen maser (BIPM ID 1405701) recorded from MJD 52769 to MJD 53034. Drift coefficient is -0.0647 ns/d^2 i.e. -7.5×10^{-16} /d.

sim	algo	N_1/d	N_2/d	N_3/d	rms/ns
1	A	1.16	5	1	6.3
2	A	1.16	15	1	5.8
3	A	1.16	15	0.8	5.7
4	A	2.16	15	2	6.6
5	B	2.16	15	5	8.2

Table 1: Simulations of UTC(CH.R) with REF clock.

is scheduled for 04:00 UTC therefore $h = 0.16$ d. Simulation 1 corresponds to variant A with the parameters of the *Autotime* algorithm. The simulation predicts a rms deviation of 6.3 ns and no bias. This is to be compared with the experimental result of Figure 3. Simulation 2 shows that by taking the optimal averaging interval $N_2 = 15$ d, i.e. minimizing the prediction error according to the criteria of the GSF-1 prediction [1], instead of 5 d, the rms deviation can be improved. Simulation 3 shows that reducing further the correction time constant N_3 improves a little more the rms deviation. If N_3 is further reduced the loop becomes unstable. The simulation with variant A and the optimal set of parameters is illustrated on Figure 6. Simulation 4 shows that if the prediction interval is increased so that the control loop tries to cancel the error predicted for the epoch of tomorrow's steering, $N_1 = 2u + h$, instead of today's steering, $N_1 = u + h$, and if the correction time constant is optimized for that condition, the rms deviation degrades. Simulation 5, finally, shows that the pure feedback variant B of the algorithm is not as good as the combined feedforward feedback variant A. Note again that both variants are predictive. A classical control loop without prediction would be unstable with comparable values of the correction time constant. In conclusion we see that the original algorithm, i.e. variant A, is the best and we see that the original parameters are quite close to the optimum so that only a small improvement can be obtained with optimization. The question that remains open is why the *Autotime* algorithm yields a rms deviation twice as large as the value predicted by the simulation with the algorithm variant A and why there is a bias. The explanation is as follows. Equations (4) and (5) which describes the prediction algorithm implemented into the *Autotime* code is an over-simplification of the steering algorithm variant A. It assumes implicitly that the moving average frequency of the REF clock does not change from day to day. More precisely, equations (4) and (5) are equivalent to (9), (10) and (11) only if

$$r_f(n-u, N_2) = r_f(n-2u, N_2) = r_f(n-3u, N_2) \quad (13)$$

which is true only in first approximation. The result was an *Autotime* steering algorithm that was functional but not optimal.

4 UTC(CH.R) from H-maser

We are looking forward to implementing the steering algorithm variant A in practice, instead of the original *Autotime* algorithm, and to verifying experimentally the improvement in performance predicted by the simulations reported above. Moreover, we now have a hydrogen maser that can be used as the free running REF clock and a DDS that can replace the late MPS for the generation of UTC(CH.R). Figure 7 shows the daily rate of our model EFOS-C hydrogen maser manufactured by Neuchâtel Observatory. This maser does not have an Automatic Cavity Tuning (ACT) system. However, the drift which is at the level of $+7.5 \times 10^{-16}$/d, as shown on Figure 7, is extremely stable. This makes the clock state highly predictable. Figure 8 represents the drift removed residual (the value of UTC-CLOCK in ns as published by BIPM, minus the best fit parabola) which varies in a range of only [-15 ns, + 15 ns] for a recording interval of 265 d. A simulation of the UTC(CH.R) generation was performed with the steering algorithm variant C and the free running maser data CLOCK-UTC(CH) used as the REF clock. Variant C of the algorithm is the same as variant A except for the addition of a drift parameter in the clock state prediction. Because of the drift the GSF-1 prediction algorithm of [1]

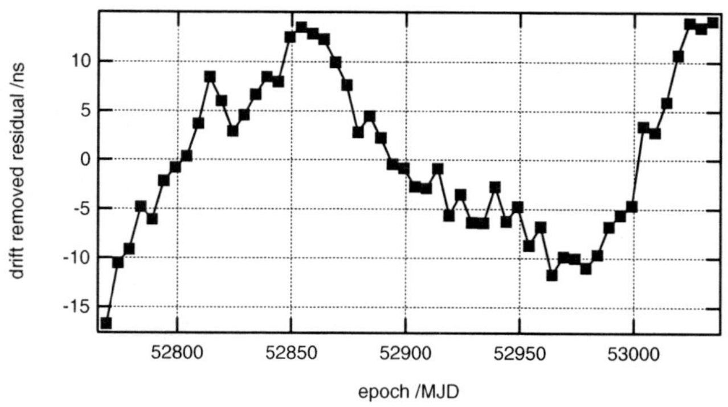

Figure 8: Drift removed residual of METAS hydrogen maser (BIPM ID 1405701) computed from MJD 52769 to MJD 53034. Residual in ns is given by (UTC-CLOCK) minus best fit parabola computed over recorded interval of 265 d.

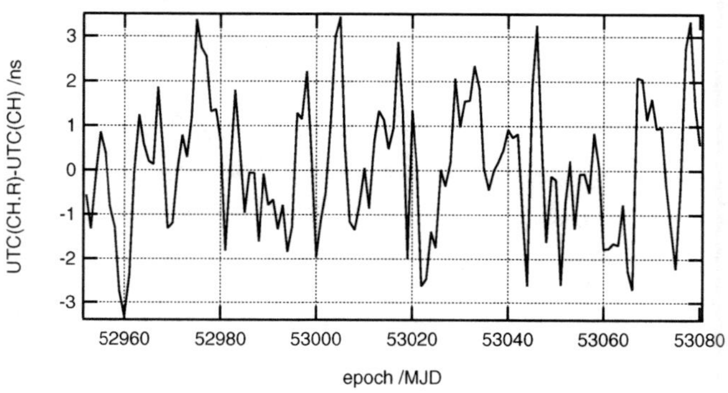

Figure 9: Simulation of steered clock UTC(CH.R) using algorithm variant C and the METAS hydrogen maser (BIPM ID 1405701) as the reference. Parameters are $N_1 = 1.16$ d, $d = 0.026$ ns/d^2, $N_2 = 20$ d, $N_3 = 0.8$ d, rms deviation from UTC(CH) = 1.9 ns.

has to be replaced by the DGSF-1 prediction algorithm of [3] which optimizes not only the frequency averaging interval N_2 but also a drift parameter d, which is half the drift coefficient. In variant C, (9) becomes

$$\hat{x}_s(n-u+N_1) = x_s(n-u)+r_sN_1+\frac{1}{2}dN_1^2(1+\frac{N_2}{N_1})+c \quad (14)$$

with r_s and c defined by (10) and (11) as before. The best results are obtained with the optimum clock prediction parameters $N_2 = 20$ d and $d = 0.026$ ns/d^2 and the same optimal control loop parameters as before, i.e. $N_1 = 1.16$ d and $N_3 = 0.8$ d. The result is an rms deviation from UTC(CH) of only 1.9 ns as shown on Figure 9.

5 Conclusion

In this paper we have reported that the predictive steering control loop implemented in the *Autotime* software in April 2002 and operated for more than a year has yielded a UTC(CH.R)-UTC(CH) rms deviation from the mean of 10.5 ns, with a bias of -9 ns. This version of the steering algorithm has demonstrated that it is possible to implement a fast control loop, with a time constant of 2 d, despite the delay of more than 1 d in the measurement arm of the control loop. The recent acquisition of a hydrogen maser and of a DDS to replace the late MPS opens new possibilities to improve the generation of UTC(CH.R). Moreover, simulations show that the *Autotime* steering control algorithm was too simple to reach optimum results. Simulations with variant A of the steering control algorithm (both feedforward and feedback terms, optimized prediction based on the moving average of the rate of the REF clock) show that an rms deviation of 6 ns with no bias could be reached using the same cesium clock as the reference. This is an expected improvement by a factor of two with respect to the *Autotime* present implementation. Simulations with variant C of the steering control algorithm (both feedforward and feedback terms, optimized prediction based on the moving average of the rate of the REF clock plus a drift parameter) show that an rms deviation of 2 ns with no bias could be reached using an hydrogen maser as the reference. This is an expected improvement by a factor of five with respect to the *Autotime* present implementation. We plan to test experimentally the combination hydrogen maser, DDS and variant C of the algorithm in the near future.

References

[1] L. G. Bernier, "Experimentation at METAS with a Simple Steering Algorithm based on Linear Prediction," in *Proc. IVth International Time Scale Algorithms Symposium*, BIPM, Sèvres, March 18-29, 2002.

[2] L. G. Bernier, "Use of the Allan Deviation and Linear Prediction for the Determination of the Uncertainty on Time Calibrations Against Predicted Timescales," in *IEEE Trans. on Instrum. and Meas.*, vol. 52, no. 2, pp. 483–486, April 2003.

[3] L. G. Bernier, "Application of the GSF-1 Algorithm to the Near-Optimal Timescale Prediction of the Hydrogen Maser," in *Proc. PTTI 2003*, San-Diego, December 2003.

THE TIME SCALE SYNCHRONIZATION ALGORITHMS AND ITS IMPLEMENTATION IN METEOR-BURST CHANNEL CASE

IVAN ANTIPOV, JURY KOVAL, VERONIKA BAVYKINA
Kharkov National University of Radioelectronics
Lenin av., 14, Kharkov, Ukraine, i_ant@mail.ru

The time scale synchronization algorithms are very important because they provide registering or removal of signal delay in the propagation medium. The algorithm choice influences the technical realization of synchronization system as a whole. Different algorithms result in different transformation of the source errors into measurement errors, as well. The preliminary analysis has shown that there is no single algorithm, which is the most preferable in all cases.

The main purpose of this work is to improve synchronization algorithms and to refine the processing of measurement results. This is necessary to improve the parameters of existing synchronization systems and for development of new ones. Some interesting results were obtained for meteor synchronization systems.

Meteor synchronization systems (MSS) are using the reflection of meter radio waves from meteor trails. The main advantage of MSS is a high reversibility of meteor-burst radio channel: the signal propagation delay is the same in both directions. To implement meteor-burst channel for time scale synchronization, special hardware complex "METKA" was designed in Kharkov National University of Radioelectronics.

We have analyzed the error sources for differ synchronization algorithms, including errors resulting from low signal-to-noise ratio, the channel errors (irreversibility and instability of propagation time and the interference influence) and the equipment errors (irreversibility and instability of delay time in receiver and in transmitter). The analysis of correlation between some errors is presented too.

The following conclusions are the result of analysis:
1. The improved analysis method, which takes into account main error sources (noise, interference, instability and irreversibility in environment and in equipment) and their correlations, allows for comparison and modernization of known algorithms and developing of a new ones.
2. The synthesis of combined and modified algorithms is needed for measurement error decorrelation in spatially separated points.
3. The mutual algorithm provides maximum accuracy among the other active algorithms. Therefore this algorithm was selected for the high accuracy phase MSS "METKA-11".
4. Among all retranslation algorithms, the reference-retranslation one provides minimum noise error, if retranslation factor n is equal to 2 for any noise situation in both points.

GPS COMMON-VIEW WITH GEODETIC RECEIVERS

T. Gotoh*, T. Otsubo†, T. Kubo'oka†, M. Imae‡, M. Fujieda*

National Institute of Information and Communications Technology*
4-2-1, Nukui-Kitamachi, Koganei, Tokyo, 184-8795, Japan
tara@nict.go.jp, fax +81 42 327 6664
Kashima Space Research Center, NICT†
893-1, Hirai, Kashima, Ibaraki, 314-8501, Japan
National Institute of Advanced Industrial Science and Technology‡
1-1-1, Umezono, Tsukuba, Ibaraki, 305-8563, Japan

Keywords: GPS time transfer, carrier phase, satellite orbit analysis

Abstract

The time transfer precision of the GPS common-view method with single-channel C/A code receivers has become inadequate for an international time comparison network because of the advances made in atomic clock technology. NICT has been developing carrier-phase time transfer software with geodetic receivers to enable more precise time transfer technology. This paper describes the details of this software and the time transfer precision attainable with this software. Especially, we discuss the influence of tropospheric delay correction models.

1 Introduction

The GPS common-view (CV) method [3] had been a mainstream technique in the international time comparison network. However, the daily averaged time transfer precision for time transfer receivers using a single-channel C/A code is a few nanoseconds, and this is being insufficient for the stability of recent atomic clocks. Therefore, a two-way time transfer method using a communication satellite is being adopted in Asia [7]. At the same time, many laboratories are investigating way to improve time transfer precision with GPS CV because GPS is easier to use than a two-way method.

Using a dual-frequency receiver makes it possible to cancel out ionospheric delay which severely degrades time transfer precision because of an increased baseline length. Using carrier-phase observations dramatically improves the observation error. Since the carrier-phase analysis with a dual-frequency receiver is the established technology in the geodetic community, most geodetic receivers can also simultaneously measure both the carrier phase and the pseudorange. However, older geodetic receivers cannot synchronize with an external reference signal. A receiver is now being developed which measures the traveling time between a GPS signal and an external reference signal, and the improvement of GPS CV precision made possible with such receivers has been investigated [8, 6].

Also, the National Institute of Information and Communication Technology (NICT) has been developing carrier-phase time transfer software. The critical problem for carrier-phase time transfer is carrier-phase ambiguity resolution. For time transfer purposes, we cannot use a double difference. Therefore, we cannot correctly fix the carrier-phase ambiguity to an integer value. On the other hand, time transfer stations have always been located at fixed points. Thus, we know the precise position of each station, and do not have to solve the position and the carrier-phase ambiguity at the same time. This advantage gives rise to a simple algorithm compared to the one for precise geodesy. In this paper, we describe the details of this ambiguity-resolution algorithm and show the time transfer results obtained when using the algorithm.

2 Overview of the software under development

We have been developing satellite analysis software which we call *concerto* [10]. The original purpose of the software was the analysis of satellite laser ranging (SLR) data. We are now extending the software for use in the analysis of a low earth orbiter with a GPS receiver. The software consists of two parts – the first part performs numerical integration of a satellite orbit, and the second part determines unknown parameters with the least squares method. We use only the second part for time transfer. Since this software can only process a pseudorange at present, we resolve carrier-phase ambiguities through other software which implements a simple algorithm.

2.1 GPS observations

The observation equations for the GPS pseudorange and carrier phase from satellite k to receiver i are written as equations (1) and (2), respectively [9].

$$\begin{aligned} P_i^k(t) &= \rho_i^k(t, t-\tau_i^k) + I_i^k + T_i^k + dm_i^k \\ &\quad + c[dt_i(t) - dt^k(t-\tau_i^k)] \\ &\quad + c[d_i(t) + d^k(t-\tau_i^k)] + e_i^k \qquad (1) \\ L_i^k(t) &= \rho_i^k(t, t-\tau_i^k) - I_i^k + T_i^k + \delta m_i^k \\ &\quad + c[dt_i(t) - dt^k(t-\tau_i^k)] \\ &\quad + c[\delta_i(t) + \delta^k(t-\tau_i^k)] \\ &\quad + \lambda[\phi_i(t_0) - \phi^k(t_0)] + \lambda N_i^k + \varepsilon_i^k \qquad (2) \end{aligned}$$

Here, P_i^k is the pseudorange observation (in meters), and L_i^k is the carrier-phase observation (in meters). ρ_i^k is the geometrical distance between a satellite and a receiver. I_i^k is the ionospheric delay and T_i^k is the tropospheric delay. dm_i^k and δm_i^k are the multipath errors of, respectively, the pseudorange and carrier phase. dt_i is the receiver clock offset and dt^k is the satellite clock offset. d_i and δ_i are the receiver internal delays of, respectively, the pseudorange and carrier phase. d^k and δ^k are the satellite internal delays of, respectively, the pseudorange and carrier phase. $\phi_i(t_0)$ and $\phi^k(t_0)$ are the non-zero initial phases for, respectively, a receiver and a satellite. N_i^k is the carrier-phase ambiguity. c is the speed of light, and λ is the wavelength of the carrier frequency. e_i^k is the pseudorange measurement error, and ε_i^k is carrier-phase measurement error.

The main differences between two observations are that the multipath error δm_i^k and the measurement error ε_i^k of the carrier phase are much smaller than the errors of the pseudorange (dm_i^k and e_i^k), and only the carrier phase is accompanied by the ambiguity terms $\phi_i(t_0)$, $\phi^k(t_0)$, and N_i^k.

To cancel out the satellite clock error and internal delay, we form a single difference observation or a common-view between two receivers and a satellite as equations (3) and (4). We also form a double difference observation between two receivers and two satellites as equations (5) and (6) to cancel out both satellite and receiver clock offsets and both internal delays. For time transfer purposes, the double difference is only used to estimate the carrier-phase ambiguity as an integer value.

$$\begin{aligned} P_{ij}^k &= \rho_{ij}^k + I_{ij}^k + T_{ij}^k + dm_{ij}^k \\ &\quad + cdt_{ij} + cd_{ij} + e_{ij}^k \qquad (3) \\ L_{ij}^k &= \rho_{ij}^k - I_{ij}^k + T_{ij}^k + \delta m_{ij}^k \\ &\quad + cdt_{ij} + c\delta_{ij} + \lambda\phi_{ij}(t_0) + \lambda N_{ij}^k + \varepsilon_{ij}^k \qquad (4) \end{aligned}$$

$$\begin{aligned} P_{ij}^{kl} &= \rho_{ij}^{kl} + I_{ij}^{kl} + T_{ij}^{kl} + dm_{ij}^{kl} + e_{ij}^{kl} \qquad (5) \\ L_{ij}^{kl} &= \rho_{ij}^{kl} - I_{ij}^{kl} + T_{ij}^{kl} + \delta m_{ij}^{kl} + \lambda N_{ij}^{kl} + \varepsilon_{ij}^{kl} \qquad (6) \end{aligned}$$

If we can fix the carrier-phase ambiguity to an integer value, we might be able to accurately measure the clock offset between two stations within the wavelength. GPS carrier frequencies are 154×10.23 MHz for the first carrier frequency and 120×10.23 MHz for the second carrier frequency. Therefore, the wavelengths of these GPS carrier frequencies are about 19 cm (or 635 ps) and 24 cm (or 815 ps), respectively. Since the uncertainty of the current GPS CV is a few nanoseconds, use of a carrier phase with the correct integer ambiguity will be dramatic improvement. However, in the process to estimate the integer ambiguity, we have to use a pseudorange. Therefore, it is impossible to fix the integer ambiguity within the wavelength. Though the accuracy of the integer ambiguity is not fixed closely, the observation error of the carrier phase is much better than that of the pseudorange. This advantage gives rise to improved short-term stability of the time transfer.

2.2 Carrier-phase ambiguity fixing

We cannot use double differences for time transfer, because double differences cancel out the clock offset between two receivers. However, since all stations are restricted to time-keeping communities which have been maintaining UTC, all station positions are known precisely. Therefore, the residual of the double difference corrected propagation delays should be equal to the carrier-phase ambiguity. If the standard deviation of the residuals after correction is smaller than the wavelength, the mean residual might indicate a correct integer ambiguity. Taking advantage of the well known station position, we have adopted a simple ambiguity-fixing algorithm.

Before describing the details of this algorithm, we will mention some linear combinations used in this algorithm. First, we define the simple double-difference equations (7) to (10), which are rewritten from equations (5) and (6). These equations also show the relation between the first and second carrier frequencies (first subscript character in the equations).

$$\begin{aligned} P_{1,ij}^{kl} &= \rho_{ij}^{kl} + \frac{f_2^2}{f_1^2 - f_2^2} I_{ij}^{kl} + e_{1,ij}^{kl} \qquad (7) \\ P_{2,ij}^{kl} &= \rho_{ij}^{kl} + \frac{f_1^2}{f_1^2 - f_2^2} I_{ij}^{kl} + e_{2,ij}^{kl} \qquad (8) \\ L_{1,ij}^{kl} &= \rho_{ij}^{kl} - \frac{f_2^2}{f_1^2 - f_2^2} I_{ij}^{kl} \\ &\quad + \lambda_1 N_{1,ij}^{kl} + \varepsilon_{1,ij}^{kl} \qquad (9) \\ L_{2,ij}^{kl} &= \rho_{ij}^{kl} - \frac{f_1^2}{f_1^2 - f_2^2} I_{ij}^{kl} \\ &\quad + \lambda_2 N_{2,ij}^{kl} + \varepsilon_{2,ij}^{kl} \qquad (10) \end{aligned}$$

All propagation delays are included in the term ρ_{ij}^{kl}, except the ionospheric delay and carrier-phase ambiguity. Using these equations, we derived equations (11) to (15). Equations (11) and (12) are called the ionosphere-free linear combination, and they cancel out the ionospheric delay. However, the carrier-phase ambiguity $N_{3,ij}^{kl}$ is no longer an integer value. Equation (15) is the Melbourne-Wübbena linear combination which includes $N_{1,ij}^{kl}-N_{2,ij}^{kl}$, and is used to fix $N_{1,ij}^{kl}$ and $N_{2,ij}^{kl}$ to integer values from $N_{3,ij}^{kl}$ which contains a real number.

$$\begin{aligned} P_{3,ij}^{kl} &= \frac{1}{f_1^2-f_2^2}(f_1^2 P_{1,ij}^{kl} - f_2^2 P_{2,ij}^{kl}) \\ &= \rho_{ij}^{kl} + e_{3,ij}^{kl} \qquad (11) \\ L_{3,ij}^{kl} &= \frac{1}{f_1^2-f_2^2}(f_1^2 L_{1,ij}^{kl} - f_2^2 L_{2,ij}^{kl}) \\ &= \rho_{ij}^{kl} + \lambda_3 N_{3,ij}^{kl} + \varepsilon_{3,ij}^{kl} \\ &= \rho_{ij}^{kl} + c\frac{f_2}{f_1^2-f_2^2}(N_{1,ij}^{kl} - N_{2,ij}^{kl}) \\ &\quad + \frac{c}{f_1+f_2} N_{1,ij}^{kl} + \varepsilon_{3,ij}^{kl} \qquad (12) \\ P_{N,ij}^{kl} &= \frac{1}{f_1+f_2}(f_1 P_{1,ij}^{kl} + f_2 P_{2,ij}^{kl}) \qquad (13) \\ L_{W,ij}^{kl} &= \frac{1}{f_1-f_2}(f_1 L_{1,ij}^{kl} - f_2 L_{2,ij}^{kl}) \qquad (14) \\ N_{\delta,ij}^{kl} &= L_{W,ij}^{kl} - P_{N,ij}^{kl} \\ &= \frac{c}{f_1-f_2}(N_{1,ij}^{kl} - N_{2,ij}^{kl}) + e_{\delta,ij}^{kl} \qquad (15) \end{aligned}$$

We fix the single-difference carrier-phase ambiguities $N_{1,ij}^{k}$ and $N_{2,ij}^{k}$ to integer values with the above equations as follows.

1. Compute the mean value of the clock offset between two stations with P_3.

2. Compute $N_{3,ij}^{k}$ with the above solution. $N_{1,ij}^{k} - N_{2,ij}^{k}$ is also computed with Melbourne-Wübbena. Using the estimated $N_{3,ij}^{k}$ and $N_{1,ij}^{k} - N_{2,ij}^{k}$, fix $N_{1,ij}^{k}$ and $N_{2,ij}^{k}$ to integer values.

3. Fix $N_{1,ij}^{kl}$ and $N_{2,ij}^{kl}$ to integer values with the same procedure as in step 2.

4. Choose a pair of appropriate $N_{1,ij}^{0}$ and $N_{2,ij}^{0}$ for the reference ambiguities. Using this pair of ambiguities, fix all single difference ambiguities through $N_{ij}^{n} = N_{ij}^{0} - N_{ij}^{0n}$.

Let's consider the accuracy of fixed integer ambiguities. From equation (12), $N_{3,ij}^{kl}$ can be divided into two terms as $N_{1,ij}^{kl} - N_{2,ij}^{kl}$ and $N_{1,ij}^{kl}$. The wavelength is about 1.2 ns with $N_{1,ij}^{kl} - N_{2,ij}^{kl}$, and 356 ps with $N_{1,ij}^{kl}$. This means that the residual has an offset in units of 356 ps corresponding to the integer value error of $N_{1,ij}^{kl}$, or 1.2 ns corresponding to that of $N_{1,ij}^{kl} - N_{2,ij}^{kl}$. The error of $N_{3,ij}^{kl}$ might be reduced to below 356 ps with precise correction models. However, since $N_{1,ij}^{kl} - N_{2,ij}^{kl}$ is computed on the basis of Melbourne-Wübbena, the error of $N_{1,ij}^{kl} - N_{2,ij}^{kl}$ is not sufficient within 356 ps. Even then, since the residual of the double difference with the correction of $N_{3,ij}^{kl}$ is unbiased, we can find out the best $N_{1,ij}^{kl}$ with the smallest standard deviation of the residuals. We adopted the same procedure for single-difference ambiguities. However, the residual of the single difference is not equal to zero because it includes biases such as station internal delays and non-zero initial phases. Moreover, the clock offset used to estimate the single-difference ambiguities is decided on the basis of the pseudorange. This means that the accuracy of the single-difference ambiguities depends on the quality of the pseudorange.

3 Time transfer results

To test our software, we performed time transfer using five IGS stations: PTBB (Braunschweig, Germany), NPLD (Teddington, UK), BRUS (Brussels, Belgium), TWTF (Taoyuan, ROC), and KGN0 (Koganei, Japan). Figures 1 and 2 show the geographical location of each station.

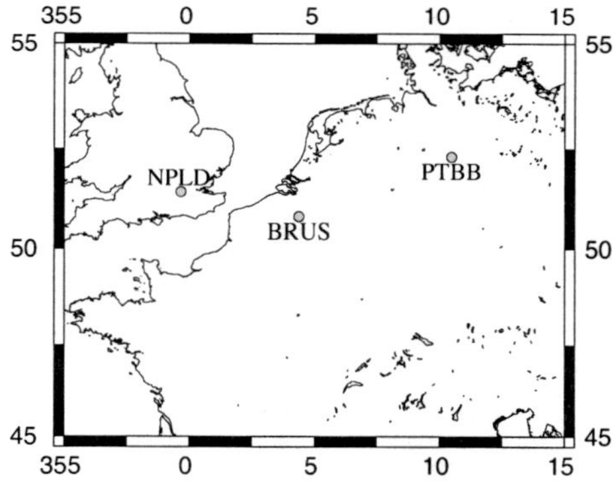

Figure 1: Locations of the IGS stations in Europe used for our time transfer experiments.

All stations used the ASHTECH Z-12T Metronome as a GPS receiver. PTBB, NPLD, and BRUS used a hydrogen maser as a reference signal, and KGN0 and TWTF used a cesium clock as a reference signal. We used the correction models shown in Table 1. We need the hydrostatic (dry) and wet zenith path delays (ZPD) for the Niell mapping function. International GPS Service (IGS) [4] provided the total (dry + wet) ZPDs of the tracking stations. We then computed the dry ZPD using a model from the IERS conventions 1996, and used the difference between the IGS total ZPD and the modeled dry ZPD as

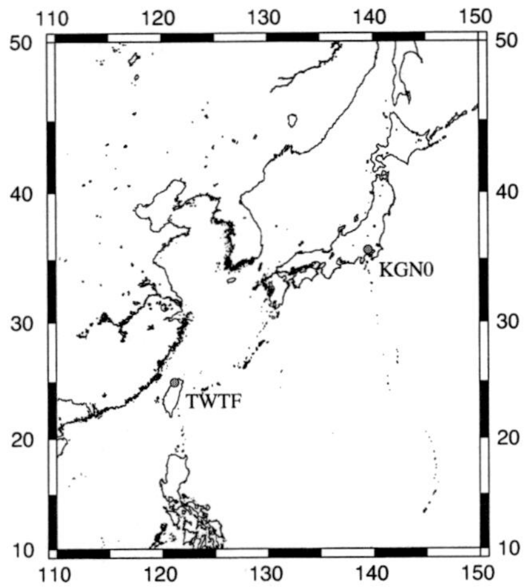

Figure 2: Locations of the IGS stations in Asia used for our time transfer experiments.

Satellite orbit	IGS final orbit
Station position	IGS final position
Earth rotation model EOP Displacement	IERS conventions 1996 [2] IERS bulletin B Solid earth tide Ocean loading with CODE product [5]
Tropospheric delay	Saastamoinen model, or Niell mapping function [1]

Table 1: Correction models

the wet ZPD. In addition, to compute the delays for both the Saastamoinen model and the modeled dry ZPD, we required meteorological data. We used a standard atmosphere model corresponding to each station's elevation for this purpose.

3.1 Residuals of ambiguity fixing

Figure 3 shows the residuals of the double-difference ambiguity fixing for the PTBB and NPLD stations on April 11, 2003. The baseline length between two stations is about 750 km. The two plots show the difference between the two tropospheric delay models – the Saastamoinen model (upper plot) and the Niell mapping function (lower plot). We applied some restrictions to these data. First, the data did not include low elevation satellites (less than 20°). Also, to avoid the influence of cycle slip we did not use a satellite path if it had Melbourne-Wübbena standard deviation of more than 0.8 cycles. Moreover, we also ignored the data where the common-view of two satellites was shorter than 50 min because the error might be increased because of a smaller number of data.

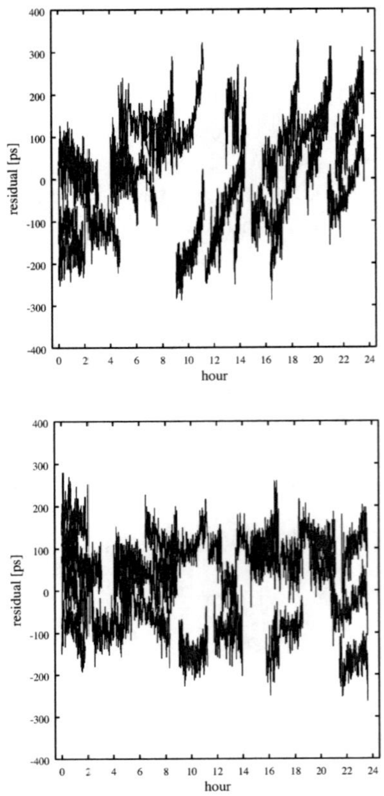

Figure 3: Residuals of double-difference ambiguity fixing on April 11, 2003. The stations were PTBB and NPLD. Tropospheric delay was corrected using the Saastamoinen model (upper) or the Niell mapping function (lower).

In Figure 3, we can clearly see drift with the Saastamoinen model which was not observed with the Niell mapping function.

3.2 Time transfer with a short baseline

Figure 4 shows the time difference between PTBB and NPLD on April 11, 2003. The tropospheric delay was corrected in the same way as the double-difference ambiguity was fixed. When the Saastamoinen model was used, we observed a large error corresponding to each satellite path. Moreover, we found inconsistency between the Saastamoinen model and the Niell mapping function in the data obtained after about 12 hours. The most likely explanation for this is that the fixed carrier-phase ambiguity with the Saastamoinen model was incorrect because of the large error in the double difference.

Figure 5 shows the time difference between PTBB and BRUS. The baseline length between two stations is about 450 km. The tropospheric delay was corrected with the same models as for PTBB vs. NPLD. In these plots, however, we cannot see a major difference between the two

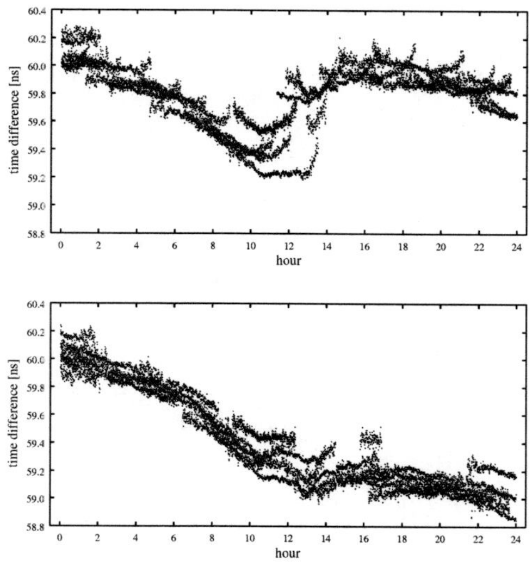

Figure 4: Time difference between PTBB and NPLD on April 11, 2003. The tropospheric delay was corrected with the Saastamoinen model (upper) or the Niell mapping function (lower).

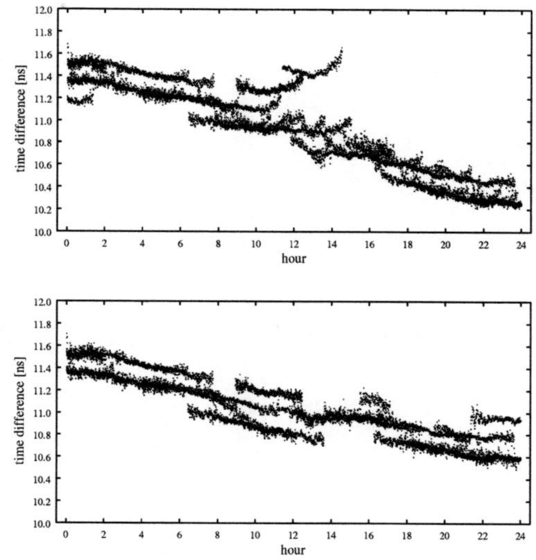

Figure 5: Time difference between PTBB and BRUS on April 11, 2003. The tropospheric delay was corrected with the Saastamoinen model (upper) or Niell mapping function (lower).

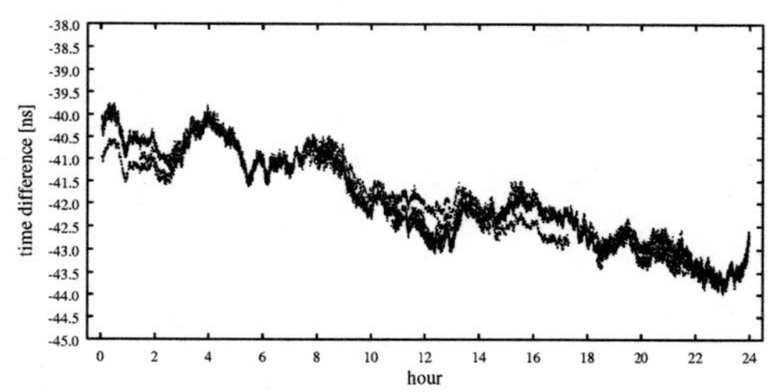

Figure 6: Time difference between KGN0 and TWTF on April 12, 2003. The tropospheric delay was corrected with the Niell mapping function.

models. However, we observed a failure of the estimated carrier-phase ambiguity after about 12 hours. Both plots, PTBB vs. NPLD and PTBB vs. BRUS, of the Niell mapping function show some discrepancies of about 100 to 200 ps. The cause of these discrepancies is not yet clear.

3.3 Time transfer with a middle baseline

Figure 6 shows the time difference between KGN0 and TWTF on April 12, 2003. The baseline length between two stations is about 2,100 km. For the tropospheric delay, we used only the Niell mapping function. We were unable to obtain a sufficient number of observations under the same restrictions as in section 3.2. Therefore, we relaxed these restrictions as follows: the minimum elevation angle 15°, and the common-view was limited to more than 30 min. We observed some offsets during the first two hours and from 10 to 18 hours. It seems that the carrier-phase ambiguity was incorrect, which was not the case with a short baseline.

Figure 7 shows the time difference between PTBB and NPLD with only SV #30, and figure 8 is shown one between KGN0 and TWTF with only SV #56. There was greater variation for KGN0 vs. TWTF than for PTBB vs. NPLD. We believe that this was due to the different reference signals of these receivers rather than the difference in the baseline length – KGN0 and TWTF both use a cesium clock, while PTBB and NPLD both use a hydrogen maser. Figure 9 shows the frequency stabilites of three baselinses: PTBB vs NPLD, PTBB vs BRUS, and KGN0 vs TWTF. We performed the arithmetic average to the data observed at the same time. The tropospheric delay was corrected the Niell mapping function. This figure clearly shows the difference between the reference signals. Carrier-phase time transfer can detect the difference in the stability between a hydrogen maser and a cesium clock, but this cannot be detected with conventional GPS CV.

4 Conclusion

We have examined the results from a GPS carrier-phase time transfer solved by the concerto software. For time transfer purposes, carrier-phase ambiguity can be fixed without difficulty since the station positions are well known. However, since the influence of tropospheric delay is dominant even if the double difference is used, we should use a mapping function and an estimated value

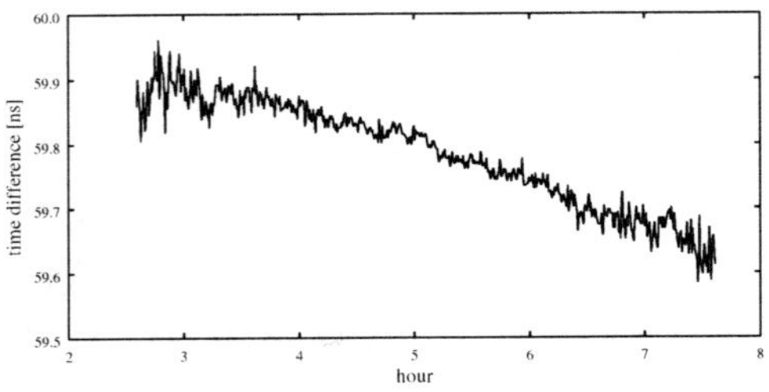

Figure 7: Time difference between PTBB and NPLD with only one satellite pass of SV #30.

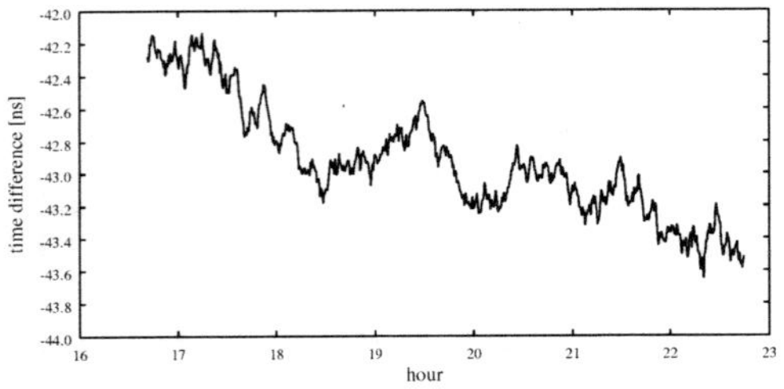

Figure 8: Time difference between KGN0 and TWTF with only one satellite pass of SV #56.

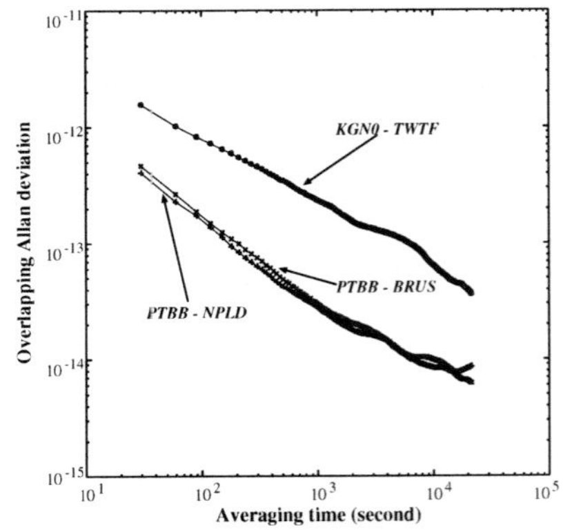

Figure 9: Frequency stabilities of three baselines: PTBB vs. NPLD, PTBB vs. BURS, and KGN0 vs. TWTF.

such as the IGS product rather than a model such as the Saastamoinen model. Furthermore, even if we use the estimated tropospheric delay, the carrier-phase ambiguity fixing might fail because the error is increased when the baseline length increases. We plan to investigate whether it is possible to estimate the correct carrier-phase ambiguity while simultaneously estimating the tropospheric delay when the baseline length is increased.

Carrier-phase time transfer can be achieved with much higher precision than when using a pseudorange. Our results have shown that it is possible to compare the clock variation with a period of less than one day, which is not possible with conventional GPS CV. However, an unknown discrepancy appeared in PTBB and BRUS baseline. We might be able to identify the error source of this discrepancy in such a way that increases the number of estimation parameters through further development of the software.

We think that carrier-phase time transfer is an efficient method for a TAI link with low observed noise and precise models provided by IGS.

References

[1] A.E. Niell, "Global mapping functions for the atmosphere delay at radio wavelengths", Journal of Geophysical Research, 101, B2, pp.3227-3246, (1996).

[2] D.D. McCarthy, *IERS Conventions (1996)*, IERS Technical Note 21, (1996).

[3] D.W. Allan, M.A. Weiss, "Accurate Time and Frequency Transfer During Common-View of a GPS Satellite", Proc. Frequency Control Symposium, pp.334-336, (1980).

[4] http://igscb.jpl.nasa.gov/.

[5] http://www.aiub.unibe.ch/.

[6] M. Fujieda, M. Imae, Y. Shibuya, "Effects of ionospheric correction on GPS time transfer throughout Asia", Metrologia, 41, pp.145-151, (2004).

[7] M. Imae, M. Hosokawa, K. Imamura, H. Yukawa, Y. Shibuya, N. Kurihara, P.T.H. Fisk, M.A. Lawn, L. Zhigang, L. Huanxin, Y. Nakadan, K. Hagimoto, "Two-way Satellite Time and Frequency Transfer Network in Pacific Rim Region", IEEE Trans. on Instrumentation and Measurement, 50, 2, pp.559-562, (2001).

[8] P. Defraigne, G. Petit, "Time transfer to TAI using geodetic receivers", Metrologia, 40, pp.184-188, (2003).

[9] P.J.G. Teunissen, A. Kleusberg, *GPS for Geodesy*, 2nd Edition, Sec. 5, Springer, (1998).

[10] T. Otsubo, T. Kubo'oka, T. Gotoh, R. Ichikawa, "Deformation of satellite laser ranging station coordinates due to atmospheric pressure loading", submitted to Geophysical Research Letters, (2004).

TIMESCALES FOR THE GPS AND GALILEO

Robert A. Nelson,[1] Dennis D. McCarthy,[2] Al Gifford,[3] and Thomas R. Bartholomew[4]

[1]Satellite Engineering Research Corporation, 7701 Woodmont Avenue, Suite 208, Bethesda, MD 20814 USA
[2]U.S. Naval Observatory ,3450 Massachusetts Avenue, NW, Washington, DC 20392 USA
[3]National Aeronautics and Space Administration , Code M-3, 300 E Street SW, Washington, DC 20546 USA
[4] Northrop Grumman IT, TASC, 2701 Technology Drive, Suite 300, Annapolis Junction, Maryland 20701 USA

Keywords: Time scale, TAI, UTC, GPS, Galileo

Abstract

GPS Time, the time scale internal to the Global Positioning System, is a uniform scale of time without leap seconds having an initial epoch of midnight, January 5/6, 1980 UTC. Galileo proposes to use a realization of TAI, another scale without leap seconds, as its internal time scale. However, TAI is ahead of GPS Time by 19 seconds, a constant value, and in 2004, GPS Time is ahead of UTC by 13 seconds. Thus GPS Time, UTC, and TAI have different epochs, which could lead to potential timing errors at the level of tens of seconds. The rationale for using either GPS Time or TAI is that it is required for those who cannot tolerate leap seconds. The international timing community is considering the adoption of a single time scale free of leap seconds to address operational concerns, as well as safety issues related to multiple time scales with different epochs. This paper provides a brief history and technical background for TAI and UTC and proposes options for the unification of all civil atomic time scales, including those of the GPS and Galileo.

1 Introduction

The international standard for civil time is an atomic time scale known as Coordinated Universal Time (UTC). The Global Positioning System (GPS) has become the *de facto* means of obtaining a realization of UTC globally. Once it becomes operational, Galileo will also disseminate a realization of UTC.

To maintain a close correspondence with solar time, UTC has occasional one-second steps called leap seconds. When the present form of UTC was adopted in 1972, it was envisioned as a convenient means of obtaining a measure of solar time for celestial navigation at sea. Today the motivation for introducing leap seconds into UTC has diminished with the increasing availability and precision of the Global Navigation Satellite System (GNSS), comprising the GPS, Glonass, their supplementary navigation systems, and new satellite systems under development, including Galileo and QZSS.

As many timekeeping systems require a timescale that is continuous, discussions are being held in the international timekeeping community to consider a redefinition of UTC by eliminating leap seconds at some date in the foreseeable future. The GPS uses its own internal system time that has no leap second steps. Similarly, Galileo will use a continuous system time. To simplify future GNSS interoperability, a means of establishing a common time reference will be needed. The present concurrent deliberations on time scales regarding the needs of both civil time and the GNSS may be viewed as a unique opportunity to bring greater unity to international timekeeping. This paper provides a background for the various time scales in use and examines alternatives for reducing the number of timescales for both civil time and the GNSS.

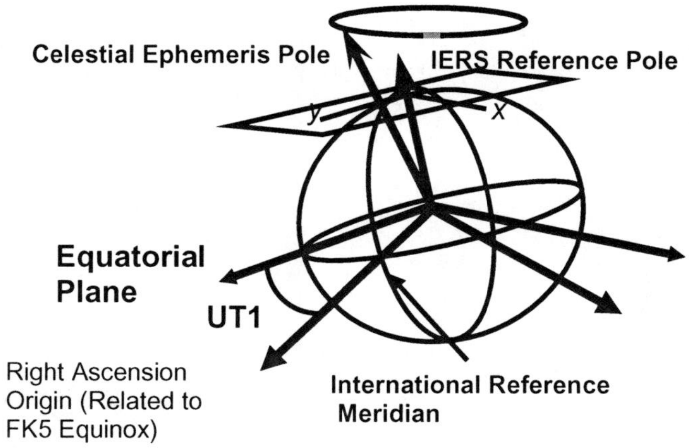

Figure 1. Definition of UT1.

2 Universal Time (UT)

Historically, time has been measured by the rotation of the Earth on its axis as reflected by the daily rising and setting of the Sun. This measure of time is called mean solar time. When referred to the meridian at Greenwich, it is called Universal Time (UT). The mean solar second is defined as 1/86 400 of a mean solar day.

The true measure of the angular orientation of the Earth is the form of UT known as UT1, as illustrated in *Figure 1.* Since the 1930s, it has been known that the Earth's rotational speed is not constant and, therefore, UT1 is not a uniform time. There are three types of variation: a steady decrease, periodic seasonal variations, and random fluctuations. The seasonally adjusted form of UT is called UT2.

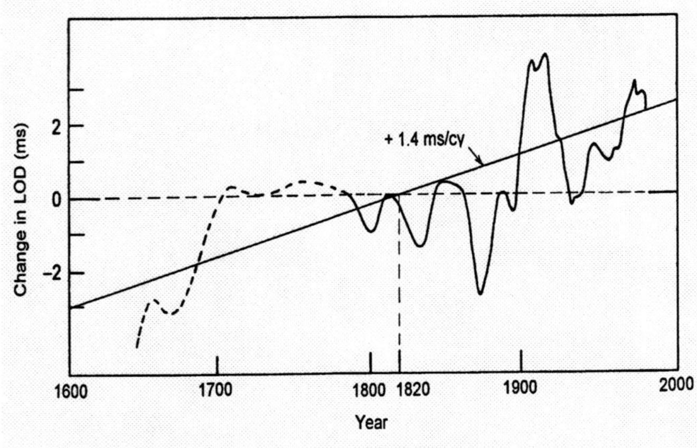

Figure 2. Length of Day (LOD) since 1600.
After F.R. Stephenson and L.V. Morrison [1].

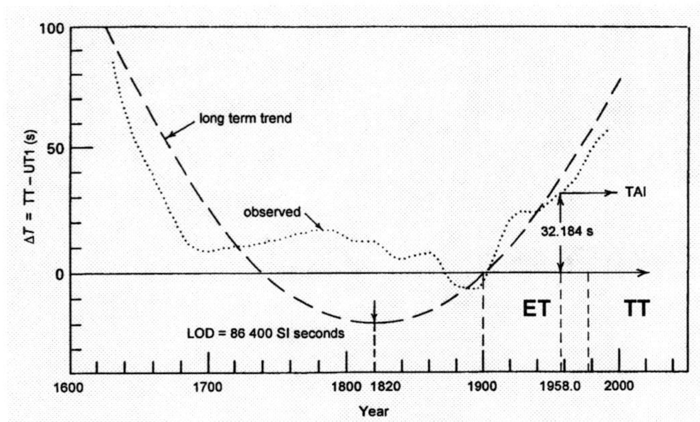

Figure 3. ΔT since 1600.
After F.R. Stephenson and L.V. Morrison [2].

The steady decrease in rate is due to tidal friction caused by the action of the Moon and corresponds to an increase in the Length of Day (LOD). Studies of the recorded times and locations of ancient solar eclipses over the past three thousand years indicate that the LOD has been increasing at the rate of about 1.4 ms per century. *Figure 2* illustrates the steady trend, together with random fluctuations having intervals of about a decade [1]. The smoothed difference between uniform time and UT1 can be represented by a parabola approximately of the form [2, 3]

$$\Delta T = (31\ \mathrm{s/cy^2})(T - 1820)^2/(100)^2 - 20\ \mathrm{s}$$

as shown in *Figure 3*. Evidence from daily and annual growth rings in coral fossils suggests that the secular rate has existed over several hundred million years. The long-term trend is maintained by conservation of angular momentum of the Earth-Moon system and will undoubtedly continue forever.

3 Ephemeris Time (ET)

Because time measured by the rotation of the Earth is not uniform, it was replaced in the 1950s by an astronomical time scale based on the Earth's revolution around the Sun. This measure of time is called Ephemeris Time (ET). In practice it is defined by Newcomb's formula for the mean longitude of the Sun, given by [4]

$$L = 279°\ 41'\ 48''.04 + 129\ 602\ 768''.13\ T + 1''.089\ T^2$$

where T is the time in Julian centuries of 36 525 days elapsed since 1900. Accordingly, the second of ET adopted by the International Committee for Weights and Measures (CIPM) in 1956 and ratified by the General Conference on Weights and Measures (CGPM) in 1960 was 1/31 556 925.9747 of the tropical year 1900. The observations analyzed by Newcomb ranged in date from 1750 to 1892 with a mean epoch of about 1820. Thus the ephemeris second is equal to a mean solar second of about the year 1820 [3].

4 International Atomic Time (TAI)

An atomic time scale based on the reference frequency of the caesium 133 hyperfine transition became possible with the creation of the first caesium atomic clock in 1955 at the National Physical Laboratory (NPL) in the United Kingdom. Soon thereafter, atomic clocks were established at several other national laboratories and observatories.

The frequency of the caesium 133 transition was measured by Essen and Parry [5] in 1955 in terms of the mean solar second as 9 192 631 830 ± 10 Hz and by Markowitz, Hall, Essen, and Parry [6] during 1955 to 1958 in terms of the second of ET as 9 192 631 770 ± 20 Hz. In 1967, only seven years after the adoption of the ephemeris second, the CGPM defined the second of the International System of Units (SI) as the duration of 9 192 631 770 periods of radiation corresponding to the hyperfine transition of the caesium 133 atom. The SI second is thus equal to the ephemeris second, which in turn is equal to the mean solar second of about 1820. Consequently, as indicated in *Figures 2* and *3*, the approximate epoch when the length of day was equal to exactly 86 400 SI seconds is 1820.

An atomic time scale has been maintained without interruption since the inception of the first atomic clock at the NPL in 1955. A statistically weighted atomic time scale with contributions from several institutions was established a short time later by the Bureau International de l'Heure (BIH). It was set equal to UT2, the seasonally adjusted form of UT, on January 1, 1958. The name International Atomic Time (TAI) was given to this scale of time in 1971. In 1988, responsibility for maintaining TAI was transferred from the BIH to the International Bureau of Weights and Measures (BIPM). TAI is defined in a geocentric reference frame with the SI second as realized on the rotating geoid as the scale unit.

The uniform atomic time scale that is continuous with ET is called Terrestrial Time (TT). A practical realization of TT is

$$\mathrm{TT} = \mathrm{TAI} + 32.184\ \mathrm{s}$$

The constant offset represents the difference between ET and UT1 at the defining epoch of TAI on January 1, 1958.

5 Coordinated Universal Time (UTC)

Coordination of the world's atomic time scales began in 1961 under the direction of the BIH. Formal rules for the system were adopted by the International Radio Consultative Committee (CCIR) of the International Telecommunication Union (ITU) in 1963. The name "Coordinated Universal Time (UTC)" was approved by Commissions 4 and 31 of the International Astronomical Union (IAU) in 1967.

From 1961 to 1972 UTC was maintained to within 0.1 second of UT2 by both frequency offsets and steps of 20, 50, or 100 ms [7]. The history of these corrections is summarized in *Figures 4* and *5*. Thus the unit interval of broadcast UTC signals was neither the SI second nor the mean solar second.

The adoption of an atomic definition of the second in 1967, coupled with dissatisfaction with the prevailing form of UTC, prompted discussions by the CIPM and CCIR. It was agreed that frequency offsets were not desirable and that the fractional steps then being used were too frequent and too small. Therefore, beginning in 1972 the current definition of UTC was implemented so that there would be no frequency offsets, but instead 1 second steps (leap seconds) would be inserted to keep UTC within 0.9 second of UT1 [8]. After further deliberation by the IAU, the name "Coordinated Universal Time (UTC)" was retained. The definition of UTC was a compromise to provide both the SI second and an approximation to astronomical time in the same radio signal.

In 1972 the difference between uniform time (ET or TT) and UT1 was 42.23 s. Thus the difference between TAI and UT1 was about 10 s. To maintain continuity with UT1, UTC was initially set behind TAI by this amount. Since 1972 there have been 22 leap seconds. Therefore, at present

$$TAI - UTC = 32 \text{ s}$$

The difference TAI – UTC since 1961 is shown in *Figure 6*.

Nearly two centuries have elapsed since 1820, the approximate epoch when the mean solar second was equal to the SI second and the LOD was exactly 86 400 seconds. Over this time span, the day has increased at the rate of 0.0014 second per century. Thus, averaged over a decade or so, the day is presently about 86 400.0025 s. (However, over the past few years it has been temporarily closer to 86 400 s due to short term fluctuations.) The difference of 2.5 ms from the day of exactly 86 400 s recorded by clocks accumulates to nearly one second in a year. The motivation for the leap second is to compensate this accumulation in time with solar time. As indicated in *Figure 6*, the rate of increase of TAI – UTC is about equal to the excess in the LOD. Thus leap seconds are inserted because the SI second is shorter than the present mean solar second by a few hundred parts in 10^{10} as a consequence of its historical derivation.

There is a growing consensus that leap seconds impose an unnecessary burden on sophisticated timekeeping systems and

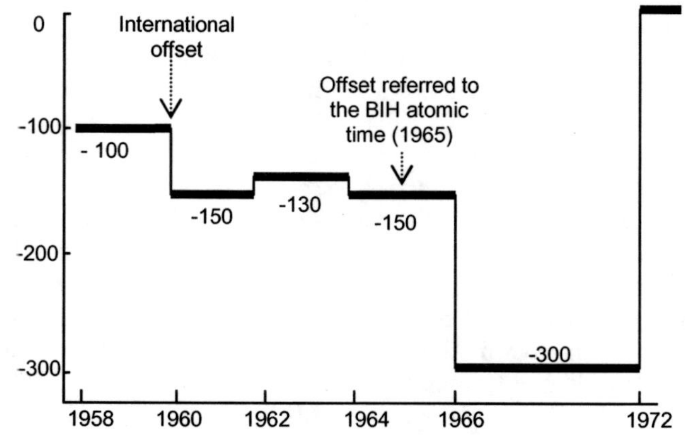

Figure 4. UTC frequency offsets (parts in 10^{-10}).
Courtesy of E.F. Arias, B. Guinot, and T.J. Quinn [7].

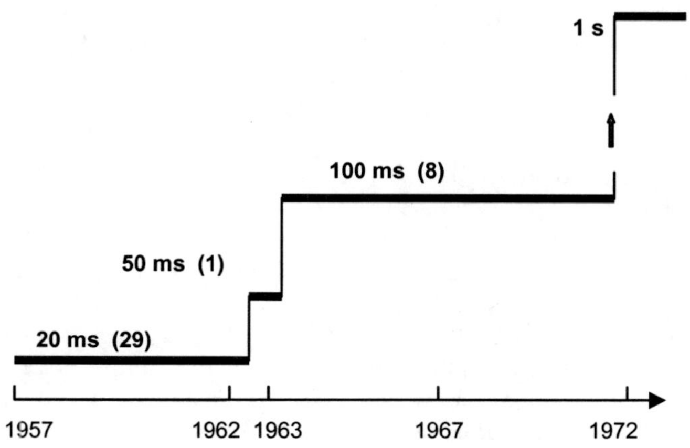

Figure 5. Evolution of UTC steps.
Courtesy of E.F. Arias, B. Guinot, and T.J. Quinn [7].

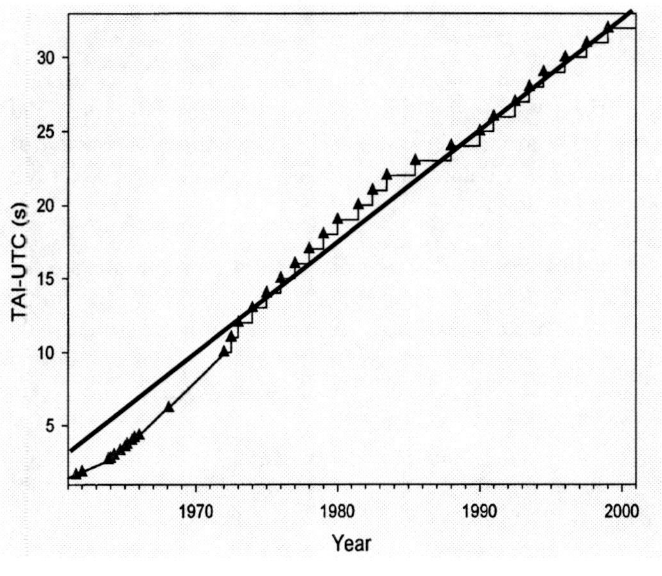

Figure 6. TAI – UTC since 1961.
The slope of the fitted line is (2.1 ± 0.05) ms per day.

that they are no longer needed or desirable. A Working Party of the IAU has been formed to study the issue. Also, in May, 2003 a colloquium on the UTC timescale was held by Special Rapporteur Group 7A of the ITU in Torino, Italy to discuss alternatives. A report is being prepared by the SRG that will be submitted to ITU Working Party 7A, which will prepare a recommendation to the full ITU. The considerations discussed by the SRG have been summarized at public forums such as the annual Precise Time and Time Interval (PTTI) conference and the Civil GPS Service Interface Committee (CGSIC) to solicit opinions. In addition, a draft report was presented to the Consultative Committee for Time and Frequency (CCTF) of the CIPM at its meeting at the BIPM on April 1 – 2, 2004.

It is expected that the SRG report will propose that leap seconds should be discontinued in the foreseeable future (possibly by 2010) and that UTC should be continuous starting with the epoch from which it is redefined. Although the possibility of a new name has been discussed, the majority of opinion appears to be that the current name UTC is well established and a new name would cause confusion. Thus it is believed, as in 1972 when the time scale was redefined once before, the name UTC should be retained.

If leap seconds are discontinued, then UTC will slowly drift from mean solar time. As shown in *Figure 7*, the difference will be about one minute after 50 years and about 2.5 minutes after 100 years [7]. By comparison, the difference between apparent solar time and mean solar time can be as much as 16 minutes, so that by the clock the afternoons are a half hour shorter than mornings in November while the mornings are a half hour shorter than the afternoons in February. Thus over the next few centuries the elimination of leap seconds would have no practical effect on customary social customs. It will take 1000 years for the difference to become one hour. At that time some future civilization might elect to change its clocks by one hour, just as we already do twice every year to convert between Standard Time and Daylight Saving Time.

6 GPS Time

The Global Positioning System (GPS) was formally chartered in 1973. It provides position with a precision on the order of a few meters or better and time with a precision on the order of 10 nanoseconds.

Each satellite carries a suite of caesium or rubidium atomic clocks. The satellite and global tracking network atomic clocks are used to form a common statistical time scale known as GPS Time, which has an initial epoch of January 5/6, 1980 UTC. GPS Time.is a continuous internal time scale that is specified to be steered to within 1 μs of UTC(USNO), except no leap seconds are inserted. (In practice the steering is much better, usually to within 10 ns.) The relationships of GPS Time with TAI and UTC are (within statistical error)

GPS Time = TAI – 19 s (constant)

GPS Time = UTC + 13 s (in 2004)

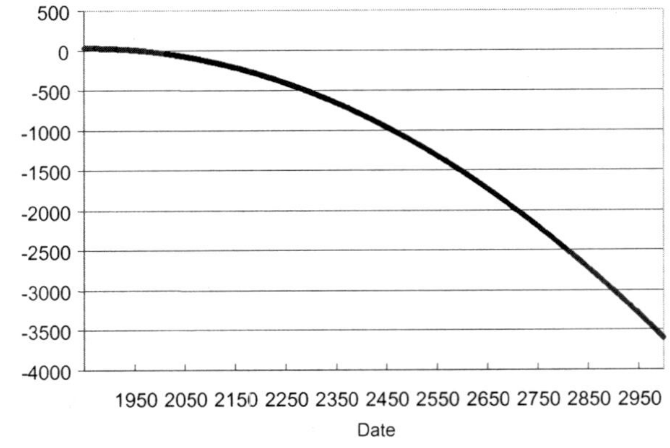

Figure 7. UT1 – uniform time without leap seconds. Courtesy of E.F. Arias, B. Guinot, and T.J. Quinn [7].

7 Galileo Time

The Galileo satellite navigation system is presently under development by the European Space Agency. The start of full operation is planned for 2008.

As the internal time scale for Galileo will be continuous without steps, it has been expected that the Galileo system time will be a realization of TAI. However, in a letter to the CCTF disseminated at its April, 2004 meeting [9], the director of the ESA Galileo Project Office expressed the hope that any decision to discontinue leap seconds should be made in the near future. The Galileo system could then adopt a realization of UTC as its internal reference time scale without facing any discontinuities.

8 Options for the GNSS

Options for the future integration of the civil time scale and the GNSS timescales may be examined in the context of several desirable objectives. These are:

(1) standardization of time scales;
(2) internal time scales without steps (no leap seconds);
(3) reduction in the number of time scales;
(4) provision for access to standard civil time.

Some of these options are identified and discussed below.

Option (1): *Status Quo*. The advantages are that no changes are required. The GPS and Galileo would provide UTC for civil use in addition to maintaining their own continuous internal time scales. The disadvantages are that separate GPS and Galileo system times add to the proliferation of time scales, may present difficulties in future interoperability, may cause confusion, and may have an impact on operations critical to safety of life services.

Option (2): *GNSS constellations move to GPS Time internally and also provide UTC.* The advantages are that the internal time scales would be unified, the systems would provide UTC for civil use, there would be no disruption to GPS operation, and Galileo could accommodate the change if a decision were made early. The disadvantages are that GPS Time is an internal system time scale that is not recognized as an international "standard" and the Galileo internal time would be artificially based on an epoch chosen by an independent system.

Option (3): *GNSS constellations move to TAI internally and also provide UTC*. The advantages are that the internal time scales would be unified and the systems would provide UTC for civil use. The disadvantages are the there would be no benefit for the GPS and there would be a disruption to GPS operation.

Option (4): *GNSS constellations move to UTC internally after UTC is redefined by international agreement to eliminate leap seconds*. This option has the advantages that it unifies all internal time scales, provides civil time, and avoids confusion between internal time scales and civil time. The significant disadvantage is that it would present a disruption to GPS operation.

9 Conclusions

A confluence of events has stimulated a far-ranging discussion on methods for international timekeeping. On one hand, there is a growing consensus on the need to revise the definition of the atomic time scale for civil use, UTC. On the other hand, there are under development new satellite navigation systems that contribute to the expanding GNSS. It is desirable that these systems be interoperable with one another.

A redefinition of UTC to eliminate leap seconds is currently under discussion at the international level. Discussions are taking place by the CCTF, the IAU, and the ITU. A possible change may be made within the foreseeable future.

The GNSS has become a *de facto* global source of time. Procedures for interoperability between the GPS and Galileo are being developed.

There now exists a unique opportunity to reduce the number of time scales and adopt an internationally recognized single continuous scale of time for both civil and GNSS use. This paper has been intended to draw attention to some of the issues and provide a brief summary of past decisions that have led to the present status. Various options for the future unification of time scales have been identified and are commended for further consideration.

References

[1] F. R. Stephenson and L. V. Morrison, "Long-Term Fluctuations in the Rotation of the Earth: 700 BC to AD 1980," *Phil. Trans. R. Soc. London* **A313**, pp. 47 – 70 (1984).

[2] F. R. Stephenson and L. V. Morrison, "Long-Term Fluctuations in the Earth's Rotation: 700 BC to AD 1990," *Phil. Trans. R. Soc. London* **A351**, pp. 165 – 202 (1995).

[3] F. R. Stephenson, *Historical Eclipses and Earth's Rotation* (Cambridge University Press, New York, 1997).

[4] S. Newcomb, *Astronomical Papers Prepared for the Use of the American Ephemeris and Nautical Almanac*, Vol. VI, Part I: *Tables of the Sun* (U.S. Govt. Printing Office, Washington, DC, 1895), p. 9.

[5] L. Essen and J. V. L. Parry, "An Atomic Standard of Frequency and Time Interval," *Nature* **176**, pp. 280 – 282 (1955).

[6] W. Markowitz, R. Hall, R. G. Essen, and J. V. L. Parry, "Frequency of Cesium in Terms of Ephemeris Time," *Phys. Rev. Lett.* **1**, pp. 105 – 107 (1958).

[7] E. F. Arias, B. Guinot, and T. J. Quinn, "Rotation of the Earth and Time Scales," *ITU-R SRG Colloquium on the UTC Time Scale* (Torino, Italy, May 28 – 29, 2003).

[8] R. A. Nelson, D. D. McCarthy, S. Malys, J. Levine, B. Guinot, H. F. Fliegel, R. L. Beard, and T. R. Bartholomew, "The Leap Second: Its History and Possible Future," *Metrologia* **38**, pp. 509 – 529 (2001).

[9] J. Hahn, European Space Agency Galileo Project Office, Letter to the CCTF (March 26, 2004).

TIME DISSEMINATION VIA TELEPHONE LINES USING COMMERCIAL MODEMS

ANCA NICULESCU, MONICA ANGELESCU
National Institute of Metrology, Sos. Vitan Barzesti 11, 042122 Bucharest, Romania

Since 1991, most time and frequency laboratories perform standard time dissemination via telephone lines using commercial modems. The time information generated by this system is referenced to the national time standard by direct telephone line and it is accessible to anyone who uses an external modem.

The dissemination service was designed to synchronise PC clocks or digital terminals with accuracy ranging from 10 ms to 100 ms.

In this paper we will present the experimental results obtain, both in laboratory conditions andconnected to the telephone network, in order to evaluate the timing capability of the system. We used a time distribution system (developed at Technische Universität Graz, Austria) which have the capability to correct the propagation delays and a commercial modem. The dissemination system used the European code, consisting of an 80 ASCII characters line transmitted each second, supplying the date, the time and some information relative to the corrections performed on the legal time or special warnings concerning time scales.

We have measured the arrival times of the coded time information in different locations all over the country and also the synchronisation errors for a PC clock under two different operation system (DOS and WINDOWS).

The results indicate errors of some tens of ms for DOS clock and of some hundreds ms for CMOS clock.

The realisation of this experimental coded time/data information dissemination system represents the first attempt to disseminate legal time, at national level, in Romania and also an attempt towards an European standardisation.

GPS C/A-CODE TIME-TRANSFER STABILITY IN ASIA

Miho Fujieda

National Institute of Information and Communications Technology
4-2-1, Nukui-Kitamachi, Koganei, Tokyo, 184-8795, Japan
miho@nict.go.jp, fax −81 42 327 6664

Keywords: GPS time transfer

Abstract

It was known that stability of time transfer by GPS C/A-code time-transfer in Asia was worse than that in Europe and the United States. Main sources to degrade the stability are the receiver-position error and the ionospheric delay. The stabilities in Asia were worse than those in the Europe and the United States when using the Total Electron Content (TEC) map by the Center for Orbit Determination in Europe (CODE), while the results compensated with measured ionospheric delay were compatible with those in other areas. We investigated the effect with the vertical position error of the receiver. It became clear that the time-difference term between the GPS time and the local time depended on the elevation and azimuth angles and the short-term stability was degraded due to the errors. When the magnitude of the error was less than a few meters, we could not discriminate the existence of the error due to other errors. However, some receiver coordinates in Asian laboratories seemed to include a error. These results encourage us to use the measured ionospheric delay and the adequate receiver coordinates in order to improve the stability of time transfer in Asia.

1 Introduction

The International Atomic Time scale (TAI) is computed by BIPM using data of atomic clocks over two hundred. In order to compare remote clocks, GPS time transfer, Two-Way Satellite Time and Frequency Transfer (TWSTFT) are used widely. It is necessary that the stability of time transfer is lower enough than that of clock. The averaging time of a few days is required to contribute for TAI as for GPS time transfer due to its stability. In order to improve the stability of GPS time transfer, the carrier phase observation[1, 2] and the use of the geodetic receiver have been tried.

In the meantime, they say that the stabilities of GPS time transfer in Asia are worse than in Europe and the United States. Figure 1 shows the differences between results of time transfers with the GPS C/A-code common-view[3] and TWSTFT[4]. Their locations of USNO, NPL, PTB, CRL and TL are presented in Figure 2. We can see

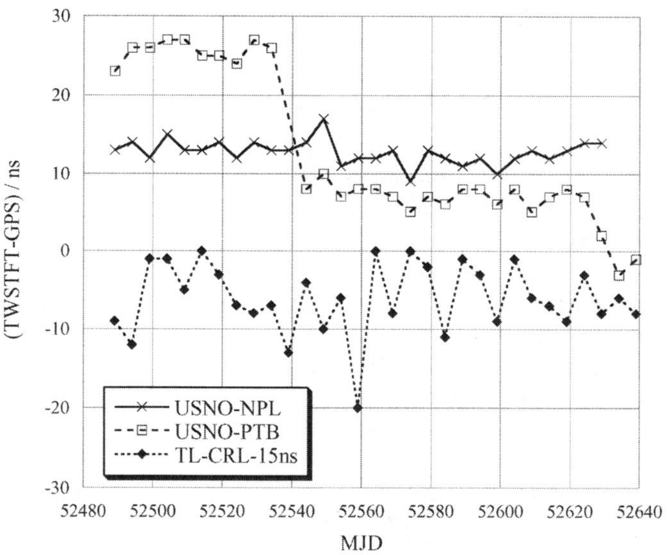

Figure 1: Differences between TWSTFT and GPS C/A-code common-view.

that the dispersion in the CRL-TL link was clearly larger than in other links. So possible causes considered. The observation equation of the satellite k for the GPS C/A-code common-view between the receivers i and j is written as

$$\begin{aligned} P_i^k - P_j^k &= \rho_i^k - \rho_j^k + I_i^k - I_j^k + T_i^k - T_j^k \\ &+ c(dt_i - dt_j) + e_i - e_j, \end{aligned} \quad (1)$$

where ρ is the geometric distance between the phase centers of the satellite antenna and the receiver antenna, I is the ionospheric refraction delay, T is the tropospheric refraction delay, dt is the difference between the local time and the satellite time and e is the measurement error including the receiver internal delay. On the other hand, the observation equation of time transfer by TWSTFT is given as[5]

$$\Delta T_i - \Delta T_j = 2(dt_i - dt_j) + (e_i - e_j) + t_s, \quad (2)$$

where ΔT is the signal-receiving time and t_s is the Sagnac effect. The time transfer by TWSTFT is more stable than GPS C/A-code common-view. Comparing between equations (1) and (2), ρ, I and T are possible to degrade the stability of GPS time transfer. It is known that T is compensated well by the Saastamoinen model[6]. Because ρ is the vector connected with the phase centers of the satellite and the receiver, both the satellite-orbit and the receiver-position coordinates are possible to cause the error. However, the satellite-orbit error does not cause the regional dependence. The stability of GPS time transfer in Asia seems to be worse because of the ionospheric delay and the receiver position error consequently.

We compare the stabilities of GPS time transfer when the ionospheric delay was compensated with a model and the measured value in Section 2. How the receiver-position error affects the stability is studied in Section 3. Section 4 concludes the paper.

2 Ionosopheric delay correction

We used the Global ionosphere map (GIM) by the Center of Orbit Determination in Europe (CODE)[7] and the measured value by the dual-frequency observation for the ionospheric correction. We call the former as "TEC" and the latter "MSIO" in this paper. The ionospheric correction by TEC is used in the computation of TAI by BIPM. It turned out that the accuracy around the equator is worse than that in higher latitudes[8].

The stabilities of the CRL-NMLS, NPL-USNO and CRL-TL links were compared. The used data in the analysis were downloaded from the BIPM website. Their locations are plotted in Figure 2. CRL, NMLS and TL have the same multi-channel receivers which can observe the dual-frequency signals. NPL and USNO use also multi-channel receivers. The baseline length of the CRL-NMLS, NPL-USNO and CRL-TL are 3800km, 5700km and 2100km, respectively. We used the period from 9/1/2003 to 11/30/2003 in the CRL-NMLS and NPL-USNO links, while the period of the used data in the CRL-TL link was from 9/1/2003 to 9/30/2003.

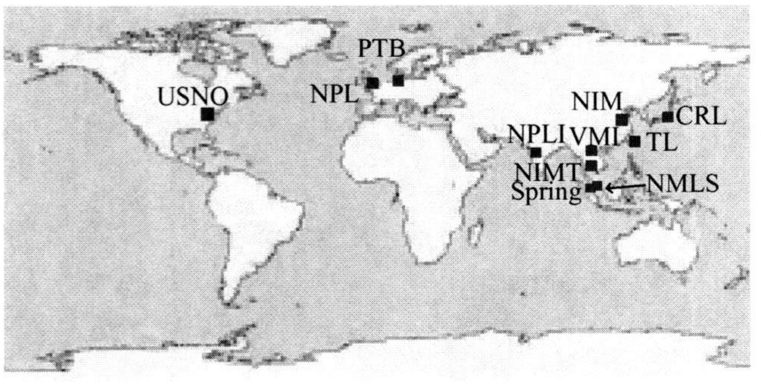

Figure 2: Location of the laboratories used in the data analysis.

Figure 3 shows the time-scale differences of the CRL-NMLS and NPL-USNO links. The absolute values are meaningless because the offset was added for better visibility. It is clearly seen that the daily variation of the NPL-USNO link was dumped but that of the CRL-NMLS link existed in the case of TEC compensation. On the other hand, when the ionospheric delay was compensated by using the MSIO, the variation almost disappeared.

The GPS observation schedule is determined by

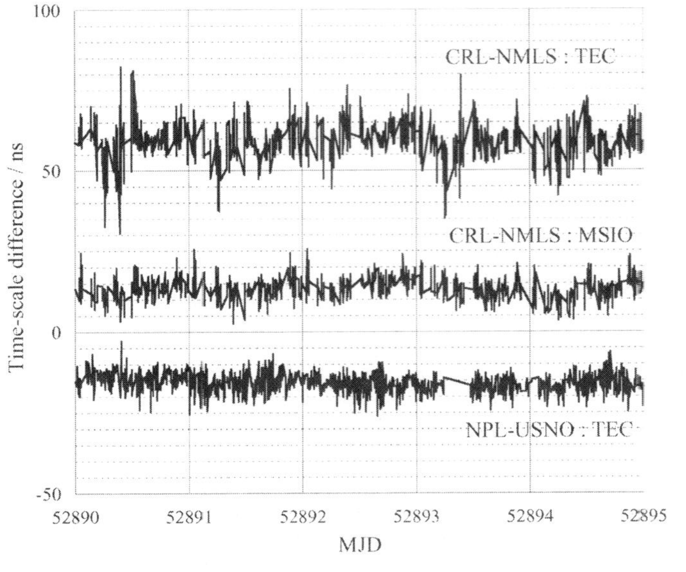

Figure 3: Time-scale differences of CRL-NMLS and NPL-USNO links.

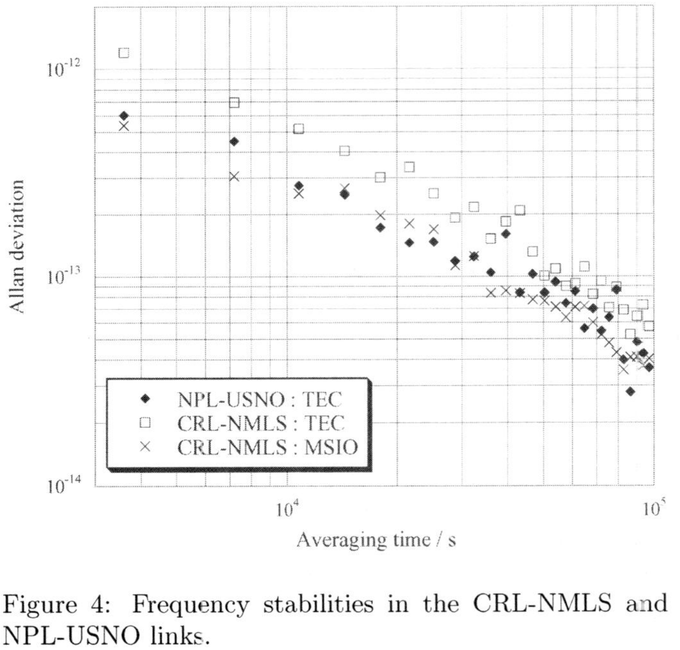

Figure 4: Frequency stabilities in the CRL-NMLS and NPL-USNO links.

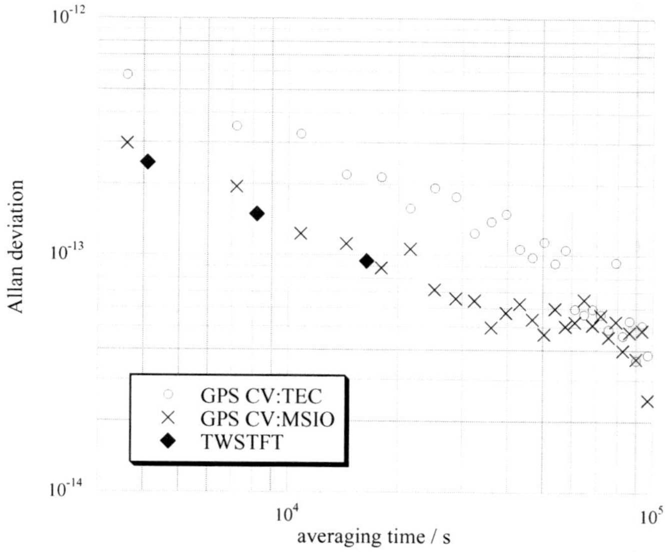

Figure 5: Stability comparison between GPS CV and TW-STFT of the CRL-TL link.

BIPM in order to be possible for the common-view with other laboratories. However, the observation is not series, which interval jumps at changing the date. So we averaged the observation values every hour and then calculated the frequency stabilities. Figure 4 shows the frequency stabilities of the CRL-NMLS and NPL-USNO links. Using the compensation by TEC for the CRL-NMLS link, the short-term stabilities under 10^5 seconds were worse than of the NPL-USNO link in spite of the shorter baselines. While the ionospheric correction was done by using the MSIO, their stabilities became equivalent to those of NPL-USNO link.

Figure 5 shows the comparison of the stabilities between GPS C/A-code time-transfer and TWSTFT in the CRL-TL link. In the case of MSIO, the stabilities became equivalent to TWSTFT. The stabilities of TWSTFT and using MSIO seemed to represent those of Cs clock. It proved that the stability using MSIO was stable enough to perform time-transfer of Cs clock. These results agree that the stability of GPS time transfer in Asia is affected by the ionospheric delay and possible to improve it by using the measured ionospheric delay.

3 Effects of receiver-position error

Because the GPS time transfer uses the time-difference term between the GPS time and the local time used by the receiver, the geometric distance and other terms should be fixed. If there is an error of the receiver position, the time-difference term also includes the error, which cannot be discriminated from errors caused by other terms. It seems that the vertical receiver-position error makes effects to not only the geometric distance but also the refraction delays by the atmosphere and seems to be more important than horizontal coordinates consequently. In this section, the influence on the GPS time transfer was studied and we tried to investigate whether the stabilities of GPS time transfer in Asia were degraded by it or not.

3.1 Observation equation

It is assumed that there is a vertical error δZ of the receiver position. A vector between the satellite antenna and the receiver antenna is written as (X, Y, Z) and the geometric distance is given as ρ. The observation equation for GPS pseudo-range of a receiver i to a satellite k is written as

$$\rho = (X^2 + Y^2 + Z^2)^{1/2}, \quad (3)$$

$$P_i^k = \rho_i^k(\delta Z) + I_i^k(\delta Z) + T_i^k(\delta Z) + c(dt_i - dt^k) + e_i^k, \quad (4)$$

where I is the ionospheric refraction delay, T is the tropospheric refraction delay, dt is the clock error and e is the measurement error including multi-path effects and the delay of the receiver. If the ionospheric delay is compensated by using TEC and the tropospheric delay is done by the Saastamoinen model, they depend on the elevation angle, therefore, depend on δZ. In the case of the measured ionospheric delay by the dual-frequency observation, I is independent of δZ.

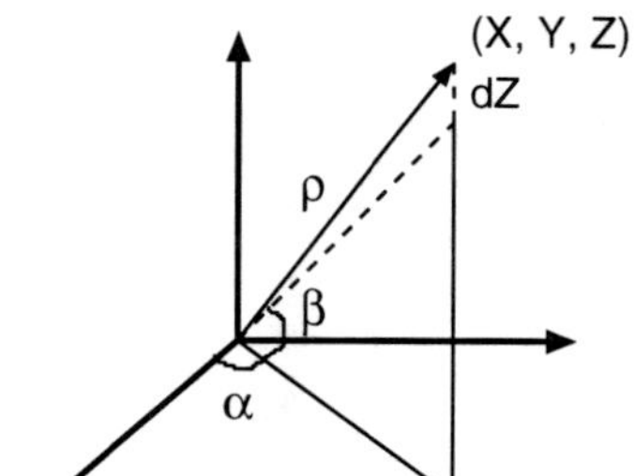

Figure 6: Vector between the satellite antenna and the receiver antenna. The original point is the phase center of the receiver antenna.

The elevation angle from the receiver to the satellite is represented as β and the azimuth angle is done as α. Using the geometric relation of $Z = \rho sin\beta$, the errors in the geometric distance and the elevation angle caused by δZ are given as

$$\delta\rho = \frac{Z\delta Z}{\rho}, \quad (5)$$

$$\delta\beta = \frac{\rho - Zsin\beta}{\rho^2 cos\beta}\delta Z. \quad (6)$$

We estimated their magnitudes of $\delta\rho$ and $\delta\beta$ in the following way. The satellite position of PRN 10 was obtained by using data observed at one of IGS stations named KGN0. The vectors between the satellite and the receiver were calculated from their positions and a vertical constant error was given in the receiver position. Figure 7 shows the estimation results of $\delta\rho$ and $\delta\beta$ when δZ is 1m, 5m and 10m, respectively. It becomes clear that the magnitude of $\delta\beta$ is quite small and negligible. The ionospheric correction using TEC is given from the coordinates of the cross point between the vector, (X, Y, Z) and the ionosphere. Because $\delta\beta$ is small and the mesh size of TEC map is larger, the coordinates of the cross point hardly change due to δZ and the ionospheric delay is not affected from δZ. The tropospheric delay is independent of δZ in the same way. Because δZ hardly cause the errors in the atmospheric refraction delays, we can see that it is not special component and the horizontal coordinates also have the same magnitudes in $\delta\rho$.

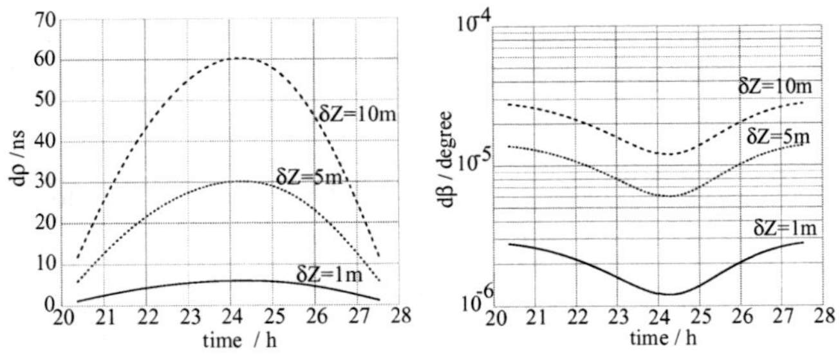

Figure 7: Magnitude of $\delta\rho$ and $\delta\beta$ when δZ is 1m, 5m and 10m, respectively.

3.2 Simulation results

In order to see how to affect the time-difference term due to δZ, a simple simulation was performed. We added the vertical constant error in the coordinate of KGN0, which observed results in RINEX format[9] were converted to the CGGTTS format[10] using R2CGGTTS program[11]. The satellite orbit was calculated by using GPS navigation message and the ionospheric delay was compensated with P3 in R2CGGTTS program. The data from 12/16/2003 to 12/31/2003 were used for the simulation.

Figure 8 and 9 show the elevation-angle and azimuth-angle dependency of the time-difference term, RefGPS, using all satellite's data. From equation (5), $\delta\rho$ is proportional to Z, which is equivalent to be proportional to the elevation angle. RefGPS values consequently show the gradient in figure 8. It seems that $\delta\rho$ causes errors in X and Y components and so that RefGPS values depend on the azimuth angle too. The results with the horizontal position error were not presented in this paper, where RefGPS showed the dependence on the elevation and azimuth angles. The effects for RefGPS values are not clearly seen when δZ is 1m. This is because other error sources, for example, the satellite orbit, are large enough so as to cover the effects due to δZ.

Figure 10 and table 1 show the frequency stabilities and the standard deviations of RefGPS in case of δZ=1m, 5m and 10m, respectively. The short-term stabilities up to about 4×10^4s are clearly degraded because of δZ, which seems to be concerned with the semi-diurnal periodicity of the satellite orbit.

These results agree that a vertical and horizontal position errors of a receiver disperse the RefGPS values and degrade the short-term stability. The dependence on the elevation and azimuth angles should be checked to know whether there is a position error larger than a few meters or not.

δZ / m	Std. Dev. / ns
none	5.3
1	5.4
5	8.9
10	15.3

Table 1: Standard deviations of RefGPS in case of δZ=1m, 5m and 10m, respectively.

3.3 Elevation-angle dependence

The dependence on the elevation angle of the RefGPS value in Asian laboratories was checked. The data between 9/1/2003 and 11/30/2003 were selected. After the drift component was removed, the offset was added to the RefGPS value for better visibility. Figure 11 shows the results. The less number of NIM and NPLI data were due to the single-channel receiver. The ionospheric delays of NMLS, CRL and TL were compensated with the measured ionospheric delay, while those of others were done by using TEC. The results of VMI, NIMT and Spring showed large gradient and dispersion. Those receiver positions seem to include errors. We cannot distinguish components of X, Y and Z from the results because the horizontal and vertical ones contribute in the same degree.

4 Conclusion

We investigated the ionospheric correction and the vertical position error which were possible to degrade the stability of GPS C/A-code time-transfer in Asia. When the ionospheric delay was compensated by using the measured value instead of the TEC, the stability became equal to that of the Europe and the United States link.

The vertical position error did not affect the elevation angle, which caused the error in the atmospheric delay negligibly. The error of the geometric distance was dominant in such case, where the horizontal and vertical co-

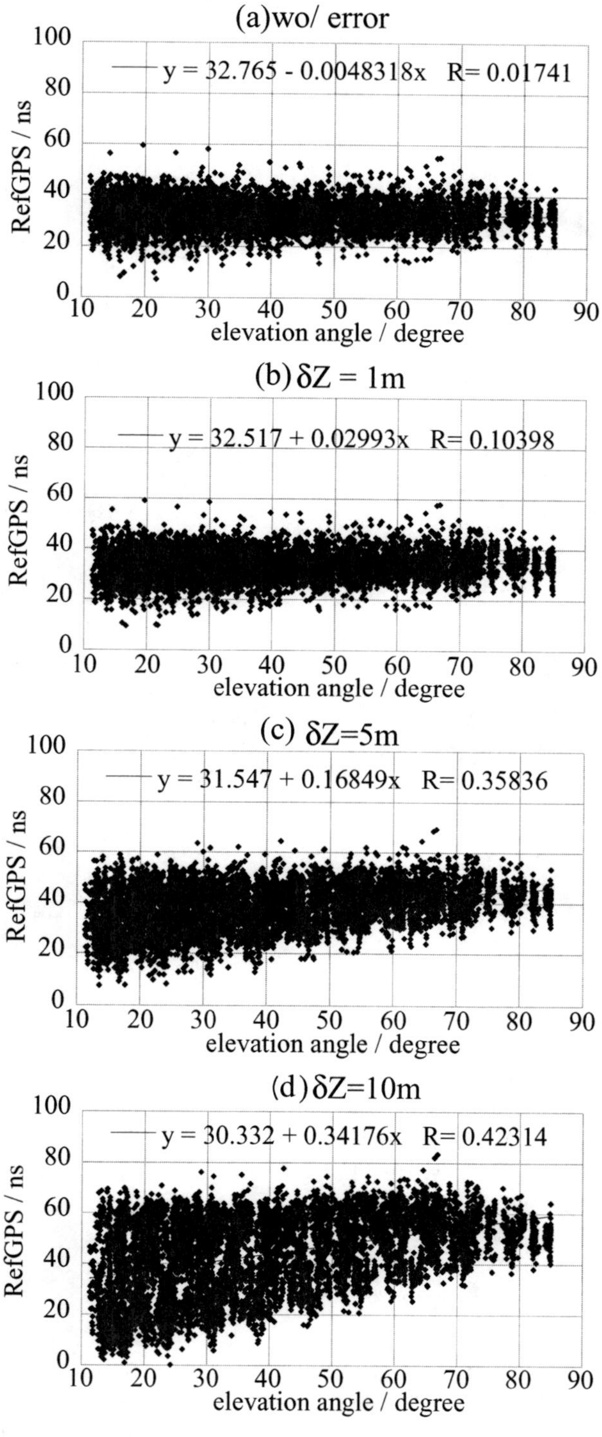

Figure 8: RefGPS in case of δZ=1m, 5m and 10m, respectively. The horizontal axes represent the elevation angle from the receiver to the satellite.

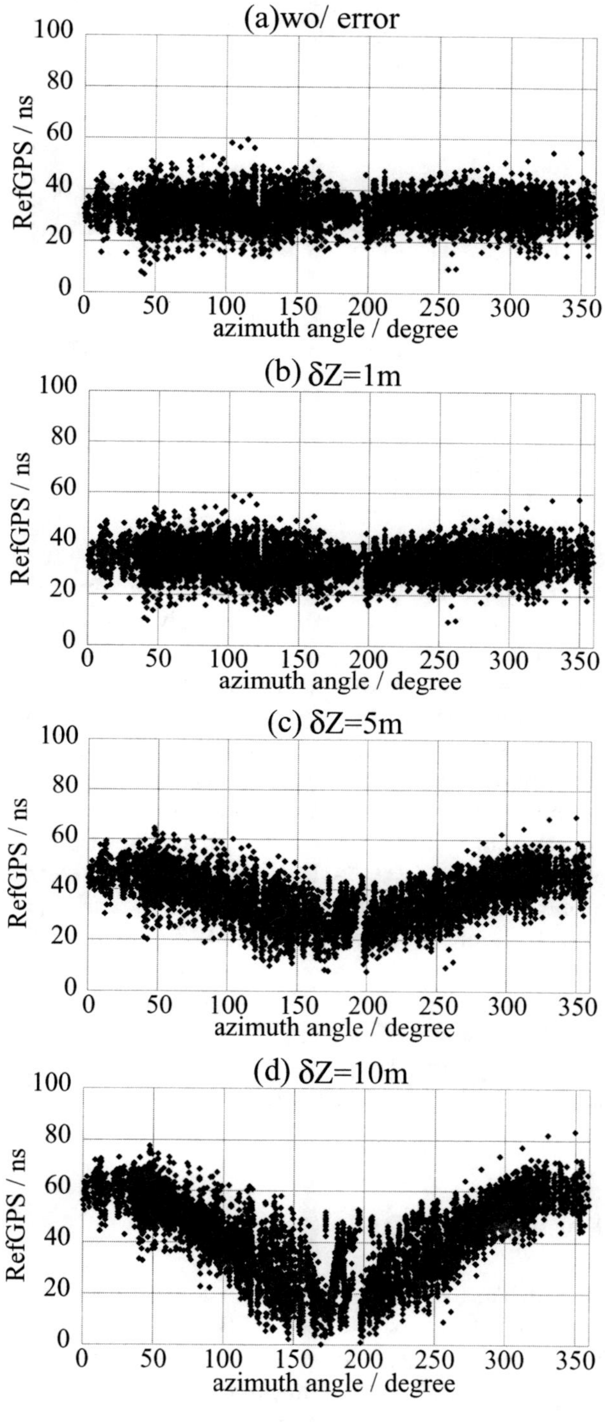

Figure 9: RefGPS in case of δZ=1m, 5m and 10m, respectively. The horizontal axes represent the azimuth angle from the receiver to the satellite.

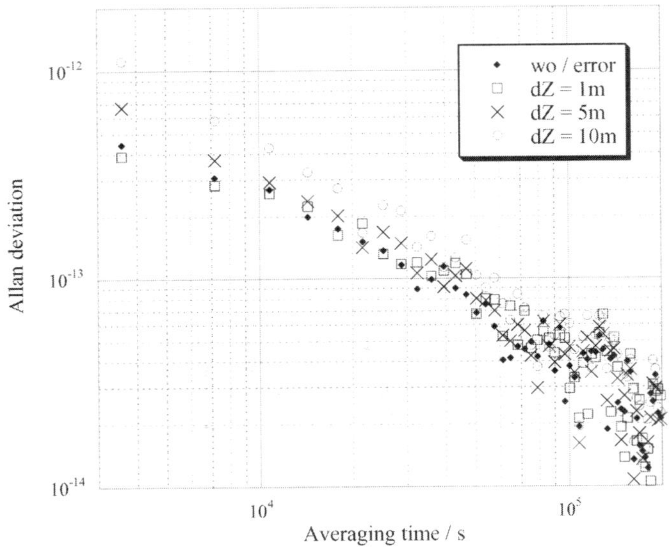

Figure 10: Frequency stability of RefGPS in case of δZ=1m, 5m and 10m, respectively.

ordinates had the same magnitudes. It turned out that the short-term stability of the time-difference between the GPS time and the local time was clearly degraded and the dispersion became large by the position error in the receiver. The influence due to the error which magnitude was less than a few meters could not be seen and distinguished from other errors. If there was a position error, the time-difference showed the dependence on the elevation and azimuth angles, which was helpful for the estimate of the presence of the error. We checked the elevation-angle dependency of data in the Asian laboratories. The receiver coordinates of some laboratories seemed to be inadequate. In order to improve the stability of GPS C/A-code time-transfer in Asia, using the measured ionospheric correction is appropriate and the receiver-position error should be less than a few meters. The discrimination of smaller position-error might be possible using the precise orbit provided by IGS. We might try to estimate the error magnitude.

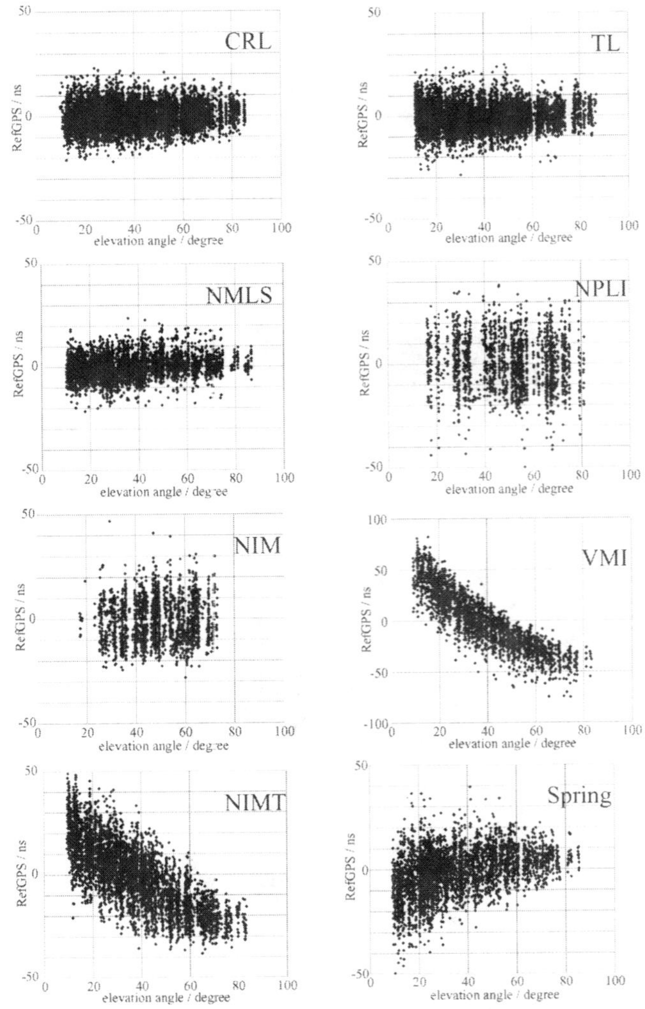

Figure 11: Dependence of the elevation angle of CRL, TL, NMLS, NPLI, NIM, VMI, NIMT and Spring

Acknowledgments

We would like to thank Mr. Gotoh for his cooperation. We would like to be grateful to Dr. Banerjee in NPL India for his fruitful advices. And we would like to thank Mr. Tisserand in BIPM for helping about the R2CGGTTS. We would like to thank Dr. Suzuyama about TWSTFT.

References

[1] R. Dach et al., "Time transfer using GPS carrier phase: error propagation and results", Journal of Geodesy, **77**, pp. 1-14.

[2] T. Gotoh et al., "GPS Common-View with Geodetic Receivers", in this proceedings.

[3] D. W. Allan, M. A. Weiss, "Accurate Time and Frequency Transfer during Common-View of a GPS Satellite", Proc. 34th Ann. Freq. Control Sympo., pp. 334-336, 1980.

[4] 23rd BIPM TWSTFT Report, 2003.

[5] H. Sun et al., "Performance of Two-way Satellite Time and Frequency Transfer in Asia-Pacific Region", Journal of Geodetic Society of Japan, **49**, No. 2, pp. 135-142, 2003.

[6] J. A. Klobuchar, J. M. Kunches, "Comparative range delay and variability of the earth's troposphere and the ionosphere", GPS Solutions, **7**, pp. 55-58, 2003.

[7] http://www.cx.unibe.ch/aiub/ionosphere.html

[8] M. Fujieda et al., "Effects of ionospheric correction on GPS time transfer throughout Asia", Metrologia, **41**, pp. 145-151, 2004.

[9] ftp://igscb.jpl.nasa.gov/igscb/data/format/rinex210.txt

[10] D. W. Allan, "Technical Directives for Standardization of GPS Time Receiver Software", Metrologia, **31**, pp. 69-79.

[11] P. Defraigne et al., "Time Transfer to TAI Using Geodetic Receivers ", Proc. of 15th EFTF, pp. 517-521, 2001.

PHASE MEASUREMENT AMBIGUITY RESOLUTION IN TIME TRANSFER BY METEOR BURST CHANNEL

VLADIMIR ALEXANDROVICH KORNEYEV, Kazan State University
420111 Kazan Kremlevskaya 18 Russia
Email: Vladimir.Kornejev@ksu.ru

A technique used in meteor synchronization equipment constructed in Kazan State University includes transmission of time marks on several carrier frequencies. Time shift between stations is calculated on secondary station by getting the differences between phases of signal carrier waves corresponding to phases of imaginary differential frequencies and resolving their ambiguity. Phase ambiguity resolution procedure goes from lowest to highest differential frequencies at each step increasing precision of the resulting measurement. The final step in the ambiguity resolution procedure—resolving ambiguity of carrier frequency phase is the hardest step requiring filtration of measurements, because for technical reasons the ratio of errors is the
highest here. The reference for carrier frequency phase ambiguity resolution is unambiguous maximal differential frequency phase.

Real-time filtration of unambiguous measurements can be performed in various ways. In the simplest case the unknown period of carrier frequency is chosen by current Kalman's estimate.

This is not always possible because due to short-term instability of frequency standards Kalman's estimate error does not necessarily reach given threshold for reliable ambiguity resolution, even on long filtration intervals. One way to deal with that is to accept whatever Kalman's estimate value is available right away and check if the resolution has been correct later, when more precise smoothing estimate becomes available. This approach is natural in meteor burst communication where automated repeat request is widely used. Other promising approach is to resolve first the ambiguity of one or more of the previous measurements where smoothing estimate is available. Thus ambiguity of measurements with better estimates gets resolved first and, if possible, it makes estimate error for later measurements small enough for reliable resolution. Second method uses history more efficiently making first resolution decisionby estimate with lower error. On the other hand actual errors of real-time ambiguity resolution in that case are hard to analyze because possible resolution errors on each step affect the estimates by which following decisions are made.

Presentation shows the possibilities of real-time unambiguous time transfer on every available meteor with errors corresponding to those of single carrier frequency phase measurement (fractions of nanosecond on the average). Both discussed methods and their combinations are tested by modeling to provide presented statistics.

Phase Correction of Reference Frequency signals transmitted by optical fiber link

J. Saalaoui[1], F. Lardet-Vieudrin[2], E. Rubiola[3]
S. Galliou[4], F. Vernotte[1], and V. Giordano[2]
[1]Laboratoire d'Astrophysique de l'Observatoire de Besançon, LAOB, UMR-6091
41bis, av. de l'Observatoire, BP1615, F-25010 Besançon, France
e-mail : saalaoui@obs-besancon.fr
[2]dept. LPMO, Institut FEMTO-ST, UMR-6174, 32 av. de l'Observatoire, F-25044 Besançon, France.
[3]Université Henri Poincaré, ESSTIN and LPMIA
2 av. Jean Lamour, F-54519 Vandoeuvre Lès Nancy, France
[4]LCEP, Institut FEMTO-ST, UMR-6174, ENSMM, 26 chemin de l'Epitaphe, F-25030 Besançon, France

***Abstract*— The time and frequency reference in the three laboratories of Besançon is based on Cesium clocks for absolute accuracy and long-term stability at the LAOB, on a H-maser for medium and long term stability at the LCEP, and on a liquid-He whispering gallery sapphire oscillator for best spectral purity and short-term stability at LPMO. These standards are located at three different sites, at a few hundreds meter from one another. In order to obtain the full accuracy and stability at least at one site, we combine these references using a frequency locked loop. The 100 MHz frequency reference is transferred via a two-way optical link. This article reports on measurement of the the phase noise and of the stability of the link.**

I. Introduction

Time and frequency research in Besançon is done in three different laboratories, LPMO, LCEP and LAOB, located in a region of a few hundreds meters. Each of these laboratories have its role, its needs, its traditions, and of course its own frequency standards. The LAOB has three HP-5071A-001 Cs clocks, which provide the highest accuracy and long-term stability ($\tau > 1$ day). These standards contribute to TAI through differential GPS comparison. The EFOS-21 H-maser at the LCEP shows high medium-term stability (parts in 10^{-15} for τ in the 10^2–10^5 range). The cryogenic sapphire oscillators in progress at the LPMO exhibits the lowest phase noise and the best short-term accuracy (10^{-14}), but drifts for τ greater than about 1 hour. Figure 1 shows the Allan deviation $\sigma_y(\tau)$ of the available standards, as a function of the measurement time τ. For future projects, each laboratory requires to extend the high-stability region of its standards. In addition, the full accuracy and stability should be available at least at one site. On the other hand the standards can not be moved for a variety of reasons, both administrative and technical. We noticed that other laboratories have similar problems.

Pioneering work in frequency transfer via optical links was done at the JPL in the late '70s [1] for the radioastronomy Deep Space Network (DNS). Later, using the phase-conjugation method [2], the phase noise of a 100 MHz two-way link was of -120 dBrad^2/Hs at the Fourier frequency $f = 1$ Hz. This method has the relevant advantage that it does not require a variable delay device, which is difficult to implement in optics. Nonetheless, the new version of the DNS frequency transfer system makes use of an optical delay unit based on a temperature-controlled fiber spool [3]. A similar solution [4] is in use at the Mizusawa Observatory, Japan, for VLBI measurements. The SYRTE (Astronomical Observatory of Paris) is studying a two-way link for metrological applications [5], yet at significantly longer distance (44 km). Surprisingly (for us) one-way frequency distribution systems through optical fiber are under study. They are aimed to transfer 10 MHz [6] and 9 GHz [7] reference signals.

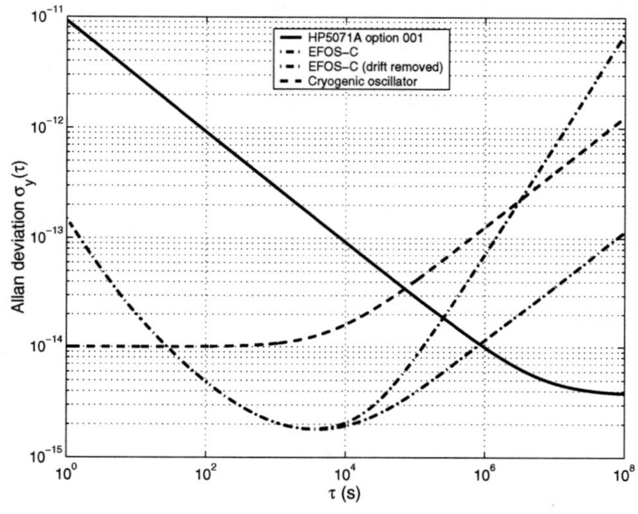

Fig. 1. Allan deviation $\sigma_y(t)$ of the fractional frequency y of the available frequency references.

Fig. 2 shows the two-way frequency distribution network in Besançon. A 100 MHz signal modulates the power of a $\lambda = 1.3$ μm carrier transmitted through a monomode optical fiber. The 100 MHz signal is restored by a conventional photodetector. The two-way link consists of two equal channels of this type. Consequently, the round-trip signal undergoes detection and modulation at half way, where it is available as the 100 MHz transferred reference.

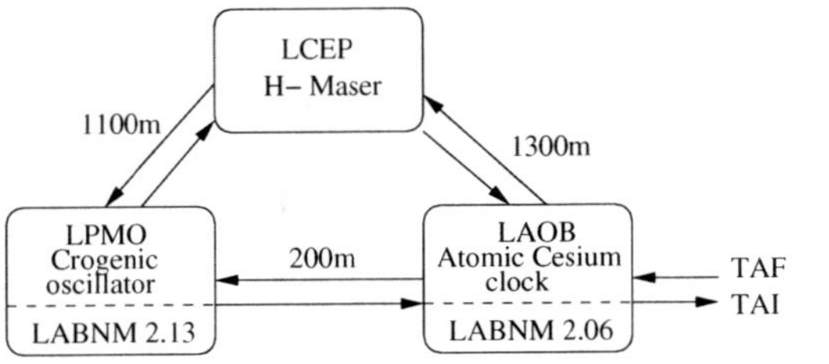

Fig. 2. Optical link between the three laboratories. Distances refer to fiber length, through underground path.

Two-way measurements rely upon channel symmmetry. Delay symmetry is guaranteed by the adoption of equal optical transmitters and receivers, and by the use of equal optical fibers in the forward and backward path. The fibers are thermally coupled. The fiber path is underground, at an average depth of 0.5 m, which reduces and filters the thermal fluctuations, and in turns improve symmetry.

In the first experiment (Fig. 3), we compare the phase of the round-trip signal to the local reference. The phase detector provides a voltage $v(t)$ proportional to the instantaneous phase fluctuation of the channel.

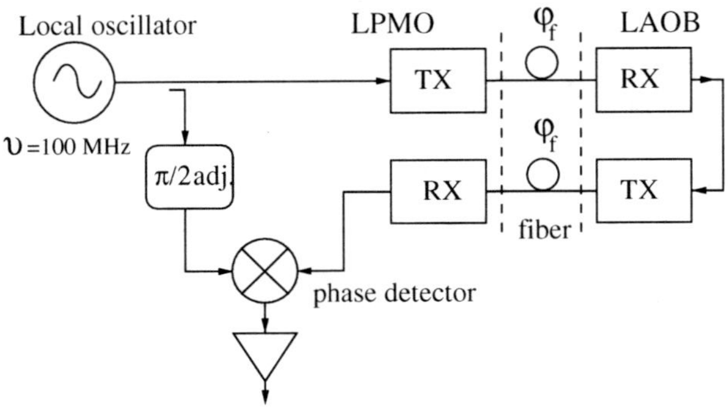

Fig. 3. Phase noise measurement of the optical link.

The measured spectrum (Fig. 4), $S_\varphi(f) = 1.6\times10^{12} f^{-1} + 2\times10^{-14}$ rad^2/Hz. In the short term, thus at Fourier frequencies higher than 1 Hz, no environmental phenomena (mainly temperature) can affect the delay of the fiber. Consequently, the measured spectrum is due to the noise of the transmitter and of the receiver.

At the present time this result is satisfactory. Thus we focus our work on medium-term stability, which in a two-way link also warrants long-term stability.

The medim-term stability of the optical link is limited by the dayly temperature fluctuation. In monomode fibers, the temperature coefficient of the refraction index, thus of the delay, is $dn/dT \simeq 6.85\times10^{-6}$ K^{-1}. Under the assumption that the dayly thermal cycle is about sinusoidal in the fiber, which is underground, a fluctuation of 1 degree K peak yields a frequency fluctuation of 3.3×10^{-15} over a 1.3 km path. We expect that the actual temperature fluctuation is of a few degrees, for the one-way frequency stability can not be better than some 1.5×10^{-14}.

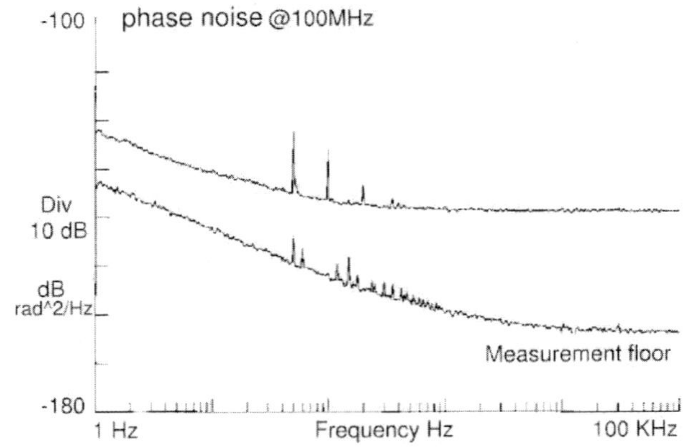

Fig. 4. Phase noise with (upper curve) or without optical link .

II. Phase compensation of the optical link

The short-term stability needed for the whispering gallery oscillators can hardly be achieved with a thermal compensation (heated fiber spool), which has a time constant of several minutes. Thus we opt for the two-way scheme shown in Fig. 5. The idea is to measure, and then to compensate half of the round-trip delay fluctuation, that is, the one-way delay fluctuation. This is accomplished by doubling frequency of the the local 100 MHz, and by up-converting to 200 MHz the received signal.

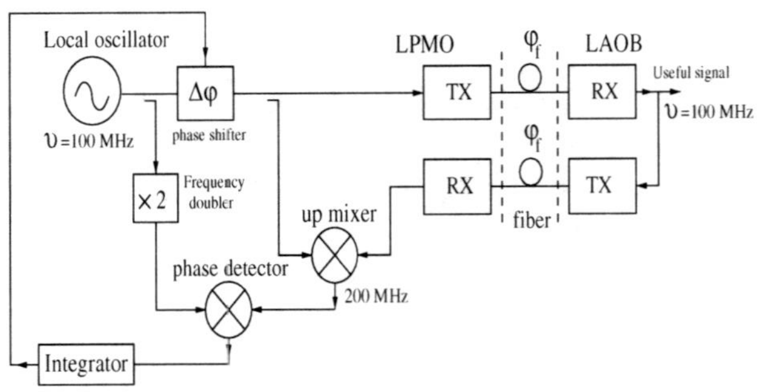

Fig. 5. Phase compensation scheme.

Let $\nu_0 = 100$ MHz the reference frequency, ϕ_f the phase fluctuation of the one-way path, and $\Delta\phi$ the correction introduced by the phase shifter. Dropping the amplitude, which is not necessary here, the signal is

$$\sin(2\pi\nu_0 t + \Delta\phi + \phi_f) \tag{1}$$

at destination (LAOB in Fig. 5), and

$$\sin(2\pi\nu_0 t + \Delta\phi + 2\phi_f) \tag{2}$$

received after the round-trip. After up-conversion, we get the 200 MHz signal

$$\sin(4\pi\nu_0 t + 2\Delta\phi + 2\phi_f) \quad . \tag{3}$$

Comparing the latter to the local reference $\sin(2\pi\nu_0 t)$ in a phase detector, we get a voltage proportional to $2\Delta\phi + 2\phi_f$.

This error is integrated and fed into the phase shifter. The steady-state closed-loop condition is

$$\Delta\phi + \phi_f = 0 \ . \tag{4}$$

Therefore the phase at destination [Eq. (1)] is compensated for the fiber fluctuation ϕ_f.

III. Results

The simplest way to measure the stability of the entire system is to bring the destination signal close to the source, as shown in Fig. 6. This configuration requires an additional TX-RX pair and two additional fibers.

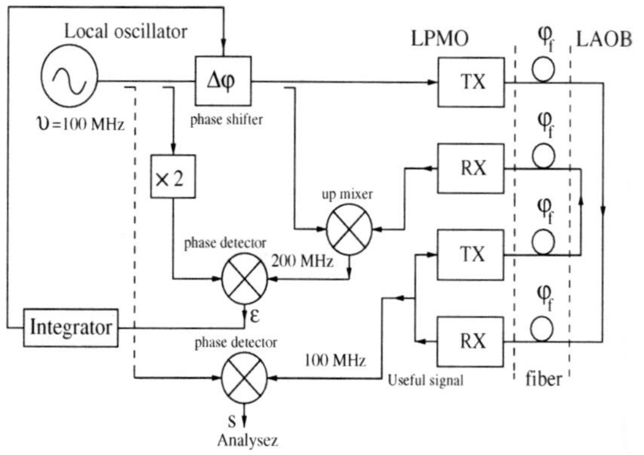

Fig. 6. Measurement of the two-way link stability.

Figure 7 shows the spectrum of the phase error, which is the phase difference betwen the reference and destination signal. The loop corrects the phase fluctuations below a cutoff frequency of some 10 kHz.

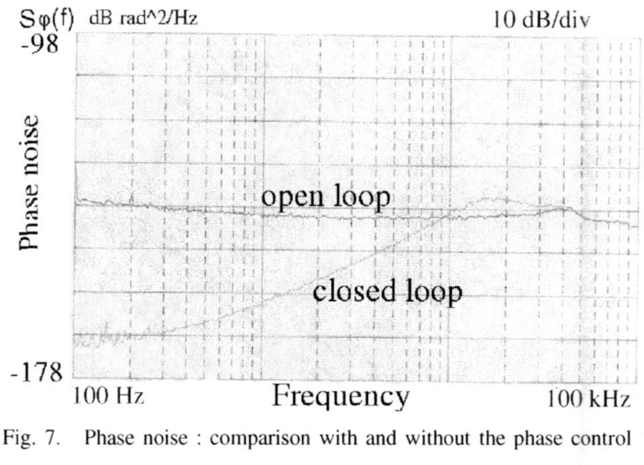

Fig. 7. Phase noise : comparison with and without the phase control

Figure 8 and 9 show the deviation $\sigma_x(t)$ and Allan deviation $\sigma_y(t)$ of the two-way link in open-loop and closed-loop conditions. The Allan deviation is sufficient to transfer all the available references, including the liquid-He whispering gallery oscillator, without stability degradation for $\tau \geq 1$ s.

Finally, we measured the response of the control to temperature changes of the optical fiber. This is done by inserting in a temperature-controlled oven two fiber spools that simulate the round-trip optical path. The results are shown in Fig. 10.

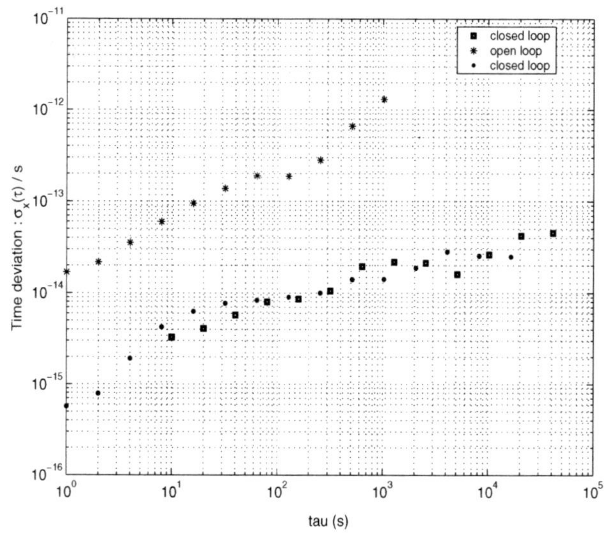

Fig. 8. Time deviation $\sigma_x(t)$ of the optical link.

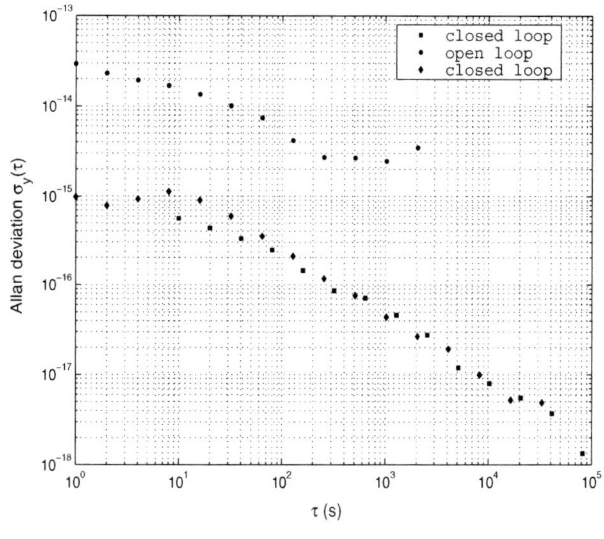

Fig. 9. (Allan deviation $\sigma_y(t)$ of the optical link.

IV. Final Remarks

The control still suffers from the presence of strong odd-order harmonics in the up-conversion mixer that feeds the 200 MHz signal into the phase detector (Fig. 5). In fact, that 200 MHz signal originates from the vector addition of several modes generated inside the mixer, each with its amplitude and phase. The main mode is the beat of two 100 MHz sinusoids. It produces most of the 200 MHz output power. Besides, the beat of each 100 MHz signal and the third harmonics of the other one produces a 200 MHz output. More generally, the beat between the i-th of each signal and the $(i+2)$-th harmonics of the other one contibute to the 200 MHz output. Although amplitudes drop rapidly as i increases, these modes are not negligible because phases are multiplied by i and by $(i+2)$. This phase multiplication is inherent in harmonic generation.

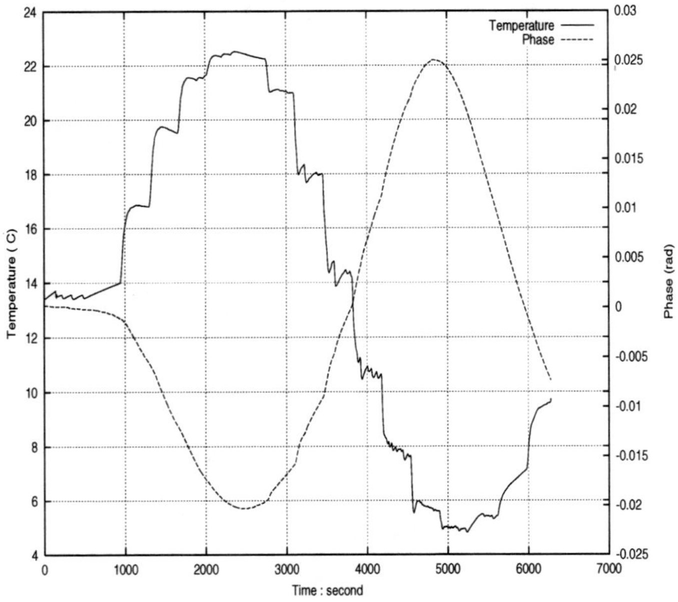

Fig. 10. Phase response to temperature sinusoidal variation.

The behavior of a mixer fed with synchronous signals of frequency ratio 1:2 was originally analyzed in Ref. [8], where the presence of the higher order modes was exploited to reduce the phase noise of regenerative dividers. In our case, the presence of higher order harmonics may cause systematic errors, depending on power and on the phase relationships in the up-converter mixer. A solution has been proposed by the SYRTE [5]. Yet this scheme shows other drawbacks, first of which complexity. The problem of harmonics is still under study.

REFERENCES

[1] G. Lutes,"Optical fibers for the distribution of frequency and timing references", Proc. 12th PTTI pp. 663–680, NASA Conference Publication CP-2175, Goddard Space Flight Center, Dec. 1980.

[2] L. Primas, G. Lutes, and D. Sydnor, "Fiber optic frequency transfer link", Proc. 42nd FCS pp. 478–484, Baltimore, Maryland, 1-3 June 1988.

[3] M. Calhoun, R. Wang, A. Kirk, G. J. Dick, and R. L. Tjoelker, "Stabilized reference frequency distribution for radio science with the Cassini spacecraft and the deep space network", Proc. 32nd PTTI, pp. 331–340, Virginia, 28–30 November 2000.

[4] K. Sato, T. Hara, S. Kuji, K. Asari, M. Nishio, and N. Kawano, "Development of an ultrastable fiber optic frequency distribution system using an optical delay control module", *IEEE Transact. Instrum. Meas.* vol. 49 no. 1 pp. 19–24, Feb. 2000.

[5] F. Narbonneau, M. Lours, C. Daussy, O.Lopez, A.Clairon, and G.Santarelli, "Ultra-stable optical links for metrological applications", Proc. *FCS and EFTF Joint Meeting* pp. 1041–1047, Tampa, FL, 5–8 May 2003.

[6] G. Quadri, H. Martinez-Reyes, B. Benazet, and O. Llopis, "A low phase noise optical link for reference oscillator signal distribution", Proc. *FCS and EFTF Joint Meeting* pp. 336–340, Tampa, FL, 5–8 May 2003.

[7] M.-B. Bibey, F. Deborgies, M. Krakowski, and D. Mongardien, "Very low phase-noise optical links — Experiments and theory", *IEEE Trans. Microw. Theory Tech.* vol. 47 no. 12 pp. 2257–2262, Dec. 1999.

[8] E. Rubiola, M. Olivier, J. Groslambert, "Phase noise in the regenerative frequency dividers", *IEEE Transact. Instrum. Meas.* vol. 41 no. 3 pp. 353–360, June 1992.

IMPROVED STEERING OF THE IGS TIMESCALES TO UTC

KEN SENIOR, U.S. Naval Research Laboratory (NRL), Washington, DC
GÉRARD PETIT, Bureau International des Poids et Mesures (BIPM), Sèvres, France
JIM RAY, Bureau International des Poids et Mesures, Sèvres, France

Currently the International GPS Service (IGS) timescales are aligned to the international timescale UTC by slowly steering (roughly consistent with a 30-40 d time constant) to observations of GPS time, which is itself steered to UTC via the UTC (USNO) realization. The instability of the IGS timescales has been determined to better than 2 parts in 10^{15} at 1 d, whereas the instability of GPS time is at least 10 times poorer. It is likely that due to its larger instability, steering to GPS time limits the stability of the IGS timescales for periods longer than a few days.

A new method (the topic of another paper at this conference) extends to geodetic systems the timing calibration from systems colocated at timing laboratories that contribute to UTC. These colocated geodetic systems and local UTC (k) realizations can have much better stabilities than GPS time. Using the empirical instrumental calibration biases together with predictions of UTC (k) offsets, suitable colocated geodetic clock data can provide a higher-quality and robust link of the IGS timescales to UTC. We discuss the techniques used to predict offsets of UTC (k) between monthly Circular T reports and assess the stability and overall quality of the IGS timescales as steered using these differentially calibrated systems instead of GPS time.

PERFORMANCE OF GPS ON BOARD CLOCKS COMPUTED BY IGS

Jérôme DELPORTE

CNES – French Space Agency – 18, avenue Belin – 31401 Toulouse Cedex 9
jerome.delporte@cnes.fr

Keywords: GPS, Galileo, IGS, clock.

Abstract

This paper presents the performances of GPS on board clocks computed by the International GPS Service (IGS) [1]. Using recently developed IGS combined clock products referenced to IGST, the identified GPS on board clocks have no longer day boundaries discontinuities, allowing extensive analysis on large integration times.

The aim of this paper is to compare the performances of the GPS on board clocks of each Block (Caesium and Rubidium clocks of Block II/IIA and Rubidium clocks of Block IIR) to the Galileo preliminary specifications.

1 Introduction

The IGS, along with a multinational membership of organizations and agencies, provides GPS orbits and clocks, tracking data, and other high-quality GPS data and data products on line to meet the objectives of a wide range of scientific and engineering applications and studies. The accuracy of the satellite and station clocks is announced to be better than 0.1 ns, while the orbits' accuracy to be less than 5 cm [1].

Historically, the internal time scale for the IGS clock products has been based upon a mere linear alignment to broadcast GPS time for each day separately. Because of the instability of GPS time, this method induces time and frequency day boundary offsets in IGS clock products.

A new strategy has recently been developed within an IGS/BIPM pilot project. A more stable reference is formed with a weighted ensemble of the IGS clocks loosely steered to GPS time [2,3]. Two timescales are generated : the IGRT corresponding to the rapid IGS combined clock products and the IGST corresponding to the final IGS combined clock products. Using the latter, we can infer the behaviour of GPS on board clocks on large integration times. In this paper, the stability and frequency drift of each GPS on board clock is presented and compared to preliminary Galileo specifications.

Inversely, the comparison with the expected performances allows us also to assess the quality of the IGS clock restitution and to compare it to other clock restitution algorithms [4].

Last, the navigation performance (1-σ extrapolation error) for each GPS clock will be given for a linear and a parabolic model and compared to theoretical results using the theory developed in [7].

2 Galileo on board clock preliminary specifications

In Global Navigation Satellite Systems, each satellite has to broadcast a certain number of time parameters so that the user can determine the satellite time offset (referenced to the phase centre of the satellite antenna) with respect to the system time. These time parameters may be a polynomial of first or second degree, which coefficients shall describe the offset for the interval of time during which they are transmitted. So, as far as timing is concerned, the system-level specification for the navigation payload is its time prediction performance.

For Galileo, the coefficients of this polynomial can be computed as follows : the raw data of the OD&TS (Orbit Determination and Time Synchronisation) process are collected with a τ_0 period during a time interval called T_m. At the end of T_m, the coefficients of the least-mean square model are computed and uploaded to the satellite to be included in the navigation message during a time interval referenced T_p.

The extrapolation error is the difference between the extrapolation of the model and the real time offset. We consider the maximum extrapolation error over T_p ; on average, this maximum is obtained at the end of this interval. The extrapolation error at the end of T_p is noted S_{ext}. The specification deals with the standard deviation of S_{ext} that is the one-sigma extrapolation error at the end of T_p. In this paper, this specification is referred to as navigation performance.

The specification deals in fact with the timing signal provided by the navigation payload. Of course this timing signal is mostly driven by the on board clock, but additional effects can degrade its performance such as the sensitivities of the on-board clock to environmental variations or the variations of the propagation delays within the equipments of the rest of the payload.

The figure below illustrates the adjustment/extrapolation process in case of a time linear model :

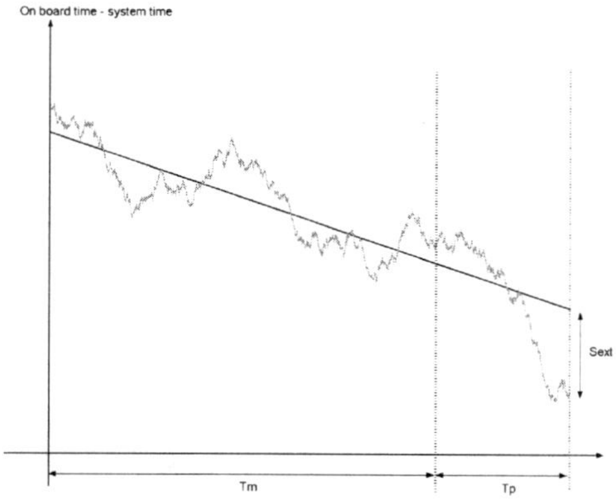

Figure 1 : Adjustment/extrapolation process

For Galileo, a preliminary specification for the Signal In Space Range Error was set to 65 cm. This SIS RE covers the orbit error and the timing error. Assuming a preliminary equal allocation between both factors, the specification for the time error is therefore 1.5 ns [9]. A preliminary study [9] had provided figures for T_m and T_p for the two foreseen Galileo on board clocks, the RAFS (Rubidium Atomic Frequency Standard) and the PHM (Passive Hydrogen Maser) :

	RAFS	PHM
T_m	10000 s	6000 s
T_p	50000 s	30000 s

Table 1 : Preliminary adjustment and extrapolation durations for foreseen Galileo on board clocks

In the following paragraph, the stability of the GPS on board clocks will be presented and their navigation performances will be computed with these values of T_m and T_p.

3 Performance of GPS on board clocks

3.1 Preliminary remarks

The analysis of GPS on board clocks behaviour is regularly performed by US organizations e.g. the National Research Laboratory (NRL). Their analysis is mainly focused on the long term behaviour and Hadamard or Allan deviation curves often start at an integration time of one day [5]. However the short term behaviour of GPS on board clocks is also very interesting. We will try to analyse in this paper whether their apparent short term stability is limited by the measurement system noise or is actually the one of the on board clock.

The analysed period consists of eight consecutive weeks : GPS weeks 1240 to 1247, that is from 12th of October to 6th of December 2003. The IGS clock products gives one point every 5 minutes. The clock products computed by IGS may present data gaps that are disturbing for stability analysis. To overcome this problem, the largest vector of continuous data is identified and processed. The table below summarizes the PRN/SVN, type of clock (as of December 2003) and maximum length of continuous data :

	SVN	PRN	Rb/Cs	Length of analyzed period (days)
Block II/IIA	13	02	Cs	19.8
	15	15	Cs	23.5
	17	17	Rb	27.7
	23	23	Cs – Rb *	-
	24	24	Cs	51
	25	25	Cs	32
	26	26	Rb	49.2
	27	27	Rb	56
	29	29	Rb	22
	30	30	Rb	56
	31	31	Rb	15.5
	32	01	Cs	56
	33	03	Cs	24
	34	04	Rb	43.8
	35	05	Cs	56
	36	06	Cs	18.2
	37	07	Rb	23.3
	38	08	Rb	53.9
	39	09	Cs	37
	40	10	Cs	56
Block IIR	41	14	Rb	56
	43	13	Rb	56
	44	28	Rb	56
	45	21	Rb	56
	46	11	Rb	56
	51	20	Rb	45
	54	18	Rb	56
	56	16	Rb	56

Table 2 : Summary of GPS constellation as of December 2003

* : the PRN 23 has undergone a clock switch in November 2003. This satellite will therefore be ignored.

3.2 Caesium clocks of Block II/IIA

The frequency stability for the different Caesium clocks is summarized below :

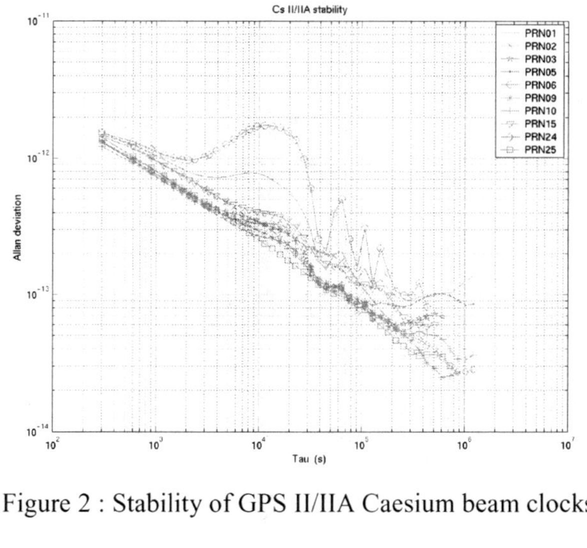

Figure 2 : Stability of GPS II/IIA Caesium beam clocks

These ten GPS Caesium beam clocks present a pretty similar stability, around $2\text{-}3.10^{-11}.\tau^{-1/2}$. Two of them, PRN 06 and 01, present significant periodic signatures which peak-to-peak are about 30 ns and 15 ns respectively and which period is the orbital period (12 h). A previous analysis [6] showed for PRN 06 that this periodic was correlated with the radial component of the Earth magnetic field and was therefore probably due to it, which would mean that the shields are shot. This systematic at the orbital period may be compensated partly by using the broadcast orbits [5].

Table 3 presents the navigation performance of the GPS on board Cs beam clocks for $T_m = 10.000$ s and $T_p = 6.000$ s :

	Standard deviation of S_{ext}	
	Linear model	Parabolic model
PRN 01	7.9 ns	10.7 ns
PRN 02	3.7 ns	7.8 ns
PRN 03	3.7 ns	6.3 ns
PRN 05	2.9 ns	5.6 ns
PRN 06	16.1 ns	16.1 ns
PRN 09	3.2 ns	6.1 ns
PRN 10	3.4 ns	5.9 ns
PRN 15	4.2 ns	7.7 ns
PRN 24	3.3 ns	5.8 ns
PRN 25	2.9 ns	6.2 ns

Table 3 : Navigation performance of GPS II/IIA Cs beam clocks for $T_m = 10.000$ s and $T_p = 6.000$ s

None of the GPS II/IIA Caesium beam clocks meet the Galileo 1.5 ns preliminary requirement.

The realizations of S_{ext} follow a normal distribution as predicted by a previous theoretical approach [7], except for PRN 06 which behaviour is dominated by the sinusoidal signature.

Here is an example of realizations of S_{ext} for PRN 25 for a linear model :

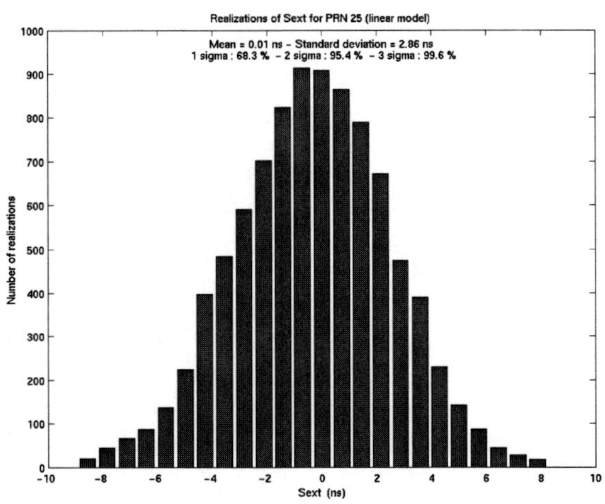

Figure 3 : Gaussian behaviour of the realizations of S_{ext}

This theoretical approach predicts a performance between 2.3 and 3.4 ns for the linear model and between 4.7 and 7.1 ns for the parabolic one for a white frequency noise comprised between 2 and $3.10^{-11}.\tau^{-1/2}$. These results are consistent with the performance shown in Table 3, except for PRN 01 and 06 that undergo a periodic.

A closer spectral analysis shows that PRN 06 on top of the 12 hours periodic exhibits also a periodic at 6 hours with a peak-to-peak of 9 ns. For this clock, the theory developed in [7] predicts the following extrapolation errors for $T_m = 10000$ s and $T_p = 6000$ s :

	Standard deviation of S_{ext}	
	Linear model	Parabolic model
Contribution of white frequency noise ($h_0 = 1.8.10^{-21}$ s)	3.4 ns	7.1 ns
Contribution of 12 hours periodic	11.7 ns	6.2 ns
Contribution of 6 hours periodic	10.5 ns	12.6 ns
Quadratic sum of the predicted contributions	16.1 ns	15.7 ns
Measured value	16.1 ns	16.1 ns

Table 4 : Theoretical and measured navigation performances of PRN 06 for $T_m = 10.000$ s and $T_p = 6.000$ s

These predicted and actual values are in good agreement for both models, which confirms the validity of the modelling developed in [7].

For the PRN 06, the extrapolation error could be greatly reduced by using a different couple (T_m,T_p). Figure 4 shows the theoretical and measured navigation performances as a function of T_m for $T_p = 6000$ s. Like for Table 4, the theoretical navigation performance is considered as the quadratic sum of the theoretical contributions of WFM and of both periodics.

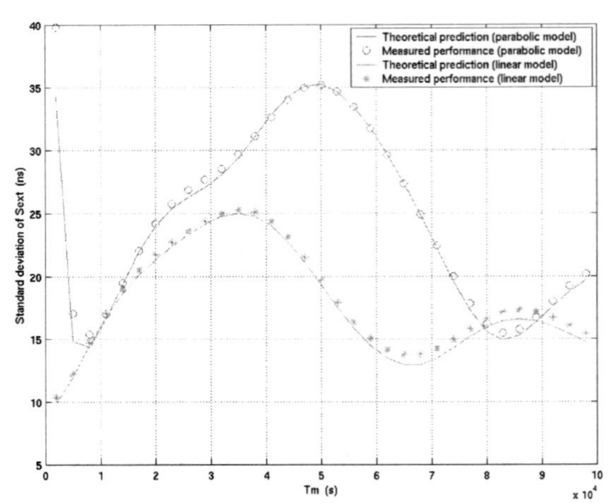

Figure 4 : Theoretical and measured navigation performances of PRN 06 for $T_p = 6.000$ s as a function of T_m

3.3 Rubidium clocks of Block II/IIA

On PRN 17, a frequency change of $+6.10^{-11}$ occurred on Tuesday, 25th of November 2003. The performances presented below are the ones before this event. The analyzed period for this clock is therefore reduced to 22 days.

The frequency drift of these clocks is given here :

	Frequency drift (/day)
PRN 04	$-3.6.10^{-14}$
PRN 07	$-8.7.10^{-14}$
PRN 08	$-6.5.10^{-14}$
PRN 17	$-8.8.10^{-14}$
PRN 26	$-6.1.10^{-14}$
PRN 27	$-1.8.10^{-13}$
PRN 29	$-8.2.10^{-14}$
PRN 30	$-1.2.10^{-13}$
PRN 31	$-1.3.10^{-13}$

Table 5 : Daily frequency drift of GPS II/IIA Rb clocks

The frequency stability drift removed for the different II/IIA Rubidium clocks is summarized in Figure 5 :

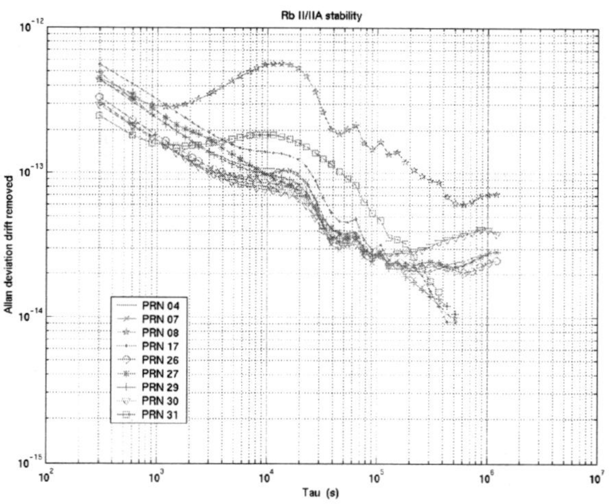

Figure 5 : Allan deviation drift removed of GPS II/IIA Rubidium clocks

Almost each clock is affected by a sinusoidal systematic at the orbital period with a different magnitude. It is particularly obvious for PRN 08. The PRN 31 undergoes also the same phenomenon but the period seems to be a bit smaller.

The Allan deviation is between 4 and $10.10^{-12}.\tau^{-1/2}$, with a Flicker floor between 2 and 3.10^{-14} for most of them.

We could remark that removing the drift on such long a period is not correct since the drift may vary a great deal. In fact, most GPS analysts [5,10] prefer to use the Hadamard deviation that will locally remove the drift (see Figure 6).

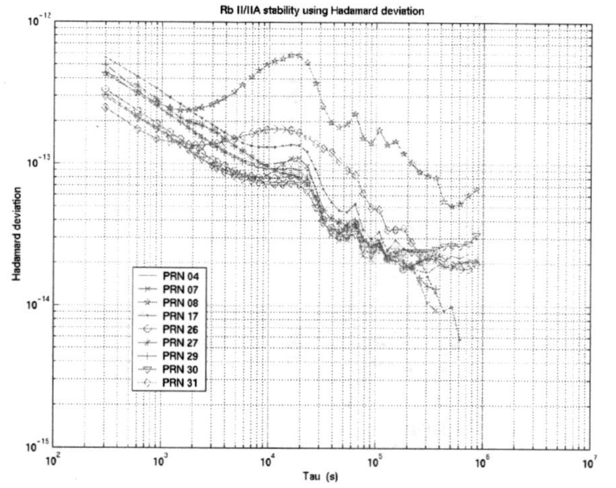

Figure 6 : Hadamard deviation of GPS II/IIA Rubidium clocks

Table 6 presents the performances of the GPS on board II/IIA Rb clocks for $T_m = 10.000$ s and $T_p = 6.000$ s :

	Standard deviation of S_{ext}	
	Linear model	Parabolic model
PRN 04	1.1 ns	1.8 ns
PRN 07	0.9 ns	1.4 ns
PRN 08	5.3 ns	5.7 ns
PRN 17	1.4 ns	2.4 ns
PRN 26	0.9 ns	1.6 ns
PRN 27	1.1 ns	2.2 ns
PRN 29	1 ns	1.8 ns
PRN 30	0.8 ns	1.3 ns
PRN 31	1.9 ns	2.6 ns

Table 6 : Navigation performance of GPS II/IIA Rb clocks for $T_m = 10.000$ s and $T_p = 6.000$ s

We conclude that the Galileo preliminary specification is met by some II/IIA Rubidium clocks when using a linear model.

The time linear model doesn't adjust a linear frequency drift, thus the latter has a deterministic contribution in the extrapolation error that affects its mean and not its standard deviation [7]. However, in this case, because T_m and T_p are short, this contribution is negligible (below 0.1 ns).

The navigation performance of Table 6 is in good agreement with the predicted values [7]. Let us take the example of PRN 17, it exhibits a $\sigma_y(\tau) = 1.10^{-11}.\tau^{-1/2}$ and a periodic at 12 hours with a peak-to-peak of about 2.2 ns. Table 7 summarizes the theoretical navigation performance and compares it to the measured 1-σ extrapolation error.

	Standard deviation of S_{ext}	
	Linear model	Parabolic model
Contribution of white frequency noise ($h_0 = 2.10^{-22}$ s)	1.1 ns	2.4 ns
Contribution of 12 hours periodic	0.9 ns	0.4 ns
Quadratic sum of the predicted contributions	1.4 ns	2.4 ns
Measured value	1.4 ns	2.4 ns

Table 7 : Theoretical and measured navigation performance of PRN 17 for $T_m = 10.000$ s and $T_p = 6.000$ s

3.4 Rubidium clocks of Block IIR

The frequency drift of these clocks is summarized here :

	Frequency drift (/day)
PRN 11	$2.9.10^{-15}$
PRN 13	3.10^{-15}
PRN 14	$-3.2.10^{-15}$
PRN 16	$-3.2.10^{-14}$
PRN 18	$-1.1.10^{-14}$
PRN 20	1.10^{-14}
PRN 21	$-5.4.10^{-14}$
PRN 28	$-5.7.10^{-15}$

Table 8 : Daily frequency drift of GPS IIR Rb clocks

One can notice that these values are quite different. We will see below that the observed drift is not that of the on board clock for Block IIR, because of additional equipment called TKS (Time Keeping System).

This equipment can correct the frequency drift, but this possibility is not used on PRN 16 and 21 and, not surprisingly, these two timing signals exhibit the most important drift.

The frequency stability drift removed for the different IIR Rubidium clocks is summarized in Figure 7.

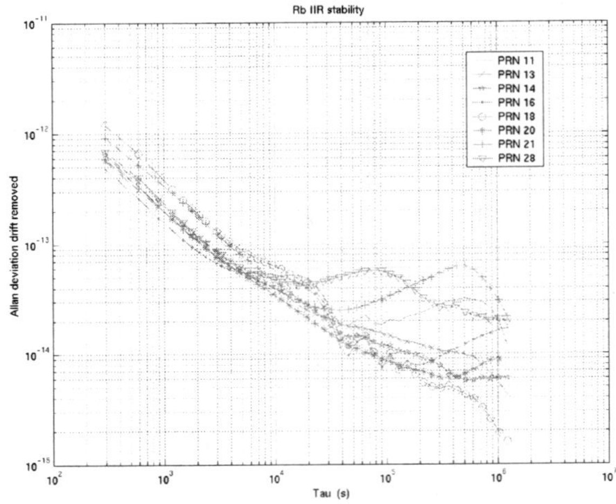

Figure 7 : Allan deviation drift removed of GPS IIR Rubidium clocks

The PRN 11, 16 and 18 are affected by periodics at the orbital period, which peak-to-peak are respectively about 1 ns, 0.8 ns and 0.8 ns.

Here is the stability of the same clocks using the Hadamard deviation :

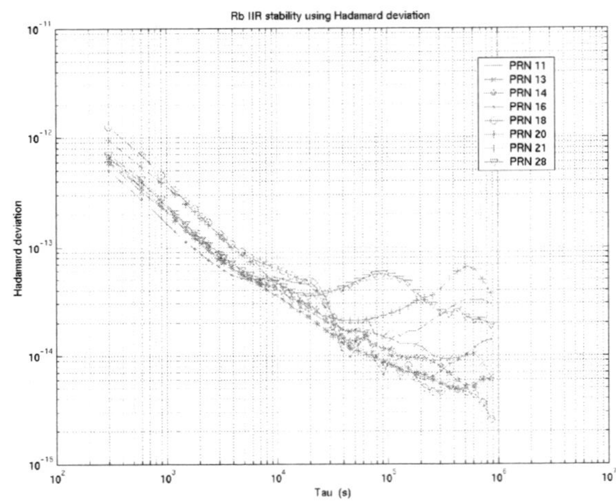

Figure 8 : Hadamard deviation of GPS IIR Rubidium clocks

Obviously the performance is better than II/IIA clocks, but the slope is no longer classically -1/2, which corresponds to a white frequency noise, but is close to -0.8. This slope is not typical of a passive atomic frequency standard.

We could firstly think that the limit of the restitution technique had been reached, in other words, this slope would be due to measurement noise. This hypothesis is put forward in [4] with a -2/3 slope. In our analysis, it is straightforward that this hypothesis is not relevant because the short term stability (between 300 and 1000 s) of several II/IIA Rb clocks is better than the one of the best IIR Rb clock. This means that we do not observe here measurement noise, otherwise it would appear exactly identically on the short term stability of II/IIA Rb clocks.

To be completely convinced, one should look at the following Allan deviations comparing the stability of IIR Rb PRN 20 and II/IIA Rb PRN 30 :

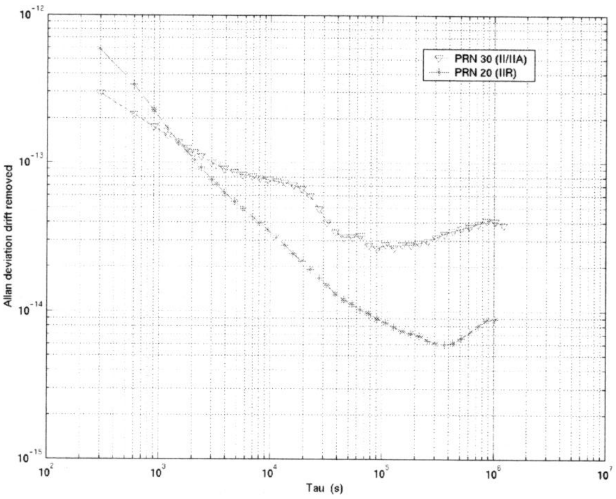

Figure 9 : Comparison of II/IIA Rb PRN 30 and IIR Rb PRN 20

The -0.8 slope is clearly not due to the measurement technique noise. Another explanation is that this noise would come from other equipments of the navigation payload. For instance, on GPS IIR payloads, there is a Time Keeping System (TKS) that enables to carry out corrections of phase, frequency and frequency drift.

This TKS is a phase lock loop between a VCXO and the Rubidium clock with software corrections. This hypothesis is confirmed by a previous article by ITT Aerospace where we can see the TKS Allan deviation when using a Rubidium clock for different loop time constants : a slope between -1 and -1/2 is obtained after 100 seconds [8].

The table below presents the performances of the GPS IIR Rb clocks for T_m = 10.000 s and T_p = 6.000 s :

	Standard deviation of S_{ext}	
	Linear model	Parabolic model
PRN 11	0.5 ns	0.6 ns
PRN 13	0.4 ns	0.7 ns
PRN 14	0.4 ns	0.7 ns
PRN 16	0.5 ns	0.7 ns
PRN 18	0.6 ns	0.9 ns
PRN 20	0.3 ns	0.7 ns
PRN 21	0.4 ns	1 ns
PRN 28	0.5 ns	0.7 ns

Table 9 : Navigation performance of GPS IIR Rb clocks for T_m = 10.000 s and T_p = 6.000 s

It is straightforward that the Galileo specification is easily met by all GPS IIR Rubidium clocks using both models, even if the TKS degrades the short term stability of these clocks. It is therefore interesting to look into longer extrapolation durations (T_p) since it means fewer operations to be performed by the ground segment. We can for instance compute the performances obtained with $T_m = 5.10^4$ s and $T_p = 3.10^4$ s, the values foreseen for the PHM (cf. Table 1) :

	RMS of S_{ext}	
	Linear model	Parabolic model
PRN 11	0.9 ns	3.5 ns
PRN 13	0.9 ns	1.8 ns
PRN 14	0.7 ns	1.9 ns
PRN 16	0.7 ns	2.9 ns
PRN 18	0.6 ns	2.9 ns
PRN 20	0.6 ns	1.4 ns
PRN 21	1.5 ns	1.7 ns
PRN 28	2.6 ns	3.1 ns

Table 10 : Navigation performance of GPS IIR Rb clocks for $T_m = 5.10^4$ s and $T_p = 3.10^4$ s

With $T_m = 5.10^4$ s and $T_p = 3.10^4$ s, the preliminary Galileo specifications are fulfilled with the linear model, except for the PRN 28.

For the linear model, the contribution of the frequency drift isn't negligible any more for the PRN 16 and 21 that do not have the TKS to correct for it. This contribution is respectively of 0.5 ns and 0.9 ns, which is in complete agreement with the theory in [7]. It has been added quadratically with the standard deviation to provide the results given in Table 10.

The periodic signature affects the navigation performance for the parabolic model for PRN 11, 16 and 18. The theory developed in [7] is unable to predict the uncertainty of the extrapolation error since the noise has not a classical slope. Yet it can foresee the following contributions of a 12 hours sinusoidal signature for $T_m = 5.10^4$ s and $T_p = 3.10^4$ s :

	Standard deviation of S_{ext}	
	Linear model	Parabolic model
Contribution of 12 hours periodic (1 ns peak-to peak)	0.1 ns	3.6 ns
Contribution of 12 hours periodic (0.8 ns peak-to peak)	0.1 ns	2.9 ns

Table 11 : Theoretical contributions of periodic signatures to navigation performance for $T_m = 5.10^4$ s and $T_p = 3.10^4$ s

The comparison of these predictions with the actual performance in Table 10 indicates that, for the parabolic model, the contribution of the periodic is completely dominating the navigation performance ; whereas, for the linear model, it is the timing signal noise contribution that drives the navigation performance. Of course these conclusions may change for other T_m and T_p.

Besides, the model developed in [7] allows to notice that, for the contribution of a 12 hours periodic, $T_m = 5.10^4$ s is nearly optimal for a given extrapolation duration $T_p = 3.10^4$ s for the linear model. For the parabolic model, the optimal T_m is about $7.5.10^4$ s, which provides a contribution of the periodic below 0.3 ns. More generally, we can say that for a given extrapolation duration, there is an optimal adjustment duration that allows to minimize the 1-sigma extrapolation error and therefore to have an optimal SIS URE. For power law noises and periodic signatures, this navigation performance can be predicted by the theory presented in [7] .

3.5 Ranking by 1-day Hadamard deviation

In order to compare with GPS constellation performance analysis [5,10] for which the measurements are collected by the National Imagery and Mapping Agency (NIMA), we ranked the GPS on board clocks by 1-day Hadamard deviation :

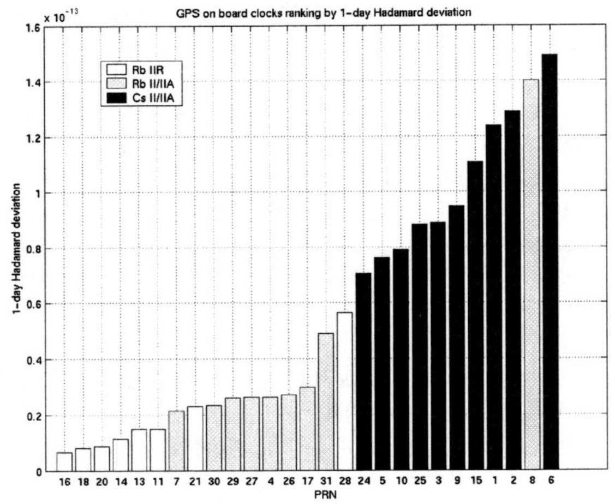

Figure 10 : Ranking by 1-day Hadamard deviation over the considered period

Our ranking is pretty consistent with [5,10]. This confirms the validity of the new IGS combined clock products. Not surprisingly, the Block IIR Rubidium clocks rank as the best in the constellation, except for the PRN 28 which is worse than Block II/IIA Rubidium clocks.

4 Conclusion

In this paper, we have analyzed the performances of the different GPS on board clocks computed by IGS with respect to classical stability analysis using variances and also with respect to the preliminary Galileo specifications.

The GPS on board clocks have very different characteristics according to the Block they belong to. It reflects an improvement of the performances with the most recent Block IIR.

Every IIR Rubidium clock Allan deviation curve presents a slope close to -0.8, instead of the expected typical -1/2 slope (WFM). We showed in this paper that this is not due to the measurement technique noise but very probably to the Time Keeping System, an on board equipment that is not present on GPS II/IIA payloads.

We also came to the conclusion that the preliminary Galileo specifications are met by some GPS II/IIA Rubidium clocks (provided a linear model is used) and every GPS IIR Rubidium clock. Moreover, it appeared very clearly that the linear model provides significantly better results than the parabolic model for the considered adjustment and extrapolation durations.

We also showed that the theory developed in [7] enables to foresee the uncertainty of the extrapolation error for random clock noises and sinusoidal signatures. It can be very useful to determine optimal adjustment and extrapolation durations for Galileo.

Acknowledgements

The author would like to thank John Vaccaro (PerkinElmer) for providing inputs on Block IIR Rubidium clocks.

References

[1] IGS web site : http://igscb.jpl.nasa.gov/

[2] K. Senior, P. Koppang, D. Matsakis, J. Ray – "Developing an IGS Time Scale", IEEE FCS, June 2001.

[3] J. Ray, K. Senior - "IGS/BIPM Pilot Project : GPS carrier-phase for time/frequency transfer and timescale formation", Metrologia 40 (2003).

[4] G. Busca and al. - "Space Clocks for Navigation Satellites", IEEE IFCS/EFTF, May 2003.

[5] J. Oaks, K Senior, M.M. Largay, W.G. Reid, H. Warren, J.A. Buisson – "Global Positioning System Constellation Clock Performance", PTTI, December 2003.

[6] M. Weiss, H. Garcia, S. Malys – "SVN#36's 12 hour periodic : Earth magnetic field", PAWG, 1999.

[7] F. Vernotte, J. Delporte, M. Brunet, T. Tournier – "Uncertainties of drift coefficients and extrapolation errors : application to clock error prediction", Metrologia, 2001, 38, 325-342.

[8] H.C. Rawicz, M.A. Epstein, J.A. Rajan – "The Time Keeping System for GPS Block IIR", PTTI, 1993.

[9] R. Lucas – "Workshop on European Clock for Galileo - System Issues", Noordwijk, 4th September 2001. See also J. Hahn – "GSTB-V1 SoW – Annex 1 : Technical Requirements", ESA-APPNG-SOW/00114-Ann1.

[10] M. Epstein, G. Feed, J. Rajan – "GPS IIR Rubidium clocks : In-Orbit Performance Aspects", PTTI, December 2003.

ULTRA-STABLE GROUND FREQUENCY DISSEMINATION VIA OPTICAL FIBRES

F. Narbonneau, M. Lours, O. Lopez*, Ch. Daussy*, D. Chambon, S. Bize, A. Amy Klein*, Ch. Chardonnet*, A. Clairon, and G. Santarelli
BNM-SYRTE, Observatoire de Paris, 61 Avenue de l'Observatoire, 75014 Paris, France
**Laboratoire de Physique des Lasers, Université Paris 13, 99 Av. Jean-Baptiste Clément, 93430 Villetaneuse, France*

Keywords - **Frequency dissemination, optical fibres, phase noise, frequency stability, Cryogenic Sapphire Oscillator, frequency standards, frequency comparisons**

Abstract - **In this paper, we demonstrate the possibility to transfer both short and long-term frequency stability of frequency standards through a standard telecom network with compensation of phase fluctuations introduced by environmental perturbations.**

I. INTRODUCTION

Ultra-stable frequency and time sources play an important role in many modern metrology and fundamental physics applications. For example, in Radio astronomy for VLBI (Very Large Base Interferometer) applications, and for navigation measurements [1] there is a strong need for, stable and ultra-low noise oscillators.

Modern cold atoms frequency standards in the microwave domain have already demonstrated an accuracy of 10^{-15} with the potential to reach the 10^{-16} level. Frequency stabilities, defined by the Allan standard Deviation (ADEV), of $10^{-13}\ \tau^{-\frac{1}{2}}$ are common for such standards and a few $10^{-14}\ \tau^{-\frac{1}{2}}$ has been demonstrated using more advanced techniques. Optical cold atom clocks have the potential to reach the 10^{-17} level of accuracy. The emergence of modern microwave-to-optical synthesizers based on mode-locked femtosecond lasers allow high resolution comparisons between microwave and optical clocks. The opportunity to compare these two types of clocks, even when the laboratories are separated by 100 km, via a ground frequency dissemination, could greatly help the evaluation of their accuracy and stability performances.

For a large number of laboratories, the access to an ultra-stable frequency reference open the way to perform new experiments in fundamental physics. Nevertheless, the development and operation of a state-of-the-art frequency standard remain a strong limitation and can be overcome by a fibre distribution system connecting reference Time and Frequency laboratory to users. The redundancy of the telecom network makes possible the use of a dedicated optical fibre for metrology applications.

The idea of disseminating high phase stability signal through an optical fibre is not new ([2], [4], [5]). Nevertheless, in our case, we intend to transfer high frequency stability signals, in a range from 0.1 to 1 GHz, by using the existing telecommunication fibre network.

II. OBJECTIVE

The scope of this system is to transfer reference signal of high performance frequency standard, at 100 MHz, (cf.: Fig. 1) via optical fibres, with a relative frequency stability $\sigma_y(\tau) \leq 2.10^{-14}$ [$\tau = 1$s] ($< 10^{-16}$ @ 1d), which implies a residual phase noise for the link of:

$$S_\phi(f) \leq 10^{-12} f^{-1} + 10^{-14} f^0 [\mathrm{rd}^2/\mathrm{Hz}] \text{ @ } 100 \text{ MHz} \quad (1)$$

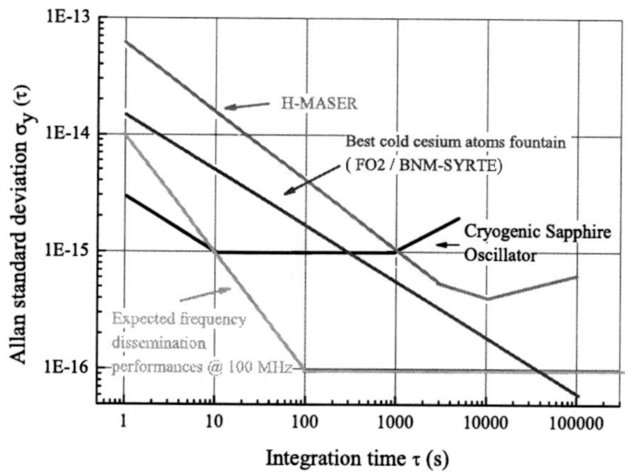

Figure 1: Allan deviation of typical frequency standards and dissemination specifications at 100 MHz

The frequency stability over one day is specified to be below 10^{-16}. A long-term phase stability of about 10 ps per day and a temperature coefficient of the order of 1 ps/°C will be necessary for the distribution device to meet these requirements.

The distribution system needs an active device to compensate signal phase variations introduced along the fibre link by external perturbations (mechanical vibrations, temperature fluctuations ...)

III. DISSEMINATION VIA URBAN METROPOLITAN FIBRE NETWORKS

Frequency dissemination via optical fibres is suitable for ultra-stable distribution over distances of a few kilometers.

Both the low attenuation (about 0.2 dB/km at 1.55 μm for a standard SMF-28 optical fibre) and high bandwidth-distance product of optical fibre allows long distance distribution by maintaining a high signal-to-noise ratio and thus outperforms the best RF cables.
Ultra-stable frequency standards are transferred via optical fibres by directly modulating a laser diode in current and detecting the RF signal on a photodiode at the link extremity. Dissemination performances are then principally limited by the signal-to-noise ratio of the detected signal and affected by the transmission medium perturbations (mechanical vibrations and temperature fluctuations).

With regard to the noise considerations, three types of noise are taking account:

the shot noise:

$$< i_s^2 >= 2 \times e \cdot I_p \cdot B \qquad (2)$$

the thermal noise:

$$< i_{th}^2 >= \frac{4 \cdot k \cdot T}{R} \cdot B \qquad (3)$$

the laser diode RIN:

$$< i_{RIN}^2 >= RIN(f) \times I_p^2 \cdot B \qquad (4)$$

where e is the electron charge (e=1.6 $\times$ 10^{-19} [C]), I_p (I_p [A] = $\eta \times P_0$ with η = 0.95 [A/W] typically at 1.55 μm and P_0 [W] the detected power) the photocurrent at the detection, B [Hz] the measurement bandwidth, k is the Boltzmann's constant (k = 1.38054 $\times$ 10^{-23} [J/K], and $RIN(f)$ the Relative Intensity Noise of the laser diode (typically -145 [dBc/Hz] for a 1.55 μm DFB laser diode).

Figure 2 is a simulation of the signal phase noise at detection, assuming an emitted optical power of 3 mW and a RF modulation level of 10 dBm, a typical detector combined to a 2 [kΩ] transimpedance low phase noise amplifier.

The previous simulation show the possibility to transfer low noise and ultra-stable signal over distance of a few ten kilometers by maintaining a white noise floor of -140 [dBrad^2/Hz].

Frequency disseminations operate in urban environment and use existing networks. Thus, fibre layout

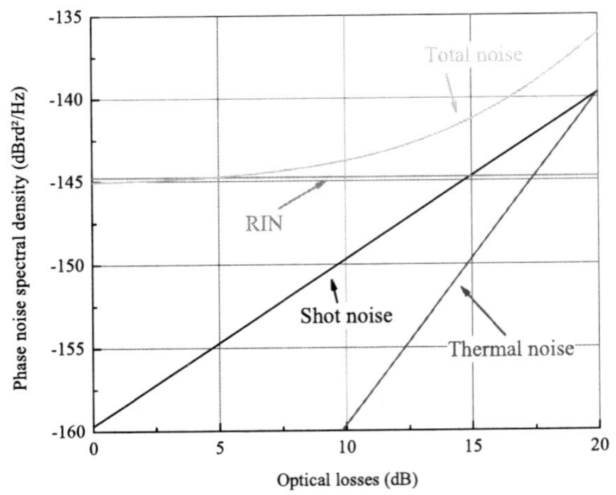

Figure 2: Phase noise spectral density simulation of each noise contribution vs optical losses, considering an emitted average optical power of 3 mW and a RF modulation level of the laser diode of about 10 dBm

and installation aspects are not ideal and the stability of the optical link can be affected by environmental effects. Optical fibre is sensitive to mechanical stress and temperature fluctuations. The first one affects phase noise and short-term frequency stability performances of the transmitted signal. The second effect, is a slowly changing phenomenon and has an impact on the long-term Allan deviation as described by the following equation (5):

$$\sigma_y(\tau) = \Delta T \times \frac{TCD \times n \times L}{c} \times \frac{\sin^2(\pi\tau\nu_0)}{\tau} \qquad (5)$$

where,

ΔT = amplitude of the temperature fluctuation [°C]

TCD = thermal coefficient of delay [ppm/°C] (typically 7 ppm/°C for standard telecom SMF28 fibre),

n = refraction index of the optical fibre,

L = optical link length [km],

c = speed velocity [3 $\times$ 10^8 m/s],

ν_0 = perturbation frequency [Hz],

τ = averaging time [s]

Two available optical links using the Paris and suburban dense France Telecom networks allow to connect BNM-SYRTE to the Laboratoire de Physique des Lasers (LPL) (about 44 km), and the Laboratoire Kastler Brossel (LKB - University Paris VI) (about 3 km).

Figure 4 presents the measured frequency instabilities (ADEV) versus integration time, on the previously

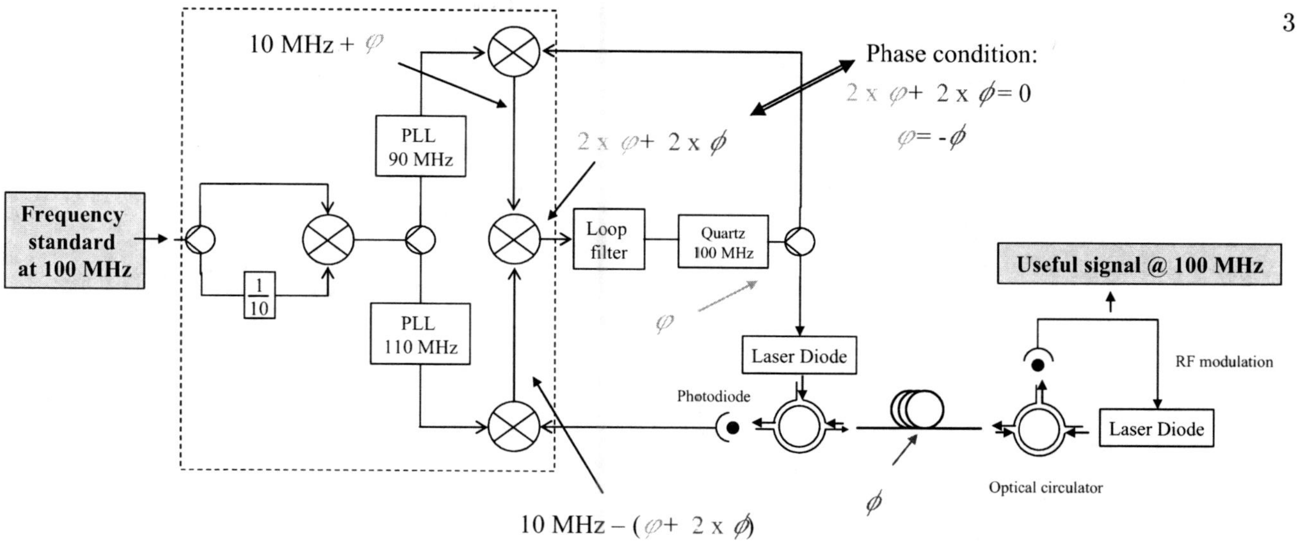

Figure 3: 100 MHz phase compensator based on the original JPL design (ϕ is the phase perturbation introduced along the link and φ the applied phase correction)

mentioned optical links. Measurements realized on the same link at different periods of the year, highlight non-stationary effects which degrade the stability of the distribution, as mechanical vibration or activities around the fibre. Periodic effects as daily temperature variations appears as a bump at the effect half period, on the ADEV .

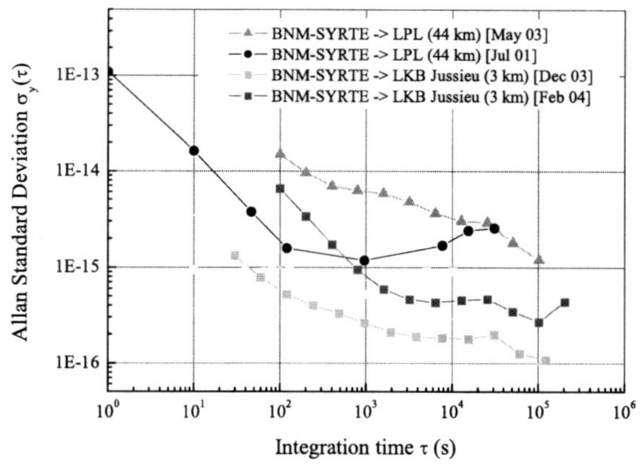

Figure 4: Frequency stability measurements of the BNM-SYRTE medium and long distance optical links

Measured frequency instabilities demonstrate the necessity to compensate for the phase fluctuations introduced along the link for reaching requirements and transferring ultra-stable signals synthesized from frequency standards.

IV. ACTIVE PHASE FLUCTUATIONS COMPENSATION SYSTEM

The general principle of the compensation system consists to measure phase fluctuation introduced by the link after one round trip and to phase shift the emitted signal of the opposite half quantity. For this we neglect amplitude modulation introduced by the distribution system and we consider that the measured phase shift after one round trip is two times the phase perturbation of the one way distribution.

Two different compensations systems have been implemented and tested.

A. Electronic phase fluctuations compensator

The first one, completely electronic, operates at 100 MHz and is based on the original JPL (Jet Propulsion Laboratory) design ([3], [6]).

Metrological signals are distributed through the optical fibre by directly modulating in current a pigtailed DFB (Distributed Feed-Back) laser diode, at 1.55 microns. Half of the RF detected signal, at the link extremity, modulates a second laser diode to generate the feedback signal, necessary for the phase fluctuations compensation. The detected signal, after one round trip, is then affected two times by the phase perturbation introduced along the link. By comparing it to the reference signal, it is possible to extract the perturbation value by a slightly complex RF process and to compensate it by phase shifting the emitted signal the opposite quantity.

For laboratory tests, two scale-down optical links of 2.5 km SMF-28 fibre has been realized and placed

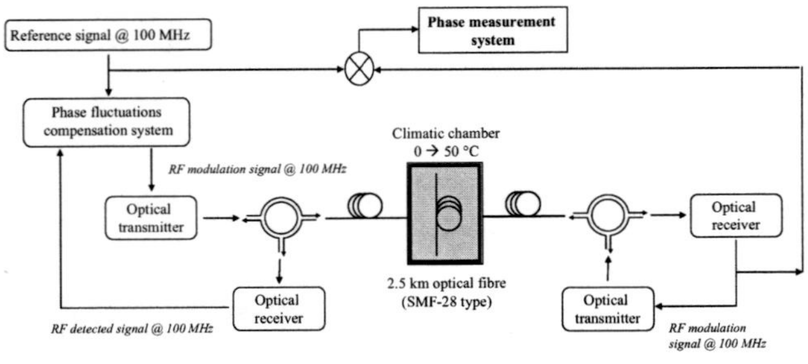

Figure 5: Laboratory tests and calibrations set-up

in a climatic chamber for simulating a periodical temperature variations in a range from 0°C to 50°C.

Phase variations between the detected and the reference signals are simultaneously measured in open and closed loop (respectively without and with phase compensation system).

Figure 6 shows phase variations measured on the scale-down optical link with and without phase fluctuations compensation for a periodical temperature variation of 2°C around an average temperature of 32°C.

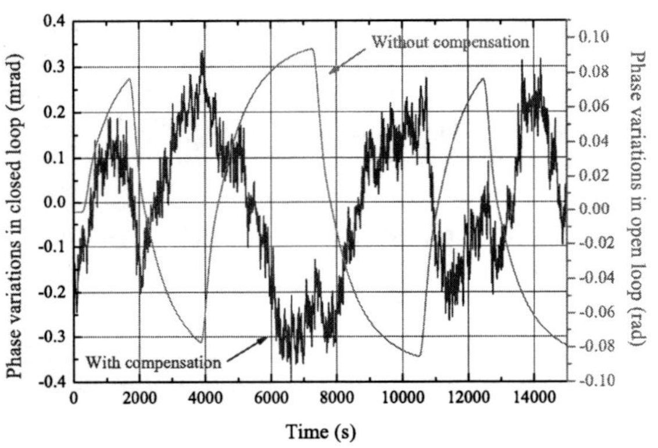

Figure 6: Phase variations in open and closed loop (respectively without and with compensation of phase fluctuations) induced by a periodical temperature variation of ±2 °C around an average temperature of 32°C

The open loop phase variation is of the order of 1.6 rad whereas we observe a 0.4 mrad variation in closed loop.With phase fluctuations correction (closed loop) a phase variation of 0.4 mrad peak-to-peak is measured. The compensation system has thus a gain better than 400 (46 dB), which is checked by calculating the correspondent ADEV in open and closed loop (cf.: Fig. 7).

Figure 8 shows the measured phase noise spectral density at 100 MHz of the complete system. Phase noise of about -125 [dBrad^2/Hz] at 100 MHz is observed, which insure the capability to transfer signal with a frequency stability $\sigma_y(\tau)$ below than 1×10^{-14} at 1 second averaging time.

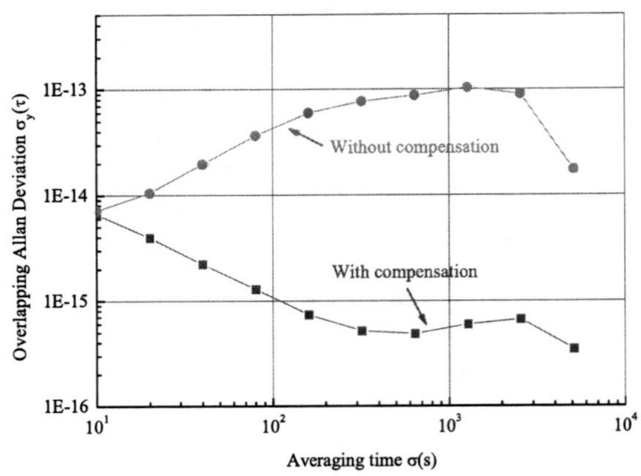

Figure 7: ADEV of the complete test laboratory distribution system calculated from the previous phase data

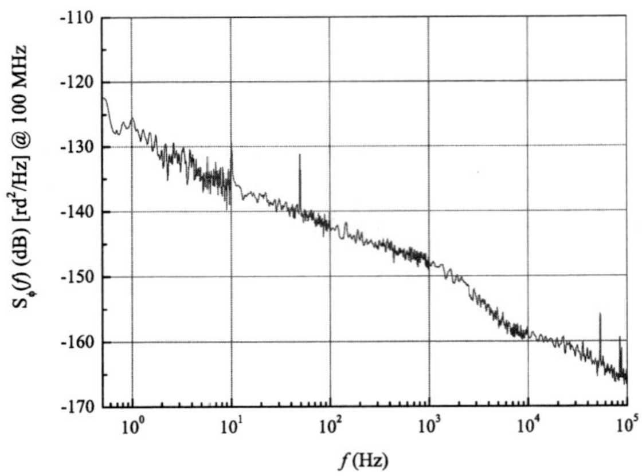

Figure 8: Phase noise spectral density of the 100 MHz phase compensator in closed loop, tested on a 2.5 km laboratory link

B. Optical compensation system

The second system operates in a slightly different way than the previous compensation system. Indeed, the phase correction is applied both on the emitted and on the feedback ways by directly acting on the optical fibre. That allows to have directly a an information on the phase perturbation terms and to cancel it via a servo loop.

The fast phase fluctuations are corrected by stretching a 15-meter optical fibre wrapped around a cylindrical piezo-electric actuator which show a 10 μm extension for a 1 kV supply voltage. That allows to compen-

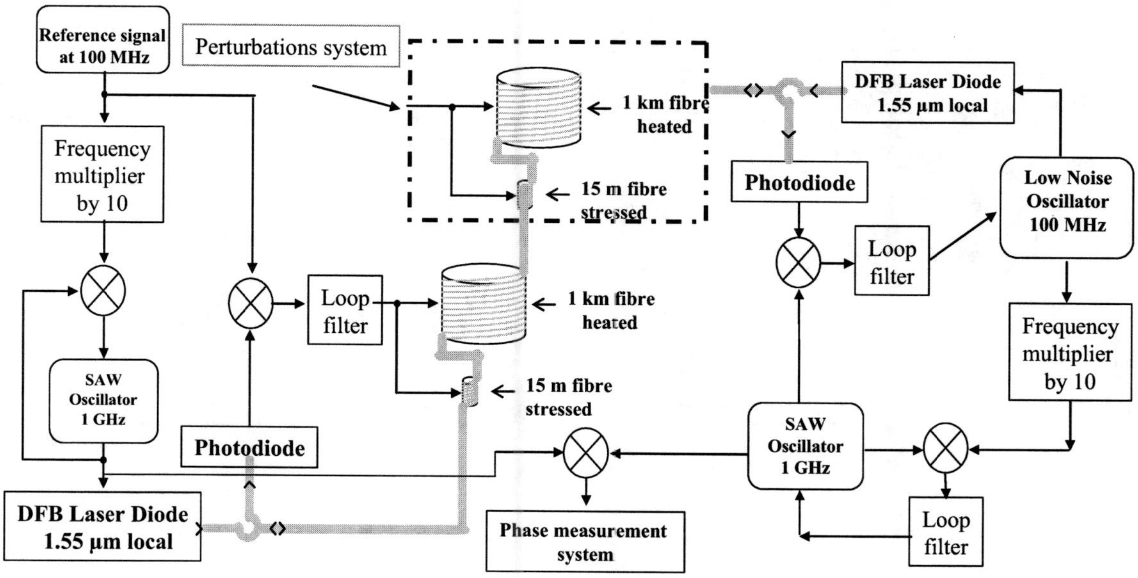

Figure 9: Set-up of the optical compensation system with phase measurement and perturbations systems

sate phase perturbations, with a bandwidth of a few hundreds hertz. The phase correction we can obtain is given by the following equation:

$$\Delta\varphi_c = N \times \pi \times \Delta D \cdot \left(\frac{c}{2\ \pi \cdot \nu_{RF} \cdot n}\right)^{-1} \quad (6)$$

$$\simeq 10\ mrad \quad (7)$$

N is the number of fibre turns (about 100 in our case) and ΔD the diameter variation of the piezo-electric actuator.

Perturbation related to slow changing phenomenon are compensated by heating a 1 km optical fibre spool:

$$\Delta\varphi_c = L \times TCD \times \left(\frac{c}{2\ \pi \cdot \nu_{RF} \cdot n}\right)^{-1} \quad (8)$$

$$\simeq 54\ mrad/°C \quad (9)$$

In order to increase isolation between channels, the emitted and the forward beams are modulated with two different frequencies (respectively 1 GHz and 100 MHz).

This system has a gain of about a factor 30 for fast fluctuations and more than 1000 for slow phase variations (Fig. 10).

Figure 11 presents phase noise spectral density measurements of the optical phase compensator, tested with a 1 km long fibre link in laboratory, in open and closed loop.

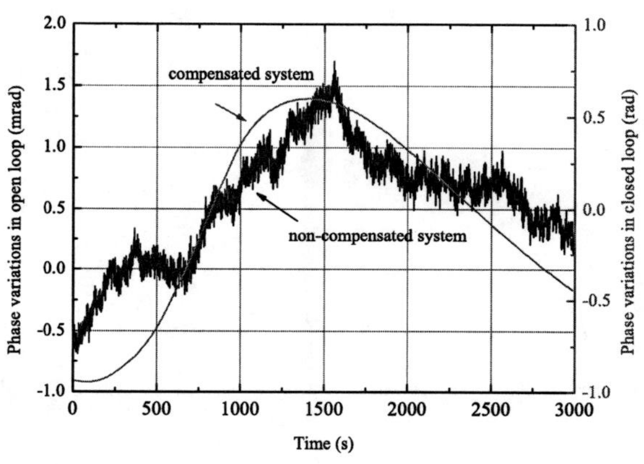

Figure 10: Phase fluctuations measurements in open loop and closed loop

V. CHARACTERIZATION OF THE TWO DIFFERENT SYSTEMS IN A FULL DIRECTIONAL LINK

A compensated full bidirectional optical link (2x44 km) has been realized by using the two fibres of the BNM-SYRTE to LPL link and implementing the previous systems at each extremity (cf Fig. 12) to be able to test independently the two systems. The use of two different systems allows highlighting systematic effects related to one of the two compensators, and thus to have a real characterization of the dissemination system.

The phase instabilities of the link are simultaneously measured by slightly modulating the transmitted signal at a non-harmonic frequency with the compensator operating frequency (e.g. 270 MHz) to be sure that

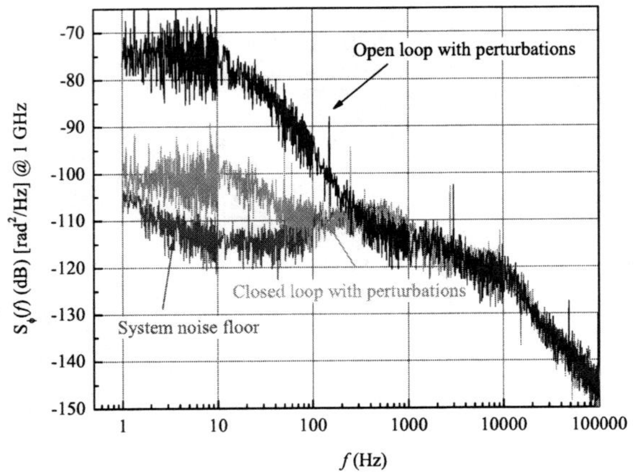

Figure 11: Measured phase noise spectral density of the optical phase compensation system

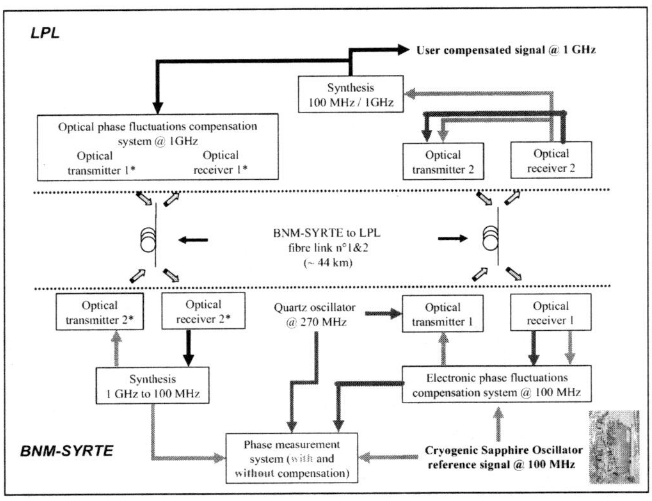

Figure 12: Full bidirectional optical link set-up using the two different compensation systems

this signal is not partially compensated by the system.

Figure 13 shows the frequency stability measurements of a full bidirectional link (2x44Km) between the BNM-SYRTE and the Laboratoire de Physique des Lasers (LPL). A fractional frequency stability of 10^{-17} at one day and 1.2×10^{-14} at one second averaging time have been measured. These results demonstrate the possibility to transfer the stability of the BNM-SYRTE atomic fountain FO2 [7] over 44 kilometers .

VI. CONCLUSION & PERSPECTIVES

For the first time, we ensure long-distance transfer of both short-term and long-term frequency stability of metrological standards, with low phase noise via

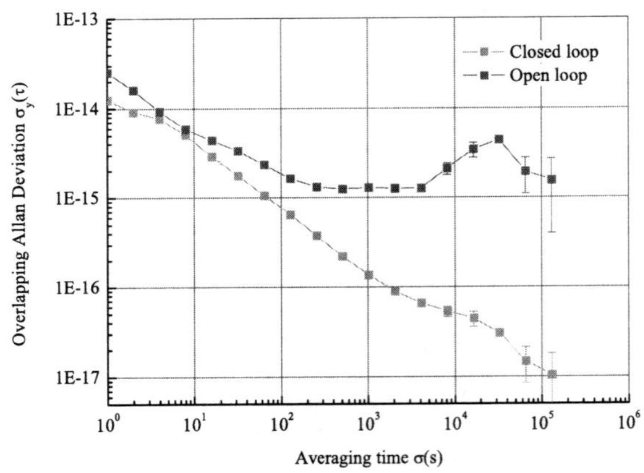

Figure 13: Simultaneous frequency stability measurements in open and closed loop on the BNM-SYRTE to LPL optical link

telecom optical fibres.

The BNM-SYRTE to LPL optical link has yet allowed to compare an optical frequency standard, in the infrared domain (CO_2 laser stabilized on a OsO_4 absorption), at LPL, and the BNM-SYRTE cryogenic sapphire oscillator [8], with an Allan deviation of 3×10^{-14} at one second.

The phase perturbations rejection is actually limited by intrinsic high-order mixing products in mixer used for phase measurements. The use of high modulation frequency, e.g. 1 GHz, allows to reduce these parasitic signals and to improve the the compensation system in terms of phase noise and insensitivity to external temperature fluctuations. We hope to have a gain of an order of magnitude on the frequency standard dissemination performances.

Preliminary laboratory measurements, realized on the laboratory breadboard of a phase compensator operating at 1 GHz, are encouraging and open the perspective to transfer signal with a fractional frequency stability of the order of 10^{-15} at one second averaging time and below 10^{-17} at one day.

VII. ACKNOWLEDGMENTS

This work was supported by the European Space Agency (ESOC contract n°17367/03/D/SW(SC)), and by the Bureau National de Métrologie.

[1] L. E. Primas, R. T. Logan and G. F. Lutes, in *Proc. of 43rd Annual Symposium on Frequency Control*, 1989, p. 202.

[2] M. Calhoun, P. Kunle and J. Law, in *Proc. of Precise Time and Time Interval*, 1987, p. 133.

[3] R.T. Logan, G.F. Lutes, in *Proc. of IEEE Frequency Control Symposium*, 1992, p. 310-316.

[4] K. Sato *et al.*, in *IEEE Transactions on Instrumentation and Measurement*, 2000, vol.49, Issue 1,p. 19-24.

[5] K. W. Holman *et al*, *Precise frequency transfer through a fiber network using 1.5-ţm mode-locked sources*, to be published

[6] F. Narbonneau, M. Lours, Ch. Daussy, O. Lopez, A. Clairon and G. Santarelli, in *Proc. of the 17th EFTF*, 2003.

[7] H. Marion *et al*, in *Proc. of the 18th EFTF*, 2004.

[8] D. Chambon *et al*, in *Proc. of the 18th EFTF*, 2004.

NEW INSIGHTS IN THE ATOMIC ORIGIN OF THE PIEZOELECTRIC EFFECT IN α-QUARTZ AND α-$GaPO_4$

ULLRICH PIETSCH, JAV DAVAASAMBUU, VASILI KOCHIN and SEMEN GORFMAN
Institute of Physics, University of Potsdam, D-14469 Potsdam, Germany
KARLHEINZ SCHWARZ and PETER BLAHA
Institute of Materials Chemistry, Vienna University of Technology, 1060 Vienna, Austria
e-mail: upietsch@rz.uni-potsdam.de

In order to understand the atomistic origin of the inverse piezoelectric effect in α-SiO_2 and α-$GaPO_4$ we have measured the changes of integrated X-ray intensities of selected Bragg reflection under influence of an external high electric field up to 8 kV/mm. To do this 300 μm thick samples were sandwiched between metallic contacts and the X-ray intensities were recorded with and without the applied field. The difference intensities were evaluated in terms of difference electron density of atoms within the respective crystallographic unit cell. Because the up to now accepted model of the field induced displacement of ionic sublattices against each other fails for the interpretation of experimental data we proposed a new model of the inverse piezoelectric effect, which considers the strong covalent bond between silicon and oxygen and gallium/phosphorous and oxygen, respectively. Here the main effect of screening the external electric field is a change in the Si-O-Si and Ga-O-P bonding angles, i.e. the rotations of nearly rigid MO_4 tetrahedra [1].

For α-SiO_2 the model has been confirmed by ab-initio calculations [2] using the FP-APW+lo method. The external electric field was modelled by a saw-like potential V_{ext} within a supercell (SC) containing 72 atoms. Applying an electric field with 550 kV/mm the atomic positions of the SC were relaxed until the forces acting on the atoms vanished. In agreement with experiment the atomic the origin of the piezoelectric effect can be described by a rotation of slightly deformed SiO_4 tetrahedra against each other. The change of the Si-O bond lengths and the tetrahedral O-Si-O angles is one order of magnitude smaller than that of the Si-O-Si angles between neighbouring tetrahedra. The calculated changes of x-ray intensities are in agreement with the experiment when the theoretical data are extrapolated down to the much field strength that was applied in the experiment.

[1] J. Davaasambuu A.Pucher, V.Kochin , U.Pietsch
"Atomistic origin of the piezoelectric effect in α-SiO_2 and α-$GaPO_4$"
Europhysics Letters **62** 834-840 (2003)
[2] V.Kochin, J. Davaasambuu, U.Pietsch, K.-H.Schwarz, P.Blaha
"The atomistic origin of the inverse piezoelectric effect in a-quartz"
J. Phys. Chem. Sol. Submitted (2003)

ORDERED $Ca_3TaGa_3Si_2O_{14}$ CRYSTALS: GROWTH, ELECTROMECHANICAL AND OPTICAL PROPERTIES

Yu.V. Pisarevsky, B.V. Mill**, N.A. Moiseeva*, A.V. Yakimov**

* Institute of Crystallography, Russia, aopt@ns.crys.ras.ru
** Moscow State University, Russia, mill@plms.phys.msu.ru

Keywords: Piezoelectric, langasite, crystals.

Abstract

Ordered $Ca_3TaGa_3Si_2O_{14}$ crystals from langasite family are grown. Full sets of elastic, piezoelectric and dielectric constant are determined. Some characteristics of elastic wave propagation are calculated.

1. Introduction

Pilot production of large diameter langasite $La_3Ga_5SiO_{14}$ and $La_3Ta_{0.5}Ga_{5.5}O_{14}$ crystals was developed in Russia, Japan, USA and various BAW and SAW devices were designed with promising characteristics.
Recently, B. Chai et al. grew 4 crystals with so-called ordered structure and some BAW and SAW characteristics were measured. ZTC existence and high QF product value point out interest for next investigations.
Here, we represent the results of growing and measurement of elastic, dielectric and piezoelectric properties of $Ca_3TaGa_3Si_2O_{14}$ (CTGS) single crystals.

2. Methods and samples

2.1 Crystal growth

CTGS single crystals were grown by the Czochralski method at temperatures near 1450°C.

2.2 Dielectric

Dielectric constants were obtained by measurement of the capacity of plates with automatic AC bridge at 1 kHz frequency.

2.3 Resonance–antiresonance

With the help of an electric field, vibrations of oriented specimens can be generated at one or another mode, which connected with a corresponding piezoelectric modulus d_{ik} or constant e_{ik}. We used resonance–antiresonance method and nonlinearity of harmonic..

2.4 Ultrasonic

Sound velocities were measured by phase-pulse method. Piezoelectric transducer was mounted to buffer line (material of buffer is fused silica) and samples were glued to opposite end of buffers. Two RF impulses with controlled delay time generated in transducer two ultrasonic impulses and returned echo pattern was displayed on oscilloscope. Time delay was adjusted to confine pulses returned from boundary buffer-sample and from free boundary of sample.
Frequencies, for which both pulse phases confined, were measured. Velocities were calculated from

$$V = 2l\Delta f,$$

where l is a sample length, Δf is a difference between neighboring frequencies. Measuring line was mounted in a temperature controlled chamber with temperature uniformity 0.01K.

2.5 Optical rotation

Rotation of polarization plane was measured on wavelength 0.6328 mkm

3. Experimental data

3.1 Crystal growth

CTGS single crystals were grown along X direction unlike [2], where pulling direction was Y axis. Melt of stoichiometric composition, Pt-crucible with active Pt-afterheater and $5O_2$–$95N_2$ gas atmosphere were used. Pull speed was 1.5 mm/h, rotation - 15 rpm. Crystal dimensions were 77x17x25 mm along X, Y, and Z direction correspondingly.

3.2 Resonance–antiresonance

Results of measurement of electromechanical properties are listed in table 1.

3.3 Ultrasonic

Results of measurement of sound velocities by phase-pulse method are given in table 2.

3.4 Optical rotation

Optical rotation along Z direction was 1.7 grad/mm right hand.

4. Constants

Full set of elastic, dielectric, piezoelectric constants at room temperature are listed in table 3.

5. Elastic wave propagation characteristic

Full set elastic, dielectric and piezoelectric constants BAW propagation characteristics at the main crystallographic planes were calculated on the base of measurements (Fig. 1, 2).

6. Conclusions

$Ca_3TaGa_3Si_2O_{14}$ (CTGS) single crystals from ordered group of langasite family were grown.

Full set elastic, dielectric, piezoelectric constants at room temperature were derived. BAW propagation characteristics at the main crystallographic planes were calculated

Acknowledgements

This paper is based upon the work supported by the European Research Office of the US Army under contract No. 62558-02-C-9013 and the Russian Foundation for Basic Research, project no. 03-02-17075.

References

[1] B.V. Mill, Yu.V .Pisarevsky, E.L.Belokoneva "Synthesis and some properties of single crystals with $Ca_3Ga_2Ge_4O_{14}$ structure", Proc. 1999 IEEE Int. Freq. Cont. .Symp., p.829

[2] S. Jen, B.C.C. Teng, M.M.C. Chou, B.H.T. Chai, T.T. Lee, J. Gwo "Experimental investigation of the BAW device potentials of singly rotated Y-cut ordered langasite-structure crystals", Proc. 2002 IEEE Int. Freq. Cont. Symp., p. 307.

[3] D. Puccio, D.C. Malocha, M.M.C. Chou, "Investigation of STGS, SNGS, CTGS & CNGS materials for use in SAW applications" Proc. 2003 IEEE Int. Freq. Cont. Symp., p.627.

Table 1. Electromechanical properties of CTGS crystal.

Vibration mode	$f_0l; f_0t$ (kHz mm)	$s_{ik}\,10^{-13}m^2N^{-1}$ $(10^{-13}\,m^2\,N^{-1})$	c_{ik} $(10^{10}Nm^{-2})$	$10^2\,k_{em}$
(XY twl) –0°/0°/0°-bar	2455,7	s_{11} = 89,75	-	10,9
(XY twl) –30°/0°/0°-bar	2435,4	s'_{22} = 91,25	-	7,7
(XY twl) –+30°/0°/0°-bar	2409,6	s'_{22} = 93,22	-	20
(XY twl) –+45°/0°/0°-bar	2486,1	s_{22} = 87,57	-	25,9
(YZ twl) –+45°/0°/0°-bar	2657,8	s'_{33} = 76,6	-	8,6
X-cut thickness-extensional mode plate	2867,2	-	c_{11} = 15,2	22
Y-cut shear-thickness mode plate	1547,1	-	c_{66} = 4,42	16
Y-cut shear-contour mode plate	1880,1	s_{44} = 239,30	-	19,7

Table 2. Phase velocities of longitudinal and transverse elastic waves in the $Ca_3TaGa_3Si_2O_{14}$ crystals

direction of propagation	type of wave	V (10 ms^{-1})	U_m, the direction cosines of the displacement vector
X – axis <1120>	longitudinal pure shear (fast) pure shear (slow)	5481 3266 2776	U_x= 1 U_y= 0, U_z= 0 U_x= 0 U_y= 0,739, U_z= 0,673 U_x= 0 U_y= -0,673, U_z= 0,739
Y – axis <1010>	quasi-longitudinal pure shear quasi-shear	5468 3093 2991	U_x= 0 U_y= 0,998, U_z= -0,071 U_x= 1 U_y= 0, U_z= 0 U_x= 0 U_y= 0,071, U_z= 0,998
Z – axis <0001>	longitudinal shear	7140 3008	$U_x = U_y = 0$, $U_z = 1$ $U_x = 1$, $U_y = U_z = 0$
+45° to the Y- and X – axes	quasi-longitudinal quasi-shear pure shear	5997 3462 3268	U_x= 0 U_y= 0,498 U_z= 0,867 U_x= 0 U_y= 0,867 U_z= -0,498 U_x= 1 U_y= 0, U_z= 0

Table 3. Piezoelectric strain d_{ik} (10^{-12} CN^{-1}) and stress e_{ik} (Cm^{-2}) constants, elastic stiffness c_{ik} (10^{10} Nm^{-2}) and compliances s_{ik} (10^{-13} m^2N^{-1}), the Young E_{ii} (10^{10} Nm^{-2}) and shear G_{ii} (10^{10} Nm^{-2}) modules, the Poisson coefficients σ_{ik} and electromechanical coupling coefficients k_{em}, permittivity constants e_{ik} of $Ca_3TaGa_3Si_2O_{14}$ crystals at 290 K.

d_{11}= -4,03•10^{-12} CN^{-1}	s_{11} = 89,75•10^{-13} m^2N^{-1}	E_{11} = 11,14•10^{10} Nm^{-2}
d_{14} = 4,8•10^{-12} CN^{-1}	s_{33} = 50,91•10^{-13} m^2N^{-1}	E_{33} = 19,64•10^{10} Nm^{-2}
e_{11} = -0,41 Cm^{-2}	s_{12} = -26,36•10^{-13} m^2N^{-1}	G_{11} = 4,24•10^{10} Nm^{-2}
e_{14} = 0,196 Cm^{-2}	s_{13} = -16,36•10^{-13} $m2N^{-1}$	G_{33} = 4,18•10^{10} Nm^{-2}
$\varepsilon_{11}/\varepsilon_0$ = 17,2 (16.7*)	s_{14} = -1,52•10^{-13} $m2N^{-1}$	σ_{12} = 0,29
$\varepsilon_{33}/\varepsilon_0$ = 24,6 (21.5*)	s_{44} = 239,3•10^{-13} $m2N^{-1}$	σ_{13} = 0,32
k_{12} = 0.109	s_{66} = 232,2•10^{-13} m2N+	σ_{31} = 0,18
k_{26} = 0.16		
k_t = 0.22		

*In brackets - data of [2]

Fig 1. Cross section of phase velocities surfaces by (100) plane in crystals a) comparati ve polar diagram CTGS, LGS; b) CTGS; Q – quasi, L – longitudinal, S – shear.

a)

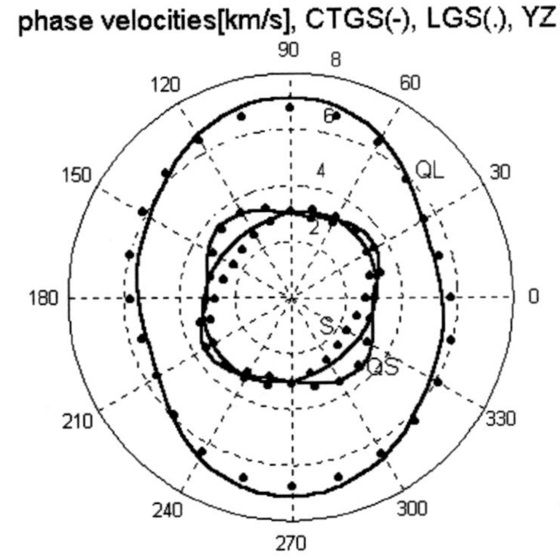

b)

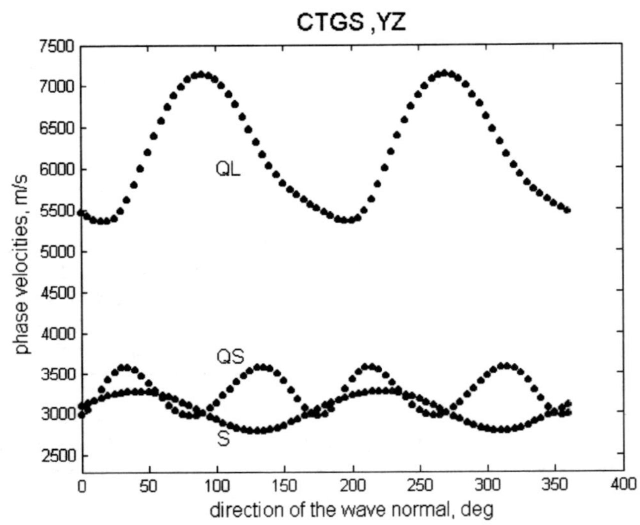

Fig 2. Direction of acoustic beam propagation (group velocities) as a function of the direction of the wave normal for CTGS

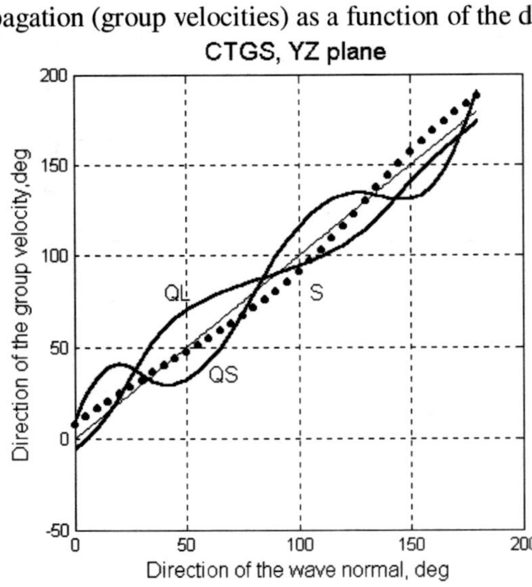

IDEALISED ANALYSIS OF SAW LONGITUDINALLY-COUPLED RESONATOR (LCR) FILTERS

David P. Morgan

Impulse Consulting, Northampton NN3 3BG, UK.
dpm@ieee.org

Keywords: SAW devices, resonators, bandpass filters.

Abstract

A longitudinally-coupled resonator (LCR) filter, also called a DMS filter, consists of either two or three interdigital transducers located between two strongly-reflecting gratings. These devices have been widely used as low-loss filters at RF frequencies. The behaviour of the devices is complicated by the fact that the transducers have strong internal reflections. This paper considers the two-transducer type and shows that, with some realistic assumptions, the device behaves much like a simple resonator. In the passband, the filter response can be expressed in a simplifed manner which gives much insight into the physical operation of the device. With appropriate electrical matching, the resonant behaviour disappears and the device behaves as a delay line. The matching required for a flat low-loss response is deduced. Equivalent circuits for two- and three-transducer devices are derived.

1. Introduction

Several special techniques can be used for SAW bandpass filters in applications which demand low insertion losses. One of the methods often used for RF devices is the longitudinally-coupled resonator filter [1, 2]. This consists of either two or three uniform interdigital transducers located between two strongly-reflecting gratings. For RF applications, the transducers are usually the single-electrode type, because this has wider electrodes than other types (for the same centre frequency), and hence gives easier fabrication. However, the use of single-electrode transducers introduces some complexity into the behaviour of the device, because of strong internal reflections from the electrodes. As a result, the operation of these devices is not very clear.

In this paper, an idealised analysis is presented in order to clarify this. The analysis makes assumptions regarding reciprocity, symmetry and power conservation. These are all valid (approximately or exactly) for practical LCR filters, for frequencies within the passband. The result is a very simple set of expressions for the device Y-matrix and for the passband response. Previous publications have explained the application to the three-transucer device [3] and to a one-port one-pole resonator [4], so the present paper concentrates on the two-transducer device. However, an equivalent circuit for the three-transducer device will also be presented for the first time.

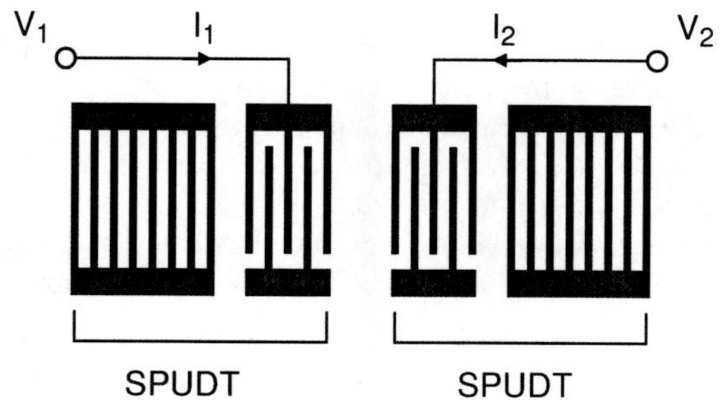

Figure 1. Structure of two-transducer LCR.

2. Analysis of two-transducer device

2.1. Basic analysis methods

The two-transducer device, shown in Fig.1, is assumed to be symmetrical. To describe a uniform transducer we use the P-matrix, defined by

$$\begin{bmatrix} A_{t1} \\ A_{t2} \\ I \end{bmatrix} = \begin{bmatrix} P_{11} & P_{12} & P_{13} \\ P_{21} & P_{22} & P_{23} \\ P_{31} & P_{32} & P_{33} \end{bmatrix} \begin{bmatrix} A_{i1} \\ A_{i2} \\ V \end{bmatrix} \quad ..(1)$$

where the A's are SAW amplitudes, with subscripts i (or t) to indicate waves incident (or leaving) the transducer. Additional subscripts 1 or 2 refer to the acoustic ports, which are located on either side of the transducer. The precise location of these ports is not significant, but they are conveniently taken to be close to the transducer edges. Also, for a symmetrical transducer the ports must be located symmetrically for the following analysis to be valid. The amplitudes are defined to be such that $|A|^2/2$ equals the SAW power, and A has the same phase as the surface potential associated with the wave. Port 3 is the electrical port, where the transducer voltage is V and the current is I. The P-matrix can also be used for a grating, in which case the P_{ij} with i or j equal to 3 can be ignored or set to zero. The components of the P-matrix can be calculated using the coupled-mode

(COM) approach [5]. For the present work, the required paramaters for the COM analysis (reflection and transduction) were obtained as in [6].

For a combination of components with the same aperture, the P-matrices can be combined to give a matrix describing the combination, and for devices such as the LCR the Y-matrix can be deduced. These familiar techniques are described by, for example, Plessky and Koskela [5]. From the Y-matrix, the device scattering matrix S_{ij} follows by standard methods. This analysis ignores some second-order effects such as attenuation, diffraction and waveguiding, but it can give results in quite good agreement with experiment, and it is adequate for present purposes.

2.2. SPUDT characteristics

We first note that, at each end of the device, the grating and its adjacent transducer must behave as a single-phase unidirectional transducer, that is, a SPUDT. Thus, in the filter passband, the SPUDT's launch SAW's preferentially towards the centre of the device. If this were not so, a significant proportion of the available applied power would be radiated (as SAW's) away from the centre, leading to substantial insertion loss. Hence it is convenient to regard the grating-transducer combination at each end as being a SPUDT. The two SPUDT's are taken to be identical, with P-matrix denoted by P_{ij}^S. Port 1 of each SPUDT is taken to be the port facing the other SPUDT. The space between the SPUDT's is assumed to be allowed for by modifying the phases of the P_{ij}^S. Hence, for both SPUDT's, port 1 is at the centre of the device. In the filter passband the SPUDT's must have high directivity, so that $|P_{13}^S| >> |P_{23}^S|$. It is shown later that this also implies high reflectivity, so that $|P_{11}^S| \approx 1$. These conditions can be assumed to be true because otherwise the filter could not have low insertion loss.

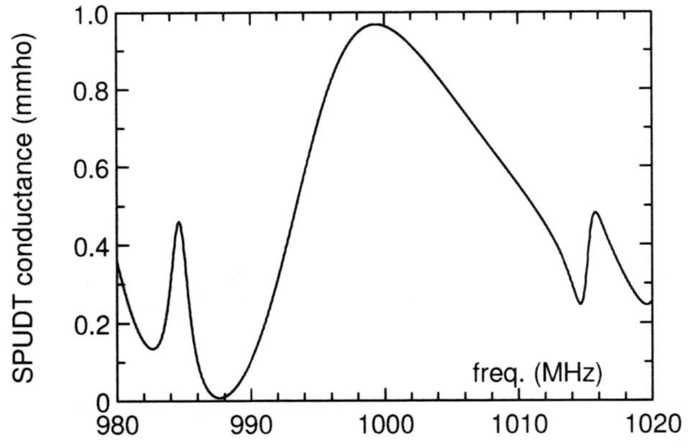

Figure 2. Conductance of SPUDT.

Some SPUDT characteristics are shown in Figs. 2 and 3. Here the substrate is taken to be Y-Z lithium niobate, with $\Delta v/v$ = 2.4% and $\varepsilon_S(\infty) = 46\ \varepsilon_0$. All components have 0.1 mm. aperture and 1000 MHz Bragg frequency, and the transducers are the single-electrode type. The electrode reflection coefficient, in both transducers and gratings, is taken as –0.04j. The transducers have 11 electrodes and the gratings have 121. The spacing between one grating and its adjacent transducer (measured to the centres of the nearest electrodes) is 0.8 λ at 1000 MHz. Figure 2 shows the SPUDT conductance $G_a^S = \mathrm{Re}\{P_{33}^S\}$, and Fig. 3 shows the directivity, $20 \log |P_{13}^S / P_{23}^S|$, dB. These functions have an important bearing on the behaviour of the filter, considered later.

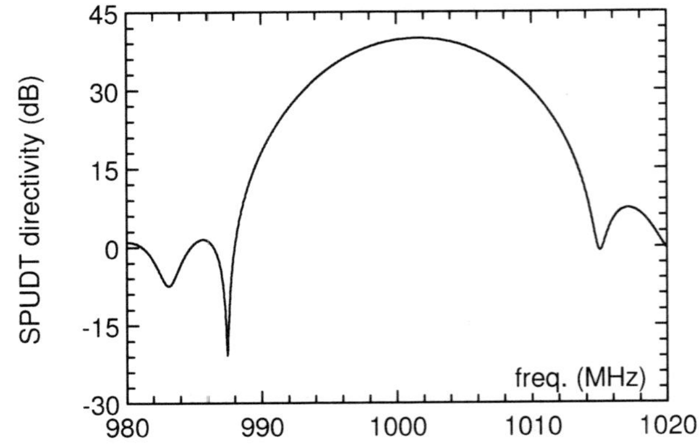

Figure 3. Directivity of SPUDT

2.3. Idealised device analysis for passband

For a device comprising two transducers with known P-matrices, the Y-matrix is easily deduced [5], and for the LCR we find

$$Y_{11} = Y_{22} = P_{33}^S - \frac{P_{11}^S (P_{13}^S)^2}{1-(P_{11}^S)^2} \qquad \text{.. (2a)}$$

$$Y_{12} = Y_{21} = -\frac{2(P_{13}^S)^2}{1-(P_{11}^S)^2} \qquad \text{.. (2b)}$$

We define θ_S such that $P_{11}^S = |P_{11}^S| \exp(j\theta_S)$. In the filter passband, $|P_{11}^S|$ must be close to unity, so the denominator in eqs.(2) is approximately $1-\exp(2j\theta_S)$. Hence, resonances occur when $\theta_S = n\pi$. We can also identify θ_S as the phase change for one transit of the resonant cavity, so that the 'round trip' phase change is $2\theta_S$, and for a resonance this equals $2n\pi$. The resonances correspond to both symmetric and antisymmetric modes.

We can also define an effective cavity length (in time units) by

$$T_c(\omega) = -d\theta_S / d\omega$$

Figure 4 shows a plot of this function, for a LCR consisting of the two SPUDT's described earlier, with a spacing of 2.8 λ.

The asymmetry in this curve is associated with the electrode reflections in the transducers. In early SAW resonators it was common to minimise these reflections by recessing the electrodes, and in that case the $T_c(\omega)$ curve becomnes a symmetrical ‘U’ shape.

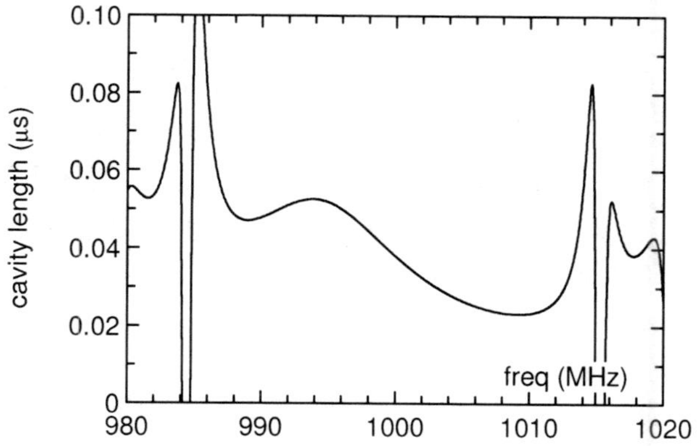

Figure 4. Cavity length of LCR, in time units.

Power conservation enables eqs.(2) to be simplified by using the relation

$$P_{11}P_{13}^* + P_{12}P_{23}^* + P_{13} = 0 \qquad .. (3)$$

This equation applies for any component with no loss; it can be derived by considering input signals to be present at ports 1 and 3, and setting the total ouput power equal to the total input power. In the passband of the filter the SPUDT needs to have high directivity, so that $|P_{13}^S| >> |P_{23}^S|$, and eq.(3) gives $P_{11}^S \approx -P_{13}^S(P_{13}^S)^*$. Hence $|P_{11}^S| \approx 1$, as stated earlier. In addition, P_{13}^S can be written as

$$P_{13}^S \approx \pm |P_{13}^S| \exp(j\theta_S/2) \qquad .. (4)$$

Substituting into eqs.(2), the Y-matrix for the passband can be approximated as

$$Y_{11} = Y_{22} \approx j\,\mathrm{Im}\{P_{33}^S\} + jG_a^S \cot\theta_S \qquad .. (5a)$$

$$Y_{12} = Y_{21} \approx jG_a^S \operatorname{cosec}\theta_S \qquad .. (5b)$$

Here $G_a^S \approx |P_{13}^S|^2$ is the acoustic conductance of the SPUDT. The term $\mathrm{Im}\{P_{33}^S\}$ is usually dominated by the transducer capacitance.

Examples of these functions are shown in Fig. 5. These are for the SPUDT parameters and spacing described above. This design is used merely to demonstrate the validity of the theory. It is not intended as an example of a practical filter. The thick solid lines in Fig. 5 are calculated using the accurate COM theory as described earlier. The thin lines are obtained from eqs.(5), using values for θ_S and G_a^S obtained from the COM theory. The agreement is very good in the region (990-1015 MHz) where the SPUDT has high directivity (Fig. 3). Oustide this region the approximate expressions fail because the assumptions are invalid. The broken lines will be explained later.

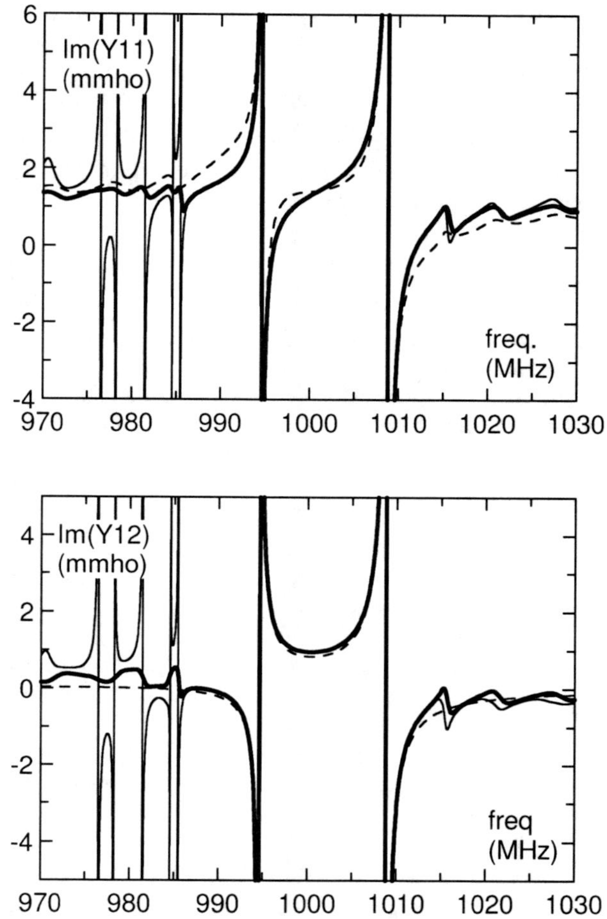

Figure 5. Y-matrix parameters for a two-transducer LCR. Thick solid line: accurate analysis. Thin solid line: approximate analysis using eqs.(5). Broken line: analysis from equivalent circuit.

2.4. Insertion Loss

The filter insertion loss, in dB, is 20 $\log|S_{12}|$, where S_{12} is given by the well-known expression

$$S_{12} = \frac{2Y_{12}\sqrt{R_{g1}R_{g2}}}{R_{g1}R_{g2}Y_{12}^2 - (1+Y_{11}R_{g1})(1+Y_{22}R_{g2})} \qquad .. (6)$$

Here R_{g1} and R_{g2} are the source or load impedances on ports 1 and 2, respectively. This expression assumes that any tuning or matching components have been accounted for by modifying the Y-matrix. If the device is tuned with parallel inductors, they can be allowed for approximately by simply ignoring the $\mathrm{Im}\{P_{33}^S\}$ term in eq. (5a).

Given the approximate Y-matrix of eqs.(5), we need to find load impedances such that the filter gives a flat low-loss response. This problem is complicated by the fact that G_a^S is a function of frequency, as seen in Fig.2. However, if the filter is to give a flat response, the SPUDT's need to be designed such that the conductance doesn't vary too much with frequency, so this condition is assumed here. Hence we can estimate the loading required by assuming G_a^S to be constant within the filter passband. Using eqs.(5) and (6), the required loading is found to be

$$R_{g1} = R_{g2} = 1/G_a^S \qquad .. (7)$$

This assumes that parallel inductor tuning is used, though the inductors may not make much difference if the SPUDT impedance is mainly resistive. Each SPUDT is just matched as if there were no waves incident. With this loading, eqs.(5) and (6) give

$$S_{12} \approx -\exp(j\theta_S) \qquad .. (8)$$

for frequencies in the filter passband. This result shows that the device is now acting simply as a delay line, with delay corresponding to the separation of the SPUDT's.

Figures 6 and 7 show the insertion loss, using parallel tuning, with load values of 1000 Ω in Fig.6 and 300 Ω in Fig.7. The line styles are as in Fig.5. Again, it can be seen that there is good agreement between the approximate and accurate theories, for frequencies in the range 990-1015 MHz where the SPUDT directivity is high.

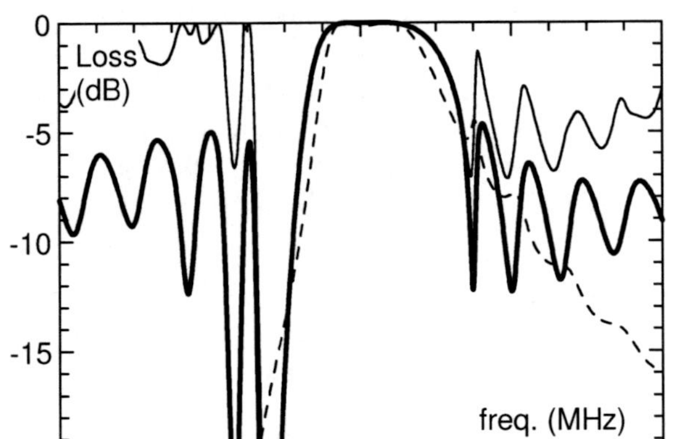

Figure 6. Insertion loss of two-transducer LCR, with loads of 1000 Ω. Line styles as on Fig.5.

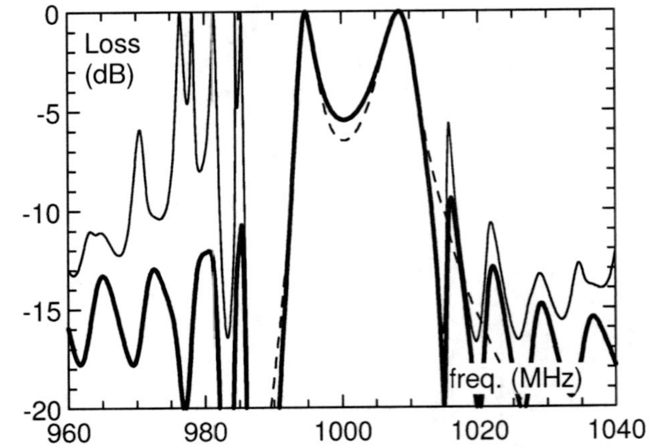

Figure 7. Insertion loss of two-transducer LCR with loads of 300 Ω. Line styles as Fig.5.

2.5. Equivalent Circuit

Figure 8 shows the well-known pi-circuit, which can be taken as an equivalent for any two-port device. For convenience, susceptances B_1 and B_2 are included to represent the $\mathrm{Im}\{P_{33}^S\}$ term explicitly; often these can be represented by the transducer capacitances. For the two-transducer LCR, the circuit will be symmetrical, so that $Z_a = Z_b$ and $B_1 = B_2$. However, an asymmetric form is presented here because it will be applied later to a three-transducer device. It is assumed here that the device has two resonance frequencies, and for each resonance the impedances Z_a, Z_b and Z_s in Fig. 8 can be represented by series resonant circuits. These circuits are simply added to give the circuit of Fig. 9, where the impedances with index 1 and 2 refer to the resonant frequencies ω_1 and ω_2.

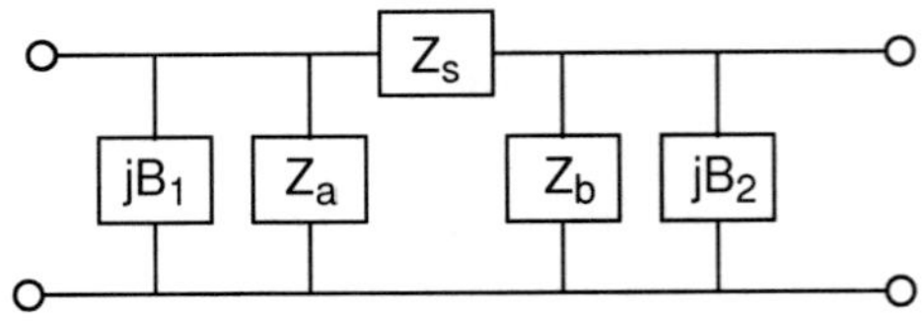

Figure 8. Pi circuit for two-port device

To deduce circuit values consider a resonance (pole of Y_{11} or Y_{12}) at frequency ω_m, where $\theta_S = n\pi$. In the vicinity of this pole we have $\theta_S \approx n\pi - T_c(\omega_m).(\omega - \omega_m)$. Here $T_c(\omega) = -d\theta_S/d\omega$ as before. Using eq. (5a), the residue of Y_{11} at the pole is found to be $(-1)^{n+1} jG_a^S(\omega_m)/T_c(\omega_m)$. A series resonant circuit with inductance L and capacitance $C = 1/(\omega_m^2 L)$ has an admittance pole at frequency ω_m with residue $1/(2jL)$. Equating these, series resonant circuits can be derived to represent Z_{am}, Z_{bm} and Z_{sm}, with resonant frequencies ω_m and m = 1, 2. For the series branch, the inductor L_{sm} is found to be

$$L_{sm} = -\frac{(-1)^n T_c(\omega_m)}{2G_a^S(\omega_m)} \quad ..(9)$$

The parallel branches (with impedance Z_{am}, Z_{bm}) are found to exist only when n is even, and then their motional inductances are $L_{am} = L_{bm} = -L_{sm}/2 > 0$. For the two resonances, n will be even for one resonance and odd for the other, so that Z_{a1} and Z_{b1} can be omitted from Fig. 9, taking these to refer to the resonance with n odd. The circuit can then be rearranged to the form of Fig. 10, where Z_1 and Z_2 refer to n odd and even, respectively [2, p. 145]. From the above analysis,

$$Z_m = j\omega L_m - j/(\omega C_m)$$

with $L_m = T_c(\omega_m)/[2G_a^S(\omega_m)]$. It is seen that Z_1 refers to the antisymmetric mode, and Z_2 to the symmetric mode.

Early resonator studies usually assumed that the transducers had no internal reflections. If this is assumed in the LCR, the inductance becomes $L_m = T_c(\omega_m)/[4G_a(\omega_m)]$, a result obtained earlier [7]. Here G_a is the transducer conductance. A factor of 2 arises because the SPUDT has high directivity, which implies that $G_a^S \approx 2G_a$.

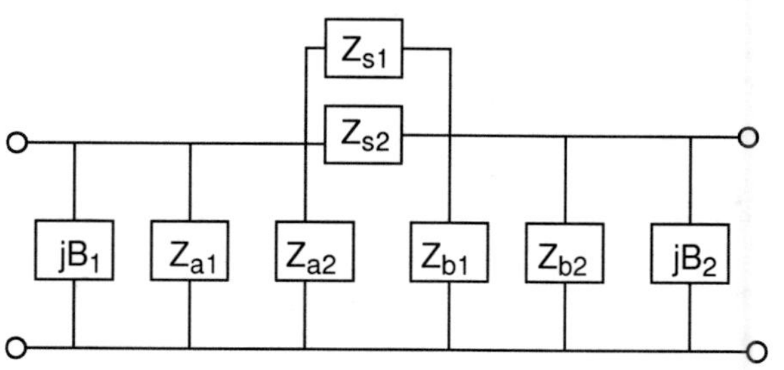

Figure 9. Equivalent circuit for two-pole LCR.

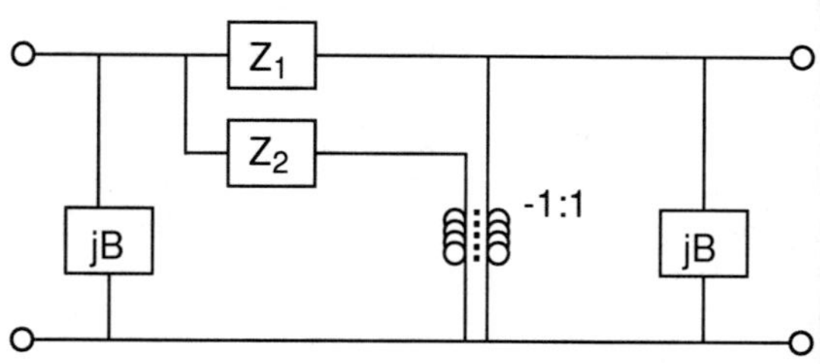

Figure 10. Equivalent circuit for symmetrical two-pole LCR

In Figs. 5-7, results obtained from the equivalent circuit are shown by broken lines. For the passband, they agree quite well with the accurate calculations. However, there are discrepancies which are due to the fact that only two poles have been included. The actual device has further poles and zeros outside the passband, and these are not represented in the equivalent circuit; their effect depends on how near they are to the passband. In addition, an equivalent circuit will usually include resistances in order to model losses, but these resistances have been ignored here. They do not have much effect on the response of a low-loss device.

3. Analysis of Three-transducer Device

Analysis of the three-transducer device (Fig. 11) has been described earlier [3], so little detail is given here except for the equivalent circuit, which was not included before. The central transducer of the 3-transducer device can be shown to behave as if it did not have internal reflections, even though it may be a single-electrode transducer with strong electrode reflectivity. This behaviour occurs because in the LCR the symmetry ensures that the transducer always has waves of equal amplitude incident from both directions. Thus, the device is essentially a simple resonator, with resonances arising from reflections at the two SPUDT's as in the two-transducer device. However, electrode reflections in the central transducer do cause some distortion because they affect the phases of the SAW's, introducing dispersion.

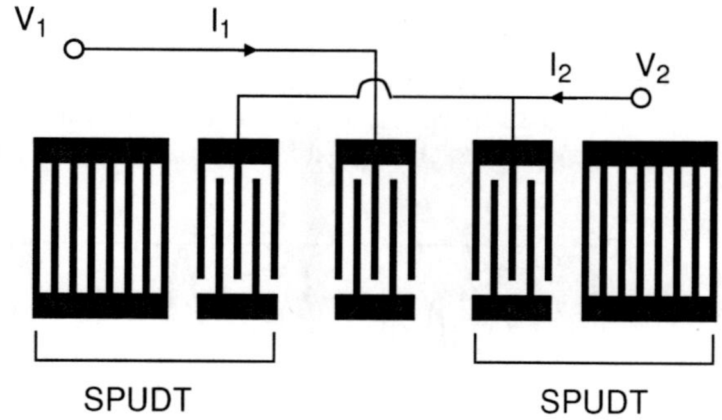

Figure 11. Three-transducer LCR

The three-transducer device is considered as two identical SPUDT's with admittance P_{33}^S and conductance G_a^S, plus a central transducer with admittance P_{33}^T and conductance G_a^T. The central transducer is port 1 of the device, and the two SPUDT's are connected together to form port 2. Using methods similar to those above, the Y-matrix for the filter passband can be approximated as [3]

$$Y_{11} \approx j\,\mathrm{Im}\{P_{33}^T\} + jG_a^T \cot(\theta_c/2) \quad ..(10a)$$

$$Y_{12} = Y_{21} \approx \pm j\sqrt{2G_a^T G_a^S}\,\mathrm{cosec}(\theta_c/2) \quad ..(10b)$$

$$Y_{22} \approx 2j\,\mathrm{Im}\{P_{33}^S\} + 2jG_a^S \cot(\theta_c/2) \quad ..(10c)$$

Here θ_c is an effective cavity phase, defined as $\theta_c = \theta_S + \theta_T$, and resonances occur when $\theta_c = 2n\pi$, these being symmetric modes. Referring to the central transducer, θ_T is defined as the phase of the expresssion $P_{11}^T + P_{12}^T$, which can be shown to have magnitude unity (assumig no losses). As before, the insertion loss is given by eq.(6). For a flat, low-loss response, the required loading (with parallel tuning) is found to be $R_{g1} = 1/G_a^T$ and $R_{g2} = 1/(2G_a^S)$. Thus the transducer and SPUDT's are again matched as if there were no incident waves. With this loading, eq.(6) gives $S_{12} \approx \pm\exp(j\theta_c/2)$ in the passband.

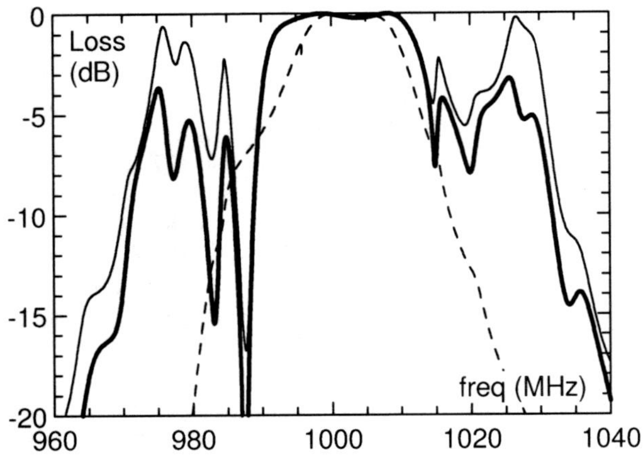

Figure 12. Insertion loss of three-transducer LCR with loads of 193 Ω (port 1) and 594 Ω (port 2). Line styles as on Fig.5, with eqs.(10) used for approximate analysis.

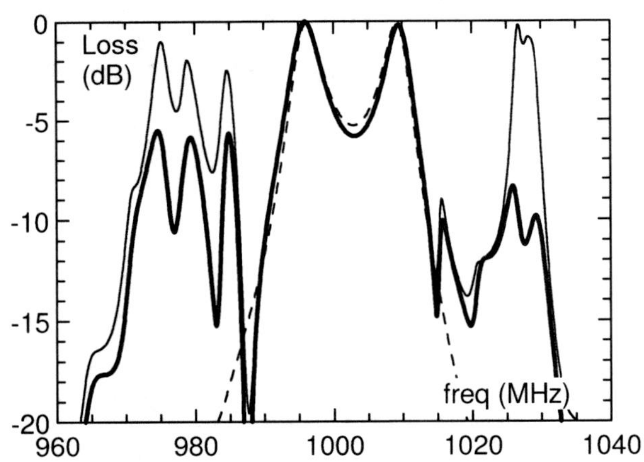

Figure 13. Insertion loss of three-transducer LCR with loads of 60 Ω (port 1) and 200 Ω (port 2). Line styles as in Fig. 12.

Insertion loss curves are shown in Figs. 12 and 13, assuming parallel tuning. Here the SPUDT design is as before. The central transducer is the single-electrode type with 51 electrodes, and the electrode pitch and reflection coefficient are as for the SPUDT transducers. The spacing between the central transducer and each SPUDT is 15.1λ at 1000 MHz. As before, the thick solid line is obtained from an accurate calculation, and the thin line is the approximate result obtained using eqs.(10). In Fig. 12 the loads on ports 1 and 2 are 193 and 594 Ω respectively, and in Fig. 13 they are 60 and 200 Ω. As before, the approximate results using eqs.(10) agree well in the device passband.

The equivalent circuit can have the form of Fig. 9, where Z_{sm} represents a series circuit resonant at frequency ω_m, with inductance L_{sm}, and for a 2-pole device we have m = 1, 2. Thus, $Z_{sm} = j\omega L_{sm} - j/(\omega C_{sm})$, with $C_{sm} = 1/(\omega_m^2 L_{sm})$. Similar comments apply to Z_{am} and Z_{bm}. Using the method of Sec. 2.5 above, the inductors are found to be [3]

$$L_{sm} = -\frac{(-1)^n T_c(\omega_m)}{4\sqrt{2G_a^T G_a^S}} \quad \text{.. (11a)}$$

$$L_{am} = \frac{T_c(\omega_m)/4}{G_a^T + (-1)^n \sqrt{2G_a^T G_a^S}} \quad \text{.. (11b)}$$

$$L_{bm} = \frac{T_c(\omega_m)/4}{2G_a^S + (-1)^n \sqrt{2G_a^T G_a^S}} \quad \text{.. (11c)}$$

Results obtained using the equivalent circuit are shown by the broken lines on Figs. 12 and 13. As for the two-transducer LCR, they show reasonable agreement with the accurate calulation, for frequencies in the filter passband.

4. Conclusions

It has been shown that the passband response of a LCR filter, with either two or three transducers, can be expressed in terms of simple approximate formulae for its Y-parameters. The formulae assume reciprocity and power conservation. It was also assumed that the gratings and their adjacent transducers behave as SPUDT's, with strong directivity in the filter passband. This condition is needed if the device is to act as a low-loss filter, but it is not necessarily valid; it needs to be obtained by design. Equivalent circuits have also been deduced, though these are only valid if the response is dominated by poles in the passband, with other poles or zeros sufficiently remote to have little effect. The method remains valid for an arbitrary number of passband poles. If there are more than two poles, the only modification necessary occurs in the equivalent circuit (Fig. 9), where resonant circuits are needed for each pole. Thus m takes values 1, 2.. M, where M is the number of passband poles.

Considering the approximate (imaginary) Y_{ij}, it is noted that $\mathrm{Im}\{Y_{11}\}$ always increases with ω, for frequencies in the passband. This is a known property of two-terminal imaginary admittances, by Foster's reactance theorem (see for example Kuo [8], p.317). Y_{11} is the admittance seen at port 1 when port 2 is shorted, so it must satisfy this condition. Hence the poles of Y_{11} all have the same sign. On the other hand, Y_{12} does not correspond to a 2-terminal admittance, and so it need not be of this form. From eq.(6) it can be seen that Y_{12} cannot be zero in the passband, because this would give $S_{12} = 0$. To satisfy this condition the poles of Y_{12} must alternate in sign.

It is also seen that poles of Y_{11} and Y_{12} occur at the same frequencies, another general requirement [8, p. 344]. There is in fact a further condition which can be deduced from eq.(6), which has 1st-order poles in the numerator and 2nd-order poles in the denominator. In order for S_{12} to be non-zero at a pole frequency, the Y-parameters need to satisfy the relation $Y_{12}^2 = Y_{11}Y_{22}$ in the vicinity of the pole. This condition is satisfied by the approximate relations of eqs.(5) and (10).

Practical LCR's often use a leaky surface wave on 42°Y-X lithium tantalate, rather than the SAW on lithium niobate assumed above. The SAW choice was made for clarity, since the leaky-wave case is more complex and gives high-frequency attenuation not exhibited by a SAW. However, it is expected that the analysis will apply approximately for a leaky-wave device, provided the attenuation is small.

It is also usual to have two tracks connected together electrically, in order to improve the sidelobe suppression, rather than the single-track device considered here. The present analysis can be extended to a two-track device straightforwardly.

References

[1]. T. Morita, Y. Watanabe, M. Tanaka and Y. Nakazawa. 'Wideband low loss double mode SAW filters'. *IEEE Ultrasonics Symp.*, Tuscon, 1992, pp. 95-104.

[2]. K.-Y. Hashimoto. *Surface Acoustic Wave Devices in Telecommunications.* Springer, Berlin, 2000.

[3]. D.P. Morgan. 'Approximate analysis of SAW longitudinally-coupled resonator (LCR) filters'. *IEEE Ultrasonics Symp.*, Honolulu, 2003 (to be published).

[4]. D.P. Morgan. 'Simplified analysis of surface acoustic wave one-port resonators'. *Electronics Lett.*, 39, 1361-1362 (2003).

[5]. V. Plessky and J. Koskela. 'Coupling-of-modes analysis of SAW devices'. In C.C.W. Ruppel and T.A. Fjeldy (eds.), *Advances in Surface Acoustic Wave Technology, Systems and Applications*, vol.2, World Scientific, Singapore 2001, pp. 1-81.

[6]. D.P. Morgan. 'Reflective array modelling for SAW transducers'. *IEEE Ultrasonics Symp.*, Seattle, 1995, pp. 215-220.

[7]. Supriyo Datta. *Surface Acoustic Wave Devices.* Prentice-Hall, Englewood Cliffs, 1986.

[8]. F. Kuo. *Network Analysis and Synthesis.* Wiley, 1966.

SCANNING WINDOW METHOD FOR SPUDT OPTIMIZATION

EVGENIY BAUSK AND RINAT TAZIEV
Institute of Semiconductor Physics of Russian Academy of Sciences
630090 Novosibirsk, Russia

Optimization of Single Phase Unidirectional Transducers (SPUDT) is main problem in thedesign of low-loss SAW filters. Here we deal with Distributed Acoustic Reflection Transducers, each elementary cell (one wavelength) of that includes not more than one "hot"electrode and/or one, two, three or four grounded electrodes [1]. For assembling the input andoutput SPUDTs we use more than ten different cell types that are considered as elementary SAW sources and SAW reflectors with different magnitudes and phases, depending on widths and positions of electrodes.

A method for SPUDT filter optimization suggested here is based on the "scanning window"technique that was earlier successfully applied to optimization of bi-directional withdrawalweighted SAW transducers [2]. The main idea of this technique consists in analysis of all possible combinations of different cells inside a short region ("window") of IDT. Checkingall parts of the filter frequency response as well as its insertion loss and the level of tripletransit echo, software chooses the best combination of sells inside the "window", memorizes it, and then changes the "window" position. Thus, the "window" scans step by step throughout both transducers of the SAW filter. The "window" scanning is repeated as long as improving of filter characteristics takes place. As a result, both SAW sources and reflectors distributions in the input and output transducers are optimized simultaneously.

Knowing transmission matrixes of IDTs, one can calculate all filter parameters [1]. Using Reflective Array Modeling [1], we form SPUDT transmission matrix by cascading transmission matrixes of cells that were calculated beforehand for each type of cell. Consideration of huge number of different IDT versions during the optimization process needs the acceleration of the filter analysis. To reduce the time for analysis of each filter modification we divide SPUDT into three regions: at the left of the "window", the "window" itself, and at the right of the "window". A transmission matrix of the short "window" is calculated for each electrode modification, while matrixes of the long IDT regions outside the "window" are calculated only once for each "window" position. This provides the analysis about 100 filter modifications per one second.

[1] D.Morgan. – IEEE 1995 Ultras. Symp. Proc., p.215.
[2] E.Bausk. – IEEE Trans. on UFFC, 1999, v.46, N5, p.1276.

A COMPACT, HIGH-PERFORMANCE LASER-PUMPED RUBIDIUM FREQUENCY STANDARD

C. Affolderbach[1], G. Mileti[1], F. Droz[2]

[1]Observatoire Cantonal de Neuchâtel, Rue de l'Observatoire 58, 2000 Neuchâtel, Switzerland
[2]Temex Neuchâtel Time, Vauseyon 29, 2000 Neuchatel, Switzerland

Keywords: rubidium frequency standard, laser optical pumping, space instrumentation.

Abstract

We present the realisation of a compact and high-performance Rubidium atomic frequency standard under development for space applications. The clock comprises a compact frequency-stabilised diode laser head for optical pumping of a lamp-removed industrial Rb clock module, in order to exploit the advantages of laser optical pumping for optimised short-term stability. We obtain typical signal contrasts around 20%, resulting in an estimated signal-to-noise limit for the short-term stability of about $1 \cdot 10^{-12}\ \tau^{-1/2}$. Modification of the buffer-gas filling in the clock resonance cell allows to reduce instabilities arising from the AC Stark effect and temperature variations, that limit the clock stability on longer timescales. The clock stability of the realised demonstrator breadboard was measured to be better than $4 \cdot 10^{-12}\ \tau^{-12}$ up to 10^4 seconds, already meeting the specifications for the GALILEO Rb clocks. Further optimisation should allow the realisation of a laser-pumped Rb clock reaching $3 \cdot 10^{-13}\ \tau^{-1/2}$ in a compact device (<1.5l, 1.5kg, 15W). Such compact and high-performance clocks can find their applications in telecommunications, science missions, and future generations of satellite navigation systems (GALILEO, GPS).

1 Introduction

Optically-pumped gas-cell atomic frequency standards (often referred to as Rubidium clocks) [16] represent a well-established type of secondary frequency references, offering competitive frequency stabilities on medium-term timescales (~days) in compact volumes around or below one litre. Industrial realisations of such clocks have their applications in timekeeping, telecommunications and satellite navigations. While commercial Rb clocks use discharge lamps for optical pumping, it has been shown that the use of tuneable, stabilised lasers with their improved spectral control can significantly improve the short-term stability of optically pumped gas-cell clocks [11,12].

Optical pumping with discharge lamps generally employs optical filtering (typically integrated or separate Rb cell filters) to produce pump light acting on one of the two ^{87}Rb D_2-line hyperfine components only. However, this filtering is rather difficult to achieve in a perfect manner, and thus spurious re-pumping on the second hyperfine component broadens the microwave resonance line to typically 1-2 kHz. In addition, the relatively broad emission spectrum of Rb lamps keeps pumping efficiencies rather low and results in typical signal contrasts below 1%. In contrast, narrow-band semiconductor laser diodes offer excellent control of the light spectrum and emission linewidths much narrower than the transition to pump (few MHz or below compared to about 500 MHz). The implementation of lasers thus offers highly selective and well-controlled optical pumping, resulting in microwave resonance lines with typical linewidths below 1 kHz and signal contrasts around 20% or more, and thus offers improved short-term stabilities.

On the other hand, the narrow spectrum of the laser pump light source also causes an increased susceptibility of the clock transition to fluctuations in laser frequency and intensity, mediated by the AC Stark effect [6,10]. While these effects tend to compromises the medium and long-term clock stability, they can nevertheless be controlled to the necessary level by careful control of the laser emission spectrum.

The integration of laser-pumping into a Rb clock thus offers the potential to realise improved high-performance gas-cell clocks while maintaining the advantageous compact design. Furthermore, the high efficiency of advanced semiconductor laser diodes offers the potential to reduce the overall power consumption, by replacing the discharge lamps with their rf excitation and high operating temperature.

Here we report recent results on our on-going project aiming to demonstrate the feasibility of a compact and high-performance laser-pumped Rb clock for space applications. The work was conducted in the frame of an ESA project within the ARTES 5 program for telecommunication technologies, in cooperation between Observatoire de Neuchâtel and Temex Neuchâtel Time as industrial partner.

2 Clock realisation

Since the main innovation regarding the clock concerns a change of the light source, a modular approach was chosen, using a modified, lamp-removed industrial Rb clock as core module and a separate laser head providing the pump light. This allows to work independently on both the laser head and the clock module for performance optimisation, while envisaging later re-integration of the laser head into the clock.

Figure 1: Experimental realisation of the demonstrator clock, comprising the stabilised laser head (left) and the modified industrial clock module (right).

For the clock module we used a RAFS type Rb clock of Temex Neuchâtel Time (TNT), showing lamp-pumped short-term stabilities of $2\text{-}3\cdot10^{-12}\ \tau^{-1/2}$ [8]. The clock modifications mainly consisted in the removal of the Rb lamp and related electronics, modifications of the electronics for easier access to control parameters, and change of the LO quartz to allow for a wider tuning range. Additional changes to the clock module during the course of the project mainly concerned modifications of the atomic resonance cell in order to adapt the clock for laser pumping, as will be discussed in sections 2.2 and 2.3 below.

As pump light source we implemented a compact, frequency stabilised laser head developed at ON for this purpose [2]. This laser head is currently used in a stand-alone configuration, but integration to the clock module is foreseen. Figure 1 gives a view of the demonstrator clock comprised of the two subunits. In the following sections, critical aspects of the two will be addressed.

2.1 Laser pump light source

In principle, today's advanced single-mode laser diodes like distributed feed-back (DBR), distributed Bragg-reflector (DBR) or vertical-cavity surface emitting lasers (VCSEL) offer sufficiently narrow spectral widths and output power for efficient optical pumping, as well as good spectral control by only few electric inputs without the need for external moving optics. However, commercial availability of such suitable lasers at the required wavelength often presents an open issue. Furthermore, the linewidths of commercially available diode lasers of these types typically are not always sufficiently small to resolve the natural homogeneous linewidth of around 6 MHz of Rb sub-Doppler saturated absorption lines, which is necessary in order to allow sufficiently precise frequency stabilisation of the pump light down to around 10^{-12} at 10^{3} seconds required for the laser clock [4,14].

For a demonstration of the laser-pumped Rb clock, the implemented laser head thus consists of an external-cavity diode laser (ECDL), spectrally narrowed and frequency-stabilised by optical feedback from a diffraction grating in Littrow configuration [2]. The physics package of the laser head (upper part on the left hand side in Fig. 1) occupies a total volume of 200 cm^3, including a compact saturated-absorption setup for laser frequency stabilisation to a thermostated and magnetically shielded Rb reference cell. This reference setup uses a simple, back-reflected beam geometry without background subtraction that assures sufficiently robust and precise stabilisation of the laser frequency to a level of $2\cdot10^{-12}\ \tau^{-12}$ up to 100 seconds and $<2\cdot10^{-12}$ between 10^4 and 10^5 seconds [4]. The laser head emission exhibits a spectral width <900 kHz and mode-hop free tuning over up to 9GHz, making it well-suited for optical pumping in the Rb clock. We also measured an FM noise of 2 $kHz\cdot Hz^{-1/2}$ and AM noise $RIN\approx10^{-13}/Hz$ at a Fourier frequency of about 300 Hz. Using the internal reference cell, the laser head frequency routinely stays locked over many days, and thus allows evaluation of the long-term clock performance.

When saturated absorption spectroscopy lines are used as reference for laser frequency stabilisation, the correct sub-Doppler resonance has to be identified among several narrow lines superimposed on a broad Doppler absorption background (Fig. 3, lower trace). While digital electronics systems for automated line detection and locking of ECDLs have already been demonstrated [5], a simpler scheme possibly based on analog electronics is clearly preferable for an envisaged space instrument. We have identified a narrow, inverted sub-Doppler line on the ^{87}Rb F=1 hyperfine component, that can be unambiguously identified even with a simple control loop. Figure 2 gives a view of the experimentally observed "inverted line" (marked by the arrow in Fig. 2), arising from competition between optical pumping processes [14]. With the laser locked to this line, the pump light frequency is also close to the centre of the F=1 transition in the clock resonator cell, where intensity related light-shift effects are small (see section 2.2). Optical pumping on the F=1 component is thus promising, in spite of its somewhat lower transition rate compared to the F=2 component.

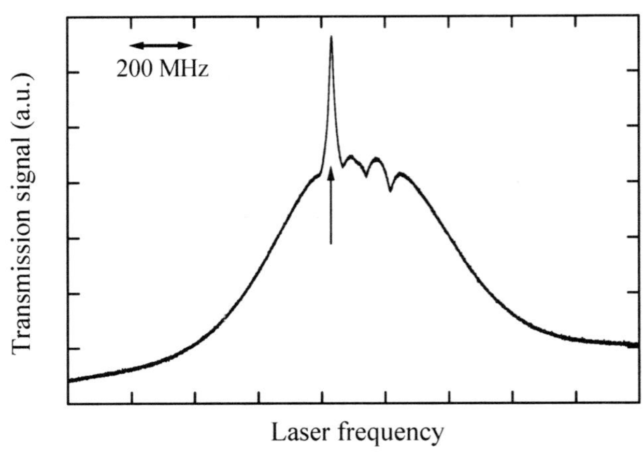

Figure 2: Spectrum of the "inverted" sub-Doppler resonance on the ^{87}Rb F=1 hyperfine transition, allowing unambiguous line identification.

2.2 Light shift effects

The AC Stark shift or "light shift" of the clock transition induced by the pump light field [6,10] constitutes a major source of instabilities in optically pumped atomic clocks. In the case of low pump light intensities present in optically pumped clocks, the interaction between the light field and the atomic levels causes the clock transition frequency f_C to exhibit a dispersive shape as function of the pump light laser frequency ν_L tuned around the pumped atomic transition ν_0 (of width Γ), given by [7]:

$$\Delta f_C = \frac{1}{4}|\Omega_R|^2 \frac{(\nu_L - \nu_0)}{(\nu_L - \nu_0)^2 + \Gamma^2/4},$$

where the amplitude of this light shift curve scales approximately linearly with the light intensity $I \sim |\Omega_R|^2$. Figure 3 illustrates this behaviour of the clock frequency, measured with the pump laser source tuned across the ^{87}Rb F=2 transition component using four different pump light intensities. At the intersection point of the four light shift curves (downward arrow in Fig. 3), variations of the clock frequency with light intensity are minimised, thus the parameter

$$\alpha = \frac{df_C}{dI}\bigg|_{\nu_L = const.}$$

(corresponding to the splitting of the light shift curves) is small and clock frequency instabilities due to fluctuations of the laser intensity are reduced. Simultaneously, fluctuations in laser frequency are translated into clock instabilities by the slope

$$\beta = \frac{df_C}{d\nu_L}\bigg|_{I = const}$$

of the light shift curves close to line centre. In a first step, we have slightly optimised the resonance cell's buffer gas composition to make the point of α=0 in the resonator cell coincide - within few MHz - with the cross-over 21-23 frequency in the evacuated Rb cell used for laser stabilisation (upwards arrow in Fig. 3) [13]. For typical clock settings and with α, β expressed in fractional stability of f_C we then have $\alpha \approx 2\cdot10^{-10}$ $\Delta I/I$ and $\beta \approx 3\cdot10^{-17}$/Hz (with almost identical values for pumping on the F=1 and F=2 transitions). For measured laser intensity stabilities around $2\cdot10^{-4}$ at 10^4 seconds one thus expects an intensity-connected stability limit around $4\cdot10^{-14}$ at 10^4s. From a laser frequency stability of $2\cdot10^{-12}$ at 10^4 seconds obtained with the simple saturated absorption stabilisation implemented in the laser head [3,4], the frequency-connected clock instability can be estimated to about $2\text{-}3\cdot10^{-14}$ at 10^4 seconds. Further reduction of these limits down to around 10^{-14} at 10^4-10^5 seconds can be expected from fine adjustments on the clock cell's buffer gas composition, on the pump light intensity and its fluctuations, as well as further optimised laser frequency stabilisation. The implementation of recently developed schemes for light shift suppression [1,7,9] could also contribute to further reductions of these limits.

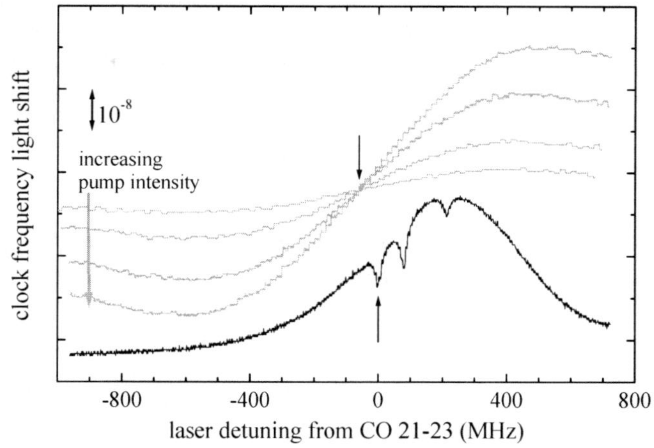

Figure 3: Measured light shift of the clock frequency f_C for different pump light intensities between 10 and 65 μW/cm^2 (upper 4 traces) and reference saturated absorption (lower trace) for the laser tuned around the ^{87}Rb F=2 transition.

2.3 Resonance cell temperature coefficient

Gas-cell based atomic frequency standards generally use a buffer-gas added to the clock resonance cell in order to reduce collisions of the Rb atoms with the cell walls, which otherwise would induce line broadening, as well as to quench spontaneous emission from the excited atomic level involved in optical pumping, which would decrease the detection signal-to-noise ratio. However, cells containing one pure buffer-gas only, as widely used in lamp-pumped Rb clocks, induce temperature coefficient of the clock transition around some 10^{-9}/K, which will contribute to the final stability limit of the clock. The choice of an appropriate mixture of buffer gases with temperature coefficients of opposite signs allows to cancel the linear dependence on temperature and thus results in a temperature coefficient of second order only [15,16]. Figure 4 gives the measured temperature dependence of a mixed buffer-gas cell specifically produced for

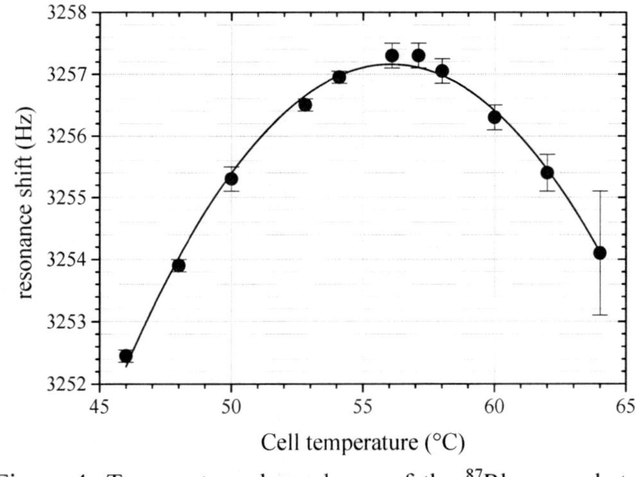

Figure 4: Temperature dependence of the ^{87}Rb ground-state hyperfine frequency in the clock resonance cell filled with a buffer-gas mixture.

implementation in the clock demonstrator. With appropriately chosen cell temperature around 57°C, the temperature coefficient is reduced to about 10^{-11}/K, resulting in a clock stability limit around 10^{-14} for cell temperature control to the millikelvin level.

3 Clock performance

Figure 5 shows examples of the demonstrator clock performance under standard laboratory conditions. First results were obtained using a pure buffer-gas cell not optimised for laser pumping, and thus are strongly degraded by the light shift and thermal effects (open circles). As expected, the introduction of a first, more appropriate buffer-gas mixture in the clock cell (filled circles) greatly improved the medium-term stability down to $4\cdot10^{-14}$ at 10^4 seconds. This stability result coincides with the performance of $3\cdot10^{-12}\ \tau^{-1/2}$ of the lamp-pumped RAFS module (solid line in Fig. 5), and already satisfies the specifications for the GALILEO Rb clocks. Depending on the laboratory conditions, the frequency drift of our clock typically amounts to around $2\text{-}6\cdot10^{-13}$ /day, probably due to the not yet fully optimised thermal stability of the laser head and residual temperature coefficients of the clock cell.

The dashed line of Figure 5 gives the short-term stability limit for the demonstrator clock of $3\cdot10^{-12}\ \tau^{-1/2}$, determined from the measured signal-to-noise ratio of the clock signal, and also presents the project objective for the clock stability up to 10^4 seconds. As the experimentally measured clock performance does not reach this signal-to-noise limit, we conclude that other contributions like the clock module electronics (LO, rf control) significantly contribute to the short-term stability encountered. We thus expect to improve the short-term stability down to its signal-to-noise-limit by optimisation of the clock electronics. With continued optimisation of the clock we envisage further improvement towards a short-term stability around $3\cdot10^{-13}\ \tau^{-1/2}$, that was already demonstrated for a laboratory laser-pumped Rb clock showing similar signal contrast and signal-to noise ratio (open squares in Fig. 5) [11].

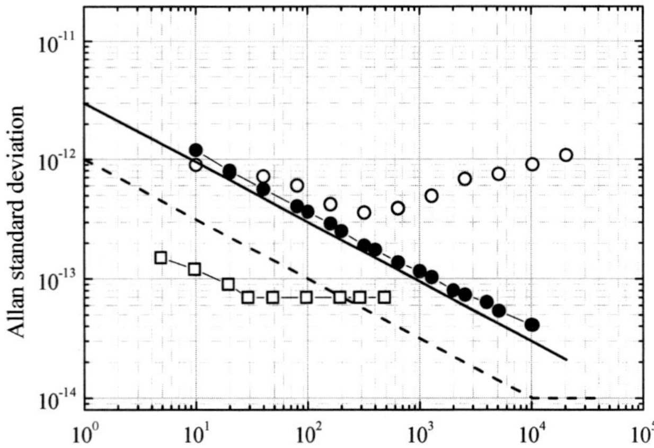

Figure 5: Performance of the demonstrator clock: with the initial cell of the lamp-pumped RAFS (open circles) and the new, optimised cell (open circles). Solid line: $3\cdot10^{-12}\ \tau^{-1/2}$; broken line: current signal-to-noise limit and project goal. The open squares give the best demonstrated short-term stability of a (much larger) laboratory laser-pumped Rb clock [11].

4 Conclusions

We have realised a compact and high-performance laser-pumped Rb gas-cell atomic clock on a breadboard demonstrator level. Under atmosphere and standard laboratory conditions, the clock shows a frequency stability of $3.8\cdot10^{-12}\ \tau^{-1/2}$, reaching $4\cdot10^{-14}$ at 10^4 seconds, within the specifications for GALILEO Rb clocks. The latter value already corresponds to the performance of other, considerably bulkier laser-pumped clocks [12] and demonstrates the performance potential of a compact, laser-pumped Rb clock. Pushing the clock subcomponent performances further towards their limits should allow for the realisation of a compact and high-performance Rb clock reaching $1\cdot10^{-14}$ at 10^4 seconds, with application potentials in satellite telecommunications, satellite navigation, space science missions, as well as similar ground applications.

Acknowledgements

This work was financed by the European Space Agency ESA/ESTEC within the ARTES 5 program under contract number 5280/01/NL/US. C.A. and G.M. also acknowledge support from the Swiss National Science Foundation (contract numbers 2100-66821, 2160-067498, and SCOPES6: JRPBUPJ062412), the Canton of Neuchâtel and the Swiss Confederation (article 16). We thank D. Slavov for his help with the spectral characterisation of the laser head.

References

[1] C. Affolderbach, G. Mileti, C. Andreeva, D. Slavov, T. Karaulanov, S. Cartaleva, "Reducing light-shift effects in optically-pumped gas-cell atomic frequency standards", *Proceedings of the 2003 IEEE International Frequency Control Symposium jointly with the 17th European Frequency and Time Forum*, pp. 27-30, (2003).

[2] C. Affolderbach, G. Mileti, "A compact, frequency stabilized laser head for optical pumping in space Rb clocks", Proceedings of the 2003 IEEE International Frequency Control Symposium jointly with the 17th European Frequency and Time Forum, pp. 109-111 (2003).

[3] C. Affolderbach, G. Mileti, D. Slavov, C. Andreeva, S. Cartaleva, "Frequency stability comparison of diode lasers locked to Doppler and sub-Doppler resonances", submitted for publication in SPIE.

[4] C. Affolderbach, G. Mileti, D. Slavov, C. Andreeva, S. Cartaleva, "Comparison of simple and compact Doppler and sub-Doppler laser frequency stabilisation schemes", Proceedings of the 18th European Frequency and Time Forum, (2004).

[5] F. Allard, I. Maksimovic, M. Abgrall, Ph. Laurent, "Automatic system to control the operation of an extended cavity diode laser", *Rev. Sci. Instrum.,* 75, pp.54-58, (2004).

[6] J. P. Barrat, C. Cohen-Tannoudji, "Etude du pompage optique dans le formalisme de la matrice densité", J. Phys. Radium, **22**, pp. 329-336, (1961).

[7] J. Deng, "Light shift compensation in a Rb gas cell frequency standard with two-laser pumping", *IEEE Trans. Ultrason. Ferroel. Freq. Control*, **48**, pp. 1657-1661, (2001).

[8] F. Droz, P. Rochat, G. Barmaverain, M. Brunet, J. Delporte, J.-F. Dutrey, F. Emma, T. Pike, U. Schmidt, "On-board Galileo RAFS, current status and performances", Proceedings of the 2003 IEEE International Frequency Control Symposium jointly with the 17th European Frequency and Time Forum, pp. 105-108 (2003).

[9] T. Karaulanov, C. Andreeva, S. Cartaleva, Y. Dancheva, B. Todorov, A. Yanev, S. Jaquet, G. Di Domenico, P. Thomann, G. Mileti, "Light shift and laser sidebands in gas-cell atomic clocks using optical pumping and coherent population trapping", *Proc. SPIE* **5137**, pp. 364-372, (2003).

[10] B. S. Mathur, H. Tang, W. Happer, "Light shift in the alkali atoms", Phys. Rev., 171, pp. 11-19, (1968).

[11] G. Mileti, J. Deng, F. L. Walls, D. A. Jennings, R. E. Drullinger, "Laser-pumped Rubidium frequency standards: new analysis and progress", *IEEE J. Quantum Electron.*, **34**, pp. 233-237, (1998).

[12] Y. Ohuchi, H. Suga, M. Fujita, T Suzuki, M. Uchino, K. Takahei, M. Tsuda, Y. Saburi, "A high-stability laser-pumped Cs gas-cell frequency standard", *Proceedings of the 2000 IEEE International Frequency Control Symposium,* pp. 651-655, (2000).

[13] Y. Saburi, Y. Koga, S. Kinugawa, T. Imamura, H. Suga,, Y. Ohuchi, "Short-term stability of laser-pumped rubidium gas cell frequency standard", *Electron. Lett.*, 30, pp. 633-635, (1994).

[14] O. Schmidt, K.-M. Knaak, R. Wynands, D. Meschede, "Cesium saturation spectroscopy revisited: How to reverse peaks and observe narrow resonances", *Appl. Phys. B*, **59**, pp. 167-178, (1994).

[15] J. Vanier, R. Kunski, N. Cyr, J. Y. Savard, M. Têtu, "On hyperfine frequency shifts caused by buffer gases: Application to the optically pumped passive rubidium frequency standard", *J. Appl. Phys.,* **53**, pp.5387-5391, (1982), and references therein.

[16] J. Vanier, C. Audoin, "The quantum physics of atomic frequency standards", Adam Hilger, Bristol and Philadelphia, 1988.

CPT MASER CLOCK EVALUATION FOR GALILEO

F. Levi, C. Calosso, S. Micalizio, A. Godone and E. K. Bertacco *

E. Detoma, P. Morsaniga and R. Zanello**

*Istituto Elettrotecnico Nazionale Galileo Ferraris, Strada delle Cacce 91, 10135 Torino, Italy
** Alenia Spazio, Corso Marche 41, 10146 Torino, Italy

Keywords: Coherent Population Trapping, light-shift, cavity pulling, short-medium term frequency stability.

1. Introduction

The possibility to apply the Λ excitation scheme to the atomic frequency standards field was considered by many authors [1]. In particular, small stable clocks based on the Coherent Population Trapping (CPT) phenomenon with interesting stability performances have recently been developed [2].
The physics involved in CPT is well known [3]. In this section we only recall the main concepts.
In the Λ interaction scheme, the atoms are excited by means of two phase coherent laser fields coupling the two hyperfine levels of the ground state with a common excited state (see Fig. 1). At the resonance, that is when the lasers frequency difference is equal to the hyperfine splitting of the ground state levels, a quantum interference phenomenon takes place: the atoms are trapped in a coherent superposition of the two ground state levels, the so called "dark state", so that they are no longer able to absorb energy from the laser fields.

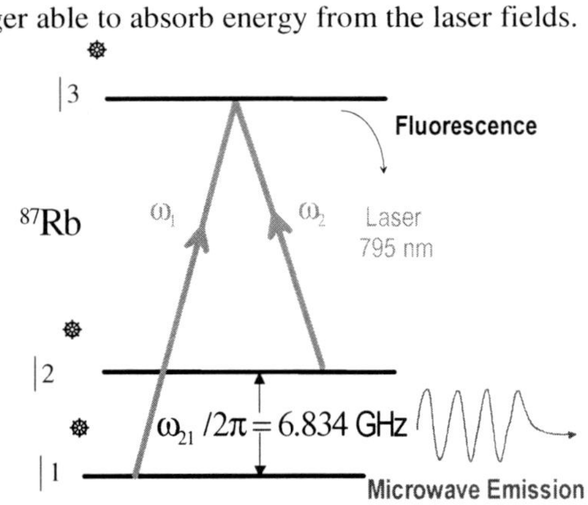

Figure 1. Three level system considered in our analysis.

This resonant mechanism can be observed both in the optical and in the microwave domain. In the first case, no fluorescence is observed from the cell and a narrow dark line appears in the fluorescence spectrum and similarly a bright line will appear in the transmitted signal. In the second case, the same CPT phenomenon creates a strong coherence in the ground state with an associated macroscopic magnetization. This magnetization is the source term of an oscillating magnetic field that can be detected as a maser emission when the atomic ensemble is placed in a suitable microwave cavity [4]. Both the approaches can be used in the implementation of a frequency standard; several studies showed that CPT standards offer some advantages relative to light shift, frequency stability and size in comparison with classical Intensity Optical Pumping technique (IOP) [5]. In particular, in the CPT maser there is no threshold regarding cavity Q and light intensity, since the CPT maser arises from the coherence generated in the ground state. Moreover, the resonance is observed without any laser background signal so that the resulting short term frequency stability is limited by the system thermal noise. Definitely, CPT frequency standards appear to be suitable for all those applications where simple working capability, small size, high stability performance are required, such as space applications.

2. Experimental set-up

A general scheme of the system we realized is shown in Fig. 2.

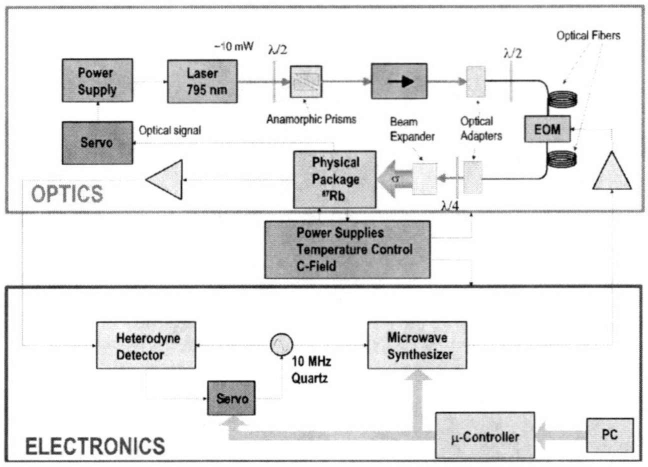

Figure 2. General scheme of the CPT maser

There are three main parts: the optical system, providing the laser beams for the Λ excitation scheme, the physical package, containing the microwave cavity and the cell, and the microwave generation and detection apparatus. In the following sub-sections we describe in more detail each part.

2.1. The optical system

With reference to Fig. 3, an external-cavity diode laser is fiber-coupled into a broad-band Electro-Optic Modulator (EOM). Applying to the modulator a microwave signal at 3.4… GHz, the first two modulation sidebands of the output spectrum provide the two phase coherent radiations fields required for the Λ excitation scheme when the laser is tuned to the D_1 center of gravity.
In order to achieve a frequency stability in the range of 10^{-14}, we realized a laser frequency servo loop. The transfer of the laser frequency instabilities to the clock transition is strongly reduced in the CPT approach in comparison with classical optically pumped frequency standards. However, this is not enough and to achieve a good laser frequency stability we used the saturated absorption technique via an external reference cell containing only ^{87}Rb. An Acousto-Optical Modulator (AOM) is placed along the laser path to allow a precise frequency tuning with respect to the optical transition; in fact, due to the presence of the buffer gas in the maser cell, the optical radiations are shifted in comparison with the reference cell.
To compensate the effects of the fiber polarization fluctuations, we realized also a laser power stabilization loop: the cavity photodiode is used as a power detector and the laser power is stabilized with a second AOM.

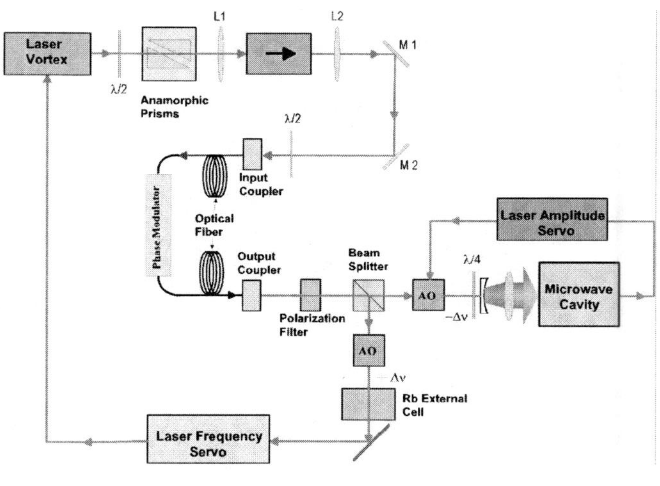

Figure 3. Optical apparatus of the CPT maser

2.2 The cavity system

The laser beam properly polarized is sent to the cavity system (see Fig. 4).
The cavity subsystem has been designed to give the best possible compromise between high output signal, high atomic quality factor and reduced sensitivity to fluctuations of environmental parameters.

The core is a λ/2 long quartz cell containing the Rb atoms and a mixture of buffer gases. The cell is placed in a copper microwave cavity that resonates at the Rb hyperfine transition frequency and operates in the TE_{011} mode. A coupling loop is used for the detection of the microwave signal.
The physical package also includes a solenoid providing the C-field, a double magnetic shield and two independent heaters to control the temperature of the system.

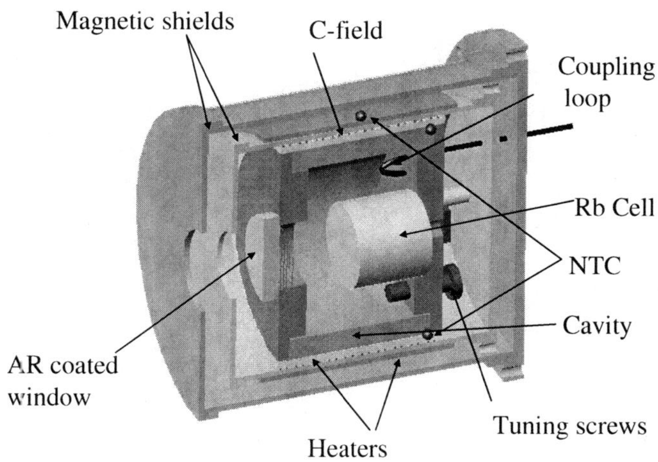

Figure 4. The cavity system

Table 1 summarizes the main features of the cavity system.

Cavity dimensions L Ø	L 40 x Ø 52 mm
Resonant frequency	6.834… GHz
Loaded quality factor	8000
Tunability	150 kHz/turn
Cavity coupling (β)	>.35
Magnetic shielding factor	100
Temperature sensitivity	100 kHz/K
Cell dimensions L Ø	L 25 x Ø 31 mm
Buffer gas mixture	N_2/Ar = 0.62
Working temperature (cavity)	60-70 °C
Heating power (cavity)	≈ 3 W
Working temperature (shield)	50-60 °C
Heating power (shield)	≈ 3 W

Table 1. Main characteristics of the physics package

3. Electronic system

3.1 Introduction

The main goal of the electronics is to lock the local oscillator to the central frequency of the atomic emission profile.
This task can be conceptually divided into three parts:
- the synthesis chain that generates the interrogating signal;
- the heterodyne receiver that detects the maser output and converts this signal into a low frequency signal;

- the servo that uses the information coming from the heterodyne receiver to correct the frequency of the local oscillator.

In the following sections we discuss in more detail each part.

3.1 Synthesis chain

The block diagram of the synthesis chain is shown in Fig. 5.

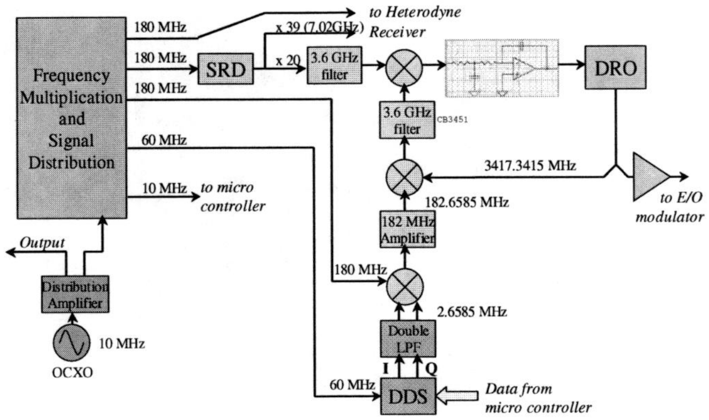

Figure 5. Block scheme of the synthesis chain.

The main goal of the synthesis chain is to provide a low noise signal at 3.4... GHz for the EOM in order to guarantee a clean and symmetric generation of the modulation sidebands. A 10 MHz low-phase noise quartz is the local reference of the microwave synthesis chain. The first frequency multiplication stages are designed to multiply the 10 MHz signal up to 180 MHz. This signal feeds a comb generator (a Step Recovery Diode, SRD) which produces a continuous comb of spectral components spaced of 180 MHz. The required components are at 7.02 GHz for the maser receiver and at 3.6 GHz for the 3.4 GHz synthesis chain. As shown in Fig. 5, the 3.6 GHz signal is mixed with the 182.6 MHz signal inside the PLL that controls the Dielectric Resonant Oscillator (DRO). The latter signal is obtained as a summation of 180 MHz and 2.6 MHz coming from a Direct Digital Synthesizer (DDS). The DRO output is then used to feed the EOM. The DDS realizes the fine-tuning of the synthesis chain, and it is also used for modulation purposes.

In Fig. 6 the two parts of the synthesis chain are shown.

(a)

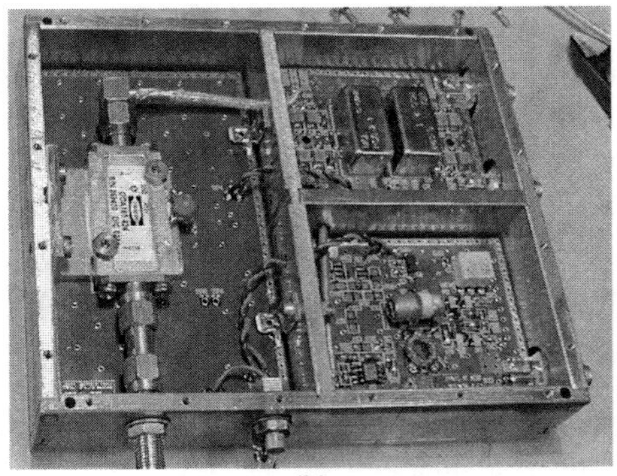

(b)

Figure 6. Synthesis chain. (a) first multiplication stages; (b) second multiplication stages; on the left it is possible to see the step recovery diode.

We tested the synthesis chain by comparing our synthesis chain with another synthesizer according to the set-up of Fig. 7.

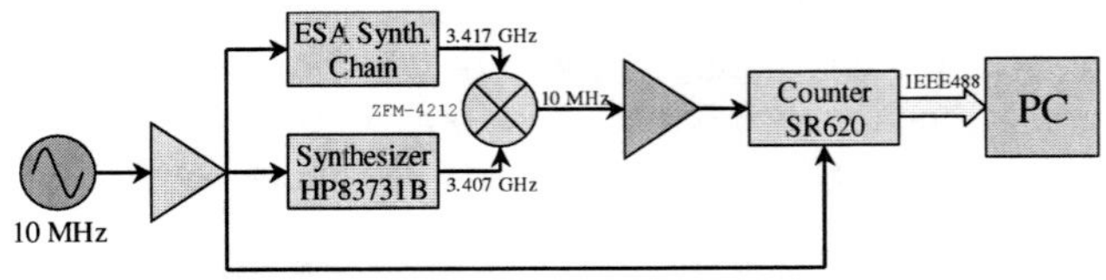

Figure 7. set-up used to test the synthesis chain.

We measured the beat note at 10 MHz between the synthesis chain and the commercial synthesizer and we obtain the Allan deviation of the relative frequency difference shown in Fig. 8.

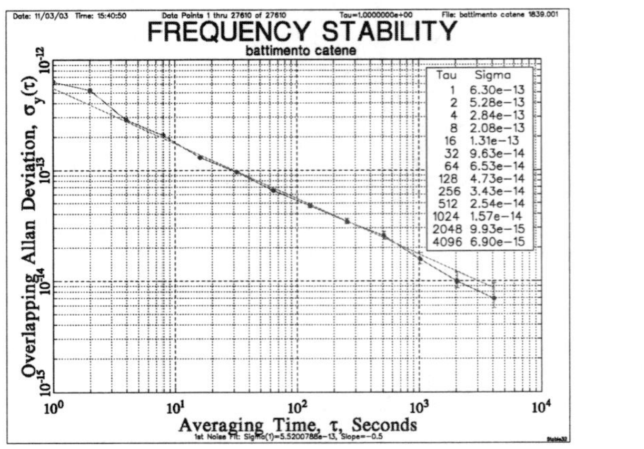

Figure 8. Allan standard deviation of the beat note between the synthesis chain and a commercial synthesizer; the $\tau^{-1/2}$ behavior is due to the gate of the counter.

At one second we have 6×10^{-13} and no flicker floor is observed up to 4000 s. This is obviously a conservative result, our synthesis chain is probably better. Anyway, the behavior shown in Fig. 8 is sufficient to affirm that our synthesis chain does not degrade both the short term and the long term stability of the CPT maser clock.

3.2. Heterodyne detector

The detector is a classical super heterodyne receiver based on a double conversion (see Fig. 9). The maser power is about 1 pW. A Low Noise Amplifier (LNA) provides a first amplification stage (27 dB amplification). Then, there are two consecutive down conversions in order to translate the atomic signal from 6.8...GHz to 5.3... MHz. In the same time, the noise bandwidth is reduced from 60 MHz to 300 kHz. The signal is then amplified again and detected with a true power detector

In Fig. 10 is shown a part of the heterodyne detector.

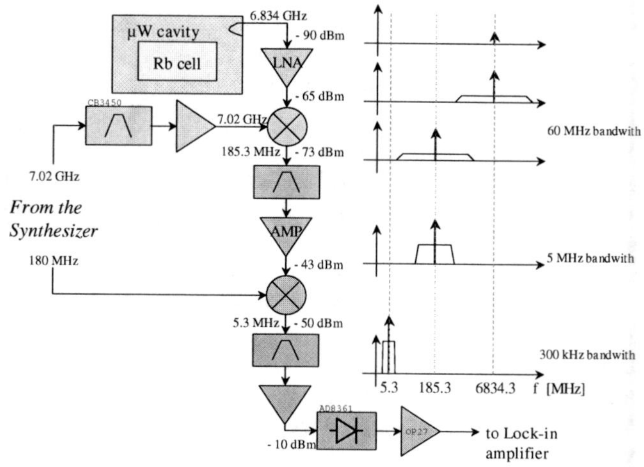

Figure 9. Block scheme of the heterodyne detector.

Figure 10. Part of the heterodyne detector. The filter at 185 MHz (left), the second down converter and the true power detector (right).

3.3 Clock Servo system

We realized a digital clock servo system (see Fig. 11). The useful information from the heterodyne receiver is acquired by an Analog to Digital Converter (ADC) and then processed by a Programmable Logic Device (PLD) that implements the lock-in and the integrator blocks. The loop gain and consequently the loop bandwidth can be adjusted by changing a multiplicative constant. All the parameters involved are set via a PC user interface. This solution strongly reduces the effect of the electronics non-idealities in comparison to an analog implementation.

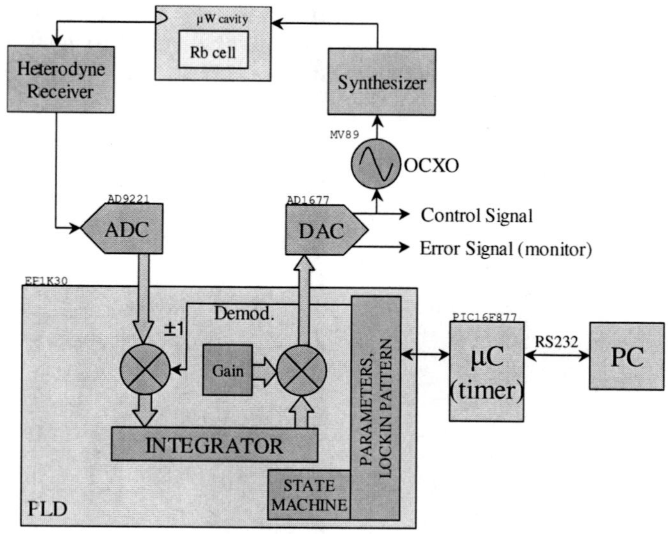

Figure 11. Digital clock servo system.

In Fig. 12 there are two pictures of the digital board we implemented.

(a)

(b)

Figure 12. Pictures of the digital board.

4 Experimental characterization of the clock

4.1. CPT maser spectrum and emission profile

Figure 13 shows the RF spectrum of the CPT maser. It is centred at the laser frequency difference and the signal to noise ratio in 1 Hz bandwidth is:

$$\left(\frac{S}{N}\right)_{1Hz} \approx 85\ dB \qquad (1)$$

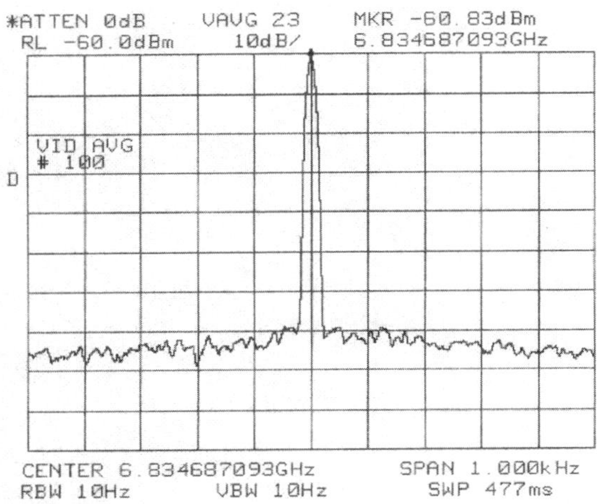

Figure 13. CPT maser RF spectrum.

The CPT maser RF spectrum is only corrupted by thermal noise as we expect when the atoms are interrogated by a low phase noise signal [4, pp. 73-74].

The maser emission profile is reported hereafter in Fig. 14. The parameters used in the registration are: optical excitation power of 300 μW, temperature of the cell 63 °C. The resulting linewidth is about 200 Hz.

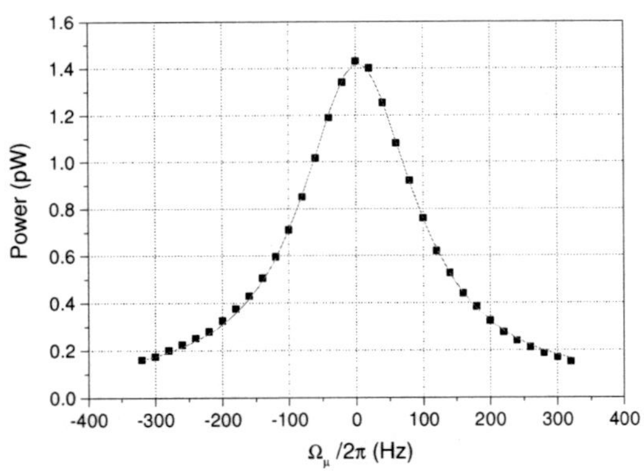

Figure 14. CPT maser emission profile

The continuous line is a fit of the experimental data with a lorentzian pattern.

4.2 Frequency stability evaluation

To evaluate the frequency stability achievable by the CPT maser, we have to consider how the different noise contributions affect the clock transition. Therefore, according to the well known theory of control loops [6], we consider the power spectral density of the relative frequency fluctuations of the quartz locked to the atomic reference via a first order loop:

$$\begin{aligned} S_y(\omega) &= \left|1 - H(j\omega)\right|^2 S_{yq}(\omega) + \left|H(j\omega)\right|^2 S_{yq}(2\omega_m) \\ &+ \left|H(j\omega)\right|^2 S_{yr}(\omega) + \frac{\left|H(j\omega)\right|^2}{\omega_{21}^2 \left|K_\varphi(j\omega)\right|^2} S_n(\omega) \end{aligned} \qquad (2)$$

In this relation, S_{yq} describes the frequency fluctuations of the free running quartz, S_{yr} describes the frequency fluctuations of the reference transition, and finally the last term is the additive noise spectrum. The first term can be made unimportant by a proper choice of the loop time constant ($H(j\omega)\approx 1$, being H the closed loop gain), the other terms are unaffected by the loop transfer function and therefore they limit the stability achievable by the standard.

In particular, the second term introduces a white frequency noise limit due to the quartz phase noise at twice the modulation frequency (Dick effect). The third term accounts for all contributions related mainly to physical effects which shift the maser emission profile. We have then to consider the effects that transfer the frequency (FM-FM) and amplitude (AM-FM) laser noises into frequency noise of the CPT maser: for the FM-FM conversion, we have the linear and the quadratic light shift and the microwave shift (due to the cavity feedback on the atoms); for the AM-FM conversion, we have the contribution due to the power light shift and to the propagation shift. Moreover, the second term accounts also for temperature dependent effects, so we have to take into account again a contribution due to power light shift and propagation shift, in addition to cavity pulling and buffer gas contributions.

Therefore, for each noise term we evaluated the power spectral density, as results from our experimental situation, and the corresponding contribution to the Allan deviation in the time domain.

As an example, we show that the noise term due to the AM-FM conversion can be strongly reduced by a proper choice of the modulation index.

In other words, there is a suitable value of the modulation index so that power light shift and propagation shift compensate each other. For this modulation index value, the CPT maser frequency does not depend from the laser power. Figure 14 shows the CPT maser frequency versus time at the wrong and the right value of the modulation index.

In table 2 we summarize all the estimated noise contributions. In particular, electronics and laser noises appear to be controlled in our apparatus, and then the ultimate stability limit is due to thermal noise only. With experimental data reported in section 4.1, we obtain an Allan devition of $1.2\times10^{-12}\ \tau^{-1/2}$. Regarding the temperature sensitivity, we expect that for a temperature stabilization at the mK level the temperature fluctuations should limit the medium term stability in the range of 10^{-14}.

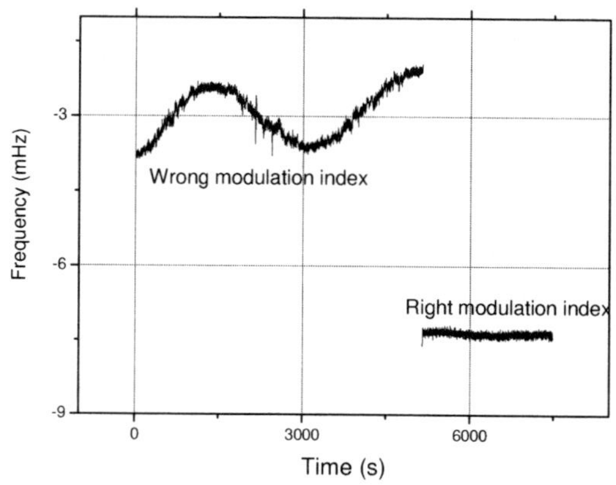

Figure 14. CPT maser frequency in the two situations described in the text.

Quartz	$\sigma_y(\tau) < 1\times10^{-13}\,\tau^{-1/2}$
2nd harmonic noise	$\sigma_y(\tau) \approx 6\times10^{-13}\,\tau^{-1/2}$
FM-FM Conversion	$\sigma_y(\tau) \le 6\times10^{-13}\,\tau^{-1/2}$
AM-FM Conversion	$\sigma_y(\tau) \le 1\times10^{-13}\,\tau^{-1/2}$
Thermal Noise	$\sigma_y(\tau) = 2.4\times10^{-12}\,\tau^{-1/2}$
Buffer Gas	$\Delta\nu/\nu \approx 10^{-11}/\mathrm{K}$
Cavity Pulling	$\Delta\nu/\nu \approx 1.8\times10^{-11}/\mathrm{K}$
Light Shift	$\Delta\nu/\nu \approx -1\times10^{-11}/\mathrm{K}$
Propagation Shift	$\Delta\nu/\nu \approx -2\times10^{-11}/\mathrm{K}$

Table 2. summary of the noise contributions to the Allan standard deviation.

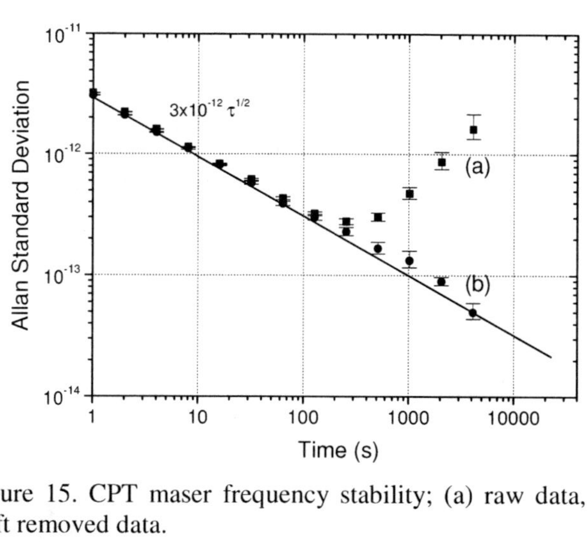

Figure 15. CPT maser frequency stability; (a) raw data, (b) drift removed data.

The measured CPT maser frequency stability in terms of the Allan standard deviation is shown in Fig. 15. In the short term the white frequency noise achieves the limit 3×10^{-12}, only a factor two worse than the expected value. For τ greater than 500 s a drift component limits the medium term frequency stability. It is important to point out that the drift removed data show that no flicker floor is observed at the level of 4×10^{-14}, so that we are confident that a better control of the temperature may allow to achieve a frequency stability in the range of 10^{-14} for τ 10000 seconds.

6 Conclusion

The research project on the CPT maser was aimed to study the feasibility of the clock and to investigate its stability capabilities. We realized a CPT maser prototype and we tested all its main components: the physical package, the optical and the electronic systems. Electronics and laser noises are well controlled in our apparatus and the measured Allan deviation at 1s is quite close to the expected value. No flicker floor is observed in the drift removed data and the medium term stability results to be limited by a thermal drift only. Therefore, a better designed temperature control allows in principle to achieve a medium term stability at the 10^{-14} level as expected.

Acknowledgments

We thank N. Beverini and E. Maccioni for their expertise in the realization of the maser quartz cells. This work was funded by the European Space Agency Contract 16104/02/NL(DS).

References

[1] N. Cyr, M. Tetu, and M. Breton, IEEE Trans. Instrum. Meas. **42**, 640 (1993); J. Vanier, A. Godone and F. Levi, Phys. Rev. A **58**, 2345 (1998); J. Kitching, S. Knappe, N. Vukicevic, L. Hollberg, R. Wynands and W. Weidmann, ibid. **49**, 1313, (2000); M. Zhu and L. Cutler, Proc. of the Precise Time and Time Interval (PTTI) Systems and Applications Meeting, Reston, Virginia, USA (2000);

[2] A. Godone, F. Levi, S. Micalizio, and C. Calosso, to appear on Phys. Rev. A; M. Merimaa, T. Lindvall, E. Tittonen, and E. Ikonen, J. Opt. Soc. Am. B **20**, 273 (2003); S. Knappe, R. Wynands, J. Kitching, H. G. Robinson, and L. Hollberg, ibid. **18**, 1545 (2001);

[3] W.E. Bell and A. L. Bloom, Phys. Rev. Lett. **6**, 280 (1961); E. Arimondo, Prog. Opt. **XXXV**, 257 (1996);

[4] A. Godone, F. Levi, S. Micalizio, "Coherent Population Trapping Maser", CLUT Editrice, Torino, (Italy), 2002;

[5] J. Vanier, M. W. Levine, D. Janssen, and M. J. Delaney, IEEE Trans. Instrum. Meas., **52**, 822 (2003).

[6] F. M. Gardner, "Phaselock Techniques", John Wiley and Sons, New York, 1979, 2nd .

STATUS OF THE ACES PROJECT: SPACE INSTRUMENTS PERFORMANCE AND GROUND SEGMENT ORGANIZATION

ACES SCIENCE TEAM : ENS/LKB, BNM-SYRTE, Neuchâtel Observatory
ACES INDUSTRIAL TEAM : Astrium, Kayser-Threde, Timetech
SPACE AGENCIES : ESA, CNES

ENS/LKB, 24 rue Lhomond, 75231 Paris cedex 5, France
BNM-SYRTE, CNRS, Observatoire de Paris, 61 avenue de l'Observatoire 75014 Paris, France
Neuchâtel Observatory, rue de l'Observatoire 58, 2000 Neuchâtel, Switzerland
ASTRIUM, 88039 Friedrichshafen, Germany and 31 avenue des Cosmonautes, ZI du Palays, 31402 Toulouse Cedex 4, France
Kayser-Threde, Wolfratshauser Str. 48, 81379 Munich, Germany
Timetech, Curiestrasse 2, 70563 Stuttgart, Germany
ESA / ESTEC, Keplerlaan, 2200 AG Noordwijk, The Netherlands
CNES, 18 Avenue Edouard Belin, 31402 Toulouse, France

Corresponding author : Noël Dimarcq, BNM-SYRTE,
Tel: 33 1 40 51 22 53 ; Fax: 33 1 43 25 55 42 ; e-mail: noel.dimarcq@obspm.fr

The ACES project aims at operating ultra stable clocks on board the International Space Station ISS and comparing them to ground clocks located all around the world. The expected performance of ACES space clocks (frequency stability and accuracy at 10^{-16} level) and of the dedicated microwave link (time stability better than 10 ps over 1 day) will be of great interest for time & frequency metrology (comparison of ground primary standards with a resolution better than 10^{-16}) and for fundamental physics (relativity tests).

The engineering models of all ACES payload sub-systems (cold atom clock PHARAO, hydrogen maser SHM, microwave link MWL, space clocks phase comparison device FCDP, ...) will be available before the end of 2004.

After a brief recall of ACES scientific objectives, the main characteristics of ACES sub-systems will be described and their contributions to global ACES performance will be pointed out. The influence of some specific contributions (thermal fluctuations, orbit determination, ...) on final performance will be discussed.

Even if the ACES payload is a European development, the ACES ground segment will have an international dimension since any ground T&F laboratory will be able to compare its clocks to ACES space clocks using a MWL terminal during dedicated sessions.

The second part of the talk will describe the organization of the ACES ground segment in order to explain how ground users will be able to participate to ACES operations.

Current Status of the Space Clock PHARAO

I. Maksimovic[1], M. Abgrall[1], A. Clairon[1], J. Grünert[1], Ph Laurent[1]*, P. Lemonde[1], G. Santarelli[1], C. Salomon[2]
C. Sirmain[3], F.Picard[3], Ch. Delaroche[3], O. Grosjean[3], M. Saccoccio[3], M. Chaubet[3], L. Guillier[3], M. Behague[3]

[1] BNM-SYRTE, 61 avenue de l'Observatoire, 75014 Paris, France
[2] ENS-LKB, 24 rue Lhomond, 75231 Paris, France
[3] CNES, 18 avenue Edouard Belin, 31401 Toulouse cedex, France
* corresponding author : philippe.laurent@obspm.fr

Keywords: PHARAO, space clock, cold atoms.

Abstract

The cold atom space clock PHARAO is under industrial construction since two years. It will be one of the two atomic clocks of the ESA supported ACES mission which will fly on an external pallet of the European module Columbus of the International Space Station. The goal of PHARAO is the study and utilization of a laser cooled cesium clock in microgravity environment.

Two models of the clock will be constructed: an engineering model (EM) and a flight model (FM). We present in this paper the current status of the EM. At present, two of the four main sub-systems have been delivered to the French Space Agency CNES: the microwave source and the UGB (On Board Data Processing Unit). The laser source is close to end and an evaluation model of the cesium tube is under tests.

1 Introduction

PHARAO, as "Projet d'Horloge Atomique à Refroidissement d'Atomes en Orbite", has started in 1993 with the objective of performing Science with a space cold atom clock [1]. As a matter of fact, the combination of atomic laser cooling techniques [2] and microgravity environment allows the development of clocks with unprecedented performances. The frequency accuracy is expected to reach 10^{-16} or better.

To demonstrate the feasibility of a compact cold atom clock, BNM-SYRTE and LKB with the support of CNES have undertaken the construction of a clock prototype in 1994. The prototype was constructed and successfully tested in 1997 on board a zeroG aircraft [3]. This work has essentially brought on the volume reduction of the clock and on the improvement of the clock operation reliability with respect to the current atomic fountains. The objective was to show that current technologies were compatible with the requirements of the space development. On the same year, ESA, the European space agency, has accepted our proposal, ACES (Atomic Clock Ensemble in Space) [4], to make fundamental physics by using a cold atom clock, a H-maser and a time and frequency transfer package. The payload is to be flown on board the space station.

After leading the usual initial phase for the development of a space instrument, CNES has decided to subcontract the construction of the different clock sub-systems by some manufacturers. The CNES is the prime contractor, the BNM-SYRTE and LKB are scientific and technical advisors. They will be also one of the users of the ACES mission.

The engineering model (EM) of the clock is now under industrial construction since 2 year. This EM is used to verify the interface compatibilities, the clock operation and its performances. Depending on the results, the design of the flight model (FM) could be slightly modified.

The PHARAO set-up is composed of four main sub-system (figure 1) developed by different companies. EADS SODERN is constructing the laser source and the cesium tube. The laser source provides all the laser tools for the atomic cooling, launching and detection. The cesium tube provides the atomic source, the controlled environment for the atomic manipulation and the interrogation process.

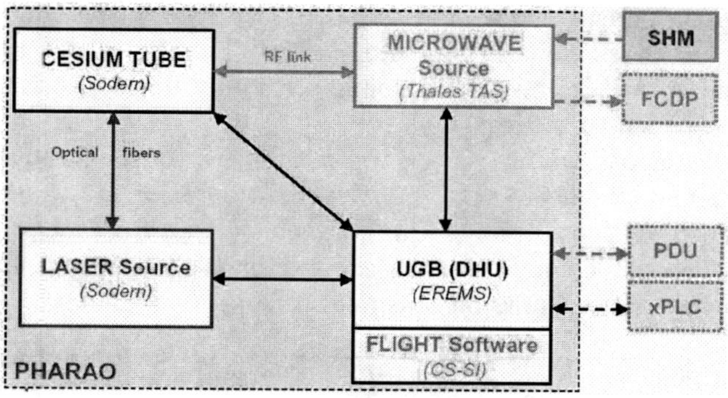

Fig 1: The PHARAO set-up.

In the cesium tube the atoms interact with the signal generated by the microwave source (THALES-TAS). All the commands and acquisition are performed by a computer, the UGB (On Board Data Processing Unit) is manufactured by EREMS and the software developed by Cs-Si. The UGB also manages the data flux between the clock and the ACES payload. All this set-up fills a volume of about 200 l for a weight of 91 kg and an electric consumption of 114 W.

The space clock operation is similar to that of an atomic fountain. It operates sequentially by playing the same successive phases: atomic capture, launching, cooling, selection, microwave interrogation and detection. The main difference with respect to atomic fountains is that in space we can change by two orders of magnitude the atomic velocity (5 to 500 cm/s).

2 Expected performances

The goal of PHARAO is to deliver a signal with a high relative level of accuracy (at the level of 10^{-16}) and a frequency stability of $10^{-13}\times\tau^{-1/2}$.

The short term frequency stability of a cold atom clock is described by the Allan variance given by:

$$\sigma_y^2(\tau)=\left(\frac{\Delta\upsilon}{\pi\ \upsilon}\right)^2\frac{T_c}{\tau}\left(\frac{1}{N_{at}}+\sigma_{dl}^2+\frac{2\sigma_{\delta N}^2}{N_{at}^2}\right)+\frac{1}{\tau}\sum_{n=1}^{\infty}\frac{g_n^2}{g_0^2}S_y^{LO}\left(n/T_c\right)$$

Where $\Delta\upsilon$ is the atomic microwave resonance linewidth, υ is the microwave frequency (9.19... GHz), T_c is the cycle time and T the integration time.

Four main sources of noise contribute to the frequency stability: the first one is the quantum projection noise [5]. It varies as the inverse of the number of detected atoms, N_{at}. This term defines the ultimate limit of the clock frequency stability. The second and the third term bring together all the technical noises such as the detection laser noise (independent of the number of detected atoms) or the detection system noise. Finally, the fourth term results from the microwave signal frequency noise. This noise level is mainly determined by the master quartz oscillator noise S_y^{LO}. The g_n are the Fourier coefficients of the atomic sensitivity function during an operating cycle [6].

By modelling the clock operation, we have obtained different frequency stability levels according to atomic velocities. These levels shown on table 1 result from a trade-off between the noise sources and also by taking into account the ACES space mission constraints (mass, volume,...).

Atomic velocities	Stability	Main noise source
< 10 cm/s	2×10^{-13}	Detection noise
10 cm/s – 20 cm/s	1.2×10^{-13}	Projection noise and Oscillator phase noise
20 cm/s – 60 cm/s	0.8×10^{-13}	Oscillator phase noise
60 cm/s – 500 cm/s	$1 - 3\times10^{-13}$	Resonance width

Table 1: Modelled noise contribution according to atomic velocity.

The main noise source is the oscillator phase noise which represents a noise floor that could not be overcome. So, we have designed the optical system in order to lead the detection and laser noise negligible in comparison with this oscillator phase noise for velocities within 20 to 60 cm/s where the oscillator phase noise reach a minimum value.

This trade-off leads to the following characteristics for each sub-system (cesium tube, microwave source, laser source and UGB)

3 The Cesium Tube

We detail the different parts of the tube (figure 2) by following the atoms along their path. First, we have the cold atoms source composed by the reservoir and the trapping zone. Then, the atoms passed into the state selection cavity and the interrogation cavity. Finally, we detect the atoms by fluorescence in the detection zone.

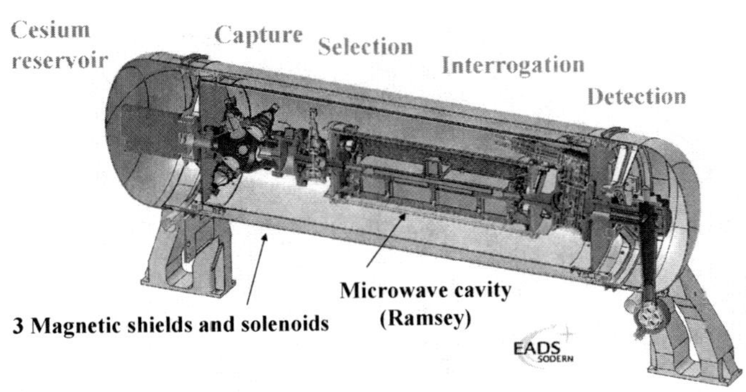

Fig 2: Section of the cesium tube. The mass is 45 kg and the volume is 70 liters.

3.1 Tube Description

Cold atoms source: reservoir and trapping zone

The cesium reservoir has a lifetime of 10 years. Its vapor pressure can be adjusted between 5×10^{-7} to 5×10^{-6} Pa to optimize the atom number with the loading time by varying the reservoir temperature (40-60°C) and valve aperture (motorized step by step). In the reservoir, a titanium matrix prevents migration of liquid cesium. A reservoir prototype has been validated.

To capture and cool a large number of atoms, while preserving a low density, we use the optical molasses configuration with six laser beams [7]. All 6 laser beams have a 13 mm beam waist radius and a 5.6 mW/cm^2 peak intensity. The capture region includes 6 collimators and 7 photodiodes for the measurements of the atomic fluorescence and for the laser power servo-loops. We expect to load about 1.5×10^8 atoms in 500ms (1.5×10^7 atoms in m_F=0 sub-level).

Selection and interrogation zones

The state selection microwave cavity induces transition from the hyperfine level F=4, m_F to the F=3, m_F. It is followed by two lasers beams. The first one induces the F=3 to F=4 transition to eventually slice the atomic cloud and the second one pushes away the F=4 atoms. 2 photodiodes are used to servo-lock the laser power.

The interrogation microwave cavity robustness is ensured by an Aluminium Silicon Carbon (AlSiC) structure (see photo1).

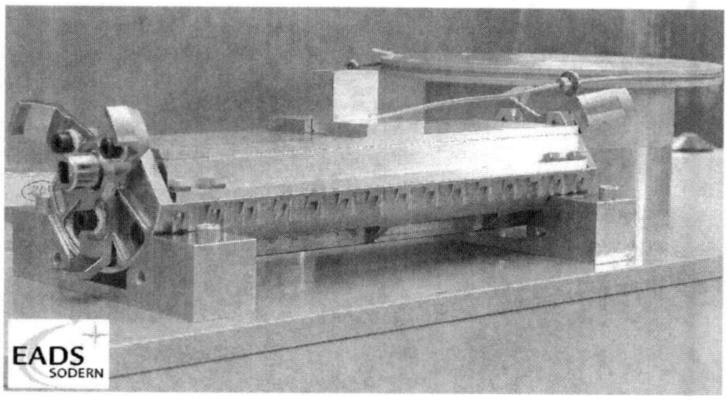

Photo 1: A model of the Ramsey cavity made by Thales TED. The cavity in pure copper is screwed on a rigid structure to avoid deformation. We can see the cut-off wave-guide with a rectangular hole for the atomic path and the central coupling upon the centre of the cavity. (Courtesy of EADS SODERN and TAS).

The Ramsey cavity we have developed is a ring resonator. One coupling system feeds two symmetrical lateral waveguides which meet at the two interaction zones. Altogether, the cavity is made of 3 brazed copper pieces. The master one includes the two interaction zones, the waveguides and, at the middle, the atom passage. A cap closes the master piece and the coupling is placed at the center (photo 1). The external dimensions are 298×100×46 mm. The advantage of this configuration is to provide large holes (8×9 mm) to let the atoms pass through with weak phase disturbances of the internal microwave field. As a precaution, at the input of the holes, a diaphragm reduces the aperture to limit atomic interactions with the cavity walls.

Five microwave cavities have been realized by Thales TED: an engineering model (EM), a flight model (FM), a back-up model, a vacuum model and a thermal model. The aim of the thermal model is to measure the cavity temperature with an absolute resolution of 0.2°C in order to evaluate the effect of the blackbody radiation on the atoms.

The best microwave cavity selected by electrical measurements will be mounted in June 2004 inside the atomic fountain FO1 of the BNM-SYRTE to study the phase shift effects before integration in the clock flight model. With the measurements and by simulations, we will be able to determine the cavity phase shift effect with an accuracy of 10^{-17}.

Detection zone

The detection region includes 4 tilted laser beams, 2 photodiodes for the laser power measurement and 2 temperature regulated photodiodes for the fluorescence detection.

The atoms are selectively detected by their fluorescence induced by 2 standing wave laser beams tuned to the F=4, F'=5 Cs line. Between the 2 beams, we have a pushing beam to remove the atoms in the F=4 level and a pumping beam to set the F=3 atoms in the F=4 state. The beams are slightly tilted (8°) with respect to the cesium tube axis. The residual cesium thermal beam is then Doppler-shifted and its contribution to stray fluorescence is reduced by an order of magnitude. Its contribution to the detection noise becomes negligible.

3.2 The Ultra-High Vacuum (UHV)

An Ultra-High Vacuum (UHV) of about 10^{-8} Pa is required for the cesium tube to reduce (to less than a few per cent) collision losses with the background gas whatever the atomic velocity is.

The UHV is ensured by different devices:

- a space qualified ion pump with a pump rate of 3 l/s to pump the rare gas,
- 5 getters near interrogation zone to complementary pump the other gases,
- graphite tube between trapping and selection zone to capture the main part of the cesium beam.

The UHV is reach in three steps: degassing individual pieces to 300°C over several days, degassing the entire tube to 140°C under high vacuum during one month and finally by getter activation.

To validate the objectives of a vacuum at the level of 2×10^{-8} Pa and to calibrate the cesium flux, an UHV evaluation model has been constructed (see photo 2). This model is representative of the FM tube in terms of material, surface, gaskets and brazed windows. It is equipped with a mass spectrometer and few gauges.

Preliminary results show that the vacuum reaches the level of 5×10^{-7} Pa without getter activation. The residual gas is mainly composed of Hydrogen (98% H_2, 2% N_2). With getter activation, the 2×10^{-8} Pa will be achieved.

3.3 The magnetic aspects

The objective is to attenuate the 25 μT amplitude modulation on ACES (due to the earth magnetic field and rotation of the space station) and get a constant and homogenous C-field. All the simulations are completed and the final version of the shields is under construction.

The cesium tube will be protected by 3 cylindrical magnetic shields (fig. 2). We have measured with a mock-up a magnetic attenuation of about 10^5. An active servo-loop, including a probe and a solenoid inside the external shield, increases the attenuation by at least a factor of ten. The residual fluctuations will then remain below 20 pT.

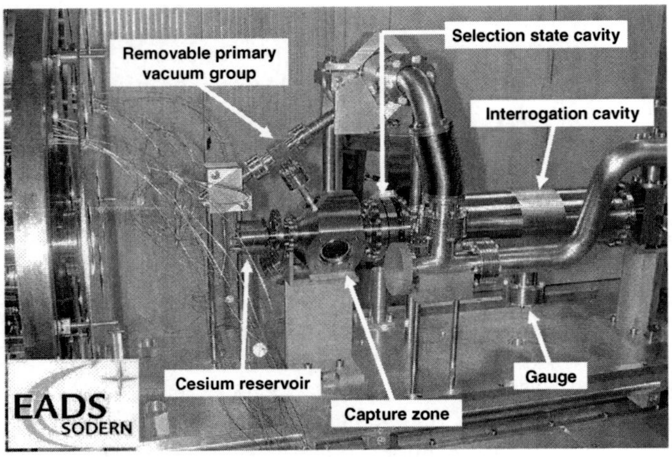

Photo 2: The UHV evaluation model. We can notice all the parts representative of the cesium tube: the cesium reservoir, the capture zone, the selection cavity and the interrogation cavity. The primary group is removed by copper tube crimping. All the cesium tube is put into the same high vacuum chamber (at the left part of the photo). This pumping system will be the same for the other models: EM and FM. (Courtesy of EADS SODERN).

4 The microwave source

The microwave chain (made by Thales TAS) synthesizes the two radiofrequency signals for the state selection cavity and the interrogation cavity.

A 100 MHz VCXO (Voltage Control Oscillator) is phase locked on an Ultra Stable Oscillator (USO) for the short term and on the Space Hydrogen Maser (SHM) for the mid term (see figure 3).

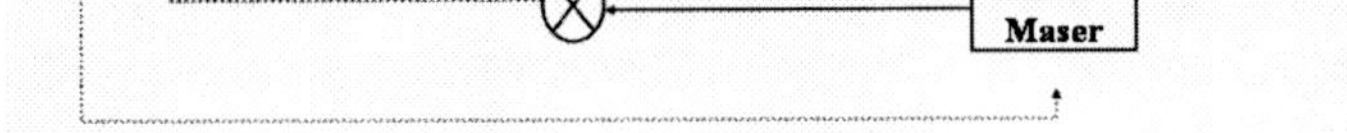

Fig 3: Schematic set-up of the microwave source.

Two USO FM were made by CMAC and have been space qualified for our application. We have compared these quartz oscillators to a cryogenic sapphire oscillator from the University of Western Australia. The stability varies from 6×10^{-14} to 7×10^{-14} from 1 to 10s integration time.

The 100 MHz and 9.192 GHz frequencies are finely tuned by two Direct Digital Synthesizers (DDS). The first one is used to lock the 100 MHz on the maser. For electrical isolation reasons a numerical error signal has been preferred to a direct analog USO correction. The second DDS is used for evaluation process to sweep or offset the microwave frequency. The microwave output levels are power controlled with a dynamic of 80 dB and can be switched off. In case of troubleshooting or during the ground tests, the 100 MHz stage can be switched off and replaced by an external signal (maser, cryogenic oscillator).

The microwave source has been qualified and delivered. During the delivery tests, a phase drift had been observed when switching the signal at selection (even though the interrogation is never pulsed). The failing circuit was identified and modified. After that, 10 ms after switching none phase drift greater than 10 µrad has been observed. Measurements show that quartz noise level is maintained after multiplication to 9.192...GHz. Presently the microwave source has the performances in agreement with the specifications.

Other measurements as electromagnetic compatibility will be performed at CNES Toulouse by using the cryogenic oscillator. Finally, the microwave source will be used by the FO2 fountain to confirm the microwave source performances.

5 The laser source

The laser source provides 10 laser beams through optical fibers to the tube. 6 are devoted to the cooling and the launching, 2 to the atomic internal state selection and 2 to the measurement of the atomic populations of the two hyperfine levels. Two extended cavity diode laser (ECDL) and two injected slave lasers provide these beams. As we have a back-up for each laser (a motor drives a half wave plate to switch from one laser to the other) we have 8 diodes on the optical bench. There are also 6 acousto-optical modulators (AOM) space-qualified to allow laser frequencies and power tuning and 7 mechanical shutters to turn off the 10 beams. The fiber coupling is optimized by mirrors mounted on piezoelectric transducers (photo 3). The optical bench is double-sided and contains the driving electronics of these devices. Its mass is 20 kg for a volume of 30 l.

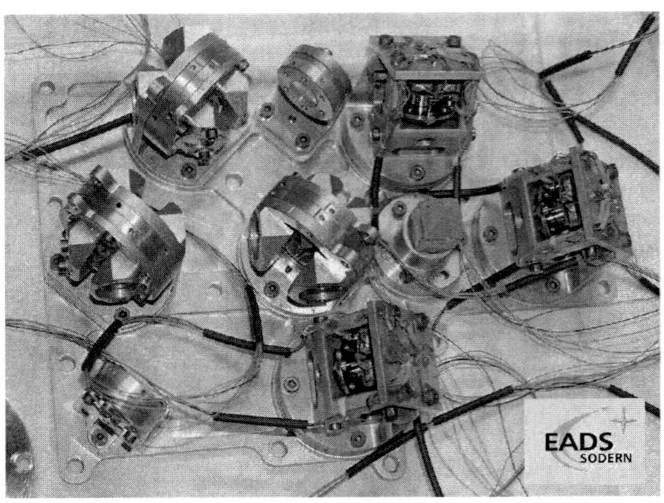

Photo 3: This photo represents a triplet cooling fiber (fibers are not present). We have four PZT on each fiber coupling mirror to optimize the coupling. (Courtesy of EADS SODERN).

We have developed a new ECDL concept [8] using an extended cavity closed by a semi-reflecting mirror and with an intra-cavity Fabry-Perot for the mode selection. With this architecture, the ECDL do not present any mode-hopping between air and vacuum operation.

The two EM ECDL have been assembled and tested (photo 4). The output power reaches 30 mW. The laser frequency is stabilized by a servo-loop to a cesium line. As an AOM is present in this servo-loop, we have checked that the laser is still locked with large variations on the AOM (0-80 MHz). The servo-loop has a bandwidth of 30 kHz and the resulting frequency noise spectral density reaches 10^4 Hz^2/Hz at 100 Hz. The laser intensity is also servo-locked by measuring the laser beam power inside the tube and by retro acting on an acousto-optic modulator of the laser source. The relative intensity noise is below $2\times10^{-10}/Hz$ at 100 Hz. With these characteristics, the laser noise contributions are negligible with respect to the projection noise.

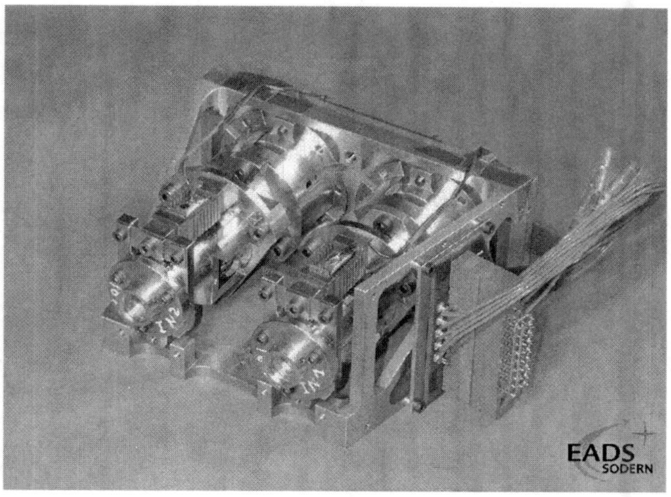

Photo 4: A picture of a redundant ECDL. (Courtesy of EADS SODERN).

6 The UGB and the software

The computer runs the real time software. It is linked to the clock sub-systems and to the ACES payload. It has a set of analog-digital converters to record all physical signals, a digital bus to configure the sub-system parameters, to receive the remote controls and to send the clock measurements. It provides the digital pulses to trigger all the phases of a clock cycle. It manages the data communication, the different servo-loops (temperature regulations, microwave frequency servo-loop, magnetic field compensation, laser relocking process…) and the clock operation.

The EM of the UGB has been delivered and fully tested at CNES (photo 5).

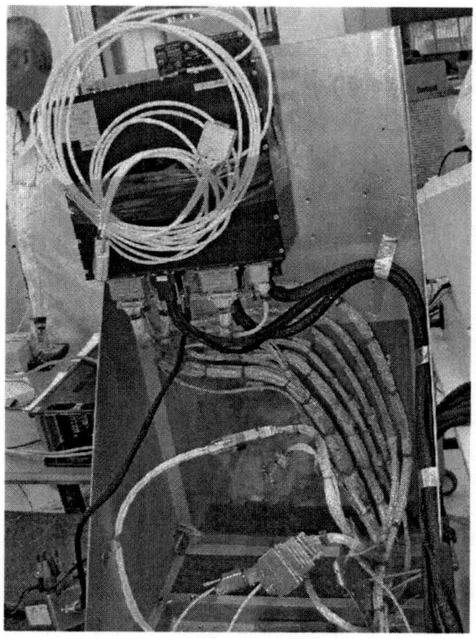

Photo 5: A picture of the EM of the UGB during the tests in Toulouse. It is fixed on the clock mounting plate and will be connected to the other sub-systems.

Conclusion and Prospects

Two models of the PHARAO clock will be constructed: an engineering model and a flight model. At present, the EM of the UGB and microwave source have been delivered to the CNES and tests are in progress. The laser source and the cesium tube are under assembly. Two decisive sub-systems performances will be measured with atomic fountain FO1 and FO2: microwave cavity and microwave source.

The EM of the clock will be assembled, tested and evaluated in CNES. The Ground Support Equipment includes a maser, the mobile fountain FOM, a cryogenic oscillator from UWA, and a common-view GPS link with the BNM-SYRTE fountains in Paris. A vacuum chamber, with base plate temperature regulation, accommodates the PHARAO clock and the space H-maser provided by Observatoire de Neuchâtel. The performance of the EM of the clock will be evaluated before assembling the FM.

References

[1] P. Laurent, G. Santarelli, S. Lea, S. Ghezali, M. Bahoura, E. Simon, A. Clairon, P. Lemonde, J. Reichel, A. Michaud, C. Salomon. "Cesium fountains and micro-gravity clocks", *25th Moriond conference on Dark matter in cosmology, clocks and tests of fundamental laws*, (June 1995).

[2] See, for instance, special issue of *JOSA* **B6**, (1989).

[3] Ph. Laurent, P. Lemonde, E. Simon, G. Santarelli, A. Clairon, N. Dimarcq, P. Petit, C. Audoin, and C. Salomon. "A cold atom clock in absence of gravity", *Eur. Phys. J. D*, **vol. 3**, pp. 201-204, (1998).

[4] C. Salomon, P. Lemonde, P. Laurent, E. Simon, G. santarelli, A. Clairon, P. Petit, N. Dimarcq, C. Audoin, F. Gonzalez and F. Jamin Changeart, *Proc. of the 1st ESA symposium on Space Station Utilization*, SP385, 295 (1996).

[5] Santarelli G., Laurent P., Lemonde P., Clairon, A., Mann A., Chang S., Luiten A., Salomon C. "Title of the article", *Phys. Rev. Lett*, **82**, pp 4619-4622, (1999).

[6] G. Santarelli, C. Audoin, A. Makdissi, Ph. Laurent, G.J. Dick and A. Clairon,. "Frequency Stability Degradation of an Oscillator Slaved to a Periodically Interrogated Atomic Resonator", *IEEE Trans. Ultra. Ferr*, **45**, nb 4, (1998).

[7] See for instance the 1997 Nobel lectures: S. Chu, *Rev. Mod. Phys.,* **70**, pp. 685, (1998); C. Cohen-Tannoudji, *ibid.,* pp. 707; W.D. Phillips, *ibid.,* pp. 721.

[8] F. Allard, M. Abgrall, I. Maksimovic and Ph. Laurent, "Automatic system to control the operation of an extended cavity diode laser", *Rev. of Scient. Instr*, **75**, nb 1, (2004).

INTRODUCTION TO OPTICAL FREQUENCY STANDARDS

PATRICK GILL
National Physical Laboratory
Teddington, Middlesex, TW11 0LW, UK
Patrick.gill@npl.co.uk

Recent progress in the development of high-accuracy optical frequency standards has resulted in some of these standards now achieving better stabilities than the caesium fountain microwave standard [1]. In parallel with this, the development of femtosecond lasers with pulse repetition rates that can be referenced to microwave standards, coupled with extensive broadening of the femtosecond pulse spectrum in microstructure fibre, has led to widespan optical frequency combs covering the visible to mid infra-red [2]. This latter technology has enabled the high-accuracy inter-relation of such optical standards to the caesium microwave standard, and provides the basis for optical clock operation. In this mode of operation, the optical standard is the anchor for the comb, which is then able to deliver highly stable microwave output.

Currently, the two main optical frequency standards technologies under research at a number of standards laboratories and research institutes worldwide are cold atoms and single cold trapped ions. The advent of laser cooling techniques a couple of decades ago led to the preparation of cold atom and single cold ion samples, where Doppler broadening could be virtually eliminated. As a result, intrinsically narrow transitions in certain cold atoms or ions, with experimental linewidths approaching a few Hz, can now be observed. The absolute frequencies of these transitions have been measured, in some cases, to better than 1 part in 10^{13}.

This talk will briefly review the state-of-the-art in respect of both cold atom and cold ion standards, contrasting the similarities and differences in each approach. Potential limiting stabilities and systematic frequency uncertainties will be outlined. It is intended to provide a backgound for the following papers within the two EFTF optical frequency standards and measurement sessions.

[1] S A Diddams, Th Udem, J C Bergquist, E A Curtis, R E Drullinger, L Hollberg, W M Itano, W D Lee, C W Oates, K R Vogel and D J Wineland 2001 Science **293** 825

[2] J Reichert, M Niering, R Holzwarth, M Weitz, Th Udem and T W Haensch 2000 Phys. Rev. Lett. **84** 3232

THE MERCURY-ION OPTICAL CLOCK: TOWARDS A MEASUREMENT OF THE QUADRUPOLE SHIFT

W. H. Oskay*, A. Bartels, S.A. Diddams, C.W. Oates, G. Wilpers, L. Hollberg, W. M. Itano, C.E. Tanner†, and J. C. Bergquist

Time and Frequency Division, National Institute of Standards and Technology
325 Broadway M.S. 847, Boulder, CO 80305 USA
*Fax: (303) 497-7375, Electronic mail: oskay@boulder.nist.gov
†Permanent address: Department of Physics, University of Notre Dame
Notre Dame Indiana 46556 USA

Keywords: optical frequency standards, quadrupole shift

Abstract

An optical clock based upon a single mercury ion in a radio-frequency trap has the potential for both high stability and high accuracy. The systematic uncertainty of this frequency standard is limited by contributions due to two types of external fields. The clock has previously been operated with a magnetic bias field that destabilizes dark states that limit the rates of laser cooling and state detection. This bias field has previously resulted in a substantial second-order Zeeman shift. We now destabilize the dark states with a polarization-scrambling technique that allows us to operate with only a minimal magnetic field. The second uncertainty that we address here is the magnitude of the quadrupole shift due to the interaction of stray electric-field gradients with the quadrupole moment of the ion. We describe the steps that we are taking to evaluate this shift in the near future.

1 Introduction

The search for more accurate and stable frequency standards has led to the development of trapped-ion optical clocks. In these systems, the ion trap provides a minimally perturbative environment that allows for high accuracy. Simultaneously, the high frequency of the clock transition, together with its narrow linewidth, leads to stability that makes it possible to realize this accuracy within a practical amount of averaging time.

The mercury-ion frequency standard has been described in detail elsewhere [6,8], and only the most important features are summarized here. In our system, a single $^{199}Hg^+$ ion is confined in a cryogenic, spherical, radio-frequency (Paul) trap and is cooled on the $^2S_{1/2}$ – $^2P_{1/2}$ dipole transition at 194 nm. The electric-quadrupole $^2S_{1/2}$ $(F=0)$ – $^2D_{5/2}$ $(F=2, m_F=0)$ transition at 282 nm serves as the reference for the frequency standard. This transition is probed by the frequency-doubled output of a highly stabilized dye laser operating at 563 nm. The short-term stability of the laser derives from an isolated, high-finesse, Fabry-Perot cavity, and the fractional instability of this local oscillator is below $5\ 10^{-16}$ (corresponding to a laser linewidth under 1 Hz at 282 nm) for times between 1 s and 10 s. The long-term stability derives from the ion, and a digital servo loop steers the laser frequency to the atomic resonance over longer time scales.

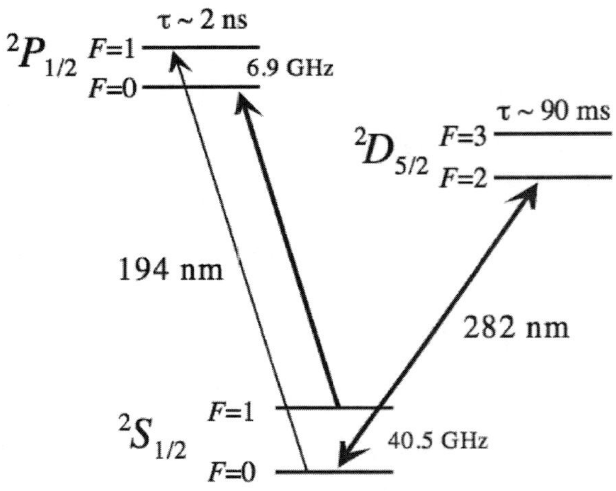

Figure 1: Partial term diagram of the $^{199}Hg^+$ ion. The $^2S_{1/2}$ – $^2P_{1/2}$ dipole transition at 194 nm is used for cooling and state detection, while the $^2S_{1/2}$ $(F=0)$ – $^2D_{5/2}$ $(F=2, m_F=0)$ "clock" transition at 282 nm serves as the reference for the optical frequency standard.

The stable light from this optical frequency standard at frequency $f_{Hg}/2$ can be phase-coherently compared with signals from other frequency standards by means of a broadband optical frequency comb [8,11]. The comb is the output of a broadband femtosecond Ti:Sapphire laser [2], which is further broadened spectrally in a short segment of microstructure fiber. The resulting spectrum consists of a series of spectral lines at frequencies $f_n = nf_r + f_0$ for integer n, which are spaced by the repetition rate f_r of the laser, near 1 GHz. Both f_r and the offset frequency f_0 common to all modes can be locked to be rational multiples of f_{Hg}, so that the entire set of spectral lines is phase coherent with f_{Hg}. The broadened spectrum typically covers a wavelength range of roughly 520 nm to 1100 nm and allows us to compare against other optical frequency standards in that range, such as the calcium optical standard at 657 nm [8, 9]. The repetition rate f_r is measured directly with a photodiode to produce a microwave-frequency output at 1 GHz or its harmonics that can be used for comparisons with frequency standards at lower frequencies. As an example, recent comparisons against a calibrated reference signal from a hydrogen maser in the NIST atomic time scale have led to constraints on the present-day variation of the fine structure constant [6].

In the past year, we have moved the clock system into an improved laboratory and resumed progress towards evaluating the uncertainties of the system. We now summarize the current status of the system and describe our recent progress towards these evaluations.

2 Destabilization of Dark States

The $^2S_{1/2}$ – $^2P_{1/2}$ transition at 194 nm is used for cooling the ion and also for determining whether or not the 282 nm probe pulse has driven the ion to the $^2D_{5/2}$ metastable state, which has a lifetime of about 90 ms. We begin our experimental cycle by cooling the ion with light tuned 35 MHz to the red of the $^2S_{1/2}$ (F=1) – $^2P_{1/2}$ (F=0) transition. A second component of the 194 nm light is phase-locked 47 GHz to the blue of the main beam, near the resonance with the $^2S_{1/2}$ (F=0) – $^2P_{1/2}$ (F=1) transition, and serves as a repumper. After cooling the ion for 3 ms, the repumper is turned off for 15 ms to pump the ion into the $^2S_{1/2}$ (F=0) state. We then extinguish the 194 nm light with a mechanical shutter and expose the ion to the probe light at 282 nm for 40 ms. Finally, we detect the state of the ion with the method of electron shelving [3,7]. Both components of the 194 nm light are then turned on, and the absence or presence of light scattered from the ion indicates that we have, or have not, made the transition to the metastable state.

For this sequence of events to be successful, it is important that the total scattering rate of the light at 194 nm is high enough that we can unambiguously determine the state of the ion after the probe period. When the ion is excited to the $^2P_{1/2}$ state, it usually decays rapidly to the $^2S_{1/2}$ ground state. The $^2P_{1/2}$ state is short-lived, with a lifetime about 2.9 ns [12]. However there are several factors that can reduce the fluorescence rate. First, off-resonant excitation can lead to the atom decaying to the "wrong" hyperfine ground state, for which we have the repumper described earlier. The ion will also occasionally decay instead to one of two metastable states ($^2D_{5/2}$ or $^2D_{3/2}$) that temporarily inhibit fluorescence.

A more important concern is that even if the ion does decay into the $^2S_{1/2}$ (F=1) state, the ion can still be pumped into dark states where it does not fluoresce. Generally speaking, an atom with a J=1 to J'=0 transition will become coherently trapped in a dark state if exposed to light of constant polarization [3]. As this happens, the fluorescence rate approaches zero, so it is critically important to destabilize these dark states. One solution is to split the atomic energy levels, usually with a magnetic field. A second solution is to modulate the polarization of the light that is causing the atom to fluoresce [5].

Previously in the operation of the mercury-ion optical clock, we have applied a magnetic bias field of typical value B = 0.3 mT to destabilize the dark states. This field strength has proven sufficient to allow us to count the scattered photons on a photomultiplier tube with a detected rate of over 10 kHz. This rate is much lower than the scattering rate of the ion (due to various inefficiencies in the imaging system), however it is more than enough fluorescence to operate the clock. The disadvantage of such a strong magnetic field is the perturbation to the frequency of the clock transition. While the $^2S_{1/2}$ (F=0) – $^2D_{5/2}$ (F=2, m_F=0) transition is magnetic-field insensitive to first order, there is a quadratic Zeeman shift $\Delta f = k^{(2)}B^2$ that must be taken into account. The magnitude of the shift is 1.7 kHz for B = 0.3 mT, where the typical fluctuations in our unshielded environment lead to an uncertainty of 2.2 Hz. The uncertainty in the coefficient $k^{(2)}$ = -18.925(28) kHz /mT^2 [10] leads to an additional uncertainty of 0.5 Hz on the clock transition frequency. Since the magnitude of the shift varies as B^2, going to a lower field makes the system less sensitive to small fluctuations in the ambient field.

We have now implemented a system of polarization modulation to destabilize the dark states so that we can operate with much lower magnetic fields. In order to fully avoid the dark states, it is necessary to have polarization components in three orthogonal directions, each with a different time dependence. The simplest way to implement this is to have two noncollinear beams intersecting the ion, where one of the two beams has its polarization modulated as a function of time. On the ion trap table, the 194 nm light is divided evenly into two beams. The first beam has linear polarization and travels in the horizontal plane to intersect the ion in the center of the rf trap. The second beam passes through an electro-optic modulator (EOM) that modulates the polarization. After the modulator, the beam is reflected off a mirror positioned below the ion trap and onto the ion. The propagation axis of the second beam is 40 degrees away from the first beam in the horizontal plane, and it is oriented 35 degrees from vertical. The polarization vector of the first beam is oriented to be in the plane spanned by the two beams. A similar system was previously implemented (using a photoelastic modulator) for a microwave mercury-ion frequency standard based upon the ground-state hyperfine transition [5].

The EOM is a transverse electro-optic phase modulator made from potassium dideuteride phosphate (KD*P). It is oriented with respect to the incident linear light polarization such that it acts as a polarization rotator. Let us write the electric field of the incident light as $\mathbf{E}_{in}(t) = E_{in}\, \boldsymbol{x} \cos(\omega t)$, where $\boldsymbol{x}$ is a unit vector orthogonal to the propagation direction $\boldsymbol{k}$. Under sinusoidal driving at frequency ω_m, the output electric field is described by

$$\mathbf{E}_{out}(t) = \eta E_{in}\,[\boldsymbol{x}\cos(\omega t)\cos(\Delta\sin(\omega_m t)) + \boldsymbol{y}\sin(\omega t)\sin(\Delta\sin(\omega_m t))]. \qquad (1)$$

Here, η is a constant and $\boldsymbol{y}$ is the unit vector orthogonal to $\boldsymbol{x}$ and $\boldsymbol{k}$. The sinusoidal modulation causes the polarization to rotate back and forth through linear, left and right elliptical, and left and right circular. The modulation index Δ must be chosen at a value high enough that there is a significant component in both the $\boldsymbol{x}$ and $\boldsymbol{y}$ directions, yet low enough that the laser power can be predominantly on the red side of the atomic resonance. The driving frequency ω_m is likewise chosen between extremes where the drive is either too slow to destabilize the dark state or high enough that there is little laser power near the atomic resonance. The parameters were chosen to maximise the final fluorescence rate, based on computational work by Berkeland and Boshier [4]. For our system, the modulation index has a value near Δ=7, and the drive frequency ω_m is near 2π x 4.7 MHz. The EOM is driven through a helical resonator attached to the EOM. The combined system of the resonator and modulator has a quality factor of approximately 230 and is directly driven by a frequency synthesizer with 20 dBm of output power.

Using this polarization modulation system, we have been able to achieve typical fluorescence count rates in excess of 6 kHz, even with vanishing magnetic fields. It is still convenient to operate with a weak magnetic bias field that defines the quantization axis and splits apart the magnetic sublevels of the $^2D_{5/2}$ (F=2) state. We determine the magnetic field magnitude by measuring the linear Zeeman splitting between the clock transition $^2S_{1/2}$ (F=0) – $^2D_{5/2}$ (F=2, m_F=0) and an auxiliary transition to the $^2D_{5/2}$ (F=2, m_F=2) state. With a bias field of 0.020 mT, the total quadratic Zeeman shift is –7.6 Hz, and typical fluctuations in the laboratory field lead to an uncertainty of 0.15 Hz on the clock transition frequency that can be further reduced with magnetic shielding. The uncertainty in the Zeeman coefficient $k^{(2)}$ contributes an additional uncertainty of about 3 mHz to the clock frequency, which could be reduced by measuring the coefficient carefully.

2 Progress Towards a Measurement of the Quadrupole Shift

The dominant contribution to the uncertainty budget of the mercury-ion optical clock is presently the quadrupole shift of the $^2D_{5/2}$ state [6]. In an ideal spherical rf trap, there are no static electric fields applied. The trap consists of a ring electrode symmetrically driven with rf about ground potential, and two grounded endcap electrodes. In this case there are no static electric field gradients and the clock transition frequency is not a function of the quantization axis orientation. In practice, there are static electric patch charges on the electrodes that we compensate with low voltages applied to the endcaps and with higher potentials applied to separate electrodes in the vicinity of the trap. The quadrupole shift is due to the interaction of the atomic electric quadrupole moment of the ion in the $^2D_{5/2}$ state with the electric field gradients arising from these patch potentials. The total magnitude of the shift is expected to be less than or of order 1 Hz [10]. However, this estimate was based upon a Hartree-Fock calculation and it is possible that there are large deviations from this value [14, 16]. A series of measurements conducted over several years with several different ion trap surfaces does not show any large variations and is consistent with this low value of the shift. However, we have not yet performed a full evaluation of this shift and so have placed a conservative 10 Hz error bar on the magnitude of this shift in the absence of this evaluation.

We are now preparing to measure the quadrupole shift. It has been shown that the quadrupole shift will average to zero if the clock transition frequency is measured over three orthogonal quantization axes [10]. In the laboratory, we are planning to orient the magnetic field at the site of the ion trap in thee orthogonal directions. The three magnet coil pairs around the ion trap have been calibrated via spectroscopy on the Zeeman-sensitive components of the $^2S_{1/2}$ (F=0) – $^2D_{5/2}$ (F=2) transition. With this calibration, magnetic fields of approximately equal magnitude but near-orthogonal orientation can be quickly exchanged by changing the currents on the magnet coil power supplies. A measurement using this method was previously performed with mercury ions in a linear rf trap, which relied upon static electric fields for confinement [13].

In order to perform this measurement, it is essential that there is a stable "flywheel" that can provide a link between successive measurements of the clock transition frequency. We would like to be able to measure the frequency at each of the three field orientations with imprecision well below 1 Hz. The best flywheel for this measurement would be a second mercury-ion optical standard, which we expect to use for this type of comparison in the long term. In the immediate future, we plan to compare against an optical frequency standard based upon neutral calcium. Figure 2 shows a plot of the Allan deviation [1] between the mercury and calcium standards, as measured with the femtosecond frequency comb. The instability averages to well below 10^{-15} (1 Hz at the 1.064 PHz Hg^+ transition frequency) with an averaging period of 1000 s.

In addition to measuring the quadrupole shift due to the stray fields in the trap, we also plan to measure the quadrupole moment by applying a known electric-field gradient to the ion. This can be achieved by symmetrically biasing the two endcap electrodes of the trap several volts above ground. A similar method has recently been employed at the PTB to measure the quadrupole moment in a Yb^+ ion trap [15]. Since it is possible to rapidly change the bias potential on the endcap electrodes, it may be possible to perform this particular measurement without using any secondary atomic references. The stable high-finesse cavities, which have a predictable linear drift for time scales under 30 s, could act as the flywheel for measurements that are conducted on a similar time scale.

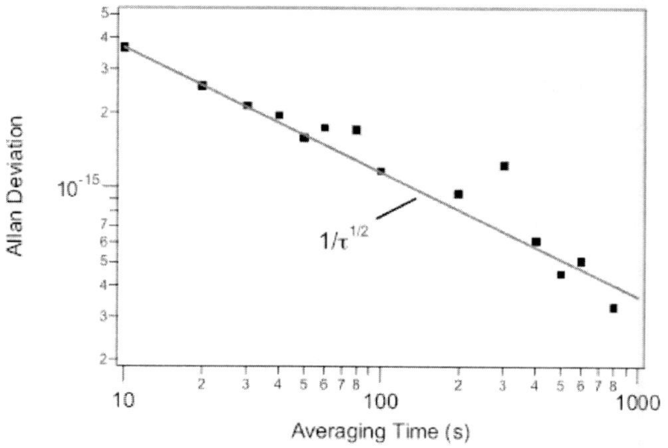

Figure 2: This Allan deviation plot shows the relative stability of the calcium and mercury optical standards as a function of the averaging period τ. This measurement is conducted by counting the effective beat frequency between the two laser sources at 657 nm and 563 nm by means of the femtosecond comb system.

Acknowledgements

We thank Roee Ozeri and Vladislav Gerginov for helpful comments in the preparation of this paper. This work was supported by the Office of Naval Research. C.E.T. was also supported by DOE and NSF. This work of an agency of the U.S. government is not subject to U.S. copyright.

References

[1] D. W. Allan, "Statistics of Atomic Frequency Standards", *Proceedings of the IEEE* **54**, pp. 221-230 (1966)

[2] A. Bartels and H. Kurz, "Generation of a broadband continuum by a Ti:sapphire femtosecond oscillator with a 1-GHz repetition rate", *Opt. Lett.* **27**, pp 1839-1841 (2002).

[3] J. C. Bergquist, W. M. Itano, and D. J. Wineland, "Recoilless optical absorption and doppler sidebands of a single trapped ion", *Phys. Rev. A* **36**, p. 428 (1987).

[4] D. J. Berkeland and M. G. Boshier, "Destabilization of dark states and optical spectroscopy in Zeeman-degenerate atomic systems", *Phys. Rev. A* **65**, 033413 (2002).

[5] D. J. Berkeland, J. D. Miller, J. C. Bergquist, W. M. Itano, and D. J. Wineland, "Laser-Cooled Mercury-Ion Frequency Standard", *Phys Rev. Lett.* **80**, p. 2089 (1998).

[6] S. Bize, S. A. Diddams, U. Tanaka, C. E. Tanner, W. H. Oskay, R. E. Drullinger, T. E. Parker, T. P. Heavner, S. R. Jefferts, L. Hollberg, W. M. Itano, and J. C. Bergquist, "Testing the stability of fundamental constants with the $^{199}Hg^+$ single-ion optical clock" *Phys. Rev. Lett.* **90**, 150802 (2003).

[7] H. Dehmelt, "Proposed 10^{14} $\delta\nu<\nu$ laser spectroscopy on Tl^+ mono-ion oscillator II (spontaneous quantum jumps)" *Bull. Am. Phys. Soc.* **20**, p. 60 (1975).

[8] S. A. Diddams, Th. Udem, J. C. Bergquist, E. A. Curtis, R. E. Drullinger, L. Hollberg, W. M. Itano, W. D. Lee, C. W. Oates, K. R. Vogel, and D. J. Wineland, "An optical clock based on a single trapped $^{199}Hg^+$ ion", Science **293**, p. 825 (2001).

[9] J. Helmcke, G. Wilpers, T. Binnewies, C. Degenhardt, U. Sterr, H. Schnatz and F. Riehle, "Optical Frequency Standard Based on Cold Ca Atoms", *IEEE Transactions on Instrumentation and Measurement*, **52**, pp. 250-254, (2003).

[10] Wayne M. Itano, "External-field shifts of the $^{199}Hg^+$ optical frequency standard", *J. Res. Natl. Inst. Stand. Technol.* **105**, p. 829 (2000).

[11] D. J. Jones, S. A. Diddams, J. K. Ranka, A. Stentz, R. S. Windeler, J. L. Hall, and S. T. Cundiff, "Carrier-envelope phase control of femtosecond mode-locked lasers and direct optical frequency synthesis", *Science* **288**, p. 635 (2000).

[12] E. H. Pinnington, W. Ansbacher, J. A. Kernahan, T. Ahmad, and Z.-Q. Ge, "Lifetime measurements for low-lying levels in Hg I and Hg II using the beam-foil technique", *Can. J. Phys.* **66**, pp. 960-962, (1988).

[13] R. J. Rafac, B.C. Young, J. A. Beall, W. M. Itano, D. J. Wineland, and J. C. Bergquist, "Sub-dekahertz Ultraviolet Spectroscopy of $^{199}Hg^+$", *Phys Rev. Lett.* **85**, p. 2462 (2000).

[14] P. G. H. Sandars and A. J. Stewart, "Beam Resonance measurements of atomic quadrupole moments", in *Atomic Physics 3*, edited by S. J. Smith and G. K. Walters (Plenum, New York) p. 429-434 (1973).

[15] T. Schneider, Chr. Tamm, E. Peik, "Comparison of Two Single-Ion Optical Frequency Standards at the Hertz Level", *Proceedings of the 2003 International Frequency Control Symposium and PDA Exhibition Jointly with the 17th European Frequency and Time Forum*, p. 72, (2003).

[16] R. M. Sternhemer, "Atomic quadrupole moments of the excited states of the rare gases", *Phys. Rev. A* **7**, pp. 887-891 (1973).

AN OPTICAL FREQUENCY STANDARD WITH ULTRACOLD CALCIUM ATOMS

U. Sterr, C. Degenhardt, H. Stoehr, H. Schnatz, J. Helmcke, F. Riehle
Physikalisch-Technische Bundesanstalt
Bundesallee 100
D-38116 Braunschweig, Germany
email: fritz.riehle@ptb.de

Keywords: Optical Ca frequency standard, optical frequency measurement, laser cooling.

Abstract

An optical frequency standard is operated at PTB where the frequency of a laser is stabilized to the intercombination transition ($\lambda \approx 657$ nm) in laser cooled Ca atoms. Previously the uncertainty resulting from the residual velocity of the cold atoms at 3 mK limited the fractional uncertainty to 2×10^{-14}. We now apply a novel method for producing ultra-cold atoms ($T \approx 10$ μK). The frequency of the intercombination line has been measured with respect to the Cs hyperfine transition over eight years and a first frequency measurement with ultracold atoms has now been performed where the contributions from the residual velocity become insignificant at that level of uncertainty. A fractional uncertainty below 10^{-15} is expected for the near future.

1 Introduction

The progress in laser manipulation of atomic absorbers and measurement of optical frequencies led to greatly improved optical frequency standards that find increasing applications in realizing basic units such as the length unit [1], to determine fundamental constants [2, 3], or to test basic theories and possible variations of fundamental constants [4, 5]. Compared to the present atomic clocks operating in the microwave region, the use of optical frequencies is advantageous since for a given linewidth, a higher quality factor $Q = \nu/\delta\nu$ is obtained with the potential to increase the stability and the accuracy. Hence today's optical frequency standards [6, 7, 8] may lead to optical clocks that eventually might become superior to the currently best microwave clocks [9, 10, 11]. Here we report on a frequency standard based on ultra-cold Ca atoms representing one of the frequency standards in the visible with the lowest uncertainty.

2 Experimental set-up

The calcium optical frequency standard makes use of the intercombination transition ($\lambda \approx 657$ nm) from the $^1S_0$ ground state to the lowest $^3P_1$ state (Fig. 1) with a linewidth of 370 Hz in ^{40}Ca. The ($m_j = 0 \rightarrow m_j = 0$) transition is well suited for a frequency standard since it is to first order insensitive to electric and magnetic fields.

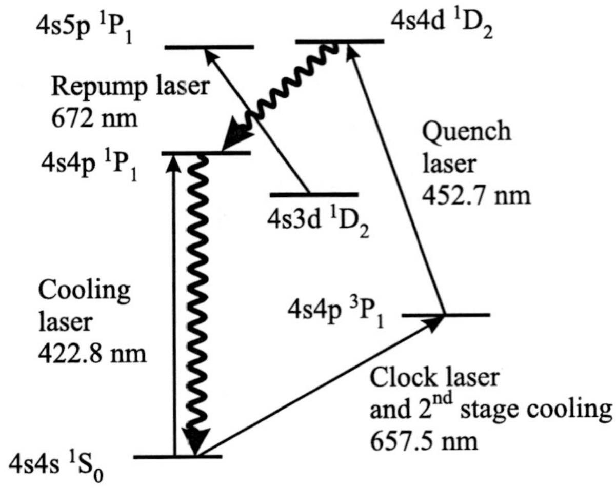

Figure 1: Energy level diagram of ^{40}Ca with the laser transitions relevant for the optical frequency standard.

2.1 Cooling the atoms

The atoms are cooled and trapped in a magneto-optical trap (MOT) using the $^1P_1$ - $^1S_0$ transition at $\lambda \approx 423$ nm. The MOT is filled by Ca atoms effusing from an oven which is located at a distance of about 15 cm from the trap center (Fig. 2). As a consequence, only a fraction of the atoms in the beam can be captured by the trap, i. e. those atoms from the Boltzmann velocity distribution ($v_{\text{prob}} \approx 600$ m/s, $T_{\text{oven}} \approx 900$ K) which are slower than the trap's maximum capture velocity of about 30 m/s. The capture velocity was increased by using two laser frequencies for the horizontal trapping beams. The additional laser frequency obtained by an acousto–optic modulator allows faster atoms to get in resonance with the trapping beams of the additional frequency.
To further increase the number of atoms a particular loss channel has to be closed originating from atoms in the $^1P_1$ decaying with a probability of 10^{-5} to the 4s3d $^1D_2$ state. From here they predominantly fall into the metastable 4s4p^{3}P states and leave the trap before

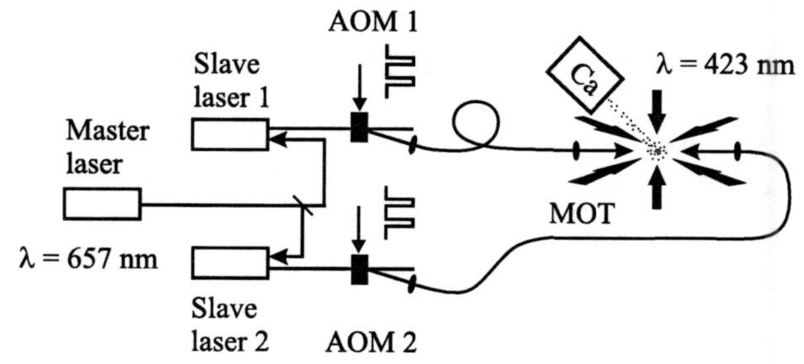

Figure 2: Experimental setup. Ca atoms are trapped in a magneto-optical trap (MOT). The clock transition is interrogated by pulses cut from two cw clock-laser beams from two slave diode lasers injection locked by a common master of extended cavity diode laser (ECDL) configuration

they can decay into the ground state. This bypass was closed by exciting the atoms on the 4s3d $^1D_2$–4s5p $^1P_1$ transition, applying 2 mW of laser light at 672 nm (repump transition in Fig. 1). From the 4s5p $^1P_1$ state the atoms can decay into the ground state and again take part in the cooling process. With these improvements the time constant of the loss of atoms from the trap was increased from 20 ms to more than 300 ms and the number of trapped atoms was increased leading to about 10^8 cold Ca atoms at a temperature of about 3 mK.

It has been shown that with cold atoms in the millikelvin range the influence of the residual first-order Doppler shift ultimately limits the fractional uncertainty of a Ca frequency standard to about 10^{-14} [16]. Lower uncertainties can be obtained by a second-stage cooling on the intercombination transition [8]. The low scattering rate on this transition, leading to a resulting force only slightly higher than the gravitational force, requires a modification of the cooling scheme [12, 13]. In this scheme, the scattering rate on the intercombination transition is increased by pumping the atoms with the help of a "quenching" laser (453 nm) back into the ground state via intermediate levels (see Fig. 1). With the quench cooling scheme we are able to achieve a temperature of the atomic cloud as low as 6 μK with up to 30 % of the atoms being transferred from the cloud at 3 mK to the ultracold cloud.

The radiation at $\lambda \approx 423$ nm used to trap the Ca atoms in the MOT was produced by the 846 nm radiation from a Ti:Sa laser ($P = 1.6$ W) which was frequency doubled in a LiB_3O_5 crystal in a buildup cavity resulting in 0.6 W of blue power.

To produce the ultracold ensemble the blue MOT is operated for typically 500 ms. After switching off the 423 nm laser beams the gradient of the magnetic field is reduced within about 0.1 ms from 0.8 T/m to about 0.04 T/m. Afterwards the laser beams for cooling (657 nm) and quenching (453 nm) are switched on for 20 ms. This time is sufficient since already after 5 to 10 ms the temperature is not reduced much any longer but the density and number of particles is reduced with a typical time constant of 210 ms. To avoid systematic frequency shifts due to the Zeeman effect of the magnetic field of the MOT and light shifts due to the 423 nm radiation, both the magnetic field and the light of the trapping laser have to be shut off before interrogating the Ca atoms. After switching off the fields the atomic cloud expands according to the root-mean-square (rms) velocity of the ballistic atoms. The rms velocity can be determined from the Doppler broadening of the intercombination line at $\lambda \approx 657$ nm (see Fig. 3) and it is in the range of $v_{rms} \approx 7$ cm/s corresponding to a temperature of about 10 μK. The velocity distributions along the three different spatial directions do not differ significantly.

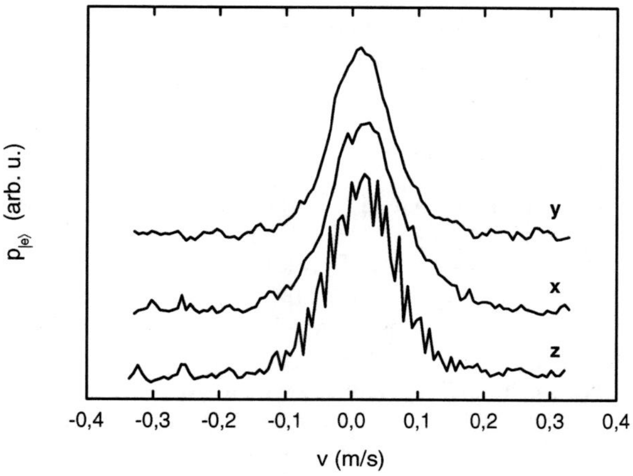

Figure 3: Velocity distributions of the laser cooled ensemble of Ca atoms released from a magneto–optical trap. The x direction denotes the direction of the interrogating laser beams. The z direction is the vertical direction parallel to the magnetic bias field. Therefore the σ transitions had to be probed, resulting in additional noise due to the fluctuations of the magnetic field.

2.2 The clock laser

The radiation to excite the Ca intercombination line and to facilitate the second-stage cooling is generated by a high-resolution laser spectrometer comprising a diode-laser system prestabilized to a suitable resonance frequency of a Fabry–Perot resonator by means of the Pound-Drever-Hall technique [14]. The diode laser system comprises an extended cavity diode laser in Littman configuration as a master laser which is essentially the same as described earlier [15]. Its laser diode is anti-reflection coated and an intracavity $LiNbO_3$ electro-optical modulator with a bandwidth of more than 1 MHz controlled the frequency of the extended cavity diode laser. To obtain a narrower linewidth as compared to [15] the Fabry–Perot resonator in its vacuum housing has been mounted on a platform which

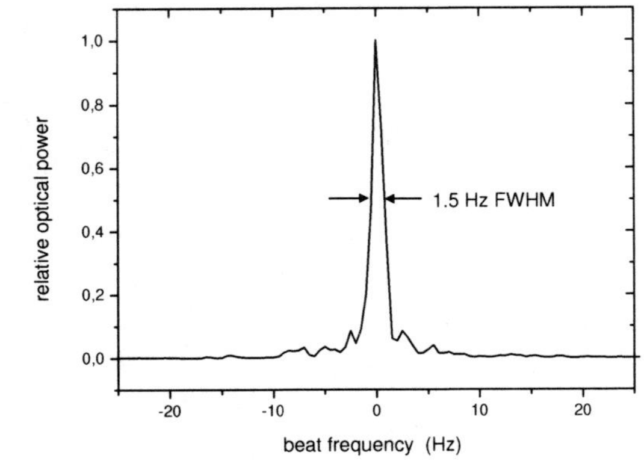

Figure 4: Beat between two diode laser systems frequency stabilized to two independent Fabry–Perot Interferometers.

is hung from a rigid framework on the optical table by four springs of 0.5 m length. The whole arrangement for the Fabry–Perot resonator is in a box which provides acoustic isolation. The light from the master laser which is located in a separate quiet room is transported to the experiment by an optical fiber whose length fluctuations resulting from thermal and acoustic noise are suppressed by means of an active stabilization. As the output power of about 6 mW was not sufficient for the interrogation and the second-stage cooling it was amplified in a series of injection-locked slave lasers comprising commercial diode lasers. The linewidth of the laser is about 1 Hz as could be inferred from a beat note measurement of two independent laser systems (Fig. 4).

3 Ramsey-Bordé excitation

In order to achieve high spectral resolution combined with a good signal-to-noise ratio (S/N), we apply the method of pulsed excitation which has similarities to the Ramsey excitation with spatially separated fields. In contrast to the Ramsey excitation originally devised for the interrogation of atomic beams by microwaves we apply the optical Ramsey excitation in the time domain. Bordé [17] has shown that this optical Ramsey excitation leads to a time-domain atom interferometer where the short pulses of $\tau \leq 1$ μs duration are used to split, redirect and recombine the wave functions of the atoms [18]. Time domain separated–field excitation was performed by two pulses of counterpropagating traveling laser fields. The laser beams used to interrogate the intercombination transition of the cold Ca atoms are derived from acousto-optic modulators (AOMs) (Fig. 2) used as "light switches" for the pulsed excitation. When the frequency of the interrogating laser is tuned close to the resonance, the probability to find the atoms in the excited state contains a contribution (Fig. 5) which varies with the cosine of the laser detuning.

The excited atoms are detected with almost unity probability using a scheme [8] similar to the shelved electron scheme [19].

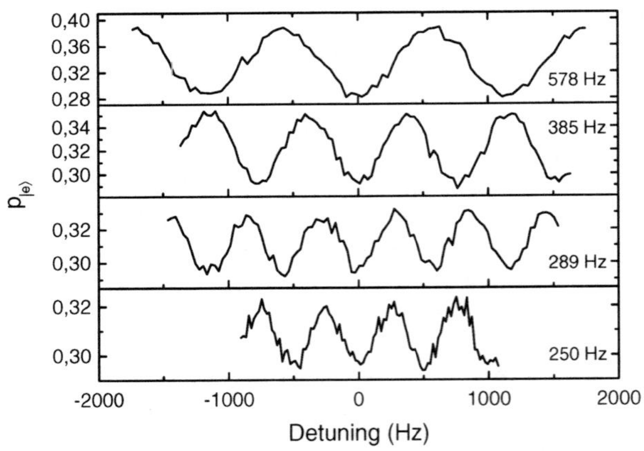

Figure 5: Optical Ramsey-Bordé resonances excited in an expanding cloud of Ca atoms released from a magneto-optical trap.

Provided that the lengths of the pulses are small compared to their separation, the width of the interference fringes $\delta\nu = 1/(4T)$ is inversely proportional to T. The narrowest fringe width obtained with trapped atoms was even below to the natural linewidth of about 370 Hz. In fact, the signals shown in Fig. 5 represent a superposition of two interference signals caused by the two recoil components which are shifted by ± 11.6 kHz. Consequently, the time T between the pulses is adjusted in such a way that the interference structures resulting from both recoil components add to give a maximum visibility of the signal. Increasing the time T reduces the period of the interference structure and consequently the linewidth of the fringes.

The central fringe was used to stabilize the frequency of the laser. The error signal for the stabilization is generated from the interference signal by modulating the laser frequency and simultaneously measuring the excitation probability. Rather than using a square-wave frequency modulation between two discrete values with the mean frequency tuned close to the center of the central fringe (1f-method) we use an analogue of the 3f–method [20].

4 Uncertainty of the Ca Standard

To convert a stable and reproducible oscillator to a frequency standard it is necessary to reference the frequency of the oscillator to the frequency of the primary standard of time and frequency, the Cs atomic clock. Hence, three independent sources contribute to the uncertainty of an optical standard. First, there is the uncertainty to realize the line center of the unperturbed Ca atom which mostly depends on the ability to reduce and correct perturbations affecting the frequency of the atomic transition. A second contribution results from the uncertainty of the primary Cs standard. Third, the comparison of the two frequencies of the Cs and Ca standards differing by five orders of magnitude may additionally contribute to the measurement uncertainty (see Tab. 1).

Effect	uncertainty achieved	expected
1st order Doppler effect	1.0 Hz	150 mHz
2nd order Doppler effect	0.02 mHz	0.02 mHz
Other phase contributions	1.6 Hz	200 mHz
Asymmetry of line shape	0.2 Hz	50 mHz
Magnetic field (64 Hz/mT^2)	0.1 Hz	100 mHz
ac Stark effect	0.1 Hz	100 mHz
Quadratic Stark effect ($E < 2$ V/cm)	0.02 Hz	20 mHz
Black - body radiation		
Oven	3.9 Hz	
Chamber (300 K)	0.07 Hz	70 mHz
Collisions of cold atoms	0.06 Hz	60 mHz
Influence of laser drift	0.1 Hz	100 mHz
Sum in quadrature	4.3 Hz	400 mHz
Stat. unc. of frequ. meas.	3 Hz	100 mHz
Cs clock (1×10^{-15})	0.5 Hz	500 mHz
Total uncertainty $\delta\nu$	5.3 Hz	700 mHz
$\delta\nu/\nu$	$1.2 \cdot 10^{-14}$	$1.6 \cdot 10^{-15}$

Table 1: Uncertainty budget for the measurement of the Ca frequency standard with ultra-cold atoms of 12 μK on 8 October 2003 (col. 2) and the achievable uncertainty (col. 3) described in the text.

We first discuss the different effects affecting the uncertainty to realize the center of the clock transition of the Ca standard. We make a distinction between the uncertainty that has been achieved during the measurement of the frequency on 8 October 2003 (second column of Tab. 1) and the uncertainty that can be achieved with this kind of standard (third column of Tab. 1). An important contribution is still caused by the residual first-order Doppler effect even though the interrogation by the Ramsey–Bordé technique is in principle a first-order Doppler-free method. However, the movement of the atoms in non-ideal (curved) wavefronts of the interogating laser beam or the acceleration due to gravity in a tilted beam leads to a phase shift between the interrogating laser pulses which increases quadratically with time. The low velocity of the atoms reduces the associated shifts and a method has been devised to determine and correct the residual shift [16] allowing one to reduce this effect in priciple below 10^{-15}. Do to the low velocities the second-order Doppler effect caused by the time dilation contributes as little as a few microhertz and can be neglected totally.

Another important contribution results from any phase chirps in the optical pulses used to interrogate the clock transition. Such chirps can occur by "ringing" in the AOM and the associated driver when it is rapidly switched on or off. The chirps in the optical pulses have been measured in an optical interferometer and in an atom interferometer where it leads to a resolution-dependent frequency shift of a few hertz which was corrected. The influences of magnetic fields by the Zeeman effect, of electric fields by the Stark effect are low. A large ac Stark shift can be caused by scattered light of the cooling laser resulting from uncomplete blocking of the AOMs. In our setup it was eliminated by the use of fast mechanical shutters. The largest contribution to the uncertainty at present results from the field of thermal radiation (so-called black-body shift) which is of particular importance since in our present setup the oven is very close to the cloud of atoms. Even though the atoms are trapped behind a circular beam stop they are exposed to the radiation from the oven during their ballistic flight or due to radiation reflected from the walls. We investigated the frequency shift when the temperature of the oven was changed by 100 K and the uncertainty given in the second column of Tab. 1 reflects the uncertainty associated with these measurements.

A particular contribution to the uncertainty may result from the interactions between the atoms in the laser cooled ensemble, i. e. cold collisions, which may shift the clock transition. The magnitude of the shift depends on the states of the two colliding atoms and their distance and results in a mean frequency shift that depends on the temperature and the density of the atomic ensemble. For the low densities of about $< 10^{16}$ m^{-3} in our trap we assume the frequency shift to be a linear function of the collision rate and, hence, of the density ρ. In order to determine the density dependence of the transition frequency we have measured the frequency shift as a function of average density differences for ultracold atoms at a temperature of about 20 μK. The density was varied by switching the repump laser at 672 nm (Fig. 1) and by changing the parameters of the second-stage cooling. All other parameters like the magnetic field, loading time, frequency of the trapping light and laser beam profile were kept constant. We have checked that slight changes of the atomic cloud in the temperature and the radius of the trapped cloud did not lead to systematic frequency shifts. The measurement of the collisional shift $\Delta\nu/\nu$ depending on the density ρ (Fig. 6) gave $\Delta\nu/\nu = (-0.4 \pm 1.2) \times 10^{-31}$ m$^3 \times \rho$. Within the uncertainty of this measurement no significant shift has been observed. From the uncertainty it can be concluded that the relative collisional shift in the optical Ca frequency standard at typical operational conditions is at least four orders of magnitude lower than in the caesium fountain [21]. Hence, the low density dependence allows one to use considerably higher densi-

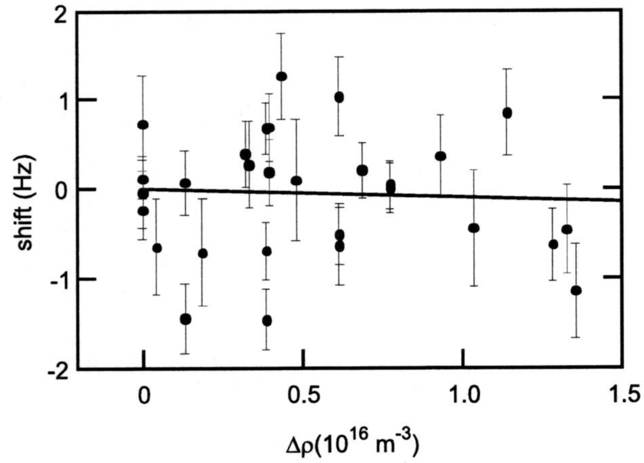

Figure 6: Measurement of the density-dependent frequency shift using ultra-cold atoms ($T \approx 20\ \mu$K). The lines represent a linear regression to the data.

ties and atom numbers compared to Cs atomic clocks. A small contribution to the uncertainty of the Ca standard in Tab. 1 results from the drift of the pre-stabilized laser during the interrogation process which is estimated to amount to less than 0.1 Hz in the present setup.

All contributions to the uncertainty of the Ca standard in the measurement on 8 October 2003 were added in quadrature and lead to a total fractional uncertainty of about 10^{-14}. Further contributions result from the measurement of the frequency and the uncertainty of the primary standard the caesium fountain.

5 Frequency measurements

The (optical) frequency of the Ca standard was measured by comparison with PTB's primary Cs atomic clocks over several years. Originally, a phase-coherent harmonic frequency measurement chain was used [22]. First measurements of the Ca standard with a femtosecond laser comb agreed within the combined uncertainties (2000) [23]. In the mean time further measurements have been performed. The uncertainties of these measurements were determined mainly by the uncertainties resulting from the corrections due to magnetic quadrupole fields of the MOT and ac Stark shifts (June 2001). Subsequent frequency measurements with reduced uncertainties employing cold atoms were performed [8].

The frequency measurement with ultra-cold atoms on 8 October 2003 gave ν_{Ca} = 455 986 240 494 144 (5.3) Hz. We emphasize that this uncertainty makes the Ca stabilized laser one of the most accurate optical frequency standards in the visible, today. Consequently, this standard has been recommended recently by the CIPM [1] for the realization of the meter with one of the lowest uncertainties.

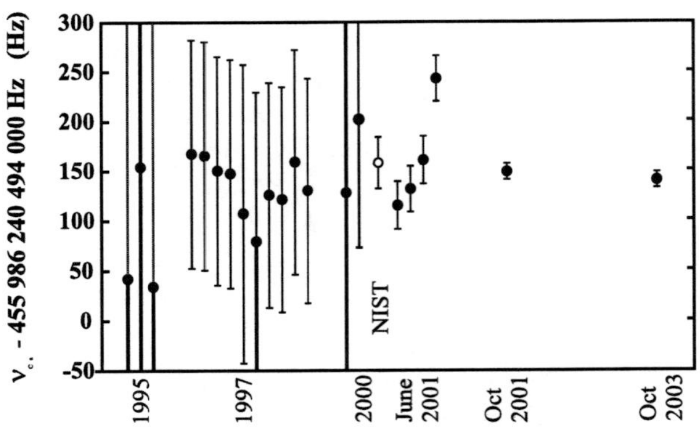

Figure 7: Measured frequency values of the Ca optical frequency standard over the years.

6 Conclusions

The accuracy of the Ca optical frequency standard can be easily improved if the contribution of the black-body shift is reduced. To this end we currently set up an apparatus where the atoms are decelerated in a Zeeman slower before being deflected by a tilted two-dimensional molasses. In this case the oven temperature will no longer contribute and one ends up with the much lower uncertainty of 0.07 Hz from the black-body shift given in the second column of Tab. 1. The influence of cold collisions is particularly low and it might even be reduced by using lower densities. The influence of the residual first-order Doppler effect seems to limit the fractional uncertainty to slightly below 10^{-15}. It remains to be explored if the use of atoms in an optical lattice operated at the magic wavelength where the first-order Doppler effect can be eliminated [24] will lead to lower uncertainties.

As has been pointed out [25, 8] the quantum projection-noise limit [26] for the Ca standard corresponds to an Allan deviation $\sigma_y(\tau) < 10^{-16}$ for $\tau = 1$ s. To approach this limit, a frequency instability of the spectroscopy laser of better than 3×10^{-16} for the duration of the atom interferometry $2T \approx 1.3$ ms is necessary. While laser instabilities better than 1 Hz in 1 s have already been demonstrated [27], to reach this value is still a challenging task. Higher stability indirectly contributes to higher accuracy because it allows one to detect systematic frequency shifts and fluctuations within short measuring times thereby identifying hitherto undetected error sources. It is expected that these improvements will allow us to reach a relative uncertainty of the optical Ca standard below 1×10^{-15} thereby competing with the best optical standards based on single ions and with the best microwave clocks.

Acknowledgements

This work was supported in part by the Deutsche Forschungsgemeinschaft (DFG) under SFB 407.

References

[1] T. J. Quinn. "Practical realization of the definition of the metre, including recommended radiations of other optical frequency standards (2001)", *Metrologia*, **40**, pp. 103–133, (2003).

[2] C. Schwob, L. Jozefowski, B. de Beauvoir, L. Hilico, F. Nez, L. Julien, F. Biraben, O. Acef, A. Clairon. "Optical Frequency Measurement of the $2S - 12D$ Transitions in Hydrogen and Deuterium: Rydberg Constant and Lamb Shift Determinations", *Phys. Rev. Lett.*, **82**, pp. 4960–4963, (1999).

[3] A. Huber, T. Udem, B. Gross, J. Reichert, M. Kourogi, K. Pachucki, M. Weitz, T. W. Hänsch. "Hydrogen-Deuterium $1S - 2S$ Isotope Shift and the Structure of the Deuteron", *Phys. Rev. Lett.*, **80**, pp. 468–471, (1998).

[4] S. Bize, S. A. Diddams, U. Tanaka, C. E. Tanner, W. H. Oskay, R. E. Drullinger, T. E. Parker, T. P. Heavner, S. R. Jefferts, L. Hollberg, W. M. Itano, J. C. Bergquist. "Testing the Stability of Fundamental Constants with the $^{199}Hg^{+}$ Single-Ion Optical Clock", *Phys. Rev. Lett.*, **90**, pp. 150802-1–4, (2003).

[5] E. Peik, B. Lipphardt, H. Schnatz, T. Schneider, Chr. Tamm, S. G. Karshenboim. "New limit on the present temporal variation of the fine structure constant", *http://arXiv.org/abs/physics/0402132*.

[6] T. Udem, S. A. Diddams, K. R. Vogel, C. W. Oates, E. A. Curtis, W. D. Lee, W. M. Itano, R. E. Drullinger, J. C. Bergquist, L. Hollberg. "Absolute Frequency Measurement of the Hg^{+} and Ca Optical Clock Transitions with a Femtosecond Laser" *Phys. Rev. Lett.*, **86**, pp. 4996–4999, (2001).

[7] J. Stenger, C. Tamm, N. Haverkamp, S. Weyers, H. R. Telle. "Absolute frequency measurement of the 435.5-nm $^{171}Yb^{+}$-clock transition with a Kerr-lens mode-locked femtosecond laser" *Opt. Lett.*, **26**, pp. 1589–1591, (2001).

[8] G. Wilpers, T. Binnewies, C. Degenhardt, U. Sterr, J. Helmcke, F. Riehle. "Optical Clock with Ultracold Neutral Atoms", *Phys. Rev. Lett.*, **89**, pp. 230801-1–4, (2002).

[9] S. Weyers, A. Bauch, R. Schröder, Chr. Tamm. 'The Atomic Caesium Fountain CSF1 of PTB'. In: *Frequency Standards and Metrology, Proceedings of the Sixth Symposium.* ed. by P. Gill (World Scientific, Singapore 2002) pp. 64–71.

[10] S. R. Jefferts, J. Shirley, T. E. Parker, T. P. Heavner, D. M. Meekhof, C. Nelson, F. Levi, G. Costanzo, A. De Marchi, R. Drullinger, L. Hollberg, W. D. Lee, F. L. Walls. "Accuracy Evaluation of NIST-F1", *Metrologia*, **39**, pp. 321–336, (2002).

[11] H. Marion, F. Pereira Dos Santos, M. Abgrall, S. Zhang, Y. Sortais, S. Bize, I. Maksimovic, D. Calonico, J. Grünert, C. Mandache, P. Lemonde, G. Santarelli, Ph. Laurent, A. Clairon, C. Salomon. "Search for Variations of Fundamental Constants using Atomic Fountain Clocks", *Phys. Rev. Lett.*, **90**, pp.150801-1–4, (2003).

[12] T. Binnewies, G. Wilpers, U. Sterr, F. Riehle, J. Helmcke, T. E. Mehlstäubler, E. M. Rasel, W. Ertmer. "Doppler cooling and trapping on forbidden transitions", *Phys. Rev. Lett.*, **87**, pp. 123002-1–4, (2001).

[13] E. A. Curtis, C. W. Oates, L. Hollberg. "Quenched Narrow-Line Laser Cooling of ^{40}Ca to Near the Photon Recoil Limit", *Phys. Rev. A*, **64**, pp. 031403(R)-1–4 (2001).

[14] R. W. P. Drever, J. L. Hall, F. V. Kowalski, J. Hough, G. M. Ford, A. J. Munley, H. Ward. "Laser Phase and Frequency Stabilization Using an Optical Resonator", *Appl. Phys. B*, **31**, pp. 97–105, (1983).

[15] V. Vassiliev, V. Velichansky, P. Kersten, T. Trebst, F. Riehle. "Subkilohertz enhanced-power diode-laser spectrometer in the visible", *Opt. Lett.*, **23**, pp. 1229–1231, (1998).

[16] G. Wilpers, C. Degenhardt, T. Binnewies, A. Chernyshov, F. Riehle, J. Helmcke, U. Sterr. "Improvement of the fractional uncertainty of a neutral atom calcium optical frequency standard to $2 \cdot 10^{-14}$", *Appl. Phys. B*, **76**, 149–156, (2003).

[17] Ch. J. Bordé. "Atomic interferometry with internal state labelling", *Phys. Lett. A*, **140**, pp. 10–12, (1989).

[18] U. Sterr, K. Sengstock, W. Ertmer, F. Riehle, J. Helmcke. "Atom interferometry based on separated light fields", in *Atom Interferometry*, P. Berman (ed.), Academic Press, San Diego, pp. 293–362, (1997).

[19] W. Nagourney, J. Sandberg, H. Dehmelt. "Shelved optical electron amplifier: Observation of quantum jumps", *Phys. Rev. Lett.*, **56**, pp. 2797–2799, (1986).

[20] F. Riehle, H. Schnatz, B. Lipphardt, G. Zinner, T. Trebst, T. Binnewies, G. Wilpers, J. Helmcke. "The optical Ca frequency standard", *Proc. of the 1999 Joint Meeting of the European Frequency and Time Forum and The IEEE International Frequency Control Symposium, Besancon*, **87**, pp. 700-705, (1999).

[21] S. Ghezali, Ph. Laurent, S. Lea, A. Clairon. "An experimental study of the spin-exchange frequency shift in a laser-cooled cesium fountain frequency standard", *Europhys. Lett.*, **36**, pp. 25–30, (1996).

[22] H. Schnatz, B. Lipphardt, J. Helmcke, F. Riehle, G. Zinner. "First Phase-Coherent Frequency Measurement of Visible Radiation", *Phys. Rev. Lett.*, **76**, pp. 18–21, (1996).

[23] J. Stenger, T. Binnewies, G. Wilpers, F. Riehle, H. R. Telle, J. K. Ranka, R. S. Windeler, A. J. Stentz. "Phase-coherent frequency measurement of the Ca intercombination line at 657 nm with a Kerr-lens mode-locked laser", *Phys. Rev. A*, **63**, pp. 021802(R), (2001).

[24] T. Ido and H. Katori. "Recoil-Free Spectroscopy of Neutral Sr Atoms in the Lamb-Dicke Regime", *Phys. Rev. Lett.*, **91**, pp. 053001-1–4, (2003).

[25] L. Hollberg, C. W. Oates, E. A. Curtis, E. N. Ivanov, S. A. Diddams, T. Udem, H. G. Robinson, J. C. Bergquist, R. J. Rafac, W. M. Itano, R. E. Drullinger, D. J. Wineland. "Optical Frequency Standards and Measurements", *IEEE J. Quantum Electron.*, **37**, pp. 1502–1513, (2001).

[26] W. M. Itano, J. C. Bergquist, J. J. Bollinger, J. M. Gilligan, D. J. Heinzen, F. L. Moore, M. G. Raizen, D. J. Wineland. "Quantum projection noise: Population fluctuations in two-level systems", *Phys. Rev. A*, **47**, pp. 3554–3570, (1993).

[27] B. C. Young, F. C. Cruz, W. M. Itano, J. C. Bergquist. "Visible Lasers with Subhertz Linewidths", *Phys. Rev. Lett.*, **82**, pp. 3799–3802, (1999).

A SINGLE INDIUM ION OPTICAL FREQUENCY STANDARD

PETRISSA ECKLE, MARIO EICHENSEER, ALEXANDER YU. NEVSKY,
CHRISTIAN SCHWEDES, JOACHIM VON ZANTHIER, HERBERT WALTHER
MPI für Quantenoptik and Sektion Physik der Ludwig-Maximilians-Universität München,
Hans-Kopfermann-Str.1, D-85748 Garching, Germany

A single indium ion stored in a radio-frequency trap and laser-cooled to a temperature below 1 mK is a promising candidate for an optical frequency standard of exceptionally high accuracy and stability. Probing the $5s^2\ ^1S_0$ - $5s5p\ ^3P_0$ transition of $^{115}In^+$ at $\lambda = 237$ nm with a natural linewidth of 0.8 Hz by a laser with a sub-Hertz linewidth can lead to an accuracy of 10^{-18} after several days of averaging. This high accuracy is possible since both states of the $^1S_0$ - $^3P_0$ transition have vanishing total electronic angular momenta; the transition is therefore free of the quadrupole shift whereas the second order Stark and Doppler shifts are at the mHz level at the temperatures to which the ion is cooled in the trap. The dominant perturbation of the transition results from a linear Zeeman shift which has been determined to 224 Hz/G for the $m_F = \pm 1/2$ $m_F = \pm 1/2$ Zeeman components; shielding the magnetic fields to within a few μG would thus lead to the anticipated 10^{-18} accuracy of the standard.
So far, a fractional resolution of 1.3 x 10^{-13} has been obtained, limited by the frequency fluctuations of the Nd:YAG laser exciting the $^1S_0$ - $^3P_0$ resonance. The frequency stability of this laser has been strongly improved recently. With a new laser setup based on a quasi-monolithic Nd:YAG ring laser, a laser linewidth < 4 Hz (FWHM) for integration times $\leq$ 26 s has been achieved. Detection and laser cooling of In^+ are performed using the $5s^2\ ^1S_0$ - $5s5p\ ^3P_1$ intercombination line at 231 nm. The dye laser formerly used for this purpose was recently replaced by a diode-based laser system allowing for long-term operation of the standard. Loading of the trap is now realized by resonant 2-photon ionization of indium at 410 nm. This strongly reduces deposition of indium on the trap electrodes so that contact potentials close to the trapped ion can be avoided.
The absolute frequency of the $^1S_0$ - $^3P_0$ transition has been measured with the frequency comb of a mode-locked femtosecond laser and a methane-stabilized He-Ne laser at 3.4 μm serving as a reference. The frequency was determined to 1 267 402 452 899.92 (0.23) kHz ($\Delta\nu / \nu = 1.8 \times 10^{-13}$), where the measurement uncertainty is dominated by the frequency uncertainty of the He-Ne laser. Further improvements in this measurement together with a comparison with other narrow atomic resonances would allow investigations of variations of fundamental constants in time.

STATUS REPORT OF THE AIUB-METAS GEODETIC TIME TRANSFER

R. Dach*, G. Dudle†, T. Schildknecht*, U. Hugentobler*

*Astronomical Institute, University of Bern, Sidlerstrasse 5. CH-3012 Bern, Switzerland
e-mail: rolf.dach@aiub.unibe.ch, phone: ++41–31–631 3802, fax: ++41–31–631 3869
†Swiss Federal Office of Metrology and Accreditation, Switzerland

Keywords: GPS carrier phase, Time transfer, Code/phase consistency, Continuous time series.

Abstract

The use of GPS Carrier Phase (CP) techniques is now widely accepted as an inexpensive and easy to use method to compare time and frequencies over long distances. A collaboration between the Astronomical Institute of the University of Bern (AIUB) and the Swiss Federal Office of Metrology and Accreditation (METAS) has produced a dedicated Geodetic Time Transfer Terminal (GeTT-Terminal) which was the subject of earlier communications.

Whereas in the past these terminals have been used as mobile units and deployed at various locations, a permanent station has now been set up at METAS and shall become a site of the International GPS Service (IGS) network. In parallel AIUB, responsible for the development of the processing software and the processing of the data, has improved the time transfer tools of its Bernese GPS Software package.

The focus of this paper is an efficient algorithm for the computation of a continuous time transfer series using the geodetic time transfer method. Based on the experience acquired over several years, special attention is paid to the long term stability of the analysis method.

1 Introduction

The Astronomical Institute of the University of Bern (AIUB) and the Swiss Federal Office of Metrology and Accreditation (METAS) have been collaborating for a long time in the field of time transfer by GPS Carrier Phase (CP) also known as geodetic time transfer. Until 2001 the focus of the collaboration was on the development of time transfer routines at the AIUB, whereas METAS was in charge of the assembly and test of the suitable hardware. Through several measurement campaigns, including a transatlantic experiment between the PTB (Braunschweig, Germany) and the USNO (Washington DC, USA) the potential of this method as tool for frequency transfer has been demonstrated [2].

In summer 2003, the collaboration between AIUB and METAS has been reactivated. Three main goals are pursued in the frame of this joint effort. First, METAS has decided to run a permanent station of the network of the International GPS Service (IGS). This step shall allow to compare the recently installed primary frequency standard FOCS-1 [6] via GPS CP with similar devices of other timing laboratories. To this end, a geodetic time transfer terminal (GeTT-terminal, description see [4]) has been installed in October 2003. Second, the collaboration shall allow to process the GPS CP data in a way that goes beyond the routinely generation of standard IGS products. And finally, it is planned to further improve the tools for time and frequency transfer in the Bernese GPS Software package [5].

The present paper concentrates on improvements on the processing software, implemented to address the well known problem of discontinuities at the day boundaries (see, e.g., [8]). Such discontinuities appear as a consequence of the noise in the code measurements when they are analysed together with the phase observations. New preprocessing features to detect and handle inconsistencies between code and phase data have been developed and implemented into the Bernese GPS Software. To test these new features experimentally, a specific campaign has been defined. The data cover a period of 150 days at the beginning of 2003. For every day, the preprocessing has been started with a set of about 70 IGS stations. The complete list of stations is given in the Appendix. Out of this set, a subset of roughly 40 stations was selected. Among the criteria for the selection was the global coverage, in order to be able to estimate also the state of the satellite clocks. 12 stations, located at timing laboratories or connected to hydrogen masers, were included in all daily solutions.

The paper is organized as follows. First, the basic equations and the corresponding notation will be introduced (Section 2). A description of identified causes for inconsistencies between code and phase measurements follows in Section 3. In the present state, the Bernese GPS Software is able to handle automatically most of these inconsistencies. Section 4 explains to what extend the remaining deficiencies represent a problem for continuous time transfer.

2 Observation equations

The basic GPS CP observation equation for code and phase measurements have the following form:

$$P_{fi}^k = |\vec{x}^k - \vec{x}_i| + \Delta_{trop} + \Delta_{ion} + c\delta_i - c\delta^k \quad (1)$$

$$L_{fi}^k = |\vec{x}^k - \vec{x}_i| + \Delta_{trop} - \Delta_{ion} + c\delta_i - c\delta^k + \lambda N_i^k \quad (2)$$

where

P_{fi}^k, L_{fi}^k	Code/phase observation of station i to satellite k on frequency f
$\vec{x}_i, \vec{x}^k$	Position vector of station i and satellite k respectively
Δ_{trop}	Signal delay in the troposphere
Δ_{ion}	Signal delay in the ionosphere
δ_i, δ^k	Clock of the receiver at the station i, transmitter of satellite k with respect to GPS time
c	Speed of light
N_i^k	Phase ambiguity
λ	Wavelength of the carrier phase

As can easily be seen in Equation 2, the phase ambiguities and the clock parameters have an one-to-one correlation. The difference, however, is that the phase ambiguities are constant over a certain time interval (e.g., for one complete satellite pass) and compensate therefore for the "mean reading" of the clocks (the mean offset of all estimated clock parameters for each receiver or each satellite clock). From the phase observations one deduces only the changes of the clock parameters $c\delta_i$ and $c\delta^k$ for all those epochs connected by the phase ambiguity parameters. To get access to the "mean reading" of the clocks, one needs to take into account the code observations as well.

As long as ambiguity parameters are connecting two epochs, the change of the clock values is available with the high precision of the phase measurements. However, if the ambiguity parameters for all satellites are lost at one epoch, the "mean reading" needs to be reestimated, adding thereby the corresponding uncertainty to the time and frequency transfer results.

For this reason, the setup of new ambiguity parameters for all satellites has to be avoided whenever possible. Fortunately, the loss of all ambiguity parameters at the same time – e.g., total loss of the signals by the receiver – is in practice very rare. Most of the discontinuities are in fact introduced artificially, mainly because individual data batches are processed independently.

In the ideal case all epochs of a computing batch (usually one day) are connected by phase ambiguities. All code observations of this time interval contribute to the "mean reading" of the clocks. From one day to the next, these parameters are newly estimated and due to the noisy code observations, discontinuities at the day boundaries of up to 1 ns are observed. This value could be decreased if longer processing batches (e.g., 3 days) were introduced.

3 Preprocessing: Code and phase consistency

Apart from the additional term for the phase ambiguity λN_i^k and from the opposite sign of the ionospheric delay Δ_{ion}, all terms are identical in the code and phase observation equations (Equations 1 and 2). Therefore, if measurements on the two frequencies are used to generate the ionosphere–free linear combination for code and phase (labeled P_{3i}^k and L_{3i}^k respectively), the change between epoch t_n and t_{n+1} is expected to be identical within the noise level mainly given by the code observations:

$$\left[P_{3i}^k(t_{n+1}) - P_{3i}^k(t_n)\right] - \left[L_{3i}^k(t_{n+1}) - L_{3i}^k(t_n)\right] = \Delta\delta_i^k(t_n) \quad (3)$$

If all the differences $\Delta\delta_i^k(t_n)$ are equal to zero for the observations from station i to all satellites in view ($k = 1 \ldots nSat$) all observations are consistent and can be accepted for the further analysis. Outliers or cycle slip cause $\Delta\delta_i^k(t_n) \neq 0$ for the affected observations whereas $\Delta\delta_i^k(t_n) = 0$ holds for the majority of the observations. The usual preprocessing tools can handle these data.

Of special interest in this paper are the cases where all differences $\Delta\delta_i^k(t_n)$ for a given station i are different from zero but identical for all satellites in view ($k = 1 \ldots nSat$). The consequence of the non-zero differences is that the receiver clock terms δ_i are not identical for code and phase measurements anymore (Equations 1 and 2) for all epochs $t \geq t_{n+1}$. This indicates an internal receiver clock event because any discontinuity of an external clock driving the GPS receiver has the same impact on code and phase data. The magnitude of the events is given by

$$\overline{\Delta\delta_i}(t_n) = \frac{1}{nSat} \sum_{k=1}^{nSat} \Delta\delta_i^k(t_n) \quad (4)$$

In order to overcome these receiver clock events, new phase ambiguities can be introduced for all satellites at epoch t_{n+1}. The inconsistency in the clock parameters may then be compensated by the new phase ambiguities. As explained earlier, the loss of the phase ambiguities for all satellites at one epoch may result in small discontinuities in the receiver clock series because of the independent "mean reading" of the clock estimated from the code data.

Example: Millisecond-jumps

If the exact size and nature of the receiver clock event is known, it can be accounted for in the preprocessing procedure avoiding thereby a discontinuity in the phase ambiguities. A well known example of known receiver

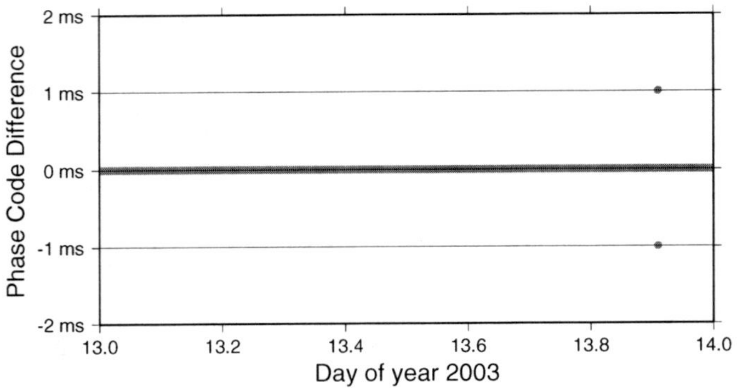

Figure 1: Differences between the changes from epoch to epoch for code and phase observations ($\overline{\Delta\delta_i}(t)$) for station USNO, day 13 of year 2003.

Epoch	Code-Phase	Remark
8:46:00	54 ns	
12:37:30	−38 μs	Different value for all satellites
12:48:30	−1.0 ms	
12:52:00	1.0 ms	
13:02:00	−1.4 ms	Only two satellites tracked

Table 1: Receiver clock events during tracking problems at station PTBB for day 148 of year 2003.

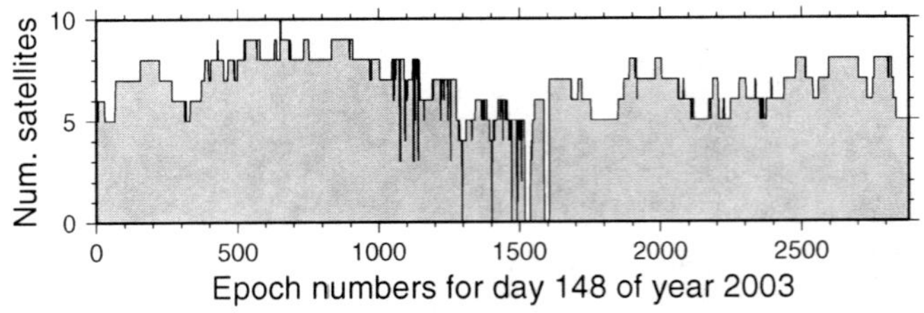

Figure 2: Number of observed satellites at station PTBB for day 148 of year 2003.

clock events are the millisecond jumps of the TRIMBLE receivers which are due to a reset of the internal receiver clock. Some of them only affect the code but not the phase data. In order to correct the phase measurements the values have to be shifted by $\overline{\Delta\delta_i}(t_n)$ which is an integer multiple of one millisecond. This type of event may also affect ASHTECH receivers.

An example of millisecond jumps is shown in Figure 1. In this graph the difference between the change in phase and code observations from epoch to epoch is plotted ($\overline{\Delta\delta_i}(t)$ in Equation 4) for the IGS-station USNO (Washington DC, USA, receiver ASHTECH Z-XII3T) for day 13 of year 2003. The noise level of 15 ns on $\overline{\Delta\delta_i}(t)$ achieved by this station is typical. Two one-millisecond-jumps can be identified in Figure 1.

An inspection of the RINEX observation file of this day shows that one observation epoch is assigned to 21:49:29.999, all other observations being assigned to the nominal measurement epoch (hh:mm:00.000 or hh:mm:30.000). The code observations correspond to the special observation epoch. They are shifted by about 300 km with respect to the measurements assigned to the nominal epochs. The phase observations do not follow this shift. They are assigned to an observation epoch of 21:49:30.000 in fact. The common analysis of code and phase data for this observation file fails because the receiver clock parameter δ_i is different for both observation types. If the phase data of the particular epoch are corrected by one millisecond, the data analysis works smoothly and the receiver clock estimate gives a perfect result for all epochs.

In this example the inconsistency did disappear after one epoch. In other examples it remains for several epochs. It is important to note that the inconsistency does not disappear in all cases. The millisecond-jumps detected and repaired in the analysis test campaign are listed in the Appendix.

In addition, ASHTECH Z-XII3 show also jumps of ± 28.75 ms and ± 57.50 ms. They are mostly found for receivers driven by their internal clock. The phase measurements are not corrected because the exact size of these events is currently not known. It can be estimated from the inconsistencies detected in the analysis campaign to $\overline{\Delta\delta_i} = 28750503.1$ ns with an uncertainty of 2.7 ns. This estimate is affected by the uncertainty of the code measurements.

Example: Tracking problems

Inconsistencies between code and phase data with an arbitrary magnitude are often accompanied by complete loss of tracking satellites or by incomplete data sets (code and phase data are not available for both frequencies for all or at least a few satellites). If such events occur repeatedly each day, environmental effects are a likely source of the problem. In contrary, isolated tracking losses are presumably due to clock or internal receiver problems. The reason for this behaviour is not fully understood. It has to be noted that not all tracking problems fall into this category. If only one satellite is lost (e.g., reduced visibility) the clock solution will not be affected.

As an illustration Table 1 summarises receiver clock events for the IGS-receiver PTBB (Braunschweig, Germany, equipped with an ASHTECH Z-XII3T receiver) for day 148 of year 2003. These events are accompanied by the loss of satellite tracking (see Figure 2). Between 08:45 and 10:00 there are several epochs where the tracking of about half of the satellites is lost. Between 10:50 and 13:20 some epochs are completely missing in the observation file. For the period of about one hour within these tracking problems the resulting time transfer series shows a shift of about 50 ns (e.g., between PTBB and NPLD – Teddington, UK, equipped with ASHTECH Z-XII3T receiver – in Figure 3). In the case of the PTBB receiver, the loss of tracking can be observed only for one day during the analysis campaign. It is, therefore, most probably not originated from environmental effects.

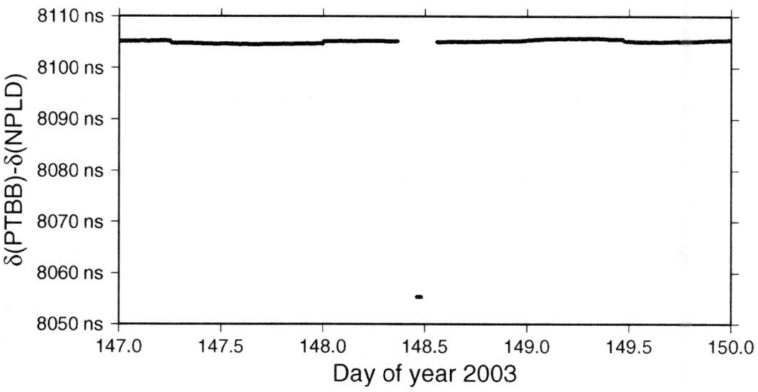

Figure 3: Receiver clock differences between PTBB and NPLD for days 147 to 149 of year 2003.

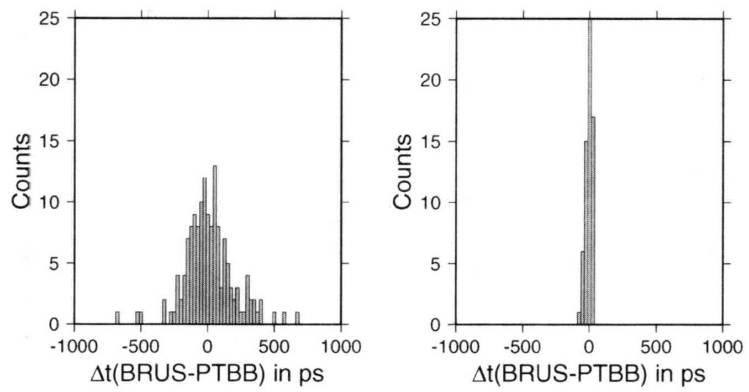

Figure 4: Histograms of the magnitude of the day boundary discontinuities for the comparison of the receiver clocks of the IGS sites PTBB and BRUS (left: independent daily solutions, right: continuous time transfer solution with parameter hand-over).

The receiver clock estimates before and after the events cannot be connected by the phase ambiguities and the clock solution will be affected by the uncertainty of an independent estimate of the "mean reading" from the code data (see Section 2).
The experience gained at the Center for Orbit Determination in Europe (CODE), one of the global analysis centres of IGS, indicates that ASHTECH Z-XII3T receivers operated permanently at time laboratories show usually at maximum two days per year with loss of tracking affecting the receiver clock solution.

Other inconsistency events

The noise level of the code measurement puts a limit on the inconsistencies between code and phase measurements detectable with the simple algorithm of Equation 3. The exact value of the noise level depends on the receiver type and the antenna environment. Values above typically 15...20 ns can be detected. To identify code/phase inconsistencies below this value, the post-fit residuals of a common analysis of the code and the phase data is necessary. An example will be discussed in Section 4.

4 Continuous time transfer

As mentioned previously, the data are normally processed in daily batches, introducing thereby artificially a discontinuity of the phase ambiguities for all satellites at the day boundaries. To overcome this, the ambiguity parameters should be connected between the independent computation batches. This is mathematically correct and technically possible but requires advanced bookkeeping of an important number of parameters to be transferred from one computation batch to the next (one parameter for each satellite observed by each station).
If the computation batches span from midnight to midnight, including the boundaries at both ends, there is one set of clock parameters that can be used to connect the computation batches in order to obtain a continuous time transfer series. In addition, the station coordinates, satellite orbits, Earth orientation parameters, and the troposphere estimates have to be continuously modelled (only very small discontinuities can be accepted). Since this approach relies on the fact that the midnight epoch connects two batches, it is essential that this particular epoch is recorded. Unfortunately this is not the case for all stations and as a result, the time transfer series for those sites are not connected. Alternatively, another epoch for the clock hand-over could also be used, but for this particular study this has not been implemented.
Independently from the method that connects batches, any approach for a continuous solution requires a better modeling of all effects for which an error could accumulate over time. In addition, any problem with the raw data that is not detected and correctly handled in the preprocessing part, affects not only the batch in which it occurs but the entire solution.
To illustrate the result of a continuous time transfer using the parameter hand-over described above, the receiver clock values of the two IGS sites PTBB and BRUS (Brussels, Belgium) are selected from the global test analysis campaign. Both stations are equipped with ASHTECH Z-XII3T receivers, connected to H-masers.
In a first step, daily solutions spanning from midnight to midnight are computed. The boundary epochs are thus included in two consecutive processing batches and two independent time transfer results are computed for this particular epoch. Figure 4 represents a histogram of the magnitude of the differences between the two consecutive batches. The histogram follows a normal distribution centered at zero. The standard deviation for this example corresponds to the results presented by other groups (e.g., [8]). Similar results can be obtained for other baselines.
In a second step, a continuous time transfer solution using the parameter hand-over method was elaborated. The right part of Figure 4 shows the histogram for the same observable as above. As expected because of the parameter hand-over the continuity of the new solution results in a very narrow histogram. It has to be noted that the

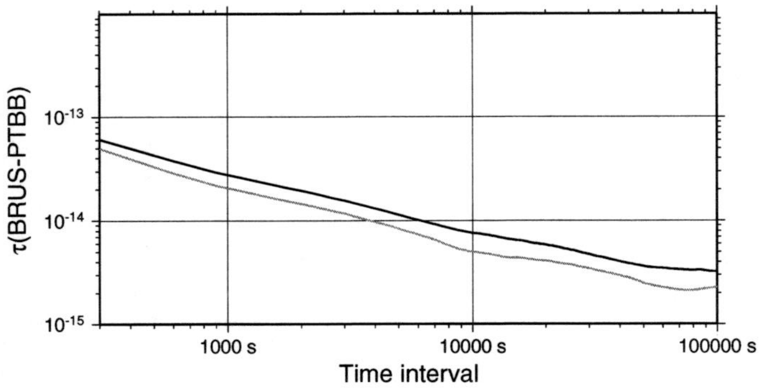

Figure 5: Allan deviation for the comparison of the receiver clocks of the IGS sites PTBB and BRUS (dark line: daily independent solutions, gray line: continuous time transfer solution with parameter hand-over).

method has also worked properly on other, longer baselines and even on intercontinental links.

Figure 5 displays the Allan deviation for the two solutions of time transfer between BRUS and PTBB. As expected, the continuous solution has a slightly better stability, since no discontinuities remain at the day boundaries.

Comparing the time transfer results from the GPS CP to UTC for three exemplary baselines in Table 2 no significant difference between the continuous solution using parameter hand-over and the daily independent solution can be found. This holds for the mean difference as well as for its standard deviation. Because UTC values are published with a sampling of 5 days only in the Circular T [1] a big number of days need to be concatenated in order to obtain a significant improvement of these statistical numbers.

On the other hand the result documents that the continuous time transfer results does not drift away. A mismodelling in the analysis procedure resulting in a drift of the daily clock estimates can be compensated at the day boundaries if daily independent solutions are generated. In the case of a continuous solution using a parameter hand-over strategy daily drifts can accumulate. That the continuous time transfer results have a quality comparable to the daily independent solutions shows that such a drift does not occur.

The offsets of the time transfer solutions to UTC can be compared, e.g., with the offsets published in [9]. They are

BRUS to	Num.	Daily indep. solution	Continuous solution	Comparison with [9]
PTBB	31	24.9 ± 0.9	24.9 ± 0.9	
USNO	29	-56.0 ± 2.0	-56.2 ± 2.0	-54.8
AMC2	27	580.7 ± 2.0	580.2 ± 2.0	581.8

Table 2: Arithmetic mean and its standard deviation for the difference between the results of the analysis campaign and UTC for an exemplary set of baselines. For comparison analogue results published in [9] are included. The values are given in nanoseconds.

derived from the combined final IGS clock products. Several independent analysis groups contributed to this solution which is based on a more stable network with more stations than in this analysis test campaign. Furthermore a longer time interval was analysed. Nevertheless the values agree very well.

An additional aspect is addressed in Figure 6. The receiver clock differences between the stations PTBB and USNO are displayed for three days (day 97 to 99 of year 2003) out of the complete time series of 150 days. Up to day 98, the difference between the continuous time transfer result (gray line) and the solution of the daily batches (black line) is well below 1 ns. But, starting on day 99, the difference increases to more than 2 ns, which is more than can be attributed to the noise of the code data used for the computation of the "mean reading" of the clocks.

Figure 7 displays a code-only-solution for clocks for the same baseline. Not surprisingly, the noise is much larger than for the geodetic solution using code and phase measurements. At the epoch 17:10 of day 98 of year 2003, the code-only-solution exhibits a jump of about 2 ns whereas the combined code and phase analysis has no discontinuity at this particular epoch.

The receiver clock difference between PTBB and USN1 (another receiver at the US Naval Observatory near to the receiver USNO) does not show a similar discontinuity at the same epoch, neither in the code-only solution nor in the code and phase combined results. The two receivers are connected to two different clocks but a discontinuity in the steering clocks can be ruled out since no anomaly can be found in published differences of the clocks [10]. The GPS receiver USNO must thus be responsible for the jump.

The discontinuity observed at 17:10 of day 98 in the code observations did not affect the phase measurements. As a consequence the clock parameters δ_i in Equations 1 and 2 are not the same for epochs after the event. If a new set of phase ambiguities is introduced for all satellites at the particular epoch, the consistency between the clock parameters derived from code and phase can be recovered. It is important to note that the inconsistency discussed here are very different from those described in Section 3. The fact that the code is the main affected observable has three major consequences:

1. A calibration for time transfer determined before the event will not be valid after.
2. The time transfer results for the clock connected to USNO cannot be used for frequency transfer over this receiver clock event because code and phase data have to be used in a combined analysis.
3. Frequency transfer using a phase-only solution is possible over this particular epoch assuming that the receiver clock event does not have any influence on the phase observations.

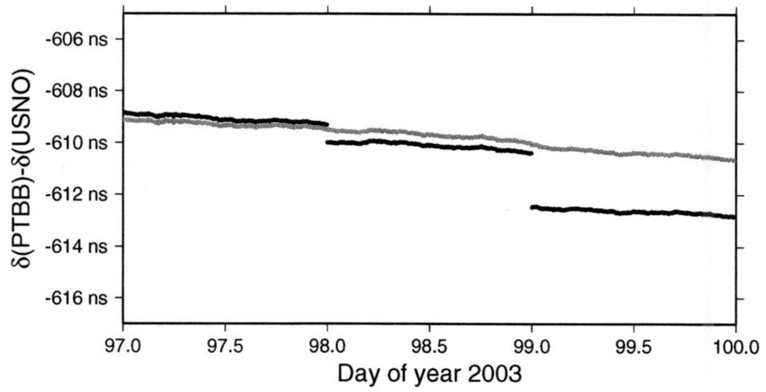

Figure 6: GPS receiver clock solution for PTBB and USNO for three days of year 2003. The dark line is the result of the daily independent solution whereas the gray line results from the continuous time transfer solution using the parameter hand-over. Code and phase measurements are used in a common analysis.

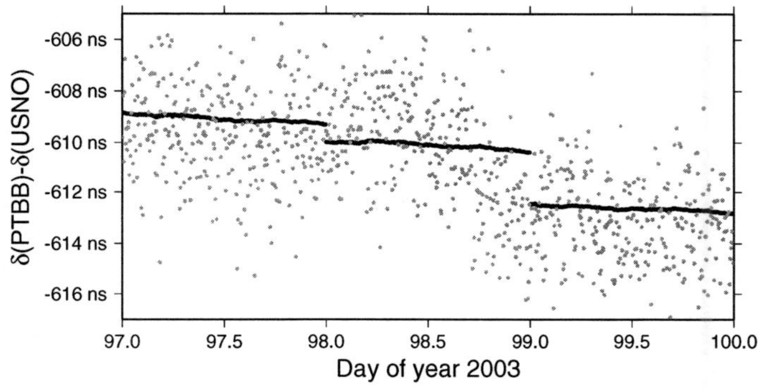

Figure 7: GPS receiver clock solution for PTBB and USNO for three days in year 2003. The dark line is the result of daily independent solutions using code and phase measurements in a common analysis. The dots are the result from a code-only solution for the clocks.

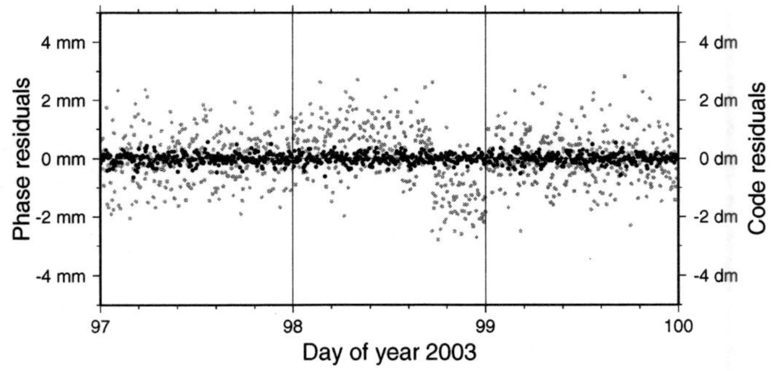

Figure 8: Code (gray dots) and phase (dark dots) residuals for the IGS site USNO for the days 97 to 99 of the year 2003. The mean residuals of each epoch for all observed satellites are plotted.

Such receiver clock events need to be detected to make the time and frequency transfer results useful. With Figure 7 it is quite clear that the receiver clock estimates obtained from the code-only solution are too noisy to detect such events reliably. Furthermore, if solely a code-only solution is available there is no way to distinguish between a step of the driving clock or an internal problem of the GPS receiver.

The most promising approach to identify such events is to check of the post-fit residuals of a combined code and phase solution. The mean residuals for all satellites of an epoch observed from station USNO for the days 97 to 99 of year 2003 are plotted separately for code and phase measurements in Figure 8. The values are expected to follow a normal distribution around zero for code and phase data. This can be seen for days 97 and 99 in Figure 8. But if the code residuals before a particular epoch are dominated by positive and after that epoch by negative values (like for day 98 in Figure 8) or vice versa, an inconsistency between the code and phase measurements may be the cause.

Comparing the signature of the event in Figure 7 and Figure 8 it seems easier to detect the event using the mean post-fit residuals of each epoch. A pattern in the code residuals can also be an indicator for multipath effects. Multipath is induced by environmental effects and the pattern in the post-fit residuals usually repeats each day as long as the environmental source of the multipath does not change. Judging from Figure 8 this is very unlikely for USNO.

To develop and implement an algorithm to distinguish reliably between a receiver clock event and multipath effects in the post-fit residuals is mandatory for an automatic processing scheme. The difficulty of this task is evident from the example for USNO illustrated in Figure 8. A continuous analysis helps any algorithm to identify events if they occur close to the boundary of a computation batch.

5 Summary and conclusions

The generation of a continuous time transfer series with parameter hand-over is an efficient method to overcome the discontinuities between daily independent solutions. The daily solutions are connected during the analysis process taking all the parameters (satellite orbits, Earth orientation parameters, station coordinates, and troposphere estimates) and their covariances into account. This is the advantage with respect to algorithms applied to the results after the processing (e.g., [3]).

The disadvantage of a continuous time transfer series is that any data or analysis problem affects the complete solution. Therefore, an improved preprocessing algorithm was implemented. The consistency of the code and phase measurements is checked in order to detect receiver clock events. If the exact size and the nature of these events is known the data can be corrected (e.g., millisecond-jumps) and their influence on the time transfer results can be elim-

inated. Otherwise new phase ambiguities for all satellites are introduced to compensate for the inconsistencies in the measurements for the time transfer solution. The drawback is that the uncertainty of the time and frequency transfer series for epochs before and after such an event is affected by the mean noise of the code data.
To check that the calibration of a GPS receiver for time transfer purposes (presented, e.g., by [7]) is valid before and after a receiver clock event a second geodetic GPS receiver may be connected to the same clock. The alternative receiver can be used to estimate the values for the common clock. The time transfer results from both receivers can be compared in order to investigate the influence of the receiver clock events on the calibration of the receivers.

Acknowledgements

The AIUB hosts the Center for Orbit Determination in Europe (CODE), which is a collaboration between the Astronomical Institute at the University of Bern (AIUB), the Federal Office of Topography (Swisstopo, Switzerland), the Federal Agency for Cartography and Geodesy (BKG, Germany), and the Institut Géographique National (IGN, France). The results presented in this paper benefit from the work of the IGS analysis centre CODE.

References

[1] Bureau International des Poids et Mesures, Pavillon de Breteuil, F-92312 Sèvres Cedex, France, *Circular T.*

[2] R. Dach, T. Schildknecht, T. Springer, G. Dudle, and L. Prost, "A Transatlantic GeTT Time Transfer Experiment – Latest Results," in *Proceedings of the* 32st *Precise Time and Time Interval Systems and Application Meeting PTTI 00*, (Reston Virginia), November 28–30 2000.

[3] R. Dach, T. Schildknecht, T. Springer, G. Dudle, and L. Prost, "Continuous Time Transfer Using GPS Carrier Phase," *IEEE Transactions on Ultrasonics, Ferroelectrics, and Frequency Control*, vol. 49(11), pp. 1480–1490, 2002.

[4] G. Dudle, F. Overney, L. Prost, T. Schildknecht, and T. Springer, "First Results on a Transatlantic Time and Frequency Transfer by GPS Carrier Phase," in *Proceedings of the* 30th *Precise Time and Time Interval Systems and Application Meeting PTTI 98*, (Reston, Virginia), December 1–3 1998.

[5] U. Hugentobler, S. Schaer, and P. Fridez, eds., *The Bernese GPS Software Version 4.2.* University of Bern: Astronomical Institute, February 2001.

[6] A. Joyet, G. Mileti, P. Thomann, and G. Dudle, "Continuous Fountain Cs Standard: Stability and Accuracy Issues," in *Proceedings of the 6th Symposium on Frequency Standards and Metrology* (P. Gill, ed.), pp. 273–280, 2002.

[7] G. Petit, Z. Jiang, J. White, R. Beard, and E. Powers, "Absolute Calibration of an Ashtech Z12-T GPS receiver," *GPS Solutions*, vol. 4(4), pp. 41–46, 2001.

[8] J. Ray and K. Senior, "IGS/BIPM pilot project: GPS carrier phase for time/frequency transfer and timescale formation," *Metrologica*, vol. 40, pp. 270–288, 2003.

[9] K. Senior and J. Ray, "Improved IGS Time Scale Stability and Tracking of UTC," Presented at the *IGS workshop and symposium*, (Bern, Switzerland), March 1–5 2004.

[10] `URL:ftp://tycho.usno.navy.mil/pub/usnodc-amc.gpscp/mc2-mc3.rtcorr.cal.`

Appendix

List of stations used in the analysis test campaign

The receiver type and the frequency base were compiled from the IGS station description files `http://igscb.jpl.nasa.gov/igscb/station/log/`. The last columns compile the number of receiver clock events detected during the preprocessing:

total: total number of days where the stations were included in the preprocessing procedure
jump: number of days where millisecond-jumps were detected and the phase data were adapted
tracking: number of days with tracking problems at the particular station
ashtech: number of days with specific ASHTECH-receiver code/phase inconsistencies (magnitude of ±28.75 ms and multiple)
other: number of days with other code/phase inconsistencies with magnitudes of >20 ns

Station ID, location	Receiver	Frequency base	total	Number of days jump	tracking	ashtech	other
ALBH : Victoria, Canada	AOA BENCHMARK ACT	H-maser	68				
ALGO : Algonquin Park, Canada	AOA BENCHMARK ACT	H-maser	68		1		
ALIC : Alice Springs, Australia	AOA ICS-4000Z ACT	internal	68				1
AMC2 : Colorado Springs, USA	ASHTECH Z-XII3T	H-maser	145		3		
ANKR : Ankara, Turkey	AOA SNR-8000 ACT	internal	119				
AREQ : Arequipa, Peru	ASHTECH UZ-12	rubidium	156			1	
ASC1 : Ascension Island	AOA SNR-8000 ACT	internal	138				
AUCK : Whangaparaoa Peninsula, New Zealand	ASHTECH Z-XII3	internal	156				
BAHR : Manama, Bahrain	ASHTECH Z-XII3	cesium	156	1			1
BOGT : Bogota, Colombia	ASHTECH Z-XII3	internal	156			31	
BRUS : Brussels, Belgium	ASHTECH Z-XII3T	H-maser	156				4
CAS1 : Casey, Antarctica	AOA ICS-4000Z ACT	internal	156				
CEDU : Ceduna, Australia	AOA ICS-4000Z ACT	internal	155		2		2
CHAT : Waitangi, New Zealand	ASHTECH Z-XII3	internal	156				
COCO : Cocos (Keeling) Island	AOA SNR-12 ACT	internal	155				
CORD : Cordoba, Argentina	ROGUE SNR-8000	internal	80				
DARW: Darwin, Australia	ASHTECH UZ-12	rubidium	136	1	8		4
DAV1 : Davis, Antarctica	AOA ICS-4000Z ACT	internal	152				3
DRAO : Penticton, Canada	AOA BENCHMARK ACT	H-maser	156				
DWH1 : Woodinville, USA	JPS LEGACY	cesium	144				
FLIN : CFS Flin Flon, Canada	AOA BENCHMARK ACT	internal	156		6		5
GLPS : Puerto Ayora, Ecuador	ASHTECH Z-XII3	internal	64			3	
GODE : Greenbelt, USA	AOA SNR-8000 ACT	H-maser	156				1
GOLD : Goldstone, USA	ASHTECH Z-XII3	internal	156			8	2
HARB : Pretoria, South Africa	TRIMBLE 4000SSI	cesium	156				1
HOB2 : Hobart, Australia	AOA ICS-4000Z ACT	H-maser	155				2
HRAO : Krugersdorp, South Africa	ASHTECH Z-XII3	internal	156			2	
HYDE : Hyderabad, India	ASHTECH UZ-12	unknown	142				
IISC : Bangalore, India	ASHTECH Z-XII3	internal	146			17	
INVK : Inuvik, Canada	ASHTECH UZ-12	internal	153	1			
KARR : Karratha, Australia	AOA ICS-4000Z ACT	internal	149				
KOSG : Kootwijk, Netherlands	AOA SNR-12 ACT	quartz	154				
MAC1 : MacQuarie Island, Southern Ocean	AOA ICS-4000Z ACT	internal	150				
MAG0 : Magadan, Russia	ASHTECH Z-XII3	internal	156	2			
MAS1 : Maspalomas, Spain	ASHTECH Z-XII3	cesium	156		2		2
MAW1 : Mawson, Antarctica	AOA ICS-4000Z ACT	internal	146				4
MKEA: Mauna Kea, USA	ASHTECH Z-XII3	internal	156			2	
NPLD : Teddington, UK	ASHTECH Z-XII3T	H-maser	129		1		
NRC1 : Ottawa, Canada	AOA SNR-12 ACT	H-maser	156				
NRIL : Norilsk, Russia	ASHTECH Z-XII3	internal	156				
NTUS : Singapore	AOA SNR-12 ACT	internal	154				
NYA1 : Ny-Alesund, Norway	AOA BENCHMARK ACT	H-maser	155				1
OHI2 : O'Higgins, Antarctica	AOA SNR-8000 ACT	internal	147				
ONSA : Onsala, Sweden	ASHTECH Z-XII3	H-maser	156				
PERT : Perth, Australia	ASHTECH Z-XII3	cesium	123	24[a]			
PTBB : Braunschweig, Germany	ASHTECH Z-XII3T	cesium	156		1		4
SFER : San Fernando, Spain	TRIMBLE 4000SSE	cesium	151		4		
SUTH : Sutherland, South Africa	ASHTECH UZ-12	internal	156				
THU3 : Thule airbase, Greenland	ASHTECH UZ-12	rubidium	156				
TIDB : Tidbinbilla, Australia	ASHTECH Z-XII3	H-maser	156				
TIXI : Tixi, Russia	ASHTECH Z-XII3	internal	142	3			
TOW2 : Cape Ferguson, Australia	ASHTECH UZ-12	rubidium	156				
TRO1 : Tromsoe, Norway	AOA BENCHMARK ACT	internal	152				
TSKB : Tsukuba, Japan	AOA BENCHMARK ACT	cesium	156				
TWTF: Taoyuan, China	ASHTECH Z-XII3T	cesium	150	7			1
USNO : Washington, USA	ASHTECH Z-XII3T	H-maser	149	1			1
VILL : Villafranca, Spain	ASHTECH Z-XII3	cesium	156	7[b]			
WILL : Williams Lake, Canada	AOA BENCHMARK ACT	cesium	138				
WSRT : Westerbork, Netherlands	AOA SNR-12 ACT	H-maser	155				
WTZR: Koetzting, Germany	AOA SNR-8000 ACT	H-maser	156				
YAKT : Yakutsk, Russia	ASHTECH Z-XII3	cesium	154				
YAR2 : Dongara, Australia	AOA ICS-4000Z ACT	rubidium	155				
YSSK : Yuzhno-Sakhalinsk, Russia	ASHTECH Z-XII3	internal	155	3			
ZAMB : Lusaka, Zambia	ROGUE SNR-8000	internal	138				

[a] more than 100 events at each of days 73 to 90 of year 2003
[b] more than 50 events per day for days 115 to 119 of year 2003

COMPARISON OF INSTRUMENTAL AND EMPIRICAL STATION TIMING BIASES FOR A SET OF ASHTECH Z12T GPS RECEIVERS

K. Senior [1], J. Ray [2,3], G. Petit [3]

[1] Naval Research Laboratory, Code 8154, 4555 Overlook Ave. SW, Washington, DC, USA, Ken.Senior@nrl.navy.mil
[2] National Geodetic Survey, Silver Spring, MD, USA
[3] Bureau International des Poids et Mesures, Sèvres, France

Keywords: GPS, calibration, time transfer, Ashtech Z12T, TAI.

Abstract

Station timing biases have been determined by two different methods for a set of Ashtech Z12T geodetic GPS receivers located at timing laboratories. The timing biases account for the internal delays within the geodetic GPS receiver and antenna hardware, including the combined effects at both the L1 and L2 frequencies, as well as the intra-laboratory offset to the local source of UTC(k). The direct method uses a standard Z12T unit from the Bureau International des Poids et Mesures (BIPM), which was previously calibrated using absolute methods with an estimated accuracy of ~3.5 ns. It is deployed to differentially calibrate laboratory receivers, each measurement having an estimated uncertainty of ~1.6 ns. The second, empirical method relies upon mutual observations of GPS time, as reported in BIPM's monthly Circular T and as observed by the International GPS Service (IGS), to extend the previously determined calibration bias of each lab's main TAI time transfer system differentially to the collocated geodetic receiver. For those stations where the geodetic system uses a frequency standard closely related to the local UTC(k) realization, we find calibration biases can be determined with repeatabilities in the range of one to a few ns. This method allows station timing biases to be monitored continuously, in near real-time, without disruption or need of any special local measurements. Results using both methods have been compared for several timing labs. We find differences of a few ns, except in two cases where there are questions related to local measurement procedures, which is roughly consistent with the overall error budgets for the calibration methods. In an operational time transfer setting, both methods can be usefully applied in complementary ways.

1 Introduction

We have previously demonstrated a method to determine empirical station timing biases for geodetic GPS receivers located at timing laboratories [8]. Mutual observations of GPS time, as reported by the BIPM in its Circular T [1] and as observed by the IGS, are used together with the Circular T offsets of the lab timescales [UTC – UTC(k)]. This method presupposes the presence of a previously calibrated time transfer system at the lab and effectively transfers that calibration differentially to the collocated geodetic receiver system without need of any additional local measurements. The empirical biases include both the internal delays within the geodetic GPS receiver and antenna hardware as well as the intra-laboratory offset to the source of UTC(k).

When applied to the subset of IGS stations located at timing labs, we previously found that useful results can be obtained provided that the geodetic system uses a frequency standard closely related to the local UTC(k) realization and that the GPS receiver does not reset its internal clock too frequently. Daily RMS repeatabilities could be observed in some cases of about 1 to 2 ns over periods up to at least a year. Given sufficient available data, the mean station bias can be estimated from a single monthly Circular T update to about the 1 ns level, which is smaller than the uncertainty in current absolute calibrations and comparable to that of other differential methods. This new method can be used, among other things, to validate laboratory calibration measurements for the geodetic systems and to continuously monitor intra-laboratory calibration stability without disrupting normal operations. It can also be exploited to improve the near real-time steering of the IGS timescale to track UTC more closely.

The direct differential calibration method uses an absolutely calibrated Ashtech Z12T receiver as a travelling standard. When data are collected simultaneously with a collocated test receiver the calibration bias of the test receiver can be determined relative to the standard Z12T [4]. Since January 2001, BIPM has circulated a Z12T calibration unit that visits participating labs roughly every year or two to differentially calibrate local receivers. The unit periodically returns to the BIPM for long-term stability checks. Each differential calibration measurement has an estimated uncertainty of ~1.6 ns. The procedure to calibrate the standard Z12T in an absolute sense was conducted at the Naval Research Laboratory (NRL) in 2000 and has been described by [5]; its accuracy is estimated to be ~3.5 ns.

In this paper, we report updated values for the empirical timing biases of those IGS stations equipped with Z12T receivers and collocated at timing labs. These empirical biases are then compared with matching determinations from direct differential calibrations by the BIPM.

2 Empirical calibration methodology

The methodology previously described by [8] to estimate empirical calibrations is first reviewed for completeness and to correct an earlier oversight in a neglected correction term.

2.1 Estimation of timing biases

Consider a geodetic GPS receiver located at a timing lab that contributes regular results to the BIPM and maintains a local realization of UTC, denoted UTC(k). Time transfers to the BIPM currently rely mostly on the common-view GPS method (using single-frequency data and coordinated observing schedules to permit common-mode cancellation of many geodetic effects) or two-way satellite time transfers (TWSTT). Dual-frequency geodetic receivers also operate at numerous timing labs and contribute data to the IGS, but are not yet used in a geodetic mode for time transfer operations. In some cases the geodetic GPS receivers are beginning to be used by the BIPM in a common-view mode that takes advantage of the dual-frequency pseudorange data, a method known as P3 common view [2].

If the geodetic receiver is driven with an external frequency standard closely related to UTC(k), then we can consider the timing bias B_k defined by

$$B_k = CLK(k) - UTC(k) \tag{1}$$

where CLK(k) is the time series of clock readings as observed using the geodetic GPS system. Clock results are obtained through a complete geodetic solution consistent with the standards and conventions used by the IGS. The accuracy of such geodetic clock determinations ranges widely, from the formal error level of about 0.12 ns for 24-hour analysis arcs up to nearly 1 ns, depending on the particular station configuration [7]. Pseudorange multipath is usually the leading source of geodetic clock error although other effects, including temperature-sensitive components and hardware problems, can also be significant.

The biases B_k include both the internal delays within the GPS receiver and antenna hardware producing CLK(k) as well as the intra-laboratory offset between the receiver clock circuit and the source of UTC(k). Direct absolute measurements of B_k are non-trivial and have typical uncertainties of several ns [5]. Alternatively, we can empirically estimate a closely related bias, B_k', for those IGS stations with all the following information available:

$$B_k' = [CLK(k) - GPST]_{IGS} - [UTC - GPST]_T + [UTC - UTC(k)]_T \tag{2}$$

where $[CLK(k) - GPST]_{IGS}$ are observed values from the IGS clock products, aligned and referenced nominally to GPS time, $[UTC - GPST]_T$ are daily monitor values for GPS time published in the BIPM Circular T [1], and $[UTC - UTC(k)]_T$ are the five-day values for the offsets of the lab k time UTC(k) from UTC as published in Circular T.

By selecting the IGS clock values at or near the midnight epochs used in Circular T and by interpolating (with a cubic spline) the [UTC – UTC(k)] time series, we can compute B_k' values at nominally daily intervals. Equation (2) reduces to

$$B_k' = [CLK(k) - UTC(k)] + [GPST_T - GPST_{IGS}] \tag{3}$$
$$= B_k + \Delta GPST \tag{4}$$

which is equivalent to B_k in Equation (1) if the GPS times observed and reported by Circular T and the IGS are identical. Of course, the empirical estimates of B_k' will only be meaningful if B_k is indeed approximately constant. If, for instance, the timing relationship between the geodetic receiver, CLK(k), and UTC(k) is not fixed or if the TAI calibration of UTC(k) is unstable, then estimates of B_k' will vary and not be useful.

2.2 Corrections for GPS time differences

Considering then only those installations with stable internal configurations, it is necessary to evaluate whether the magnitudes of the differences in observed GPS time, ΔGPST(t), are significant. In fact, differences exist in the IGS and BIPM methods. The BIPM follows the GPS Interface Control Document (ICD), which describes the use of the broadcast navigation message, whereas the IGS produces its own products using its own models and conventions. This gives rise to slowly-varing, systematic differences. In addition, the IGS and BIPM monitor GPS time from very different perspectives and may use different subsets of the full constellation. The BIPM observes the satellites only while they are visible above Paris whereas the global IGS tracking network allows the full constellation to be continuously in view. This causes smaller day-to-day differences. On infrequent occasions, the IGS linear alignment of its clocks to GPS time has malfunctioned in its Rapid products (those issued ~17 hours after each observed day). This effect compounds the quasi-random differences.

There are two effects that dominate the slowly-varying systematic GPS time differences. 1. The BIPM refers its observations of the satellite clocks to the GPS system values (approximately) for the transmitter antenna to spacecraft center-of-mass offsets, while the IGS uses its own set of satellite offsets. The differences are negligible for the older, more numerous Block II/IIA spacecraft, but exceed 1 m for the newer Block IIR satellites. 2. The BIPM single-frequency data apply the broadcast group delay correction, T_{GD}, to compensate for the inter-frequency pseudorange bias inherent in the broadcast clocks, as recommended in the GPS ICD. The average T_{GD} correction over the full constellation is roughly –5.5 ns. The IGS, on the other hand, uses a datum convention that sets the average differential code bias (DCB) to zero.

Taking these considerations into account, we can then regard ΔGPST(t) as consisting of three components:

$$\Delta GPST = \Delta_{ANT} + \Delta_{DCB} + \delta \tag{5}$$

The first two terms vary slowly as the constellation evolves and can be well determined from known system parameters. The last, quasi-random term accounts for the day-to-day variations in specific satellite usage and similar effects (e.g., possible drifts in the calibration of the Paris GPS receiver used by the BIPM). The GPS time difference due to differing antenna offsets is given by

$$\Delta_{ANT} = \sum [dZ_j(IGS) - dZ_j(BIPM)] / (c * N) \quad (6)$$

where the sum runs over each satellite j in the constellation of N spacecraft, c is the speed of light, and dZ_j are the respective center-of-mass to antenna phase center offsets, in the Earth direction, used by the IGS and BIPM. For the IGS, dZ_j are 1.023 m for the Block II/IIA satellites and 0.0 m for the newer IIR satellites. The BIPM used 1.03 m for all satellites until 1 January 2001 when a value of 1.17 m was adopted for the IIRs. Consequently, based on the actual numbers of satellites of each block aloft at any given time, the Δ_{ANT} correction term has gradually decreased from -0.15 ns in early 1999 to -1.41 ns currently.

The GPS time difference due to different pseudorange bias conventions is simply

$$\Delta_{DCB} = - \sum T_{GD}(j) / N \quad (7)$$

since the average is zero by design for the IGS. In terms of the equivalent differential code bias (DCB) values used by the IGS and others, note that

$$DCB_j = [1 - (77/60)^2] * T_{GD}(j) \quad (8)$$

or

$$T_{GD}(j) = -1.54573 * DCB_j \quad (9)$$

Monthly values for the broadcast DCB values have been archived by S. Schaer (Astronomical Institute of the University of Berne; see website at http://www.aiub.unibe.ch/ionosphere.html); those have been used to evaluate Equation (7). Since the start of 2000, the Δ_{DCB} correction term has gradually increased from 4.30 ns to 6.70 ns currently. Taken together with the Δ_{ANT} contribution, the trends partially offset one another so that the total systematic corrections to ΔGPST have only ranged over 1.14 ns during the past four years, from 4.15 ns to 5.29 ns.

Other differences in the two methods of measuring GPS time do not contribute significant systematic components to ΔGPST. For instance, the BIPM uses L1 C/A pseudorange data while the IGS uses P1 and P2 codeless pseudoranges. Due to distinct delays in the signal processing hardware in the transmit systems, small satellite-specific biases occur between the P1 and C/A observables which can manifest themselves in receiver-specific tracking variations [6]. However, this has no net effect on ΔGPST because the (P1 - C/A) bias convention adopted by the IGS has zero mean over the full constellation.

The effects of other procedural differences between the IGS and BIPM methods will mainly cause quasi-random variability on short time scales. These can be compensated by estimation from the $B_k'(t)$ time series themselves provided that simultaneous results are available from more than one station. The procedure assumes that the underlying station biases B_k are constant and relies on the fact that deviations due to different GPS times will affect each station bias estimate, B_k', by the same amount. That is, Equation (4) can be rewritten as

$$B_k = B_k'(t) - \Delta_{ANT} - \Delta_{DCB} - \delta(t) \quad (10)$$

where all the terms on the righthand side are known except the quasi-random variations δ(t) that are common to the $B_k'(t)$ observations from all labs at time t. When time series of empirical station timing biases are available for several independent stations, the δ(t) variations can be determined by an iterative process that minimizes the variance of each B_k for all the stations simultaneously. We have implemented such a procedure to give improved estimates of B_k.

3 Empirical calibration results

Table 1 summarizes our basic empirical station timing results, including mean and standard deviations for the observed biases before the systematic and quasi-random corrections have been applied, B_k', as well as the estimated biases after those adjustments, B_k. The time ranges for some stations are discontinuous because of changes in station configuration or because of resets of the internal receiver clocks. In nearly all cases, the bias ranges shown in Table 1 have been detected by an automated procedure that checks for discontinuities larger than three sigmas as computed by a maximum likelihood operator over the 20 epochs before and after the break. One unnecessary break at BRUS and two at NPLD have been manually suppressed. In addition, an extra break was inserted for a small discontinuity at PTBB known from a written report to the BIPM from the station operator. No results are given for the IGS stations IENG, MDVO, MIZU, NRC1, OBE2, PENC, SFER, SPT0, TLSE, or WTZR either because of insufficient data or because the geodetic receiver is not locked to the local UTC(k) standard.

Stations equipped with Ashtech Z12T receivers are indicated in Table 1. Some non-Z12T stations, particularly BOR1, can have frequent bias changes due to resets of the internal receiver clock. The AMC2, BRUS, NPLD, USNO, and USN1 receivers all use external hydrogen-maser frequency standards, while PTBB uses a primary cesium (Cs) standard and BOR1, KGN0, and TWTF use industrial Cs standards. Judging from the bias scatters, there seems to be a general tendency for labs with Cs standards to show somewhat poorer bias stabilities, which is probably because of inherently larger variations in the [UTC – UTC(k)] offsets for those labs.

We have used the IGS Rapid and Final clock products independently as a confidence check. However, since both series should yield the same bias estimates, more robust results would be possible by combining them into a single calibration bias estimate for each station. Such a procedure should be followed in any operational use of this method. Of the 38 station-intervals listed in Table 1 having Rapid and Final results, all agree much better than the root-sum-of-squares (RSS) of their respective standard deviations and only two disagree by more than 1.79 ns. The exceptions are for two brief, poorly sampled periods at BOR1 with especially high internal scatters. Overall excellent agreement between the bias estimates from the Rapid and Final clocks provides some assurance that the method is sound. The global mean Rapid-Final bias difference, weighted by the individual RSS

IGS site	MJD range	# Values		Mean B_k’		STD B_k’		Mean B_k		STD B_k	
		Rapid	Final	Rapid (ns)	Final (ns)	Rapid (ns)	Final (ns)	Rapid (ns)	Final (ns)	Rapid (ns)	Final (ns)
AMC2	52118-52437	305	41	-452.45	-454.15	3.28	2.50	-457.19	-458.49	2.49	1.52
‘’ A	52485-53093	593	565	3.92	3.84	3.07	3.04	-0.51	-0.68	2.15	2.28
BOR1	51855-52020	97	72	-2956.61	-2954.66	12.00	12.92	-2960.32	-2959.14	8.48	9.09
‘’	52251-52293	37	41	-352.89	-353.17	2.40	2.17	-358.26	-357.91	1.16	0.85
‘’	52342-52363	18	16	-438.83	-437.94	10.17	7.57	-445.21	-443.98	6.89	6.50
‘’	52365-52399	33	17	-262.95	-263.26	2.76	3.09	-268.23	-268.15	1.24	1.04
‘’	52407-52431	24	11	-43.75	-45.49	5.99	2.04	-47.14	-48.93	5.25	1.22
‘’	52460-52481	21	11	-220.76	-212.62	12.26	10.62	-225.40	-217.30	8.58	5.56
‘’	52504-52531	26	27	30.96	31.34	3.82	2.98	28.14	27.56	2.66	2.08
‘’	52584-52607	22	23	156.17	156.30	2.26	2.96	151.86	151.76	1.90	1.94
‘’	52647-52678	31	29	121.20	121.50	3.35	3.14	114.77	115.31	1.48	1.53
‘’	52693-52724	31	30	-26.41	-26.41	2.30	2.33	-30.59	-30.34	1.37	1.51
‘’	52923-52947	20	24	34.38	34.29	3.01	3.05	30.76	30.09	2.37	2.29
‘’	52948-52976	24	28	60.20	59.80	6.19	6.03	56.17	55.87	4.36	3.94
‘’	53063-53090	26	14	-5.97	-1.64	10.79	13.14	-12.80	-8.30	8.43	10.57
BRUS A	52213-52354	126	4	452.99	452.10	2.19	0.97	447.51	448.13	1.52	0.99
‘’ A	52383-52419	35	0	598.27		3.92		593.24		2.17	
‘’ A	52420-52883	445	377	585.03	584.92	3.54	2.85	580.63	580.34	2.46	2.16
‘’ A	52884-53093	206	191	574.63	574.47	2.29	2.45	570.06	569.94	1.11	1.17
KGN0 A	52422-53093	457	296	-29.20	-29.29	3.06	3.03	-33.67	-33.99	1.99	2.07
NPLD A	52134-53093	530	279	-8074.99	-8075.09	3.38	2.85	-8079.64	-8079.96	1.93	1.88
PTBB A	52922-52987	12	0	518.66		2.10		514.93		1.27	
‘’ A	52991-53093	59	79	525.16	524.64	2.34	2.11	518.45	518.41	1.31	1.32
TWTF A	52415-52453	32	33	300.59	299.81	3.59	4.23	297.38	295.70	2.98	2.95
‘’ A	52483-52561	66	78	381243.39	381243.75	4.02	4.10	381240.61	381240.17	3.05	2.85
‘’ A	52563-52605	39	42	296.33	296.34	3.51	3.52	291.51	291.74	2.31	2.15
‘’ A	52606-52739	127	133	283.94	283.95	3.35	3.44	278.49	278.56	2.53	2.87
‘’ A	52751-52788	33	34	328.04	327.75	3.12	3.01	324.38	323.97	2.31	2.25
‘’ A	52789-53093	273	287	315.41	315.42	2.87	2.88	311.12	311.04	2.32	2.13
USNO	51922-51993	69	71	-70.78	-70.67	3.44	2.85	-74.55	-75.21	5.63	5.10
‘’	51994-52018	23	24	-83.29	-83.26	3.53	3.49	-86.76	-87.41	3.54	5.25
‘’	52027-52065	35	38	157.12	157.10	4.38	3.35	152.74	152.72	4.38	3.35
‘’	52069-52129	60	60	624.11	624.17	3.21	3.18	619.35	619.77	2.94	3.18
‘’	52131-52158	25	27	-157.69	-157.98	2.06	2.03	-160.47	-161.99	0.53	1.96
‘’	52180-52255	71	71	-70.05	-69.83	2.69	2.46	-74.76	-74.30	0.84	2.37
‘’	52257-52343	81	84	7.34	7.45	2.35	2.27	2.06	2.57	1.07	1.75
‘’ A	52357-52529	161	165	85287.80	85287.65	2.53	2.04	85283.43	85283.19	1.94	1.99
‘’ A	52531-52890	346	349	640.06	640.04	2.90	2.86	635.50	635.44	2.08	2.12
‘’ A	52891-53093	197	185	656.11	656.12	3.13	3.33	651.58	651.62	1.43	1.62
USN1 A	52491-53093	251	63	2.37	3.17	2.51	2.56	-2.75	-1.79	1.49	1.88

Table 1: Empirical station timing biases are summarized here; mean and standard deviation (STD) estimates have units of nanoseconds. No results are given for IENG, MDVO, MIZU, NRC1, OBE2, PENC, SFER, SPT0, TLSE, or WTZR – other IGS stations located at timing labs – because of insufficient data or because the geodetic receiver is not locked to the local UTC(k) standard. Site names marked with “A” indicate those equipped with Ashtech Z12T receivers. AMC2, BRUS, NPLD, USNO, and USN1 use external hydrogen-maser frequency standards; BOR1, KGN0, PTBB, and TWTF use cesium standards.

scatters, is 0.005 ns with a standard deviation of 0.70 ns. These results suggest that the appropriate uncertainty for the empirical mean bias estimates is about 1 ns rather than the measured standard deviations, which are too pessimistic by about a factor of nearly 4, or the standard error of the means, which would usually be far too optimistic.

The effectiveness of accounting for small temporal variations in the observed GPS time differences, $\delta(t)$, can be judged by comparing the reduction in standard deviations for B_k in the rightmost two columns in Table 1 with those for B_k'. In most cases the reduction in scatter is substantial, especially during more recent times when a larger number of labs is available. So we judge the procedure to be justified. However, since the actual bias values can be affected only slightly, including this processing step is not critical. Figure 1 shows the $\delta(t)$ adjustments applied to derive the values in Table 1.

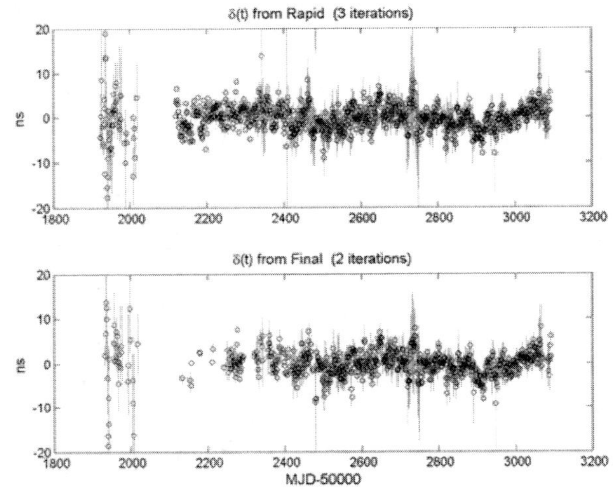

Figure 1: Plots of the adjustments $\delta(t)$ applied in computing the bias values in Table 1 to account for time-varying differences in the observation of GPS time by the BIPM and the IGS. The upper panel is for the IGS Rapid clock series and the lower panel is for the Final series.

4 Direct differential calibration

Calibration of a GPS receiver amounts to the determination of the total delay incurred by the signal from the antenna phase center to a fiducial point, the "internal reference", to which the measurements are referred. The total delay may be broken down into as many as six independent parts (see Figure 2 for a typical set-up). The receiver internal delay for each frequency, denoted $X_{R1/2}$, and the antenna internal delay for each frequency, denoted $X_{S1/2}$, are the components to be determined by differential calibration. The other delays are directly measured. X_C is the delay of the cable between the antenna and the receiver; when necessary, we distinguish the antenna cable itself, which delay is denoted X_C, from the set of splitter and short cable linking the splitter to the receiver, which delay is denoted X_D. By convention, the internal reference is taken as lagging by 15.8 ns the positive zero crossing of the 20 MHz signal entering the receiver. The offset between the 20 MHz and the 1-PPS signals entering the receiver is measured with a digital oscilloscope and, after adding 15.8 ns, provides the delay between 1PPS-in and internal reference, denoted X_O. Finally X_p is the cable delay between the 1PPS-in and the laboratory reference.

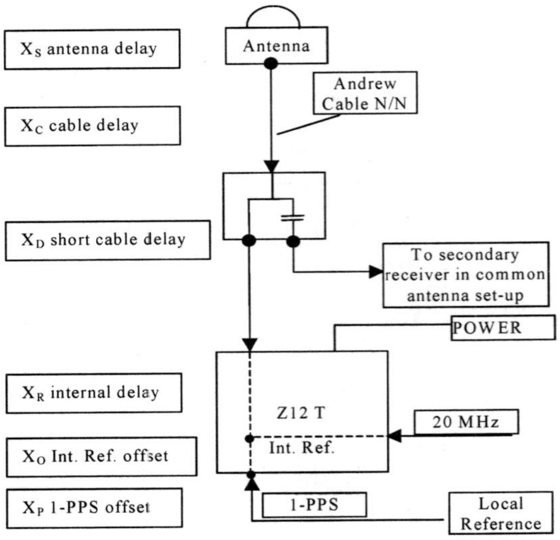

Figure 2: Definition of the different delays used in the typical experimental set-up for differential calibration of an Ashtech Z12T GPS receiver.

The "directly measured" calibration delay of such a receiver is then $C_k = X_{R3} + X_{S3} + X_C - X_O - X_P$, where subscript 3 refers to the ionosphere-free linear combination of the delays at the two frequencies. (Note that we assume a single cable linking antenna and receiver, a configuration which is advised in routine operation.) At each change in the set-up, the last three values should be re-measured, while the X_{R3} and X_{S3} values should be periodically checked by direct comparison with a travelling receiver of known delay values. For a number of receivers operating at time laboratories, these measurements have been carried out and the available results are summarized in Table 2.

IGS site	Date of calibration	X_{R3} + X_{S3} (ns)	Interval of fixed set-up	C_k (ns)
BRUS	July 2003	289.1	52420-52882	582.8
"	"	"	52883-.....	574.3
KGN0	Oct. 2003	295.4	52422-.....	-14.2
PTBB	June 2003	282.3	52923-52990	518.5
"	"	"	52991-.....	522.8
TWTF	Jan. 2002	291.6	52606-52739	304.7
"	"	"	52740-52788	354.7
"	"	"	52789-.....	358.0
USN1	Dec. 2002	-4.6	52491-.....	-4.6

Table 2: Direct differential biases C_k for a set of receivers calibrated using the BIPM travelling Z12T standard and the operational intervals of fixed set-up. The raw instrumental biases for the GPS equipment, $X_{R3} + X_{S3}$, are indicated with the date of their determination. Note that the USN1 value is relative to another absolute calibration applied to the raw observables (see section 5). The C_k values for TWTF before MJD 52789 are suspect due to a discrepancy of about 15 ns in the X_O value used with later measurements. The 50 ns change at TWTF on MJD 52740 is due to a slip of 1 cycle in the 20 MHz input frequency with respect to the 1-PPS reference.

5 Comparison of calibration results

Our empirical timing biases are ultimately based on the absolute calibration of the GPS equipment used to refer GPS time to UTC for Circular T. A C/A-code, single-channel receiver located at the Paris Observatory is used for this purpose. It has been repeatedly calibrated over the past 20 years with an observed stability at the 5 ns level over long times and much better during the past few years [9].

Table 3 compares our empirical timing biases (from Table 1) with those from differential instrumental calibration (from Table 2) for those stations and periods with simultaneous results. For BRUS (Royal Observatory of Belgium) during two periods, PTBB (Physikalisch-Technische Bundesanstalt, Germany) during two periods, and USN1 (U.S. Naval Observatory, USA) the two calibration methods all agree within -4.4 to 2.2 ns. It should be noted that the BRUS to PTBB time transfer link for TAI uses data from the same Z12T geodetic receivers in the P3 common view mode. So the B_k bias transfer for BRUS, in particular, represents a closure check of sorts on the internal accuracy within the Circular T values.

IGS site	MJD range	Mean B_k (ns)	C_k (ns)	B_k - C_k (ns)
BRUS	52420-52882	580.5	582.8	-2.3
‘’	52883-.....	570.0	574.3	-4.3
KGN0	52422-.....	-33.8	-14.2	-19.6
PTBB	52923-52990	514.9	518.5	-3.6
‘’	52991-.....	518.4	522.8	-4.4
TWTF	52606-52739	278.5	304.7	-26.2
‘’	52740-52788	324.2	354.7	-30.5
‘’	52789-.....	311.1	358.0	-48.0
USN1	52491-.....	-2.4	-4.6	2.2

Table 3: Comparison of empirical timing biases (B_k), using a weighted average of the Rapid and Final estimates, with direct differential biases (C_k); units are nanoseconds. The direct determinations shown include the raw instrumental biases for the GPS equipment as well as the corrected biases including cable connections to the local UTC(k) realization.

On the other hand, the agreement for TWTF (Telecommunication Laboratories, National Standard Time and Frequency, Taiwan) and KGN0 (Communications Research Laboratory, Koganei, Japan) is very poor. The KGN0 empirical bias is very stable and well determined over nearly two years. The middle of the two spans for TWTF has little data for B_k but the other two periods are also well determined. For TWTF, the equipment used for the TAI link has been known for years to be poorly calibrated. It has been re-measured recently but the results have not yet been introduced into the TAI computation. The new calibration indicates that the value $[UTC - UTC(k)]_T$, and thus also B_k, should be increased by 46.3 ns [3], which is consistent with the results in Table 3 for the period since 52789. For KGN0, the question is open and may be linked to a particular mode of operation of the Z12T, which, to our knowledge, only applies to this receiver.

Another comparison is also possible between our empirical bias estimates and absolute instrumental calibration values for the two stations shown in Table 4, AMC2 and USN1. In both cases the absolute delay calibrations were performed similarly to the BIPM travelling Z12T, using a GPS signal simulator in a chamber at NRL. All delays to the local realizations of UTC(k) have been applied to the raw pseudorange observations so that the nominal clock biases should be zero (see IGS station log files at http://igscb.jpl.nasa.gov/network/list.html). The agreement of these two empirical calibrations with the absolute determinations is even better than with the differential measurements in Table 3. The differential and absolute instrumental calibrations for USN1 differ by 4.6 ns; the empirical bias estimate is midway between them.

IGS site	MJD range	Mean B_k (ns)	A_k (ns)	$B_k - A_k$ (ns)
AMC2	52485-.....	-0.6	0.0	-0.6
USN1	52491-.....	-2.4	0.0	-2.4

Table 4: Comparison of empirical timing biases (B_k) with absolute calibrations using a GPS signal simulator (A_k); units are nanoseconds. For these two stations the raw GPS data have been modified at the receiver to remove the absolute biases (including cable delays) so that the empirical biases should be nominally zero.

6 Discussion and conclusions

Any process that relies on comparisons among independently running clocks or dissemination of time from a central source requires that the enabling time transfer equipment be adequately calibrated for internal delays and other biases. The time transfer process itself is generally not self-calibrating. Instead, dedicated calibration procedures are usually conducted that interrupt normal time transfer operations and are consequently performed infrequently. Even when performed, the calibration procedures themselves can lead to subtle calibration shifts due to effects such as impedance mismatches at couplings that are changed. Obviously, any errors in the assumed values for timing biases enter directly into the results of time transfer users.

The empirical method of determining station timing biases we have demonstrated here does not require any hardware changes nor any additional measurements. It can be used to validate independent direct measurements of hardware biases, absolute or differential. Indeed, such checks are recommended in order to avoid potential discontinuities in UTC/TAI when time transfer operations are switched between techniques. For two stations (TWTF and KGN0) our results have identified potential problems in differential calibration results.

Moreover, the empirical method has the distinct advantage of permitting continuous monitoring of intra-laboratory timing stability in a completely non-disruptive manner. Figure 3 illustrates an example of such a possible use. Empirical timing biases for the three IGS stations linked to UTC(USNO) – namely AMC2, USNO, and USN1 – are compared for a period of nearly 600 days. A common systematic variation, possibly an annual trend, is evident in the AMC2 and USNO data. It may also be present in the much sparser USN1 series. Matching behavior is not observed in simultaneous data from other stations, such as BRUS and TWTF. So we believe it more likely to be related to variations within the USNO laboratory than with any defect in the empirical calibration method. A closely correlated trend in the AMC2 and USNO geodetic receiver data, such as due to environmental effects, seems very unlikely since these stations are separated by 2360 km and both are environmentally controlled. Instead, we suggest that the UTC(USNO) calibration for TAI has varied slowly over a few-ns range during the period.

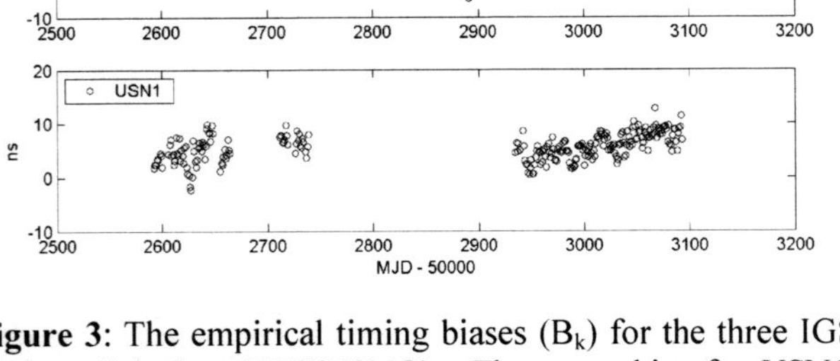

Figure 3: The empirical timing biases (B_k) for the three IGS stations linked to UTC(USNO). The mean bias for USNO has been adjusted to match AMC2 and a bias shift on 52891 has been removed for easier comparison.

Based on the results presented here, the empirical calibration method has been validated to the few-ns level. Differences with two absolutely calibrated Ashtech Z12T receivers agree well within the estimated accuracies of ~3.5 ns. Comparisons with direct differential calibrations are less clear in two cases, probably due to unresolved problems in [UTC – UTC(k)] offsets and local delay measurements. For the remaining stations the agreement is –4.4 to 2.2 ns. We estimate the nominal uncertainty of the empirical calibrations to be near the 1 ns level provided that sufficient data are available (at least one month). The leading errors in the empirical bias determinations are probably caused by instabilities in the local UTC(k) calibrations and in Circular T [UTC – UTC(k)] offset measurements.

Acknowledgements

The assistance of Stefan Schaer (Astronomical Institute of the University of Berne) in providing broadcast DCB values is gratefully acknowledged. The BIPM Circular T values used here can be obtained at the website http://www.bipm.org/en/scientific/tai/time_ftp.html. All information concerning the IGS is available via http://igscb.jpl.nasa.gov/ and the IGS clock products can be accessed via https://goby.nrl.navy.mil/IGStime/.

References

[1] Bureau International des Poids et Mesures. "Circular T", Pavillon de Breteuil, F-92312 Sèvres Cedex, France.

[2] P. Defraigne, G. Petit. "Time transfer to TAI using geodetic receivers", *Metrologia*, **40**, pp. 184-188, (2003).

[3] W. Lewandowski, P. Moussay. "Determination of the differential time corrections for GPS time equipment located at the OP, NTSC, CRL, NMIJ, TL, and NML", Bureau International des Poids et Mesures, Rapport BIPM-2003/05, Pavillon de Breteuil, F-92312 Sèvres Cedex, France, (2003).

[4] G. Petit, Z. Jiang, P. Uhrich, F. Taris. "Differential calibration of Ashtech ZXII-T receivers for accurate time comparisons", *Proc. 15th European Frequency and Time Forum*, **15**, pp. 164-166, (2001).

[5] G. Petit, Z. Jiang, J. White, R. Beard, E. Powers. "Absolute calibration of an Ashtech Z12-T GPS receiver", *GPS Solutions* , **4**(4), pp. 41-46, (2001).

[6] J.R. Ray, H. Dragert, J. Kouba. "Recommendations for handling non-Rogue data", in *IGS 1999 Technical Reports*, Jet Propulsion Laboratory Publication, Pasadena, California, pp. 445-451, (2000).

[7] J. Ray, K. Senior. "IGS/BIPM pilot project: GPS carrier phase for time/frequency transfer and timescale formation", *Metrologia*, **40**(3), pp. S270-S288, (2003).

[8] K. Senior, J. Ray. "Estimating empirical station timing biases using IGS clock products", Frequency Control Symposium & PDA Exhibition Jointly with the 17th European Frequency and Time Forum, *Proc. 2003 IEEE International*, pp. 250–257, (2003).

[9] M.A. Weiss, V. Zhang, W. Lewandowski, P. Uhrich, D. Valat. "NIST and OP GPS receiver calibrations spanning twenty years: 1983-2003", this volume, (2004).

COMPARISON OF REMOTE CESIUM FOUNTAINS USING GPS P3 AND TWSTFT LINKS

. . Richard, P. Uhrich, A. Clairon, . Achkar, S. Bi e, H. Marion, P. Laurent, M. Abgrall, I. Maksimovic, . Gr nert,
Bureau National de M trologie Syst me de R f rence Temps Espace
OBSERVATOIRE de PARIS, 61 av. de l Observatoire, 75014 PARIS, France
Corresponding author: J.- . Richard, Tel. + 33 (0)1 40 51 22 23, Fax + 33 (0)1 43 25 55 42,
Jean-yves.richard@obspm.fr

A. Bauch, S. Weyers, R. Wynands
Physikalisch-Technische Bundesanstalt
POB 3345, D-38023 Braunschweig, GERMAN
Stefan.Weyers@ptb.de

Keywords : Time transfer, Cesium Fountain, Comparison

Abstract

The cesium fountain primary frequency standards FO2 and FOM of BNM-S RTE (Bureau National de M trologie Syst me de R f rence Temps Espace) and CSF1 of PTB (Physikalisch-Technische Bundesanstalt) have been compared using two different time transfer links: GPS P3 CV (Global Positioning System Precise Code, Common View) and TWSTFT (Two-Way Satellite Time and Frequency Transfer) during 10 days of almost continuous fountain operation. The resulting relative frequency differences between the remote fountains amount to several 10^{-15}, whereas the total uncertainty of each clock (including statistical and systematic uncertainty) was = 10^{-15} for the 10-days comparison. An experimental investigation of the noise on the two links between PTB and BNM-S RTE was performed later during two months. This allows us to deduce a link noise for a 10-day period of about $2x10^{-15}$.

Introduction

Over the course of the last years atomic cesium fountains have established themselves as the primary frequency standards contributing to TAI (Temps Atomique International) with the lowest uncertainty achieved so far (around 10^{-15}) [1-4]. As in metrological laboratories worldwide the number of locally available primary frequency standards is generally limited, which is particularly true for cesium fountains, there is a strong interest in remote comparisons using state-of-the-art comparison means [5-8]. A first remote comparison between the fountains CSF1 of PTB and F1 of NIST (National Institute of Standards and Technology) was reported in [9]. This comparison exhibited good agreement between the two fountains when taking into account the respective uncertainties. At the same time a comparison of the techniques TWSTFT, GPS carrier phase, and GPS common view could be performed.

Here we present the results of a first remote comparison between the fountains FO2 and FOM of BNM-S RTE and CSF1 of PTB by the techniques of GPS common view and TWSTFT.

The GPS receivers used are multi-channel geodetic models with software developed by P. Defraigne (ORB) [10] to compute the ionosphere-free multi-channel GPS P3 common-views that BNM-S RTE has implemented in 2002. At PTB the GPS receiver is on loan from the BKG (Bundesamt f r Kartographie und Geodäsie), which also takes care of the data evaluation. The TWSTFT link between European laboratories is realized by the geostationary satellite Intelsat 706 @ 307°E in the Ku-band. The receiver equipment at BNM-S RTE is composed of a Vertex antenna, a Miteq RF transceiver and a SATRE v.4 modem [11], and at PTB a Vertex antenna, a Dornier transceiver, a SATRE v.3 modem and a high resolution time interval counter.

The GPS P3 CV Ashtech receivers, as well as the TWSTFT SATRE modems, were driven by H-masers in both laboratories. During the comparison, the H-maser frequencies were continuously monitored by the cesium fountains with a frequency resolution of the order of 10^{-15} per day. The H-maser used at BNM-S RTE is denoted H805 and the H-maser used at PTB is denoted H2.

Frequency Measurements of the Fountains

The fractional frequency differences of H805 relative to the fountains FO2 and FOM corrected for physical effects are plotted in Figure 1. These measurements had durations of about one day, interrupted by short idle periods where bias evaluations were performed. There were larger dead times of FO2 between MJD 52823,876 and MJD 52824,301 and of FOM between MJD 52823,871 and MJD 52824,337. Since the TWSTFT measurements were performed each day at 14:20 UTC (i.e., 0,597 in MJD decimal units) we have chosen to remove data of fountain measurements from MJD 52823,597 up to MJD 52824,597. Other cut-offs are applied at the date of the beginning of measurements, the 1st July 2003 at 14:20 UTC or MJD 52821,597, and at the date of the end of measurements, the 11th July 2003 at 14:20 or MJD 52831,597. These restrictions were applied in order to compare only synchronous periods of operation of the three fountains within the TWSTFT schedule.

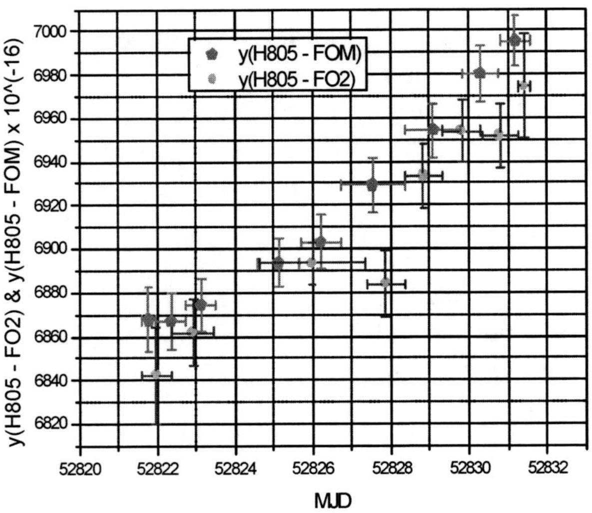

Figure 1: Fractional frequency differences of H-Maser H805 relative to FO2 and FOM

The mean frequency differences between the respective fountain and H805 are evaluated by weighted linear fitting of the data using the respective statistical uncertainties of the single measurements as weights. The results of the ten days of simultaneous measurements are reported in Table 1.

For the PTB-CSF1 fountain the recorded fractional frequency differences relative to the local hydrogen maser H2 were corrected for systematic effects and are presented in Figure 2 with the dead time MJD 52823,597 - 52824,597 removed. The mean frequency of CSF1 is calculated from these 7745 frequency differences and is given in Table 1.

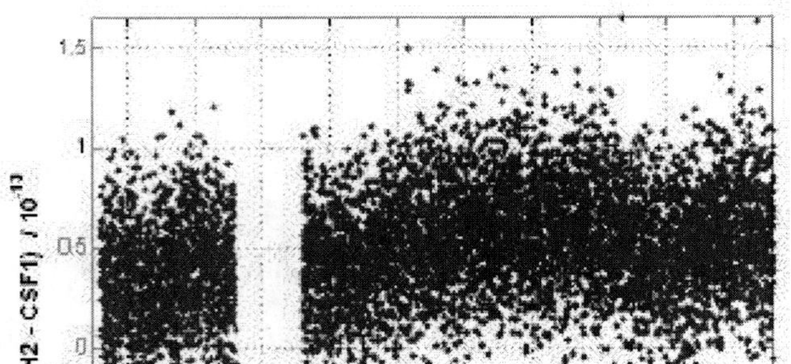

Figure 2: Fractional frequency differences of H-maser H2 relative to CSF1. The data points taken during the dead time MJD 52823,597 - 52824,597 have been removed. Each point represents the result of a 100-s measurement time

In order to estimate the statistical uncertainty of the frequency averages we have analyzed the frequency stability of each fountain. Figure 3 gives the behavior of the instability of the three fountains with respect to their local H-maser.

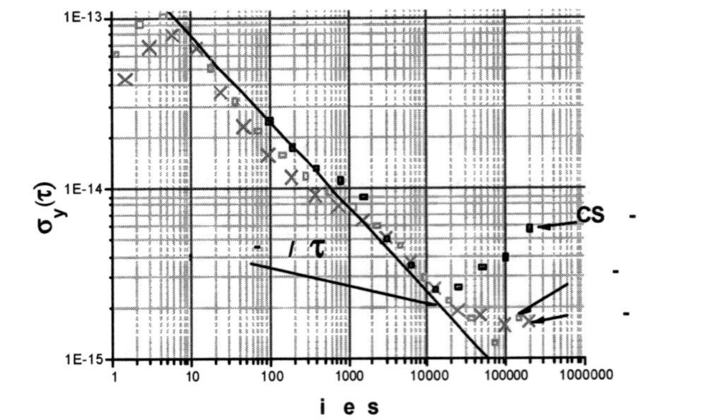

Figure 3: Overlapping Allan deviation of PTB-CSF1, S RTE-FO2, and S RTE-FOM

By extrapolating the white noise level of the frequency standard to 1s we obtain an instability of CSF1 of $_y(\tau) = 2{,}5 \times 10^{-13} \times (\tau/s)^{-1/2}$ for comparing the three fountains. The long-term behavior is probably due to the H-maser frequency drift. With the assumption that the CSF1 frequency stability is determined by white frequency noise up to 10 days it reaches an extrapolated instability of $_{y\ CSF1}(10d) = 3 \times 10^{-16}$.

For FO2 and FOM the stability we obtain by analyzing the fractional frequency differences between S RTE-FO2 and S RTE-FOM over 100 s is also determined mainly by white frequency noise to $_y(\tau) = 2{,}3 \times 10^{-13} \times (\tau/s)^{-1/2}$, as indicated in Figure 4.

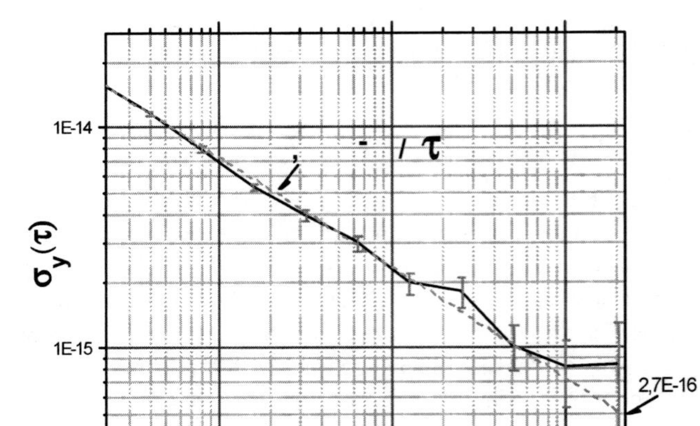

Figure 4: Overlapping Allan deviation of synchronized fractional frequency differences S RTE-FO2 - S RTE-FOM

With the hypothesis of white frequency noise up to 10 days the frequency instability of FO2 relative to FOM reaches

$y_{FO2\ FOM}(10\ d) = 2,5 \times 10^{-16}$, thus representing an estimate of the statistical uncertainty of the mean frequency difference of FO2 and FOM.

During the period of comparison the systematic uncertainties of FO2, FOM and CSF1 were evaluated as indicated in Table 1.

Clocks	Mean frequency	Statistical uncertainty	Systematic uncertainty
H805 - FO2	+6900,1	2,5	5,4
H805 - FOM	+6921,1	2,5	10,1
H2 - CSF1	-485,5	3	9

Table 1: Mean frequencies of the fountains relative to their local H-maser (in units of 10^{-16})

Frequency Transfer

Both the Two-Way Satellite Time and Frequency Transfer and the GPS P3 common-view techniques were used for the remote fountain comparison in order to assess the relative qualities of the two techniques. The TWSTFT measurements followed the standard one-measurement-per-day schedule and the GPS CV measurements followed the BIPM tracking schedule.

From TWSTFT the phase differences between the remote hydrogen masers, H805 at S RTE and H2 at PTB, are extracted at MJD 52821,597 and MJD 52823,597 for calculating the mean hydrogen maser frequency difference for the first period (effectively 2 days) of the comparison and at MJD 52824,597 and MJD 52831,597 for the calculating the mean of the second period (7 days) of comparison. By taking these data only, the common dead time MJD 52823,597-52824,597 is omitted and the frequency difference between the remote H-masers can be calculated as the weighted average of the 2-day and 7-day measurement periods. The resulting frequency difference is given in Table 2. The determination of the uncertainty of this result is described in the next section.

Link	Clocks	Mean frequency	Statistical uncertainty
TWSTFT	y(H2 H805)	-7362,3	19,2
GPS	y(H2 H805)	-7373,6	19,2

Table 2: Mean frequencies of the fountains relative to their local H-maser (in units of 10^{-16})

For GPS P3 CV the phase difference measurements are collected for the interval MJD 52821,597 - MJD 52831,597 with a gap in the data between MJD 52823 and 52824 due to operational reasons of the GPS comparison. The number of data points is 797 and a second-order polynomial fit was computed.

The mean frequency difference between the H-masers evaluated by GPS P3 CV during the 10 days is given in Table 2. The determination of the uncertainty of this result is described in the next section.

Comparison of the two time transfer links

During two months, from 1st July to 31st August (MJD 52821 to 52882), the phase differences between the hydrogen masers were continuously extracted from TWSTFT and GPS P3 CV time transfers. After removing a second order polynomial fit independently for the two data sets of phase differences, and after removing outliers (3 away) from the GPS data, we obtain the graph in Figure 5.

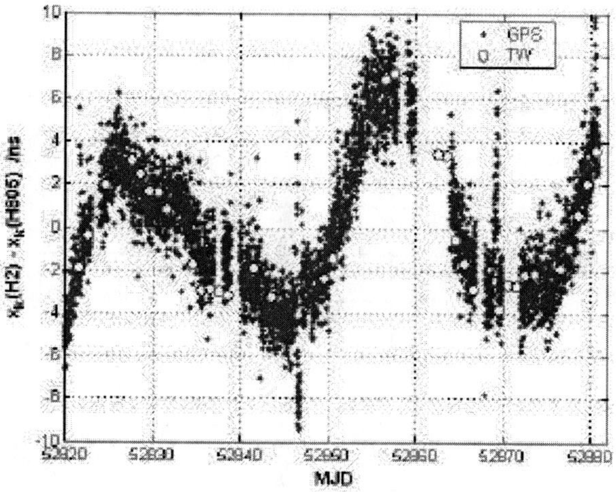

Figure 5: TWSTFT and GPS P3 CV phase differences x_k(H2 - H805) between MJD 52821 to 52882, linear & quadratic fit removed

In this plot 4427 GPS phase difference data points have been plotted together with the 35 TWSTFT data points that are available. From the study of the GPS data, it seems that the uncertainty of the H-Masers is reached for an averaging period of two days. Therefore the GPS mean phase difference is computed using a linear least-square fit of the GPS data averaged over two-day intervals centered around each TWSTFT date of measurement. Figure 6 shows the result of that phase-difference comparison between the two time transfer techniques.

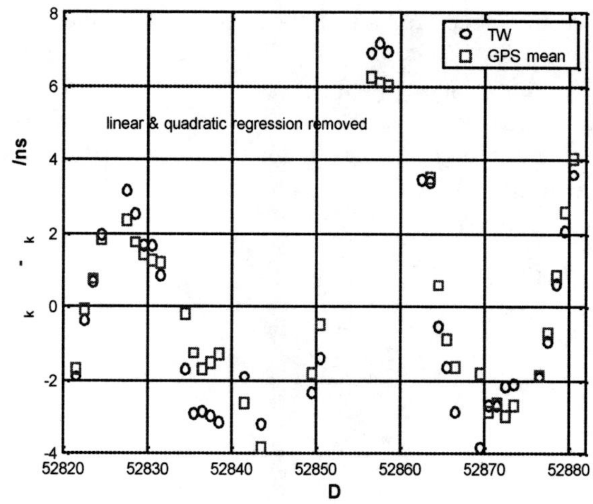

Figure 6: TWSTFT and GPS P3 CV phase differences between MJD 52821 and 52882. GPS phase data are averaged over ±1 days around each TWSTFT date.

The differences between TWSTFT and GPS P3 CV phase data measurements on each of the 35 common dates can be computed. They are plotted in Figure 7. As no deterministic trend can be seen, the computation of a RMS value is justified. One obtains 1ns.

From the data points in Figure 7 one can compute the time deviation and therefore an estimate for the link uncertainty in the following way.

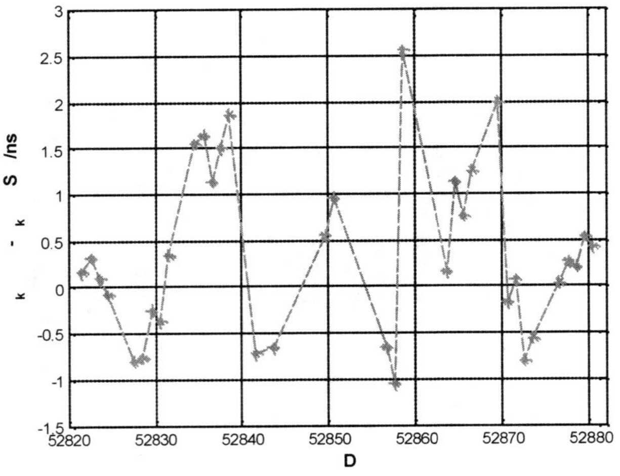

Figure 7: Time differences between phase data measured by TWSTFT and GPS P3 CV between MJD 52821 and 52882

In a first step, the data points are distributed evenly over the 62-day interval, giving an averaged time separation of τ = 153051 s between two data points. For these epoch-adjusted time values the Time Allan standard deviation is calculated for averaging times $\tau = \tau_0$, $2\tau_0$, $4\tau_0$, etc. This Time Allan deviation is plotted in figure 8 and, despite large statistical uncertainties, shows that the phase differences stability is close to a flicker phase noise modulation (slope 0).

A conservative value of $_x(10\ d)$ = 0,8 ns is chosen and the frequency stability in the case of a flicker phase noise modulation is $_{Link}(10\ d) = 27{,}1 \times 10^{-16}$. For simplicity, we attribute this uncertainty equally to both techniques, so that for a single link a factor of $2^{1/2}$ has to be included, resulting in a link uncertainty $_{link} = 19{,}2 \times 10^{-16}$.
The agreement between the two time transfer techniques is given by the mean frequency difference between the two links for the original 10 days of fountain comparison, within the uncertainty of $_{Link}$ (τ = 10d):

$$y(H2\quad H805)_{TW} - y(H2\quad H805)_{GPS} = 11{,}3 \times 10^{-16} \pm 27{,}1 \times 10^{-16}.$$

This result shows that both time transfer links give qualitatively the same phase-difference fluctuations between the remote Hmasers at S RTE and at PTB, and that the frequency difference lies well within the one-standard-deviation interval for this 10-day comparison.

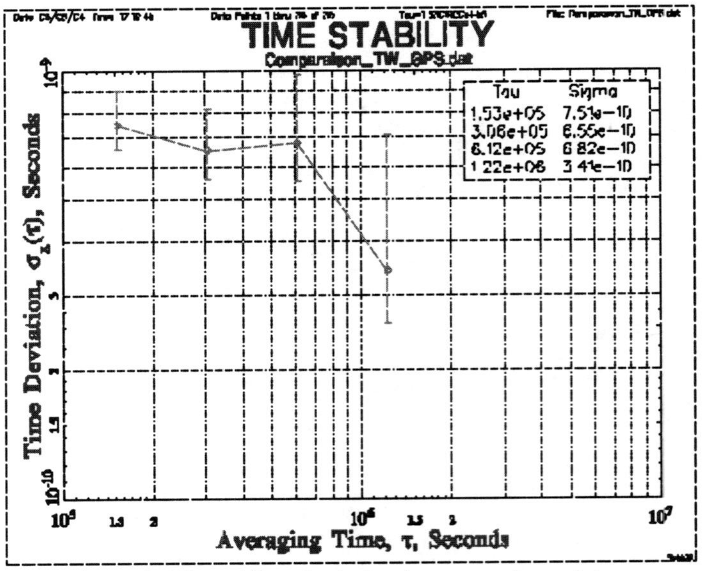

Figure 8: Time Allan deviation between phase data differences measured by TWSTFT and GPS P3 CV between MJD 52821 and 52882

Frequency Comparison Results

The results of the remote frequency comparison between the fountains S RTE-FO2 and S RTE-FOM with respect to PTB-CSF1 by the two time transfer links TWSTFT and GPS P3 CV and the frequency difference between the local fountains FO2 and FOM are summarized in Table 3. For the local comparison, only the root-quadratic-sum of the cable link uncertainties of 1×10^{-16} each is considered. The statistical uncertainty of each mean frequency difference is computed as the root-quadratic-sum of the respective instabilities of the fountains. Likewise, the root-quadratic-sum of the individual systematic uncertainties is formed to obtain an overall systematic uncertainty. Finally, the total uncertainty is calculated as the square root of the quadratic sum of link uncertainty, statistical uncertainty, and systematic uncertainty. All these uncertainties are evaluated at 1 . A graphical representation of these numbers is shown in Figure 9.

Link	Clocks	Mean frequency Over 10 days	Link uncertainty	Statistical uncertainty	Systematic uncertainty	Total uncertainty
GPS	y(CSF1 FO2)	+12,0	19,2	3,9	10,5	22,2
	y(CSF1 FOM)	+33,0	19,2	3,9	13,5	23,8
TWSTFT	y(CSF1 FO2)	+23,3	19,2	3,9	10,5	22,2
	y(CSF1 FOM)	+44,3	19,2	3,9	13,5	23,8
Local	y(FO2 FOM)	+21,0	1,4	2,5	11,5	11,9

Table 3: Summary of the comparison results of the fountains S RTE-FO2, S RTE-FOM, and PTB-CSF1 during 10 days in July 2003 (in units of 10^{-16})

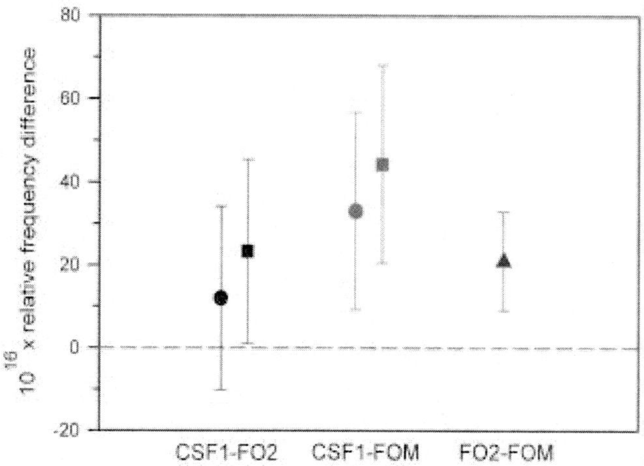

Figure 9: Graphical representation of the relative frequency differences between the fountain clocks CSF1, FO2 and FOM. The circles represent the GPS comparison results, the squares the TWSTFT comparison results and the triangle the local comparison result. The error bars reflect the total uncertainty of the remote comparisons and of the local comparison, respectively.

Conclusions Outlook

For the rather short and incomplete data set of only ten days we find that link uncertainty dominates the uncertainty budget for the remote frequency comparison. Of course, it would be interesting to repeat the experiment with a longer data run, so that the contribution by the link noise would be negligible compared to the systematic uncertainties of the individual fountains. The evaluation of the two-month remote comparison of maser frequencies indicates that this should be the case for a data run of about two months.

Furthermore, it appears that CSF1 is running faster than FO2 by roughly the same amount that FO2 is running faster than FOM. These differences lie roughly within the 1 -interval of the overall uncertainty stated for the respective fountains. So although the difference is not statistically significant in the strict sense it is nevertheless large enough to make further remote frequency comparisons advisable, in order to be able to discover potential problems with the evaluation of systematic frequency shifts in one or more of the cesium fountain clocks.

References

[1] H. Marion, F. Pereira Dos Santos, M. Abgrall, S. hang, . Sortais, S. Bize, I. Maksimovic, D. Calonico, J. Gr nert, C. Mandache, P. Lemonde. G. Santarelli, Ph. Laurent, A. Clairon, C. Salomon, Search for Variations of Fundamental Constants using Atomic Fountain Clocks, Phys. Rev. Lett. (15), 150801, 2003.

[2] S. Weyers, U. H bner, R. Schr der, Chr. Tamm and A. Bauch, Uncertainty evaluation of the atomic cesium fountain CSF1 of PTB, Metrologia **38**, 343-352, 2001 S. Weyers, A. Bauch, R. Schr der, Chr. Tamm, The atomic caesium fountain CSF1 of PTB, Proc. 6th Symposium on Frequency Standards and Metrology, St. Andrews, pp. 64-71, World Scientific, 2002.

[3] S. R. Jefferts, J. H. Shirley, T. E. Parker, T. P. Heffner, D. M. Meekhof, C. W. Nelson, F. Levi, G. Constanzo, A. DeMarchi, R. E. Drullinger, L. Hollberg, W. D. Lee, and F. L. Walls, Accuracy evaluation of NIST-F1, Metrologia **3** , 321-336, 2002.

[4] F. Levi, L. Lorini, D. Calonico, and A. Godone, Systematic Shift Uncertainty of IEN CSF1 Primary Frequency Standard, IEEE Trans. Instrum. Meas. **52** (2), 267-271, 2003.

[5] G. Petit and . Jiang, "Long-term comparison of GPS P3 links with other techniques used for TAI", BIPM report GP/TM.126, May 2003

[6] G. Petit and . Jiang, "Computation of TAIP3 links: December 2002", BIPM report TAIP3 comp 0212, January 2003.

[7] S. L. Shemar, J. D. Clarke, J. A. Davis, A. J. Lowe and P. J. Chapman, "Comparisons of The Two-Way, GPS Carrier-Phase and GPS Common-View Time Transfer Methods", in Proc. 2000 14th EFTF-2000, Torino, Italy, pp. 46-50, 2000.

[8] F. Levi, L. Lorini, D. Calonico, E.K. Bertacco and A. Godone, "IEN-CSF1 Accuracy Evaluation & Two Way Frequency Comparison", Proc. 2003 IEEE International Freq. Control Symp., pp. 199-204, 2003

[9] T. Parker, P. Hetzel, S. Jefferts, S. Weyers, L. Nelson, A. Bauch, J. Levine, "First Comparison of Remote Cesium Fountains," in Proc. 2001 IEEE International Freq. Control Symp., pp. 63-68, 2001 and Proc. EFTF 2001, Neuch tel, pp. 57-61.

[10] P. Defraigne, G. Petit, "Time transfer to TAI using geodetic receivers", Metrologia **4** (2003) 184-188.

[11] J. Achkar, P. Merck, "Development of a TWSTFT station at BNM-S RTE ". Proceedings of the 2003 IEEE International Freq. Control Symp. Jointly with the 17th EFTF & Time Forum, pp. 1021-1024, Tampa, Florida, USA 2003.

PERFORMANCE COMPARISON AND STABILITY CHARACTERIZATION OF TIMING AND GEODETIC GPS RECEIVERS AT IEN

R. Costa, D. Orgiazzi, V. Pettiti, I. Sesia, P. Tavella

Istituto Elettrotecnico Nazionale "Galileo Ferraris", Strada delle Cacce 91, Torino, Italy, metf@ien.it

Keywords: geodetic GPS receivers, stability, outliers, precise point positioning, time scale comparison.

Abstract

In this paper the noise characterization of the GPS receivers (both timing and geodetic) hosted at the Time and Frequency Laboratory of the Istituto Elettrotecnico Nazionale (IEN) "Galileo Ferraris" was taken into account. In particular, a specific algorithm as well as two different approaches (a classical GPS "common view" technique and a geodetic-based Precise Point Positioning, PPP) employed to process the data coming from such receivers were described.

Moreover, the paper focused on the preliminary timing performance of the PPP geodetic method by comparing them with the International GPS Service (IGS) estimates on the link between two European metrological institutes, such as IEN and PTB (Physikalisch-Technische Bundesanstalt), Braunschweig, Germany.

1 Introduction

The Time and Frequency Laboratory of the Istituto Elettrotecnico Nazionale (IEN) "Galileo Ferraris", Turin (Italy) is currently operating a 3S Navigation (in the following 3SN) multi-channel single-frequency GPS/GLONASS timing receiver performing "common view" measurements [1], allowing time scale comparisons with other remote UTC(k) laboratories. A second timing receiver of the same type, a TTS-2 with temperature stabilized antenna (in the following named TTS), is also working since mid December 2003.

In addition, two dual-frequency geodetic GPS receivers are available in a very short baseline set up. An Ashtech Z-12T "Metronome" receiver (hereafter named ASH) has been working since 2001, with external time and frequency reference signals directly connected to the national time scale UTC(IEN). This receiver, differentially calibrated in the context of an international campaign arranged in 2001 by BIPM, is now involved in the BIPM TAI P3 activities [5],[6].

Starting from March 2003 a second geodetic GPS receiver, a Javad Legacy with timing option (in the following JAV), is also operated. In the framework of the Galileo System Test Bed V1 (GSTB V1) [7], the first experimental phase of the Galileo project supported by the European Space Agency, this receiver is part of the Experimental Precise Timing Station (E-PTS), implemented at IEN in collaboration with Alenia Spazio, Roma (Italy), with the aim to generate and disseminate the Experimental Galileo System Time for the whole experimentation period planned for 2004.

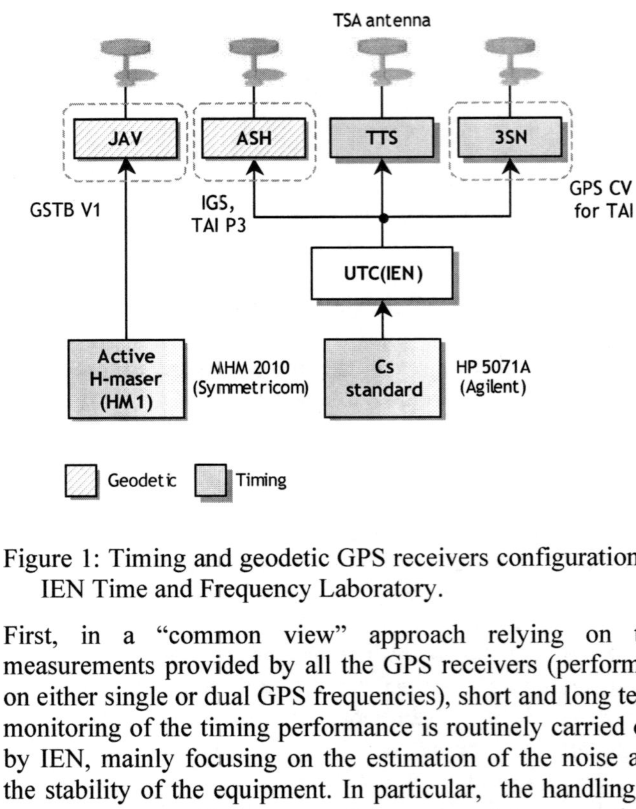

Figure 1: Timing and geodetic GPS receivers configuration at IEN Time and Frequency Laboratory.

First, in a "common view" approach relying on the measurements provided by all the GPS receivers (performed on either single or dual GPS frequencies), short and long term monitoring of the timing performance is routinely carried out by IEN, mainly focusing on the estimation of the noise and the stability of the equipment. In particular, the handling of measurement outliers as well as the data smoothing techniques are addressed with the aim of improving the robustness of the data processing. Real data sets were used, dealing with the influence of outliers and discussing the effectiveness of the Hampel filter (based on outlier-resistant median) as an outlier identification method.

Moreover, some additional efforts have been devoted to test the clock offset estimates obtained by Precise Point Positioning (PPP), as provided from the service kindly supplied by the Geodetic Survey Division (GSD) of Natural Resources, Canada (NRCan) [10]. This geodetic approach is based on high-quality GPS products (precise satellite orbits and clocks) made available in near real time by the International GPS Service (IGS) [9]. From a first evaluation, this method seems to provide stability performance even better than that provided by a "common view" approach, mainly due to the use of both carrier phase measurements and high quality IGS products. Preliminary "timing" performance of this method are presented and discussed in this paper.

Besides, since January 2004 the Ashtech Z-12T receiver is qualified as an IGS station and the clock driving the receiver is thus estimated by the IGS via a carrier phase technique [15],[16]. A preliminary comparison of such estimates with respect to those coming from a PPP reduction is also reported in the paper.

2 Robust outliers detection algorithm

The computation of the time offset between the local UTC(k) and both GPS time and other remote time scales is one of the major duty of each metrological institute.

At the Time and Frequency Laboratory of IEN, this task is carried out by processing the GPS receivers data through a specific three-stages algorithm, which has the main aim to detect and remove any outliers and to provide robustly smoothed estimates.

At first, starting from the CGGTTS-formatted (Common GPS GLONASS Time Transfer Standard) [2], measurement data collected at IEN (3SN receiver), a MAD (Median of the Absolute Deviation) based filtering procedure is performed with the aim of removing outliers caused mainly by unhealthy satellites or less favourable geometric conditions. Such a filter is applied on the set of all the measurements concerning the satellites in view at each epoch, in order to detect outliers and to make use of all information available at that epoch.

In particular, the MAD-based filter is obtained by applying the Hampel identifier [4],[13], which replaces the outlier-sensitive classical statistical indexes (mean and standard deviation), with the outlier-resistant median and median absolute deviation from the median, respectively.

As documented in [14], the median x^C of a generic data sequence is obtained by first rank-ordering it from smallest to largest, i.e.:

$$x_{(1)} \leq x_{(2)} \leq \ldots \leq x_{(N-1)} \leq x_{(N)} \qquad (1)$$

and then taking x^C as either the middle value (if N is odd) or the average of the middle two values (if N is even). The MAD scale estimate is then defined as:

$$S = 1.4826 \cdot median\{|x_k - x^C|\}, \qquad (2)$$

where the factor 1.4826 in Equation (2) is chosen so that the expected value of S is equal to the standard deviation σ for normally distributed data.

A rejection threshold t (with t as an integer positive value) is then introduced and if

$$|x_k - x^C| > t \cdot S \qquad (3)$$

x_k is declared an outlier and it is removed from the dataset.

Figure 2 shows an example of the general behaviour of both a classical 3σ dynamic filter and a MAD-based filter with a rejection threshold t =3 (hereafter named 3S MAD-based filter) for a small real dataset containing a great outlier.

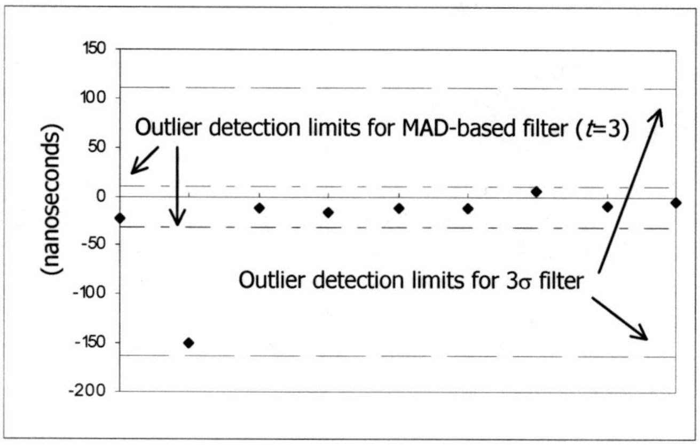

Figure 2: Behaviour of different filtering on a small real dataset containing a great outlier (≅ 4σ).

As it comes out observing the example reported in Figure 2, in general the 3S MAD-based filter seems to be a better choice than the classical 3σ filter, concerning the detection and removal of outliers especially when the parameters are estimated from a small data set (as in the case of epoch measurements), as also documented in literature [18],[14]. On the other hand, it is worth to notice that, for large datasets, the 3σ dynamic filter and the 3S MAD-based filter seem to provide the same performance, as confirmed by the values in Table 1, concerning the estimates of the mean and the standard deviation values.

Data set length	*Filter*	*Estimated*	
		Mean (ns)	*Std. Dev. (ns)*
89 samples	3σ	5.6	3.4
	3S MAD	5.6	3.4
9 samples	3σ	-25.9	47.1
	3S MAD	-10.4	7.9

Table 1: Effect of 3σ and 3S MAD filters on both a small dataset (epoch measurements of the example reported in Figure 2) and a large dataset (GPS daily measurements at IEN for MJD 52988).

Then, after the epoch-based filtering procedure, the average of all validated values is computed and only one datum per each epoch results. With the aim of removing any residuals anomalies caused mainly by the receiver along with its external reference (the UTC(k) time scale in case of metrological laboratories), a further 3σ filtering procedure is then applied on the 24 hours interval of the considered day. As mentioned above, since the receiver could collect up to 90 measures on each day, both the MAD-based and the 3σ filter are comparable in term of outlier detection performance. The latter was selected for software implementation purposes.

Finally, the linear regression of all the mean-epoch values is computed for each 24 hours interval centred at 00h UTC of the considered day. This smoothing procedure has the capability to filter out the short-term instabilities noticed on the measures. Only one datum per day is then produced by the algorithm as the estimate of the mean offset between the local reference UTC(IEN) and either GPS time or other remote UTC(k) laboratories.

The estimated UTC(IEN) versus GPS time offset provided by the described algorithm is plotted in Figure 3 for the 2 months period from MJD 52640 (January 1st, 2003) to MJD 52698 (February 28th, 2003). As showed in the plot, the results coming from the processing algorithm are very close to those coming from the BIPM Circular T data, having residuals with 2.2 ns rms.

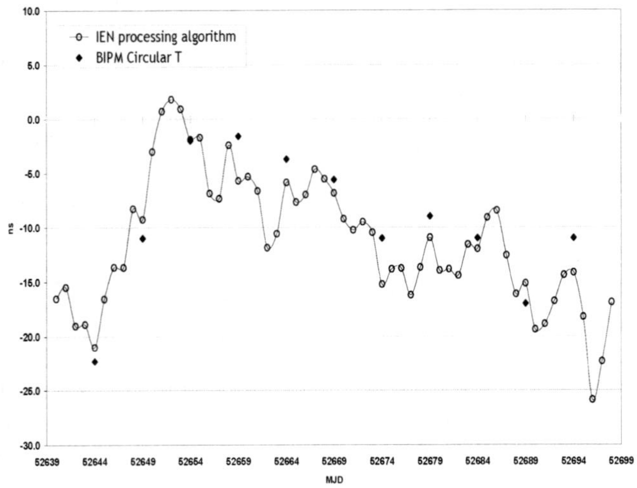

Figure 3: Behaviour of UTC(IEN) versus GPS time as estimated by the processing algorithm on the 3SN receiver data (January 2003 – February 2003). The same quantity as estimated by the BIPM Circular T data (5 days spaced) are also reported.

In case of time scale comparison with remote UTC(k) laboratories, the processing algorithm is quite different , even if it is based on three-stages too. In particular, the *3S* MAD-based epoch filtering is now applied on the set of raw differences between the measurements of satellites in strict common view by the two laboratories.

As a consequence, the differences between raw measurements lead to remove all the "common" source of uncertainties. Besides, the further filtering on each epoch allows coping with anomalous asymmetrical common views between the two laboratories.

3 Receivers noise characterization

With the aim to characterize the noise and the stability of the IEN GPS receivers, the measurements provided by the two timing and the two geodetic GPS receivers have been processed using the MAD-based filter algorithm described in the previous section.

For a 3 months period from MJD 52988 (December 15th, 2003) to MJD 53079 (March 15th, 2004), the offset between UTC(IEN) and GPS time has been estimated from the data of each receiver. In particular, for the 3SN and the TTS timing receivers, weekly CGGTTS-formatted data files [2] have been considered. On the other hand, for the geodetic receivers ASH and JAV, the daily RINEX files [8] (containing phase and pseudorange dual-frequency measurements with 30 seconds sampling rate) have first been converted in 5-days CGGTTS-formatted files using specific software [6], as required in BIPM TAI P3 activities.

Figure 4 shows the time series of the offset UTC(IEN) versus GPS time, as one datum per day at the 0h UTC standard epoch.

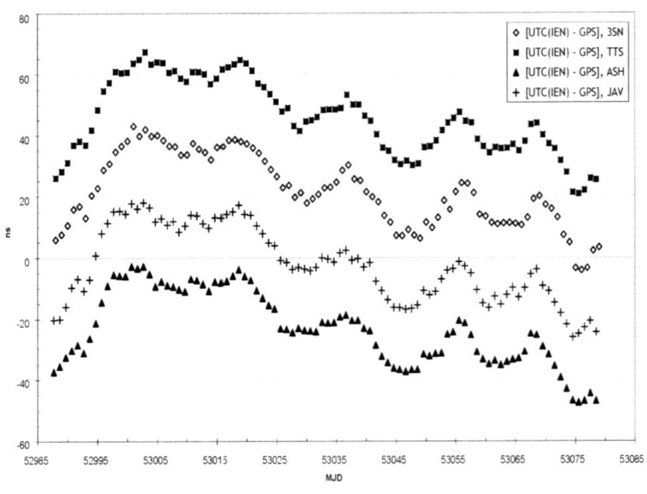

Figure 4: Behaviour of UTC(IEN) versus GPS time as estimated by different receivers (December 2003 – March 2004). A 20 ns offset has been added to JAV for purpose of plotting.

Since the JAV receiver was referenced by the free-running H-maser #1 operated by IEN (in the following HM1), the [HM1–GPS] offset resulting from the algorithm has been post-processed using the daily [UTC(IEN)–HM1] internal measurements, in order to remove the HM1 contribution. A costant bias equal to 140627 ns is also removed.

Apart from not compensated calibration biases (discussed later), the time series coming from different receivers seem to be very similar and consistent with the BIPM Circular T data. This is also confirmed by the day-by-day residuals reported in Figure 5, where some pairs of receivers have been taken into account.

Please note that such residuals have been estimated by a strict common view approach, instead of compare the daily values, achieving a small improvement in term of data spreading. In this case, as mentioned in section 2, the processing algorithm jointly treats the two receivers and only the differences between the measurements of the same satellite at the same epoch are computed.

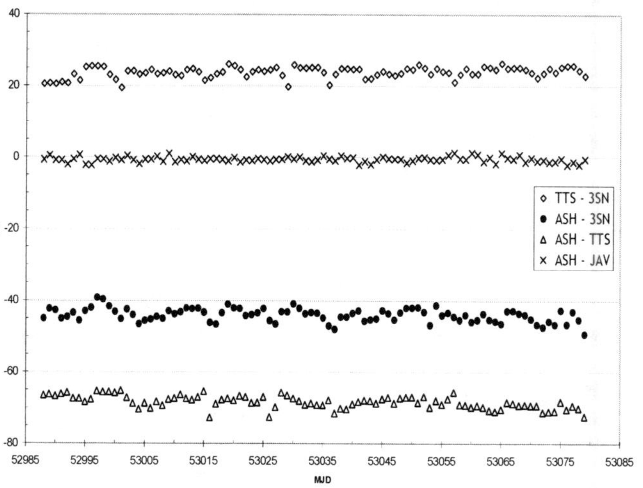

Figure 5: Time residuals between pairs of GPS receivers (December 2003 - March 2004)

Due to the UTC(IEN) "common clock" configuration of the receivers (native for 3SN, TTS and ASH, but also applicable for JAV after removing the HM1 contribution), the resulting residuals are intended to be the differential delay of pair of receivers. Table 2 reports the statistical indexes of such residuals, where the mean values are not significant in performance evaluation since there are calibration biases to be compensated. The corresponding frequency stabilities, in term of Allan deviations, are also given in Figure 6.

Receivers	*Mean (ns)*	*σ (ns)*
TTS - 3SN	23.8	1.6
ASH - 3SN	-44.0	1.8
ASH - TTS	-68.3	1.7
ASH - JAV	-0.6	0.8

Table 2: Statistical indexes for the time residuals between pairs of GPS receivers (depicted in Figure 5).

It is worth to notice that the results in Table 2 concern a pair of receivers, since the residuals can be computed only comparing the measurements provided by two receivers. If the receivers could be assumed identical and independent, the noise figure of each receiver could be estimated to be a factor $\sqrt{2}$ lower.

Looking at the plots in Figure 6, the Allan deviations show a behaviour close to a noise with a τ^{-1} slope: as expected, the residuals are only affected by the typical additive thermal noise of electronic devices (the receiver), since both the reference time scale and the GPS contributions have been removed by the processing procedure. Actually, an additional investigation using the modified Allan deviation detects both white phase (WP) and flicker phase (FP) noise components at all the averaging times, with different level, but only for the residuals involving timing receivers.

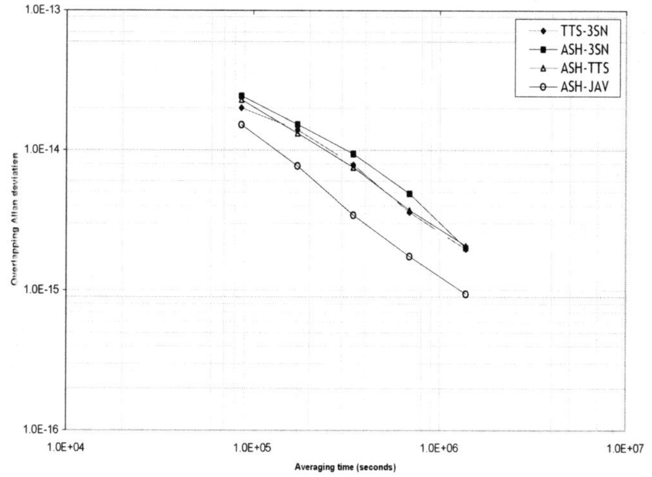

Figure 6: Allan deviation of the time residuals between pairs of GPS receivers (depicted in Figure 5).

In the last row of Table 2, the statistical indexes concerning the residuals for the two geodetic GPS receivers are also indicated, after removing the HM1 contribution. As well showed in Figure 5, both the classical variance and the Allan deviation in Figure 6 are a factor close to 2 lower than for either timing/timing (TTS-3SN) or timing/geodetic (ASH-TTS) pair of receivers. For instance, over 1 day, the Allan deviation decreases from ($2.3 \cdot 10^{-14} \cdot \tau^{-1}$) for ASH-TTS pair to ($1.5 \cdot 10^{-14} \cdot \tau^{-1}$) for ASH-JAV.

Furthermore, in case of geodetic/geodetic pair, the Allan deviation shows a pure WP noise for all the averaging times considered (up to 2 weeks) and no FP component is noticed. Then, it is possible to infer that using the P3 iono-free combination of the dual frequency P-code measurements provided by geodetic GPS receivers, the receiver noise could be estimated with higher accuracy, filtering out any other source of noise (such as, the measurement method itself).

4 A "geodetic-based" approach

The precise point positioning (PPP) is a post-processing approach using un-differenced dual frequency pseudorange and carrier phase observations coming from a single geodetic GPS receiver, together with the high-quality GPS products such as those provided by the International GPS Service (IGS) [9].

Taking advantage of the precise satellite clock estimates as well as the precise satellite coordinates in the IGS products, the PPP allows stand-alone point positioning (kinematic and static) with centimetre precision in term of station coordinates and receiver clock offset estimates with sub-ns precision.

It means that the PPP eliminates the need to acquire simultaneous observations from a reference station or a network of tracking stations. As a consequence, the stand-alone position of a single geodetic GPS receiver can be autonomously estimated with cm precision, even if the receiver is not part of a network of stations (such as, the world wide distributed IGS network).

For a number of years, PPP algorithms have been available in the GIPSY [12] GPS analysis software and, more recently, in the BERNESE software [17].

In particular, since the mid-90's the PPP approach has been developed by the Geodetic Survey Division (GSD) of Natural Resources Canada (NRCan), which is one of the IGS Analysis Centers contributing daily predicted, rapid and final GPS orbits and clocks to the IGS for combination. The current NRCan PPP implementation [10] is available as an on-line application [3] and only requires the submission of a valid observation RINEX file to obtain a stand-alone position estimate, as well as station clock estimates and tropospheric zenith path delays.

4.1 PPP clock offset estimates evaluation

With the aim to validate the timing-oriented performance of the NRCan PPP software, an experimental activity is carried out at IEN since few months using the on-line service kindly provided by GSD NRCan.

Preliminary results are reported and discussed in the following, using a 9 days period from MJD 53026 (January 22nd, 2004) to MJD 53034 (January 30th, 2004). Since this time window is fully included in the 3 months period considered in the receivers noise analysis reported in Section 3, a direct comparison of classical GPS common view (based on 3SN and TTS timing receivers), iono-free P3 (using ASH and JAV geodetic receiver data, as converted in CGGTTS format) and geodetic-based (PPP) approaches is then available.

In our analysis, phase and pseudorange dual-frequency measurements have been collected from both geodetic receivers and formatted in daily RINEX files. These RINEX files have been then processed with the PPP software using IGS final products (available with two weeks latency), providing receiver clock offset estimates every 30 seconds. It is worth to mention that the precise satellite orbits and clocks estimates in the IGS final products are given with 15 minutes and 5 minutes interval, respectively, with an accuracy less than 5 cm/0.1 ns.

Let ASH_{PPP} and JAV_{PPP} these estimates for Ashtech and for Javad, respectively:

$$ASH_{PPP} = \{GPS - [UTC(IEN) + k_{ASH}]\} \quad (1)$$

$$JAV_{PPP} = \{GPS - [HM1 + k_{JAV}]\}, \quad (2)$$

where k_{ASH} and k_{JAV} are the delays of the receiver internal references versus UTC(IEN) and HM1 external reference signals, respectively.

Subtracting the two estimates (1) and (2), at the same epochs, results in the offset of UTC(IEN) versus the free-running HM1, except for the differential delay (k_{ASH} - k_{JAV}) of the two receivers:

$$JAV_{PPP} - ASH_{PPP} = [UTC(IEN) - HM1] + (k_{ASH} - k_{JAV}). \quad (3)$$

The time series of the offset (3) is reported in Figure 7, after removing a quadratic trend (a = 0.3 ns/day^2; b = 51.2 ns/day; c = 143 609 ns). As showed by the Allan deviation in Figure 8, the residuals are very similar to a typical behaviour of a commercial Cs standard, which UTC(IEN) is based on.

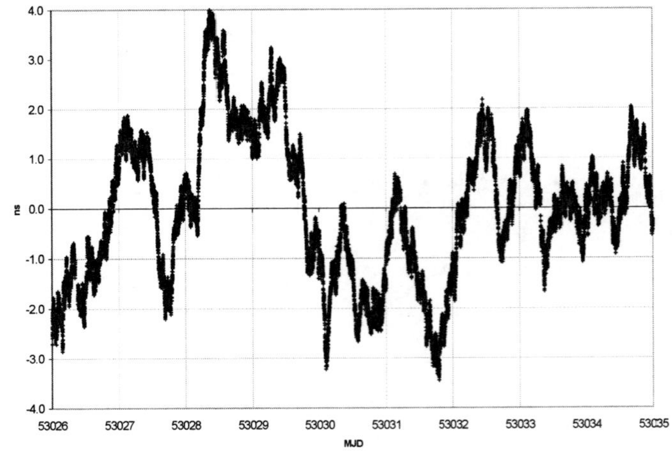

Figure 7: Time residuals between the Ashtech and the Javad receivers (January 22nd to January 30th, 2004), representing UTC(IEN)–HM1. Quadratic fit removed: a=0.3 ns/day^2; b=51.2 ns/day; c=143 609 ns.

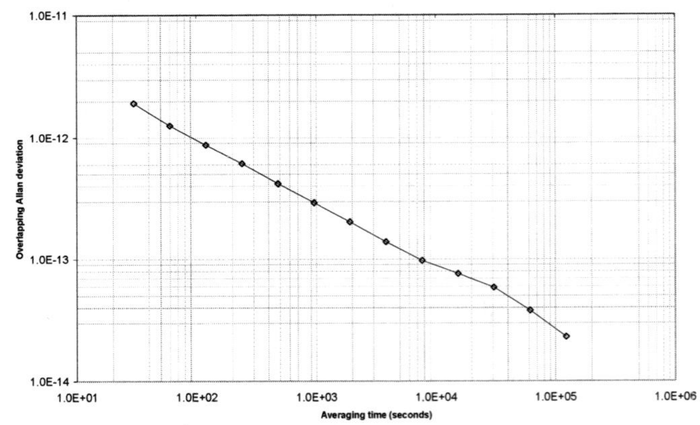

Figure 8: Allan deviation of the time residuals in Figure 7 between the PPP solutions for the Ashtech and the Javad receivers (January 22nd to January 30th, 2004), representing UTC(IEN)–HM1.

Moreover, in order to extract from Equation (3) the differential delay of the two receivers, the hourly [UTC(IEN)–HM1] values provided by the measurement system of IEN Time and Frequency Laboratory and the (JAV_{PPP} – ASH_{PPP}) estimates at each hour have been subtracted at the same epochs. A constant bias equal to 140603.3 ns is further removed, assuming it as the mean differential delay of the two receivers.

As plotted in Figure 9, the resulting hourly residuals are all in the range of –1.0 ns and 1.5 ns with a zero mean value (being removed the constant bias) and a 0.35 ns standard deviation.

Looking at the results in Table 3, the PPP geodetic approach seems to provide a stability performance better than that provided by both a classical common view approach on timing GPS receivers and a P3-based common view approach using the same geodetic GPS receivers.

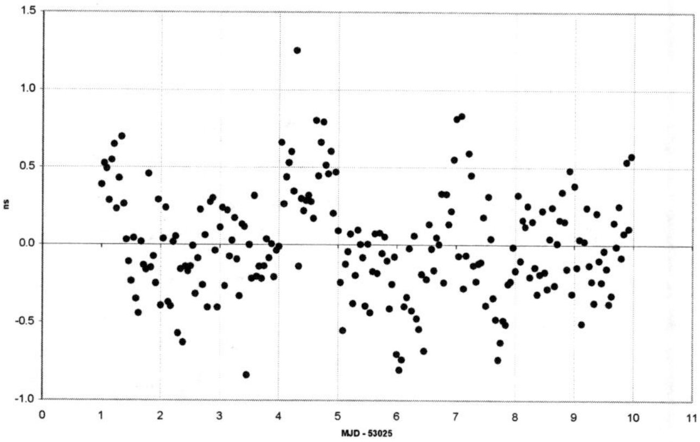

Figure 9: Time residuals of the differential delay between Ashtech and Javad receivers as estimated by PPP (January 22nd to January 30th, 2004). Constant bias removed: 140603.3 ns (mean differential delay).

Receivers	σ *(ns)*	$\sigma_y(\tau)$ @ $\tau = 1$ *day*
3SN – TTS via GPS CV	1.6	$2.0 \cdot 10^{-14}$
ASH - JAV via P3 (CGGTTS)	0.8	$1.5 \cdot 10^{-14}$
ASH - JAV via PPP	0.3	$0.8 \cdot 10^{-14}$

Table 3: Statistical index (classical standard deviation) and Allan deviation over 1 day for the time residuals of the differential delay between Ashtech and Javad receivers (depicted in Figure 9), compared with similar index previously reported in Table 2.

Moreover, the following Figure 10 reports the Allan deviation of the residuals of the two geodetic receivers as computed both via P3 (τ_0 = 1 day) and via PPP (τ_0 = 1 hour), where the stability performance improvement is well depicted.

Such result is mainly due to the use of dual frequency carrier phase measurements in the PPP algorithm, as well as the availability of high-quality GPS products (precise satellite clocks and orbit information) as provided in near real time by IGS.

However, it is worth to notice that the noise affecting PPP residuals has not the expected τ^{-1} slope (WP/FP) for all the averaging times. In effect, looking at the bottom line in Figure 10, a significant change in slope is observed for averaging times close to half a day. This issue has been already noticed in [11] and it is currently under inspection with the helpful support of the Canadian GSD.

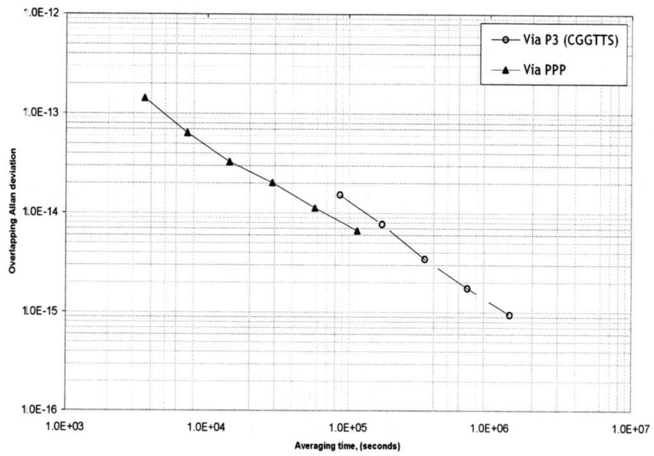

Figure 10: Allan deviation of the time residuals of the differential delay between Ashtech and Javad receivers as estimated by both PPP and P3 (January 22nd to January 30th, 2004).

4 Time transfer link between IEN and PTB using PPP

Relying on the promising results, the PPP method has then been used in the frame of a European link between the two metrological institutes, IEN and the Physikalisch-Technische Bundesanstalt (PTB), Braunschweig, Germany, in order to test its time transfer capability.

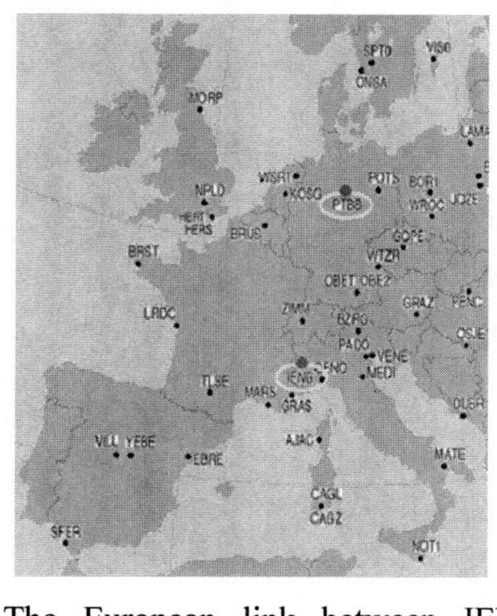

Figure 11: The European link between IEN and PTB metrological institutes.

Over such a link, two other time transfer methods (namely the TWSTFT and the GPS Common View) are regularly operated and a direct comparison of different synchronisation systems can then be achieved. In addition, the Ashtech Z12-T receivers of both laboratories are part of the IGS network (named IENG and PTBB, respectively).

For a 2 months period from MJD 53005 (January 1st, 2004) to MJD 53064 (February 29th, 2004), the daily RINEX files of the receivers of the two laboratories have been reduced with the PPP software using IGS final products of precise satellite clocks and orbits. The resulting 30 seconds estimates have been compared, at the same epoch, and the differences have been then linear fitted (over 1 day) to have the best daily estimates at 0h UTC standard epoch.

It is worth to notice that the data coming from the above processing represent [UTC(IEN) – UTC(PTB)] except for the differential delay of the configuration of the two receivers hosted by IEN and PTB. Such a bias has been estimated with respect to the recent calibrated TWSTFT link between the two laboratories and a calibration value of –514.4 ns (σ = 1.1 ns) has then to be considered in the PPP link between IEN and PTB.

The time series of the calibrated PPP estimates is shown in Figure 12, together with the daily [UTC(IEN) – UTC(PTB)] offset as provided by the TWSTFT and the GPS CV. In addition, the same offset as computed using the BIPM Circular T data are also reported.

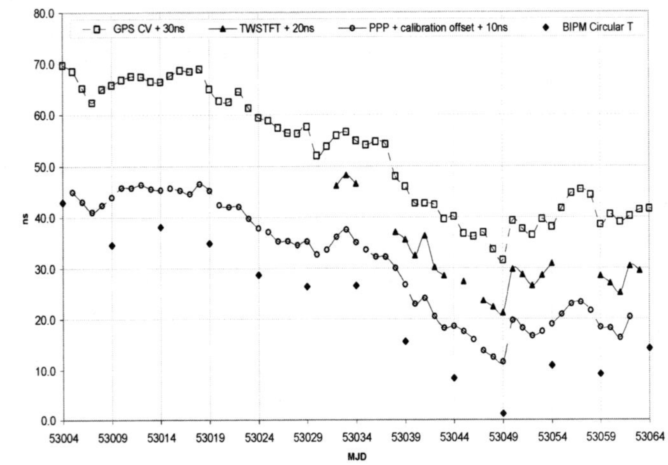

Figure 12: Behaviour of UTC(IEN) versus UTC(PTB) (January 2004 – February 2004) as provided by TWSTFT, GPS CV and PPP. The BIPM Circular T data (5 days spaced) are also depicted. An arbitrary offset is added to each series for purpose of plotting.

The residuals of the PPP estimates with respect to both TWSTFT and GPS CV are plotted in the following Figure 13.

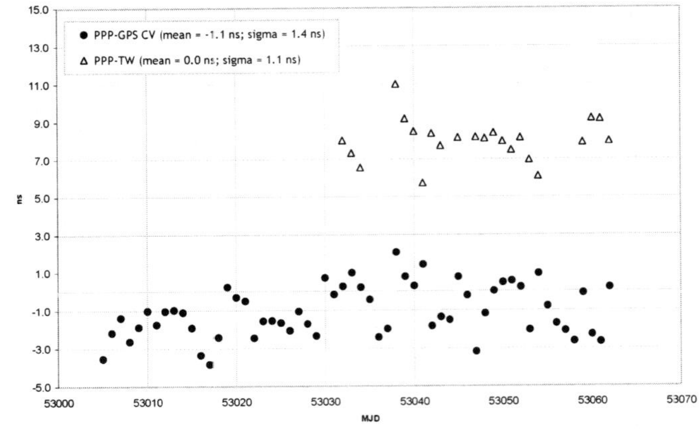

Figure 13: Time residuals of the PPP estimates (as corrected for the calibration bias computed using TWSTFT) with respect to both TWSTFT and GPS CV. An offset of +8 ns is added to the PPP-TW series for purpose of plotting.

As mentioned before, since January 2004 the ASH receiver is qualified as a permanent station of both EPN (EUREF Permanent Network) and IGS networks. The RINEX data from the receiver are regularly processed by some of the IGS Analysis Centers and they are also contributing to the IGS time scale [16].

In this frame, the clock products are daily available reporting the estimated behaviour of the clock driving the receiver (that is, UTC(k) in our case). Figure 14 shows a first comparison between IEN and PTB time scales (over a 5 days period), as estimated using the PPP approach and the IGS clock products (realigned to both rapid and final IGS time scale). Please note that, for purpose of plotting, an offset is added to each series.

Looking at the actual differences of the three time series, in term of accuracy, the PPP processing results are consistent at sub-ns level with the estimates from IGS which are based on a "network solution". In detail, the residuals show σ = 0.2 ns for PPP versus rapid IGS time scale, IGRT, and σ = 0.1 ns versus the final IGST, over a 5 days period.

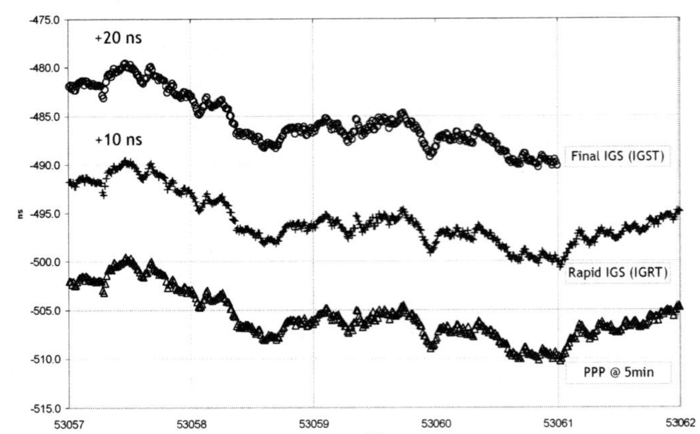

Figure 14: Behaviour of UTC(IEN) versus UTC(PTB) (22nd to 27th February, 2004) as estimated using PPP, Rapid IGS (IGRT) and Final IGS (IGST). The offsets shown in the graph are added to the series for purpose of plotting.

6 Conclusions

The operation and the noise characterization of the IEN GPS receivers (both timing and geodetic) have been addressed in this paper. The GPS measures are firstly treated by a IEN processing algorithm aiming to detect the outliers and to provide a robust, smoothed, daily estimates. From the analysis of the experimental results, it appears that the noise affecting GPS measures is close to $\sigma_y(1 \text{ day}) \cong 2 \cdot 10^{-14}$ when using timing receivers but it can be halved when using geodetic receivers with a TAI P3 like approach.

In addition, preliminary timing performance of the NRCan PPP geodetic approach have been presented, mainly focused on the estimation of the differential delay of two co-located geodetic GPS receivers (an Ashtech Z-12T and a Javad Legacy). Looking at the results, the PPP seems to provide a good stability performance over short/medium term, even better than that provided by both a classical common view approach on timing GPS receivers and a P3-based common view approach using the same geodetic GPS receivers.

The PPP has been also tested on the European link between IEN and PTB, where other time transfer methods (such as the TWSTFT and the GPS CV) are operated. After performing a calibration with respect to the recent calibrated TWSTFT link, the estimates provided by PPP are consistent with both the BIPM Circular T and the other time transfer methods with accuracy better than 2 nanoseconds.

Finally, concerning the same time transfer link between IEN and PTB, the PPP results are consistent (at sub-ns level) with the estimates carried out using the IGS clock products, which are available since the two metrological institutes are part of the IGS world wide network.

Acknowledgements

On behalf of the Istituto Elettrotecnico Nazionale "Galileo Ferraris", the authors wish to thank the Geodetic Survey Division, Natural Resources Canada (NRCan) for providing the GPS Precise Point Positioning (PPP) software. Special thanks to François Lahaye (NRCan) for kind support and helpful advices provided on the usage of PPP software.

The authors are also grateful to Andreas Bauch (PTB, Germany) and his staff for making available GPS CV data as well as RINEX data for PTBB IGS station. Special thanks to our colleague L. Lorini for providing TWSTFT data as well as to the students with the Time and Frequency Laboratory for the helpful support in data retrieving and processing.

References

[1] D.W. Allan, C. Thomas, "Technical Directives for Standardization of GPS Time Receiver Software", *Metrologia*, **31**, pp. 69-79, (1994).

[2] J. Azoubib, W. Lewandowski, "CGGTTS GPS/GLONASS Data Format Version 01", *VII CGGTTS Meeting*, (1998).

[3] Canadian Space Reference System (CSRS) – Precise Point Positioning (PPP) on-line service, *Geodetic Survey Division (GSD) of Natural Resources Canada (NRCan)*, http://www.geod.nrcan.gc.ca/index_e/products_e/services_e/ppp_e.html.

[4] L. Davies, U. Gather, "The identification of multiple outliers", *J. Amer. Statist. Assoc.*, pp. 782-801, **vol. 88**, (1993)

[5] P. Defraigne, G. Petit, C. Bruyninx, "Use of Geodetic Receivers for TAI", *Proceedings of the 33rd Annual Precise Time and Time Interval (PTTI) Meeting*, Long Beach, California, pp. 341-348, (2001).

[6] P. Defraigne, C. Bruyninx, "Time Transfer for TAI Using a Geodetic Receiver; An Example with the Ashtech ZXII-T", *GPS Solutions*, pp. 43-50, **vol. 5 n. 2**, (2001).

[7] G. Graglia, E. Detoma, F. Cordara, R. Costa, L. Lorini, D. Orgiazzi, V. Pettiti, P. Tavella, J. Hahn, "The Experimental Precise Timing Station and the Galileo System Time generation in the Galileo System Test Bed Phase V1", *Proceedings of GNSS 2003 – The European Navigation Conference*, Graz, Austria, (2003).

[8] W. Gurtner, "RINEX: The Receiver Independent Exchange Format Version 2.10", (2002).

[9] J. Kouba, "A Guide to Using International GPS Service (IGS) Products", *IGS Central Bureau*, ftp://igscb.jpl.nasa.gov/igscb/resource/pubs/GuidetoUsingIGSProducts.pdf, (2003).

[10] J. Kouba, P. Héroux. "Precise Point Positioning Using IGS Orbits and Clock Products", *GPS Solutions*, **vol. 5 n. 2**, (2001).

[11] F. Lahaye, P. Collins, J. Popelar, "The Virtual Reference Clock (VRC) Time Scale of the NRCan Real Time Wide Area GPS Correction Service (GPS·C)", *IV International Time Scales Algorithm Symposium*, Sèvres, France, (2002).

[12] S. M. Lichten, Y. E. Bar-Sever, E. I. Bertiger, M. Heflin, K. Hurst, R. J. Muellerschoen, S. C. Wu, T. P. Yunck, J. F. Zumberge, "GIPSY-OASIS II: A High precision GPS Data processing System and general orbit analysis tool", *Technology 2006, NASA Technology Transfer Conference*, Chicago, Il., (1995).

[13] R.K. Pearson, "Exploring process data", *J. Process Contr.*, pp. 179-194, **vol.11**, (2001)

[14] R.K. Pearson, "Outliers in Process Modelling and Identification", *IEEE Transactions on Control Systems Technology*, **vol. 10**, pp. 55-63, (2002).

[15] J.R. Ray, "IGS/BIPM Time Transfer Project", *GPS Solutions*, pp. 37-40, **vol. 2 n. 3**, (1999).

[16] J. Ray, K. Senior, "IGS/BIPM pilot project: GPS carrier phase for time/frequency transfer and timescale formation", *Metrologia*, **40**, pp. S270-S288, (2003).

[17] M. Rothacher, L. Mervart, "The Bernese GPS Software Version 4.0", *Astronomical Institute, University of Berne*, Berne, Switzerland, (1996).

[18] D.F. Vecchia, J.D. Splett, "Outlier - Resistant Methods for Estimation and Model Fitting", *Proceedings of the Workshop Advanced Mathematical Tools in Metrology*, Torino, Italy, pp. 143-154, (1993).

STABILITY AND ACCURACY OF GPS P3 TAI TIME LINKS

G. Petit and Z. Jiang

Bureau International des Poids et Mesures, 92312 Sèvres France

Keywords: GPS, P-code, time link, TAI.

Abstract

In April 2002, a pilot experiment was initiated by the BIPM in order to study the use of GPS P3 code measurements obtained with geodetic type dual frequency receivers to compute time links. For several time links, independent computations using two or three different techniques may be carried out and compared to assess the stability of each technique. This paper presents several such comparisons that show that the stability of GPS P3 is below 1 ns (one standard deviation) for averaging times between a few hours and several months. After more than one year of operation, such time links were introduced into TAI computation starting June 2003.

1 Introduction: the TAIP3 experiment

In recent years, most time links used for International Atomic Time TAI have been based on C/A code, single frequency, GPS receivers and on Ku-band Two way time transfer (hereafter TW) using geostationary satellites. In April 2002, the TAIP3 pilot experiment was proposed by the BIPM to laboratories participating to TAI. Its goal is to study the use of GPS P3 code measurements obtained with dual frequency receivers of the type Ashtech Z12-T to compute time links [5]. A dozen laboratories participate to this experiment and provided the data reported in this study.

Statistical analysis allows to estimate the short term stability of the P3 technique. On the other hand, the long term stability is estimated by comparison with independent techniques that are available for several of the links. This paper presents all possible comparisons between GPS-P3 and TWTT, and some triple comparisons including also GPS-CA. Results concerning the stability of each technique are presented in section 2. Because the receivers have been differentially calibrated, the accuracy of GPS-P3 links may also be assessed by comparison to other differentially calibrated links, as presented in section 3. Finally section 4 concludes and discusses the introduction of such time links into TAI computation, that was initiated in June 2003 for three links.

2 Stability of GPS P3 time links

In the following, we report results for several P3 links, including all those which can be compared to the TW technique, plus some links for which the only comparison is C/A single channel. See Table 1 for a list of equipment, compiled from GPS data files and from [3], and Table 2 for a list of the computed links.

Laboratory City	GPS P3 equipment	TW equipment	GPS C/A equipment
METAS (CH) Bern	GeTT (Ashtech)		AOA TTR5A (SC)
IEN Torino	Ashtech Z12T	MITREX 2500A	3S Nav. GNSS-300T (MC)
SYRTE (OP) Paris	Ashtech Z12T	TimeTech /SATRE	NBS TTR5 (SC)
PTB Braunschweig	Ashtech Z12T	TimeTech /SATRE	AOA TTR5 (SC) AOS SRC TTS-2 (MC)
USNO Washington DC	Ashtech Z12T	MITREX 2500	AOS SRC TTS-2 (MC)
CRL Tokyo	Ashtech Z12T	AOA/Atlantis	3S Nav. R100(MC) EURO-80 (MC) AOA TTR6 (SC)
NMIJ Tsukuba	Ashtech Z12T	AOA/Atlantis	AOA TTR6 (SC)
TL Chung Li	Ashtech Z12T	AOA/Atlantis	AOA (SC)

Table 1: List of laboratories considered in this study and their equipment (SC = Single Channel, MC = Multi Channel)

Link	Distance	Techniques
CH-PTB	650 km	P3, C/A SC
IEN-PTB	800 km	P3, TW, C/A SC-MC
OP-PTB	700 km	P3, TW, C/A SC
CRL-PTB	8300 km	P3, C/A SC-MC
USNO-ORB	5900 km	P3
USNO-PTB	6300 km	P3, TW, C/A MC
NMIJ-CRL	70 km	P3, TW, C/A SC
TL-CRL	2100 km	P3, TW, C/A SC

Table 2: List of the links considered in this study.

All GPS data presented in this section are measurements of the difference between the local time reference UTC(k) and GPS time provided directly by C/A receivers or generated by an external software for P3 [2]. These data were corrected for GPS precise ephemerides from the International GPS Service using the standard TAI procedure. In addition the C/A data were corrected using ionosphere maps from the IGS. Strict common-view measurements were differenced and a Vondrak smoothing was applied following the standard TAI procedure: The smoothing used for the P3 data is equivalent to a low pass filter with a cut-off period of about 0.6 day while the one used for the C/A data corresponds to a cut-off period of about 1.3 day. This difference is to account for the lower noise of the P3 data, mostly due to the larger number of measurements.

In sections 2.1 we study the stability of a P3 link by a statistical analysis. On the other hand, in section 2.2, all results are differences between two techniques, i.e. all should be constant with time. To evaluate a "long term instability" in section 2.2, we therefore compute the standard deviation of a time series over a long interval, which typically includes several tens of points over several months. In section 2.3, we consider cases where three nearly continuous series are available for three techniques and we infer the instability of each technique by a three-corner hat analysis.

2.1 Short-term stability

The "standard" processing of GPS common view measurements used for TAI consists in simple averaging of all simultaneous measurements (for multi-channel receivers) after correction for IGS ephemerides. By averaging such measurements for a few hours, measurement noise is reduced to a level below a few hundred ps. It is known that several physical effects which are not accounted for in this "standard" processing have a similar order of magnitude, and several of them may be taken into account.

- By using a weighted average of simultaneous measurements, according to the lowest elevation at the two stations, we expect an improvement of measurement noise (short term).
- By applying a model for solid Earth tides, we expect an improvement at 0.5 day averaging time
- By using troposphere zenith delay values obtained from the IGS, we expect an improvement at several days averaging, and possibly short term.

Here we use such "enhanced" processing (which would be "standard" in e.g. geodetic GPS analysis) to estimate the short-term stability achievable with GPS-P3. Stability is estimated by computing the time deviation on the link data (averaged to one point per epoch, every 16 minutes). Figure 1 shows such results for USNO-ORB (two masers) where the enhanced processing provides 0.2-0.3 ns time deviation from

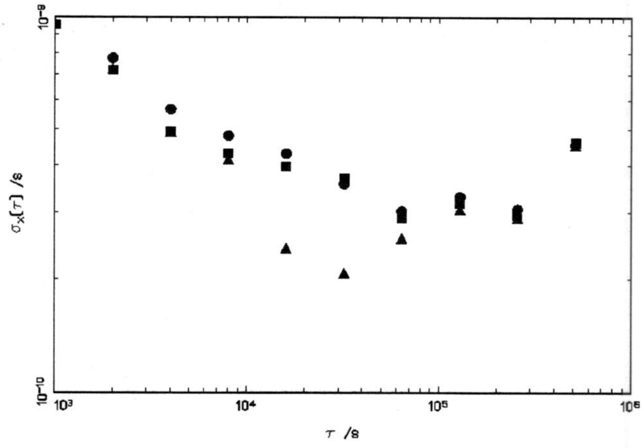

Figure 1: Time deviation for USNO-ORB GPS P3 link (circle = standard, square = weighted, triangle = weighted and tidal correction)

3 hours to 3 days averaging time. The last value corresponds to a 3-day frequency instability of 1.5×10^{-15} and clearly reveals the clock behavior for this averaging time, without limitation from the time transfer technique.

We acknowledge that the effect of propagation multipaths is the important remaining effect, that cannot easily be accounted for. Because it cannot be averaged out and can have long-term variations, it could be a dominant factor in the long term comparisons presented in the section below.

2.2 Long-term stability: comparison to other techniques

The following links for which both P3 and TW(Ku) are available have been computed: IEN-PTB, USNO-PTB, OP-PTB, NMIJ-CRL, TL-CRL. The difference [P3-TW] is plotted in the Figures. For the Europe-America links, TW data are 3 points per week until about MJD 52800, then 1 point per day until about MJD 53030, then 4 points per day. For the Asia-Pacific links, 2 sessions per week are conducted, each providing 2 or 3 measurements separated by a few minutes (i.e. that can be considered as simultaneous for our purpose).

For IEN-PTB (Figure 2) over several months, the RMS of the difference is between 1.0 ns and 1.2 ns. Note that the 2-month gap around MJD 52640, as well as smaller ones, are due to missing TW data, while a 15-day gap around MJD 52580 is due to P3. The two discontinuities between different intervals are due to undocumented changes in the P3 set-up at IEN.

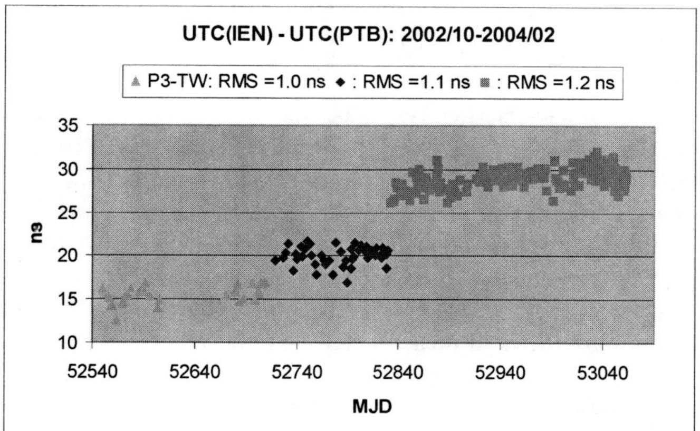

Figure 2: Difference [P3-TW] for the link IEN-PTB

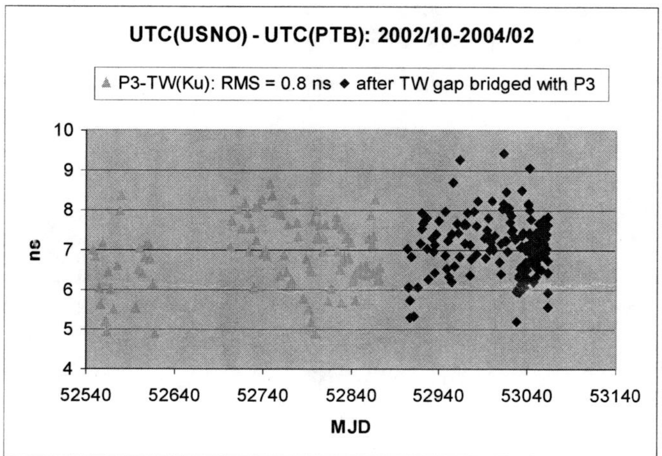

Figure 3: Difference [P3-TW] for the link USNO-PTB

For USNO-PTB (Figure 3), the RMS of the difference over 17 months is 0.8 ns. However there are two distinct intervals in this comparison due to a change in the TW set-up, and the corresponding discontinuity has been bridged using P3 data themselves, so that the very long-term stability could be aliased by this procedure. Note also that the 3-month gap is due to the interruption of P3 data transmission at USNO and that this data could be recovered.

For OP-PTB (Figure 4), the RMS over one month is 1.2 ns and is dominated by short-term or diurnal noise, mostly attributed to a problem in the TW data processing.

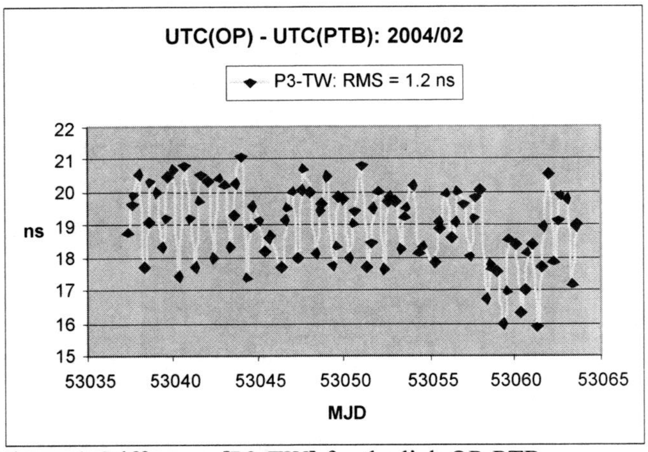

Figure 4: Difference [P3-TW] for the link OP-PTB

For the Asian-Pacific TW links, the RMS of the difference is 1.2 ns for NMIJ-CRL over 15 months (Figure 5) and between 1.1 and 1.3 ns for CRL-TL over several months (Figure 6). Note that the TW data used for NMIJ-CRL in Figure 5 has been edited for more than 10 outliers. In such cases, the 2-3 nearly simultaneous measurements differ significantly, by up to several ns. This indicates a problem because the uncertainty of each measurement should be well below 1 ns, but the TW data themselves do not allow to determine which measurements are wrong. Outliers are identified by comparison of the TW link with the P3 or C/A link over an extended period. Note also that the gap in theses links around MJD 52920 are due to change in the P3 set-up during a calibration exercise at CRL.

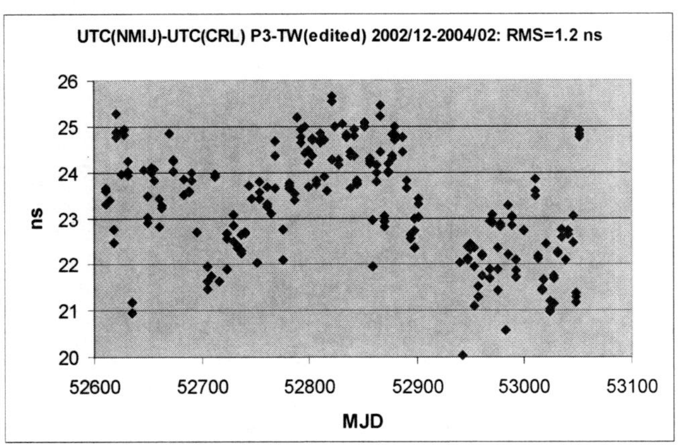

Figure 5: Difference [P3-TW] for the link NMIJ-CRL

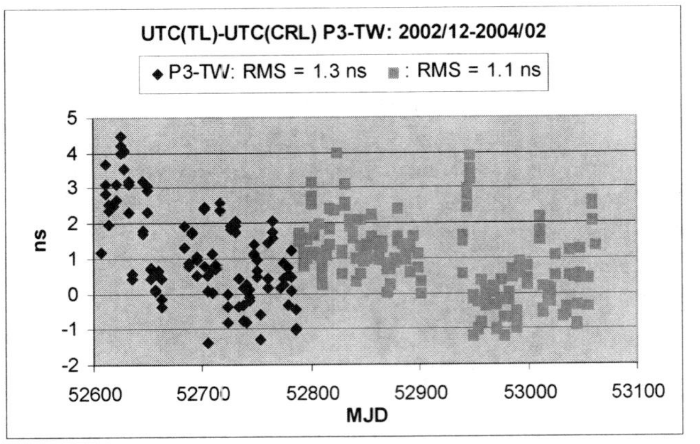

Figure 6: Difference [P3-TW] for the link TL-CRL

These comparisons indicate that, for a given link computed over several months, the standard deviation of the differenced data [P3-TW] is typically 1 ns, and slightly better in the best case (USNO-PTB).

For some links for which TW is not available, the comparison with GPS C/A, computed at the standard TAI dates, is presented: For CRL-PTB the RMS of the difference over 18 months is 1.7 ns (Figure 7) and for CH-PTB the RMS of the difference over 10 months is 1.9 ns (Figure 8). In addition, the results for OP-PTB are presented in Figure 9, with a RMS of the difference over 7 months of 1.6 ns. For the links with TW, the GPS C/A link has also been computed at the dates of the TW measurements over the longest and most recent continuous interval. For these comparisons, the RMS of the differenced data [P3-C/A] is typically at or below 2 ns, e.g. 1.5 ns for IEN-PTB over 5 months, 1.1 ns for USNO-PTB over 5 months, 1.6 ns for OP-PTB over 1 month, 0.8 ns for NMIJ-CRL over 3 months, 2.0 ns for TL-CRL over 4 months.

All [P3-C/A] comparisons display some long-term variations at the level of a few ns (see Figures 7 to 9), except in the very short link NMIJ-CRL (70 km) not shown here. It seems that these long-term variations, rather than short term noise, set the level of the RMS of the difference [P3-C/A] and that they are attributable to the C/A link (see below).

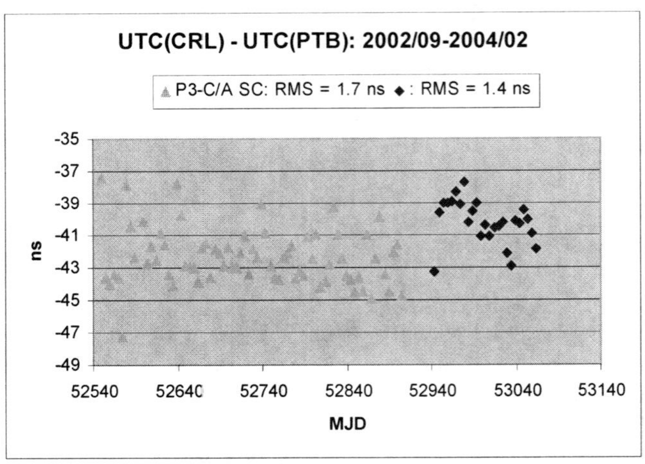

Figure 7: Difference [P3-C/A] for the link CRL-PTB

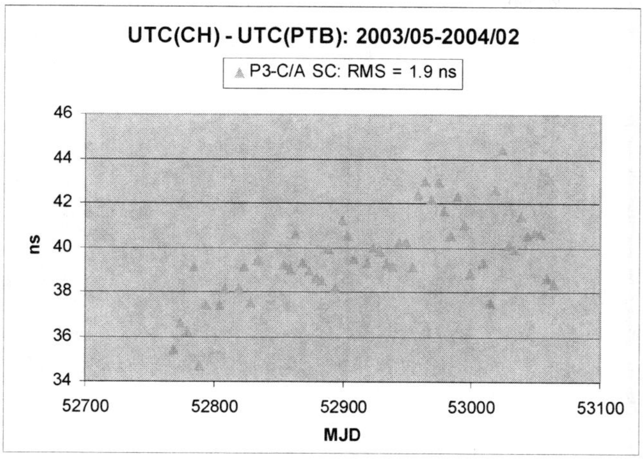

Figure 8: Difference [P3-C/A] for the link CH-PTB

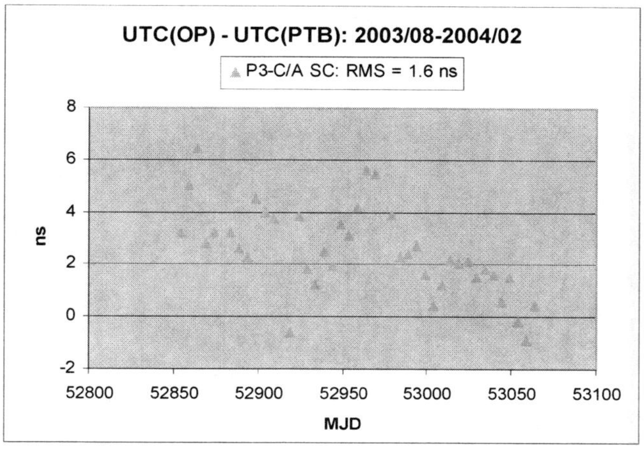

Figure 9: Difference [P3-C/A] for the link OP-PTB

2.3 Long-term stability: three corner hat analysis

When three independent techniques are available for a long period without any significant gap, the three corner hat technique can retrieve the intrinsic stability of each techniques from the three pair-differences. This has been estimated for (IEN-PTB) over a 7-month interval and for (USNO-PTB) over a 5-month interval. The time variance of the three pair-differences were used as input and the results of the 3-corner hat analysis are shown in Figures 10 and 11.

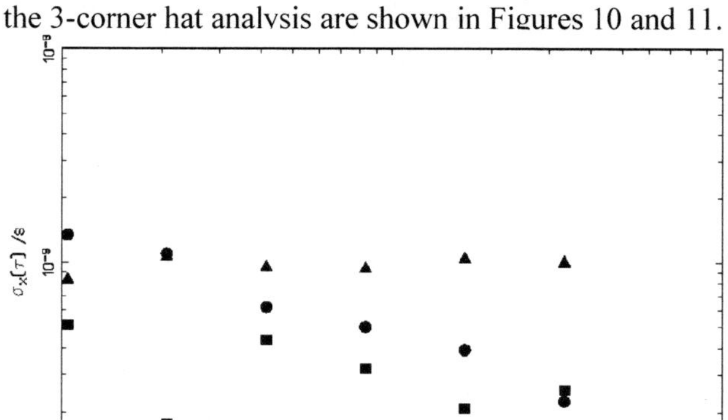

Figure 10: Time stability from 3-corner hat analysis for IEN-PTB (square = P3, circle = TW, triangle = C/A).

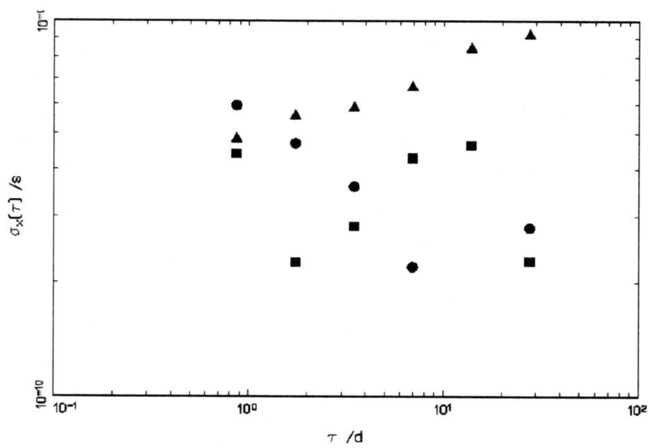

Figure 11: Time stability from 3-corner hat analysis for USNO-PTB (square = P3, circle = TW, triangle = C/A).

We observe that the time stability obtained for the P3 technique never exceeds 0.6 ns for an averaging time up to one month. That obtained for the TW technique is about equivalent except for a larger level of short term noise for IEN-PTB, at about 1 ns. The time stability for the C/A technique is consistently larger, around 1 ns, noting that the smaller values obtained for short averaging times are biased by the smoothing applied to the C/A data (see top of section 2) and are too optimistic.

3 Accuracy of GPS-P3 time links

In 2001, calibration of P3 equipment has been started using a travelling receiver from the BIPM that has been previously absolutely calibrated [6]. Each visited system is differentially calibrated against the travelling system and all links between two calibrated systems may thus be considered as calibrated. The uncertainty in the differential calibration of a P3 link has been stated at 3 ns [4]. Nevertheless discrete variations of up to a few ns in the (P1-P2) delay have sometimes been observed for some receivers and have been associated with changes in the "environment" in a general meaning, e.g. change in cable connections. These discrete events are detectable in the comparisons between techniques and thus do not affect the comparisons presented in section 2. However we cannot exclude that such events occur between calibrations or even that the results of a calibration exercise depend on the precise set-up of the travelling receiver. Until further information is obtained, we use a more conservative value of 5 ns for the uncertainty in the differential calibration of a P3 link.

All P3 equipment considered in this study have been differentially calibrated. However the results for the CRL P3 receiver are still under review and have not been included here. Therefore, out of the P3 links considered in this study, three can be compared with other independently calibrated techniques and the results are detailed below.

For OP-PTB, the C/A link is regularly calibrated. The average difference [P3-C/A] is of order 2.5 ns (Figure 9), while the stated uncertainty of a C/A differential calibration is 3 ns [3] so that the agreement between the C/A and P3 calibrations is very good. Note that long-term variations at the level of a few ns over several months are apparent and attributed to the C/A link. For USNO-PTB (Figure 3) a TW calibration was available until reconfiguration of the TW link around MJD 52880. The average difference [P3-TW] is about 7 ns, with no significant long-term variations, while the uncertainty of the TW calibration is estimated to be 1 ns [D. Matsakis, USNO, pers. comm.] The difference between the two calibrations is thus slightly larger than expected. For CH-PTB (Figure 8), the two C/A receivers have been calibrated in two different calibration trips, therefore the uncertainty of the C/A calibration is somewhat larger than the usual 3 ns value. Accounting for the result of the C/A calibration, not included in the results of Figure 8, the average difference [P3-C/A] is about 6 ns, so that the agreement between the C/A and P3 calibrations is satisfactory. Again long-term variations at the level of a few ns over several months are apparent.

4 Conclusions: introducing P3 links into TAI

Through a study of several time links with different techniques, we show that the time instability of GPS P3 time links is below 1.0 ns (one standard deviation) for averaging times up to one month. This number is supported by the results of statistical analysis (sections 2.1 and 2.3) and by comparisons with independent techniques (section 2.2). This performance is maintained at longer averaging times because the long comparisons with TW(Ku) links all show a standard deviation of about 1 ns. This performance is typically at least twice better than GPS C/A links which consistently display some long-term instabilities, and probably about equivalent to that of the TW(Ku) links.

A few P3 time links have already been introduced into TAI computation: DLR-PTB, IFAG-PTB, ORB-PTB in July 2003. In addition, all other P3 time links are computed as backup links and have occasionally been used. Several P3 links are considered for introduction into TAI mostly in replacement of present C/A links, e.g. CH-PTB, CRL-PTB, OP-PTB, NRC-USNO. In addition several TW links presently used in the TAI computation display various forms of short-term instabilities, such as diurnal variations or outliers. It is therefore expected that improvement would result from complementing these links by the use of P3 links.

Acknowledgements

We thank all colleagues participating to the TAIP3 experiment for their collaboration.

References

[1] J. Azoubib, W. Lewandowski, "24th BIPM TWSTFT Report" *BIPM TM.124*, (September 2003).

[2] P. Defraigne, G. Petit, "Use of geodetic receivers for TAI", *Metrologia* **40**, 184, (2003).

[3] W. Lewandowski, L. Tisserand., "Determination of the differential time corrections for GPS time equipment located at the OP, PTB, AOS, KRIS, CRL, NIST, USNO and APL", BIPM report 2004/06 (2004).

[4] G. Petit, "Estimation of the values and uncertainties of the BIPM Z12-T receiver and antenna delays, for use in differential calibration exercises", *BIPM TM.116*, (July 2002).

[5] G. Petit, Z. Jiang, P. Moussay, "TAI time links with geodetic GPS receivers: a progress report", *Proc. PTTI*, p. 19, (2002).

[6] G. Petit, Z. Jiang, J. White, R. Beard, E. Powers, "Absolute calibration of Ashtech Z12-T GPS receiver", *GPS Solutions* 4 (4), p. 41, (2001).

PM AND AM NOISE IN VOLTAGE CONTROLLED PHASE SHIFTERS

V.N.Kuleshov, A.E.Mordvinov

Moscow Power Engineering Institute (Technical University), Moscow, Russia, kuleshov@srv-vmss.mpei.ac.ru

Keywords: voltage controlled phase shifters, phase control characteristic, wideband noise sources, 1/f noise sources, PM and AM noise.

Abstract

PM and AM noise analysis in voltage controlled phase shifter, consisting of two amplifiers with voltage controlled gain, phase splitter and summing amplifier, is presented. It is based on the theory of PM and AM noise in bipolar junction transistor (BJT) amplifiers. Both 1/f and wideband noise sources in BJT amplifiers are taken into consideration. It is shown that conversion of AM noises of the amplifiers with voltage controlled gain to PM noise of the phase shifter gives the main contribution to total 1/f PM noise. The formulae for power spectral densities of PM and AM noises are obtained. An influence of feedback through unbypassed and bypassed emitter resistances in the voltage controlled amplifiers on the phase shift control characteristic and PM noise is investigated. Opportunities to decrease PM noise using this feedback are analyzed and discussed. An examples of PM noise calculation are considered.

1 Introduction

Voltage controlled phase shifters (VCPS) (phase modulators) based on the transformation of amplitude modulation (AM) to phase modulation (PM) were offered by E.M. Armstrong [1]. Other types of such circuits were analyzed by A.A.Kulikovsky [6] and I.S.Gonorovsky [2]. As such circuits can be manufactured using integral technologies they are convenient for some applications. But such modulators bring PM and AM noise in output signal. It is important for applications to know power spectral densities (PSD) of this noise caused by internal noise sources of the VCPS.

In this paper an approach to PM and AM noise calculation is developed. It is applied to particular example of the VCPS circuit.

2 The phase shifter block diagram

A block diagram of the VCPS that is considered in this paper is shown in Fig.1.

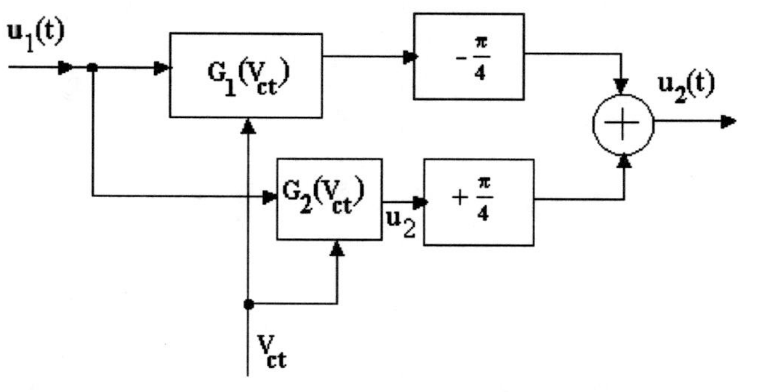

Fig. 1. A block diagram of the voltage controlled phase shifter (VCPS)

In this diagram $G_1(V_C)$ and $G_2(V_C)$ are voltage controlled amplifiers of input harmonic signal

$$u_1(t) = U_1 \cos \omega_S t. \tag{1}$$

The control voltage is V_{ct}. The blocks "$-\frac{\pi}{4}$" and "$+\frac{\pi}{4}$" provide constant phase shifts equal to $\left(-\frac{\pi}{4}\right)$ and $\left(\frac{\pi}{4}\right)$. The output voltage of this VCPS is

$$u_2(t) = U_1 G_1(V_{ct}) \cos\left(\omega_S t - \frac{\pi}{4}\right) + U_1 G_2(V_{ct}) \cos\left(\omega_S t + \frac{\pi}{4}\right) \tag{2}$$

The oscillation (2) can be expressed as follows

$$u_2(t) = K_{US} U_1 \cos(\omega_S t + \varphi_S), \tag{3}$$

where
$$K_{US} = \sqrt{G_1^2(V_{ct}) + G_2^2(V_{ct})}, \tag{4}$$

$$\varphi_S = -\frac{\pi}{4} + \operatorname{arctg}\frac{G_1(V_{ct})}{G_2(V_{ct})}. \tag{5}$$

From (4), (5) it is evident that, when $(-V_{ctm}) \le V_{ct} \le V_{ctm}$, and

$$G_1(V_{ct}) = \sin\frac{\pi}{4}\left(\frac{V_{ct}}{V_{ctm}} + 1\right);\ G_2(V_{ct}) = \cos\frac{\pi}{4}\left(\frac{V_{ct}}{V_{ctm}} + 1\right), \tag{6}$$

we obtain
$$K_{US} = 1, \qquad \varphi_S = \frac{\pi}{4}\frac{V_{ct}}{V_{ctm}}. \tag{7}$$

In this case dependence of phase shift φ_S on the control voltage V_{ct} is linear and an amplitude of the output signal doesn't depend on V_{ct} (Fig.2). Let us consider this VCPS as an ideal one.

Real functions $G_1(V_{ct})$ and $G_2(V_{ct})$ differ from the ones of the ideal VCPS. Real voltage controlled amplifiers bring their

AM and PM noise in the signal. That gives rise to PM and AM noise of the VCPS. To investigate it we consider particular circuit diagram of the VCPS corresponding to the block diagram Fig.1.

3 Circuit diagram of the VCPS

We consider here the simple circuit diagram of the VCPS, using transformation of AM to PM (Fig.3). Only the stages that give main contribution to total PM and AM noise will be analyzed in detail.

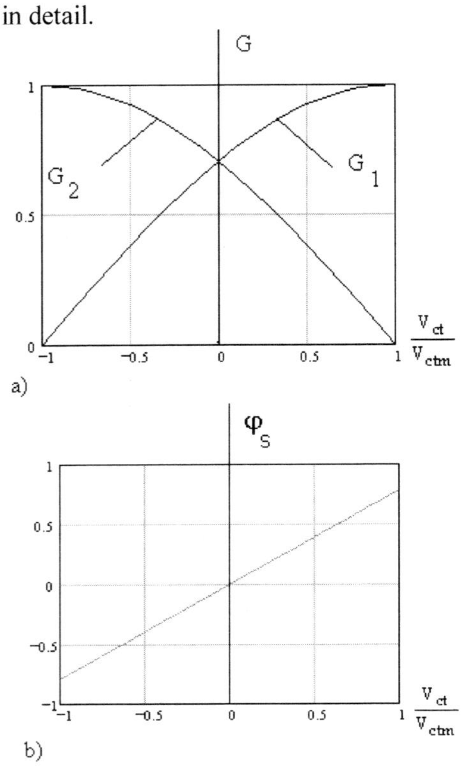

Fig.2. Characteristics (6) of voltage controlled amplifiers (a) and dependence of φ_S on the control voltage (7) in ideal VCPS.

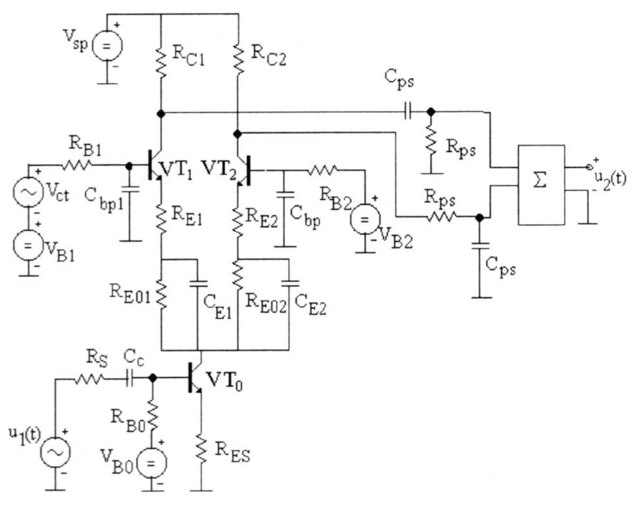

Fig.3. A simplified circuit diagram of the VCPS.

A high frequency signal $u_1(t)$ is applied to the input of the transistor VT_0. This transistor is a source of the current

$$I_{0S}(t) = I_0 + i_S(t), \tag{6}$$

where I_0 is a constant current and $i_S(t)$ is a high frequency current caused by $u_1(t)$.

A control voltage V_{ct} is applied to transistor VT_1 of the differential pair (VT_1, VT_2). This voltage changes average collector currents I_{C1}, I_{C2} of the transistors VT_1 and VT_2, and their incremental current gains, that are defined as follows

$$G_1 = \frac{i_{C1}(t)}{\alpha i_S(t)}\ ; \qquad G_2 = \frac{i_{C2}(t)}{\alpha i_S(t)}, \tag{7}$$

where α – common base current gain of the transistors (it is supposed constant and equal for both of them), $i_{C1}(t)$ and $i_{C2}(t)$ – high frequency components of the collector currents.

It is supposed that capacitors C_{bp1}, C_{bp2} provide very small impedance for high frequencies by comparison to R_{B1} and R_{B2}. Resistors R_{E1} and R_{E2} provide current feedback for high frequencies, and resistors R_{E10} and R_{E20} provide current feedback for modulation frequencies that are assumed much lower than signal frequencies. Bypassing capacitors C_{E1} and C_{E2} provide small impedance's for high frequencies by comparison to R_{E10} and R_{E20}.

4 A circuit model for control frequencies

As the current gains G_1 and G_2 depend on the currents I_{C1} and I_{C2} in the operating points of the transistors it is necessary to find the currents I_{C1} and I_{C2} as functions of V_{ct}. To find these currents we use nonlinear models of the transistors VT_1 and VT_2 shown in Fig.4. In this consideration we supposed that $V_{b1} = V_{b2} = V_b$; $R_{b1} = R_{b2} = R_b$; $r_{b1} = r_{b2} = r_b$; $R_{E1} = R_{E2} = R_E$; $R_{E10} = R_{E20} = R_{E0}$. We suppose also that collector currents depend on emitter junction voltages in accordance with Ebers-Moll model [3].

$$I_{C1} = I_{sat}\left(\exp\frac{V_{b'e1}}{V_T} - 1\right), \quad I_{C2} = I_{sat}\left(\exp\frac{V_{b'e2}}{V_T} - 1\right), \tag{8}$$

where V_T – is a thermal voltage, I_{sat} – saturation currents of the BJT, and base currents can be presented as follows

$$I_{b1} = [1+\mu_1(t)]\frac{1}{\beta}I_{C1}, \quad I_{b2} = [1+\mu_2(t)]\frac{1}{\beta}I_{C2}, \tag{9}$$

where β – common emitter current gain of the transistors, and $\mu_1(t)$ and $\mu_2(t)$ – relative 1/f fluctuations of the recombination rates in VT_1 and VT_2 that present sources of 1/f noise in these transistors. We suppose that fluctuations $\mu_1(t)$ and $\mu_2(t)$ are uncorrelated and have equal power spectral densities $S\mu_1(\omega) = S\mu_2(\omega) = S\mu(\omega)$. Such presentation corresponds to 1/f noise model discussed in the papers [4,5].

From the model Fig.4 and (8), (9) the next equations follow:

$$R_{F0}(I_{C1} - I_{C2}) + V_T \ln\frac{I_{C1}}{I_{C2}} = V_{ct} - R_{N0}(I_{C1}\mu_1 - I_{C2}\mu_2), \tag{10}$$

$$I_{C1} + I_{C2} = \alpha I_0 - (\beta+1)^{-1}(I_{C1}\mu_1 + I_{C2}\mu_2), \tag{11}$$

where

$$R_{N0} = \beta^{-1}(R_B + r_b + R_E + R_{E0}), \quad (12)$$

$$R_{F0} = (R_{N0} + R_E + R_{E0}). \quad (13)$$

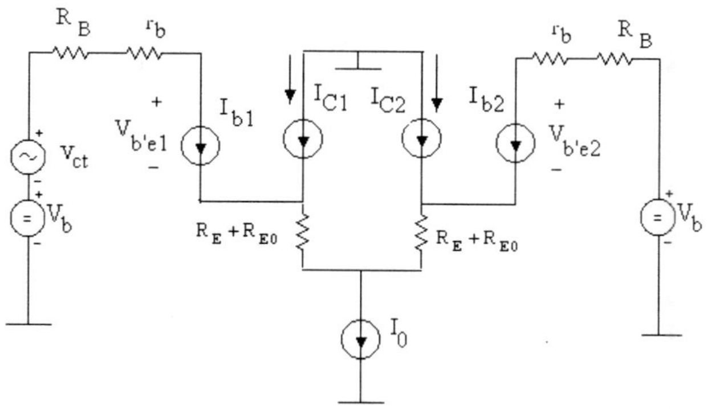

Fig.4. A circuit model of the differential transistor pair for low control frequencies

If one puts $\mu_1 = \mu_2 = 0$ the famous dependencies of $I_{C1}(V_{ct})$ and $I_{C2}(V_{ct})$ can be easily obtained. But in addition to them we obtained from (10), (11) the values $\delta_1(t)$, $\delta_2(t)$ of relative fluctuations of I_{C1} and I_{C2} that are caused by $\mu_1(t)$ and $\mu_2(t)$. These fluctuations are necessary to calculate 1/f PM and AM noise of the VCPS.

5 An incremental model of the VCPS with 1/f noise source

To calculate current gains G_1 and G_2 we used an incremental model shown in Fig.5. We didn't take into consideration the internal capacitances of the transistor incremental model. Their influence adds some corrections, but they are not discussed in this paper. Here we discuss only fundamental mechanisms of 1/f noise influence on PM and AM noise of the VCPS.

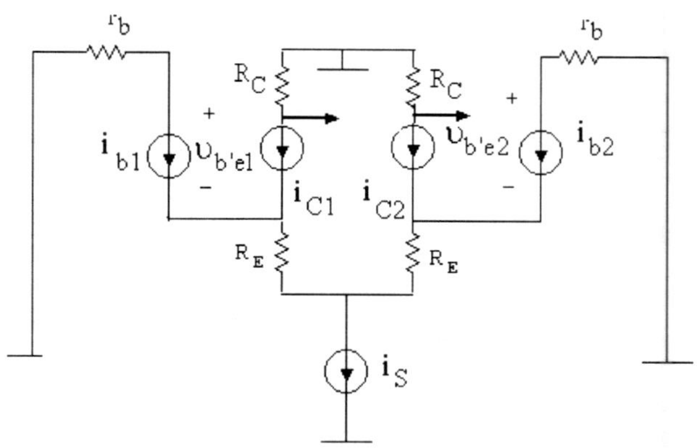

Fig.5. An incremental model of the differential pair for signal frequency

From the circuit Fig.5, taking into consideration that

$$g_{m1} = I_{C1} / V_T, \; g_{m2} = I_{C2} / V_T, \quad (14)$$

$$i_{C1} = (1+\delta_1) g_{m1} \upsilon_{b'e1}, \; i_{C2} = (1+\delta_2) g_{m2} \upsilon_{b'e2}, \quad (15)$$

$$i_{b1} = (1+\mu_1)\frac{1}{\beta} i_{C1}, \quad i_{b2} = (1+\mu_2)\frac{1}{\beta} i_{C2}, \quad (16)$$

where $\delta_1(t)$ and $\delta_2(t)$ are relative fluctuations of g_{m1} and g_{m2} that are equal to relative fluctuations of I_{C1} and I_{C2}, we obtain the equations

$$\left(R_F + \frac{V_T}{I_{C1}}\right) i_{C1} - \left(R_F + \frac{V_T}{I_{C2}}\right) i_{C2} = = -R_N (i_{C1}\mu_1 - i_{C2}\mu_2) + \frac{V_T}{I_{C1}} (i_{C1}\delta_1 - i_{C2}\delta_2), \quad (17)$$

$$i_{C1} + i_{C2} = \alpha i_S - (\beta+1)^{-1} (i_{C1}\mu_1 + i_{C2}\mu_2), \quad (18)$$

where

$$R_F = \beta^{-1}(r_b + R_E) + R_E, \quad (19)$$

$$R_N = \beta^{-1}(r_b + R_E). \quad (20)$$

6 An algorithm of the phase control characteristic calculation

Assuming that $\mu_1 = \mu_2 = 0$ and as follows $\delta_1 = \delta_2 = 0$, we obtain from equations (17), (18) the formulae for current gains G_1 and G_2:

$$G_1 = \frac{J_1(\lambda J_2 + 1)}{J_1(\lambda J_2 + 1) + J_2(\lambda J_1 + 1)}, \quad (21)$$

$$G_2 = 1 - G_1, \quad (22)$$

where $J_1 = I_{C1} / \alpha I_0, \; J_2 = I_{C2} / \alpha I_0,$ (23)

$$\lambda = \alpha I_0 R_F / V_T. \quad (24)$$

Using equations (10), (11) with $\mu_1 = \mu_2 = 0$ we express V_{ct} in terms of J_1, J_2:

$$(V_{ct}/V_T) = \lambda_0 (2J_1 - 1) + \ln \frac{J_1}{1-J_1}, \quad (25)$$

$$J_2 = 1 - J_1, \quad (26)$$

Where

$$\lambda_0 = \alpha I_0 R_{F0} / V_T. \quad (27)$$

From (21), (22), (25) one can obtain the functions $G_1(V_{ct})$ and $G_2(V_{ct})$. Using them and (4), (5), the dependencies $\varphi_S(V_{ct})$ and $K_{US}(V_{ct})$ can be calculated.

7 An algorithm of 1/f PM and AM noise calculation

To calculate 1/f PM and AM noise we have found from (17), (18), (10), (11) relative fluctuations of G_1 and G_2, and then from (4), (5) expressed fluctuations $\Delta\varphi_S(t)$ and $\Delta K_{US}(t)$ in terms of $\mu_1(t)$ and $\mu_2(t)$.

$$\Delta\varphi_S = T_{\varphi 1}\mu_1(t) + T_{\varphi 2}\mu_2(t), \quad (28)$$

$$\Delta K_{US} = T_{a1}\mu_1(t) + T_{a2}\mu_2(t), \quad (29)$$

where $T_{\varphi 1}$, $T_{\varphi 2}$, T_{a1}, T_{a2} – partial transformation coefficients. As we assumed that $\mu_1(t)$ and $\mu_2(t)$ are uncorrelated and have equal power spectral densities (PSD) $S\mu_1(\omega) = S\mu_2(\omega) = S\mu(\omega)$, the PSD's $S\varphi_S(\omega)$ and $S_{KUS}(\omega)$ of $\Delta\varphi_S(t)$ and $\Delta K_{US}(t)$ can be expressed as follows

$$S_{\varphi S}(\omega) = T_\varphi^2 S_\mu(\omega), \quad (30)$$

$$S_{KUS}(\omega) = T_a^2 S_\mu(\omega), \quad (31)$$

where

$$T_\varphi^2 = T_{\varphi 1}^2 + T_{\varphi 2}^2, \quad (32)$$

$$T_a^2 = T_{a1}^2 + T_{a2}^2. \quad (33)$$

The values of T_φ^2 and T_a^2 are transformation coefficients that characterize an influence of both 1/f noise sources of the transistors VT_1 and VT_2 on PSD of phase and amplitude fluctuations of the VCPS output signal.

These coefficients can be calculated as functions of the parameters β, r_b, R_B, R_E, R_{E0} and control voltage V_{ct}. Some results of these calculations are given further.

8 An algorithm of wideband PM and AM noise calculation

To calculate wideband PM and AM noise we use an incremental model of the VCPS different pair with incremental sources of wideband noise of the transistor VT_1 and VT_2. It is shown in Fig.6.

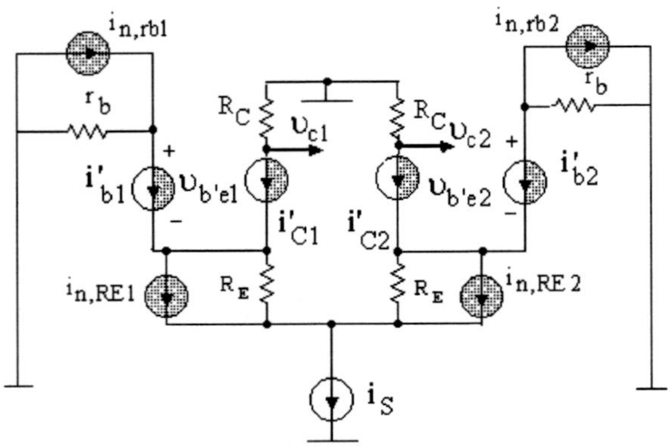

Fig.6. An incremental model of the differetial pair with wideband noise sources

In this model: $i_{n,rb1}$, $i_{n,rb2}$ - thermal noise currents of base resistances; $i_{n,RE1}$, $i_{n,RE2}$ - thermal noise currents of emitter resistances;

$$i'_{C1} = g_{m1}\upsilon_{b'e1} + i_{n,C1}, \quad i'_{C2} = g_{m2}\upsilon_{b'e2} + i_{n,C2} \quad (34)$$

- collector currents with shot noise sources $i_{n,C1}$, $i_{n,C2;}$

$$i'_{b1} = \frac{g_{m1}}{\beta}\upsilon_{b'e1} + i_{n,b1}, \; i'_{b2} = \frac{g_{m1}}{\beta}\upsilon_{b'e2} + i_{n,b2} \quad (35)$$

- base currents with recombination noise sources $i_{n,b1}$, $i_{n,b2}$.

To simplify resulting formulae we introduce the next designations:

$$\left.\begin{array}{l} i_{w,1} = i_{n,rb1}, i_{w,2} = i_{n,rb2}, i_{w,3} = i_{n,b1}, i_{w,4} = i_{n,b2}, \\ i_{w51} = i_{n,C1}, i_{w,6} = i_{n,C2}, i_{w,7} = i_{n,RE1}, i_{w,8} = i_{n,RE2}, \end{array}\right\} (36)$$

We fulfill at first PM and AM noise calculation for one source of wideband noise $i_{w,k}$ (k=1,2... 8).

Let us present signal current $i_S(t)$, noise current $i_{w,k}$, and collector currents $i'_{C1}(t)$, $i'_{C2}(t)$, using complex form [7]:

$$i_S = \mathrm{Re}\left(I_S e^{j\omega_s t}\right), \quad (37)$$

$$i_{w,k} = \mathrm{Re}\left[\left(I_{ws,k} + jI_{wq,k}\right)e^{j\omega_s t}\right], \quad (38)$$

$$i'_{C1,k} = \mathrm{Re}\left[\left(\dot{I}'_{C1,k}\right)e^{j\omega_s t}\right], \; i'_{C2,k} = \mathrm{Re}\left[\left(\dot{I}'_{C2,k}\right)e^{j\omega_s t}\right]. (39)$$

The currents I'_{C1}, I'_{C2} can be expressed as follows

$$I'_{C1,k} = \alpha G_1 I_S + H_{1k} I_{ws,k} + jH_{1k} I_{wq,k}, \quad (40)$$

$$I'_{C2,k} = \alpha G_2 I_S + H_{2k} I_{ws,k} + jH_{2k} I_{wq,k}. \quad (41)$$

In these equations current gains G_1, G_2 (7) are defined by (21), (22), and H_{1k}, H_{2k} – transfer functions connecting noise components of $i'_{C1,k}$, $i'_{C2,k}$ with noise current $i_{w,k}$. They are real functions of V_{ct}, that have to be calculated from the circuit model Fig.6.

In accordance with block diagram Fig.1 and circuit model Fig.6 complex amplitude of the output voltage can be written in the form:

$$U_{2,k} = R_T\left(I'_{C1,k} e^{-j\frac{\pi}{4}} + I'_{C2,k} e^{j\frac{\pi}{4}}\right), \quad (42)$$

where R_T is some constant value.

Using (38), (39), we obtain from (40):

$$U_{2,k} = \alpha R_T\left(G_1 e^{-j\frac{\pi}{4}} + G_2 e^{j\frac{\pi}{4}}\right)(1 + a_{w,k} + j\phi_{w,k}), \quad (43)$$

where

$$a_{w,k} = W_{1,k} I_{ws,k} - W_{2,k} I_{wq,k}, \quad (44)$$

$$\phi_{w,k} = W_{2,k} I_{ws,k} + W_{1,k} I_{wq,k}, \quad (45)$$

$$W_{1k} = \frac{H_{1k}G_1 + H_{2k}G_2}{G_1^2 + G_2^2}, \; W_{2k} = \frac{H_{2k}G_1 - H_{1k}G_2}{G_1^2 + G_2^2}. \quad (46)$$

From (43) it is clear that $a_{w,k}$ and $\phi_{w,k}$ are relative amplitude fluctuations and phase fluctuations caused by noise source $i_{w,k}$. Fluctuations, caused by all sources of wideband noise

$$a_w = \sum_{k=1}^{8} a_{w,k}, \quad \phi_w = \sum_{k=1}^{8} \phi_{w,k}, \quad (47)$$

can be found from (47), (44), (45)

$$a_w = \sum_{k=1}^{8} \left(W_{1k} I_{ws,k} - W_{2k} I_{wq,k}\right)$$
$$\phi_w = \sum_{k=1}^{8} \left(W_{2k} I_{ws,k} + W_{1k} I_{wq,k}\right) \quad (48)$$

It is known that for stationary wideband noises power spectral densities (PSD) $S_{IS,k}$, $S_{Iq,k}$ of components $I_{ws,k}$, $I_{wq,k}$ are equal and they can be expressed in terms of PSD $S_{i,k}$ of the current $i_{w,k}$ as follows [7]

$$S_{IS,k} = S_{Iq,k} = 2S_{i,k} \,. \quad (49)$$

It is known also that $I_{ws,k}$ and $I_{wq,k}$ are not correlated. That is why from (48) it follows

$$S_w = S_{a,w} = S_{\phi,w} = 2 \sum_{k=1}^{8} \left(W_{1k}^2 + W_{2k}^2\right) S_{i,k} \,. \quad (50)$$

From (36) and noise theory of the BJT one can find:

$$S_{i,1} = S_{1,2} = 4k_B T r_b^{-1},$$
$$S_{i,3} = 2q_e \alpha \beta^{-1} I_0 J_1, S_{i,4} = 2q_e \alpha \beta^{-1} I_0 (1 - J_1),$$
$$S_{i,5} = 2q_e \alpha I_0 J_1, \quad S_{i,6} = 2q_e \alpha I_0 (1 - J_1),$$
$$S_{i,1} = S_{1,2} = 4k_B T R_E^{-1}, \quad (51)$$

where k_B – Boltzmann constant, T- absolute temperature, q_e – electron charge.

Taking in consideration (46), (50) and (51) one can see that the problem of calculation of wideband PM and AM noise is reduced to calculation of transfer functions H_{1k}, H_{2k} (k=1,2…8) using circuit model shown in Fig.6.

Using (50), (51) it is possible to find contribution of each noise source (or group of noise sources) to total wideband noise level. One can analyze also dependence of wideband noise on the control voltage V_{ct}, circuit parameters and BJT parameters.

9 Total PM and AM noise calculation

As 1/f noise sources and wideband noise sources are statistically independent, total PSD's of PM and AM noise can be calculated as sum of PSD's (30), (31) and (50)

$$S_\phi(\omega) = S_w + T_\phi^2 S_\mu(\omega), \quad (52)$$

$$S_a(\omega) = S_w + T_a^2 S_\mu(\omega). \quad (53)$$

Supposing that $S_\mu(\omega)$ can be presented by formula

$$S_\mu(\omega) = S_\mu(\omega_m)\frac{\omega_m}{\omega}, \quad (54)$$

where ω_m – measurement frequency at which the PSD of the BJT 1/f noise was found experimentally, we obtain from (52), (53):

$$S_\phi(\omega) = S_w + T_\phi^2 S_\mu(\omega_m)\frac{\omega_m}{\omega}, \quad (55)$$

$$S_a(\omega) = S_w + T_a^2 S_\mu(\omega_m)\frac{\omega_m}{\omega} \quad (56)$$

Using these formulae one can find corner frequency of PM noise at which 1/f component is equal to wideband component:

$$\omega_{c,\phi}(\omega) = T_\phi^2 \frac{S_\mu(\omega_m)}{S_w}\omega_m \,. \quad (57)$$

Similar corner frequency can be found for AM noise

$$\omega_{c,a}(\omega) = T_a^2 \frac{S_\mu(\omega_m)}{S_w}\omega_m \,. \quad (58)$$

Using an approach and formulae that were obtained above, we consider a particular example of VCPS.

10 Case study

In this part we analyze a particular example of VCPS. We suppose that BJT's VT_1 and VT_2 have the next parameters

$$r_b = 60 \text{ Ohm}, \beta = 50, \alpha = 0{,}98 \quad (59)$$

and we set average emitter current

$$I_0 = 5 \text{ mA}. \quad (60)$$

We suppose also that

$$R_B = 50 \text{ Ohm}. \quad (61)$$

Let us consider first an influence of resistors R_E and R_{E0} on the VCPS characteristics $\varphi_S(V_{ct})$ and $K_{US}(V_{ct})$. Some examples of dependencies $\varphi_S(V_{ct})$ and $K_{US}(V_{ct})$ are given in Fig.7,8,9.

Dimensionless parameters of high frequency feedback λ (24) and low frequency feedback λ_0 (27) for which the curves Fig.7,8,9 were calculated are given in Tables 1,2. The values of R_E and R_{E0} corresponding to these parameters are also given in the Tables 1,2. They were calculated from (24), (27), (19), (13).

λ	0,235	1	4
R_E	0	3,8 Ohm	18,8 Ohm

Table 1: High frequency feedback parameters used in the case study

λ_0	0,431	10	100
$R_{E0} + R_E$	0	47,8 Ohm	497,8 Ohm
R_{E0} with λ =1	0	44 Ohm	494 Ohm
V_{ct} range	±0,15 V	±0,35 V	±2,6 V

Table 2: Low frequency feedback parameters used in the case study

One can see from Fig.7-9 that increasing λ_0 leads to increasing of control voltage range, providing available variation of the phase shift. It leads also to decreasing of phase shift sensitivity to variations of control voltage. An influence of λ on phase control characteristics is shown in Fig.8,9. It is clear from Fig.8,9 that increasing λ from minimal value to some optimal value improves linearity of the control characteristic $\varphi_S(V_{ct})$. When we increase λ further the linearity becomes worse. The optimal value of λ is close to 1. That is why PM and AM noise is calculated further for the case when λ=1.

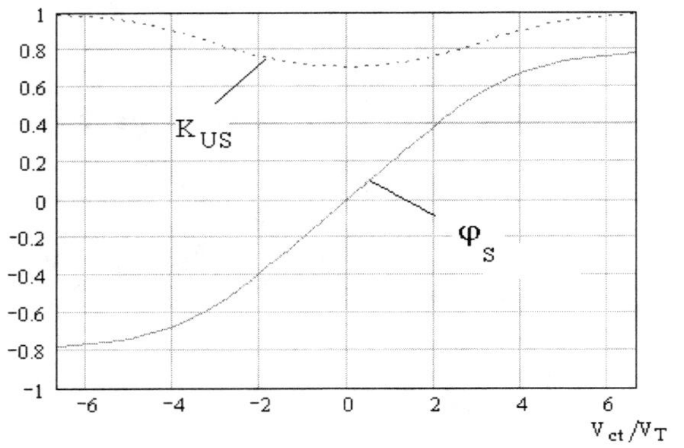

Fig.7. The phase control characteristic and amplitude variation characteristic of the VCPS with λ =1 and λ_0 =1,2

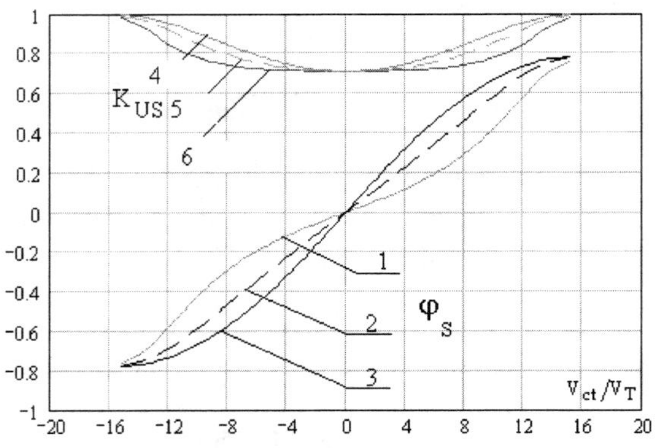

Fig.8. The phase control characteristic (curves 1,2,3) and amplitude variation characteristic (curves 4,5,6) of the VCPS with λ_0 =10 and λ =0,235 (curves 1,4), λ = 1 (curves 2,5), λ =4 (curves 3,6).

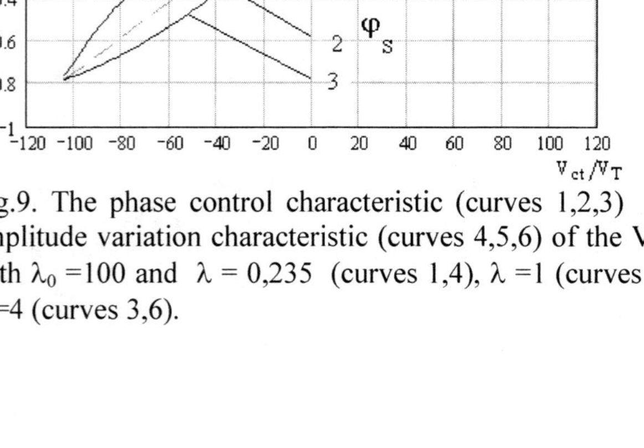

Fig.9. The phase control characteristic (curves 1,2,3) and amplitude variation characteristic (curves 4,5,6) of the VCPS with λ_0 =100 and λ = 0,235 (curves 1,4), λ =1 (curves 2,5), λ =4 (curves 3,6).

Results of calculation of 1/f noise transformation coefficients for λ = 1 and λ_0 given in the Table 2 are presented in Fig.10. One can see that the curves $T_\varphi^2(V_{ct})$ have maximum at the point V_{ct} = 0. The maximum values of T_φ^2 become less when λ_0 becomes more (–23 dB when λ_0 = 1,2; -26 dB when λ_0 = 10; -28 dB when λ_0 = 100).

The curves $T_a^2(V_{ct})$ have minimum at the point V_{ct} = 0. The values of T_a^2 at this point are equal to –40 dB.

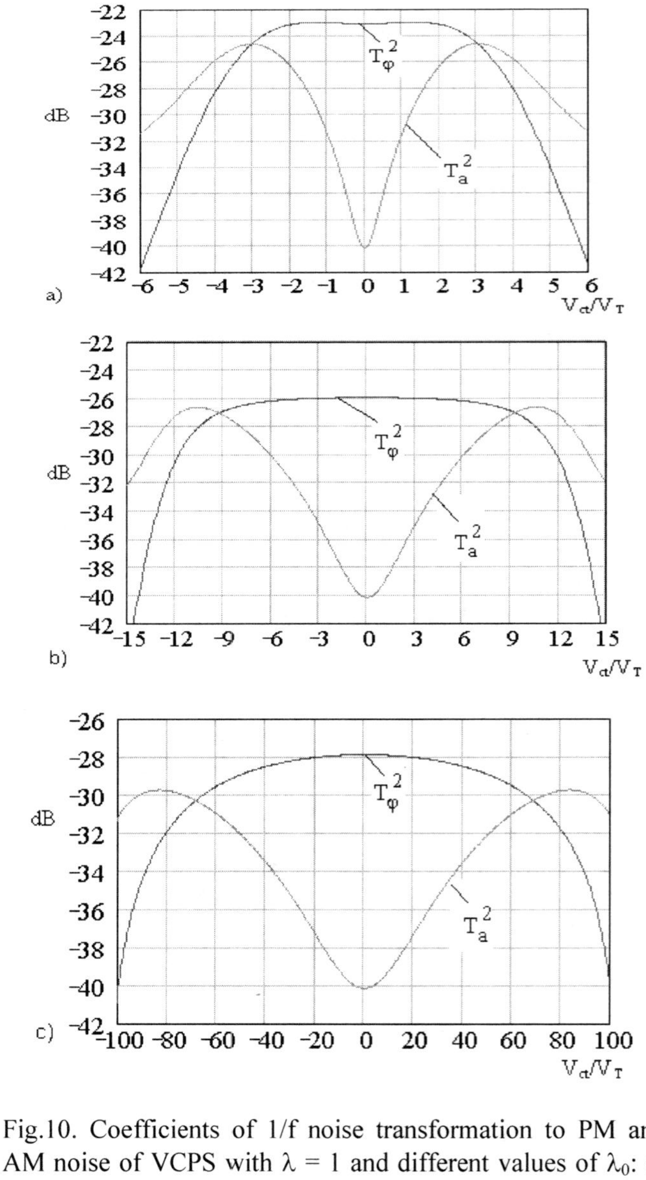

Fig.10. Coefficients of 1/f noise transformation to PM and AM noise of VCPS with λ = 1 and different values of λ_0: a) λ_0 =1,2; b) λ_0 =10; c) λ_0 =100

Results of total wideband PM and AM noise calculation are shown in Fig.11. Maximum value of S_w with I_S = 3 mA is about (-146) dBrad/Hz. It is located at V_{ct}=0. Its value does not depend on λ_0. It depends on the amplitude of signal current as I_S^{-2}. Contributions of different sources of wideband noise in transistors are shown in Fig.12. One can see that thermal noise of base resistances r_b (noise currents $i_{n,rb1}$, $i_{n,rb2}$ in Fig.6) give the biggest contribution. If one takes the BJT's with smaller r_b, total wideband noise can be decreased.

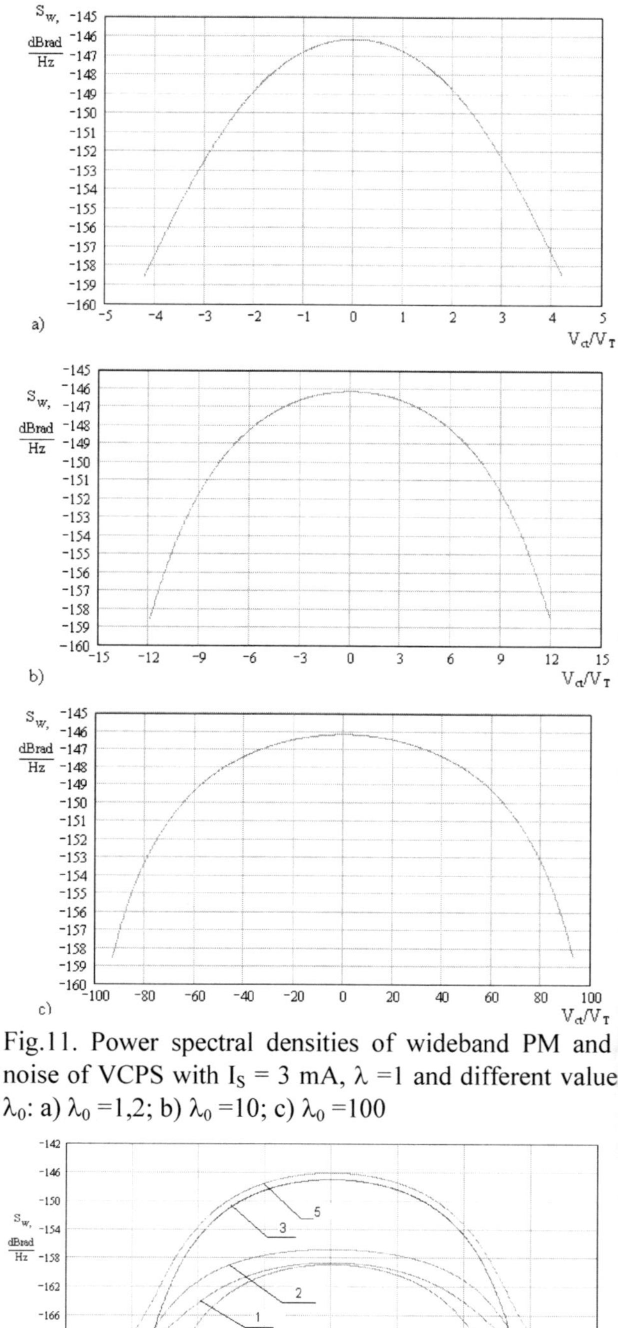

Fig.11. Power spectral densities of wideband PM and AM noise of VCPS with I_S = 3 mA, λ =1 and different values of λ_0: a) λ_0 =1,2; b) λ_0 =10; c) λ_0 =100

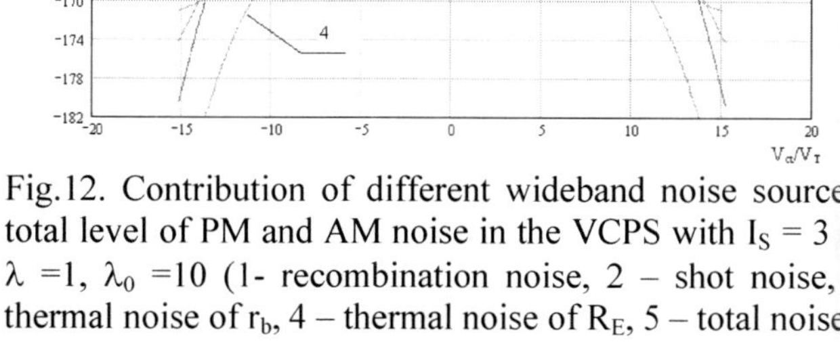

Fig.12. Contribution of different wideband noise sources to total level of PM and AM noise in the VCPS with I_S = 3 mA, λ =1, λ_0 =10 (1- recombination noise, 2 – shot noise, 3 – thermal noise of r_b, 4 – thermal noise of R_E, 5 – total noise).

To calculate total PM and AM noise of the VCPS using (55), (56), one needs to know PSD $S_\mu(\omega)$. Using experimental data given in the paper [5]

$$S_\mu(\omega) = \frac{4{,}8 \cdot 10^{-12}}{I_0/2}\omega^{-1} \qquad [1/\text{Hz}]$$

and supposing that I_0 = 5 mA, $\omega_m/2\pi = 10^3$ Hz, we find

$$S_\mu(\omega_m) = 3 \cdot 10^{-13} \quad [1/\text{Hz}] \tag{59}$$

Using (55) and assuming V_{ct}=0, I_S = 3 mA, λ =1 we obtain the next formula for PSD of PM noise:

$$S_\phi = 2{,}5 \cdot 10^{-15} + T_\phi^2 \cdot 3 \cdot 10^{-10} f^{-1}. \tag{60}$$

These PSD's are shown in Fig.13 for values λ_0 =1,2; λ_0 =10 and λ_0 =100, corresponding to Fig.10. One can see from Fig.13 that corner frequency of PM noise PSD decreases when λ_0 goes up, because T_ϕ^2 becomes less. From (57), (59), (60) we obtain

$$f_{c,\phi} = T_\phi^2 \cdot 1{,}2 \cdot 10^5 \quad \text{Hz}.$$

For the curves shown in Fig.13 the calculation of $f_{c,\phi}$ gives the next values: λ_0 =1,2 → $f_{c,\phi}$ = 240 Hz; λ_0 =10 → $f_{c,\phi}$ = 380 Hz; λ_0 =100 → $f_{c,\phi}$ = 600 Hz.

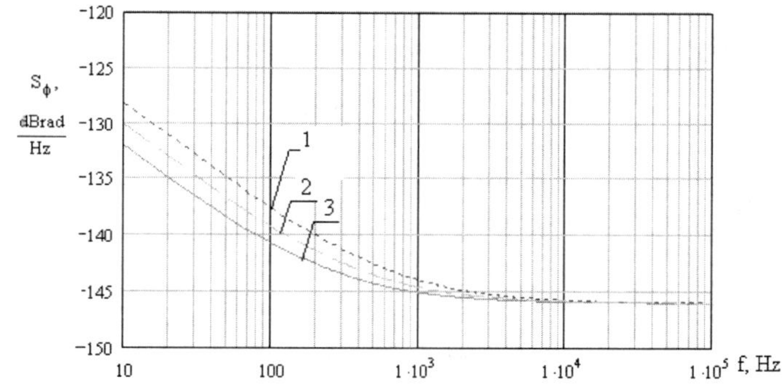

Fig.13. Power spectral densities of PM noise of VCPS with I_S = 3 mA, I_0 = 5 mA, λ =1 at the point V_{ct} = 0. (Parameters: curve 1 - λ_0 =1,2; curve 2 - λ_0 =10; curve 2 - λ_0 =100).

11 Conclusion

The analysis of PM and AM noise in voltage controlled phase shifters presented in this paper was fulfilled under several assumptions.

We used linear models of BJT's when we calculated high frequency signal and noise transformation. We did not take into consideration an influence of internal capacitances in BJT models.

We took into consideration only noise sources of transistors VT_1 and VT_2 that provide voltage controlled gain in the circuit shown in Fig.3. It can be proved that contribution of other stages to total PM noise can be done much less than contribution of voltage controlled stages.

The analysis of PM noise has shown that to decrease 1/f component of this noise it is necessary to use as small value of R_B as possible and choose BJT with small 1/f noise and small r_b. To decrease wideband component of PM noise one has to use the largest possible amplitude of signal current I_S and choose BJT with small r_b.

Using high frequency emitter current feedback through R_{E1}, R_{E2} (Fig.3), that helps to reduce PM noise in amplifiers [5], is limited in VCPS, as increasing $R_E = R_{E1} = R_{E2}$ leads to nonlinearity of phase control characteristic (Fig.8,9).

Analysis opportunities to improve phase control characteristics and to decrease PM and AM noise of VCPS's, that can be manufactured using integral technology seems interesting for many applications.

References

[1] E.M.Armstrong. "A method of reducing disturbance in radio signaling by a system of frequency modulation". *Proc. IRE, 1936,* volume 24, *N5*, pp.89, (1936)

[2] I.S.Gonorovsky. " Frequency modulation and its application". M.: State publishing house on communication and radio, 1948 (In Russian).

[3] P.E.Gray and C.L.Searle. "Electronic principles. Physics, models and circuits. John Wiley, NY. (1969).

[4] V.N.Kuleshov and T.I.Boldyreva, «1/F AM and PM noise in bipolar transistor amplifiers: sources, ways of influence, techniques of reduction», *Proc. 1997 IEEE Int. Freq. Control Symp.,* pp.446-455, (1997).

[5]. V.N.Kuleshov, "1/F models of bipolar junction transistor and their application to PM and AM noise calculation", *Proc. 1998 IEEE Int. Freq. Contr. Symp.* pp.167-171, (1998).

[6] A.A.Kulikovsky. "Frequency modulation in broadcasting and wireless communication". M-L. Gosenergoisdat, 1947 (In Russian).

[7] V.Zalud, V.N.Kuleshov, "Noise in semiconductor devices". Soviet Radio, Moscow, (in Russian), (1977).

Novel Jitter and Phase Noise Reduction Circuit

S. Bunnjaweht, M. J. Underhill and I. D. Robertson

Microwave Systems Research Group
School of Electronics and Physical Sciences, University of Surrey,
Guildford, Surrey GU2 7XH, UK
Tel: +44 (0) 1483 689862; Fax: +44 (0) 1483 686081
E-mail: sawat@mut.ac.th

Keywords: Phase noise, jitter, noise suppression.

ABSTRACT

This paper presents a novel jitter reduction circuit. The reduction technique is based on the time domain jitter averaging. The technique is a simple phase noise reduction method, which can be applied to a signal path as a drop-in building block. The jitter averaging technique allows the circuit to be applied in cascade for higher phase noise reduction.

1. INTRODUCTION

It is phase noise or jitter that limits operation range of radar, degrades the quality of television pictures, limits the precision of satellite positioning and spoils the quality of digital communication systems. For data link systems, deficiencies in transmission and propagation can seriously damage the frequency stability of originally high stable clock signal. Using jitter or phase noise reduction techniques can significantly improve system's bit error rate.

2. JITTER REDUCTION TECHNIQUE

The phase fluctuations, $\phi(t)$, give rise to fluctuations of the zero crossing of the carrier signal and can be expressed as:-

$$v(t) = \sin[2\pi f_o t + \phi(t)] = \sin[2\pi f_o (\tfrac{t+\phi(t)}{2\pi f_o})]$$

or

$$v(t) = \sin[2\pi f_o (t + \Delta T(t))]\,. \qquad (1)$$

However, it is more convenient to view the zero crossing when the carrier signal is converted into a square wave or clock signal.

Phase jitter is defined as the standard deviation, $\sigma\Delta_\phi$, of the phase difference between the first cycle and *i*th cycle of the clock waveform. Timing jitter can be expressed in terms of phase jitter by $\sigma_{\Delta T} = (T/2\pi)\sigma\Delta\phi = (1/\omega_o)\,\sigma\Delta\phi$ where the clock period, T, is $2\pi/\omega_o$. From equation (1), $\Delta T(t)$ is a random function that cause the jitter phenomena. The square wave signal with uniform distribution time jitter is shown in Figure 1.

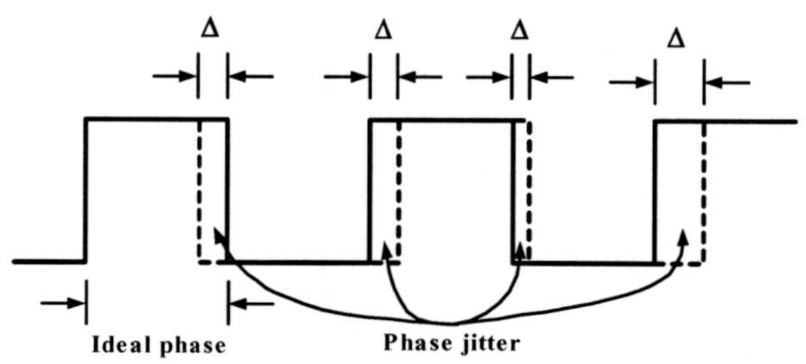

Figure 1 Time jitter with uniform distribution.

The uniform distribution jitter can be defined as [1]:-

$$T_{jitter} = T_{ideal} + \sum_{j=0}^{1} \Delta_j \qquad (2)$$

where Δ is a random variable distributed uniformly from α to $-\alpha$, α is equal to the peak-to-peak of the jitter of the clock signal, T_{ideal} is the period of the jitterless clock. It is assumed that all jitters are independent and identical in statistics. There are algorithms applied to compensate for timing distortions due to jitter [2]. However, time domain averaging process offers simplicity. The time-domain average of periods of equation (1) is [1]:-

$$T_{avearage} = T_{ideal} + \frac{\sum_{i=0}^{n-1}\sum_{j=0}^{1}\Delta_{ij}}{n} \qquad (3)$$

The second term in (3) is an error after average. By the central limit theorem, it is shown that the variance after averaging process in (3) is less than the original signal. This means that jitters have been reduced by average. It is possible to minimize the second term of equation (3), but it cannot be reduced to zero.

3. JITTER REDUCTION CIRCUIT

The functional block diagram of jitter reduction circuit is shown in Figure 2. The circuit operates as follows; the sinusoidal input signal is first converted into a square wave in order to drive pulse splitter circuit. Then the integrators convert the positive and negative pulse train into sawtooth waveforms. These sawtooth waveforms have a constant ramp and its amplitude is modulated by the FM (PM) noise of the input signal. The integrators average the leading and trailing edge jitters of the pulse splitter's output signals. The comparators reference voltage is equal to the mean value of the sawtooth waveform. The output signal of the jitter reduction circuit is reconstructed by the phase comparison of the comparators output waveforms.

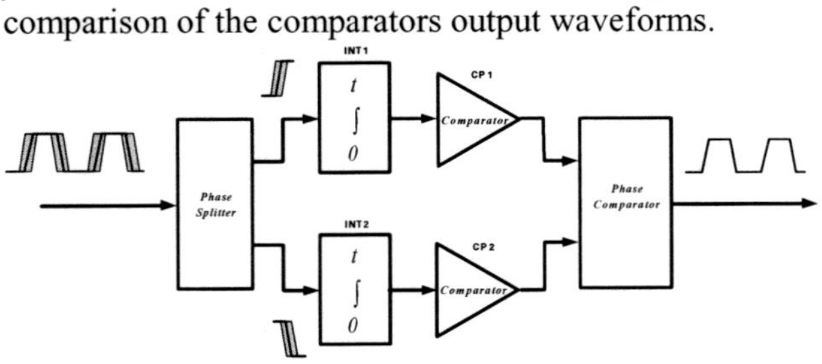

Figure 2 Jitter reduction block diagram.

4. IMPLEMENTATION

The pulse splitter can be implemented by conventional monostable or differential output direct differentiator circuits [3]. Since it is not easy to design a very high gain integrator band pass circuits are used as integrators. The stop band response of 6 dB per octave and a 90 degree constant phase shift in the stop band region of the band pass circuit can performed an integration function.

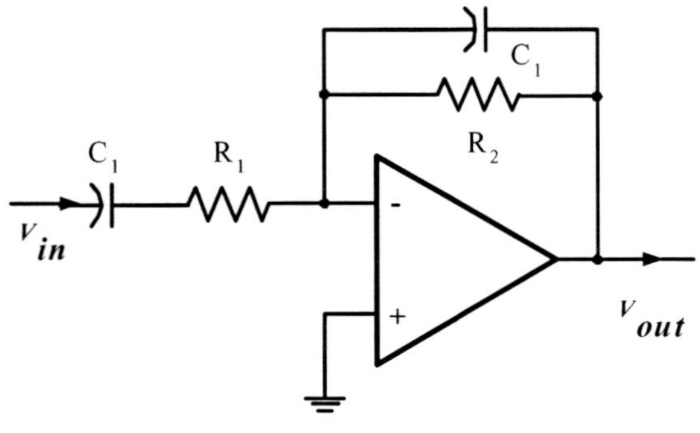

Figure 3 Band pass network as an integrator.

Furthermore, band pass circuit gives less mean-square output noise voltage than ideal integrator due to its smaller noise bandwidth [4] and has no DC response. A simple bilinear transfer function band pass circuit is shown in Figure 3.The transfer function of the band pass network is given by [5]:-

$$\frac{V_{out}}{V_{in}}(s) = \frac{SC_1R_2}{(SC_2R_{2+1})(SC_1R_1+1)}\,. \tag{4}$$

The zero at DC offers the advantage of removing the DC component of the input waveform, which varies with the pulse width of the input signal. Equation (4) shows 2 poles. In order to achieve a correct integration, the operating frequency must be at least 10 times higher than the highest pole's frequency.

5. SIMULATIONS

The block diagram in Figure 2 is use for investigation of the jitter reduction performance. A behavioral model is employed as a building block. ADS2002® has been used to analyse the behavior of the jitter reduction circuit. The suppression achievable on the noise sidebands is measured in the frequency domain by comparison of the input and output spectra as observed by fast Fourier transformation of the time domain signals.

The input test signal is 2 MHz FM signal with a modulation index of 0.5. Figure 4 shows input and output waveform of the comparator 2. The phase comparators must be triggered by the correct transition ramp, otherwise the jitter is not reduced. The correct transition is the same as the AJC circuit [6-7]. The input and output signals of the phase comparator is shown in Figure 5. As can be seen, the upper trace shows constant pulse width after the process.

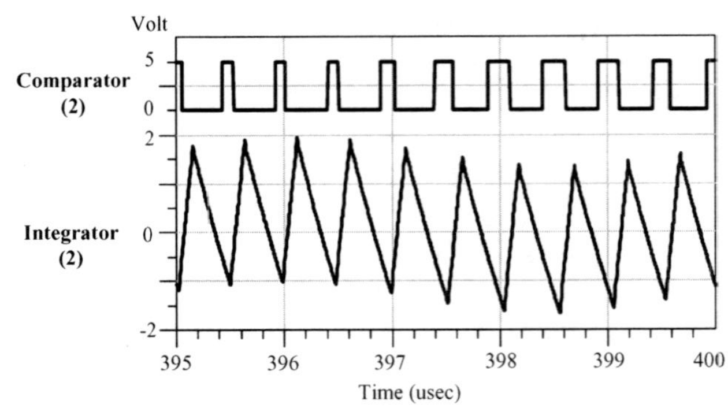

Figure 4 Comparator 2 input-output signal.

Figure 6 shows the simulated input and output power spectrum when the modulating signal, *fm*, is equal to 200

KHz. Comparing the power spectrum of the input signal and the output signal, we can see that the first sideband is reduced by 7 dB. However, the suppression ratio of the other sidebands is obscured by the additional effect due to square wave operation [8] and the dynamic range of the simulator's FFT operation.

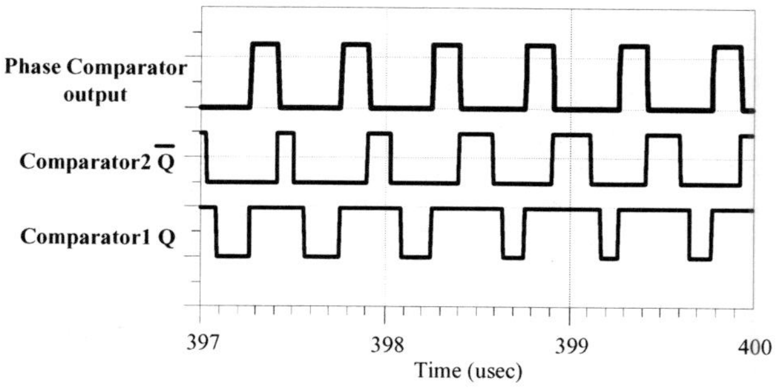

Figure 5 Phase comparator input-output signals.

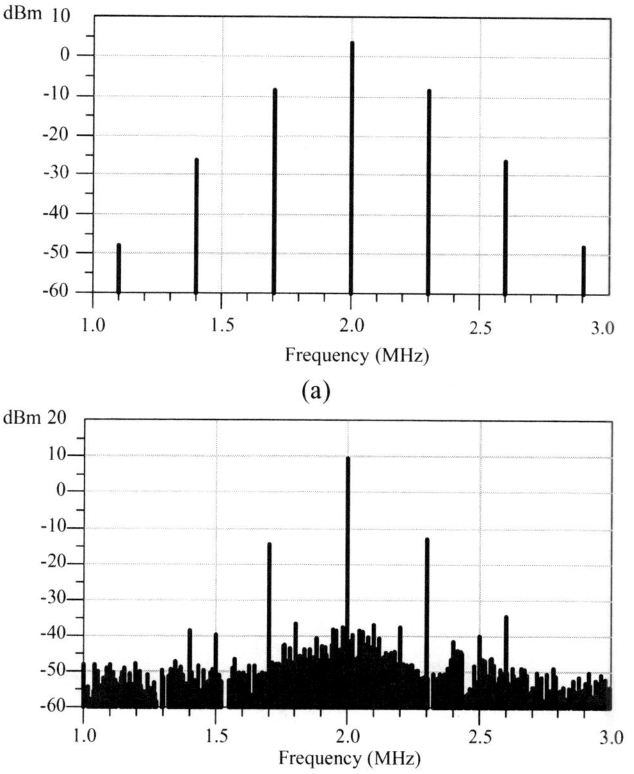

(b)

Figure 6 (a) Simulated input power spectrum and (b) output power spectrum of the jitter and phase noise reduction circuit.

The simulated jitter reduction response in the frequency domain is shown in Figure 7. By inspection, above 100 KHz, the sideband suppression is around 6 dB per octave of the offset frequency.

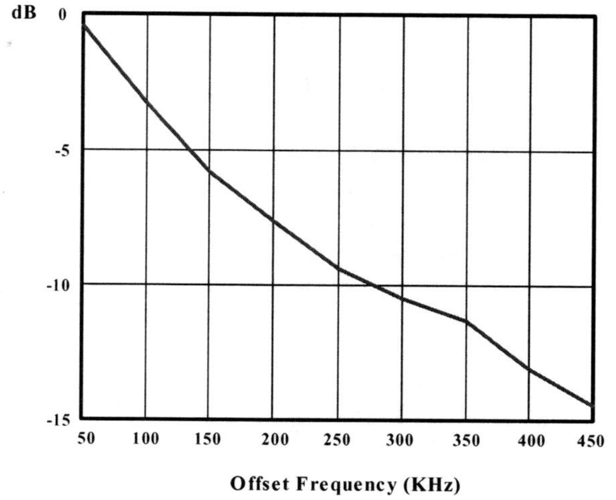

Figure 7 Phase noise suppression response.

6. CONCLUSIONS

The simulation results indicate that a simple jitter and phase noise reduction circuit can improve the phase noise or timing jitter of signal sources. The structure of the circuit is cascadable for higher phase noise reduction.

ACKNOWLEDGMENTS

The authors wish to acknowledge Mahanakorn University of Technology for financial support.

REFERENCES

[1] T. Y. Hus, et al.," Design of a Wide-Band Frequency Synthesizer Base on TDC and DVC Techniques", *IEEE J. Solid State Circuits*, vol. SC-37, No. 10 pp1244-1255, Oct. 2002.

[2] H. Meyr, M. Moeneclaey and S. A. Fechtel, *Digital Communication Receivers-Synchronization, Channel Estimation and Signal Processing*, New York: Wiley, 1998.

[3] K. K. Clark and D. T. Hess, *Communication Circuits: Analysis and Design,* Addison-Wesley Publishing Inc., 1971.

[4] R. S. Carson, *Radio Communications Concepts: Analog*, John Wiley & Sons, New York, 1990.

[5] M. E. van Valkenburg, *Analog Filter Design*, Holt-Saunders, Japan, 1982.

[6] M. J. Underhill and M. J. Blewett, "Spectral Improvement of Direct Digital Frequency Synthesisers and Others Frequency Sources," *Proc. 10th EFTF*, Brighton, pp. 452-460, March 1996.

[7] M. J. Underhill and M. J. Blewett, "Performance of a Delay Compensation Phase Noise and Time Jitter Reduction Method," *Proc. 11th EFTF*, Neuchatel, pp. 364-368, March 1997.

[8] B. Razavi, "A Study of Phase Noise in CMOS Oscillators," *IEEE J. Solid-State Circuits*, vol. Sc-31, pp. 331-343, March 1996.

PM AND AM NOISE IN VOLTAGE CONTROLLED OSCILLATORS

V.N.Kuleshov, T.I.Boldyreva, A.A.Perfilyev

Moscow Power Engineering Institute (Technical University), Moscow, Russia, kuleshov@srv-vmss.mpei.ac.ru

Keywords: voltage controlled oscillators, PM and AM noise, simple nonlinear analysis, amplitude-frequency conversion.

Abstract

An approach to PM and AM noise calculation in oscillator with voltage controlled oscillations frequency (VCO) is developed. It is based on quasi-steady state equations of the oscillator. Both wideband and 1/f noise sources are taken into consideration. Under several assumptions simple formula for PM noise in VCO is obtained and discussed. An influence of feedback through unbypassed emitter resistance R_e on PM and AM noise of the oscillator is investigated and optimal values of R_e, providing minimal values of PM noise due to wideband and 1/f noise sources are found. Dependencies of PM noise in VCO on the oscillations amplitude are discussed theoretically and investigated in experimental VCO. The ways to decrease PM noise of the VCO that follow from main results of the paper are discussed.

1 Introduction

Voltage controlled oscillators (VCO) are widely used in electronic systems. They are produced by many manufacturers. Some of the manufacturers issue designers handbooks, that contain useful information about basic VCO circuits and their characteristics [3,9]. They often discuss qualitatively how VCO performance parameters affect each other. This issue is very important for VCO development. It is especially important to analyze quantitatively how the frequency tuning range affects the PM noise of the VCO. It helps to understand the ways to decrease PM noise in VCO with fixed tuning range.

Recent results of theory of PM and AM noise in bipolar junction transistor (BJT) amplifiers [1,4,6,10] can be used to fulfill this analysis. An approach to such analysis and some of its results are presented in this paper.

2 The VCO model. Assumptions

We consider here the VCO that has a block diagram shown in Fig.1. It this block diagram NA is a nonlinear amplifier and FC is a resonant feedback circuit. A resonant frequency of the FC ω_{00} is controlled by voltage V_{ct}, that usually has a noise component $V_{n,ct}$. We assume that FC open circuit output voltage v(t) is harmonic:

$$v(t) = \mathrm{Re}[\mathbf{V}\exp(j\omega_0 t)], \qquad (1)$$

where $\mathbf{V} = V\exp(j\varphi_v)$ – a complex amplitude of the v(t), and ω_0 – oscillations frequency that can differ from ω_{00}.

We suppose also, that only the first harmonic of output current of the NA

$$i_1(t) = \mathrm{Re}[\mathbf{I}\exp(j\omega_0 t)], \qquad (2)$$

where

$$\mathbf{I}(V,\omega_0) = I(V,\omega_0)\exp(j\varphi_v + j\varphi_I(V,\omega_0)) \qquad (3)$$

- a complex amplitude of the $i_1(t)$, has significant influence on the VCO performance.

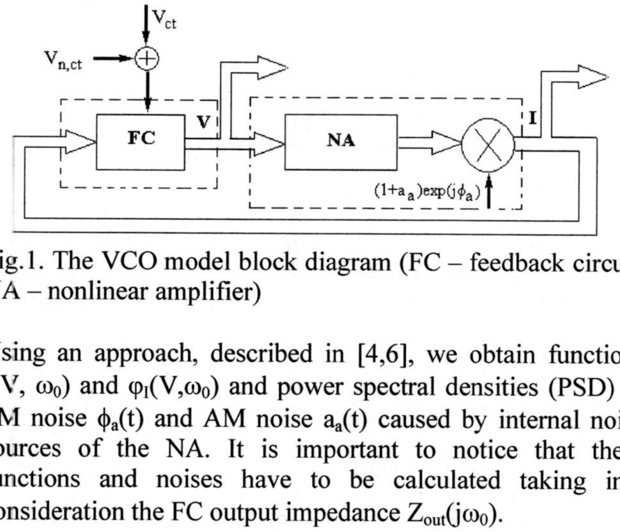

Fig.1. The VCO model block diagram (FC – feedback circuit, NA – nonlinear amplifier)

Using an approach, described in [4,6], we obtain functions $I(V, \omega_0)$ and $\varphi_I(V,\omega_0)$ and power spectral densities (PSD) of PM noise $\phi_a(t)$ and AM noise $a_a(t)$ caused by internal noise sources of the NA. It is important to notice that these functions and noises have to be calculated taking into consideration the FC output impedance $Z_{out}(j\omega_0)$.

3 Basic equations

If there is no sources of noise a steady-state operation of the VCO is described by complex equation

$$\mathbf{V_0} = Z(j\omega_0)\mathbf{I}(V_0, \omega_0), \qquad (4)$$

where $Z(j\omega_0)$ – is open circuit mutual impedance, that describes an influence of $i_1(t)$ on the output voltage of the FC. Here

$$\mathbf{V_0} = V_0\exp(j\varphi_{v0}),\ \mathbf{I_0} = I(V_0, \omega_0)\exp(j\varphi_{v0}+j\varphi_I(V_0, \omega_0)) \qquad (5)$$

- steady-state complex amplitudes.

In the VCO a resonant frequency of the FC ω_{00} depends on control voltage V_{ct} and oscillations amplitude V_0:

$$\omega_{00} = \omega_{00}(V_{ct},V_0). \qquad (6)$$

The steady-state oscillation frequency ω_0 can be presented as follows

$$\omega_0 = \omega_{00}(V_{at},V_0) + \Delta\omega_0, \qquad (7)$$

where $\Delta\omega_0$ is a frequency shift due to constant phase shift in the amplifier.

A complex equation (4) is equivalent to two equations:

$$V_0 = |Z(j\omega_0)| I(V_0, \omega_0)\ , \qquad (8)$$

$$\varphi_Z(\omega_0) + \varphi_I(V_0, \omega_0) = 0\ , \qquad (9)$$

from which one can find the values of V_0 and ω_0.

An influence of the noise sources $a_a(t)$ and $\phi_a(t)$ and $V_{n,ct}(t)$ gives rise to the VCO AM and PM noise $a_0(t)$ and $\phi_0(t)$. Then an amplitude of the oscillation $V_f(t)$ can be expressed in terms of V_0 and $a_0(t)$:

$$V_f(t) = V_0(1+a_0(t)). \qquad (10)$$

A resonant frequency (6) can be written as follows

$$\omega_{00,f}(t) = \omega_{00}(V_{ct} + V_{n,ct}, V_0(1+a_0(t)) = = \omega_{00}(V_{ct}, V_0) + \frac{d\phi_{n,ct}}{dt} + \frac{d\phi_{0,v}}{dt}, \qquad (11)$$

where

$$\frac{d\phi_{n,ct}}{dt} = \left(\frac{\partial\omega_{00}}{\partial V_{ct}}\right) V_{n,ct}(t), \qquad (12)$$

$$\frac{d\phi_{0,v}}{dt} = \left(\frac{\partial\omega_{00}}{\partial V_0}\right) V_0 a_0(t) \qquad (13)$$

- are angular frequency fluctuations due to noise in the control voltage circuit, and due to amplitude – frequency conversion. A frequency shift is also fluctuating and it can be presented in the form

$$\Delta\omega_{0f} = \Delta\omega_0 + \frac{d\phi_{0,a}}{dt}, \qquad (14)$$

where $\frac{d\phi_{0,a}}{dt}$ - fluctuations of the angular frequency due to phase shift in the oscillators feedback loop fluctuations.

Supposing that fluctuations are slow, we can write quasi steady state equations:

$$V_0(1+a_0) = = |Z(j\omega_{00,f} + j\Delta\omega_{0f})| I(V_0(1+a_0), \omega_{00,f} + \Delta\omega_{0f})(1+a_a)\ , \qquad (15)$$

$$\varphi_Z(\omega_{00,f} + \Delta\omega_{0f}) + \varphi_I(V_0(1+a_0), \omega_{00,f} + \Delta\omega_{0f}) + \phi_a = 0. \qquad (16)$$

As both amplitude and frequency fluctuations are very small, we can linearize these equations in the vicinity of steady-state values of V_0, ω_{00} and $\Delta\omega_0$. Then, taking (8), (9) into consideration, we obtain basic equations for AM and PM noise in the VCO.

$$D_a a_0 + D_{a\omega}\frac{d\phi_{0a}}{dt} = a_a + W_{a,v}V_{n,ct}\ , \qquad (17)$$

$$D_{\omega a} a_0 + D_\omega \frac{d\phi_{0a}}{dt} = -\phi_a + W_{\varphi,v}V_{n,ct}\ , \qquad (18)$$

where

$$D_a = 1 - \frac{\partial I/\partial V_0}{I/V_0} - \left(\frac{\partial \ln|Z|}{\partial\omega_{00}} + \frac{\partial \ln I}{\partial\omega_0}\right)\frac{\partial\omega_{00}}{\partial V_0}V_0, \qquad (19)$$

$$D_{a\omega} = -\left(\frac{\partial \ln|Z|}{\partial(\Delta\omega_0)} + \frac{\partial \ln I}{\partial\omega_0}\right), \qquad (20)$$

$$D_{\omega a} = \left(\frac{\partial\varphi_I}{\partial V_0} + \frac{\partial\varphi_I}{\partial\omega_0}\frac{\partial\omega_{00}}{\partial V_0}\right)V_0\ , \qquad (21)$$

$$D_\omega = \left(\frac{\partial\varphi_Z}{\partial(\Delta\omega_0)} + \frac{\partial\varphi_I}{\partial\omega_0}\right), \qquad (22)$$

$$W_{a,v} = \left(\frac{\partial \ln|Z|}{\partial\omega_{00}} + \frac{\partial \ln I}{\partial\omega_0}\right)\frac{\partial\omega_{00}}{\partial V_{ct}}, \qquad (23)$$

$$W_{\varphi,v} = \frac{\partial\varphi_I}{\partial\omega_0}\frac{\partial\omega_{00}}{\partial V_{ct}}. \qquad (24)$$

These equations have to be solved with respect to $\frac{d\phi_{0a}}{dt}$ and a_0. Total PM noise

$$\frac{d\phi_0}{dt} = \frac{d\phi_{0a}}{dt} + \frac{d\phi_{0v}}{dt} + \frac{d\phi_{n,ct}}{dt} \qquad (25)$$

can be found from (25), (12), (13):

$$\frac{d\phi_0}{dt} = \frac{d\phi_{0a}}{dt} + \left(\frac{\partial\omega_{00}}{\partial V_0}\right)V_0 a_0 + \left(\frac{\partial\omega_{00}}{\partial V_{ct}}\right)V_{n,ct}. \qquad (26)$$

Although it is not difficult to solve equations (17), (18) and find total PM noise from (26) for some particular VCO, it seems useful for applications to simplify these calculations using several additional assumptions.

4 Simplified equations

If the amplifier in the block diagram Fig.1 gives the phase shift $|\varphi_I| < \pi/8$ and its dependence on ω_0 is rather smooth it is possible to neglect in (17), (18) the terms, containing coefficients $D_{a\omega}$ and $D_{\omega a}$ (20), (21). Then

$$D_a a_0 = a_a + W_{a,v}V_{n,ct}\ , \qquad (27)$$

$$D_\omega \frac{d\phi_{0a}}{dt} = -\phi_a + W_{\varphi,v}V_{n,ct}. \qquad (28)$$

Using (26), (27), (28), we obtain

$$\frac{d\phi_0}{dt} = -D_\omega^{-1}\phi_a + \left(\frac{\partial\omega_{00}}{\partial V_0}\right)V_0 D_a^{-1}a_a + + \left(\frac{\partial\omega_{00}}{\partial V_{ct}} - D_\omega^{-1}W_{\varphi,v} + D_a^{-1}W_{a,v}\right)V_{n,ct}\ . \qquad (29)$$

As a rule

$$\left|\frac{\partial\omega_{00}}{\partial V_{ct}}\right| >> \left|D_{\omega}^{-1} W_{\varphi,v}\right| \text{ and } \left|\frac{\partial\omega_{00}}{\partial V_{ct}}\right| >> \left|D_{a}^{-1} W_{a,v}\right|, \quad (30)$$

because the tuning sensitivity to direct influence of the control voltage is much more than indirect sensitivities. Under assumptions (30) equation (29) can be simplified and presented as follows:

$$\frac{d\phi_0}{dt} = -D_{\omega}^{-1}\phi_a + \left(\frac{\partial\omega_{00}}{\partial V_0}\right) V_0 D_a^{-1} a_a + \left(\frac{\partial\omega_{00}}{\partial V_{ct}}\right) V_{n,ct}. \quad (31)$$

This equation could be written just by inspection. The first term in the left part of (31) is PM noise component of the oscillator frequency fluctuations due to phase shift fluctuations in the amplifier. It exists also in oscillators without voltage control [2]. The second term arises due to conversion of AM noise of the oscillator to frequency fluctuations because of nonlinearity of varactor. The third term is directly connected with noise that arises in the control circuit.

5 PM and AM noise power spectral densities in oscillators without voltage controlled frequency

Let us consider PM noise in the oscillator with linear FC without voltage controlled (VC) frequency. In this oscillator in accordance with (31) and (27) with $W_{a,v} = 0$ we have

$$\frac{d\phi_0}{dt} = -D_{\omega}^{-1}\phi_a, \quad (32)$$

$$a_0 = D_a^{-1} a_a. \quad (33)$$

It is well known that PM and AM noise of the amplifier can be presented as a sum of wideband components ϕ_{aw}, a_{aw} and flicker components ϕ_{af}, a_{af}, having 1/f type PSD's

$$\phi_a = \phi_{aw} + \phi_{af}, \quad (34)$$

$$a_a = a_{aw} + a_{af}. \quad (35)$$

From (32), (33), (34), (35) PSD's $S_\phi(\omega)$ and $S_a(\omega)$ of PM and AM noise of this oscillator can be presented by formulae

$$S_\phi(\omega) = \frac{1}{D_\omega^2 \omega^2}\left[S_{\phi a,w}(\omega) + S_{\phi a,f}(\omega)\right], \quad (36)$$

$$S_a(\omega) = \frac{1}{D_a^2}\left[S_{aa,w}(\omega) + S_{aa,f}(\omega)\right], \quad (37)$$

where $S_{\phi a,w}(\omega)$, $S_{aa,w}(\omega)$ – PSD'S of the amplifier wideband PM and AM noise ϕ_{aw}, a_{aw} and $S_{\phi a,f}(\omega)$, $S_{aa,f}(\omega)$ – PSD's of the amplifier 1/f PM and AM noise ϕ_{af}, a_{af}.

As it was found in [6] wideband noise PSD's can be written as follows

$$S_{\phi a,w} = F_\phi \frac{k_B T}{P_{av}}, \quad (38)$$

$$S_{aa,w} = F_a \frac{k_B T}{P_{av}}, \quad (39)$$

where

$$P_{av} = V_0^2 / 8\,\mathrm{Re}\, Z_{out}(j\omega_0) \quad (40)$$

- available power of the oscillation source at the input of nonlinear amplifier,

F_ϕ – "phase" noise factor of the NA, F_a – "amplitude" noise factor [6], k_B – Boltzmann constant, T – absolute temperature.

Suppose that ω_0 coincides with resonant frequency. Than $Z_{out}(j\omega_0) = R_{out}$, $\mathrm{Im}(Z_{out}(j\omega_0)) = 0$, and

$$P_{av} = V_0^2 / 8R_{out}. \quad (41)$$

Assume also that input impedance of the amplifier $R_{A,in}$ is active ($\mathrm{Im}\, Z_{A,in}(j\omega_0)) = 0$. Than full power of oscillations dissipated in the loaded feedback circuit can be presented by formula

$$P_{osc} = \frac{1}{2}\left(\frac{V_0}{R_{out}}\right)^2 \frac{R_{out} R_{A,in}}{R_{out} + R_{A,in}}. \quad (42)$$

(This formula is valid under assumption that output current of the NA is saturated).

Using (41), (42), we express P_{av} in terms of P_{osc}, R_{out} and $R_{A,in}$:

$$P_{av} = \frac{1}{4}\left(1 + \frac{R_{out}}{R_{A,in}}\right) P_{osc}. \quad (43)$$

The 1/f PM and AM noise components in accordance with [2,4] can be presented in the form

$$S_{\phi a,f}(\omega) = T_\varphi^2 S_\mu(\omega_m)\left(\frac{\omega_m}{\omega}\right), \quad (44)$$

$$S_{aa,f}(\omega) = T_a^2 S_\mu(\omega_m)\left(\frac{\omega_m}{\omega}\right), \quad (45)$$

where

$S_\mu(\omega)$ – PSD of relative fluctuations of recombination rate $\mu(t)$ [5], expressed in terms of its value $S_\mu(\omega_m)$ at the frequency of measurements ω_m, supposing that $S_\mu(\omega)$ depends on Fourier frequency as $1/\omega$,

T_φ, T_a – transformation coefficients that characterize an influence of $\mu(t)$ on phase and amplitude of the NA output oscillations with signal source having frequency ω_0, open circuit voltage V_0 and output impedance R_{out} at the NA input. An approach to calculation T_φ and T_a was discussed in the paper [4].

From (36-39), (44,45), taking into consideration (43), (19), (22) and supposing that an influence of the third term in the sum (19) is negligible, we obtain the formulae for PSD's of PM and AM noise of the oscillator without voltage controlled frequency

$$S_\phi(\omega) = \frac{1}{\left[\frac{\partial\varphi_Z}{\partial(\Delta\omega_0)} + \frac{\partial\varphi_I}{\partial(\Delta\omega_0)}\right]^2 \omega^2} \cdot \left[F_\phi \frac{4k_B T}{\left[1 + \frac{R_{out}}{R_{A,in}}\right] P_{osc}} + T_\varphi^2 S_\mu(\omega_m)\left(\frac{\omega_m}{\omega}\right)\right], \quad (46)$$

$$S_a(\omega)=\frac{1}{\left[1-\left(\frac{\partial I}{\partial V_0}\Big/\frac{I}{V_0}\right)\right]^2}\cdot\left[F_a\frac{4k_BT}{\left[1+\frac{R_{out}}{R_{A,in}}\right]P_{osc}}+T_a^2S_\mu(\omega_m)\left(\frac{\omega_m}{\omega}\right)\right].(47)$$

For single resonance feedback circuit oscillating at the resonant frequency

$$\frac{d\varphi_Z}{d(\Delta\omega_0)}=-\frac{2Q_0}{\omega_{00}},\qquad \frac{d\varphi_Z}{d(\Delta\omega_0)}+\frac{d\varphi_I}{d(\Delta\omega_0)}=-\frac{2Q_L}{\omega_{00}},\quad (48)$$

where Q_0 – quality factor measured with open-circuit operation at the output of the FC,
Q_L – loaded quality factor measured with input impedance of the NA at the output of the FC.

In this case instead of (46) we can use the next formula:

$$S_\phi(\omega)=\frac{(\omega_{00}/2Q_L)^2}{\omega^2}\left[F_\phi\frac{4k_BT}{\left[1+\frac{R_{out}}{R_{A,in}}\right]P_{osc}}+T_\varphi^2S_\mu(\omega_m)\left(\frac{\omega_m}{\omega}\right)\right].\ (49)$$

This formula is similar to the Leeson's one [7]. But all values in this formula can be calculated theoretically under definite conditions. For example the phase noise factor F_ϕ, that replaces a noise factor of linear amplifier in [7], has to be calculated for nonlinear amplifier with signal source output impedance $Z_{out}(j\omega_0)$ and taking into consideration an influence of a system of output amplitude limitation (self-biasing or automatic gain control) [6]. The transformation coefficient T_φ has to be calculated under the same conditions. Using the approach developed in this paper we obtain from (47) the PSD of AM noise of the oscillator. It is not very important for an oscillator without voltage controlled frequency. But as one can see from (31) it is very important for VCO's.

6 PM and AM noise power spectral densities in VCO

We calculate PSD's of PM and AM noise in VCO using assumptions (30) that lead to equations (31), (33). Under these assumptions the PSD of AM noise is presented by formula (47).

To calculate PSD of PM noise we substitute (34), (35) in (31), taking into consideration, that

$$a_a = T_a\,\mu(t),\qquad \phi_a = T_\varphi\,\mu(t).\quad (50)$$

Then we obtain

$$\frac{d\phi_0}{dt}=-D_\omega^{-1}\phi_{aw}+\frac{\partial\omega_{00}}{\partial V_0}V_0D_a^{-1}a_{aw}+\left(-D_\omega^{-1}T_\varphi+\frac{\partial\omega_{00}}{\partial V_0}V_0D_a^{-1}T_a\right)\mu(t)+\left(\frac{\partial\omega_{00}}{\partial V_{ct}}\right)V_{n,ct}\quad (51)$$

From (51) taking into consideration that $\phi_{a,w}(t)$ and $a_{a,w}(t)$ are not correlated [11] we get an expression for $S_\phi(\omega)$:

$$S_\phi(\omega)=\frac{\Omega_w^2}{\omega^2}\frac{4k_BT}{\left(1+\frac{R_{out}}{R_{a,in}}\right)P_{osc}}+\frac{\Omega_f^2}{\omega^2}S_\mu(\omega_m)\left(\frac{\omega_m}{\omega}\right)+\frac{1}{\omega^2\left(\frac{\partial\omega_{00}}{\partial V_{ct}}\right)^2}S_{v,ct}(\omega),\ (52)$$

where

$$\Omega_w^2=\left(\frac{\omega_{00}}{2Q_L}\right)^2F_\phi+\frac{\left(\frac{\partial\omega_{00}}{\partial V_0}V_0\right)^2}{\left(1-\frac{\partial I}{\partial V_0}\Big/\frac{I}{V_0}\right)^2}F_a,\quad (53)$$

$$\Omega_f^2=\left(\left(\frac{\omega_{00}}{2Q_L}\right)T_\varphi+\frac{\left(\frac{\partial\omega_{00}}{\partial V_0}V_0\right)T_a}{\left(1-\frac{\partial I}{\partial V_0}\Big/\frac{I}{V_0}\right)}\right)^2,\quad (54)$$

One can see from (52) that an influence of AM noise through amplitude-frequency conversion due to nonlinearity of the tank circuit with varactors leads to increasing of wideband component of PM noise. An influence of 1/f fluctuations $\mu(t)$ through phase shift in the amplifier is correlated with influence through amplitude-frequency conversion. That is why it is necessary to take into consideration the signs of T_a, T_φ and $(\partial\omega_{00}/dV_0)$. If the signs of $(\partial\omega_{00}/dV_0)T_a$ and T_φ are the same an influence of amplitude-frequency conversion monotonously increases 1/f PM noise of the VCO.

Using formula (52) one can quantitatively analyze an influence of different characteristics both the nonlinear amplifier and feedback circuit with voltage controlled resonant frequency on the level of PM noise. This analysis helps to look for the opportunities to decrease PM noise.

7 An influence of emitter current feedback on PM noise in oscillator without voltage controlled frequency

It is reasonably to begin development of low noise VCO from development of the low noise oscillator without VC frequency. One of the ways to decrease PM noise of the BJT oscillator is connected with using emitter current feedback that decreases PM noise of BJT amplifiers [1,2,4,6,11].

Let us consider an influence of emitter feedback resistor on PM noise of the Clapp oscillator shown in Fig.2.

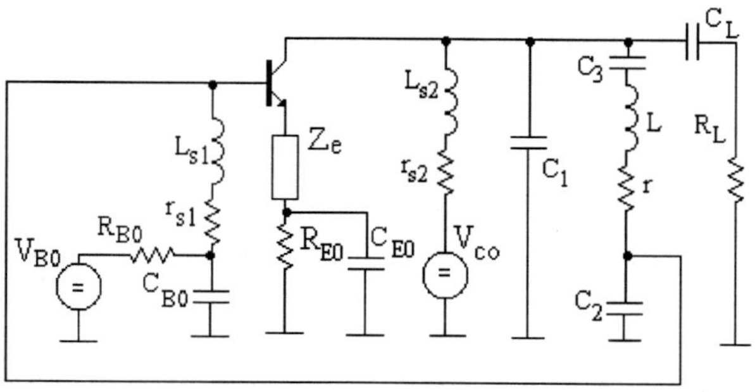

Fig.2. Circiut diagram of a Clapp oscillator with emitter current feedback

The BJT used in this oscillator has the next parameters of the hybrid pi model: $r_{bb'}$ =30 Ohm, β=50, $(\omega_T/2\pi)$=500 MHz, C_{ie} = 5 pF. In this model we assume that C_{ic} = 0. We assume also that at the point of oscillations self-exiting the BJT has collector current I_{C0} = 5 mA. At this operating point the BJT incremental model parameters are: g_m = 0.2 A/V, $r_\pi = \beta/g_m$=250 Ohm, $C_\pi = (g_m/\omega_T)+C_{ie}$ = 64 pF.

The oscillator under consideration has $(\omega_0/2\pi)$=20 MHz, collector load resonant resistance R_{ce}=1 kOhm. The feedback resistance is varied from R_e = 0 to R_e = 500 Ohm. For each value of R_e parameters of positive feedback circuit are chosen to provide constant values of $(\omega_0/2\pi)$=20 MHz, Q_0 =50, I_{C0} = 5 mA, $I_{C1}=I_{C0}*(\pi/2)$=7,85 mA. It means that we support the loop gain compression ratio equal to 2. We also supported R_{B0}=2,5kOhm and $R_{E0} = 10\cdot\left(g_m^{-1} + (R_{B0}/\beta) + R_e\right)$. In this oscillator $P_{osc,0}$ = 30,8 mW is equal to the power delivered by BJT in the resonant load when R_e = 0.

Under these conditions we found the values of $R_{A,in}$ and R_{out}, corresponding to each value of R_e. Than we presented $S_\phi(\omega)$ (49) as a sum of PM noise PSD caused by wideband noise sources $S_{\phi w}(\omega)$ and PM noise due to 1/f noise sources $S_{\phi f}(\omega)$, and calculated the ratio:

$$\frac{S_{\phi w}(\omega)}{S_{\phi w,0}(\omega)} = \left(\frac{Q_0}{Q_L}\right)^2 \frac{F_\phi P_{osc,0}}{\left(1+\frac{R_{out}}{R_{A,in}}\right)P_{osc}}, \quad (55)$$

where

$$S_{\phi w,0}(\omega) = \frac{(\omega_{00}/2Q_0)^2}{\omega^2}\frac{4k_B T}{P_{osc,0}} \quad (56)$$

- the wideband PM noise PSD of the oscillator with $Q_L = Q_0$, $F_\phi = 1$, $P_{osc} = P_{osc,0}$. Results of these calculations are shown in Fig.3. They show that there is an optimal value of $g_m R_e$ that provides the PSD $S_{\phi w}(\omega)$ about 10 dB higher than the value $S_{\phi w,0}(\omega)$ that is calculated using a very simple formula (56).

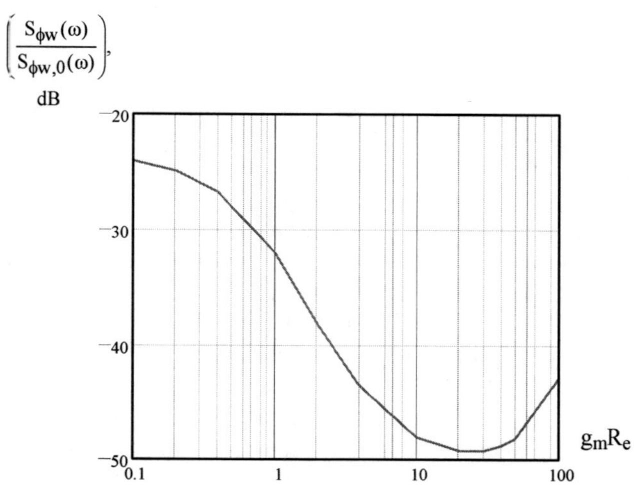

Fig.3. Dependence of wideband PM noise PSD on emitter current feedback parameter

The PM noise PSD due to 1/f noise source in accordance with (49) can be presented in the form

$$S_{\phi f}(\omega) = \left(\frac{Q_0}{Q_L}\right)^2 T_\varphi^2 \frac{(\omega_{00}/2Q_0)^2}{\omega^2} S_\mu(\omega_m)\left(\frac{\omega_m}{\omega}\right). \quad (57)$$

In this formula only the multiplicator $\left(\frac{Q}{Q_L}\right)^2 T_\varphi^2$ depends on $g_m R_e$. Dependence of this multiplicator on $g_m R_e$ calculated at the same conditions as the curve Fig.3 is shown in Fig.4. This curve shows that there is an optimal value of $g_m R_e$, providing minimal value of PM noise due to 1/f noise source. This curve can be easily explained. When $g_m R_e$ grows the PM noise of the amplifier goes down and its power gain goes down also. If $g_m R_e$ is less than the optimal value then decreasing of the amplifier PM noise predominates. When $g_m R_e$ is more than the optimal value the loaded quality factor Q_L goes down very quickly and PM noise of the oscillator goes up.

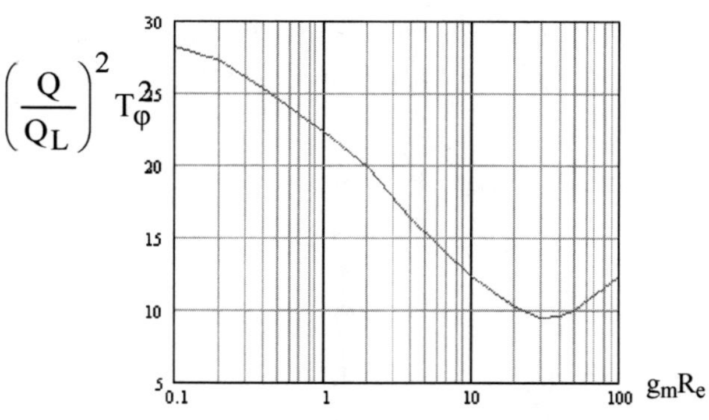

Fig.4. Dependence of 1/f PM noise PSD on emitter feedback parameter

8 The VCO experimental study

The aim of the experimental study reported here was to identify an influence of different sources of PM noise and different ways of their transformation to PM noise.

The circuit diagram of the VCO is shown in Fig.5 [8]. We used three types of frequency controlling devices (FCD) shown in Fig.6: variable capacitor (a), 2-varactor FCD (b) and 8-varactor FCD (c). The BJT KT371A and varactor 2B110B were used in this VCO. The oscillation frequency was changed from 100 MHz to 110 MHz. The BJT parameters $r_{bb'}$ =21 Ohm, β=80, $(\omega_T/2\pi)$ = 4200 MHz, C_c = 0,72 pF. An opportunity to control average value of emitter current I_{E0} by variable resistor R_{var} (Fig.5) was provided. The basic circuit parameters: C_1 = 15 pF, C_2 = 75 pF, C_3 = 8 pF, C_4 = 3,3 pF, L = 0,3 μH, Q_0 = 100, R_{E0} = 300 Ohm, R_{C0} = 300 Ohm, L_{sh} = 140 μH, C_{bp} = 33 nF.

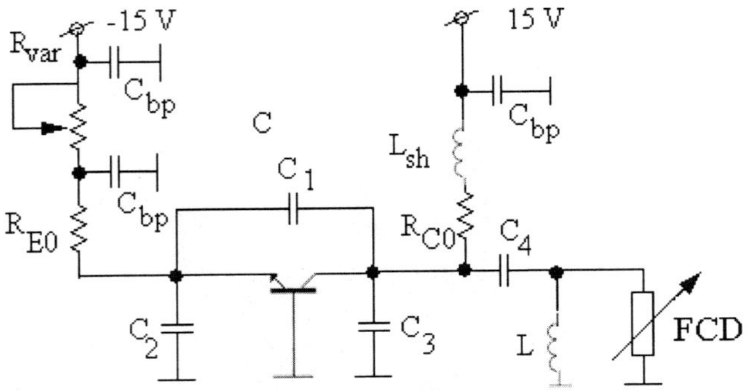

Fig.5. Circuit diagram of the experimental VCO

Results of measurements of dependencies of voltage U_L across the inductance L (Fig.5) and signle sideband PM noise at the offset frequency $(\omega/2\pi)$ = 20 kHz on emitter current I_{EO} are given in Fig.7,a,b. These curves show an influence of amplitude-frequency conversion on PM noise in the VCO with 2-varactor FCD (Fig.6,b). One can see that in the oscillator without varactor increasing I_{E0} and U_L leads to decreasing of PM noise. In the VCO dependencies of £ = 0,5 S_ϕ on I_{E0} have points of minima. The cause of PM noise growth with increasing I_{E0} and U_L in the VCO is conversion of AM noise to PM noise.

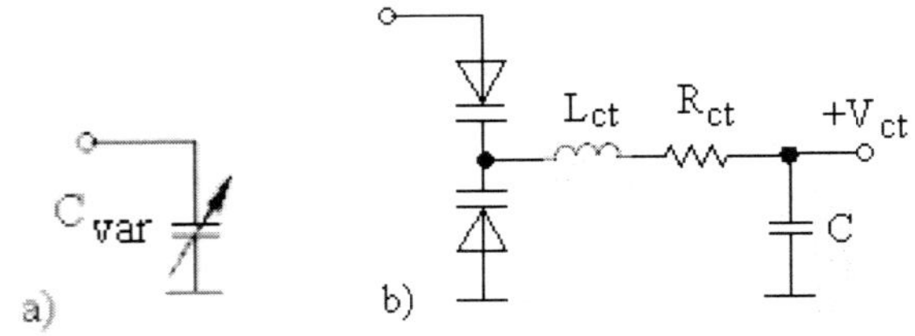

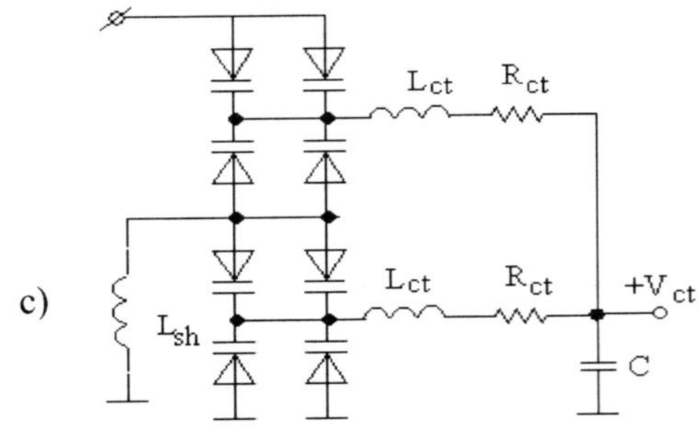

Fig.6. Frequency control devices (FCD), used in the VCO experimental study

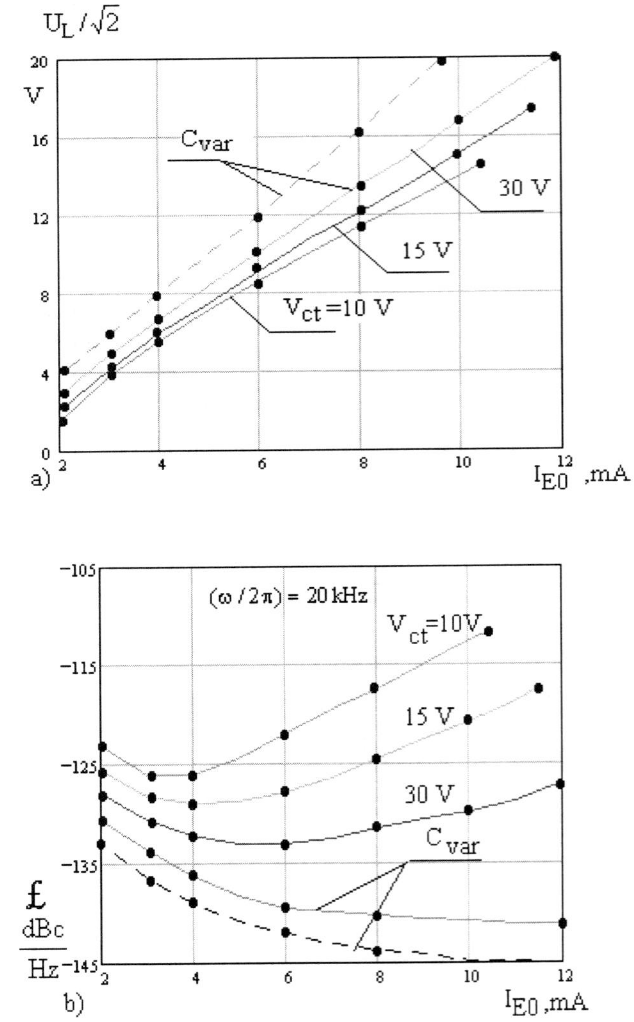

Fig.7. Dependencies of amplitudes of oscillations across the inductance L (a) and single sideband PM noise at the offset frequency $(\omega/2\pi)$ = 20 kHz (b) on average emitter current of the BJT. (Dotted curves correspond to the tank circuit with higher quality factor, than in the circuit with varactor).

The dependencies of £ on the offset frequency $(\omega/2\pi)$ presented in Fig.8 show that increasing of oscillations amplitude U_L leads to more intensive growth the PM noise component caused by 1/f AM noise conversion. That is why the slope of the curve $\pounds(\omega/2\pi)$ becomes more with U_L increase.

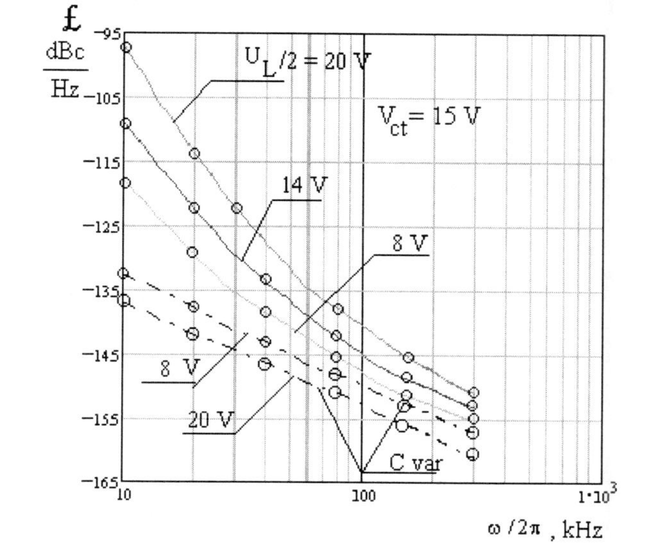

Fig.8. Single side band PM noise of the VCO's with different values of the oscillation amplitude

An influence of thermal noise of the output resistor of the control circuit (R_{ct} in Fig.6,b) is illustrated by Fig.9. One can see that insertion of R_{ct} = 10 kOhm leads to 3 dB increasing of the minimal value of PM noise PSD.

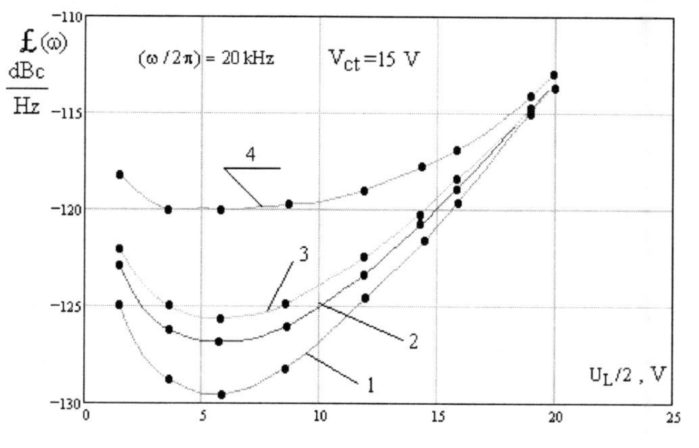

Fig.9. Contribution of thermal noise of the resistor R_{ct} in voltage control circuit Fig.6,b to PM noise of the VCO: curve 1 - R_{ct} = 0; curve 2 - R_{ct} = 10 kOhm; curve 3 - R_{ct} = 20 kOhm; curve 4 - R_{ct} = 100 kOhm

Finally we compared two types of FCD shown in Fig. 6,b,c. Dependencies of U_L on oscillation frequency given in Fig.10,a show that these FCD's insert almost the same losses in the tank circuit. But conversion of AM noise to PM noise has to be less in the FCD with 8 varactors. One can see it in Fig.10,b. The conversion influence becomes less when oscillation frequency becomes higher because of increasing of V_{ct}.

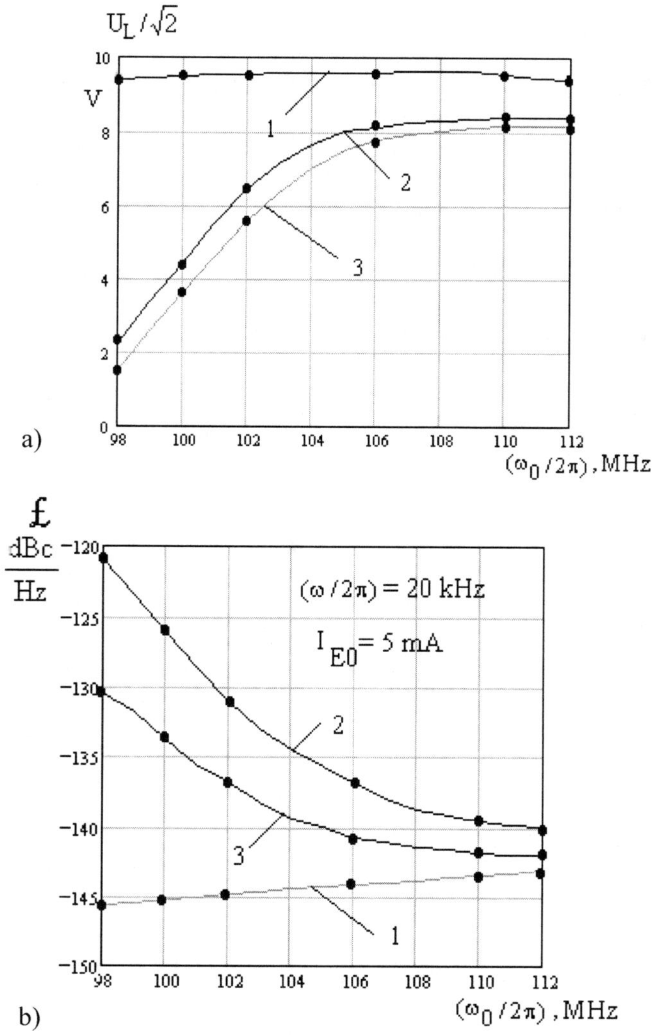

Fig.10. Oscillation voltage $U_L/\sqrt{2}$ (a) and single sideband PM noise £ at the offset frequency 20 kHz (b) as functions of oscillation frequency that is controlled by variable capacitor (curves 1), by 2 varactors (curves 2) and by 8 varactors (curves 3)

9 Conclusion

An approach to VCO PM and AM noise analysis developed in this paper presents an opportunity to understand a contribution of all main sources of this noise by means of basic ways of their influence on PM noise. A simplified formulae (52-54) contain the values that can be found basing on the results of PM and AM noise analysis in BJT amplifiers [1,4,6] and calculations of amplitude-frequency conversion in the tank circuit with nonlinear capacitance. The assumptions that were made to obtain the formulae (52-54) are explained in the paper. They can be checked if it is necessary to estimate the accuracy of the simplified formulae.

Basing on this approach different algorithms of low PM noise VCO's development can be offered.

References

[1] T.I.Boldyreva, "Comparative study of phase and amplitude noise of CE, CE-CB and CC-CB bipolar transistor amplifiers", *Proc. 1998 IEEE Int. Freq. Contr. Symp.* pp.218-255, (1998).

[2] T.I.Boldyreva, V.N.Kuleshov. "PM and AM noise in oscillators based on wideband BJT amplifiers". *Proc. of EFTF* 2002, St.Peterburg, Russia, pp. D-029-D-032, (2002).

[3] Designers handbook. Synergy Microwave Corporation. USA.Paterson. N.J. (2001).

[4] V.N.Kuleshov. "An influence of nonlinearities on PM and AM noise in BJT amplifiers". *Proc. 2001 IEEE Int. Freq. Contr. Symp.,* pp.175-182, (2001).

[5] V.N.Kuleshov, "1/F models of bipolar junction transistor and their application to PM and AM noise calculation", *Proc. 1998 IEEE Int. Freq. Contr. Symp.* pp.167-171, (2001).

[6] V.N.Kuleshov. "PM and AM noise in nonlinear BJT amplifier". *Proc. of the 1999 Joint Meeting of the Time Forum and the IEEE International Frequency Control Symposium.* 13-16 April. Besancon. France, pp. 1180-1183, (1999).

[7] D.B.Leeson. A Simple Model of Feedback Oscillator Noise Spectrum. Proccedings of IEEE, volume 54, N.2, pp. 329-330, (1966).

[8] M.P.Savchenko, V.N.Kuleshov. "Experimental study of noise in bipolar junction transistor voltage controlled oscillator". № 2, Elektrosviaz, (1984).

[9] VCO Designers handbook. Minicircuit. USA.NY, (2001).

[10] F.L.Walls, E.S.Ferre-Pikal, and S.R.Jefferts. "The origin of 1/F PM and AM noise in bipolar junction transistor amplifiers". *Proc. 1995 IEEE Int. Freq. Contr. Symp.,* pp.294-304, (1995)

[11] V.Zalud, V.N.Kuleshov, "Noise in semiconductor devices". Soviet Radio, Moscow, (in Russian), (1977).

AN ASYMPTOTICAL NOISE MODEL OF A CRYSTAL OSCILLATOR WITH A NONLINEAR RESONATOR

Yu. Shmaliy* and R. Brendel**

*Guanajuato University, FIMEE, Salamanca, 36730, Mexico, shmaliy@salamanca.ugto.mx
**Laboratoire de Physique et Métrologie des Oscillateurs, FEMTO–ST LPMO, Besançon, France, brendel@lpmo.edu

Keywords: Crystal oscillator, noise, differential equation.

Abstract

A generalized noise model is discussed of a crystal oscillator, in which the resonator equivalent parameters of the fundamental modes, overtones, and anharmonic branches are assumed to be drive level dependent (DLD) and flicker-noisy. An additive thermal noise of the resonator losses is also taken into account. A feedback amplifier is presented by the generalized nonlinear noise model, in which the flicker fluctuations of the amplitude and phase are involved along with the thermal noise. The model is designed in a frame of the asymptotical methods allowing studying the crystal oscillator nonlinear effects, such as "sleeping sickness", "second level of drive", DLDs with low and high drive levels, transients in the amplitude and phase, and others. An example is given of the model use in the low drive levels sub range.

1 Introduction

Fluctuations in crystal oscillators have been under peer investigation during several decades owing to a number of reasons. In electronics, optoelectronics, and radio electronic systems, the oscillator phase noise influences selectivity of channels and purity of signals. In the precise positioning and measuring systems, it is ordinarily a prime limiter of accuracy. During decades, the crystal oscillator theory has passed over several stages depending upon improvement of the resonator model. First, it had been noticed a multi-modal nature of a crystal resonator. Accordingly, the single-mode net was complicated by the anharmonic series branches. Next, investigations of the anisochronisms have forced to think that the resonator motional capacity, inductance and losses are DLD. Finally, the noise studies in the time and frequency domains had required complicating the model by the noise sources and thinking that the resonator motional parameters are flicker-noisy.

2 Crystal oscillator model

Generalizing the above-given observation, we combine the model of a single mode crystal oscillator by the piezoelectric resonator and feedback amplifier as it is shown in Fig. 1.

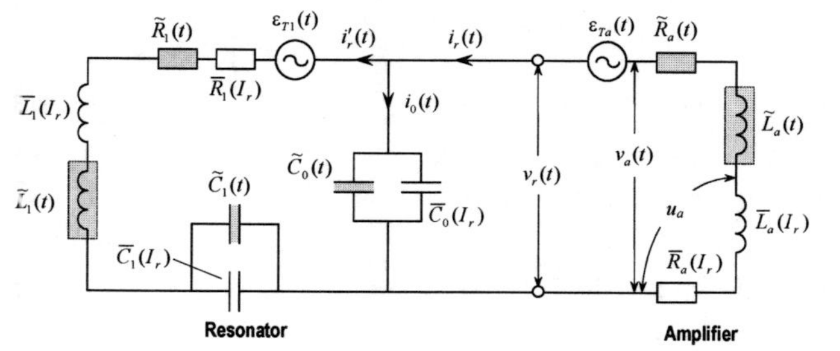

Figure 1: The model of a single-mode crystal oscillator

The resonator motional inductance, capacity, and losses, as well as the static capacity are modeled by, respectively[1],

$$L_1 = \overline{L}_1 + \widetilde{L}_1 = L_{10} + \Delta L_1 + \widetilde{L}_1 = L_{10}(1+\overline{\delta}_L + \widetilde{\delta}_L), \quad (1)$$

$$C_1 = \overline{C}_1 + \widetilde{C}_1 = C_{10} - \Delta C_1 + \widetilde{C}_1 = C_{10}(1-\overline{\delta}_C + \widetilde{\delta}_C), \quad (2)$$

$$R_1 = \overline{R}_1 + \widetilde{R}_1 = R_{10} + \Delta R_1 + \widetilde{R}_1 = R_{10}(1+\overline{\delta}_R + \widetilde{\delta}_R), \quad (3)$$

$$C_0 = \overline{C}_0 + \widetilde{C}_0 = C_{00} + \Delta C_0 + \widetilde{C}_0 = C_{00}(1+\overline{\delta}_{C0} + \widetilde{\delta}_{C0}), \quad (4)$$

where $\overline{L}_1 = L_{10} + \Delta L_{10}$, $\overline{C}_1 = C_{10} + \Delta C_{10}$, $\overline{R}_1 = R_{10} + \Delta R_{10}$, and $\overline{C}_0 = C_{00} + \Delta C_0$ are their deterministic part; $\Delta L_1(I_r,t)$, $\Delta C_1(I_r,t)$, $\Delta R_1(I_r,t)$, and $\Delta C_0(I_r,t)$ are the relevant DLD terms, which are also time dependent in a general case; $\overline{\delta}_L(I_r,t) = \Delta L_1(I_r,t)/L_{10}$, $\overline{\delta}_C(I_r,t) = \Delta C_1(I_r,t)/C_{10}$, $\overline{\delta}_R(I_r,t) = \Delta R_1(I_r,t)/R_{10}$, and $\overline{\delta}_{C0}(I_r,t) = \Delta C_0(I_r,t)/C_{00}$ are the relevant fractional DLD terms, which are also time dependent in a general case; $\widetilde{L}_1(t,I_r)$, $\widetilde{C}_1(t,I_r)$, $\widetilde{R}_1(t,I_r)$, and $\widetilde{C}_0(t,I_r)$ are zero-mean flicker-noise increments of the resonator parameters that may also be DLD; $\widetilde{\delta}_L(t,I_r) = \widetilde{L}_1(t,I_r)/L_{10}$, $\widetilde{\delta}_C(t,I_r) = \widetilde{C}_1(t,I_r)/C_{10}$, $\widetilde{\delta}_R(t,I_r) = \widetilde{R}_1(t,I_r)/R_{10}$, and $\widetilde{\delta}_{C0}(t,I_r) = \widetilde{C}_0(t,I_r)/C_{00}$ are the relevant fractional instabilities; and L_{10}, C_{10}, R_{10},

[1] Here, we save the Gagnepain's definition of the DLD motional capacity: $C_1^{-1} = C_{10}^{-1}(1+\overline{\delta}_C)$ for $\overline{\delta}_C << 1$.

and C_{00} are values of the resonator parameters with normal drives.

The concept of a generalized amplifier for the crystal oscillator assumes substituting it with the equivalent negative resistance and inductance, $R_a = \overline{R}_a + \widetilde{R}_a$ and $L_a = \overline{L}_a + \widetilde{L}_a$, respectively, that are combined with the nonlinear deterministic [$\overline{R}_a(i_r)$ and $\overline{L}_a(i_r)$] and flicker-noisy [$\widetilde{R}_a(t)$ and $\widetilde{L}_a(t)$] components. Accordingly, the deterministic amplifier voltage $u_a(t)$ is supposed to be induced on $\overline{R}_a(i_r)$ and $\overline{L}_a(i_r)$ by the piezoelectric current $i_r(t)$. Yet, $v_a(t)$ is the feedback amplifier flicker-noisy voltage.

The resonator Johnson (thermal) noise $\varepsilon_{T1}(t,i_r)$ inherently becomes multiplicative owing to the loss's DLD and the amplifier thermal noise $\varepsilon_{Ta}(t)$ is assumed to be additive.

Most commonly, both the resonator and amplifier are nonlinear and flicker noisy. Accordingly, the differential equations system regarding the resonator current i_r and feedback voltage v_r takes the form of

$$\begin{cases} \dfrac{d}{dt}L_1 i'_r + i'_r R_1 + \displaystyle\int \frac{i'_r}{C_1}dt = v_r + \varepsilon_{T1} \\ v_r = i_r R_a + \dfrac{d}{dt}L_a i_r - \varepsilon_{Ta} \\ i'_r = i_r - \dfrac{d}{dt}C_0 v_r \end{cases}, \tag{5}$$

where $v_r(i_r) = v_a(i_r) + \varepsilon_{Ta}(t)$ is combined by the nonlinear amplifier voltage v_a and its additive thermal noise ε_{Ta}. The canonic form of (5) is then

$$\frac{d^2}{dt^2}i_r + \omega_{r0}^2 i_r = \mu F\left(i_r, \frac{d}{dt}i_r\right) + \omega_{r0}^2 E(t, I_r), \tag{6}$$

in which $\omega_{r0} = 1/\sqrt{L_{10}C_{10}}$ is a natural frequency of the resonator series branch; $\mu = Q_{r0}^{-1}$ is a small parameter; $Q_{r0} = \omega_{r0}/\Pi_{r0}$ is the resonator quality factor; and $\Pi_{r0} = R_{10}/L_{10}$ is the resonator unloaded bandwidth.

For the following convenience, we now separate the function F into the deterministic and random subfunctions, F and E, namely we write $F = F_0 + F_1$, where F_0 is called the *equilibrium function* and F_1 is called the *anisochronisms function*. Reasoning similarly, we write $E = E_0 + E_1$, where E_0 is called an *additive thermal noise function* and E_1 is said to be the *flicker noise function*. Neglecting products of small values, the above-defined functions become as in the following.

2.1 Equilibrium function

This function is mostly responsible for the oscillator equilibrium and reads

$$F_0 = -\omega_{r0}\frac{d}{dt}i_r - \frac{\omega_{r0}\Pi_{r0}}{p_r R_{10}}u_a + \frac{\omega_{r0}}{R_{10}}\dot{u}_a, \tag{7}$$

where $p_r = C_{10}/C_{00} << 1$ is a resonator's capacities ratio and $u_a(i_a)$ is a nonlinear "deterministic" feedback voltage.

2.2 Anisochronisms function

The nonlinear resonator effects are described by the function

$$F_1 = -Q_{r0}\omega_{r0}^2 D_1 i_r - \omega_{r0}D_2\frac{di_r}{dt} + \frac{\omega_{r0}\Pi_{r0}}{p_r R_{10}}D_3 u_a + \frac{\omega_{r0}}{p_r R_{10}}D_4 \dot{u}_a, \tag{8}$$

in which small coefficients D_1, D_2, D_3, and D_4 are:

$$D_1 = \overline{\delta}_C - \overline{\delta}_L + \frac{1}{\omega_{r0}^2}\left(\ddot{\overline{\delta}}_L + \Pi_{r0}\dot{\overline{\delta}}_R\right) \cong 2\overline{\delta}_\omega, \tag{9}$$

$$D_2 \cong \overline{\delta}_R - \overline{\delta}_L + \frac{2}{\Pi_{r0}}\dot{\overline{\delta}}_L = \overline{\delta}_\Pi + \frac{2}{\Pi_{r0}}\dot{\overline{\delta}}_L, \tag{10}$$

$$D_3 \cong \overline{\delta}_L - \overline{\delta}_{C0} - \overline{\delta}_R - \frac{2}{\Pi_{r0}}\left(\dot{\overline{\delta}}_{C0} + \dot{\overline{\delta}}_L\right) + \frac{1}{\omega_{r0}^2}\left(\ddot{\overline{\delta}}_{C0} + \frac{1}{\Pi_{r0}}\dddot{\overline{\delta}}_{C0}\right) \cong -D_2, \tag{11}$$

$$D_4 \cong \overline{\delta}_C - \overline{\delta}_L - p_r\overline{\delta}_L + \frac{1}{\omega_{r0}^2}\left[\Pi_{r0}\left(2\dot{\overline{\delta}}_{C0} + \dot{\overline{\delta}}_R\right) + 3\ddot{\overline{\delta}}_{C0} + \ddot{\overline{\delta}}_L\right] \cong 2\overline{\delta}_\omega - p_r\overline{\delta}_L, \tag{12}$$

where $\overline{\delta}_\omega = \frac{1}{2}(\overline{\delta}_C - \overline{\delta}_L)$ and $\overline{\delta}_\Pi = \overline{\delta}_R - \overline{\delta}_L$ are fractional DLD coefficients of the resonator resonance frequency and bandwidth, respectively. Function (8) covers all the permitted range of drive levels and simplifies for each of the three practically distinguished cases:

Case 1—Low drive levels. Here, only DLD of the resonator losses is usually of interest. If so, then one sets $D_1 = D_4 = 0$ and $D_2 = -D_3 = \overline{\delta}_R$. Accordingly, (8) degenerates to

$$F_1 = -\omega_{r0}\left(\frac{\Pi_{r0}}{p_r R_{10}}u_a + \frac{di_r}{dt}\right)\overline{\delta}_R \cong -\omega_{r0}\overline{\delta}_R\frac{di_r}{dt}, \tag{13}$$

Case 2—Normal drives. Here all DLD effects are assumed to be zero: $D_1 = D_2 = D_3 = D_4 = 0$. Thus, $F_1 = 0$.

Case 3—High drive levels. In this case, the following simple relation may efficiently be used:

$$F_1 \cong -2Q_{r0}\omega_{r0}^2\tilde{\delta}_\omega i_r . \tag{14}$$

2.3. Additive thermal noise function

The function associated with additive thermal noise is

$$E_0(t,I_r) = C_{00}\left[\dot{\varepsilon}_a(1+p_r) + \omega_{r0}^{-2}(\ddot{\varepsilon}_a\Pi_{r0} + \dddot{\varepsilon}_a) + \dot{\varepsilon}_{T1}(t,I_r)p_r\right], \tag{15}$$

where the noise terms may be substituted by the quasi harmonic models, namely by

$$\varepsilon_a(t) = e_{ac}(t)\cos\omega_{r0}t - e_{as}(t)\sin\omega_{r0}t , \tag{16}$$

$$\varepsilon_{T1}(t) = e_{Tc}(t)\cos\omega_{r0}t - e_{Ts}(t)\sin\omega_{r0}t . \tag{17}$$

The quadrature components $e_{ac}(t)$ and $e_{as}(t)$ are jointly coupled the zero-mean stationary in a wide sense random processes, $\langle e_{ac}(t)\rangle = \langle e_{as}(t)\rangle = 0$, with known power spectral densities. Thus, (15) becomes

$$E_0 = C_{00}\left(e_1\cos\omega_{r0}t - e_2\sin\omega_{r0}t\right), \tag{18}$$

where $e_1 \cong -\omega_{r0}p_r\left(e_{as} + e_{Ts}\right)$ and $e_2 \cong \omega_{r0}p_r\left(e_{ac} + e_{Tc}\right)$.

2.4. Flicker noise function

The flicker (colored) noise function is

$$E_1 = -B_1 i_r - \frac{B_2}{Q_{r0}\omega_{r0}}\frac{di_r}{dt} + \frac{B_3}{p_r R_{10} Q_{r0}^2}u_a + \frac{B_4}{p_r R_{10} Q_{r0}\omega_{r0}}\dot{u}_a , \tag{19}$$

where B_1, B_2, B_3, and B_4 are derived in the same manner as (9)—(12) to be

$$B_1 = \frac{1}{\omega_{r0}^2}\ddot{\tilde{\delta}}_L + \frac{\Pi_{r0}}{\omega_{r0}^2}\dot{\tilde{\delta}}_R + \tilde{\delta}_C - \tilde{\delta}_L \cong \tilde{\delta}_C - \tilde{\delta}_L , \tag{20}$$

$$B_2 \cong \tilde{\delta}_R - \tilde{\delta}_L + \frac{2}{\Pi_{r0}}\dot{\tilde{\delta}}_L , \tag{21}$$

$$B_3 \cong -\frac{2}{\Pi_{r0}}\left(\dot{\tilde{\delta}}_{C0} + \dot{\tilde{\delta}}_L + \frac{R_{10}}{\overline{Z}_-}\dot{\tilde{\delta}}_A + \dot{\tilde{\delta}}_W\right) + \tilde{\delta}_L - \tilde{\delta}_{C0} - \tilde{\delta}_R - \frac{R_{10}}{\overline{Z}_-}\tilde{\delta}_A - \tilde{\delta}_W , \tag{22}$$

$$B_4 \cong \frac{3}{\omega_{r0}^2}\ddot{\tilde{\delta}}_{C0} + \frac{\Pi_{r0}}{\omega_{r0}^2}\left(2\dot{\tilde{\delta}}_{C0} + \dot{\tilde{\delta}}_R\right) + \tilde{\delta}_C - \tilde{\delta}_L(1+p_r) + p_r\frac{R_{10}}{\overline{Z}_-}\tilde{\delta}_A + p_r\tilde{\delta}_W , \tag{23}$$

where $\tilde{\delta}_W$ and $\tilde{\delta}_A$ are fractional fluctuations of the feedback amplifier's frequency and amplitude, respectively. A simplification of (19) is available if to take into consideration the relationship between the resonator flicker noises, this is $\tilde{\delta}_L \sim \tilde{\delta}_C <<< \tilde{\delta}_{C0} << \tilde{\delta}_R$, and the typical relation for the amplifier flicker noises $\tilde{\delta}_A \sim \Delta\varphi_- \cong f_{r0}\delta_W \sim \tilde{\delta}_R$. This leads to $B_1 = \tilde{\delta}_C - \tilde{\delta}_L$, $B_2 \cong \tilde{\delta}_R$, $B_3 \cong -\tilde{\delta}_R - \frac{R_{10}}{\overline{Z}_-}\left(\tilde{\delta}_A + \frac{2}{\Pi_{r0}}\dot{\tilde{\delta}}_A\right)$, and $B_4 \cong p_r\frac{R_{10}}{\overline{Z}_-}\tilde{\delta}_A$. However, the next step may be done with some lack in accuracy. We first notice that even being gained by the Q-factor, the coefficient $Q_{r0}B_1$ is much smaller then each other, and we therefore may neglect the first term in (19). Second, since $B_2 \sim B_3$ then the second term is about $p_r Q_{r0}$ times lesser then the third one and the firth term seems close to the second term. So that finally

$$E_1 \cong \frac{1}{Q_{r0}\omega_{r0}}\left(\frac{1}{\overline{Z}_-}\tilde{\delta}_A\dot{u}_a - \tilde{\delta}_R\frac{di_r}{dt}\right), \tag{24}$$

where $\overline{Z}_-$ is a total negative impedance of the amplifier.

The model (Fig. 1) is now performed enough, by (6)—(24), to transfer to the shortened equations for the random amplitude and phase (frequency) in each of the particular cases associated with low, normal, or high drive levels. Below, we show how this model works for low drives to describe statistically the effect termed "sleeping sickness" in crystal oscillators.

3 Low drive levels

The effect of "sleeping sickness" in crystal oscillators is associated with low drive excitation after a long storage when the crystal resonator losses become large and nonlinear. Gerber has shown in 1954 that the resonator may be reworked by removing its electrodes, cleaning them, and then etching them just prior to replating them [1]. It then was supposed that such a DLD is due to poor adhesion of the electrodes to the quartz surface (electrode flap). The other cause of DLD, the surface contamination, was studied by Bernstein [2]. In 1971, Nonaka, Yuuki, and Hara have experimentally shown that DLD may be caused by scratches in the electrode area [3]. In 1974, the last known DLD effect was indicated by Knowles [4] that is the loosely attached particles on the resonator surface. In further, several models were examined for the supposed physical causes by Dybwad [5], Dworsky and Kinsman [6], and EerNisse [7]. In [8], it was proposed to describe the DLD effect in the resonator losses by the inverse dependence on the piezoelectric current amplitude. It was also shown in

[8] that the following approximation fits reasonably well the toward DLD of R_1 if to start measuring from the noise floor and then gradually increase the piezoelectric current up to the normal drives, with which the losses are R_{10}. In the fractional term $\bar{\delta}_R(I_r) = \Delta R_1(I_r)/R_{10}$ [9], this function becomes

$$\bar{\delta}_R = \frac{\gamma_R}{I_r^n}, \tag{25}$$

where γ_R is a DLD coefficient, I_r is a peak amplitude of the piezoelectric current, and an integer n ranges from 1 to 4, depending on the crystal resonator quality factor Q. It is important to notice that when coming backward, from the normal drives to the noise floor, the DLD coefficient measures almost zero and only after a long time the effect recurs. However, this hysteresis phenomenon is not well studied yet.

By low drives, the model (Fig. 1) becomes as in Fig. 2.

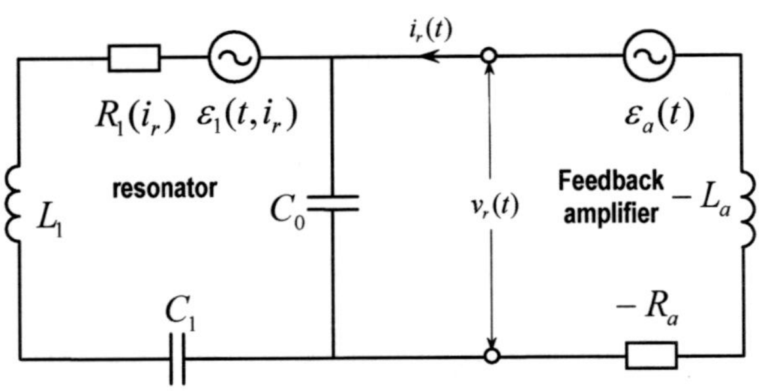

Fig. 2. Electric equivalent of a crystal oscillator with a nonlinear resonator and low drives

Here, the resonator parameters L_1, C_1, and R_1 are responsible for the natural frequency $\omega_s \cong \omega_0 = 1/\sqrt{L_1C_1}$ and bandwidth $2\eta = R_1/L_1$ of a series resonance. The static capacity C_0 induces a parallel resonance at the natural frequency $\omega_p \cong 1/\sqrt{L_1C_1C_0/(C_1+C_0)}$ that tends toward infinity if C_0 becomes zero. The amplitude equilibrium is obtained by the amplifier nonlinear negative losses $-R_a(I_r)$. Finally, the phase equilibrium is provided with an equivalent negative inductance $-L_a$ to make possible for the oscillator to operate at the desired frequency ω that corresponds to some positive phase angle φ between the feedback voltage $v_r(t)$ and the piezoelectric current $i_r(t)$.

3.1. Motion equation

It may be shown that for $\bar{\delta}_R \neq 0$, that is given by (25), and for $\bar{\delta}_L = \bar{\delta}_C = \bar{\delta}_{C0} = 0$ and $\tilde{\delta}_R = \tilde{\delta}_L = \tilde{\delta}_C = \tilde{\delta}_{C0} = 0$, the motion equation (6) of a crystal oscillator (Fig. 2) with the nonlinear resonator losses $R_1(I_r) = R_{10}[1+\bar{\delta}_R(I_r)]$, can be written as follows

$$\ddot{\upsilon} + \omega_0^2\upsilon = 2\eta\left[\dot{\nu} - \left(\dot{\upsilon} + \frac{2\eta}{p}\nu\right)\left(1+\bar{\delta}_R\right)\right] - 2\eta\omega_0\varsigma, \tag{26}$$

where $\upsilon = i/\Delta$, $\dot{\upsilon} = d\upsilon/dt$, $\nu(i) = v(i)/R_{10}\Delta$, $p = C_1/C_0 << 1$, η is a linear damping coefficient, and Δ is a normalizing constant. The DLD noise $\varsigma(t,\upsilon)$ is combined by the additive and multiplicative terms, which are of additive and flicker origin. However, the latter one does not contribute substantially to excitation dynamics, so that the function may be performed by the narrowband model

$$\varsigma(t,\upsilon) = \left(e_{as} + e_{Ts}\right)\cos\omega_0 t + \left(e_{ac} + e_{Tc}\right)\sin\omega_0 t, \tag{27}$$

where $e_{Ts}(\bar{\delta}_R)$, $e_{Tc}(\bar{\delta}_R)$ and e_{as}, e_{ac} are quadrature components of the normalized multiplicative and additive noises of a resonator and a feedback, respectively.

3.2. Stochastic differential equation

We now substitute Δ by the noise intensity induced by the crystal resonator with normal drives within its bandwidth $\Delta f = \eta/\pi$, namely by $\Delta = \sqrt{2kT\eta/\pi R_0}$. Thereafter, the Markov approach applied by Stratonovich in [9] to the oscillator fluctuation differential equation transforms (26) to the stochastic differential equation (SDE) for the normalized envelope $x = I_r/\Delta > 0$ of the piezoelectric current

$$\dot{x} = \left(\alpha - \frac{\gamma}{x^n}\right)x + \frac{\alpha_1}{x}\left(1 + \frac{\gamma_1}{x^n}\right) + \sqrt{\left(1+\frac{\gamma}{x^n}\right)^2 + g^2}\,\xi, \tag{28a}$$

$$dx = f(x)dt + q(x)dw, \tag{28b}$$

where $\dot{x} = dx/d\tau$; $\tau = \eta t$; $\alpha = G_L - 1$ is a linear friction coefficient; $\alpha_1 = (1+g^2)/2$; $\gamma = \gamma_R\left(\pi R_0/2kT\eta\right)^{n/2}$; $\gamma_1 = \gamma/2\alpha_1$; $g = \sqrt{\beta/R_0}$; and $f(x)$ and $q(x)$ are readily defined by comparing (28a) and (28b). The noise $\xi(\tau)$ is here zero-mean, $\langle\xi(\tau)\rangle = 0$, white Gaussian, having a uniform power spectral density $N_0/2 = 1$. Finally, $w(t)$ is a mean-zero, $\langle w(t)\rangle = 0$, Wiener process with a variance $\langle w^2(t)\rangle = t$. It may also be shown that the linear gain of a closed system is calculated by

$$G_L = g^2\left(\cos\varphi - \kappa\sin\varphi\right), \tag{29}$$

where $\kappa = \omega_0 RC_0$ is a resonator κ-factor. We notice that, by $\gamma = 0$, (28a) inherently degenerates to the Langevin equation describing the *Rayleigh* process:

$$\dot{x} = \alpha x + \frac{\alpha_1}{x} + \sqrt{2\alpha_1}\,\xi,\ x > 0 \tag{30}$$

The SDE (28a) is not soluble. Its approximate solution may be found either in a sense of Ito or Stratonovich. Noting that the solution in a sense of Ito is positive valued, we take into account that the increments of the Wiener noise do not correlate any function $\Im(x)$ of the amplitude $x(t)$ and then average the left-hand and right-hand sides of (28a). In doing so, we set $\langle \Im[x(t)]dw(t)\rangle = 0$, and arrive at the nonlinear ordinarily differential equation (ODE) for the oscillator mean envelope

$$\langle \dot{x}\rangle = \left(\alpha - \frac{\gamma}{\langle x\rangle^n}\right)\langle x\rangle + \frac{\alpha_1}{\langle x\rangle}\left(1+\frac{\gamma_1}{\langle x\rangle^n}\right), \; x>0. \quad (31)$$

Noting that, with arbitrary n, a solution of (31) entails difficulties, we consider below its particular solutions.

3.2.1. Approximate solutions of the SDE

Having incidentally large excitation SNR x_0, one may be satisfied with dropping the last term in (31) and consider instead the Bernoulli equation with the constant coefficients

$$\langle \dot{x}\rangle = \alpha\langle x\rangle - \gamma\langle x\rangle^{1-n}, \quad (32)$$

which solution is

$$\langle x\rangle = x_0\sqrt[n]{e^{n\alpha\tau} + \frac{\gamma}{\alpha x_0^n}\left(1-e^{n\alpha\tau}\right)}. \quad (33)$$

Even though (33) does not fit (31) accurately it leads to the important finding. It follows that the linear friction ($\gamma = 0$) induces the linear transient $\langle x\rangle = x_0 e^{\alpha\tau}$ that inherently starts with x_0 and then moves up without bounds ($\alpha > 0$), stays at x_0 ($\alpha = 0$), or goes down to zero ($\alpha < 0$). Contrary, setting $\langle x\rangle = x_0$ into (33) and then solving it produces a critical value γ_b, separating the development and attenuation processes, namely

$$\gamma_b = \alpha x_0^n. \quad (34)$$

It means that, in contrast to the linear case, the excitation occurs if $\alpha > \gamma / x_0^n$, the amplitude stays at x_0 if $\alpha = \gamma / x_0^n$, and the attenuation starts when $\alpha < \gamma / x_0^n$. In other words, the linear friction produces a zero critical level, whereas the nonlinear friction elevates it up to (10). We therefore deduce that if an oscillator has low strength of a limit cycle (α ranges very close to zero) then even a negligible DLD coefficient $\gamma > 0$ may enable attenuation leading to "sleeping sickness". A critical value γ_{b1} is derived for an arbitrary n, supposing $\langle \dot{x}\rangle = 0$ and setting $x = x_0$, that is,

$$\gamma_{b1} = 2x_0^n \frac{\alpha x_0^2 + \alpha_1}{2x_0^2 - 1}. \quad (35)$$

And as a matter of notation we say that, with $\gamma < \gamma_{b1}$, the mean amplitude develops and, with $\gamma > \gamma_{b1}$, it attenuates. Setting $n = 2$ and $\langle \dot{x}\rangle = 0$, we calculate two available stationary levels:

$$\langle x\rangle_{\pm} = \sqrt{\frac{\gamma - \alpha_1 \pm \sqrt{(\gamma-\alpha_1)^2 - 2\alpha\gamma}}{2\alpha}}, \quad (36)$$

where the sign "+" holds true for excitation and "–" is for attenuation. The calculus (36) is unconditional if $\alpha < 0$. With $\alpha > 0$, the amplitude attenuates down to $\langle x\rangle_-$ only if $\gamma > \alpha_1 + \alpha + \sqrt{\alpha(2\alpha_1 + \alpha)}$. And, if $\gamma = \alpha_1 + \alpha + \sqrt{\alpha(2\alpha_1 + \alpha)}$, the only level is $\langle x\rangle = \sqrt{(\gamma - \alpha_1)/2\alpha}$.

The development and attenuation processes, calculated by (31), are shown in Fig. 3.

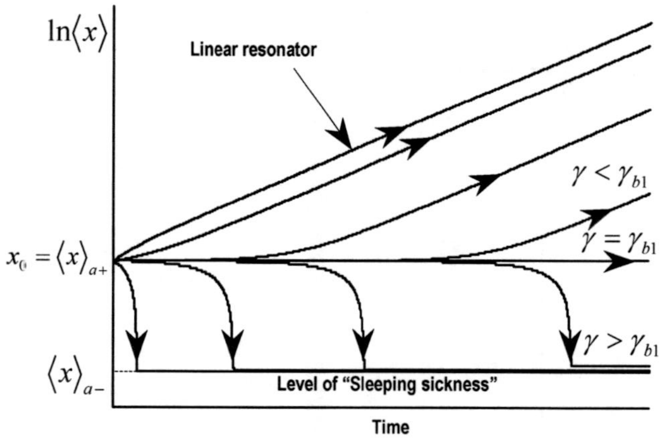

Fig 3. Dynamics of the crystal oscillator mean amplitude with low drives

We note that the linear case of $\gamma = 0$ reduces (31) to the Bernoulli ODE with its solution

$$\langle x\rangle = \sqrt{x_0^2 e^{2\alpha t} - \frac{\alpha_1}{\alpha}(1 - e^{2\alpha t})} \quad (37)$$

and one watches for the exponent-shaped excitation process of van der Pol's oscillator, starting from x_0. Assuming $\gamma < \gamma_{b1}$, the amplitude still develops, however, the time delay in the transient occurs. At the crucial point (35), $\gamma = \gamma_{b1}$, the oscillations sustain at the start point x_0. Finally, a stronger nonlinearity, $\gamma > \gamma_{b1}$, attenuates the amplitude down to some steady state. Afterwards, it may easily be deduced that the attenuation process is characterized by the three major stages:

- the amplitude keeps near the constant value $\langle x\rangle_{a+}$,
- it then falls dawn rapidly and, finally,
- it conserves at the stationary level $\langle x\rangle_{a-}$.

It is now just a matter of notation to say that $\langle x\rangle_{a-}$ means nothing more than the level of the crystal oscillator "*sleeping sickness*". A recipe follows instantly: *to rework such a "sleeping" oscillator the initial value of x_0 should somehow be increased to achieve $\gamma < \gamma_{b1}$*.

A typical picture of the amplitude development with low drives is sketched in Fig. 4*a*.

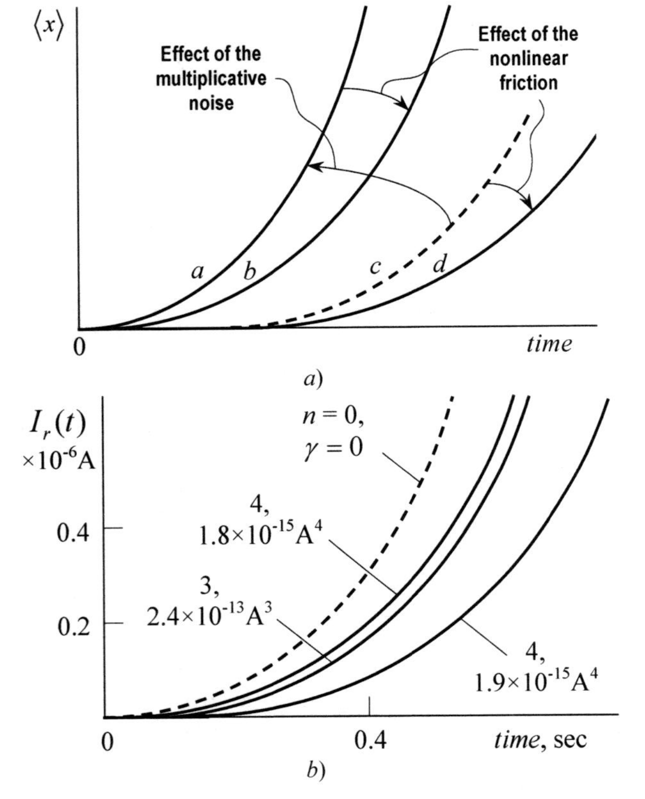

Fig. 4. Development of the crystal oscillator amplitude with low drives: a) *c* is linear noiseless self-excitation; *a* involves a multiplicative noise; *b* and *d* show effects of the nonlinear friction; b) amplitude development caused by three different resonators: dashed curve corresponds to a linear resonator.

The curve "*c*" in Fig. 4a is basic, corresponding to the linear noiseless self-excitation, by setting $\gamma = \alpha_1 = 0$ in (31) and thereafter arriving at $\langle \dot{x}\rangle = \alpha\langle x\rangle$. Accounting the noise and setting $\gamma = \gamma_1 = 0$ in (31) leads to the Rayleigh process "*a*" and we indicate that noise accelerates the amplitude development. Contrary, accounting γ results in a delay in both curves "*c*" and "*a*" to mean that the nonlinear friction in the resonator increases the quality factor of the closed loop up to the limiting case of $\gamma = \gamma_{b1}$ when the dissipated energy is fully compensated and the oscillations conserve. As an example, we show in Fig. 4*b* the simulated excitation processes for three different crystal resonators along with the linear case.

4 Conclusions

Is summary, we indicate that the above-discussed asymptotical model of a crystal oscillator incorporates the thermal and flicker noise sources as well as the DLD effects of a crystal resonator. Therefore it seems challenging for investigating many of the problems associated with crystal oscillator excitation and operation. An example of the model use in the low drive levels sub range is given here in section 3. Other applications are currently under investigation.

References

1. E. A. Gerber, VHF Crystal Grinding, Electronics, **27**, no. 3, pp. 161—163, (1954).
2. M. Bernstein, "Increased crystal unit resistance at oscillator noise levels," *Proc. of 21st Ann. Freq. Control Symp.*, pp. 244—258, (1967).
3. S. Nonaka, T. Yuuki, and K. Hara, "The current dependency of crystal resistance at low drive level," *Proc. of 25th Ann. Freq. Control Symp.*, pp. 139—147, (1971).
4. J. E. Knowles, "On the origin of the 'second level of drive effect' in quartz oscillators," *Proc. of 29th Ann. Freq. Control Symp.*, pp. 230—236, (1971).
5. G. L. Dybwad, "A sensitive new method for the determination of adhesive bonding between a particle and a substrate," *J. Appl. Phys.*, **58**, p. 2789, (1985).
6. L. Dworsky and R. G. Kinsman, "A simple single model for quartz crystal resonator low level drive sensitivity and monolithic filter intermodulation," *IEEE Trans. On Ultrason., Ferroel., and Frequency Control*, **41**, no. 2, pp. 261—268, (1994).
7. E. P. EerNisse, "An analysis of drive level sensitivity in thickness shear quartz resonators," *Proc. of 1996 IEEE Int. Freq. Control Symp.*, pp. 346—356, (1996).
8. S. A. Trushliakov and Yu. S. Shmaliy, Influence of the crystal resonator losses on its excitation in oscillator, *Radiotekhnika*, **44**, no .7, pp.41—44, (1989).
9. R. L. Stratonovich, *Topics in the Theory of Random Noise*, Gordon and Breach, New York, vol. 1, 1963; vol. 2, 1967.

PM AND AM NOISE IN BJT AMPLIFIERS DUE TO FLUCTUATIONS OF SUPPLY VOLTAGES

T.I.Boldyreva

Moscow Power Engineering Institute (Technical University), Moscow, Russia, BoldyrevaTI@mpei.ru

Keywords: bipolar transistor amplifiers, supply voltages fluctuations, PM and AM noise, 1/f noise, nonlinear collector capacitance.

Abstract

A theoretical analysis of output signal phase and amplitude sensitivities of the common emitter BJT amplifier to variations in supply (or biasing) voltages is presented. The dependence of collector junction capacitance on biasing voltage applied to this junction is taken into consideration. A separate influence of base biasing voltage noise and collector biasing voltage noise on PM and AM noise of the amplifier is investigated. An influence of internal 1/f noise sources of the BJT is also analyzed taking into consideration collector capacitance nonlinearity. The approach, developed in this paper, is applied to the CE amplifier with one source of supply voltage and balance circuit of the base biasing. Conditions of compression of the supply voltage contribution to total PM noise are analyzed and discussed. Results of this analysis can be used to set demands to biasing voltages fluctuations in low PM noise amplifiers.

1 Introduction

Development of the amplifiers with low PM and AM noise at low Fourier frequencies demands more deep understanding of biasing voltages fluctuations influence on amplitude and phase of the amplified harmonic signal. A theoretical analysis of sensitivities of phase shift and gain of the amplifier to biasing voltages variations gives an opportunity to find out basic mechanisms of the biasing voltages influence on phase and amplitude of the output signal and investigate methods of this influence reduction.

In this paper such analysis is done for bipolar junction transistor (BJT) common emitter (CE) amplifier. This analysis is based on the approach developed in [1,2]. But in this paper a dependence of collector junction capacitance on biasing voltage applied to this junction is taken into consideration. It is shown that in some cases an influence of biasing voltages on the phase shift due to collector capacitance variations gives significant contribution to total phase shift variations.

An opportunity to decrease PM noise of the amplifier using the biasing circuit offered in the paper [4] was also analyzed.

2 The amplifier circuit and model

The circuit under consideration is shown in Fig.1. It is CE amplifier with feedback impedance Z_E in the emitter circuit. The biasing voltage V_B is applied to the base through resistor R_{B0}. The biasing voltage V_C is applied to the collector through the load resistor R_L. The signal source voltage v_{SS} is harmonic

$$v_{SS}(t) = \mathrm{Re}[V_{SS}\exp(j\omega_s t)] \tag{1}$$

with angular frequency ω_S and amplitude V_{SS}. The source output impedance R_{SS} is real. The impedance Z_E consists of self-biasing circuit R_{E0}, C_{E0} with large bypassing capacitor C_{E0} and element R_E, providing the necessary feedback at the signal frequency.

We suppose that voltages V_B and V_C can be presented as follows

$$V_B = V_{B0}(1+\mu_B(t)), \tag{2}$$

$$V_C = V_{C0}(1+\mu_C(t)), \tag{3}$$

where V_{B0}, V_{C0} – average values of the voltages and $\mu_B(t)$, $\mu_C(t)$ – relative fluctuations (or ripples) of these voltages.
To calculate an influence of biasing voltages fluctuations and internal 1/f noise source of the BJT on its operating point we used a low frequency circuit model of the amplifier shown in Fig.2. In this figure $V_{b'e}$ is the voltage across emitter junction,

$$I_C(V_{b'e}) = I_{CS}\left(\exp\frac{V_{b'e}}{V_T} - 1\right), \tag{4}$$

$$I_B(V_{b'e}) = \beta^{-1} I_C(V_{b'e})(1+\mu_f(t)), \tag{5}$$

V_T – thermal voltage, I_{CS} – the saturation current, β - common emitter current gain of the BJT, μ_f – relative fluctuations of the recombination current, that is a source of internal 1/f noise of the BJT [1-3]. The bypassing capacitance C_{E0} can be taken into consideration in the model Fig.2. But in this paper we suppose that it is not so large to change the power spectral density of 1/f fluctuations of the emitter current. That is why it is connected to R_{E0} by dotted line.

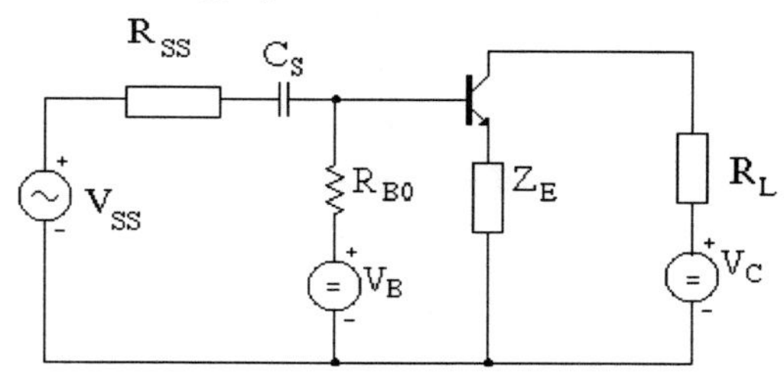

Fig.1. Circuit diagram of the CE amplifier

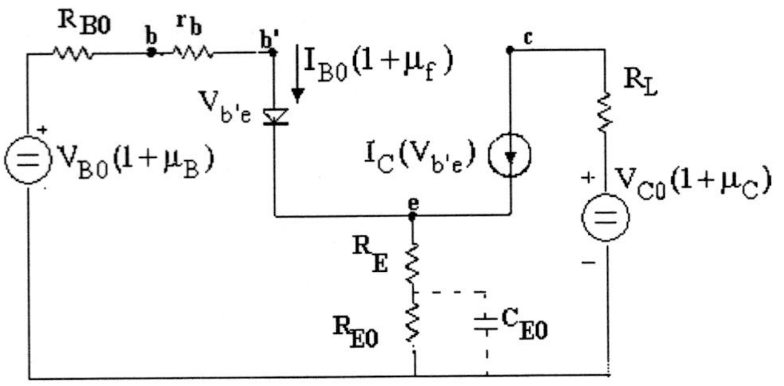

Fig.2. Low frequency circuit model of the CE amplifier

To calculate the signal transformation by the amplifier Fig.1 we use an incremental high frequency model of the BJT amplifier shown in Fig.3. In this circuit the BJT is replaced by its incremental hybrid pi model. In this model: $g_m = I_C/V_T$ - mutual conductance (I_C – collector current in the operating point), $g_\pi = g_m/\beta$ - recombination conductance, $C_b = g_m/\omega_T$ - incremental charging capacitance, $\omega_T = 2\pi f_T$ – the frequency at which the magnitude of short circuit CE current gain extrapolates to unity (it is supposed that C_b is much more than emitter junctions capacitance). The dimensionless variables δ_π, δ_{Cb}, δ_{Cc}, δ_m – are relative low frequency fluctuations caused both by biasing voltage fluctuations and by internal 1/f noise of the BJT. They will be expressed in terms of μ_B, μ_C and μ_f from the circuit model Fig.2 and used to find the amplifier transfer function sensitivity to the fluctuations of μ_B, μ_C and μ_f.

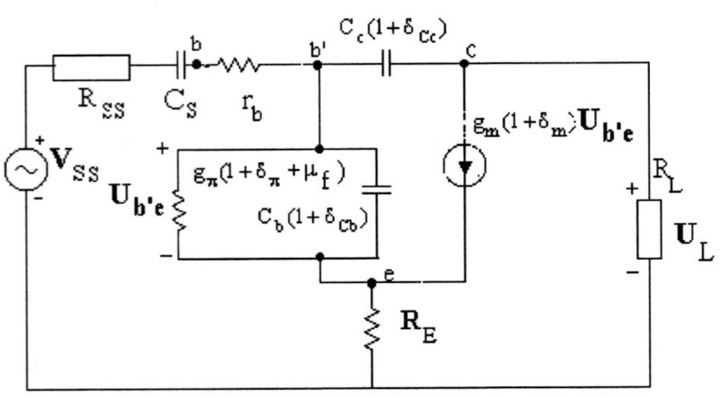

Fig.3. An incremental high frequency model of CE amplifier

3 Algorithms of PM and AM noise calculation

An approach to calculation of low frequency PM and AM noise due to biasing voltages fluctuations is similar to the one developed in the paper [1]. We express a complex transfer function (gain) of the amplifier in terms of parameters g_m, g_π, C_b and C_c:

$$\mathbf{G}(j\omega_s; g_m, g_\pi, C_b, C_c) = \frac{\mathbf{U}_L}{\mathbf{V}_{SS}} \tag{6}$$

Than we write an increment of this function in the form

$$\Delta\mathbf{G}(j\omega; g_m, g_\pi, C_b, C_c) = g_m \frac{\partial \mathbf{G}}{\partial g_m}\delta_m + g_\pi \frac{\partial \mathbf{G}}{\partial g_\pi}(\mu_f + \delta_\pi) + C_b \frac{\partial \mathbf{G}}{\partial C_b}\delta_{Cb} + C_c \frac{\partial \mathbf{G}}{\partial C_c}\delta_{Cc}. \tag{7}$$

As g_m, g_π and C_b are proportional to I_C in the operating point δ_m, δ_π and δ_{Cb} can be expressed in terms of μ_f and μ_B, causing collector current fluctuations:

$$\delta_m = \delta_\pi = \delta_{Cb} = A_f\mu_f + A_B\mu_B\,, \tag{8}$$

where

$$A_f = \frac{(R_{B0} + r_b + R_{E0} + R_E)(g_m/\beta)}{1 + (R_{B0} + R_{E0} + r_b + R_E)(g_m/\beta) + g_m(R_{E0} + R_E)}\,, \tag{9}$$

$$A_B = \frac{(V^0_{b'e}/V_T) + (R_{B0} + r_b + R_{E0} + R_E)(g_m/\beta)}{1 + (R_{B0} + R_{E0} + r_b + R_E)(g_m/\beta) + g_m(R_{E0} + R_E)}\,, \tag{10}$$

- are coefficients of I_C sensitivity to relative fluctuations of recombination current and biasing voltage respectively. The formulae (9), (10) can be easily obtained from the circuit Fig.2 analysis, taking into consideration (4), (5) and formulae for g_m, g_π and C_b given above.

To obtain an expression for δ_{Cc} we suppose that

$$C_c = C_{c0}\left(1 + \frac{V_{cb'}}{\psi_0}\right)^{-1/3}\,, \tag{11}$$

where ψ_0 – contact potential of the collector junction, C_{c0} – the junction capacitance when $V_{cb'} = 0$. From (11) and circuit model Fig.2 one obtains

$$\delta_{Cc} = \frac{N_c V_T}{3(\psi_0 + V_{cb'})} A_f\mu_f + \frac{N_c V_T}{3(\psi_0 + V_{cb'})} A_B\mu_B - \frac{V_{c0}}{3(\psi_0 + V_{cb'})}\mu_C \tag{12}$$

where

$$N_c = 1 + (R_E + R_{E0})\left(1 + \frac{1}{\beta} g_m\right) + R_L g_m\,. \tag{13}$$

It was shown in [2] that the relative gain fluctuations a(t) and phase fluctuations $\phi(t)$ are connected with $\Delta\mathbf{G}(j\omega_s; g_m, g_\pi, C_b, C_c)$ by formula:

$$a(t) + j\phi(t) = \frac{\Delta\mathbf{G}(j\omega_s; g_m, g_\pi, C_b, C_c)}{\mathbf{G}(j\omega_s; g_m, g_\pi, C_b, C_c)}\,. \tag{14}$$

From (14), using (7), (8), (12) one obtains

$$a + j\phi = \mathbf{T}_f(j\omega_s)\mu_f + \mathbf{T}_B(j\omega_s)\mu_B + \mathbf{T}_C(j\omega_s)\mu_C\,, \tag{15}$$

where

$$\mathbf{T}_f(j\omega_s) = -\left(\frac{g_m}{\mathbf{G}}\frac{\partial \mathbf{G}}{\partial g_m} + \frac{C_b}{\mathbf{G}}\frac{\partial \mathbf{G}}{\partial C_b}\right)A_f + \frac{g_\pi}{\mathbf{G}}\frac{\partial \mathbf{G}}{\partial g_\pi}(1 - A_f) + \frac{C_c}{\mathbf{G}}\frac{\partial \mathbf{G}}{\partial C_c}\frac{N_c V_T}{3(\psi_0 + V_{cb'})}A_f\,, \tag{16}$$

$$\mathbf{T}_B(j\omega_s) = \left(\frac{g_m}{\mathbf{G}}\frac{\partial \mathbf{G}}{\partial g_m} + \frac{C_b}{\mathbf{G}}\frac{\partial \mathbf{G}}{\partial C_b} + \frac{g_\pi}{\mathbf{G}}\frac{\partial \mathbf{G}}{\partial g_\pi} + \frac{C_c}{\mathbf{G}}\frac{\partial \mathbf{G}}{\partial C_c}\frac{N_c V_T}{3(\psi_0 + V_{cb'})}\right)A_B\,, \tag{17}$$

$$\mathbf{T}_C(j\omega_s) = -\frac{C_c}{\mathbf{G}}\frac{\partial \mathbf{G}}{\partial C_c}\frac{N_c V_T}{3(\psi_0 + V_{cb'})}\,. \tag{18}$$

Introducing transformation coefficients

$$T_{fa}(\omega_s) = \mathrm{Re}[\mathbf{T}_f(j\omega_s)], \qquad T_{f\phi}(\omega_s) = \mathrm{Im}[\mathbf{T}_f(j\omega_s)], \quad (19)$$

$$T_{Ba}(\omega_s) = \mathrm{Re}[\mathbf{T}_B(j\omega_s)], \qquad T_{B\phi}(\omega_s) = \mathrm{Im}[\mathbf{T}_B(j\omega_s)], \quad (20)$$

$$T_{Ca}(\omega_s) = \mathrm{Re}[\mathbf{T}_C(j\omega_s)], \qquad T_{C\phi}(\omega_s) = \mathrm{Im}[\mathbf{T}_C(j\omega_s)], \quad (21)$$

we obtain from (15)

$$a = T_{fa}\mu_f + T_{Ba}\mu_B + T_{Ca}\mu_C, \quad (22)$$

$$\phi = T_{f\phi}\mu_f + T_{B\phi}\mu_B + T_{C\phi}\mu_C. \quad (23)$$

Using (22), (23), it is easy to find power spectral densities (PSD) of PM noise $\phi(t)$ and AM noise $a(t)$, if one knows PSD's of $\mu_f(t)$, $\mu_B(t)$, $\mu_C(t)$ and transformation coefficients (TC) (19-21).

To calculate TC one has to find the complex transfer function (6), using circuit model Fig.3, and sensitivities $\left(\frac{g_m}{G}\frac{\partial G}{\partial g_m}\right)$, $\left(\frac{g_\pi}{G}\frac{\partial G}{\partial g_\pi}\right)$, $\left(\frac{C_b}{G}\frac{\partial G}{\partial C_b}\right)$, $\left(\frac{C_c}{G}\frac{\partial G}{\partial C_c}\right)$. These sensitivities were found numerically.

4 The amplifier, used for particular case study

To provide particular case study we consider the amplifier with BJT KT306A. The data of particular BJT example: r_b = 30 Ohm, f_T = 500 MHz, β = 50. The collector junction capacitance data (11): C_{c0} = 5,4 pF, ψ_0 = 0,8 V. The BJT operating point data: I_C = 10 mA, $V_{CE}=V_C - V_E$ = 4,5V. The BJT model parameters in the operating point g_m = 0,4 A/V, C_b = 127 pF >> C_{je} = 3 pF, g_π = 0,008 A/V.

The amplifier and signal source parameters: R_L =350 Ohm, R_{E0} = 200 Ohm, R_{B0} = 2,500 kOhm, R_{SS} = 50 Ohm. The emitter feedback resistor R_E will be changed from 1 Ohm to 25 Ohm.

5 PM and AM noise of the amplifier due to internal 1/F noise source of the BJT

First of all we consider PM and AM noise in the BJT amplifier caused by internal source of 1/f noise. There are two reasons why this consideration is interesting.

In the first place the analysis presented here differs from [1,2], because we have taken into consideration the C_c nonlinearity. It seems interesting to estimate contribution of this channel of BJT 1/f noise influence on PM and AM noise of the amplifier.

Secondly the results of this AM and PM noise calculation can be used for comparison with contribution of biasing voltages fluctuations.

Let us suppose that there is no fluctuations of biasing voltages: $\mu_B=\mu_C=0$. Then using (22), (23) one presents PM and AM noises as follows

$$\phi(t)=T_{f\phi}(\omega_S)\,\mu_f(t), \qquad a(t)=T_{fa}(\omega_S)\,\mu_f(t). \quad (24)$$

Using (24) one can express power spectral densities (PSD) $S_{\phi,f}(\omega)$ and $S_{a,f}(\omega)$ of PM and AM noise in terms of PSD of the BJT 1/f noise $S_{\mu f}(\omega)$:

$$S_{\phi,f}(\omega) = T^2_{f\phi}(\omega_s)S_{\mu f}(\omega); \quad S_{a,f}(\omega) = T^2_{fa}(\omega_s)S_{\mu a}(\omega). \quad (25)$$

One can see from (25) that an influence of the BJT internal 1/f noise on PM and AM noise is described by transformation coefficients (TC) $T^2_{f\phi}(\omega_s)$ and $T^2_{fa}(\omega_s)$ that can be found from (16), (19). Using these formulae the dependencies $T^2_{f\phi}(\omega_s)$ and $T^2_{fa}(\omega_s)$ on signal frequency were calculated for the amplifier described above. Results of the calculations are shown in Fig.4. The curves in Fig.4 were calculated for two values of R_E (1 Ohm and 25 Ohm). Dotted lines correspond to TC's calculated for the case of constant C_C. In this case the last term in (16) is not taken into consideration. Solid lines were calculated taking into consideration an influence of collector-base voltage fluctuations induced by the BJT 1/f noise on the value of nonlinear C_C.

One can see that the difference between these two cases is very small. Fig.5 illustrates an influence of feedback resistance R_E on $T^2_{f\phi}(\omega_s)$.

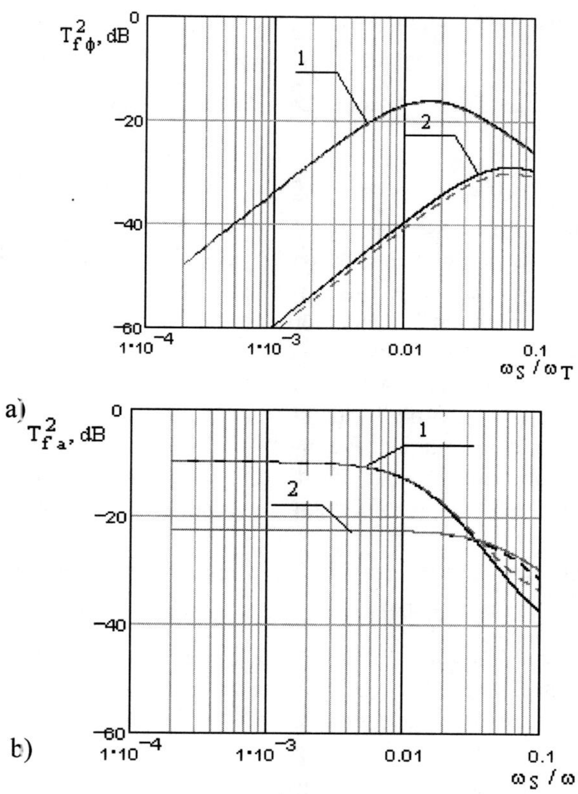

Fig.4. Transformation coefficients "1/f noise – PM noise" (a) and "1/f noise – AM noise" (b) as functions of the relative signal frequency ω_S/ω_T calculated for the cases of nonlinear C_C (solid lines) and constant C_C (dotted lines). Case 1 – R_E = 1 Ohm, Case 2 – R_E = 25 Ohm.

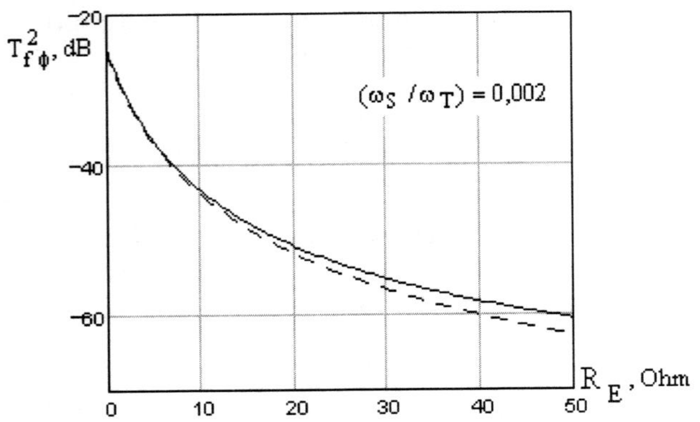

Fig.5. Dependence of transformation coefficient "1/f noise – PM noise" on the value of emitter feedback resistor at constant signal frequency. (Dotted line corresponds to the model with constant C_C).

6 PM and AM noise due to base biasing voltage fluctuations

Suppose that μ_f = 0 and μ_C = 0. Than from (22), (23) it follows that PSD's of PM and AM noise $S_{\phi,B}(\omega)$ and $S_{a,B}(\omega)$ can be presented by formulae

$$S_{\phi B}(\omega) = T^2_{B\phi}(\omega_s) S_{\mu B}(\omega)\,;\, S_{aB}(\omega) = T^2_{Ba}(\omega_s) S_{\mu B}(\omega)\,. \quad (26)$$

where $S_{\mu B}(\omega)$ – PSD of $\mu_B(t)$. Transformation coefficients $T^2_{B\phi}(\omega_s)$ and $T^2_{Ba}(\omega_s)$ can be calculated from (20), (17).

Results of of $T^2_{B\phi}(\omega_s)$ and $T^2_{Ba}(\omega_s)$ calculations for same amplifier that was considered in the previous part are presented in Fig.6. Comparing the curves, calculated under assumption that C_C=const (dotted lines), and curves, calculated taking into consideration the C_C dependence on the voltage $V_{b'c}$ (solid lines) one can see that contribution of mechanism connected with dependence of C_C on $V_{b'c}$ to total PM noise level caused by $\mu_B(t)$ is significant. This influence increases total PM noise level by 5-10 dB. AM noise (Fig.6,b) almost doesn't change inside the amplifier band and decreases, when ω_S is more than the amplifier bandwidth.

Comparing dependencies Fig.7of transformation coefficient $T^2_{B\phi}$ on feedback resistance R_E, calculated for the cases when C_C = const and when C_C is defined by (11) (in) one can see that contribution of the C_C nonlinearly increases PM noises for all values of R_E. If one compares the curves shown in Fig.6,a and Fig.4,a it is evident that when R_E = 10 Ohm $T^2_{B\phi}(\omega_s)$ is almost equal to $T^2_{f\phi}(\omega_s)$. When R_E = 25 Ohm $T^2_{B\phi}(\omega_s)$ is more than $T^2_{f\phi}(\omega_s)$. The difference is about 10 dB.

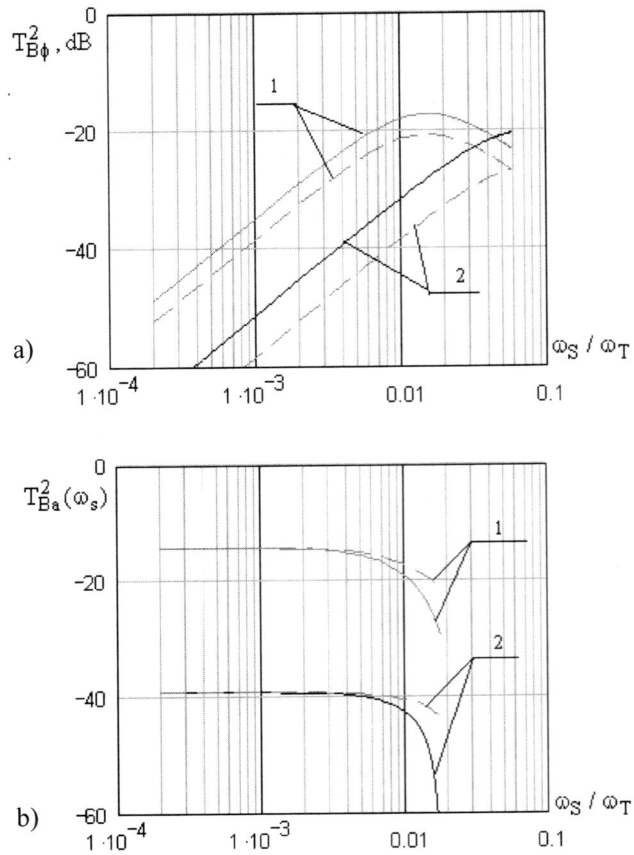

Fig.6. Transformation coefficients "V_B fluctuations – PM noise" (a) and "V_B fluctuations – AM noise" (b) as functions of relative signal frequency ω_S / ω_T calculated for the cases of nonlinear C_C (solid lines) and constant C_C (dotted lines). Case 1 – R_E = 1 Ohm, Case 2 – R_E = 25 Ohm.

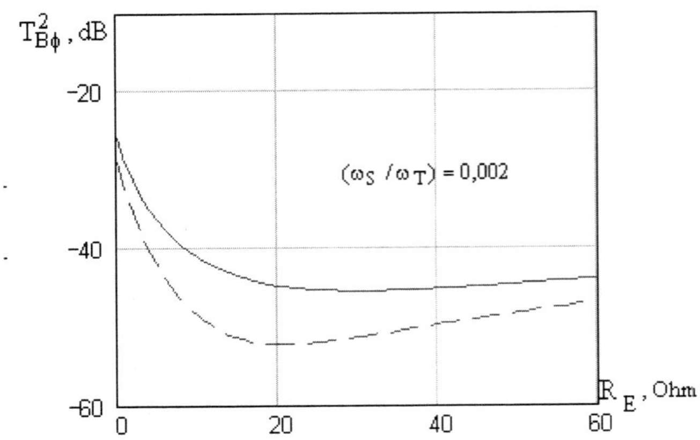

Fig.7. Dependence of transformation coefficient "V_B fluctuations – PM noise" on value of emitter feedback resistor at the constant signal frequency. (Dotted line corresponds to the model with constant C_C).

7 PM and AM noise due to collector biasing voltage fluctuations

Let us consider the case when $\mu_f = 0$ and $\mu_B = 0$. Then from (22), (23) we obtain the next formulae for PSD's of PM and AM noise $S_{\phi,C}(\omega)$ and $S_{a,C}(\omega)$:

$$S_{\phi C}(\omega) = T^2_{C\phi}(\omega_s) S_{\mu C}(\omega),\ S_{aC}(\omega) = T^2_{Ca}(\omega_s) S_{\mu C}(\omega). \quad (27)$$

where $S_{\mu C}(\omega)$ – PSD of $\mu_C(t)$. Transformation coefficients $T^2_{C\phi}(\omega_s)$ and $T^2_{Ca}(\omega_s)$ can be calculated from (21), (18).

The functions $T^2_{C\phi}(\omega_s)$ and $T^2_{Ca}(\omega_s)$ are shown in Fig.8. One can see from Fig.6,a and Fig.8,a that transformation coefficient $T^2_{C\phi}(\omega_s)$ is less than $T^2_{B\phi}(\omega_s)$ by 10-12 dB.

One can also notice that dependence T^2_{Ca} on ω_S differs qualitatively from dependence T^2_{Ba} on ω_S. Transformation "V_C fluctuations – AM noise" is connected with variations of C_C, and their influence on the amplifier gain decreases when ω_S becomes less. An influence of emitter feedback resistance R_E on $T^2_{C\phi}(\omega_s)$ is shown in Fig.9.

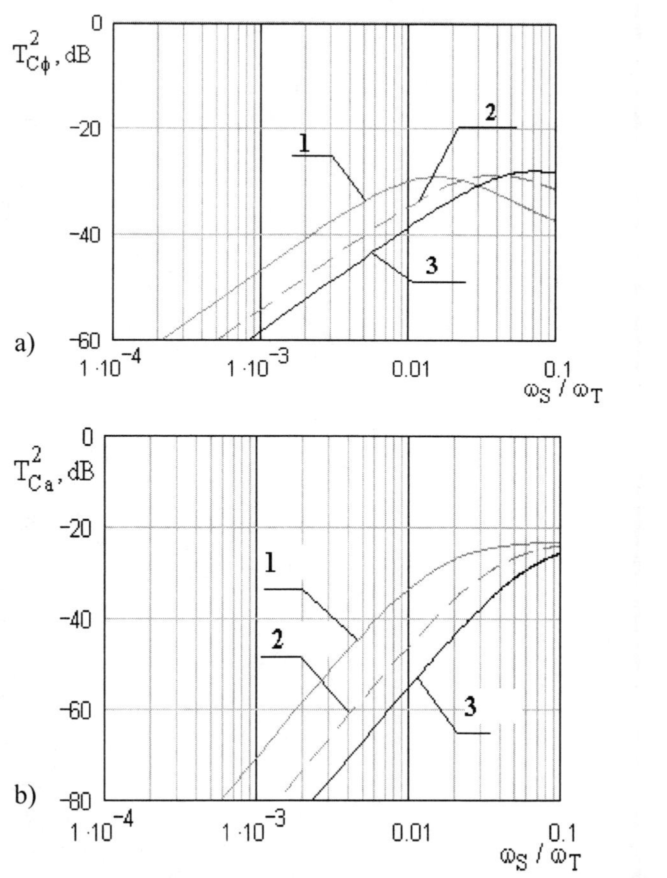

Fig.8. Transformation coefficients "V_C fluctuations – PM noise" (a), and "V_C fluctuations – AM noise" (b) as functions of relative signal frequency ω_S/ω_T. Case 1 – R_E = 1 Ohm, Case 2 – R_E = 10 Ohm, Case 3 – R_E = 25 Ohm.

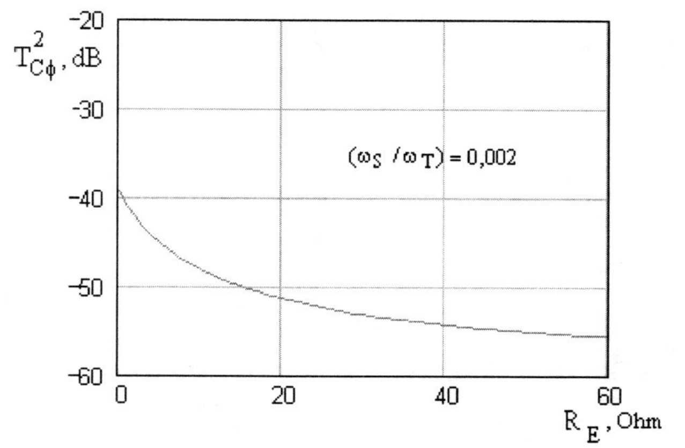

Fig.9. Dependence of transformation coefficient "V_C fluctuations – PM noise" on the value of emitter feedback resistor at the constant signal frequency.

It is interesting to notice from (17), (18), (20), (21) that the sign of $T_{C\phi}(\omega_s)$ is opposite to the sign of the part of $T_{B\phi}(\omega_s)$ connected with C_C nonlinearity.

It means that there is an opportunity to compensate an influence of biasing voltage on phase shift of the amplifier if one uses the same bias voltage source for both collector and for base. Investigation of this opportunity (that was first noticed and analyzed by F.Walls and co-authors [3]) is presented below.

8 PM and AM noise due to biasing voltage fluctuations in DC-balance amplifier circuit

The circuit diagram of the amplifier with one source of biasing voltage, providing an opportunity of DC-balance for V_{cb}, is shown in Fig.10.

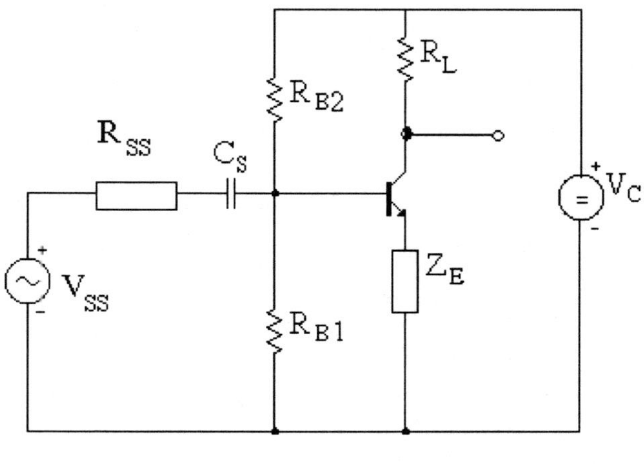

Fig.10. CE amplifier with one source of biasing voltage

The formulae for PM and AM noise calculation in this amplifier follow from (20-22). The circuit Fig.1 becomes equivalent to the circuit Fig.10 if we take

$$V_B = \frac{R_{B1}}{R_{B1}+R_{B2}} V_C, \quad R_{B0} = \frac{R_{B1}R_{B2}}{R_{B1}+R_{B2}}. \qquad (28)$$

It is evident that in this case

$$\mu_B(t) = \mu_C(t) \qquad (29)$$

and from (22), (23) with $\mu_f(t) = 0$ it follows

$$\phi(t)=T_{V\phi}(\omega_S)\,\mu_C(t), \qquad a(t)=T_{Va}(\omega_S)\,\mu_C(t), \qquad (30)$$

where

$$T_{V\phi}(\omega_S)=T_{B\phi}(\omega_S)+T_{C\phi}(\omega_S),\ T_{Va}(\omega_S)=T_{Ba}(\omega_S)+T_{Ca}(\omega_S). \quad (31)$$

From (30) we obtain the formulae for PSD's:

$$S_{\phi V}(\omega) = T^2_{V\phi}(\omega_s)S_{\mu V}(\omega)\,;\, S_{aV}(\omega) = T^2_{Ca}(\omega_s)S_{\mu V}(\omega). \quad (32)$$

The equation of DC-balance for $V_{cb'}$ can be obtained from(31), (20), (21), (17), (18):

$$A_B N_C V_T = V_{C0}. \qquad (33)$$

Using (10), (13) and supposing that $R_{B0} << \beta R_{E0}$, $R_{B0} << \beta R_L$ we obtain from (33) an approximate balance equation

$$\frac{R_{B1}}{R_{B2}} = \frac{R_{E0}+R_E}{R_L} \qquad (34)$$

It is similar to the one presented in [4].

To investigate the opportunities of PM noise decreasing using compensation effect we analyzed the amplifier Fig.10, where BJT was at the same operating point as in the amplifier Fig.1 and had the same values of R_L = 350 Ohm and R_{E0} = 200 Ohm. We have chosen R_{B1} = 1,35 kOhm.

Results of transformation coefficients $T^2_{V\phi}(\omega_s)$ and $T^2_{Va}(\omega_s)$ calculation are shown in Fig.11. The curve 1 corresponds to the case when R_E = 1 Ohm and R_{B1} = 1,4 R_{B1} kOhm (chosen in accordance with (33)). The curve 2 corresponds to the case when R_E = 1 ohm and R_{B2} = 2,7 R_{B1}. One can see that in these cases $T^2_{V\phi}(\omega_s)$ and $T^2_{Va}(\omega_s)$ go down when R_{B2} goes up. The curves 3,4 correspond to R_E = 25 ohm and R_{B2} = 1,4 R_{B1} (curves 3) and R_{B2} = 2,7 R_{B1} (curves 4). In these cases $T^2_{Va}(\omega_s)$ goes down when R_{B2} goes up, but $T^2_{V\phi}(\omega_s)$ goes up.

To understand these results we calculated dependencies $T^2_{V\phi}(\omega_s)$ on R_E with $\omega_S = 0.002\omega_T$ = const for the cases when R_{B2} = 1,4 R_{B1} (curve 1 in Fig.12) and R_{B2} = 2,7 R_{B1} (curves 2 in Fig.12). One can notice a compensation notch at the curve 2. It is similar to the one that was described in [4]. The authors of [4] obtained such curve both by simulation, and by experiment.

It turned out that to obtain the compensation one has to take the ratio R_{B2}/R_{B1} more then it follows from (34). When R_E is small the PM noise that exists in the model with C_C = const (dotted line in Fig.12) is dominating. The part of PM noise that arises due to C_C nonlinearity is less. So it compensates only part of the PM noise. When R_E goes up the part, corresponding to the model with C_C = const, decreases more quickly then the part that arises due to nonlinearity of C_C. At the compensation point these parts are equal (in the Fig.12 it corresponds to $R_{Ecomp} \approx$ 6,7 Ohm). When $R_E > R_{Ecomp}$ PM noise that arises due to nonlinearity C_C becomes dominating. From this consideration it follows that one can obtain compensation at different values R_E choosing the proper ratio R_{B2}/R_{B1}. It is confirmed by Fig.13 where compensation point moves from R_E = 2,5 Ohm to R_E = 44 Ohm.

It is important to notice that the values of R_E and R_{B2}/R_{B1} at the compensation point doesn't depend on the signal frequency. One can see it from Fig.14.

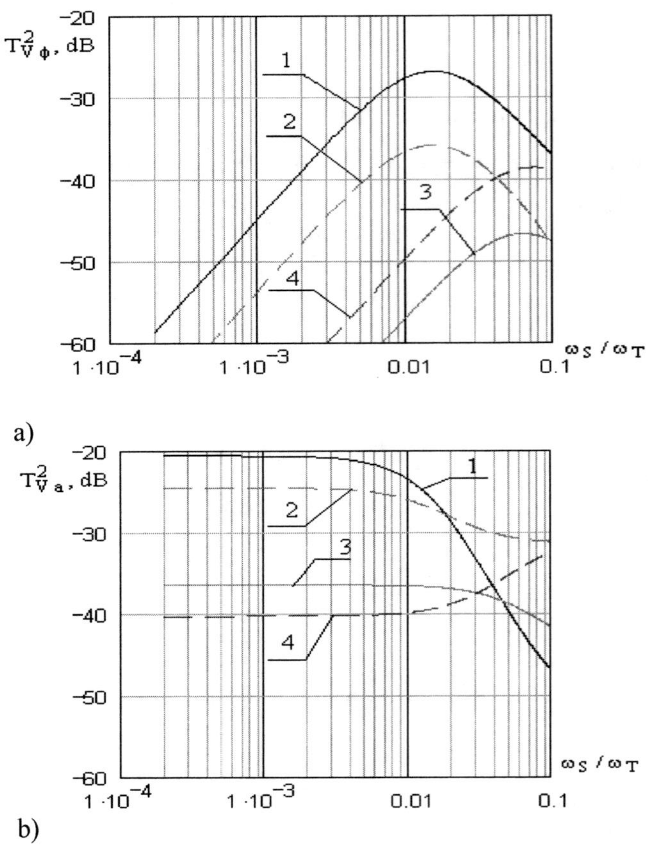

Fig.11. Transformation coefficients $T^2_{V\phi}$ (a) and T^2_{Va} (b) as functions of relative signal frequency ω_S/ω_T for several examples of the amplifier Fig.10: Case 1 – R_E = 1 Ohm, R_{B2}/R_{B1} = 1,4; Case 2 – R_E = 1 Ohm, R_{B2}/R_{B1} = 2.7; Case 3 – R_E = 25 Ohm R_{B2}/R_{B1} = 1,4; Case 4 – R_E = 25 Ohm, R_{B2}/R_{B1} = 2.7

9 Conclusion

An approach developed in this paper gives an opportunities to formulate demands to biasing voltages fluctuations to provide the level of PM and AM noise of the amplifier that is close to the one caused by internal noise sources. The opportunities to decrease an influence of biasing voltage fluctuations on PM and AM noise of the amplifier are also analyzed.

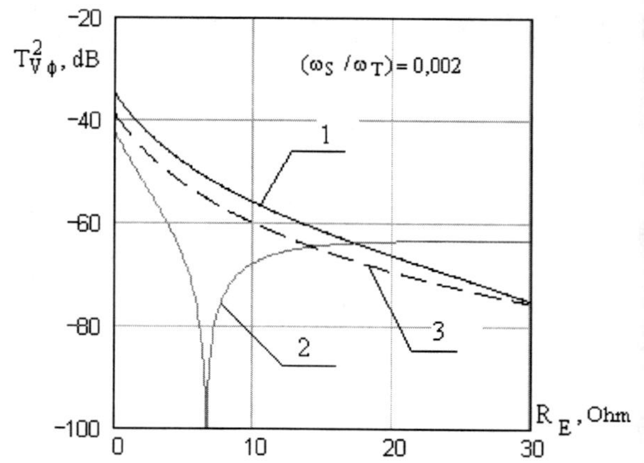

Fig.12. Dependencies of $T^2_{V\phi}$ on R_E at the constant signal frequency: Case 1 – R_{B2}/R_{B1} = 1,4; Case 2 – R_{B2}/R_{B1} = 2,7; Case 3 – R_{B2}/R_{B1} = 2,7 (model with C_C =const).

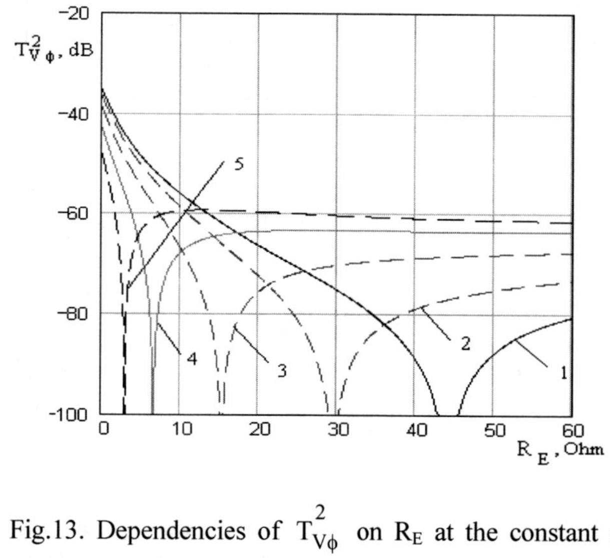

Fig.13. Dependencies of $T^2_{V\phi}$ on R_E at the constant signal frequency: curve 1 - R_{B2}/R_{B1} = 1,4; curve 2 - R_{B2}/R_{B1} = 1,6; curve 3 - R_{B2}/R_{B1} = 2; curve 4 - R_{B2}/R_{B1} = 2,7; curve 5 - R_{B2}/R_{B1} = 3,5.

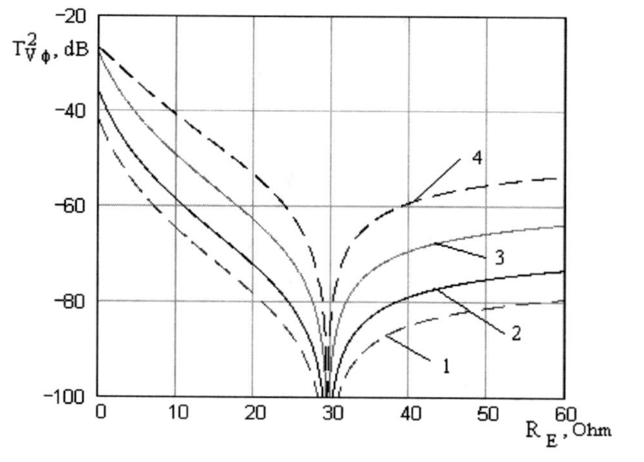

Fig.14. Dependencies of $T^2_{V\phi}$ on R_E at the constant R_{B2}/R_{B1} = 1,6: curve 1 – (ω_S/ω_T)= 0.001; curve 2 – (ω_S/ω_T)= 0.002; curve 3 – (ω_S/ω_T)= 0.006; curve 4 – (ω_S/ω_T)= 0.02.

Acknowledgment

The author would like to thank Prof. V.Kuleshov under guidance of whom this research was fulfilled and Dr.F.Walls who drew the author's attention to this problem.

References

[1] T.I.Boldyreva, "Comparative study of phase and amplitude noise of CE, CE-CB and CC-CB bipolar transistor amplifiers", *Proc. 1998 IEEE Int. Freq. Contr. Symp.* pp.218-255. (1998)

[2] V.N.Kuleshov and T.I.Boldyreva, «1/F AM and PM noise in bipolar transistor amplifiers: sources, ways, of influence, techniques of reduction», *Proc. 1997 IEEE Int. Freq. Control Symp.*, pp.446-455. (1997)

[3] V.N.Kuleshov, "1/F models of bipolar junction transistor and their application to PM and AM noise calculation", *Proc. 1998 IEEE Int. Freq. Contr. Symp.* pp.167-171.(1998)

[4] J.Li, E.S.Ferre-Pikal, F.L.Walls, D.Forbes and S.H.Talisa. "Predicting 1/f PM and AM noise in Si BJT amplifiers: a new computer program". *Proc. 1998 IEEE Int. Freq. Control Symp.*, pp.204-209. (1998)

ROM-LESS DIRECT DIGITAL FREQUENCY SYNTHESIS USING THE 3rd-ORDER PARABOLIC APPROXIMATION

Mohammad Mehrmohammadi, Amir M. Sodagar

Electrical Engineering Dept., Iran University of Science and Technology,Tehran, Iran.
E-mail: m_mehrmohammadi@ee.iust.ac.ir

Electrical Engineering Dept., K.N.Toosi University of Technology,Tehran, Iran.
E-mail: amsodagar@eetd.kntu.ac.ir

Keywords: Direct Digital Frequency Synthesis, ROM-Less DDS, Parabolic Approximation.

Abstract

This paper deals with the introduction of "the third-order parabolic approximation (PB3)" as a new method for sine computation in direct digital frequency synthesizers (DDSs). PB3 is, in fact, an improvement to the second-order parabolic approximation (PB2) [1] that leads to more accurately computing a sine function. The possibility of being implemented using standard logic gates (not even a bit of memory) is one of most important advantages of the third-order parabolic approximation. Another interesting property of the proposed digital implementation for PB3 is that the additional hardware needed for developing a PB2-based DDS architecture to a PB3-based one is quite simple. This is because the hardware utilizes some internal signals that had already developed for PB2 computation.

1 Introduction

Recently, direct digital frequency synthesizers (DDSs or DDFSs) are preferred in precise electronic equipment, and some modern communication systems because of their significant advantages over PLL-based synthesizers. A simplified block diagram of a sine-output DDS is shown in

Fig. 1. The M-bit phase accumulator acts as a slope-controlled digital sweep generator. The phase accumulator content m, which is an M-bit digital sweep, and its slope is determined by the value of frequency control word F, which varies between zero and 2^M-1. This digital sweep can be corresponded to a quantized phase sweep, $\phi=2\pi.(m/2^M)$, varying between zero and 2π. To avoid an extra-large read-only memory, the digital phase sweep is truncated to N bits and then delivered to the sine-computation block to be converted into a digital representation of the output waveform as $y(n)=sin(2\pi.n/2^N)$. The synthesized digital signal can be either directly used in digital systems or converted to analog form and then filtered in analog applications. There are several methods for phase to sine-amplitude conversion, but due to difficulties of sine-function computation in direct digital frequency synthesizers, it is typically performed by

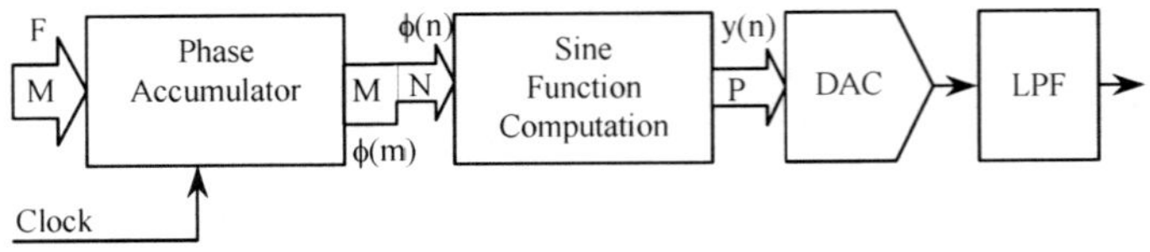

Fig. 1 Simplified block diagram of a DDS

ROM lookup tables. However, using ROM causes some limitations in the synthesizer's performance in terms of frequency synthesis bandwidth and power dissipation. This has led to a wide variety of memory reduction techniques and also some attempts to introduce ROM-less DDS architectures [1].

The first- and the second-order parabolic approximations proved successful initial guesses in memory reduction [2-4]. The former was able to significantly reduce the required ROM and the latter was close enough to an ideal sinusoid that could be used to develop a ROM-less DDS architecture with 10-bit output amplitude resolution.

This paper introduces the third-order parabolic approximation, which presents a more accurate approximation for sine function than the second-order one. Employing this approximation in a DDS system as the sine computation method makes it possible to have a ROM-less DDS architecture with one more bit of amplitude resolution than that was based on the second-order parabolic approximation.

2 The 3rd-Order Parabolic Approximation

The first-order parabolic approximation (PB1) was defined in [2] as:

$$p1(\phi) = (4\phi/\pi)(1-\phi/\pi) \qquad (1)$$

where ϕ was the phase between 0 and 2π. In this method the maximum difference between the initial guess (PB1) and the desired sinusoid is 5.6%. Substituting the truncated digital phase sweep, n, in (1) results in

$$p1(n)=(4/2^{2N}).n.(2^N-n) \qquad (2)$$

and eliminating the constant multiplier $4/2^{2N}$ we have:

$$P1(n)=n.(2^N-n). \quad (3)$$

As was introduced in [2], a *Parabola Generator*, which consists of a 2's complementor and a digital multiplier, can simply convert the truncated digital phase sweep to the required parabola.

The second-order parabolic approximation (PB2), proposed in [4], introduces an improved approximation for the sine function in which, a signal similar to the error between PB1 and an equivalent ideal sinusoid is generated by digital hardware and used to correct PB1. It was described in [4] that the above-mentioned error can be reproduced by using a second parabola generator whose input is P1. The final relationship for the reproduced error was obtained there as:

$$e1(n) =(4K/2^{4N-4}).P1(n).\{ 2^{2N-2} - P1(n)\} \quad (4)$$

and:

$$E1(n) = P1(n).\{ 2^{2N-2} - P1(n)\} \quad (5)$$

As the final relationship for the second-order parabola, the reproduced error was to be scaled down and subtracted from the first-order parabola:

$$P2(n) = 2^{2N}.P1(n) - K'.E1(n) \quad (6)$$

where the scaling factor K' was 0.896. Maximum error between P2 given by (6) and an ideal sinusoid was in the order of 9e-4, which is equivalent to a quantized sinusoid with 10-bit amplitude resolution.

The errors in approximating the first quarter period of a sinusoid with PB1 and PB2 are shown in Fig. 2.

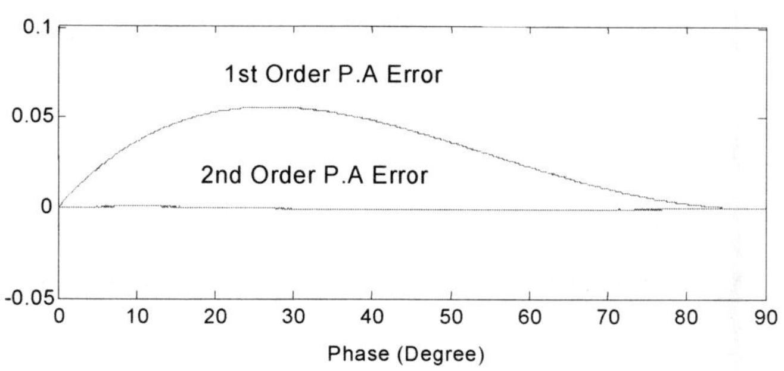

Fig. 2 PB1 and PB2 approximation errors

After the above background review, now we are ready to deal with the third-order parabolic approximation (PB3). The essence of the work is similar to that of PB2, i.e., using a reproduced error to further reduce the approximation error. To find a suitable solution, let's take a closer look at PB2 error, shown in Fig. 3.

The main idea is derived from the fact that PB2 error of Fig. 3 is dividable into two separate parts as shown. These two parts can be reconstructed separately, and be used whenever needed. A phase comparator can detect the zero-

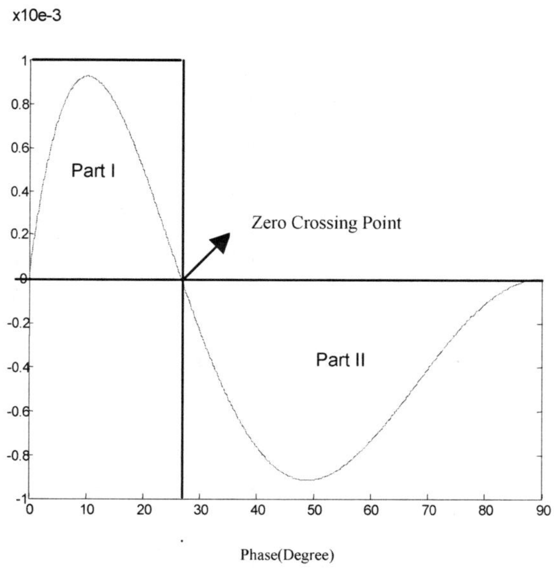

Fig. 3 Plot of the PB2 error

crossing point by comparing the digital phase at the output of the phase accumulator with a fixed value equivalent to the zero-crossing phase point and decide on whichever part is to be used for error reduction.

There are two different ways for generating either part of PB2 error:

1. A simple look at each part of PB2 error shows that it is possible to reconstruct each part by using additional parabola generators. In other words, we can produce each part by using a first or even second order parabola generator. Simulation results show that the first part can be fitted to a first-order parabola between zero and the zero-crossing phase point, ϕ_C. But for the second part a second-order parabola matches much better than a first order parabola between ϕ_C and 90°. This method can be used to accurately compensate for PB2 error but in the expense of increased circuit complexity. Also adding two or three parabola generators increases the power dissipation.
2. The second approach, which adds no significant complexity to a PB2-based architecture, uses a simple mathematical concept. The concept is that the subtraction of a straight line from our first-order parabola can generate another parabola with different zero-crossing and peak points. In other words, we can generate each part of PB2 error by scaling the digital phase and subtracting it from the first order parabola. The advantage of this approach is that both the digital phase and the first-order parabola are already available in the system and there is no additional hardware needed to generate them. This concept is illustrated in Fig. 4.

The zero-crossing point occurs at approximately 26.72° and choosing the second approach, the error of the third-

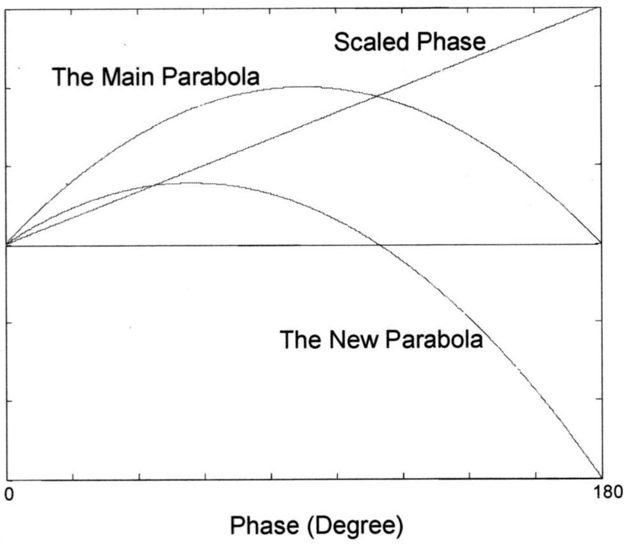

Fig. 4 Generating a new parabola

order parabolic approximation is half that of PB2 and hence, one additional bit of amplitude resolution will be achieved. This way, PB3 can be used to develop a ROM-less DDS system with 11-bit output amplitude resolution.

Frequency Domain analyses

Fig. 5 shows the normalized power spectrum for a 1-KHz quasi-sinusoid waveform synthesized by PB3. The harmonic levels in this spectrum are small enough that confirm in frequency domain that PB3 is an accurate method for approximating a sinusoid.

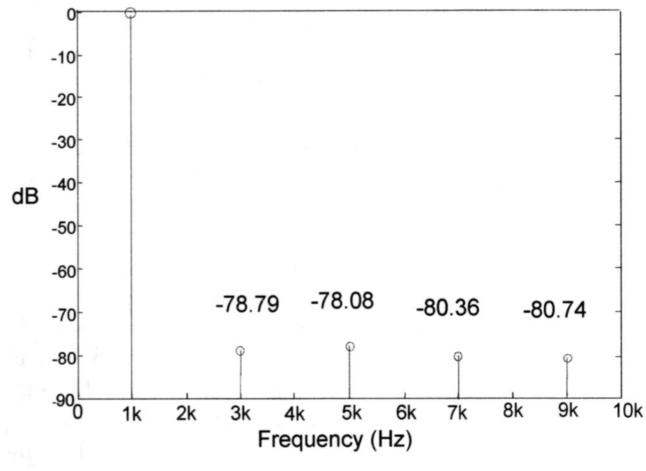

Fig. 5 Power spectrum of a 1-KHz quasi-sinusoid waveform synthesized by PB3 (Normalized to the fundamental component)

To compare the harmonic levels of the second- and the third-order parabolic approximation, Table I summarizes the more important harmonic levels for PB2 and PB3. It shows

Table I PB2 and PB3 harmonic levels with respect to the fundamental component (all in dB_C)

	3rd	5th	7th	9th
2nd Order P.A.	-85.72	-64.12	70.82	-76.38
3rd Order P.A.	-78.79	-78.08	-80.36	-80.74

that except for the third harmonic in PB2 that is lower than that of PB3, the other harmonics are much better in PB3.

3 Digital Realization

Fig. **6** shows the block diagram that is responsible for reproducing PB2 error in order to be properly combined with PB2 and produce PB3.

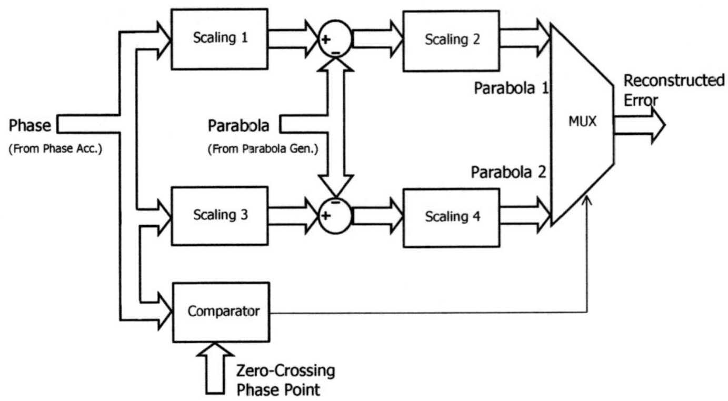

Fig. 6 Proposed block diagram for reproducing PB2 error

As is shown in this block diagram, the truncated digital phase that comes from the output of the phase accumulator should be scaled (Scaling 1 and Scaling 3) and then subtracted from the main parabola that has been already provided for the second-order parabolic approximation. Then two other scaling factors are required to match the generated waveforms with each part of the second order P.A error (specified in Fig. 3). Finally a digital phase comparator compares the value of the digital phase with a constant digital value equivalent to the zero-crossing point (ϕ_C) and generates a selection control signal for the 2:1 multiplexer. In fact, this part of the circuit controls the application of the appropriate error part (I or II) by applying an appropriate command to the multiplexer. If the phase value is smaller than or equal to the zero-crossing phase, then the multiplexer output will be the reconstructed waveform of the part I. Otherwise, the multiplexer output will be the second part reconstructed waveform.

Canonic Signed-Digit Implementation

There are well-known and accurate implementations for all the blocks in the block diagram of Fig. 6 except for the scaling blocks. This is because the scaling factors are not powers of two or even integer numbers, so they are not easily implementable. A good solution for implementing such scaling factors is to represent them in Canonic Signed Digit (CSD) form. In this method all the coefficients has the general form of:

$$C = \sum_{j=0}^{I} d_j 2^{-p_j} \qquad (7)$$

where d_j and p_j are belonging to $\{-1,0,1\}$ and $\{0,1,2,3,....N\}$ sets, respectively. For example, one of the scaling factors,k_1 (Scaling I), is represented in CSD format as follows:

$$K1(CSD) = 4\times\left(1-\frac{1}{4}+\frac{1}{8}-\frac{1}{64}-\frac{1}{128}\right) = 4\times(0.8515625) = 3.40625$$

So, if we accept converting $k_1 = 3.4062 \Rightarrow 3.40625$, multiplying the phase by k_1 will be realized by simple *shift* and *add/subtract* operations rather than using digital multipliers. However, this may increase the 3rd-order parabolic approximation error. The most important point is that these added errors shouldn't exceed the maximum permitted error for 11-bits of amplitude resolution:

$$error \le \left(\frac{1}{2^{11}}\right) \Rightarrow error \le 4.882\times10^{-4} \qquad (8)$$

After implementing the whole PB3 block diagram (including the four scaling factors) in digital world, simulation results show that the digitized 3rd order parabolic approximation error satisfies (8). Fig. 7 shows the approximation error for the full-digital implementation of PB2 and PB3.

4 Conclusion

"The third-order parabolic approximation (PB3)" as a new method for sine computation in direct digital frequency synthesizers was introduced. PB3 is, in fact, an improvement to the second-order parabolic approximation that leads to more accurately computing a sine function. The possibility of being implemented using standard logic gates (not even a bit of memory) is of the most important advantages of the proposed approximation. Another interesting property of the proposed digital implementation for PB3 is that the additional hardware needed for developing a PB2-based DDS architecture to a PB3-based one is quite simple. This is because the hardware utilizes some internal signals that had already developed for PB2 computation. Based on PB3, a ROM-less DDS system with 11-bit of amplitude resolution has been designed in digital form. The SFDR of the synthesized waveform is about –78dBc.

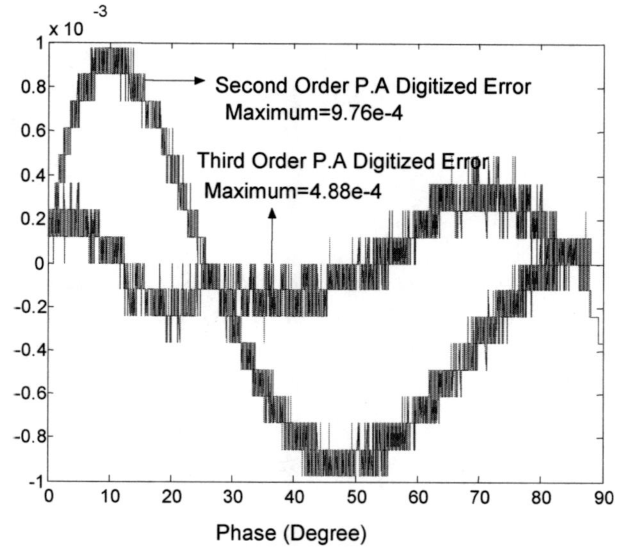

Fig. 7 PB2 and PB3 errors

References

[1] J.vankka, "Methods of mapping from phase to sine amplitude in direct digital synthesis", Proceedings of the 1996 IEEE International Frequency Control Symposium, pp.946-950.

[2] Amir M. Sodagar, G. Roientan Lahiji, "Mapping From Phase to Sine-Amplitude in Direct Digital Frequency Synthesizers By Using Parabolic Approximation", IEEE Transactions on Circuits and Systems, Part II: Analog and Digital Signal Processing, Vol. 47, No. 12, December 2000, pp. 1452 -1457.

[3] Amir M. Sodagar, G. Roientan Lahiji, Ali Azarpeyvand, "Reduced-Memory Direct Digital Frequency Synthesizer Using Parabolic Initial Guess," Analog Integrated Circuits and Signal Processing, An International Journal, Kluwer Academic Publishers, Vol. 34, No. 2, pp. 89–96, February 2003.

[4] Amir M. Sodagar, G. Roientan Lahiji, "A Pipelined ROM-Less Architecture for Sine-Output Direct Digital Frequency Synthesizers Using the Second-Order Parabolic Approximation", IEEE Transactions on Circuits and Systems, Part II: Analog and Digital Signal Processing, Vol. 48, No. 9, September 2001, pp. 850-857.

L-BAND FM TELEMETRY MODULATOR (TWO-POINT MODULATION)

SIGHAKOLLI SATYANARAYANA
VISWANATHAN A.N.
Vikram Sarabhai Space Centre(VSSC)
Veli,RFAFD,MSB
IN 695 022 Trivandrum
India

Conventional FM transmitters lack DC response to the modulation input. Two point modulation is a solution to get around this problem and achieves dc response. In this method, both the VCO in the PLL and the reference oscillator are modulated by the input modulation signal. Modulation of the reference oscillator provides the low frequency response and modulation of the VCO provides high frequency response. But low sensitivity of VCXO's used in the reference channel is a problem for getting uniform response from dc to the required bandwidth. With the advent of inverted mesa crystals, high frequency VCXO's generating in 100MHz region (in fundamental mode) can be easily realized and this problem can be solved. Availability of Fractional-N Synthesizer, will reduce carrier phase noise, near the carrier. A block diagram of the proposed scheme is given below.

In the main paper, loop design criteria to achieve constant frequency response of the deviation w.r.t modulation signal for all loop-bandwidths, will be presented.

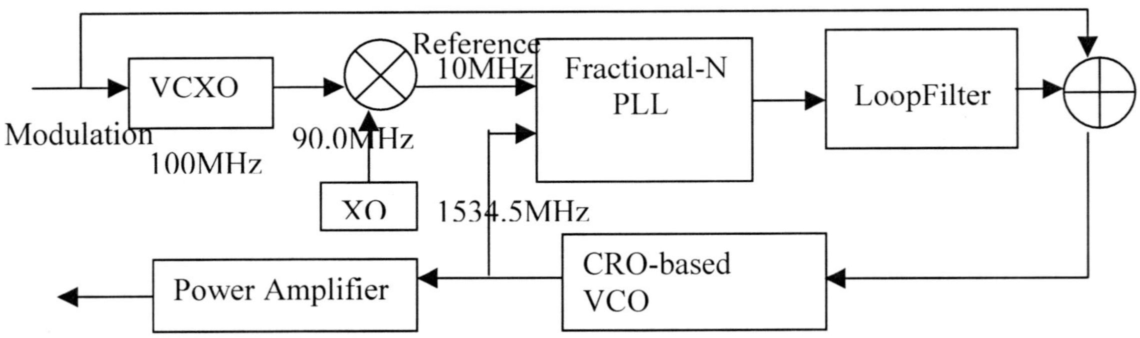

Specifications Achieved:

Frequency = 1534.50MHz
Modulation = TRUE FM, sense positive
Frequency response = DC to 5.0MHz (with ± 1.0dB)
Maximum Frequency deviation = 4MHz
Frequency stability = ± 20ppm (0°C to 70°C)
Incidental FM $\leq$ 4 KHz

Abbreviations used

CRO = Coaxial Resonator Oscillator
VCXO = Voltage Controlled Crystal Oscillator
PLL= Phase Locked Loop
XO = Crystal Oscillator

EMPLOYING ARTIFITIAL NEURAL NETWORKS FOR MAPPING FROM PHASE TO SINE APLITUDE IN DIRECT DIGITAL FREQUENCY SYNTHESIZERS

Amir M. Sodagar

Electrical Engineering Dept., K.N.Toosi University of Technology,Tehran, Iran.
E-mail: amsodagar@ieee.org

Keywords: Direct Digital Frequency Synthesis, ROM-Less DDS, Artificial Neural Networks.

Abstract

This paper presents novel approaches for mapping from phase to sine amplitude in direct digital frequency synthesizers (DDSs) based on the function-approximation capability of artificial neural networks (ANNs). Feasibility, advantages, and some implementation issues associated with the proposed neural DDS family are discussed.

1. INTRODUCTION

Direct digital frequency synthesizers are of increasingly interest in modern digital communication systems and high-precision function generation. Simplified block diagram of a sine-output DDS is shown in Fig. 1. There are a wide variety of approaches for mapping from phase to sine amplitude in DDSs, which can be classified into two major categories: methods based on lookup tables, and those utilizing computational techniques. DDS architectures introduced by Nicholas, et al [1], Sunderland, et al [2], Yamagishi, et al [3], and by the author [4,5] are examples of lookup-table-based methods, and the architectures based on CORDIC algorithm by G. Gielis, et al [6], and the second-order parabolic approximation by the author [7,8] can be named as examples of the computational methods. However the sine function computation block in Fig. 1 converts the digital phase samples to the desired digital sine amplitude samples as is desired, there are some other factors associated with the analog parts of the D/A converter (e.g. temperature, aging, and micro-fabrication tolerances), that may lower the quality of the synthesized waveform.

This paper deals with utilizing the *function approximation* capability of feed-forward neural networks (FFNNs) for phase-to-amplitude conversion in DDSs. Since function approximation in FFNNs is not limited to any specific function, the resulted DDS will be capable of synthesizing arbitrary waveforms as well as the sine function.

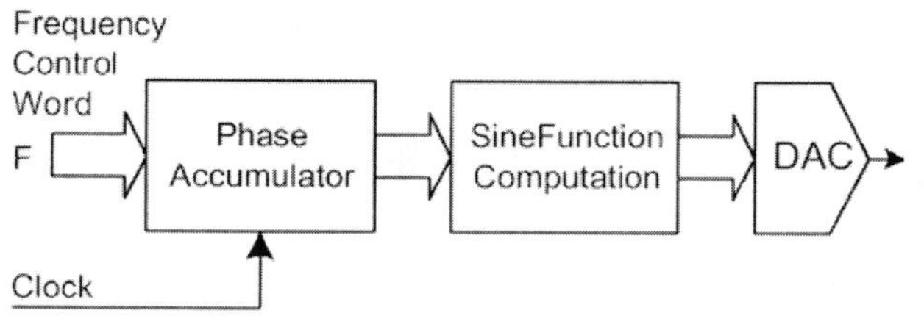

Fig. 1 Simplified block diagram of a sine-output DDS

2. IDEA DESCRIPTION

Fig. 2 shows two direct digital frequency synthesis architectures proposed for synthesizing sine waves and arbitrary waveforms as well, using FFNNs. These architectures are generally called Neural DDS (NDDS) family [9].

The first architecture, shown in Fig. 2(a), exactly follows the traditional DDS architecture in which, the sine computation block has been replaced with a digital-in digital-out FFNN. This architecture, called Digital NDDS (DNDDS) architecture hereafter, is capable of learning any arbitrary periodic waveform to be synthesized. The most important feature of DNDDS is its capability of synthesizing arbitrary waveforms without any architectural complexity compared to that is required for sine-wave synthesis. The only thing that may be imposed to the system in the case of some complex waveforms is the increased number of neurons in the FFNN. In the second architecture, as is shown in Fig. 2(b), the D/A converter brings the digital phase sweep into analog domain and then the FFNN converts the analog phase sweep to the desired waveform. Since in this architecture phase-to-amplitude conversion is entirely performed in analog domain, it is called Analog NDDS (ANDDS) architecture. Such an architecture will then be capable of compensating some sources of non-linearity in the input-output characteristic of the D/A converter. When placed in the architecture of Fig. 2(b), the FFNN can be trained to convert the non-ideal analog ramp at the output of DAC to a sinusoid.

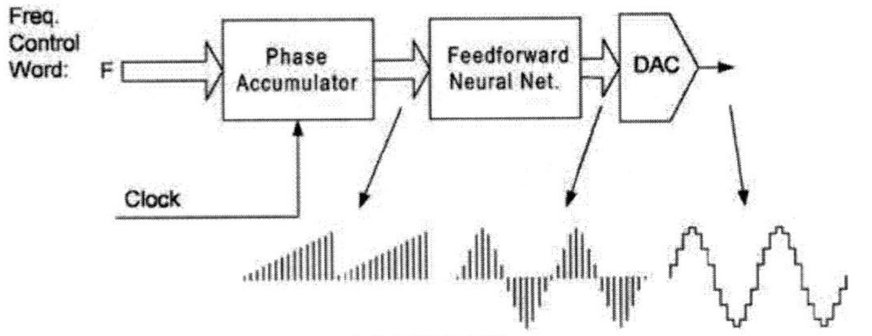

(a) DNDDS

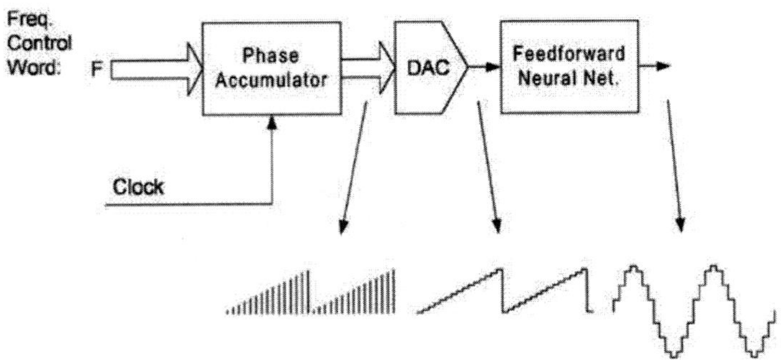

(b) ANDDS

Fig. 2 Proposed Neural DDS Architectures

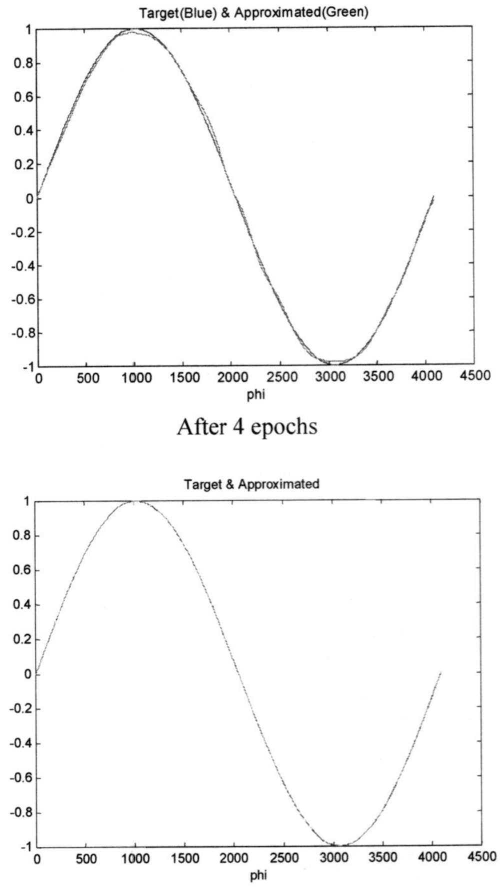

Fig. 3 Sine function learned in the ANDDS architecture during training phase

However the FFNNs in both proposed NDDS architectures can be implemented with predetermined fixed weights in order to get rid of the complexities associated with on-chip learning, it is possible to design the ANDDS architecture with on-chip learning capability in order to be calibrated and compensate for the errors caused by fabrication tolerances, probable changes in environmental conditions, and aging.

In both proposed NDDS architectures, the frequency resolution is still determined by the phase accumulator, and the amplitude resolution in ANDDS architecture is determined by the maximum acceptable error considered in the FFNN learning phase.

3. DESIGN & SIMULATION

An analog-input analog-output (AIAO) FFNN, with 10 neurons in the hidden layer and one output neuron, is trained to learn a complete period of a sinusoid. Characteristic functions for the hidden layer and the output neuron are hyperbolic tangent sigmoid and linear, respectively.

The synthesized sinusoid is supposed to be equivalent to the output of a DDS with 10-bit amplitude resolution. This means that the maximum between the synthesized waveform and an ideal sinusoid can be $1/2^{10} \simeq 9.76e\text{-}4$. As is recommended, the phase is considered with two more bits of resolution. So, the input to the D/A converter is a 12-bit digital phase sweep in the architecture of Fig. 2(b).

Back-propagation algorithm with momentum is used to train the FFNN. The network is trained by the back-propagation (with momentum) algorithm with: LearningRate=0.01 and MomentumConstant=0.9. Due to our application, the FFNN should be trained once, and its main task is to operate with fixed weights and biases. Hence, learning speed is not the matter and the learning rate and momentum constant should be so chosen that only satisfy convergence of the learning algorithm to the desired states with acceptable final approximation error.

In order to have an acceptable accuracy in sine function computation (after training), a long training phase (594 epochs*) has been considered so that the maximum error between the learned and the target functions is acceptable, as is shown in Fig. 3.

- **The FFNN in the ANDDS architecture can compensate some non-linearities of the DAC:**

One of the most important points in ANDDS architecture is the ability of the FFNN to compensate some of the unavoidable non-linearity behaviors of the D/A converter.

* One epoch of training is defined as a single presentation of all input vectors to the neural network. The network is then updated according to the results of all those presentations.

When placed in ANDDS architecture, the FFNN can be trained to convert the non-ideal analog ramp at the output of DAC to a sinusoid. Fig. 4 shows the perfect sine computation performed by the AIAO FFNN in the case of an ADC with non-linear characteristic curve.

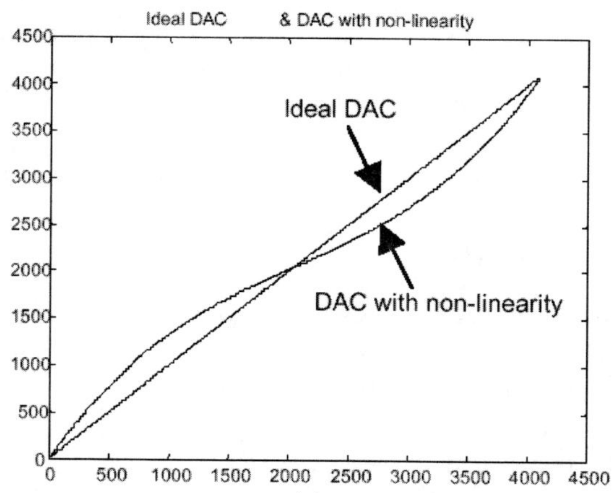

(a)

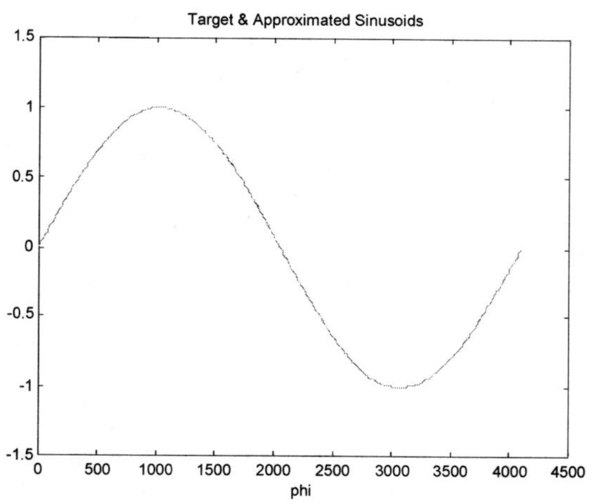

(b)

Fig. 4 Sine wave synthesis in the presence of DAC non-linearity

- **Resolution**

At the first step of this work, the neural networks was designed, simulated, and trained in system level. The sine computation accuracies have been obtained in MATLAB by using "long" numeric format (i.e. scaled fixed point format with 15 digits). Despite this excellent numerical resolution in system-level simulations, high resolution of weights and biases of a neural network cannot be easily obtained in the real world, in neither digital nor analog implementations. Hence, the resolution of computations should be lowered as much as possible. In order to determine the minimum required resolution in realizing wieghts and biases within the network, all the numeric values are truncated in the AIAO FFNN. By each decrement in the number of digits, the maximum error between the approximated sinusoid and an ideal sinuoid is studied. Fig. 5 shows the error between the approximated and ideal sinusoids in a complete period, in the

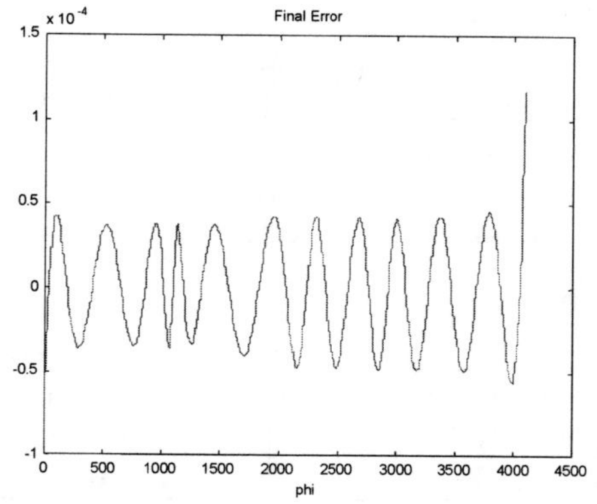

No truncation

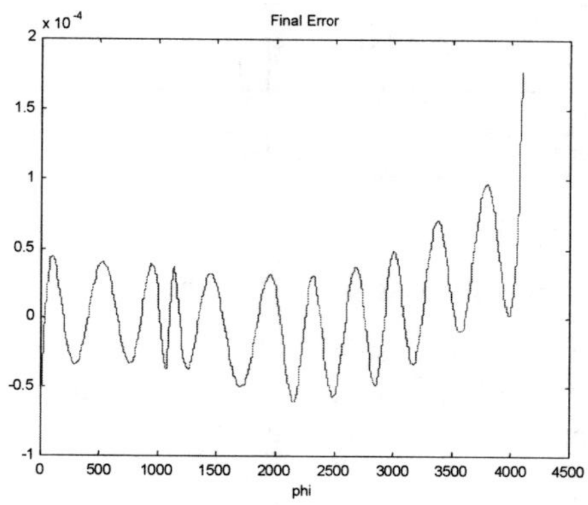

8 digits after decimal point

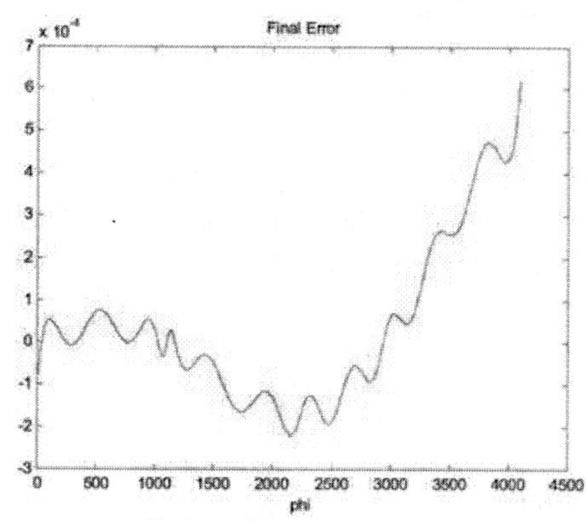

7 digits after decimal point

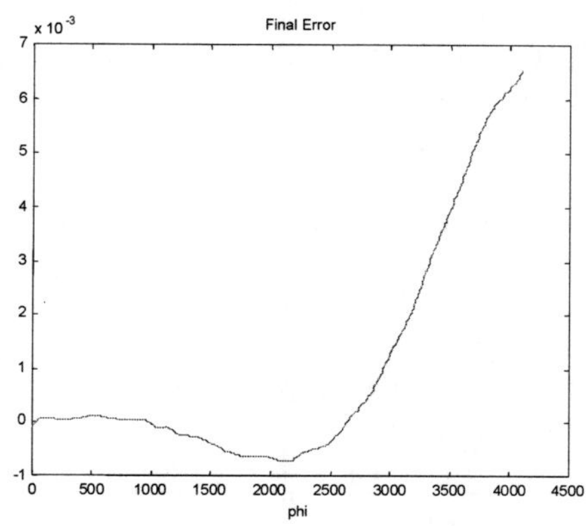

6 digits after decimal point

Fig. 5 Impact of the reduction in the number of digits in the function computation error

case of 'no truncation', and truncation to 10, 7, and 6 digits. The maximum difference between the approximated and ideal sinusoids during the process of digit reduction is summarized in Table I. As is marked, weight and bias values can be maximally truncated to 7 digits after the decimal point and the error is still below that is associated with output quantization to 10 bits.

Table I Function approximation error for different cases of weight/bias truncation for AIAO FFNN

No. of digits after decimal point	Approximation Error
No Truncation	1.17e-4
12	1.17e-4
10	1.18e-4
8	1.77e-4
7	6.19e-4*
6	6.5e-3

- **The smallest possible network**

However the designed AIAO feed-forward neural network successfully computes the required sine function with 10 neurons in its hidden layer, it can be designed with less number of neurons. Simulations show that the hidden layer can contain minimally 3 neurons and the AIAO FFNN is still capable of perfectly operating. After 200 epochs, the difference between the approximated and the ideal sinusoids has a maximum value of 1.9e-4. However this excellent accuracy has been obtained with "long" numeric format, it can still be quite low for truncation up to 7 digits after the decimal point, as is demonstrated in Table II.

Some of weights and bias components can be further truncated without any considerable effect on the computation accuracy. Maximally-truncated weights and biases are:

```
net.iw{1} = [0.00079677  0.0006987  -0.0007965]';
net.lw{2,1} = [2.56  -3.2893  -2.5614];
net.b{1} = [-3.1128  -1.4308  0.1508]';
net.b{2} = -0.00054;
```

which will lead to the maximum error of 2.92e-4 with respect to an ideal sinusoid. As can be seen, some of the weights and biases can be simply scaled up by a proper scaling factor, so the number of digits is easily reduced further.

Table II Function approximation error for different cases of weight/bias truncation for the AIAO FFNN with 3 hidden neurons

No. of digits after decimal point	Approximation Error
No Truncation	1.8986e-4
10	1.8986e-4
8	2.010e-4
7	5.7956e-4*
6	0.0059796

4. CONCLUSION

Employing artificial neural networks in direct digital frequency synthesis was proposed. Two different neural DDS (NDDS) architectures were introduced, ANDDS and DNDDS, in which the neural network performs phase to amplitude conversion, in analog and digital domains respectively. Some implementation issues such as resolution and hardware complexity were discussed.

REFERENCES

[1] Nicholas, H. T., Samueli, H. and Kim, B., "The optimization of direct digital frequency synthesizer performance in the presence of finite word length effects," in Proceedings of the 42nd Annual Frequency Control Symposium, pp. 357–363, 1988.

[2] Sunderland, D.A., et al., "CMOS/SOS frequency synthesizer LSI circuit for spread spectrum communications," IEEE Journal of Solid-State Circuits, 19(4), pp. 497–506, August 1984.

[3] Yamagishi, A. et al., "A 2-V, 2-GHz low-power direct digital frequency synthesizer chip-set for wireless communication." IEEE Journal of Solid-State Circuits 33(2), pp. 210–217, February 1998.

[4] Amir M. Sodagar, G. Roientan Lahiji, Ali Azarpeyvand, "Reduced-Memory Direct Digital Frequency Synthesizer Using Parabolic Initial Guess," *Analog Integrated Circuits and Signal Processing, An International Journal*, Kluwer Academic Publishers, Vol. 34, No. 2, pp. 89–96, February 2003.

[5] Amir M. Sodagar, G. Roientan Lahiji, Ali Azarpeyvand, "A Novel Architecture for Sine-Output Direct Digital Frequency Synthesizers By Using Parabolic Approximation", *Proceedings of the 2000 IEEE International Conference on Electronics, Circuits, and Systems (ICECS'2K)*, Vol. 1, pp. 256-259.

[6] G. Gielis *et al.*, "A 540 MHz 10 b polar-to-Cartesian converter," *Proc. 1991 IEEE Int. Solid-State Circuits Conf. (ISSCC'91)*, pp. 160–161.

[7] Amir M. Sodagar, G. Roientan Lahiji, "A Pipelined ROM-Less Architecture for Sine-Output Direct Digital Frequency Synthesizers Using the Second-Order Parabolic Approximation", *IEEE Transactions on Circuits and Systems, Part II: Analog and Digital Signal Processing*, Vol. 48, No. 9, September 2001, pp. 850-857.

[8] Amir M. Sodagar, G. Roientan Lahiji, "A Novel Architecture for ROM-Less Sine-Output Direct Digital Frequency Synthesizers By Using the 2nd-Order Parabolic Approximation", *Proceedings of the 2000 IEEE International Frequency Control Symposium*, pp. 284-289.

[9] Amir M. Sodagar, "New Techniques for Mapping from Phase to Amplitude in Sine-Output Direct Digital Frequency Synthesizers," Ph.D. Dissertation, Iran University of Science & Technology (IUST), 2000.

Analysis of the PM and AM associated with the natural eigenfrequencies of spinning metallic and dielectric cylinders illuminated at high frequencies

. K. Christensen *, M. . Underhill †

* MIEE, Defence Science and Technology Laboratory, Fort Halstead, Sevenoaks, Kent, TN14 7BP, UK

† FIEE, Advanced Technology Institute, University of Surrey, Guildford, Surrey, GU2 7 H, UK

Keywords AM, Ambiguity, Doppler, PM, Sideband.

Abstract

Doppler spectra from broadside 77GHz plane wave radar illuminated dielectric and metallic cylinders are investigated under static and dynamic target conditions. The classical formulae of modulation theory are reviewed and verified empirically using measured data containing AM and PM components manifested as discrete sideband imbalances. Particular emphasis is placed on the phasor representation of the Doppler and on the practical understanding of the intrinsic AM / PM ambiguities. Techniques used in the analysis could be applicable to the problem of AM and PM conversion in oscillators and sources. In particular ambiguities are represented and bound using a vector in a box approach.

1 Introduction

Contemporary interest in digital communications for broadband terrestrial and satellite applications has stimulated a surge of research into optimal and sub-optimal multi-position (multi-phase / multi-amplitude) modulation and demodulation schemes [1]. Extant standard techniques such as phase modulation (PM), phase difference modulation (PDM), and quadrature amplitude (amplitude-phase) modulation (AM) are being widely adopted for sophisticated phase shift key (PSK) applications such as binary PSK (BPSK), and quadrature PSK (PSK), and for variants of these such as offset SPK (O PSK) and differential SPK (D SPK). A common feature of all of these modulation methods is the need to understand the inherent phase and amplitude characteristics of the signals under consideration, to quantify error rates reliably, and to identify and eliminate sources of uncertainty in the physical system design. Perhaps understandably research activities have coalesced around those issues of immediate commercial interest, e.g. efficient transmission methodologies, new circuit design processes and component technologies (oscillators and sources etc), to the exclusion of other fields of enquiry such as phase noise and jitter.

In this paper the AM and PM Doppler spectra from broadside plane wave radar illuminated dielectric and metallic cylinders are investigated under static and dynamic target conditions. The study emphasis is not that of understanding target phenomenology, although that will be a natural outcome, but rather, given the availability of coherent data containing novel and interesting spectral features, to review and develop techniques to analyse the intrinsic AM / PM ambiguities present in the data.

Complex Fourier analyses are applied to data from targets with discrete rotation rates from 0 to 480rpm. Typical results for nominally axi-symmetric spinning cylinders indicate that an asymmetric and discrete line Doppler spectrum will be evident. Such a spectrum can be represented as correlated AM and PM sideband components.

A subsidiary objective is to show that the classical formulae of modulation theory can be verified empirically using the measured data. It will be shown that for a given discrete sideband imbalance, at least two ratios of AM and PM will be present. Particular emphasis is placed on the phasor representation of the Doppler where the techniques used could be applicable to the problem of AM and PM conversion in oscillators and sources.

2 Theoretical review

Mathematically it can be seen that the formula for an AM signal, shown equation (1)

$$(\ ) = \sqrt{2}\left[\sin\omega + \frac{}{2}\cos(\omega - p) - \frac{}{2}\cos(\omega + p)\right] \quad (1)$$

has a synergistic relationship with the formula for a PM signal, shown equation (2),

$$(\ ) = \sqrt{2}\left[\sin\omega + \frac{\theta}{2}\sin(\omega + p) - \frac{\theta}{2}\sin(\omega - p)\right] \quad (2)$$

given small angular sideband to carrier ratio [2], i.e. the phase deviation is small and where = carrier power, = modulation depth, and θ = peak angular deviation.

The phasor representation of these formulae, shown figures 1 and 2, when exaggerated for visual effect reveals fundamentally important information describing the nature of the complex AM / PM vector relationships. In the simple scenario described the amplitudes of each sideband pair will always be of equal magnitude and will have a definite phase relationship. It is noted that the respective vectors for the AM and PM cases rotate in opposite directions with the sideband pairs summing on an amplitude basis.

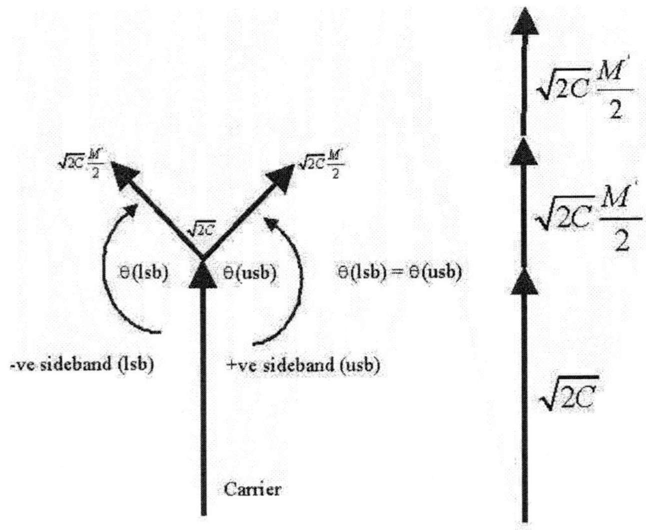

Figure 1: Amplitude modulation vector representation

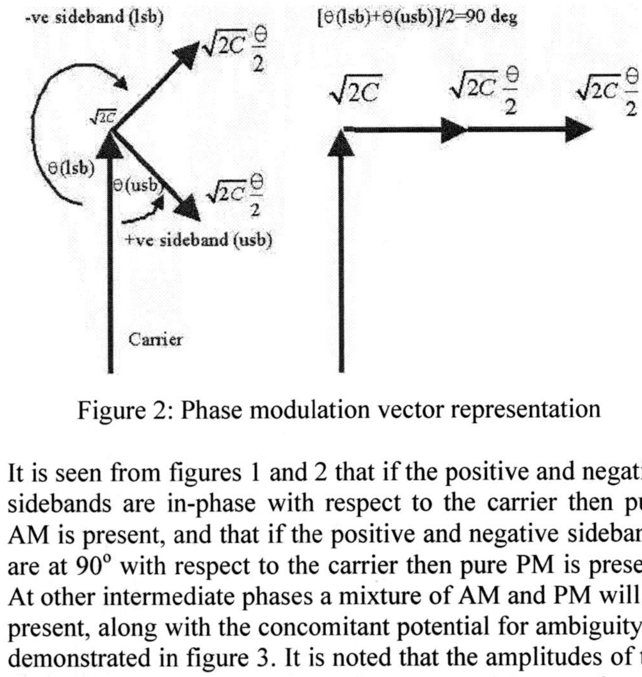

Figure 2: Phase modulation vector representation

It is seen from figures 1 and 2 that if the positive and negative sidebands are in-phase with respect to the carrier then pure AM is present, and that if the positive and negative sidebands are at 90° with respect to the carrier then pure PM is present. At other intermediate phases a mixture of AM and PM will be present, along with the concomitant potential for ambiguity as demonstrated in figure 3. It is noted that the amplitudes of the respective vector components, figures 1 and 2, are of equal magnitude. The AM vector pair could also be pointing downwards or the PM vector pair pointing to the left with similar effect.

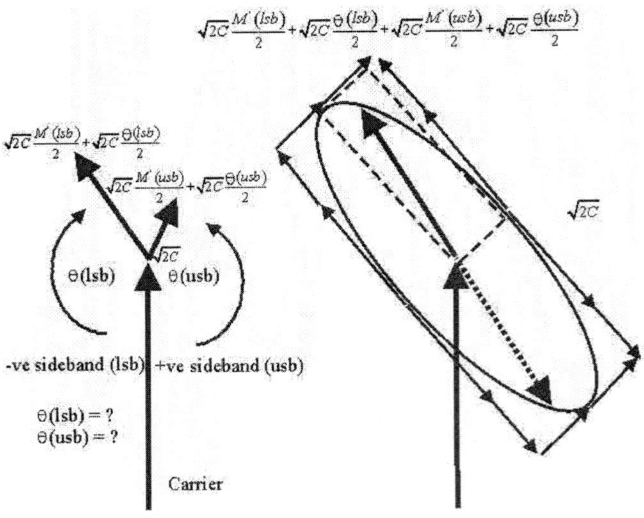

Figure 3: Vector representation illustrating AM / PM ambiguity

In figure 3 it is assumed that phase information is unavailable and that the modulation components are of unequal magnitude. A number of phasor representations could therefore be drawn. In such cases the potential vectors can be only be characterised within certain limits of ambiguity as shown. The limits of the modulation vector can be represented and bound within a rectangular box that is defined by the worst case AM / PM ratios (that are determined by both the ratio of amplitudes of upper and lower sidebands and arbitrary phase relative to the carrier).

From figure 3 it can be seen that at least two ratios of AM and PM are present that cannot be resolved without knowledge of the respective phases of the original modulating waveforms. The resultant vector in the box could be represented by an elliptically rotating vector, linearly fluctuating vector (for example across a diagonal), or some other vector trajectory function. A suitable analogy being that of the lissajous figures [3] that can be produced when the horizontal and vertical plates of an oscilloscope are independently driven by two independent sinewave sources of differing frequencies.

Phasor representation of AM and PM components in signals has mainly found application in the assessment of the noise performance of oscillators [4]. The technique has also been recently used to analyse the spectral components of a signal containing an unwanted parasitic AM component [5]. In this paper the theory will be extended and applied to data containing complex asymmetrical spectral features. The paper will demonstrate the key point that whenever an imbalance in a complex frequency spectrum is observed there will always be at least two solutions describing the ratio of AM to PM signal content. This fact has not, to the authors knowledge, been previously stated in such explicit form although many commendable texts [2, 4, and 5] have alluded to the phenomenon and have addressed the underpinning concepts. A pertinent analogy to support this precept is taken from transmission line theory [6] where for a standing wave ratio of S:1 the ratio of impedance to characteristic impedance can be either S or the reciprocal of S. The study therefore advocates that by using coherent data it is possible to process a signal differently such that AM and PM can be resolved unambiguously.

3 Empirical approach and measurement data

The coherent radar data set under investigation here was recorded using a polarimetric bi-phase coded pulse Doppler 77GHz radar. As such the measured data will contain an associated error uncertainty particular to the system implementation. Subsequent frequency domain analyses will therefore contain spectral components attributable to these unwanted phenomena. It has been established that timing jitter will manifest itself in a pulse Doppler system as both pulse position jitter and pulse width jitter [7]. Similarly bi-phase modulation schemes using pseudo random coded waveforms will also introduce errors. Noise artefacts will be produced due to amplitude imbalance, phase imbalance, rise and fall times, and timing errors associated with the bi-phase modulation process [8]. It is suggested that the total sum of these effects will not significantly impact on the fidelity of the results derived in this paper as the error contributions will be of small magnitude with respect to the modulation effects being studied.

Data sets for target rotation rates from 0 to 480rpm (nine cardinal points in 60rpm contiguous steps) are available for dielectric and metallic cylinders. Details of the radar and experimental test set up will be discussed elsewhere.

4 Time frequency domain analyses

Measurement results for the metallic cylinder and dielectric cylinder are shown in figure 4.

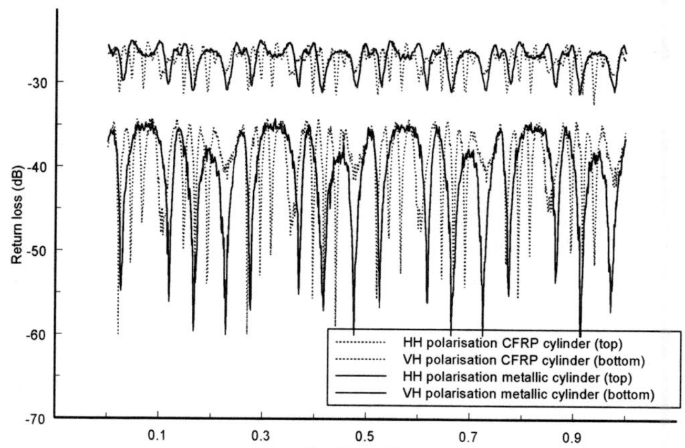

Figure 4: Time-domain response for nominally axi-symmetric spinning steel and CFRP cylinders: 240rpm clockwise

The magnitude responses shown in figure 4 reveal some interesting features. It is immediately recognisable that non-linear modulation effects are present, and that these modulation effects are at twice the rotation frequency rather than mirroring the rotation speed. It is also seen that the modulation fluctuates about a quiescent level, that the target returns from the CFRP cylinder have greater amplitude instabilities, and that distinct nulls occur for the cross-polar responses. It is hypothesised that positive and negative Doppler components from the cylinders are constructively and destructively mixing together causing these null responses, i.e. the spectral components are beating together. The manifestations of which are also producing frequency deviations that are a function of time.

The actual measured data in complex form has been improved using standard phase unwrapping techniques based on successive differences [9]. Typical data sets showing phase against time are respectively shown for the metallic and dielectric cylinders in figures 5 and 6.

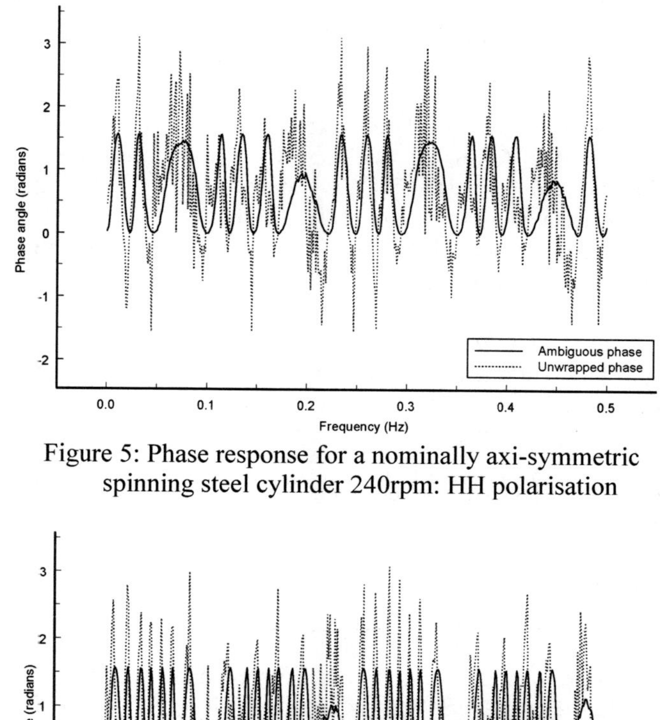

Figure 5: Phase response for a nominally axi-symmetric spinning steel cylinder 240rpm: HH polarisation

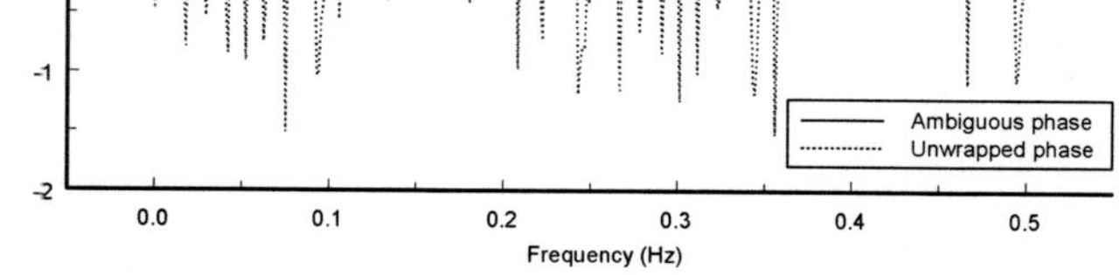

Figure 6: Phase response for a nominally axi-symmetric spinning CFRP cylinder 240rpm: HH polarisation

It is seen from the phase responses shown in figures 5 and 6 that the CFRP cylinder generates greater phase instabilities than the steel cylinder. Distinct phase reversals are noted for both data sets (and especially for the cross-polar plots not shown here). It is speculated that surface scattering effects are responsible for these differences in behaviour between the steel and CFRP cylinders the steel cylinder is electrically smooth whereas the CFRP whilst appearing to be electrically smooth may actually be rough due to the CFRP manufacturing process used. The CFRP cylinder has axial and longitudinal filament windings that are impregnated with resin. It is therefore conceivable that the actual carbon fibre surface is rough. Hence it is hypothesised that the phase variations will be correlated for the steel cylinder but partially de-correlated for the CFRP cylinder.

On the basis of the preceding time domain analyses it would seem plausible that the rotating cylinders are introducing non-linear amplitude-frequency modulation type effects into the radar return signals. Whilst further more detailed time domain analyses are available, such as the Allan deviation technique for statistically characterising the frequency deviation [10], these approaches are not pursued in favour of using frequency domain techniques that are more quantitative in nature. Thus complex Fourier analyses are performed to identify the constituent Doppler / spectral components present in the backscattered target data.

It is stressed that the Fourier technique is a very powerful tool. In the following analyses 1575 point complex fast Fourier transform (FFT) s have been performed providing 0.5Hz frequency resolution [11]. A high confidence level can therefore be placed on the results, i.e. the results contain real features that are not artefacts of the Fourier process.

Complex Fourier analyses allow the upper and lower sideband components to be extracted from the I- data without the loss of information that is associated with using standard magnitude only Fourier algorithms.

Complex FFTs are applied to the data sets with typical results for the steel and CFRP cylinders respectively shown in figures 7 and 8. In figure 7 it is seen that the sideband frequency spectra contains strong components (10dB with respect to the zero Doppler frequency) out to 49Hz. This is contrasted to the CFRP case shown figure 8 where strong components exist out to 105Hz. This trend is repeated within the full data set, and for differing polarisations, and will be the subject of a further publication.

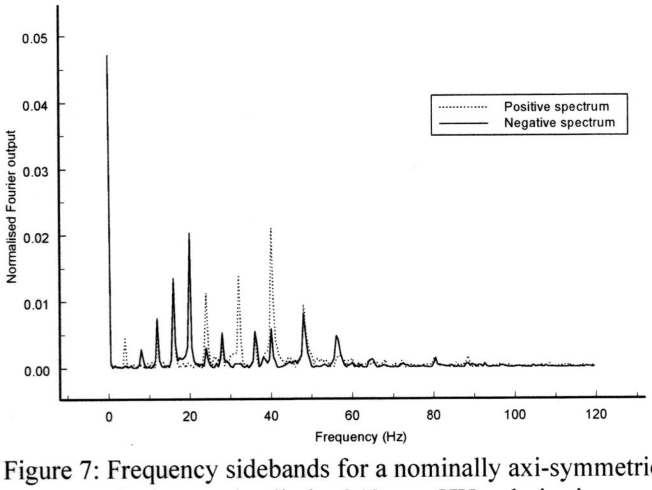

Figure 7: Frequency sidebands for a nominally axi-symmetric spinning steel cylinder 240rpm: HH polarisation

Detailed observation of figure 7 reveals that

- in total there are 12 significant sideband frequency pairs (greater than -10dB with respect to the zero Doppler carrier),
- at 20Hz there is a large negative sideband but no corresponding positive sideband suggesting the presence of both AM and PM,
- at 24Hz and 32Hz there are large positive sidebands but no corresponding negative sidebands suggesting the presence of both AM and PM,
- at 16Hz and 48Hz there are large positive and negative sidebands respectively of equal amplitude suggesting the presence of AM only, and
- at 40Hz there are large negative and positive sidebands, but with the negative component 6dB lower than the positive component suggesting the presence of both AM and PM.

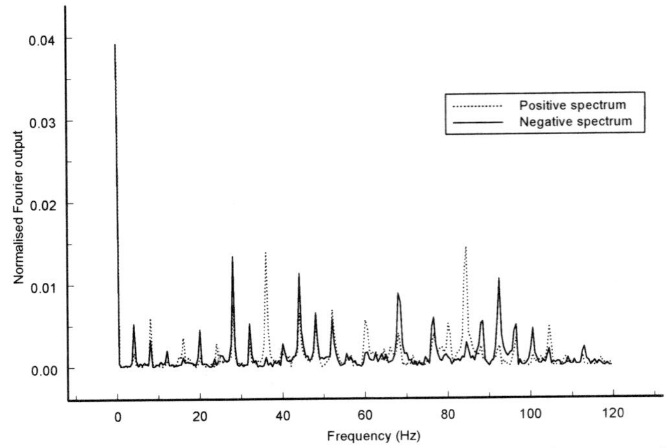

Figure 8: Frequency sidebands for a nominally axi-symmetric spinning CFRP cylinder 240rpm: HH polarisation

Detailed observation of figure 8 reveals that

- in total there are 29 significant sideband frequency pairs (greater than -10dB with respect to the zero Doppler carrier),
- the peak magnitude of the maximum sideband is lower than that seen in figure 7,
- at 69Hz there is a large negative sideband but no corresponding positive sideband suggesting the presence of both AM and PM,
- at 36Hz and 85Hz there are large positive sidebands but no corresponding negative sidebands suggesting the presence of both AM and PM,
- at 48Hz and 52Hz there are large positive and negative sidebands respectively of equal amplitude suggesting the presence of AM only, and
- at 28Hz there are large negative and positive sidebands, but with the positive component 3dB lower than the negative component suggesting the presence of both AM and PM

It is confirmed from analysis of the I- data for the steel cylinder at 16Hz and 48Hz, and the CFRP cylinder at 48Hz and 52Hz, that the frequency components are in phase and hence consistent with the AM-PM theory discussed earlier. Similarly results are consistent with expected theory where there is an imbalance in the magnitudes of a particular sideband pair and / or where the respective components are out of phase.

Visual analyses of the full data set have shown similar imbalances in the resultant power density spectra, with respect to the positive and negative sidebands, and about the zero Doppler component. In every case an asymmetric and discrete line Doppler spectrum was evident.

5 Conclusions

It has been demonstrated that there are always at least two ratios of AM to PM, when amplitude only information is available, and that this ambiguity can be resolved by using coherent radar processing. Complex Fourier techniques were found to be extremely useful in this context since the ratio of AM to PM depends on the relative phases of the modulation sidebands (for any given AM / PM ratio).

Other techniques used in the analysis, such as the vector in a box concept could be applicable to the problem of AM and PM conversion in oscillators and sources. The vector in the box concept allows ambiguities to be represented and bound in an easily understandable form. Statistical methods based on the concept could also be developed, for example, to improve the fidelity of incoherent polarisation data.

The results indicate that for nominally axi-symmetric rotating cylinders an asymmetric and discrete line Doppler spectrum will always be present that can be characterised as AM and PM sideband components. The pure AM content appears to be similar for both the steel and CFRP cylinders. However, more AM / PM components are exhibited by the CRFP cylinder.

For a 4Hz rotation rate the presence of sidebands out to 49Hz (for the steel cylinder) and 105Hz (for the CFRP cylinder) implies that significant phase modulation is present for both cases. The causes of these modulation effects will be discussed elsewhere. Similar results were observed across the whole data set providing confidence in the fidelity of the measured results.

6 Recommendations

The current emphasis on vector modulation techniques should be extended to encompass alternative fields of enquiry and especially for understanding the dynamic signatures of spinning targets. In particular the causes of the modulation effects exhibited by the steel and CFRP cylinders should be studied further. Methods of applying the vector in a box concept to incoherent data sets should also be explored.

Acknowledgements

The following are thanked Dstl for sanctioning the release of this Letter, for professional support and permission to use the experimental radar, Thales Missile Electronics (TME) for technical support, and ineti for the loan of the cylinder rotation rig assembly and control hardware.

References

[1] OKUNEV, ., Phase and phase-difference modulation in digital communications, Artech House, Norwood, MA, USA, 1997.

[2] ROBINS, Phase noise in signal sources, IEE Telecommunications Series 9, Peter Peregrinus Limited, Stevenage, United Kingdom, 1982.

[3] HENNE , K, Radio engineering handbook, 5th edition, McGraw-Hill Book Company, Inc, USA, 1959.

[4] EVERARD, J., Fundamentals of RF circuit design with low noise oscillators, Wiley, Chichester, West Sussex, England, 2001.

[5] KOMAROV, I.V., and SMOLSKI , S.M., Fundamentals of short-range FM radar, Artech House, Norwood, MA, USA, 2003.

[6] DAVIDSON, C.W., Transmission lines for communications with CAD programs, Page 74, Macmillan, 2nd edition, London, UK, 1989.

[7] BUDGE Jnr, M.C., and GILBERT, S.M., Timing jitter spectrum in pulsed and pulsed Doppler radars, IEEE, 1993.

[8] LEVETT, R., Millimeter-wave PRC radar seeker studies, IEE colloquium on high time-bandwidth product waveforms in radar and sonar, Pages 4/1 to 4/6, 1st May 1991.

[9] GHIGLIA, D.C., and PRITT, M.D., Two-dimensional phase unwrapping theory, algorithms, and software, John Wiley and Sons, N , USA, 1998.

[10] IEEE STANDARDS COORDINATING COMMITTEE 27 ON TIME AND FRE UENC , IEEE standard definitions of physical quantities for fundamental frequency and time metrology random instabilities, IEEE Std 1139-1999, IEEE, N , USA, 26 March 1999.

[11] COOLE , J.W., and TUKE , J.W., An algorithm for the machine calculation of complex fourier series, Mathematics of Computation, Volume 19, Page 297, 1965.

1/F NOISE IN CRYSTAL RESONATORS

V. F. KROUPA
The Institute of Radio Engineering and Electronics of the Academy of Sciences of the Czech Republic

It has been found that factional frequency 1/f noise in crystal oscillators is generated by the energy dissipated in the resonator [1]. Earlier investigations found the same origin of 1/f noise also in quartz resonators [2]. In the present contribution we shall show that the losses connected with the current flow through the resonator generate frequency fluctuations which are changed into 1/f phase fluctuations due to the transportation delay. Further we shall prove that the magnitude of the corresponding Power Spectral Density (PSD) is independent of the resonator resonance frequency, f_o, since it is function of the material constant Qf_o.

In addition, we will present a theoretical PSD model of the phase noise at the output of the resonator with a discussion on the $1/f^2$ and $1/f^3$ slopes [2,3]. Finally, we shall investigate contribution of the white noise on the measurement and shall discuss the Allan variance for the resonator and an eventual crystal oscillator [4].

[1] V.F. Kroupa, “Origin of 1/f noise in crystal oscillators,” 2003 IEEE International Ultrasonics Symposium, Honolulu, 5-8 October 2003.

[2] V.F. Kroupa, "Flicker Frequency Noise in BAW and SAW Quartz Resonators," IEEE Tr. UFFC-35 (1988), pp. 406-420.

[3] J.J. Gagnepain, “Fundamental noise studies of quartz crystal resonators”, Proc. 37th Freq. Contr. Symposium , pp. 84-91, 1976.

[4] F.Sthal et al., “Characterization of quartz crystal resonators on phase modulation noise without an oscillator”, 2000 IEEE International Freq. Contr. Symposium, pp. 393-6, 2000.

DUAL-MODE SAPPHIRE LOADED CAVITY MICROWAVE OSCILLATOR WITH A FREQUENCY DIVIDER IN TEMPERATURE CONTROL LOOP

D.P. Tsarapkin

Moscow Power Engineering Institute (Technical University), Moscow, 111250, Russia. E-mail: tsarap@srv-vmss.mpei.ac.ru

Keywords: Microwave oscillator, sapphire loaded cavity, whispering gallery mode, dual-mode frequency stabilization

Abstract

This paper develops the effective method of sapphire loaded cavity (SLC) microwave oscillator frequency stabilization proposed two years ago [1]. The method uses the dual-mode temperature frequency control based on two orthogonal "whispering gallery" (WG) modes and differs from similar approaches by applying of output microwave frequency division to form a control signal. The desired resonant spectrum is provided by selecting a WGE-WGH modes' pair via 3D FE electrodynamic simulation. Main attention is paid to the case of the 1:2 SLC resonant frequencies' ratio.

1 Introduction

The problem of middle-time frequency stability of advanced SLC-based room-temperature low phase noise microwave sources arose immediately after their invention in 1980 [2]. Working for better temperature performance, specialists use direct thermostabilization, supply sapphire "whispering gallery" mode resonators (SWGR) with additional dielectric elements having inverse temperature dependences from that of sapphire and apply thermomechanically compensated SWGRs [3, 4]. However, the best way here, as professional practice shows, lies in using of some intrinsic SWGR's properties, particularly, resonant mode temperature coefficient of frequency (TCF) dependences on electromagnetic field polarization, sapphire disk format and characteristic indices [5, 6]. It follows, a frequency gap between some particular pair of orthogonal WGE- and WGH-modes, $f_{EH} = |f_E - f_H|$, is a sensitive function of temperature. Hence, the problem of precise SWGR's temperature stabilization can be reduced to a simpler task of fixing the intermode frequency difference f_{EH}. Practical implementation of that basic idea was performed successfully in a dual-mode SWGR oscillator at the end of the 80th [7]. Now this direction is under intensive investigation [8].

Published technical solutions exploiting the described method differ in particular ways used to keep f_{EH} constant. For proper operation the most of them demand to introduce or a frequency discriminator (FD) with extremely stable tuning at a chosen f_{EH} [7] or a high quality reference RF oscillator [8]. Internal instabilities of the auxiliary device limit final system parameters in both cases.

The novel dual-mode oscillator approach appeared in 2002 permits to reduce this drawback [1]. The main advantage of the proposed temperature frequency stabilization method is that it does not require any additional frequency reference, which is used in the competitive oscillator design as a control signal now is formed by output microwave frequency division (Figure 1).

The role of the WGE and WGH modes is reversed in Figure 1 in comparison with [1] to improve oscillator performance. It can be shown that if two resonant modes, having fractional TCFs k_{y1} and k_{y2} accordingly, are used in the dual-mode scheme and namely k_{y1} characterizes the "main" oscillation then the loss in automatic control loop sensitivity, χ_2, is

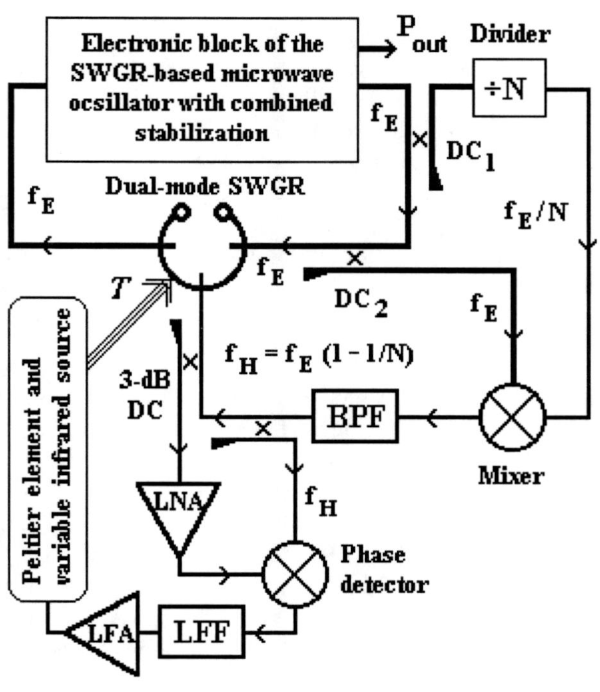

Figure 1. Block diagram of a dual-mode oscillator with the 1:N frequency divider in a microwave part of the temperature control loop [1].

$$\chi_2 = |k_{y1}/(k_{y1} - k_{y2})|. \qquad (1)$$

The loss arises due to both frequencies change in the same direction when SWGR temperature T_r varies. Thus, it is profitable to choose a mode with less $|k_y|$, i. e. WGE one in the case of SWGR, to decrease the loss.

As assumed in [1], the dual-mode oscillator having 10 GHz output utilizes the $WGE_{7,1,1}$–$WGH_{8,1,1}$ mode pair and a divide ratio $N = 64$. So high N demands a multistage divider that can be considered like some drawback of this design. This paper represents further development of the dual-mode oscillator with a frequency divider in the temperature control loop. The idea now is to use a pair of the modes with far spaced resonant frequencies. Particularly, the auxiliary operational mode is supposed to have twice lower frequency than the main one.

Two principal advantages can be seen here. First, the frequency divider providing transfer a part of oscillator output to a frequency discriminator formed on a base of the auxiliary SWGR mode becomes simpler significantly. Second, a twice-lower frequency of the auxiliary mode doubles its quality factor in comparison with a case of about the same f_E, f_H. As expected, these two points can provide lower oscillator fractional frequency deviation.

2 Oscillator description

A block diagram of the microwave SWGR-based dual-mode oscillator having a 1:2 frequency divider into a microwave section of the temperature control loop (called further as the “doubly-stabilized microwave oscillator” (DSMO)) is presented in Figure 2.

Like an oscillator in Figure 1, DSMO incorporates a “usual” microwave SWGR-based oscillator with combined frequency stabilization [9, 10] and an additional temperature control. The first stage of a microwave section of the latter network comprises the directional coupler DC1, the 1:2 frequency divider, the microwave amplifier (MA) working at a half frequency and the SWGR WGH-mode input. A reflected signal selected with the 3-dB directional coupler (3-dB DC) goes to a phase detector (PD) via the low noise microwave amplifier (LNA).

The other PD input is excited through the directional coupler DC2. The mentioned elements together constitute a highly effective interferometric frequency discriminator (FD) sensitive to temperature fluctuations [11]. FD output is filtered and amplified by a low-frequency filter (LFF) and a low-frequency amplifier (LFA) accordingly after what the formed control signal comes to thermal actuators [1]. Thus, the completed automatic temperature control (ATC) repeats basically that one in [1] except the FD operates now at a half frequency. Notice, the network in Figure 2 is somewhat simpler in comparison with its analog in Figure 1.

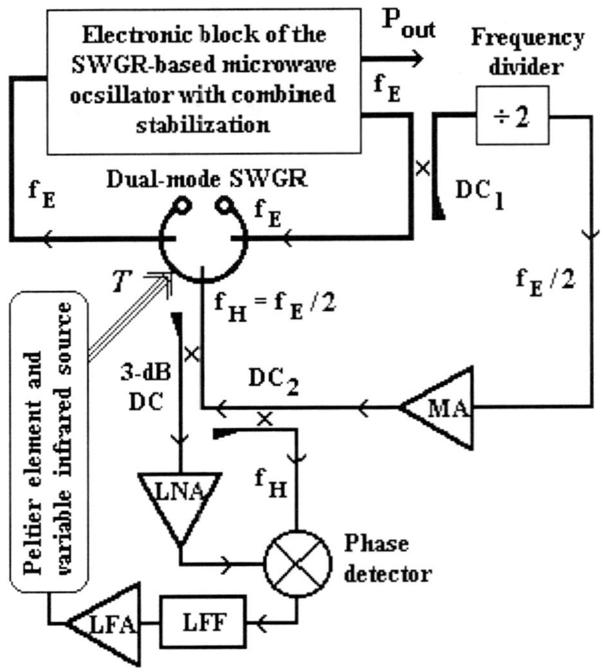

Figure 2. Block diagram of the microwave SWGR-based dual-mode oscillator having the 1:2 frequency divider into a microwave section of the temperature control loop.

3 Dual-mode resonator design

Dual-mode SWGR oscillator operation requires to excite two orthogonal polarized WG modes. In our particular design SWGR is supposed to operate at the main WGE-type mode having resonant frequency $f = f_E$ and at the auxiliary WGH one having the twice-lower resonant frequency $f_H = f/2$. Thus, the operational modes are assumed to have octave splitting between their resonant frequencies.

The lower mode should have relatively high azimuthal index to maintain low radiation losses. The higher mode, on the contrary, is to be chosen with the lowest possible azimuthal index to reduce SWGR size from one side and to guaranty mode frequencies matching from the other. As found by computer simulation, the compromising solution resolving these contradictions could be to choose the $WGH_{6,1,1}$ or $WGH_{7,1,1}$ mode as the lower mode and the $WGE_{14,1,1}$ or $WGE_{15,1,1}$ mode as the higher one.

Mode’s resonant frequencies matching can be obtained by selection of a particular disk format $\Phi = R/H$ (R — the sapphire disk radius, H — its thickness). The fR-product vs. Φ responses for the supposed WGE-modes and those ones doubled for the supposed WGH-modes are drawn in Figure 3. It is easy to see there are three possible SWGR format values for which modes resonant frequencies can be matched.

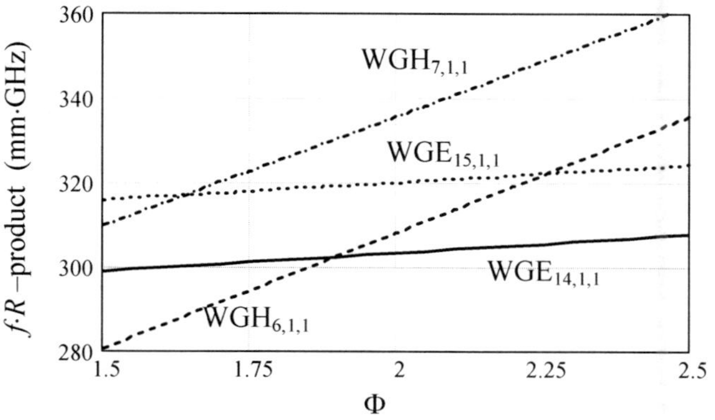

Figure 3. $f{\cdot}R$ vs. Φ plots for the used WGE modes and twice valued plots for the WGH ones.

Another important design aspect is to avoid operational and spurious modes intersection. The spurious modes become a great problem especially when resonant WG modes having high azimuthal indexes have to be used.

The problem of spurious modes has been analyzed for each suitable SWGR format. The spurious modes around the operational (higher) WGE-modes were identified using 3-D finite element electromagnetic analysis software. It was revealed the purest, as far as the spurious modes are concerned, spectrum in vicinity of the operational frequency takes place for the $WGE_{15,1,1}$ mode and disk format Φ = 1.64. Spectrum around 10 GHz calculated for these conditions is shown in Figure 4. The nearest dangerous spurious WGE mode is located about 45 MHz higher the operational one. Thus, the mode pair $WGE_{15,1,1}$-$WGH_{7,1,1}$ can be used to provide the dual-mode SWGR operation.

TCF for the chosen mode set consists of $k_f^E = -48.0$ ppm/K for $WGE_{15,1,1}$ and $k_f^H = -68.7$ ppm/K for $WGH_{7,1,1}$ [6].

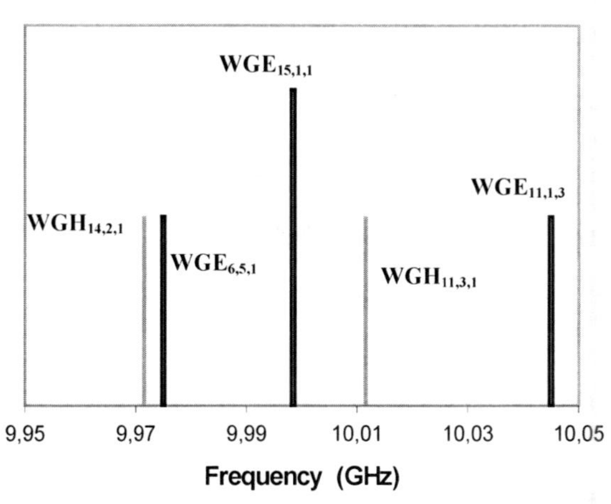

Figure 4. SWGR's modes spectrum around the operational $WGE_{15,1,1}$ mode at 10 GHz in the case of Φ = 1.64.

Figure 4 does not include the most unpredictable spurious modes due to disk enclosure resonances (so called "volume parasites"). My practice shows the only reliable way to eliminate volume parasites in a dual-mode SWGR oscillator is to replace the metal lids of a usual cylindrical enclosure (at least, one of them) with the lids cowered with (or made from) microwave absorber. A sketch of possible SWGR design is drawn in Figure 5. Notice that following this way, one needs to increase the distance between disk faces and the adjacent lids to keep SWGR modes Q-factor constant.

A method of coupling with operational modes is extremely important as it, besides all, must prevent modes' interaction. Capacitive and inductive coaxial probes usually do not work properly in absorbent surrounding. A convenient way here is to use filled rectangular waveguides as described in [3] or wide strip lines.

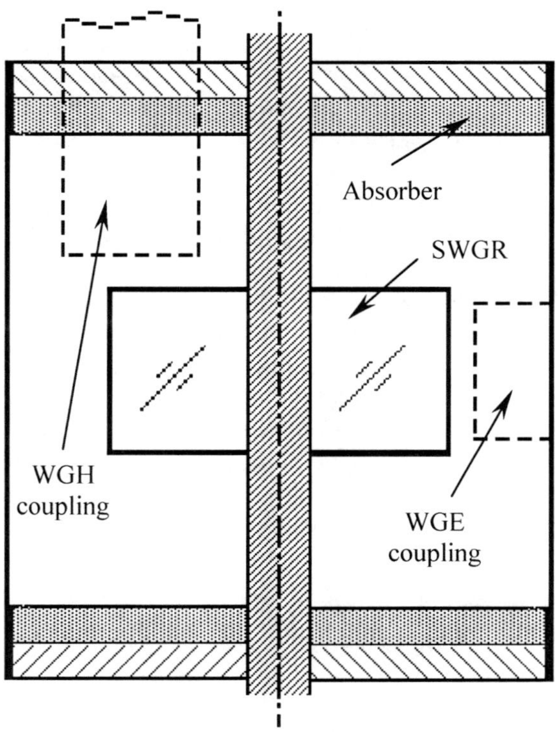

Figure 5. A sketch of possible dual-mode SWGR design with suppressed volume parasites.

4 Oscillator noise floor analysis

A typical transfer function of an arbitrary automatic control system (ACS) with respect to being suppressed distortions has a form

$$\pounds(F) = \pounds_0(F) \,/\, |\, 1 + \boldsymbol{G}_{ol}(F) \,|^2 + \pounds_{fl}(F). \qquad (2)$$

Here $\pounds(F)$ — a final single sideband (SSB) power spectrum density (PSD) of the parameter under analysis; F — a Fourier frequency; $\pounds_0(F)$ — an initial SSB PSD of the controlled parameter; $\boldsymbol{G}_{ol}(F)$ — a frequency response of an open loop gain (a negative feedback is supposed normally); $\pounds_{fl}(F)$ — a

ACS's noise floor. Thus, in principle, one can get about the same total result if starting from rather noisy hardware and suppressing its noise very deeply or taking better equipment and applying more moderate negative feedback.

To define a fractional frequency deviation that can be achieved in this dual-mode oscillator, the ATC noise floor should be estimated. Noise floor analysis of the SWGR oscillator with an additional temperature control can be accomplished like in [8-11] as the ATC in Figure 2 incorporates all typical features of the frequency control with carrier suppressing except the ferrite circulator that is changed with a 3-dB hybrid to eliminate impact of its excess noise [9].

The frequency divider in Figure 2 can be considered as a noiseless device [12, 13]. At first approach it is possible to ignore the phase noise of MA and PD as well. Then the LNA in Figure 2 can be supposed to be a main noise contributor. Following [11] and bearing in mind that: (*a*) the reflected wave power is 6 dB lower now in comparison with a circulator based interferometer; (*b*) in a dual-mode case the automatic control sensitivity should be additionally reduced by χ_2 times, the system noise floor due to LNA phase noise (in rad^2/Hz) can be defined as

$$£_{fl}^{LNA}(F) = [2kT_K N_{LNA}(F) / P_{inc}^H] \cdot (\chi_2 / T_o)^2 \cdot (B_o / F)^2 . \quad (3)$$

Here k — the Boltzmann constant; T_K — current temperature in Kelvin scale; $N_{LNA}(F)$ — LNA noise figure; P_{inc}^H — the microwave power incident on the 3-dB DC placed in front of the resonator; $B_o = f_H / (2Q_{oH})$ — a half bandwidth of an unloaded SWGR at the auxiliary mode; F — an offset frequency; T_o — a value of the figure of merit T describing microwave interferometer efficiency at a central frequency [11]. As shown in [11], T_o reaches 0.5 near exact balance for the conventional phase noise suppression scheme.

Put, for example, $N_{NLA}(1 \text{ kHz}) = 2.5$ dB, $P_{inc}^H = 20$ mW, $f_H = 5$ GHz, $Q_{oH} = 3.5 \cdot 10^5$. Then $£_{fl}^{LNA}(1 \text{ kHz}) = -151.1$ dBc/Hz from Equation (3).

The AFC noise floor in rad^2/Hz corresponds with additional phase noise of the oscillator output. Thus, translation of $£_{fl}^{AFC}(F)$ into Allan deviation $\sigma_y^2(\tau)$ can be obtained by

$$\sigma_y^2(\tau) = \int_0^\infty 2 \cdot \frac{\sin^4(\pi F \tau)}{(\pi F \tau)^2} \cdot \left(\frac{F}{f}\right)^2 \cdot £_{fl}^{AFC}(F) dF . \quad (4)$$

A signal frequency $f \equiv f_E$ here. As supposed, $f_E = 10$ GHz.

To estimate the oscillator fractional frequency deviation $\sigma_y(\tau)$ that arises due to LNA phase noise impact let $£_{fl}^{LNA}(F)$ be

$$£_{fl}^{LNA}(F) = N_\infty \cdot (1 + F_{fc}/F). \quad (5)$$

Here N_∞ — the LNA noise figure far from carrier; F_{fc} — a flicker corner frequency.

The simulation results are presented in Figure 6 with curves 1...4.

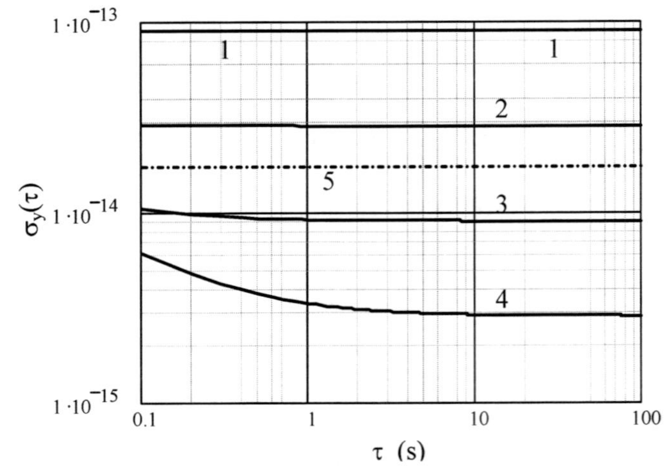

Figure 6. Plots of the fractional frequency deviation due to LNA phase noise (curves 1...4) and due to the frequency divider phase noise (curve 5). Simulation conditions as far as LNA are concerned: $P_{inc}^H = 20$ mW, $N_\infty = 1.25$ dB; $F_{fc} = 1$ kHz (curve 1), 100 Hz (curve 2), 10 Hz (curve 3), 1 Hz (curve 4). The microwave frequency divider phase noise is characterized with $£_{div}(F) = 10^{-17}/F$ rad^2/Hz.

Curve 1 corresponds to SiGe HBT specifications by CEL (California Eastern Laboratories) and other producers given for typical high power regimes. In this case the $\sigma_y(\tau)$ plot looks like a horizontal line drawn at a level close to $1 \cdot 10^{-13}$. The next three curves describe possible middle-time frequency stability improvement when LNA input power decreasing shifts the amplifier to so called "a micro regime." Thus, SiGe technology looks very promising for this particular application also.

Evaluating oscillator $\sigma_y(\tau)$, one has to take into account other noisy components inside the AFC loop. In this context curve 5 in Figure 6 is a useful illustration of the frequency divider noise impact on dual-mode oscillator frequency stability. The supposed divider phase noise level (-140 dBc/Hz at 1 kHz offset frequency that yields $£_{div}(F) = 10^{-17}/F$ rad^2/Hz) is rather typical for commercially available GaAs ICs dividers. One can see even less divider noise in [13].

More thorough investigation might include phase noise impact of such electronic components as a phase detector and auxiliary microwave amplifier (MA in Figure 2). But it does not affect substantially the got results.

5 Conclusion

Novel SLC oscillator approach with an improved frequency temperature control based on the dual-mode operation with the 1:2 frequency divider into a microwave section of the automatic temperature control loop has been proposed and simulated. It is found that the oscillator fractional frequency deviation due to electronics phase noise impact can be as low as $1 \cdot 10^{-13}$ for measurement intervals from one tenth to some hundred seconds.

Acknowledgements

This work is sponsored by Russian Ministry on Education under a grant #209.04.01.020.

References

[1] D. P. Tsarapkin, N. A. Shtin. "Sapphire loaded cavity microwave oscillator with improved temperature stability," in *Proc. 2002 IEEE Int. Freq. Contr. Symp. & PDA Exhibition*, pp. 572-579, (2002).

[2] D. P. Tsarapkin, E. N. Ivanov. "Microwave oscillator," *USSR Patent*, #995,675, int. cl. H 03 B 7/14, Oct 8, 1982, prior. Mar 6, 1981.

[3] S. L. Abramov, E. N. Ivanov, D. P. Tsarapkin, "Low noise microwave oscillator with a thermo-compensated disk dielectric resonator," *Radiotechnika,* no. 11, pp. 81-83, (1988) (in Russian; this magazine is translated in English under name *Telecommunications and Radio Engineering*).

[4] G. J. Dick, "Introduction to sapphire microwave frequency sources," in *Tutorials 2001 IEEE Int. Freq. Contr. Symp. & PDA Exhibition,* Part 1b, (2001).

[5] D. P. Tsarapkin, S. L. Abramov, E. N. Ivanov. "On difference of temperature coefficients of frequency of leucosapphire disk dielectric resonator different modes," *Radiotechnika i Electronika,* vol. 35, no. 2, pp. 431-432, (1990) (in Russian).

[6] D. P. Tsarapkin. "Sapphire disk dielectric resonator temperature coefficient of frequency dependence on temperature, disk configuration and resonant mode," in *Proc. 48th Annu. Freq. Contr. Symp.*, pp. 447-450, (1994).

[7] D. P. Tsarapkin, S. L. Abramov, E. N. Ivanov. "Microwave source," *USSR Patent*, #1,605,890, int. cl. H 03 B 7/14, July 8, 1990, prior. Aug. 22, 1988.

[8] M. E. Tobar, E. N. Ivanov, J. G. Hartnett, C. R. Locke. "Novel temperature control of a sapphire loaded cavity oscillator from the difference frequency of WGE and WGH modes", in *Proc. 2001 IEEE Int. Freq. Contr. Symp. & PDA Exhibition*, pp. 699-702, (2001).

[9] D. P. Tsarapkin. "Low phase noise sapphire disk dielectric resonator oscillator with combined stabilization", in *Proc. 48th Annu. Freq. Contr. Symp.*, pp. 451-458, (1994).

[10] E. N. Ivanov, M. E. Tobar, R. A. Woode. "Advanced phase noise suppression technique for next generation of ultra low-noise microwave oscillators," in *Proc. 49th Annu. Freq. Contr. Symp*, pp. 314-320, (1995).

[11] D. P. Tsarapkin, N. A. Shtin. "Performance limits of microwave oscillators with combined stabilization," in *Proc. 16th European Frequency and Time Forum,* 12-13 March 2002, St.-Peterburg, Russia, pp. C028-C031, (2002).

[12] O. Llopis, M. Regis, S. Desgrez, J. Graffeuil. "Phase noise performance of microwave analog frequency dividers application to the characterization of oscillators up to the millimeter-wave range," *IEEE Trans. on UFFC,* vol. 46, no. 4, pp. 935-940, (1999).

[13] M. Mossammaparast, C. McNeilage, P. Stockwell, J. H. Searls, M. E. Suddaby. "Low phase noise division from X-band to 640 MHz," in *Proc. 2002 IEEE Int. Freq. Contr. Symp. & PDA Exhibition*, pp. 685-689, (2002).

INTRINSIC NOISE IN FERRITE CIRCULATORS AND ITS IMPACT ON MICROWAVE OSCILLATOR PHASE NOISE

D.P. Tsarapkin

Moscow Power Engineering Institute (Technical University), Moscow, 111250, Russia. E-mail: tsarap@srv-vmss.mpei.ac.ru

Keywords: Ferrite circulator, intrinsic noise, microwave oscillator, phase noise

Abstract

This paper deals with intrinsic noise in ferrite circulators and its impact on microwave oscillator phase noise. An approximate formula for spectral density of ferrite circulator intrinsic phase noise with respect to the isolation port is derived and discussed. The derivation is based on a view of a ferrite circulator like a kind of an interferometer with intrinsically noisy arms. It is found that the relative noise with respect to the isolation port might be much more than the noise added to a signal passing between adjacent circulator ports in forward direction.

Some examples demonstrate the use of the derived formula in oscillator design. As shown, namely intrinsic circulator noise limits in many cases output phase noise of the advanced ultra-low phase noise microwave sources. The only radical way to overcome that drawback is to replace a ferrite circulator with a suitable hybrid junction.

1 Introduction

For many years people engaged in design of low phase noise microwave sources did not pay any attention to noise characteristics of ferrite elements. That situation changed only in 1993 when D. P. Tsarapkin revealed an important role of ferrite circulator intrinsic modulation noise in microwave oscillators having £(1 kHz) below -130 dBc/Hz [1]. Two years later E. N. Ivanov et al. published the first experimental results [2]. The latest experimental data are placed in [3]. Nevertheless, until now specialists mistake often as far as the effect of circulator noise on microwave oscillators is concerned.

This paper, dealing with intrinsic noise in ferrite circulators and its impact on microwave oscillator phase noise, is aimed to improve understanding. In its first part the paper contains development and discussion of the approximate formula for spectral density of ferrite circulator intrinsic phase noise with respect to the isolation port postulated in [1]. The second half of the paper includes some examples demonstrating the use of the mentioned formula in oscillator design.

2 Circulator noise model

Advanced microwave sources reach their superb phase noise due to use of precise microwave interferometers. A typical way to suppress carrier and, thus, increase automatic control loop sensitivity is an application of a ferrite circulator (FC) for separating out a wave reflected from a stabilizing resonator (SR) (Figure 1).

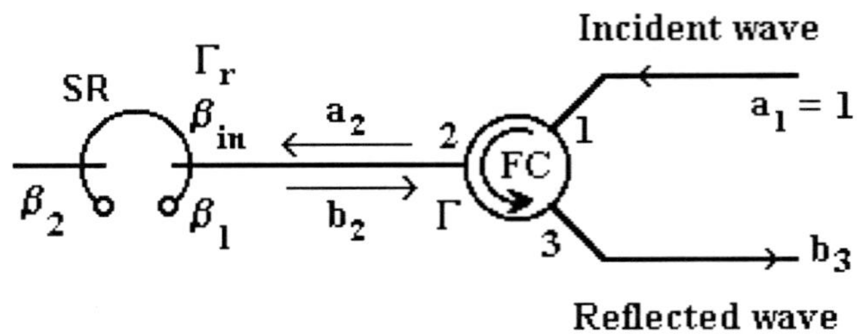

Figure1. Typical using of a ferrite circulator in a microwave interferometer.

An ideal FC is described with Equation (1)

$$\boldsymbol{R} = [\boldsymbol{S}_{cir}] \times \boldsymbol{V}_{ex}. \quad (1)$$

Here $[\boldsymbol{S}_{cir}] = \mathrm{e}^{-j\varphi}\begin{bmatrix}0 & 0 & 1\\ 1 & 0 & 0\\ 0 & 1 & 0\end{bmatrix}$ — a FC's scattering matrix,

$\boldsymbol{V}_{ex}$ — an excitation column vector; $\boldsymbol{R}$ — a reaction column vector.

Let us consider a classical Y-circulator consisting of an axially symmetrical ferrite cylinder with three ports at 120°, biased by an axial dc magnetic field. To describe the wave propagation in such a system one needs to take into account *three* waves [4]:

- a fast wave — $\boldsymbol{V}_{ex1} = ⅓\{1, \exp(120^{\circ}), \exp(-120^{\circ})\}$;
- a slow wave — $\boldsymbol{V}_{ex2} = ⅓\{1, \exp(-120^{\circ}), \exp(120^{\circ})\}$;
- a standing wave — $\boldsymbol{V}_{ex3} = ⅓\{1, 1, 1\}$.

The fast wave rotates clockwise in Figure 1, the slow one – counter-wise.

Thus, the total excitation vector is

$$V_{ex}=\frac{1}{3}\begin{bmatrix}1\\ \exp(120^\circ)\\ \exp(-120^\circ)\end{bmatrix}+\frac{1}{3}\begin{bmatrix}1\\ \exp(-120^\circ)\\ \exp(120^\circ)\end{bmatrix}+\frac{1}{3}\begin{bmatrix}1\\1\\1\end{bmatrix}. \quad (2)$$

A total reaction is a sum of partial reactions. For example, if FC is ideal, the total reaction $\boldsymbol{R}_{\Sigma 1}$ at port 1 is

$$\boldsymbol{R}_{\Sigma 1} = ⅓(\exp(-120^\circ) + \exp(120^\circ) + 1) = 0,$$

while $\boldsymbol{R}_{\Sigma 2} = ⅓(1 + 1 + 1) = 1$,

and $\boldsymbol{R}_{\Sigma 3} = ⅓(\exp(120^\circ) + \exp(-120^\circ) + 1) = 0$

in accordance with the nominal phase shifts among FC ports values $\varphi_1 = -120^\circ$, $\varphi_2 = 120^\circ$, $\varphi_3 = 0^\circ$.

The ferrite cylinder represents the noisy medium which modulates the waves spreading through it in amplitude and phase. Most probably, this intrinsic noise arises due to elementary magnetic domain boundaries fluctuations [5]. If so, the partial reaction in respect of a port "k" due to a vibration "i" takes a form

$$\boldsymbol{R}_{ik} \approx [(1-\Delta_i+m_i)+\mathrm{j}(\zeta_i+\phi_i)]\exp(\mathrm{j}\varphi_{ik})\ (k=1,2,3). \quad (3)$$

Here Δ_i — relative losses; ζ_i — a phase deviation in respect of its nominal value;
$m_i(t)$, $\phi_i(t)$ — amplitude and phase fluctuations of $\boldsymbol{R}_{ik}$.

Summarizing Equations (3), one comes to

$$\begin{aligned}\boldsymbol{R}_{\Sigma 2} &= 1+\{[-(\Delta_1+\Delta_2+\Delta_3)+(m_1+m_2+m_3)]\\ &+\mathrm{j}[(\zeta_1+\zeta_2)+(\phi_1+\phi_2+\phi_3)]\}/3.\end{aligned} \quad (4)$$

for a transmission port and

$$\begin{aligned}\boldsymbol{R}_{\Sigma 3} &= \{[(\Delta_1+\Delta_2-2\Delta_3)-(m_1+m_2-2m_3)+(\phi_2-\phi_1)\sqrt{3}]\\ &+\mathrm{j}[(\Delta_2-\Delta_1)\sqrt{3}+(m_1-m_2)\sqrt{3}-(\zeta_1+\zeta_2)\\ &-(\phi_1+\phi_2-2\phi_3)]/6\}.\end{aligned} \quad (5)$$

for an isolation one.

Equation (5) is especially interesting. Firstly, it follows, the FC isolation (and, hence, leakage) is restricted with

$$\begin{aligned}L_{13} &= -10\log\{|(\Delta_1+\Delta_2-2\Delta_3)\\ &+\mathrm{j}[(\Delta_2-\Delta_1)\sqrt{3}-(\zeta_1+\zeta_2)]|/6\}\ \text{(dB)}.\end{aligned} \quad (6)$$

It seems possible to tune FC at some particular frequency in such a way that $(\zeta_1 + \zeta_2) = 0$. But it is hardly possible to provide simultaneously $\Delta_1 = \Delta_2 = \Delta_3$. So, some leakage is inevitable. Secondly, though it is a main result, Equation (5) gives us ability to analyze the noise, induced by FC.

The "null" at the isolation port appears as a result of interference. Formally, an instantaneous voltage here is a sum of three big vectors. This situation exists constantly and does not depend on FC ports matching. Equation (5) is derived with perfectly matched ports, be the way. Thus, the FC isolation port is a source of some additive noise with a point of view of a following network. A peculiar feature of the situation discussed is that additive noise (AN) appeared due to random processes in ferrite medium and so the 1/f-type in nature.

Quite obviously, AN power is directly proportional to FC incoming power while FC properties does not depend on a signal level. This fact was proved experimentally [6: look at p. 591 & Fig. 8].

A relative phase of the noise vector could be defined in respect of the residual leakage or a signal, passing through FC. The latter looks somewhat more reasonable when one is interested to calculate phase noise. In any case, a principal moment here is the resulting noise defined both amplitude and phase random modulation into FC.

There are all sorts of imponderables when trying to estimate phase noise at the isolation port as it depends on mutual levels of amplitude and phase components and even on a delay of a regular reference signal that positions the noise in respect of the reference. Many attempts to do it came the author to a conclusion that at the moment we need to put up with it and take the effect into account via an auxiliary coefficient. It appears later as $k_{cir} \approx 0.25\ldots1$.

Equations (4) and (5), considering together, give essentially lower noise at the isolation port in comparison with transmission one if $(m_1 - m_2) = 0$ and ϕ_3 is negligible. It coincides with some observations in [3].

To simplify subsequent analysis let

$$\langle m_3^2\rangle < \langle m_1^2\rangle << \langle m_2^2\rangle,\ \langle \phi_3^2\rangle < \langle \phi_1^2\rangle << \langle \phi_2^2\rangle. \quad (7)$$

The physical reason for such assumption lies in the fact that the slow wave is "slow" due to a resonant phenomenon into a ferrite sample and, hence, its parameters are more sensitive to medium disturbances.

When a FC transmission port is coupled with a resonator like in Figure 1 the transfer function for the network output is

$$b_3 = \boldsymbol{S}_{21}\,\boldsymbol{S}_{32}\,\Gamma + \boldsymbol{R}_{\Sigma 3} \quad (8)$$

where $\Gamma = \Gamma_r \cdot \exp(-\mathrm{j}2\varphi_{c\text{-}r})$ — a reflection coefficient in the isolation port cross-section; Γ_r — a resonator reflection coefficient; $\varphi_{c\text{-}r}$ — a phase delay between FC and SR.

Put a current frequency f equal f_o — the SR's resonant frequency. Then

$$\Gamma_r(f_o) \equiv \Gamma_{ro} = |\beta_{in}-1|(\beta_{in}+1)^{-1} \quad (9)$$

and $\beta_{in} = \beta_1/(1+\beta_2)$ — the SR's resonant input impedance.

The terms $\boldsymbol{S}_{21}$, $\boldsymbol{S}_{32}$ describe signal passing through FC including noise impact. Denote as $\pounds_{co}(F)$ [rad^2/Hz] the SSB phase noise imposing on a signal passing between adjacent circulator ports. According to [2, 6]

$$£_{co}(F) \approx -150 - 12 \cdot \lg(F), \text{ dBc/Hz} \quad (10)$$

if the incoming microwave power, $P_{inc,}$ is moderate enough. Then, sorting noise terms in Equation (8), bearing in mind the intermediate results and performing another tedious procedures, one arrives to a final expression describing phase noise at the output:

$$£_{cir}(F) \approx £_{co}(F) \cdot (3 + k_{cir}/\Gamma_{ro}^2) \ [\text{rad}^2/\text{Hz}]. \quad (11)$$

Two terms in brackets reflect here impact of two different mechanisms forming the output noise: direct modulation of a passing signal (*a*) and the interferometric additive noise (*b*).

A value of an integer in the brackets (three instead two) arises due to a double pass of the signal through FC (direct and reflected waves) and mutual correlation of low frequency phase perturbations within a part of a signal path. This estimate may change when more experimental results become available.

The second term in brackets demonstrates noise growth along with the regular signal decreasing. The physical picture is just the same as in one-stage transistor amplifier whose phase noise in inversely proportional to input power.

3 Impact on oscillator phase noise

The phase noise floor of a stabilized microwave oscillator arising due to FC intrinsic noise can be described as [1, 7]

$$£_{fl}^{cir}(F) = £_{co}(F) \cdot (3 + k_{cir}/\Gamma^2) \cdot (B_o / q_e F)^2 \quad (12)$$

where $B_o = f_o / (2Q_o)$ — a half bandwidth of an unloaded SR;

$q_e = Q_e / Q_o$ — an effective Q-factor. In many cases

$$q_e = [(1-\beta_{in})(1+\beta_2)]^{-1} \ (\beta_{in} < 1) \quad (13)$$

or this value approximately. Then from Equations (9, 13) it follows

$$q_e \times \Gamma_{ro} \approx \text{const} \quad (14)$$

if deep carrier suppression is in use.

To check the theory it is reasonable to compare the calculated results with published experimental data.

Start from [8] where experimentally $£$ = -148 dBc/Hz at 1 kHz offset. One can reveal Γ_{ro} = 0.21. Using (10) and putting this value and other data from [8] to (12) one can find $£_{fl}^{cir}$ (1 kHz) = -164.4 + (9.4…14) dBc/Hz within the possible range of k_{cir}. Hence, it is reasonable to choose $k_{cir} \approx 0.5$.

The most reliable proof follows from [9]. Using data [9] and Equation (12) one can calculate the phase noise floor and compare it with the experiment (Figure 2). No doubt, our estimate is in good agreement with the experimental curve

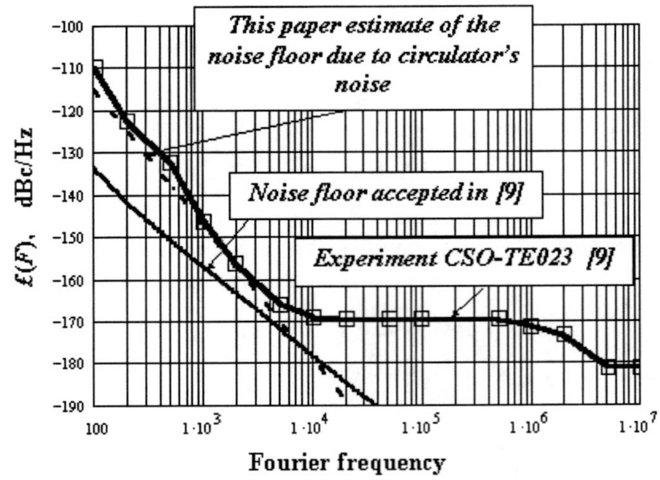

Figure 2. Published experimental and theoretical data [9] as compared to this paper phase noise floor estimate. k_{cir} = 0.25.

while the original variant deviates 24 dB at 100 Hz offset and has another gradient.

Some additional theoretical estimates of microwave oscillator phase noise floor due to intrinsic ferrite circulator noise is shown in Figure 3 where we put: f = 10 GHz, $Q_o = 2 \cdot 10^5$, k_{cir} = 0.5. The SR's transition losses influencing loaded Q-factor are used as a parameter.

The curves demonstrate rather flat minima in respect of SR input impedance ranging within -(149…155) dBc/Hz. It reflects a sense of the problem: the more a degree of carrier suppression the less signal-to-noise ratio and, hence, more noise influence. It does not matter even whether one suppresses a carrier "by one stage" trying to match the resonator almost ideally or makes it step-by-step like in [2, 9] using the two- or multi-stage suppression schemes.

It is no joy to increase P_{inc} trying to improve the noise floor caused by intrinsic circulator noise as that one grows with the same rate.

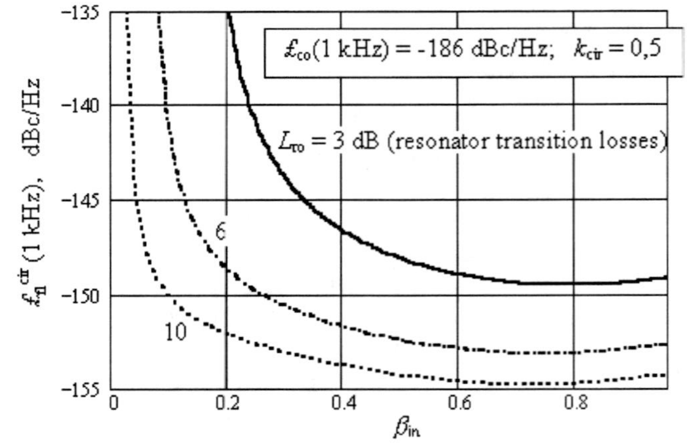

Figure 3. Theoretical estimates of microwave oscillator phase noise floor due to intrinsic ferrite circulator noise.

Conclusion

In many cases namely the intrinsic circulator noise limits a noise floor of the advanced ultra-low phase noise microwave source using additional automatic control loop to lower phase noise. The only radical way to solve this problem is to replace a ferrite circulator with a suitable noiseless hybrid junction.

Acknowledgements

This work is sponsored by Russian Ministry on Education under a grant #209.04.01.020.

References

[1] D. P. Tsarapkin, "Low phase noise sapphire disk dielectric resonator oscillator with combined stabilization," in *Proc. 48th Annu. Freq. Contr. Symp.*, pp. 451-458 (1994).

[2] E. N. Ivanov, M. E. Tobar and R. A. Woode, "Experimental study of the noise phenomena in microwave components," in *Proc. 50th Annu. Freq. Contr. Symp.*, pp. 858-865 (1996).

[3] E. Rubiola, Y. Gruson, and V. Giordano, "On the flicker noise of ferrite circulators for ultra-stable oscillators," in *Proc. 2003 IEEE Freq. Contr. Symp. & PDA Exhibition Jointly with 17th EFTF,* 5-8 May 2003, Tampa, USA, pp. 395-400 (2003).

[4] D. M. Sazonov, A. N. Gridin, and B. A. Mishustin, "Microwave circuits," – Moscow: Vysshaya shkola, 1981. – 295 p. (in Russian; there is an English edition).

[5] N. N. Kolachevsky, "Fluctuation phenomena in ferromagnetic materials," – Moscow: Nauka, 1985. – 184 p. (in Russian).

[6] Tobar M. E., Ivanov E. N., Blondy P., Cros D., Guilon P., "High-Q whispering gallery travelling wave resonators for oscillator frequency stabilization," *in Proc. 1999 Joint Meeting EFTF-IEEE IFCS*, 13-16 April 1999, Besancon, France. – pp. 589-592 (1999).

[7] Tsarapkin D. P., Shtin N. A., "Performance limits of microwave oscillators with combined stabilization," in *Proc. 16th European Frequency & Time Forum*, 12-14 March 2002, St.-Peterburg, Russia. – pp. C028-C031.

[8] Hartnett J. G., Tobar M. E., Ivanov E. N., "Novel interferometric frequency discriminators for low noise microwave applications," *IEEE Trans. UFFC*, vol. 45, no. 6, pp. 1526-1536, (1998).

[9] A. Sen Gupta, D. A. Howe, C. Nelson, A. Hati, F. L. Walls, and J. F. Nava, "High-spectral-purity microwave oscillator: design using conventional air-dielectric cavity," in *Proc. 2003 IEEE Freq. Contr. Symp. & PDA Exhibition Jointly with 17th EFTF,* 5-8 May 2003, Tampa, USA , pp. 423-429 (2003).

MINIATURISED OC O PROFESSIONAL COMPONENTS FOR SPACE APPLICATIONS

Philippe Guillemot*, r me Delporte*, Ale is Morin†, Guy Richard†

*CNES, 18 avenue E. Belin, 31401 Toulouse Cedex 9, France
THALES Microelectronics, .I. Bellevue, 35221 Ch teaubourg Cedex, France

Keywords OC O, Highly Accelerated Tests, Radiation, Space Applications, Stability.

Abstract

For a few years, THALES and CNES have been developing new miniaturised OC Os for space applications. These "micro-OC O" and "mini-OC O Light Version" (respectively referenced EWOS 513 and EWOS 810) are based on professional components instead of space qualified ones. Because standard space qualification rules cannot be used with such components, a new method have been developed to evaluate their ability to fulfil space missions. The retained method derives from the Highly Accelerated Tests.

This paper first describes the Highly Accelerated Tests approach and how it has been applied to miniaturised OC O. Then specifications and typical OC Os performances are given, under conditions such as thermal vacuum, magnetic field or radiation.

1 Introduction

The Nineties were the occasion of a notable change in the space industry with a renewed interest for smaller satellites (in opposition to the large telecommunication payloads). That led all contributors of this sector to a great effort in terms of saving weight, power consumption and cost.

The principal explored way was the use of professional components rather than space qualified ones. That gave access to more integrated technologies authorising significant profits mainly on costs and delays of development. That also brought us to reconsider the methods of qualification of space equipment. Traditional methods, based on the individual qualification of each component or process and on the traceability of the suppliers were not applicable any more. It was thus necessary to imagine some new ways, by taking as a starting point what was made in industry, in particular aeronautical and automobile.

First products to be developed at CNES from such an approach were miniature OC O derived from those produced on a large scale for ARGOS/SARSAT beacons by THALES Microelectronics (Formerly SOREP).

2 Highly Accelerated Tests

2.1 The method

The Highly Accelerated Tests gather a whole of methods (like HALT, HASS [1]) which main objective is seeing where and how failures occur, investigating real design margin and using that information to improve product reliability or make better component selections within the shortest delay and at the lowest cost. These methods were developed by industry to reduce cycles of development, and thus costs, to get a reliable product to market in the shortest period of time. They are based on processes that quickly stress the product, stresses of temperature, rapid thermal cycle, vibration , and allow to rapidly discover design weaknesses and margins. These methods dissociate from traditional methods of accelerated tests [2] by the fact that they do not make it possible to estimate the level of reliability but only the margins of operation. Indeed, the factors of equivalence calling upon the law of acceleration of phenomena of degradation under constraint (Arrhenius, Eyring...) cannot apply in the vicinity of the technological limits.

The method of the Highly Accelerated Tests made its entry in the space field with the introduction of professional components on payloads of satellites. With such components, traditional approaches of qualification of equipment are not applicable any more. First, the individual qualification of each component, such as it is normally considered, would reduce to nothing the profits hoped on costs. Second, available data are insufficient to carry out relevant calculations of reliability. In this context, the method of the Highly Accelerated Tests must answer, at lower cost and within the shortest deadlines, with the double objective [3]:

- To explore the limits imposed by technologies, which makes it possible to determine the limits of operation of the product (i.e. beyond the theoretical design margins Figure 1) and, if necessary, to push back them by new technological choices when the margins compared to the specifications appear insufficient

- To detect as soon as possible (so that they can be corrected) the chargeable causes of defects, these causes being inherent in errors of design or in an insufficient control of the manufacturing processes, and not with technologies themselves.

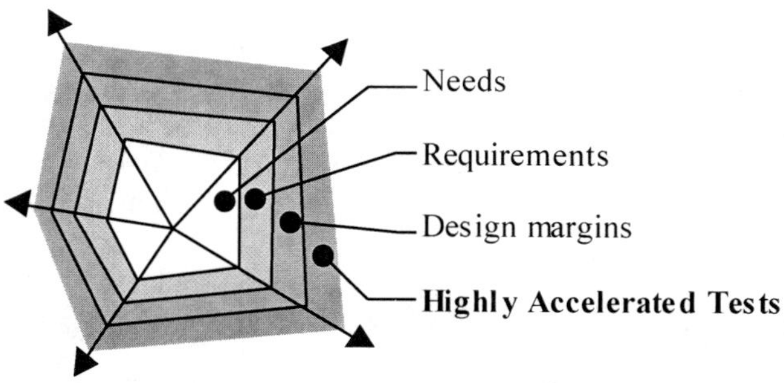

Figure 1 The Highly Accelerated Tests and operating domain.

These objectives are achieved while applying to the product increasing levels of constraint, up to the failure or nearly (Figure 2). Constraints are determined thanks to failure mode analysis and in regards with the use of the product and its environment. They can be divided into 4 main groups :

- Mechanical constraints : To stress the structure,
- Electrical constraints : To stress components,
- Thermal constraints : To stress the assembly,
- Life test : To test the reliability.

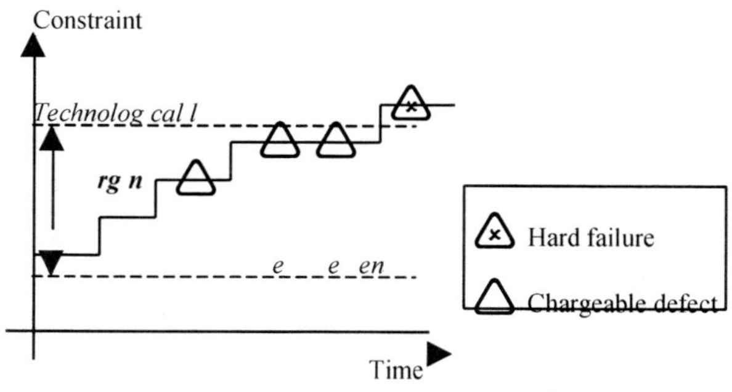

Figure 2 Highly Accelerated Tests approach.

In the particular case of space applications it is also advisable to take into account the radiation aspect. The use of commercial components imposes to us an approach on two levels :

- First of all at the component level : Because of a lack of radiation data and of an insufficient traceability for active components (Diodes, transistors, integrated circuits), we have to decide to apply a specific policy of procurement. Each active component will have to come from a single manufacturing batch, batch which is stored by the provider of the equipment. Thus, components from the same batch will be used for the Highly Accelerated Tests and for the future flying models
- Then at the equipment level: Radiation tests are added to the standard tests set.

2.2 Application to space OC O

The analysis of the specification of the OC O, their internal structure and their operating mode led us to define a test routine as presented Figure 3. The main guidelines are :

- Each main family of tests must involved 3 to 4 oscillators. At least two oscillators will have to undergo each main family of tests, but with only half of the constraint. And two oscillators will be put of as reference parts
- The oscillators being oven controlled, their power consumption (and thus the stress of components) is maximum at cold temperature. Consequently, tests normally carried out at hot temperature will be also held at cold temperature
- The types of tests will be selected among the tests carried out in a traditional way of a standard space qualification or specifically defined according to the constraint related on the experiment or the oscillator.

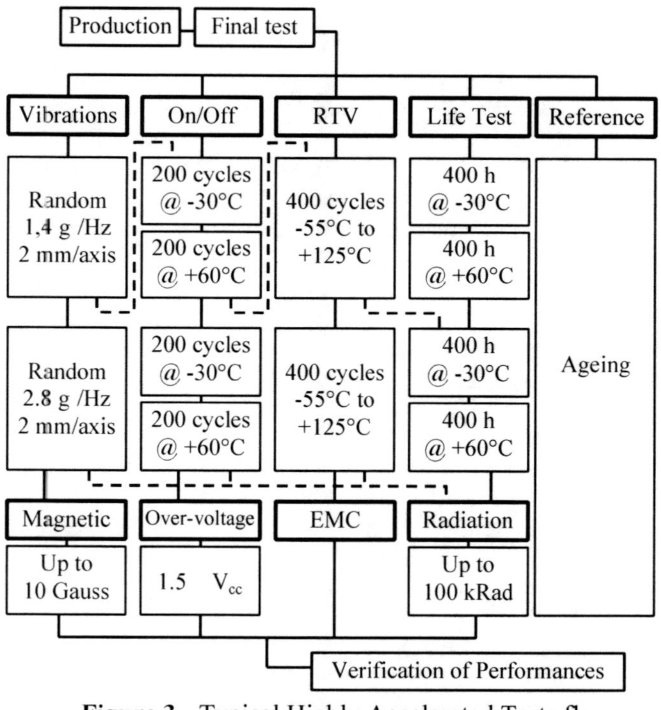

Figure 3 Typical Highly Accelerated Tests flow.

The levels of constraints were defined according to the missions requirements, without necessarily arriving up to the failure of the oscillator. It is sufficient to demonstrate the existence of sufficient margins.

echan cal con a n
The stress of the components is caused by random vibrations up to 4 times the requirements in level and duration. Tests are carried out with power switched off.

lec cal con a n
The stress of the components is caused by series of On/Off. The duration of the On periods corresponds to the duration necessary to the oscillator to reach his steady state (Warm up). The duration of the period Off is the same as the On one. For the reasons mentioned above, cycles will be carried out at the two ends of the temperature range of use of the oscillator. The number of cycles, a few hundred at each temperature, shall be large enough in regards with the number of On/Off that we estimate that the oscillator should undergo during its life (around a hundred).

The al con a n
The stress of assemblies is carried out by application of Rapid Temperature Variation (RTV), with the oscillator s power switched off. The duration of the plates of temperature was estimated from the thermal time-constant of the oscillator.

e e
As for On/Off, the life test is carried out at hot and cold temperatures.

a a on
The radiation tests were evolutionary during various series of measurements. The first tests, 50 kRad with high dose rate (1 kRad /h) without intermediate measurements, quickly showed their limits: No milestone in the case of any problem. That led us to imagine tests with dose rates closer to reality (while keeping duration of tests reasonable), with continuous recording of the frequency and the short term stability of the oscillators (see 3). We also extended our tests up to 100 kRad at the request of some potential customers. For the last test (in progress, see 4), we have to also add measurements of supply voltage and current and series of On/Off. The profile of the tests is thus the following :

- Simulation of a low orbit : A few days with radiation of 40 minutes every 120 minutes with a radiation rate of 5 Rad/h. Fast measurement rate of frequency (1 measurement with $\tau = 1$ s every 40 s) without measurement of stability
- Accumulation of radiation dose followed by 25 On/Off : 25 kRad with a slow measurement rate of frequency (30 consecutive samples over 1 second every 8 minutes) and calculation of short term stability over 1 second
- Simulation of a low orbit
- Accumulation of radiation dose followed by 25 On/Off : 25 kRad with a radiation rate of 50 rad/h
- Simulation of a low orbit
- Accumulation of radiation dose followed by 25 On/Off : 25 kRad with a radiation rate of 100 rad/h
- Accumulation of radiation dose followed by 25 On/Off : 25 kRad with a radiation rate of 100 rad/h for a total dose of 100 kRad.

3 The micro OC O EWOS 513

The EWOS 0513 is a very small size and low power OC O especially designed for space application [4]. A specific integrated circuit is used in the device in order to heat and to control with accuracy the temperature of the crystal unit and the electronic. The EWOS 0513 is housed in an hermetic metallic DIL package.

Figure 4 Micro-OC O EWOS 513

The EWOS 513 is the first OC O to have sudden the Highly Accelerated Tests. Twelve oscillators were implied, principal results are presented Table 1 and Figures 5 to 7.

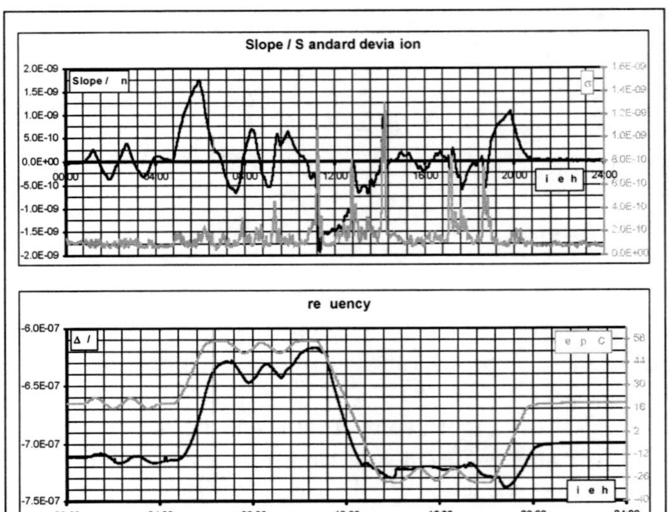

Figure 5 Mid term stability of EWOS 513 (vacuum)

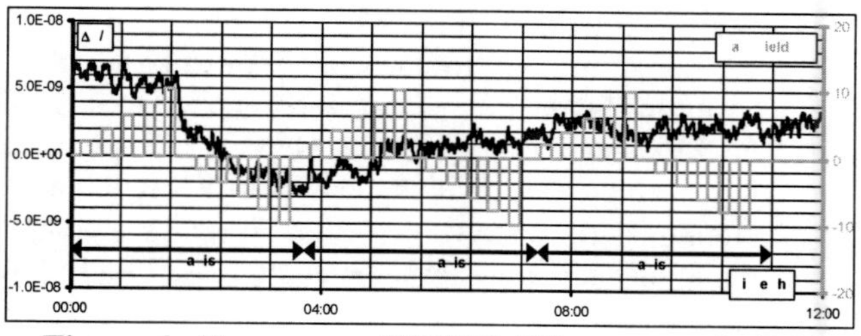

Figure 6 Frequency of EWOS 513 vs magnetic field

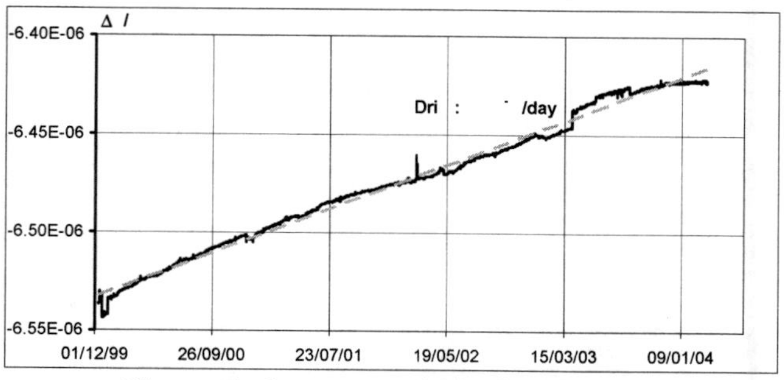

Figure 7 Long term drift of EWOS 513

Item	EWOS 513		
	Req.	Avg.	Units
Frequency Tuning	10 ±5	10 ±5	MHz ppm
Power supply Supply current (vacuum, -30°C)	5 150	5 140	V mW
Stability (vacuum) Allan (1s) In [-30 / +60 °C] Slope / mn (with 0,5 °C/mn) Per day Per year	 2 0.09 ±4 1 0.25	 1.7 0.03 ±3 1,4 0.5	 10^{-11} ppm 10^{-10}/mn 10^{-9} ppm
Phase noise @ 10 MHz 10 Hz 100 Hz 1000 Hz 10000 Hz	 - - -150 -	 -110 -140 -150 -150	 dBc/Hz dBc/Hz dBc/Hz dBc/Hz
Mass	4.2		g
Volume	1.6		cm^3
Operating Conditions Temperature Random vibrations Radiation	 [-30, +60] 0,732 100		 °C g /Hz kRad

Table 1 EWOS 513 Requirements and performances (average of the 12 OC O) after Highly Accelerated Tests (before radiation).

Highly Accelerated Tests, but radiation, did not highlight any failure of the oscillator. During a specific tests, we have even checked its operation with a bored cap although the EWOS 513 is normally hermetic.

Concerning radiation, whereas first tests up to 1 kRad were perfectly conclusive, following tests up to 50 kRad with high dose rate showed a strong degradation of the short term noise. This led us to carry out tests with a lower dose rate and continuous measurement of frequency and short term stability. Unfortunately, several series of tests with different rates confirmed this degradation of the stability (Figure 8). The origin of this degradation was identified as being related to the technology used for the ASIC. This degradation is however compatible with applications concerned and a batch of 100 flying models was carried out. First EWOS 513 units now fly with the 11 years long ROSETTA mission. The EWOS 513 are also retained for the receivers / transmitters of remote control and telemetry of the family of micro-satellites M RIAD.

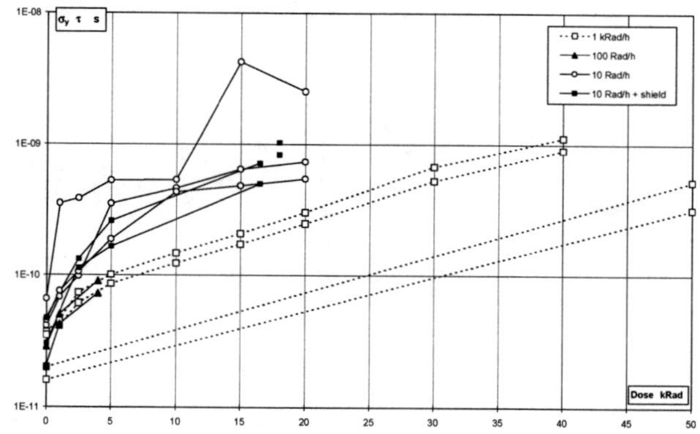

Figure 8 Short term stability degradation of the first version of EWOS 513 versus radiation

Moreover, a new version of the EWOS is being studied (availability envisaged at the beginning of 2005) integrating a new ASIC less sensitive to radiation. This ASIC already was the subject of an evaluation which shows on the one hand that the stability of the oscillator is not degraded any more and on the other hand that the frequency shift due to the radiation remains around 1 to 2 10^{-11} per Rad (Figure 9 and 10).

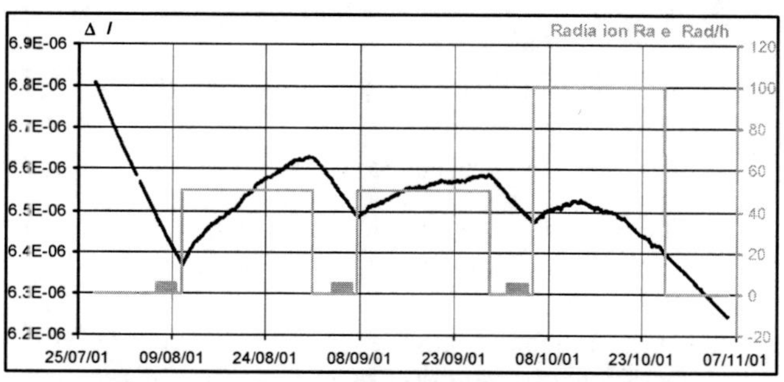

Figure Frequency of EWOS 513 during radiation (100 kRad)

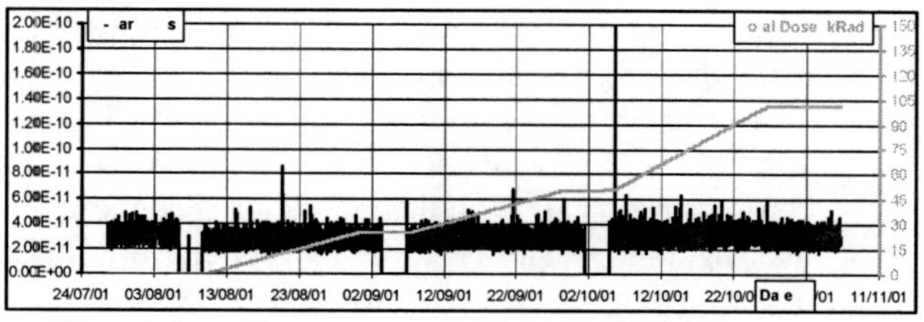

Figure 1 Short term stability of EWOS 513 during radiation (100 kRad)

4 The Mini OC O VL EWOS 81

The EWOS 0810 is a small size high stability OC O especially designed for space application [5]. It uses a 10 MHz SC P3 resonator set in an aluminium oven. It includes a voltage reference and a telemetry (heating current) outputs. The EWOS 0810 is housed in an metallic sealed package.

Figure 11 : Mini-OC O-VL EWOS 810

Highly Accelerated Tests (i.e. Rapid Temperature Variation tests) highlighted a weakness at the level of the assembly of the heating elements which fell apart of the oven (Figure 12). That led us to modify the process of assembly of heaters in order to eliminate this defect.

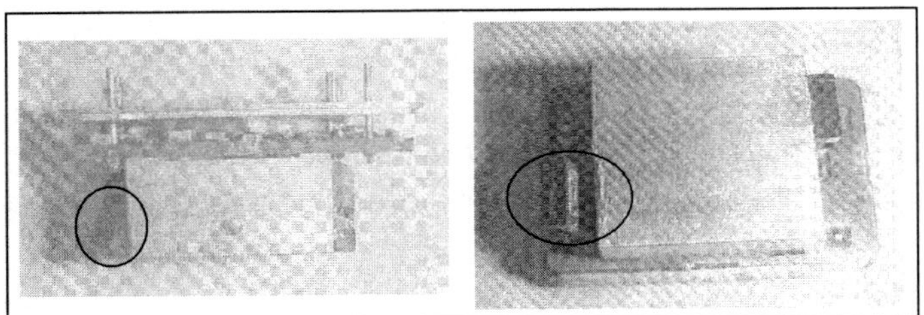

Figure 12 EWOS 810 after first RTV

The tests in radiation proceeded in 3 times :

- The first run was concluded by a quasi simultaneous breakdown from all the oscillators in test around 40 kRad. The origin was a voltage regulator. The total dose reached having been considered to be insufficient, a solution have consisted of replacement for this regulator.
- During a new test, oscillators thus modified could function up to 100 kRad, but it was impossible to make them start again at the end of the tests. However, a cure was observed after a storage of a few days with 125 °C. An analysis of the breakdown and complementary tests on the regulation made it possible once again to propose a corrective action (modification of the settings of the regulation of voltage).
- A new batch of oscillator was thus carried out and has undergone a complete series of Highly Accelerated Tests. Tests in radiation, which are still running, were to arrange in order to introduce series of On/Off each 25 kRad (Figure 13 and 14). Intermediate results, after 50 kRad and two series of On/Off, are completely convincing. Sensitivities noted on 5 oscillators are around 1 to 2 10^{-11} per Rad at the beginning of radiation and non-measurable after 50 kRad (lower than 1 10^{-12}).

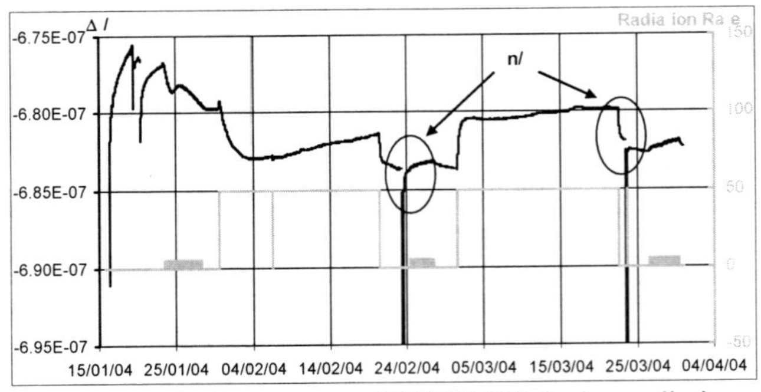

Figure 13 Frequency of EWOS 810 during radiation

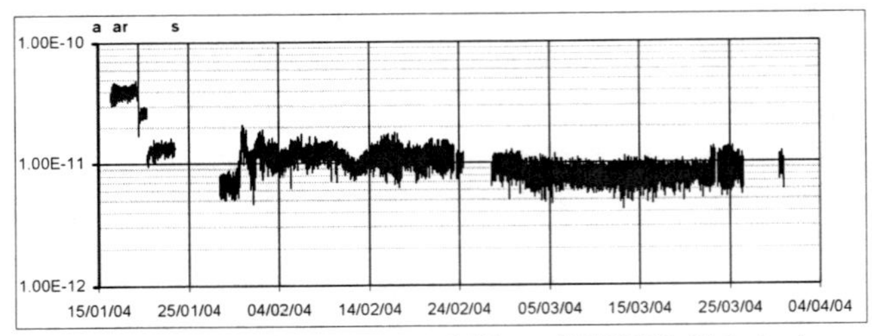

Figure 14 Short term stability of EWOS 810 during radiation (Measurements are limited by the performances of the test bench).

At the end of the tests in radiation, the performances of all oscillators will be checked. The final validation of the EWOS 810 should then be marked mid-2004. Table 2 points out the expected performances while some typical results are presented Figures 15 and 16.

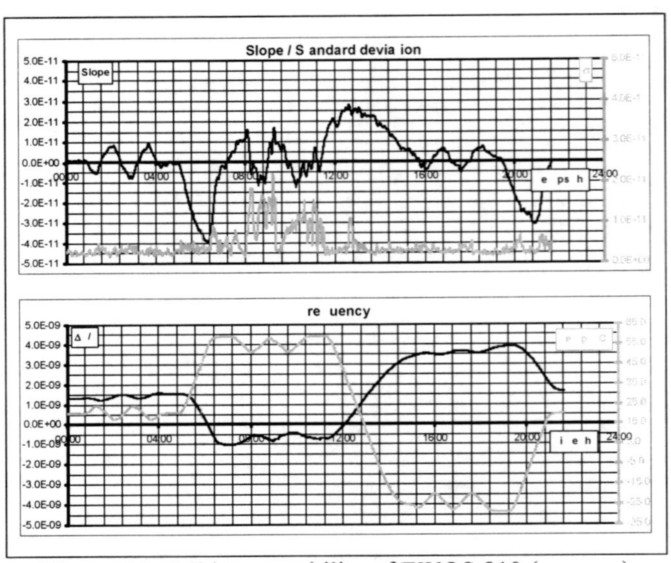

Figure 15 Mid term stability of EWOS 810 (vacuum)

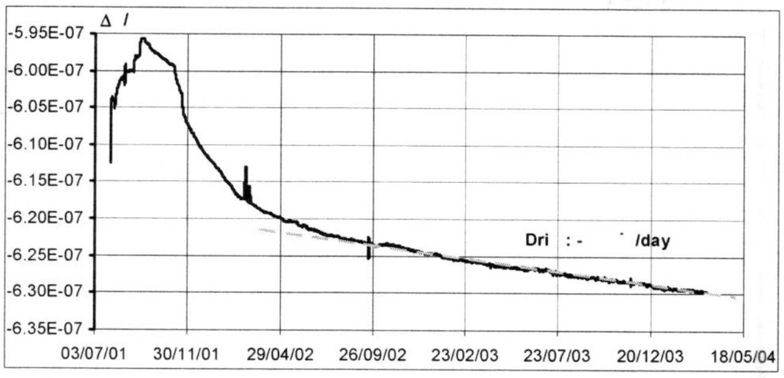

Figure 16 long term drift of EWOS 810

Item	EWOS 81	
	Typical Perf.	Units
Frequency Tuning	10 ±0.5	MHz ppm
Power supply Supply current (vacuum, -30°C)	12 720	V mW
Stability (vacuum) Allan (1s) In [-30 / +60 °C] Slope / mn (with 0,5 °C/mn) Per day Per year	 2 5 ±2 ±1 ±2	 10^{-12} 10^{-9} 10^{-11}/mn 10^{-10} 10^{-8}
Phase noise @ 10 MHz 10 Hz 100 Hz 1000 Hz	 -133 -149 -153	 dBc/Hz dBc/Hz dBc/Hz
Mass	36	g
Volume	20	cm^3
Operating Conditions Temperature Random vibrations Radiation	 [-30, +60] 0,732 100	 °C g /Hz kRad

Table 2 EWOS 810 Typical performances [5]

5 Conclusion

The demonstration was made that Highly Accelerated Tests can offer a powerful alternative when they are applied to new developments using professional components. They make it possible to reach lower costs advanced technologies allowing significant profits on the mass, consumption, volumes and of course cost. Nevertheless, they must be applied with understanding. It is advisable to take into account specificity of space applications. In particular, radiation must be treated upstream. Indeed, if the taking into account of constraints relative to vacuum operations or severe mechanical environment is relatively easy, the lack of radiation data on active components makes any prediction on their behaviour in the face of such type of aggression relatively hazardous.

Finally, this method enables us to have today two oscillators for space applications, and that at the end of relatively fast and not very expensive developments and for an attractive cost. The EWOS 513 is already a success, with 100 flying models sold, it is retained on ROSETTA and on micro-satellites and a new rad-hard version (guaranteed without degradation up to 100 kRad) planned for the end of the year.

References

[1] NIST/SEMATECH, "e-Handbook of Statistical Methods", http://www.itl.nist.gov/div898/handbook/, March 2004.
[2] "START Accelerated testing", DoD Reliability Analysis Center, Vol 6 Num 4, 1999.
[3] "Guide for defining and performing highly accelerated tests", RG.AERO.000.029, BNAE, August 2003
[4] "EWOS 513 Data sheet", THALES, 1998
[5] "EWOS 810 Data sheet", THALES, 2000

High performance flywheel source for atomic fountains and advanced metrology applications

D. Chambon, S. Bize, H. Marion, L. Cacciapuotti, F. Narbonneau, M. Lours, A. Clairon, and G. Santarelli
BNM-SYRTE, Observatoire de Paris, 61 Avenue de l'Observatoire, 75014 Paris, France

A. Luiten and M. Tobar
The University of Western Australia, School of Physics, 35 Stirling Hwy, Crawley, Western Australia.

Keywords - **cryogenic oscillator, atomic fountain, frequency synthesis chain, phase noise, frequency stability**

Abstract - **Atomic fountain frequency standard have the potential to reach a frequency stability of about $10^{-14}\tau^{-1/2}$, if the interrogation oscillator frequency noise which degrades the stability via the "Dick effect" ([2, 7]) is made negligible. In this aim a Sapphire Cryogenic Oscillator (SCO) is used as very low noise frequency source $\sigma_y(\tau) \leq 1 - 2 \times 10^{-15}$ $1\ s < \tau < 1000\ s$. In this paper, we describe a low phase/frequency noise flywheel source for atomic fountains and other demanding metrological applications (cold atom inertial sensors, optical frequency standards, optical synthesizer based on femto-second laser).**

I. INTRODUCTION

The frequency stability plays a fundamental role for atomic frequency standard and high resolution frequency measurements. For example, the measurement of a frequency shift in a atomic fountain at the 10^{-16} resolution needs at least 10 days of integration with a stability of 10^{-13} at 1s. This is the typical stability obtained by driving the fountain with a low phase noise quartz oscillator. The fountain have the potential to reach 10^{-14} level stability with a negligible phase noise interrogation oscillator. The combination of an ultra low noise Sapphire Cryogenic Oscillator SCO ($\sigma_y(\tau) \leq 1 - 2 \times 10^{-15}$ for $1\ s < \tau < 1000\ s$) with an H-Maser (pivot oscillator of the laboratory) makes an ideal flywheel oscillator for driving atomic fountains and high performance metrological experiments. We will describe this flywheel source with all the synthesizers used to obtain the Rb/Cs interrogation signals.

II. CLOCK ENSEMBLE AT BNM-SYRTE.

A. Overview.

At BNM-SYRTE , there is a set of ultra precise metrology experiments. Figure 1 shows the principle of reference signal distribution, obtained from a combination of Maser and SCO, to the experiments: three atomics fountains: FO1 ([6]), FOM (the transportable fountain) both based on ^{113}Cs and a dual fountain (FO2)[1, 4] operating with both ^{133}Cs and ^{87}Rb. The reference signal is also distributed to inertial sensors experiments (gyroscope and gravimeter) and to the optical synthesizer based on a femto second laser. Distant experiments, such as CO_2 stabilized on O_sO_4 and SF_6 at Laboratoire physique des Lasers (LPL) , is served using an 44 km optical fiber [5].

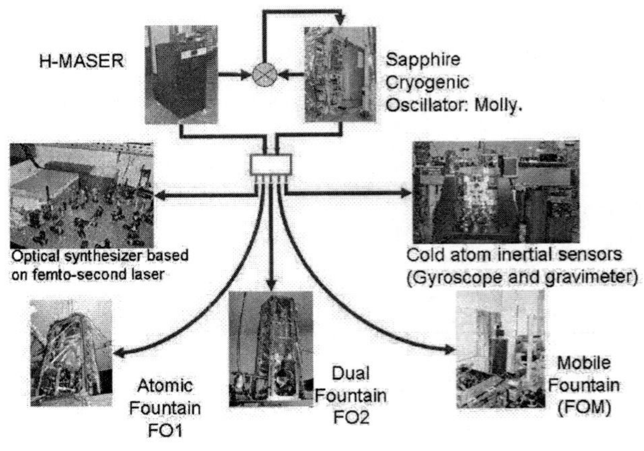

FIG. 1: Distribution schematic of the metrological signal

B. Flywheel oscillator synthesis.

The sapphire cryogenic oscillator operates continuously for 25 days between each LHe refilling, and delivers an ultra low phase noise microwave signal at 11.932GHz. The frequency stability (FIG. 3) of this oscillator is estimated to be $\sigma_y(\tau) = 1 - 2 \times 10^{-15}$ for $1\ s < \tau < 1000\ s$ with a drift of about 1.5×10^{-13} per day [3, 9]. The H-Maser reaches the 10^{-15} level stability at about 1000 s. The combination of these two oscillators makes an ideal source of our metrological applications. From the free-running SCO output, we produce a microwave signal at 11.98GHz (see FIG. 2) by offset phase locking a Dielectric Resonator Oscillator (DRO) via a Direct Digital Synthesizer (DDS). This signal is slightly tunable and nearly reproduces the SCO phase noise performances. The H-Maser delivers a metrological signal at 100MHz. This signal is already distributed to all the experiments via high phase stability RF cables. In order to transfer the low phase noise and high short term frequency stability of the SCO, we need to bridge the gap between

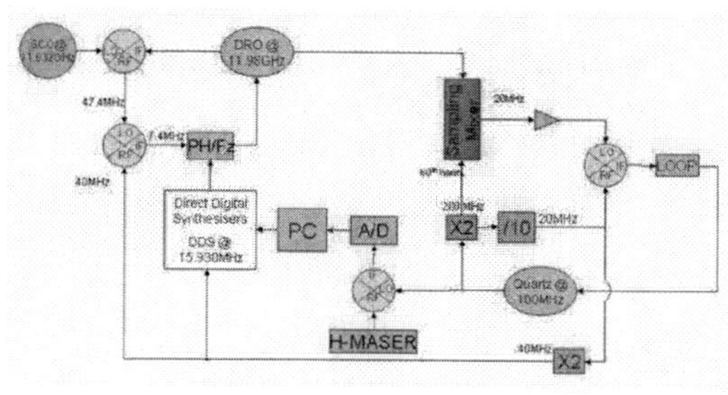

FIG. 2: A synthesis chain to convert the SCO's microwave frequency from 11.932GHz to 11.98Gh100MHz

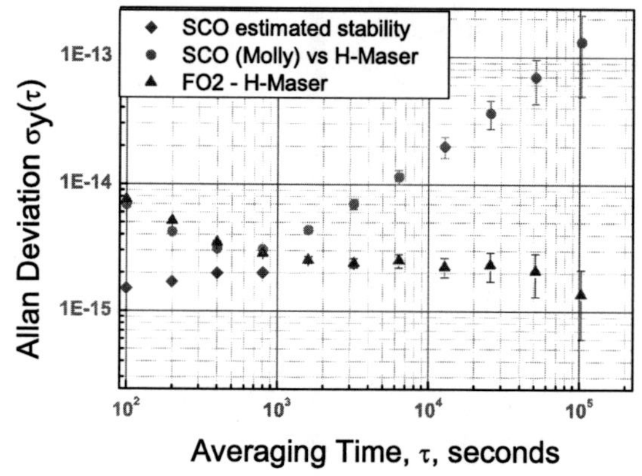

FIG. 3: Allan deviation between the SCO, the H-Maser and the fountain FO2

11.980 GHz down to 100 MHz. Figure 2 describes the synthesizer we use for this conversion. A 100MHz low phase noise VCXO is frequency doubled and drives the LO port of a sampling mixer. The RF input is driven by the 11.98GHz. The IF signal at 20MHz signal is produced by mixing the 60th harmonics of the LO signal with the microwave RF input. This output signal is phase compared to a 20MHz signal, obtained by dividing the VCXO output. The zero beat signal is used to phase lock the quartz oscillator on the 11.98GHZ signal(BW=400Hz). The baseband beat signal of the VCXO and the H-Maser output is used to perform, via a computer, a slow digital phase lock loop ($\tau = 1000s$) controlling the DDS frequency.

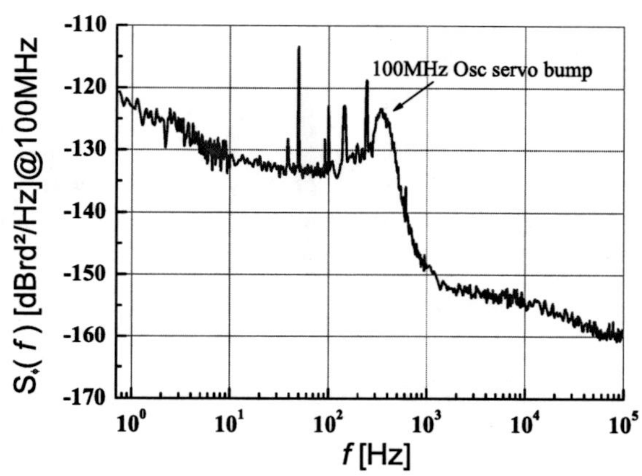

FIG. 4: Phase noise of two identical down converters driving by the same reference signal at 11.98GHz

C. Experimental results

The resulting phase noise of the 11.932GHz to 11.98GHz conversion is $S_\phi(f) = 10^{-10} f^{-1} + 10^{-12} f^0 (rad^2/Hz)$ at 11.98GHz, probably limited by the phase/frequency comparator. That is equivalent in term of frequency stability to $\sigma_y(\tau) = 1 \times 10^{-15}$ for $\tau = 1\ s$ integration time.

The phase noise of the 11.98GHz to 100MHz synthesizer (FIG. 4) has been measured by beating two identical systems at 100MHz; we obtain $S_\phi(f)$=-125dBrad^2 at 1Hz with a white phase floor of -160dBrad^2; that is equivalent in term of frequency stability at $\sigma_y(\tau) = 5 \times 10^{-15}$ for $\tau = 1\ s$ integration time. We can also observe a bump at 300 Hz due to the 100 MHz quartz oscillator phase lock. The sampling mixer of this synthesizer degrades the SCO phase noise by about 20dB at 1Hz from the carrier.

III. A HIGH PERFORMANCE CHAIN FOR THE FO1 CESIUM FOUNTAIN FROM 100MHZ TO ν_{Cs}

A. Hardware description

Figure 5 represents a schematic block diagram of a "traditional" synthesis . A low phase noise 100MHz VCXO is phase locked on the input metrological signal (BW=300Hz) to have a constant power level of the synthesizers internal signals and filter out spurious and harmonic sidebands beyond the loop bandwidth. A sampling mixer is used to phase lock a Dielectric Resonator Oscillator (DRO) at 8.792 GHz. The RF input is driven by the microwave signal and the LO port by the 100 MHz frequency doubled signal; the filtered IF beat-note at 7.36 MHz is phase compared to the DDS signal via a phase/frequency detector to perform the phase lock. The frequency of the DRO is shifted by 400 MHz from 9.192 GHz in order to avoid the microwave leakage towards the Cesium fountain. The VCXO output signal at 100 MHz is multiplied by 4 to provide a 400 MHz signal, which is mixed with the DRO signal, for generating the Cs hyperfine transition frequency (ν_{Cs}).

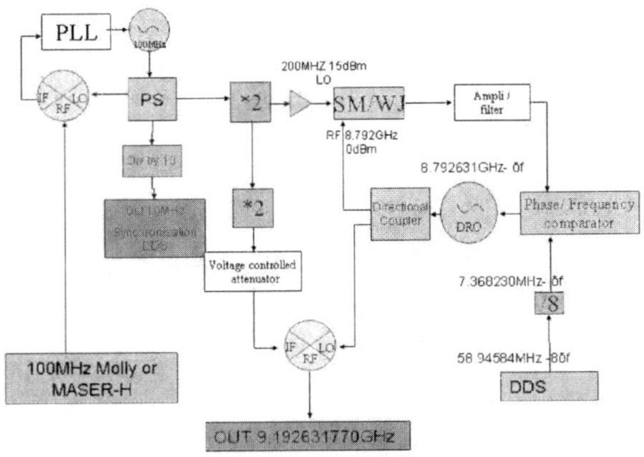

FIG. 5: Complete layout of the FO1 synthesizer (from 100 MHz to 9.192631770 GHz)

The level of this output microwave signal is adjusted using a voltage controlled attenuator at 400 MHz, and frequency-tuneable with a resolution of 0.25μHz (2.5×10^{-17}).

B. Experimental results

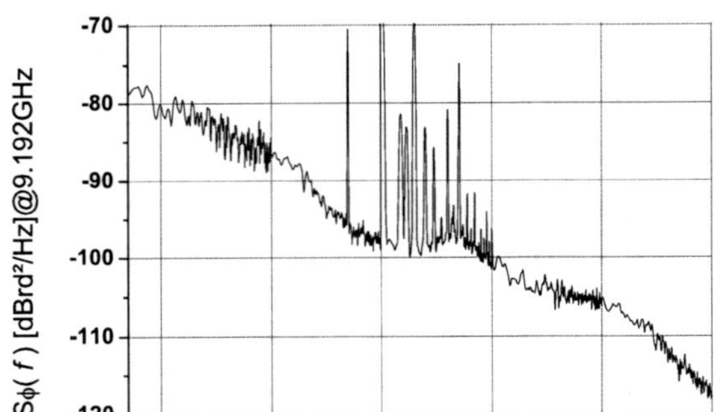

FIG. 6: Phase noise of the combination of the down converter and FO1 synthesis chain

To measure the phase noise of this synthesis chain, we mix it with a very low phase noise reference signal. Figure 6) is the phase noise spectral density measurement showing a noise of -81 dBrad^2/Hz at 1Hz at 9.2GHz with a white phase floor of -160dBrad^2/Hz; that's equivalent in term of frequency stability at $\sigma_y(\tau) = 1.10^{-14}$ for $\tau = 1$ s. We can also observe a small bump at 500 Hz due to the 100 MHz oscillator quartz phase lock. Let's notice that this phase noise measurement is the quadratic sum of this synthesis chain and the down converter phase noises.

As indicated in Figure 7, we obtain, with this synthesis chain, on the fountain FO2, a very good stability of 2.2×10^{-14} at 1 s . The stability is still limited by the synthesizers phase noise via the "Dick effect". To overcome this limitation, we designed a new frequency synthesis chain operating in the microwave domain.

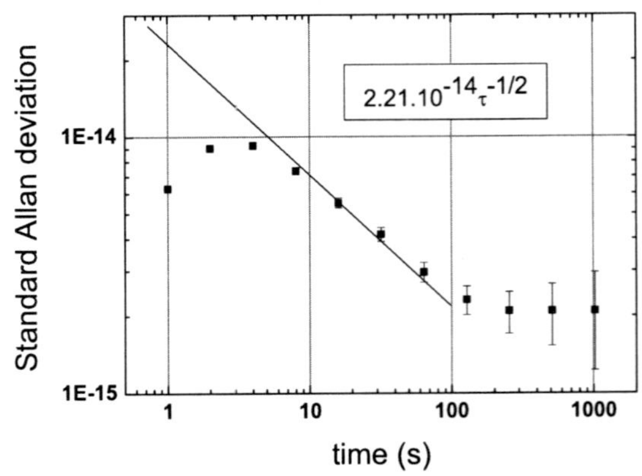

FIG. 7: Frequency stability of the fountain with the FO1 chain driven by the SCO

IV. A DIRECT FREQUENCY CONVERSION DEVICE OF THE SCO SIGNAL TO THE CLOCK TRANSITION FREQUENCIES

The proximity of the fountain FO2 with respect to the SCO allows us to transfer the signal at 11.98 GHz through a microwave cable with a small insertion loss. Working directly in the microwave domain has two advantages: the spurious phase modulation at 50 Hz, due to powerline, is substantially reduced and

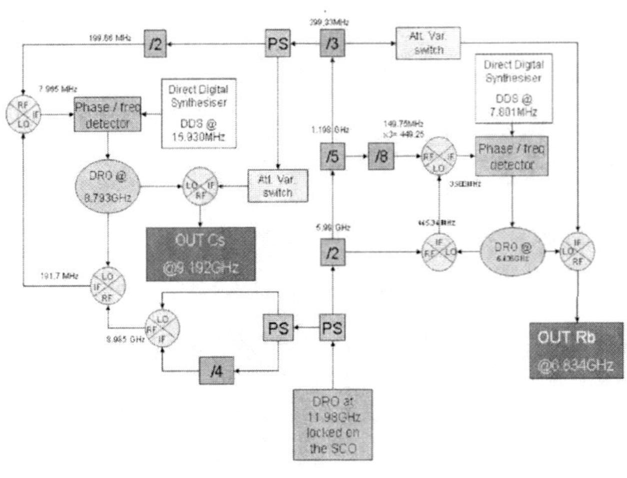

FIG. 8: Principle of the double Rb/Cs synthesis chain

the phase noise is improved by 15dB with respect to the previous synthesis chain. The synthesis chain is represented in figure 8. The reference signal is the signal at 11.98 GHz from the SCO synthesis chain. A voltage controlled DRO at 8.792 GHz is offset phase locked to the 11.980GHz by several dividing and mixing stages which include a DDS at 7.8MHz to provide tunability. The Cs clock transition frequency is generated by mixing the DRO output signal with the 11.98GHz signal divided by 30 (about 400MHz). A similar design is used to generate the Rubidium clock frequency at 6.834GHz.

This chain has been fully characterized in term of phase noise, phase stability and temperature to phase sensitivity. All the test, are performed by using a second low phase noise simplified synthesizer. For the phase noise, we obtain a very good flicker noise of -95dBrad^{2}Hz^{-1} at 1 Hz from the carrier (9.2GHz) which is 15dB better than the noise obtained with the previous synthesizer and a white phase noise floor at about -120dBad$^2Hz^{-1}$ (FIG. 9).

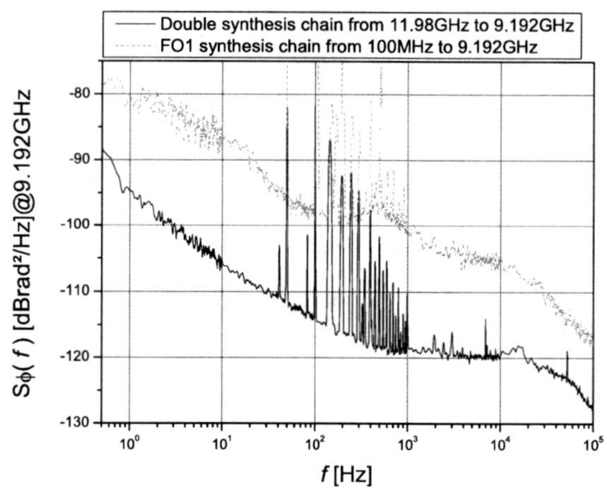

FIG. 9: Phase noise of the double synthesis chain compared to the FO1 synthesis chain

The frequency stability is below 10^{-17} beyond 1000s. To determine the temperature to phase coefficient, we apply a sinusoidal thermal profile of 1.5°C. peak-to-peak with a period of 3600 s. The measured coefficient is 70fs/K.

With this low noise synthesizer driven by the cryogenic sapphire oscillator, we obtain, with the FO2 fountain, a record frequency stability of $1.6 \times 10^{-14}\tau^{-1/2}$ (FIG. 10) limited by the atomic quantum projection noise [[8]].

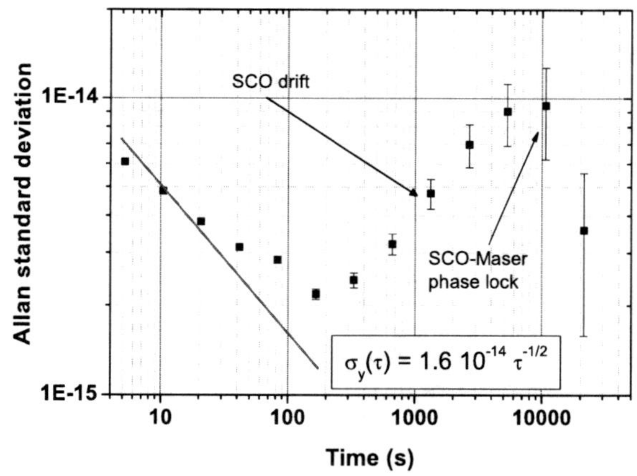

FIG. 10: Stability of the Cs fountain FO2 using the double synthesis chain driving by the SCO

V. CONCLUSION

We have described three different synthesis chains, one to transfer the excellent performances of the SCO and the H-Maser to 11.98GHz and 100MHz; and two frequency synthesizers to obtain the cesium interrogation frequency at 9.192 GHz. The first one starts at 100 MHz signal which is multiplied up to 9.192 GHz. With this kind of synthesis chain, we obtain an excellent stability with FO2 of 2.2×10^{-14} at 1s which is very good but still not limited by the quantum projection noise. So we decide to build a second one with a new approach: a direct synthesis to the Cs frequency from 11.98 GHz staying in the microwave domain. The stability of this double Rb/Cs synthesis chain is 5×10^{-16} at 10s and below 10^{-17} beyond 1000s. This allows a $10^{-14}\tau^{-1/2}$ frequency stability for a fountain without significant degradation due to the phase noise of the interrogation signal.

---

[1] S. Bize *et al.*, in *Proc. of the 6th Symposium on Frequency Standards and Metrology* (World Scientific, Singapore, 2001), p 53.

[2] G.J. Dick *et al.*, "Local oscillator induced instabilities in trapped ion frequency standards" in Proc. Precise Time and Time Interval,Redondo Beach, 1987

[3] A. Luiten *et al.*, IEEE Trans. on Instr. and Meas. **44**, 132 (1995).

[4] H. Marion *et al.*, Proceeding EFTF 2004.

[5] F. Narbonneau *et al.*, Proceeding EFTF 2004.

[6] P. Rosenbusch *et al.*, Proceeding EFTF 2004.

[7] G.Santarelli *et al.*, "Frequency stability degradation of an oscillator slaved to a periodically interrogated atomic resonator" in IEEE Transactions on Ultrasonics, Ferro., and frequency control July 1998

[8] G.Santarelli *et al.*, "Quantum projection noise in an Atomic Foutain: A high stability Cesium Frequency Standard" in Physical Review Letters June 1999

[9] P. wolf *et al.*, Phys. Rev. Lett. **90**, 060402 (2003).

ULTRA-STABLE 5 MHz Xtal RESONATOR WITH VERY LOW G SENSITIVITY AND NOISE

J. LAMBOLEY*, V.CANDELIER*
F.COURTADE**, Ph.BAVIERE**, M.BRUNET**
F.DEYZAC***

* **C-MAC Frequency Products**, 44, Avenue de la Glacière –95100 Argenteuil – FRANCE
** **CNES**, 18, Avenue Edouard Belin –31055 Toulouse – FRANCE
*** **ONERA**, 29, Avenue de la Division Leclerc-92320 Châtillon – FRANCE

ABSTRACT

New demanding targets have been specified for on board Ultra Stable Oscillators (USO) in the frame of future programs. The aim is to reach the highest feasible stability performance for phase noise with the USO operating in all space environmental conditions. Immediate applications are dedicated to accurate orbitography (DORIS program) and frequency metrology (PHARAO/ACES project)

The required performance in the presence of severe micro-vibrations is :

- Phase noise : L(f) in dBc/Hz =
  -131 @ 1 Hz
  -147 @ 10 Hz
  -156 @ 100 Hz
- Allan deviation: $\sigma_Y(\tau) = 7 \times 10^{-14}$ @ 1s and 7×10^{-14} @ 10s
- G sensitivity: 2×10^{-10}/G all axes or less

(Micro-vibration spectrum : 1×10^{-9} to 1×10^{-7} G^2/Hz from 0.1 Hz to 100 Hz).

A new 5 MHz QAS type resonator was developed to greatly reduce the G sensitivity. Several analyses were performed on the resonator process to detect the noise limitation determined by the finished Xtal resonator.

This paper describes :

- the performance respectively needed for the USOs and for the quartz resonators,
- the G sensitivity analysis
- the description and results of several analyses in the process
- data synthesis of the new 5 MHz QAS resonator.

INTRODUCTION

A new 5 MHz quartz crystal resonator was studied to reduce the acceleration sensitivity without short-term degradation. The need is generated by new space programs like DORIS, PHARAO and navigation O.B. clocks, proposed by CNES. This paper describes the needs, the study and the first batch performance, of the resonator.

1. NEEDS IN VERY ULTRA STABLE SPACE USOs

1.1/ DORIS Precise Orbit Determination

DORIS (Doppler Orbitography and Radiopositioning Integrated In Space) is a CNES, IGN and GRGS French space program able to determine with very high precision (1-2 cm) the position of satellites and ground beacons. DORIS O.B. payload has been integrated into several satellites since 1990, SPOT 2 to 5 series, TOPEX, Jason 1, Envisat, and will be embarked on Cryosat, Jason 2,...Spot series and Envisat are at 830 kms, TOPEX and Jason 1 are at 1300 kms. Presently, 6 DORIS O.B. payloads are in orbit. Based on one way bi-frequency Doppler techniques with O.B. very accurate measurements from a network of 55 ground transmitted beacons. Localization and orbitography are performed by several entities (CNES-SOD, IGN, NASA-JPL, IERS,...). The Doppler measurement precision needed to achieve the 1 cm radial orbit error is less than 0.3 mm/s or 1.10-12, including short term noise over 10s, medium term over 10 minutes, and contributions from all instruments (OB and Ground USO, transmitter and receiver) This budget allows to have short and medium stability in the 10^{-14} to 10^{-13} range taking into account the space environment.

1.2/ PHARAO Cold Atom O.B. clock

PHARAO is a cold cesium atom space clock developed by CNES from the experience of BNM-SYRTE and ENS-LKB laboratories in PARIS. The Pharao frequency stability expected is to be better than $\sigma_Y(\tau) = 10^{-13} .\tau^{-1/2}$ for τ = 1s to 10^6 s. To achieve this performance, it is necessary to have a very stable USO in the 5 to 7 $.10^{-14}$ range from 1 to 10s, and a very low phase noise in presence of micro-vibrations given by ISS.

1.3/ Navigation O.B. clocks

In navigation systems, each satellite should broadcast all OB parameters (clocks,...) to determine the position of users. In Galileo, the User Range Error (URE) specification is < 65 cm 1σ, composed of 45 cm for ephemeris error, and 45 cm (1.5 ns) for synchronization error. The 1.5 ns should be considered as the extrapolation error over TP= 6000 seconds following TM= 10^4 seconds used to determined the clock coefficient model. This necessitates a TDEV less than 1.5 ns, or $\sigma_Y(\tau) = 2.6 .10^{-13}$ over 10^4 seconds.

2. SPACE USOs SPECIFICATIONS

The main specifications for DORIS-PHARAO O.B. USOs are given in the figure 1.

Parameters	DORIS USOs specs	PHARAO USOs specs
Frequency	5 or 10 MHz	5 MHz
Short term stability	$\leq 1.\ 10^{-13}$ $\tau = 10s$	$\leq 1.\ 10^{-13}$ $\tau = 1$ to 10s goal: 7. 10^{-14} $\tau = 1$ to 10s
Medium term stability	$\leq 1.\ 10^{-13}$ / min over 10 min pass including noise, temperature, magnetic field and radiation	$\leq 2.\ 10^{-11}$ pp over 90 min orbit including noise, temperature, magnetic field and radiation
Long term aging	1. 10^{-10}/day	1. 10^{-10}/day
Phase noise L(f) in dBc/Hz	• 1 Hz N.A. • 10 Hz - 120 • 100 Hz N.A. • 1 KHz - 140 • 10 kHz - 150	• 1 Hz - 131 • 10 Hz - 147 • 100 Hz - 156 • 1 KHz - 156 • 10 kHz - 156
Temperature sensitivity	$\leq 4.\ 10^{-10}$ pp in -15 to + 50°C	$\leq 2.5\times10^{-10}$ pp over 0 to 48°C
Magnetic sensitivity	$\leq 5.\ 10^{-13}$ / gauss	$\leq 5.\ 10^{-13}$ / gauss
Radiation sensitivity	$\leq 1.\ 10^{-12}$ / rad	$\leq 1.\ 10^{-12}$ / rad
Acceleration sensitivity	$\leq 1.\ 10^{-9}$ / g	$\leq 2.\ 10^{-10}$ / g
DC power	3.5W @ 0°C in vacuum	3.5W @ 0°C in vacuum
Random vibration	20 G rms	11 G rms
Sine vibration	20 G	N.A.
Shock	300 G, 11ms	200G, 0.5 ms
Radiations	50 krads	1 krads
Reliability	300 fits	300 fits
Masse; Volume	1 kg; 1 liter	1 kg; 1 liter

Figure 1 : DORIS and PHARAO O.B. USOs specifications

The phase noise specified for Pharao USO should be compliant with the micro vibration spectrum generated by the ISS, and given in the figure 2 below.

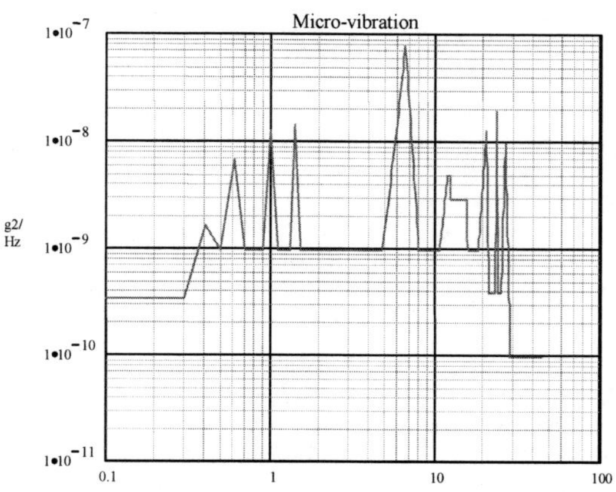

Figure 2 : ISS micro-vibration spectrum applied to Pharao Clock.

The simulation of phase noise spectral density L(f) with micro-vibration gives, by using the Filler law [1], the required acceleration sensitivity of the crystal resonator frequency. It must be lower than 2.10^{-10}/g to obtain the specification. The figure 3 gives a simulation of phase noise spectral density with and without micro-vibration.

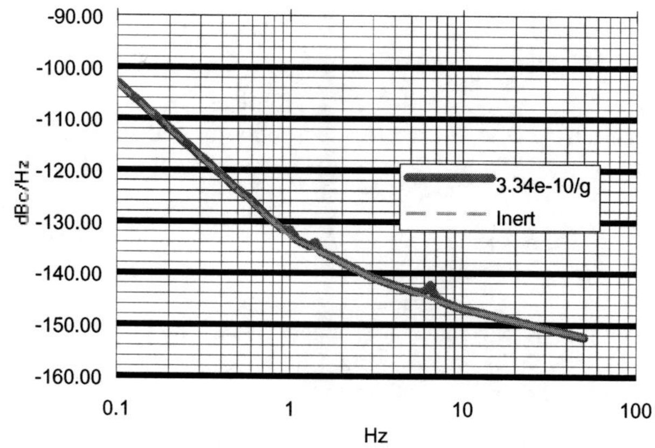

Figure 3 :simulated phase noise spectral density L(f) with and without micro-vibration as in figure 2 . The acceleration sensitivity is taken as 3.3E-10/g for the simulation.

3. SPACE QUALIFIED NEW QAS 5 MHz RESONATOR

3.1/ Resonator specifications

The USO main specifications are obtained with the resonator characteristics seen in figure 4. The frequency 5 MHz is chosen for the best performance given in short-term stability [2].

N°	Characteristic	Symbol	Values min	Values max	Unit	Remarks
1	Nominal Frequency	F0	5	5	MHz	
2	Reference Temperature	T0	70	80	°C	turn over point
3	Overtone order		3			SC cut
4	Load capacitance	CL	16	30	pF	
5	Drive Level	P0		0,1	mW	
6	Frequency Adjustment Tolerance	DF/F	NA	NA		
7	Resonance resistance	R		100	Ω	over Top
8	Frequency variation with temperature	DF/F	NA	NA		
9	Resistance variation with temperature	DR/R	-20	20	%	from 25°C
10	Operating temperature range	Top	-30	80	°C	
11	Variation of frequency with drive level	DR/R		1.5	10-9/µW	
12	Variation of resistance with drive level	DR/R	NA	NA		
13	Motional inductance	L1	6.5	7.5	H	
14	Motional capacitance	C1	NA	NA	pF	
15	Static capacitance	C0		4	pF	
16	Q factor	Q	2		E+06	
17	Ratio of unwanted response resistance		6		dB	
18	Aging	DF/F	-3	3	E-11	after 4 months
19	Storage temperature	Tstg		85	°C	
20	Barometric sensitivity	SB		1	E-8/bar	
21	Short term stability ALLAN	ST		7	E-14	τ = 1 to 10s
22	Gravitation sensitivity	SA		2	E-10/g	worst case 3axis
23	Magnetic field sensitivity	SMa		3	E-12/ Gauss	
24	Radiation sensitivity	SR		1	E-11/rad	
25	Mechanical sensitivity	SMe		2	E-08	after shocks, and vibr.
26	case		HC40DM			

Figure 4 : Resonator characteristics

3.2/ Design definition of QAS 5 MHz resonator

The resonator is a 5 MHz QAS, SC cut, 3rd overtone, quartz crystal.

The QAS structure is an appropriated response to minimise the accelerometric sensitivity of quartz crystal frequency. That is a self-suspended resonator with a sticky electrode (BVA 4 version).

Old versions of the 5 MHz QAS or BVA structures were in bigger cases [3], [4]. The need of HC40DM case (height = 8 mm , diameter = 21 mm) (see figure 5) required us to reduce the dimensions of the vibrating crystal with its support ring without degradation of Q factor and short-term stability.

The frequency range for random vibration is 2000 Hz. That implies a four mounts configuration of the 5 MHz crystal.

Figure 5 : 5 MHz QAS HC 40 DM case

The first batch motional parameters is given in figure 6. The motional values are similar to the 5 MHz QHS quartz resonator space qualified in 2003 [5].

	R (ohms)	self (mHy)	quality factor Q (10^6)	turn over point (°C)	adjustment capacity (pF)	mechanical resonance (Hz)
mean	98	7044	2,1	71	22	2300
standard deviation	18	148	0,6	2	3	100

Figure 6 : QAS 5 MHz batch motional parameters

3.3/ G sensitivity analysis

Accelerometric sensitivity of the QAS resonator was firstly studied in the case of a **two-point mount** configuration [6]. In the ideal case , the mounting clips are positioned orthogonally with respect to the bridges axe. It was shown that accelerometric sensitivity was mainly due to extension-compression and shear stresses appearing in the vibrating crystal. On the other hand, these stresses are due to back strengths of mounts under the effect of acceleration applied to the crystal mass.

This study has also tried to evaluate the impact of assembly angular defect (between crystal blank and mounting clips) on g-sensitivity of QAS. The sensitivity of QAS to diametric compression was experimentally measured for different values of angle α (figure 7).

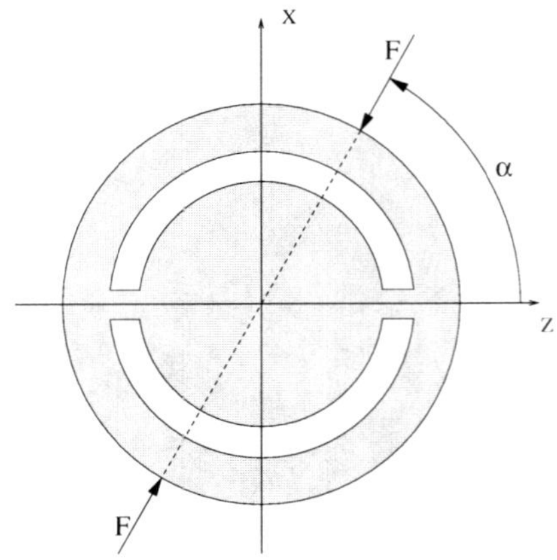

Figure 7 : Angular convention for experimental study of QAS sensitivity to diametric compression.

The results of this study lead to consider that this diametric compression with F strength induces two kinds of strength at the level of bridges – F_C compression strength and F_Σ shear strength such as:

$$F_C(\alpha) = F \cos 2\alpha$$
$$F_\Sigma = F/7 \sin 2\alpha$$

The present resonator is a **four point mount** QAS structure . In this case , mounts are attached to the ring along two orthogonal axes respectively such that $\alpha = 45°$ and $\alpha = 135°$. In this way the compression forces are zero and shear forces are compensated. If the resonator is perfectly realized a very low g-sensitivity may be guaranteed. For an industrial process two kinds of defects have to be taken into account: crystal orientation defect and mount positions defects. A theoretical approach has shown that mount positions are more critical than crystal orientation in fabrication of a low g-sensitivity resonator.

But there are other defects which have predominant influence on the g-sensitivity of the **QAS**, independently of the mounting clips number. Before, let us remind: the sensitivity of a resonator to applied forces may be obtained by summation of stresses effects in each point weighted by its vibratory energy level. The distribution of energy levels, called acoustic spot, is mainly determined by the curvature of the blank. Some of stresses lead to zero effect when their distribution (which is antisymmetric) is centred on the middle of acoustic spot. In the case of a perfectly realized resonator, these two centres are confused. During industrial process two defects may lead to separate these centres :

1) curvature defect :
If the convexity is not well-positioned in the middle of the blank (figure 8), then the acoustic spot (drawn under the curvature) will be out of the blank centre.

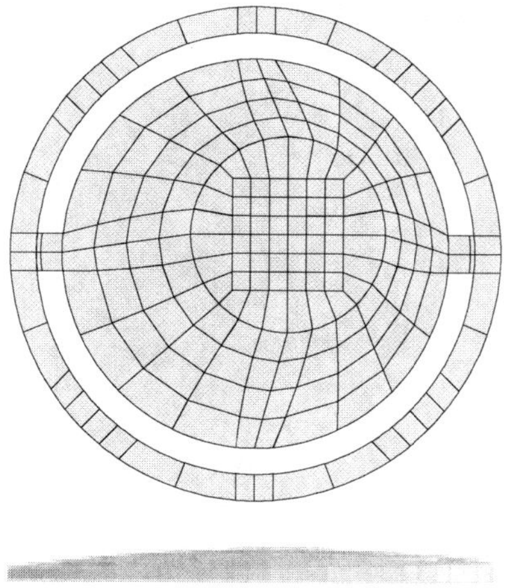

Figure 8 : Curvature defect –> acoustic spot is shifted

2) machining defect :
QAS resonator is realized by ultrasonic machining of a quartz crystal blank. This machining digs two semi-circular grooves which isolate the vibrating part from the ring (only two bridges link it to the ring). If the machining is not well-centred (figure 9) then stresses field will be shifted from the blank centre.

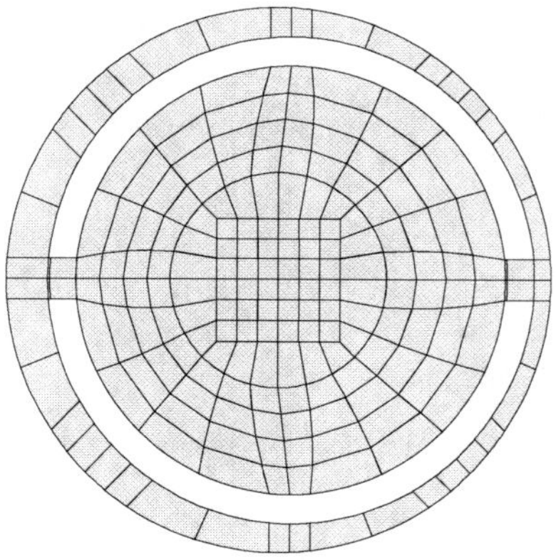

Figure 9 : Machining defect –> stress distribution is shifted

Whatever the origin of shift between acoustic spot and stresses distribution, the effect on g-sensitivity is very significant.

We have taken account of these findings for the realization of the first resonator batch.

3.4/ Noise analysis

Some technological analysis (granulometric, optical, chemical, residual gas) were done and correlated with the short-term performance of different production batches. Different parts of the process were modified.

An improvement of the crystal polishing quality has been firstly performed, most precisely the mechanical step has been reviewed to get a roughness improvement of the bare crystal. A selection of abrasives, slurries and cloths has allowed to get a same polishing quality on the both sides of the crystal. Figure 10 presents the difference of final polishing quality before process improvement.

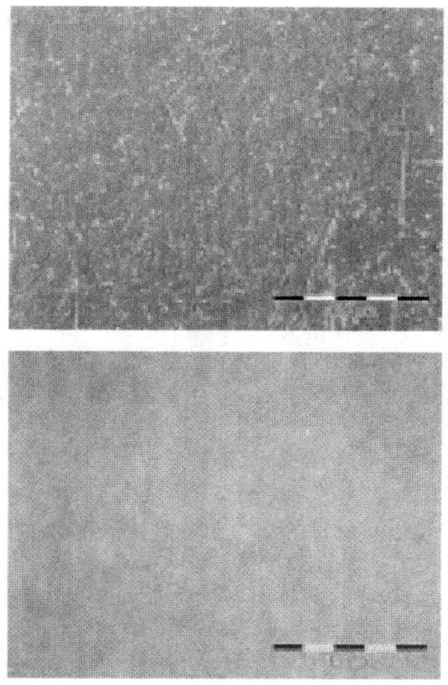

Figure 10 : optical view done by nomarsky mode allow to show the difference on quality polishing between the both sides of the bare crystal before process improvement

Metallization step has been pinpointed as a major contributor and many improvements on tie layer thickness management, metal deposition rate, annealing condition have been done in order to get the best uniformity on thickness and grain boundary distribution and to minimise residual stress. Figure 11 exhibits a SEM (Scanning Electronic Microscope) view of the achieved aspect and grain boundary distribution of the typical metallization layer.

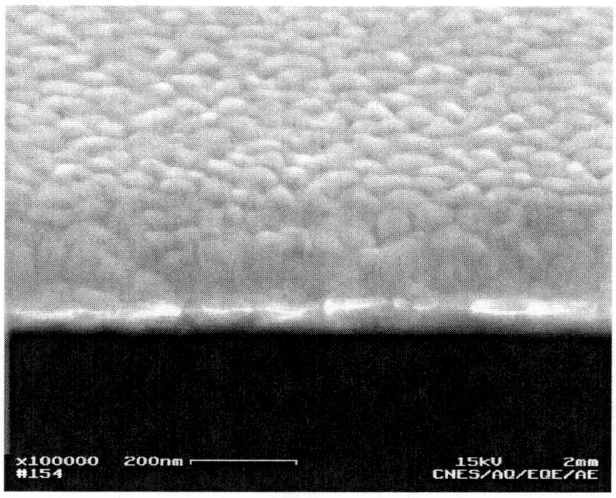

Figure 11 : SEM view of the gold metal layer after metallization step improvement.

The result is an improvement of the short-term stability yield in the range lower than 1.10^{-13}.

The 5 MHz QAS resonator batch is realised with a production process which has been improved in polishing, metallization and cleaning.

3.5/ 5 MHz QAS resonator batch performances

The figures 12 to 14 give the resonator acceleration sensitivity in the three orthogonal axis OX, OY,OZ, described in figure 5. The performance is three times better than that obtained from the 5 MHz QHS resonator [5]. The figures 15 to 17 give the resonator magnetic field sensitivity. It is ten times better than that obtained from the 5 MHz QHS resonator. The same improvement has been observed for ageing and barometric sensitivity (figures 18 and 19).

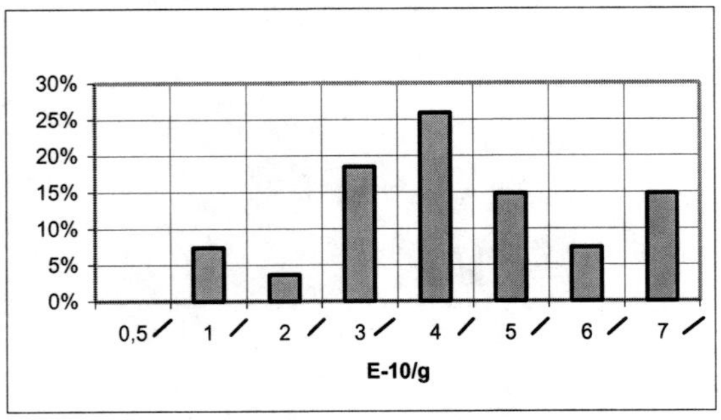

Figure 12 : QAS 5MHz acceleration sensibility distribution with acceleration axis in OX OY plan. The mean value is 3.3e-10/g.The standard deviation is 2e-10/g.

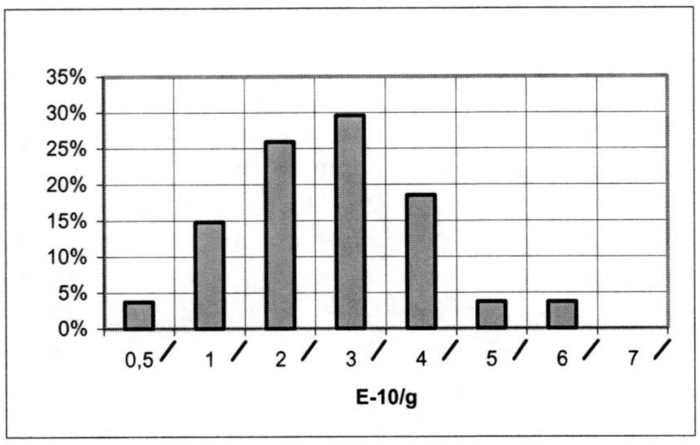

Figure 13 : QAS 5MHz OZ axis acceleration sensibility distribution .The mean value is 2.2e-10/g.The standard deviation is 1.1e-10/g.

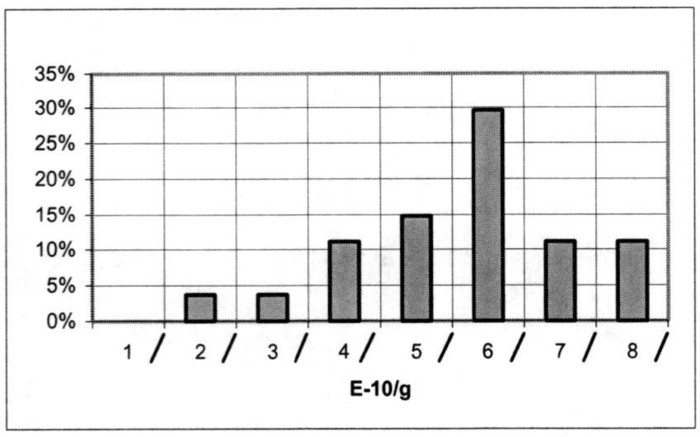

Figure 14 : QAS 5MHz worst case axis acceleration sensibility distribution. The mean value is 5.6e-10/g.The standard deviation is 2e-10/g.

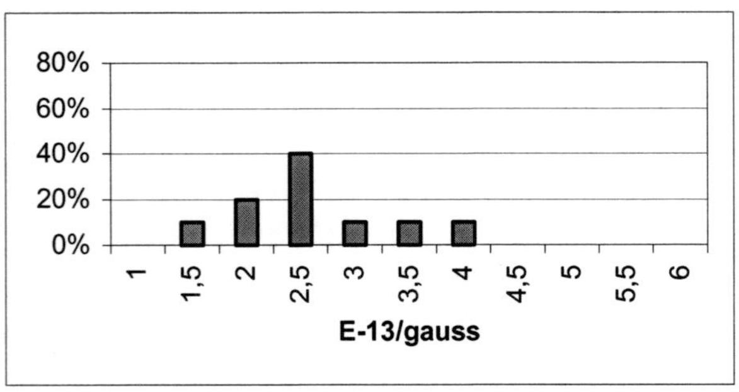

Figure 15 : QAS 5 MHz X axis magnetic field frequency sensitivity distribution . The mean value is 2.2E-13/gauss.

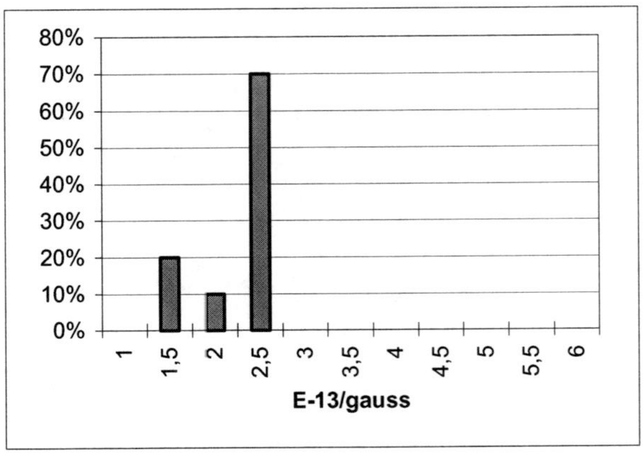

Figure 16 : QAS 5 MHz Y axis magnetic field frequency sensitivity distribution. The mean value is 2.5E-13/gauss.

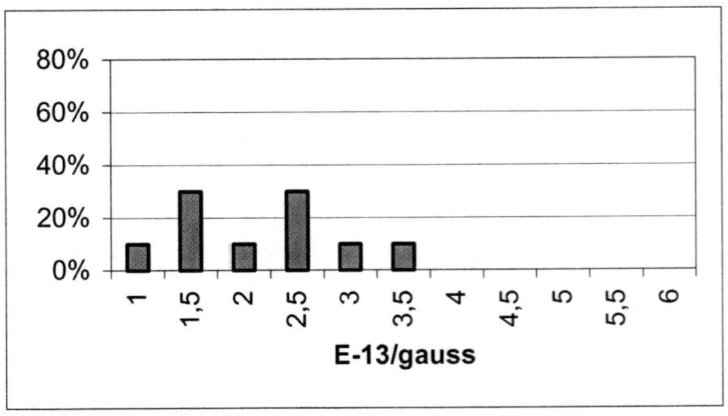

Figure 17 : QAS 5 MHz Z axis magnetic field frequency sensitivity distribution. The mean value is 2.1E-13/gauss.

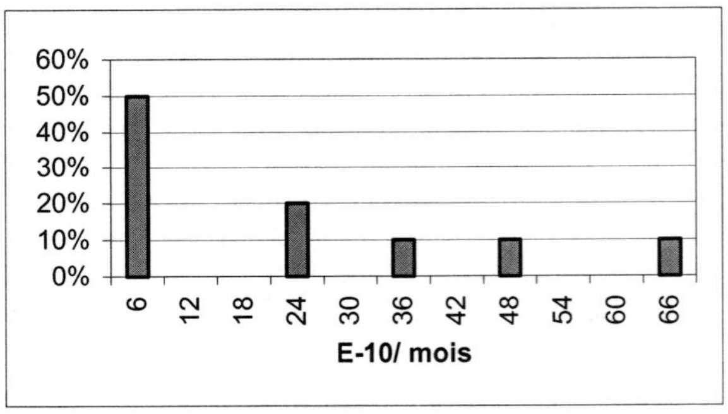

Figure 18 : QAS 5 MHz first month frequency ageing distribution. The mean value is 2E-9/month.

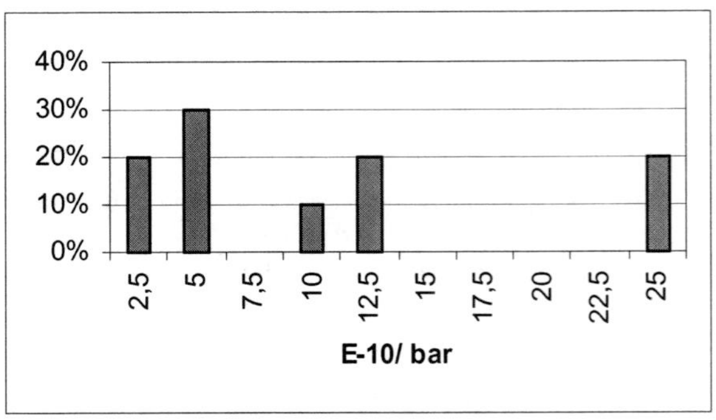

Figure 19 : QAS 5MHz barometric sensibility distribution. The mean value is 10E-10/bar. The standard deviation is 8E-10/bar.

In the batch, ten samples were tested in short-term frequency stability with an integration time of 10s. The

Allan standard deviation mean value was $1.3.10^{-13}$ and some parts were below 1.10^{-13} (figure 20). The three best samples were tested with an integration time of 1s, with the saphir frequency reference of the SYRTE laboratory at PARIS. These Allan standard deviations are also lower than 1.10^{-13} (figure 21 for one of them).

A summary of the environmental performance of the three best 5 MHz QAS resonators is given in figure 22.

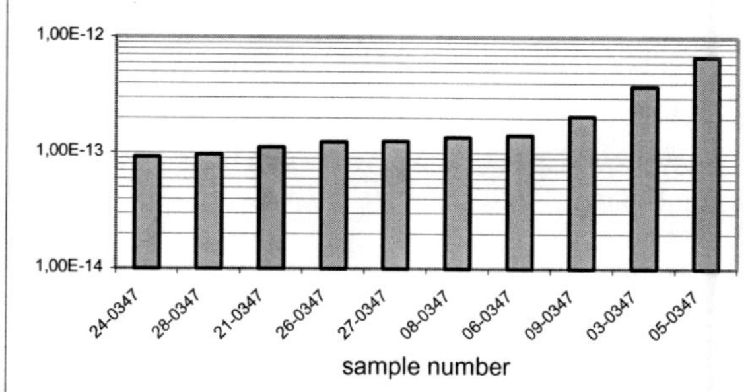

Figure 20 : QAS 5MHz Allan standard deviation with integration time of 10s.The mean value is 1.3E-13.

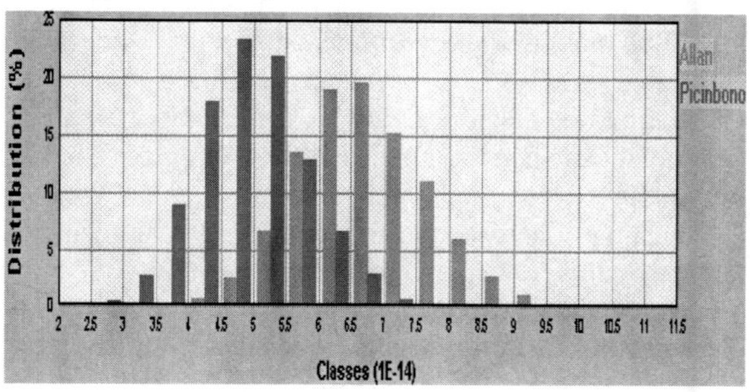

Figure 21 : Distribution of 9000 Allan and Picinbono [7] *standard deviations of 30 frequency samples with time integration of 1s for the 5MHz QAS number 24-0347. The mean value for the Allan standard deviation is 6.5E-14.*

resonator number	O34721	O34724	O34728	unit
OX accelerometric sensibility	0,8	1,5	2,6	E-10 /g
OY accelerometric sensibility	1,4	2,0	2,2	E-10 /g
OZ accelerometric sensibility	0,9	2,3	0,9	E-10 /g
module accelerometric sensibility	1,9	3,3	3,5	E-10 /g
Allan standard deviation (1s)	7,5	6,5	9,5	E-14
Allan standard deviation (10s)	9,4	7,7	0,8	E-14
Ox magnetic field sensitivity	2,5	2,5	2,5	E-13 /Gauss
Oy magnetic field sensitivity	1,6	3,8	2,5	E-13 /Gauss
Oz magnetic field sensitivity	2,5	1,0	1,5	E-13 /Gauss
retrace 24h off DF/F	-3,0	1,0	0,5	E-10
first month ageing	5,0	3,3	62,0	E-10 /month
barometric sensitivity	-12,5	25,0	-8,0	E-10 /bar

Figure 22 : *Environmental performance of the three best 5MHz QAS resonators*

CONCLUSIONS

We have proved by samples that QAS structure can improve the acceleration frequency sensitivity of the 5 MHz resonator without degradation of its short-term stability and in the same holder as the 10 MHz QAS resonator. Others parameters as ageing , barometric and magnetic field sensibilities are also improved . The 5 MHz QAS resonator complies fully with the space application objectives. Some adjustments of the process are being done to minimise the acceleration frequency sensitivity.

ACKNOWLEDGMENTS

We would like to thank Mr. SANTARELLI of BNM-SYRTE for his help in the frequency short-term stability measurement.

REFERENCES

[1] Raymond L. Filler, John A. Kosinski and John R. Vig - F*urther Studies on the Acceleration Sensitivity of Quartz Resonators* - 37th Annual Frequency Control Symposium (1983).

[2] Vincent Candelier, Jacques Lamboley, Gérard Marotel, Patrice Canzian, Pierre Poulain – *Recent progress in ultra stable oscillators for space on board and ground applications* – 15th EFTF proceedings 2001.

[3] JP. Aubry et A. Debaisieux – *Further results on 5 and 10 MHz Resonators of BVA and QAS Designs* – 38th Annual Symposium on Frequency Control (1984), p. 240

[4] Gilles Robichon – *A New 5 MHz Mini BVA OCXO* – Proceedings of 12th EFTF - 10 ÷12 March 1998 – Warsaw – Poland, p. 358

[5] Vincent Candelier, Patrice Canzian, Jacques Lamboley, Michel Brunet, Giorgio Santarelli – *Space Qualified 5 MHz Ultra Stable Oscillators* - IEEE EFTF Frequency Symposium Proceedings 2003.

[6] Deyzac François - *Finite Elements method modelling of accelerometric and barometric sensitivity of quartz resonators* - Proc. 7th European Frequency and Time Forum (1993), pp. 291-295

[7] V. Candelier, J. Chauvin, C. Gellé, G. Marotel, M. Brunet, R. Petit – *Ultra Stable Oscillators* – Proceedings of the 12th EFTF - March 1998.

QUARTZLOCK MODEL A7A FREQUENCY/PHASE DIFFERENCE COMPARATOR

Clive Green

Quartzlock UK Ltd, cgreen@quartzlock.com

Abstract

This paper describes various future avenues for improvement and extension of the Quartzlock A7 frequency/phase difference comparator.

Initially a brief description of the instrument will be given, which will include outlining the current modes of operation. Along with this, a detailed description of the current methodology for achieving the stated time domain resolution will be given

At present the A7A frequency/phase difference comparator is for measuring a wide range of frequency standards, isolation amplifiers, frequency multipliers and dividers, and passive devices such as cables. The instrument includes a moving coil meter for rapid, unambiguous display of fractional frequency difference or relative phase difference between two sources. Outputs are also provided for an external counter to provide higher resolution analysis of the time domain stability of a source or amplifier.

The instrument combines the production oriented capability of rapidly adjusting a source to within a certain tolerance using the panel meter, along with the metrology capability of a full time domain analysis of a source or passive component using data acquisition from the frequency counter.

Quartzlock are presently investigating two potential avenues for improving the performance, and widening the potential usage for this instrument. These are improving the pc-based frequency counter and broadening the acceptable input frequency range.

At present the frequency counter runs under DOS, which leads to numerous measurement and usability problems. Quartzlock are researching a windows native frequency counter, utilising a multi channel phase meter with better than 100ps resolution and capable of storing 30000 time interval measurements. It would designed to enable a measurement run lasting maybe weeks and would be much more secure from data loss or interruption.

The extension of the A7A to input frequencies other than the present 5 or 10MHz involves numerous technical challenges, which are discussed in detail within the paper. A discussion of possible methods for implementing a broader input frequency range, along with potential drawbacks is also highlighted.

1 Introduction

The A7A frequency/phase difference comparator is a Quartzlock product for measuring a wide range of frequency standards, isolation amplifiers, frequency multipliers and dividers, and passive devices such as cables. The instrument includes a moving coil meter for rapid, unambiguous display of fractional frequency difference or relative phase difference between two sources. Outputs are also provided for an external counter to provide higher resolution analysis of the time domain stability of a source or amplifier. The instrument combines the production oriented capability of rapidly adjusting a source to within a certain tolerance using the panel meter, along with the metrology capability of a full time domain analysis of a source or passive component using data acquisition from the frequency counter.

Quartzlock can supply a suitable PC card based counter, which can accumulate readings at 1000 readings per second. The data file can then be analysed using the Industry Standard software package Stable 32. Stable 32 can plot all the variants of time domain statistics, and can convert to the frequency domain to provide close in phase noise analysis. The rapid data acquisition rate of 1000 readings / second enables phase noise to be calculated to a Fourier frequency of 500Hz from the carrier. Close in phase noise analysis may be made to within 1mHz of the carrier by extending the length of the data acquisition to several thousand seconds. The very low drift of the A7A makes such measurements valid. Stable 32 can also extract frequency offset and drift information.

The A7A comparator has state of the art noise floor and drift characteristics. Its technique of frequency multiplication followed by down conversion provides lower noise floors than the simpler dual mix down convert system. The very low drift is achieved by providing identical multiplier/mixing chains for the reference and measurement channels. When the multiplied signals are finally mixed together (subtracted), any drift in the multiplier chains is cancelled.

The automatic battery backup facility fitted as standard enables very long measurement runs to be undertaken without concerns over line power failures. An external 24V car battery will power the instrument for at least 24 hours.

Measurements are made in the time domain and consist of time difference measurements between a reference source and a measurement source. Measurements may be made on passive devices such as amplifiers by splitting a source output and comparing the time delay through the item under test with the direct path. In this way the time or phase stability of the amplifier may be measured. Unlike a general-purpose time interval meter the inputs must be substantially sine wave and at either 5MHz or 10MHz. The resolution is much better than even the fastest counters, being around 50fs for a single measurement with filter off, and around 15fs with 10Hz filter.

The comparator has two modes of operation, frequency measurement mode and phase difference mode. In frequency mode the moving coil meter indicates fractional frequency difference and the external counter is configured as a frequency counter. Meter full-scale ranges are selectable from the front panel in the range +/-10^{-7} to +/-10^{-12}. The external counter is configured as a frequency counter with gate times selected on the counter as usual for a frequency measurement. The frequency measured is actually the sum of the multiplied frequency difference at the inputs of the A7A plus 100kHz. It is therefore desirable that the external counter has a maths facility to remove the offset and allow for the multiplication. The RMS resolution is typically better than 5 parts in 10^{14} for a 1 second gate.

In phase mode, the meter is configured to read phase difference between the reference and the measurement inputs. The full-scale range is selectable between +/-10us to +/-100ps. An extended range phase detector is used so phase roll over will be between +10 and 0 on the meter if the frequency is increasing, and between -10 and 0 on the meter if the frequency is decreasing. The meter shows relative phase difference between the reference and measurement inputs. Because of the multiplication process in the comparator, the absolute phase difference is not available. A phase-reset key is provided that zeros the indicated phase to within +/- 100ps.

The external counter is configured as a time interval meter and measures the time difference between pulses on its A channel and B channel. The pulse rate is set from the front panel of the A7A. The time difference between the pulses is the multiplied time difference between the inputs to the A7A. Thus if the counter has 1ns time interval resolution, the effective resolution (multiplication factor 10^5) at the input of the A7A is 10fs. In practice this resolution is not achievable due to instrument noise. Single shot time resolution has been measured at 15fs (with 10Hz filter).

2 Principles of operation

The principle behind the A7A is to increase the resolution of a frequency counter, which is essentially a time interval measurement device. This is achieved by multiplying the frequency to be measured to a higher frequency, and then mixing it down to a lower frequency using a local oscillator derived from the frequency reference. The principle is illustrated in FIG 3.1, and has been made the basis of a number of instruments in the past. The relationship is shown for signals down the mix/multiply chain for an input signal with a difference of delta f from the reference, and also for a signal with no frequency difference, but with a phase difference of delta t. (An important clarification is that "phase" difference between two signals can either be measured either in time units or angle units. A measurement in time units does not specify or imply the frequency of the signals. A measurement in angle units (radians) needs a prior knowledge of the frequency. Throughout this manual, phase will be measured in time units) It should be noted that a frequency multiplication multiplies a frequency difference but leaves a phase difference unchanged. Conversely, a mixing process leaves a frequency difference unchanged, but multiplies a phase difference. When the frequency differences are converted to fractional frequency differences by dividing by the nominal frequency, it will be seen that the multiplication factors for frequency and phase are the same.

The big disadvantage in the simple approach shown in FIG 3.1 is that phase drift with temperature will be excessive. As rate of phase drift is equal to the fractional frequency difference, the measurement of the frequency of an unknown device will be in error. For example, a drift rate of 10ps per second in the first multiplier in the FIG 3.1 diagram will be multiplied to 1ns per second at the output. This is equivalent to a 1 x 10^{-12} frequency error due to drift. Phase drift may occur in mixers and multipliers, but more especially in multipliers. If harmonic multipliers are used, drift will occur in the analogue filters that are used to separate the wanted harmonic from the sub harmonics and unwanted mixer products. If phase lock multipliers are used, phase drift will occur in the digital dividers.

To overcome the drift problem, the multiplier/mixer chain is made differential, ie the reference signal is processed in an identical way to the unknown. When the two channels are subtracted, any drift in the multipliers will cancel. The method of doing this can be seen from the functional block diagram of the A7A, FIG 3.2. The first stage of the processing for both the reference and measurement channels is a multiplication by 10 (20 for 5MHz inputs). The multipliers are phase locked loops with a VCXO of 100MHz locked to the input by dividing by 10 (20 for 5MHz inputs). The phase detectors used are double balanced diode mixer type phase detectors. These exhibit the lowest phase drift with temperature. The dividers used are ECL types with very small propagation delays. The outputs of the dividers are reclocked using a D type flip-flop clocked by the 100MHz VCXO signal. In this way the divider delay is made equal to the propagation delay of one D type, approx 500ps. As a further refinement, the reclocking D types for the reference and measurement channels share a dual D type chip. As the divider propagation delays are equal to the reclocking flip-flop delays, the tracking between the reference and measurement channels is exceptionally good.

The VCXO signals at 100MHz also drive double balanced diode mixers for the first down conversion to 1MHz. The 99MHz LO is common to both the reference and measurement channels, and is obtained from a 4 way passive inductive type power splitter. The remaining two LO outputs feed the second down convert mixers. The output from the mixers is filtered by diplexer type filters to remove the image at 199MHz and the signal and LO feed through at 100MHz and 99MHz respectively. The wanted IFs at 1MHz are passed without further processing to the second multipliers. The avoidance of IF amplifiers at this point avoids drift which could be substantial as the propagation delay of the IF amplifier could be several 100 nanoseconds. IF amplifiers are used for the first IF take off points to the IF processing board. The first IFs are used when a multiplication of 10^3 is selected.

The second multipliers are nearly identical to the first multipliers with the difference that the phase lock loop dividers divide by 100. This multiplies the first IF of 1MHz to the second VCXO frequency of 100MHz. The second downconvert is identical to the first, with the second IFs being passed to the IF processing board.

The first and second multipliers/mixers for the reference and measurement channels are built symmetrically on one PCB (Printed Circuit Board). In order to ensure the best possible temperature tracking beween the channels, the PCB is in good thermal contact with a thick metal base plate. This minimises rapid temperature changes between the channels.

The two pairs of IF signals (sine wave) are passed to the IF processing PCB. The two pairs are the outputs from the first and second downconvertors. They correspond to final multiplication factors of 10^3 and 10^5. Also on the IF processing board is the 99MHz LO generation and phase lock. A 10MHz unmultiplied signal is passed to the IF processing board from the reference channel on the Multiplier board.

The 1MHz IFs could be divided down and measured directly by the frequency counter, which would make a time difference measurement between the measurement and reference IF signals. In this way the difference between the channels would be measured and any drift would cancel. Although this would work for a phase measurement, there would be no way of making a conventional frequency measurement. The IFs cannot be directly subtracted in a mixer as they are both nominally 1MHz, and the nominal difference frequency would be zero. In order to avoid this problem, the multiplied reference IF is frequency shifted to 900kHz using an LO of 100kHz derived from the unmultiplied reference. The 900kHz is then mixed with the 1MHz measurement channel IF to give a final IF of 100kHz. This final IF contains the multiplied frequency difference, but drift in the multipliers and phase noise in the common 99MHz LO will have been canceled out.

This IF is now further processed to provide the counter outputs as will be described in the next paragraphs.

The measurement bandwidth of the system has been defined up to this point by the loop bandwidths of the phase lock multipliers and the bandwidth of the 100kHz LC filter. The 3dB bandwidth is about 10kHz. This means that Fourier frequencies further displaced from the carrier of greater than 5kHz will be attenuated. The phase measurement process essentially samples the phase of the unknown signal relative to the reference at a rate determined by the selected tau (selectable from 1ms to 1000sec). As with any sampling process, aliasing of higher frequency noise into the baseband will occur. Thus further band limiting of the 100kHz IF is desirable before measurement takes place. It is important that the eventual calculation of the Allen variance is not in error due to excessive band limiting in the frequency domain. The effect of band limiting typical phase noise spectra on the calculation of Allen variance can be investigated by integrating the phase noise spectrum using the standard formula for phase spectral density to Allen variance conversion. This has been done for various types of noise band limited by a single pole filter. The general rule is that for a 3dB bandwidth of f Hz, the Allen variance calculation is accurate for taus greater than 1/f seconds. Thus if a 200Hz bandpass filter is used to filter the IF, then a value of tau less than 5ms should not be used. The A7A has a crystal filter following the LC filter with selectable bandwidths of nominally 10Hz, 60Hz, and 200Hz. It can be seen that for most Allen variance plots at least the 200Hz filter should be used. The use of a filter will reduce the noise floor of the instrument, which is desirable when measuring very stable active sources and most passive devices.

After the crystal filter the 100kHz IF is limited to a square wave by a zero crossing detector. This output is made available to the counter A channel when frequency mode is selected. Both the 100kHz IF containing the multiplied frequency difference information and the 100kHz unmultiplied reference are divided in identical divider chains down to 1kHz to 1mHz in selectable decade steps. The output of the dividers trigger digital (clocked) monostables to generate 10us pulses which are routed to the counter A and B channels when phase mode is selected.

3 Future improvements

a) Frequency counter

The computer card frequency counter currently supplied by Quartzlock for use with the A7A has many excellent characteristics, mainly that it will make time interval measurements at a rate of at least 1000/s, and will store the results as a formatted text file suitable for reading by Stable32 or other analysis program. Its one big disadvantage is that it must run under DOS to do this. Any attempt to run it as a window, or even as a DOS application under Windows results in random communication errors. The manufacturers of this card have stated that they do not intend to write a Windows driver for the card. The fundamental problem is that Windows is no use as a real time operating system as it is not possible to bypass its time scheduling.

In order to address the problem that the DOS operating system is now nearly extinct, Quartzlock commissioned the design of a new frequency counter that would address all the problems of the existing computer card. K+K Messtechnik GmbH who already had a multi channel phase meter with better than 100ps resolution that could be used as a time interval meter undertook this work. In order to avoid the problems with the windows interface, a special handshake RS232 interface would be used to ensure no data would be lost, however slow Windows became. The counter would have sufficient storage for 30000 time interval measurements. Thus even if the computer link was lost, or the computer crashed the readings resulting from a measurement run would be intact and could be read out from the counter at any time. The counter would be built in to the A7A case, and would have a simple RS232 cable to the computer. This would no longer have to house the counter card, and therefore could be a notebook type. The counter would take advantage of the A7A battery backup, and a measurement run lasting maybe weeks would be much more secure from data loss or interruption.

A full windows application would be provided that could operate with other applications running such as Stable32. This would provide a "virtual front panel" type of counter display which could offer variable gate times in frequency mode, variable sampling rates in phase mode, digital filtering to complement and extend the analogue filter provided on the A7A, statistics, data storage, and graphing facilities that updated as the run progressed. Basic Allen variance could be provided, although Stable32 would remain the preferred analysis tool.

All maths functions need to correct the A7A multiplication factors would be automatically applied, and various display formats such as frequency, fractional frequency, time, and phase angle would be provided. The counter module would read the front panel controls on the A7A, and would apply corrections and set modes automatically.

b) Extension to other frequencies

The A7A operates with reference and measurements of 5MHz or 10MHz and automatically switches between the two. The first multipliers operate with fixed dividers of 10 or 20, dividing down from a VCXO at 100MHz. Key to the low drift of the A7A is the reclocking of the divider outputs by the direct signal from the VCXO. In this way the temperature dependant divider delay is replaced by the delay in one D type flipflop. By making the reclock D type for the reference and measurement channel part of the same chip, good temperature tracking is assured.

The simplest way of making the A7A work with variable frequency inputs is to replace the fixed ECL dividers with a DDS synthesiser used as a divider, and clocked by the 100MHz VCXO. The stepped sine wave from the DAC in the DDS chip would be used as one input to the phase comparator, and the input signal as the other. In order to get sufficient frequency resolution, a 48 bit DDS is required. This would give a maximum fixed frequency error at 10MHz of 1.8×10^{-14}. One such DDS device seems to be available, with a built in DAC. However the problem is to achieve the same equalisation of divider delays (in this case the delay in the DDS and DAC) as in the case of fixed dividers. If a DDS chip without a built in DAC was available, then a dual DAC could be used to generate the stepped sine references for the two channels, with good tracking between the channels. Unfortunately such a device is not available.

A possible solution to this problem was eventually conceived. This is a track and hold device that is clocked by the 100MHz signal from the VCXO, and has as its input the stepped sine wave from the DDS/DAC combination. Provided that the "hold" period overlaps the period when changes to the sine wave may be expected, which may be temperature dependent, then the output sine wave phase will be determined only by the phase of the 100MHz clock. The only temperature dependence will be variations in the track/hold delay of the device.

By building two track and holds on a high thermal conductivity substrate such as aluminium nitride, good thermal tracking would be obtained between the channels. Good electrical isolation will be important as the major limitation to the performance of the current A7A is caused by crosstalk between the measurement and reference channels.

It is believed that a track and hold may be constructed using enhancement mode microwave GASFETS. These will provide very rapid transitions, and are physically very small. Only very small hold capacitors are required, as the device will operate at a clock frequency of 100MHz. The complete thick film hybrid need only be about 15mm square.

Extension of the A7A to variable input frequencies may be done on either just the measurement channel, or on both measurement and reference channels. The first case is sufficient to measure devices against a house standard on 5 or 10 MHz. However a customer may have two identical devices of which he wishes to measure the combined time domain stability. This is quite a usual technique, which avoids the use of a much better reference standard. If these devices are not on 5 or 10MHz, then the A7A will need variable frequency capability on both channels.

Practical implementation of variable frequency involves many changes to the existing design. The front panel will need a digital display and keyboard in order to enter new frequencies. However if the new counter option goes ahead, an alternative would be to dispense with all front panel controls and set up the A7A exclusively from the computer. This would then mean that the stand-alone A7A with just the meter display would not be possible.

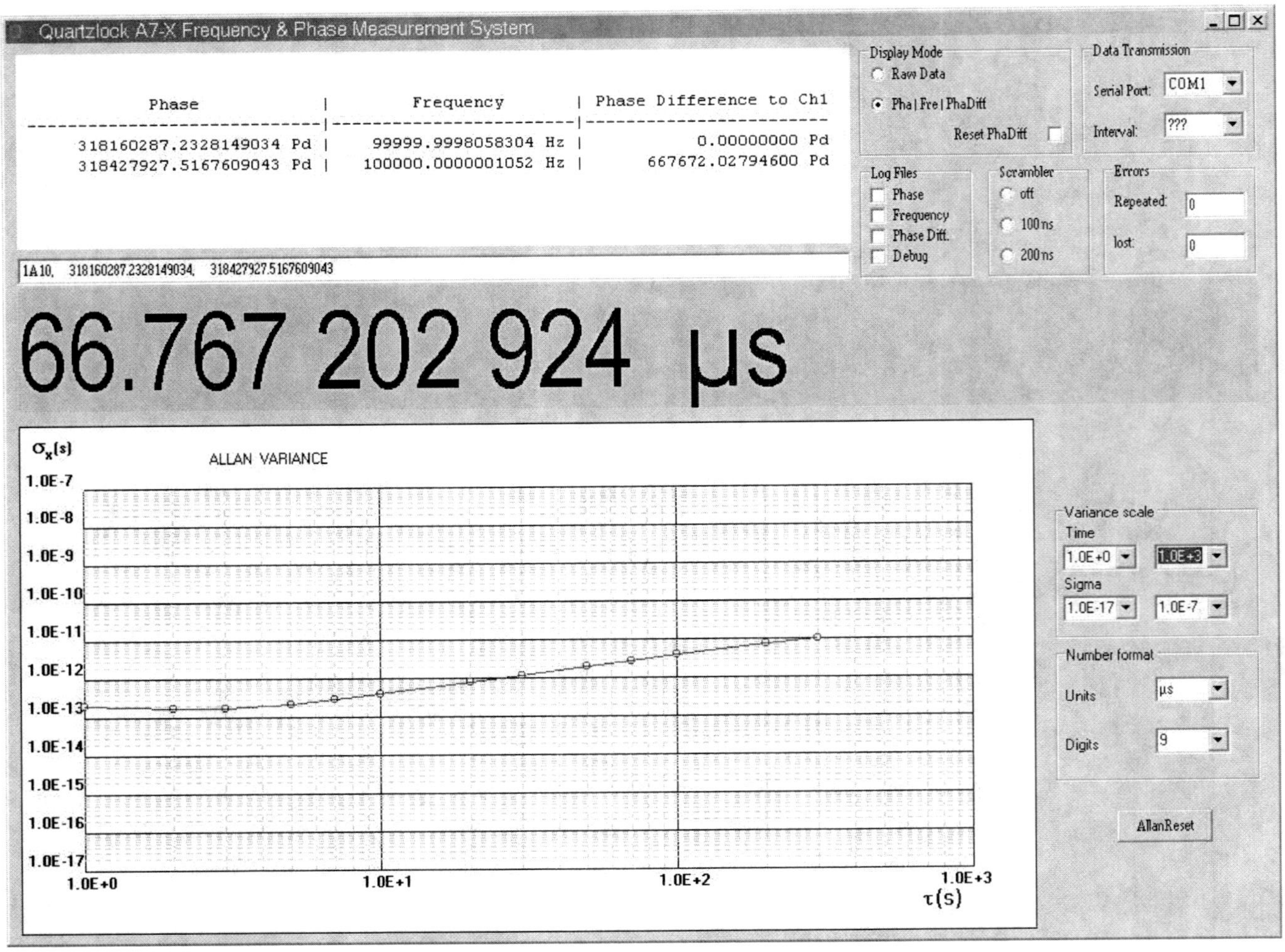
Quartzlock A7-X Frequency & Phase Measurement System
Phase | Frequency | Phase Difference to Ch1
318160287.2328149034 Pd | 99999.9998058304 Hz | 0.00000000 Pd
318427927.5167609043 Pd | 100000.0000001052 Hz | 667672.02794600 Pd
1A10, 318160287.2328149034, 318427927.5167609043
66.767 202 924 µs
Display Mode
Raw Data
Pha | Fre | PhaDiff
Reset PhaDiff
Data Transmission
Serial Port: COM1
Interval: ???
Log Files
Phase
Frequency
Phase Diff.
Debug
Scrambler
off
100 ns
200 ns
Errors
Repeated: 0
lost: 0
ALLAN VARIANCE
$\sigma_x(s)$
1.0E-7
1.0E-8
1.0E-9
1.0E-10
1.0E-11
1.0E-12
1.0E-13
1.0E-14
1.0E-15
1.0E-16
1.0E-17
1.0E+0
1.0E+1
1.0E+2
1.0E+3
$\tau(s)$
Variance scale
Time
1.0E+0
1.0E+3
Sigma
1.0E-17
1.0E-7
Number format
Units µs
Digits 9
AllanReset

NARROW LINES IN MOLECULAR IODINE NEAR THE DISSOCIATION LIMIT

F. du Burck, J.-P. Wallerand*, A. N. Goncharov†, O. Lopez, C. Daussy,

A. Amy-Klein, C. Chardonnet

Laboratoire de Physique des Lasers, 99 av. J.-B. Clément, 93430 Villetaneuse, France
*BNM-INM/CNAM, 292 rue St-Martin, 75003 Paris, France
†Institute of Laser Physics, Siberian Branch of the Russian Academy of Sciences, Novosibirsk, Russia.

Keywords: molecular iodine, high-resolution spectroscopy, FM spectroscopy, absolute frequency measurements.

Abstract

The experimental set-up for high-resolution spectroscopy of iodine at 501.7 nm, near the dissociation limit, is described. Line shapes of about 30 kHz wide are obtained with hyperfine components of the R(26)62-0 transition. Preliminary results on the absolute frequency measurements of the Ar^+ laser locked onto these transitions are given.

1 Introduction

Molecular iodine is of great interest for metrology both for the materialisation of the metre [17] and for realisation of optical frequency standards [19]. For instance, the high metrological qualities of the frequency doubled diode pumped Nd:YAG laser at 532 nm stabilised onto a hyperfine component of the R(56)32-0 transition are well-known [13]. Simple portable optical frequency standards are produced with this iodine-stabilised laser [14]. The width of the transition used as frequency reference gives a limitation to the metrological performances of an optical frequency standard: the error onto the line centre frequency decreases with the line width and so do the distortions and the shifts of the line due to pressure, wave front curvature, residual amplitude modulation, ... In the case of the above 532 nm frequency standard, the reference line is about 250 kHz wide.

Transitions between hyperfine levels of rovibronic states of the ground electronic state X and the excited one B in iodine have been extensively studied. Their natural width is defined by their radiative width and by predissociation effects. The latter result from the gyroscopic and hyperfine couplings between the excited state $^3\Pi_{0_u^+}$ and the dissociative state $^1\Pi_{1_u}$ [1]. Both radiative and predissociative widths decrease near the dissociation limit. In particular, direct lifetime measurements by fluorescence decay have shown that the hyperfine components of the R(26) 62-0 transition, accessible with an Ar^+ laser at 501.7 nm should have a natural line width around 10 kHz [16]. Such transitions are clearly of particular interest for the realisation of optical frequency standards, especially with the development of compact sources around 500 nm realised with doubled diode-pumped solid-state lasers based on Yb^{3+}-doped materials [15]. However, in a recent study of the transitions in the region 523-498 nm, the narrowest natural line width was estimated to be 43 kHz (HWHM) [3]. In a previous work, we had obtained 30 kHz wide (HWHM) lines for the R(26)62-0 transition at 501.7 nm in a molecular beam [4] and we present here line shapes with similar widths obtained in a cell.

In Sect. 2, we describe our experimental set-up for high-resolution spectroscopy of iodine at 501.7 nm. In particular, we have developed an original narrow-band stabilisation of the beam power, which is applied both to increase the signal-to-noise ratio (SNR) of the error signal for the long-term stabilisation of the laser and to reject the residual amplitude modulation (RAM) in the detection scheme of the narrow lines in frequency modulation (FM) spectroscopy. We present the principle of the controller and we describe these applications. In Sect. 3, we present one of the narrow lines obtained with hyperfine components of the R(26)62-0 transition. In Sect. 4, we give the preliminary results on the absolute frequency measurements of the Ar^+ laser locked onto these transitions. We conclude in Sect. 5.

2 Experimental set-up

2.1 The spectrometer

A monomode Ar^+ laser is prestabilised on a Fabry-Perot resonator mode by a classic Pound-Drever technique, in order to reduce the frequency jitter. It is then locked onto a hyperfine component of the R(26)62-0 transition of iodine at 501.7 nm detected with a saturation spectroscopy technique in a 50 cm long sealed cell (Fig. 1). The iodine pressure in the cell, of about 3.3 Pa (25 mTorr), is controlled by the thermostabilisation of the cold finger. Probe and saturating beams frequencies are shifted by 250 MHz and 80 MHz respectively by the acousto-optic modulators (AOM) AOM1 and AOM2. The error signal for the long-term stabilisation of the laser is obtained by a modulation transfer technique: the saturating beam is frequency modulated at 125 kHz by AOM2 and the first harmonic of the saturated absorption signal is detected on the probe beam. The correction signal for frequency stabilisation is applied to a piezoceramic transducer that controls the length of the Fabry-Perot resonator. AOM1 is used to control the probe beam intensity around the frequency detection, as it will be seen in Sect. 2.3. In order to record narrow lines, the spectroscopy is performed in a low-pressure iodine cell with FM spectroscopy technique (Fig. 2). The laser beam is split into the probe beam and the saturating beam which are frequency shifted by AOM3 and AOM4 respectively, by the same frequency shifts introduced in the stabilisation scheme. The probe beam is either frequency modulated at $\omega_m/2\pi$ = 2.5 MHz by AOM3 or phase modulated by an electro-optic modulator (EOM) located after AOM3. To record the line shape, a frequency scanning is applied to the saturating beam. Consequently, noting Γ the line width of the molecular resonance, the detected signal is composed of 2Γ wide resonances located at $\pm p\omega_m$ with p = 0, ±1, ±2, … [2,12]. In order to improve the wave fronts quality, the beams with crossed linear polarizations are superimposed before crossing a telescope with a pinhole. The diameter of the beams at the output of the telescope is 6 mm. Both polarizations are then split by a polarizing beam splitter cube (BSC), and the probe and saturating beams are contra-propagating through the cell. The probe beam intensity is measured at the output of the cell by photodiode PD4. The saturated absorption signal is obtained from a first-harmonic detection using a RF lock-in amplifier (SR844 Stanford Research System).

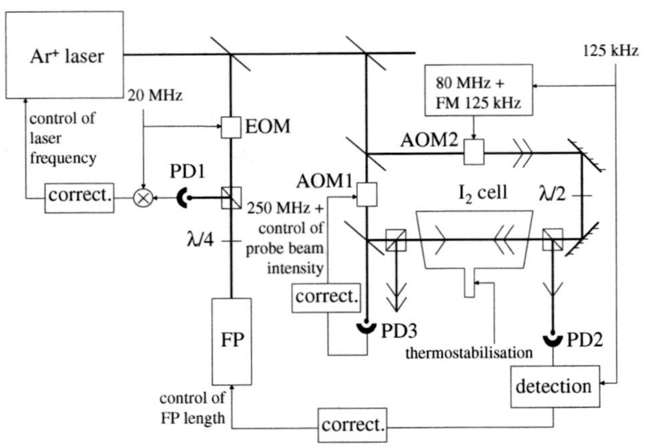

Fig 1: Ar⁺ laser long-term stabilisation scheme (FP: Fabry-Perot resonator; correct.: controller; PD: photodiode; AOM: acousto-optic modulator; EOM: electro-optic modulator).

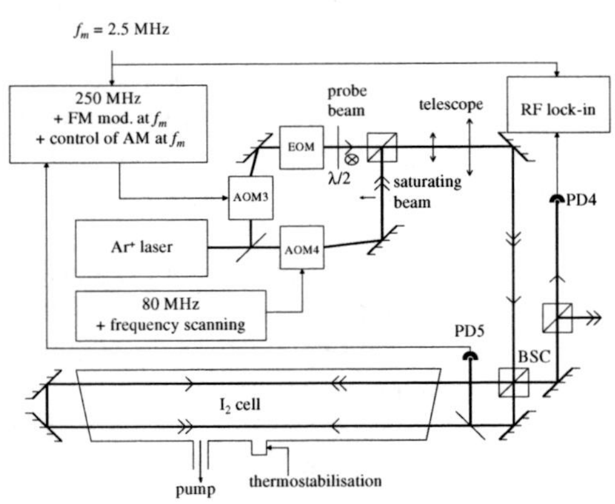

Fig 2: The spectrometer at 501 nm (PD: photodiode; AOM: acousto-optic modulator; EOM: electro-optic modulator; BSC: beam splitter cube).

2.2 Principle of the narrow-band controller

The narrow-band controller is derived from the adaptive noise canceling techniques developed in the low-frequency region. The disturbance corrupting a primary signal is estimated by an adaptive filter from a correlated auxiliary signal (reference input). The estimate is then subtracted from the primary signal [18]. For the rejection of a sinusoidal interference, the reference signal is a sinusoid at the same frequency. The conditions under which the system can be analysed as a servo loop including a time-invariant linear controller are shown in [10]. In this case, the block diagram of Fig. 3 including a continuous-time model of the controller can represent the structure of the servo loop. The in-phase and quadrature components at frequency f_m of the error signal are obtained at the output of the integrator on each channel. Then, these components are modulated by the sine and cosine references and both channels are added to produce the cancelling signal. This latter is subtracted from the primary signal to form the error signal. The gain α fixes the rejection bandwidth. In practice, a time delay τ in the loop may minimize the phase margin and limit the stability. Its effect is compensated by introducing a phase shift ϕ between the reference signals used for the modulation at the input and the output of the controller.

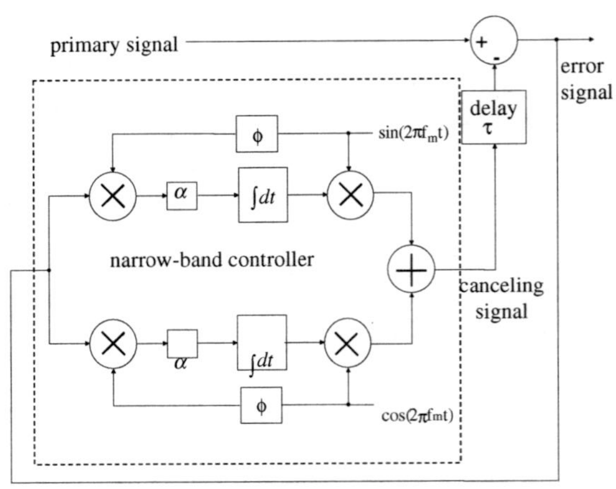

Fig 3: Principle of the narrow-band controller.

We have developed a controller based on a field programmable gate array (Xilinx XC2S200) [7]. The cancelling signal drives the RF input power of an AOM which is used as the actuator. The error signal is obtained from a photodiode monitoring the beam intensity at the output of the modulator.

We describe in the following two applications of the narrow-band controller in our set-up. Firstly to control the intensity of the probe beam at the detection frequency in the saturation spectroscopy system for the laser long-term stabilisation (Sect. 2.3). Secondly, to reject the RAM in the detection scheme of the narrow lines in FM spectroscopy [8] (Sect. 2.4).

2.3 Long-term stabilisation of Ar^+ laser: noise rejection at 125 kHz

The laser is long-term stabilised onto a hyperfine transition of iodine detected in saturated absorption. The error signal is obtained from a transfer modulation technique with a frequency modulation at 125 kHz applied to the saturating beam. The possible correction bandwidth is limited by the SNR of the detected signal. The narrow-band controller is used to reject the intensity noise of the probe beam at 125 kHz. The cancelling signal generated by the controller drives the RF input of AOM1 that controls the probe beam intensity monitored by photodiode PD3 (Fig. 1) at the input of the iodine cell.

Noise rejection efficiency in a narrow-band centred at 125 kHz is illustrated in Fig. 4. It shows the spectral density of the signal of photodiode PD2 located at the output of the cell. A hole in the intensity noise near 125 kHz is clearly seen. The noise is rejected by 27 dB, only 9 dB above the electronic noise level.

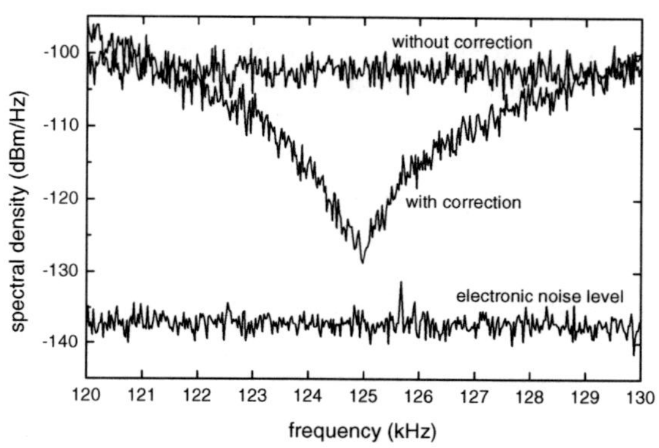

Fig 4: Noise spectral density of the probe beam near 125 kHz for the Ar^+ laser long-term stabilisation.

2.4 RAM rejection at 2.5 MHz

In FM spectroscopy, the probe beam is frequency or phase modulated at a frequency much greater than the width of the studied line. During the interaction with the sample, each Fourier component of the modulated beam experiences attenuation and phase shift. The beat notes between them on the photodetector give rise to an amplitude modulation which is detected. This technique is of particular interest for high-resolution and high-sensitivity spectroscopy (no modulation broadening, high modulation frequency). However, it is well known that these possibilities are limited by the residual amplitude modulation (RAM) generated by the modulator. As a result, a non-zero baseline appears in the detected signal and, in the case of an external modulator, a part of the laser low frequency technical noise is transferred to the detection frequency. Moreover, the detected line shape is distorted.

We use the narrow-band controller to operate an active correction of the RAM generated by the modulator at 2.5 MHz. The probe beam intensity is controlled by AOM3 (Fig. 2). In the simplest scheme, this AOM is also used for the frequency modulation of the probe beam. In this case, the fluctuations of the frequency response of the AOM around 2.5 MHz result in a few % amplitude modulation of the beam intensity. It was shown that the narrow-band controller may achieve a 70 dB rejection of the RAM [7]. However, the rejection level of the RAM on the probe beam is only 40-50 dB and is very sensitive to the photodiode location on the beam. This is attributed to the angular splitting of the Fourier components of the modulated beam at the AOM output. Any occultation of a part of the beam gives rise to an additional amplitude modulation. In particular, a photodiode may act as a diaphragm if it is not perfectly aligned on the beam and it must be carefully located to receive the entire beam.

Fig. 5 shows the rejection of the RAM on the probe beam. The transferred noise, clearly seen around the 2.5 MHz carrier without correction, is rejected by 50 dB with the correction and the noise is decreased at 10 dB above the electronic noise level.

As noted above, it is also possible to use an EOM, located after the AOM, to produce a phase modulation of the probe beam, without the specific difficulties of AOM. In this case, the RAM obtained with the correction is about the same than in the previous case.

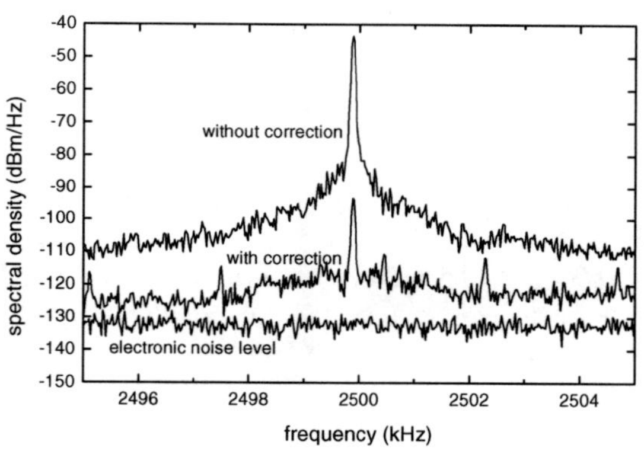

Fig 5: RAM rejection of the probe beam near 2.5 MHz.

2.5 The low-pressure iodine cell

An other specificity of our set-up is the 4 m long low pressure iodine cell which gives the possibility to observe high contrast saturation resonances with iodine pressure in the range 0.07 - 0.7 Pa (0.5 - 5 mTorr). The cell is not sealed and may be pumped during the experiment to minimize buffer gases and impurities. Indeed, the width of the studied transitions at 501.7 nm is known to be very sensitive to the impurities present in the cell [9]. An iodine crystal is formed in the cell before the experiment and the iodine pressure is controlled by thermo-stabilisation of a cold finger.

In previous works on transitions between hyperfine levels of the ground vibrational state of iodine, resonance widths HWHM = 2.6 kHz were obtained with this cell in resonant Raman spectroscopy and HWHM = 2.1 kHz in Rayleigh spectroscopy [5,6]. This indicates an upper limit of residual buffer gases and impurity pressure less than 0.04 Pa (0.28

mTorr) and demonstrates that the cell itself does not limit the resolution of our experiment.

3 Detected lines

The saturation signal detected in FM spectroscopy corresponding to hyperfine component a_7 of the R(26)62-0 transition is shown in Fig. 6. The probe beam was phase modulated at 2.5 MHz by the EOM. The solid curve is the result of the fit of experimental data with the imaginary part of a complex Lorentzian with HWHM = 33 kHz.
Such a line width is to our knowledge the narrowest one ever observed in iodine spectroscopy in cell. These narrow transitions in iodine close to the dissociation limit are very promising for realization of frequency standards.

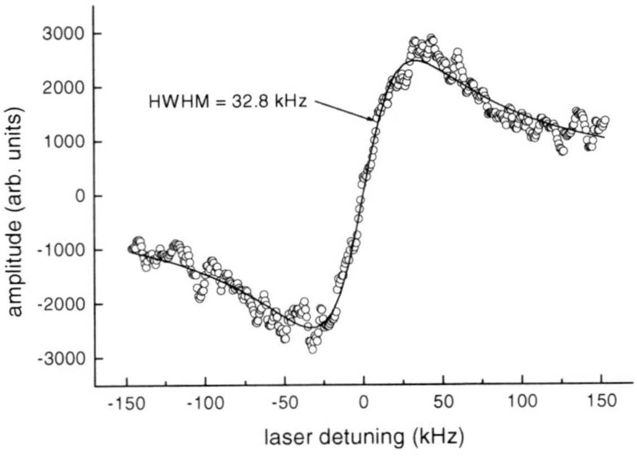

Fig 6: Hyperfine component a_7 of R(26)62-0 transition in iodine at 501.7 nm detected in FM spectroscopy (iodine pressure: 0.066 Pa (0.5 mTorr); saturating beam power: 0.2 mW; probe beam power: 0.5 mW; beams diameter: 6 mm; time constant: 100 ms; two sweeps of 60 s).

4 Absolute frequency measurement of these transitions

From the large number of iodine spectroscopic data, the identification of a particular iodine line and the calculation of the hyperfine structure are possible. Nevertheless, in order to increase the accuracy of the hyperfine components frequencies, especially in the most interesting and less explored range 515 - 500 nm, where narrow lines are encountered, absolute frequency measurements of iodine lines are required. Recently, we measured the frequency of iodine stabilized Ar^+ laser at 514.6 nm using a femtosecond optical frequency comb [11]. The first absolute frequency measurements of hyperfine components of transitions near 501.7 nm are now in progress. Moreover, these measurements give the possibility to study the reproducibility of the source.
The Ar^+ laser, prestabilised onto the Fabry-Perot resonator mode, as was seen above, is locked onto a hyperfine component of the R(26)62-0 transition of iodine detected in FM spectroscopy in the 4 m long continuously pumped cell. The line width is HWHM = 45 kHz. In order to cancel the residual offset of the error signal due to the Doppler background, an additional amplitude modulation of the saturation beam at 200 Hz and a second phase sensitive detection are used. A small amount of Ar^+ laser power (approx. 5 mW) is directed along a 20 m polarization maintaining single mode fibre to the experimental room where the femtosecond comb generator set-up is located.
We have measured the a_7 hyperfine component of the R26(62-0) transition several times during a period of a few days, with different configurations of the set-up: the probe beam was frequency modulated with the AOM or phase modulated with the EOM, the optical scheme producing the beams was modified. The mean value was found to be 597 366 498 654.7 kHz with a standard deviation of 1.3 kHz.
To estimate the reproducibility of a source stabilised onto such a reference line, 5 measurements were operated the same day during a few hours, the laser being unlocked and the optical system being re-aligned between each measurement. The standard deviation was found to be 65 Hz which demonstrates the high metrological potentiality of these narrow lines.

5 Conclusion

We have described our ultra-high resolution spectrometer based on an Ar^+ laser and a continuously pumped cell at 501.7 nm. We have obtained 33 kHz wide line shapes in saturation spectroscopy with a hyperfine component of the R(26)62-0 transition. The preliminary results of frequency measurements of Ar^+ laser locked onto such sharp transitions demonstrate the metrological potentialities of iodine transitions close to the dissociation limit for the realisation of frequency standards near 500 nm.

References

[1] M. Broyer, J. Vigué, J.-C. Lehmann. "Hyperfine predissociation of molecular iodine", *J. Chem. Phys.*, **64**, pp. 4793-4794, (1976).

[2] G. Camy, Ch. J. Bordé, M. Ducloy. "Heterodyne saturation spectroscopy through frequency modulation of the saturating beam", *Opt. Comm.*, **41**, pp. 325-330, (1982).

[3] W.-Y. Cheng, L. Chen, T.H. Yoon, J.L. Hall, J. Ye. "Sub-Doppler molecular-iodine transitions near the dissociation limit (523-498 nm)", *Opt.Lett.*, **27**, pp. 571-573, (2002).

[4] F. du Burck, J.-P. Wallerand, A.N. Goncharov, M. Himbert, Ch.J. Bordé. "Recent progress in ultra-high resolution spectroscopy of molecular iodine", *in Laser Spectroscopy XII, eds. M. Inguscio, M. Allegrini, A. Sasso (World Scientific : London, 1996)*, pp 220-221.

[5] F. du Burck, J.-P. Wallerand, B. Mercier, A. N. Goncharov, M. Himbert, Ch. J. Bordé. "Resolving power and sensitivity in modulation transfer stimulated resonant Raman spectroscopy", *IEEE J. Quantum Electron.*, **37**, pp. 837-843, (2001).

[6] F. du Burck, J.-P. Wallerand, A. N. Goncharov, M. Himbert. "High-frequency modulation transfer technique for ultra-high resolution spectroscopy of I_2", *IEEE Trans. Instrum. Meas* , **50**, pp. 493-496, (2001).

[7] F. du Burck, O. Lopez. 'Stabilisation of laser beam intensity at 2.5 MHz", *Electron. Lett.*, **38**, pp.14471449, (2002).

[8] F. du Burck, O. Lopez, A. El Basri. 'Narrow-band correction of the residual amplitude modulation in FM spectroscopy", *IEEE Trans. Instrum. Meas.*, **52**, pp. 288-291, (2003).

[9] G. Flory, M. Broyer, J. Vigué, J.-C. Lehmann. 'Perturbation de l'iode moléculaire en phase vapeur par la présence de gaz étrangers, conséquence pour la stabilisation du laser He-Ne", *Revue Phys. Appl.*, **12**, pp. 901-904, (1997).

[10] J. R. Glover, JR. "Adaptive noise canceling applied to sinusoidal interferences", *IEEE Trans. Acoust., Speech, Signal Processing*, **25**, pp. 484-491, (1977).

[11] A. Goncharov, A. Amy-Klein, O. Lopez, F. du Burck, C. Chardonnet. "Absolute frequency measurement of the iodine-stabilized Ar^+ laser at 514.6 nm using a femtosecond optical frequency comb", *to be published in Appl. Phys. B.*

[12] J. L. Hall, H. G. Robinson, T. Baer, L. Hollberg. 'The lineshapes of subdoppler resonances observable with FM side-band (optical heterodyne) laser techniques" *in Advances in Laser Spectroscopy, eds F. T. Arecchi, F. Strumia and H. Walther (Plenum Press : New York and London, 1983)*, pp. 99-126.

[13] J. L. Hall, L.-S. Ma, M. Taubman, B. Tiemann, F.-L. Hong, O. Pfister, J. Ye. 'Stabilization and frequency measurement of the I2-stabilized Nd:Yag laser", *IEEE Trans. Instrum. Meas.*, **48**, pp. 583-586, (1999).

[14] F.-L. Hong, J. Ishikaya, Z.-Y. Bi, J. Zhang, A. Onae, J. Yoda. "A portable I_2-stabilized Nd:Yag laser for international comparaisons", *IEEE Trans. Instrum. Meas.*, **50**, pp. 486-489, (2001).

[15] M. Jacquemet, F. Balembois, S. Chénais, F. Druon, P. Georges, R. Gaumé, B. Ferrand. 'First diode-pumped Yb-dopped solid-state laser continuously tunable between 1000 and 1010 nm", *Appl. Phys. B*, **78**, pp. 13-18, (2004).

[16] J.-C. Keller, M. Broyer, J.-C. Lehmann. 'Mesure directe de la durée de vie et des facteurs de Landé du niveau $^3\Pi_{0_u^+}$, $v' = 62$, $J' = 27$ de la molécule $^{127}I_2$", *C. R. Acad. Sci. Paris*, **277** série B, pp. 369-372, (1973).

[17] T. J. Quinn. 'Practical realization of the definition of the metre (1997)", *Metrologia*, **36**, pp. 211-244, (1999).

[18] B. Widrow, J. R. Glover, JR., J. M. McCool, J. Kaunitz, C. S. Williams, R. H. Hearn, J. R. Zeidler, E. Dong, JR., R. C. Goodlin. "Adaptive noise cancelling: principles and applications", *Proc. IEEE*, **63**, pp. 1692-1716, (1975).

[19] J. Ye, L.-S. Ma and J. L.Hall. 'Molecular iodine clock", *Phys. Rev. Lett.*, **13**, pp. 2708-01, (2001).

COMPARISON OF SIMPLE AND COMPACT "DOPPLER" AND "SUB-DOPPLER" LASER FREQUENCY STABILISATION SCHEMES

C. Affolderbach[1], G. Mileti[1], D. Slavov[2], C. Andreeva[2], S. Cartaleva[2]

[1]Observatoire Cantonal de Neuchâtel, Rue de l'Observatoire 58, 2000 Neuchâtel, Switzerland
[2]Institute of Electronics, Bulgarian Academy of Sciences, boul. Tsarigradsko Shosse 72, 1784 Sofia, Bulgaria

Keywords: laser stabilisation, atomic spectroscopy.

Abstract

The availability of frequency stabilised laser sources is essential for the realisation of laser-pumped atomic clocks and other applications. Here we present a comparison of two simple spectroscopic schemes for laser stabilisation to the Rb D_2 line transitions, based on linear (Doppler-broadened) and non-linear (saturated absorption) spectroscopy in thermal vapour cells, covering time scales up to 10^5 seconds. We measured short-term stabilities of $2\cdot10^{-11}\ \tau^{-1/2}$ and $2\cdot10^{-12}\ \tau^{-1/2}$ up to 100 seconds and long-term stabilities at one day around $1\cdot10^{-10}$ and $2\cdot10^{-12}$ for the Doppler and sub-Doppler scheme, respectively. From a drift analysis of different experimental parameters we find the frequency stability beyond 10^4 seconds to be limited essentially by variations of the reference cell temperature for both schemes, and estimate limits around $9\cdot10^{-12}$ and $2\cdot10^{-12}$ for improved experimental realisations. The performance and simplicity of both stabilisation schemes studied makes them interesting for applications in compact atomic clocks and other precision instruments.

1 Introduction

While the short-term frequency stability of free-running diode laser modules may be sufficient in some cases, most high-precision applications in metrology demand improved laser frequency stability over an extended range of time, and thus stabilisation of the laser emission to an atomic or molecular reference line is required. For example, laser-pumped gas-cell atomic frequency standards [10,11,15], commonly referred to as Rb type clocks, offer improved short-term stabilities down to $3\cdot10^{-13}\ \tau^{-1/2}$ [10] compared to conventional lamp pumped devices, but have to fight light shift effects introduced by the narrow-band laser pump light on longer timescales. In order to maintain the compact size of such Rb clocks while optimizing their performance, a compact realisation of a high-performance frequency stabilisation scheme for the pump light laser is of major importance [4].

Here we report on our study regarding the long-term frequency stability of grating-stabilised diode lasers locked to the D_2 line of ^{87}Rb, with focus on the limitations imposed by environmental parameters. Compact and robust realisations of such frequency-stabilised laser heads could find applications not only for optical pumping in compact, laser-pumped Rb atomic clocks [4,5], but also in high-performance atomic magnetometers [6,7], and as secondary frequency references [14].

We have studied in detail two selected schemes that offer the potential to realise compact and robust stabilised laser heads for precision instruments: sub-Doppler stabilisation by the so-called saturated absorption spectroscopy [12] and stabilisation to simple Doppler-broadened absorption lines. While the sub-Doppler scheme has the advantage of resonance linewidths almost a factor 100 narrower than the Doppler absorption, the Doppler scheme could nevertheless be suitable for instruments where less frequency stability is required but the physics package needs to remain the simplest possible. Furthermore, automated identification of the correct line to lock to is much simpler in the Doppler case due to the lower total number of lines to be distinguished.

2 Experimental study

For a systematic and quantitative comparison of the two schemes, we have performed heterodyne frequency stability measurements of two extended-cavity diode lasers (ECDL), stabilized independently to different reference cells (Figure 1). The home-made lasers use Hitachi HL7851G diodes emitting around 780 nm, spectrally narrowed to below 300 kHz by optical feedback from a 1200 lines/mm diffraction grating in Littrow configuration. The laser frequencies are stabilized to absorption spectroscopic lines from evacuated reference cells containing atomic Rubidium in natural isotope mixture. Error signals for laser stabilisation are obtained by FM modulation of the laser frequency via direct modulation of the diode current at 50 kHz and subsequent 1f lock-in detection. The fast and slow correction signals obtained are applied to the diode current and a piezo element controlling the grating position, respectively.

The first laser was stabilised to a sub-Doppler absorption signal of the ^{87}Rb D_2 transition from a thermal Rb vapour cell and was used as a frequency reference. The physical parameters of this stabilisation setup like, e.g., cell temperature and the magnetic field, remained unchanged under the measurements. The second laser was stabilised using either a second sub-Doppler or a Doppler stabilisation

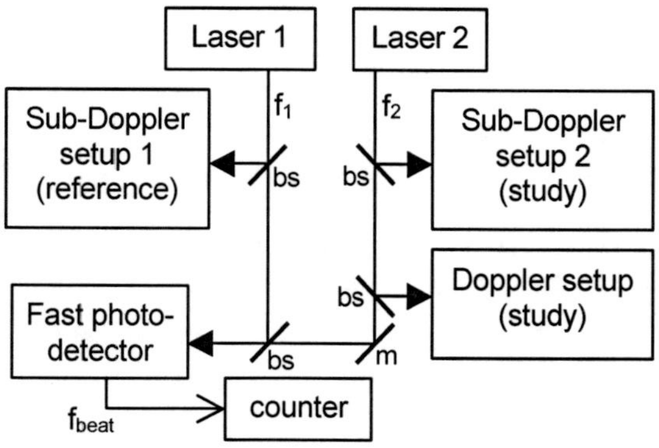

Figure 1: Sketch of the experiment. Two lasers are stabilised to independent Doppler or Sub-Doppler Rb spectroscopy references. The relative frequency stability is measured via the beat note of the two laser frequencies. bs: beam splitters, m: mirror.

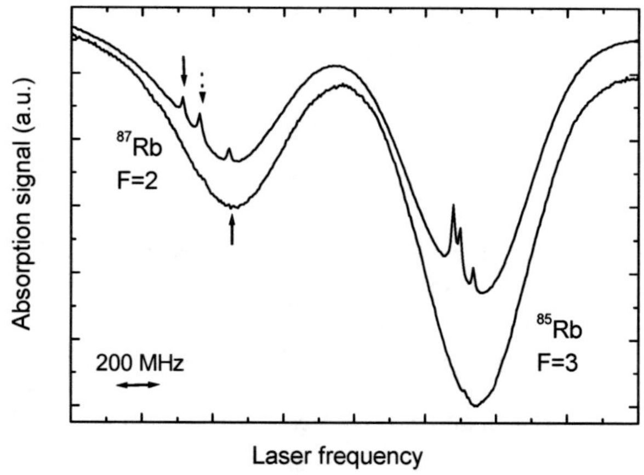

Figure 2: Measured reference absorption signals from the sub-Doppler (upper trace) and Doppler (lower trace) setups. Solid arrows indicate the frequency positions to which the test laser was stabilised, the dashed arrow gives the position of the reference laser.

setup where environmental parameters could be changed in a well-controlled way, and thus provided the stabilisation scheme under test. In order to keep the spectroscopic setups simple and compact, the sub-Doppler configurations use the pump beam retro-reflected onto itself to act as probe beam, and no subtraction of the Doppler background [12] was implemented. The frequency difference between the two laser frequencies was measured by the heterodyne beat signal from a fast photodetector, compared to a stable in-house hydrogen maser reference using a frequency counter. Both laser stabilisation setups remained routinely locked over several days and thus allowed long-term studies of the frequency stability.

We concentrated our studies on the ^{87}Rb D_2 line F=2 component, as this is a convenient transition for optical pumping in Rb atomic clocks. In the case of sub-Doppler stabilisation, the cross-over resonance 21-23 was chosen which combines a high reference line contrast with the possibility to achieve low light shift in a Rb laser pumped clock [3,8]. The reference laser was always stabilised to the cross-over resonance 22-23. Figure 2 gives examples of the absorption spectra and illustrates the laser frequency positions.

Let us note that the background slopes present in the spectroscopic signals due to the Doppler background and the presence of the ^{85}Rb isotope in the sub-Doppler and Doppler scheme, respectively, cause offsets in the error signals and thus result in small systematic shifts of the laser frequencies from the centre of the atomic transitions. While the absolute values of the laser frequencies are not explicitly studied here, their changes are nevertheless implicitly included in the shift coefficients of the experimental parameters discussed in section 3.2 that influence the background slope or line centre positions.

2.1 Experimental results

Using the setup described in section 2, we have performed measurements of the relative laser frequency stability over the course of several days while monitoring a variety of experimental parameters that influence the absolute laser frequency value [2] and thus by their drifts relate to the obtainable frequency stability. Figure 3 summarises the experimentally measured laser frequency stabilities obtained using the Doppler and sub-Doppler schemes in terms of the Allan standard deviation for the two schemes studied. As can be seen, the sub-Doppler scheme offers about one order of magnitude better stabilities, but even the Doppler scheme meets the requirements for the application example of an atomic clock at short integration times up to 100 seconds (solid lines).

Previous work reported flicker floor stabilities of $2.4\cdot10^{-10}$ for 3 to 10^4 s integration time for Doppler stabilisation [13], which we were able to improve by 1-2 orders of magnitude. Our measurements using the sub-Doppler scheme also show

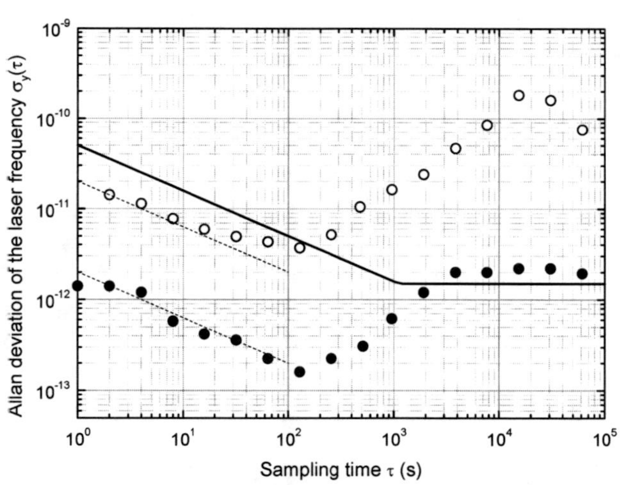

Figure 3: Experimental frequency stabilities in terms of Allan standard deviation for the Doppler (open circles) and sub-Doppler (bullets) stabilisation. Dashed lines correspond to short-term stabilities of $2\cdot10^{-11}\ \tau^{-1/2}$ and $2\cdot10^{-12}\ \tau^{-1/2}$. The solid line illustrates a typical laser frequency stability required in Rb atomic clocks as given by light shift effects.

improved stability on the short-term time scales up to about 500 s compared to the $3.1\cdot10^{-12}\ \tau^{-0.36}$ obtained in earlier results [16], but with a considerably simpler and more compact setup.

3 Result analysis

For an analysis of the measured stabilities we distinguish two regimes: On the short-term time scales up to about 100 seconds, the frequency stability is essentially limited by the signal-to-noise ratio of the reference line used. In contrast, for longer integration times frequency drifts are related to drifts of the experimental and environmental parameters.

3.1 Short-term stability

On short time scales the laser frequency stability [1], discussed here in terms of the Allan standard deviation σ_f, is governed by the noise level of the error signal and the width and contrast of the atomic reference line. We take into account the implemented detection scheme by expressing σ_f in terms of the detection noise power spectral density N_{PSD} and the discriminator slope D of the error signal, as obtained, e.g., by frequency modulation and subsequent lock-in detection [9,10]:

$$\sigma_f = \frac{N_{PSD}}{\sqrt{2}\cdot D\cdot f_L}\cdot\tau^{-1/2} \qquad (1)$$

With the FM modulation amplitude optimised to the width Γ of the reference line, the discriminator slope roughly depends on the resonance amplitude like $D\sim A/\Gamma$. In thermal Rb vapour, the Doppler broadened widths of the D_2 lines are almost 100 times larger than the natural widths of the Doppler-free saturated absorption features. However, the latter possess signal amplitudes almost one order of magnitude less than for the Doppler lines. One thus expects a difference in short-term laser stability of about one order of magnitude between the two schemes. Table 1 summarises the stability limits calculated according to Eq. (1) for the two stabilisation schemes implemented, as well as the experimental results.

	Doppler	Sub-Doppler
Calculated limit	$1\text{-}3\cdot10^{-11}\ \tau^{-1/2}$	$1\text{-}4\cdot10^{-12}\ \tau^{-1/2}$
experimental	$2\cdot10^{-11}\ \tau^{-1/2}$	$2\cdot10^{-12}\ \tau^{-1/2}$

Table 1: Calculated (S/N limit) and experimental short-term laser frequency stabilities for timescales of 1-100 seconds.

3.2 Long-term stability

On medium and long-term timescales the laser frequency stability is limited by the stability of the atomic reference, as well as by drifts in the implemented stabilisation setup and control electronics. We have measured the variations of several experimental parameters of relevance, most of which were not actively controlled during measurements, and obtained the following Allan deviations at 10^4 seconds (about one day):

Magnetic field: magnetic flux density variations in a laboratory environment typically exceed the variations of the earth's magnetic field due to working day activities, elevators, and others. We used μ-metal shields to reduce the field fluctuations on the reference vapour cells to an estimated level of about 1 mG, taking into account measured laboratory magnetic noise levels.

Optical power: We observed laser frequency changes with varying optical intensity, which we attribute mainly to population redistribution by optical pumping, and which also depend on cell temperature [12]. No active stabilisation of the laser power was implemented and measured drifts amount to about $3\cdot10^{-4}$ or total variations of 3-5 nW in incident power.

Reference cell temperature: The cell temperature not only influences the reference signal contrast via the atomic density probed, but also shifts the stabilised laser frequency, with much stronger impact in the Doppler case [2,14]. Our measurements for the Doppler scheme on the ^{87}Rb D_2-line F=2 component show a minimum shift rate of 300 kHz/K around 35°C cell temperature, at least 3 times larger than values reported elsewhere [13,14]. Reference cell temperatures during our experiments were about 35°C and 24°C in the Doppler and sub-Doppler scheme, respectively, with a moderate stability of 0.1 K over 1 day.

Beam angle: While in the sub-Doppler scheme the angle between pump and probe beams shifts the resonance position via the Doppler effect, the angle of incidence of the only existing laser beam on to the Doppler reference cell was also found to shift the laser frequency. We attribute these shifts to slight changes in the laser polarisation caused by the cell entrance window and residual magnetic fields across the resonance cell. Variations in beam angle due to movements of the laser grating amount to about 0.9 μrad over one day.

FM modulation amplitude: The laser frequency modulation is derived by direct modulation of the diode's injection current, and this modulation signal represents a major source of drifts due to the control electronics. We measured significant shifts of the stabilised laser frequency with the amplitude of this FM modulation signal, probably due to residual AM modulation contributions that shift the zero crossing of the error signal. Thermal drifts of the electronics result in FM amplitude variations amount to about 0.02 mV and 0.12 mV in the sub-Doppler and Doppler schemes, respectively.

An overview of the parameter drifts and resulting frequency stabilities is given in Table 2, where the shift rates of different Rb spectroscopic lines under variations of experimental parameters from a previous communication were used [2]. Note, that the shift coefficients for other sub-Doppler resonances than the cross-over 21-23 studied here can be 3-8 times smaller.

The total drift calculated as the sum of the single parameter contributions of Table 2 reproduces very well the measured frequency stabilities around 10^4 seconds (cf. Figure 3).

Obviously, for both stabilisation schemes, the influence of reference cell temperature changes presents the dominating contribution to the total laser frequency drift. Table 2 also gives an optimised stability limit for optimised control of experimental parameters, from which a laser stability improved by about a factor of 10 can be expected. The considered optimisations mainly concern the implementation of improved reference cell temperature control to within 0.01 K, and optimisation of the reference cell temperature to reduce the shift coefficient connected to the FM modulation amplitude. The implementation of isotopically pure Rb cells or 3f lock-in detection might further help to reduce the susceptibility related to cell temperature change and AM modulation effects.

	Doppler		Sub-Doppler	
	Coefficient	Frequency drift [kHz]	Coefficient	Frequency drift [kHz]
Magnetic field	35 Hz/mG	0.004	120 kHz/mG	0.012
Optical power	8 kHz/mW	0.04	10 kHz/mW	0.05
Cell temperature	300 kHz/K	30	4 kHz/K	0.4
Beam angle	5.73 MHz/rad	0.005	80.2 MHz/rad	0.07
FM amplitude	11 kHz/mV	1.3	1 kHz/mV	0.02
Total drift		31 kHz		0.5 kHz
Total stability		$8\cdot10^{-11}$		$1.3\cdot10^{-12}$
Optimised limit		$9\cdot10^{-12}$		$2\cdot10^{-13}$

Table 2: Measured variations of experimental parameters and corresponding laser frequency drifts over 10^4 seconds. The estimated limit gives the laser stability expected for optimised control of the experimental parameters.

3 Application to laser pumping in a Rb clock

We have implemented both frequency stabilisation schemes studied above for the realisation of optical pumping in a demonstrator laser-pumped Rb atomic clock. This demonstrator clock is currently being developed at Observatoire de Neuchâtel in cooperation with Temex Neuchâtel Time as industrial partner in the frame of an ESA project within the ARTES 5 program [4,5]. The goal of this project is the development of a compact (<1.5 kg, <1.5 l, < 15 W) and high-performance (10^{-12} at 1s, 10^{-14} at 1 day) laser pumped Rb clock for satellite telecommunications and other space applications. The demonstrator clock consists of a modified commercial lamp-pumped Rb clock (TNT RAFS type) whose discharge lamp was removed, and a compact laser head including frequency stabilisation to a sub-Doppler Rb reference line [4].

For an evaluation of the clock performance using the Doppler scheme, the clock pump laser was stabilised to the same Doppler stabilisation setup used in the experiments described above. Figure 4 shows the resulting clock stabilities using both stabilisation schemes. The stability of the Doppler-stabilised clock is found to be about one order of magnitude worse than for the case of sub-Doppler stabilisation and stays rather constant at a level of $2\text{-}3\cdot10^{-12}$ over all integration times measured. This can, however, not be attributed to the difference in laser frequency stability alone, because for a clock only limited by the laser frequency stability, one would expect the clock to perform equally well up to 100 seconds for both stabilisation schemes (cf. Fig. 3). In fact, in the Doppler case the laser frequency is offset much further (about 200 MHz, cf. Fig. 2) from the laser frequency corresponding to zero light shift, and thus the clock stability is additionally degraded by light shift effects stemming from pump light intensity variations [8]. Furthermore, the significantly larger laser FM modulation amplitude applied for the Doppler stabilisation might introduce additional effects, not yet taken into consideration. In view of existing strategies for the reduction of light shift effects in Rb clocks [3,8], the Doppler approach remains nevertheless interesting for the development of a simple and robust laser pumped Rb clock.

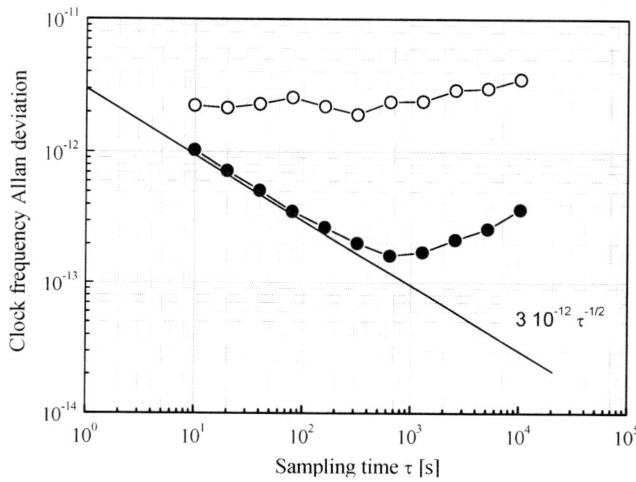

Figure 4: Demonstrator Rb laser clock stabilities using laser stabilisation to Doppler (open circles) and sub-Doppler (bullets) frequency stabilisation.

4 Conclusion

We have investigated the long-term stability limits of two simple and compact schemes for laser frequency stabilisation to Rb D_2 reference lines, implementing both simple linear Doppler absorption and a simple realisation of "saturated absorption" sub-Doppler signals obtained from thermal vapour cells. On all time scales considered, the Doppler scheme offers 1-2 orders of magnitude less stability than the sub-Doppler scheme. Nevertheless, with stabilities of $2\text{-}3\cdot10^{-12}\ \tau^{-1/2}$ up to 100 seconds it can be interesting in applications where a simple and robust laser stabilisation is required. On time scales beyond one day both schemes are mainly limited by variations in the reference cell temperature. The measured relative laser frequency stabilities at 10^4-10^5 seconds are in agreement with the limits derived from measured drifts of experimental parameters, reaching $2\cdot10^{-10}$ and $2\cdot10^{-12}$ for the Doppler and sub-Doppler scheme, respectively. For optimised parameter control we expect stabilities around $9 \cdot 10^{-12}$ and $2 \cdot 10^{-13}$, respectively, for the two schemes. Compact realisations of laser heads implementing one of these stabilisation schemes [4] offer interesting frequency-stabilised laser sources for compact atomic clocks and other precision applications like, e.g., wavelength references or atomic magnetometers.

Acknowledgements

This work was financed by the European Space Agency (ESA) under contract numbers 5280/01/NL/US, the Swiss National Science Foundation under project numbers 2100-066821.01, SCOPES 7BUPJ062412, and R'Equip 2160-067498, the Canton de Neuchâtel, and the Swiss Federation (Art. 16). We acknowledge fruitful collaborations with the atomic physics group of A. Weis at the University of Fribourg, Switzerland, and with Temex Neuchâtel Time on the ESA clock project.

References

[1] Strictly speaking, the Allan standard deviation σ_f denotes a measure of fluctuation and thus of frequency *in*stability rather than stability. For reasons of simplicity, we will use the two terms as synonyms, understanding a high frequency stability to mean a low Allan standard deviation.

[2] C. Affolderbach, G. Mileti, D. Slavov, C. Andreeva, S. Cartaleva, "Frequency stability comparison of diode lasers locked to Doppler and sub-Doppler resonances", accepted for publication in SPIE.

[3] C. Affolderbach, G. Mileti, C. Andreeva, D. Slavov, T. Karaulanov, S. Cartaleva, "Reducing light-shift effects in optically-pumped gas-cell atomic frequency standards", *Proceedings of the 2003 IEEE International Frequency Control Symposium jointly with the 17th European Frequency and Time Forum*, pp. 27-30, (2003).

[4] C. Affolderbach, G. Mileti, "A compact, frequency stabilized laser head for optical pumping in space Rb clocks", *Proceedings of the 2003 IEEE International Frequency Control Symposium jointly with the 17th European Frequency and Time Forum*, pp. 109-111, (2003).

[5] C. Affolderbach, G. Mileti, "A compact, high-performance laser-pumped rubidium frequency standard", *Proceedings of the 18th European Frequency and Time Forum*, (2004).

[6] G. Bison, R. Wynands, A. Weis, "Dynamical mapping of the human cardiomagnetic field with a room-temperature, laser-optical sensor", *Optics Express,* **11**, pp. 904-909, (2003).

[7] D. Budker, D. F. Kimball, S. M. Rochester, V. V. Yashchuk, M. Zolotorev, "Sensitive magnetometry based on nonlinear magneto-optical rotation", *Phys. Rev. A* **62,** 043403 (2000).

[8] T. Karaulanov, C. Andreeva, S. Cartaleva, Y. Dancheva, B. Todorov, A. Yanev, S. Jaquet, G. Di Domenico, P. Thomann, G. Mileti, "Light shift and laser sidebands in gas-cell atomic clocks using optical pumping and coherent population trapping", *Proc. SPIE* **5137,** pp. 364-372, (2003).

[9] G. Mileti, P. Thomann, "Study of the S/N performance of passive atomic clocks using a laser pumped vapour", *Proceedings of the 9th European Frequency and Time Forum*, pp. 271-276, (1995).

[10] G. Mileti, J. Deng, F. L. Walls, D. A. Jennings, R. E. Drullinger, "Laser-pumped Rubidium frequency standards: new analysis and progress", *IEEE J. Quantum Electron.*, **34**, pp. 233-237, (1998).

[11] Y. Ohuchi, H. Suga, M. Fujita, T Suzuki, M. Uchino, K. Takahei, M. Tsuda, Y. Saburi, "A high-stability laser-pumped Cs gas-cell frequency standard", *Proceedings of the 2000 IEEE International Frequency Control Symposium,* pp. 651-655, (2000).

[12] O. Schmidt, K.-M. Knaak, R. Wynands, D. Meschede, "Cesium saturation spectroscopy revisited: How to reverse peaks and observe narrow resonances", *Appl. Phys. B*, **59**, pp. 167-178, (1994).

[13] M. Têtu, B. Villeneuve, N. Cyr, P. Tremblay, S. Thériault, M. Breton. "Multiwavelength sources using laser diodes frequency locked to atomic resonances", *IEEE J. Lightwave Technol.*, **7**, pp. 1540-1548 (1989).

[14] M. Têtu, N. Cyr, B. Villeneuve, S. Thériault, M. Breton, P. Tremblay, "Towards the realization of a wavelength standard at 780 nm based on a laser diode frequency locked to Rubidium vapor", *IEEE Trans. Instr. Meas.*, **40**, pp. 191-195, (1991).

[15] J. Vanier, C. Audoin, "The quantum physics of atomic frequency standards", Adam Hilger, Bristol and Philadelphia, 1988.

[16] J. Ye, S. Swartz, P. Jungner, and J. L. Hall., "Hyperfine structure and absolute frequency of the $^{87}Rb5P_{3/2}$ state", Opt. Lett., **21**, pp. 1280-1282, (1996).

3D COMPUTER MODELING OF CONVENTIONAL CESIUM BEAM TUBE

K.W. Radecki[1], D. Wasiak[2], A. Moryc[3]

[1]Institute of Radioelectronics, Nowowiejska 15/19, 00-665 Warsaw, Poland
e-mail: radecki@ire.pw.edu.pl, fax: +48 22 825 3759
[2] Telecommunications Research Institute, Poligonowa 30, 04-051 Warsaw, Poland
e-mail: wasiak@pit.edu.pl, fax:=+48 22 810 2570
[3]Institute of Radioelectronics, Nowowiejska 15/19, 00-665 Warsaw, Poland
e-mail: amoryc@elka.pw.edu.pl, fax: +48 22 825 5248

Keywords: Atomic clocks, Computer modelling.

Abstract

We are elaborating 3D computer model of conventional cesium beam tube. The atomic source, selection magnets and detector are modeled in Matlab environment. Model of the source describe intensity and angular distribution of atoms for multichannel array nozzle geometry. Conventional "two-wire" model of A and B magnets has been assumed. Microwave Ramsey type cavity has been modeled by using QuickWave-3D electromagnetic simulator.

In this paper the calculations of intensity distributions of atoms in various cross section of a cesium tube are also presented.

1 Introduction

Most of all portable cesium beam frequency standards hold conventional cesium tube, with magnetic selectors of atoms. In conventional cesium beam tube (Fig.1), atoms effusing out of the source are deflected by magnet A before and magnet B after excitation in microwave cavity at the presence of weak homogeneous C field. The C field separates clock transition ($F=4$, $m_F=0$)$\leftrightarrow$($F=3$, $m_F=0$) from other magnetic sensitive transitions.

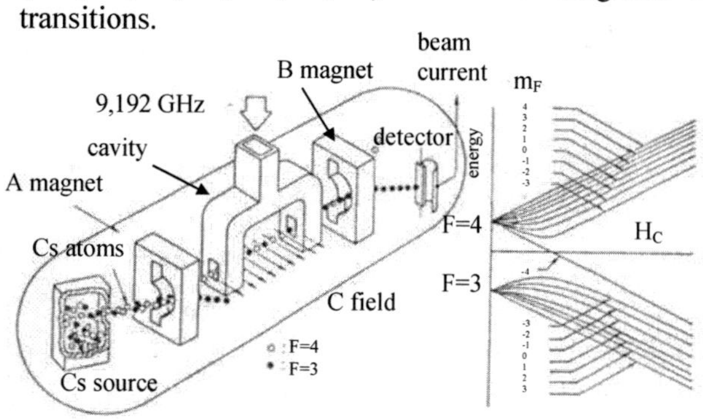

Fig.1. Schematic block diagram of conventional cesium beam tube and Cs^{133} hyperfine energy levels [4].

Magnet A selects atoms which are in desired hyperfine state and steers towards the entry of resonator. Magnet B selects excited atoms and steers towards detector with surface ionization. Ion current, after amplification in electron multiplier, is processed to automatically regulate frequency of voltage controlled quartz oscillator from which the microwave frequency is synthesized. Intensity of atoms at the detector depends on atomic beam geometry and will determine output signal to noise ratio and short term stability of frequency standard.

Because of angular and velocity distributions of atoms in the beam and divergences between C field and magnetic microwave field directions, the Ramsey and Rabi frequency pulling of atomic resonance line is observed [10,11,2]. At present a conventional commercially available cesium clocks have advanced tube design with reducing the Ramsey and Rabi pulling [2]. Using computer modeling, the atomic beam geometry in the tube may be investigated to achieve high signal to noise ratio and some frequency pulling effects reduction.

2 Cesium beam tube model

At present only the atomic source, selection magnets and detector has been modeled in Matlab environment. Microwave Ramsey type cavity is modeled by using QuickWave-3D electromagnetic simulator.

2.1 Atomic beam source

The cesium beam is formed by effusion of atoms through an array of channels [1,3,5]. The angular intensity distribution of atoms formed by one channel and emitted per second, m^2, and steradian can be approximated by [6,7,8]:

$$I(\theta) = I_0(1 - \mathrm{tg}\theta / 2\eta), \quad \text{for } 0 \le \theta \le \eta \qquad (1)$$

$$I(\vartheta) = I_0(\sin\eta\cos^2\theta)/(2\sin\theta\cos^2\eta) \quad \text{for } \eta \le \theta \le \pi/2 \qquad (2)$$

where: $\eta=2r_k/L_k$, $I_0=p(2\pi^3 mkT)^{-1/2}$ is the intensity of cesium atoms effusing trough thin edge aperture, m- atomic mass, k is Boltzmann's constant, T is thermodynamic temperature of cesium vapour, p is the saturating vapour pressure of cesium. For T>302K [14]:

$$\log p[\mathrm{Tr}]=11{,}0531-1{,}35\log T-4041/T \qquad (3)$$

Examples of calculated angular intensity distributions of Cs atoms effusing through long channels are shown in Fig.2.

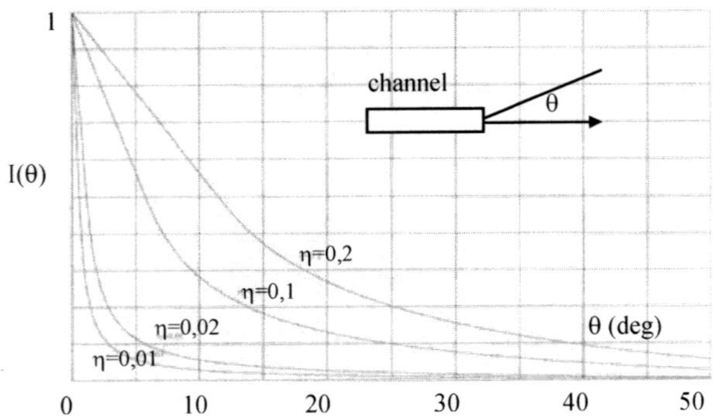

Fig.2: Angular distributions of atoms effusing through long channels

The model of the source channels array is shown on Fig.3. The atomic beam is formed by two hundred long cylindrical channels. Position of the source may be oriented in any direction towards to A magnet. A modified Maxwell velocity distribution of atoms in the beam has been assumed [13,9]:

$$p(\upsilon)=2(\upsilon^3/\upsilon_p^4)\exp(-\upsilon^2/\upsilon_p^2) \qquad (4)$$

where $\upsilon_p=(2kT/m)^{1/2}$ is the most probable velocity in the beam. Additional modification of this distribution is caused by A and B magnets and slots in microwave cavity.

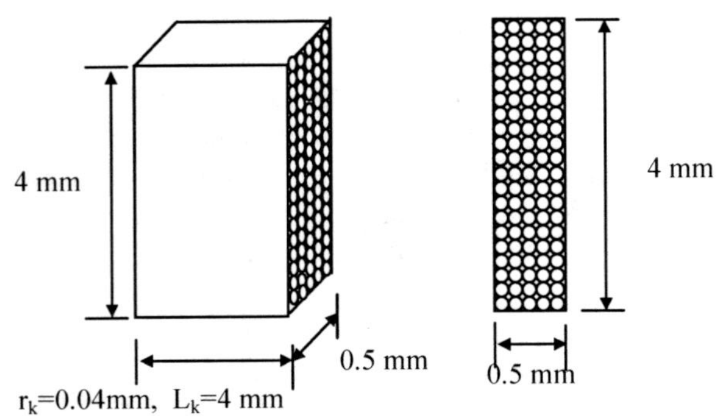

Fig.3: Model of the source channels array

The computer program "Source" has been elaborated in Matlab environment to calculate:

1. the coordinates and intensity distributions of atoms on a given surface
2. the components of atoms velocities leaving the source

for the given source temperature, geometry of the channels array, range of the angle θ, step Δθ and the position of the measurement plane.

2.2 State selectors

In a conventional cesium tube two-pole magnet state selectors are commonly used. Fig.4 shows the pole pieces of A and B magnets. Magnetic fields in the gap of both magnets correspond to within a good approximation with two-wire magnetic field configuration (points A and B on Fig.4).

Magnetic field between pole pieces has a gradient **∇B** and Cs atoms are deflected by the force

$$\mathbf{F}=\mu_{eff}\nabla\mathbf{B} \qquad (5)$$

where

$$\mu_{eff} = \mp\frac{(\xi + m_F/4)g_J\mu_B}{2(1+m_F\xi/2+\xi^2)^{1/2}} \qquad (6)$$

is the effective magnetic moment of atom, $\xi = g_J\mu_B B/hf_0$, f_0=9192631770Hz, μ_B is the Bohr magneton, g_J = 2,002 is the Lande factor, plus sign is for the (3, m_F) and (4,-4) states, minus sign is for other (4,m_F) states [13,14].

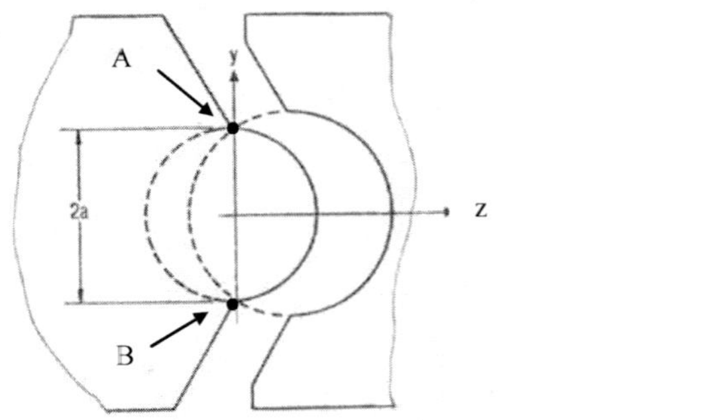

Fig.4: Schematic view of the cross section of a "two wire" type A and B magnets

Fig.5 shows variation of the effective magnetic moment of Cs atoms against magnetic induction B.

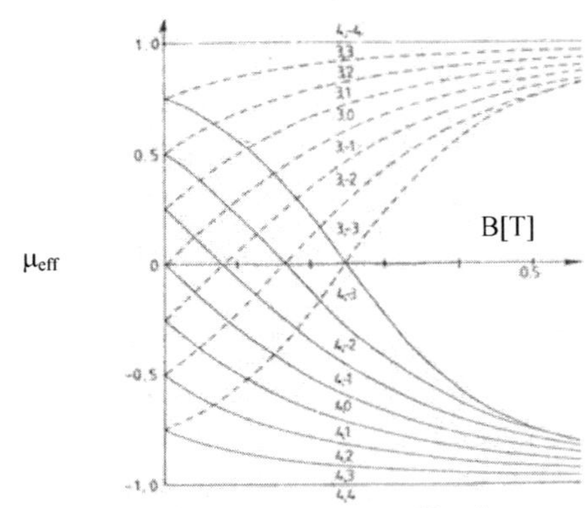

Fig.5: variation of the Cs atoms effective magnetic moment against magnetic induction

At high magnitude of magnetic induction B>0.5T the effective magnetic moment of atoms is negative or positive. The magnetic field induction and field gradients in the gap of two poles magnets are given by the following equations:

$$B=2a^2B_0/(r_1 r_2) \qquad (7)$$

$$\frac{\partial B}{dz} = -B_0\frac{2a^2z(r_1^2+r_2^2)}{r_1^3r_2^3} \qquad (8)$$

$$\frac{\partial B}{dy} = -B_0\frac{2a^2y(r_1^2+r_2^2)}{r_1^3r_2^3} - B_0\frac{2a^3(r_1^2-r_2^2)}{r_1^3r_2^3} \qquad (9)$$

where $r_1=z^2+(y-a)^2$, $r_2=z^2+(y+a)^2$, B_0 is the magnetic induction at the convex pole tip, 2a is the diameter of the convex pole. Fig.6 and Fig.7 shows variation of the magnetic induction B and field gradients between pole tips. Since μ_{eff} and gradients ∂B/dz , ∂B/dy are functions of position, then the force acting on atoms will also depends on atomic position inside the magnetic gap.

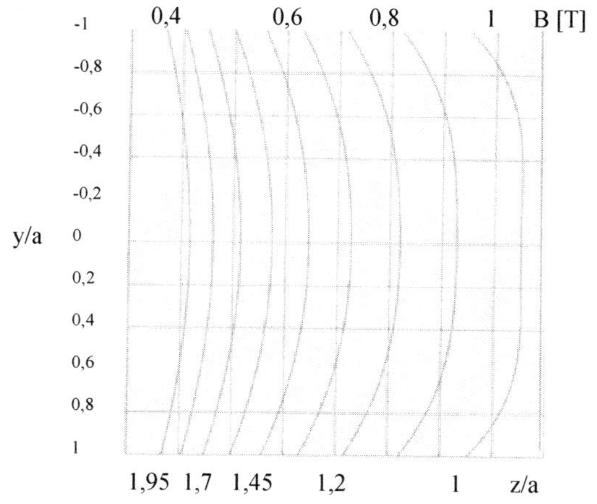

Fig.6: Variation of the magnetic induction B between pole pieces of selection magnets.

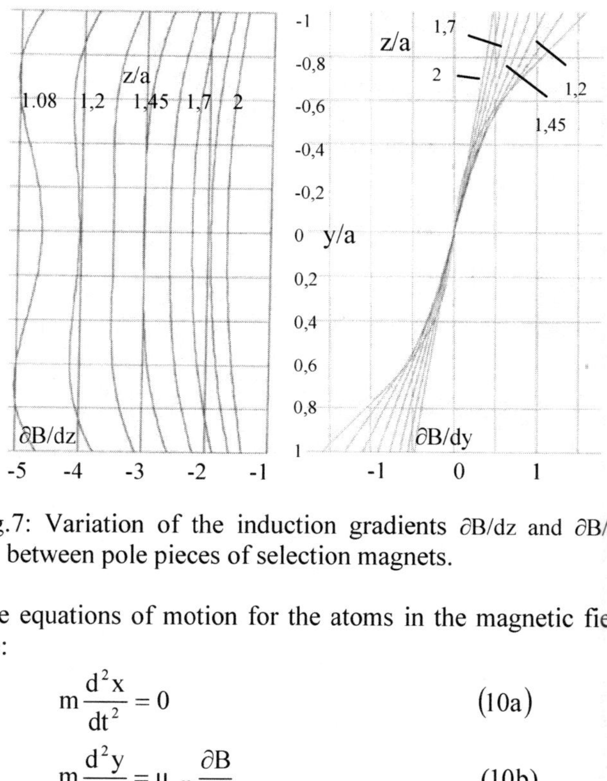

Fig.7: Variation of the induction gradients ∂B/dz and ∂B/dy between pole pieces of selection magnets.

The equations of motion for the atoms in the magnetic field are:

$$m\frac{d^2x}{dt^2}=0 \qquad (10a)$$

$$m\frac{d^2y}{dt^2}=\mu_{eff}\frac{\partial B}{dy} \qquad (10b)$$

$$m\frac{d^2z}{dt^2}=\mu_{eff}\frac{\partial B}{dz} \qquad (10c)$$

The computer program "Magnets" has been elaborated in Matlab environment to calculate:

1. the coordinates and intensity distributions of atoms on a given surface at the output of the magnet
2. the components of velocity of atoms leaving the magnet

for a given magnet geometry and magnetic induction B_0, the coordinates and intensity distributions of atoms which entry to the magnet, dimensions of the field lattice.

This program calculates the atomic trajectories for 3D field distribution inside the magnet gape and take into account variation of the induction magnitude, gradients and effective atomic moment with the position.

Fig.8 illustrates the trajectories of cesium atoms going through the gape of magnet.

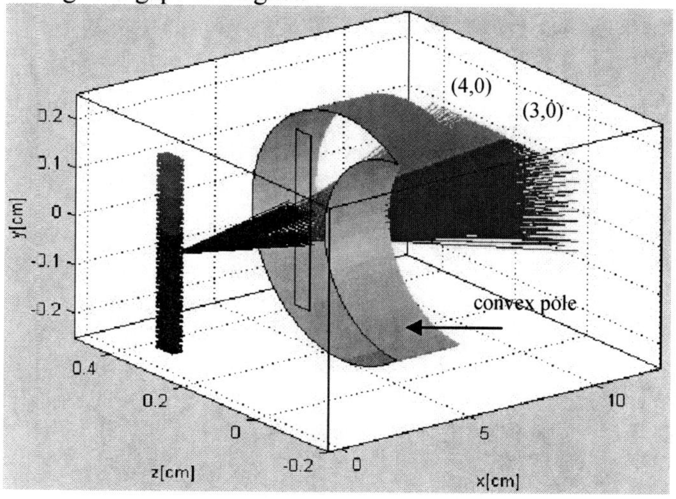

Fig.8: Example of the trajectories of cesium atoms in (4,0) and (3,0) states and velocity υ=240m/s, going through the gape of magnet of 2.5cm length and B_0=1T, 2a=4mm.

2.2 C field region

Hyperfine transition occurs in a "C field" region with small value of static magnetic induction and high homogeneity. This region is enclosed in set of magnetic shields. Several precautions are made to avoid leakage and uncontrolled magnetic fields. In our computer simulation perfectly homogeneous the C field and transverse magnetic induction with respect to the beam tube Ox axis has been assumed.

2.2 Microwave cavity

In a manufactured cesium beam tubes the Ramsey type, E-plane microwave cavity is usually used. In our simulation the following assumptions for a model of this cavity has been made:

1. inside the cavity only standing TE wave exist
2. the slots in the cavity arms for atomic beam passing through have a dimension of $1\text{x}6\text{mm}^2$.
3. the distance between the cavity arms is 10cm.

The magnitude of the microwave field components acting to atoms are:

$$H_x=0$$

$$H_y=H_{ym}\cos\frac{\pi z}{a}\sin\frac{2\pi y}{\lambda_g} \qquad (11)$$

$$H_z=-H_{ym}\frac{2a}{\lambda_g}\sin\frac{\pi z}{a}\cos\frac{2\pi y}{\lambda_g}$$

where a=2.286cm and b=1.016cm are width and height of the waveguide respectively, H_{ym} is the amplitude of H_y component, λ_g is the wavelength limit.

The Quick–Wave simulator [12] has been used in the computation of 3D field distribution inside the Ramsey cavity. This tool is based on the Finite Difference Time Domain method (FDTD). This method is one of the common numerical methods due to being very useful and effective in

simulations. The FDTD method requires discretization of time and space. So the time t is divided into *n* steps $t=n\Delta t$ and the space is divided into cells $(x, y, z) \rightarrow (i\Delta x, j\Delta y, k\Delta z)$. The physical size of the cells is dictated by the need for accuracy.
In the simulation the H_{101} mode has been set up to verify resonant frequency and to investigate the distribution of the electromagnetic fields. At the beginning a wide band of excitation frequency has been used. Through the S parameters analysis the resonant frequency has been tested and was found to be in good agreement with theoretical cesium resonant frequency f_0 Next, the resonant structure has been excited with a sinusoidal signal of the resonant frequency in order to obtain the distribution of the electric and magnetic field components. In this simulation it was needed to check if the H maximum is located exactly at the slot junction of the cesium beam. In the Fig.9 and Fig.10 the distribution of magnetic field H in the whole cavity is presented.

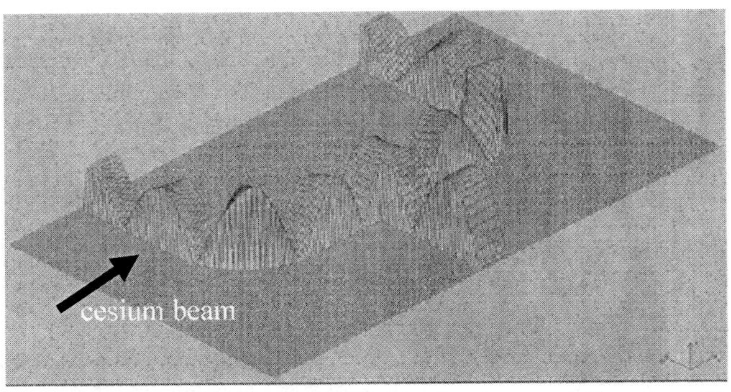

Fig.9: The distribution of the resultant magnetic strength inside the Ramsey cavity

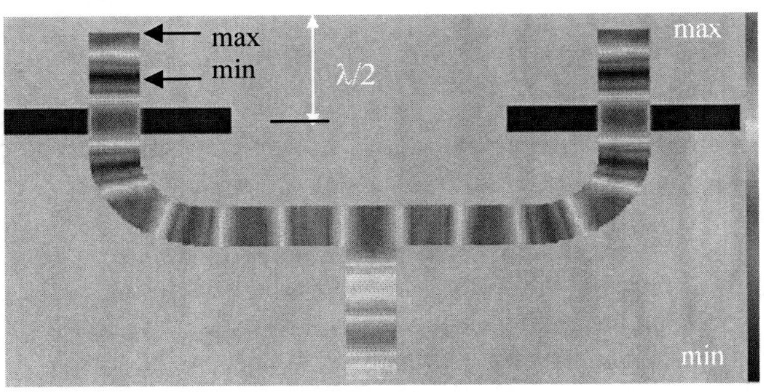

Fig.10: The distribution of the magnitude of H_z field component inside the Ramsey cavity and field shields at the entry and exit of the cavity.

The microwave magnetic field components (inside the Ramsey cavity) calculated by the Quick–Wave simulator will be used to calculate the probability of the clock transition and resonant frequency pulling effects. The computer program for this purpose is under preparation.
The probability of the clock transition between two energy levels of cesium atom can be derived from matrix density formalism and is given by equation [14]:

$$P(\upsilon, \Phi_0, \theta_0) = 1 - \frac{a_3(\tau_1, T, \tau_2)}{a_3(0)} \qquad (12)$$

where:

$a_3(\tau_1,T,\tau_2)$ describe the fractional population difference between quantum states of cesium atom at the output of the second arm of the Ramsey cavity. The quantity a_3 depends on the Rabi angular frequency in each microwave region, the durations of the atom interaction with microwaves τ_1,τ_2 and time of flight of atoms between two arms of cavity. $a_3(0)=1$ or $a_3(0)=-1$ when atoms in states (4,0) or (3,0) are detected.

2.2 Detector

The cesium beam is detected by surface ionization on a hot wire. The efficiency of ionization is determined by equation:

$$\frac{I^+}{I_d} = \frac{1}{1+2\exp[-(E_w - E_I)/kT]} \qquad (13)$$

where: I^+ is the intensity of ions, $I_d=\iint I_{mn}(y,z)dydz$ is the intensity of atoms at the ionizer surface, E_w is the work function of the ionizer, E_I is the ionization energy of cesium atom, T is the ionizer temperature, k is Boltzmann's constant.
In manufactured cesium standards an electron multiplier is often installed to have better signal to noise ratio.
In our computer model we simply compute the efficiency ionization of hot wire surface with a width of 1mm and length of 5mm.

3 Atomic beam geometry

Fig. 10 shows schematic diagram of the elementary atomic beam trajectory inside the cesium tube.

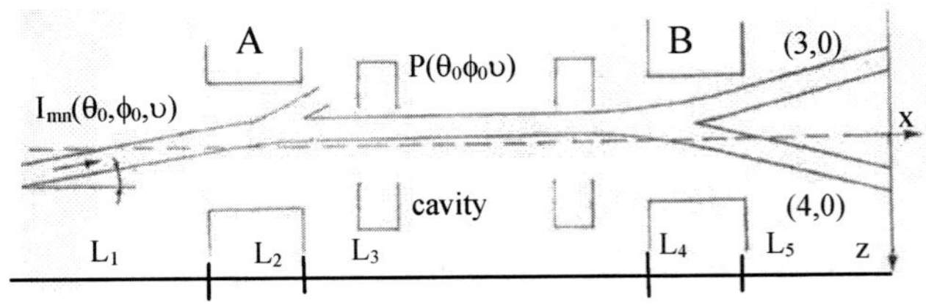

Fig.11: Elementary atomic beam trajectory

This beam has a circular cross section and leaves the source at the set of angles Φ_0 and θ_0. The intensity of this beam depends on the channel directivity diagram and the velocity of atoms in the beam. The coordinates of the beam centre z_d and $\tan\Phi_d$ on the detector plane Ozx are given by equation:

$$\begin{vmatrix} z_d \\ tg(\phi_d) \\ 1 \end{vmatrix} = \begin{vmatrix} 1 & L_5 & 0 \\ 0 & 1 & 0 \\ 0 & 0 & 1 \end{vmatrix} \cdot \begin{vmatrix} 1 & L_4 & z_B \\ 0 & 1 & tg\Phi_B \\ 0 & 0 & 1 \end{vmatrix} \cdot \begin{vmatrix} 1 & L_3 & 0 \\ 0 & 1 & 0 \\ 0 & 0 & 1 \end{vmatrix} \cdot \begin{vmatrix} 1 & L_2 & z_A \\ 0 & 1 & tg\Phi_A \\ 0 & 0 & 1 \end{vmatrix} \cdot \begin{vmatrix} 1 & L_1 & 0 \\ 0 & 1 & 0 \\ 0 & 0 & 1 \end{vmatrix} \cdot \begin{vmatrix} z_0 \\ tg(\phi_0) \\ 1 \end{vmatrix} \qquad (14)$$

where z_A, $tg\phi_A$ and z_B, $tg\Phi_B$ are the coordinates of the elementary beam at the A and B magnets outputs respectively. Similarly we can calculate y_d, $tg\theta_d$. coordinates.
The intensity distribution of atoms on the plane of detector when the microwaves are applied is

$$I_{mn} = \sum_{\upsilon}\sum_{\theta_0}\sum_{\Phi_0} I_{mn}(\upsilon) I_{mn}(\theta_0, \Phi_0) P_{mn}(\theta_0, \Phi_0, \upsilon) \quad (15)$$

where $I_{de}(\upsilon)$ is the velocity distribution of atoms which reach the detector. I_{mn} depends on microwave excitation frequency and also represents the Ramsey resonance line for the elementary beam.

The resultant intensity distribution at the detector plane is the superposition of all elementary beams:

$$I_{de}(y,z) = \Sigma I_{mn} \qquad (16)$$

where m and n are the numbers of source channels along Oz and Oy axis respectively.

Fig.12 shows the example of the intensity distribution of atoms on the plane of cavity.

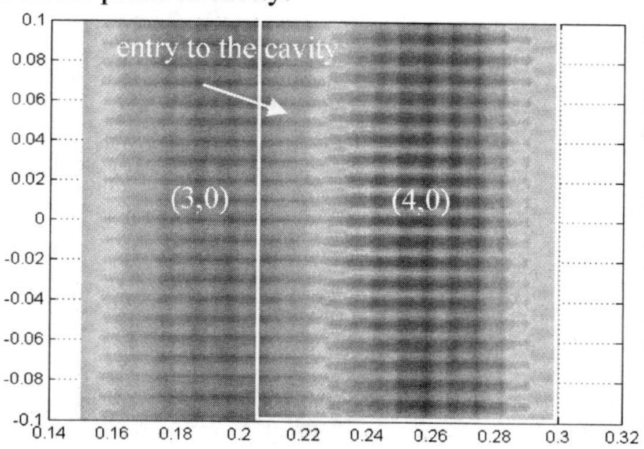

Fig.12: Intensity distribution of atoms on the plane of cavity. (50 channels with too small steps $\Delta\theta_0$ and $\Delta\Phi_0$)

Fig.13a and Fig.13b shows the examples of the intensity distribution of non excited and excited atoms on the plane of the detector at the distance 35cm from the source.

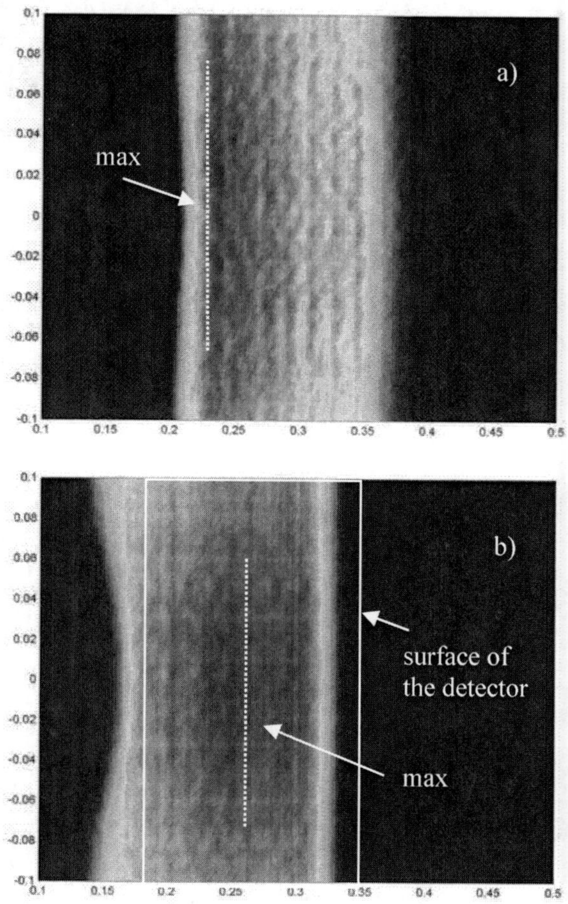

Fig.13: Intensity distribution of atoms on the plane of detector (50 channels) for a) non excited atoms b) excited atoms in the cavity

4 Conclusions

In the paper 3D computer model of conventional cesium beam tube has been described and tested. Also the method of atomic beam intensity calculation has been presented.

Preliminary simulations shows that we can accurate calculate the atoms intensity and velocity distribution on the detector plane and to set optimal position of the detector. Using our model we can also investigate the influence of the position of cesium tube components on the atomic trajectories.

We plan to realize more advanced simulations (e.g. to calculate the resonance lines and resonant frequency offsets) when the computer program "Resonator" will be finished.

References

[1] A. Chachulski, P. Gajewski, R. Herczyński, K. Radecki. "Free molecular effusion of cesium atomic beam", *Prace IPPT PAN*, vol.59, (1970) (in Polish)

[2] L. S. Cutler, C. A. Flory, R. P. Giffard, A. De Marchi. "Frequency pulling by hyperfine σ transition in cesium beam atomic frequency standards", *J. Appl. Phys.*, 69, pp.2780-2792, (1991).

[3] J. A. Giordmaine, T. C. Wang. "Molecular Beam Formation by Long Parallel Tubes", *J. Appl. Phys.*, vol.34, pp.463-471, (1960).

[4] T. Hashi, K. Kariya, H. Chyba. "Analysis, Design and Performance of cesium Beam frequency Standard", Fujitsu Sci. Tech. J., vol.20, pp.467-512, (1984).

[5]. Ikegami. "Angular Distribution Measurement of Cesium Atomic Beam from Long Tube Collimators". *Jpn. J. Appl. Phys.* vol.33, pp. 4795, (1994).

[6] B. Jaduszliwer. "A cesium beam atomic clock computer model", *IEEE Int. Frequency Control Symp.*, pp.1097-1105, (1996).

[7] B. Jaduszliwer. "A cesium beam atomic clock computer model: II. The clock signal". *Freq. Control. Symp.*, pp. 240-244, (1997).

[8] B. Jaduszliwer. "Atomic Trajectories in Compact Cesium-Beam Clocks". *IEEE Trans. On Ultasonics, Ferroelectrics and Frequency Control,* vol.37, pp.121-125, (1990).

[9] S. K. Karuza, W. A. Johnson, J. P. Hurell, M. F. Botjer, F. J. Voit, "An Automated System for Measuring Velocity Distribution in Cesium Beam Tubes". *4th EFTF, Conf. Proc.,* Neuchatel, pp.555-563, (1990).

[10] S. K. Karuza, W. A. Johnson, F. J. Voit, "The interrelatioonship of output frequency, micowave power, and C-field for cesium beam tubes from four manufacturers". *CPEM Conf. Proc.,* pp.165-166, (1992).

[11] A. De Marchi, G. P.Bava, "On cavity phase shift in commercial caesium beam frequency standards". *Metrologia*, vol. 20, pp.33-36, (1984).

[12] QuickWave-3D. "A general purpose electromagnetic simulator based on conformal finite-difference time-domain method, version 2.1"

[13] N. F. Ramsey, "Molecular beams". Oxford Univ. Press, Oxford, (1956).

[14] J. Vanier, C. Audoin, "The Quantum Physics of Atomic Frequency Standards", Adam Hilger, pp. 855-858, (1989).

DESIGN STATUS OF THE ACES MICROWAVE LINK

S. Föckersperger*, S. Bedrich*, W. Schäfer†

* Kayser-Threde GmbH, Wolfratshauser Str. 48, 81379 Munich, Germany
spacetech@kayser-threde.de, phone +49(89)72495-186, fax +49-89-72495-104

† TimeTech GmbH, Curiestr. 2, 70563 Stuttgart, Germany
wolfgang.schaefer@timetech.de

Keywords: Time & Frequency, International Space Station, High Performance Clock

Abstract

The ACES Microwave Link (MWL) is a cross-link microwave signal transfer system consisting of a Flight Segment, which shall be installed on the ACES payload aboard the International Space Station (ISS), and dedicated Ground Terminals. These Ground Terminals are intended to be installed world-wide at major time and frequency (T&F) laboratories participating in the ACES experiment programme.
The system will allow to compare the phase stability of the master clock signal generated by the ACES space clocks (PHARAO and SHM) with the most precise frequency standards on ground, operated e.g. at LPTF, NIST, PTB with unprecedented accuracy.
The ACES MWL project successfully concluded Phase C0 (preliminary design review) and is presently in Phase C1/D. The paper will report on the current status of design and performance results of the electronics breadboards.

1 Introduction

The ACES project, one of the most exciting microgravity physics experiments ever, will test a new generation of atomic clock in space. In order to compare those clocks to primary standards world-wide a specially designed Microwave Link (MWL) is used.
The MWL is using the well known 2-way principle. 2-way clock comparison via satellite (TWSTFT) is regularly performed by metrological institutes since more than 20 [1]. The ACES MWL incorporates the same basic technique, but cancels the satellite transponder and allows, therefore, better precision and less noise. In addition, it uses much higher carrier and code frequencies than the traditional TWSTFT and includes improved H/W monitoring [2].
In order to perform space-ground clock comparisons at the level of the involved clocks, the MWL must exhibit very low phase noise compatible with a clock stability of 10^{-16}.

Figure 1: ACES MWL Mission Concept

2 ACES MWL Mission Description

The ACES Microwave Link (MWL) is a two-way crosslink microwave signal transfer system consisting of:

- MWL Flight Segment, installed on the ACES payload aboard the International Space Station (ISS)
- MWL Ground Terminals installed world-wide at major time and frequency (T&F) laboratories participating in the ACES experiment programme.

The objectives of the ACES microwave link are to:

- Compare the ACES space clock ensemble (PHARAO and SHM) with high-performance ground clocks operated e.g. at LPTF, NIST, PTB
- Support evaluation and determination of short- and long-term stability of the on-board clocks
- Disseminate highly accurate on-board T&F information
- Contribute to universal time scale computation (UTC).

3 ACES MWL System Performance

During times of visibility (passage of the ISS over an ACES Ground Terminal) the MWL Flight Segment and the Ground Terminals communicate by pn-coded microwave signals (Figure2):

- Downlinks signals at Ku-band (14.703 GHz) and S-band (2.248 GHz) for ionospheric refraction correction
- Uplink signals at Ku-band (13.475 GHz).

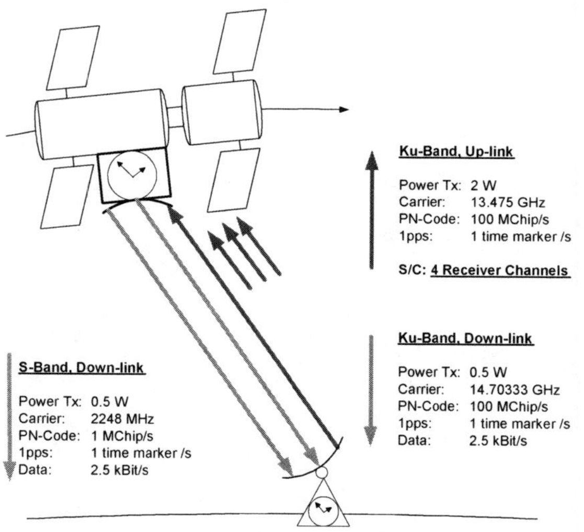

Figure 2: ACES MWL Signals

Synchronously on ground and in space the following measurements are carried out:

- Unambiguous high-precision phase measurements, performed directly on the RF carriers
- Identification of single carrier cycles through coherent pn-signal code
- Correlation of the space and ground measurements through time tagging of all data w.r.t. local clock
- Data correlation and distinct error correction.

This finally leads to the determination of the frequency stability of the space master clock w.r.t. the local primary standard on ground with the accuracy shown in Table 1.

Short Term		*Long Term*	
10s	1.30 ps	1 d	5.10 ps
300s	0.24 ps	10 d	16 ps
1000s *	0.60 ps	*: only for ground test	

Table 1: MWL Performance Specification (TDEV)

4 ACES MWL Science Data Flow

The MWL Flight Segment can communicate to up to four Ground Terminals in parallel which allows mutual ground clock phase determination. Data obtained at the Ground Terminals are transferred via Internet directly to the ACES Science and Operations Centre (ASOC) for central processing. On board data are sent via serial link to the ACES computer and forwarded through ISS communication to Columbus control centre and to ASOC subsequently. The MWL system is independent from the absolute link delay and from the location of the clocks under consideration.

5 ACES MWL Design Drivers

The major challenge for the ACES MWL system design and development process is the harsh ISS operational environment:

- Signal dynamics (high Doppler ±400 kHz , Doppler Rate ±6.5 kHz/s)
- Thermal environment (-10°C to 50 °C)
- Thermal slopes (1K/300s)
- Multipath reflection (0 to -40 dBm).

High design effort is needed caused by system specific constraints like:

- Strict mass and power requirements
- Antenna gain characteristic
- Antenna phase centre stability.

6 ACES MWL System Composition

6.1 ACES MWL Flight Segment

The Flight Segment consists of one electronic unit (about 20 boards assembled on two levels) and two antenna assemblies (Ku- & S-band). The two separately mounted turnstile antennae are covered by a radome each. The technical characteristics of the sub-assemblies are shown below.

<u>Flight Segment Electronics</u>

- Envelope: 260x220x200 mm
- Mass: 9 kg
- Power consumption: 31 W
- Transmit output power: 0.5 W (Ku,S)
- Power interface: 28 VDC
- Data interfaces: RS422 & Digital I/O
- RF interfaces: 100 MHz input, Ku Rx/Tx, S Tx
- Test interfaces: Ku&S-band monitor, 100 MHz monitor, RS 232 data I/F
- Data rate: 4 kbit/s.

Figure 3: ACES MWL Flight Segment Mock-up

Flight Segment Antennae

- Envelope: S-band 320x320x90 mm, Ku-band 130x130x90 mm
- Field of View (nadir): ± 70°
- Phase centre stability: < 0.1 mm (after calibration)
- Radome material: Tecapeek GF30
- Antenna unit mass: ca. 700g.

Figure 4: ACES MWL S-band Antenna Assembly

6.2 ACES MWL Ground Terminal

The Ground Terminal consists of an offset dish antenna with Ku- and S-band front-ends (helical feeds), RF electronics unit, both mounted onto an Azimuth/Elevation steering unit, a support electronics unit, support structure & cables and a laptop for system monitoring. As an option a meteorological station can be added.

Ground Terminal Characteristics

- Envelope: 1200x700x1400 mm
- Mass: ca. 100 kg
- Power consumption: ca. 200 W (w/o PC)
- Tx output power: 2 W
- Power interface: 220 VAC
- RF interfaces: 100MHz input, Ku Rx/Tx, S Rx.
- Data interface: Ethernet (TCP/IP)
- Field of View: ± 90° (entire hemisphere)
- Offset antenna dish: ca. 0.4m
- Pointing accuracy: ± 0.5°

Figure 5: Preliminary Set-up of ACES MWL Ground Terminal

6.3 Ground Terminal Key User Requirements

In the following table the key user requirements of the ACES MWL Ground Terminal are summarized.

Interface
High precision clock must be available (primary standard)
Licenses
User has to apply for a local RF transmission license (13.475GHz)
Signal Interference
The site shall be free from interference at receive frequencies (14.7 and 2.25 GHz)
Antenna Field of View
User has to take care that no obstacles are in the Field-of-View of the antenna
Segment masking of the antenna pattern is part of the Ground Terminal SW

Table 2: Key User Requirements

7 Project Implementation

7.1 Current Development Status

The ACES MWL project successfully concluded the initial design and development phases, as well as breadboard testing. It is now in the detailed design and development phase, which results in an EM board design. Special emphasis is laid to evaluate key implementation aspects, including:

- Detailed link budget
- Removal of 1st order Doppler effect by use of the TWSTFT principle
- Thorough error assessment, including signal path-induced and orbit-induced effects
- Determination of the ionospheric delay and ionospheric delay variations
- Sensitivity to thermal variations and thermal slopes
- Development of a suitable signal coding and modulation scheme with CDMA capability
- Minimisation of Multipath induced errors
- Strictly synchronous measurement of Code- and Carrier phase.

A block diagram of the Flight Segment is presented in Figure 6. Crucial hardware elements have been developed and tested, including:

- Phase-stable amplifiers
- Phase measurement sub-system
- Antenna pattern and phase centre.

Other design activities include:

- Detailed mechanical design
- Thermal model
- Mechanical model
- Parts analysis and parts selection.

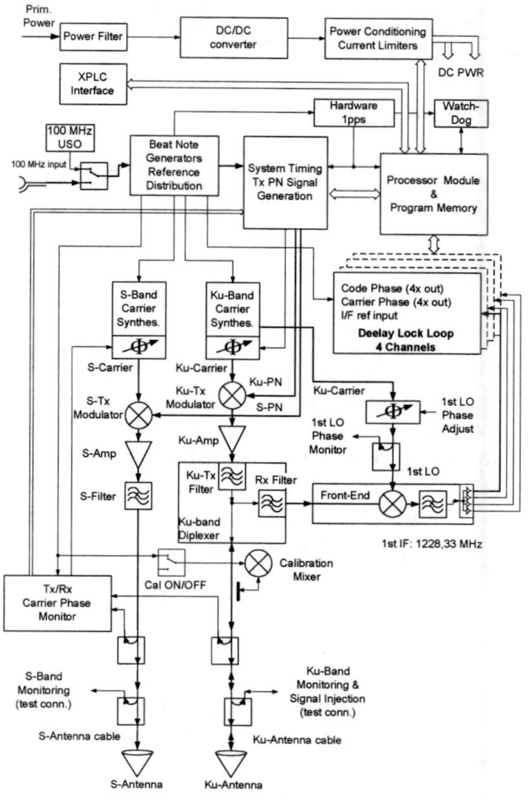

Figure 6: Block Diagram of the ACES MWL Flight Segment

7.2 Phase Measurements

The ACES MWL relies on 2-way phase measurements at Ku-band, using code- and carrier phase data. The ground terminal determines in addition the differential delay between the received S-band and Ku-band signals as well as the differential Doppler between S-band and Ku-Band carriers. The performance of the phase measurement sub-system (breadboard results) is as follows:

- Carrier, Phase Ku-Band: 7 fs
- Code Phase, Ku-Band: 380 fs
- Carrier Phase, S-Band: 14 fs
- Code Phase, S-band 21 ps.

Time tagging of measurements:

- Real-time 50 ns
- After post-processing 5 ns

(all values precision, 1-sigma, for $\tau = 1$s, excluding thermal noise).

7.3 Verification Approach

The ACES MWL shall be verified prior to launch to demonstrate, meeting all performance requirements and operating under the harsh space environment.
The orbit-induced signal variations are verified with a programmable signal generator capable to simulate:

- Distance: 0..2000 km (Time Delay)
- Relative Velocity: 0..7 km/s (Doppler)
- Acceleration according to ISS orbit
- Orbit-induced amplitude variations
- Interference (narrow- and wide-band).

The verification includes:

- Short term stability (1 pass, max 300s)
- Medium term stability (up to one day)
- 10 day stability
- Stability of built-in delay calibration
- Thermal sensitivity
- Performance under thermal vacuum conditions
- Reception of simulated Doppler-shifted signals including removal of 1st order Doppler effect
- Reception of multiple ground signals
- Differential delay between Ground Terminals
- Environmental effects on Ground Terminals
- Ground Terminal Delay Monitoring.

7.4 ACES MWL Key Milestones

The following key milestones are foreseen for the ACES MWL project:

- Critical Design Review 09/05
- Delivery of Flight Segment EM 10/05
- Delivery of Ground Terminal 1 10/05
- Final Acceptance Review 12/06
- Delivery of Flight Hardware 02/07.

8 Conclusion

The ACES Microwave Link (MWL) is a cross-link microwave signal transfer system consisting of a flight segment, which shall be installed on the ACES payload aboard the International Space Station (ISS), and dedicated MWL Ground Terminals. The system will allow to compare the phase stability of the master clock signal generated by the ACES space clocks (PHARAO and SHM) with the most precise frequency standards on ground, operated e.g. at LPTF, NIST, PTB with unprecedented accuracy. The project has passed System PDR successfully. Critical hardware elements have been developed and tested. The detailed design phase (Phase C1) was started beginning of December 2003 and is well under way. Assuming final commitments of all involved parties within the next months, manufacturing of the engineering model will start shortly.

Acknowledgements

The work is funded by the European Space Agency (ESA).

References

[1] "Microwave Link for ACES, Phase A Study, Final Report", ESTEC Contract 13671/99/NL/JS, (Dec. 1999).

[2] Bedrich, S., W. Schäfer. "ACES Crosslink - A New Method for Ultra-Accurate Time and Frequency Synchronization of Satcom Terminals", *Space Technology* **22 (1)**, pp. 1-8, (2002).

ACES:
A TIME AND FREQUENCY MISSION FOR THE INTERNATIONAL SPACE STATION

S. Feltham[1], G. Gianfiglio[2] and F. Reina[3]

European Space Agency,
ESTEC, 2200 AG Noordwijk, The Netherlands
Contact: Stephen.Feltham@esa.int
Ref: MSM-GF/306.2004/sf

Keywords: ACES, PHARAO, SHM, ISS, ESA.

1 Abstract

ACES (Atomic Clock Ensemble in Space) will test a new generation of micro-gravity or cold atom Caesium clock with frequency stability in the range 10^{-16} to 10^{-17}. This clock will be compared to a Space Hydrogen Maser in orbit. Both clocks will also be compared to primary ground standards world-wide with unprecedented accuracy via a specially designed Micro-Wave Link (MWL). The ACES payload has recently undergone its Preliminary Design Review (PDR) and is preparing for the Mission PDR. A description of the ACES mission is given and status of the ACES payload presented.

2 Introduction

The fact that time can be measured very precisely, far better than any other physical parameter is a technological asset of great importance. This statement defines the essence of the ACES mission. One of the most exciting microgravity physics experiments ever conceived, the ACES project is accredited with the highest scientific merit. ACES will test a new generation of atomic clock in space. The manipulation of cold atoms under microgravity conditions allows longer interaction between the atoms and the microwave radiation because the perturbation from the Earth's gravitational force is removed. The microgravity or Caesium cold atom clock PHARAO and the SHM will be characterised and compared to each other and to national frequency standards world wide using a specially developed MWL. The ultimate performance of PHARAO in a microgravity environment will be explored and a number of fundamental physics experiments will be performed.

3 ACES Organisation

ACES is an ESA Mission. An industrial consortium led by EADS Friedrichshafen[4] (D) is developing the payload. EADS is responsible for the design, development, integration and verification of the payload. Kayser-Threde[5] and TIME TECH[6] (D) are developing the MWL and Frequency Comparison and Distribution Package (FCDP) under EADS sub-contract.

The two atomic clocks (or instruments) are national developments. The French National Space Agency (CNES) is responsible for the funding and development of PHARAO. The SHM is sponsored by Switzerland via the ESA PRODEX Program and developed by the Observatory of Neuchâtel (ON). The ACES Principal Investigator[7] (PI) defines the scientific objectives and mission goals supported by the ACES Project Scientist[8]. The Instrument Investigators[9,10] oversee the development of their specific instruments under the guidance of the Instrument Principal Investigators.[11,12,13]

4 Main Scientific Objectives

A number of important experiments will be performed in several scientific domains including a demonstration of high precision time and frequency transfer, atmospheric propagation, high precision geodesy and global network synchronisation. The fundamental physics experiments are briefly described below.

4.1 Gravitational Frequency Shift (Einstein effect)

A source of radiation in a gravitational potential appears to an observer in a different gravitational potential to be shifted in frequency. The Einstein effect will be determined with a relative uncertainty of 3 10^{-6} representing a factor 25 improvement.

4.2 Time Variation of the Fine Structure Constant

The fine structure constant 'alpha' characterises the strength of the electromagnetic interaction. According to Einstein's equivalence principle, fundamental constants should not vary with time. However some modern theories predict the existence of new interactions that violate Einstein's equivalence principle. The frequencies from clocks using different elements as a function of time will be compared via ACES. An improvement of a factor 100 / year is anticipated.

4.3 Test of Special Relativity

A number of theories predict violations of special relativity not yet observed. The ACES experiment is expected to improve previous observation limits by about one order of magnitude. The experiment will compare the space and ground clocks continuously during several passages of ISS.

5 The ACES Mission Scenario

The ACES mission consists of a number of elements as out lined below. The payload will be launched with a Space Shuttle using the Lightweight Integrated Cargo Carrier (ICC-Lite). The on-orbit transfer to the Columbus External Payload Facilities (CEPF), Nadir location, will be performed using the ISS robotic systems i.e. the SSRMS (Space Station Robotic Manipulator System) provided with the SPDM (Special Purpose Dexterous Manipulator). An artist's impression of Columbus and ACES location on the CEPF is shown in fig. 1.

Figure 1: The Columbus Laboratory External Payload Facility showing the ACES location.

The End-to-End Mission Operations and Ground Segment (GS) consist of the NASA GS, Columbus GS including the Columbus Control Centre (COL-CC) and the Facility Responsible Centre (FRC) installed at CNES Toulouse. The Interconnection Ground Sub-network (IGS) links the FRC to the COL-CC.

6 The ACES Payload

ACES is a complex payload. Both clocks are extremely sensitive to their operating environment. In particular thermal and magnetic sensitivity place severe design constraints on the payload. A sketch of the ACES payload is shown in fig.2 The ACES Payload consists of the following key elements:

- The cold Caesium atom clock (PHARAO)
- The Space Hydrogen Maser (SHM).
- The ESA developed subsystems:
  - *Micro Wave Link (MWL) for T&F dissemination.*
  - *Frequency Comparison & Distribution Package (FCDP) for phase noise comparison and signal distribution.*
  - *eXternal PayLoad Computer (XPLC) for command, control and data handling.*
  - *Power Distribution Unit (PDU)*
  - *Dedicated mission operation software,*
  - *Mechanical and thermal subsystems.*
- The ESA-provided Columbus External Payload Adapter (CEPA), developed by NASA/Boeing, the interface between the payload and mounting platforms.

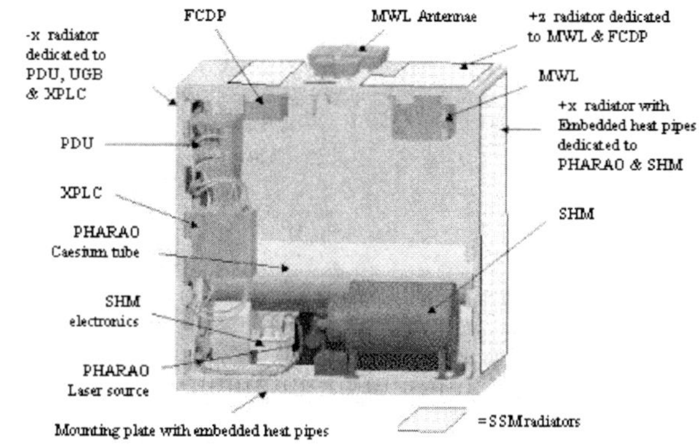

Figure 2: The ACES Payload (without CEPA). Volume ~ 1m^3, mass 225 kg.

Mechanical and thermal analyses were performed as part of the baseline design to ensure the design was capable of surviving the environment whilst maintaining satisfactory performance over the mission duration.

6.1 The Instruments

6.1.1 PHARAO

Projet d'Horloge Atomique par Refroidissement d'Atomes en Orbite (PHARAO) is a cold atom Caesium device with six elements as shown in fig. 3.

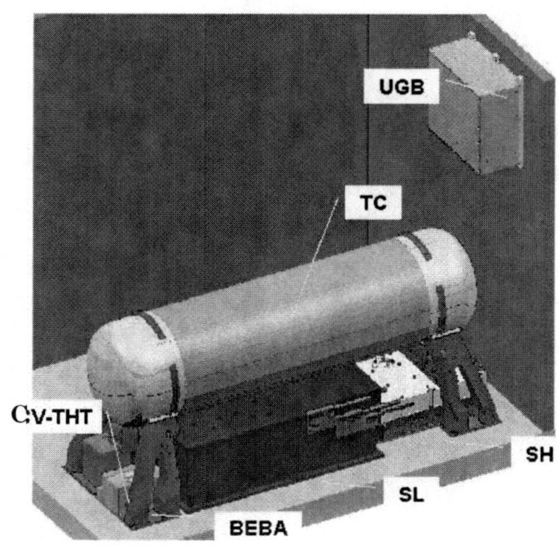

Figure 3: The Component parts of PHARAO

The Cesium Tube (TC) in which atoms are captured, launched and cooled. The Laser Source (SL) provides the six laser beams necessary for capture, launch, cool down, atomic selection and detection of atoms.

The Microwave Source (SH) supplies the signal to drive the cavity interaction. The Electronic Command and Control System (UGB) processes the detection signal to command SH frequency corrections. The BEBA electronic unit controls the Caesium Tube magnetic coils and acquires the Caesium Tube analog signal. The power converter (CV-THT) drives the ion pumps.

6.1.2 SHM

The Space Hydrogen Maser (SHM) instrument is a miniaturised 35 kg active Hydrogen Maser. It consists of two parts: the Physics Unit (PU) and the Electronics Unit (EU), see fig. 4.
The heart of the maser is an atomic storage bulb made from a single a crystal of sapphire. The bulb dielectrically loads the microwave cavity to reduce the overall dimensions. Automatic cavity tuning is required to compensate for the increased temperature sensitivity.
The SHM possesses extremely good medium term frequency stability. It will be frequency-locked to PHARAO for long-term accuracy and T&F distribution.

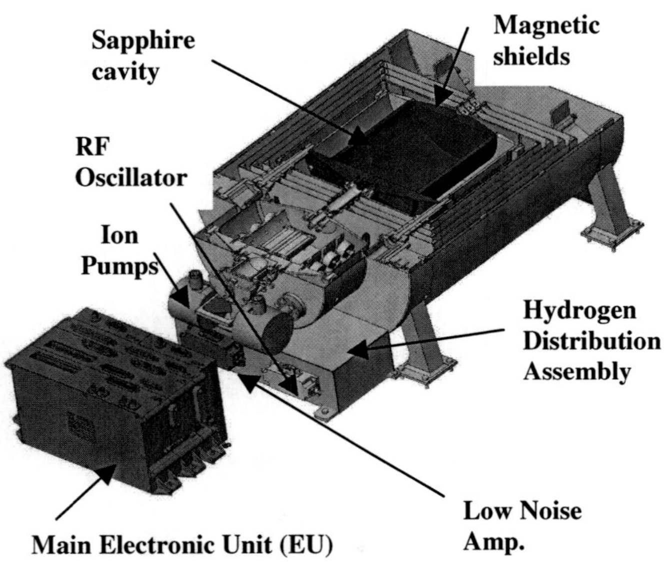

Figure 4: A Cross Section of the SHM

6.2 The Payload Structure and Thermal Control

The ACES structure is made from Aluminium sandwich honeycomb. The base plate contains 8 embedded "L-shaped" heat pipes for thermal control of PHARAO and SHM. The side panels provide the thermal rejection sink for the sub-system units via an internal cavity radiative coupling and external Secondary Surface Mirror (SSM) act as radiators to deep space. The design allows easy access to the internal payload volume for hardware integration. Areas other than the radiators are covered with Multi Layer Insulation (MLI). The XPLC monitors the system temperature via an array of sensors and maintains the equipments/instruments within their safe operational ranges using heater pads. The parameters of the thermal control loop can be reconfigured by telecommand. PHARAO and SHM also have their own internal thermal control.

6.3 Electrical architecture

The Power Distribution Unit (PDU) provides the power conversion, conditioning, switching and protection functions for ACES during transportation and operation. For ground storage and transportation the PHARAO and SHM Ion pumps are operated via ground support equipment using a special connector.

6.4 Data Handling Architecture

A MIL 1553 Bus is used to transfer caution and warning data and an Ethernet is used to transfer scientific telemetry data to the Columbus Data Management System (DMS) at a max rate of 50 kbits per second and for reception of commanding. RS422 interfaces are used for ACES internal transfer of standard housekeeping and scientific telemetry data. Each instrument and subsystem is interfaced by its dedicated RS422 interface with the XPLC. A mass memory unit allows the storage of up to 50 Mbytes of data locally. ISS High data rate tape Recorders provide back up in case of a data flow outage. A Schematic of the ACES layout is shown in Fig. 5.

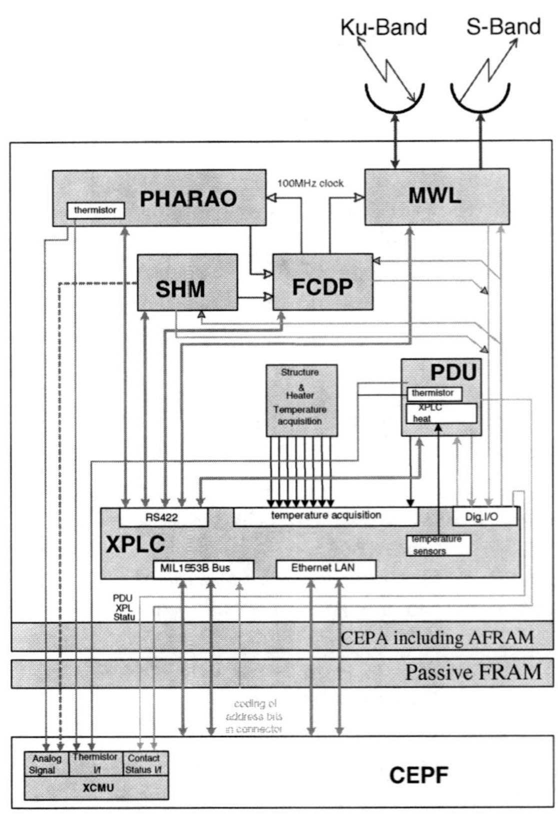

Figure 5: ACES payload Communication Block Diagram

6.5 FCDP

The ACES mission objectives are mainly based on clock comparisons, either between PHARAO and SHM or between ACES and ground clocks via the MWL. The FCDP is the central node ensuring the distribution of 100 MHz clock reference signals. The phase comparison function is the main operating mode of the FCDP and allows a mutual steering of the two atomic clocks. A block diagram of the RF signal circuitry is shown in fig. 6.
A short-term servo-loop allows the PHARAO local oscillator to be locked to the SHM thus ensuring a better mid-term stability. The long-term servo loop allows steering of the SHM drift by the PHARAO Caesium resonator thus ensuring the long-term stability and accuracy of the ACES signal.

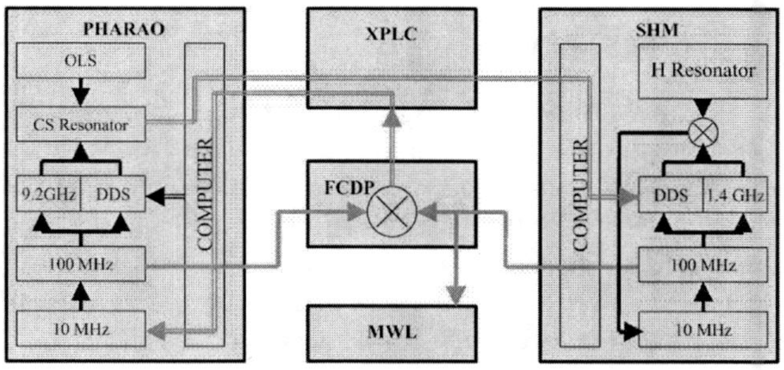

Figure 6: RF circuit Block diagram

6.6 MWL

The MicroWave Link (MWL) provides the means for accurate comparison of the on-board clock ensemble with ground-based clocks. In order to perform space-ground clock comparisons, the MWL must exhibit very low phase noise compatible with a clock stability of 10^{-16}. Two characteristic times are taken into account:

- The mean duration of an ISS pass over a ground user terminal where the T&F link must exhibit rms time fluctuations below 0.2 ps over 300 s.
- The minimal dead time between two successive comparison sessions will be 90 min. (ISS orbital period). Larger dead times (as large as 1 day) may occur requiring a T&F link with a very good long term time stability, better than 6 ps over 1 day.

A time comparison system using the 2-way principle is used in order to cancel to first order a number of the errors. Key design perimeters take into account error modes such as ionosphere, troposphere, multi-path, Doppler rejection etc. A conceptual drawing of the MWL system is shown in fig. 7. Common view comparison with up to three simultaneous ground clocks can be performed. Three simultaneous users can be accommodated. When two ground stations are not in common view, the on-board clock stability determines the system capability to 'transport' the time from one ground location to the other.

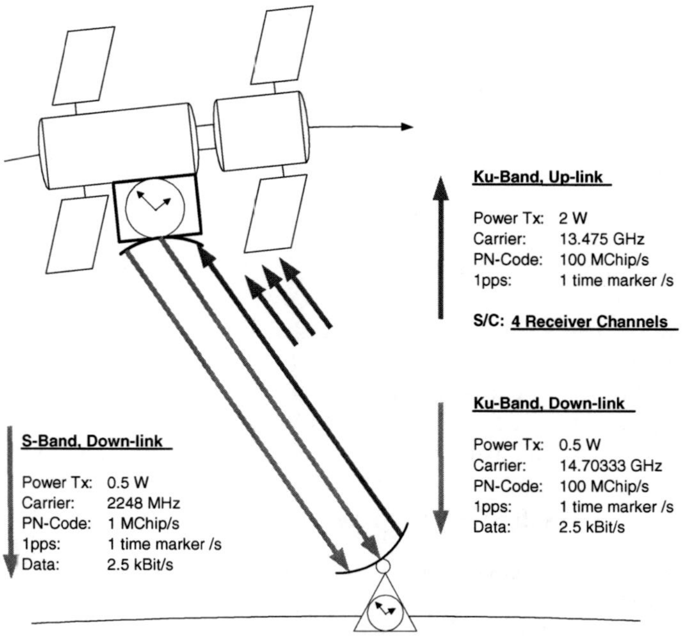

Figure 7: The MWL Concept

6.7 CEPA

The Columbus External Payload Adapter (CEPA) is an integral part of the ACES payload. It uses an Active Flight Releasable Attachment Mechanical (AFRAM) to attach the payload to sites fitted with passive FRAM interface such as the CEPF. The CEPA also provides the robotic interface for on-orbit transportation and manipulation. The CEPA is shown in fig. 8.

Figure 8: The Columbus External Payload Adapter.

7 PA and Safety

For human space, safety is of paramount importance. Product Assurance (PA) and safety plans are implemented according to ESA and NASA standards. All safety-related activity in the first instance is under ESA responsibility.

Formal Safety Reviews are performed at ACES payload level. The ISS Safety Authority, the Payload Safety Review Panel and Ground Safety Review Panel, however, provide final safety approval of ACES System Design and Operations. As part of the Payload PDR, the ACES payload successfully underwent the level 0/1 Safety Review

8 Mission Operations

The nominal duration of the ACES mission is 18 months. Following in-orbit commissioning two operational scenarios are planned. The first part of the ACES mission will last 6 months and will be dedicated to the characterisation and evaluation of the clocks and T&F link. This will be followed by an utilisation phase that will last 12 to 30 months, depending on the duration of the mission.

9 ACES Development History and Status

A mission feasibility study (Phase A) was made by CNES in 1997 and a 'peer evaluation' was performed. The 'peer group' accredited ACES with the highest scientific merit. The Manned Space Program Board unanimously endorsed the peer group recommendation. Matra Marconi Space (MMS, now EADS-Astrium-Space) performed an industrial feasibility phase (phase A/B) from 9/98–3/99. A preliminary payload layout was proposed and system resources allocated. A consolidation phase (Phase B2) was performed from 7/99 to 2/00. The ACES payload design was consolidated. The general feasibility of ACES was confirmed. A bridging phase spanned the period up to placement of the flight equipment procurement contract (Phase C/D). During this period the accommodation of ACES on the CEPF was confirmed.

10 Current Development Status

The industrial contract for the development, manufacture, integration and verification was signed in July 2003. Phase C0 established a baseline design and was concluded by a successful payload PDR in Feb 04. Phase C1will complete the detailed design and addresses the manufacture, integration and verification of the sub-system Engineering Model (EM) units and the EM payload. The EM units and payload will undergo full EMC testing. The payload Critical Design Review (CDR) will conclude phase C1.

Phase D will focus on the manufacture, integration and verification of the Flight Model (FM). The FM sub-systems & the payload will be submitted to functional, performance, electrical and environmental tests as necessary to finally demonstrate the health of the design, its ability to fulfil the mission requirements and the workmanship quality. Phase D will be completed by a Flight Model Acceptance Review (FAR). Both the EM and FM payloads will be subject to scientific verification tests under ACES PI responsibility.

11 Key Milestones

The following key milestones are foreseen for the mission:

- Payload SRR Dec 2002
- Safety 0/1 Oct 2003
- Payload PDR Feb 2004
- Mission PDR Jun 2004
- Safety level 2 Sep 2005
- Payload CDR Oct 2005
- Mission CDR mid 2006
- Safety level 3 end 2006
- Payload FAR end 2006
- Mission FAR mid 2007
- Delivery to KSC end 2007
- Ready for launch mid 2008

12 Conclusion

The ACES mission is one of the most exciting microgravity physics experiments ever conceived and is accredited with the highest scientific merit. The industrial development contract for the payload Phase C/D leading to EM and FM manufacture is placed. The Payload PDR was recently completed and preparations are now underway for the mission PDR, which will be held in June this year.

13 Acknowledgements

The contributions to ACES from the engineers at ESTEC, CNES, ON, EADS, sub-contractors and all the scientific and technical institutes are sincerely acknowledged.

14 Contacts

1 **S.Feltham,** ACES Payload Manager, ESA/ESTEC, Keplerlaan, 2200AG Noordwijk, Holland, stephen.feltham@esa.int

2 **G. Gianfiglio**, Head of ISS External Payloads Section, ESA/ESTEC, Keplerlaan, 2200AG Noordwijk, Holland, giacinto.gianfiglio@esa.int

3 **F. Reina,** ACES Support System Engineer, Vitrociset, ESTEC, Keplerlaan, 2200AG Noordwijk, Holland, Francisco.Reina@esa.int

4 **J. Winter,** ACES Project Manager, EADS (D) josef.winter@space.eads.net

5 **S. Foeckersperger**, MWL & FCDP Project Manager, Kayser-Threde (D), Stefan.Foeckersperger@kayser-threde.de

6 **W. Schaefer**, MWL & FCDP Design, TIMETECH (D) wolfgang.schaefer@timetech.de

7 **Ch. Salomon**, ACES Principle Investigator, Christophe.Salomon@physique.ens.fr

8 **N. Dimarcq**, ACES Project Scientist, noel.dimarcq@obspm.fr

9 **C. Sirmain,** PHARAO Project Manager, CNES, sirmain@cnes.fr

10 **W. Faedo ;** SHM Project Manager, Walter.Faedo@ne.ch

11 **Ch. Salomon**, PHARAO co Principle Investigator, Christophe.Salomon@physique.ens.fr

12 **A, Clarion**, PHARAO co Principle Investigator, andre.clairon@obspm.fr

13 **A. Jornod,** SHM Principle Investigator, Alain.Jornod@NE.CH

Human Spaceflight

SPACE FOR LIFE

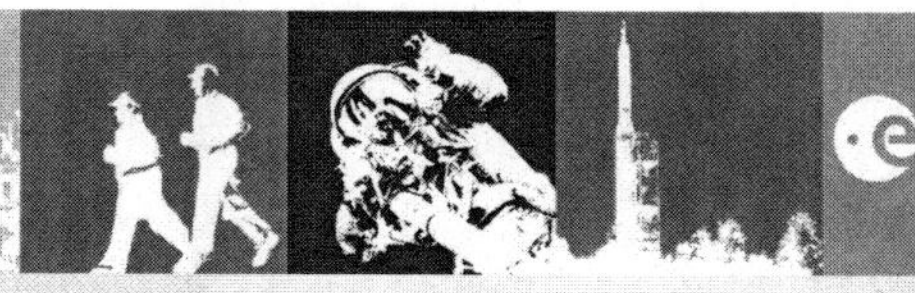

A Time & Frequency Mission for the International Space Station

S. Feltham, G. Gianfiglio and F. Reina

ACES
A Time & Frequency Mission for the International Space Station

S. Feltham, G. Gianfiglio and F. Reina[1]
European Space Agency
[1]Vitrociset
ESTEC, 2200 AG Noordwijk, The Netherlands
Contact: stephen.feltham@esa.int

ABSTRACT

ACES (Atomic Clock Ensemble in Space) will test a new generation of micro-gravity or cold atom Caesium clock with frequency stability in the range 10^{-16} to 10^{-17}. This clock will be compared to a Space Hydrogen Maser in orbit. Both clocks will also be compared to primary ground standards world-wide with unprecedented accuracy via a specially designed MicroWave Link. The ACES payload has recently undergone its Preliminary Design Review (PDR) and is preparing for the Mission PDR. A description of the ACES payload is given and latest status of the ACES payload presented.

ACES — Time Flies on The International Space Station

ACES

A Atomic
C Clock
E Ensemble in
S Space

Objectives

- Characterisation and comparison of atomic clocks in space.
- Exploration of their ultimate performance in microgravity.
- Demonstration of high precision T&F transfer.

Peer Evaluation

One of the most exciting microgravity physics experiments ever conceived.
The ACES project has the highest scientific merit

Applications

- Fundamental physics.
- Global clock comparison
- Conceptual demonstration of cold atom devices
- Global network synchronisation
- Advanced clock technology, eg future navigation applications

ACES Organisation

ACES is an ESA Mission. The payload is being developed by an industrial consortium led by EADS Friedrichshafen (D). EADS is responsible for the design, development, integration and verification of the payload. The MicroWave Link (MWL) and Frequency Comparison and Distribution Package (FCDP) are being developed under EADS sub-contract by Kayser-Threde and TIME TECH (D).
The two atomic clocks (or instruments) are national developments. The French National Space Agency (CNES) is responsible for the funding and development of PHARAO. The SHM is sponsored by Switzerland via the ESA PRODEX Program and developed by the Observatory of Neuchâtel (ON). The Scientific objectives and mission goals are defined by the ACES Principal Investigator (C. Salomon) supported by the ACES Project Scientist (N. Dimarcq). The Instrument Investigators oversee the development of their specific instruments under the guidance of the Instrument Principal Investigators.

Mission Scenario

The ACES payload will be launched with a Space Shuttle using the Lightweight Integrated Cargo Carrier (ICC-Lite). The on-orbit transfer to the Columbus External Payload Facilities (CEPF), Nadir location, will be performed using the ISS robotic systems i.e. the SSRMS (Space Station Robotic Manipulator System) provided with the SPDM (Special Purpose Dexterous Manipulator).
The ACES mission consists of the following key elements:

- The ACES Payload
- The Launcher/Carrier and the ISS On-orbit Transportation
- The ISS/Columbus with the Columbus External Payload Facility
- The End-to-End Mission Operations and Ground Segment (GS) consisting of the NASA GS, Columbus GS including the Columbus Control Centre (COL-CC) and the Facility Responsible Centre (FRC) installed at CNES Toulouse

The Interconnection Ground Sub-network links the FRC to the COL-CC.

The Instruments: PHARAO

Projet d'Horloge Atomique par Refroidissement d'Atomes en Orbite (PHARAO) is a cold atom Caesium device with six sub-systems.
The Caesium Tube (TC) in which atoms are captured, launched and cooled.
The Laser Source (SL) provides the six laser beams necessary for capture, launch, cool down, atomic selection and detection of atoms.
The Microwave Source (SH) supplies the signal to drive the cavity interaction.
The Electronic Command and Control System (UGB) processes the detection signal to command SH frequency corrections.
Ion pump power converter (CV-THT)

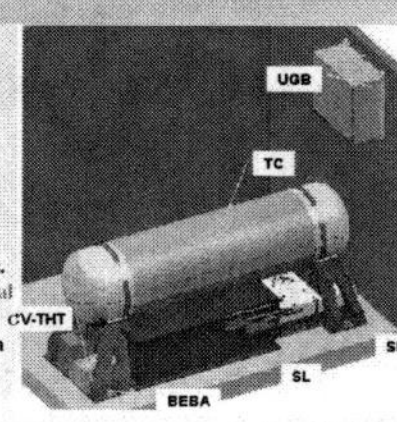

The BEBA electronic unit controls the Caesium Tube magnetic coils and acquires the Caesium Tube analog signal.

The Instruments: SHM

The Space Hydrogen Maser (SHM) instrument is a miniaturised 35 kg active Hydrogen Maser. It consists of two parts: The Physics Unit (PU) and the Electronics Unit (EU). The heart of the maser is an atomic storage bulb made from a single a crystal of sapphire. The bulb dielectrically loads the microwave cavity to reduce the overall dimensions. Automatic cavity tuning is required to compensate for the increased temperature sensitivity.

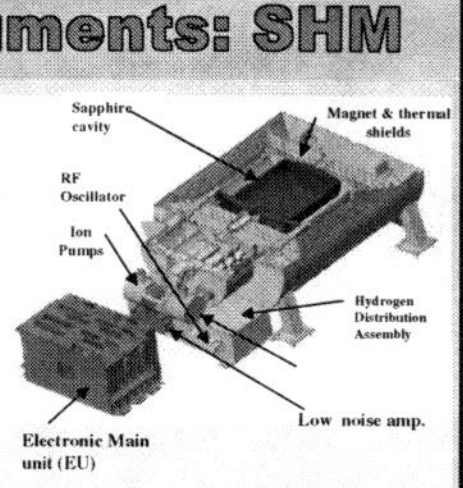

The SHM possesses extremely good medium term frequency stability. It will be frequency-locked to PHARAO for long-term accuracy and Time & Frequency (T&F) distribution.

The ACES Payload

The ACES Payload consists of the following key elements:

- The ESA developed subsystems:
  - Micro Wave Link (MWL) for T&F dissemination
  - Frequency Comparison & Distribution Package (FCDP) for phase noise comparison and signal distribution
  - eXternal PayLoad Computer (XPLC) for command, control and data handling
  - Power Distribution Unit (PDU)
  - Dedicated mission operation software
  - Mechanical and thermal subsystems
- The ESA provided Columbus External Payload Adapter (CEPA), developed by NASA/Boeing, the interface between the payload and mounting platforms
- The cold Caesium atom clock (PHARAO) instrument
- The Space Hydrogen Maser (SHM) instrument

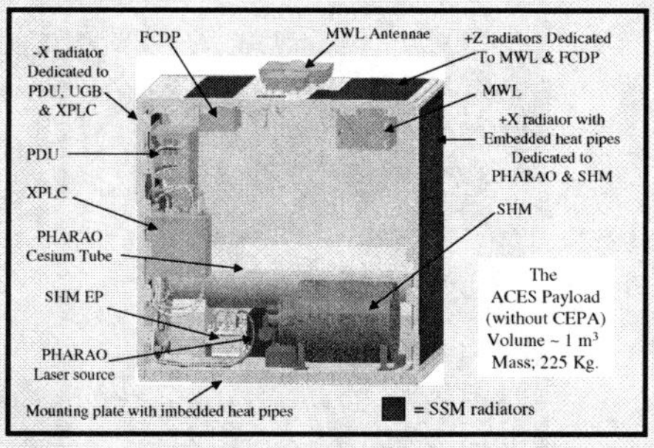

The ACES Payload (without CEPA) Volume ~ 1 m^3 Mass; 225 Kg.

Structure & Thermal Control

The ACES structure is made from Aluminium sandwich honeycomb. The base plate contains 8 embedded "L-shaped" heat pipes for thermal control of PHARAO and SHM. The side panels provide the thermal rejection sink for the sub-system units via an internal cavity radiative coupling and external Secondary Surface Mirror (SSM) act as a radiators to deep space. The design allows easy access to the internal payload volume for hardware integration. Areas other than the radiators are covered with Multi Layer Insulation (MLI). The XPLC monitors the system temperature via an array of sensors and maintains the equipments/instruments within their safe operational ranges using heater pads. The parameters of the thermal control loop can be reconfigured by telecommand. PHARAO and SHM also have their own internal thermal control.

Electrical & Data Handling

Electrical architecture

The Power Distribution Unit (PDU) provides the power conversion, switching and protection functions of ACES during transportation and operation. For ground storage and transportation the PHARAO and SHM Ion pumps are operated via a special connector.

Data Handling Architecture

A MIL 1553 Bus is used to transfer caution and warning data and an Ethernet is used to transfer scientific telemetry data to the Columbus at a max rate of 50 kbits per second and for reception of commanding. RS422 interfaces are used for ACES internal transfer of standard housekeeping and scientific telemetry data. Each instrument and subsystem is interfaced by its dedicated RS422 interface with the XPLC. A mass memory unit allows the storage of up to 50 Mbytes of data locally. ISS High data rate tape recorders provide back up in case of a data flow outage.

Freq. Control and Distrubution Package

FCDP

The ACES mission objectives are mainly based on clock comparisons, either between PHARAO and SHM or between ACES and ground clocks via the MWL. The FCDP is the central node ensuring the distribution of 100 MHz clock reference signals towards the MWL. The phase comparison function is the main operating mode of the FCDP and allows a mutual steering of the two atomic clocks. A short term servo-loop allows the PHARAO local oscillator to be locked to the SHM thus ensuring a better mid-term stability.

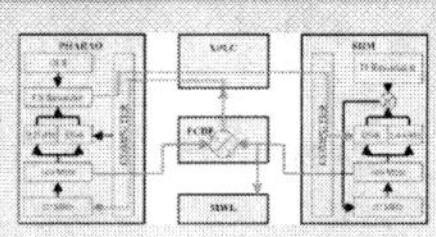

Block Diagram of the 100 MHz RF signals

The long term servo-loop allows to steer the SHM drift thanks to the PHARAO Caesium resonator thus ensuring the long term stability and accuracy of the ACES signal.

MicroWave Link

The MicroWave Link (MWL) provides the means for accurate comparison of the on-board clock ensemble with ground-based clocks. A time comparison system using the 2-way principle is used in order to cancel to first order a considerable number of errors.
Common view comparison with 3 simultaneous users can be performed.
When two ground stations are not in common view, the on-board clock stability determines the system capability to 'transport' the time from one ground location to the other.

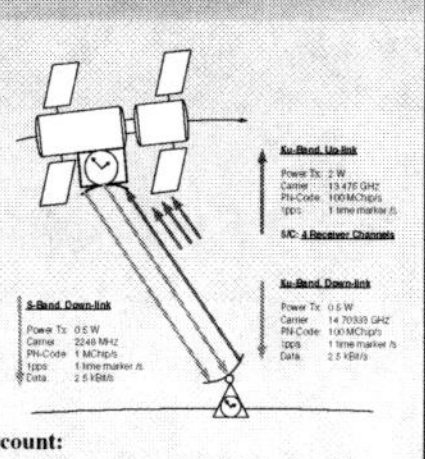

Two characteristic times are taken into account:
The mean duration of an ISS pass over a ground user terminal, 0.2 ps over 300 s.
The dead time between two successive comparison, 6 ps over 1 day.

CEPA

The Columbus External Payload Adapter (CEPA) is an integral part of the ACES payload. It uses an Active Flight Releasable Attachment Mechanical (AFRAM) to attach the payload to FRAM compatible sites like the CEPF. The CEPA also provides the robotic interface for on-orbit transportation

ACES Communication Block Diagram

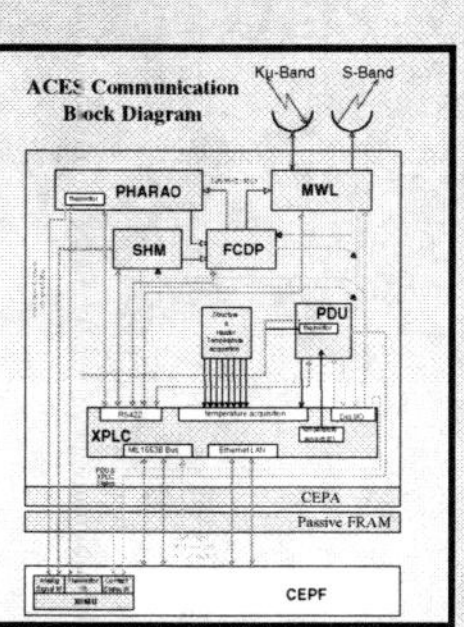

Payload Status

Development Status

The ACES payload industrial contract is signed. Phase C0 established a baseline design and was concluded by a successful payload PDR in Feb 04. Phase C1completes the detailed design and addresses the manufacture, integration and verification of the Sub-system Engineering Model (EM) units and the EM payload. The EM units and payload will undergo full EMC testing. The phase will be concluded by the Payload Critical Design Review (CDR). Phase D focuses on the manufacture, integration and verification of the Flight Model (FM). The FM units & payload will be submitted to functional, performance, electrical and environmental tests as necessary to finally demonstrate the health of the design, its ability to fulfil the mission requirements and the workmanship quality. The phase will be completed by a Flight Model Acceptance Review (FAR). Both the EM and FM payloads will be subject to scientific verification tests under ACES PI responsibility.

Key Milestones

Payload SRR	Dec 2002
Safety 0/1	Oct 2003
Payload PDR	Feb 2004
Mission PDR	Jun 2004
Safety level 2	Sep 2005
Payload CDR	Oct 2005
Mission CDR	mid 2006
Safety level 3	mid 2006
Payload FAR	end 2006
Mission FAR	mid 2007
Delivery to KSC	end 2007
Launch Ready	mid 2008

18th European Frequency and Time Forum, Guildford, England 5-7 April 2004

MSM-GF/307.2004/sf
Issue 1; 31/3/04

HIGH STABILIT PERSPECTIVES WITH PULSED OPTICALL PUMPED FRE UENC STANDARD

S. Micali io, A. Godone, F. Levi, and C. Calosso

Istituto Elettrotecnico Nazionale Galileo Ferraris, Strada delle Cacce 91, 10135 Torino, Italy

Keywords Pulsed optical pumping, free induction decay, light-shift, cavity pulling, short-medium term frequency stability.

Abstract

The use of pulsed coherent technique in optically pumped vapor cell frequency standards was proposed in sixties by Arditi and Carver in order to eliminate light shift and to increase the stability performances. In this paper, we re-consider the pulsed-scheme idea taking into account the present technology: semiconductor lasers and fast pulsed electronics are nowadays available.
We prove here that the microwave detection allows to achieve an atomic quality factor that is twice that obtained with the usual optical detection approach. Moreover, light shift and cavity pulling effects are studied both from the theoretical and the experimental point of view. A very promising preliminary measurement of the frequency stability of our prototype is also reported.

1 Introduction

Local oscillators with high frequency stability in the short and medium term are required in different application fields such as radio-navigation, telecommunication and high resolution spectroscopy. In this contest, we re-propose the idea of Pulsed Optically Pumped (POP) frequency standard conceived mainly to eliminate the light-shift effect [1, 2]. In particular, we consider the case of pulsed optical pumping performed by a semiconductor laser on a vapor cell containing buffer gas. The optical pumping phase is followed by a Ramsey interaction scheme defined by a double microwave pulse. The clock transition can be detected at the end of the second Ramsey pulse in two ways. The first one is the commonly used: the transparency of the atomic medium (proportional to the population inversion) is observed via a probe laser pulse. In the second one, the free induction decay (related to the atomic coherence) is detected via a high quality factor microwave cavity. We show both theoretically and experimentally that the component of the Bloch vector related to the microwave coherence rotates at a double speed with respect to the population inversion one when submitted to /2 Ramsey pulses, resulting in a higher atomic quality factor. Moreover, the microwave detection allows to reach the thermal noise limit without any laser background signal. The theoretical stability limit achievable by the POP using the microwave signal has been calculated as $_y(\tau)$ 3 10 ^{13} τ $^{1/2}$, a value not far from the typical performances of H-masers.

As regards the shifts affecting the atomic resonance, light shift, which mainly impairs the short term frequency stability of vapor cell frequency standards, is strongly reduced in the pulsed operation scheme moreover, there is a suitable value of microwave power that makes negligible the cavity pulling. Experimental characterization of these effects are reported as well as a preliminary frequency stability measurement of our laboratory prototype operating at the moment with the optical transmitted signal.

2 Theory

In the theoretical analysis, we consider the pure three-level atomic system shown in Fig. 1.

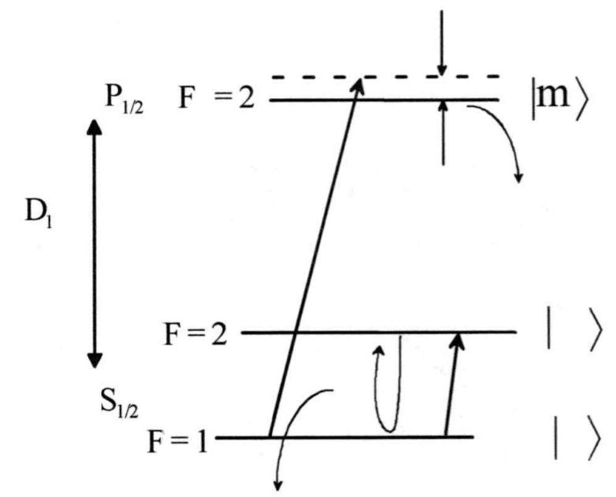

Figure 1: Three-level system described in the theory. $_0$ and $_1$ are the angular frequencies of the applied microwave field and of the laser field respectively. The other parameters are described in the text.

In particular, we consider a sample a ^{87}Rb atoms excited by means of a laser pulse tuned to the D_1 transition. The atoms are contained in a cell and diluted in a proper buffer gas. The

cell is placed in a microwave cavity operating in the TE_{011} mode and tuned to 6.834 GHz, the ground state hyperfine transition frequency. The rates $_1$ and $_2$ characterize the relaxation of the population difference and of the ground state coherence respectively, while * is the relaxation rate of the excited state [3] $_0$ is the laser detuning from the optical resonance.
A detailed description of such a three level atomic system can be found in [4] here we just recall the main results.

2.1 Ramsey fringes

Figure 2 shows the timing sequence of the pulsed-operation scheme.

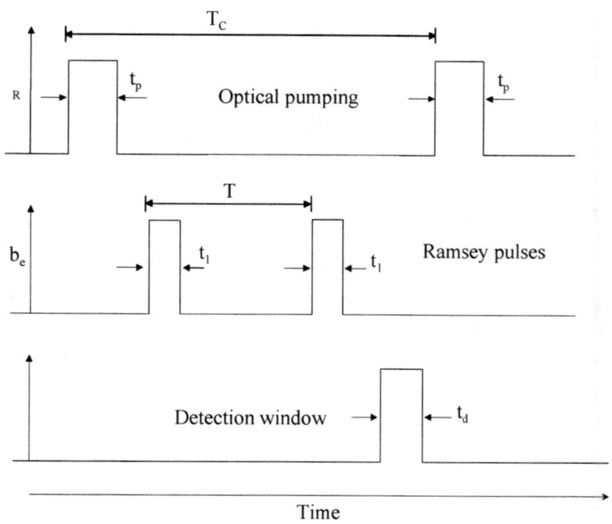

Figure 2: POP timing sequence T_c cycle period, $_p$ optical pumping time, $_1$ Rabi time, T Ramsey time and $_d$ detection time.

In Fig. 2, $_R$ is the optical Rabi frequency proprotional to the laser field amplitude and b_e is the microwave Rabi frequency proportional to the amplitude of the microwave signal applied to the cavity.
In the first phase, the atoms are submitted to a laser pulse in order to create a large population difference between levels $|\mu\rangle$ and $|\mu\rangle$. Afterwards, the atoms are interrogated by a Ramsey interaction scheme defined by two microwave pulses and finally, in the third phase, the clock transition can be detected either in the optical domain or in the microwave domain, as explained in section 1. We point out that in the POP operation the three phases (preparation, interrogation and detection) are separated in time so that the atoms behave, in the ideal case, as a pure two-level system. No coupling between microwave and optical coherence takes place and, in principle, light shift effect and noise conversion from the laser to the clock signal are avoided.
The behavior of an atomic system submitted to such an interaction scheme is well known [5] however, here we report our analysis based on the unified theory of optically pumped frequency standards developed in [6].

First of all, we observe that the timing sequence of Fig. 2, with period T_c, is in practice periodic. Obviously, we are interested to the steady-state condition of the atomic ensemble or, in other words, after many cycles of optical pumping and Ramsey pulses.
Let us analyse now each phase in the framework of the density matrix formalism.
During the optical pumping process, the atomic system is described by the following set of equations:

$$\gamma_1 \quad p \quad p \qquad \delta_{\mu\mu} \quad \gamma_2 \quad p \quad i \quad \mu \quad p\delta_0 \quad \delta_{\mu\mu} \quad 0 \tag{1}$$

In (1), is the ground state population difference and is the microwave coherence in the rotating wave approximation we defined $p \; \frac{\omega^2}{2\;^*}$ the longitudinal pumping rate, $\mu \;\; \omega_0 \;\; \omega_{\mu\mu}$ the microwave detuning and $\delta_0 \;\; \frac{2\;\;_0}{^*}$ the normalized laser detuning. The hypothesis we used to obtain Eq. (1) are discussed in [4].
The second equation of (1) gives:

$$\delta_{\mu\mu} \;\; p \qquad \delta_{\mu\mu} \qquad 0 \; e^{\;\gamma_2 \;\; p \;\; p} e^{i \;\; \mu \;\; p\delta_0 \;\; p} \tag{2}$$

where (=0) is the residual microwave coherence at the beginning of the laser pulse and $_p$ is the duration of the laser pulse. It is evident from (2) that owing to the laser pulse application, the residual coherence is reduced by a factor $e^{\;p\;p}$ and consequently also the light shift contribution is strongly lowered. Therefore, the following condition should be satisfied:

$$_p \; _p \quad 1 \tag{3}$$

Actually, due to the atomic absorption, the pumping rate is a decreasing function of the path covered by the laser pulse in the cell: $_p = \;_p(\;)$, being the propagation axis. Taking this point into account, it can be demonstrated that at the input of the cell (=0) we should have:

$$p \quad 0 \quad \zeta\gamma_1 \quad \frac{1}{p} \tag{4}$$

where is the optical length of the cell [4]. If (4) is satisfied, the two phases of optical pumping and interrogation are really separated.
In the second phase, the atoms are submitted to two microwave pulses whose duration is $_1$ (Rabi time), separated by the Ramsey time T ($T \;\; _1$). In the time between the two pulses the atoms decay freely in the cavity. The atomic state at the end of the second microwave pulse is described by the well known matrix product:

$$(\;_f) = \;_1(\;_1) \quad _D(T) \quad _1(\;_1) \quad (\;_i) \tag{5}$$

where $_{I}(\ )$ and $_{D}(\ )$ are the usual interaction and decay matrices respectively [3, 4], and R(t) is the Bloch vector whose components are:

$$\mathrm{Re}\,\delta_{\mu\mu} \qquad \mathrm{Im}\,\delta_{\mu\mu} \qquad (6)$$

In (5), t= $_{i}$ ($_{f}$) corresponds to the beginning (end) of the first (second) microwave pulse. We point out that to write Eq. (5) we make the following hypothesis:

(i) $_{1}$ $_{1}$ ^{1}, $_{2}$ ^{1}, in this way the decay rates can be neglected during the interaction with the microwave pulse

(ii) in the free decay region between the two Rabi pulses, the cavity pulling is not considered. We will see later on what happens when the cavity pulling effect is taken into account.

Equation (5) allows us to obtain Ramsey patterns as observed in the optical signal (proportional to the population inversion) and in the free induction decay signal (proportional to the microwave coherence).

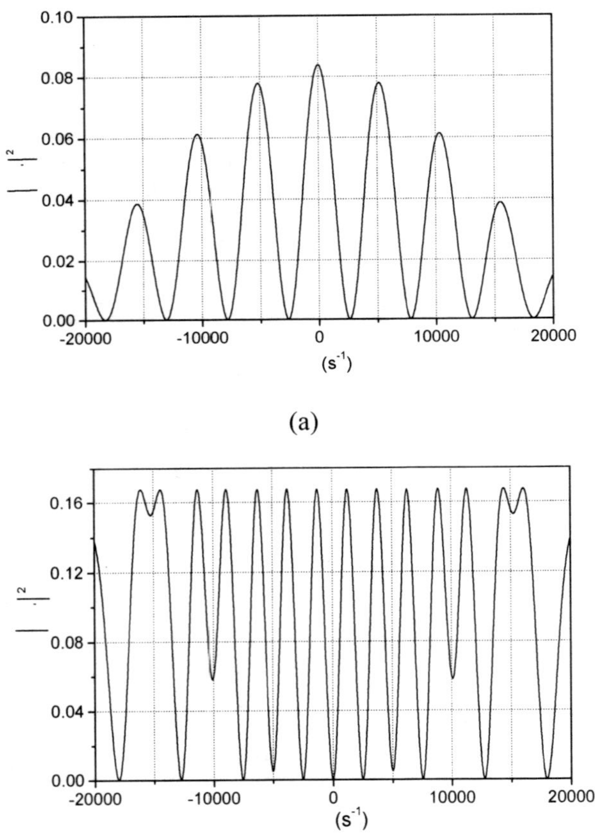

(a)

(b)

Figure 3. Computed Ramsey fringes as observed in the free induction decay signal: $_{1}$ = $_{2}$= 200 s ^{1}, $_{0}$=0, $_{1}$= 200 s, T= 1 ms (a) = /8 (b) = /2

Figure 3 shows the computed Ramsey fringes observed through the microwave signal for two values of the microwave pulse area = b$_{e}$ $_{1}$. It is evident that for = /2 the clock signal oscillates with a double speed with respect to the case of = /8 near = 0.

This feature, very interesting from a metrological point of view, is not present in the optical domain where, in the same situation of Fig. 3, we observe an increase of the contrast for a = /2 pulse but not a double atomic quality factor. This is also evident from the analytic expressions describing the central Ramsey fringe for the optical transmitted signal and for the microwave signal in the case of = /2:

$$e^{\ \gamma_2 T}\cos\ _{\mu}T \qquad \left|\delta_{\mu\mu}\right|^2\ \ \frac{1}{4}e^{\ 2\gamma_2 T}\sin^2\ _{\mu}T \qquad (7)$$

Therefore, the microwave detection appears very promising for a proper implementation of the POP frequency standard.

2.2 Residual Light shift

In the previous section we demonstrated that, provided relation (4) is satisfied, any residual microwave coherence is destroyed before the beginning of a new cycle and no light shift is observed in the clock signal. However, due to the finite laser intensity, a "phase memory" between consecutive cycles survives [5]. In particular, it is possible to show that the light shift due to the residual coherence $\delta_{\mu\mu}$ at the beginning of the first Ramsey pulse can be expressed as:

$$\frac{\omega}{\omega_{\mu\mu}}\ \ \frac{2}{\pi\ _{a}}\mathrm{Im}\,\delta_{\mu\mu} \qquad (8)$$

being $_{a}$ the atomic quality factor.

To evaluate the light shift, we computed the frequency shift of the central fringe solving with a numerical recursive approach Eqs. (2) and (5). The results are shown in Figs. 4 and 5.

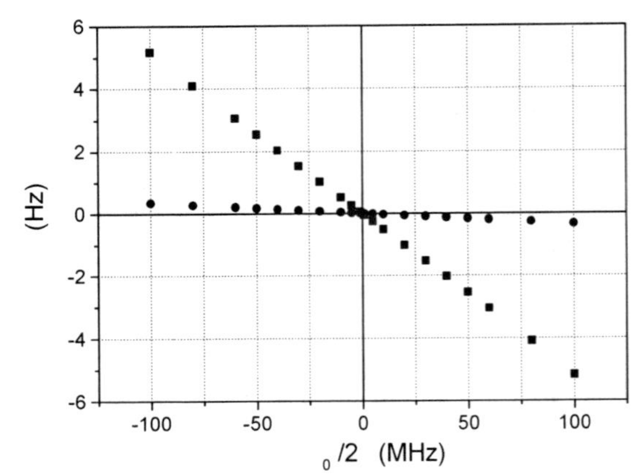

Figure 4. Computed residual light shift versus the laser detuning $_{p}$=10000 s ^{1}, $_{1}$= $_{2}$= 200 s ^{1}, $_{1}$=0.2 ms, T=1 ms: squares: = /4, circles: =12 /25.

In particular, in Fig. 4 we report the absolute frequency shift versus the laser detuning for two values of , while in Fig. 5 the frequency shift is plotted versus for two values of the pumping rate. The values of the parameters refer to typical experimental conditions.

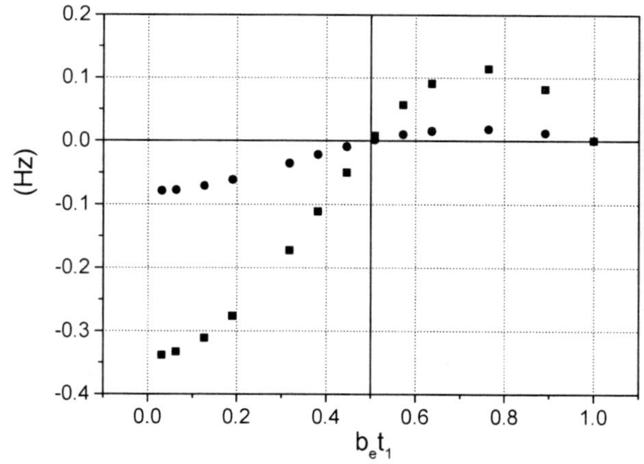

Figure 5. Computed residual light shift versus the microwave pulse area $_0/2$ = 1 MHz. Squares: $_p$=5000 s^{-1}, circles: $_p$=10000 s^{-1}.

The physical meaning of the previous figures is evident the more the laser power in increased the more the residual microwave coherence is destroyed and the resulting light shift is strongly reduced. Moreover, for = /2 light shift becomes completely negligible and no PM-PM conversion takes place.

2.3 Cavity Pulling

During the time between the two Ramsey pulses, the atoms decay freely, being submitted only to the electromagnetic field excited in the cavity by the atoms themselves. If the cavity is not exactly tuned to the hyperfine transition frequency, a dephasing in the free evolution of the coherence is introduced, resulting in a detuning $_{cp}$ of the atomic resonance. The cavity pulling effect is evaluated taking into account in the equations the terms describing the coupling between the atoms and the cavity we have precisely [4, 6]:

$$\dot{} \quad \gamma_1 \quad 4 \left|\delta_{\mu\mu}\right| \qquad \dot{\delta}_{\mu\mu} \quad \gamma_2 \quad \mathrm{i} \quad _{\mu} \ \delta_{\mu\mu} \qquad (1 \quad \psi) \ \delta_{\mu\mu} \tag{9}$$

where is the gain factor (number of microwave photons emitted by each atom in 1 s) and ψ is the cavity detuning parameter defined as ψ 2 ν / ν_0 [6], being ν the cavity detuning. We have found the following analytical solution for the cavity pulling of the central Ramsey fringe:

$$\omega_{cp} \quad \frac{\psi}{T} \ln \ \cosh \ \frac{}{\gamma} 1 \quad e^{\ \gamma T} \qquad \cos\theta \sinh \ \frac{}{\gamma} 1 \quad e^{\ \gamma T} \tag{10}$$

where we assumed = $_1$ $_2$, a typical situation for alkali atoms in buffer gas at the usual operating temperatures. In Fig. 6 we show the calculated cavity pulling versus the microwave pulse area for different values of the parameter $/\gamma$. From Eq. (10), the value of that makes the cavity pulling effect negligible is readily obtained:

$$\theta_0 \quad \pi \quad \arccos \ \tanh \ \frac{}{2} 1 \quad e^{\ \gamma T} \tag{11}$$

It is important to point out that in the low cavity feedback situation (1), $_0$ is not far from /2 and in this case both residual light shift and cavity pulling are minimized.

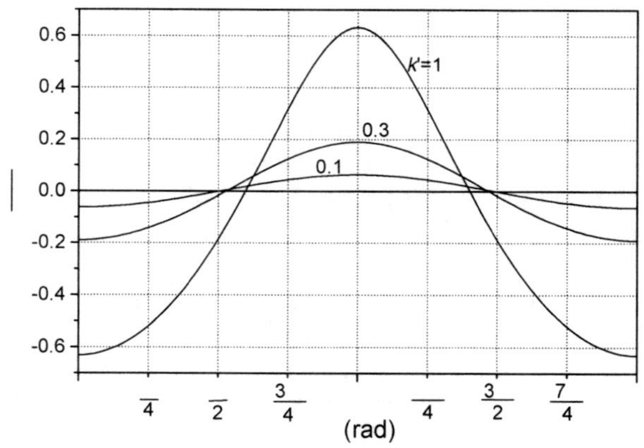

Figure 6. Calculated cavity pulling as a function of T=1.

2.4 Frequency stability

Since under suitable operating conditions the laser noise is not transferred to the clock signal, we expect that the main contribution to the short term frequency stability of the POP frequency standard is due to the thermal detection noise. We consider then an additive Gaussian noise at the output of the microwave cavity described by the following spectral power density:

$$S_n \ \omega \qquad T_0 \tag{12}$$

where is the noise figure of the detection system and is the Boltzman constant. It is possible to show [4] that assuming a square-wave modulation of period T_m = 2 T_C the Allan standard deviation can be written as:

$$\sigma_y \ \tau \qquad \frac{\sqrt{2\pi}}{a} \sqrt{\frac{\quad T_0}{}} \sqrt{\frac{T}{}} \tau^{\ 1/2} \tag{13}$$

being τ the average time and the microwave power detected at the end of the second Ramsey pulse.

For an operating temperature of T_0 around 60 °C and for a cell of length = 40 mm and diameter 2 = 40 mm with 15 Torr of buffer gas (N_2 and Ar), we have *=1.9 10^9 s^{-1} and $_1$ $_2$ 200 s^{-1}. Moreover, for the timing sequence T = 4 ms, $_d$

= 4 ms, $_p$ = 2 ms and $_1$ = 0.5 ms we have T_C = 11 ms and $_a$ = 1.1 10[8]. Finally, for a cavity operating in the TE_{011} we may obtain an output power at end of the second Ramsey pulse around $_0$ = 100 pW [4]. For these values of the parameters, Eq. (14) gives:

$$\sigma_y \tau \quad 3 \quad 10^{13} \tau^{1/2} \tag{14}$$

a very interesting result for a vapor cell frequency standard. Since the cavity pulling effect can be made negligible with a proper choice of , we expect that in the medium term the main instability source is due to the buffer gas temperature sensitivity. The latter can be reduced by a proper mixture of buffer gases and with a high performance temperature control.

3 E periment

The experimental set-up used to test the theoretical predictions is shown in Fig. 7. The core of the physical package is a quartz cell (length = 18 mm, diameter 2 =32 mm) containing the ^{87}Rb (98) atoms and a mixture of buffer gases (Ar and N_2). The total pressure of the buffer gases is 25 Torr. The cell is placed inside an aluminium microwave cavity operating in TE_{011} mode. The cavity is thermally stabilized by means of a preliminary PID analogue controller with the loop time constant of about 20 s. It allows to achieve a temperature stability of 1 mK for an integration time of 10000 s.
The temperature of the cell is about 63 °C at this temperature we have *=3 10^9 s ^{1}, $_1$ $_2$ 350 s ^{1}, = 4.5 and n 4 10^{11} cm ^{3}, being n the atomic density.

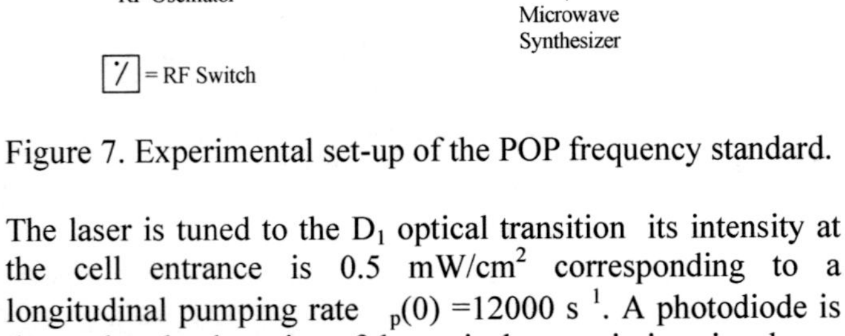

Figure 7. Experimental set-up of the POP frequency standard.

The laser is tuned to the D_1 optical transition its intensity at the cell entrance is 0.5 mW/cm^2 corresponding to a longitudinal pumping rate $_p$(0) =12000 s ^{1}. A photodiode is devoted to the detection of the optical transmission signal.
A commercial synthesizer, driven by a 10 MHz quartz oscillator, provides the interrogation microwave and a spectrum analyzer operating in the video mode and synchronized to the timing pattern is used as heterodyne detector.

3.1 Ramsey fringes

We obtain the Ramsey patterns shown in Fig. 8 by measuring the amplitude of the free induction decay signal versus the microwave detuning /2 , as observed at the end of the second microwave pulse.

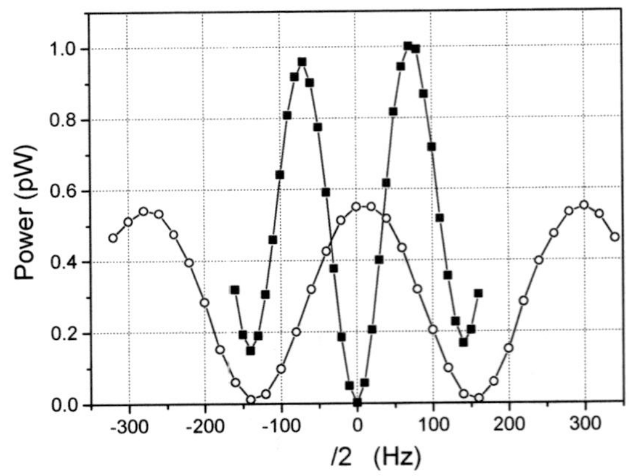

Figure 8. Ramsey fringes observed in the microwave domain open circle: = /8 squares = = /2.

With reference to Fig. 2, we have $_p$ = 2 ms, $_1$ = 200 s, T =3.5 ms and T_c = 8 ms. For these values of the parameters, the inequality (4) becomes $_p$(0) 2000 s ^{1}, widely satisfied in our experimental situation.
For a microwave power of about 100 nW, no signal is observed at the end of the second microwave pulse when /2 = 0 this power value corresponds then to = /2.
A double atomic quality factor is clearly observed in the case of = /2 and the theoretical behaviors of Fig. 3 are then confirmed. In this case the atomic quality factor turns out to be $_a$=1 10^8.
The detected output power at the end of the second Ramsey pulse is = 1 pW and the signal-to-noise ratio in a 1 Hz bandwidth is 10^4.

3.2 Residual light shift

In section 2.1 we demonstrated that a high intensity laser pulse can destroy the residual microwave coherence, resulting in a no transfer of the laser fluctuations to the clock transition. The coherence destruction after the application of the laser pulse is shown in Fig. 9 recorded at the spectrum analyzer. It is possible to observe the Ramsey pattern and free induced decay of the coherence between the two Ramsey pulses.
The figure refers to = /2. Therefore, to make more evident the effect of the laser pulse, the interrogation microwave field is detuned with respect to the atomic resonance, so that at the end of the second Ramsey pulse there is a non-null microwave coherence (see Fig. 8). In Fig. 9 we superimposed the laser pulse window during its application the microwave coherence is destroyed according to Eq. (2). The timing pattern for this experiment is $_p$ = 2.16 ms, $_1$ = 200 s, T=2.3 ms and T_c = 6.72 ms.

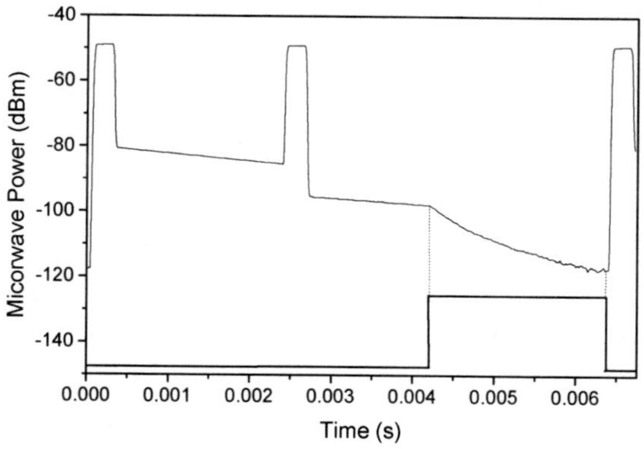

Figure 9. Atomic coherence behavior in the cavity when submitted to a Ramsey interaction scheme and to a laser pulse. The bold step refers to the laser pulse. Resolution Bandwidth 30 kHz Video Bandwidth 30 kHz Center 6.834686670 GHz /2 = 10 Hz laser power 1.1 mW.

To measure the residual light shift we lock the laser frequency to the profile of the D_1 line using an external cell containing only ^{87}Rb. The nominal detuning of the laser from the optical resonance is around 100 MHz, widely inside the linear part of the light shift curve. Then, for a fixed laser power, we finely change the laser frequency by means of an acousto-optical modulator (AOM). For a 10 MHz of laser frequency variation we measure the differential POP frequency by locking the quartz oscillator to the optical signal via a first order control loop. The frequency of the modulating sine-wave signal is $_m$= 39 Hz and the loop time constant is τ = 40 ms. For this experiment the timing sequence is T = 2.33 ms, corresponding to a linewidth of $_a$=214 Hz and to an atomic quality factor of $_a$=3.2 10^7, $_1$= 322 s, $_p$= 2 ms and T_c=5 ms. The results are shown in Fig. 10. In particular, we point out that there is a suitable value of the microwave power making negligible the POP frequency from the laser power. Moreover, in correspondence of this microwave power value, we measure the relative variation of the POP frequency versus the laser frequency we find (1.6 0.5) 10 ^{13}/MHz.

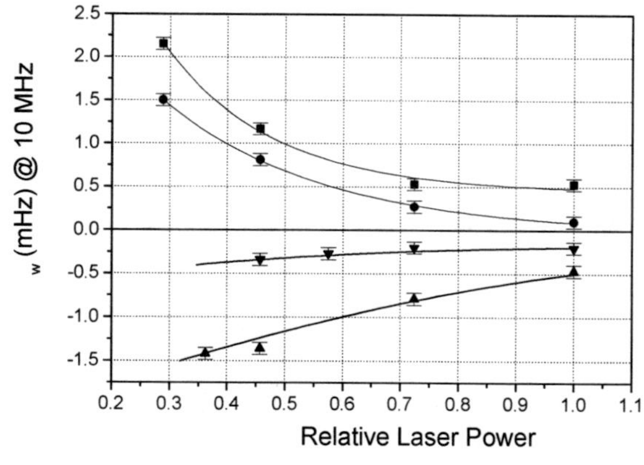

Figure 10. Differential frequency shift of the POP versus the laser power for four values of the applied microwave power: squares: 0 dB circles: 2 dB down triangle: 4 dB up triangle: 6 dB

3.3 Cavity pulling

In the same conditions of the previous paragraph, we also measure the cavity pulling shift. The microwave cavity has a linewidth of $_{1/2}$ = 400 kHz, corresponding to a quality factor of $_L$=17000. The gain factor , that characterizes the cavity feedback, turns out to be 450 s ^{1} moreover, we have 1.5 and $_2 T$ = 0.7.

To measure the cavity pulling we detune the cavity by means of the tuning screw. The following figure refers to three values of the detuning parameter the laser power is about 0.8 mW.

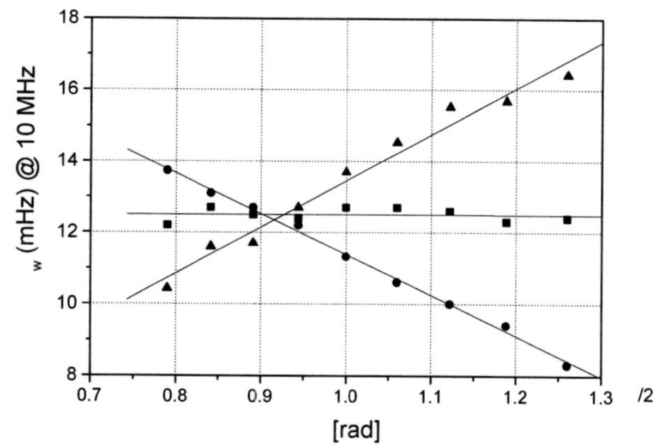

Figure 11. Cavity pulling versus the microwave pulse area. Squares: $_C$= 10 kHz (= 0.05) circles: $_C$= 140 kHz (= 0.69) up triangle: $_C$= 140 kHz (=0.69).

Around θ_0 $0.9\frac{\pi}{2}$ the output frequency hardly depends on the cavity detuning, as expected from the theoretical computations. However, for the previous values of the parameters Eq. (11) foresees θ_0 $1.23\frac{\pi}{2}$. This discrepancy can be understood taking into account that inside the cavity the TE_{011} mode has a spatial distribution so that the real power felt by the atoms depends on their position inside the cell. This effect is not considered in the theory where all the atoms are submitted to the same microwave power, that is to the same θ.

3.4 Frequency stability

Finally we also perform a measurement of the frequency stability of our prototype. As mentioned in section 3.2, the local oscillator is locked to the optical transmission signal detected by the photodiode at the end of the cell. In particular, in our present structure there is not a pure detection phase, since the Ramsey fringes are superimposed to the same signal used in the optical pumping process. Therefore, there is not a probe laser pulse devoted to the detection phase.

The behavior of Fig. 11 refers to θ $\frac{\pi}{2}$ and laser power 0.8 mW. The other physical operation parameters are the same of the previous paragraphs. We obtain:

$$\sigma_y \tau \quad 3 \quad 10^{11}\tau^{1/2} \qquad (15)$$

for integration times up to 4000 s. What is important in Fig. 11 is that no flicker floor is observed, i.e. no laser fluctuations are transferred to the clock transition.
The short term stability is deteriorated at the moment by the adopted detection scheme and cannot be compared with (14).

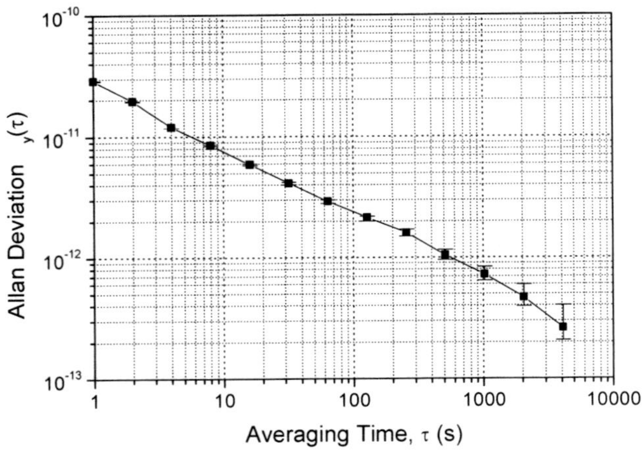

Figure 11. ^{87}Rb POP frequency stability.

3 Conclusions

In this paper we presented a theoretical review of the POP frequency standard and some preliminary experimental results. From a theoretical point of view, we have demonstrated that the detection of the clock transition observed through the free induction decay signal shows some advantages with respect to the more common optical detection: the clock signal is observed without any laser background and the atomic quality factor is twice that obtained in the optical domain when the clock transition is observed through the Ramsey interaction scheme. Light shift and cavity pulling can be cancelled out or at least strongly reduced. We performed a first experimental characterization of light shift and cavity pulling in the optical domain and the results are in good agreement with the theoretical predictions.
As mentioned in the previous section, in our present experimental apparatus the clock transition is detected via the transmitted optical signal used for optical pumping purpose, there is not any real detection phase. In spite of this, the clock signal turns out to be affected by white frequency noise up to integration times of 4000 s, a result that clearly shows the potential of the POP frequency standard.
Work is in progress to realize a low noise synthesis chain and a suitable pulsed electronics in order to use the microwave output as a clock signal. From the preliminary experimental data here reported, we are confident to approach the result of Eq. (14) in an optimised system.

Acknowledgments

This work has been partially supported by ASI (Italian Space Agency). The authors would like also to thank M. Siccardi for useful and stimulating discussions.

References

[1] C. O. Alley "Coherent pulse techniques in the optical detection of the 0-0 ground state hyperfine resonance in rubidium 87", in *an elec on c* , C. H. Townes Editor, Columbia university Press, N. ., pp. 146-155, (1960).

[2] M. Arditi and T. R. Carver "Atomic clock using microwave pulse-coherent techniques", IEEE Trans. Instrum. Meas., **13**, pp. 146-152, (1964).

[3] J. Vanier and C. Audoin, *The an hy c o o c e ency S an a* , Adam-Hilger, Bristol, England (1989).

[4] *o one S cal o an e l e op cally p pe e ency an a hy cal e e*

[5] E. I. Alekseyev, . N. Bazarov, and A. E. Levishin, Radio Eng. Electron. Phys., **1** , pp. 77-83, (1974) E. I. Alekseyev, . N. Bazarov, and G. I. Telegin, ibid., **2** , pp. 73-80, (1975).

[6] A. Godone, F. Levi, and S. Micalizio, Coherent Population Trapping Maser, CLUT Editrice, Torino, Italy (2002).

A NEW MODEL OF AN ACTIVE HYDROGEN MASER WITH A FREQUENCY-SWITCHING CAVITY AUTOTUNING

N. DEMIDOV, V. GAVRILOV, V. GOREV, V. VORONTSOV
KVARZ Institute of Electronic Measurements
176, Prospect Gagarina
RU 603009 Nizhny Novgorod
Russia
E-mail: kvarz@kvarz.ru

One of the most recent developments to improve the performance of the CH1-75 Active Hydrogen Maser is an advanced cavity-autotuning system, which ensures that the H-Maser remains centered on the atomic hydrogen line over a long period of time. The paper describes the new model of an Active Hydrogen Maser using a well-known method of cavity tuning for its frequency switching.

Due to low extraneous noise of the tuning system it is possible to set an autotuning loop time constant of the order of a few thousands of seconds with the result that even with auto-tuning engaged for improved long term stability, the effect on short and medium term stability characteristics is negligible ($\sigma_y(\tau) \sim 7 \cdot 10^{-16}$ at $\tau = 10^4$ s).

A high enough modulation frequency (~87 Hz) allows to achieve a low phase noise at modulation frequency in the Maser output signal. The new Maser environmental sensitivity measurement results are given, in particular the Maser temperature sensitivity below $1 \cdot 10^{-15}$/ °C.

HIGHER-ORDER LIGHT SHIFTS IN OPTICAL FREQUENCY STANDARDS

V.G. Pal'chikov [1], V.D. Ovsiannikov [2], M. Takomoto [3], H. Katori [3]

[1] Institute of Metrology for Time and Space at National Research Institute for Physical-Technical and Radiotechnical Measurements, Mendeleevo, Moscow Region 141570, Russia ; E-mail: vitpal@mail.ru , Fax: +007-095-534-0609
[2] Department of Physics, Voronezh State University, Voronezh 394006, Russia
[3] Engineering Research Institute, the University of Tokyo, Bunkyo-ku, Tokyo 113-8656, Japan

Keywords: Light shift, ac Stark effect.

Abstract

Higher-order perturbation theory is used to study frequency shifts of the spin-forbidden $^3P_J - {}^1S_0$ transitions for Mg, Sr and Yb atoms in an external laser field. Calculations of the higher-order dynamic Stark effect have been performed in connection with precise measurements on cooled and trapped atoms and the design of ultra stable optical clock in an engineered light shift trap. The form of the model potential of the valence electrons suggested by Simons is used for the calculation of atomic dynamic characteristics given by second- and fourth-order perturbation theory for the interaction of the atoms with the electromagnetic field. Dispersion properties of the polarizability and hyperpolarizability components and their behavior in the vicinity of both resonances and so-called "magic wavelengths" for the upper and lower atomic states, are analyzed. Calculations of ac multipole polarizabilities and dipole hyperpolarizabilities for the $^3P_0 - {}^1S_0$ clock transition in ^{87}Sr indicate that the contribution of the higher-order light shifts can be reduced to less than 1 mHz, allowing for a projected accuracy of better than 10^{-17}.

1 Introduction

The dynamic (or ac) Stark shift has played an important role in many recent advances in atomic physics, including precision measurements of atomic polarizabilities, laser cooling of atoms and molecules down to ultracold temperatures, optical frequency standards, atomic interferometry, etc.

As a prime candidate for an optical frequency standards in neutral atoms, laser cooling of alkaline earth atoms have been studied in Mg, Ca, Sr and Yb atoms (see, for example [1-5]). It is important to note, that optical frequency standards can be developed based on narrow transitions that allow exceptionally high Q-factor, that are orders of magnitude higher than in Cs-fountains. The atomic configuration of two outer electrons introduces a unique complexity in their energy structure, specifically, the spin-singlet ground state and the spin-forbidden triplet excited states. For example, by consideration of the prospects for an optical clock on the $^1S_0 - {}^3P_0$ transition, it was shown both experimentally [6] and theoretically, that hyperfine structure (hfs) effects can mix significantly levels which cross for some atomic isotopes with nonzero nuclear spin ($I \neq 0$). In the case of ^{97}Sr atoms, the hfs mixing of the $^3P_0$ ($I = 9/2$) state with the $^{1,3}P_1$ states give rise to an electric-dipole transition.

In our recent paper [3], we have discussed the feasibility of an "optical lattice clock", which utilises millions of neutral atoms separately confined in an optical lattice that is designed to adjust the dipole polarizabilities for the probed electronic states in order to cancel light field perturbations on the measured spectrum. This scheme permits an exceptionally low instability of $\sigma_y(\tau) \approx 10^{-18}$ with an interrogation time of only $\tau = 1s$, which may open up new applications of ultra precise metrology, such as the search for the time variation of fundamental constants and the real time monitoring of the gravitational frequency shift. Figure 1 illustrates the proposed scheme.

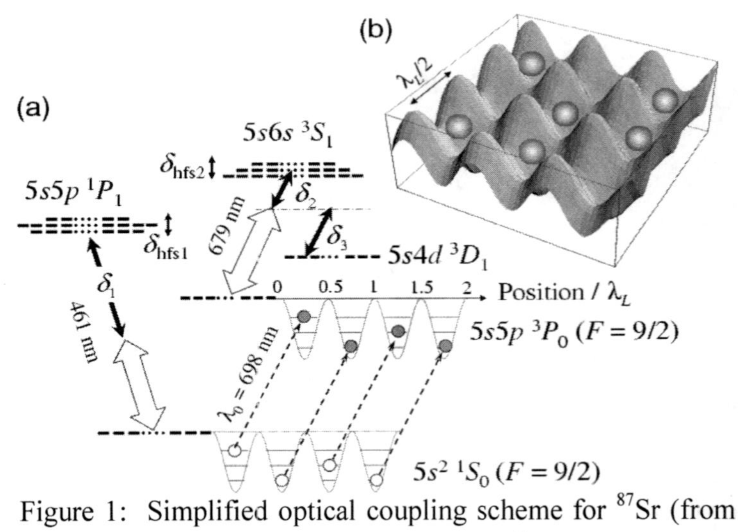

Figure 1: Simplified optical coupling scheme for ^{87}Sr (from [3]). (a) In the limit of large detunings δ_i of the coupling laser compared to the hyperfine splittings δ_{hfs}, the squared transition dipole moment of the upper J manifold can be simply added up, resulting in a quasi-scalar light-shift. (b) 3D optical lattice provides Lamb-Dicke confinement while it prevents atom-atom interactions.

Sub-wavelength confinement provided by the optical lattice localizes atoms in the Lamb-Dicke regime (LDR), where the first order Doppler shift as well as the photon recoil shift disappears and the second order Doppler shift can be made negligibly small by sideband-cooling atoms down to the vibrational ground state . Therefore this scheme simulates a system where millions of single-ion clocks operate simultaneously.

Here we report our theoretical results for the light shift in Mg, Sr and Yb atoms as a function of the trapping laser wavelength at fixed intensity in the vicinity of the crossing points. The wavefunctions are generated using semiempirical Fues model potential theory [7-9] which have proven to be well suited for accurate calculating ac polarizabilities and ac hyperpolarizabilities. The details of the calculations of angular and radial matrix elements have been described elsewhere [10-12] and are omitted here for brevity. Atomic units are used throughout.

2 General formulations

The spectral properties of an atom in a field F of a light wave are determined by the quasienergies, corresponding to the nonperturbed atomic spectrum F_n. Except for some special cases (for example, single-photon resonance or excited states of the hydrogen atom in a low-frequency field) the perturbation of spectra in the first nonvanishing order turns out to be quadratic with respect to the electric field amplitude F of the wave (dynamic or ac Stark effect).In the general case, the energy shift $\Delta E = E - E_{nJ}$ of an atomic level under the action of monochromatic radiation with electric-field strength

$$\vec{F}(t) = F\,\mathrm{Re}\left\{\vec{e}\,\exp\left[(\vec{k}\cdot\vec{r} - \omega t)\right]\right\} \qquad (1)$$

(where $\vec{e}$ is the unit polarization vector and $\vec{k}$ is the wavevector) is the root of the secular equation

$$\det\left|\left|\,\Delta E\,\delta_{MM'} - U_{MM'}\right|\right| = 0, \qquad (2)$$

where $U_{MM'}$ is the matrix element of the atom-field interaction operator.

The ac polarizabilities and ac hyperpolarizabilities are calculated starting from formal expressions for the second- and fourth-order quasienergy in terms of the field-free wave functions $|\,0\rangle$ and the reduced quasi-energy Green's functions G [7]:

$$\Delta E^{(2)} = U^{(2)}_{MM} = -\frac{F^2}{4}\,\alpha(\vec{e},\omega) = -\langle\langle 0|\,HGH|0\rangle\rangle, \qquad (3)$$

$$\Delta E^{(4)} = U^{(4)}_{MM} = -\frac{F^4}{64}\,\gamma(\vec{e},\omega) = -\langle\langle 0|\,HGHGHGH|0\rangle\rangle + \langle\langle 0|\,HGH|0\rangle\rangle\langle\langle 0|\,HG^2H|0\rangle\rangle, \qquad (4)$$

$$G_E = \sum_m \frac{|\,m\rangle\rangle\langle\langle m\,|}{\varepsilon_m^{(0)} - E}, \qquad (5)$$

where the double brackets indicate the integration over the spatial variables and averaging over time, $\varepsilon_m^{(0)} = E_m^{(0)} + \omega$ is the quasi-energy. In order to calculate higher-order contributions to the polarizability, the magnetic-dipole and electric quadrupole atom-field interactions should be taken into account together with the electric-dipole term in the interaction Hamiltonian [7]

$$\hat{H}(\vec{r},t) = \hat{V}(\vec{r})e^{-i\omega t} + \hat{V}^{+}(\vec{r})e^{i\omega t}, \qquad (6)$$
$$\hat{V}(\vec{r}) = \hat{V}_{E1} + \hat{V}_{M1} + \hat{V}_{E2}.$$

In the dipole approximation $\hat{V}(\vec{r}) = -\frac{1}{2}F(\vec{e}\cdot\vec{d})$, where $\vec{d}$ is the atomic dipole moment.

Finally, the transition frequency ν of an atom exposed to the trapping laser field of F is described as [3]

$$\hbar\nu = \hbar\nu^{(0)} - \frac{1}{4}\Delta\alpha(\vec{e},\omega)F^2 - \frac{1}{64}\Delta\gamma(\vec{e},\omega)F^4 - \ldots,$$

where $\nu^{(0)}$ is the transition frequency at zero-field condition, $\Delta\alpha(\vec{e},\omega)$ and $\Delta\gamma(\vec{e},\omega)$ are the differences between the ac polarizabilities and ac hyperpolarizabilities, respectively.

2.1. Polarization dependence

To consider the polarization properties of the laser wave we shall use the normalization for which the light intensity

$$I = \frac{c}{8\pi}\left(\vec{F}\cdot\vec{F}^{*}\right) \qquad (7)$$

does not depend on the type of polarization . Applying the Wigner-Eckart theorem the second order matrix element may be obtained in an explicit form

$$\Delta E^{(2)} = -\frac{F^2}{4}\sum_{p=0,\lambda}^{2} C^{JM}_{JMp\lambda}\left(\frac{6}{(p+2)}\frac{(2J+2)_p}{(2J+1-p)_p}\right)^{1/2} \cdot \left(\vec{e}\times\vec{e}^{*}\right)_{p\lambda}\alpha_p(\omega) \qquad (8)$$

Here $(a)_p = a(a+1)\ldots(a+p-1)$ is the Pochhammer symbol and $C^{JM}_{J_1M_1J_2M_2}$ is Clebsch-Gordan coefficient.

Using standard methods of quantum theory of the angular momentum it is convenient to introduce the "scalar" α^s, "asymmetric" α^a and "tensor" α^t polarizabilities for linear (l) and circular (c) polarization of the electromagnetic wave [7]:

$$\Delta E^{(2)} = -\frac{F^2}{4}\sum_{p=0}^{2} C^{JM}_{JMp0}\left(\frac{(2J+2)_p}{(2J+1-p)_p}\right)^{1/2}\alpha_p^{l,c}(\omega), \quad (9)$$

$$\alpha_0^l(\omega) = \alpha_0^c(\omega) = \alpha^s(\omega)$$

$$= \frac{1}{3(2J+1)}\sum_{J'=J,J\pm1}(-1)^{J-J'}\left(R_{J'}(\omega) + R_{J'}(-\omega)\right), \quad (10)$$

$$\alpha_1^c(\omega) = \frac{1}{2}\alpha^a(\omega)$$

$$= -\frac{1}{3(2J+1)_2}\sum_{J'=J,J\pm1}(-1)^{J-J'}X\left(R_{J'}(\omega) - R_{J'}(-\omega)\right), \quad (11)$$

$$\alpha_2^l(\omega) = -2\alpha_2^c(\omega) = \alpha^t(\omega)$$

$$= -\frac{1}{3(2J+1)_3}\sum_{J'=J,J\pm1}(-1)^{J-J'}\left[3X(X-1) - 8J(J+1)\right] \times \left(R_{J'}(\omega) + R_{J'}(-\omega)\right). \quad (12)$$

Here $X = J(J+1) + 2 - J'(J'+1)$ and R is the reduced matrix elements of the atomic dipole moment d:

$$R_J(\omega) = \left\langle J\left\| dG_J(E_J^0 + \omega)d\right\| J\right\rangle \quad (13)$$

Finally, with the help of these parameters the quadratic light shift may be expressed in a simple form

$$\Delta E^{(2)} = -\frac{F^2}{4}\left[\alpha_J^s(\omega) + \alpha_J^t(\omega)\frac{3M^2 - J(J+1)}{J(2J-1)}\right] \quad (14)$$

for the linearly polarized wave (here M is the total-momentum projection onto the direction of the polarization vector $\vec{e}$), and

$$\Delta E^{(2)} =$$

$$= -\frac{F^2}{4}\left[\alpha_J^s(\omega) + A\,\alpha_J^a(\omega)\frac{M}{2J} - \alpha_J^t(\omega)\frac{3M^2 - J(J+1)}{2J(2J-1)}\right] \quad (15)$$

for the circularly polarized ($A = 1$ for the right-hand; $A = -1$ for the left-hand polarization) laser radiation: M is the projection onto the propagation axis of the light wave.

2.2. Analytical expression for Green's function.

The main difficulty in the exact computation of the amplitudes of second to fourth orders in Equations.(3) -(4) is related to the infinity sums over the total atomic spectrum including continuum, which represent the Green's function (Equation 5).

In the present work the frequency dependence of the polarizability (α) and the hyperpolarizability (γ) is calculated using Sturmian Green's functiom method [7-11]. It provides a reliable control for the accuracy of the calculation with the finite sums of discrete states. The angular part of the Green's function $G_E(\vec{r}_1, \vec{r}_2)$ is simply the product of spherical harmonics, while for the radial part $g_l(E; r_1, r_2)$ we have taken an expansion in Sturmian functions [7,9-10]:

$$G_E(\vec{r}_1, \vec{r}_2) = \sum_{lm} g_l(E; r_1, r_2)\, Y_{lm}(\vec{n}_1)\, Y^*_{lm}(\vec{n}_1),$$

$$g_l(E; r_1, r_2) = \frac{4Z}{\nu}\sum_{k=0}^{\infty}\frac{U_{kl}\left(\frac{2Z r_1}{\nu}\right)U_{kl}\left(\frac{2Z r_2}{\nu}\right)}{k + \lambda_l + 1 - \nu}, \quad (16)$$

where $\nu = \frac{Z}{\sqrt{-2E}}$ and

$$U_{kl}(x) = \sqrt{\frac{k!}{\Gamma(k + 2 + 2\lambda_l)}}\, x^{\lambda_l}\exp\left(-\frac{x}{2}\right)L_k^{2\lambda+1}(x). \quad (17)$$

Here $L_m^l(x)$ is the Laguerre polynomial and $\Gamma(x)$ is the gamma function. The radial functions $R_{nl}(r)$ are obtained from the residues of the Green's function at the poles of $g_l(E; r_1, r_2)$: $\nu = \nu_{nl} = n_r + \lambda_l + 1$, with $n_r = 0,1,2,\ldots$ the radial quantum number, λ_l the effective orbital momentum and ν_{nl} the effective prinzipal quantum number. These effective quantum numbers for the ground and metastable levels, are discussed in [9-10]. As a result, all the radial integrals were presented in the form of absolutely converging series, well suited for numerical computations.

3 . Results and discussions

The quantitative relation between shift and splitting of atomic levels with respect to the external field is thus determined by relations between the set of polarizabilities and hyperpolarizabilities, presented above. Therefore, it is important to know their numerical values for the correct examination of the atomic spectra in the field of the light wave.

The hyperpolarizability (γ) may be also resolved into five irreducible parts (similarly to the irreducible parts of the polarizability tensor (Equation 8)), $\gamma_q (q = 0,1,2,3,4)$ of which only scalar term γ_0 contribute in a state with the total momentum $J = 0$ [7]. The general expressions for the tensor components of γ_q have quite complicated forms [7], but they can be simplified significantly in the case of particular atomic states. For example, the hyperpolarizability of the ground state with two equivalent electrons can be presented as [3] (fine structure effects are omitted here):

$$\gamma^{l}(\omega)=\gamma^{c}(\omega)+\frac{8}{9}\left[\sigma_{101}(\omega,2\omega,\omega)+\frac{3}{5}\Sigma_{121}-\frac{2}{5}\sigma_{121}(\omega,2\omega,\omega)\right] \quad (18)$$

$$\gamma^{c}(\omega)=\frac{8}{9}\left[\Sigma_{101}+\frac{1}{5}\Sigma_{121}+\frac{6}{5}\sigma_{123}(\omega,2\omega,\omega)\right]-2\alpha_0(\omega)S_{-3}(\omega) \quad (19)$$

where

$$\alpha_0(\omega)=\frac{2}{3}\left\langle 0\left| r\left(g_1^{\omega}+g_1^{-\omega}\right) r\right| 0\right\rangle$$

is the polarizability,

$$S_{-3}(\omega)=\frac{2}{3}\left\langle 0\left| r\left(g_1^{\omega}g_1^{\omega}+g_1^{-\omega}g_1^{-\omega}\right) r\right| 0\right\rangle$$

is the so-called frequency-dependent oscillator strength moment; g_l^{ω} is the radial Green's function in the subspace of the jumping electron's states with the angular momentum l; the following notations for the radial matrix elements and their combinations were used above:

$$\Sigma_{l_1l_2l_3}=\sigma_{l_1l_2l_3}(\omega,0,\omega)+\sigma_{l_1l_2l_3}(\omega,0,-\omega),$$

$$\sigma_{l_1l_2l_3}(\omega_1,\omega_1,\omega_3)=R_{l_1l_2l_3}(\omega_1,\omega_1,\omega_3)+R_{l_1l_2l_3}(-\omega_1,-\omega_1,-\omega_3)$$

$$R_{l_1l_2l_3}(\omega_1,\omega_1,\omega_3)=\left\langle 0\left| r\, g_{l_1}^{\omega_1} r\, g_{l_2}^{\omega_2} r\, g_{l_3}^{\omega_3} r\right| 0\right\rangle$$

Results of numerical calculations are presented in Figures 2-4.

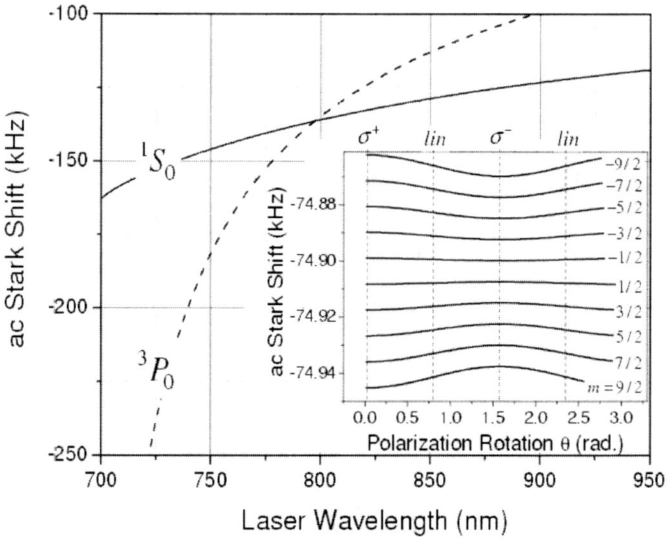

Figure 2. Light shift of Sr atom as a function of the trapping laser wavelength at intensity $10kW / cm^2$ in the vicinity of the crossing point (from [3]). The inset illustrates the insignificance of the polarization-dependent light shifts of the $^3P_0(F = 9/2)$ state by taking into account the dipole coupling to the $^3S_1$ and $^3D_1$ state in the presence of a magnetic field $B_0 = 30mG$.

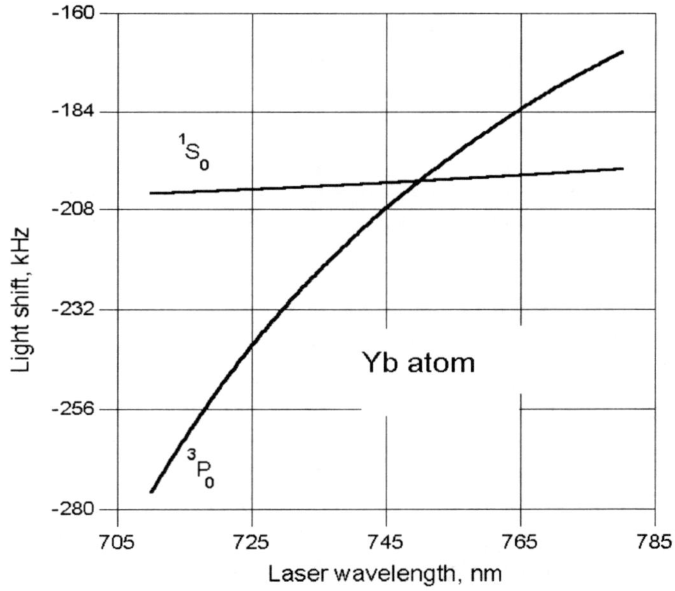

Figure 3. Light shift of Yb atom as a function of the trapping laser wavelength at intensity $10kW / cm^2$ in the vicinity of the crossing point for linearly polarized beam.

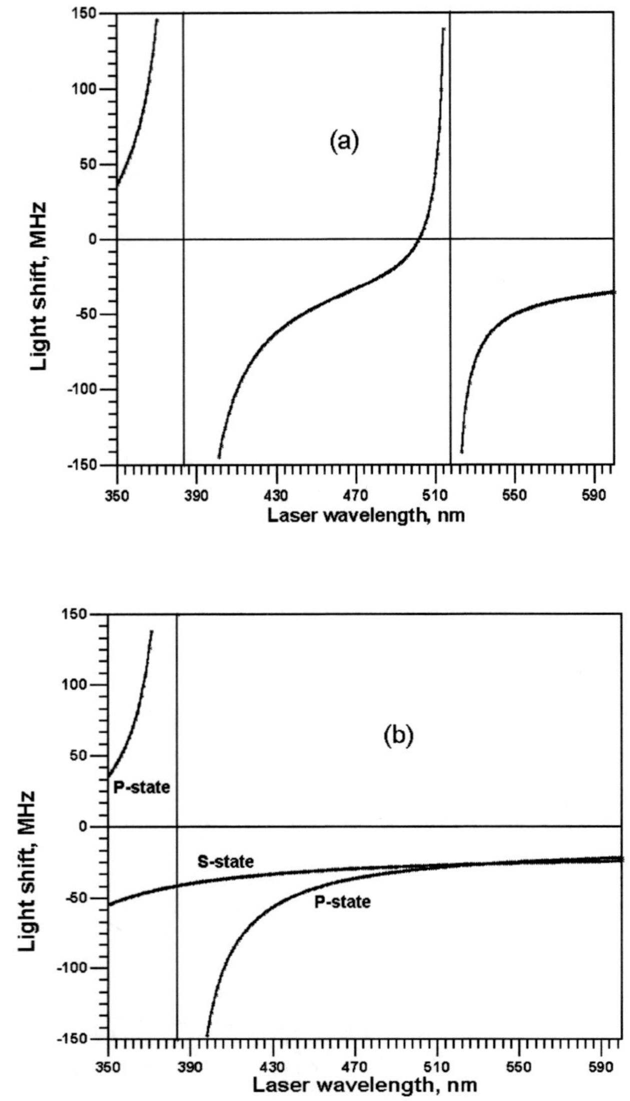

Figure 2. Dispersion properties of the second order light shift of Mg atom as a function of the trapping laser wavelength for a laser intensity of $45W\ /\ (40\mu m)^2$ for linear type of polarization: (a) light shift for the $^3P_1$ $(M = 0)$ state; (b) light shift for the $^1S_0$ $(M = 0)$ and $^3P_1$ $(M = 1)$ states.

The computation is performed by summing terms up to $n = 10$, in which we employed a multiconfiguration Dirac-Fock (GRASP) program for the determination of the dipole matrix elements, together with experimental transition energies. Alternatively, we have also carried out similar calculations with the Fues model potential method. These calculations have reproduced Figure 2. to within 2%-3% accuracy, confirming the validity of these two independent approaches. Thus, the crossing point for both the $5s5p\ ^3P_0$ ealized at the wavelengths of about 795 nm.

the E1 polarizability, we have taken into account two terms ($\hat{V}_{M1}$ and $\hat{V}_{E2}$) in the interaction Hamiltonian (Equation 6). Numerical estimates for the crossing point in Figure 2 give $\alpha_{M1} \approx \alpha_{E2} \approx 10^{-7}\alpha_{E1}$ for the both levels.

The numerical results for hyperpolarizability (γ) in the $5s^2\ ^1S_0$ and $5s5p\ ^3P_0$ states at the crossing point (Figure 2) are 6.3×10^6 a.u. and 2.7×10^8 a.u., respectively. Thus, as seen from these data and Equation (4) the fourth-order Stark contribution ($\Delta E^{(4)}_{5s^2\ ^1S_0} \approx -10^{-4}$ Hz and $\Delta E^{(4)}_{5s5p\ ^3P_0} \approx -3\times10^{-3}$ Hz) is not significant for the employed trapping laser intensity of $10kW\ /\ cm^2$. The present study may also be important for further experimental investigations on the Yb atoms (Figure 3) and Mg atoms (Figure 4).

4. Conclusions

To support efforts on cooling and trapping measurements of Mg, Sr and Yb atoms and designs of atomic clocks in optical domain, we perform the second-, and fourth-order ac susceptibilities for these atoms in the ground and excited ^3P_J-states. In particular, in order to provide the theoretical analysis for precise measurements employing a "light shift cancellation technique" [2-3] we have used the general theory of [7-9] for consecutive estimates for the higher-order light-shift contributions (described by ac hyperpolarizabilities, magnetic dipole $_{M1}$ and electric quadrupole $_{E2}$ terms in addition to the electric-dipole polarizability α_{E1}). For the accurate calculations of these quantities we have used the Fues' model potential method [7-12]. The key point is that it ia consistent to use analytical wave functions for the evaluation of radial matrix elements, together with experimental ionization energies. Quite simple presentation for the Green's function enables us to evaluate the contribution of the higher excited states (including continuum) which are neglected in the finite-sum approach.

As a result, calculations of ac multipole polarizabilities for the $5s^2\ ^1S_0 - 5s5p\ ^3P_J$ clock transition in Sr atom indicate that the contribution of the higher-order light shifts can be reduced less than 1 mHz, allowing for a projected accuracy of better than 10^{-17}. The present study may also be important for further experimental investigation on the other alkaline-earth atoms, where the experimental accuracy can be improved considerably [5]. The theoretical approach presented in this work is sufficiently universal and allows one to obtain not only a full set of second- and fourth-order light shift corrections in the spectrum of atoms considered, but also to outline a method for systematic calculations of the ac Stark effect in higher-orders of perturbation theory that are of interest from the viewpoint of modern experiments.

Acknowledgements

This work was supported in part by the Deutsche Forschungsgemeinschaft under project SFB 407, by the US Civil Research and Development Foundation (US CRDF), by Japan Society for the Promotion of Science under Grand-in-Aid for Young Scientists (A) KAKENHI 14702013 and by the Ministry of Education of the Russian Federation (Grant No VZ-010-0). Valuable discussions with Dr. P. Lemonde (BNM-SYRTE, Observatoire de Paris, France) and Professor T.H. Yoon (Kriss, South Korea) are acknowledged.

References

[1] H. Katori, T. Ido, Y. Isoya, M. Kuwata-Gonokami. "Laser cooling of strontium toward quantum degeneracy", *Proceedings of the 17th International Conference ICAP 2000*, edited by E. Arimondo, P. De Natale and M. Inguscio (AIP Conference Proceedings, Vol.551, Melville, New York, 2001) pp.382-396.

[2] H. Katori. "Spectroscopy of strontium atoms in the Lamb-Dicke confinement", *Proceedings of the 6th Symposium on Frequency Standards and Metrology*, edited by P. Gill (World Scientific, Singapore, 2002) pp.323-330.

[3] H.Katori, M. Takamoto, V.G. Pal'chikov, and V.D. Ovsiannikov. "Ultrastable optical clocks with neutral atoms in an engineered light shift trap", *Phys. Rev. Lett.,* **91**, 173005, (2003).

[4] S.G. Porsev, A. Derevianko, E.N. Fortson. "Possibility of an optical clock using the 6^1S_0 - 6^3P_0 transition in ^{171,173}Yb atoms held in an optical lattice", *Phys. Rev.,* **69A**, 021403(R), (2004).

[5] T. Binnewies, G.T. Wilpers, U. Sterr, F. Riehle, J. Helmcke, T.E. Mehlstraubler, E.M. Rasel, and W.Ertmer. "Doppler cooling and trapping on forbidden transitions", *Phys. Rev. Lett.,* **87**, 123002, (2001).

[6] H. Bergström, C. Levinson, H. Lundberg, S. Svanberg, C.G. Wahlström, and Zhao You Yuan. "Hyperfine-dependent lifetimes induced by singlet-triplet mixing", *Phys. Rev.,* **33A**, pp. 2387-2390, (1986).

[7] N.L. Manakov, V.D. Ovsiannikov, and L.P. Rapoport. "Atoms in a laser field", *Phys. Reports* **141**, pp. 319-433, (1986).

[8] N.L. Manakov, and V.D. Ovsiannikov. "The use of a model potential for the calculation of dynamic polarizabilities, dispersion forces and the light shifts of atomic levels", *J.Phys.* **10B**, pp. 569-581, (1977).

[9] A. Derevianko, W.R. Johnson, V.D. Ovsiannikov, V.G. Pal'chikov, D.R. Plante, and G. von Oppen. "Higher-order Stark effect on an excited helium atom", *Phys. Rev.,* **60A**, pp. 986 -995, (1999).

[10] A. Derevianko, W.R. Johnson, V.D. Ovsiannikov, V.G. Pal'chikov, and D.R. Plante, and G. von Oppen. "Fine-structure effects in relativistic calculations of the static polarizability of the helium atom", *Zh. Eksp.Teor.Fiz.* **115**, pp. 494-504, (1999) [*JETP* **88**, pp. 272-277, (1999)].

[11] I.L. Bolgova, V.D. Ovsiannikov, V.G. Pal'chikov, A.I. Magunov, and G. von Oppen. "Higher orders of perturbation theory for the Stark effect on an atomic multiplet", *Zh. Eksp.Teor.Fiz.* **123**, pp.1145-1159, (2003) [*JETP* **96**, pp. 1006-1018 (2003)].

[12] V.G. Pal'chikov, Yu.S. Domnin, and A.N. Novoselov. "Black-body radiation effects and light shifts in atomic frequency standards", *J.Opt.*, **5B**, pp. S131-S135, (2003).

SHORT AND LONG TERM STABILITY OF HP CS STANDARDS INCLUDING ENVIRONMENTAL EFFECTS

K. Kalliomäki [1], T. Mansten [1], J. Mannermaa [2]

[1] MIKES, Electricity Group, Otakaari 7B, FI-02150, Espoo, Finland
[2] Nokia Mobile Phones, P.O.Box 68, FI-33721, Tampere, Finland

Keywords: Cesium clock, Allan stability, Environmental effects

Abstract

The Finnish Centre for Metrology and Accreditation (MIKES) has two HP 5071A Cs-beam standards (standard version from 1997 and high stability version from 2000) and one CH1-76 Passive hydrogen maser (since February 2003). A special four-channel phase comparator with better than 0.3 ps single shot resolution is used to study the short-term stability of the Cs-standards, with hydrogen maser as a reference. Long term stability is measured with GPS as a reference. Allan stability plots indicate that the high stability version 5071A is approximately three times better than the standard version. The high stability version has only slightly smaller long term drift (ageing) than standard version but frequency retrace after clock failure or power blackout is significantly better in high stability version.

1 Introduction

Short and long term stabilities are of utmost importance in keeping time and frequency. Rubidium oscillators and -clocks are becoming more popular and their testing requires highly stable references. HP 5071A Cs-clocks are quite common as primary standards in calibration laboratories but sometimes the stability of a standard tube clock is not good enough.

Approximately half of industry reference oscillators to be calibrated are rubidium oscillators. As a routine test of oscillators in MIKES short-term stability, i.e. Allan FDEV is determined. "Time is money"; measurements and tests in industry are done faster and faster. This means that the importance of short-term stability is enhanced.

We checked both long and short-term stability of our Cs-clocks. Due to our new hydrogen masers and corresponding phase comparator femtosecond range instabilities are easily measured. In addition, frequency or phase shifts caused by problems in air conditioning can be measured and evaluated with specifications of standards. Earlier we have studied Oscilloquarz Cs-standards /2/.

As a conclusion the performance of HP 5071A Cs-clocks is much better than their specifications. The only problem encountered with the clocks is that they both have once suffered similar failures for unknown reason.

2 Measurement system

An automatic clock and environmental monitoring system has been in operation since 1989 /1/. Sixteen analogue and sixteen digital channels operate at ten minutes intervals. We use GPS-receivers as long-term phase standards and Cs-clocks as flywheels. Due to excellent stability of our new hydrogen masers, one of them will replace Cs-clock as the Master clock in the near future. The phase comparator has four channels and supports 10 fs time interval resolution.

When studying environmental effects we simply waited for problems in air conditioning to happen. In normal ambient conditions changes in temperature and humidity are so small that we cannot resolve the frequency changes caused by them. Air pressure seems to vary too fast to produce any detectable phase or frequency changes.

During the last seven years we have had once or twice per year problems with air conditioning. They have produced valuable information about the environmental stability of clocks. In addition, our UPS-power system was once unintentionally switched off by removal staff late in the evening. All our clocks were turned off for several hours giving important information e.g. from reproducibility after restart.

3 Results

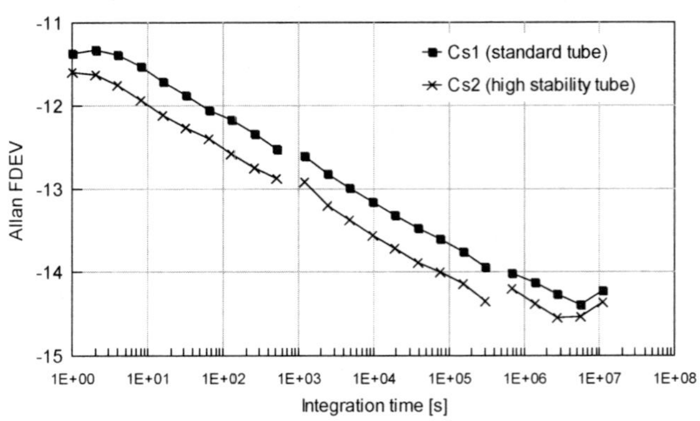

Fig. 1: Allan FDEVs for the two HP 5071A Cs clocks

Fig. 1 shows Allan FDEV from 1 s up to 100 days integration time for both Cs-clocks. Due to extremely wide time range the FDEV curves were calculated from three data groups with different sampling intervals (1 s, 10 min and one day).

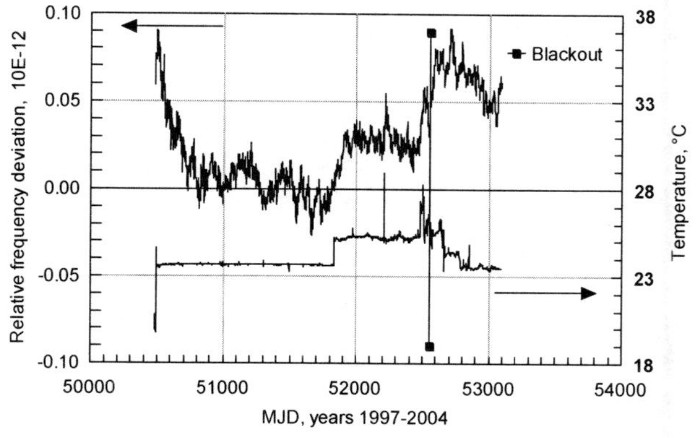

Fig. 2: Relative frequency deviation of Cs1

Fig. 2 shows relative frequency deviation of Cs1 (standard tube) during 7 years of continuous operation. Smoothing time is 30 days corresponding the flicker floor. In the same picture the temperature of Cs-cabinet is shown and blackout time is marked.

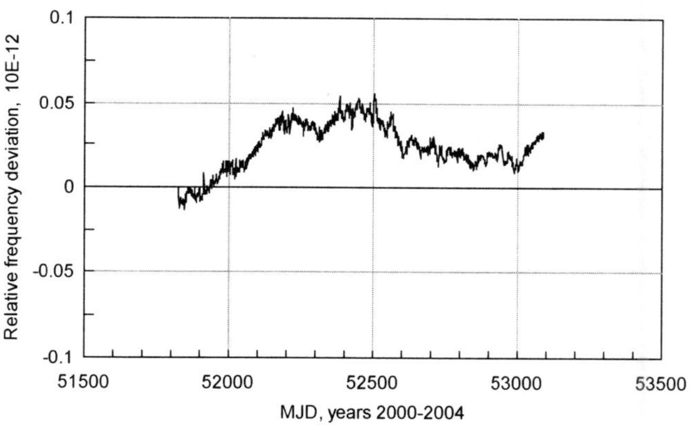

Fig. 3: Relative frequency deviation of Cs2

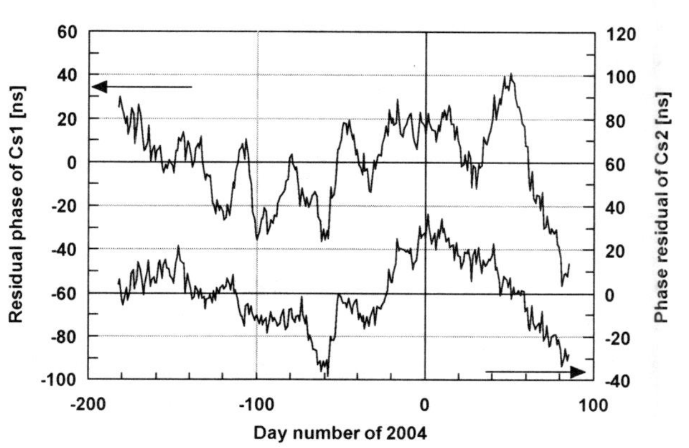

Fig. 4: Residual phase fluctuation of both Cs-clocks

Fig. 3 shows the relative frequency of Cs2 (high stability) during 4 years of continuous operation (excluding the blackout mentioned earlier). Smoothing time is 30 days corresponding to the flicker floor.

Fig. 4 shows the residual phase fluctuation of both Cs-clocks during the last 300 days.

The effects of environmental factors (temperature, pressure and humidity) on our standards seem to be clearly within specifications.

4 Discussion

The Allan FDEV plots in Fig. 1 indicate that the stability of the high stability version of HP 5071A is more than two times better than the stability of the standard version. When comparing to the specifications, the performances seem to be nearly one decade better than quality guarantees. Statistically thinking, accuracy in mind, one high stability version replaces at least four standard versions!

The frequency of Cs1 has varied approximately $1 \cdot 10^{-13}$ (peak-to-peak) during 7 years operation. After first start-up an initial aging lasting one year is clearly visible. Later we can see some environmental effects. At MJD 51850 the temperature of Cs-cabinet was increased by 1.5 °C and the relative frequency increased by $3 \cdot 10^{-14}$. Frequency retrace after power blackout at MJD 52550 is interesting, too. The effect seems to be roughly $2 \cdot 10^{-14}$.

The frequency of Cs2 has varied approximately $0.5 \cdot 10^{-13}$ (peak-to-peak) during 4 years of operation. In this case we cannot see any "initial aging", because Cs2 had already operated for 12 months in another institution. Neither can we see any temperature or retrace effects. Cs2, high stability version, is clearly better than the standard version Cs1.

Slow phase fluctuation during the last 300 days is about 90 ns peak-to-peak at Cs1 and 60 ns peak-to-peak at Cs2. One interesting observation is that the random phase variation is more abrupt in Cs1 than in Cs2.

References

[1] A Computerized Time and Frequency Standards Monitoring System, Kalliomäki, Kalevi; Mansten, Tapio, proceedings of the 4th European Frequency and Time Forum/EFTF 90, Neuchatel, Switzerland, 13 - 15 March 1990. (1990), pp. 177 – 179

[2] Aging and environmental stability of Oscilloquarz Cs-standards, Kalliomäki, Kalevi; Mansten, Tapio; Rautiainen, Anssi. Proceedings of 12th European Frequency and Time Forum, Varsaw, PL, 10 - 12 March, 1998 Tele & Radio Research Institute. Varsaw. (1998), pp. 132-133

IMPROVEMENTS OF THE FIRST CS FOUNTAIN CLOCK, FO1

P. Rosenbusch, C. Vian, S. Zhang, G. Santarelli, and A. Clairon
BNM-SYRTE, Observatoire de Paris, 61 Avenue de l'Observatoire, 75014 Paris, France

C. Salomon
Laboratoire Kastler Brossel, Ecole Normale Suprieure, 24 rue Lhomond, 75231 Paris, France

Keywords: Atomic fountain, frequency measurement, cold collisions, adiabatic passage

ABSTRACT

To date atomic fountain clocks using laser cooled Cs atoms prove to be the most accurate frequency standards. Among the three set-ups at BNM-SYRTE, the historically first Cs fountain clock, FO1, has recently undergone substantial changes. The cold atom preparation and selection have been rebuilt. Selection of the $F = 3$, $m_F = 0$ state can be carried out by adiabatic passage allowing the preparation of 100% or 50% of the atom density with a precision of 10^{-3}. Thereby, we can quantify the collisional frequency shift at a level of 10^{-16}. Furthermore, a cryogenic sapphire oscillator is used for the synthesis of the microwave interrogation signal. With these improvements FO1 is currently operating at a stability of $2.8 \times 10^{-14}\text{s}^{-1/2}$ and an accuracy of a few 10^{-16}.

INTRODUCTION

Since the first operation of an atomic fountain clock using laser cooled Cs in 1993 [1], this kind of device has become the most precise primary standard with a handful of set-ups around the world currently contributing to the International Atomic Time (TAI). Following the historic set-up FO1, the BNM-SYRTE has developed 2 further devices: the double fountain FO2 which can operate with Cs or Rb, and the mobile fountain FOM. Recently, considerable changes have been completed on FO1 and operation has been resumed with a good stability making it the second primary standard to explore the low $10^{-14}\tau^{-1/2}$ region. The changes concern mainly the cold atom preparation, state selection and detection.

COLD ATOM PREPARATION

The cold atom preparation no-longer employs a magneto-optical trap, instead an optical molasses is directly loaded from a decelerated atomic beam. The atomic beam is created by diffusing Cs vapour through a bundle of 420 capillary tubes (100 μm inner diameter, 200 μm outer diameter, 8 mm length) which are heated to about 120°C. The beam travels with a mean velocity of $\sim$ 300m/s and a divergence angle of 21 mrad and is decelerated to a few m/s by a counter-propagating laser beam. The laser is frequency chirped by 350 MHz in 4 ms to account for the Doppler shift.
The slowed atoms are captured in an optical molasses of linear $\perp$ linear polarisation, where its beams of 26 mm diameter are aligned in the (111) geometry. The beams, no-longer limited in diameter by the interrogation cavity, are provided from custom made optical collimators which expand the light emerging from an optical fibre. The collimators are mounted directly onto the precision machined vacuum chamber as to be aligned to better than 1 mrad. During 400 ms we load about 3×10^8 atoms. After switching to a moving molasses the atoms are accelerated to typically 4 m/s and post-cooled in the moving frame to 1-2μK.

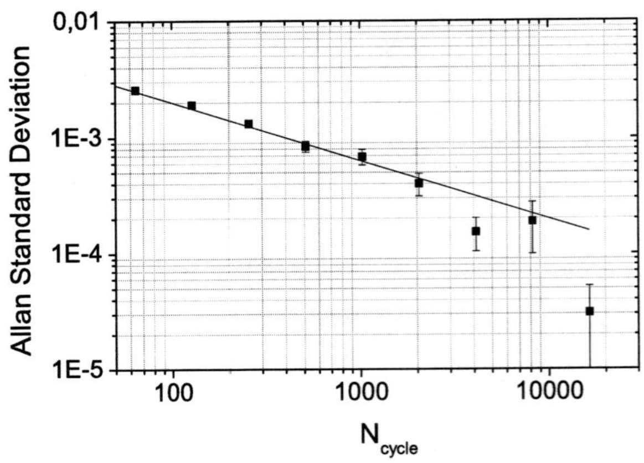

FIG. 1: Allan standard deviation of the ratio $N^{(50)}_{atom}/N^{(100)}_{atom}$ as a function of the number of experimental cycles. After 500 cycles the ratio is known to better than 10^{-3}.

STATE SELECTION

8 cm above the capture zone the atoms pass through the state selection cavity. By the method of adiabatic passage [2] either 50 or 100% of the atoms in $F = 4$, $m_F = 0$ are transferred into $F = 3$, $m_F = 0$. A successive push beam resonant with $F = 4$, eliminates all other atoms. The technique of adiabatic passage has the advantage of being robust against variations in the experimental parameters and thereby allows us to realize the ration 50% in atom number and density to within 10^{-3} after 500 cycles. Figure 1 shows the Allan

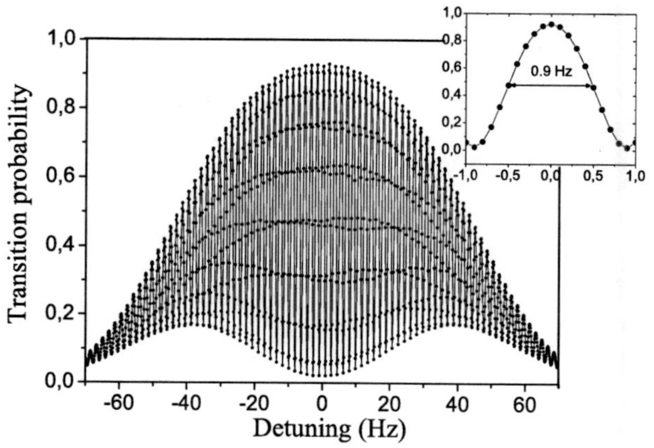

FIG. 2: Ramsey finges obtained with the modified set-up of FO1. The transition probability from the state $F = 3$, $m_F = 0$ to $F = 4$, $m_F = 0$ is plotted as a function of the detuning from the atomic resonance. The inset shows a zoom of the central finge. Its FWHM is 0.9 Hz.

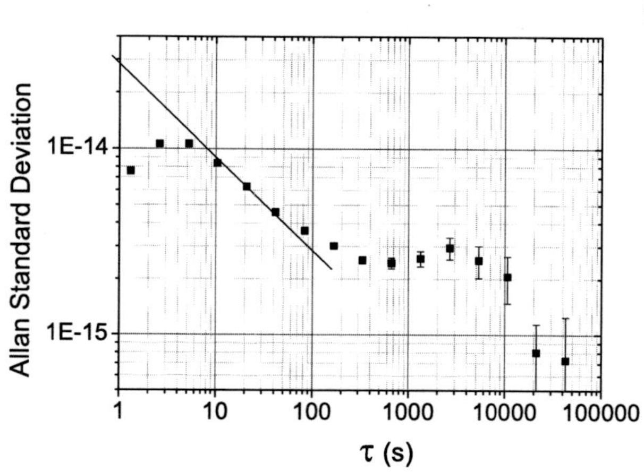

FIG. 3: Relative Allan standard deviation of the interrogation frequency as measured by the atomic fountain as a function of the duration of integration τ

deviation of the experimental ratio $N^{(50)}_{atom}/N^{(100)}_{atom}$ as a function of the number of cycles, N_{cycle}. The experimental data follow the line representing the function $0.02 \times N_{cycle}^{-1/2}$.
The selection by adiabatic passage allows us to prepare two atom samples with identical properties apart from a precise ratio of 1/2 in the atom density. Thereby we can measure systematic effects such as the collisional shift and the cavity pulling to a level of 10^{-16} and extrapolate to the case of an isolated atom, as it is employed in the definition of the second.

DETECTION

The Ramsey zone of interrogation has not been modified with respect to the initial set-up [1]. Only the microwave interrogation signal now is referenced to a cryogenic sapphire oscillator [3] locked to a hydrogen maser.
After interrogation the atoms are detected 14 cm above the capture zone (5,5 cm above the push beam). With respect to the old set-up where the detection was below the capture zone, this has the advantage of a shorter cycle and a more symmetric atom trajectory, diminishing the effect of microwave leakage.
The detection system is composed of two laser beams of 7×10 mm^2 and practically flat intensity distribution. Apart from a 2 mm region at the bottom of the upper beam, the light is retro-reflected to balance radiation pressure on the atoms. The upper beam detects the atoms in F=4, the lower thanks to mixed-in repump light, those in F=3. The fluorescence from the atoms is collected with an efficiency of $\sim 3\%$ onto two photodiodes. All detection beams can be power stabilized using photodiodes which are illuminated by un-used parts of the beams.

STABILITY PERFORMANCE

Figure 2 displays the typical Ramsey fringes recently obtained with the modified set-up of FO1. The width of the central fringe is 0.9 Hz and the signal-to-noise ratio 3000, with typically 3×10^6 detected atoms. Figure 3 shows the relative Allan standard deviation σ_y of the interrogation frequency as measured by the atomic fountain as a function of the duration of integration τ. The line indicates the dependence $\sigma_y = 2.8 \times 10^{-14}\tau^{-1/2}$. For $\tau > 100$ s the deviation increases due to a combination of the drift of the cryogenic oscillator and the hydrogen maser phase noise. At 20000 s, the measured deviation decreases to below 10^{-15} reaching the frequency flicker floor of the maser. The data given in Figure 3 prove FO1 to be, after FO2 [4], the second realization of a frequency standard with a stability in the low $10^{-14}\text{s}^{-1/2}$ region.

CONCLUSION

The historically first Cs atomic fountain clock, FO1 has undergone substantial changes concerning the cold atom preparation, selection and detection. Thanks to the method of adiabatic passage the collisional shift and cavity pulling can be determined to a few 10^{-16}. The new set-up displays a relative frequency stability of $2.8 \times 10^{-14}\tau^{-1/2}$ when compared to a cryogenic sapphire oscillator. Proposing of two fountain clocks with such a stability an unprecedented comparison between two primary frequency standard becomes possible at BNM-SYRTE.

ACKNOWLEDGMENTS

The authors wish to thank H. Marion and S. Bize for fruitful discussions and technical support. This work was financed in part by BNM and CNRS. BNM-

SYRTE and Laboratoire Kastler-Brossel are Unités Associées au CNRS, UMR 8630 and 8552.

[1] A. Clairon *et al* Proceedings of the Fifth Symposium on Frequency Standards and Metrology, p. 49-59, World Scientific (1995)

[2] F. Pereira Dos Santos *et al.*, Phys. Rev. Lett. **89**, 233004 (2002).

[3] A. N. Luiten *et al*, IEEE Trans. Instrum. Meas. **44**, 132, (1995)

[4] Paper by H. Marion *et al* of these proceedings

ADIABATIC INVERSION AND INTERFEROMETRY OF ATOMIC STATES ON ALKALINE ATOMS

V.G. Pal'chikov [1],Yu.S. Domnin [1],G.A. Yolkin [1], A.V. Novoselov [1], V.N. Barychev [1], Yu.L. Sokolov [2]

[1] Institute of Metrology for Time and Space at National Research Institute for Physical-Technical and Radiotechnical Measurements, Mendeleevo, Moscow Region 141570, Russia ; E-mail: vitpal@mail.ru , Fax: +007-095-534-0609
[2] Kurchatov Institute Russian Research Centre, Moscow 123182, Russia

Keywords: Adiabatic passage , population inversion.

Abstract

We consider a two-level atomic system where a pair of discrete quantum states are coupled by a general time-dependent electromagnetic field. In the first part of the paper, effects of stationary interference 2S-2P states of hydrogen-like atom embedded in the electric field are studied. With decreasing of the effective rise time of the field, the measured and calculated population inversion exhibits a transition from adiabatic to sudden behavior. In the second part, the adiabatic transition probability for the Cs atom to be excited following the pulse is studied as a function of the field strength and pulse shape.

1 Introduction

The adiabatic rapid passage (ARP) technique has been widely used in magnetic resonance to achieve population inversion of spin system, in measurements of field-atom decay on the emission spectrum in a cavity, in observing self-stepping of optical pulses in atomic vapors, etc. Very recently, this method is used to perform of the "clock's" atomic states for cold Cs atoms with a good defined of atomic density in connection with precise measurements of cold collision frequency shift in atomic fountain [1,2]. In particular, this approach is successfully used to perform a measurement of the cold collision frequency shift in a laser cooled cesium clock for the evaluation of the cesium fountain accuracy on the level 10^{-16}. In the present work this approach has been developed in several ways.

First, we present and discuss results of an experimental and theoretical investigation in which the field-induced Lyman-α radiation in hydrogen is for the first time systematically studied in the ARP regime at small and large field strengths. Specifically, the present study is important for precise Lamb shift measurements in hydrogen by using atomic interferometer method [3-5].

Second, we performed a numerical simulation based on the use of exact solutions of time dependent Bloch equations for two-level alkali atoms, coupled by a general ARP microwave field with a pulse-like form: Blackman (which minimize off-resonance excitation), hyperbolic secant, Lorentzian, Hyperbolic secant squared, Lorentzian squared, and Gaussian. The corresponding transition probabilities as a function of the Rabi frequency, are presented. (including Blackman pulses). These results are useful for precise estimation of the collisional frequency shift in ^{133}Cs fountain clocks.

2 The interference of atomic states for hydrogen-like system

We recall the basic principles [3-5] of observation of the interference of $2\ _{1\ 2}$ (or $2P_{1/2}$) states of atomic hydrogen states, as illustrated in Figure 1.

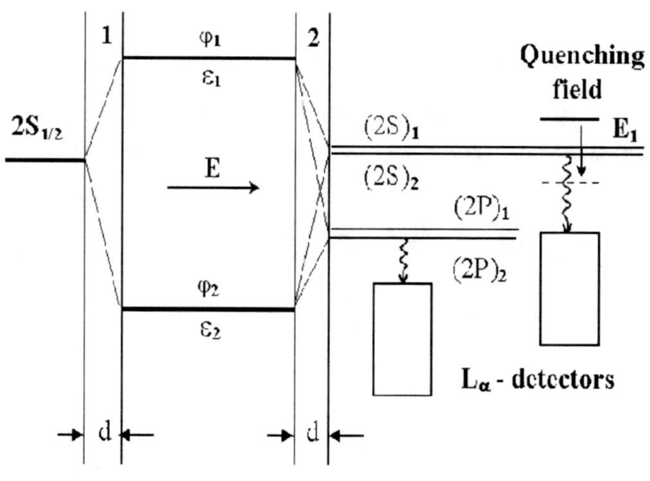

Figure 1: Scheme of the atomic interferometer

A beam of metastable $2S_{1/2}$ hydrogen atoms crosses the electric field E, which sharply (non-adiabatically) drops at the boundaries 1 and 2. After crossing boundary 1, the atoms pass into the superposition of eigenstates ϕ_1 and ϕ_2 with energies ε_1 and ε_2, respectively, which essentially depend on the field strength due to the Stark effect. At the boundary 2, where the field vanishes, each of the states ϕ_1 and ϕ_2 should be considered as a superposition of eigenstates $2S_{1/2}$

and $2P_{1/2}$ in the field-free space. So the resulting amplitude of these states will contain phase- shifted contributions $(2S)_1 - (2S)_2$ and $(2P)_1 - (2P)_2$ from the states ϕ_1 and ϕ_2, respectively, which evolved along different "paths" before. The value of the shift $(\varepsilon_1 - \varepsilon_2)T / \hbar$ depends both on the time-of-flight T and on the frequency $(\varepsilon_1 - \varepsilon_2) / \hbar$ of the transition between terms ε_1 and ε_2. The latter, in turn, is determined by the field strength E. As a result, the monotonic change in the field strength will be accompanied by periodic counter-phase oscillations in the flux intensity of $2S_{1/2}$ and $2P_{1/2}$ atoms, caused by the interference of the $(2S)_1 - (2S)_2$ contribution and, similarly, of the $(2P)_1 - (2P)_2$ contribution. The same picture will be observed at gradual change of time-of-flight, i.e. the distance between boundaries 1 and 2.

The interferometer based on the above principle is similar to a two-beam optical interferometer in which an individual photon interferes with itself. Two “channels ”, ϕ_1 and ϕ_2, appear here due to the electric field playing the role of the semi-transparent mirror that splits the evolutionary “paths”. It mixes the states with opposite parity, so the initial $2S_{1/2}$ state receives a coherent addition of the $2P_{1/2}$ state.

The interference pattern of the short-living $2P_{1/2}$-state ($\tau = 1.59 \times 10^{-9} s$) is observed by measuring the flux of L_α-quanta, which result from one-photon transitions to $1S_{1/2}$-state. To observe the interference of the metastable $2S_{1/2}$-state, the quenching field E_1 is used.

Note that for comparatively weak fields (when the Stark shift is of the order of the Lamb shift), the analysis can be simplified by considering only the two-level system $2S_{1/2}$-$2P_{1/2}$ and by introducing small corrections to account for the effect of the $2P_{3/2}$-level. All experiments described below were performed with a beam from which the $2S_{1/2}$-component with the total angular momentum $F = 1$ was removed by a high-frequency field. In this case the observed interference pattern corresponds to the transition between the states $|1\rangle \equiv |2S_{1/2}, F = 0, F_z = 0\rangle$ and $|2\rangle \equiv |2P_{1/2}, F = 1, F_z = 0\rangle$ with the frequency $\upsilon = 909.9 MHz$.

Figure 2 shows the scheme of the simplest atomic interferometer consisting of two plane electrodes producing the field E and separated by a variable distance L. On the electrodes, there are two slits for the beam passage. The width of the slits meets the condition of the non-adiabatic change of the field E at the boundaries.

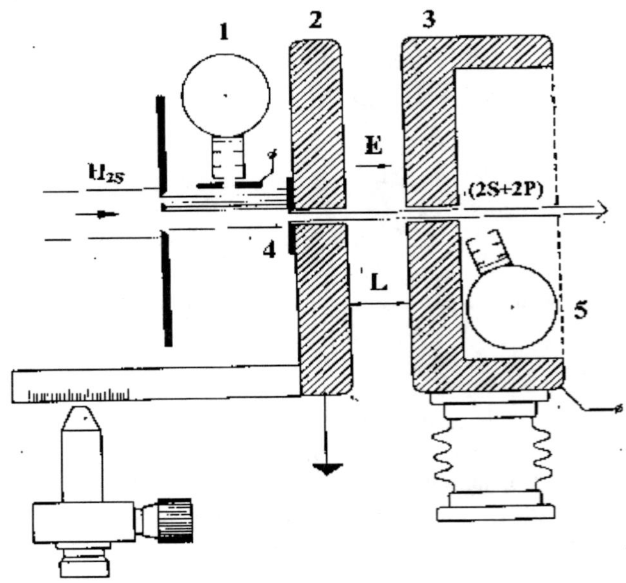

Figure 2: Scheme of the two-electrode interferometer with a longitudinal field: 1 - detector of the $2S_{1/2}$-monitor; 2 and 3 -flat electrodes with slits for passing the beam; 4 - collimator slit; 5 - L_α detector.

2 Nonadiabatic behavior of electric-field-induced Lyman-α radiation

The interference is possible only when the change of the fields at the interferometer boundaries meets the non-adiabaticity condition expressed as $v / d \geq 2\pi\hbar\delta_L$, where v is the beam velocity and d is the scale of the electric field change (Figure 1). Usually the length of the slits (along the particle trajectory) in the input and output electrodes of the interferometer exceeds the slit width (Figure 2). In this case it is the slit width that determines the scale d.

Below we restrict to two-level approximation in describing excited $|\alpha\rangle \equiv |2S_{1/2}\rangle$ and $|\beta\rangle \equiv |2P_{1/2}\rangle$ atomic states.

We investigate the non-stationary solution of the Schrödinger equation

$$i\hbar \frac{\partial}{\partial t} \Psi(t) = \hat{H}\Psi(t) \tag{1}$$

for a two-level atom interacting in the electric-dipole approximation with the static field E. The Hamiltonian $\hat{H}$ is written in two parts:

$$\hat{H} = \hat{H}_0 + \hat{H}_1, \quad (2)$$

where $\hat{H}_0$ is the Hamiltonian of the atom and $\hat{H}_1$ represents the interaction of the atom with the field E. For generality, we shall also assume that the unperturbed Hamiltonian $\hat{H}_0$ contains the anti-Hermitian term first introduced by Weiskopf and Wigner to account for effects of the spontaneous decay of the bound states in the atom in the absence of the field, i.e.

$$\varepsilon_\alpha = \varepsilon_\alpha^{(0)} - i\Gamma_\alpha \hbar / 2 \quad (3)$$

In this formula $\varepsilon_\alpha^{(0)}$ determines the energy of the bound state and Γ_α corresponds to the radiative width of the quasi-stationary state $|\alpha\rangle$. The solution of Equation (1) is obtained by expanding the wave function $\Psi(t)$ in terms of the eigenstates of the Hamiltonian $\hat{H}_0$. We write

$$\Psi(t) = C_\alpha(t)e^{-i\varepsilon_\alpha t/\hbar}|\alpha\rangle + C_\beta(t)e^{-i\varepsilon_\beta t/\hbar}|\beta\rangle \quad (4)$$

and substitution of Equation (4) into Equation (1) yields the following set of coupled linear differential equations for the amplitudes C_α and C_β

$$i\hbar \frac{\partial}{\partial t}\begin{pmatrix} C_\alpha(t) \\ C_\beta(t) \end{pmatrix} = \begin{pmatrix} 0 & e^{\frac{it\varepsilon_{\alpha\beta}}{\hbar}} V_{\alpha\beta} \\ e^{-\frac{it\varepsilon_{\alpha\beta}}{\hbar}} V_{\beta\alpha} & 0 \end{pmatrix} \times \begin{pmatrix} C_\alpha(t) \\ C_\beta(t) \end{pmatrix} = \hat{H}' \Psi(t) . \quad (5)$$

In Equation (5) stands for the time-depended matrix element $\langle\alpha|\hat{H}_1|\beta\rangle = \langle\beta|\hat{H}_1|\alpha\rangle$. The coefficients C_α and C_β of this superposition are proportional to the probability amplitudes for atoms being in states $|\alpha\rangle$ and $|\beta\rangle$, respectively.

In the nonadiabatic approximation the probability of the output atoms to be in the $2P_{1/2}$-state is proportional to the value [3]

$$I_{2P}(E) = \frac{2|V|^2 e^{t(\lambda_1 - \lambda_2)}}{\hbar^2\left[\left(\lambda_1 - \lambda_2\right)^2 + \left(\mu_1 - \mu_2\right)^2\right]} \times \left\{\coth t(\lambda_1 - \lambda_2) - \cos t(\mu_1 - \mu_2)\right\}. \quad (6)$$

Here λ_j and $\mu_j (j = 1,2)$ represent real and imaginary parts of the field dependent frequencies, i.e.

$$\lambda_j + i\mu_j = -i\frac{\varepsilon_1 - \varepsilon_2}{2\hbar} \pm \sqrt{\left(i\frac{\varepsilon_1 - \varepsilon_2}{2\hbar}\right)^2 - \left(\frac{V}{\hbar}\right)^2}, \quad (7)$$

where

$$\varepsilon_1 = \varepsilon_{2S}^{(0)}, \varepsilon_2 = \varepsilon_{2P}^{(0)} - i\hbar / 2\tau_{2P}, \quad V = DE, D \equiv \langle 2S|\hat{D}_z|2P\rangle. \quad (8)$$

Figure 3 shows the typical plot illustrating the interference of $(2P)_1 - (2P)_2$ atomic states in hydrogen at the beam energy 18 keV and a distance of 1 cm between the interferometer plates.

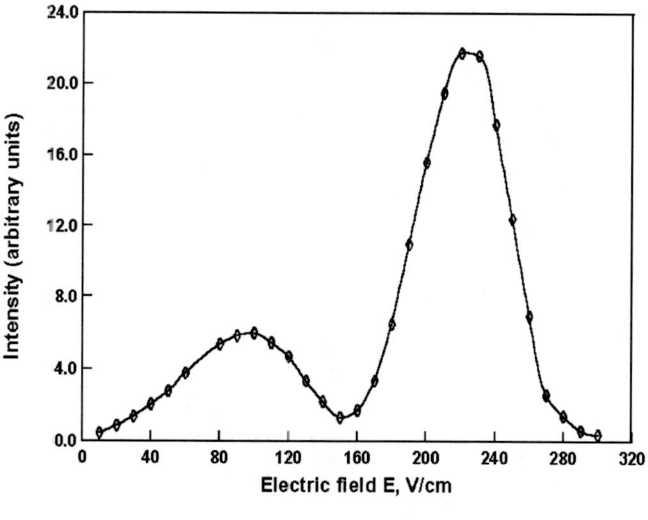

Figure 3: Experimental and theoretical interference curves of $2P_{1/2}$ -atomic states in hydrogen at the beam energy 18 keV. The solid line corresponds to the theoretical calculation with account of the hyperfine structure effects.

This figure shows the results of measurements of the output $(2P_{1/2}, F = 1)$-component of the atomic beam as a function of the longitudinal field strength in the interferometer space. The initial $(2S_{1/2}, F = 1)$-state was prepared beforehand from $2_{1/2}$-atoms by removing the hyperfine $F = 1$ components using UHF-fields localized across the beam trajectory. The corresponding theoretical dependence of the $2P_{1/2}$-atom output I_{2P} on E (the solid curve in Figure 3) was calculated with account of the hyperfine structure effects.

2 Adiabatic passage regime for hydrogen-like atoms

The conditions of the previous investigations were characterised by a fast entrance of the metastable hydrogen atoms into the field region. As it was mentioned before, the effective rise time of the field as seen by the atoms in their rest frame was small as compared with the characteristic times associated with the $2S - 2P$ Lamb splitting in the unperturbed hydrogen system. Here we present results of the investigations in which the interference of atomic states is for the first time studied in the adiabatic passage regime at both small and large field strengths.

The geometry shown in Figures 1,2 can be easily varied to yield the interference pictures ranging from the adiabatic to the impulsive behaviour by changing the slit width. The electric field E at points along the z-axis is given by

$$E(z) = \frac{1}{2} E_0\left[1 + z\left(z^2 + s^2\right)^{-1/2}\right], \qquad (9)$$

where z is the displacement measured from the slit , s is one-half of the slit width, and E_0 is the electric field inside the capacitor far from the slit. Results of measurements are presented in Figure 4.

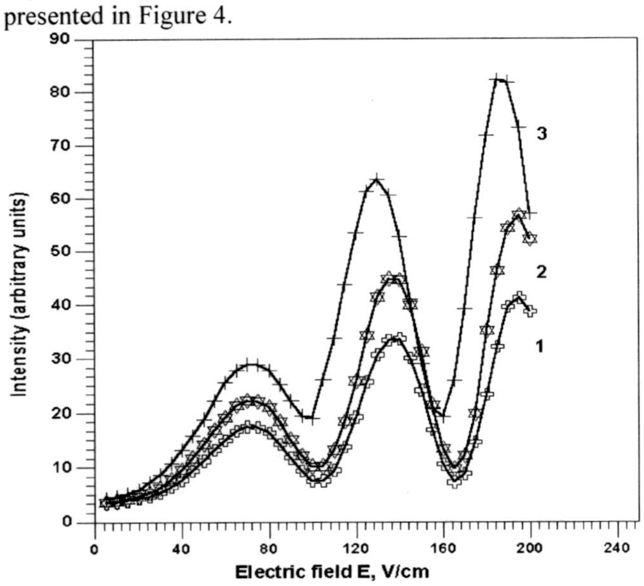

Figure 4: Experimental interference curves of hydrogen at the beam energy 22 keV. All three plots (1)-(3) are for s=0.01, 0.02 and 0.03 cm, respectively.

The similarity between the interference curves is due to the fact, that in these cases the nonadiabatic approximation works quite well.

Figures (5)-(7) show the interference curves I_{2} and I_{2S} for different values of the $^4He^+$ beam velocity and the fixed distance between the interferometer plates 1 and 2. As in the case of neutral hydrogen, the dependence I_{2P} shows well pronounced interference picture of atomic states in an external static field. For that purpose, the solutions for the transition amplitudes in Equation (4) are determined numerically using the above approximation for the electric field (Equation (9)).

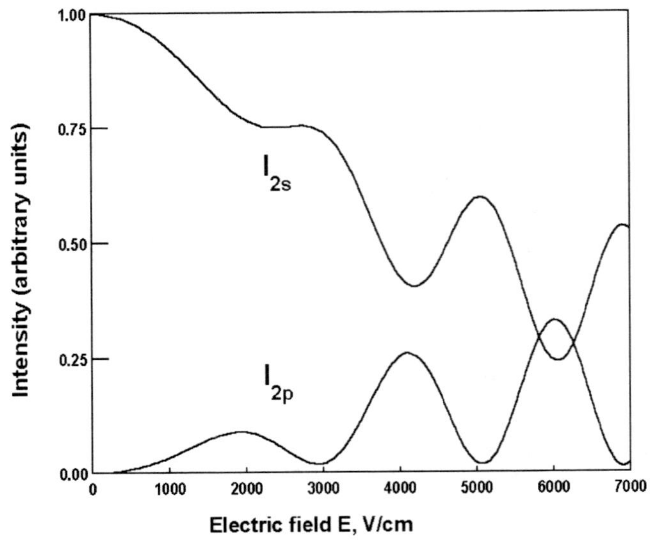

Figure 5: Interference of the $2S$- and $2P$- atomic states in $^4He^+$ for the $v / c = 0.1$ velocity.

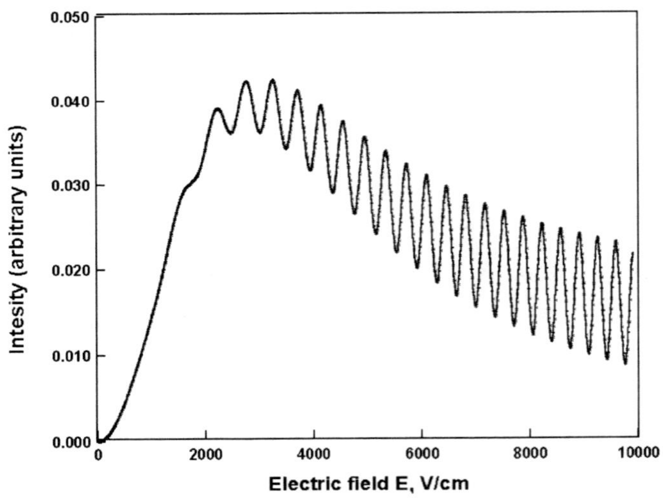

Figure 6: I_{2P}-dependence in $^4He^+$ for the $v / c = 0.05$ velocity.

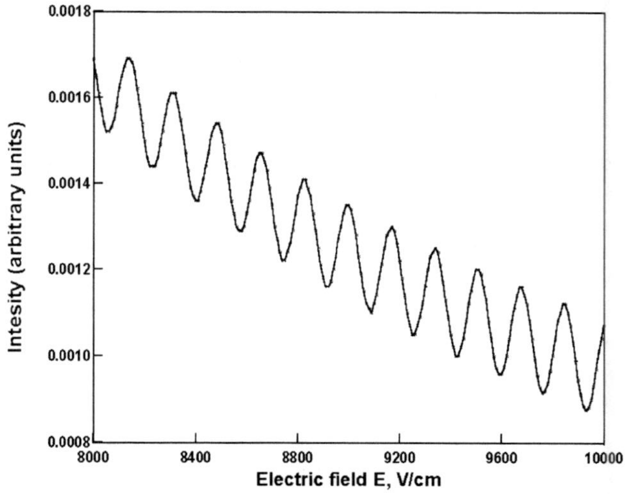

Figure 7: I_{2P}-dependence in $^4He^+$ for the $v / c = 0.01$ velocity.

Thus, if the entry time is larger than the Lamb splitting period, then the interference of atomic states will be near its adiabatic value. Oscillations indicative of impulsive entry will appear when the entry time is less than or nearly equal to the Lamb period. If the entry time is larger than the Lamb period then there are very small oscillations since this is clearly adiabatic (Figure 7).

2 Coherent population transfer in Cs atom based on adiabatic rapid passage method

An application of cold atoms in atomic fountain produce sizable frequency shifts in this kind of frequency standards since atomic collisions are different at the low temperature achievable with laser cooling. This frequency shift results from unequal collision-induced energy shifts of the two hyperfine states of a clock transition. Recently, it has been demonstrated that the population of the atomic states in Cs fountain can change considerably if the atom are contained in a selection cavity by applying the microwave Blackman pulses in an ARP regime [1,2]. In this way, there is an unique possibility to prepare cold atomic samples with a well-defined ratio of atomic density and atom number. In particular, this new approach is successively used to perform a precise measurement for the determination of the cold collision frequency shift at the 10^{-16} level of accuracy.

In connection with these measurements on cold Cs atoms employing ARP technique [1,2] we report here the theoretical results for transition probability of atoms to be excited following the six different pulse shapes: Blackman, Rosen-Zener, Hyperbolic secant squared, Lorentzian, Lorentzian Squared and Gaussian. These calculations are performed in the framework of the adiabatic or semiclassical dressed-state representation by using the rotating-wave approximation. In addition, the spontaneous decay during the pulse duration is assumed to be negligible.

The general time-dependent electromagnetic field vector has the form:

$$E = E_0 f(t) \cos\omega t\,, \qquad (10)$$

where E_0 is the field amplitude and the smooth field envelope function $f(t)$ has a temporal width of order T. It is also assumed that $f(t)$ and all its deviations are continuous functions and

$$\int_{-\infty}^{\infty} f(t)dt = 1 \qquad (11)$$

In the framework of the rotating-wave or resonance approximation, the atomic state amplitude (Equation (4)), written in an interaction representation, evolve as

$$\begin{aligned} \frac{\partial C_\alpha}{\partial t} &= -iBf(t)\exp(iat)C_\beta\,, \\ \frac{\partial C_\beta}{\partial t} &= -iBf(t)\exp(-iat)C_\alpha\,. \end{aligned} \qquad (12)$$

Here C_α and C_β represent the probability amplitudes for the two level considered (ground and excited), $B = -\frac{\mu E_0 T}{2\hbar}$ is a coupling strength (μ is a dipole moment matrix element), and $a = (\omega - \omega_0)T$ is an atom-field detuning. Results of numerical calculations are summarized in Figure 8 and Table 1.

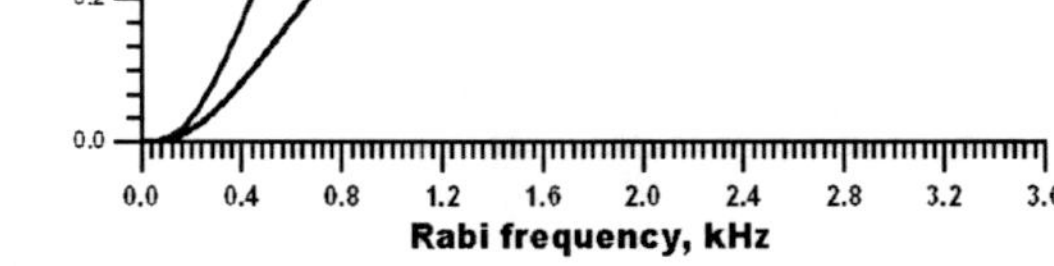

Figure 8: Transition probabilities $|F = 4, m_F = 0\rangle \rightarrow |F = 3, m_F = 0\rangle$ as a function of the Rabi frequency for a Blackman pulse with a duration 4 ms .

The maximal value for the frequency detuning from resonance is 4 kHz.

Pulse	Formula	W1	W2
Rosen-Zener	$f(t)=(1/2)\mathrm{sech}(\pi t/2)$	0.5004	1.0009,
Hyperbolic secant squared	$f(t)=(\pi/4)\mathrm{sech}^2(\pi t/2)$	0.5001	1.0000
Lorentzian	$f(t)=1/\left[\pi\left(1+t^2\right)\right]$	0.5008	1.001
Lorentzian Squared	$f(t)=2/\left[\pi\left(1+t^2\right)^2\right]$	0.5005	1.001
Gaussian	$f(t)=1/\sqrt{\pi}\,\exp\left(-t^2\right)$	0.5004	1.0008

Table 1: Transition probabilities for the full and ½ pulse for the various envelope function $f(t)$ at Rabi frequency in 2.5 kHz.

In our calculations, the detuning a is determined from the equation $\frac{\partial a}{\partial t}=\Omega^2(t)$, where $\Omega/2\pi$ is the Rabi frequency.

For particular case of the Blackman pulse, the presented numerical data are in good agreement with results of the independent calculations and measurement, published in [1,2].

4. Conclusions

We study interference of the $2S_{1/2}$ and $2P_{1/2}$ atomic states due to passage of metastable hydrogen-like atoms through external electromagnetic field. Basic theory of atomic state interference is presented and experimental conditions which allow observation of interference picture with a fast atomic beam are analyzed. Specifically, our study attempts to answer the question related to the relevant time that controls the adiabatic (nonadiabatic) behaviour of the atomic interference. Our numerical simulations for an adiabatic transfer of population for the Cs atomic states, which takes into account the transition probabilities in atomic fountain, are important for the precise determination of the cold collision shift on the "clock transition ".

References

[1] F. Pereira Dos Santos, H. Marion, S. Bize, Y. Sortais, A. Clairon and C. Salomon. "Controlling the cold collision shift in high precision atomic interferometry", *Phys.Rev.Lett.*, **89**, 233004, (2002).

[2] F. Pereira Dos Santos, H. Marion, M. Abgrall, S. Zhang, Y. Sortais, S. Bize, I. Maksimovic, D. Calonico, J.Grunert, C. Mandache, C. Vian, P. Rosenbuch, P. Lemonde, G. Santarelli and A. Clairon, "^{87}Rb and ^{133}Cs laser cooled clocks: testing the stability of fundamental constants ",*Proc. of the 2003 IEEE Int. Frequency Control Symposium and PDA Exhibition jointly with the 17th European Frequency and Time Forum* (Tampa, USA 2003) pp 55-67.

[3] Yu.L. Sokolov and V.P. Yakovlev. "Measurement of the Lamb shift in the hydrogen atom (n=2)", Zh. Eksp. Teor. Fiz. **83,** pp. 15-27, (1982) [Sov.Phys. JETP **56,** pp.7-16, (1982)]

[3] V.G. Pal'chikov, Yu.L. Sokolov and V.P. Yakovlev. "Measurement of the Lamb shift in H, n=2", Metrologia **21,** pp. 99-105, (1985).

[4] Yu.L. Sokolov. "An interference method for measuring atomic state parameters", Physics-Uspekhi **42,** pp. 481-503, (1999).

ATOMIC HYDROGEN MASER GROUP FOR CHINESE POSITIONING SYSTEM

WEI-QUN ZHANG, GUAN-ZHONG WANG, CHUAN-FU LIN, YI-PING ZHANG and ZAO-CHENG ZHAI
Shanghai Observatory, Chinese Academy of Sciences, Shanghai200030
Email: zwq64@sina.com.cn

Three hydrogen masers were put into operation at the earth station of Chinese positioning system four years ago. We have made many physical and electronics improvement on these hydrogen masers for improving the long-term performance. The hydrogen maser's size is 110cm(high)*54cm(width)*73cm(depth). It weighs 250kg.By performance comparing with each other. we may choose the best one as a frequency reference for this system. A kind of cavity auto-tuning system was applied for improving of the long-term performance. Performance evaluation showed the frequency stability is in the order of E-15 for time interval from 80s to 1day and the phase noise (SSB) is -150 dBc/100Hz. The environmental sensitivities are also given in this paper.

Key words: Atomic hydrogen maser Cavity auto-tuning Frequency stability Frequency accuracy

Microwave interrogation in a compact atomic clock

S. Trémine, S. Guérandel, D. Holleville, N. Dimarcq and A. Clairon
BNM-SYRTE – CNRS UMR8630 – Observatoire de Paris
61 avenue de l'Observatoire, 750014 Paris, France
Email: stephane.tremine@obspm.fr

***Abstract*— HORACE is a compact cold atom clock where the atoms are cooled inside the microwave interrogation cavity. About 10^8 atoms can be cooled at kinetic temperatures as low as 2.5 μK. We report for the first time Ramsey pattern observed with a 14 Hz linewidth and fringe contrast better than 80%. Since this clock is designed for Space applications, some properties are extrapolated to micro-gravity operation.**

I. Introduction

Isotropic laser cooling of cesium atoms in Horace has been studied and reported previously [3] [4] [5]. Temperatures as low as 2.5 μK have been obtained. The time cooling sequence and cloud properties are summarized in table I.

	Capture	Sisyphus I	Sisyphus II
Δt (ms)	1000	80	2
N_{at}	1.6 10^8	8 10^7	8 10^7
T (μK)	35	9	2.5
Φ (σ) (mm)	8	4.6	4.6

TABLE I

The duration Δt of each phase is given on the first line. Other lines are relatives to the properties of the cloud at the end of the corresponding phase. N_{at} is the number of cold atoms in the cavity deduced from time of flight signals, taking into account the device's geometry owing to a MonteCarlo simulation. T is the kinetic temperature after correction by the cloud "size". Φ is the cloud diameter given at σ for a gaussian distribution.

Fig. 1. Inside the microwave cavity, one can see the quartz bulb containing the Cs vapor, in which the vacuum is about 2 10^{-9} Torr. The "isotropic" cooling is performed carrying the light to the cavity using multimode optical fibres.

Fig. 2. Experimental setup for the Horace clock project. The microwave cavity where all interactions take place is located in the front end.

Figure 1 shows a section of the physics package. Figure 2 shows the whole experiment including the optical bench.

A clock cycle is divided into three steps : cooling, atomic interrogation, and clock signal detection. If the repetition rate is fast enough the same cold atoms can be used from one cycle to the next. Experiments and results about cold atoms recapture are presented in section II.

Ramsey microwave interrogation has been performed. The atomic response detected with time of flight measurements are presented in section III.

As gravity accelerates the atoms, the interrogation time inside the cavity cannot exceed 80 ms on Earth, but can be increased during Space operation. A comparison of the cold atoms main contribution to systematic effects is made in section IV, with the objective of a short term frequency stability lower than $10^{-12}\tau^{-1/2}$ in Space.

II. Recapture of cold atoms

In a pulsed operating clock, the quartz oscillator is not continuously locked to the atomic signal. The resulting frequency correction sampling is the source of the Dick effect [1]. In our conditions, where the optimal cooling phase is as long as one second, this effect becomes rapidly a limitation for the short-term stability. To avoid this limitation, it is necessary to increase the cyclic ratio given by $R_c = T_{int}/(T_{int} + T_{cool})$, where T_{int} is the interrogation time and T_{cool} the cooling time.

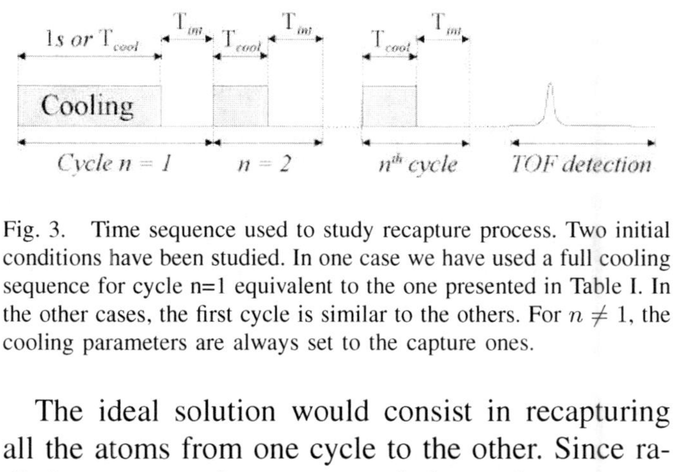

Fig. 3. Time sequence used to study recapture process. Two initial conditions have been studied. In one case we have used a full cooling sequence for cycle n=1 equivalent to the one presented in Table I. In the other cases, the first cycle is similar to the others. For $n \neq 1$, the cooling parameters are always set to the capture ones.

The ideal solution would consist in recapturing all the atoms from one cycle to the other. Since radiation pressure forces are needed to make recapture possible, the laser cooling power and detuning for this process are those used in the capture phase ($\Delta \sim -1.8\Gamma$ and P $\sim$ 35 mW). To test the efficiency of such a process, we have measured the number of atoms still inside the cavity after n cycles by TOF method. The sequence followed is presented on figure 3, and results are presented on figure 4. For 20 ms interrogation time, four different cooling times have been used : 180 ms, 80 ms, 30 ms, 20

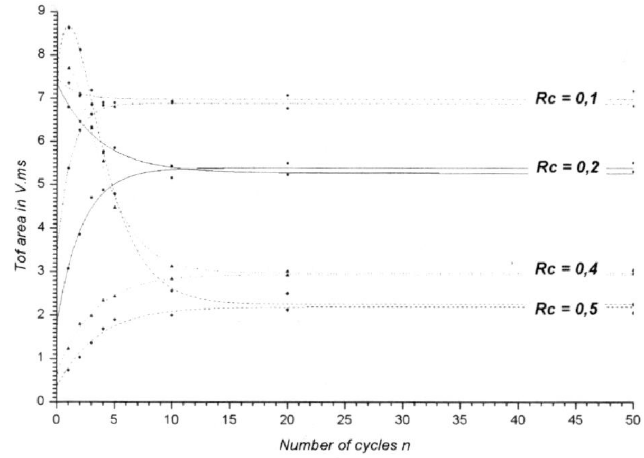

Fig. 4. Time of flight areas versus number of cycles. Each curve is obtained for a given cyclic ratio R_c. T_{int} is set to 20 ms and T_{cool} is the adjustebal parameter in this measurement. For example, given our present cyclic ratio R_c = 0.2, the stationary number of atoms is about 70% of the initial loaded cloud.

ms. For each R_c, we measure the number of atoms recaptured, with two opposite initial conditions. Starting from no cold atom, the recapture process is given by the increasing curves. On the contrary, the decreasing curves correspond to the optimal loading sequence ($\sim$ one second, see table I), in which the recapture process starts from about 10^8 atoms.

In both cases, the more efficient is the recapture process, the worse is the clock cyclic ratio. Moreover, a stationary number of atoms is reached after about 15 cycles, and does not depend on the load of the initial cloud. This is easily understood since the losses occur during the interrogation time. The losses are due to gravity, natural cloud expansion, and collisions with atoms from the thermal vapor. On Earth, gravity is the main loss factor. Indeed, a cold atom needs about 60 ms to fall from the centre and leave the cavity, whereas it would take about 350 ms only limited by its rms velocity (about 4.7 $cm.s^{-1}$).

Furthermore, the gravity changes the cloud shape. On figure 5.b, time of flight signals are recorded after one recapture only, for several dead times. The distorted distributions appearing for long falling times are interpreted as double spatial distributions of the atoms in the cavity, one after the capture phase and the other one after recapture, just where the atoms are trapped when cooling light is back on again. This proves recapture process is not efficient

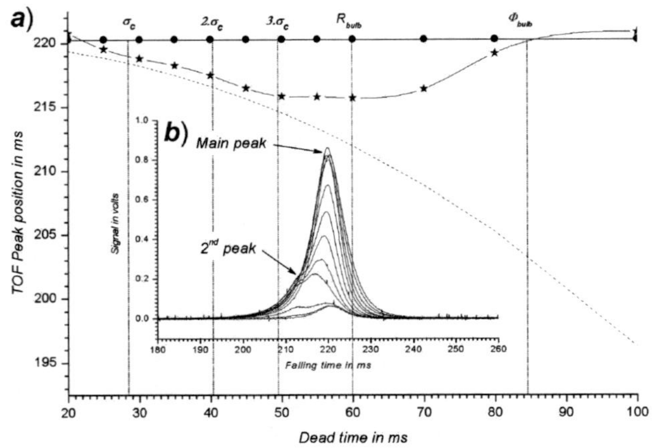

Fig. 5. a) Assuming double Gauss distribution for the time of flight signals, arrival times of the two maxima are plotted against T_{int}. "– • –" : main peak position set to 220.3 ms (expected arrival time for a free fall from the center of the cavity). "– ⋆ –" : second peak position. "..." : expected position of the second peak, resulting of the free fall of the cloud during T_{int}, which is simply given by $\sqrt{\frac{2(H-Z_0)}{g}}$ with $Z_0 = \frac{1}{2}gT_{int}^2$. H = 238 mm, is the distance between the center of the cavity and the TOF beam. b) Time of flight signals obtained for those measurements. In all those measurements, T_{cool} is set to 20 ms.

enough to push back the atoms towards the centre of the cavity. The behaviour of the two clouds is described on figure 5.a where the TOF distribution peaks originate from initial cloud positions. Obviously, the comparison between the movement of the recaptured cloud and the gravity law only makes sense as long as the atoms are still in the capture region. This region has a typical radius of $4mm$ (cloud size) equivalent to $T_{int} < 50ms$.

Finally, the choice of the cyclic ratio will be given by the compromise between the limitation of the short term stability by Dick effect, and the signal to noise ratio depending of the number of atoms recaptured.

III. Microwave interrogation

Microwave interrogation have been performed on cold atoms following the time sequence presented on figure 6. Just after cooling, all the atoms are evenly distributed in all $(4, m_F)$ levels. Next, they are interrogated by two Ramsey pulses for the $(4,0)-(3,0)$ clock transition. The atoms $(4, m_F \neq 0)$ are then blasted using the cooling light blue-shifted by 0.9 Γ from the optical cycling transition. Atoms having made the clock transition are then optically pumped, and $(4,0)$ atomic populations detected 238 mm downwards with the TOF beams.

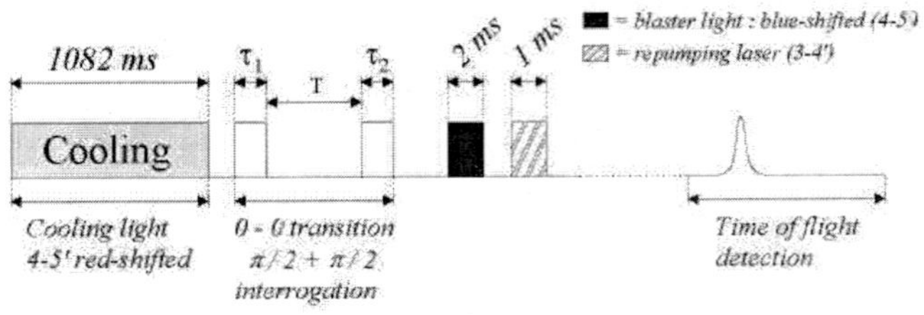

Fig. 6. Time sequence followed for microwave interrogation.

Figure 7 is a typical scan of the microwave frequency across the clock resonance. Experimental data are averaged over four measurements. The central fringe width is 25 Hz corresponding to the 20 ms Ramsey pulse separation. The contrast is presently better than 90%. Experimental data are in good agreement with calculated curve shown on this graph from a simple model describing the time evolution of the atomic population during atom-microwave interaction. The loss of fringe contrast can be attributed to several causes : Rabi frequency pulses unbalance, phase differences induced when switching, stray external magnetic field fluctuations, or any other hyperfine coherence relaxation causes, like collisions. Investigation is still in progress. Figure 8 is a scan around the central fringe obtained for 35 ms Ramsey pulse separation time. The frequency shift from $\nu_0 = 9.192631770$ GHz corresponding to second order zeeman shift is attributed to strong magnetic field inhomogeneities in our setup. Further work will definitely require to improve the experiment. However we can still forecast some performance operation in Space where the atoms are no longer submitted to gravity.

IV. A clock for operation in Space

Gravity is the main loss factor in the recapture process, and limits the number of cold atoms contributing to the clock signal. The resulting loss of signal affects the short term stability :

$$\sigma_y(\tau) \propto \frac{1}{\nu/\Delta\nu} \frac{1}{S/N} \sqrt{\frac{T_c}{\tau}} \tag{1}$$

Gravity also affects clock accuracy because of cloud motion. Figure 9 compares the losses of cold atoms submitted to various gravity accelerations (Earth, reduced gravity and micro-gravity) after one recapture

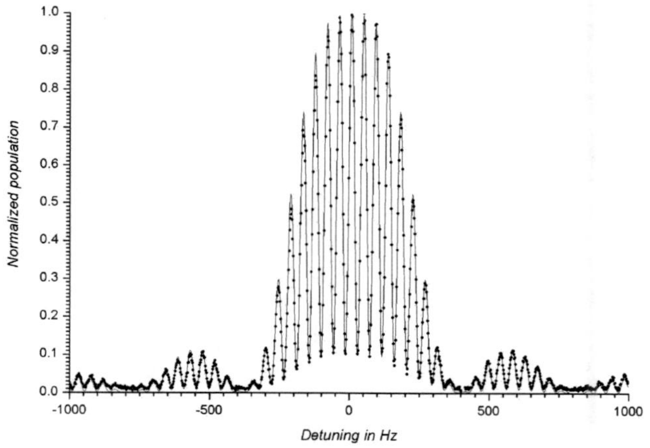

Fig. 7. Ramsey interrogation is performed in the cavity by two microwave pulses of τ = 2.5 ms each and separated by T = 20 ms. Consequently, the linewidth of the central fringe at middle height is 25 Hz. The associated contrast is better than 90%. Experimental points, from time of flight detection, are superposed to calculation represented by the continuous line. The theoretical curve is obtained for coherence lifetime of 110 ms, in order to fit the lost of contrast. Experimental and theoritical signals are both multiplied by a normalisation factor to be compared.

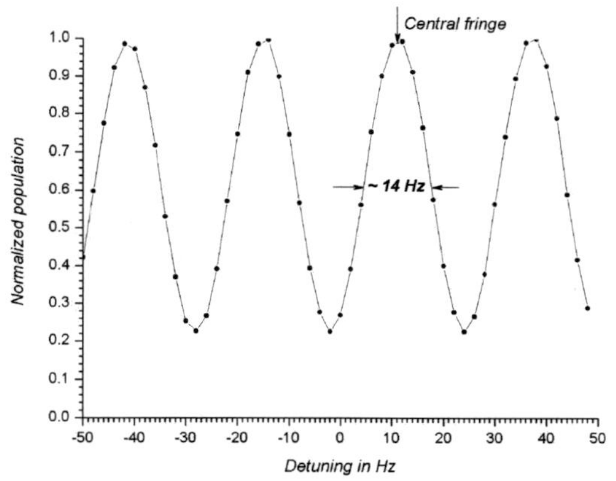

Fig. 8. Scan over the central Ramsey fringe for 35 ms dead time giving a 14 Hz linewith with 80% contrast. The frequency deviation from the origin is due to applied static magnetic field.

only. The curves are computed using Monte-Carlo simulations [2]. For example the interrogation time is as twice as long between Earth and space for 90% recapture efficiency. These curves can be used with TOF areas from figure 5 and combined into table II. For example, taking 90% recapture efficiency, we extrapolate the position shifts of the cloud from operation on Earth to Space.

Table II shows interest in using cold atoms at 2.5 μK, which suggests to add a Sisyphus cooling

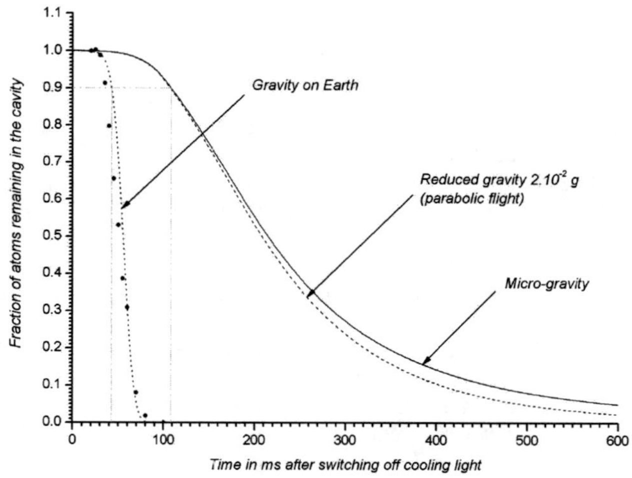

Fig. 9. Fraction of atoms remaining in the quartz bulb (diameter = 33 mm) as a function of the atom free flight time [2]. "•" : experimental points on earth from time of flight measurements, with T_{cool} = 20 ms. Lines are calculated decay from a MonteCarlo simulation with a temperature of 35 μK (i.e. a rms velocity of 4.7 cm.s^{-1}), with a spatial distribution characterized by σ = 4 mm, and for respectively g (dotted line), 2 10^{-2} g (dashed line), and 0 g (continuous line).

	On Earth	In space
Interrogation time T_{int} (ms)	40	110
Cyclic ratio R_c	0.67	0.85
Central Fringe Linewidth (Hz)	12.5	4.5
Position shift due to gravity (mm)	7.8	1.2
Position shift at T = 35 μK (mm)	1.9	5.2
Position shift at T = 2.5 μK (mm)	0.5	1.4

TABLE II

Assuming 90% recapture efficiency, we extract from figure 9 the two dead times associated respectively to Earth and Space operation of the clock. Then linewidth and position shifts are deduced.

phase in recapture described in section II.

V. Conclusion

We managed to cool 10^8 atoms at 2.5 μK using isotropic light in a clock cavity. About 10^6 atoms have been successfully interrogated with microwave and detected by time of flight method. We next plan to detect the clock signal with a laser beam across the cavity itself. Clock performances will be ultimately limited by gravity on Earth and kinetic temperature in Space.

VI. Acknowledgment

This work is supported in part by CNES. S. Trémine is grateful to CNES and SODERN for providing research studentship.

REFERENCES

[1] G. Santarelli et al., *Laser cooling of atoms in micro-gravity*, C. R. Acad. Sci. Paris, t. 316, Série II, p 739-744 (1993).

[2] B. Lounis, J. Reichel, C. Salomon, *Laser cooling of atoms in micro-gravity*, C. R. Acad. Sci. Paris, t. 316, Série II, p 739-744 (1993).

[3] E. Guillot, P.-E. Pottie, N. Dimarcq, *Three-dimensional cooling of cesium atoms in a reflecting copper cylinder*, Opt. Lett., vol 26, p 1639, (2001).

[4] P.-E. Pottie, *Conception et réalisation d'un démonstrateur d'horloge atomique miniature à atomes refroidis en cellule*, Thesis from the Université Paris-VI, Paris (2003).

[5] P.-E. Pottie, S. Trémine, S. Guérandel, N. Dimarck, A. Clairon, *3D speckle cooling in a microwave clock*, In Proc. of the 17th EFTF, Tampa, Florida (2003).

COLD SR ATOMS FOR AN OPTICAL FREQUENCY STANDARD

P.E. POTTIE, T. BINNEWIES, H. STOEHR, U. STERR, J. HELMCKE, F. RIEHLE
Physikalisch-Technische Bundesanstalt,
Bundesallee 100, 38116 Braunschweig, Germany, e-mail: paul-eric.pottie@ptb.de

Sr-atoms are of special interest for an optical frequency standard since a large number of atoms can be cooled down efficiently to ultra-low temperatures in a magneto-optical trap (MOT) [1]. The ultra-narrow line on the forbidden $^1S_0 \rightarrow {}^3P_0$ transition of ^{87}Sr ($^3P_0$ natural lifetime is approx. 100 s), and the ability to interrogate Sr atoms in a perturbation free dipole trap [2] promise un-precedented low inaccuracy level [2] and relative frequency instability, well below $10^{-15}\tau^{-1/2}$ [3]. We will show the realization of an efficient atomic source of cold Sr atoms and the construction of an ultra-stable laser for the interrogation of the atoms.

The spectral density of noise of the laser of interrogation and its spectral aliasing in the Fourier space by the Dick effect is a key parameter for the frequency instability of the future frequency standard. Our ultra-stable laser for the Ca-frequency standard exhibits a line-width (FWHM) smaller than 1.5 Hz, with a linear drift of 0.2 Hz/s, and a relative frequency instability of $3 \cdot 10^{-15}$ from 0.4 s to a few seconds. To reach the frequency instability that would be possible with long interaction times in a dipole trap, especially the low frequency fluctuations of the interrogation laser frequency have to be reduced. One possible strategy is to pre-stabilize the interrogation laser to a broaden atomic line. We will discuss possible schemes to realize such a stabilization.

For this pre-stabilization it is essential to get as many cold atoms as possible in a time as short as possible. Our experimental set-up is optimized for that purpose. Sr atoms are first slowed in a Zeeman slower, then deflected with the help of a 2-D molasses, and trapped in a MOT. We will present the parameters of our MOT and its implication on an optical frequency standard.

References

[1] H. Katori, T. Ido, Y Isoya, and M. Kuwata-Gonokami, Phys. Rev. Lett. **82**, 6, pp.1116, 1999

[2] H. Katori, M. Takamoto, V.G. Palchikov, and V.D. Ovsiannikov, Phys. Rev. Lett. **91**, 17,
173005, 2003

[3] A. Quessada, R.P. Kovacich, I. Courtillot, A. Clairon, G. Santarelli and P. Lemonde, J. Opt.B 5, S150 (2003).

PROGRESS TOWARDS AN OPERATIONAL RUBIDIUM FOUNTAIN

Scott Crane, Steven Peil, Thomas B. Swanson, and Christopher R. Ekstrom

U. S. Naval Observatory, Washington D. C. USA
crane@atom.usno.navy.mil

Keywords: Atomic Fountain, Rubidium

Abstract

The U.S. Naval Observatory has an ongoing program to develop atomic fountains for eventual incorporation into the USNO Master Clock. The fountains are being constructed to provide a stable frequency reference which will allow for rapid characterization of drift of masers in our clock ensemble. In support of this goal we are using rubidium to exploit the small cold-collision frequency shift. We report on the progress made in the construction and testing of our first rubidium device.

1 Introduction

The USNO has undertaken a program to build several atomic fountains to upgrade our timing ensemble. The current ensemble is based on commercial hydrogen masers and cesium beam clocks. The atomic fountains will be used to provide a stable frequency reference and to allow rapid drift characterizations within the ensemble.

The Observatory will need three operational fountain clocks, with an additional two housed at our alternate facility at Schriever Air Force Base. These devices will need to run unattended for roughly 10 years. We are designing the devices to operate with a short term stability of $1\text{-}2\times10^{-13}$ at one second and with a long term frequency reproducibility of 10^{-16}.

2 Overall Layout

The physics package is contained within a frame of $0.8\text{x}0.8\text{x}1.8\ \text{m}^3$. In addition, there are two equipment racks to house all of the lasers, optical tables, frequency synthesis, and control systems.

3 Vacuum Chamber

The vacuum chamber is assembled with welded connections wherever practical. Most of the vacuum connections along the main vertical bore are made with conflat joints. The vacuum chamber is housed inside of magnetic shields and is designed with machined flats to control the alignment of all optical systems.

3.1 Atom Collection

The collection region is a monolithic stainless steel sphere. The flats that house optical windows have machined weld preparations. The optical windows are 1.5" diameter sapphire blanks mounted in copper weld collars with a high temperature braze. The windows are electron beam welded into the chamber. We have had mixed results with the welds due to thermal shock to the sapphire causing small cracks. In the future, we are moving to titanium collars and chambers. Our welding tests show dramatically lower weld temperatures with titanium.

Atoms will be collected from background vapor and cooled in a lin⊥lin molasses. The atoms will be launched vertically along the <1,1,1> axis.

3.2 State Selection

A state preparation cavity is located above the atom collection region. The cavity is built from stainless steel and is e-beam welded. There is a single microwave feed through a ceramic window. The ceramic window is brazed to a weld collar and welded into the cavity feed structure at the equator of the cavity.

Atoms are transferred from the $|2,0\rangle$ state to the $|1,0\rangle$ state in this cavity. The remaining $|2,m\rangle$ atoms are removed in the detection region.

3.3 Detection Region

The detection region is a copper vacuum chamber with two fluorescence collection regions. In addition there is an optical pumping region between these two regions. The optical windows are the same as used in the collection region. There are conflat flanges made from copper-titanium explosion bonded material welded to the top and bottom of the detection region.

3.4 Spectroscopy Region

The spectroscopy region is a copper drift tube with an integrated microwave cavity. The microwave cavity is fed symmetrically from two sides at its equator through vertical slits. There are two ceramic microwave feedthroughs of the same type used in the state preparation cavity. The cavity has a loaded Q of 6,500.

4 Microwave Frequency Chain

The chain starts with a low phase noise 5MHz BVA crystal oscillator. This oscillator is phase locked to an active hydrogen maser with a time constant of roughly 3 seconds. The 5MHz is multiplied to 40 and 100 MHz. The 100 MHz is used to phase lock a 6.8GHz Dielectric Resonator Oscillator.

The final frequency is generated by mixing the output of a synthesizer (34.6x MHz) with the 6.8 GHz from the DRO. The synthesizer is clocked from the 40 MHz that is generated from the 5MHz crystal. The synthesizer allows easy phase modulation for interrogation of the atomic line.

We will use the same strategy as in our previous work to produce a steered output that reflects the stability of the fountain. The fountain will monitor the frequency of an active hydrogen maser and provide corrections to a high resolution synthesizer that is driven by the same maser. The steering values are generated by a critically damped Kalman filter with outlier rejection and holdover to reduce the impact of any missed fountain cycles.

5 Lasers and Optics

5.1 Laser Source

The laser source is based on frequency doubling of 1560nm light to get the 780nm light required for trapping, cooling, launching, and detection of the rubidium atoms. After performing several experiments ourselves, we contracted with a commercial company (IRE Polus Group*) to provide a high-power 780nm source of this type.

The laser is housed in a 4U high enclosure and supplies up to 1.5W of 780nm light in a single mode fiber. We are currently evaluating the laser linewidth and stability.

5.2 Optical tables

The majority of the optics are rack-mounted on a single 38x56 cm^2 optical table. The table is built around two custom "fiber bench" tables from OFR*. The high power (1.5 W) laser source comes to the table on a single mode (PM) optical fiber. The shifted laser frequencies are generated in double-pass through AOMs and returned to the fiber bench for splitting and coupling. An EOM provides a sideband on one beam for optical pumping and repumping that is tuned to the F=1 to F'=2 transition.

This main optical table provides 11 fiber outputs: 3 each for the upward and downward trapping/launching beams, 2 detection beams, 2 pumping/repumping beams, and an output that goes to a frequency locking setup. All fibers for the physics package are single mode and polarizing.

The laser frequency is offset locked to a rubidium vapor cell. The spectroscopy setup is in a small rack-mounted enclosure and receives a fiber from the main optical table.

5.3 Monolithic Optics

The optics for producing the atom collection and detection are all housed in monolithic assemblies. The collection beam couplers form 32mm diameter beams and have integrated polarizers. The detection beam couplers form 25mm x 4 mm elliptical beams. All of the beam shaping and fluorescence collection optics bolt directly to the vacuum chamber.

6 Magnetic Shielding

The magnetic shielding has three layers over the entire apparatus and a fourth around the spectroscopy region. Our models indicate a shielding factor exceeding 100,000. Measurements of individual layers have met our modeling expectations. Because the shielding of the combination of the four layers is so large, evaluation of the full shielding will have to be performed with the atoms.

7 Current Status

With most of the components in hand, we foresee preliminary operation this summer. The vacuum chamber is under vacuum and the shielding subsystems have been assembled.

Acknowledgements

This work was supported by the Office of Naval Research.

References

* Mention of specific company names does not constitute an endorsement by the U.S. Government, the Navy or USNO.

Optical frequency measurements at BNM-SYRTE

A. Brusch*, I. Courtillot, A. Quessada, D. B. Kolker, G. Santarelli, A. Clairon

P. Lemonde, J. J. Zondy, O. Acef, G. D. Rovera

BNM-SYRTE, France

* Email: anders.brusch@obspm.fr

Keywords: femtosecond laser, optical frequency standards, CO_2/OsO_4.

Abstract

We report on recent progress toward high stability transfer of microwave frequencies to optical frequencies using a femtosecond laser.

1 Introduction

The development in recent years of Kerr-lens modelocked femtosecond lasers and their use in converting oscillations in the radio-frequency domain to optical frequencies, has made these lasers indispensable tools in frequency metrology as a means to measure optical frequencies [1,2,3]. Conversely the femtosecond laser is the only possibility for a realistic clockwork to transfer the stability of optical frequency standards down to microwave frequencies [4,5,6]. For both these applications it is important to test the ultimate stability that can be achieved. To provide tests of stability transfer several frequency standards are available at the BNM-SYRTE. Figure 1 shows the Allan standard deviation of some of these frequency standards.

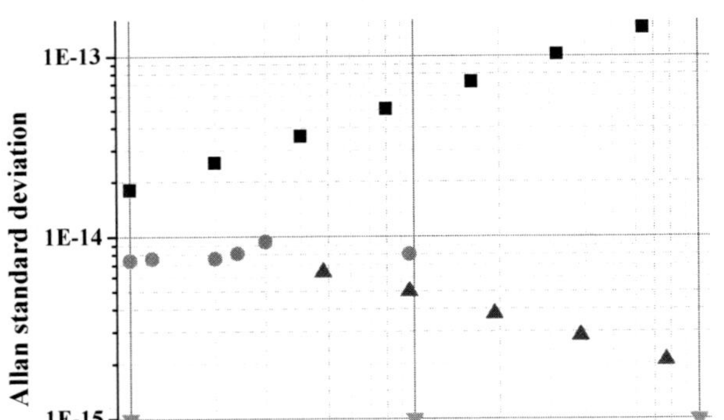

Figure 1. Allan standard deviation of some frequency standards at the BNM-SYRTE. ■ shows the ultra stable cavities (without temperature stab.), ▲ shows an atomic fountain, ● is the CO_2OsO_4 frequency standard and ▼ is the cryogenic sapphire oscillator.

In this paper we report recent progress toward obtaining high stability in transferring microwave frequencies to optical frequencies. In the experiment discussed here we phaselock the femtosecond laser to the 1 GHz signal from a cryogenic sapphire oscillator with an inferred stability below $1*10^{-15}$ with 1 second of averaging. The femtosecond laser is used to measure a CO_2 laser at 10 μm stabilized to a transition in OsO_4 with a stability of $7,4*10^{-15}$ with 1 second of averaging.

2 Experimental setup

For our experiment we use a commercial femtosecond laser developed by Giga optics, with a repetition rate around 840 MHz. The output power of the laser is approximately 500 mW with a pump power of 5 W.

The signal from the cryogenic oscillator is delivered in an optical fibre using amplitude modulation. The stability of this transfer has been shown to have a stability below $1*10^{-14}$ at 1 second. The 1 GHz signal is multiplied to 9 GHz by a microwave frequency chain, and divided down to 200 MHz and 10 MHz. To lock the femtosecond laser we detect the 11 th. harmonic of the repetition rate on a photoconductive detector. After mixing with the 9 GHz and 200 MHz from the microwave chain, the resulting 40 MHz beatnote is phase locked to a synthesizer. To achieve as high a bandwidth (BW) in locking as possible we use both an AOM with a BW of 100 kHz and a mirror mounted on a piezo with a BW of 30 kHz. The AOM is usually only used for locking of the carrier envelope offset frequency, but in our measurements it is not necessary to control or detect the offset frequency.

The measurement of the CO_2/OsO_4 standard is performed by measuring two laser diodes at 778 and 841 nm. The difference between the two lasers is locked by phaselocking the sum frequency of the 841 nm laser and a CO_2 laser phaselocked to the CO_2/OsO_4 at 10 μm, to the laser at 778 nm. To remove the drift of the free running laser diodes, the 778 nm laser is locked to the $5S_{1/2} - 5D_{5/2}$ two photon transition in Rb with a low BW.

A great advantage in performing the measurement in this way is that the two laser diodes we measure are within the frequency spectrum of the femtosecond laser. This means that we do not have to use a photonic crystal fibre to broaden the spectrum, leading to ease of operation and no extra noise due to noise amplification in the fibre.

To detect the beat notes between the femtosecond laser comb and the stabilized laser diodes, the light from the two lasers is superimposed in the same polarization using a dichroic mirror. The overlapped beam is superimposed with the femtosecond laser in a broadband polarizing beamsplitter. A Glan-Thompson prism is then used to project the orthogonal polarizations onto a common axis, and the light is launched into a polarization maintaining fibre. The use of a fibre at this point ensures perfect spatial overlapping between the beams and greatly simplifies the alignment of the three beams. The reduction in intensities due to the fibre is compensated for by these advantages. The two beatnotes between the diode lasers and the femtosecond laser comb is detected on a single avalanche photodiode. The signal is then split and the two beatnotes are filtered and amplified separately. Finally we mix the two signals and use the sum beatnote for subsequent filtering, tracking and counting. By choosing the correct sum beatnote we remove the influence of the changes in offset frequency due to both long and short term drifts, and also due to the locking with the AOM of the repetition rate, due to the high BW of mixing.

3 Experimental results

Figure 2 shows the standard Allan deviation for two measurements performed with 1 and 10 second gate time. The stability for both gate times is $6.6*10^{-14}\ \tau^{-1/2}$.

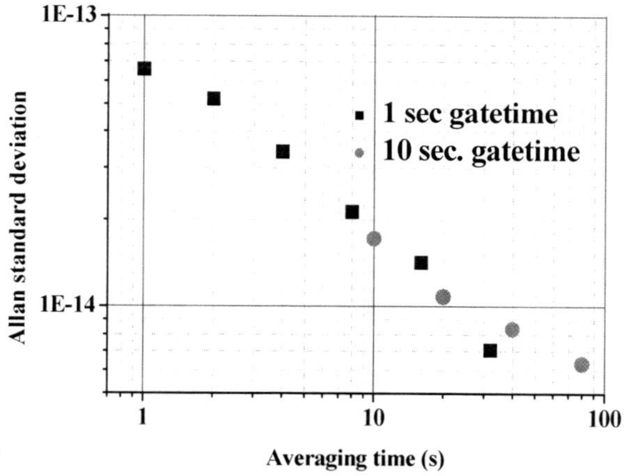

Figure 2: Allan standard deviation for measurement of the CO_2/OsO_4.

These measurements have been performed for a variation of locking bandwidths on the lock of the repetition rate and little or no variation has been found, suggesting other fundamental limitations to our transfer stability.
The presented result is a first result, and we are not currently aware of where the limitations are in our systems.

4 Conclusion

We have shown the use of a femtosecond laser to phase coherently transfer frequencies from the microwave to the near infrared region. Although the measured stability of this transfer is still a factor of 5 below what can ultimately be expected, we have nevertheless shown an improvement of the transfer to optical frequencies. In the near future we expect perform the down conversion of the CO_2/OsO_4 to microwave for comparison with the cryogenic oscillator. We also plan to perform the same measurements on light from an ultra stable laser at 698 nm. used in the strontium experiment in our laboratory [2].

Acknowledgements

Anders Brusch acknowledges his grant under the CAUAC network. Furthermore we acknowledge J.C. Knight, P. St. J. Russel and W.J. Wadsworth from the University of Bath for supplying us with photonic crystal fibres.

References

[1] J. Stenger et al. PRA **63**, 021802 (R), (2001)
[2] I. Courtillot et al. PRA **68,** 030501(R), (2003)
[3] Th. Udem et al. PRL **86**, 4996, (2001)
[4] Jun Ye et al. PRL **87**, 270801, (2001)
[5] S. A. Diddams et al. Opt. Lett. **27**, 58, (2002)
[6] A. Bartels et al. Opt. Lett. **28**, 664, (2003)

COLLIMATION OF A CONTINUOUS COLD ATOMIC BEAM USING RAMAN SIDEBAND LASER COOLING

N. Castagna*, G. Di Domenico*, G. Dudle♣, G. Mileti*, M. Plimmer*, A.V. Taichenachev†, P. Thomann*, V.I. Yudin†,

* Observatoire cantonal, 58 rue de l'Observatoire, CH-2000 Neuchâtel, Switzerland
♣ Swiss Federal Office of Metrology and Accreditation (METAS), CH-3003 Bern-Wabern, Switzerland
† Novosibirsk State University, Pirogova 2, Novosibirsk 630090, Russia and
Institute of Laser Physics SB RAS, Lavrent'eva 13/3, Novosibirsk 630090, Russia

E-mail address: Natascia.Castagna@ne.ch; Fax Number: +41 32 722 04 20

Keywords: Cold atoms, frequency standards, continuous atomic fountain clock, laser cooling.

Abstract

In this article we report on the first use of degenerate Raman sideband cooling for the collimation of a continuous beam of cold cesium atoms in a fountain geometry. Using this cooling technique we have reduced the atomic beam transverse temperature from 60 K to 1.6 K in a few milliseconds. We present experimental results and future directions that we will investigate.

1 Introduction

Since the discovery of laser cooling [1], beams of slow and cold atoms have played an ever more important role in high precision experiments, e.g. in atom interferometry [2,3] and in atomic fountain clocks [4]. In this context, the continuous beam approach [5] is interesting because it dramatically reduces all undesirable effects of atomic density (collisions, cavity pulling) [4] and the Dick effect, in pulsed beams, imposes much more severe requirements on the local oscillator [6,7]. A continuous fountain can combine low density for accuracy and high flux for stability. However, to take full advantage of the continuous beam approach, one needs to increase the useful flux. One method is to collimate the atomic beam.

In this paper we present a laser cooling experiment for the collimation of a continuous beam of cold cesium atoms in a fountain geometry. This approach is directly related to the development at the Observatoire Cantonal de Neuchâtel, in collaboration with the Swiss Federal Office of Metrology and Accreditation (METAS), of the continuous atomic fountain clock FOCS-1 [8], whose stability we wish to improve. The technique we use is Zeeman shift degenerate Raman sideband cooling. Here we report on the first demonstration of this mechanism for two dimensional cooling of a continuous atomic beam, on the efficiency of the process and on possibilities to optimise this technique.

2 FOCS-1

FOCS-1 is the first continuous fountain clock and results from a collaboration between the Observatoire Cantonal in Neuchâtel and METAS in Bern. A schematic view of FOCS-1 is shown in Fig.1.

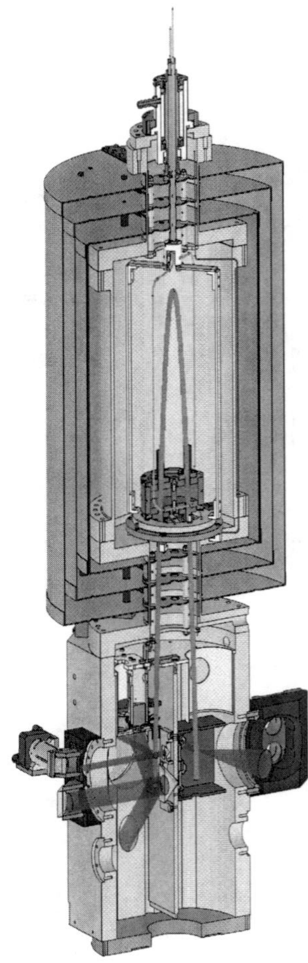

Fig.1: Overview of the continuous atomic fountain clock FOCS-1.

A partial state of the accuracy budget of FOCS-1 is presented in Table 2; we can remark that collisional shift does not represent a limit in a continuous fountain. On the other hand the strong light shift needs to be attenuated: for this purpose we have introduced a light trap between the source of the beam and the microwave cavity [5, 9]. Simulation models and measurements of the stray light confirm a reduction of the light shift by a factor 10^4.

Shift	Value ($\times 10^{-16}$)	Uncertainty ($\times 10^{-16}$)
Zeeman second order (*)	244	3
Light shift without trap (*)	14000	-
Light shift with trap (***)	< 10	< 1
Collisional shift (***)	0.07	< 1
End-to-End phase shift (**)	< 5	< 10
Distributed phase shift (**)	0.4	< 1

Table 1: Partial state of the accuracy budget; symbol (*) indicates the measured values, (**) the calculated values and (***) stands for calculated from measured values.

The stability, which is proportional to the square root of the atomic flux density, is at present equal to 2.5 10^{-13} after one second (relative Allan deviation). An improved short term stability could be achieved with a more efficient collimation of the atomic beam. For this reason we are experimenting with Raman sideband laser cooling in a separate modular system; the technique adopted is the Zeeman shift induced degeneracy.

3 The Raman sideband laser cooling

Sideband cooling was first applied to trapped ions [10-12]. Later, following a theoretical proposal by Taïeb *et al.*[13], it was adapted to neutral atoms in optical lattices by several groups [14-16]. Recently, the group of Chu developed an improved scheme [17] for 3D cooling of a pulsed cesium beam [18]. Here we report on the first demonstration of this mechanism for two-dimensional cooling of a continuous atomic beam [19].

3.1 Cooling Principle

The scheme of the sideband cooling cycle, similar to the one used by Chu and his group [17, 18], is depicted in Fig. 2.

Cesium atoms are trapped in a far off-resonance optical lattice; vibrational levels of potential wells associated with different values of m_F are shifted and brought into degeneracy by a static magnetic field B (Zeeman effect). Degenerate Raman transitions can then take place (double arrows in Fig. 2) stimulated by the lattice laser light itself.

In this context we cool the atoms with a succession of cycles decreasing their vibrational energy level n until they reach the ground state $n = 0$.

For $n > 1$, each cycle consists of two Raman transitions $|F=3,m_F=3,n$ $|3,2,n-1$ $|3,1,n-2$ followed by an optical pumping cycle towards $|3,3,n-2$. Each Raman transition removes one vibrational quantum but the optical pumping conserves n with high probability because the atoms are in the Lamb-Dicke regime. Notice that for $n = 1$ we need a cycle made up of only one Raman transition $|3,3,n=1$ $|3,2,0$ followed by optical pumping to bring the atom to the ground state $|3,3,0$. The Raman cooling process ends there since the ground state is dark to both the lattice and pump beams. However, as the atoms leave the lattice one can cool them further by adiabatic expansion of the potential well as in Ref. [20].

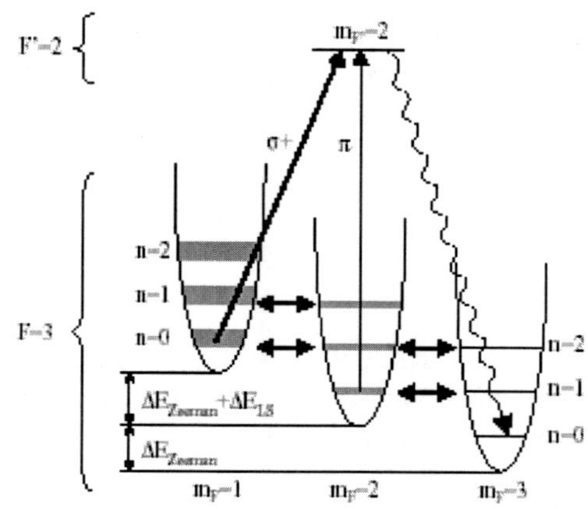

Fig.2: Degenerate Raman sideband cooling scheme. E_{Zeeman} is the shift due to the magnetic field; E_{LS} is the light shift induced by the pump laser beam. Double arrows indicate the Raman transitions; optical pumping pumps atoms to F' = 2.

3.2 Optical lattice geometry

A laser locked on the F=4 F'=4 transition of the D_2 cesium line creates the optical lattice. Figure 3 shows the chosen geometry: a 2D configuration with four laser beams that guarantees a symmetric optical lattice. Phase stability of the lattice is ensured, despite the fact that four beams are used instead of the usual three, by the fact that the single laser beam is folded onto itself to create the lattice [21]. An additional advantage of this folded configuration is that the laser power is recycled.

There are no laser beams in the vertical direction (perpendicular to the plane of the picture) which avoids any problem of light shift due to the strong lattice beams.

The peak intensity of the gaussian laser beam is equal to 380 mW cm^{-2} and the polarization is linear; the red detuning with respect to the F = 3 to F' transitions is = 2 9 GHz.

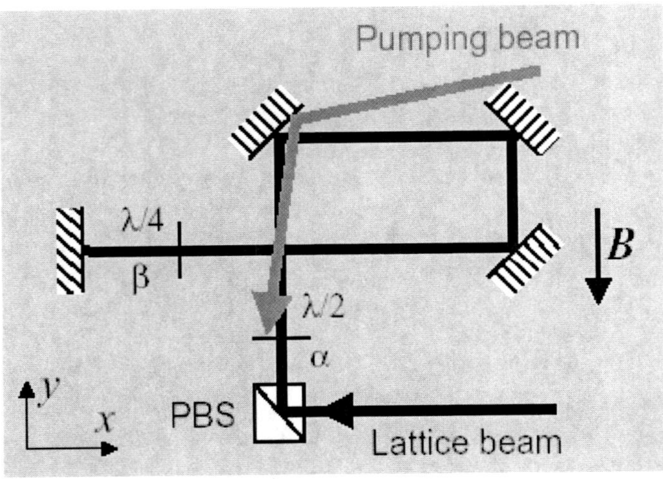

Fig. 3: Optical lattice geometry. Two wave plates allow us to control the linear polarization (and   are the tilt angles with respect to the vertical). PBS: Polarizing Beam Splitter; **B** is the static magnetic field.

The static magnetic field B acts along the y axis and its direction also determines the quantization axis. Its intensity of 60 mG is calculated to match the Zeeman frequency with the average vibrational frequency, in order to bring into degeneracy different vibrational states of adjacent potential wells.

3.3 Experimental results

The experimental scheme is presented in Figure 4. The source of the continuous beam of cold atoms was presented in detail in [22]. It is an optical molasses loaded by a thermal cesium vapour $p = 10^{-8}$ mbar. The atoms are continuously cooled and launched upwards by the moving molasses technique. We thus obtain a continuous beam of cesium atoms with a flux of the order of 10^8 at/s, a temperature between 50 and 100 K and an adjustable velocity of around 4 m/s [22]. Collimation is carried out 22 cm above the source in a plane tilted at 3° with respect to horizontal in such a way that the atoms arrive at detection after a parabolic flight.

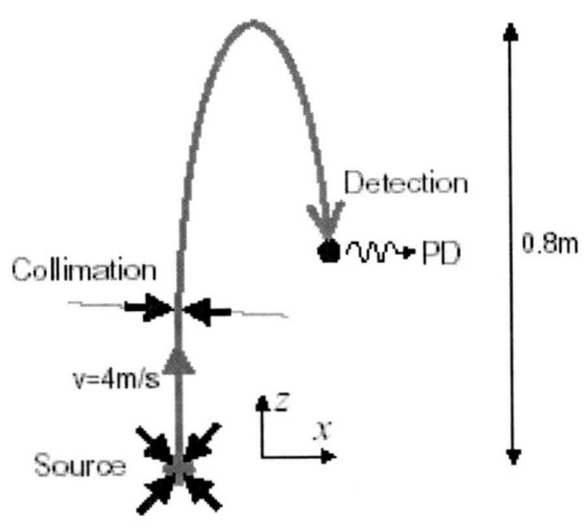

Fig. 4: Scheme of the experiment. The source of the atomic beam is a six beam moving molasses, four of which are shown (horizontal beams in and out of the page); PD: Photo-Detector.

The atoms are detected by fluorescence induced by a retro-reflected probe laser propagating along the y axis. We can move the detection system in a horizontal plane to obtain the profile of the atomic beam, whose velocity distribution is related to the transverse temperature by the Maxwell-Boltzmann statistics:

$$\rho\ v \qquad \exp \frac{M\ v^2}{2\ k_B\ T} \tag{1}$$

where M is the mass of the cesium atom, k_B the Boltzmann constant and T the transverse temperature.
We started our study by identifying situations where sideband cooling takes place. The first situation we investigated is reported in Figure 5. In this graph we plot the maximum flux density and the transverse temperature as a function of the static magnetic field.

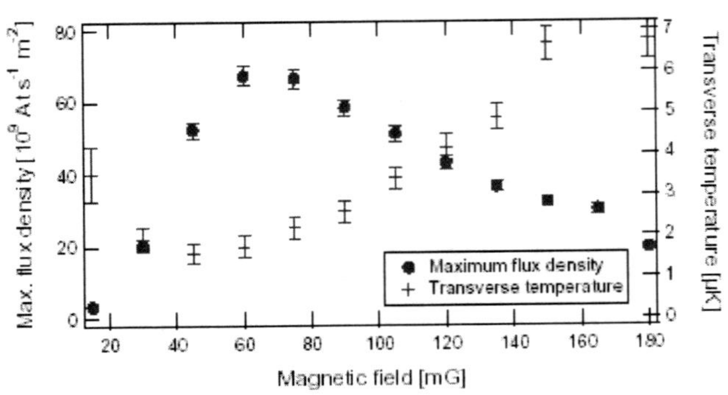

Fig. 5: Atomic flux density and transverse temperature as a function of magnetic field along the y axis.

The result is a peak in the detected flux in correspondence with the magnetic field necessary to bring into degeneracy the vibrational levels of different potential wells. The transverse temperature has a dip in the same region, which is consistent with a minimum value and so a better collimation. The value of the magnetic field for which we observe this behaviour is about 60 mG, in good agreement with theoretical expectations. The flux density is measured in the centre of the atomic beam where it reaches its maximum value. The experimental conditions are the following: = 22.5°, lattice locked on F = 4 F' = 4 transition, pumping beam locked a few megahertz above the F = 3 F' = 2 transition, lattice power 190 mW per beam and pump power 0.6 mW; the polarization of the pump laser beam is almost entirely $^+$, with a weak component (see fig.2).
A second interesting result is reported in Fig. 6, namely a plot of the fluorescence signal versus the pump laser frequency. We scan over all F = 3 F' transitions of cesium D_2 line in 20 s. The polarization of the pumping beam is $^+$ and **B** is nearly parallel to the pumper. Other conditions: = 22.5°, optical lattice locked on F = 4 F' = 4, B = 60 mG, lattice power 190 mW per beam and pump power 0.6 mW. Some

remarks deserve the reader's attention. On the two hyperfine pumping transitions F = 3   F' = 3,4 we observe large dips because atoms spend more time in the F = 4 hyperfine ground state where they are heated by the strong lattice light.

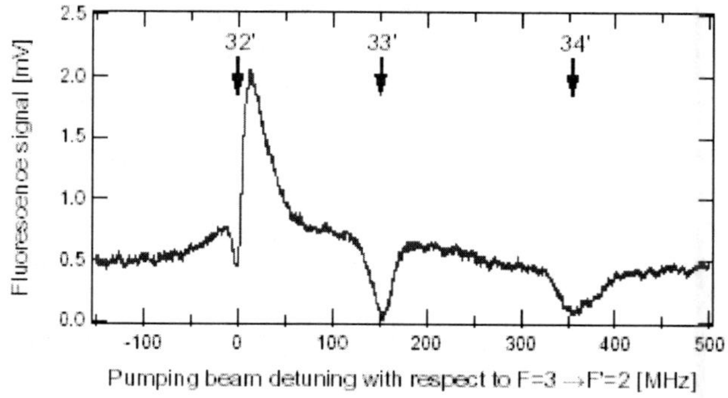

Fig. 6: Fluorescence signal detected in the centre of the atomic beam as a function of pumping beam frequency. For its calibration, we used saturated absorption spectroscopy.

The 34' dip is wider than the 33' dip because of the power broadening induced by the lattice laser which is locked to the F = 4   F' = 4 transition. We observe a very strong sideband cooling peak on the blue side of the F = 3   F' = 2 transition, as well as a narrow dip on the red side. We can explain this result as follows: because of the $^{+}$ polarization of the pumping beam, the F = 3, m_F = 1 sub-level is shifted, but the F = 3, m_F = 2 and F = 3, m_F = 3 sub-levels are not as they are dark to the $^{+}$ pumper light. If the light-shift is negative, which is the case for a red detuning, this will lead to a heating, instead of a cooling cycle if the light shift is comparable with the Zeeman shift. This argument explains the dip observed in Fig. 6 and the fact that the cooling efficiency is much higher on the blue side than on the red side of the F = 3   F' = 2 transition. This behaviour is indeed observed in Figure 7, which is a zoom of the previous Fig. 6.

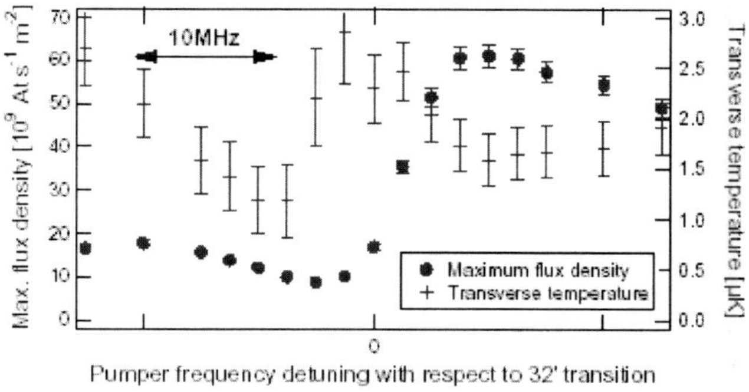

Fig. 7: Atomic flux density and transverse temperature as a function of pumping laser frequency, in a range of a few tens of megahertz around the F = 3   F' = 2 transition.

We have observed that only 15% of the atoms are trapped and collimated. This statement is confirmed by a calculation of the percentage of atoms trapped when entering the potential wells, whose depth is equal to 20 K.

The transverse temperature measured without the collimation is 64 4 K, which implies a capture efficiency of about 30% and a uniform population distribution among the different vibrational levels of 3%. After sideband cooling collimation, the lowest transverse temperature measured is 1.6 0.3 K, which indicates a redistribution of the bound atoms among the vibrational levels and a population of the ground vibrational level of 15%. The detected flux increases from 2 10^{10} at s^{-1} m^{-2} without collimation to 7 10^{10} at s^{-1} m^{-2} with sideband cooling.

In conclusion, we have observed collimation with Raman sideband cooling: we decreased the transverse temperature by a factor 40 and we increased the flux by a factor 3.5. However it would be useful to improve the capture efficiency still further and find a way of increasing the initial flux in the source region. In the next paragraph we present some ideas to achieve these aims.

4 Future directions

Among the possibilities to improve our experimental results we are working on the following:

- A 2D-MOT as pre-source to increase the flux of the atomic beam while lowing the cesium pressure in the source region [23].
- A combination of Sisyphus and sideband cooling in the same lattice, to improve the cooling efficiency and to collimate atoms still further.
- Realization of sideband Raman cooling using the Stark effect instead of the Zeeman effect, to accumulate atoms directly in the clock state m_F = 0 (*Oz*).

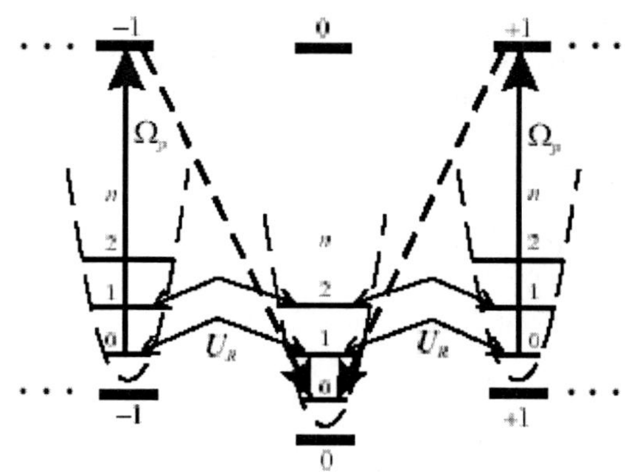

Fig. 8: Sideband cooling cycle using Stark effect. U_R indicates the rate of the Raman transitions; $_P$ is Rabi coupling for the transitions induced by the pump laser.

The last point is particularly interesting because it allows one to accumulate atoms in the desired Zeeman state $m_F = 0$ (*Oz*) required for the clock transition whereas the technique described in the present paper populates $m_F = +3$ (*Oy*). The Stark sideband cooling cycle is depicted in Figure 8.
The vibrational levels of different potential wells would be brought into degeneracy by the light shift due to the pump laser beam. This shift is proportional to m_F^2, so states with symmetric value of m_F are shifted by the same amount and they both participate in the cooling cycle towards the ground vibrational level of the state $m_F = 0$. A detailed theoretical description of this process can be found in [24].

Acknowledgements

This work was supported by the Swiss National Science Foundation, the Swiss Federal Office of Metrology and Accreditation (METAS), the canton of Neuchâtel, and the Swiss Confederation. AVT and VIYu were partially supported by a grant INTAS-01-0855 and by RFBR through grants 04-02-16428, 04-02-16488, and 04-02-16525.

References

[1] H. J. Metcalf, P. van der Straten. "Laser cooling and trapping", (Springer, 1999).
[2] P.R. Berman. "Atom Interferometry", (Academic Press, 1997).
[3] S. C. A. Peters, K.Y. Chung, *Nature*, **400**, p.849, (1999).
[4] Y. Sortais *et al.* , *Physica Scripta,* **50**, (2001).
[5] A. Joyet, G. Mileti, P. Thomann, G. Dudle, "Continuous fountain CS standard: stability and accuracy issues", in *"Frequency standards and metrology", Proceedings of the 6th Symposium*, St. Andrews, Scotland, pp. 273-280, 9-14 Sept. 2001, P. Gill editor (World Scientific 2002).
[6] G. J. Dick, J. D. Prestage, C. A. Greenhall, L. Maleki, *Proceedings of the 22nd Annual PTTI Meeting*, pp. 487-508, (1990).
[7] A. Joyet, G. Mileti, P. Thomann, G. Dudle, *IEEE Trans. Instr. Meas.*, **50**, p. 150, (2001).
[8] G. Dudle, G. Mileti, A. Joyet, E. Fretel, P. Berthoud, P. Thomann, "An alternative cold cesium frequency standard: the continuous fountain", *IEEE Transactions on Ultrasonics, Ferroelectrics and Frequency Control*, **47**, pp. 438-442, (2000).
[9] A. Joyet, "Aspects métrologiques d'une fontaine continue à atoms froids", *PhD Thesis,* University of Neuchâtel, Switzerland, (2003).
[10] D. J. Wineland, H. Dehmelt., *Bull. Am. Phys. Soc.*, **20**, p.637, (1975).
[11] F. Diedrich, J. C. Bergquist, W. M. Itano, D. J. Wineland, *Phys. Rev. Lett*, **62**, p.403, (1989).
[12] D.J. Heinzen, D.J. Wineland. *Phys. Rev. A*, **42**, p. 2977, (1990).
[13] R. Taïeb, R. Dum , J. I. Cirac, P. Marte, P. Zoller, *Phys. Rev. A*, **49**, p. 4876, (1994).
[14] S. E. Hamann, D. L. Haycock, G. Klose, P. H. Pax, I. H. Deutsch, P. S. Jessen, *Phys. Rev. Lett*, **80**, p. 4149, (1998).
[15] H. Perrin, A. Kuhn, I. Bouchoule, C. Salomon, *Europhysics Letters*, **42**, p. 395, (1998).
[16] V. Vuletic, C. Chin, A. J. Kerman, S. Chu, *Phys. Rev. Lett*, **81**, p. 5768,)1998).
[17] A. J. Kerman, V. Vuletic, C. Chin, S. Chu, *Phys. Rev. Lett*, **84**, p. 439, (2000).
[18] P. Treutlein, K. Y. Chung, S. Chu, *Phys. Rev. A*, **63**, p. 051401(R), (2001).
[19] G. Di Domenico *et al.*, to be published in *Phys. Rev. A.*
[20] A. Kastberg, W. D. Phillips, S. L. Rolston, R. J. C. Spreeuw, P. S. Jessen, *Phys. Rev. Lett.*, **74**, p.1542, (1995).
[21] A. Rauschenbeutel, H. Schadwinkel, V. Gomer, D. Meschede, *Opt. Commun.*, p. 148, **45**, (1998).
[22] P. Berthoud, E. Fretel, P. Thomann, *Phys. Rev. A*, **60**, p. R4241, (1999).
[23] S. Chelkowski, *Diplomarbeit*, Institut für Quantenoptik Universität Hannover, (2002).
[24] A.V. Taichenachev, A. M. Tumaikin, V. I. Yudin, L. Hollberg, *Phys. Rev. A*, **63**, p. 033402, (2001).

RECENT PROGRESS ON CS ATOMIC FOUNTAIN AT NICT

M. Kumagai, H. Ito, K. Fukuda, M. Kajita, M. Hosokawa, T. Morikawa
Atomic Frequency Standards Group
National Institute of Information and Communications Technology, JAPAN
mkumagai@nict.go.jp

Keywords atomic fountain, primary frequency standard, moving molasses, saturation effect

Abstract

National Institute of Information Communications Technology (NICT) has been developing a Cesium atomic fountain primary frequency standard. In order to achieve the high stability as the frequency standard, we have made efforts to improve the signal to noise (SN) ratio of the signal. The frequency stability is 2 $10^{-12}/\tau^{1/2}$ presently. This report presents the current status of the development of NICT atomic fountain.

1 Introduction

NICT has been developing a Cs atomic fountain frequency standard to aim at an operational primary frequency standard, whose frequency uncertainty is 1 10^{-15}[1,2]. Cs atoms are cooled below 2 K by both magneto-optical trap (MOT) and polarization gradient cooling (PGC), and launched vertically by a moving molasses method. The launched atoms pass through a microwave cavity twice, on the way upward and downward. In 2002, we succeeded to observe a Ramsey signal, which is narrower than 1Hz. We also locked the microwave frequency to the atomic resonance however, the frequency stability was not high enough at that time [3]. We judged that the low frequency stability is due to the low SN ratio of the Ramsey signal, and several efforts have been done to improve SN ratio. One is the reduction of the noise level of the photo detector and its amplifiers, which detects the fluorescence emitted by the falling atoms. Another is to increase the number of the launched atoms by optimising the condition of the moving molasses with taking account of the saturation effect. As a result, the frequency stability is one-order of magnitude improved.

2 NICT Atomic Fountain

2.1 System Description

The design of the NICT atomic fountain is shown in Fig.1 Our fountain consists of three parts laser cooling area, microwave interaction region and detection zone. The detection zone is set between the laser cooling area and the microwave interaction region. The height of total system is about 2m. The laser cooling area consists of a trap chamber with a pair of anti-helmholtz coils. This area is surrounded by one layer magnetic shield. The microwave interaction region has a C-field coil and a TE_{011} microwave cavity, whose loaded quality factor is about 15000. The whole of this region is surrounded by four-layers magnetic shield, which produce the shielding of order 10^5. The C-field coil and the most inner layer of the shields are set inside the high-vacuum chamber. The detection chamber has three laser interaction zones where the detection beams irradiate the falling atoms. Two photodiodes, which have a large active area ($1cm^2$), are set at the highest and the lowest zones to observe the fluorescence the falling atoms emit. The middle zone is used for an optical pumping and blasting. At present, the selection cavity is not installed. By using two ion pumps and two Ti-getter pumps, very high vacuum (<3 10^{-8} Pa) of the whole system has been achieved.

We use three extended-cavity diode lasers (ECDLs) for

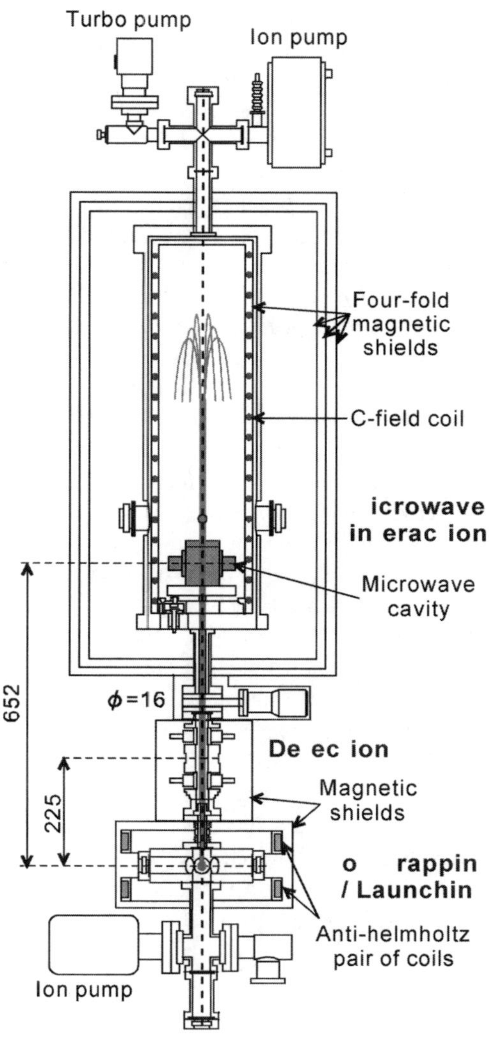

Fig.1 NICT Atomic Fountain

cooling, repump and detection. All lasers are locked to Cs D_2 absorption line by modulation transfer spectroscopy technique. To amplify the laser power for the laser cooling, two 150mW laser diodes (SDL5422) are injection-locked to one of the ECDLs. The frequency and the power of each laser beam are controlled with acousto-optical modulator (AOM). A direct digital synthesizer (DDS) synchronizes all AOMs. Each laser beam is delivered to the vacuum chamber through polarization maintaining fiber (PMF). The power density of each cooling laser beam is at least 10mW/cm^2.

For the source of 9.2GHz microwave interrogation, we use an ANRITSU Corporation synthesizer. All crystal oscillators and DDS are phase-locked to the 5MHz of the hydrogen maser, which is linked to UTC(CRL). Controlling DDS by PC, we tune the microwave frequency.

2.2 Fountain operation

The process to lock the microwave frequency to the Ramsey signal is as follows. Cs atoms are captured in a MOT, and then cooled in the optical molasses. To use the full region of the cooling beams for the launch, the cooled atoms are at first push down to the lower edge of the horizontal cooling beams by (0,0,1) moving molasses. Then, the atoms are launched upward by (0,0,1) moving molasses. In our system, in order to observe the Ramsey fringe narrower than 1Hz, the atoms must be launched with the initial velocity of over 4.4m/s. After giving the launch velocity to the atoms, the launched atoms are post-cooled by the PGC. All the lasers are turned off by AOM and mechanical shutters just before the atoms jump out from the horizontal beams.

All launched atoms are pumped to the =3 state by the optical pumping before the microwave interaction. Twice /2-pulse microwave interactions move the atom in the =3 state to the =4 state. Both the number of the =4 atoms ($_4$) and the number of the =3 atoms ($_3$) are detected independently at the detection zone. The signal is normalized to avoid the cycle-to-cycle fluctuation of the number of the launched atoms. The fraction = $_4$/($_4$+ $_3$) indicates the normalized transition probability. With the initial velocity of 4.4m/s, the atoms are launched to a height of 30cm above the microwave cavity, which gives the drift time of 520ms. The total launching height from the loading point is nearly 1m. As a result, the Ramsey signal of 0.96Hz is observed.

We lock the microwave frequency to the narrow atomic resonance. At present, we adopt a simple locking method, which gives a constant frequency step by comparing the signal intensities at the frequencies of $_0$- and where $_0$ is the microwave frequency, and is the linewidth of the Ramsey fringe. The operation is, when the signal at the lower frequency is stronger than that at the upper frequency, $_0$ is adjusted lower by the amount of the constant frequency step, and vice versa.

3 Recent Progress

3.1 Noise reduction of the detector

In order to improve the SN ratio of the Ramsey signal, we made efforts to reduce the noise. At first, we checked over the photo detector and its amplifiers. By refining the electric circuits and choosing the ultra-low noise operational amplifier, we have reduced the noise of the detector. Although the transimpedance gain is 700MV/A, the noise level without any light can be restricted to less than 2mV.

3.2 Optimise the power of the hori ontal beams

In order to increase the number of the launched atoms, we optimised the laser power of the horizontal beams. As mentioned in Section 4, the high power of the horizontal beams makes the time to reach the equilibrium velocity longer at (0,0,1) moving molasses process. This is due to the saturation effect caused by the horizontal beams. In our case, that the horizontal beams are turned off is the best condition for the moving molasses. By this optimisation, the more atoms are launched vertically, and the signal intensity obtained from the falling atoms increases consequently.

3.3 Frequency stability

Because the SN ratio is getting higher by the various trials, we tried to lock the microwave to the atomic resonance again. As mentioned in Section2.2, we adopt the simple method of the frequency stabilization at present. Fig. 2 shows the frequency stability $_y(\tau)$ of the present NICT fountain. The short-term stability (less than 100s) looks better than the actual because of the constant frequency step. However, the stability at the averaging time longer than 100s is independent of the constant frequency step. Also, $1/\tau^{1/2}$ trend is seen in the stability for τ 100s. These results imply that the obtained stability is reliable. We will not adopt this simple locking method at the accuracy evaluation because there may remain some frequency offset in this method. Nevertheless, it is useful for just preliminary check of the frequency stability. A linear fitting shows the frequency stability of about 2 10$^{-12}/\tau^{1/2}$. It is getting better than the previous stability [3],

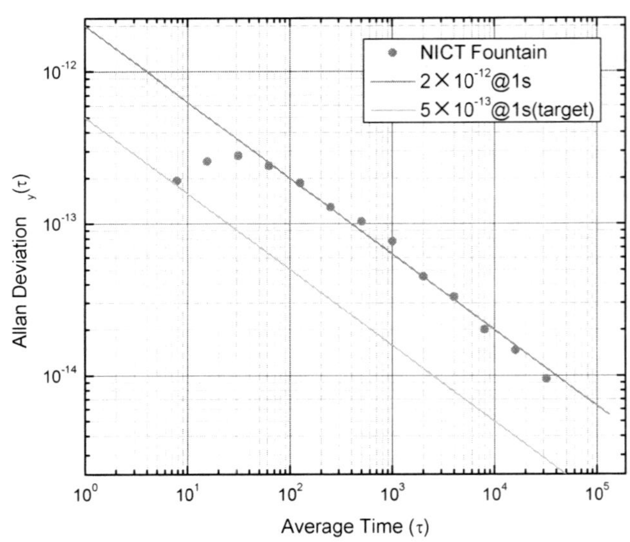

Fig.2 Frequency Stability

however, 3 or 4 times improvement is still required. We guess that this low stability is due to the normalization and the synthesizer itself. The improvements of the normalization scheme, for example, to improve the stability of the probe laser, to change the diameter of the probe beams and so on, are required. Furthermore, the frequency stability of the synthesizer itself should be checked.

4 Saturation effect at the atom launching

4.1 Laser power dependence of the hori ontal beams

In order to increase the number of the launched atoms, we take account of the saturation effect caused by the horizontal beams at the atom launching. At present, the cooled atoms are launched by (0,0,1) moving molasses formed by two vertical beams and four horizontal beams. The launch velocity is determined by the frequency detune of two vertical laser beams. In this process, we evaluate the influence of four other horizontal lasers. We measured the variation of the TOF signal intensity with changing the power density of the horizontal laser beams. Fig.3.a shows the result with the vertical laser intensity of 8mW/cm^2. -axis indicates the duration of the moving molasses process. -axis indicates the number of the launched atoms, which is obtained by Gaussian fitting to the TOF signal. Fig.3.a shows the time to reach to the equilibrium launch velocity. The peak of each line indicates the essential duration of moving molasses to give the launch velocity to the most atoms. The larger the horizontal beam intensity, the longer duration is necessary to give the launch velocity to many atoms efficiently. In fact, when the horizontal lasers are turned off while the moving molasses process, the initial launch velocity is given to the cooled atoms within the shortest period. It is because the horizontal lasers influence the acceleration force of the atomic launching. The high power of the horizontal beams disturbs the acceleration induced by the vertical beams because of the saturation effect. Next, we investigate the dependence of the vertical laser power intensity. Fig.3.b shows the case of 4mW/cm^2 vertical lasers. In comparison with the Fig.3.a,

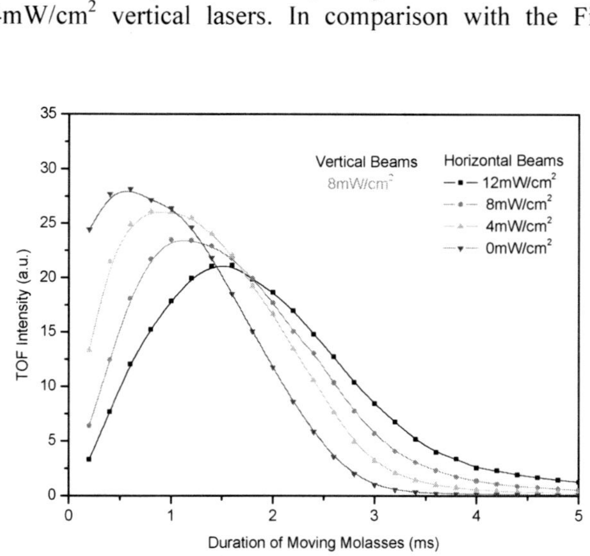

Fig.3.a Laser Power Dependence of the Horizontal Beams at the Vertical Beams of 8mW/cm^2

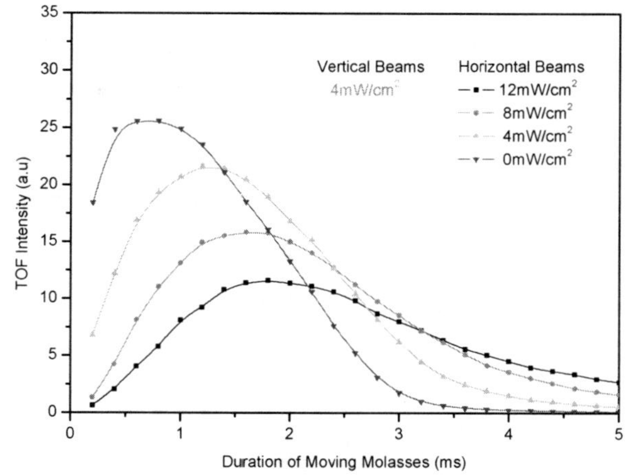

Fig.3.b Laser Power Dependence of the Horizontal Beams at the Vertical Beams of 4mW/cm^2

when the horizontal laser intensity is stronger, it takes a longer time to reach to the equilibrium velocity. We are planning to analyse these results theoretically.

4.2 Making the most use of the PGC effect

From Fig.3.a and Fig.3.b, it can be seen that the TOF signal intensity decreases suddenly after the duration of about 2ms. This is due to the insufficiency of the cooling by the PGC. At present, the atoms launched by the moving molasses are post-cooled to less than 2 K by the PGC. The PGC is carried out until the launched atoms jump out from the horizontal beams. In fact, because the diameter of the horizontal beams is limited (about 30mm), the longer the duration of the moving molasses, the shorter the time for PGC is. Consequently, the pre-cooled atoms are not further cooled sufficiently and most of the launched atoms cannot reach at the detection zone because of the dispersion. In other words, at the atom launching in the atomic fountain, it is essential to give the launch velocity to the many atoms as rapidly as possible to make the most use of the PGC effect. In our case, at (0,0,1) moving molasses process, the horizontal laser beams should be turned off to give the launch velocity to the atoms rapidly and obtain PGC effect sufficiently.

5 Frequency Stability of Synthesi ers

In order to investigate the reason why the frequency stability of our fountain is still low, we checked the frequency stability of the microwave synthesizers themselves. We have prepared three synthesizers, ANRITSU corporation synthesizer, NIST synthesizer and Spectradynamics Incorporation (SDI) synthesizer. At present, we use the synthesizer of ANRITSU Corporation. All synthesizers are locked to the 5MHz output of the hydrogen maser. The frequency stability of 100MHz output from each synthesizer is measured by dual-mixer time difference (DMTD) method [4]. The 100MHz output of each synthesizer is compared with the 100MHz output of the hydrogen maser. The obtained frequency stability is shown in Fig.4. In our measurement, the frequency stabilities of ANRITSU synthesizer and NIST synthesizer are not high enough to achieve less than 1 10^{-12} at

1 second. Presently, these local oscillators do not set limit to the frequency stability of our fountain. However, it will be a critical problem as the frequency stability of our fountain get better.

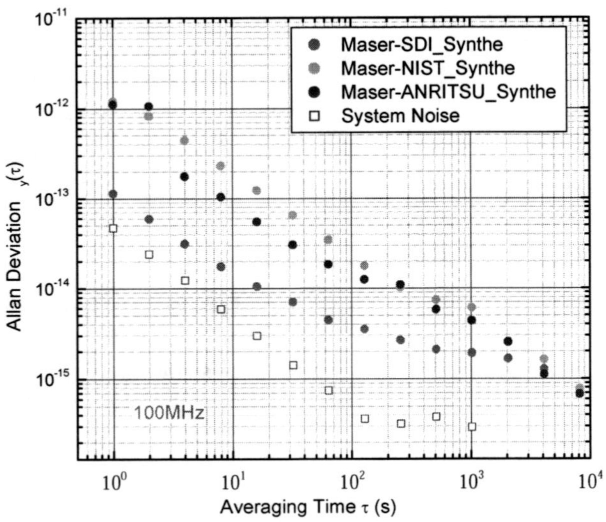

Fig.4 Frequency Stability of 100MHz output from several synthesizers. All synthesizers are locked to the hydrogen maser.

6. Summary and Future Plan

NICT have been developing the atomic fountain frequency standard. So far, we have reduced the noise of the photo detector and its electric circuit and optimise the moving molasses condition to increase the number of the launched atoms. Several efforts have been done to increase the SN ratio of the Ramsey signal. As a result, the frequency stability is presently 2 $10^{-12}/\tau^{1/2}$, which is about tem times better than the previous one.

It is known that the selection cavity is useful to reduce the collision shift and improve the SN ratio of the normalized signal. We will install the selection cavity into the present fountain. And, the SDI synthesizer will be installed as the microwave source. Furthermore, we plan to construct second atomic fountain with making the most use of the experiments and the results obtained from the development of the first fountain. It will be more compact in the size and sophisticated in the design.

References

[1] K. Fukuda, H. Ito, M. Kumagai, M. Kajita, A. Hasegawa, M. Hosokawa, T. Morikawa, "Cesium Fountain Development at CRL" *ocee ng o T* , pp. 217-223, (2000)

[2] M. Kumagai, H. Ito, K. Fukuda, M. Kajita, M. Hosokawa and T. Morikawa: "Development of Cs Atomic Fountain Frequency Standard at CRL" *oc o T* , pp. 150-155, (2002)

[3] M. Kumagai, H. Ito, K. Fukuda, M. Kajita, M. Hosokawa and T. Morikawa: "Development of Atomic Fountain Frequency Standard at CRL" *oc o h T* , pp. 120-122, (2003)

[4] L.Sojdr, J. Cermak, and G. Brida, "Comparison of high-precision frequency-stability measurement systems" *oc o h T* , pp.317-325

CRYOGENIC ALL-SAPPHIRE OPTICAL-AND-MICROWAVE FLYWHEEL OSCILLATOR

Sarika Pugla[+], M. Oxborrow[#], P. Gill[‡] and D. Segal[*]

[+ # ‡]National Physical Laboratory, Teddington, UK, [+ ‡ *]Imperial College, London, UK
sarika.pugla@npl.co.uk

Keywords: Cavity-stabilized lasers, ultra-stable oscillators, frequency stability, cryogenic sapphire, whispering gallery

Abstract

Microwave and optical oscillators both based on cryogenic sapphire resonators offer advantageous frequency stabilities over time intervals ranging from 0.1 to 10000 seconds. We present progress on the development of a combined optical-and-microwave oscillator, whose component optical and microwave resonators form the same physical package and, furthermore, whose resonance frequencies are defined by the same piece of mono-crystalline sapphire.

1 Introduction

Ultrastable 'flywheel' oscillators are essential tools for measurements demanding the highest levels of sensitivity and stability, which include the realization of new atomic frequency standards, the study of astrophysical processes, tests of special [1,2] and general relativity (gravitational red-shift, local positional invariance [3,4,5]), and measurements of small forces [6]. At optical frequencies, the most suitable such oscillators are made from frequency-tunable lasers locked to the modes of high-finesse optical reference cavities. Frequency stability demands that the cavity's length be held constant. Changes in a cavity's temperature, the circulating power within it, and the mechanical forces supporting and/or restraining it can all cause its length to change. These perturbations must be suppressed and/or the cavity itself made less sensitive to them.

At liquid-helium temperature (4.2 K) or below, high-purity mono-crystalline sapphire exhibits a number of properties that favour it as a material both for the spacer and mirror-substrates of linear, vacuum-pathed, optical cavities and for the dielectric puck defining whispering gallery modes in microwave resonators. These properties include (i) low thermal-expansion coefficient ($\sim 5\times10^{-11}$ K^{-1} along its c-axis); (ii) absence of isothermal creep (in contrast to ULE or Zerodur at near-room temperature), (iii) low dielectric loss (and thus high Qs) at microwave frequencies, and (iv) high thermal diffusivity.

Cryogenic sapphire cavities, subject to independent temperature control, have attained stabilities of 2.4×10^{-16} at 32s time intervals in the microwave regime [7] and stabilities of 2.3×10^{-15} at 20s time intervals in the optical regime [8]. Since our optical cavity and microwave resonator are made from the same, and moreover isothermal, piece of cryogenic sapphire, we expect to be able to measure correlations between optical and microwave frequencies at these or greater levels of stability.

2 Experimental setup

Our current experimental setup is shown in Fig. 1. Two all-sapphire optical cavities are mounted horizontally inside the same cryostat. Each cavity comprises a spacer to which a pair of ultra-high-reflectivity mirrors are optically contacted; all three components are made from c-axis-aligned, mono-crystalline sapphire. One cavity is 90mm long and enclosed within a microwave can –see Fig. 2. The other cavity is 70mm long and held in a simple copper clamp. They are oriented parallel to one another and positioned within the cryostat along a common axis.

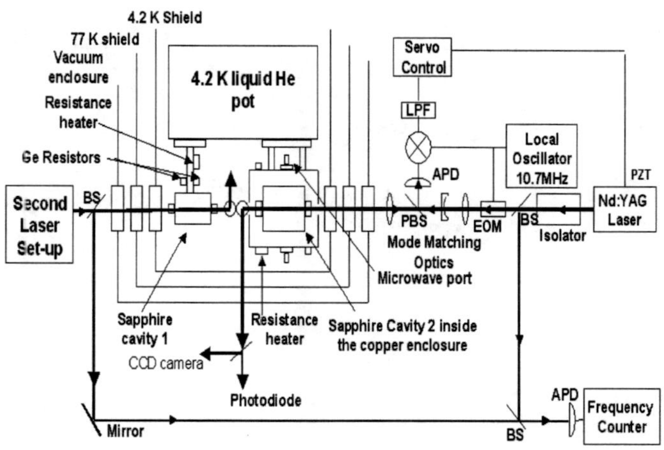

Fig. 1: Schematic of our set-up. Only one PDH subsystems is shown. BS: beam splitter, PBS: polarizing beam splitter, APD: avalanche photo detector, EOM: electro-optic modulator, LPF: low-pass filter, PZT: frequency-controlling piezo actuator. The PDH 'Servo Control' incorporates a double integrator.

Our single laser is a diode-pumped monolithic Nd:YAG non-planar ring oscillator (NPRO), at 1064 nm. The Pound-Drever-Hall (PDH) technique [9] is used to lock, in frequency, two independent laser beams (namely 'Laser 1' and 'Laser 2' in Fig. 3) to the two optical cavities. The first beam is derived directly from the laser and passes through a set of PDH optics mounted on an optical breadboard to one side of the cryostat. The PDH circuit locks the output of the laser to the 90-mm cavity by adjusting the voltage applied to a piezo element attached to the laser's NPRO crystal. The second laser beam is split off from the first and routed to a separate optical breadboard (not shown) on the opposite side of the cryostat. Here, it is first frequency-shifted by 'double-passing' it through an acousto-optic modulator (AOM), before passing through a separate set of PDH optics. The frequency-shifted beam is locked to the 70mm cavity, by adjusting the frequency (100-200 MHz) of the rf signal that is applied to the AOM. The relative frequency stability of the two cavity-locked laser beams is measured by frequency-counting the rf beat note produced by overlapping picked-off samples of each on an avalanche photon detector. Due to their different lengths (90mm and 70mm), TEM_{00} longitudinal modes supported by each optical cavity can always be found whose beat frequency lies within the range (300-MHz) of an SRS 620 counter without recourse to UHF prescaling.

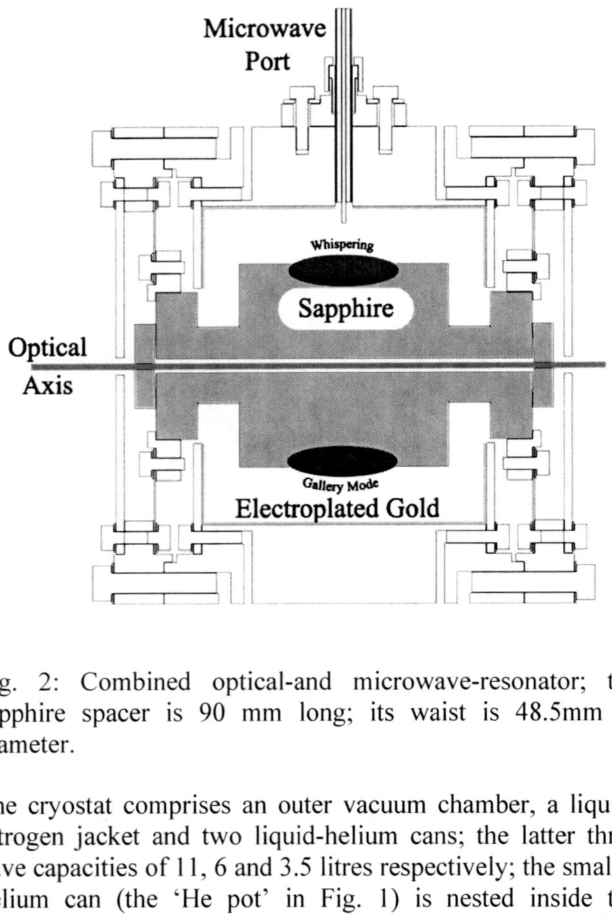

Fig. 2: Combined optical-and microwave-resonator; the sapphire spacer is 90 mm long; its waist is 48.5mm in diameter.

The cryostat comprises an outer vacuum chamber, a liquid-nitrogen jacket and two liquid-helium cans; the latter three have capacities of 11, 6 and 3.5 litres respectively; the smaller helium can (the 'He pot' in Fig. 1) is nested inside the radiation shield of the larger, upper helium can. Each optical cavity and its supporting OFHC copper structure(s) are thermally anchored (weakly) to the inner helium can's 'cold face' through copper straps of designed thickness and length. Including the cryostat's outer vacuum chamber, there are three radiation shields, each holding four optical windows. Fig. 3 shows a schematic of the optical routing within the cryostat.

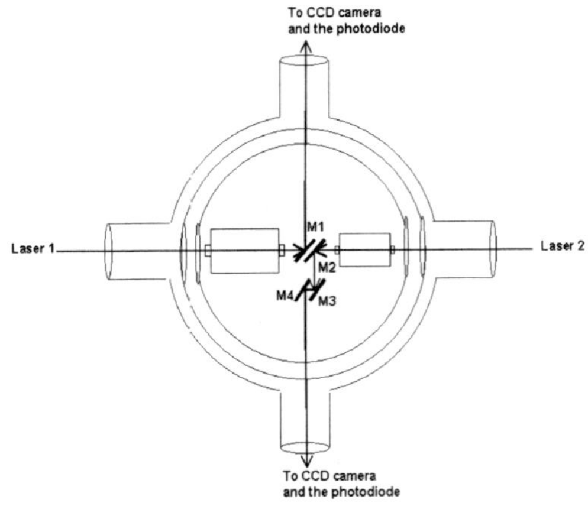

Fig. 3: Bottom-view of our four-port optical cryostat. Mirrors M1, and {M2, M3, M4} are used to route the light transmitted by each cavity out of the cryostat.

To get a genuine measure of the frequency stability, we provide independent temperature control to each of our cavities. The temperature deviation is measured using a four-wire ac-bridge circuit (Fig. 4) wired to a germanium thermistor (Lakeshore GR-200A-2500) mounted on the copper structure that supports the cavity inside the cryostat. An EG&G 7260 lock-in amplifier supplies an ac drive voltage to the bridge circuit and measures the amplitude of the synchronous error voltage returning from it; the latter indicates the temperature deviation from the bridge's set-point. A PID control program written in LabVIEW reads the error signal from the lock-in amplifier via a GPIB link. This program also sends instructions (again via GPIB) to a programmable current source that controls the power dissipated by a resistance heater mounted to the same copper structure as supports the cavity, thus closing the temperature-control loop.

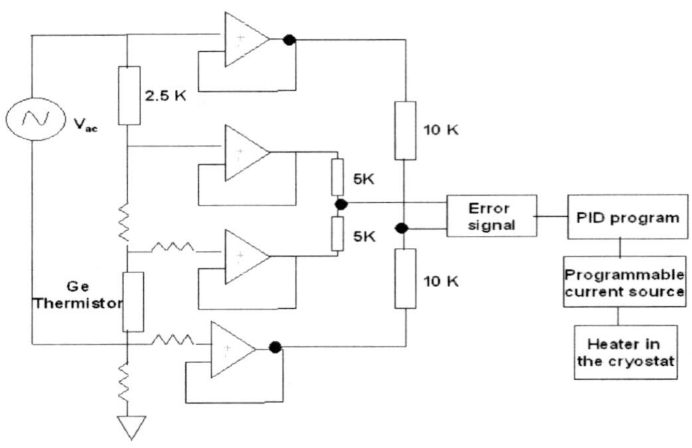

Fig. 4: Four-wire ac-bridge temperature control circuit used to control the temperature of both cavities.

There are in fact two germanium resistors mounted on each copper support; one is used for controlling the temperature while the other is used for monitoring it. The resistance of the latter was measured using a digital multimeter; when the temperature control was working and Ziegler-Nichols optimised at 4.2 K, the multimeter's reading fluctuated in the fourth significant decimal place every few seconds, corresponding to a temperature stability of ~2 μK; the bandwidth of the PID control loop was about ~1 Hz. Furthermore, our 90 mm optical cavity can be regarded as having a supplementary 'thermometer' provided by the frequency of a high-order whispering-gallery mode localized around the belly of its sapphire spacer –see Figs. 2 and 5.

As already mentioned, the laser frequency is locked to the frequency of a TEM_{00} longitudinal mode supported by each optical cavity through the Pound-Drever-Hall (PDH) technique [9]. The technique's phase modulation is provided using an electro-optic modulator (EOM), driven by a local oscillator at 10.7 MHz –see Fig. 1. A disadvantage of this set-up, as compared to directly modulating the laser crystal with the piezo element attached to it, is the presence of a polarization-dependent dc offset in the PDH error signal, due to residual amplitude modulation introduced by the EOM. To minimize this dc offset (and hence its absolute drift), we use a Glan-Taylor (calcite) polarizer before the EOM to select a polarization that nulls out the dc offset.

3 Results so far

Some initial measurements at room temperature were performed with a beat frequency of ~180 MHz. At 4.2 K, the optical finesse of our 90-mm optical cavity, was measured to be approx 108,000; the finesse of our shorter 70-mm cavity was also $>10^5$. Unfortunately, before any beat data could be taken at 4.2K (with both cavities under full temperature control), the mirror at one end of the 90mm cavity became detached from the spacer during a slow warm-up from liquid-He to liquid-N_2 temperature.

Whilst the 90mm cavity was still near liquid-He temperature, an as yet unidentified microwave whispering-gallery mode was observed at 9.237 GHz whose linewidth was just 81 Hz, corresponding to an unloaded Q of 114 million.

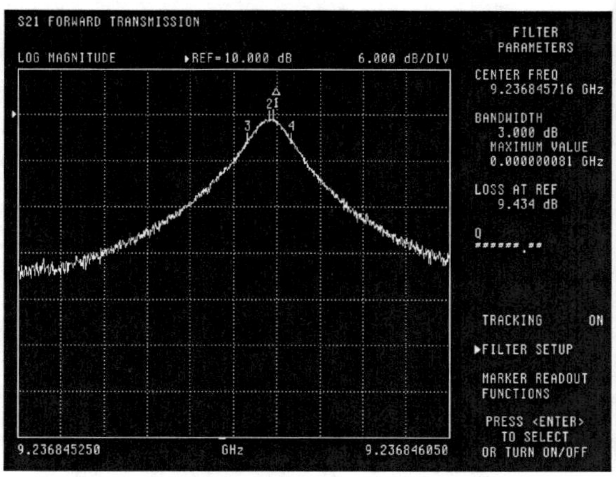

Fig. 5: Transmission (S21) across our microwave can showing a high-Q resonance at 9.237 GHz.

4 Outlook

The values of finesse and Q reported above are extremely promising for the 90mm cavity's ultimate performance as either an optical or microwave flywheel oscillator, and for its ability to transfer frequency stability between the microwave and optical frequency domains. Recent interferometric tests on the individual elements of the 90mm cavity have indicated that neither end of the cavity's spacer, nor the contacting annuli of the mirrors (once) optically contacted to it, were flat enough to realise optical contacts capable of surviving repeated thermal cycling to and from cryogenic temperatures. We have thus elected to have all of the relevant optical-contacting surfaces, including those forming the contact that did not fail, re-polished to higher levels of flatness. This re-polishing work is underway.

Acknowledgements

This work was supported through the UK government's NMS Quantum Metrology Programme

References

[1] A. Brillet and J. L Hall. "Improved Laser Test of the Isotropy of Space", Phys. Rev. Lett, **42**, pp. 549-552 (1979).

[2] D. Hils and J. L. Hall. "Improved Kennedy-Thorndike experiment to test of relativity", Phys. Rev. Lett., **64**, pp. 1697-1700 (1990).

[3] R.F.C Vesset *et.al.*, "Test of Relativistic Gravitation with a Space-Borne Hydrogen Maser", Phys. Rev Lett. **45**, pp. 2081,2084, (1980); C. J. Berglund *et.al.*, "New Limits on Local Lorentz Invariance from Hg and Cs Magnetometers", Phys. Rev. Lett. **75**, pp. 1879-1882 (1995).

[4] J. P. Turneaure *et al.*, "Test of the principle of equivalence by a null gravitational red-shift experiment" Phys. Rev. D **27**, 1705-1714 (1983); A. Godone *et al.*, "Null gravitational redshift experiment with nonidentical atomic clocks" Phys. Rev. D **51**, pp. 319-323 (1995).

[5] H. Muller, S. Herrmann, C. Braxmaier, S. Schiller, A. Peters, "Modern Michelson-Morley experiment using cryogenic optical resonators" Phys. Rev. Lett. **91**, pp. 020401 (2003)

[6] H. Walesch *et al.*, "Determination of the gravitational constant G using a Fabry-Perot pendulum resonator", Precision Electromagnetic Measurements Digest, pp. 144 -145 (1998).

[7] S. Chang, A. Mann, and A. Luiten, "Improved cryogenic sapphire oscillator with exceptionally high frequency stability," Electron. Lett., **36**, pp. 480-481 (2000).

[8] S. Seel, R. Storz, G. Ruoso, J. Mlynek and S. Schiller, "Cryogenic Optical resonators: A New Tool for Laser Frequency Stabilization at the 1Hz Level", Phys. Rev. Lett. **78**, pp.4741-4744 (1997).

[9] E. D. Black, "An introduction to Pound-Drever-Hall laser frequency stabilization", Am.J. Phys. **69**, pp. 79-87 (2001).

SPACE WORTHY ELECTRONICS PACKAGE FOR THE 35KG SPACE ACTIVE HYDROGEN MASER ON ACES

B. Thieme, Dr. D. Zöschg and G. Baister

Contraves Space AG, Schaffhauserstrasse 580, 8052 Zurich, Switzerland

Keywords: ACES, SHM-35, Maser, Space, Electronics

Abstract

The 35kg active Space Hydrogen Maser (SHM-35) developed at the Observatory of Neuchâtel is one core element of the ACES experiment (Atomic Clock Ensemble in Space). ACES is an European experiment on the International Space Station (ISS) comprising the SHM-35 and the cold atom caesium clock PHARAO locked to each other. The resulting ultra-stable clock provides an optimised time scale and allows for fundamental physics tests. SHM-35 consists of the physical package (PP) and of the electronics package (EP).

The operation outside the ISS and thus in space environment together with the demanding performance needs of the SHM-35 puts challenging requirements on the EP design. The development of the EP is on-going at Contraves Space with the flight design underway together with the build and test of an elegant breadboard model (EBB) of this EP. The EP EBB incorporates the same architecture, design and functionality as the flight hardware but using commercial EEE parts. With this approach the performance of the EP and SHM can be verified at an early stage in the flight development programme.

This paper presents the architecture of the space worthy electronics for SHM-35 with emphasis on the radio frequency part. The control part and special problems arising from the operation in the space environment will also be discussed. The required performance is compared with measurement results at the EBB EP. Furthermore first results of the integrated SHM consisting of the EBB EP and a ground Maser physical package will be reported.

1. Introduction

This paper presents an overview of the Electronics Package (EP) needed for the control and operation of the 35kg space hydrogen maser (SHM-35). The design and the performance of the Physical Package (PP) of the SHM-35 has been reported previously in [1]. All electronics functions required to operate the SHM-35 are implemented in the EP. In particular the EP is to be the interface to the ACES experiment, will operate all control loops and monitor channels and will generate the 100 MHz frequency standard output from the 1.420405751 GHz signal generated by the PP.

Based on the design of the laboratory electronics by the Observatory of Neuchâtel (ON) Contraves Space AG (CSAG) is responsible for the development of the flight SHM electronics. The development of the flight electronics follows a three model philosophy:

- Elegant Breadboard (EBB) Model to demonstrate the performance and functionality of a flightworthy design.
- Engineering Model (EM) - fully representative of the flight hardware and will undergo qualification tests.
- Proto-Flight Model (PFM).

2 Architecture of Electronics

The general architecture of the SHM-35 is depicted in the block diagram in figure 1. The two main parts of the SHM-35 are the Physics Package (described in [1]) and the electronics package (EP).

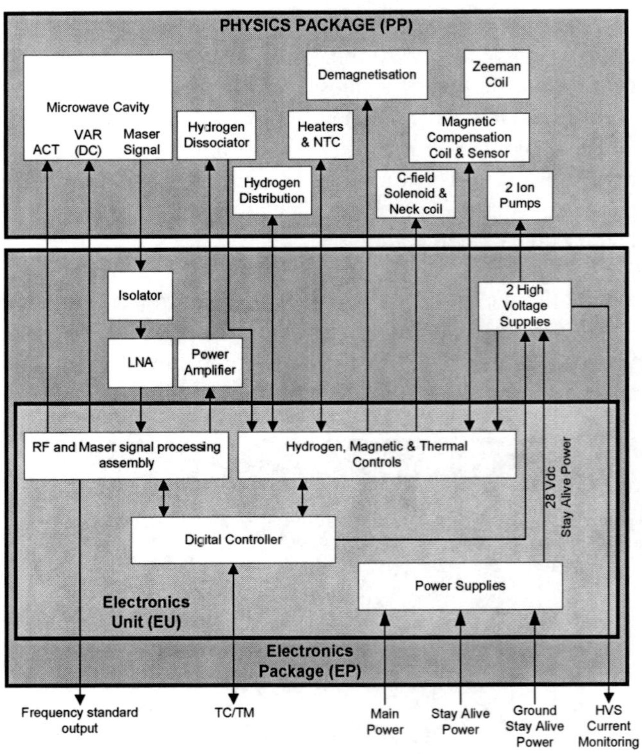

Figure 1: Block diagram of SHM

The EP itself comprises the following:

- An Electronics Unit which itself is made up of:
  - RF and Maser Signal Processing Assembly – the main function of which is to generate the 100 MHz frequency standard signal from the maser signal.
  - Hydrogen, magnetic and thermal controller.

- Digital Controller.
- Power supplies for SHM-35.

- A 1.4 GHz isolator and a Low Noise Amplifier (LNA) which are used to isolate and amplify the output signal from the PP Microwave cavity.
- A power amplifier amplifying a tunable 120 to 150 MHz signal capable to provide up to 10W to a 50Ω load for the hydrogen dissociation.
- High voltage supplies for the ion pumps.

The remainder of this section describes the design of the Electronics Unit.

2.1 RF and Maser signal processing assembly

The RF and Maser signal processing assembly is part of the EP and performs the following functions:

- Synthesis of a tuneable 100 MHz frequency standard output signal out of the amplified atomic signal input (1.420405751 GHz).
- Synthesis and switching of two tuneable interrogation signals, 15±5 kHz above and below the atomic signal frequency. The interrogation signals are periodically and alternately fed into the microwave cavity of the PP for a short time, to perform the in [1] reported Automatic Cavity Tuning (ACT) which guarantees that the resonance frequency of the microwave cavity is stabilised over all working conditions at an adjustable frequency closed to the hydrogen atomic frequency. The ACT consists of generation of injection frequencies locked to the Maser frequency, detection of the cavity response and derivation of a DC-voltage for tuning of the cavity resonance frequency via a varactor.
- Synthesis of a 120 MHz – 150 MHz signal with a selectable power level up to 10 W. This signal is applied to the hydrogen dissociator of the PP.

Figure 2 shows the functional block diagram of the RF and Maser signal processing assembly.

The down-converter block performs a triple down-conversion of the atomic signal input coming from the LNA with intermediate frequencies of 20.405751 MHz, 4.594249 MHz, and 405.751 kHz. The required local oscillator signals of 1400 MHz, 25 MHz, 5 MHz are derived from the master oscillator (MO), which operates at 10 MHz and is locked to the Maser.

The MO depicted in figure 3 is a special designed 10 MHz OCXO based on the Datum 9600 model (now Symmetricom). It has been especially adapted by Symmetricom for the combined requirements of low g-sensitivity, phase noise, short term stability as well as the use in space environment. The short term stability of the MO is $\leq 2 \cdot 10^{-12}$ at 1s and the phase noise is –110 dBc/Hz at 1 Hz offset.

The synthesiser block takes the same 25 MHz signal as input as the down-converter stage and the 10 MHz MO signal and generates a digitally tuneable 405.751 kHz signal, two digitally tuneable ACT signals nominally at 20.39 MHz and 20.42 MHz, and a hydrogen dissociator signal digitally tuneable between 120 MHz and 150 MHz. All generated frequencies have a tuning resolution smaller or equal 1 Hz. The frequencies in the synthesiser block are generated by Direct Digital Synthesis (DDS). DDS is also used to lock all generated signals to the 10 MHz MO as well as to adjust frequency without phase step and step sizes down to 1.4 μHz. To achieve spectral purity and because frequency range of DDS is limited to half of clock frequency (10 MHz clock) the required frequencies (20.39 MHz, 20.42 MHz, 120…150 MHz, 405.751 kHz) are generated from the DDS by digital PLLs. The generated DDS frequencies are 1.27 MHz, 1.28 MHz, 1.8…2.4 MHz and 1.9 MHz. The synthesiser (DDS) for the 405.751 kHz signal includes a 36 bit counter which allows for a relative stepsize of the 10 MHz signal of $<1 \cdot 10^{-15}$ (Δf/f). The synthesised 405.751 kHz signal and the downconverted 405.751 kHz signal are used to lock the 10 MHz signal to the Maser frequency.

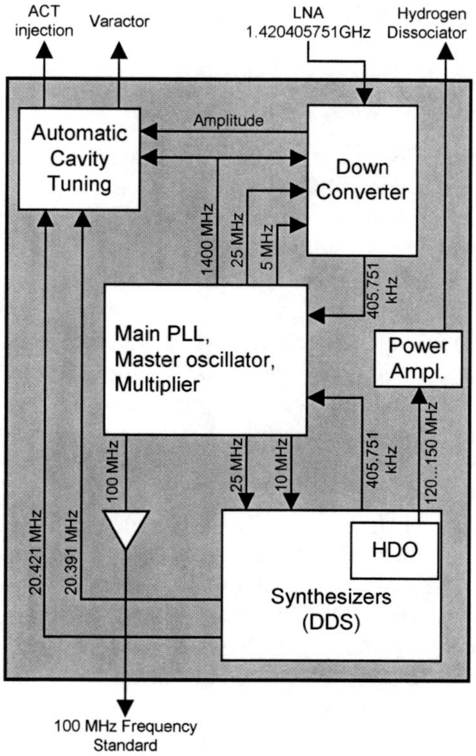

Figure 2: Block diagram of RF and Maser signal processing assembly

A PLL and a 1:10 frequency divider generate the 100 MHz frequency standard output from the MO. From the 100 MHz signal, frequency division generates the 25 MHz signal. The 1400 MHz signal is generated by filtering the 14th harmonic out of non-linear multiplication of the 100 MHz signal by a step recovery diode.

Finally, in the ACT block, the two synthesised ACT signals in the 20 MHz band are up-converted to 1420 MHz and alternately injected into the microwave cavity. The main PLL is open during the ACT injection cycle. The ACT detection circuit will be implemented in digital logic which is standard PC electronics for the EBB EP.

Figure 3: 10 MHz OCXO Master Oscillator

2.2 Digital Controller and Hydrogen, magnetic and thermal controller

The Digital Controller comprises the actual micro-processor circuit including RAM/ROM, the digital to analogue converter (DAC) channels and the analogue to digital converter (ADC). With exception of the demagnetisation signal, which is generated directly by the micro-processor, all hydrogen, magnetic and thermal control circuits are implemented as analogue control loops. The set-point of the control loops is given via a DAC-channel by the micro-processor. A multiplexed analogue signal which allows the ADC on the digital controller to measure the control and housekeeping parameters is generated from the sensor signals in the PP.

The design of the Digital Controller unit will incorporate a Leon2 processor (SPARC V8 architecture) implemented in a FPGA. The operational clock frequency of the FPGA is selected to be 20MHz. Five 4Mbit SRAMs will be implemented, providing 512Kword x 32 bits. The additional 512Kword x 8 bits are reserved in order to be able to implement an EDAC protected RAM memory in the EM/PFM model. This memory is installed as a single bank. 4Mbits of Flash EPROM are implemented as four 1Mbit devices. The PROM memory is implemented in two banks, the first bank providing 128Kword x 32 bits wide memory, and the second bank 128Kword x 8 bits. For the lower memory bank, the ability to erase and write to this memory bank is only enabled when a link in the test connector is installed. Typically, this section of the memory will hold the program, and will be programmable on ground only. For the final configuration before flight, the link will be removed and further erasure or re-writing of the memory disabled. However, for the upper memory bank Read/Write functionality will be allowed in flight. This area of memory will be used to store and change program constants and parameters, or for non-critical sections of the program code.

A significant number (ca. 20) of digitally controlled analogue outputs are required. In principle, this could be simply achieved by implementing a large number (ca. 20) of DAC IC's, or with a single DAC and a large number of Sample and Hold circuits. However, the impact of this on the design is significant in terms of control interconnections required, number of components required, and the corresponding impact on the PCB area, mass and power. Therefore, bearing in mind that only relatively low bandwidth control is required for the Controller Unit (<< 100Hz), a Pseudo DAC scheme will be implemented, based on digitally controlled PWM signals, which are low pass filtered to generate the necessary analogue control signals. Externally to the Digital Controller, all drive signals are analogue. The generation of the Pulse Width Modulation (PWM) signals is implemented in VHDL in the logic of the FPGA. The PWM ratio is programmable by the software, by writing to memory mapped registers in the I/O memory of the processor. This circuit is implemented 20 times, for the various control functions.

The selected ADC is an AD9042. This device is known to be radiation hard and used for space applications. Standard multiplexing and buffer amplifier circuits are implemented. The sequencing of the multiplexing and clock of the ADC is performed by the hardware logic of the FPGA.

2.3 Power Supply Unit

The Power Supply Unit generates isolated secondary voltages for the SHM from the Primary (+28V) power supply input from the ACES experiment. To reduce the complexity of the system, only the main voltages are generated in the Power Supply Unit. Specific voltage required by other units are generated locally within to those units (e.g. the +3.3V logic voltage required by the FPGA and digital circuits in the Controller Unit). The high voltage of 3500V for operation of the ion pumps will be generated by two separate high voltage power supply units. These units are independently powered by a stay-alive power line from the ACES experiment.

The EP contains also a 1.4 GHz isolator and a LNA for the conditioning of the output signal of the cavity. These elements are installed directly at the PP as well as the power amplifier for the generation of a 120...150 MHz signal with selectable power level up to 10W for hydrogen dissociation.

3 Implementation

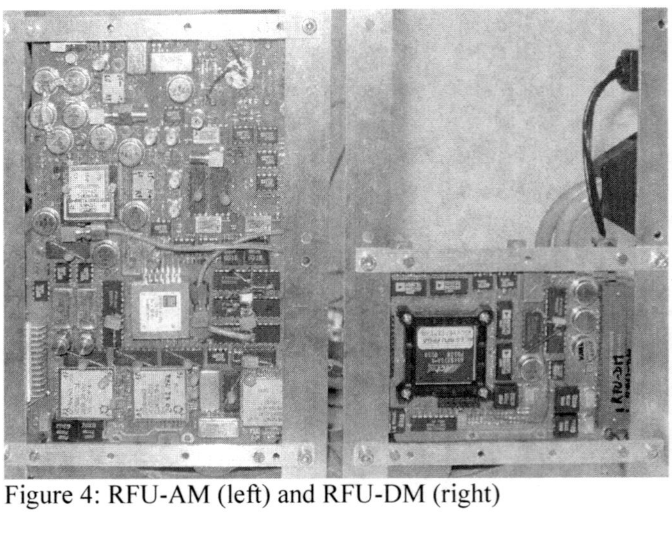

Figure 4: RFU-AM (left) and RFU-DM (right)

The functions are implemented on different Printed Circuit Boards (PCB) and grouped into units. The RF and signal processing assembly is implemented in the RF Unit (RFU). Conveniently, the partitioning of the functions has resulted in three parts with only a few interconnections in-between. One PCB the RFU Analogue Module (RFU-AM) hosts mainly high frequency-parts such as the downconverter chain. The second PCB (RFU-ACT) incorporates the automatic cavity tuning detection circuit. The third PCB, the RFU Digital Module (RFU-DM) comprises a FPGA that generates the programmable reference frequencies for all synthesisers using DDS techniques including the frequencies for the ACT. Figure 4 depicts the RFU-AM and the RFU-DM integrated into the EBB EP.

The Controller Unit (CU) comprising the Digital Controller and the hydrogen, magnetic and thermal control functions will be implemented on two PCB's. The partitioning of the functions has resulted in one PCB containing mainly digital functions and interfaces, the Processor Module (CU-PM). The CU-PM PCB is depicted in figure 5. A second PCB, the Sensor Module (CU-SM), contains mainly analogue signal conditioning for the PP sensors and actuators.

Figure 5: Controller Unit Processor Module PCB

The Power Supply Unit (PSU) comprises three DC/DC modules. A fourth module provides an EMI input filter. A simple PCB with mainly passive components will be incorporated to enable the interconnection of the connectors and signals.

At EBB design stage all PCB's are mounted on a frame construction to allow easy access to all circuits during the integration process with the PP. The EBB EP is depicted in figure 6.

The EU at later EM and PFM design stages will be implemented as a package of electronics as represented in figure 7, consisting of three mechanical frames into which the electronics are mounted. The detailing of each frame is tailored to the necessary connector cut-outs and mounting points required by the PCB design. The package of electronics is held together with four stainless steel threaded rods. The RFU-AM, the RFU-DM and the RFU-ACT are mounted into one frame forming the RFU module. The CU module comprises the CU-SM and the CU-PM PCB mounted back to back into a second frame. The PSU module is integrated into the third frame where the DC/DC modules are bolted directly to a 4mm thick aluminium wall for thermal reasons.

The frames are machined from Aluminium, and provide a rigid structure into which the PCB's are mounted. Analysis and experience indicates that the frame design and PCB support can meet the design strength and eigen-frequency requirements.

Figure 6: Integrated EBB model of EP

The aluminium structure provides the thermal path for the heat dissipation from the electronics, with the main transfer being by conduction via the baseplate of the unit. To this end, highly dissipating components are preferably placed towards the bottom of the PCB, and the highest dissipating components, e.g. power transistors, are mounted with a bolt connection to the frame itself.

All functional connectors are mounted on the top of the unit. Interconnections between the units, and from the EU to PP are made with conventional connectors and cabling. This provides the simplest connection, yet allows also the units to be easily tested individually without requiring complex adaptation. A simple backplane (PCB) is foreseen to connect between the RFU and CU and the backplane connectors will be enclosed by a metal cover. A standard type of space qualified backplane connector has been defined for this connection. Test connectors are installed on the side of the unit, and will be covered with a mechanical cap when in flight.

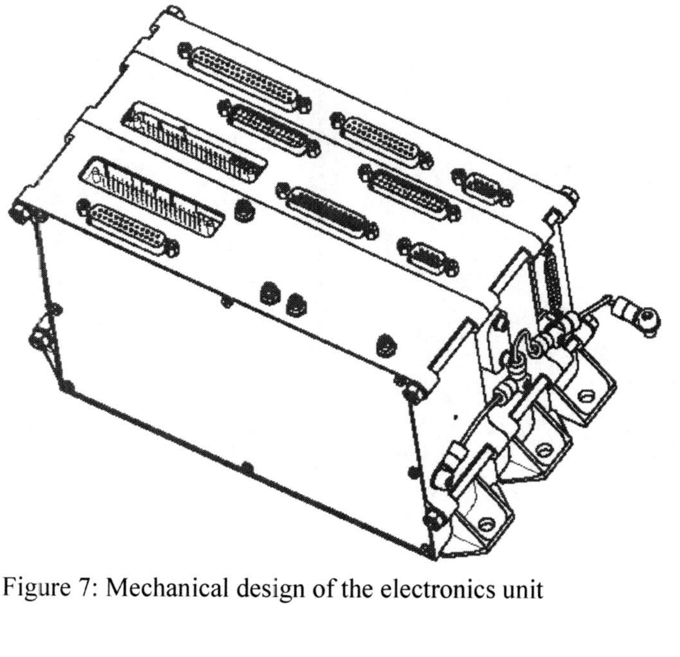

Figure 7: Mechanical design of the electronics unit

The aluminium structure of the frames provides an overall conductive enclosure for the EU, and will be provided with a suitably conductive surface finish. The Controller Unit, being in the middle of the 'sandwich' has two covers installed to enclose it. The two outside frames, are provided with a cover on the outside only, but interlock with the controller unit, with a small overlap, to ensure that the package provides an overall shielded enclosure for EMC tightness. It is required that the EU has to be bonded to the ground with a bonding strap. For the location the PSU Frame has been chosen, since the local secondary grounding point for the EU is in the PSU.
Beside the thermal and mechanical design of the EP additional effort has been taken to make the PCB's compliant with the requirements of the space environment. All components are selected to be radiation tolerant allowing the operation in the radiation environment of the International Space Station. With at least 2mm aluminium shielding the total accumulated dose over 3 years lifetime is 3.4 krad. Therefore as example less radiation sensitive technologies such as ECL or Bipolare are preferably used. The FPGA implementing the DDS in the RFU-DM is an ACTEL radiation tolerant device. All PCBs made from FR4 substrate are designed according appropriate IPC layout standards to guarantee the required reliability over lifetime. For this reason all used components are at least qualified according military standards. Components in circuits interfacing other ACES or Columbus Module sub-systems are qualified according space standards.

4 Performance

After successful verification of the function of the stand-alone EBB EP at CSAG premises the EP is currently under test with an ON ground hydrogen maser (EFOS) and the 50kg prototype EM PP (PEM-PP). The ground hydrogen maser is used to verify the performance of the electronics package in terms of the Allan variance and phase noise. The measurement set-up comprising EBB EP and ground hydrogen maser is depicted in figure 8. Control and monitor functions will be verified with the PEM-PP.

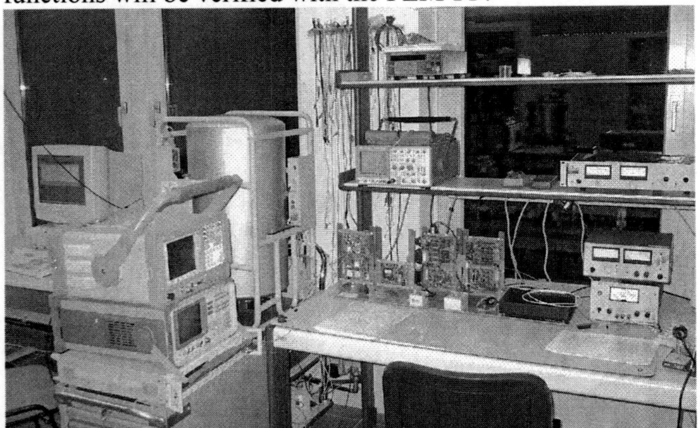

Figure 8: Laboratory set-up with EBB EP and ground maser

Tests have been performed with the EBB EP and the PEM-PP to verify the performance of the thermal and magnetic control and monitor functions. The performance of the hydrogen distribution assembly (HDA) was tested at a special breadboard. The performance of the EBB EP could be verified by comparison with the recorded performance of the ON breadboard electronics.
Tests of the RF and maser signal processing assembly as well as a verification of the EP in a thermal chamber are currently on-going.

5 Conclusion

The electronics package for the 35kg space hydrogen maser has been developed to EBB stage and is currently verified with a ground hydrogen maser and the PEM-PP at ON premises. This paper has reported the architecture and the design of the Electronics Package with main emphasis on the RF and maser signal processing assembly. Special design solutions required by the operation in space environment have been depicted. The operation of the hydrogen, magnetic and thermal controls has been tested with the PEM PP. Tests of the RF and maser signal processing assembly with an ON ground hydrogen maser are currently on-going.

Acknowledgements

The authors would like to acknowledge the work and efforts of the project team at Contraves Space. In addition, we gratefully acknowledge the contributions and support from Observatoire Cantonal de Neuchâtel. The development of the Electronics Package by Contraves Space for the SHM instrument is under subcontract to Observatoire Cantonal de Neuchâtel and is funded by ESA through the Prodex programme.

References

[1] A. Jornod, D. Goujon, D. Gritti and L.G. Bernier. "The 35kg Space Active Hydrogen Maser (SHM-35) for ACES", *Frequency Control Symposium and PDA Exhibition Jointly with the 17th European Frequency and Time Forum, 2003. Proceedings of the 2003 IEEE International*, pp. 82-85, (2003).

FIFTY YEARS OF ADVANCE IN FREQUENCY STANDARDS

D. J. E. Knight

DK Research, 110 Strawberry Vale, Twickenham, TW1 4SH, UK
djek@cix.co.uk

Keywords: frequency standards, accurate, laser, list, history

Abstract

Fifty years ago next October, the Ve Congrès International de Chronométrie was held in Paris. In its proceedings there is a paper by L. Essen which shows a drawing of a caesium-beam frequency standard then under construction. Measurements in the paper were referred to quartz ring oscillators. The caesium apparatus was still for the future. Here I draw comparisons with that time. I also collate new data and present a list of precise frequency measurements, primarily at laser frequencies, following previous presentations of the list in 1997 and 2001. Recent advances have been accentuated by the introduction of femtosecond lasers for measuring optical frequencies against the caesium standard, and by a laser-cooled fountain version of that standard. Measurements noted include narrow transitions in atoms or ions susceptible to cooling by optical techniques, and so capable of observation with reduced second-order Doppler effect. Applications include searches for change with time of the fundamental constants.

1 Introduction

Over the past decade or so, substantial advances have come to fruition in the interrogation of atoms, molecules and ions as frequency standards, particularly as regards the use of lasers for first reducing their motion in the laboratory frame. I have at intervals collected and listed precise measurements of spectroscopic transitions capable of use as frequency standards to indicate new progress [1,2]. Recent advances have been accentuated by the introduction of femtosecond lasers for measuring optical frequencies against the caesium standard [3], and by a laser-cooled fountain version of that standard [4].

At this point, just as the measurement of optical frequencies has become much simplified, one can look back 50 years to an era with quite different measurement problems, just before the caesium clock itself was demonstrated to work. The Ve Congrès International de Chronométrie was held in Paris fifty years ago this October. In its proceedings [5] there is a paper by L. Essen [6] which shows a drawing of a caesium-beam frequency standard then under construction. Measurements in the paper however were referred to quartz ring oscillators, and the caesium apparatus was still for the future [7].

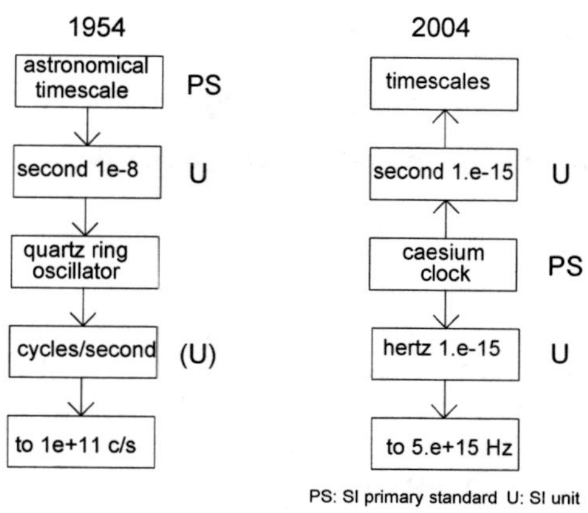

Figure 1: T & F Standards and Units 1954 and 2004

2 Historical view

The essential differences for frequency metrology between 50 years ago and the present are indicated in Figure 1. At that time the SI unit of time interval, the second, was realised through astronomical measurements, and that of frequency, the cycle per second, its inverse, was maintained through the use of precise quartz crystal oscillators such as the Essen ring oscillator. These were capable of intercomparison to 1 part in 10^9, but they tended to drift in frequency by about 1 part in 10^8 per year, leading to realisation of the second to about 1 part in 10^8. At that time Rabi, Ramsey and others working on nuclear magnetic resonance had laid the foundations for the successful caesium frequency standard [8,9]. It was clear to microwave spectroscopists such as Gordy and colleagues [10] already working to 130 GHz frequencies that there would be a new primary standard given a sufficiently high-Q resonance to use, although techniques to eliminate the first-order Doppler effect were in infancy. Lyons and colleagues actively studied the strong ammonia resonance at 23 GHz in a gas cell [11] but the work was limited by Doppler broadening.

The parallel situation today, Figure 1, shows that we have a separate unit, the hertz, for frequency, and that the caesium resonance utilised in 1955 still provides a satisfactory primary standard for both that and the second, and time scales derived from the second. The range of frequencies accessible to precise measurement essentially includes all those for which sufficiently spectrally-pure lasers can be constructed, to beyond 10^{15} Hz, the laser of course being entirely new since1954.

1955	caesium clock (beam, separated field)
1961	laser, gas (optical coherent source)
1967	definition of second on caesium
1967	first laser frequency measurement
1969	stabilised lasers, 1st-order Doppler free
1973	definitive laser speed of light, at 88THz
1983	definition of metre on fixed speed of light (visible frequencies measured)
1995	laser-cooled atom standard (Cs fountain)
1999	measurement of lasers by femtosecond comb

Table 1: Events for Frequency Standards 1954-2004

Some historical developments that contributed to the accuracy and range of present-day frequency and time standards are summarised in Table 1. From the caesium standard to redefinition of the second based on it took 12 years. The first measurement of a laser frequency the same year was followed 16 years later by a new definition of the metre based on a fixed value for the speed of light. Then 12 years to the laser-cooled caesium standard and a further 4 years to the unexpected discovery of the extreme usefulness of the femtosecond laser for measuring optical frequencies. Also important were developments of techniques for first-order Doppler-free stabilisation of lasers, by saturated absorption in gases, in methane[12] and CO_2[13] for the speed of light measurements, and in I_2 for realisation of the metre[14].

3 Table of frequency standards

The updated table of frequency standards is shown in Table 2. This follows two earlier versions [1,2] but is the same in format, on which more follows. The list ranges in frequency from the ultra-violet at the top to RF at the bottom. A table extracted from current CIPM recommendations [Qu03] is at the foot, and indicates working metrological uncertainties. Entries of uncertainty less than 1 part in 10^{12} are in bold type, and new entries have displaced certain older ones, within the single-page length.

4 The list format

The list in Table 2 is intended for transmission as a text file, or for use as a computer data file, with space to accommodate the frequency to 16 or more significant figures, together with its uncertainty. Exponential notation is used. There is space for up to 3 references, also a brief note on the frequency measurement route used. There is not room however for the wavelengths in addition, except that for multiple gas reference lines, such as in I_2 or Rb, the lines are given a wavelength-based label placed in the `reference' column, and which is used to identify it in the `notes' column. Apart from the `system' column at the left, for which transition information can appear as the first item in the `notes' column at the right, there is a single-letter column `app.', for `apparatus' which indicates how the system was interrogated: B: beam; G: gas; L: gas laser Lamb dip; or T: trap. (`~B' is entered for thermal evaporation from a nearby surface into vacuum.) This broadly correlates with increasing potential accuracy in the order L, G, B and T. All standards listed, except those using traps, use a method such as saturated absorption or two-photon excitation (TP) to eliminate the Doppler effect to first order. The Table references have codes in the form of two or three letters for the start of the first author's name plus two digits for the year, and are listed separately.

5 Improvement of frequency-standard accuracy of over 50 years.

The way in which accuracy has increased over the past 50 years is indicated in Figure 2. This figure and the one following were originally used to survey optical measurements going back 30 years [15]. The vertical scale spans 10 decades and the horizontal the past 50 years. The lines link measurements mostly of optical frequency

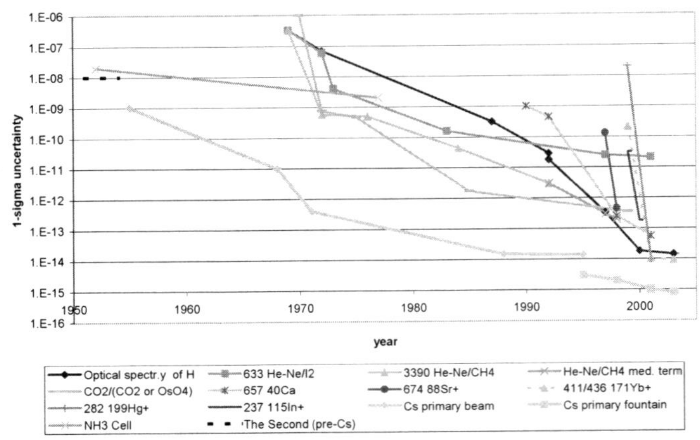

Figure 2: Improvement of frequency accuracy over 50 years

standards, and the two regions of steep drop represent accession to frequency- rather than wavelength- measurement against the primary standard. The lowest lines show the accuracy of the best realisation of the primary standard, the base for measurement. Up to 1954 this was the astronomically-realised second, but from 1955 *de facto* the caesium standard, although this was not fully adopted until 1967. A line from the left, 1952 to 1977 represents the early work by Lyons *et al.* [11] and a 1977 re-examination by Wineland *et al.* of the ammonia transition at 23 GHz, limited by first order Doppler broadening. The discontinuity in the bottom curve for caesium indicates the introduction of the caesium laser-cooled fountain by Clairon et al. [4]. A key indication of improvement in optical measurement is that for the best-measured of the hydrogen transitions, shown by a dark line, for which uncertainty has decreased from 3.3 parts in 10^7 in 1969 to about 2 parts in 10^{14} at present. Some steeply declining lines about 2000 result from the introduction of measurement using the femtosecond laser technique, so

TABLE 2: FREQUENCY STANDARDS FOR LASERS

system	app*	reference(s)	label	frequency/Hz		uncert.	notes (line;)std used; etc
He	B	Ei97		5130.495083	+12	8.8 -09	1S-2P;I2
H	**B**	**Ud97Re00Ni00**		**2466.061413187103**	**+12**	**1.8 -14**	**1S-2S TP;Cs fountain**
muonium	~B	Sch95Me00		2455.5289410	+12	4.0 -09	1S-2S TP;I2
115In+	**T**	**Pe99Za00**		**1267.40245289992**	**+12**	**1.8 -13**	**5:^1S-^3P0coldion;CH4**
Ps#	~B	Ch84Da89		1233.6072189	+12	8.7 -09	1S-2STP;D beta viaTe2
199Hg+	**T**	**Ud01**		**1064.721609899143**	**+12**	**9.4 -15**	**cold ion;Cs fountain**
107Ag	B	Ba04		1040.7060380	+12	3. -09	4d..S1/2-D3/2 TP; I2
H	B	Sc99		799.1917274028	+12	8.4 -12	2S-12D5/2TP;Cs/778Rb
D	B	Sc99		799.4091849676	+12	8.1 -12	2S-12D5/2TP;Cs/778Rb
D	B	deB97Sc99		770.8592528483	+12	7.1 -12	2S-8D5/2TP;Cs/778Rb
4He	G	Pav94		770.732839058	+12	2.4 -10	2\3S1-3\3P0;778Rb
H	B	deB97Sc99		770.6495615811	+12	7.6 -12	2S-8D5/2TP;Cs/778Rb
172Yb+	T	Gi95Tay97		729.4688689	+12	5.5 -10	633I2 via 1m FP
171Yb+	**T**	**St01bPe03**		**688.358979309310**	**+12**	**9. -15**	**cold ion;Cs fountain**
171Yb+	**T**	**Bly03**		**642.12149677126**	**+12**	**3.6 -13**	**octupole,cold ion;Cs**
172Yb+	T	Gi95Ro97		642.1167853	+12	1.1 -09	octopole;633I2/467Te2
127I2	G	Ed96	637I2	470.76645546	+12	2.2 -10	a4;633I2
40Ca	**T**	**St01aRi04**	**657Ca**	**455.986240494143**	**+12**	**1.3 -14**	**cooled; Cs**
40Ca	**T**	**Ud01**	**657Ca**	**455.986240494158**	**+12**	**5.7 -14**	**cooled; Cs fountain**
88Sr+	**T**	**Mar03**		**444.77904409552**	**+12**	**2.2 -13**	**cold ion; Cs**
88Sr+	**T**	**Be99**		**444.7790440954**	**+12**	**4.5 -13**	**S-D, cold ion; Cs**
87Sr	T	Cou03		429.228004235	+12	4.7 -11	cold atom w=.001Hz;Cs
85Rb	G	To97	778Rb	385.28514237828	+12	5.2 -12	5S-5D F=3-5 TP; OsO4
87Rb	G	Ye96	780Rb	384.2279818773	+12	1.4 -11	D2,d/f; 778'Rb
85Rb	G	Ba93	795Rb	377.10627149	+12	1.5 -10	D1,c'; 633I2
Cs	B	Av86	852Cs	351.7257682	+12	3.5 -08	D2;633I2;l-m,hyp.res.
Cs	G	Ud99	895Cs	335.116048807	+12	1.2 -10	D1;CH4
C2HD	G	Ye296		281.635363961	+12	1.2 -10	532I2; 1.3Pa
20Ne	L	Po83		260.10324926	+12	3.1 -10	CH4 F
12C2H2	G	Nak96		197.75046656	+12	7.6 -10	v1+v3 R(18)high;778Rb
13C2H2	G	Nak96		196.92974592	+12	7.6 -10	v1+v3 R(23)high;778Rb
12C2H2	G	Nak96		194.91619955	+12	7.7 -10	v1+v3 P(21)low;778Rb
13C2H2	G	Ed03		194.3695693859	+12	1.5 -11	P(16) v1+v3; Cs viaH
13C2H2	G	Nak95Nak96		193.22000472	+12	5.7 -10	v1+v3 P(30)low;778Rb
HCN	G	Nak95		192.6224469	+12	5. -10	2v1 P(27); 778Rb
CO	G	Po283		129.019381761	+12	5.4 -10	R12; CH4 F
CO	G	Po283		128.360848969	+12	5.5 -10	R05; CH4 F
CO	G	Po283		127.598179703	+12	6.3 -10	P01; CH4 F
CO	G	Po283		126.990495584	+12	6.3 -10	P06; CH4 F
CH4 E	G	Er98		88.373149028553	+12	2.3 -12	Cs; <1.4Pa
OsO4	G	Ac93		29.37081407841	+12	2.7 -12	CO2 R26; OsO4 R10
OsO4	**G**	**Ac93Du99**		**29.054057446579**	**+12**	**3.5 -13**	**CO2 R10; Cs**
192OsO4	G	Do87		28.464676938787	+12	3.3 -12	CO2 P14; Cs
138Ba+	T	Mad96		24.01204831717	+12	1.8 -11	Cs
26Mg	B	Go93		601.278870352	+09	5. -12	
24Mg	**B**	**Go87Go93**		**601.2771578691**	**+09**	**1. -12**	**Ramsey; Cs**
199Hg+	**T**	**Poi96Ber98**		**40.50734799684159**	**+09**	**1.1 -14**	**cold ions; Cs**
113Cd+	T	Tan96		15.199562858	+09	1.3 -10	rf trap; Cs
171Yb+	**T**	**Bau96**		**12.642812118471**	**+09**	**7. -13**	**Cs**
171Yb+	**T**	**Se95Fi96**		**12.642812118468**	**+09**	**1.3 -13**	**Cs**
87Rb	**F**	**Bi99Le01Ma03**		**6.834682610904324**	**+09**	**1.2 -15**	**fountain;two Cs fount**
9Be+	**T**	**Bo85Bo91**		**303.016377265070**	**+06**	**1.9 -13**	**Cs**
Some CIPM Recommendations**							
H	**B**	**Qu99Qu03**	**243H**	**1233.03070659355**	**+12**	**2.0 -13**	**1S-2S TP CIPM2001;Cs**
127I2	G	Qu99Qu03	532I2	563.260223513	+12	8.9 -12	(a10)2fYAG CIPM2001
127I2	G	Qu03	633I2	473.612353604	+12	2.1 -11	(f),HeNe,CIPM2001;Cs
40Ca	**T,B**	**Qu99Qu03**	**657Ca**	**455.986240494150**	**+12**	**1.1 -13**	**station.y atomCIPM01**
88Sr+	**T**	**Qu99Qu03**	**674Sr+**	**444.7790440955**	**+12**	**7.9 -13**	**centreZee.mult.tCIPM01**
85Rb	G	Qu99Qu03	778Rb	385.285142375	+12	1.3 -11	F=3-5TPdiode CIPM2001
CH4 F	G,B	Qu99Qu03		88.3761816005	+12	2.3 -11	hyp.unres.CIPM2001;Cs
CH4 F	G,B	Qu99Qu03		88.37618160018	+12	3.0 -12	hyp.res. CIPM2001;Cs
OsO4	**G**	**Qu03**		**29.054057446579**	**+12**	**1.4 -13**	**R(10)<0.2Pa CIPM01;Cs**

Notes
(i) All uncertainties are 1-sigma.
(ii) Labels of I2 etc. lines have 3-digit wavelength in nm: e.g. 633I2.
**: omitting here 515/543,612,640 nm Ar+/HeNe stabilised lasers
*: app: apparatus
#: positronium (Ps)
1m FP: 1 metre Fabry Perot (resonator)
B: in beam
F: fountain, after laser cooling
F-t: Fourier transform (spectroscopy)
G: in gas
hyp.res.: hyperfine resolved
L: in gas-laser Lamb dip
l-m: lambda meter
T: in trap
TP: two photon;f/2 for 778Rb,CIPM01

that some previously-developed laser-cooled trapped-ion standards are now measured to 1 part in 10^{14}. The past few years show however a more gentle rate of improvement, similar to that for the caesium standard over the years, indicating the difficulties involved in assessing primarily systematic contributions to uncertainty at very high levels of accuracy.

6 Accuracy over the frequency range

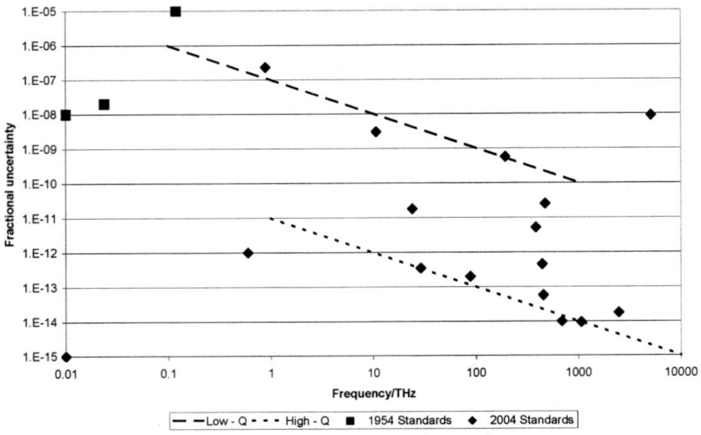

Figure 3: accuracy vs frequency 1954 and 2004

A perspective on how accuracies have changed across the frequency range in 50 years is provided in Figure 3. There, 10 decades of uncertainty are plotted against 6 decades of frequency, from that of the caesium standard to the highest precise measurement reported, near 5000 THz [Ei97]. A few points dating from 1954 are plotted as squares at the top left, reflecting both the great increase in accuracy that has come with the caesium standard, 7 decades, and the limited range of frequencies then accessible. However, microwave harmonic generation had by then reached 130 GHz [10], before laser radiation was discovered. Two dotted lines are given to guide the eye through the scatter of points, the upper indicating broadly first-order Doppler-limited standards and the lower, transitions narrowed by some form of cooling. The less-accurate point at the highest frequency results from having to use a pulsed laser at so high a frequency in the ultra-violet.

7 Discussion and conclusion

The table in Table 2 of precisely-measured frequencies in atomic, molecular and ionic systems is limited in length to one page, but indicates continuing progress in accuracy, more noticeable at the extremes of the frequency range, in the microwave and optical regions. A limitation is in the stability of local oscillators available for interrogating transitions.

The comparison with 50 years ago shows that the adoption and improvement of the caesium primary standard has resulted in an improvement of 7 orders of magnitude, fairly gradually achieved over the time interval, Figure 2. Other measurements have undergone generally a 6-order improvement. What is not shown is that the time and frequency standard is by far the best of the primary standards used to realise the S. I. measurement system, and that other measurements are increasingly referred to it. The high resolution of frequency measurement in the optical region brings hope of a yet-better standard, and supports evermore precise measurements of fundamental interest to physics, such as measuring the Rydberg constant, testing field theories, and investigation of any time variation of the fundamental constants.

8 Acknowledgements

The author is grateful to the Bureau National de Métrologie and the Institut National de Métrologie at the Conservatoire National des Arts et Métiers, Paris, for support whilst writing this paper.

9. Table 2 References

Ac93: (OsO4) O. Acef, J. J. Zondy, M. Abed, D. G. Rovera, A. H. Gérard, A. Clairon, Ph. Laurent, Y. Millerioux and P. Juncar, "A CO2 to visible optical frequency synthesis chain: accurate measurement of the 473 THz HeNe/I2 laser", Optics Commun. 97, 1993, 29-34.

Av86: (Cs D2) G. Avila, P. Gain, E. de Clercq and P. Cerez, "New absolute wavenumber measurement of the D2 line of caesium", Metrologia 22, 1986, 111-114.

Ba93: (85Rb, 87Rb) G. P. Barwood, P. Gill and W. R. C. Rowley, "Optically narrowed Rb-stabilized laser GaAlAs diode laser frequency standards with 1.5 x 10-10 absolute accuracy", Proc. SPIE - Int. Soc. Opt. Eng. 1837, 1993, 262-270.

Ba04: (107Ag) T. Badr, M. D. Plimmer, P. Juncar, M. E. Himbert, G. D. Rovera and J. D. Silver "First observation of the 4d10 5s 2S1/2-4d9 5s2 2D3/2 transition by Doppler-free two-photon spectroscopy in a siver atomic beam", Proc. EFTF04, Guildford, Surrey, UK, 5-7 April 2004.

Bau96: (171Yb+) A. Bauch, D. Schnier and Chr. Tamm, "Microwave spectroscopy of 171Yb+ stored in a Paul trap", Proc. 5th Symp. on Frequency Standards and Metrology, Ed. J. C. Bergquist (Singapore: World Scientific) 1996, 387-388.

Be99: (88Sr+) J. E. Bernard, A. A. Madej, L. Marmet, B. G. Whitford, K. J. Siemsen and S. Cundy "Cs-based frequency measurement of a single, trapped ion transition in the visible region of the spectrum", Phys. Rev. Lett. 82 (16), 1999, 3228-3231.

Ber98: (199Hg+) D. J. Berkeland, J. D. Miller, J. C. Bergquist, W. M. Itano and D.J. Wineland, "Laser-cooled mercury ion frequency standard", Phys. Rev. Lett. 80 (10), 1998, 2089-2092.

Bi99: (87Rb) S. Bize, Y. Sortais, M. S. Santos, C. Mandache, A. Clairon and C. Salomon "High-accuracy measurement of the ^87Rb ground-state hyperfine splitting in an atomic fountain", Europhys. Lett. 45 (5), 1999, 554-564.

Bly03: (171Yb+) P. J. Blythe, S. A. Webster, H. S. Margolis, S. N. Lea, G. Huang, S.-K. Choi, W. R. C. Rowley, P. Gill and R. S. Windeler, "Subkilohertz absolute-frequency measurement of the 467-nm electric octupole transition in 171Yb+", Phys. Rev. A67, 020501(R) 2003, 1-4.

Bo85: (9Be+) J. J. Bollinger, J. D. Prestage, W. M. Itano and D. J. Wineland, "Laser-cooled atomic frequency standard", Phys. Rev. Lett. 54 (10), 11 March 1985, 1000-1003.

Bo91: (9Be+) J. J. Bollinger, D. J. Heinzen, W. M. Itano, S. L. Gilbert and D. J. Wineland, "A 303-MHz frequency standard based on trapped Be+ ions", IEEE Trans. Instrum. Meas. 40 (2), April 1991, 126-128.

Ch84: (positronium (Ps), e+e-) S. Chu, A. P. Mills Jr. and J. L. Hall, "Measurements of the positronium 13S1-23S1 interval by Doppler-free two-photon spectroscopy", Phys. Rev. Lett. 52 (19), 7 May 1984, 1689-1692.

Cou03: (87Sr) I. Courtillot, A. Quessada, R. P. Kovacich, A. Brusch, D. Kolker, J-J. Zondy, G. D. Rovera and P. Lemonde, "Clock transition for a future optical frequency standard with trapped atoms", Phys. Rev. A 68, 2003, 030501(R)-1-4.

deB97: (H and D) B. de Beauvoir, F. Nez, L. Julien, B. Cagnac, F. Biraben, D. Touahri, L. Hilico, O. Acef, A, Clairon and J. J. Zondy, "Absolute frequency measurement of the 2S-8S/D transitions in hydrogen and deuterium: new determination of the Rydberg constant", Phys. Rev. Lett. 78 (3), 20 January 1997, 440-443.

Do87: (OsO4) Yu. S. Domnin, N. B. Koshelyaevskii, A. N. Malimon, V. M. Tatarenkov and P. S. Shumyatskii, "Infrared frequency standards based on osmium tetroxide", Sov. J. Quant. Electron. 17 (6), June 1987, 801-803; also Kvantovaya Elektron. (Moscow) 14, June 1987, 1260-1263.

Du99: (OsO4) F. Ducos, G. D. Rovera, C. Daussy and O. Acef "Performance of OsO4 stabilized CO2 lasers as optical standards near 29 THz", Proc. 1999 Joint Meeting EFTF-IEEE IFCS, 13-16 April 1999, Besançon, France,(IEEE Cat. No. 99CH36313), pp714-717.

Ed96: (127I2) C. S. Edwards, G. P. Barwood, P. Gill, F. Rodríguez-Llorente and W. R. C. Rowley, "Frequency stabilised diode lasers in the visible region using Doppler-free iodine spectra", Optics Commun. 132, 1996, 94-100.

Ed03: (13C2H2) C. S. Edwards, H. S. Margolis, G. P. Barwood, S. N. Lea, P. Gill, G. Huang and W. R. C. Rowley, "Absolute frequency measurement of a 1.5 um acetylene standard using a combined frequency chain and femtosecond comb", Optics Lett. 15 March 2004, 566-568.

Ei97: (He) K S Eikema, W Ubachs, W Vassen and W Hogervorst "Lamb shift measurement in the 1^1S ground state of helium", Phys. Rev. A 55 (3), 1997, 1866-1884.

Er98: (CH4 E line) P. S. Ering, D. A. Tyurikov, G. Kramer and B. Lipphardt "Measurement of the absolute frequency of the methane E-line at 88 THz", Optics Commun. 151, 1998, 229-234.

Fi96: (171Yb+) P. T. H. Fisk, M. J. Sellars, M. A. Lawn and C. Coles, "A microwave frequency standard based on trapped, buffer gas-cooled 171Yb+ ions - some accuracy issues", Proc. European Frequency Time Forum 5-7 March 1996, (London:IEE) Conf. Publ. 418, 1996, 212-217.

Gi95:(172Yb+) P. Gill, H. A. Klein, A. P. Levick, M. Roberts, W. R. C. Rowley and P. Taylor, "Measurement of the 2S1/2-2D5/2 411-nm interval in laser-cooled trapped 172Yb+ ions", Phys. Rev. A 52 (2), August 1995, R909-R912.

Go87: (24Mg) A. Godone, E. Bava and C. Novera, "High-resolution frequency measurement of the 3P1-3P0 24Mg transition", Metrologia 24, 1987, 133-138.

Go93: (26Mg) A. Godone, C. Novero "The magnesium frequency standard", Metrologia 30(3), 1993 163-181.

Le01: (87Rb fountain): P. Lemonde, Ph. Laurent, G. Santarelli, M. Abgrall, Y. Sortais, S. Bize, C.Nicolas, S. Zhang, A. Clairon, N. Dimarcq, P. Petit, A. G. Mann, A. N. Luiten, S. Chang and C. Salomon "Cold-atom clocks on earth and in space", in A. N. Luiten, (Ed.): *Frequency measurement and control*, Topics Appl. Phys. **79**, 131-152 (2001)

Ma03: (133Cs, 87Rb) H. Marion, F. Pereira Dos Santos, M. Abgrall, S. Zhang, Y. Sortais, S. Bize, I. Maksimovic, D. Calonico, J. Grünert, C. Mandache, P. Lemonde, G. Santarelli, Ph. Laurent, A. Clairon and C. Salomon, "Search for variations of fundamental constants using atomic fountain clocks", Phys. Rev. Lett. 90(15) 18 April 2003, 150801-1-4.

Mad96: (Ba+, Sr+) A. A. Madej, K. J. Siemsen, B. G. Whitford, J. E. Bernard and L. Marmet, "Precision absolute frequency measurements with single atoms of Ba+ and Sr+", Proc. 5th Symp. on Frequency Standards and Metrology, Ed. J. Bergquist (Singapore: World Scientific) 1996, 165-170, and A. A. Madej and K. J. Siemsen, "Absolute heterodyne frequency measurement of the 88Sr+ 455-THz S-D single ion transition", Optics Lett. 21 (11), 1 June 1996, 824-826.

Mar03: (88Sr+) H. S. Margolis, G. Huang, G. P. Barwood, S. N. Lea, H. A. Klein, W. R. C. Rowley and P. Gill, "Absolute frequency measurement of the 674-nm 88Sr+ clock transition using a femtosecond optical frequency comb", Phy. Rev. A 67, 032501, 2003, 1-5.

Me00: (Muonium μ+e-) V Meyer + 28, "Measurement of the 1s-2s energy interval in muonium", Phys. Rev. Lett. 84 (60), 2000, 1136-1139.

Nak95: (HCN, C2H2, 13C2H2) K. Nakagawa, M. de Labachelerie, Y. Awaji and M. Kourogi, "Precise optical frequency atlas of acetylene lines in the 1.5 μm region", Proc. Laser Spectoscopy XII, Isle of Capri, Italy, 11-16 June 1995, Ed. M. Inguscio, M. Allegrini and A. Sasso (Singapore: World Scientific) 1996, 228-229.

Nak96: (C2H2, 13C2H2) K. Nakagawa, M. de Labachelerie, Y. Awaji and M. Kourogi, "Accurate optical frequency atlas of the 1.5 μm bands of acetylene", J. Opt. Soc. Am. B 13 (12), December 1996, 2708-2714. Note: 36 lines in 12C2H2 and 54 lines in 13C2H2 are listed: the high and low extremes are given.

Ni00: (H 1S-2S) M Niering, R Holzwarth, J Reichert, P Pokasov, Th Udem, M Weitz, T W Hänsch, P Lemonde, G Santarelli, M Abgrall, P Laurent, C Salomon and A Clairon "Measurement of the hydrogen 1S-2S transition frequency by phase coherent comparison with a microwave cesium fountain clock", Phys. Rev. Lett. 84 (24) 12 June 2000, 5496-5499.

Pav94: (4He) F. S. Pavone, F. Marin, P. De Natale, M. Inguscio and F. Biraben, "First pure frequency measurement of an optical transition in helium: Lamb shift of the 23S1 metastable level", Phys. Rev. Lett. 73 (1) 4 July 1994, 42-45.

Pe99: (155In+) E. Peik, J. Abel, Th. Becker, M. Fries, J. von Zanthier and H. Walther "Towards an indium single-ion optical clock", Proc. 1999 Joint Meeting EFTF-IEEE IFCS, 13-16 April 1999, Besançon, France,(IEEE Cat. No. 99CH36313), pp682-685.

Pe03: (171Yb+) E. Peik, B. Lipphart, H. Schnatz, T. Schneider, Chr. Tamm, and S. G. Karschenboim "New limit on the present temporal variation of the fine structure constant", arXiv:physics/0402132 v1 27 Feb 2004, 4pp.

Po83: (20Ne) C. R. Pollock, D. A. Jennings, F. R. Petersen, J. S. Wells, R. E. Drullinger, E. C. Beaty and K. M. Evenson, "Direct frequency measurements of transitions at 520 THz (576 nm) in iodine and 260 THz (1.15 μm) in neon", Optics Lett. 8 (3), March 1983, 133-135.

Po283: (CO) C. R. Pollock, F. R. Petersen, D. A. Jennings and J. S. Wells, "Absolute frequency measurements on the 2-0 band of CO at 2.3 μm; calibration standard frequencies from high resolution color center laser spectroscopy", J. Mol. Spectrosc. 99, 1983, 357-368.

Poi96: (199Hg+) M. E. Poitzsch, J. C. Bergquist, W. M. Itano and D. J. Wineland, "Cryogenic linear ion trap for accurate spectroscopy", Rev. Sci. Instrum. 67 (1), January 1996, 129-134.

Qu99: (CIPM 1997 revision of "Mise en pratique" of the definition of the metre) T. J. Quinn, "Practical realisation of the definition of the metre (1997), Metrologia 36, 1999, 211-244.

Qu03: (CIPM 2001) T. J. Quinn, "Practical realisation of the definition of the metre, including recommended radiations of other optical frequency standards (2001), Metrologia 40, 2003, 103-133

Re00: (H 1s-2s) J Reichert, M Niering, R Holzwarth, M Weitz, Th Udem and T W Hänsch "Phase coherent vacuum-ultraviolet to radio frequency comparison with a mode-locked laser", Phys. Rev. Lett. 84 (15), 2000, 3232-3235.

Ri04: (Ca) F. Riehle, communication, Guildford April 2004: value in the PTB EFTF04 paper.

Ro97: (172Yb+) M. Roberts, P. Taylor, G. P. Barwood, P. Gill, H. A. Klein and W. R. C. Rowley, "Observation of an electric octupole transition in a single ion", Phys. Rev. Lett. 78, 1997, 1876-1879.

Sc99: (H,D 2S-12D) C Schwob, L Jozefowski, B de Beauvoir, L Hilico, F Nez, L Julien and F Biraben, O Acef and A Clairon "Optical frequency measurement of the 2S-12D transitions in hydrogen and deuterium: Rydberg constant and Lamb shift determinations", Phys. Rev. Lett. 83 (25), 1999, 4960-4963.

Sch95: (muonium, μ+e-) W. Schwarz + 24, "Spectroscopy of the 1S-2S energy splitting in muonium", IEEE Trans. Instrum. Meas. 44 (2), April 1995, 505-509.

Se95: (171Yb+) M. J. Sellars, P. T. H. Fisk, M. A. Lawn and C. Coles, "Further investigation of a prototype microwave frequency standard based on trapped 171Yb+ ions", in Proc. 1995 IEEE International Symposium on Frequency Control (49th Annual Symposium) (New York: IEEE) 1995, 66-73.

St01a: (Ca) J. Stenger, T. Binnewies, G. Wilpers, F. Riehle, H. R. Telle, J. K. Ranka, R. S. Windeler and A. J. Stentz, "Phase-coherent frequency measurement of the Ca intercombinatioon line at 657 nm with a Kerr-lens mode-locked femtosecond laser", Phys. Rev. A 63, 17 January 2001, 021802 (R).

St01b: (171Yb+) J. Stenger, Chr. Tamm, N. Haverkamp, S. Weyers and H. R. Telle, Absolute frequency measurement of the 435.5 nm 171Yb+ clcck transition with a Kerr-lens mode-locked femtosecond laser", preprint, 19 March 2001.

Tan96: (113Cd+) U. Tanaka, H. Imajo, K. Hayasaka, R. Ohmukai, M. Watanabe and S. Urabe, "Laser microwave double-resonance experiment on trapped 113Cd+ ions", CPEM96 Digest, TH1B-4, 515-6; see also Phys. Rev. A, 53(6), June 1996, 3982-3985.

Tay97: (172Yb+) P. Taylor, M. Roberts, S. V. Gateva-Kostova, R. B. M. Clarke, G. P. Barwood, W. R. C. Rowley and P. Gill, Investigation of the ^2S1/2-^2D5/2 clock transition in a single ytterbium ion", Phys. Rev. A, 1997, 56, pp.2699-2704.

To97: (85Rb two-photon) D. Touahri, O. Acef, A. Clairon, J.-J. Zondy, R. Felder, L. Hilico, B. de Beauvoir, F. Biraben and F. Nez, "Frequency measurement of the 5S1/2(F=3)-5D5/2(F=5) two-photon transition in rubidium", Optics Commun. 133, 1 January 1997, 471-478.

Ud97: (H) Th. Udem, A Huber, M. Weitz, D. Leibfried, W. König, M. Prevedelli, S. Dimitriev, H. Geiger and T. W. Hänsch, "Phase-coherent measurement of the hydrogen 1S-2S frequency with an optical frequency interval divider chain", IEEE Trans. Instrum. Meas. 46(2) April 1997, 166-168; also Phys. Rev. Lett. 79(14), 6 October 1997, 2646-2649.

Ud99: (Cs D) Th. Udem, J. Reichert, R. Holzwarth and T. W. Hänsch, "Absolute optical frequency measurement of the cesium D1 line with a mode-locked laser" Phys. Rev. Lett., 82(18), 3 May 1999, 3568-3571.

Ud01: (199Hg+, 40Ca) Th. Udem, S. A. Diddams, K. R. Vogel, C. W. Oates, E. A. Curtis, W. D. Lee, W. M. Itano, R. E. Drullinger, J. C. Bergquist and L. Hollberg, "Absolute frequency measurements of the Hg+ and Ca optical clock transitions with a femtosecond laser", electronic preprint (arXiv:physics/0101029) 4 January 2001.

Ye96: (87Rb) J. Ye, S. Swartz, P. Jungner and J. L. Hall, "Hyperfine structure and absolute frequency of the 87Rb 5P3/2 state", Optics Lett., 21 (16), 15 August 1996, 1280-1282.

Ye296: (C2HD) J. Ye, L-S. Ma and J. L. Hall, "Ultra-stable optical frequency reference at 1.064 μm using a C2HD molecular overtone transition", CPEM96 Digest, WE2A-6, Braunschweig, Germany, 17-20 June 1996, Supplement, 19-20.

Za00: (115In+) J. von Zanthier, Th. Becker, M. Eichenseer, A. Yu. Nevsky, Ch. Schwedes, E. Peik, H. Walther, R. Holzwarth, J. Reichert, Th. Udem and T. W. Hänsch, "Absolute frequency measurement of the In+ clock transition with a mode-locked laser", Optics Lett., 25 (23), 1 December 2000, 1729-1731.

10. Text References

[1] D. J. E. Knight, Proc. EFTF'97, Neuchâtel, Switzerland, 4-6 March 1997, pp. 623-627.

[2] D. J .E. Knight, Proc. 15th EFTF, Neuchâtel, Switzerland, 6-8 March 2001, pp. 373-377.

[3] Th. Udem, J. Reichert, R. Holzwarth and T. W. Hänsch, Phys. Rev. Lett., **82** (18), pp. 3568-3571 (1999).

[4] A. Clairon *et al.*, in Proc. 5th Symposium on Frequency Standards & Metrology 1995, ed. J. C. Bergquist (World Scientific, Singapore, 1996) pp. 49-59.

[5] Proc. Ve Congrès International de Chronométrie, Paris, 1-5 Oct. 1954, Ed. P. Libessart, R. Baillaud (Observatoire National de Besançon 1955).

[6] L. Essen, ref. [5] pp 343-353.

[7] L. Essen and J. V. L. Parry, Phil. Trans. Roy. Soc. Lond. **250** (A973), pp. 45-69 (1957).

[8] N. F. Ramsey, *Nuclear moments*, Wiley, NY 1953; pp37-52 on molecular beam resonance methods.

[9] N. F. Ramsey "Measurements with separated oscillatory fields and hydrogen masers" in Proc. Atomic Physics 12, Ann Arbor, Michigan 1990, ed. J. C. Zorn, R. R. Lewis (Amer. Inst. Phys., New York, 1991, Conf. Proc. 233) pp. 1-27.

[10] W. Gordy , W. V. Smith and R. F. Trambarulo, *Microwave Spectroscopy* (New York: Dover 1966) Originally publ. New York:Wiley 1953.

[11] H. Lyons Proc. NY Acad. Sci. 55, pp. 831- (1952) and NBS Report No.1848, 1952.

[12] R. L. Barger and J. L. Hall, Phys. Rev. Lett. **22**, pp. 4-8 (1969).

[13] C. Freed and A. Javan "Standing wave saturation resonances in the CO_2 10.6 μm transitions observed in a low-pressure room-temperature absorber gas" Appl. Phys. Lett. **17** pp. 53-56 and 541 (1970)

[14] G. R. Hanes and C. E. Dahlstrom, "Iodine hyperfine structure observed in saturated absorption at 633 nm," Appl. Phys. Lett., **14**, pp. 362-364 (1969).

[15] D. J. E. Knight, "Advances in optical frequency measurement over 30 years", in Proc. 6th Symp. on Frequency Standards and Metrology, St. Andrews, Scotland 9-14 September 2001, ed. P. Gill (Singapore: World Scientific, 2002) pp. 557-559.

Reduction of residual amplitude modulation in Pound-locking schemes for ultra-stable oscillators

Giuseppe Marra, Mark Oxborrow and Dale Henderson

National Physical Laboratory, Queens Road, Teddington, UK
giuseppe.marra@npl.co.uk

Keywords: Pound locking, residual amplitude modulation, zero-crossing detection, IQ mixer, sapphire resonator

Abstract

Residual Amplitude Modulation (RAM) in Pound-locking schemes is known to limit the long-term frequency stability of ultra-stable oscillators. In this paper, we propose two simple yet effective techniques for reducing it. We have implemented both at X-band using commercially available components. We report experimental results for the levels of RAM reduction achieved.

1 Residual amplitude modulation (RAM) –an explanation of the problem

The most frequency-stable microwave oscillators over time intervals of 1-1000s [13,3,2] incorporate ultra-high-Q resonators within loop oscillators that implement Pound locking [12,4,1]. At NPL, we have constructed two such, nominally identical, ultra-stable oscillators. The frequency of each is Pound-locked to the same whispering-gallery-mode resonance of separate cryogenic sapphire-ring resonators. This resonance is located at approx 9.204 GHz, and exhibits a loaded Q of approx 800 million at 4.2 K. A minimal version of our (quite generic) Pound-locking scheme is shown in Fig. 1; we proceed to consider its salient details and functionality.

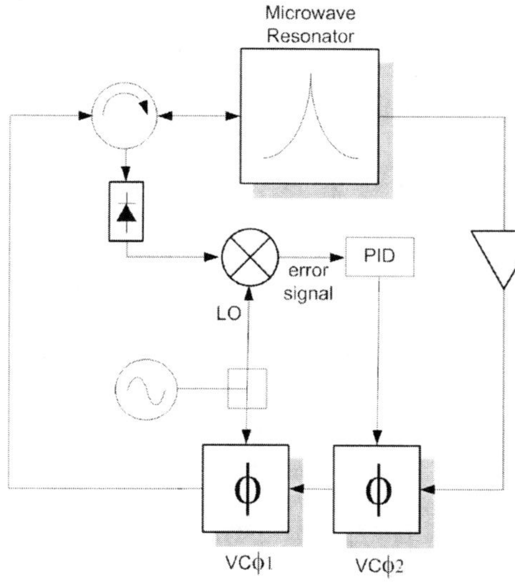

Fig. 1: Minimal Pound-locking scheme

Pound locking involves phase-modulating the oscillator's carrier at a particular offset or 'intermediate' frequency (IF) such that the signal incident on the resonator comprises a carrier accompanied by discrete PM sidebands. The requisite phase modulation is typically implemented with a voltage-controlled analogue phase shifter (VCØ1 in Fig. 1) incorporating abrupt varactor diodes. Unfortunately, the insertion loss of such a phase shifter invariably varies slightly though significantly with the applied modulation voltage, such that the carrier is modulated in amplitude as well as in phase; here lies the origin of the residual amplitude modulation or RAM.

The error signal of the Pound-locking scheme is the low-passed output from a synchronous detector or mixer, whose reference (LO) input receives the same modulation signal as is applied to the phase shifter VCØ1. The Pound error signal is thus the 'in-phase' synchronous component of the power level of the microwave signal that is reflected back from the resonator, where this power level is detected with a diode, ideally operating in its square-law region. Any RAM introduced by the phase shifter will cause a voltage offset in the error signal that is indistinguishable from that caused by a deviation in the loop oscillator's phase/frequency from that defined by the resonator itself. Any fluctuations in the depth of the RAM will thus cause corresponding fluctuations in the loop oscillator's frequency.

Previous works confronting the processes that limit the long-term frequency stability of ultra-stable oscillators [6,5,7] indicate that RAM (or rather uncontrolled fluctuations in its depth) is a significant problem, whose reduction, either at source (*i.e.* 'avoidance') or subsequently (*i.e.* 'suppression', 'cancellation' or 'removal') is necessary to reach the ultimate limits of frequency stability defined by the resonator.

2 Known approaches to RAM reduction

RAM can be avoided (to first order) by biasing the voltage-controlling phase shifter at a turning point in its loss-versus-voltage characteristic [6]; but not all devices exhibit a turning point. The approach only works if the voltage traversed by the applied modulation signal is small, corresponding to a small phase-modulation depth. From the standpoint of maximizing the size of the Pound error signal, for a given microwave power incident on the resonator and a given phase/frequency error, the approach is thus not wholly satisfactory.

RAM can be cancelled with a voltage-controlled attenuator (VCA), whose control signal is a suitably phase-shifted and amplitude-adjusted version of that applied to the modulating phase shifter [8]. But good cancellation requires constant manual or feedback-controlled adjustment.

RAM can be suppressed by inserting a saturating amplifier or limiter after the modulating phase-shifter; we pursue this idea, to its logical extreme, in section 4. From typical output power versus input power characteristics published in the datasheets of different commercial pin-diode limiters, we can infer that the maximum suppression of the AM sidebands achievable is no more than 40 dB when the limiter works in its hard-limiting zone and the amplitude modulation index is lower than 20 %.

We are also aware of several 'modulation-feedback' techniques, as have been applied to the linearization of rf power amplifiers [14], and also to phase-noise metrology [9], which can be adapted to effect active cancellation of the RAM; these techniques come under the names of (in order of increasing complexity and general performance) 'Envelope feedback' (*i.e.* power levelling), 'Polar loop' [10] and 'Cartesian loop' [11,9]. The degree of cancellation afforded by envelope feedback is limited by the ratio between the control bandwidth of its VCA to the chosen modulation IF adopted by the Pound servo (and this ratio is often small). Though promising excellent performance and stability, the circuits associated with the latter two techniques are relatively complicated; we have yet to fully implement either in our laboratory.

In the following two sections, we motivate and describe two alternative –and we believe reasonably novel– techniques for reducing the RAM, which are straightforward to implement yet effective.

3 RAM avoidance with an IQ mixer

IQ mixers are widely used as phase modulators. If the input to such a mixer is a sinusoidal carrier, $\sin(\omega t)$, then we can write its output as

$$V_0(t) = i(t)\cdot\sin(\omega t) + q(t)\cdot\cos(\omega t),$$

where $i(t)$ and $q(t)$ are respectively the mixer's in-phase and quadrature modulation inputs, which in general will alter as functions of the time t. The mixer's output can be seen as a phasor, whose phase and modulus are functions of $i(t)$ and $q(t)$. If

$$i(t)^2 + q(t)^2 = const,$$

then the modulus of the phasor, and hence the amplitude and power at the mixer's output, will remain constant while the phase can vary. By synthesizing suitable modulation inputs $i(t)$ and $q(t)$ that satisfy the above 'circle' condition, RAM is completely avoided –at least in the case of an ideal IQ mixer.

But there is an easier solution for the specific application of a Pound servo, whose synchronous detector is, in general, only sensitive to RAM at odd harmonics of the Pound servo's IF. As pictured in Fig. 2, a constant offset (in practice, a fixed bias current) is applied to one of the two modulation inputs of the IQ mixer, while a sine wave (or, more generally, any symmetric periodic function) is applied to the other; the sine wave's frequency is the Pound servo's IF. The signal power at the IQ mixer's output will undergo two identical periods of oscillation for every period of the applied sine wave, such that RAM is generated only at even harmonics of the Pound servo's IF, to which the Pound's synchronous detector is insensitive.

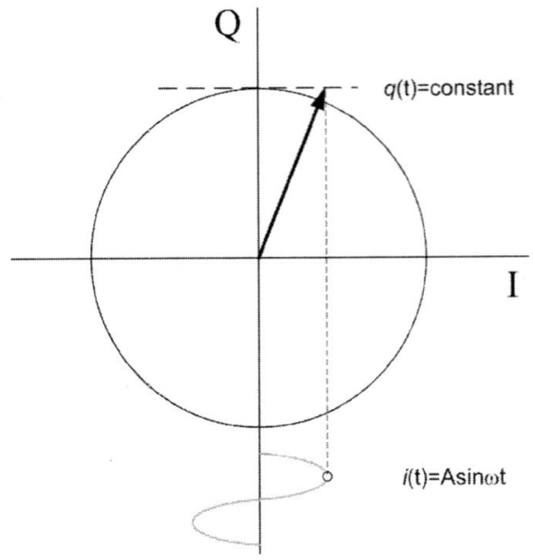

Fig. 2: Generation of amplitude modulation at only the even harmonics of the fundamental modulation frequency

When tested experimentally, the imperfect balance among the diodes inside the IQ mixer requires an additional bias current to be superposed on the sine wave. By fine-tuning this current, we have achieved a PM-to-AM sideband ratio of >50dB. The maximum achievable absolute depth of phase modulation was –6dBc (single PM sideband power relative to carrier).

4 RAM removal with a zero-crossing detector

The positions in time at which an ac-coupled periodic signal crosses zero (*i.e.* its 'zero crossings') are insensitive to fluctuations in the signal's amplitude or power level. This insensitivity, though compromised by noise and the dc-bias fluctuations found within real circuits, inspires the following extremely simple approach, as shown in Fig. 3. A carrier modulated in both phase and amplitude is sent through a zero-crossing detector. The detector's output is a square wave carrying no information about the amplitude of the signal at its input. The phase information is, in contrast, entirely preserved.

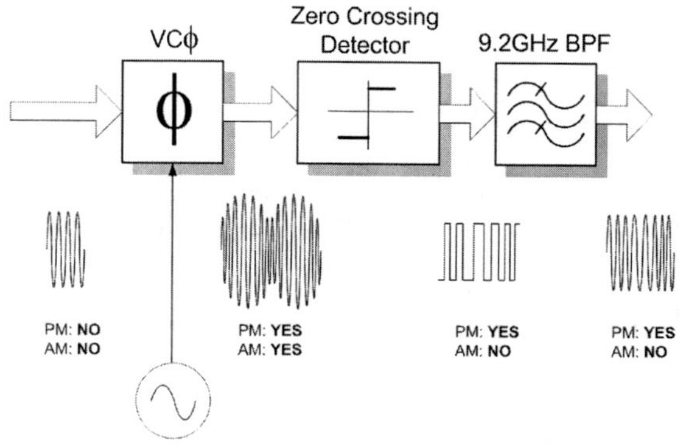

Fig. 3: Principle of the RAM removal using a zero-crossing detector

We are not aware of any commercial zero-crossing detector that functions at X-band. The closest equivalent to such a device, made from components that we had available in our laboratory, comprised a digital divider-by-4 prescaler, whose output at 2.3 GHz was multiplied by 4 to recover the 9.2 GHz carrier. The circuit's performance has been tested with an input provided by a commercial microwave CW source followed by the same voltage-controlled phase shifter as used in our Pound servos. Test signals, all at a total input power of 4.5 dBm, with various phase-modulation depths and thus different amounts of RAM, were applied to the input of the zero-crossing detector. Their AM sidebands had powers at the fundamental offset (IF) ranging from –43 dBc to –65 dBc. The corresponding AM sideband powers at the output were all suppressed by at least 60 dB. The phase noise introduced by the prescaler and multiplier could of course be significant in certain applications. The manufacturer quotes –125 dBc at 100 Hz offset and –140 dBc at 10 kHz offset for both components. In our application (Fig. 1), the phase noise that they introduce is defeated, within the bandwidth of the Pound loop, by the phase correction added by VCØ2. A significant advantage of this technique over the RAM avoidance afforded by an IQ mixer (section 3), is that no pre-tuning is required. When incorporated within a loop oscillator, it additionally provides power regulation.

6 Conclusions

We have explored two techniques for reducing residual amplitude modulation (RAM) in Pound locking schemes. An IQ mixer has been configured to avoid the production of RAM (at those harmonics that matter) and offers an AM sideband power at the fundamental harmonic >50 dB below that of the intended PM sideband. A zero-crossing detector for removing RAM at X-band has been assembled and achieves a reduction of 60 dB in the incident AM sideband power. These results compares to a suppression of just 40 dB in the AM sideband power with a pin-diode limiter. The zero-crossing-detector technique is generic and can be extended to almost any application where phase (or even frequency) modulation without amplitude modulation is required.

Acknowledgements

We would like to thank Enrico Rubiola for useful discussions.

References

[1] E. D. Black, "An Introduction to Pound-Drever-Hall laser frequency stabilization", Am. J. Phys., **69**, pp. 79-87 (2001)

[2] S. Buchman, M. Dong, W. Moeur, S. Wang, J. A. Lippa, and J. P. Turneaure, "A space-based superconducting microwave oscillator clock", Adv. Space Res. **25**, pp. 1251-1254 (2000).

[3] S. Chang, A. Mann, and A. Luiten, "Improved cryogenic sapphire oscillator with exceptionally high frequency stability," Electron. Lett., **36**, pp. 480-481 (2000).

[4] A. J. Giles, A. G. Mann, S. K. Jones, D. G. Blair, and M. J. Buckingham, "A very high stability sapphire loaded superconducting cavity oscillator", Physica B **165**, pp. 145-146 (1990).

[5] C. R. Locke, S. Munro, M. E. Tobar, E. N. Ivanov, G. Santarelli, "Constructing the next generation cryogenic sapphire oscillator", Proc. 2003 IEEE Int. Freq. Control Symp. and 17th EFTF, pp. 350-354 (2003).

[6] A. N. Luiten, A. G. Mann, N. J. McDonald and D. G. Blair, "Latest Results of the U. W. A. Cryogenic Sapphire Oscillator", Proc. 1995 Int. Freq. Control Symp., pp. 433-437 (1995).

[7] H. Müller, S. Herrmann, C. Braxmaier, S. Schiller, and A. Peters, "Modern Michelson Morley Experiment using Cryogenic Optical Resonators", Phys, Rev, Lett. **91**, 020401 (2003).

[8] M. Oxborrow, "Phase Modulation Without Amplitude Modulation", unpublished (2001).

[9] E. Rubiola, "Advanced interferometric phase and amplitude noise meansurements", Rev. Sci. Instr. **73**, pp. 2445-2457 (2002).

[10] V. Petrovic and W. Gosling, "Polar Loop Transmitter", Electron. Lett. **15**, pp 1706-1712 (1979).

[11] V. Petrovic and A. N. Brown, "Application of Cartesian feedback to HF SSB transmitters", Proc. 3rd Internat. Conf. on HF Communication Systems and Techniques, London, February 26-28, pp. 81-85 (1985).

[12] R. V. Pound, "Electronic Frequency Stabilization of Microwave Oscillators", Rev. Sci. Instr. **17**, pp. 490-505 (1946).

[13] R. T. Wang and G. J. Dick, "Cryocooled Sapphire Oscillator with Ultrahigh Stability", IEEE Trans. Instr. & Meas. **48**, pp. 528-531 (1999).

[14] S. Whittle et al, "A review of rf power amplifier linearization techniques", Toracomm Ltd, UK (2000).

HIGH-Q FREQUENCY-TEMPERATURE COMPENSATED SOLID–NITROGEN–COOLED RESONATOR-OSCILLATORS: FIRST RESULTS

J.G. Hartnett*, P.Y. Bourgeois†, J.D. Anstie*, M.E. Tobar*
N. Bazin†, E.N. Ivanov*, D. Cros‡ and V. Giordano†

*School of Physics, University of Western Australia
Crawley, 6009, WA, Australia
†Dpt. LPMO Institut FEMTO-ST UMR 6174 CNRS
32, avenue de l'Observatoire, 25000 Besançon, FRANCE
‡ IRCOM, UMR 6615 CNRS
123 avenue A. Thomas, 87060 Limoges Cedex, FRANCE

Keywords: Microwave oscillators, sapphire, whispering gallery modes.

Abstract

Experimental results for two solid-nitrogen-cooled microwave resonator oscillators are presented. The Allan Deviation was measured to be just less than 2×10^{-13} for $10\text{s} \leq \tau \leq 100$ s in the dual-mode oscillator and $2 - 3 \times 10^{-13}$ for $1\text{s} \leq \tau \leq 6\text{s}$ in the sapphire-rutile composite design.

1 Introduction

The frequency-temperature dependence of Whispering Gallery (WG) modes in sapphire is quite strong at temperatures accessible to liquid and solid nitrogen (approximately 50 K to 77 K). In recent years efforts have been made to annul this dependence [1-6], and hence build a microwave oscillator with state-of-the-art frequency stability for use as a local oscillator for the new generation of atomic fountain clocks. This paper reports and compares the first oscillator stability measurements for solid nitrogen cooled oscillators based on two different techniques. In one, the resonator is a single crystal cylindrical sapphire with thin rutile rings attached to the end face and in the other the resonator is a dual mode single–crystal cylindrical sapphire. The resonator frequency–temperature dependence was annulled in both cases, at 52.8 K and 50.6 K respectively.

2 Sapphire-rutile composite resonator oscillator:

A sapphire-rutile composite resonator was constructed from a 31.65 mm diameter, 30.01 mm high cylindrical single-crystal sapphire-dielectric resonator with rutile rings (of the same diameter) held to the end faces of the sapphire and supported also by one of its spindles in a copper cavity (see Fig.1). The copper cavity has an internal diameter of 50.00 mm and an internal height of 48.29 mm. The dimensions of the Rutile rings were determined by finite element analysis and were measured to be: internal diameter = 23.64 mm, outer diameter = 31.68 mm, and thickness = 0.42 mm [5]. The sapphire was held in the cavity by a copper support for the bottom sapphire spindle, held in place by a brass nut. The operational design mode was the $WGE_{8,0,0}$ mode at 12.031 GHz. Its turnover temperature is 52.8 K. This mode was coupled to the oscillator coaxial lines by straight antenna probes. The effective coupling was set on the input port of the resonator to be 0.83 and approximately 0.05 on the other port. The unloaded Q-factor was measured to be approximately 6×10^6.

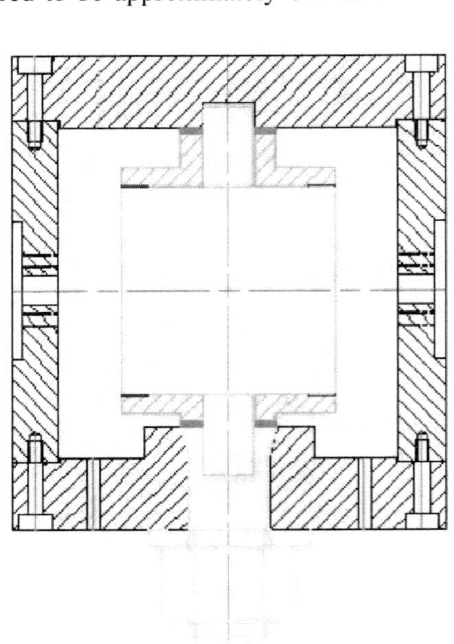

Fig. 1. *Schematic diagram of the composite sapphire-rutile resonator*

A Pound-servo-controlled loop oscillator was implemented. The carrier is phase modulated by a voltage controled phase shifter (VCPS). The signal reflected by the resonator is sent to a backward diode placed outside the cryogenic system. The signal delivered by the backward diode then demodulated by synchronous detection (SR830 lock–in amplifier). The resulting error signal is filtered and sent back to the VCPS bias stage where it is added to the bias voltage and to the modulation signal. A schematic of the microwave circuit is shown is shown in Fig 2.

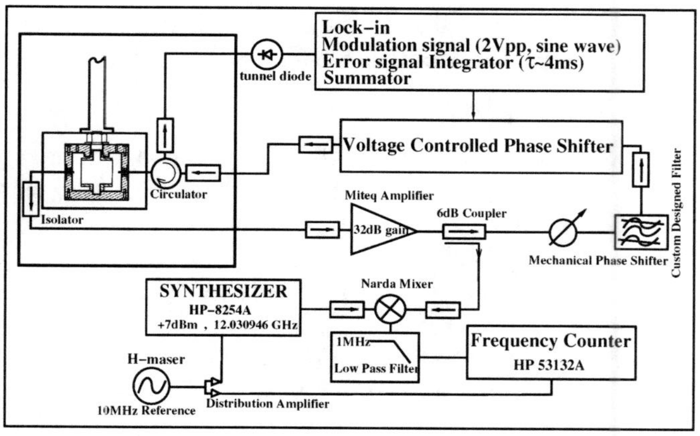

Fig. 2. *Schematic diagram for the Pound loop oscillator based on the sapphire-rutile resonator*

The 12 GHz oscillator output signal was mixed with the signal of a microwave synthesizer referenced to a Hydrogen Maser. The resulting beat note was analysed by a frequency counter interfaced to a computer.

3 Dual-Mode Sapphire Resonator Oscillator:

A sapphire resonator approximately 50.00 mm in diameter with a tapered height ranging from 24.50 mm in the center to 20.27 mm at the diameter. It was also supported by its bottom spindle in a silver-plated copper cavity, with a similar mechanism to that shown in Fig. 1. The dual-mode resonator cavity is of a similar design but with an 80.0 mm internal diameter and 50.0 mm high.

The output of the oscillator was the beat frequency between a chosen pair of Whispering Gallery modes. In this case they were the $WGH_{12,0,0}$ mode (9.086 GHz with an unloaded Q-factor of 1.3×10^8) and the $WGE_{15,0,0}$ mode (12.605 GHz with an unloaded Q-factor of 0.94×10^8) at the turnover temperature of 50.6 K. The beat frequency was 3.5188 GHz. The effective port couplings for the WGH mode were 0.36 and 0.04, and for the WGE mode they were 0.58 and 0.11.

The two modes were coupled to the oscillator coaxial lines by straight antenna probes (see figure 3). The probes were placed in the side walls for the WGE mode but for the WGH mode in the top lid of the resonator and positioned on opposite sides on the cavity at 90 degrees around from the probes for the WGE mode. A straight antenna probe is made when the central conductor of semi-rigid coaxial cable is cut away. The probes are positioned so that they as sensitive to the dominant component of the electric field for each mode. This design is fully explained in [2, 6].

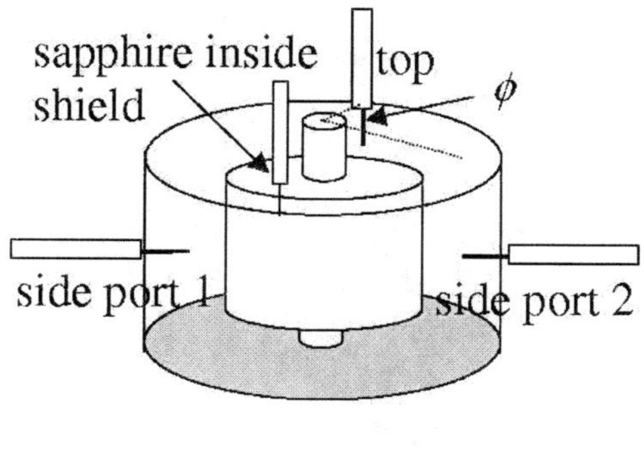

Fig. 3. *Schematic diagram for the dual mode resonator cavity*

Two Pound servo-controlled loop oscillators were implemented based on the single resonator described above. These loop oscillators were of a similar design to the preceding oscillator but separate electronic phase shifters were used as modulators and error correctors. The beat frequency of the two loop oscillators was used as the output signal.

The figure 4 shows the frequency variation of each mode and of the beatnote. This data was recorded during the preliminary phase of the experiment. The temperature controller had not yet stabilised on the control point. Then we observed large variations in the frequency of the two modes due to transient effect in the thermal control. The beatnote remained stable which demonstrates the common mode rejection.

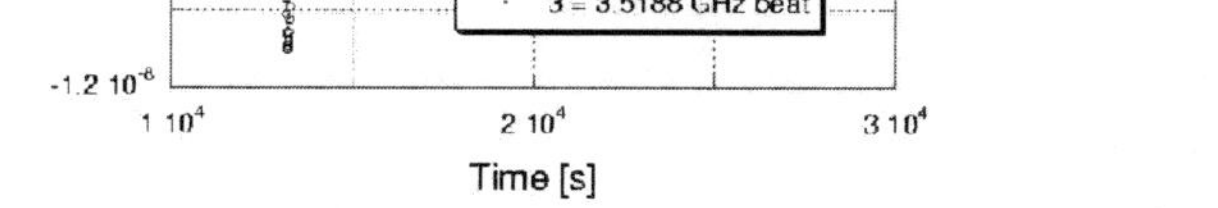

Fig. 4. *Frequency variations of the two modes and of the beatnote*

4 Frequency Stability:

Both resonators were temperature stabilized at what was believed to be the frequency-temperature turnover points of the resonators, with temperature control of at least 1 mK using platinum resistance thermometers.

As the two experiments were realized in two different places (UWA in Perth for the dual–mode system and LPMO in Besançon for the composite resonator) different measurement procedures were used :

*Sapphire-rutile resonator oscillator:*A 50 kHz beat frequency was obtained with an HP8254-A synthesizer, referenced to a 10 MHz reference from an H-maser and measured. From this the Allan Deviation [7] was calculated and is shown in Fig. 5 as a function of integration time τ. The dashed line represents the noise floor resulting from the frequency instability of the Hydrogen Maser and from the phase noise of the microwave synthetizer.

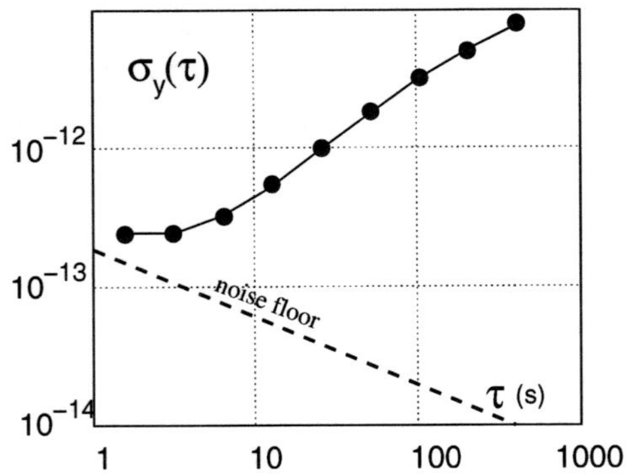

Fig. 5. *Frequency satbility obtained with the sapphire-rutile resonator oscillator*

Dual mode resonator oscillator Using an HP8673 synthesizer referenced to a very stable (8600 Oscilloquartz OCXO) reference oscillator we were able to measure a 20 kHz difference frequency between the dual mode oscillator beat signal and the synthesizer. The fractional frequency Allan Deviation of the beat signal is shown in Fig. 6. The Allan deviation for the noise floor of this measurement system is shown as the dashed line. It was obtained by measuring our helium cooled oscillator with the same measurement system before it was sent to France. That data didn't extend beyond $\tau > 50$s.

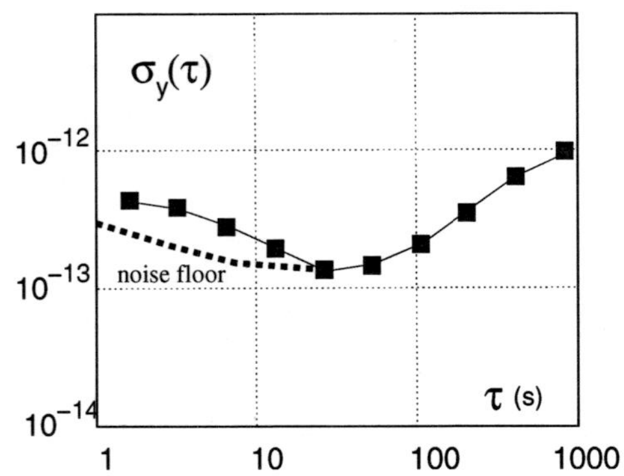

Fig. 6. *Frequency satbility obtained with the dual–mode resonator oscillator*

A fractional frequency drift of 10^{-9}/day was measured in the data for the sapphire-rutile oscillator (from which curve 1 was calculated) and 10^{-10}/day for the dual-mode oscillator (from which curve 2 was calculated).

It is not clear why the electronics noise floor of the dual-mode oscillator is a few times higher than that of the oscillator based on the compensated resonator. Due to the almost order magnitude higher Q-factor of the pure sapphire resonator, in the dual mode design, we would expect the reverse. The higher Q-factor of the dual-mode resonator is an advantage over the compensated resonator but to some extent that is reduced by the ratio of microwave frequency of the individual loop oscillators (9.086 GHz and 12.605 GHz) to the beat frequency (3.5188 GHz). However we expect that the dual-mode oscillator should exhibit better short term stability than the compensated oscillator. This needs further investigation.

5 Conclusion:

The fractional frequency Allan deviations of two microwave oscillators, based on novel resonator designs, are compared. The oscillators were operated with resonator temperatures stabilized so that their frequency-temperature dependence was annulled slightly above 50 K. The best Allan Deviations were measured to be between 2 and 3×10^{-13} for 1s<τ<6s in the sapphire-rutile oscillator, and slightly less than or equal to 2×10^{-13} for 10s<τ<100s in the dual mode oscillator.

6 References:

[1] J. G. Hartnett, M. E. Tobar, E. N. Ivanov, and D. Cros, "High-Q frequency-temperature compensated sapphire/rutile resonator," Electron. Lett., vol. 36, pp. 726-727, 2000.

[2] J. G. Hartnett, J. D. Anstie, M. E. Tobar, and E. N. Ivanov, "The progress in the development of a solid nitrogen cooled dual-mode frequency-temperature compensated sapphire-resonator oscillator," Proceedings of the 2003 IEEE International Frequency Control Symposium and PDA Exhibition Jointly with the 17th European Frequency and Time Forum, pp. 376-379, 2003.

[3] Y. Kersalé, N. Boubeker, P. Y. Bourgeois, N. Bazin, S. Vives, C. Meunier, and V. Giordano, "Comparison of WGE and WGH modes for temperature compensated sapphire-rutile resonator," Proceedings of the 2003 IEEE International Frequency Control Symposium and PDA Exhibition Jointly with the 17th European Frequency and Time Forum, pp. 360-364, 2003.

[4] M. E. Tobar, J. Krupka, J. G. Hartnett, E. N. Ivanov, and R. A. Woode, "High-Q sapphire-rutile frequency-temperature compensated microwave dielectric resonators," IEEE Trans. on UFFC, vol. 45, pp. 830-836, 1998.

[5] M. E. Tobar, D. Cros, P. Blondy, J. G. Hartnett, and P. Guillon, "Finite Element Realization of Ultra-High Quality Factor Frequency-Temperature Compensated Sapphire-Rutile Whispering Gallery Mode Resonators," Proc 1999 Int. Microwave Symp., pp. 1323-1326, 1999.

[6] M. E. Tobar, G. L. Hamilton, E. N. Ivanov, and J. G. Hartnett, "New method to build a high stability sapphire oscillator from the temperature compensation of the difference frequency between modes of orthogonal polarization," IEEE Trans. UFFC, vol. 50, 2003.

[7] D. W. Allan, "Statistics of atomic frequency standards," Proceedings of the IEEE, vol. 54, pp. 221-230, 1966.

A LIGHT WEIGHT BAND MILLIMETER WAVE RADAR MASTER OSCILLATOR

. Ruoskanen*, P. Eskelinen†, H. Heikkilä**

*Defense Force's Technical Research Centre, FINLAND, jukka.ruoskanen@pvtt.mil.fi
†Helsinki University of Technology, FINLAND
**Armoured Brigade, FINLAND

Keywords Millimeter wave radar, frequency stability.

Abstract

The usability of millimeter wave frequencies in mobile radar systems is largely dictated by the stability characteristics of the system's master oscillator. The oscillator also has to meet the system specific tactical requirements in terms of the tunable band and tuning speed. In this paper the demands the desired system performance sets on the radar's master oscillator are discussed. In the discussion the emphasis is on how the oscillator instabilities restrict radar's signal processing capabilities. The paper also describes one possible solution suitable as the radar's millimeter wave carrier oscillator. The practical prototype comprises of a Ku-band oscillator and diode multiplier that generates the desired Q-band output. The oscillator uses dielectric resonator stabilization and it has only one GaAs FET as an active element.

1 Introduction

In a requirement specification of a moder radar system mobility is often emphasized. Radars are installed on moving platforms that usually allow very little room for antenna and other radar hardware. Ideally, mobility has to be achieved without any degradation in system performance. Millimeter wave carrier frequencies are an attractive choice for lightweight transportable radar systems due to the available reduction in antenna size and system overall weight. However, certain difficulties must be overcome [1]. One of the great challenges is to achieve adequate frequency tuning and stability characteristics with electronics having comparable size and power supply requirements. Both fundamental operational and tactical needs must be fulfilled [2]. The frequency stability has to enable sensible coherent Doppler processing of the returned echoes, at least over the pulse repetition interval [3]. Because transmitter power is limited at these high frequencies, we may want to use narrow band IF filtering, which again sets frequency stability requirements, but here the time scale is much longer [4]. Some frequency agility is desirable in order to be able to overcome, to some extent, multipath problems and clutter [5], to get tools for compression, and most notably due to demands for effective electronic counter-counter measures. Tuning speed and available span are both of interest [4]. Finally, one would like to avoid heavy and expensive synthesizers, see [6] and [7] and their power supplies, and if possible, use direct millimeter wave oscillators [8] instead.

2 Moving target processing in radar systems

Target detection in radar systems is based on the amplitude and phase characteristics of the return echo signal. In a non-coherent radar system the detection is based on the amplitude of the signal alone. Therefore the requirement for the short-trem frequency stability of non-coherent radar system oscillator mainly emanates from other than signal processing needs, since the backscattering coefficient of typical targets and surface/volume scatterers does not vary strongly over minuscule changes in frequency and thus the amplitude variations of received signals are not related to the oscillator instabilities. The reliability of target detection is considerably improved in coherent radar systems, where the phase information can be utilized. A moving target will give rise to a Doppler effect – a slight shift in the frequency of the echo signal relative to the frequency of the transmitted signal. This well-known phenomenon provides means for detecting a moving target in the presence of clutter whose amplitude exceeds that of the target.

There are several moving target processing methods in which the phase information of the echo signals is applied. In majority of radar systems the extraction of moving targets from other echoes is carried out by MTI- (moving target indication) or Doppler processing. These methods are similar in nature, but there is a distinctive difference between the applications they are used in. In immovable radars the MTI-processing involves using a high-pass filter, which eliminates echoes from the vicinity of zero frequency in the Doppler spectrum. Stationary targets as well as clutter echoes with very small radial velocities are filtered out and the moving targets are passed through.

When the radar is installed on a moving platform, the Doppler frequency of stationary targets is shifted in Doppler spectrum according to the radial velocity component of the platform in the direction of the scanning antenna main beam. Therefore the stopband of the MTI filter has to be constantly shifted accordingly to compensate for the platform movement. With high antenna scan rates and rapidly varying platform velocities this is a demanding task, and therefore the moving target processing in non-stationary radar systems is usually realized using Doppler processing technique. This method is based on spectral analysis of the return echoes in each range cell, with the radar's pulse repetition frequency being the sampling rate. The radar signal is analyzed by applying the fast Fourier transform, which performs a sophisticated coherent integration of successive echoes. Contrary to MTI-processing, in Doppler processing the entire frequency information is preserved, which allows target detection also in circumstances where the spatial resolution cell is occupied by both the target and volume clutter with non-zero radial velocity. The compensation for the platform movement can be done in subsequent signal processing stages based on the center frequency of a Doppler bin where the heavy clutter concentrations arising from ground echoes are located.

2.1 Effect of oscillator phase noise in radar detection

The stopband of an MTI-filter should be carefully selected so, that the clutter power from ground echoes is attenuated to desired level and on the other hand so that the elimination of slow velocity targets is minimized. The width of ground clutter spectrum is narrow by nature but the phase noise of a radar system will spread it causing power leakage to the passband region and thereby degrading the clutter cancellation ability of the MTI-processor.

In Doppler processing technique the spreading of target and clutter echoes in frequency domain due to system phase noise results in power leakage to the adjacent Doppler bins. This may deteriorate the theoretical velocity resolution obtainable by the given combination of illumination time and pulse repetition frequency. In addition to the degradation of velocity resolution the phase noise sidebands may also completely prevent target detection if the target's true or aliased radial velocity is close to that of clutter or another target. Figure (1) depicts a case where a weak target is obscured by rain clutter spectrum.

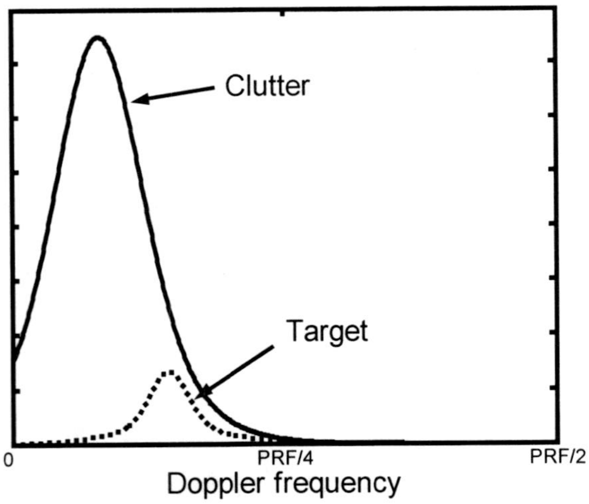

Figure 1. *An example of a situation where a weak target echo is buried under a strong clutter return.*

In order to be able to evaluate the performance of MTI- or Doppler processing under various frequency stability conditions of radar's master oscillator, one needs the knowledge of the spectral distributions of targets and different clutter types. Literature shows that clutter measurements at millimeter wave frequencies have mostly been noncoherent [10]. In a radar capable of Doppler processing it is advantageous to preserve the sign of the Doppler frequency, and therefore a coherent spectral shape is required.

In the following, the clutter and target echoes spectral shapes are expected to be described by Gaussian distribution function. The spectrum of an echo from a resolution cell containing both clutter and target is

$$S(f) = \frac{C}{\sigma_c\sqrt{2\pi}}\exp\left(-\frac{(f-f_c)^2}{2\sigma_c^2}\right) + \frac{1}{\sigma_t\sqrt{2\pi}}\exp\left(-\frac{(f-f_t)^2}{2\sigma_t^2}\right) \tag{1}$$

where C is the clutter power relative to the target power, σ_c and σ_t are the standard deviations of clutter and target echoes, and f_c and f_t are the mean Doppler frequencies of clutter and target echo spectra, respectively. The effect of phase noise on the return signal can be taken into account in the equation (1). If the phase noise of an oscillator is approximated with Gaussian spectral shape, equation (1) can be written as

$$S(f) = Ce^{-2P_n^2}\sum_{k=0}^{\infty}\frac{2^k P_n^{2k}}{\sqrt{2\pi}\left(\sigma_c^2 + k\frac{\Delta f_n^2}{2}\right)^{1/2}}\exp\left(-\frac{(f-f_c)^2}{2\left(\sigma_c^2 + k\frac{\Delta f_n^2}{2}\right)}\right) + e^{-2P_n^2}\sum_{k=0}^{\infty}\frac{2^k P_n^{2k}}{\sqrt{2\pi}\left(\sigma_t^2 + k\frac{\Delta f_n^2}{2}\right)^{1/2}}\exp\left(-\frac{(f-f_t)^2}{2\left(\sigma_t^2 + k\frac{\Delta f_n^2}{2}\right)}\right) \tag{2}$$

where the oscillator's phase noise power relative to the received target echo level is P^2_n and Δf_n is the standard deviation of the phase noise spectrum [11]. Figure (2) illustrates the degradation in signal-to-clutter ratio, caused by phase noise sidebands, compared to noiseless situation as a function of Doppler frequency offset between a target and clutter echo for various phase noise power levels (relative to the target echo). The spectral width of clutter is 180 Hz [10] and that of target significantly narrower. The spectral width of oscillator phase noise is proportional to short-term stability of 6 x 10^{-9}. Figure (2) shows that the phase noise sidebands cause considerable reduction in the signal-to-clutter ratio in clutter echo's vicinity. The relative loss becomes larger as the frequency offset increases. This is due to the fact that the noiseless spectrum of clutter (and target) echo falls off more rapidly with frequency than the phase noise sidebands. The maximum reduction is attained at the offset frequency where the noiseless clutter echo spectrum reaches the thermal noise floor and is therefore system dependent. After this point the effect of the phase noise sidebands decreases proportionally to the decline slope of phase noise spectrum.

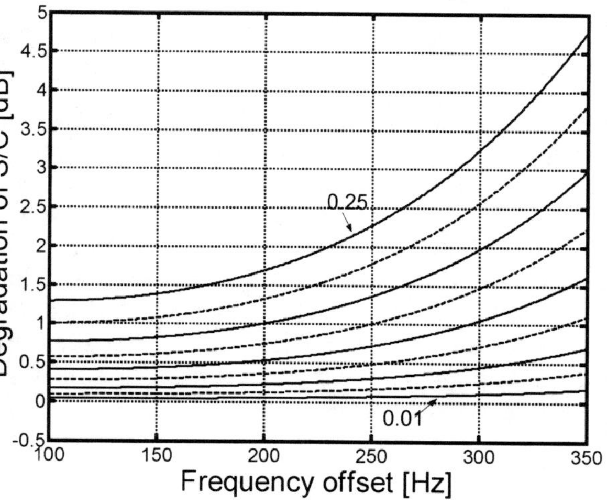

Figure 2. *The degradation of signal-to-clutter ratio as a result of phase noise versus clutter and target echo separation when compared to noiseless situation. The curves represent different phase noise power levels normalized to target echo level.*

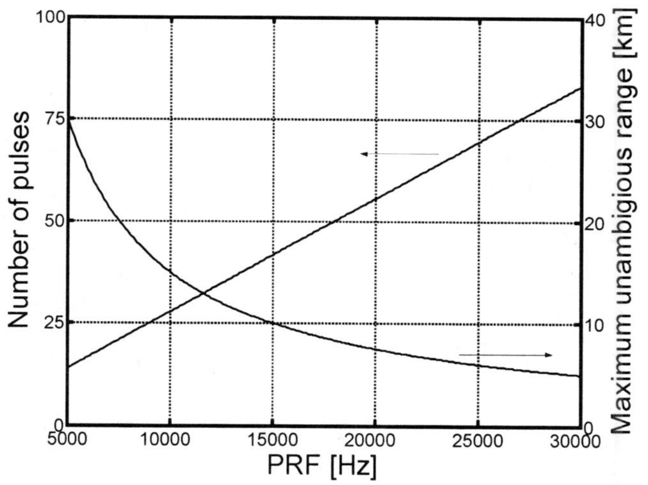

Figure 3. *The selection of pulse repetition frequency is a trade-off between the desired ambiguous range and the amount of pulses needed for integration.*

If we consider a surveillance radar with antenna beam width of 1° and scanning rate 60 rpm, the illumination time becomes t_d = 2,78 ms. If a short term stability of 6 x 10^{-9} is assumed, an achievable spectral resolution at Q-band is around 200 Hz which represents a radial velocity change in the order of 1 m/s. From the oscillator stability, desired unambiguous range and the available illumination time one can derive the suitable pulse repetition frequency. Figure (3) contains two curves. One curve links the PRF to the desired unambiguous range and the other presents the number of pulses within the given illumination time with different PRF-values. If the radar's surveillance range is up to 20 km, the prf is 7,5 kHz and the amount of integrable pulses within the illumination time will be 21. This leads to frequency resolution of 180 Hz, which matches well with the resolution constraint set by the oscillator phase noise

3 The prototype oscillator

In order to evaluate and study the performance of a millimeter wave radar in practice a test system was built. The master oscillator for the instrumentation radar was constructed of a Ku-band oscillator and diode multiplier to produce the desired carrier frequency at Q-band (figure 4). The oscillator uses dielectric resonator stabilization and it has only one GaAs FET as an active element. A ferrite circulator is connected at the output to reduce load-pulling effects. Frequency control is arranged through two paths. The drain voltage of the GaAs FET is adjustable through an SMA interface and additionally the bias of the multiplier PIN diode is tunable as well. All the electronics is housed for better stability in vibrating environments into a milled 100 x 50 x 30 mm^3 aluminium frame (figure 5) [9].

Preliminary experiments that have been carried out show, that with the present test arrangement the frequency tuning range is up to 950 MHz at Q-band if output power level variation is required to stay smaller than 1 dB. The respective adjustment scale of tuning voltage is roughly 3 V. A control current of 90 mA is needed at the FET drain connector but just 20-30 μA is enough for the reverse-biased multiplier. In case the output power level variation needs to be reduced to 0.5 dB or less, a 350 MHz span can still be achieved. The video bandwidth for the tuning input has been measured to exceed 100 MHz.

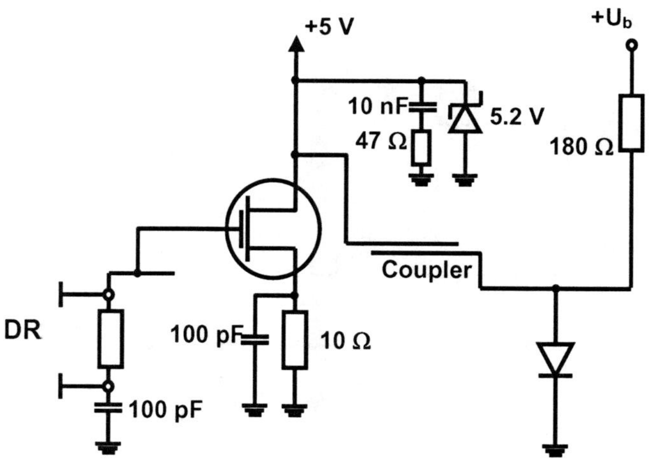

Figure 4. *Schematic diagram of the prototype Q-band oscillator for a millimeter wave instrumentation radar.*

Because two separate control inputs are at hand, the DRO's temperature characteristics can be further enhanced through NTC-type control. At present we are able to keep fluctuations less than 1 MHz / 10°C. The measured short-term stability is around 6 x 10^{-9} over 1 ms processing time. Desired spectral purity close to the carrier is comparable e.g. to commercial Gunn-units.

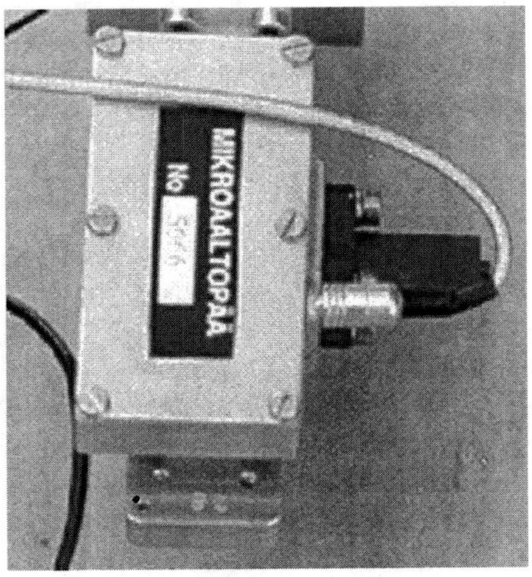

Figure 5. *Photograph of the oscillator aluminium frame.*

4 Measurement results

The short-term stability of the test instrumentation radar's Q-band master oscillator measured in laboratory conditions is around 6 x 10^{-9}. Preliminary measurements of the oscillator stability behaviour in field circumstances have been carried out. In order to evaluate the degradation of oscillator's frequency stability when onboard a moving platform the instrumentation radar was installed on a ground vehicle. First a 30 second measurement was conducted when the vehicle was not moving. After that the measurement was repeated for a moving vehicle, and the Doppler shift was removed from the data. The normalized distribution-curves are shown in figure (6). The standard deviations of frequency in both measurements are exactly the same within the measurement accuracy.

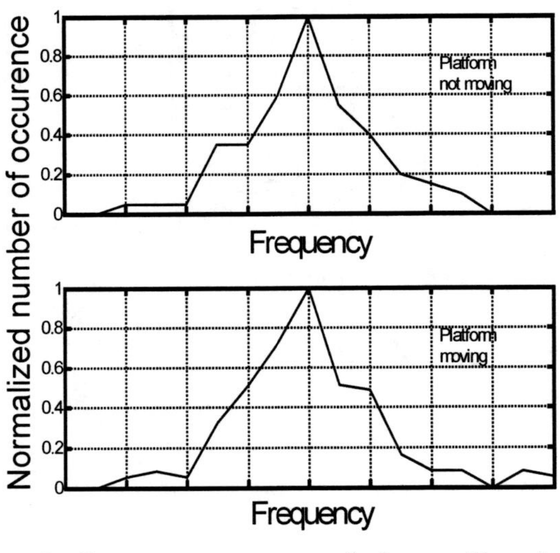

Figure 6. *Two measurements of the oscillator's output frequency. The one on top is of a situation when the platform was not moving. The lower curve shows the frequency distribution in a case of moving platform.*

As a result of the field measurements one can conclude, that a moving platform does not have a strong effect on the frequency stability of an oscillator, and an installation of a millimetre wave radar to a moving platform is realizable option. To confirm this more stability measurements will be carried out for a moving platform in different terrain and at various platform speeds.

5 Conclusions

A prototype Q-band master oscillator for a light-weight millimeter wave radar was presented. The short-term frequency stability of the oscillator in laboratory conditions is 6 x 10^{-9}. This is in a good agreement with a typically attainable Doppler frequency resolution of a radar system. The field measurements showed no degradation in oscillator's frequency stability when installed onboard a moving platform. Based on the measured performance it is shown that the constructed oscillator prototype is suitable as a millimeter wave radar's master oscillator.

References

[1] W. Delaney, W. Ward: "Radar Development at Lincoln Laboratory: An Overview of The First Fifty Years", *MIT Lincoln Laboratory Journal*, **vol 12**, No 2, pp. 147-166 (2000).

[2] M. Kihara, S. Ono, P. Eskelinen: *Digital Clocks for Synchronization and Communications*, Artech House, Norwood, 2002.

[3] J. Ruoskanen, P. Eskelinen, H. Heikkilä: "Target Detection Trials with A Millimeter Wave Radar System", *IEEE Aerospace and Electronic Systems Magazine*, **vol 18**, No 11, pp. 26-30, (2003).

[4] P. Lacomme et al: *Air and Spaceborne Radar Systems*, William Andrew Publishing, Norwich, 2001.

[5] P. Eskelinen et al: "Test arrangements for a mobile millimeter wave battlefield radar", *Proc. URSI General Assembly 2002*, Maastricht, August 19-24, 2002.

[6] J. Crawford: *Frequency Synthesizer Design Handbook*, Artech House, Norwood, 1994.

[7] G. Tan: "A highly stable, phase coherent, wideband microwave synthesizer for radio astronomy applications", *Proc. IEEE Intl. Frequency Control Symposium*, pp. 607–610, April 1999.

[8] H. Eskelinen, P. Eskelinen: "An experimental arrangement for injection locking Ka-band oscillators", *Proc. Joint IEEE International Frequency Control Symposium and EFTF 2003*, Tampa, Florida, May 2003.

[9] H. Eskelinen, P. Eskelinen: *Microwave Component Mechanics*, Artech House, Norwood, 2003.

[10] N. Currie et al.: *Millimeter-Wave Radar Clutter*, Artech House, Norwood, 1992.

[11] J. Lee, Y. Byun: "Analysis of Clutter Cancellation Capability Considering Radar System Phase Noise Effects", *Proc. 1997 IEE AP-S International Symposium*, paper 132.9.

METAL-SEMICONDUCTOR POINT-CONTACT DIODES FOR FREQUENCY MIXING IN THE NEAR INFRARED

E. Bava[1], C. Svelto[1], G. Galzerano[2], N. Beverini [3], G. Carelli[3], A. De Michele[3], E. Maccioni[3], F. Sorrentino[3], A. Moretti[3,4], M. Prevedelli[5]

[1]INFM and IEIIT-CNR - Department of Elettronica e Informazione - Polytechnic of Milan, Via Ponzio 34/5, 20133 Milano (MI), Italy, Phone: +39 02 23993610, Fax: +39 02 23993413, E-mail: svelto@elet.polimi.it
[2]IFN-CNR, Department of Physics, Polytechnic of Milan, Piazza Leonardo da Vinci 32, 20133 Milano (MI), Italy, Phone: +39 02 23996115, Fax: +39 02 23996126, E-mail: galzeran@fisi.polimi.it
[3]Department of Physics, University of Pisa and INFM, via Buonarroti, 2 56127 Pisa, Italy, Phone: +39 050 2214285, Fax: +39 050 2214333, E-mail: demich@df.unipi.it
[4]IPCF-CNR, via G. Moruzzi,1 56124 Pisa, Italy
[5]Department of Chemical Physics, University of Bologna, Via del Risorgimento 4, 40136 Bologna, Italy, Phone: +39 051 2093703, Fax: +39 051 2093690, E-mail: mrp@ms.fci.unibo.it

Keywords: mixer, near infrared, phase lock.

Abstract

We tested the performance of metal-semiconductor point-contact diodes as frequency mixers in the near infrared region. We performed preliminary experiments in order to phase-lock two diode laser at 850 nm some hundred GHz apart. We used GaSb, InAs and InSb as semiconductor layer. The frequency bridge between the two lasers was covered by a Gunn diode frequency locked to a 1 GHz SAW oscillator. The optical phase-lock loop is implemented using a novel analog+digital phase detector.

1 Introduction

In the fields of frequency metrology and high-resolution spectroscopy, it is usual to measure wide frequency differences (up to several terahertz) between optical standards. Ultra broadband non-linear devices like Metal Insulator Metal (MIM) point contact diodes and Schottky barrier diodes are usually employed for this purpose. Also the Metal-semiconductor point contact diodes have proved to be good detectors and mixers for radiation from the far-infrared to the visible [1,5,6,7], they present performances that are similar to the usually employed MIM and Schottky barrier diodes and even better in the near infrared region [2,10]. They are also cheaper and easier to use. In a Schottky diode the barrier region is usually between the semiconductor (normally GaAs) and a metal layer (usually gold) deposited on it. In our case the metal-semiconductor barrier region is reduced to the very small region of the contact between a semiconductor platelet and the sharp point of a tungsten whisker. Thus in these devices both electron tunnelling, as in Metal Insulator Metal (MIM) point contact diodes, and thermionic emission, as in Schottky diodes, are probably important for their rectifying and mixing properties. The purpose of this work is to investigate the capabilities of a metal semiconductor point contact diode to phase-lock two diode laser in the visible or near infrared at frequency differences around 100 GHz or larger. The frequency down-conversion was obtained by beating the two optical frequencies with a Gunn diode, phase-locked to a Surface Acoustic Wave (SAW) oscillator. The optical phase-lock loop is based on a novel analog+digital phase detector.

2 Experimental set up

The experimental set up is shown in Figure 1.

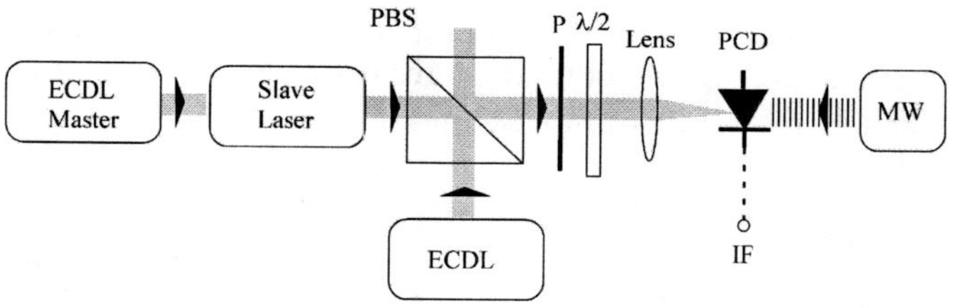

Figure 1: Scheme of the apparatus. ECDL: extended cavity diode laser; PBS: polarizing beam splitter; P: polarizer; PCD: point contact diode; MW: microwave source; IF: intermediate frequency.

Two diode lasers at 850 nm were mounted in extended cavity configuration in order to narrow their linewidth at about 1 MHz. The output power is few mW. One of the two lasers is amplified by optical injection of a second more powerful laser diode (master-slave configuration). At the output of this second laser we obtained 40 mW with the same linewidth of the master ECDL laser. The two beams are focused together on the junction of the point contact diode with a lens of 45 mm of focal length. The effective light power impinging on the diode is 1mW for the first laser and 10mW for the second one in master-slave configuration. By using a pre-amplifier the DC sensitivity of our arrangement was as low as a few microvolts. Due to the non linearity of the device, on the point-contact diode output is present a signal at the frequency difference between the two lasers ν_{IF}, that is amplified by a low noise +40 dB amplifier and finally displayed on a spectrum analyser. A microwave radiation can be coupled to the point-contact diode in order to down convert the IF in the case that it is greater than the detection bandwidth. In the presence of the MW the intermediate frequency is given by:

$$\nu_{IF} = \nu_{ecdl1} - \nu_{ecdl2} - \nu_{mw} \qquad (1)$$

We used a microwave sweeper as MW source for frequency until 18 GHz, or a frequency stabilized Gunn oscillator at 72 GHz. The microwave was coupled to the point-contact diode with some suitable horns. The output power was 100 mW for the MW sweeper and 60 mW for the Gunn oscillator.
The signal at IF was sent at the input of a optical phase-lock loop in order to stabilize the frequency difference between the two lasers.

2.1 Point-contact diodes

The metal-semiconductor diode consists of a tungsten cat whisker pressed onto the surface of a semiconductor platelet.
The whisker is electrochemically etched to the desired tip in a 2N-NaOH solution. A detailed description of the diode can be found in [10]. We tested four different semiconducting samples: intrinsic InAs and InSb, n-doped GaSb and InSb. The n-doped (Te) GaSb and InSb crystal have a carrier concentration of 6-8 10^{17} cm^{-3} and 4-8 10^{14} cm^{-3}, respectively. Undoped InAs and InSb wafer have a carrier concentration of 1.8-2.0 10^{16} cm^{-3} and 3 10^{13} cm^{-3}, respectively. As described in [8,10] no dramatic changes in video detection or in mixing properties were noted changing the diameter of the tungsten wire, nor changing the radii of the tip. We used wires of 127 μm in diameter, with tip radius of about 100 nm. This choice allowed us to obtain very stable and long lived contacts, also owing to a deep penetration of the tip into the semiconductor substrate [2]. The open structure of the diode allows us to change the resistance of the contact in order to optimise the rectified ($R > 3\text{-}4$ kΩ) signal or the mixing one ($R \approx 30\text{-}300$ Ω).

2.2 Frequency stabilized Gunn oscillator

The Gunn oscillator is frequency stabilized [9] with respect to a low phase-noise SAW oscillator at a frequency of 1 GHz. The 9 GHz component of the comb, generated from the 1 GHz signal through a step recovery diode, was selected, amplified, and mixed with the Gunn radiation producing an IF frequency of 10 MHz.
The rectified signal was fed into an integrating servo which controls the current, and hence the frequency, of the 72 GHz Gunn diode. With a control bandwidth of 10 kHz, the medium term relative frequency stability reaches 2 10^{-9} on an integration times of 1 s.

2.3 Optical phase-locked loop

For the optical phase-locked loop (OPLL) we employed a dual analog+digital phase-frequency detector [3]. For stable phase locking the residual phase fluctuations must be kept well below the monotonic range of the phase detector. This requires either a very fast loop or a broad-range detector. While the monotonic range of an analog mixer is limited to π radians, there are several ways to build a broad-range digital phase detector. The most convenient one is to send the beat-note signal to an upward counter and the local oscillator signal to a downward counter. The outputs of the counters are finally sent to a digital adder, whose output, after D/A conversion and low-pass filtering, gives a linear response to the phase difference. The linear range only depends on the number of bits in the counters and the adder. In order to avoid a periodic response, saturation is obtained by blocking the appropriate counter when the adder reaches the minimum or maximum value. In this way when the input frequencies are different the detector gives a constant value indicating the sign of the frequency difference. Thus the system acts as a phase-frequency detector (PFD). Compared to analog phase detectors, digital PFDs offer several advantages in the construction of an OPLL. The broader operating phase range allows slip-free phase locking even in presence of a large residual phase noise. This can be a frequent condition when operating with broad sources as diode lasers, or when the signal-to-noise ratio in the beat note is low. Moreover, digital PFDs provide larger capture range and sideband selectivity. On the other hand they are intrinsically noisier and slower than analog mixers. In order to combine the advantages of both kinds of phase detector, the output from the analog mixer is used only when the phase error is below one cycle, while the output from the digital PFD is employed to extend the phase detection range and to allow for frequency detection when the loop is out of lock. While using both phase detectors together, care is taken to avoid that the analog one works when the phase error is more than one cycle; on the other hand, when the phase error is less than one cycle the digital PFD is disabled. These features are obtained by adding few extra logics to the scheme described above. The local oscillator input to the analog phase detector is switched by a logic signal revealing whether the phase error is below one cycle. In this way the input of the analog detector is disabled and its output is zero when the phase error is larger than one cycle. On the other hand, by using a synchronous adder clocked by the local oscillator signal, the PFD response becomes a staircase function of the phase difference and exhibits a ``dead zone" in the range where the analog detector is enabled. Finally, the local oscillator signal is sent to the analog detector after an appropriate delay, so that the analog locking point is in the exact center of the dead zone.
In our system the functions of counters and adder are all performed by a single Complex Programmable Logic Device (CLPD), which also controls the signal enabling the analog detector. The CLPD (EPM7064, from the Altera MAX7000S family), can be easily re-programmed by connecting the circuit to a computer. In this way it is possible to change the bit number of the counters and of the adder. It is also possible to add frequency scalers to the local oscillator or to the beat note signal. In our experiment we used a 7-bit adder. Thus the monotonic phase range of our PFD is 128 cycles (about 804 radians). The scheme of our circuit is shown in Figure 2.

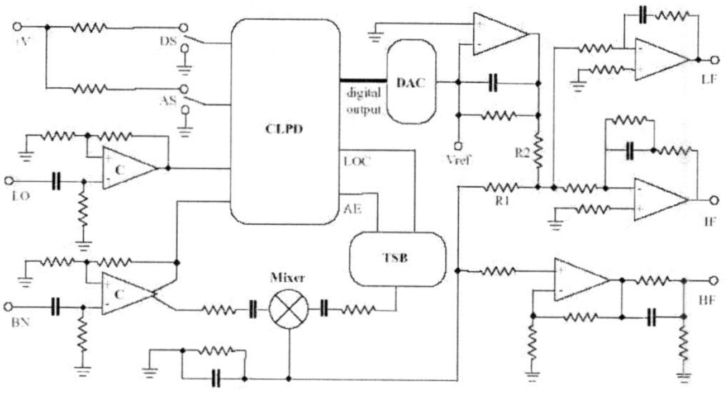

Figure2: Electrical scheme of our analog+digital phase-lock loop. DAC: digital/analog converter. TSB: three-state buffer. C: comparator. DS (AS): switch for digital (analog) operation. LO: local oscillator. BN: beat note. LOC: copy of the local oscillator signal enabled by AS. AE: control signal enabling LOC input to the mixer. Vref: voltage reference for the bias in the digital output. LF: output for PZT driver. IF: output for diode laser current supply. HF: direct output to diode laser injection current.

The signals of local oscillator and beat note are squared by two fast comparators (LT1016) and sent as inputs to the CLPD together with two control signals, one for enabling the digital detector and the other for enabling the analog one. In this way the system allows four different operating modes: digital PFD alone, analog phase detector alone, mixed PFD and zero output. The squared beat note is also sent to the analog phase detector (Minicircuits RPD-1). The CLPD has three outputs: the counter signal is sent to a D/A converter. A copy of the squared local oscillator, together with the signal enabling analog detection, is sent to a three-state buffer (74ACT245), used as both a time delay and a switch. The output of the buffer goes to the analog phase detector as local oscillator input. The signals from the analog phase detector and the D/A converter are low-pass filtered and summed. The phase sensitivity should be the same in the two operating conditions (analog and digital phase detection) for closed loop stability. The ratio R1/R2 takes into account the different sensitivities of the two detectors.The feedback loop is composed of three paths, to control three actuators in different Fourier frequency regions. The fastest path acts on the diode laser injection current directly. An RC phase-lead filter is used to balance the roll-off in the diode laser response and therefore to extend the loop bandwidth (to about 1.7 MHz). For this path we use the analog phase detector output alone, as in tight lock conditions the digital PFD is not active however. The intermediate-frequency correction is sent to the current supply of the diode laser, providing high gain in the 10^1-10^4 Hz region. The unity-gain point for this path is around 100 kHz. In order to keep the injection current around a constant value (the current-to-frequency characteristic of a diode laser, and therefore the loop gain, changes with the injection current), long-term corrections are performed by changing the cavity length of the ECDL through a PZT. An integrator increases the low-frequency gain.

3 Results

We made measurements of the signal to noise ratio of the beat signal at the output of the different kinds of diodes in three different conditions. The results are plotted in Figure 3 and summarized in Table 1, where in column 1, Δν is the frequency distance between the two ECDL lasers.

Δν (GHz)	GaSb S/N (dB)	InAs S/N (dB)	InSb S/N (dB)	InSb (Te) S/N (dB)
0.5	65	75	55	65
8	40	35	30	40
72	25	20	15	30

Table 1: Signal to noise ratios obtained in different situations with a resolution bandwidth of 300 kHz.

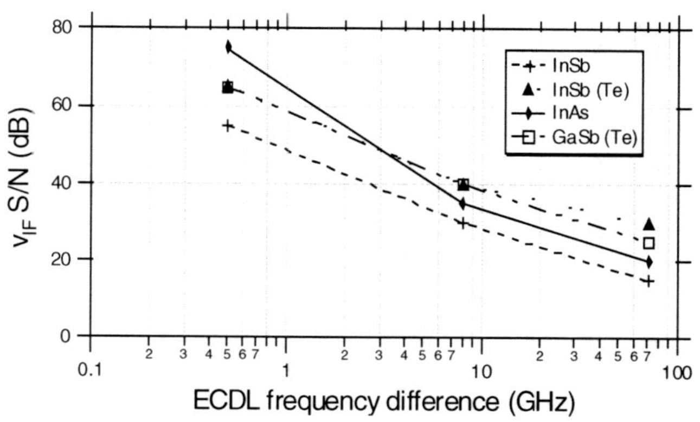

Figure 3: Signal to noise ratios obtained in different situations with a resolution bandwidth of 300 kHz.

When Δν is 0.5 GHz we observe the direct beat note between the two lasers while when Δν is 8 and 72 GHz we bridge the gap with the microwave sweeper or the Gunn oscillator, respectively. Without microwave InAs, InSb and InSb (Te) demonstrate better performances with high forward bias current (≈30 mA). On the other hand they do not need bias in presence of microwave radiation. On the contrary GaSb never needs bias. The GaSb diode has also the advantage to generate a video signal two orders of magnitude higher than the other diodes [4], and so it is easier to align. The decrease of the S/N ratio between the case with Δν=0.5 GHz and the case with Δν=8 GHz (at least of 25 dB) is mainly due to the higher order of mixing (three instead of two). There is also the problem that the optimal operating condition of the diode as mixer are not the same for the microwave or infrared region respectively, as proved by the different optimal bias current in the two cases. Between Δν=8 GHz and Δν=72 GHz the S/N ratio decrease of 15 dB. This can be due to the frequency response of the diode or to the different coupling of the microwave radiation on the diode. From Figure 3 it is possible to expect values of the signal to noise ratio of 10 dB when the two ECDL frequencies are distant several hundreds of GHz. Further investigation using the harmonics of the Gunn oscillator are necessary to explore the limit of this technique.

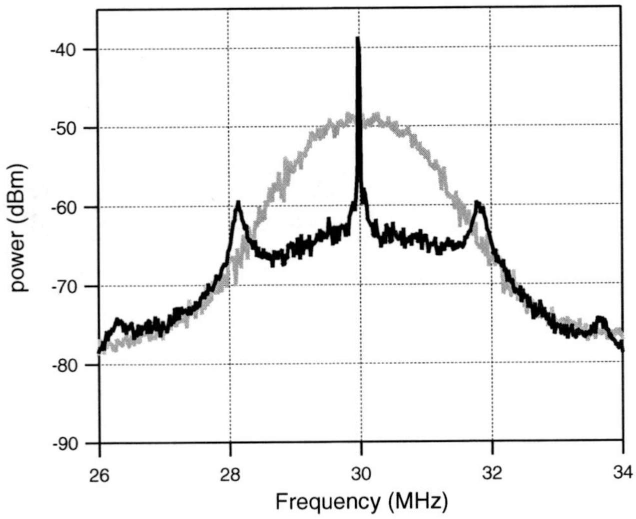

Figure 4: Spectrum of the beat signal between the two ECDLs and the 72 GHz Gunn diode, in two different phase-lock conditions. Grey line: only digital phase-frequency detector; black line: analog+digital detector. Resolution bandwidth: 10 kHz.

To test the feasibility of phase-locking the two lasers using our point-contact diode as mixer, we phase-lock the two lasers at a frequency distance of 72 GHz. Figure 4 shows the beat signal in two conditions with only the digital servo loop closed and with the analog+digital servo loop closed. This phase-lock-loop was obtained using the InSb (Te) diode that has shown the better performances in the mixing between the two diode lasers and the Gunn.

4. Conclusions

Using metal-semiconductor point-contact diodes it is possible to phase-lock two diode lasers with large frequency distance using microwave oscillators to bridge the gap. As a preliminary test we phase-locked the lasers 72 GHz apart. We are still working to improve the experimental set-up in order to detect the mixing between the lasers when they are beyond 72 GHz apart, using the harmonics of the Gunn diode. Since the phase noise of the 72 GHz signal can be approximated with a white contribution at a level of ≈75 dBc/Hz for Fourier frequencies in the range from 1 Hz to 10 kHz the corresponding collapse frequency of the oscillator is ≈4 THz, which is the theoretical limit for the frequency difference of the diode lasers.

References

[1] A. Bertolini, G. Carelli, C.A. Massa, A. Moretti, F. Strumia, "Detection and mixing properties of an InSb metal-semiconductor point contact diode", *Int. J. Infrared Millimeter Waves*, **20**, pp. 1121-1127, (1999).

[2] A. Bertolini, G. Carelli, A. Moretti, G. Moruzzi, "Performance of W-InSb metal-semiconductor point contact diodes as detectors and mixers in the near infrared", *Int. J. Infrared Millimeter Waves*, **22**, pp. 1761-1768, (2001).

[3] N. Beverini, Beverini, M. Prevedelli, F. Sorrentino, B. Nyushkov, A. Ruffini, "Phase locking of a diode laser to a Ti:Sa comb by means of an analog + digital phase-frequency detector", *Proc. Laser Optics 2003 Conf.*, **SPIE Proc. vol. 5478**, in press, St. Petersburg (2003).

[4] G. Carelli, A. De Michele, M. Finotti, K. Bousbahi, N. Ioli, A. Moretti, "GaSb and InAs: new materials for metal-semiconductor point-contact diodes", *Int. J. Infrared Millimeter Waves*, **24**, pp. 799-811, (2003).

[5] H. U. Daniel, , B. Maurer, M. Steiner, "A broadband Schottky point contact mixer for visible laser light and microwave harmonics", *Appl. Phys. B*, **30**, pp. 189-193, (1983).

[6] H. U. Daniel, M. Steiner, H. Walter, "Response of metal-insulator-metal point contact diodes to visible laser light", *Appl. Phys. B*, **25**, pp. 7-12, (1981).

[7] R. E. Drullinger, K. M. Evenson, D. A. Jennings, F. R. Petersen, J. C. Berquist, L. Burkins, H.U. Daniel, "2.5-THz frequency difference measurements in the visible using metal-insulator-metal diodes", *Appl. Phys. Lett.*, **42**, pp. 137-138, (1982).

[8] K. M. Evenson, M. Inguscio, D. A. Jennings, "Point contact diode at laser frequencies", *J. Appl. Phys.*, **57**, pp. 956-960, (1985).

[9] G. Galzerano, C. Svelto, E. Bava, G. Carelli, M. Finotti, A. Moretti, N. Beverini, "High-stability 72-GHz Gunn oscillator for the characterization of ultra-high-speed optical receivers based on InP and InSb Schottky diodes", *IEEE Trans. Instrum. Meas.*, **52**, pp. 1190-1194, (2003).

[10] A. Moretti, E. Maccioni, M. Nannizzi, "A W-InSb point contact diode for harmonic generation and mixing in the visible", *Rev. Sci. Instr.*, **71**, pp. 585--586, (2000).

INTEROPERABILITY OF THE GPS AND GALILEO TIMESCALES FOR POSITIONING AND METROLOGY

W. M. Forrest

Thales, UK. email: william.forrest@thalesatm.com, fax: +44 208 391 6455

Keywords: GPS, Galileo, Timescale, Interoperability.

Abstract

Today, GPS plays an important role in positioning and the dissemination of time and frequency on a global scale. In Europe, Galileo is under development and will enter service some time after 2008. Both systems will have their own timescales: GPS Time for GPS and Galileo System Time for Galileo. There are enormous benefits to be gained if GPS and Galileo can be used together as one "super-constellation" of over 50 satellites. There is therefore a need for a clear understanding of the factors that will contribute to, and ultimately limit interoperability of the GPS Time and Galileo System Time.

In an ideal world, both GPS and Galileo timescales would be aligned with TAI/UTC and interoperability would be assured, however, the systems each have their own timescale and TAI/UTC is not available in real time; offsets between the two are inevitable.

Section 2 of this paper details the GNSS interoperability considerations with respect to system time. Section 3 outlines the timing concept and requirements for Galileo, and Section 4 presents the implications for positioning receivers of various time offsets for a mixed constellation in an urban environment. Section 5 presents a GPS / Galileo traceability model from a user receiver to TAI/UTC. In Section 6, a range of conclusions are drawn on how Galileo could interact with GPS and World Timekeeping, and recommendations are made for avenues of research that may further reduce the problem and so enhance GNSS interoperability.

1 Background to this paper

This paper summarises the main results and conclusions from a dissertation [1] submitted to the University of Nottingham in August 2003 in partial fulfilment for the authors degree of Master of Science in Satellite Positioning Technology. The full results and analysis is contained in the dissertation.

2 Introduction

There are two main service objectives of Global Satellite Navigation Systems:

i) Positioning
ii) Dissemination of Time and Frequency

Both services are made possible by a synchronised network of highly stable atomic clocks. This network of clocks provides: a) a reference timescale used throughout the system for positioning, b) a reference for dissemination of UTC and c) a stable frequency source. These "time" products are highly interdependent, and satisfy many different user applications.

GPS today is a well-established system with a baseline of 24 operational satellites providing high quality services on a global basis. It is essentially a military system transmitting both open signals in the L1 band, and encrypted signals in L1 and L2 bands. The GPS system is synchronised to GPS System Time which is based on UTC(USNO). Within the GPS program and at the United Stated Naval Observatory (UNSO), considerable experience in clock technology and operations has been accumulated which will contribute to further improvements in the planned GPS modernisation program.

Europe is developing Galileo, a new satellite navigation system under civil control that is planned to be operational some time after 2008. Galileo's services will be similar to those of modernised GPS, and will also be based upon a synchronised network of highly stable atomic clocks. The Galileo constellation will have 27 satellites, complemented by three in-orbit spares to reduce replenishment time in case of failure. Galileo will transmit ten individual navigation signals in the upper and lower L-Bands, these signals will combine to provide a range of services for public, commercial and regulated applications. A fundamental requirement is that Galileo is to be interoperable with GPS, but independent from it. This independence requires the creation of a new timescale known as Galileo System Time (GST), and a new Geodetic Reference Frame known as Galileo Terrestrial Reference Frame (GTRF).

In an ideal world, both the GPS and Galileo timescales would be perfectly aligned with International Atomic Time (TAI), but two physically separate systems results in two timescales and each will have different offsets and fractionally different drift rates. They will therefore deviate from TAI/UTC, and from each other.

There are enormous benefits to be gained if GPS and Galileo can be used together as one "super-constellation" of over 50 satellites and these benefits are well reported. Their frequencies and navigation messages are highly compatible, the Geodetic Reference Systems will be within acceptable limits, and therefore the major hurdle for true interoperability is the alignment of their timescales. Perfect alignment would be practically impossible, so a workable fallback solution is necessary where the instantaneous difference between the two must be either delivered to or calculable at the user receiver to acceptable degree of error. There is therefore a need for a clear understanding of the factors that will contribute to, and ultimately limit interoperability of the GPS Time and Galileo System Time.

Three fundamental questions need to be answered to determine the level of time interoperability between GPS and Galileo:

1. For a future GNSS receiver using stand-alone positioning in an urban canyon environment, to determine a solution of single-system-equivalent accuracy from say 2 GPS and 3 Galileo satellites, what degree of time synchronisation is needed between the two constellations
2. Does Galileo need to transmit Galileo-GPS time offset in the navigation message and if so to what accuracy, or, with emerging technology (clock and receiver) can this be avoided
3. For time and frequency users, how can traceability from the GNSS receiver through the GPS / Galileo constellations to International Atomic Time and the SI units be assured

This paper examines the magnitude of the problems caused by offsets between the GPS and Galileo timescales for the positioning, time and frequency users, and examines existing plans to minimise these effects. Conclusions are drawn and recommendations are made for avenues of research that may further reduce the problem and so enhance GNSS interoperability and performance.

Since GPS selective availability was set to zero on 2nd May 2002, and would appear unlikely ever to be reintroduced and there is no complementary feature on Galileo, selective availability will not be considered.

3 Navigation and Metrology concept for Galileo

There are two service objectives for Galileo:

i) **Navigation Services** – requires all satellite clocks to be synchronised (with offset parameters broadcast), which is also needed for orbit determination. The need is short-term stability for the navigation function and long stability in order to estimate clock parameters. Navigation, however will also need a link to UTC to relate the orbits with an earth reference frame, UT1/UTC, although the accuracy here is not demanding as with metrology

ii) **Metrology Service** – to disseminate TAI / UTC with high accuracy to support time user applications. This service is not critical to the navigation function.

In 2003, it had not been finally decided how the metrological interface is to be implemented. Three main options were considered:

Independent GST - all navigation and metrological functions and interface with BIPM performed inside of the Galileo system with a link to a Time Services Provider (TSP) for comparison with UTC(K)

GST dependant on UTC(K) - An external Time Service Provider (TSP) performs [GST - UTC(K)] computation, TAI prediction and GST steering corrections and the link to BIPM

Mixed solution - A mix of the above two approaches where the independent scenario is modified for the TSP to provide fine steering corrections for GST.

The current strategy is actually neither of the above, but one called the *final mixed solution* whereby the navigation critical functions are retained inside Galileo and metrological functions outsourced to the Time Service Provider. Although outsourced, the functions of the TSP are fundamental to operations. Functions include the [GST - UTC(K)] analysis, TAI prediction and provision of [TAI_P – GST] time and frequency offsets and steering corrections on a daily basis. The TSP will take responsibility for all TWSTFT and CV time transfers with the Control Segment, and similarly provide the interface with BIPM.

Within Galileo, a Precise Timing Facility (PTF) maintains a stable ensemble of clocks and measures offsets for all clocks in the ensemble against a master clock by use of a measurement system. From these measurements, GST is computed. The PTF will receive the daily steering corrections to the master clock(s), and provide GST to the Orbit Determination and Time Synchronisation process. In addition, all Galileo ground and satellite clocks are monitored.

Galileo Timing Requirements

The specifications upon which the Galileo system design is based are the European Commission High Level Document (HLD) and Mission Requirement Document (MRD), which are endorsed by Member States. It is these two documents that elaborated by the GalileoSat program in the ESA System Requirement Document (SRD). Interpreting the above requirements will result in GST being constrained as per clock once every 5-days, a 0h on MJDs that end with a 4 and 9.

below. Note the 5-day periods have been added to align with measurements that reported by the NMIs for each clock once every 5-days, a 0h on MJDs that end with a 4 and 9.

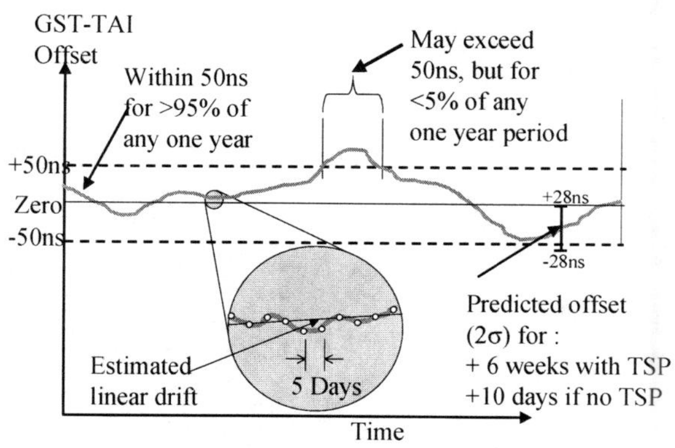

Figure 1: Galileo System Time in relation to TAI

4 Timescale Implications for Stand-alone Positioning Receivers

4.1 Overview of the Issues

Knowledge of the system time and individual satellite clock performance is vital for the determination of a position solution in stand-alone positioning. For a single GPS or Galileo constellation, system availability is generally degraded in an urban canyon environment, and, if one is to gain the ultimate benefit from interoperable systems, then there are three considerations with regard to system time:

1. GPS and Galileo use the same timescale – this option is rejected by nature of the requirement for independence and would introduce a common failure mode between the two systems
2. Galileo transmits the difference between GST and GPS time such that the user receiver is able to take this into account in the position solution
3. GPS and Galileo remain truly independent and do not attempt to inform the user of the offset between the two GNSS timescales

Having rejected option 1 above, then detailed consideration must be given to options 2 and 3 and the implications for a receiver to derive a position solution. Two important parameters that need to be discussed in relation to time are availability and accuracy.

Availability - For GPS alone there is a problem in poor locations where visibility of four satellites is intermittent or perhaps not possible. The EC Gala study indicates that in some urban canyon environments, only two or three GPS satellites will be visible at any one time and the expected availability can be in the region of 30 to 70%. This will be approximately the same for Galileo alone, but together and with known clock parameters for the two systems, then the availability will increase to a figure in the region of 95%.

Accuracy - In the urban canyon environment, high satellite angles directly translate into higher GDOP values and hence poorer positioning accuracy. Determining a position solution with a mixed GPS/Galileo constellation will not improve this situation but will degrade it further if the two timescales cannot be related by the transmission of time offset parameters.

Given that accuracy requirements are different for the various applications, there is no single figure that can be stated as an acceptable measure of accuracy, so the methodology of analysis using specifically designed simulation results was adopted which allow various system time offset parameters can be introduced to the simulations.

4.2 Simulation Results

For each simulation, the individual satellite azimuth, elevation, Right Ascension of the Ascending Node (RAAN) and UERE were recorded. Each simulation was run a total of seven times, each with a different bias added to the GPS satellites for zero, 5ns, 10ns, 20ns, 30ns 40ns and finally 50ns to create a bias between the GPS and Galileo timescales. Each run creates a file with the 10,000 vertical and horizontal position errors with respect to the "truth", least squares residuals, and test statistics.

i) Simulation 1 – 15 satellites (8 GPS and 7 Galileo), with optimised GPS/Galileo plane spacing and receiver in open view used as a baseline for comparison
ii) Simulation 2 – 5 GPS/Galileo satellites, urban canyon with optimised plane spacing between the GPS and Galileo orbit planes
iii) Simulation 3 – 5 GPS/Galileo satellites, urban canyon with one GPS and Galileo plane aligned

The receiver co-ordinates were set to the London area. London was selected because it is a mid-latitude city with a large population of potential mass-market users of applications such as Location Based Services, and secondly it has an asymmetric view of the GNSS constellation that gives slightly worse geometry than equatorial locations.

Simulation 1 using 8 GPS and 7 Galileo satellites with optimised plane spacing and an unobstructed receiver. In this simulation, the two constellations were configured such that the Right Ascension of the Ascending Node (RAAN) of the orbit planes is optimised at 30° apart. It can be seen in Figure 2 below that even with a 50ns offset between GPS and Galileo, a good position solution is achieved.

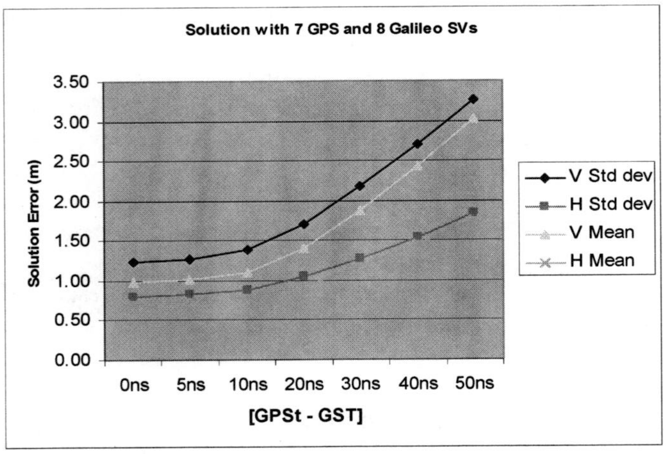

Figure 2: Position accuracy, 15 satellites

The conclusion which can be drawn is that in an open environment, and providing the [GPSt - GST] time values are within 50ns (which would appear entirely reasonable), then the transmission of time offset would not be essential for the stand-alone positioning market.

Simulation 2 – 2 GPS and 3 Galileo Satellites *with optim*ised plane spacing, and a receiver urban canyon environment. The second simulation is a typical urban canyon environment. A value of 40° cut-off angle was used to represent satellite signal obstruction. The result was a poor geometry – all five satellites have medium or high elevations, two of those satellites in the region of 070° and three satellites in a southerly direction.

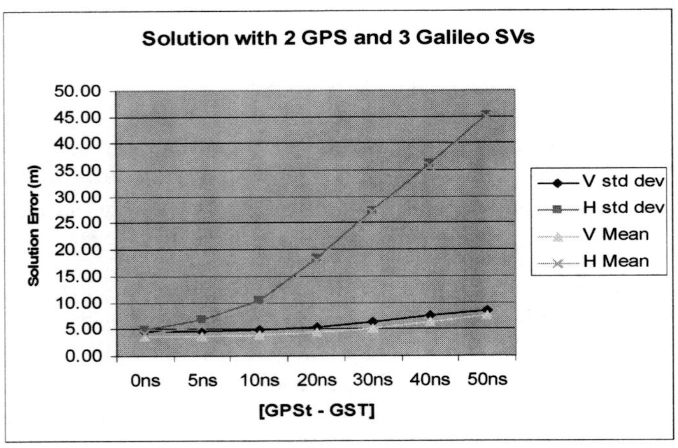

Figure 3 : Vertical and horizontal position chart, 5 SVs and optimised plane spacing

Figure 3 above shows that the vertical and horizontal standard deviations of the position errors increase significantly once the timescales of the two constellations separate by more than 5 ns.

Simulation 3 – 5 Satellites, urban canyon with GPS and Galileo plane alignment. The third simulation used the same parameters as simulation 2 with the only change being a reconfiguration of the Geometry such that the RAAN of the GPS and Galileo constellations were aligned, i.e., the three Galileo orbits aligned with the alternate GPS orbits. Five satellites were still visible but the geometry was significantly improved with two satellites to the west, two to the northeast and one in the south. Accordingly, both horizontal and vertical showed improved accuracy.

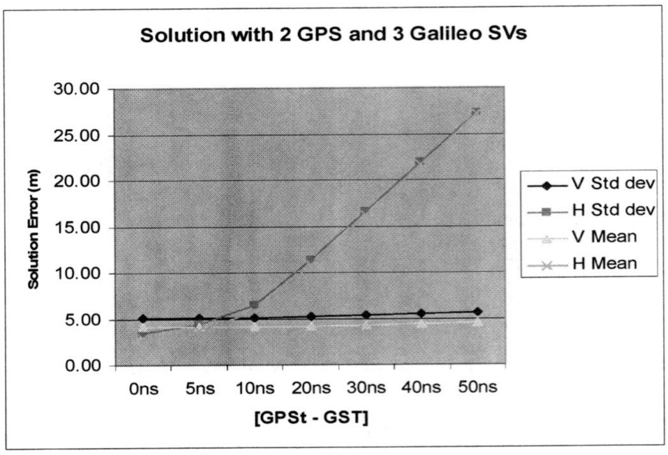

Figure 4 : Vertical and horizontal position chart, 5 SVs and zero plane spacing

Figure 4 gives improvements due to the geometry, but similarly shows that the standard deviations of the position errors increase significantly once the timescales of the two constellations separate by more than 10 ns.

4.3 Analysis

The relationship between time offset and a position error is non-linear, with larger offsets creating progressively larger errors. This has been demonstrated both statistically and by examination of the vertical and horizontal error distributions.

The simulations indicate that for stand-alone positioning in an open environment, that the [GPSt - GST] time offset need not be transmitted. However, in an a typical city environment with limited satellite visibility, then the offsets will need to be broadcast for the receiver to determine a position with adequate accuracy for mass market applications. Two points should be noted:

i) that the most mass-market users are in this category and that the forecast growth in GNSS products and applications will be dependant on ease of use and confidence in the service
ii) if the offset is not broadcast in the navigation message, the market will find an alternative mechanism such as assisted-GPS but this will add cost and complexity

The [GPSt - GST] time offset should therefore be included in the navigation messages for GPS and Galileo not exceed +/-5ns (2σ).

The satellite geometry will change with time and has a significant effect on the results. The best results for the selected location and time were achieved with the GPS and Galileo constellations aligned (which on a global perspective is actually the "worst"). As the results are position/time dependent, these simulations can only be regarded as snapshots, however, the actual results for an urban canyon environment are valid.

5. Traceability issues for GNSS

For positioning and navigation applications, time synchronisation within the system is essential: the system time could be hours or days apart from UTC and positioning is still achieved so it could be argued that traceability is not important. However, GNSS is an ideal, low-cost tool for distribution of time and frequency allowing time and frequency calibrations and synchronisation on a global scale.

In order to establish GNSS traceability for measurements made by a dual mode (GPS/Galileo) receiver, it is necessary to produce a time traceability model showing unbroken measurements with known uncertainties through both GNSS systems at all stages back to the SI units at BIPM. It is then possible to certify the GPS/Galileo receiver measurements. This is, in principle, the same methodology for traceability used for other SI units. The metre for example where a national standard is compared to the SI meter at BIPM, and lower-order calibrations are performed from the national standard and so-on down the chain.

For GPS, the GPS timescale is referenced to UTC(USNO) and in the US the national standard is UTC(NIST). For Galileo, the time interface will be provided by the Time Service Provider. It follows therefore, that the USNO / Galileo timescales will require frequent comparison if traceability is to be achieved. This is already done for GPS by NIST/USNO comparisons and an equivalent mechanism will be required for Galileo.

For Galileo, the situation is less clear: the system is in the design stage and no field experience is available. The analysis for Galileo traceability will need to incorporate an analysis of the specifications, and also the way in which the time services are provided and the operational concept for the Time Service Provider. In this context one can then consider the validity of those specifications with current achievable capabilities.

Traceability Model

In the *International Vocabulary of Basic and General Terms in Metrology* [3], traceability is defined as:

> *The property of a result of a measurement or the value of a standard whereby it can be related to stated references, usually national or international standards, through an unbroken chain of comparisons all having stated uncertainties.*

For calibration laboratories, measurements must be traceable to the SI units of measurement (ISO Guide 17025), and this is normally achieved via the national standards. It follows that for GNSS traceability on a global scale, documented evidence of all measurements is required on a continuous basis.

For GPS, the GPS timescale is referenced to UTC(USNO) and in the US the national standard is UTC(NIST). For Galileo, the Time Service Provider will provide the time interface. It follows therefore, that the USNO / Galileo timescales will require frequent comparison if traceability is to be achieved. This is already done for GPS by NIST/USNO but for Galileo, the situation is less clear: the system is in the design stage and no field experience is available. The analysis for Galileo traceability will need to incorporate an analysis of the specifications, and also the way in which the time services are provided and the operational concept for the Time Service Provider. In this context one can then consider the validity of those specifications with current achievable capabilities.

The traceability model shown in Figure 5 below details the links that need to be considered for GNSS. For each line, a series of measurements or comparisons are needed and the uncertainties stated. In generating the traceability model, the assumption has been made that three European metrology laboratories (EU1..3) will contribute to the provision the Galileo Time Interface (GTI). It must be stressed that this is a hypothesis for the purpose of this paper only to enable the issues can be explored.

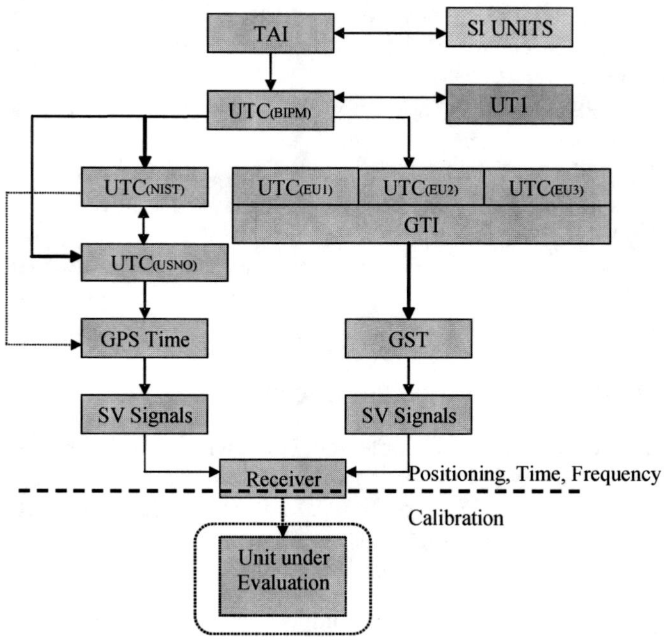

Figure 5 : GPS/Galileo Traceability Model

6 Conclusions and Recommendations

6.1 Galileo Operational Concept

The distinction between navigation and metrology for Galileo is appropriate, i.e., the navigation function performed by the Galileo system and the metrological services provided by a Time Service Provider. This concept, if correctly managed should allow the Galileo system optimal navigation accuracy whilst benefiting from the technical advances that will be made by the time and frequency community. The Time Service Provider should be a combination of at least three European metrology labs realising UTC(EU1..n), and in this way GST prediction to UTC will be assured.

6.2 Direct Transmission of TAI and UTC

GST should be steered towards TAI and GALILEO should disseminate UTC – this is clear, however, due to restrictions in the release of Galileo documents from ESA, it has not been established within this paper if the dissemination of UTC is by offset parameters to GST (as is the case with GPS) of if the timescale is to be transmitted independent of GST. Transmission of two timescales would be a benefit:

- UTC for Civil use (with prospects for international / EU law?)
- TAI (International Atomic Time) with transmission of the number of leap seconds between UTC and TAI

6.3 Galileo contribution to UTC

It has been suggested in various fora, that Galileo System Time could contribute to UTC once the system is stable. Having analysed the approach to world timekeeping to determine TAI and UTC, and the approach used by USNO and 2 SOPS for GPS TIME, the logical conclusion is that GST should not directly contribute to UTC. The rational being the objectives of the two timescales will be differently optimised:

- GPS Time is a short-term diffuse timescale, optimised for stability over 1 to 2 days
- Galileo System Time will be optimised in the short/medium term, 10 days (using a master clock)
- UTC is long term, > 6 weeks

However, the raw clock measurements for the individual clocks within the Galileo Precise Timing Facility will be of value to BIMP in the computation of TAI.

6.4 GPS and Galileo Co-operation

The technology is available today to synchronise the two timescales to within 1 or 2 ns. If the steering strategies for GPS Time and GST are shared, and with the anticipated technology advances then this figure could be reduced to 0.5 ns by the time Galileo is operational.

By having combined GPS/Galileo receivers at selected monitor stations in the GPS and Galileo ground segments, then the GPS and Galileo systems could provide the highest level of geodetic reference precision and time synchronisation. However, this is in direct conflict with the Galileo autonomy requirements. A case could be made that common-mode failure points do not exist provided the systems do not depend on this co-operation to achieve the performance specifications of the individual systems.

Although Multi-Channel Common View time transfer is not a real time comparison, developments are taking place to send data over the Internet in File Transfer Protocol, and have processed results soon after the end of a 10-minute average. If this were to be done on a 24-hour basis, then one laboratory could know the uncertainty of their frequency standard compared to another with a 10-minute latency, and within 5ns. This could be a useful mechanism for the Galileo and GPS operators to know the time offset of the two systems in near real time, and is certainly within the 100 minutes uplink period for Galileo. This could avoid the expense incurred in the setting up and operation of the two-way frequency and time transfer links.

GPS/GST Offset Determination/Dissemination

It is clear that the transmission of the GPS/Galileo time offsets is a major interoperability opportunity for both GPS and Galileo, and if implemented, will substantially benefit the user community. The conclusion is that for maximum interoperability, the time offset value should be mutually agreed between GPS (US) and Galileo (Europe) and included in the navigation messages of both systems and should be kept within +/- 5ns (2σ).

To eliminate single point failure modes, the offset should be calculated independently by say the USNO and the Galileo Time Service Provider using multi-channel GPS/Galileo receivers at precisely surveyed sites (perhaps VLBI/SLR sites), the result compared and a single value agreed before dissemination. The period of validity should be more than double the navigation data uplink period to cover the case of navigation message uplink failure. Other dissemination means should also be considered such as the Internet and mobile telephone services.

As the Galileo and modernised GPS L2/L5 navigation messages are not yet finalised, it should be possible to find accommodation for the Galileo/GPS time offset parameters. The accuracy of the offset value should be commensurate with the calibration limit values with a safety margin. Today's practical limit would suggest 5ns (2σ) over any 24-hour period, but this could be reduced with technological developments to perhaps 1 to 3ns. Further investigation would be needed to determine the stability and the impact of the more frequent Galileo system uploads. As well as the time offset update at an agreed epoch, the rate of change of the offset would give users the ability to improve the accuracies in the user terminals in a similar manner to that currently implemented in GPS for satellite clock corrections to GPS. From analysis of the clock performances, the drifts will be predictable for 24 hours so simple polynomial terms should suffice. If a resolution of say 0.1ns was provisioned, then a single 24-bit word would allow for 12-bits for the offset (range 204.8ns) at the reference epoch plus a first order polynomial term.

The first GPS IIR-M launch is scheduled for late 2004 and will have the capability for testing the new L2 signal, however, the debate continues to decide if the new L2 civil signal is transmitted or the legacy C/A code: both options have merits and drawbacks. GSTB V2 should therefore have the ability to transmit the GPS offset for initial interoperability trials with the GPS system for test purposes and to have results before IIR-F and Galileo IOV launches.

The first GPS IIR-F launch will be in late 2006 and transmit both L2 and L5 in a similar timeframe to Galileo in-orbit validation. By this time interoperability trials would be possible with user equipment.

Regarding the legacy GPS L1 C/A navigation message, ICD-GPS-200 indicates that many words are spare or reserved for system use: the use of these bits are not public and would need to be investigated to see if the GPS/Galileo time offset could be included in GPS L1 SPS. This would allow mass-market users with single frequency (L1) receivers to benefit from time synchronisation.

Week Number and Time of Week

Week number and time of week need to be considered for GPS/Galileo interoperability, especially in relation to week rollovers and the simultaneous introduction of leap seconds.

It must be recognised that there is a large GPS community with equipment that will operate for many years, and that this community will need to be served even when GPS is modernised - the term legacy L1 SPS is used. New receivers will be developed for GPS L2/L5 and for Galileo, however, they will still need to accommodate the legacy L1 SPS navigation message for many years.

Constellation Design

From the urban canyon analysis, GPS/Galileo constellation design would appear to be worthy of further consideration to reduce the effects of performance change due to RAAN drift.

The conclusions of Skinner [2] are that the optimisation of the GPS and Galileo orbit planes will improve PDOP / TDOP. In his paper, Skinner goes further and concludes that if GPS were to change to share the Galileo orbits (3 planes at 56°), then further performance improvements will result. The PDOP (worst-case) would increase from 1.992 to1.663 to and the TDOP would improve from 1.335 down to 0.829. It is clear from both sets of analysis that there is further scope for consideration in the constellation designs of the two systems.

6.5 Galileo Time Interface

UTC Prediction

Investigations have identified a key success factor for GPS is the ability of UTC(USNO) to predict UTC and to steer the ensemble towards this prediction. Galileo must be therefore be able to predict forward UTC, however, to do this Galileo will need to maintain a timescale with sufficient stability. As we have seen, for a single clock to contribute to UTC, six to nine months of history are needed before BIPM will consider its introduction. To create an ensemble with adequate stability and confidence will take a significant time (several years), and if Galileo is to be deployed and operational in 5 years time, then the construction of the Galileo Time Interface and PTF's cannot wait until the deployment phase.

Work is in progress at NPL to develop time transfer combining and steering algorithms to maintain a real time UTC(NPL) within 10 ns of UTC, if Galileo were to use this figure we have the situation where both UTC and USNO would be maintained within 10ns, giving a worst case 20ns offset from each other, however, as USNO contributes significantly to UTC, then this would maintained be much tighter.

Time Service Provider Roles and Responsibilities

For Galileo, through discussions with JU and ESA staff, the institutional arrangements and responsibilities for GTI and provision metrological timekeeping are far from clear. Conclusions from this project reveal that the role and responsibilities will change as the Galileo system is realised:

- For the Galileo System Test Bed, Version 2 (GSTBV2) and full system development, the Precise Timing Stations, satellite and sensor station clocks and their performances must be the responsibility of ESA. The GSTBV2 and In Orbit Validation satellites should both have the capability to transmit the offset between their respective timescales and GPS time, and this must be determined within the Galileo Phase C/D development contract
- The concessionaire will be responsible for system deployment and operations but the concept of the Time Service Provider requires further elaboration. Due to the critical nature of GST and the consequences of failure (commercial, operations, risk, liability...), the control of GST should be with the concessionaire, either directly if technical competence can be gained, or if commercial arrangements permit it could be contracted to at least three NMIs to ensure adequate skills and operational redundancy. Whilst it may not politically acceptable, co-operation with USNO could be a major de-risking approach for the concessionaire in the early years of Galileo, especially in the deployment phase and perhaps into operations with either GST or USNO operating as a backup solution depending on confidence and EU/US co-operation.

6.6 Time Services Roadmap

A recommendation is that a clear *time-service roadmap* needs to be developed very soon to detail the evolution of the Galileo system through its various phases of the program, both from technical, institutional and commercial viewpoints. Such a roadmap will de-risk the program and ensure that the necessary timescales and co-operations can be effective to create a stable Galileo Timescale well in advance of the system deployment. In addition, it will further engage the time and frequency community in the Galileo program. System performance and validation will depend heavily on a stable GST.

6.7 Calculation Period for UTC

At present, Circular T is calculated monthly and up to 6 weeks in arrears. This is a limiting factor in the predictions towards UTC for the GNSS Systems. There would appear to be no technical limitation, with modern communications and processing capability, why this period should not be calculated at two-weekly or even weekly intervals. Three important benefits would be:

i) the NMIs would be able to know their own performance against UTC much sooner and could introduce more frequent and more gentle steering corrections and hence achieve tighter maintenance of [UTC - UTC(k)]. With more frequent, smaller corrections, the predictions for UTC will also improve.
ii) Many NMI's go to great lengths in developing prediction algorithms to compensate for the latency of the "monthly block" calculation. Predictability is extremely good over periods of days but deteriorates more rapidly thereafter, and more frequent calculations of UTC would significantly reduce, or even remove this complexity
iii) The Galileo specification calls for UTC and GST to be kept within 50ns (95%) over any one year period and <28ns offset uncertainty (wrt UTC) when predicted forward 6 weeks. The implication is that GST must be able to predict this offset for 6-weeks in advance to ensure it is achieved in retrospect. Reducing the Circular T latency could therefore reduce the clock ensemble complexity and tighten the uncertainty figures in the Galileo and GPS specifications

A second conclusion is that Circular T will need the addition of the [UTC – GST] figures and the user community will benefit from [GPS – GST]. This will need an interface to and co-operation with BIPM.

6.8 Legal Time

GPS today is the most suitable mechanism for UTC dissemination of time on a global basis and its use will expand and further improve, but although widely used it has no legal standing. With the introduction of Galilleo, there is an ideal opportunity to establish a new legal time in the European Union, however, it should be recognised that a change of national law will be needed in most countries. On a global basis, this would be impractical. It would be impossible to change US Law for example to depend on a European system under industrial control.

Galileo could however, be regarded as the official mechanism to distribute UTC, but will it be stable enough in the early stages of operation? The answer could depend on the Time Service Providers ability to predict UTC.

6.8 GPS/Galileo Traceability

A GPS/Galileo traceability model has been produced between UTC (BIPM) / TAI and the Dual Mode GNSS Receiver.

If commercial GPS receivers are correctly configured, traceability is achieved through the various GPS data archives described above and the uncertainty can be estimated through manufactures figures or by calibration. Traceability for GPS is therefore secure, although this may be applicable to the US only. Some European NMIs publish the GPS timescale and satellite clock offsets against their UTC(K) which extends this traceability. For example, NPL do this in the UK.

The documentation studied for Galileo does not specifically address traceability, and areas of omission have been identified. The issue of traceability for Galileo should be included in the next phase of the Galileo workplan. Traceability for the deployed signals must be a co-operation between the Galileo Supervisory Authority, the Galileo Concessionaire, Time Service Provider, EU Labs and BIPM.

Future GNSS Developments to Improve Timing and Time Transfers

With the introduction of Galileo, the modernisation of GPS, and improved time and frequency equipment and transfer techniques, modernised GPS and Galileo will further contribute towards our control of time and frequency in numerous ways:

- Implementation of a second civilian frequency for GPS C/A code time transfers. A third frequency (L5) will help long baseline geodetic time transfers
- Higher chip rates on Galileo signals and GPS L2 and L5 will improve common view time transfer accuracy
- Use of SBAS (Satellite Based Augmentation Systems) systems (e.g. EGNOS, WAAS) can help stand-alone GPS time transfer to give the IGS final orbit accuracies, with the advantage of the SBAS orbits is that they are available in "real-time"

The limits will continue to be stretched in the time and frequency domain and the benefits will be global.

Acknowledgements

The author would like to thank the following for the contributions they have made that has helped enormously in writing this paper:

Dr. John Davies and Dr John Laverty UK's National Physical Laboratory, for their support in preparing the original MSc dissertation, and to Peter Whibberly, also of NPL for presenting this paper at EFTF'04.

John Loizou (Vega L^{td}) for providing the Navigation Engine Simulator and GPS/Galileo constellation files which were used in the analysis of GPS/Galileo time offsets.

References

[1] Dissertation for the Degree of Master of Science: Interoperability of the GPS and Galileo System Timescales for Positioning and Metrology, W M Forrest, August 2003

[2] GNSS II: Imperative for Galileo-GPS Rationalization and Harmonisation, R W Skinner, Lockheed Martin Space Systems Company

[3] *International Vocabulary of Basic and General Terms in Metrology*, second edition, International Organisation for Standardisation (ISO), 1993

DESIGN OF THE GALILEO PRECISE TIME FACILITY (PTF)

S. BEDRICH[1], A. BAUCH[2], J. LAVERTY[3], A. MOUDRAK[4], W. SCHÄFER[5]

[1]Kayser-Threde GmbH, Munich, Germany
email: spacetech@kayser-threde.de, phone: +49-89-72495-231, fax: +49-89-72495-104

[2]Physikalisch-Technische Bundesanstalt (PTB), Braunschweig, Germany

[3]National Physical Laboratory (NPL), Teddington, UK

[4]Deutsches Zentrum für Luft- und Raumfahrt (DLR), Oberpfaffenhofen, Germany

[5]TimeTech GmbH, Stuttgart, Germany

Keywords: Precise Time Facility, Galileo System Time, Galileo/GPS Time Offset, Failure Handling.

Abstract

The current paper gives an overview on the present design baseline of the Precise Time Facility (PTF) for Galileo. The major task of the PTF is to generate and maintain Galileo System Time (GST). The current design has been worked out during Galileo Phase C0 by a European team of experts and provides viable solutions for most of the requirements of this key element of Galileo.

1 Introduction

Since November 2003, Kayser-Threde GmbH together with its sub-contractors TimeTech GmbH, the UK National Physical Laboratory (NPL), the German Aerospace Research Centre (DLR) and the German Physikalisch-Technische Bundesanstalt (PTB) are designing the Precise Time Facility (PTF) for Galileo. The major purpose of the PTF is to generate, maintain and distribute Galileo System Time (GST). The PTF will consist of an ensemble of atomic clocks with appropriate time and frequency (t&f) measurement equipment and S/W filtering algorithms. During the Galileo In-Orbit Verification (IOV) phase, characterised by a reduced number of spacecraft, one PTF is planned to be physically implemented, whereas for Galileo Final Operational Capability (FOC) two identical and synchronized PTFs at two different sites are foreseen.

A preliminary PTF turn-key architecture has been proposed in 2003 and further detailed during the Galileo Phase C0 study in 2004 by the team. The architecture is based on the results of the earlier Galileo system studies and trade-offs, in particular those of Phase B2, and the segment requirements derived thereof. In addition, our turn-key design proposal covers a number of features which are deemed key to successful and timely procurement, installation, and operations of the PTF during IOV and FOC.

Concerning the technical PTF requirements and its present design baseline, the PTF redundancy mechanisms are considered more critical than the GST generation and steering algorithms, where valuable experience exists world-wide at National Metrological Institutes (NMI). Several outstanding critical issues refer to parametrization of the GST algorithms and their operational implementation (e.g. ensemble algorithm, redundancy switching). The paper reports about the team's PTF turn-key design concept and on current results of the critical algorithm assessment activities.

2 Technical Requirements

The ground time system for Galileo which will generate Galileo System Time (GST) has two primary functions:

1. Navigation timekeeping: this function is critical for fulfilling the navigation mission and is needed for Galileo spacecraft orbit determination and time synchronisation (ODTS);
2. Metrological timekeeping: this function is uncritical to the mission, but needed to steer GST towards International Atomic Time (TAI) and to provide the UTC timing dissemination service to the user.

Navigation timekeeping is the core task of the PTF, whereas the metrological timekeeping shall be performed by an external Time Service Provider (TSP).

2.1 PTF Tasks

The PTF is considered a key element of the Galileo Mission Segment (GMS) and shall, therefore, be physically implemented at the Galileo Control Centre (GCC). Based on this concept, all PTF tasks and top-level performance requirements are summarized in a hierarchical order in the following list:

1. maintain ensemble of ground clocks in well controlled environment, whereof one clock is referred to as Master Clock (MC)
2. measure all individual clock offsets w.r.t. the MC incl. covariance matrix through a local t&f measurement system
3. monitor and survey ground clock stability
4. compute Galileo System Time (GST)
5. provide GST parameters and clock data to the TSP
6. steer GST towards TAI by applying frequency steering corrections provided by the TSP
7. provide physical realization of GST (1pps, 10 MHz)
8. operate Galileo and GPS Time Receiver(s) driven by physical representation of GST
9. determine and provide GST/GPS Time Offset (GGTO)
10. link to the orbit determination and time synchronisation process (ODTS) by providing GST and by forwarding raw measurements and calibration data from Time Receiver(s)

2.2 PTF Requirements

In terms of performance requirements, the following list provides a comprehensive overview of all relevant figures as specified today.

1. GST Frequency Stability shall be optimised on short-term (τ = 1 day):

τ [sec]	ADEV
1	1.5 E-13
10	2.0 E-14
100	5.0 E-15
1000	2.0 E-15
10.000	2.0 E-15
> 86.400	1.0 E-14

Table 1: GST frequency stability requirements

2. GST Time Stability shall be optimised to medium-/ long-term (τ = 30 days)
3. Based on above two requirements, the design baseline after [1] assumes 1+1 Active H-Maser clocks (AHM) for short-term timing plus 3+1 high-performance Caesium (Cs) standards for long-term drift removal. This would result in a typical GST frequency stability performance as expressed in the following figure. The extra AHM and Cs clocks shall be included as hot spares.

   We propose 2+1 AHM which would allow immediate isolation of any faulty AHM.

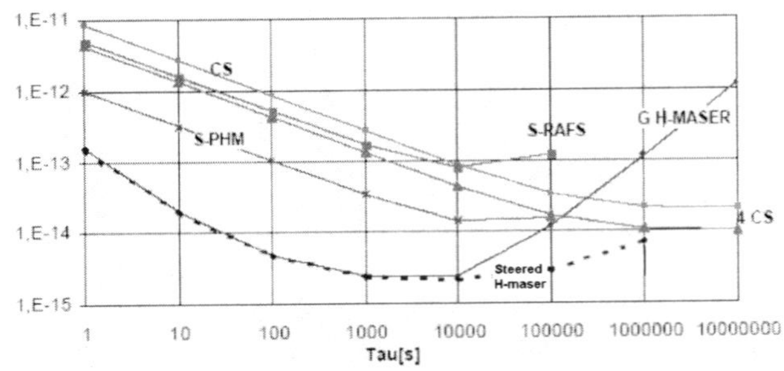

Figure 1: Specified GST time stability (dotted curve at bottom) based on clock ensemble of 1 AHM + 3 Cs (G- = ground, S- = space)

4. GST Frequency Offset (normalized to TAI): <5.5 E-14 (1 day)
5. GST Limits: accuracy of GST – TAI offset <50 ns, 2σ (95% of any yearly interval)
6. GST Uncertainty: precision of GST – TAI offset <28 ns, 2σ (95% confidence level)
7. GST Autonomy: accumulation of less than 28 ns (2σ) uncertainty over 10 days (95% conf. level)

To achieve above specified performance, a dedicated PTF environment is necessary which basically should consist of a separate clock operations room with the following characteristics: room temperature 25+/-1°C, temperature slope <1°C/h, humidity 50+/-10%, dedicated means for magnetic shielding, air filtering and corrosion prevention, uninterruptable power supply for at least two consecutive days.

2.3 TSP Tasks

The main reason to involve an external TSP is simplification of the PTF core timekeeping system by outsourcing the metrological functions and relying on an – already existing – infrastructure.

It will enable institutional national timing labs UTC(k) and commercial providers to research and improve GST generation independently and to find the best approach for UTC dissemination via Galileo with the FOC system.

This will most probably result in considerable cost savings for the PTF itself. Moreover, such shared task allocation will enable to continuously improve the timing service of Galileo without frequent changes of the core infrastructure, e.g. by involvement of new and better (external) clocks or better TAI prediction algorithms.

This in turn will support the inevitable comparison with the ongoing and planned GPS modernisation initiatives in the US and will help to stabilise and promote GST in the long-term. It is assumed that the challenge of an ever improved GPS time in the coming years can be faced best by such work share between PTF and TSP and still create long-term confidence to the Galileo user in the quality and performance of the Galileo time product and services.

Based on these assumptions, all TSP tasks are summarized in the following list:

1. install and operate two-way satellite t&f transfer (TWSTFT) and common-view (CV) equipment at the PTF, which shall be driven by the physical realization of GST
2. operate daily links to n UTC(k) labs (at least three; required for reliable determination of TAI offset) incl. periodic calibration of equipment as well as provision and/or leasing of external infrastructure (e.g. satellite transponder time)
3. receive GST parameters, individual clock data Ck_i and clock ensembling information from PTF
4. perform data analysis of all measurements GST – Ck_i and GST – UTC(k)
5. develop and operate TAI_p prediction algorithm
6. provide daily predicted value of (TAI_p – GST) time and frequency offset and daily frequency steering correction to PTF
7. provide current value of UTC – TAI time offset (leap seconds) to PTF
8. interface with BIPM by exchange of all relevant clock data
9. support extended scientific activity

The following figure summarizes the most important PTF and TSP tasks and gives an overview of the relations between PTF and TSP.

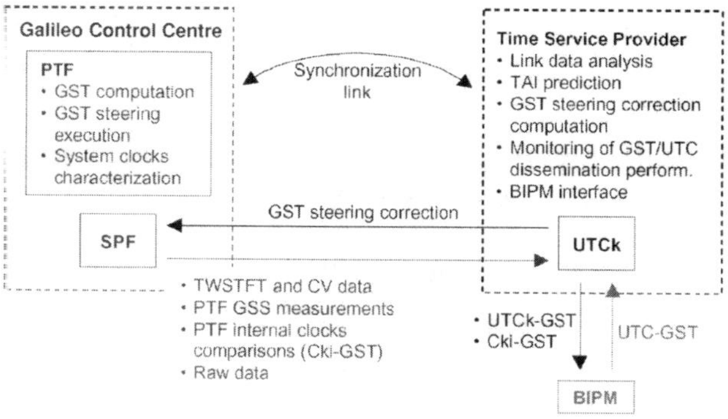

Figure 2: Task allocations of PTF (left) and TSP (right) after [1]. Links between PTF and TSP. SPF = "Service Products Facility". I/F to GPS (USNO) not shown here.

2.4 Requirements for IOV

Concerning the requirements and the design baseline as defined in Phase B2 for Galileo IOV, [2] states that „due to the equivalent number and type of operative clocks in the PTF (i.e. 3 Cs and 1 AHM), which will be used both in FOC and IOV, the same algorithm and the same stability performances are expected both in FOC and IOV ... The following table summarises the functions that can be tested during IOV and the level of verification achievable."

Function	Performance level achievable at IOV
GST Generation	FOC
Steering to TAI	To be agreed with TSP
Determination of GGTO	~ FOC
Dissemination of GST to fixed user	~ FOC
Dissemination of UTC	Depending on TSP
Switching from PTF1 to PTF2	Depending on UTC(k) lab to be used as PTF2

Table 2: PTF functions testable during IOV after [2]

The above statements are valid as far as GST performance is concerned. As far as important other functions are concerned, in particular redundancy and master clock switching mechanisms, the PTF architecture baseline for IOV is deemed not sufficient. Assuming an adequately equipped and experienced UTC(k) lab with NMI functions is involved as TSP, it would have to burden the three-fold work load of acting as TSP with all related tasks, providing and operating back-up master clock(s), and still providing national metrological service in the t&f area.

Since, moreover, the PTF redundancy mechanisms are considered more critical than the GST generation and steering algorithms, where valuable experience exists world-wide, the presently defined focus of the PTF related tasks during IOV should be reconsidered carefully.

Alternatively, facilities, equipment and operational expertise available in non-NMI time labs and industry incl. the ESA deep space tracking facilities in New Norcia or Cebreros (AHM, CV/TWSTFT equipment, etc.) could be re-used to test:

a) phase-synchronous switching between the master and the hot-spare AHM within the master PTF,

b) steering of the slave to the master PTF and switching from the master to the slave PTF (will have non-neglegible procedure/effect on ODTS).

In addition, it is considered inevitable to evaluate operational PTF procedures and to gain hands-on experience with clocks and timing equipment at the earliest possible state of implementation. According to NMI experience, such processes generally last several years, in particular when covering detection of yearly changes to clocks, facilities, etc. It is, therefore, recommended to put in operation both the master PTF and the TSP already for IOV.

2.5 Architecture Baseline

To fulfill the above objectives, the Galileo PTF should consist of the following major components/sub-systems:

1. Clock ensemble;
2. Local t&f measurement subsystem, consisting of:
   - core measurement equipment;
   - support measurement equipment.
3. GST/GGTO data processing subsystem;
4. GST realization subsystem;
5. T&F comparison subsystem (synchronization links);
6. Time data interfaces subsystem, made up of:
   - external communication I/F and
   - data storage/archiving facilities;
7. Monitoring & Control (M&C) subsystem, incl. environmental sensors.

The following figure shows a block diagram of the PTF architecture covering any of the above mentioned sub-systems/major equipment.

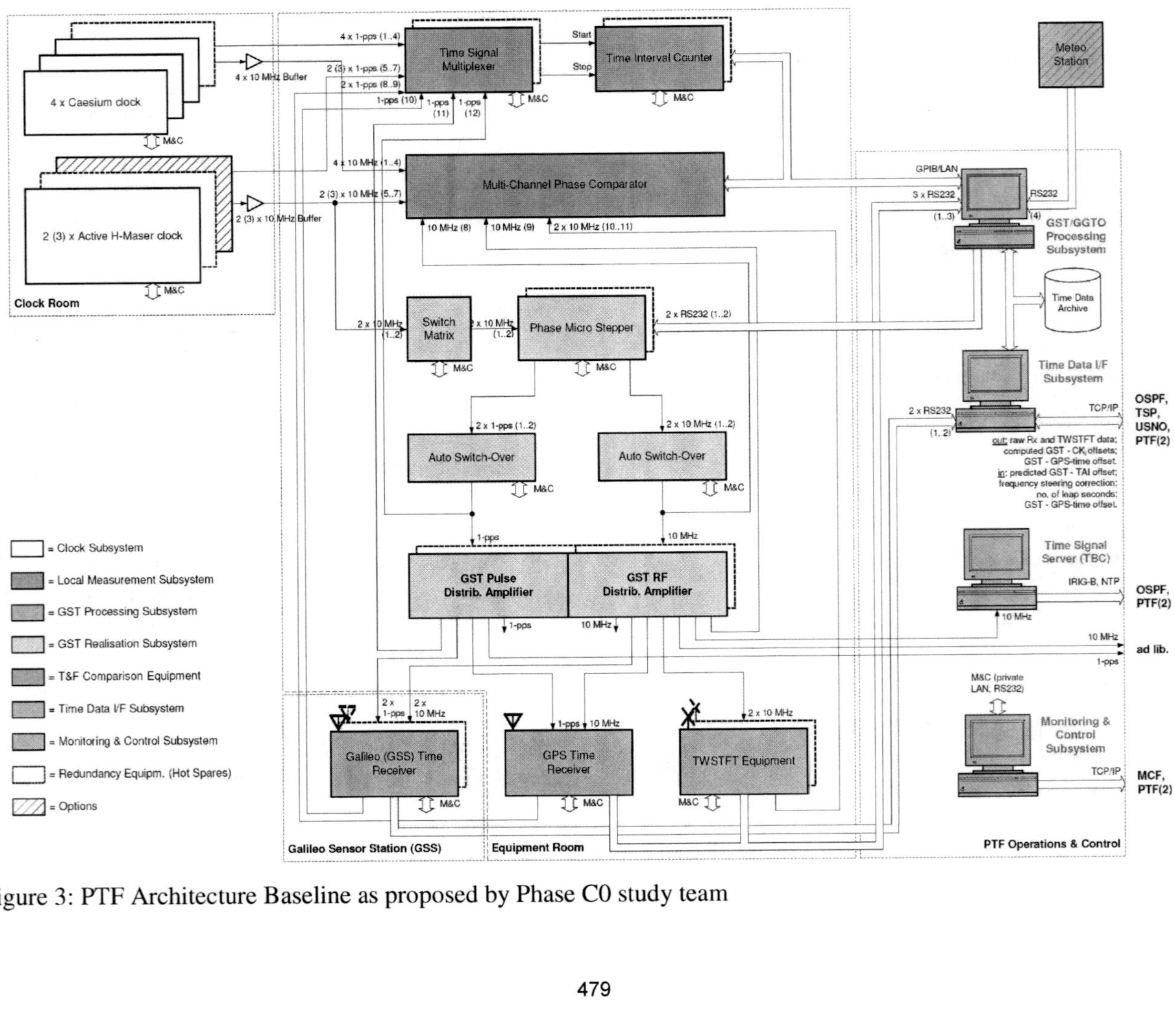

Figure 3: PTF Architecture Baseline as proposed by Phase C0 study team

3 Algorithms Baseline

3.1 GST Generation and Steering

Preliminary analysis of the GST generation and steering algorithms worked out in Phase B2 and GSTB-V1 has lead to the conclusion that the solution proposed in the baseline is sound and feasible.

GST generation is described in detail in [1] and [3]. The basic approach is to establish a physical realisation of GST by steering the output of an AHM operated at PTF to

a) TAI with the help of a steering correction provided by the TSP, and/or

b) the free-running ensemble timescale (GSTR) computed from the Cs clocks operated at the PTF.

[1] defines that in IOV configuration only one PTF should be operational, equipped with 1 AHM and 3 Cs clocks. Considering this configuration, the generation of GST can be represented at the conceptual and physical levels as illustrated in Figure 4 and Figure 5. Note that neither cabling, networks, processing units nor the Monitoring and Control Subsystem are shown.

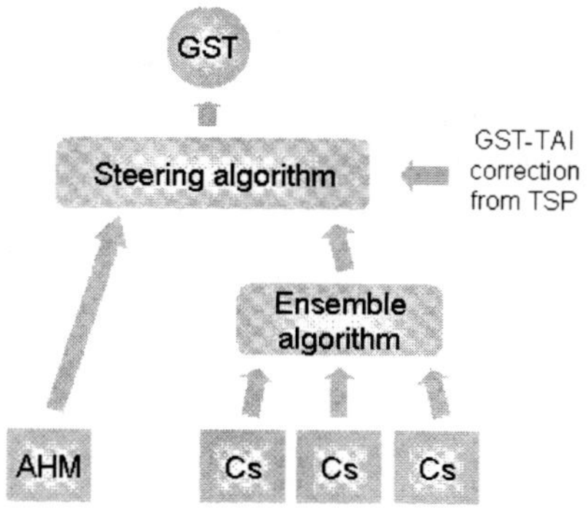

Figure 4: GST generation/steering chain: conceptual level

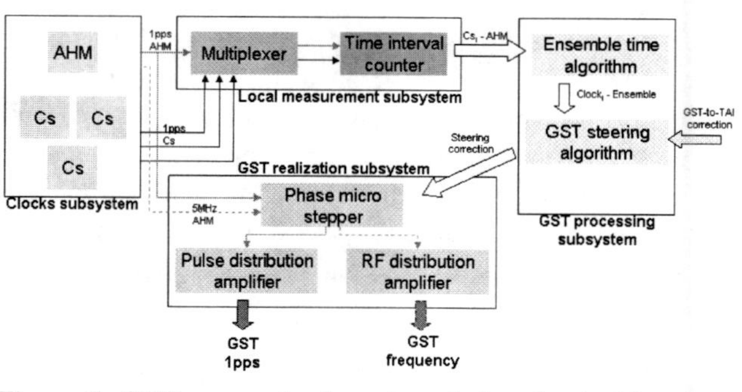

Figure 5: GST generation/steering chain: physical level

GST generation in FOC will follow the same scheme as in IOV, but will involve redundant elements [1; 3]. The most important of these elements are: second AHM clock, fourth Cs clock, second micro phase stepper, redundant cabling. We propose 3 AHMs which would allow immediate isolation of any faulty AHM. Since two PTFs shall be operated, dedicated algorithms and procedures to manage the redundancies (failure detection, switching procedures etc.) will be required for. These are discussed in chapter 4.

3.2 GGTO Determination

Since Galileo will rely on its own internal time scale GST, but future users most likely will (have to) handle a combination of both GPS and Galileo data, it has been decided to include a common pre-determined Galileo/GPS Time Offset (GGTO) parameter in the navigation messages of both systems. This will enable users with limited satellite visibility (e.g. two Galileo, two GPS) to process valid solutions without the need to solve for the time unknown.

A basic GGTO determination scheme as worked out in Phase C0 is shown in the figure below. It uses data of a GPS time receiver and TWSTFT equipment both installed at the PTF and driven by GST. More details about the GGTO algorithm can be found in [5] and [6].

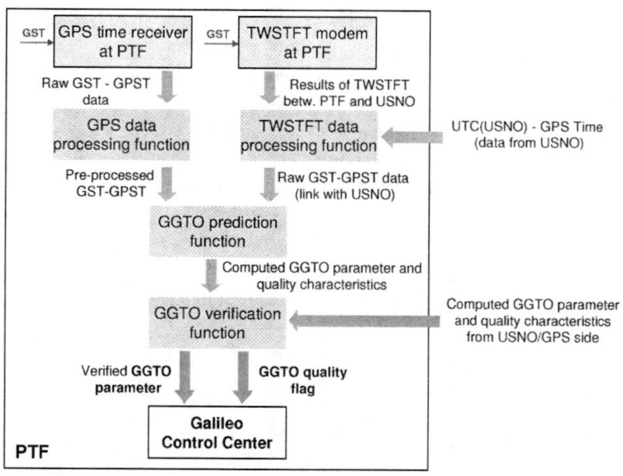

Figure 6: GGTO determination scheme as proposed by Phase C0 study team

4 Failures and Redundancies

4.1 Failure Identification

The following table illustrates possible failure events on PTF component level relevant to the GST and GGTO generation chains in the context of the present PTF redundancy concepts defined for IOV and FOC. The events were identified based on operational experience in NMIs and/or time laboratories.

The probability of events is designated as follows: "M" – medium (in order of a few times a year), "L" – low (in order of one or two times during the component lifetime). The probability of certain algorithm failures can presently not be estimated due to the lack of knowledge on failure detection procedures in these algorithms and their verification. Moreover, the failures/malfunctions of the following components are presently not considered:

- cables (which typically do not produce long-lead failures if they were initially properly tested);
- data collection/device control S/W (needs to be included in further detailed analysis).

Component	Pr	Event
AHM clock	M	Temporary malfunction (phase or frequency step) affecting both 1pps and RF outputs
	M	Degradation of performance (increase of frequency drift, increase of ADEV) affecting both 1pps and RF outputs
	L	Failure of single output (either 1pps or RF output is not available)
	L	Device failure (all 1pps and/or RF outputs are not available)
Cs clock	L	Temporary malfunction (phase or frequency step) affecting both 1pps and RF outputs
	L	Degradation of performance (increase of frequency drift, increase of ADEV), affecting both 1pps and RF outputs
	L	Failure of single output (either 1pps or RF output is not available)
	L	Device failure (all 1pps and/or RF outputs are not available)
Multiplexer	M	False switching (wrong input is connected to the device output)
	L	Degradation of performance of single input (e.g. loose contact on relay)
	L	Failure of single input (one 1pps or RF output is not available)
	L	Device failure (multiplexer not working)
TIC	L	Degradation of performance
	L	Device failure (TIC not working)
Micro phase stepper	L	Degradation of MPS output performance (e.g. increase of noise due to MPS H/W problems)
	L	Failure of one MPS output
	L	Device failure (MPS not working)
GST Pulse distributor	L	Degradation ofperformance of one pulse distributor output (e.g. increase of noise in terms of ADEV)
	L	Failure of one distributor output
	L	Device failure (distributor not working)
GST RF distributor	L	Degradation of performance of one RF distributor output (e.g. increase of noise in terms of ADEV)
	L	Failure of one RF distributor output
	L	Device failure (distributor not working)

Table 3: Possible failure events on PTF components (list not exhaustive)

4.2 Failure Handling

Failure handling covers high-level measures that should preferably be undertaken automatically at the PTF following detection of a failure in the GST generation chain (maser-phase stepper-distributor-receiver) to restore generation of GST and to eliminate any impact on GST performance. As mentioned above, we consider redundancy handling issues to be a critical point in GST generation, therefore we propose to test the corresponding mechanism already during IOV to ensure their proper functioning at and after reaching FOC. Thus, the PTF configuration would be in IOV basically the same as in FOC (only one Cs clock less). The switching between the PTFs could be tested in collaboration with an existing time laboratory or NMI.

The current baseline [1] foresees the failure handling to be executed at the level of individual PTF components. This strategy implies the ability to detect a failed component and to execute the switching or other response measures within such a short interval of time that would make the failure non-critical for the overall system performance.

It is understood that this interval of time (which is basically the requirement of maximum Time-To-Repair for GST) is set mainly by the ODTS processing. Assuming that the navigation processing is executed in ODTS in 10 minutes batches, and the update (upload) rate of satellite navigation messages is 90 minutes, we consider the Time-To-Repair of GST to be equal to the interval that would lead in the worst case to a loss of two processing batches. Note that in the current baseline the PTF is not included in the integrity chain, therefore we do not consider here requirements w.r.t. the integrity processing which would lead to Time-To-Repair of GST of less than 1 s.

Figure 7 illustrates the logic of the switching and steering mechanism as proposed in Phase B2. It is foreseen that one of the H-masers (AHM1a in Figure 7) represents GST, thus being the Master Clock (MC). The other AHM is working as hot spare MC, and the second PTF is operated on a master-slave basis with the primary one. Note that titles as "MC" or "primary" refer only to the role of a certain element and does not constitute a permanent designator; these roles may change during Galileo operations as a result of switching between individual elements or components.

We found that in real operations it will be hardly possible to make all switching operations automatic. It would be possible to some extent to automate the detection and corresponding switching for failures related to the lack of output signals on PTF components. However, such failures as performance degradation of a component or situations when a planned maintenance should be executed would involve certain manual operations which would require trained personnel.

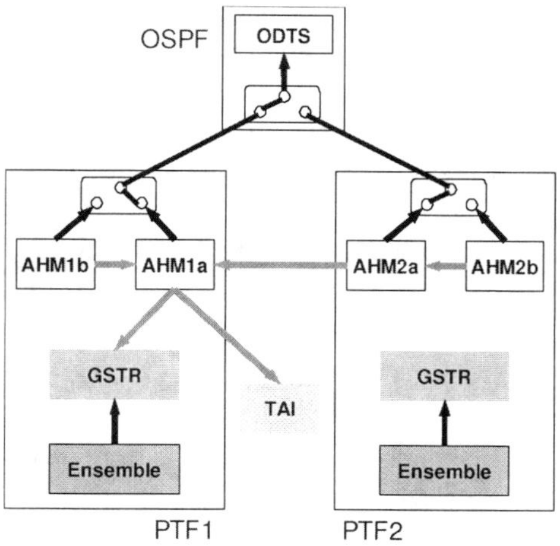

Figure 7: Baseline PTF switching and steering mechanism in FOC after [1]

Following these guidelines, our Phase C0 approach is deemed to be more reliable and much simpler in operation than the Phase B2 baseline. We propose to place the switch inside the ODTS functions (Figure 8). This way, the switching process would be reduced to solely informing ODTS that switching has occurred and which AHM is the (current) MC. ODTS should then derive GST from the corresponding data set.

In fact, each of the AHM could be connected to its "own" Galileo Time Receiver (collocated within each PTF) and the data from all four receivers (two at each of the PTFs) could be delivered to ODTS via standard data transfer interfaces.

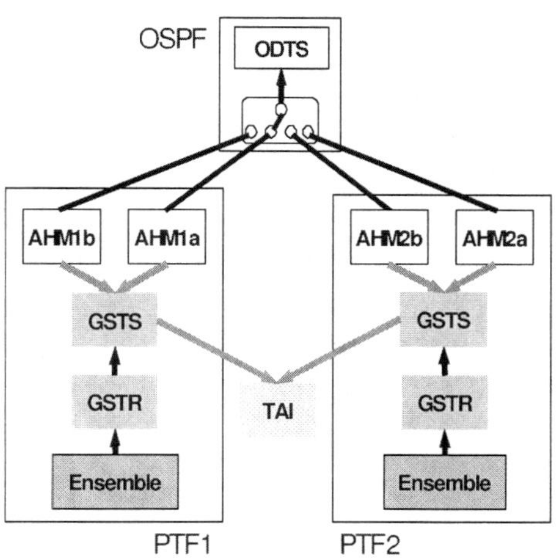

Figure 8: Alternative PTF switching and steering mechanism as proposed by study team

The overall switching logic is rather simple leaving individual modules and steering schemes unaffected by failures in other modules. Note that all AHM are steered to GSTR (which is the free-running ensemble time) as produced at the PTF, and that each GSTR is individually steered to TAI. Moreover, this concept allows a much higher flexibility and a simple, but powerful scalability throughout the various Galileo implementation phases.

To avoid phase and frequency steps in GST following any switching, it is advisable to use a GST model (e.g. a linear polynomial) within ODTS. The model should be activated following a switching to compensate the GST step. Its coefficients (time and frequency offsets between the AHMs) can be determined from PTF measurements and ODTS processing results.

Hardware realisation of such switching scheme within one PTF is illustrated in Figure 9 showing the individual GST generation chains (AHM-stepper-distributor-receiver). Note that there is no H/W switching element on PTF side, thus following a failure in one of the GST generation chains, personnel will have rather moderate time constraints for organizing repair or replacement of failed components. The system could always run without impact on GST performance (but with reduced reliability until the failed chain will have been repaired). Absence of H/W switching elements simplifies PTF operation and increases system reliability. This design has also benefits from the AIV point of view as each module can be built and tested separately without disturbance.

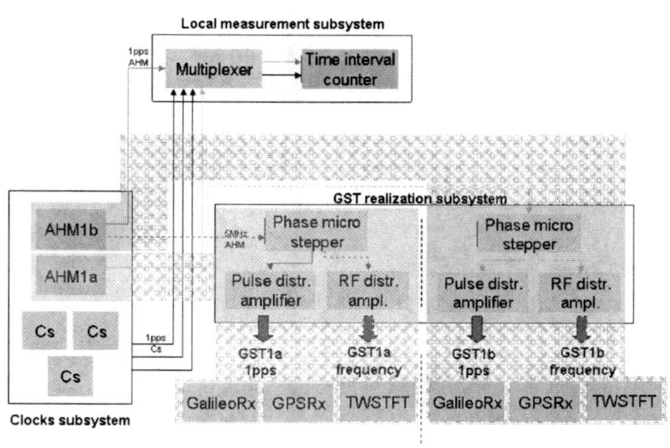

Figure 9: Independent hardware chains for each AHM

There is a need to define where the decision on selection of the master clock and on switching occurs. PTF is not a good candidate, since a) the two PTFs may come to contradictious decisions, and b) it is desirable to have a human supervision of the decision making. Thus, it is advisable to allocate the responsibility for GST definition and switching to one of the manned facilities of the GCC. PTF should provide comprehensive clock monitoring data and warning messages to support the decision making.

5 Summary and Conclusions

The main recommendations worked out in the PTF Phase C0 design study can be summarized as follows: due to limited experience on implementation of automatic redundancy/failure management systems, it is advisable to implement and test the corresponding mechanisms and algorithms already during IOV. Otherwise, there is a danger to enter FOC with a not thoroughly tested PTF which may fail to meet the requirements on availability and reliability of GST.

The mission of IOV critically depends on the availability and performance of GST. From practical experience it is known that the most critical PTF components are at the same time long-lead items (6 months or more). Moreover, they are produced only by very few manufactures world-wide. In particular the work horse of GST – the active hydrogen maser – is known to be more a scientific than a commercial instrument, whose failure modes are not very well studied. Therefore, it is advisable to have hot spares for critical PTF components including the AHM already in IOV. The impact of the procurement of PTF hot spares on the overall IOV costs is estimated to be rather moderate.

The Phase B2 baseline concerning PTF redundancies and switching which foresees physical steering of all back-up AHMs to the MC is rather complex in implementation and operation. Switching would in addition require skilled personnel to undertake certain activities within a rather short reaction time.

Our alternative approach seems to be more attractive in terms of reliability and simplicity of operation. In this approach each GST generation chain (maser-phase stepper-distributor-Galileo receiver) will be operated separately and deliver its data continuously to the ODTS function. ODTS will then derive GST from the data relating to the AHM currently assigned as the MC. No hardware switching is needed and no immediate actions on PTF hardware are required to be undertaken by the GCC personnel to realize the switching. The switching itself is limited to some actions within ODTS which would have to derive GST from another set of observations corresponding to the new MC.

The Phase B2 baseline architecture foresees physical steering of all AHMs to the (current) master one. The stringent steering requirements can be hardly met due to intrinsic calibration errors of the measurement equipment. An alternative solution is to steer all AHMs to individual free-running ensemble time scales GSTRs produced at corresponding PTFs, and steer these GSTRs in turn to TAI. As a result, all AHMs will be kept very close to each other. To compensate for steps following any switching between the AHMs, a GST model (e.g. a linear polynomial) might be used in ODTS. Coefficients of this model can be determined using both the results of ODTS processing and the clock measurements made at the PTFs.

In any case it is desirable that ODTS provides a feed-back to PTF about its estimates of the quality of the PTF raw data and the quality of GST, since these effects are not visible to PTF. This information would complement clock monitoring made at PTF and assist a reliable real-time assessment of the quality of GST.

Acknowledgments

The work described in this paper has been funded by the European Space Agency (ESA).

References

[1] PTF Concept File (Phase B2 Consolidation), ESA document, Nov. 2002.

[2] Anlysis of IOV System Configuration (Phase B2 Consolidation), ESA document, Oct. 2002.

[3] Galileo System Time Generation (Phase B2 Consolidation), GSTGENALG ID.48, Iss. 7, ESA document, Oct. 2002.

[4] Galileo Mission Segment Requirements (Phase C0), GMSREQ, Iss. 9, ESA document, March 2004.

[5] Moudrak, A., et al., Determination of GPS/ Galileo Time Offset to Support the System Operability, accepted for publ. in Proc. ENC-GNSS 2004, Rotterdam, May 2004.

[6] Bauch, A., et al., Time Comparisons between USNO and PTB: A Model for Determining the Time Offset between GPS Time and the future Galileo System Time, accepted for publ. in Proc. IEEE Freq. Contr. Symp., Montreal, Aug. 2004.

GPS and Galileo UTC Time Distribution

Edward Powers*, Jörg Hahn †

*United States Naval Observatory (GPS)
† European Space Agency (Galileo)

Keywords: Galileo, GPS, UTC, Time, Navigation, GGTO,

Biography

Edward D. Powers received his B.S. and M.S in Electronic Engineering and Instrumental Science from the University of Arkansas in 1984 and 1987 respectively. In 1987 he joined the Naval Research Laboratory as an engineer working on GPS clock development program. In 1997 he joined the USNO and now works as the senior engineer responsible for development of improved precise time synchronization and GPS timing.

Jörg H. Hahn received his M.Sc. in Physics and Mathematics from the Belorussian State University, Minsk in 1993 and his Ph.D. in Engineering Sciences from the University FAF, Munich in 1999. In the year 2000 he joined ESA's Galileo Project Office as a system engineer and is now responsible for all Galileo End-to-end Performances.

Abstract

Precise timing is an inherent part of Global Navigation Satellite Systems (GNSS) like the Global Positioning System (GPS), Glonass and the future Galileo. GPS has been providing a UTC (Coordinated Universal Time) time distribution service since its inception. This paper will outline present performance of GPS UTC time distribution and show projections for Galileo and the planned modernized GPS UTC distribution performance. This paper will also outline efforts to harmonize the underlying navigation time scales of both GPS and Galileo to better facilitate a combined Navigation solution. A Preliminary Galileo to GPS Time Offset (GGTO) Interface Control Document (ICD) has been drafted and will be outlined as part of this presentation.

1 Introduction

Existing Radio Navigation Services (RNS) like Loran, Global Positioning System (GPS), and Glonass are based on providing precise time and frequency synchronized ranging signals. Because of the exploitation of very precise timing signals these RNS systems are used to provide both navigation and time distribution services. Typically the navigation mission only requires relative timing but by providing an absolute timing reference a RNS system may also act as a time distribution system. Interoperability between RNS systems can be enhanced by coordinating the timing references of radio navigation systems.

2 Coordinated Universal Time (UTC)

The International Bureau of Weights and Measures, "Bureau International des Poids et Mesures" (BIPM) in France is charged with providing the standard UTC for the World. The BIPM receives data contributed from more than 200 atomic clocks and a few primary ("absolute") frequency standards, from over 50 institutions around the world. Once a month these data are used to produce the standard international references for frequency and time, International Atomic Time (TAI) and UTC. UTC is equal in rate to TAI, but adjusted by an integer number of leap seconds to account for variations in the rotation of the Earth. Real time realizations of TAI and UTC are produced at most of the contributing laboratories.

3 GPS Timing Service

The GPS is not only a high accuracy navigation system, but it also delivers time globally with unprecedented accuracy. Free to the user, GPS is presently the world's most accurate globally available one-way source of time and frequency. The GPS system was declared fully operational in 1995 and has revolutionized both navigation and timing. Dual use of GPS (military and civilian) was announced as a goal in the wake of the 1983 downing of the Korean Airlines Flight 007, and civilian users now far outnumber military users. In addition to navigation, the GPS system has also had a huge impact on the use of precise time.

Each GPS satellite contains several atomic clocks and continually broadcasts its position in orbit above the Earth and timing corrections relative to a common time scale (GPS Time). A GPS receiver tracking at least four GPS satellites can solve for the receiver's unknown position (x, y, z) and time (t) at virtually any location on the globe with a precision of a few meters and a time error of a few tens of nanoseconds (excluding receiver calibration errors). A timing user operating from a known fixed location can derive time from GPS using as few as one satellite and with local averaging of GPS errors, a timing accuracy of a few nanoseconds are possible. Time from GPS is now used for many civilian purposes, including synchronization of communications systems, cell phones, and power grids, and also for many commercial applications where accurate time tagging is becoming increasingly important.

The U.S. Naval Observatory (USNO) has been designated [1] to provide the external coordinating time scale for all navigation systems operated by the United States and produces a real time realization of UTC and TAI as a contributing laboratory to the BIPM. This time scale is identified as UTC(USNO) and Figure (1) shows the accuracy over the previous eight years. UTC(USNO) is the external UTC timing reference for GPS, and consequently the UTC time derived from GPS is considered fully traceable to all international standards for civil and legal time, subject to user equipment errors.

A GPS timing user can obtain UTC time directly from GPS by using the UTC timing corrections broadcast in the navigation message of the GPS satellites (Subframe 4, page 18). The largest uncontrolled error for an L1-only C/A code user is the remaining ionosphere propagation error that cannot be completely removed using the simple Klobuchar Model. Currently only military users and a small group of civilian codeless users make use of the second GPS frequency at L2 (1227 MHz), which allows for the direct estimation of the user ionosphere error. This ionosphere error limits timing accuracy to a few tens of nanoseconds and positioning accuracies to around ten meters.

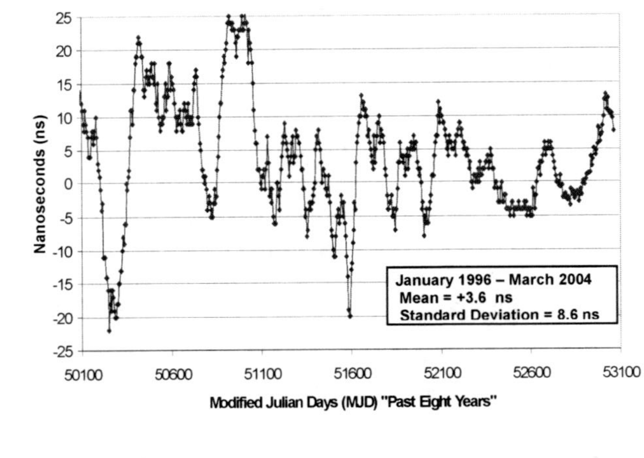

Figure 1 : UTC - UTC(USNO)

In early 2005 GPS will introduce its first IIR-M (modernized) GPS satellite that will broadcast a new civil L2 signal which will not be encrypted and is available to all users free of charge (like the present GPS L1 C/A code signal). Then in 2006, with the first GPS IIF satellite launch, a third civil signal called L5 will be added. This new L5 signal has higher power and will provide a third redundant signal. Since a timing user at a fixed location only needs a single GPS satellite to obtain time, the benefits of these new signals will be immediate.

Each GPS satellite includes in its navigation message a prediction of its position, an estimate of the difference between its internal clock and GPS system time, and an estimate of the difference between GPS system time and UTC. The GPS Master Control Station updates these predictions approximately once a day. GPS Time is the internal GPS navigation time scale, which is not adjusted for leap seconds, and which is very gently steered to UTC(USNO) modulo 1 second. GPS Time is specified to be maintained to within one microsecond modulo integral seconds; and for the past eight years it has been maintained to within (+/-25 ns) of this goal. A GPS timing user obtains precise time by first computing the user position and time offset relative to GPS Time, and then applying the UTC correction terms including the leap seconds offset. The performance of both timescales is shown in Figures 2 and 3. The daily averages of broadcast estimates of UTC(USNO) are generally precise to a few nanoseconds (ns) RMS relative to UTC(USNO) and UTC(USNO) has been within 9.3 ns RMS of UTC(BIPM) for the past eight years. Thus the UTC time obtained from GPS has been accurate to better than 9.4 ns RMS over this period.

Once GPS modernization is completed, more rapid updates of the GPS navigation message will be possible which will improve both timing and navigation accuracy. Additional improvements are expected with the implementation of an expanded network of ground monitor stations and improvements to the GPS Operational and Control Segment's orbit and clock determination software. Improvements to UTC(USNO) are also planned with RMS accuracies expected to be less than 5 nanoseconds. In fact for the last three years UTC(USNO) has already achieved an RMS of 5 nanoseconds.

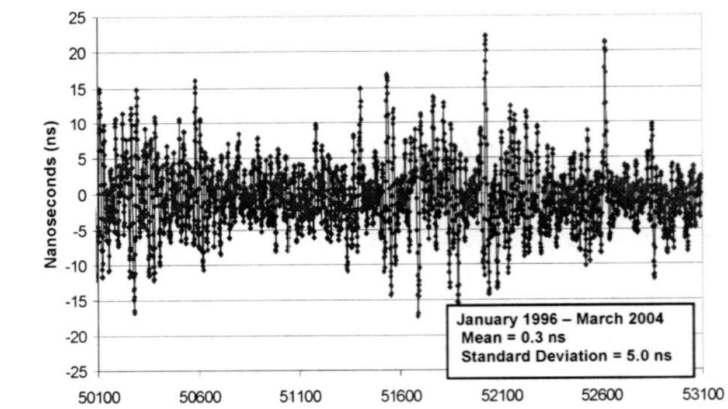

Figure 2: Daily averages of UTC(USNO) – GPS Time (module one second)

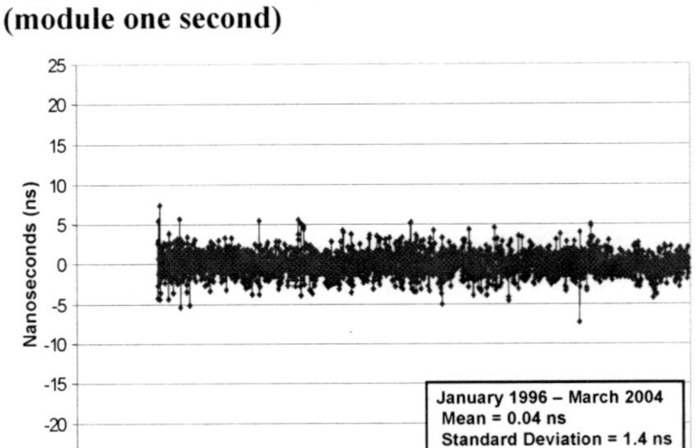

Figure 3: Daily averages UTC(USNO) via GPS – UTC(USNO)

4 Galileo and Galileo Timing

In partnership the European Commission (EC) and European Space Agency (ESA) are now in the Development and In-Orbit Validation Phase of a new Radio Navigation and Timing Service called Galileo. Galileo is specified as a high-accuracy navigation system providing integrity to Safety-of-Life critical users and accurate time information on a global scale. Together with GPS it will become a second source of accurate time. Galileo's final operational capability is envisioned for 2010.

Each Galileo satellite will contain several atomic clocks (2 Rubidium, 2 passive Hydrogen masers) and will continually broadcast its position in orbit above the Earth and timing corrections relative to Galileo System Time (GST). The principle application of Galileo's navigation signals is to calculate a position and accurate time identical to the one used with GPS.

Two fundamental system functions are concerned with timing. The "Navigation Timing" is critical for the entire navigation mission. It will be needed for orbit determination/ prediction and internal satellite clock synchronization. This function is not directly intended for timing applications.

The "Metrological Timekeeping" may not be critical for navigation, but is required to provide TAI and UTC timing services (time dissemination) in support of communication systems, banking, power grid management, etc.

Galileo System Time (GST), modulo one second, will be steered to a prediction of TAI obtained through an external timing service provider. Galileo System Time will be kept to within 50 ns (95%) of TAI over any one-year time interval. The offset between TAI and GST will be known with a maximum uncertainty of 28 ns (2-sigma), assuming the estimation of TAI six weeks in advance. See Figure 4.

Users equipped with a Galileo timing receiver will be able to predict UTC to 30 ns for 95% of any 24 hours of operation.

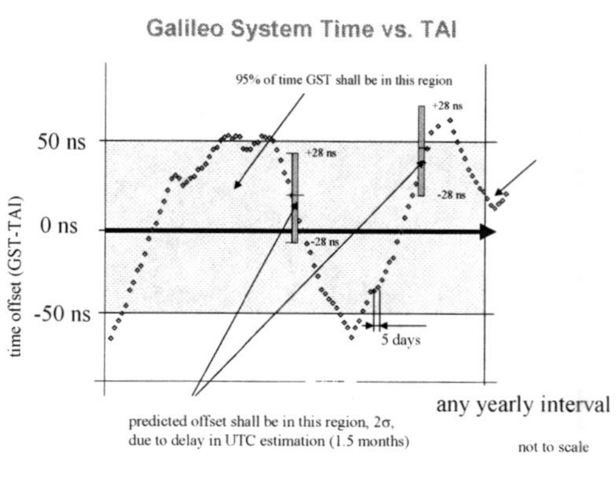

Figure 4: Illustration of the GST vs. TAI specifications

The key timing element in the Galileo system architecture will be the Precision Timing facility (PTF). This element is in charge of the navigation timekeeping and will:

- Maintain a stable ensemble of clocks in a well controlled environment;
- Measure the time offsets of all the clocks compared to the master clock through the local measurement system;
- Compute Galileo System Time (GST);
- Perform Space and ground clock monitoring and surveillance;
- Steer GST towards TAI through the steering correction provided by the external Time Service (capability of being autonomous from TSP for 10 days);
- Provide GST to the orbit determination and time synchronization process.

The metrological function and associated performance will be provided by an external Timing Service and will:

- Operate the daily links to UTC(k) laboratories required for the determination of TAI, the periodic calibration of the equipment and remote control facilities, etc.;
- Perform the data analysis of all the measurements GST-UTC(k);
- Develop and operate the TAI_p prediction algorithm;
- Provide Galileo with the daily predicted value of (TAI_p-GST) time and frequency offset and the daily steering correction;
- Interface with the BIPM by sending the internal clock data and GST-UTC(k) and receive from BIPM the Circular T (GST-$UTC(k)_{old}$);
- Provide data exchange under request from Galileo Control Segment (Two-Way Satellite Time Transfer (TWSTT) and GPS Common View);
- Provide an extended scientific activity, in collaboration with the leading European laboratories, to improve accuracy of GST (which is expected to exceed present specifications) and match the present and future accuracy of the USNO time scale.

An experimental version of the PTF called E-PTS is available through ESA's Galileo System Test Bed V1. First results are reported in [2].

5 GPS/Galileo Time Offset Message (GGTO)

During the US/EU GPS/Galileo Technical Working Group (TWG) discussions, both parties identified the dissemination of the GPS/Galileo Time Offset (GGTO) through the navigation signals of each respective system as a significant interoperability enhancement of the two systems. This led to the development of a Preliminary Interface Control Document

(ICD) to provide a starting point for developing an ICD for the GGTO.

The broadcast of the GGTO would enhance interoperability between the GPS and Galileo system. It can be anticipated that even with full GPS and Galileo constellations some users will experience limited visibility, for example when located in urban canyons. In such situations it is quite likely that as few as four satellites, not necessarily all of the same system, may only be visible to this user. Even if the geometry of the satellites were poor (high GDOP), the availability of any position estimate could be more important for these users than its accuracy. Thus the TWG set an early goal for the GGTO message to enable a navigation user equipped with a combined GPS/Galileo receiver to determine position with only a mixed combination of four GPS and Galileo satellites.

Benefits may be obtained from a combined GPS/Galileo solution as early as the Galileo In-Orbit Validation Phase (IOV) which will provide three to four additional navigation ranging satellites. During the IOV phase, the Galileo IOV satellites could broadcast the GGTO in the navigation message. All users of GNSS equipment would benefit from the new Galileo signals processed together with GPS signals. Receiver manufacturers and users all around the world (and not only in Europe) could take advantage of the Galileo satellites they have in view in the computation process of their position together with the GPS satellites.

Once Galileo is fully operational, the broadcast of the GGTO could be made by both the GPS and Galileo satellites, or by other communication means. Indeed, users needing this GGTO information are more likely to be located in areas where communications networks are already available. In such cases, the GGTO could also be disseminated by local GNSS augmentation services.

5.1 Key GGTO related Specifications

The following specifications have been agreed to among GPS and Galileo representatives:

Performances:

- The accuracy of the GGTO modulo 1 second shall be less than 5 ns with 2-sigma confidence interval over any 24 hour period. The proposed GGTO broadcast accuracy of 5 ns (2 sigma) is appropriate today. However, by the time Galileo becomes operational it is anticipated that an accuracy of 1 ns (2 sigma) will readily be achieved. Hence, the accuracy requirement will need to be revised in the future.
- The stability of the GGTO, expressed as an Allan deviation, shall be better than 8×10^{-14} over any one day. This implies an appropriate level of stability for the two system times and for the time transfer comparisons.
- The validity period of the GGTO shall be one day as a minimum.

Implementation:

- The GGTO shall be broadcast on all frequencies provided by GPS and Galileo navigation signals where feasible and appropriate. The Legacy GPS L1 navigation message is defined in GPS ICD-200. Because of backward compatibility concerns it might not be possible to broadcast additional information from the GPS L1 signal. The new GPS civil L2 and L5 signals provided as part of GPS modernization have new navigation messages that are now under construction. It is suggested that GPS and Galileo attempt to broadcast these corrections on all frequencies deemed feasible and appropriate.
- The GGTO shall be openly available to all users.
- The GPS "Improved Clock and Ephemeris" initiative (ICE) has a defined goal of improved navigation and timing clock message resolution of 0.03 ns Thus the GGTO message shall accommodate a range of $\pm 1\ \mu s$ with a resolution of 0.03 ns, and a rate of change of the GGTO $\pm 1 \times 10^{-12}$ s/s with a resolution of 4×10^{-16} s/s [TBC]. A quadratic term A2 is also being explored which can be used to account for frequency changes within a day.
- Preliminary sizing:
    - A0: 16 bits (~30 ps LSB) which allows for +/- 1 microsecond range
    - A1: 13 bits (~4E-16 sec/sec LSB) which allows for +/- 1E-12 range or 100 ns per day rate difference
    - A2: 7 bits (3E-16/day drift rate) which allows for ~30 ps/day drift rate difference.
    - T0: 8 bits to 14 bits, WN: 8 bits

- The determination of the GGTO shall be introduced:
    - for Galileo - starting with the first satellite launched;
    - for GPS - starting with the modernized GPS satellites. The first modernized GPS IIR satellite will be launched early in 2005 and initially may be broadcasting a C/A code like test signal with a legacy navigation message format. The GPS IIR satellites will later be switched to an improved civil code with a new navigation message after initial testing. The exact schedule is to be determined. The first fully modernized GPS IIF satellite will be launched in 2006 and will broadcast the new GPS L5. The exact schedule for the inclusion of the GGTO in the GPS IIF navigation message is not known, but it should be ready by the time Galileo is fully operational.

GGTO Estimation technique

- A GGTO measurement infrastructure is still being investigated with two approaches being explored.

1. *Physical measurement of the underlying timing references for GPS and Galileo using time transfer techniques such as TWSTT and GPS "Common-View".*

USNO maintains an estimate of the timing bias between GPS time and UTC(USNO). The timing difference between UTC(USNO) and most major National Metrological Institutes (NMI) in Europe is already routinely estimated. Through the Galileo timing service a traceable set of physical estimates of the timing bias between Galileo system time and UTC(k) would be maintained. With exchange of these measurements, the timing difference between GPS and Galileo could be estimated.

2. *Estimation of the GGTO within a combined GPS/Galileo monitor station receiver.*

A special dual tracking GPS/Galileo monitor station receiver would be developed to estimate the GGTO. The time offset between GPS and Galileo satellites would be measured within this special dual monitor receiver. Any biases between the GPS and Galileo tracking channels would need to be calibrated. Either broadcast clock corrections or corrections obtained through a cooperative interface would be used to derive an estimate of both GPS Time and GST.

It is anticipated that this equipment would first be installed as part of the Galileo test bed/IOV phase. It is important to make use of this existing infrastructure to validate the GGTO Signal in Space (SIS) Interface, which is anticipated to be the preferred GGTO interface in the long run.

- High-accuracy calibrated combined GPS/Galileo timing receivers shall be available to enable the GGTO to be measured from the SIS's. Assuming that computation of the GGTO will take place in GPS and Galileo before coordination, the proposed baseline method relies on the SIS of both GPS and Galileo. From that, the GGTO can be generated.
- An operational coordination interface between GPS and Galileo shall be implemented to agree on a unique value for up-linking to all the satellites. This implies a dedicated communication link [TBD] between GPS and Galileo. In the future, the process for determining the GGTO will be automated. A plan for developing this automated process will be required.

6 Conclusion

Today GPS serves the world as both the most widely available and the most accurate source of UTC time. For the past eight years UTC accuracies of 10 ns RMS (daily averages) has been demonstrated. It is anticipated that by the year 2010 when Galileo becomes operational Galileo will provide a UTC timing service equal to that of GPS.

Once Galileo is operational it is anticipated by many that most users will use a combined GPS and Galileo positioning, navigation and timing service. Interoperability between Galileo and GPS is enhanced by providing the user information about the navigation time scale offset between GPS and Galileo. Thus the US and EU have agreed in principle to provide this information through each system's navigation signals.

The opinions discussed in this paper are those of the authors and do not necessarily represent those of USNO and ESA and other agencies in charge of the GPS and Galileo programs.

References

[1] *2001 Federal Radionavigation System Document:* DOT-VNTSC-RSPA-01-3.1/DOD-4650.5, published jointly by the Department of Transportation and Department of Defense, December 2001, p.1-3. http://www.navcen.uscg.gov/pubs/frp2001/FRS2001.pdf

[2] *F. Cordara et al., The Galileo System Test Bed V1: First Experimental Results,* Proc. of the 18th EFTF 2004 Meeting, *Guildford, UK, April 2004, to be published*

THE GALILEO S STEM TEST BED V1 FIRST E PERIMENTAL RESULTS

F. Cordara*, R. Costa*, L. Lorini*, D. Orgia i*, V. Pettiti*, I. Sesia*, P. Tavella*, G. Graglia†, M. Mascarello†, M. Falcone, . Hahn****

*Istituto Elettrotecnico Nazionale (IEN) G. Ferraris , Torino, Italy, metf@ien.it
Alenia Spazio, Roma, Torino, Italy
**ESA/ESTEC Noordwijk, The Netherlands

Keywords Satellite Navigation, Atomic clock, Timekeeping, Time scale algorithms.

Abstract

The Experimental Precise Timing Station generating the reference time scale for the first experimental phase of the Galileo system is described, together with the first experimental results.

1 Introduction

The Galileo System Test Bed (GSTB) V1 is the first experimental phase in the Galileo project supported by the European Space Agency (ESA). The aim is testing the Galileo algorithms in a mixed configuration where the space segment is given by the existing GPS constellation, while the ground segment is an experimental set-up as close as possible to the Galileo architecture. In this framework the realization of a reference time scale, the Experimental Galileo System Time (E-GST), plays an important role since it provides the time reference inside the GSTB V1. In addition, since E-GST is steered to the International Atomic Time (TAI), also the possibility to disseminate accurate time is tested.
The infrastructure that allows the realization of E-GST is the Experimental Precise Timing Station (E-PTS), that is implemented at the Istituto Elettrotecnico Nazionale (IEN), Turin, Italy using the facilities and personnel of the IEN Time & Frequency laboratory in collaboration with Alenia Spazio. The experiment is carried on also with the participation of the PTB (Germany) and the NPL (UK) time laboratories.
The experiment aims at testing the real-time E-GST generation and steering algorithms in a real environmental situation. In addition, off-line, alternative time scale algorithms as well as different clocks ensembles, measurement systems, remote synchronization techniques are used to test the stability and accuracy performances of the resulting time scale.
The project started in mid 2002 and the complete installation, calibration, and verification of the E-PTS is scheduled for the end of 2003. The experimental period is planned to last for the whole 2004.
Till now, the E-PTS activities were mostly focusing on the installation and calibration activities, including the calibration of the TWSTFT IEN-PTB link and on the detailed definition of the E-GST generation and steering algorithms.
The first experimental results are reported and discussed in the paper.

2 The Galileo System Test Bed V1 (GSTBV1)

2.1 The GSTB V1 Purpose

During the GalileoSat Definition Phase, an early experimental stage, the GSTB, has been defined as an integral part of the Galileo Design Development and Validation Phase in order to mitigate program risks [1,2,3].
The GSTB first stage (GSTB V1) main goal is to verify the Galileo concepts for the algorithms. For this purpose the GPS constellation is used as space segment while the ground one integrates present infrastructure with new elements installed for the GSTB V1.

2.2 The GSTB V1 activity

The GSTB V1 experimentation phase, started in February 2004, will last for 12 months. The European Space Agency has specified a list of experiments (the so called Test Cases, Fig. 1) to be performed in GSTB V1:

- Orbit Determination and Time Synchronization (OD&TS) algorithms performance assessment and comparison of alternative methods.
- E-GST generation algorithms.
- E-GST steering to the TAI algorithms.
- User Equivalent Range Error (UERE) budget characterization.
- Signal In Space Accuracy (SISA) computation algorithm and Integrity Flag (IF) computation performance assessment.

The GSTB V1 core infrastructure provides the following functions (Figure 1):

- GPS raw measurements in the sensor stations and data exchange to the GSTB V1 Processing Center (GPC).
- Data processing with OD&TS and SISA/IF algorithms generating the experimental results (core products).
- Precision timing including in particular the generation and maintenance of E-GST. This function includes the E-GST steering to TAI and the generation of timing core products.

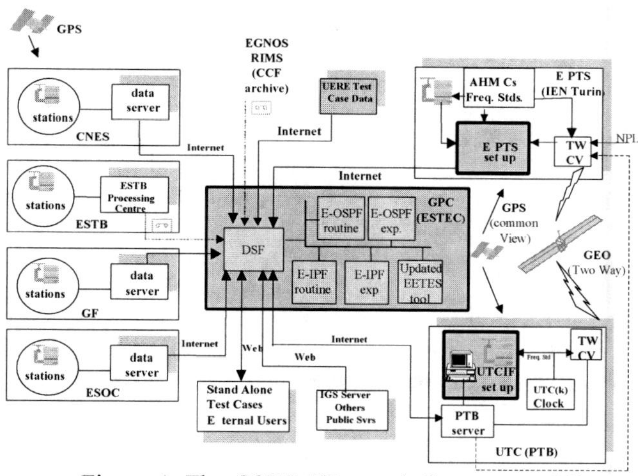

Figure 1. The GSTB V1 core infrastructure [1]

2.3 Timing activity inside the GSTB V1

As a part of the GSTB V1, the E-GST is intended both as a experimental version of Galileo System Time (GST) and as the reference time for the whole GSTB V1 experiment. These two scopes define both the infrastructure required to the E-GST generation and the experimental scenarios in which the E-GST algorithms will be tested. As a prototype of the final GST, the E-GST needs to have optimal characteristics for navigation and time dissemination purposes.

The use of E-GST as a reference time of GSTB V1 activities requires that E-GST is realized in real time and with an hardware representation. In fact a GPS geodetic receiver, which is part of the GSTB sensor stations network, has to be referenced to a signal representing the E-GST. In this way the E-GST is injected in the OD&TS algorithms.

In addition, the E-GST algorithm has to be robust and flexible to fulfill the GST aims and allow further modifications.

To accomplish to these different scopes the reference time scale will be realized in several versions:

- *a eal e e on* giving the time reference to the OD&TS process via the E-PTS sensor station,

- *o l ne e on* that are mainly intended to test the various time scale and steering algorithms.

3. THE E PERIMENTAL PRECISE TIMING STATION (E PTS)

The infrastructure dedicated to the E-GST generation and steering is the Experimental Precise Timing Station (E-PTS). The E-PTS architecture is designed in such a way to give a complete support to the GSTB V1 experimentation phase (Figure 2).

The E-PTS was installed during the year 2003 at the Istituto Elettrotecnico Nazionale "G. Ferraris" (IEN), Turin, Italy. The E-PTS partially re-uses the infrastructure of the IEN Time and Frequency laboratory, in particular the IEN clock ensemble and the remote transfer equipments, limiting the new hardware installation at the GSTB-V1 experimental stage [4, 5].

The equipments especially procured for E-PTS are all located in dedicated rack hosted in the IEN time and frequency laboratory.

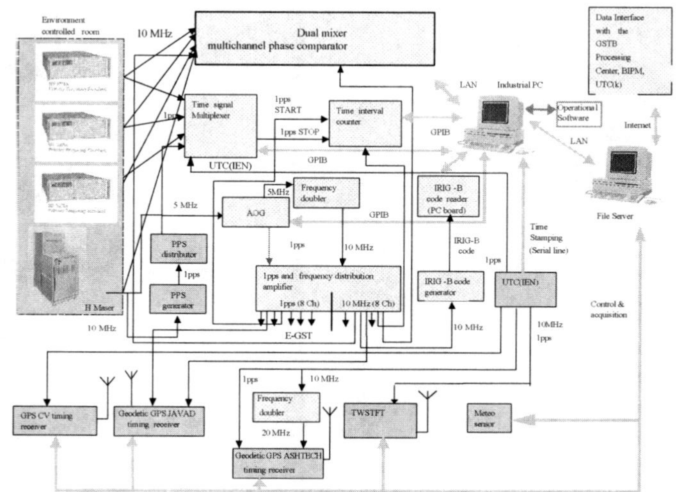

Figure 2. E-PTS architecture

3.1 The E GST generation hardware

The E-PTS clock ensemble is composed by 1 Hydrogen maser and 3 Cesium (Cs) clocks. They are part of the IEN clock ensemble.

The frequency output from the H maser is fed to a phase microstepper which has the capability not to degrade the short-medium term stability of the H maser. The microstepper offset frequency is programmable via serial port, allowing a closed-loop control of the E-GST physical realization by the E-PTS software and the generation of the 5 MHz and 1 PPS signals representing the E-GST. The E-GST time scale originates from one of the 1PPS outputs of the distribution amplifier. EGST is then a physical real time signal as recommended [12].

A time scale generation algorithm requires measurements of time difference between the ensemble clocks. The primary measurement system relies on a 1 PPS signal multiplexer followed by a high-resolution Time Interval Counter (TIC).

An auxiliary system, based on a Dual Mixer Multi-channel Phase Comparator, compares all the clocks simultaneously, with a frequency noise around 4 10^{-13} at 1 second. The simultaneity of the measurements and the continuity of the data acquisition at a relatively high data rate allows real time monitoring of the clock ensemble, frequency stability computation and anomalies detection.

3.2 Automated control system

The E-GST generation and clock monitoring functions is performed by a local computer, which drives the acquisition acquires measurements both from the multiplexed TIC and from the Multi-channel Phase Comparator.

In the operational configuration primary measurement system performs measurements every hour. Data coming from these measurements are used to compute E-GST. Secondary chain acquisition software is devoted to the real time clock analysis. A continuous running module acquires phase differences of every clock with respect to E-GST a real time phase and frequency anomalies detection tool is active. The software calculates also the short term (up to 1 hour) stability of all the cesium clock with respect to E-GST as a further anomaly check

The operational software, developed by Alenia Spazio with the scientific support of IEN, is in charge of the following main functions:

- automated schedule and management of the E-PTS operations
- generation, steering and physical realization of the E-GST time scale.
- Algorithm testing in different scenarios
- Core product generation and formatting

3.3 Remote synchroni ation hardware

The link with the external UTC laboratories (PTB, NPL), which is needed for the TAI steering of E-GST, is realized with the GPS Common View (CV) and the Two-Way Satellite Time and Frequency Transfer (TWSTFT) comparison systems. Both systems are shared with IEN laboratory and are driven by the UTC(IEN) time scale and their measures are related to E-GST through a supplementary internal measurement.

The calibration of such time transfer tools is fundamental for an accurate remote time comparison. In fact the instrumentation involved in the GPS CV and TWSTFT systems (receivers, transceivers, modems) introduces some unknown delays that must be calibrated. Calibration of the GPS CV equipment is performed using the periodical circulations between the UTC labs of an absolutely calibrated receiver from BIPM. For the TWSTFT a calibration trip between IEN and PTB of a transportable station is organized as a GSTB V1 activity under a contract with Joanneum Research G.m.b.H. and the Technical University of Graz (Austria).

3.4 E PTS sensor station

The E-PTS provides a complete sensor station, based on GPS geodetic receiver, which is part of the GSTB V1 sensor stations network. The GPS geodetic receiver is directly connected with the physical representation of E-GST (via the 10 MHz and 1 PPS signals) and this allows the OD&TS functions to be referenced to E-GST itself. A second geodetic receiver, connected to the UTC(IEN) time scale, supports, on a campaign basis, the OD&TS Test Case "clock performance assessment" experimental activity.

3.5 E PTS data e change

Data exchange at E-PTS is based on the IEN Time and Frequency laboratory file server. The server retrieves and hosts the input data required by the E-GST generation algorithm coming from IEN and external labs (PTB, NPL, BIPM).

The E-PTS raw measurements (RINE files from sensor station) and status (log file) are made available to the GPC via an Internet connection supported by the server. Experimental results (Core Products) are transferred to the GPC through the same connection.

All the produced Core Product are now available on the ESA data server at www.gstb-v1.esa.int. A list of the EPTS Core Products corresponding to the on line EGST generation is reported in Table 1

Core roduc	crony
Time series of the offset of EGST vs TAI/UTC	TSTA1
Stability of EGST vs TAI/UTC	STTA1
Stability of EGST vs TAI(NPL)/UTC(NPL), via GPS CV	STGPN
Stability of EGST vs TAI(NPL)/UTC(NPL), via TWSTFT	STTWN
Stability of EGST vs TAI(PTB)/UTC(PTB), via GPS CV	STGPP
Stability of EGST vs TAI(PTB)/UTC(PTB), via TWSTFT	STTWP
Time series of the offset of EGST vs TAI(NPL)/UTC(NPL), via GPS CV	TSGPN
Time series of the offset of EGST vs TAI(NPL)/UTC(NPL), via TWSTFT	TSTWN
Time series of the offset of EGST vs TAI(PTB)/UTC(PTB), via GPS CV	TSGPP
Time series of the offset of EGST vs TAI(PTB)/UTC(PTB), via TWSTFT	TSTWP
Time series of the offset of EGST vs TAI(IEN)/UTC(IEN)	TSIEN
Time series of the offset of EGST vs GPS time	TSGPS

Table 1:list of EPTS Core Product

4 Installation and test of the E PTS

During the year 2003 was carried out the installation, testing and calibration activity of E-PTS. The installation mainly concerned the integration of new hardware with the IEN existing equipment in particular the signal distribution inside the IEN Time and Frequency laboratory was redesigned to provide the required clock signals to the E-PTS rack. Moreover the signal lines connecting the E-GST to the remote time transfer equipments and the remote links were absolutely calibrated to provide the required accuracy to the E-GST steering activity.

4.1 E PTS early versions

During the installation phase of the GSTB V1 core infrastructure, E-PTS provided support to the GSTB V1 early operations with reduced functionality. Since April 2003 a first version of the E-PTS, called E-PTS V1, has been operated, providing raw data from the IEN geodetic GPS receiver referred to the UTC(IEN) time scale. Since July 2003, after the installation of the E-PTS GPS geodetic receiver, a second provisional version has been provided raw data (1 Hz sampling) from an additional geodetic GPS receiver referred to a free running active H maser.
These early versions the E-PTS were provided as very stable time reference to be used during the OD&TS preliminary assessment.

4.2 E PTS functional tests

Many functional and performance tests are performed after the E-PTS installation with the aim of testing the compliance to the specified requirements.
Performance tests were carried out on the clocks stability, maser frequency drift, microstepper accuracy and stability, clock room temperature stability. Functional tests were carried out on the integrity of frequency and 1 PPS signals through the distribution and measurement chain and on the geodetic GPS receiver. Test have been performed also on the remote communication through the file server and on data security.

4.3 Internal delay calibration

E-GST TAI steering needs for accurate calibration of remote links with UTC(NPL) and UTC(PTB). As the GPS common view and TWSTFT measurement systems are driven by the UTC(IEN) time scale, a 1PPS signal representing UTC(IEN) is sent to E-PTS measurement chain, where the difference E-GST-UTC(IEN) is measured. The absolute time delay of the signal chain connecting E-GST to UTC(IEN) and UTC(IEN) to both the GPS CV and TWSTFT receivers was accurately calibrated the quoted uncertainty for this internal calibration was below 500 ps.

4.4 TWSTFT calibration

The TWSTFT time transfer link has better performances with respect to a GPS CV link [6] however the TWSTFT link has to be calibrated. With respect to the calibration with Circular T, which relies mainly on calibrated GPS CV links, best accuracies are achievable calibrating the TWSTFT link with a portable reference TWSTFT station, but such an activity is a very challenging one. In the framework of GSTB V1 IEN received funding for such a calibration the activity was conducted under a contract with Johanneum Research G.m.b.H. and the Technical University of Graz (Austria), which provided a rented TWSTF portable station (Figure 3) its operation, and data analysis during the activity.

Figure 3. The TUG02 portable station installed at IEN

The calibration with portable station was limited to the TWSTFT IEN-PTB link, with an estimated uncertainty of 1 ns (1) the TWSTFT IEN-NPL link was calibrated with Circular T, with an estimated uncertainty of 5 ns (1).
During September 2003, the satellite link provider (Intelsat) moved the TWSTFT service from the satellite I706 to the satellite I903. The new satellite has been easily pointed from both the IEN and PTB location, but a re-evaluation of earth station delays was needed to the transmit and receive frequency change (PTB had to change part of the receive hardware). The delays re-evaluation (details are reported in another work presented at this conference [7]) did not add relevant uncertainty to the link.

4.5 Geodetic GPS calibration

The “Clock performance assessment” test case, part of the OD&TS activity in charge of Alenia Spazio, required the time delay calibration of the E-PTS GPS geodetic receiver. This calibration has been derived by difference with the IEN GPS receiver that was calibrated in the framework of the BIPM TAIP3 experiment.

4.6 GPS CV calibration

The GPS common view calibration was made in the framework of a reference receiver circulation between European laboratories. This circulation is organized by the BIPM every year.

The IEN receiver involved in GSTB V1 was calibrated in 2002 and 2003 calibration uncertainty was quoted by BIPM at 5 ns in both cases. The 2003 calibration was performed during the TWSTFT calibration activity a discussion of this contemporary calibration is reported in an another work presented at this conference [8].

5 E perimental EGST activity

The experimental activity to be carried out for the whole year of experiment consists in generating the Experimental Galileo System Time (E-GST) in two different context [9]:

- Real time E-GST obtained from an ensemble of 3 Cesium clocks and 1 Hydrogen maser, steered versus TAI and monitored by the comparison versus UTC(PTB) and UTC(NPL) (Fig.4),
- Offline test using different ensembles of clocks, measurement rate, steering and averaging algorithms.

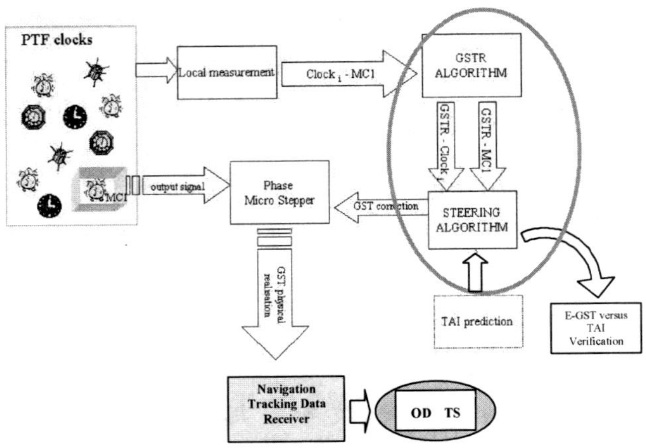

Figure 4 - Scheme of the generation of E-GST

5.1 Algorithms implementation

Till now the activity was mainly devoted to the real time E-GST generation by the definition of a complete time scale algorithm that proceeds along the following steps:

1. Weighted average of the Cesium clocks providing medium-long term stability
2. Steering of the weighted average versus TAI to ensure accuracy and long term stability
3. Real time steering of the H maser output versus the above reference to obtain a real time signal with the short term stability of the maser and with the medium long term stability of the Cesium reference.

The algorithm is based on different submodules which perform different activities on different epoch. He time line and the main submodules of the EGST algorithm are depicted in the Figure 5.

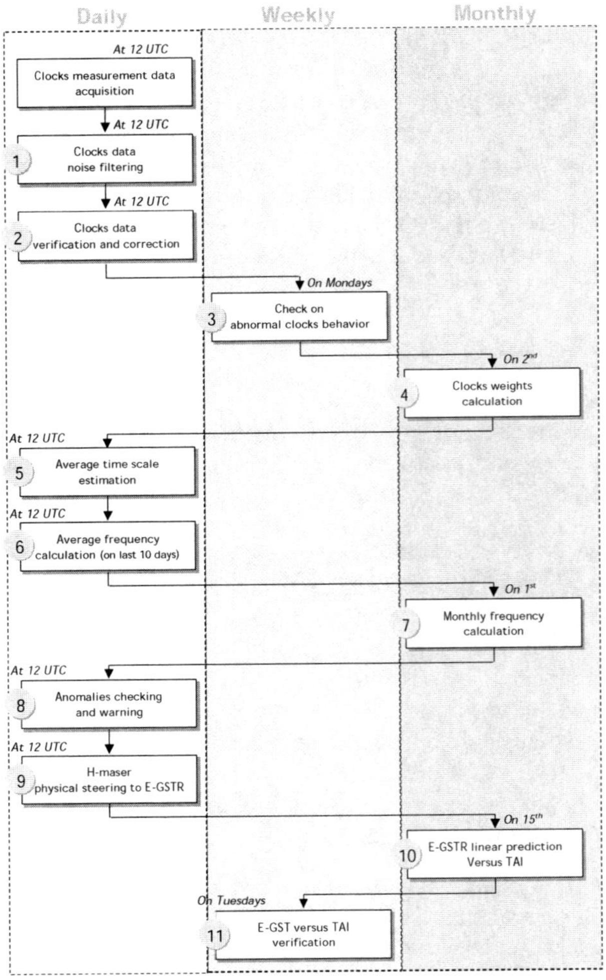

Figure 5 Timing schedule of the EGST generation algorithm

5.2 Algorithm test

The time scale algorithm was coded inside the main software managing the EPTS activities and it was repeatedly tested to evaluate its fundamental functions as well as additional routines appositely devised to ensure robustness to the time scale in case of clock anomalies.

The tests of the real time GST generating software was performed in three steps:

a) test offline on IEN historical data (clock measurements, GPS CV and TWSTFT data, BIPM circulars T)
b) test on line with EPTS clock data without a physical realisation of EGST to gain confidence on the obtained results
c) test on line with results used to steer the H maser and therefore generating the hardware real time E-GST

These validation steps were carried out in December 03 / February 04 and on March 5th the E-GST was switched on by applying the estimated frequency correction on a frequency microstepper (AOG) (Fig.6).

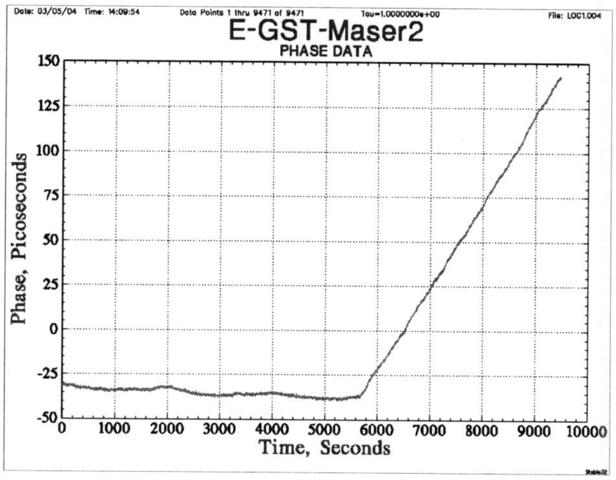

Figure 6 Switching on of the H maser frequency steering to produce EGST

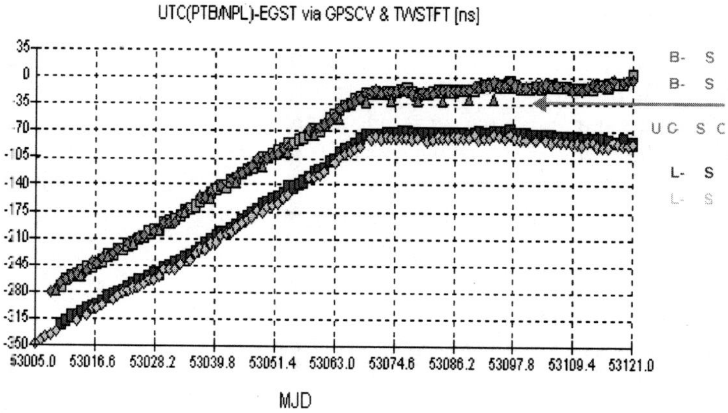

Figure 8 UTC(k) EGST from January 2004 till April 2004

5.3 Real Time E GST. First results

The main requirements on E-GST concern its accuracy, stability, and predictability versus TAI. The required performance for the final Galileo system, GSTB V1, as well as the results obtained in the first month of operation are reported in the Figure 7 .

	Galileo target	GSTB target	Preliminary results
Accuracy |TAI - E-GST|	<50 ns (95%) over any yearly interval	< 1 μs	~ 30 ns
Stability @1day |TAI - E-GST|	$< 5.5 \cdot 10^{-14}$	$< 5.5 \cdot 10^{-14}$	$\sim 7.2 \cdot 10^{-15}$
Uncertainty on the prediction |TAI - E-GST|	28 ns (2 σ)	33 ns (2 σ)	~ 40 ns (2 σ)

Figure 7 Requirement and preliminary results for EGST performance

As the E-GST is on line only since March 5th, its behaviour versus TAI is currently known only for the first month of operation (March 2004), therefore its statistical characterisation is poor at the moment. Nevertheless, the results of the first month of operation are quite encouraging as depicted in the Figure 8.

5.4 E perimental activity. Ne t stages

Next step in the experimental activity, besides the real time EGST generation, concern a series of off-line tests that will evaluate the following issues:

- E-GST algorithm with clocks presenting anomalous behaviour
- E-GST computed with an enlarged set of Cesium clocks
- E-GST computed with alternative algorithm, namely
  a. The NIST AT1 algorithm [10]
  b. A GPS Composite Clock-like algorithm [11]

All the tests will help in understanding the feasibility of E-GST expected behaviour and the feedback on the architectural or algorithm choices.

6 Conclusions

GSTB V1 is the first experimental test bed in the Galileo program context. The architecture and Test Cases are designed to evaluate the criticalities of the algorithms for the Galileo system.
Concerning time and frequency activities, the GSTB timing test cases are expected to give important results on the use of different clocks, measurement systems and algorithms for the generation of a reference time scale, adding some value to the entire time and frequency community.
The operation of the E-PTS at IEN has just started and now it will continue for one year carrying out many different test cases. All timing core products will be accessible through the GSTB-V1 Web-server at http://www.gstb-v1.esa.int/

Acknowledgements

The authors wish to thank all the colleagues from ESA, GalIn, Alenia, and IEN for the invaluable help that they are receiving every day in setting up this project.
The opinions discussed in this paper are those of the authors and do not necessarily represent those of ESA and other agencies in charge of the Galileo program.

References

[1] M. Falcone, F. Amarillo-Fernandez, J. Hahn, H. Favin-Leveque, The GALILEO System Test Bed for Early Prototyping and Experimentation of GALILEO Algorithms , *oc o he S ee ng* Portland, USA, September 2002

[2] M. Falcone, R. Lucas Rodriguez, F. Emma, J. Hahn, The GALILEO System Test Bed (GSTB) , *oc o he SS on e ence*, Sevilla, Spain, (2001)

[3] E. Breeuwer, R. Franco, M. Falcone, J.C. De Mateo, P. Erhard, NM. Hollreiser, Simulation and Experimentation in the GALILEO System Design and Verification , *oc o he SS on e ence* (Copenhagen, May 2002)

[4] G. Graglia et al., "The Experimental Precise Timing Station (E-PTS) for the Galileo System Test Bed (Phase V1): Architectural Design and Experimental Goals", *oc o h TT ee ng*, Reston (USA), 2002

[5] G. Graglia, E. Detoma, F. Cordara, R. Costa, L. Lorini, D. Orgiazzi, V. Pettiti, P. Tavella, J. Hahn, "The Experimental Precise Timing Station and the Galileo System Time generation in the Galileo System Test Bed Phase V1", *oc o he SS on e ence* (Copenhagen, May 2002)

[6] Kirchner, D., Two-Way Satellite Time and Frequency Transfer (TWSTFT): "Principle, Implementation and Current Performance", in *e e o a o Sc ence* , Editor W. Ross Stone, Oxford University Press, Oxford, UK. (1999)

[7] F. Cordara, L. Lorini, V. Pettiti, P. Tavella, D. Piester, J. Becker, T. Polewka, G. De Jong, O. Koudelka, H. Ressler, B. Blanzano, C. Karel, "Calibration of the IEN-PTB TWSTFT link with a portable reference station" To appear in the *ocee ng o he h opean e ency an T e o* Guildford, UK, 2004

[8] W Lewandowski, F Cordara, L Lorini, V Pettiti, A Bauch, D Piester, O Koudelka, "A simultaneous calibration of the IEN/PTB time link by GPS CV and TWSTFT portable equipment" *ocee ng o he h opean e ency an T e o* Guildford, UK, 2004

[9] F. Cordara, R. Costa, L. Lorini, D. Orgiazzi, V. Pettiti, P. Tavella, G. Graglia, E. Detoma, J. Hahn, "The generation of the Experimental Galileo System Time in the Galileo System Test bed V1 *ocee ng o he n e na onal e ency on ol Sy po o n ly h he h opean e ency an T e o* , Tampa, USA (2003), pp 306-311

[10] F.B. Varnum, et al.," Comparison of time scales generated with the NBS ensembling algorithm" *oc h ec e T e an T e n e al ee ng*, pp. 13-24, Dec. 1997.

[11] Brown, "The theory of GPS Composite clock", *S on e ence*, 1991.

[12] J. Laverty, P. Tavella, et al., "Report and recommendations of the GalileoSat Working Group on the Galileo Time Interface", ESA STR-243, ISSN 0379-4067, 2001

CLOCK MODEL FOR GSTB V1 CLOCK PREDICTION PERFORMANCE ASSESSMENT EXPERIMENTATION

M. Gotta, G. Graglia*, M. Falcone, J. Hahn [†]

* Satellite System and Navigation Division, Alenia Spazio, Roma, Italy,
Tel: +39 06 4151 521, Fax: +39 06 4151 506, m.gotta@roma.alespazio.it
[†] Galileo Project Office, ESA /ESTEC, Noordwijk, The Netherlands
Tel: +31 71 565 5832, Fax: +31 71 565 4369, Marco.Falcone@esa.int

Keywords: Allan Variance, Time Standards, Error Modelling, Power Law, Flicker Noise

Abstract

The Galileo System Test Bed (GSTB) V1 aims to investigate and to make early experimentation on those elements of the Galileo System, which are considered critical. The GSTB V1 is then a great added value to the final Galileo, in terms of design consolidation, key technology qualification and gained confidence and experience.
In the frame of GSTB V1 the experimentation activities will be organised and conducted in different Test Cases related to the different critical areas of Galileo Ground Processing. One of the project area to be investigated is related to modelling of atomic clocks in the frame of Orbit Determination and Time Synchronisation and verification of clock estimation performance. In fact in navigation systems, models are used to simulate the process, to estimate the clock parameters based on the measurements and to predict the clock behaviour for the user navigation/positioning function. In the frame of Clock Prediction Accuracy Assessment Test Case Alenia Spazio is in charge to develop a clock model algorithm able to represent the different atomic clock types in order to support the experimentation. Both the systematic and stochastic errors are considered and modelled in the clock simulator tool.

This paper presents the mathematical model, its implementation in a dedicated environment and the first validation results both with respect to theoretical expected behaviour in terms of clock stability evaluation and by comparison with real clock data.

1 Introduction

One of the first and fundamental activities foreseen in the Clock Prediction Accuracy Assessment Test Case is the implementation of a mathematical model representing the atomic clock oscillator inside the Orbit Determination & Time Synchronisation (OD&TS) Filter.
In fact one-way ranging measurements provide the observable for OD&TS that are solved by this last to produce new estimates for the orbit elements and the clock parameters. The Orbit Determination & Time Synchronisation process baseline does not include a model for the atomic clock but estimates the orbit starting from the pseudorange measurements and the clocks are extracted in a second time in snapshot. This baseline will results in the impossibility to solve for orbits and clocks when not enough pseudorange measurements are available, for example in the case of a reduced constellation or ground sensor station network. On the contrary, if a model for the atomic clock is implemented in the Orbit Determination & Time Synchronisation filter, the 'holes' due to lack of measurements can be avoided and its robustness improved.

2 Clock model in navigation systems

Clock models differs as far as their domain of validity and their end use is concerned. In navigation systems, models are used to simulate the process, to estimate the clock parameters based on the measurements and to predict the clock behaviour for the user navigation/positioning function. Their domain of applicability varies from a few seconds to several days.
In particular, once the clock model has been defined, two steps are required to maintain a timing system when the data is acquired in batches and the clock state vector is estimated at regular intervals (as in the case of navigation systems):

1. characterisation of the clocks behaviour, i.e., estimation via measurements of the relevant parameters entering clock model equation, namely:
   - time offset Δt at some (initial) time t_0, t_0 can be selected as completely arbitrary;
   - fractional frequency offset $\frac{\Delta f(t)}{f_0}$ at t_0;
   - frequency drift $\frac{\Delta^2 f}{f_0}$,assuming that this term is constant to avoid a further integration in the model;
   - frequency stability ,expressed as the Allan variance $\sigma_y^2(\tau)$ over the time interval of interest (in this case the time required to carry on the measurements used to characterise the device);
   - sensitivity parameters p_k of the clock to the various environmental factors.

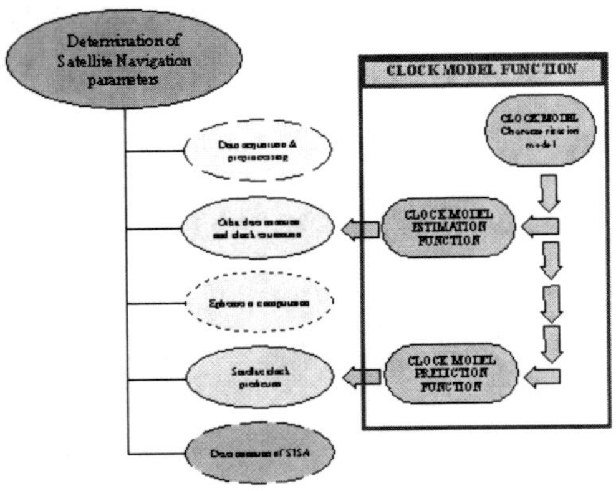

Figure 1 1 : Clock models in the OD&TS overall process

2. prediction of the clock behaviour, i.e. forecasting of the clock error $\Delta t(t)$ in time over some time interval in the future, based again on the clock model

Figure 1 1 shows in which way and at which level the clock activities are involved in the OD&TS process.

In the following section the mathematical representation of the atomic clock model to be used both for estimation and prediction functions is provided.

3 Clock Model mathematical representation

Stable atomic clocks are affected by systematic and stochastic errors that must be accurately modelled and characterised to insure optimum performance in the use of the device.

3.1 Clock systematic model

Generally, the basic model for a stable clock includes three parameters:

1. an initial time offset, $a_0=\Delta t_0$, which represents the inaccuracy in the setting of the initial time at t_0;
2. an initial frequency offset, a_1, which represents the inaccuracy in setting the frequency of the clock at the time t_0; this parameter is customarily represented as normalised with respect to the nominal frequency of the oscillator f_0, and in this case it is referred to as the initial fractional frequency offset of the oscillator, $\frac{\Delta f_0}{f_0}$
3. a frequency drift a_2 (a parameter referred to as ageing in crystal oscillators), which represents the change in time of the initial frequency offset, i.e., is the first derivative of the frequency $\dot{f}(t_0)$; this parameter is generally assumed constant (or slowly varying with time) for all practical purposes.

Considering the clock as a numerical integrator, the above three terms can be regarded as the integration constants for the cascaded integration that made up the oscillators and the clock, resulting in the following equation:

$$\Delta t(t) = \Delta t(t_0) + \int_{t_0}^{t} \frac{\Delta f(t_0)}{f_0} + \left(\int_{t_0}^{t} \frac{\Delta^2 f(t_0)}{f_0} \cdot dt \right) \cdot dt = \\ = \Delta t(t_0) + \frac{\Delta f(t_0)}{f_0} \cdot (t - t_0) + \frac{1}{2} \cdot \frac{\Delta^2 f(t_0)}{f_0} \cdot (t - t_0) \quad (1)$$

It should be noted that this model includes only the systematic terms intrinsic to the oscillator and clock, but not external systematic effects that affect any additional departure from the nominal frequency and phase settings.

3.2 Complete clock model

In order to completely model the atomic clock the stochastic part should be added to the systematic one.
The noise processes affecting the output frequency of the oscillator dominate the stochastic behaviour of the clock.

These processes have been intensively studied during the past 40 years, and are characterised by a measurement of frequency instability expressed as spectral density of the phase noise (in the frequency domain) and Allan variance (in the time domain).

Since for a clock we are usually interested in the time domain aspects of frequency instabilities, the next equation is completed to account for stochastic effects as follows:

$$\Delta t(t) = \Delta t(t_0) + \frac{\Delta f(t_0)}{f_0} \cdot (t - t_0) + \frac{1}{2} \cdot \frac{\Delta^2 f(t_0)}{f_0} \cdot (t - t_0)^2 + \\ + \sigma_y[(t - t_0)] \cdot (t - t_0) \quad (2)$$

In particular, for the aim of the present study the 3 typical frequency stochastic noises are considered, namely White Frequency (WFM), Flicker Frequency (FFM) and Random Walk Frequency (RWFM).
The White and Random Walk Frequency are easily modelled through the theory of the stochastic process. In fact WFM (also called Phase Random Walk) is a process well known (Wiener process) for which the mathematical model is represented by the integration of a gaussian variable with variance equal to the q_{WFM} coefficient characterising the spectral level of the WFM:

$$v_{xk} \approx N(0, q_{WFM}) \quad (3)$$

In the same way the RWFM is modelled through the integration of a Phase Random Walk (to obtain a Frequency Random Walk):

$$Z(t) = \int_{0}^{t} W(s) ds \quad (4)$$

where W(s) is the Wiener Process. Therefore the RWFM mathematical model is represented by the double integration of a gaussian variable with variance equal to q_{RWFM} coefficient characterising the spectral level of the RWFM:

$$v_{yk} \approx N(0, q_{RWFM}) \qquad (5)$$

Let's denote x(t) the instantaneous time error of a clock run from an oscillator with instantaneous normalised frequency deviation y(t), so that:

$$\begin{cases} \dot{x} = y + n_x \\ \dot{y} = z + n_y \\ \dot{z} = 0 \end{cases} \qquad (6)$$

where n_x, n_y are white noises on frequency and drift respectively and drive the random walk on phase (or WFM) and the random walk on frequency (RWFM) respectively.

Through the integration and discretization of Eq. (6), with Δ representing the time step, t=kΔ, the following system equations can be easily found [1]:

$$\begin{cases} x_k = x_{k-1} + y_{k-1}\Delta + \frac{1}{2}d_0\Delta^2 + \Delta^2 v_{yk} + \Delta v_{xk} \\ y_k = y_{k-1} + d_0\Delta + \Delta v_{yk} \\ z_k = z_{k-1} \end{cases} \qquad (7)$$

where the first equation represents the discrete time error equation corresponding to Eq. (2), where WFM and RWFM noises are considered, as per Eq. (3) and (5).

The relationship between the model noises and the common clock characterisation in terms of the Allan variance is expressed by the typical relation [2] provided in Table 1 1 where:

$$q_\alpha = \frac{h_\alpha}{2(2\pi)^\alpha} \qquad (8)$$

α	*Allan Deviation $\sigma y(\tau)$*
0 WFM	$\sqrt{\frac{h_{WFM}}{2}\tau^{-1}}$
-1 FFM	$\sqrt{2h_{FFM}\ln 2}$
-2 RWFM	$\sqrt{\frac{2\pi^2}{3}h_{RWFM}\tau^{+1}}$

Table 1 1: Relation among $\sigma_y(\tau)$ and h_α

In order to have a complete clock model with all the frequency noises included, it remains to add the flicker noise part.

The flicker noise FFM, on the other hand, is not easily modelled. Infact it's not possible to reproduce the 1/f noise with a physical circuit, as any circuit has a transfer function that can be decomposed at the best in polynomial with a single pole and therefore with spectral power multiple of 2. However in literature studies can be found on the subject, trying to find an approximated representation for the flicker noise. We adopt the theory of the Barnes-Jarvis Generator [3] that is able to model the flicker at least in a range of validity.

The flicker noise samples are generated by feeding the output of a Gaussian white noise generator to a n-sections low-pass filter.

The discrete-time n stage Barnes-Jarvis flicker-noise generator [4] is represented by the overall transfer function:

$$G(j\omega) = G_0 \prod_{i=1}^{n} \frac{j\omega + \alpha_i}{j(\omega/\beta) + \alpha_i} \quad \infty \quad \frac{h_{FFM}}{|\varpi|^{1/2}} \qquad (9)$$

with a range of validity represented by :

$$\omega_{\inf} = \alpha_1 \beta^{3/2} < \omega < \alpha_1 \beta^{2n-3/2} = \omega_{\sup} \qquad (10)$$

The difference equation [4] that implements the generator is therefore represented by:

$$y_j(t+1) = a_j y_j(t) + y_{j-1}(t+1) - b_j y_{j-1}(t), \qquad j = 1,...n \qquad (11)$$

where:

$$\gamma_j = \frac{1}{6 \cdot 9^{j-1}}, a_j = 1 - \gamma_j, b_j = 1 - 3\gamma_j \qquad (12)$$

and where $y_0(t)$ is the cascade filter input consisting of white noise (mean zero, variance one),and should be opportunely initialised, and $y_n(t)$ is the output, i.e. the vector of flicker sample.

To obtain the complete Barnes-Jarvis process y(t), it is necessary to set up the initial vector $y_0(0)$, $y_1(0)$, $y_2(0)$.... $y_n(0)$, with random value having the correct joint distribution, through and ad hoc initialisation. The mathematical demonstration of the initialisation routine can be found in [4] and is based on the generation of a set of new random variable $Z_j(0)$, j=1,...n each one orthogonal to $y_0(t)$ with covariance:

$$R_Z(i,j) = \oint_{|z|=1} K_i(z) K_j(1/z) \frac{dz}{2\pi i z} \qquad (13)$$

so that $y_j(0)=y_{j-1}(0)+Z_j(0)$.

Note that the j-th component Z_j is obtained by acting on y_0 with a filter whose transfer function is:

$$K_j(z) = G_1(z)....G_{j-1}(z)\lfloor G_j(z) - 1 \rfloor \quad (14)$$

The $y_n(t)$ is than generated for $t \geq 1$, starting from Eq.(11), from the white noise inputs $y_0(t)$ ($t \geq 1$).

In order to choose the number of stages to be used, one can observe from Eq.(10) that the approximation interval varies as β^{-2n} while the precision of the approximation increases as β^2 tends toward unity. Choice of n=6 assures a flicker modelization over 12 decads; as we are interested in a flicker for t_{inf} = 1 s, n=6 allows to model a flicker correctly up to 10^12 s, that is largely acceptable.

Finally, in order to obtain the correct noise spectral level amplitude of the flicker, the term G_o adjust the magnitude of the transfer function, H(z). It is possible to demonstrate (eq.7 of [5]) that the white noise driving the n-state flicker generator ($y_0(t)$) should be N(0, q_{FFM}), where:

$$q_{FFM} = h_{FFM} |G_0|^2 = h_{FFM} \cdot 3.0942 \qquad (15)$$

that is found coherent with Eq.(8) that specificies $q_{FFM}=\pi h_{FFM}$.
In order to obtain the complete clock model, we therefore add in frequency (Eq. (7)) the flicker model, as obtained from Eq. (11), by obtaining the complete clock model:

$$\begin{cases} x_k = x_{k-1} + y_{k-1}\Delta + \frac{1}{2} d_0 \Delta^2 + \Delta^2 v_{yk} + \Delta v_{xk} \\ y_k = a_k y_{k-1} + y_k(5) - b_k y_{k-1}(5) + d_0 \Delta + \Delta v_{yk} \\ z_k = z_{k-1} \end{cases} \qquad (16)$$

where $y_k(5)$ is the 5-th Barnes-Jarvis stage output (at t=kΔ) and $y_{k-1}(5)$ is 5-th Barnes-Jarvis stage output (at previous instant, t=(k-1)Δ).

The clock model as expressed by Eq.(16) has been validated versus the theoretical expected results. Infact, once identified the noise levels of each composing noises (between WFM, FFM and RWFM), it is possible to express the theoretical expected clock stability behaviour in terms of Allan Variance as the quadratic sum of the Allan Variance of each noise (as expressed in Table 1 1):

$$\sigma_y(\tau) = \sqrt{\frac{h_{WFM}}{2}\tau^{-1} + 2h_{FFM}\ln 2 + \frac{2\pi^2}{3}h_{RWFM}\tau^{+1}} \quad (17)$$

The clock model has been run over t=1e4 s with the initial condition:

$a_0=x_0=-4.9e-12$, $a_1=y_0=1.0e-11$, $a_2=d_0=z_k=1.1e-14$

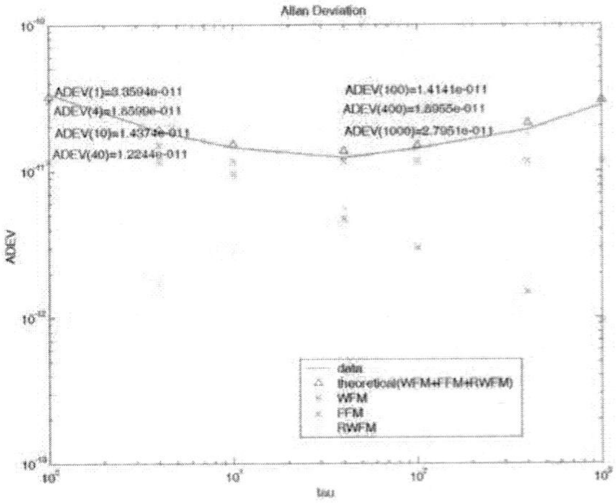

: Figure 1 2 : Clock model validation versus theoretical expected results

and with the following noise levels:

q_{WFM}=9e-22, q_{FFM}=2.9e-22, q_{RWFM}=2.1e-24

h_{WFM}=1.8e-21, h_{FFM}=9.6e-23, h_{RWFM}=1.1e-25

and the clock error (phase) has been evaluated through the Allan Deviation analysis, by comparing the Allan Deviation of the data (as obtained by implementing the clock model) and the theoretical expected values (as per Eq. (17)). The clock model shows to correctly behave, as demonstrated in Figure 1 2.

Moreover a validation of the clock model versus real data is currently on-going to verify over which domain of applicability and at which extent the theoretical model represents the clock behaviour. It is experimentally found that a real clock data can exhibit behaviour in phase and frequency, that are not taken into account in the theoretical model as specified in Eq.(16). Infact the fundamental hypothesis that is behind the developed model is that the deterministic parameters a_0, a_1 and a_2 should not change over the interval of applicability of the model. Moreover it assumes that the estimation of the deterministic parameters (with a least square fitting for example) is done with a negligible error over this interval and that the remaining contribution after detrending is due to the stochastic noises only. Therefore it is easy to understand that a particular clock model is necessarily limited to be within a certain error budget over a particular time interval. Extension or reduction of the said interval does not guarantee an equivalent performance of the model: reduction of the interval may result in a smaller number of measurements and a corresponding increase in the uncertainty of the model parameters, while an increase in the time interval may result into taking into consideration additional effects that are unimportant over shorter time spans, such as drift. On the other end, modelling effects, such as drift, which are

negligible over short time intervals results in a worsening of the results in prediction and estimation using the model. It has been experimentally proven that, for this reason, linear models produce better results than quadratic models over medium-term intervals even if these clocks are known to exhibit frequency drift; drift will become an issue for time intervals in the order of one day or greater.
Recent studies [6] provides the means to determines the best timeline in terms of T_m, measurements time over which the deterministic parameters are estimated, and T_p, prediction time or applicability time for the clock model).

We consider a set of 1 sample per second time offset measurements of an EM1 RAFS, measured versus EFOS 16 (AHM) as reference with a Timing Solutions 5110A.
We choose to estimate the deterministic parameters of the model over Tm=10^4 s that is the baseline measurement time for Galileo RAFS clock parameters estimation. Moreover we choose Tp=6000 s as time interval over which to test the model as it is the baseline prediction time to keep the extrapolation error below a threshold of 1.5 ns (1σ).
Due to the fact that in the first samples the real clock data suffers an increase of the frequency instability as it can be seen in Figure 1 3, we select an interval inside vector 2 of Figure 1 3, where the clock doesn't show abnormal behaviour. For example starting from t=20000, the following parameters are found over the first 10^4 samples (linear model):

a_0= -9.09 e-007, a_1= -4.44e-011

while the noise levels are found to be:

h_{WFM}=2.3e-23, h_{FFM}=1.1e-28, h_{RWFM}=1.38e-31

The clock model (complete: deterministic + stochastic) is run and the resulting clock error compared with the real data. From a stability analysis we can find the correct stability behaviour (the same of the real data), as shown in Figure 1 4. This is a confirmation that the clock model stochastic part correctly represents the noise of the real data.
However the clock error as obtained through the clock model (complete) exhibit a not-negligible departure from real data already after 4000 samples as shown in Figure 1 5.
This can be interpreted in the following way: the stochastic part, as modelled in Eq.(16) is just one realisation of the non-stationary process affecting the clock and therefore the fact to add it to the deterministic part it's worsening the clock modelling other than improve it. This can be easily seen from Figure 1 7 where different realisations of a WFM stochastic process are presented. Each realisation represents a WFM with correct stability behaviour (characterising the stochastic process itself). However single realisations can differ punctually for an offset that is equal, in the worst case, to 2 sigma uncertainty and that is completely random.

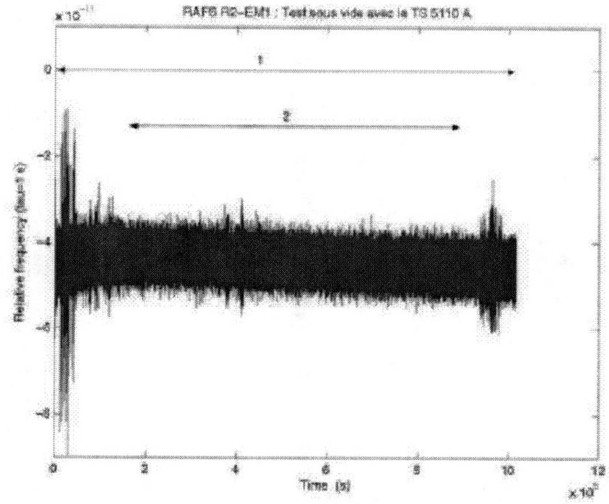

Figure 1 3: RAFS frequency behaviour

This means that the best way to proceed is simply to keep the stochastic error, rather then to choose a single realisation of it.

The preliminary conclusion that can be drawn is that the clock model, including only the deterministic part (through propagation of a_0, a_1 and eventually a_2), is suitable (as can be seen in Figure 1 6) to the purpose of modelling the clock inside the OD&TS filter. On the contrary, the complete clock model as presented in Eq. (16) is suitable for simulation purposes where the designer, assuming the same deterministic part, is interested to study the clock extrapolation accuracy (in terms of mean and standard deviation of the error) due to clock instability as covered in [7].

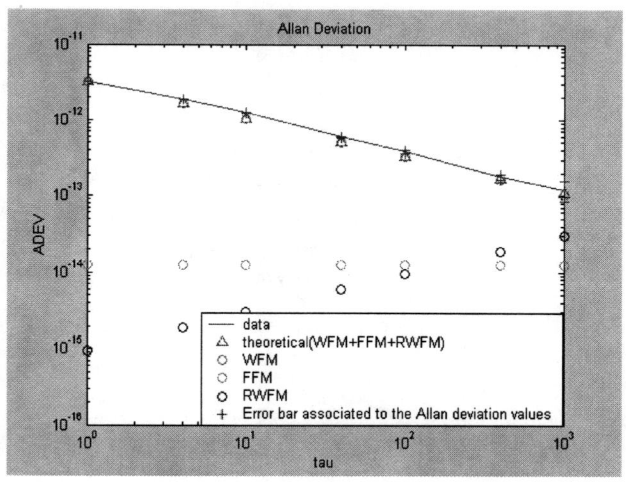

Figure 1 4: Stability of data as obtained from Clock model

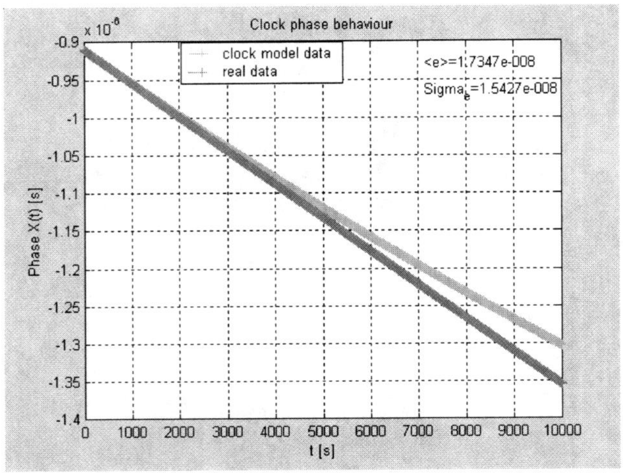

Figure 1 5: Clock model (complete) representativeness of RAFS data

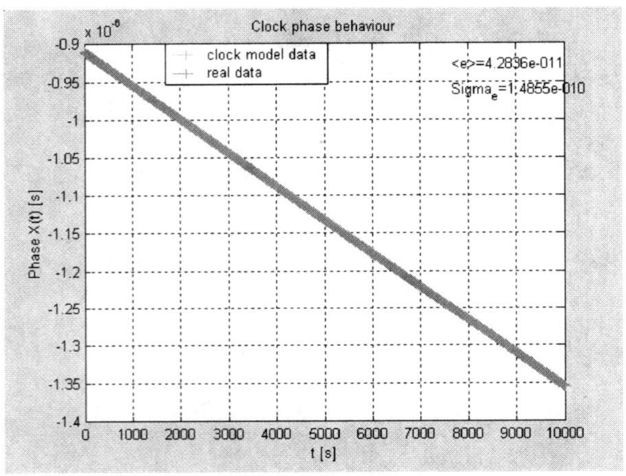

Figure 1 6: Clock model (deterministic-only) representativeness of RAFS data

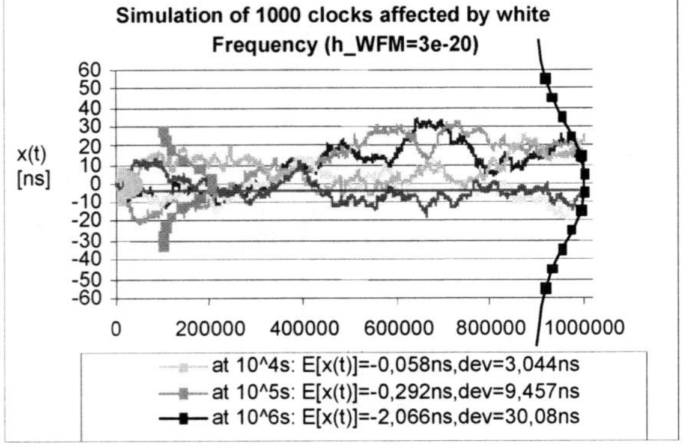

Figure 1 7: Realisations of a stochastic process

Acknowledgements

The authors are thankful to Jerome Delporte and Michel Brunet of CNES for their kind support during the preparation of this work.

References

[1] Antonella Albanese, Gianluca Graglia, Patrizia Tavella. "Clock error modelling and frequency prediction in the Galileo GNSS2 System.", Proceeding of GNSS 2002.

[2] J.W. Chaffee, "Relating the Allan Variance to the Diffusion Coefficients of a Linear Stochastic Differential Equation Model for Precision Oscillators", *IEEE Trans. Ultras. Ferroel. Freq. Control*, vol. UFF C-34, no.6, pp.655-658, November 1987

[3] Barnes J., Jarvis S., "Efficient Numerical and analog modeling of flicker noise processes", *NBS Tech. Note 604, National Bureau of Standards*

[4] C.A.Greenhall, "Time Interval error of a Flicker Noise Generator", *TDA Progess Report 42-77,* January-March 1984

[5] G.L. Mealy, D. R. Vander Stoep, "*Time Standard Error Modeling with applications to satellite navigation*", Air Force Avionics Laboratory under Contract No.F33615-72-C-1787

[6] J.Delporte, F.Vernotte, M.Brunet, T.Tournier, "*Uncertainties of drift coefficients and extrapolation errors: application to clock error prediction*", Metrologia, January 2001

[7] M.Gotta, P.Tavella, "*Uncertainty and Prediction of Clock Errors in Space and Ground Applications*", Proceeding of 14th EFTF, p.77-81, March 2000

ESTIMATION OF TOTAL JITTER AND JITTER PROBABILITY DENSITY FUNCTION FROM THE SIGNAL SPECTRUM

M.J. Underhill *[†], P.J. Brown [†]

* University of Surrey, UK. Email: m.underhill@toric.co.uk
† Toric Ltd, UK

Keywords: Phase noise spectrum, total jitter, jitter PDF.

Abstract

The total jitter of a signal can be found directly by integrating the sideband spectrum as it appears on a spectrum analyser. If there are any significant spur components, these can be included in the calculation of total RMS jitter if they are identified and corrected in power level according to the spectrum analyser resolution bandwidth. The jitter amplitude PDF (Probability Density Function) can be calculated on the assumption that the jitter amplitude has a Gaussian PDF provided that there are no sufficiently strong spurs. In the limiting case of a single dominant sine wave spur with a power that exceeds the rest of the sideband noise the PDF is the central $\pm\pi$ region of the $1/\cos x = \sec x$ function. This has a well-defined width equal to the sine-wave peak-to-peak amplitude. When Gaussian noise is added to a single spur the PDF becomes 'Rician' representing a convolution of the PDFs of the spur and the Gaussian noise. The convolution process may be extended to include further spur components (of unrelated frequencies) to give the final PDF. To find such total jitter measures as 'Time Interval Error jitter', 'Period jitter', 'Cycle-to-Cycle jitter' or any other total jitter definition, the appropriate frequency weighting functions should be applied before the spectrum is integrated and then it can be converted to a PDF. It is the 'tails' of a jitter PDF that mainly determine the Bit Error Rate (BER); in general these can be reasonably well estimated from spectrum analyser measurements, even though the exact shape of the main central part of the PDF is not easily computable. A simple approximation is proposed.

1 Introduction

In a previous paper [1] a total jitter estimate for the output of an Anti-Jitter Circuit was examined. The point was made that the total jitter power, and therefore the RMS jitter amplitude, can always be calculated from the phase noise spectrum $L(f)$.

The direct relationship between the sideband spectrum $L(f)$ and the time jitter spectrum $\tau(f)$ is $L(f)=2\pi f_o \tau(f)$, where f_o is the carrier frequency.

If the spectrum is noise-like, with no dominant spur (no spurs having a significant portion of the total sideband power), total jitter can be found directly by integrating the sideband spectrum as it appears on a spectrum analyser. In fact any spur components can be included in the calculation of total rms jitter if they are identified and corrected in power level according to the spectrum analyser resolution bandwidth.

In analogue communications and radar systems it is the phase noise spectrum and the total (RMS) jitter that are the ultimate determinants of system performance. In this case a measurement of the total system phase noise spectrum will suffice to determine the system limiting performance.

In digital and data systems it is the system Bit Error Rate (BER) that determines the final system performance. For no bit errors the clock must always sample the bit steam of a data signal correctly within a time window that is a defined fraction of the bit period. The (raw) bit error rate is defined by how often this fails to happen; and this is ultimately determined by how much of the Probability Density Function (PDF) of the time jitter amplitude falls outside the prescribed time window.

One aim of this paper is to examine to what extent the time jitter amplitude PDF may be computed from measurements of phase noise on a standard spectrum analyser. The emphasis is on the effect that strong discrete components may have on the 'tails' of an overall PDF.

Another overall aim is to find how far the time jitter 'seen' and measured on an oscilloscope or time jitter analyser, can be 'manufactured' from a single spectrum analyser frequency plot.

2 Types of jitter

In the frequency domain time jitter and phase jitter have exactly the same Power Spectral Density (PSD) functions and are therefore equivalent. The direct relationship between the sideband spectrum $L(f)$ and the time jitter spectrum $\tau(f)$ is $L(f)=2\pi f_o \tau(f)$, where f_o is the carrier frequency. This relationship holds at all frequencies down to the frequency below which the total modulation index approaches or exceeds unity.

Jitter can be separated into 'random' and 'deterministic' noise types [2]. 'Random' noise has a smoothly varying and continuous PSD for $L(f)$ as can be seen on a spectrum analyser. 'Deterministic' noise in general consists of a number of discrete frequency components that are usually harmonically (or sub-harmonically) related to one or more frequency components and their mixing products. Discrete components have a fixed (or slowly varying) measured power. Often one, or at the most two or three of these components will visibly dominate in power. This will be seen to be useful.

2.1 Measurement of total jitter and frequency weighting functions

There are three main types of Total Jitter [2] and each of these has an associated frequency weighting or 'filter function'.

1 **TIE = Time Interval Error Jitter**. This is the time error between transitions of the signal waveform and an ideal clock. This is un-weighted phase noise. Total jitter is direct integration of the phase noise spectrum from a chosen lower frequency to half carrier frequency.

2 **Period Jitter.** This is time fluctuations of the period of the waveform. The frequency weighting function is a first order filter function of $e^{-\pi fT} \sin(\pi fT)$, where T is the period.

3 **Cycle-to-Cycle Jitter** This is the time fluctuations of the period and the function is a second order filter of $\{e^{-\pi fT} \sin(\pi fT)\}^2$

Allan variance is a much-used measure of frequency stability in the low frequency long time-constant region. It plots the 'fractional frequency stability' against the time τ taken for each measurement. It is more suitable for addressing the 'phase noise' or the 'time jitter spectrum' at frequencies below about 10Hz. It is not practical to obtain direct phase noise plots in this region and so this where Allan variance comes into its own. Random noise in this region is now usually called 'wander' rather than 'jitter' (ITU definition [2]). Allan variance has its own (second order difference) filter function there is also Modified Allan Variance (MVAR) that has a third order difference filter function.

In fact every total jitter measurement method imposes a specific frequency weighting function on the time jitter or phase noise spectrum. The total time jitter 'power' is then the integral of the weighted spectrum, and the total time jitter RMS 'amplitude' is the square root of this integral.

All frequency weighting functions, to a greater or lesser extent, act to exclude the low frequency region. For example the measurement time T_{meas} automatically limits creates a 'high-pass filter' with a low frequency cut-off at $f_{lf} = 1/(2\pi T_{meas})$. Thus the measurement time can be chosen to exclude the region where the phase and time jitter are not directly related.

The measurement time of a single measurement also has a frequency weighting or filtering function that has a 'comb' of 'notches' that extends over all frequencies. The notches appear at frequencies that are integer multiples n of the reciprocal of the measurement time: $f_{notch} = n/(T_{meas})$. A problem then arises if there is a discrete 'spur' component with a frequency that falls into one of these notches. Its energy will be excluded from the measurement energy, which will not then be a correct measurement of total time jitter. This problem can be overcome by taking multiple sample measurements at random phases or by making measurements to a fraction of a clock or signal cycle.

Fortunately measurements taken with a spectrum analyser do not suffer from this particular problem, because any detectable spurs are included in the measurements. However the finite bandwidth of any spectrum analyser creates a different problem. The spurs (and any carrier, if present) are measured as total power, while the (random) noise is measured as a Power Spectral Density (PSD). The measured noise level changes as the analyser resolution bandwidth is changed, but the strength of the spurs does not. How this problem can be solved 'automatically' is treated in section 4.

As already mentioned, 'deterministic jitter' appears on a spectrum analyser as a set of spurs which are all sub-harmonics of the system clock that is coupling noise onto the 'wanted' signal. As will be seen, deterministic jitter can be included in the total jitter estimates, and the effect that it can have on the system Bit Error Rate (BER) can also be estimated.

3 Spectrum analyser jitter measurement

There are various different definitions of total jitter that have been offered as 'standards'. Each 'standard' is in general focussed on the different system requirements of different applications. As already mentioned some examples are, 'Time Interval Error', 'Period Jitter' and Cycle-to-cycle Jitter. Any standard can be interpreted as the application of a 'filter' with an implied frequency weighting function. For a given spectrum, the appropriate 'filter' weighting function may be applied and the modified spectrum integrated to give the relevant total time jitter figure. For real signals a spectrum analyser with a suitable data I/O port may be connected to a PC that performs the required filter function and integration processes. We have used a HP8560A spectrum analyser with a GPIB connection to a standard desk-top PC. Our PC algorithms can also be used on the spectra predicted from PLL (Phase Locked Loop) or AJC (Anti Jitter Circuit) simulations.

For certain applications the peak or 'peak-to-peak' jitter amplitude or the actual jitter amplitude PDF (Probability Density Function) is required. In the above we have assumed that the jitter amplitude has a Gaussian PDF. In the limiting case of a single dominant sine wave spur with a power that exceeds the rest of the sideband noise the PDF is found to be the secant (reciprocal of a cosine) function; and this has a well-defined peak excursion equal to the sine-wave amplitude. When Gaussian noise is added to a single spur the PDF becomes 'Rician' representing a convolution of the PDFs of the spur and the Gaussian noise [3]. The convolution process may be extended to include further spur components (of unrelated frequencies) to give the final PDF. This final PDF contains the information from which the total jitter measures may be obtained for any total jitter definition.

We have developed a process for identifying in the data those spur signals that need to be taken into account in the calculation of the final PDF. We are exploring ways of identifying harmonically related and correlated components so that more accurate PDFs may be obtained in these cases, as for example for 'pattern noise'.

4 The jitter estimation method

The steps required to extract the total time jitter from a spectrum are:

1 Obtain spectrum.

2 Extract spurs and weight them.

3 Apply frequency-weighting function, including spectrum limits according to type of jitter.

4 Integrate the spectrum remaining after the spurs have been extracted.

5 Add in the extracted spur energy to get total RMS jitter.

6 Assume all jitter components are random.

7 Convolve to get a PDF estimate.

8 Use approximation to get an estimated BER.

4.1 The initial spectrum

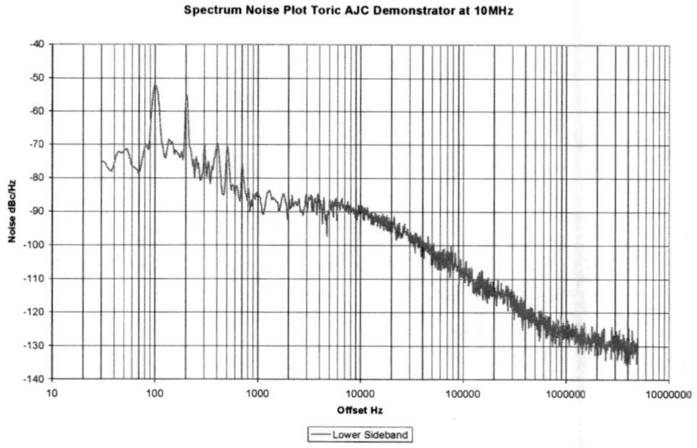

Figure 1: Measured Spectrum from HP 8642 to be converted to Time Jitter PDF and Total Jitter Estimate

Figure 1 shows the spectrum of the output of an Anti-Jitter Circuit (AJC) as measured on an HP8642 Spectrum Analyser. We find the jitter amplitude Probability Density Function (PDF) and the required Total Jitter Measure from this.

4.2 Probability Density Function of Random Noise

If all the noise and jitter components of a signal are random, the Probability Density Function (PDF) is by definition Gaussian. The standard deviation is then the RMS phase error as obtained above. Figure 2 shows the Gaussian PDF of any random noise. The amplitude deviation function σ (shown as d in the following graphs) is equal to the total RMS amplitude of all the random noise signals combined.

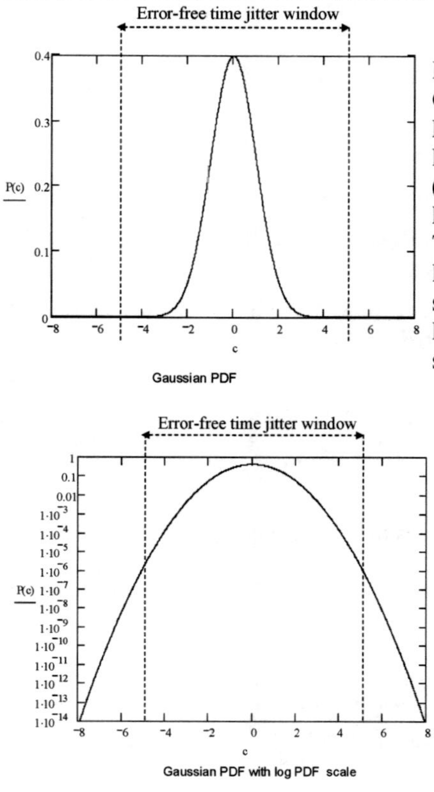

Figure 2: The Gaussian Probability Density Function (PDF) of Random Noise Time Jitter
Left: = linear scale
Below left: = log scale

4.3 Bit Error Rate (BER) estimation

To find the Bit Error Rate (BER) of any Gaussian signal we assume that there is a time jitter window within which no bit errors are generated. This is shown in Figure 2.

The BER is then the total area of the probability curve remaining outside the window.

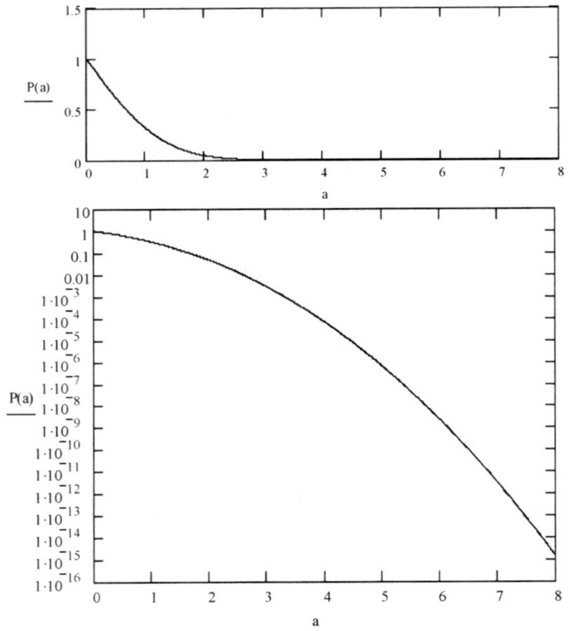

Figure 3: BER as a function of (symmetrical) time window half-width *a*

Figure 3 shows the area under the curve on both sides and this is therefore the BER. P(a) is the BER (Bit Error Rate).

$$P(a)= 2/\sqrt{(2\sigma\pi)} \times \int_a^{\infty} exp(-a^2/2\sigma)\, da \quad (1)$$

$$= 2\int_a^{\infty} P(c) \quad (2)$$

where a = c is the time window half-width.

In the upper graph P(a) is on a linear scale and in the lower graph P(a) is on a log scale

The BER is the value of the PDF P(c) at the window edge × 2c (=2a) – typically to within a few percent. Note that at a = 7 the BER is 2.8×10^{-12}. This is when the total window width is 14 times the RMS total jitter. This point is often quoted as an industry standard of reliable system performance.

4.4 Probability Density Function of a sine wave

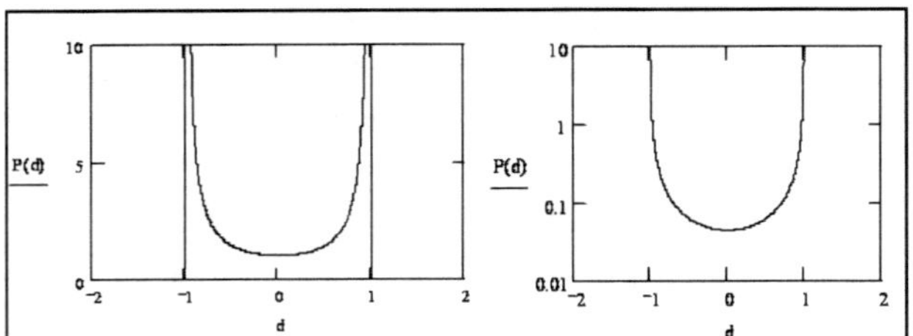

Figure 4: The PDF of a Pure Sine wave with no added noise (linear and log scales)

Figure 4 is the PDF of a sine wave of unit amplitude d = 1. Note that the BER is zero until the sine-wave peak-to-peak amplitude just exceeds the time jitter window width. At that point there is a very rapid increase in BER.

The PDF is $P(d) = 0.04332\ sec(d\pi/2)$ *for* $-1 < d < 1$, and elsewhere $P(d) = 0$. The normalisation factor 0.04332 is the integral of sec(dπ/2) over the range.

Also note that the PDF can be usefully approximated as two half-unit impulses at d = -1 and +1. This allows a simplified convolution calculation and a simpler estimation of the BER, as mentioned below.

4. 5 Combined Probability Density Function of a sine wave and random noise

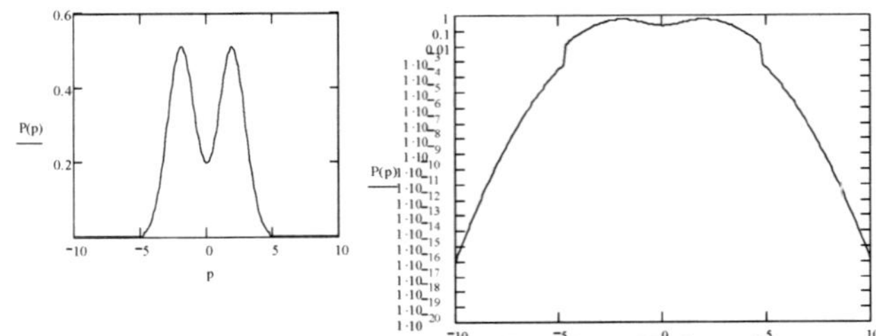

Figure 5 Combined Spur and Random Noise: Spur amplitude =2, Noise SD = RMS amp, $\sigma_d = 1$

The joint Probability Density Function P(p) is the (normalised) convolution of the Random Noise and Sine-wave PDFs:-

$$P(p) = \int_{-\infty}^{+\infty} [\sqrt{(2\sigma\pi)}^{-1} \times exp\{-(x-p)^2/2\sigma\} \times 0.1085\ sec(x\pi/2d)\ (|x/d| < 1)]\, dx \quad (3)$$

This is shown in Figure 5 with a linear P(p) scale on the left-hand graph and a log P(p) scale on the right-hand graph.

In Figure 5 the total RMS random noise jitter amplitude is unity, σ = 1, and the sine wave 'spur' peak jitter amplitude is d = 2 units.

The 'kinks' in the curves around P(p) = 10^{-2} to 10^{-3} are thought to be from (unavoidable?) granularity or numerical error in the Mathcad 11 simulation. The kinks move and change shape if the integral limits or the parameters are fine-tuned. However it could also be that a single convolution is not sufficient to represent the combined PDF with better accuracy than shown. In spite of the 'kinks', the accuracy of the 'skirts' beyond this is found to be very good.

4. 6 The BER of a sine wave and random noise

As already mentioned the PDF of a sine -wave can be usefully approximated as two half-unit impulses at d = -1 and +1. As a consequence the BER can be calculated on the assumption that the skirts have been reduced in amplitude by 1/√2 and move apart by the peak-to-peak amplitude of the sine wave.

4. 7 Examples of the joint Probability Density Function of a sine wave and random noise

The following jitter amplitude PDF examples were also obtained from a Mathcad implementation of Equation 3 giving the combined or joint probability of various proportions of a single sine wave spur of deterministic jitter together with a proportion of random Gaussian jitter.

The important point of note, seen on the logarithmic plots, is that the shape of the cures at low probability values is well approximated by:

$$P(a) = 2/\sqrt{(2\sigma\pi)} \times exp\{-(a^2 - \sigma_d{}^2)/2\sigma\} \qquad (4)$$

This represents the splitting of the random noise Gaussian curve by an amount equal to the peak-to-peak amplitude of the sine wave signal = $2\sigma_d$.

The Bit Error Rate BER is then given by the area of the PDF outside the error-free time window and it is:

$$P(a) = 2/\sqrt{(2\sigma\pi)} \times \int_a^\infty exp\{-(a^2 - \sigma_d{}^2)/2\sigma)\}\, da \qquad (5)$$

The values of random and deterministic noise in Figures 5 to 9 are chosen to be illustrative of the general time jitter histogram shapes as might be found from a jitter analyser.

Note that the random noise amplitude Standard Deviation (SD) is taken to be the same as the RMS amplitude of the random noise.

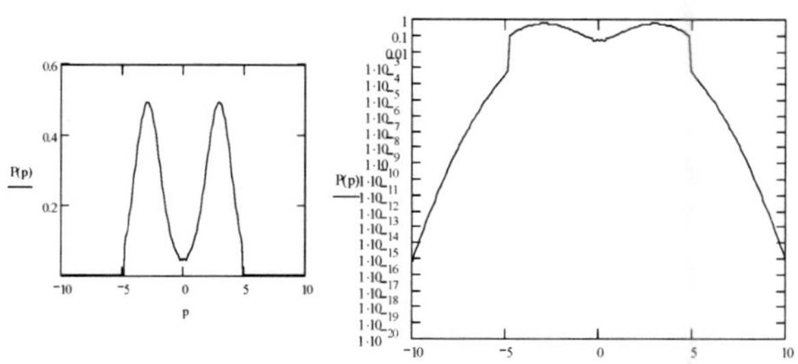

Figure 6 Combined Spur and Random Noise PDF: Spur amplitude =3.001, Noise SD = RMS amp, σ_d = 1

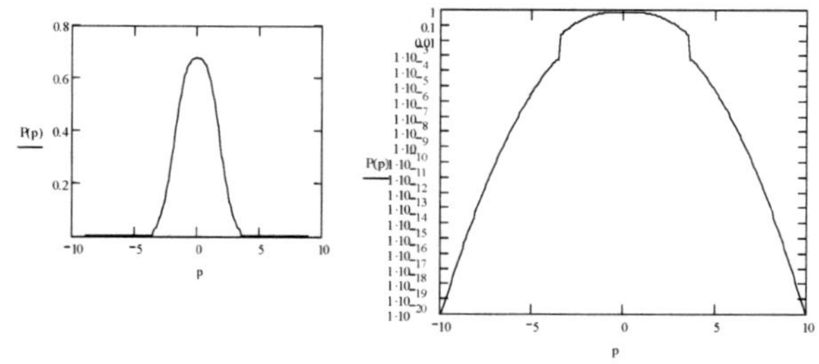

Figure 7 Combined Spur and Random Noise: Spur amplitude =3.001, Noise SD = RMS amp, σ_d = 1

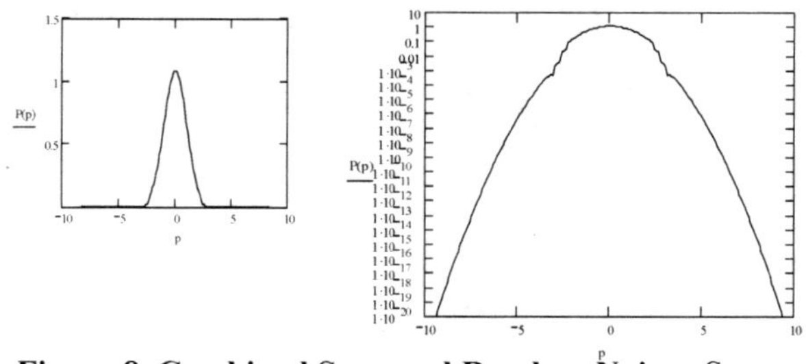

Figure 8 Combined Spur and Random Noise: Spur amplitude =3.001, Noise SD = RMS amp, σ_d = 1

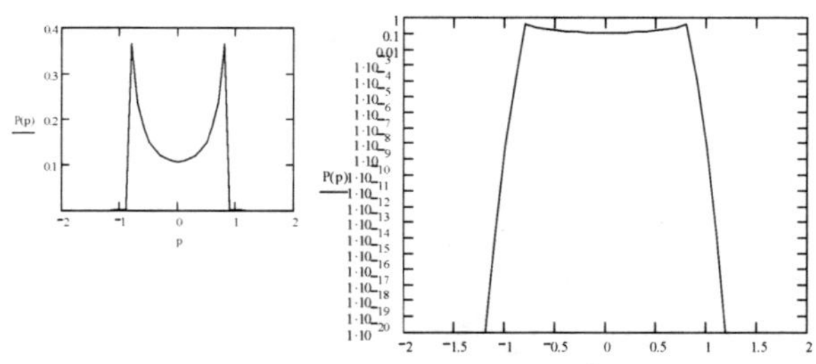

Figure 9 Combined Spur and Random Noise: Spur amplitude =1.01, Noise SD = RMS amp, σ_d = 0003

4. 7 Bit Error Rate Estimation

In summary from these figures we can see that the BER can be well estimated from the assumption of the splitting og the Gaussian random noise PDF by the peak-to-peak sine wave amplitude.

If the deterministic noise consists of several components then the BER for low error rates is well represented on the assumption that the deterministic noise components are combined according to the 'RSS' (Root of the Sum of the Squares) into a single sine wave deterministic noise component.

5 Detail of Spectrum to Time Jitter Conversion

The Toric Integration Algorithm is used as follows:-

1. Output Power Spectrum is measured using a spectrum analyser :
2. This is translated into Phase Noise by applying the following corrections for each data point :
   a. Log Scale under-response, add 2.51dB
   b. 3dB BW over-response, subtract 0.52dB
   c. Normalise to 1Hz BW subtract 10*log(RBW)
3. Data is measured for both the Upper and Lower sidebands;-
   a. Each measured data point will have an associated offset frequency
   b. One sideband will have m frequency points
   c. The frequency closest to carrier is called f_1,
   d. The frequency furthest from carrier is f_m
   e. At each frequency offset f there is a measured noise N.
   f. For the lower sideband the noise measurements are $NL_1..NL_m$, corresponding to offset frequencies $f_1..f_m$.
   g. For the upper sideband the noise measurements are $NU_1..NU_m$, corresponding to offset frequencies $f_1..f_m$.
4. To find the noise between two adjacent frequency points f_1 and f_2 the following assumptions are made:
   a. The noise is assumed to be a straight line between f_1 and f_2 at the level given by N_2 (for lower sideband use NL_2, for upper sideband use NU_2)
   b. The power in the band f_2-f_1(as a ratio) is given by $(f_2-f_1)\times(10^{(N2/10)})$
5. To find the noise for frequency points $f_1..f_m$ the following calculation is performed:
   $\Sigma(f_k - f_k\text{-}1)^{Nk/10}$ from k=2 to k=m
   a. Take noise power (as a ratio)
   b. This can then be performed for both the upper and lower sidebands.
   c. These are added together to give the Double Sided noise power from $f_1..f_m$.
6. The Noise power (as a ratio) may now be converted into dBc
   a. (10×log (DSB Noise Power ratio))
   b. or into radians √(DSB Noise power Ratio) which gives the RMS jitter.
7. One period is 2π radians. The value in UI is given by (RMS Jitter in Radians) / (2π)
8. The value in UI (Unit Interval) can be multiplied by the Carrier period (T) to give the time jitter RMS

6 Results from spectrum measurements

The following three examples compare the results obtained from a LeCroy SDA600 Jitter Analyser with those obtained from the Toric Spectrum Integration Method described above.

6.1 Example 1

The TIE result from a SDA6000 measurement of a Mk2 AJC was as follows:

Mean: -7.45ps
Min: -293ps
Max: 359ps
SDEV: 97.56ps

If the FFT data, used in Figure 1 (and repeated as background in Figures ? and ?) is integrated using the Toric Calculation (with integration limits 1.192kHz – 5MHz), the RMS jitter is yielded as 99.7ps. This compares extremely well with the SDEV measurement given by the TIE.

6.2 Example 2

In Table 1 the Time Jitter Time Interval Error (TIE) results were taken with the LeCroy SDA6000. The Toric 'Results' are calculated from the phase noise spectrum as already shown. In using the TIE method the Golden PLL of the SDA was turned OFF. The lower limit for the Toric integration was taken from the inverse of the SDA record length, i.e. 200us/div = 2ms of data, inverse is 500Hz. The upper limit was 5MHz in each case.

The following results were yielded from the dataset:

Time Base	TIE SDEV	Results	Int B/W
200us/div	131ps	129ps	500Hz – 5MHz
100us/div	134ps	121.3ps	1kHz – 5MHz
500us/div	125.6ps	116.6ps	2kHz – 5MHz

Table 1: Comparison of Jitter Results.

Again the agreement can be seen to be good.

6.3 Example 3

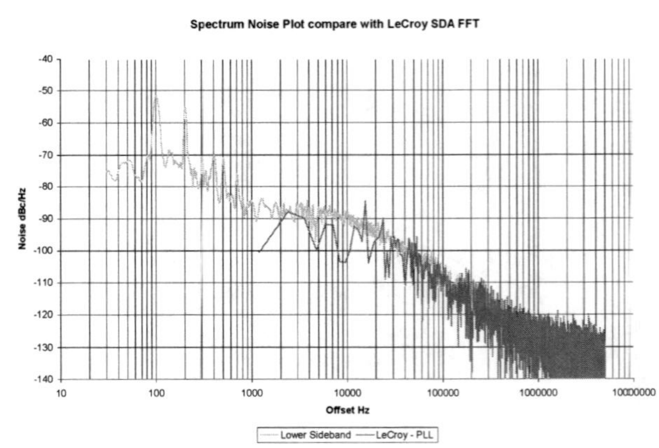

Figure 9: Spectrum Comparison with FFT of Time Jitter

If the data used to produce the plot is anti-logged, then smoothed with a 10point moving average, and then re-logged, this gives the curve shown in 9 . The noise plot taken using the Toric equipment has also been shown for comparison.

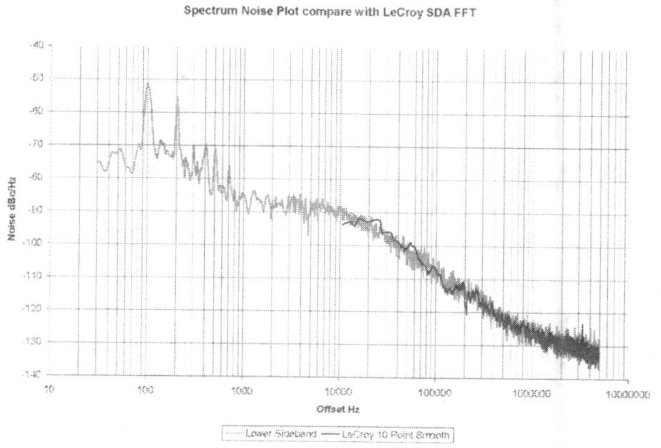

Figure 10: Spectrum Comparison with FFT of Time Jitter after Smoothing of Jitter Data

The averaged noise profile shown in Figure 3 demonstrates that the noise profile given by the SDA6000 follows closely to that measured using a spectrum analyser.

7 Discussion and conclusions

For *random Gaussian noise*, a spectrum measurement can give:-

1. Any type of total time jitter to high accuracy (1%) – an appropriate frequency weighting of the spectrum is used for each type of total jitter.
2. The shape of the time jitter PDF or its 'histogram' to high accuracy.
3. An accurate prediction of BER for a given allowable time window width, or 'clock skew'. Note that the BER projection is obtainable in much less time than from a set of time jitter measurements. Also note that BER prediction depends on the (good?) assumption that the tails of the noise PDFs are exactly Gaussian.

'Deterministic' noise can usefully be modelled as a single dominant spur in random noise, and then:

1. Any type of total time jitter can be found to high accuracy (1%) – an appropriate frequency weighting of the spectrum is used for each type of total jitter.
2. The shape of the time jitter PDF or its 'histogram' can be found to a high accuracy. It is approximated by two half power PDFs, $1/\sqrt{2}$ in amplitude, spaced apart by the peak-to-peak value of the spur sine wave.
3. The BER can then be found from the area of the two 'outer' Gaussian 'tails' of the combined PDF.

Total Jitter of any type can much more quickly be computed from the Spectrum and not from the Jitter Histogram. The type of Jitter determines the 'filter' that has to be applied prior to computing. (Some commercial companies are starting to do this, e.g. Aeroflex PN9000).

In summary most, but not quite all of the outputs of a jitter analyser can be 'constructed' from a single spectrum measurement (and almost always in much less time). The exception is the exact shape of the central part of the PDF if the Deterministic Noise is, for example, a triangular wave. A simple oscilloscope measurement can then be used to identify the wave shape to find PDF central part of the PDF. 'Higher Order Convolution', to produce higher order PDFs, should in principle be able to improve the accuracy of representation of this central part.

8 References

1. M.J. Underhill. "The Noise and Suppression Transfer Functions of the Anti-Jitter Circuit", Proc.2003 Int'l IEEE FCS and 17th EFTF, Tampa, pp 490-498
2. www.tektronix.com/jitter "Understanding and Characterizing Jitter"
3. Simon Haykin, "Communication Systems", 3rd Ed. Section 4.15, pp 295-298, ISBN 0-471-30584 -7

IF-LO Delay Mismatch and Noise Reduction in Transposed Gain Oscillators

S. Bunnjaweht*, M. J. Underhill ** and I. D. Robertson**

*Research Centre of Applied Electromagnetic Waves
Mahanakorn University of Technology, Bangkok 10530, THAILAND
**Microwave Systems Research Group
School of Electronics and Physical Sciences, University of Surrey,
Guildford, Surrey GU2 7XH, UK
Tel: +44 (0) 1483 689862; Fax: +44 (0) 1483 686081
E-mail: sawat@mut.ac.th

Keywords: Oscillators, transposed-gain, sideband noise.

ABSTRACT

This paper describes delay mismatch effects in transposed gain oscillators. Unequal time delay between IF and LO signal paths give rise to sub-optimum local oscillator's noise suppression. It is also shown that the phase response of the transposed gain amplifier is a function of LO signal path and optimum delay matching may not give a condition for oscillation.

1. INTRODUCTION

The block diagram of the transposed gain oscillator, which was introduced by Pan, et al. [1], is shown in Figure 1. It is also known as double frequency conversion oscillator, is so-called because the oscillation-sustaining loop gain is achieved by a transposition of low frequency IF amplifier's gain to higher RF frequencies on both sides of the local oscillator signal. The local oscillator is applied at a frequency close to that required for the gain. By utilizing two mixers, the downconverted high frequency RF signal can be amplified by low frequency amplifiers and then upconverted back to the resonator's frequency using the same LO signal. The upconverted signal is then fed back to a downconversion mixer, mixer1, via the resonator. This resonator determines the oscillation frequency of the oscillator circuit.

The output frequency is stabilized by the operation of both feedforward and feedback paths. If the frequency f1 from the local oscillator begins to rise, then the frequency f2 at the out put of the second mixer decrease proportionally to compensate for this rise in the LO frequency. A similar compensation is also gained when the LO frequency begins to decrease. Thus, the double frequency conversion oscillator achieves precise control of the output frequency, f1+f2, even though the circuit is driven by a non-stable or noisy LO. Very low phase noise microwave oscillators using Sapphire dielectric resonators have been reported based on the transposed gain design [2], [3].

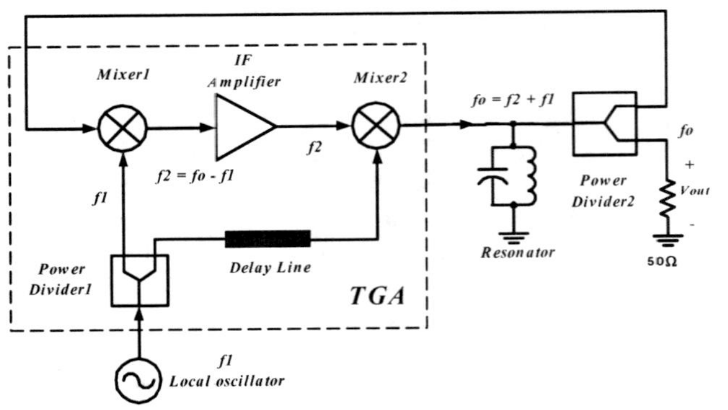

Figure 1Transposed gain oscillator.

2. DELAY EFFECTS AND LO SIDEBAND NOISE SUPPRESSION

Beside the resonator loaded Q, the path delays are found to play an important role in determining LO noise suppression.

If the LO signal with phase noise is represented by an FM (PM) signal as follows [4]:-

$$f_1(t) = \cos[\omega_1 t + \beta \sin \omega_m(t)] \quad (2)$$

where β is a modulation index and ω_m is a modulating signal ($2\pi f_m$). The carrier to single sideband noise power can be calculated from [5]:-

$$\frac{C}{SB} = \frac{1}{(\beta\Delta\tau)^2} \tag{2}$$

where $\Delta\tau$ is a sideband suppression factor and the suppression factor is defined as:-

$$\Delta\tau = \sin\omega_m\left(\frac{T_2 - T_1}{2}\right). \tag{3}$$

T1 and T2 are group delay of LO and IF signal paths respectively. This calculation assumed that both signal paths are linear and have the same dispersion characteristics.

3. TRANSPOSED GAIN AMPLIFIER

Figure 2 shows a block diagram of the transposed gain amplifier or TGA. AS can be seen, transposed gain is achieved using two mixers, an IF amplifier and a local oscillator. The low frequency gain of the IF amplifier is converted to microwave frequencies on both side of the reference LO. Mixer 2 determines the saturated output power of the TGA as same as the output power of the transposed gain oscillator or TGO.

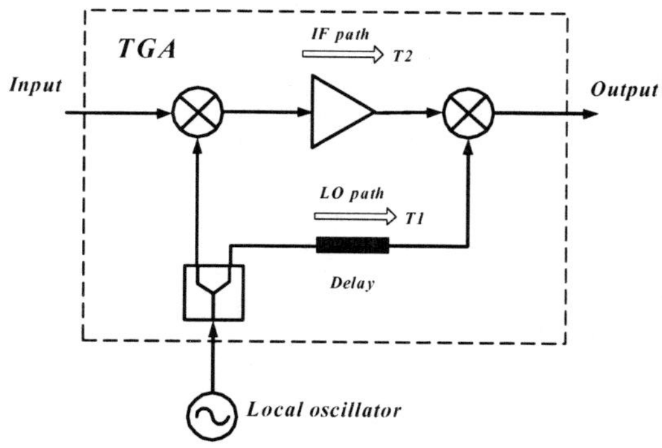

Figure 2 Transposed gain amplifier.

Since the oscillator waveforms play important role in the transformation of *1/f* and the other sources of low-frequency noise into an oscillator's phase noise spectrum [6-8], thus designing schemes based on optimum waveforms are supposed to give good oscillator performance. One of the waveforms of this class is the half-wave symmetric co-sinusoids signal in which contains only odd harmonics. There is no DC Fourier component in this waveform, thus LO leakage is minimized. Because of these requirements, a push-pull amplifier is the circuit of choice.

The push-pull cascode amplifier comprises two identical feedback cascode amplifiers and two baluns is shown in Figure 3. The cascode push-pull amplifier has 24 dB power gain and the noise figure of 4 dB. As the mixer conversion losses are 7dB, this makes a total noise figure of the transposed gain amplifier of 11.03 dB.

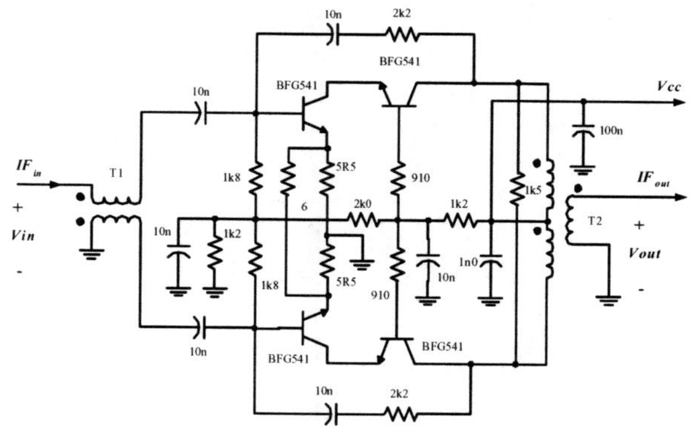

Figure 3 Push-pull cascode IF amplifier.

4. TGA FREQUENCY RESPONSE

A 200 MHz bandwidth TGA has been built with a pair of SME1400-17B WJ electronics high level diode mixers, a push-pull cascode IF amplifier, wideband Wilkinson power divider and a delay line made from PTFE coaxial cable.

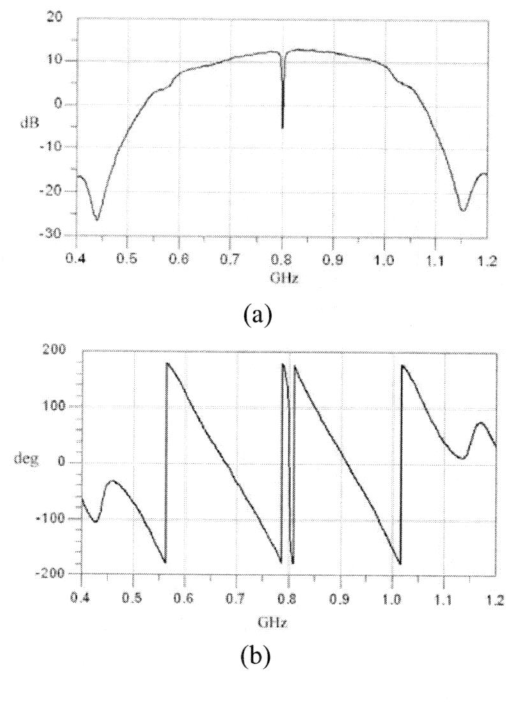

Figure 4 Magnitude (a) and phase (b) response of TGA.

Using LO signal of 800 MHz, typical performance of this amplifier across the double-side bandwidth is shown in figure 4.

Figure 4 (b) shows zero-phase at 675 and 915 MHz. Thus, TGO can be made around these frequencies. This is the main object when the circuit was invented [1]. If the input frequency is fixed and the equaliser time delay is varied, a strong connection of TGA phase response and delay is displayed. A simulated phase response due to time delay variation is depicted in Figure 5.

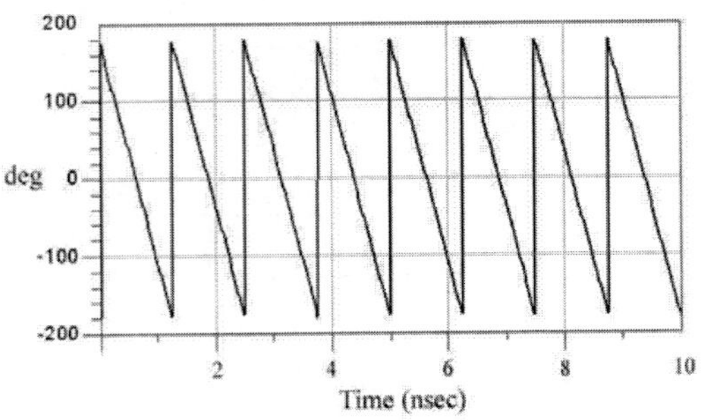

Figure 5 Phase response as a function of equaliser's delay time.

In case of the IF-LO group delay matching does not give a condition for oscillation, an addition phase shifter is needed in the feedback path as shown in Figure 6.

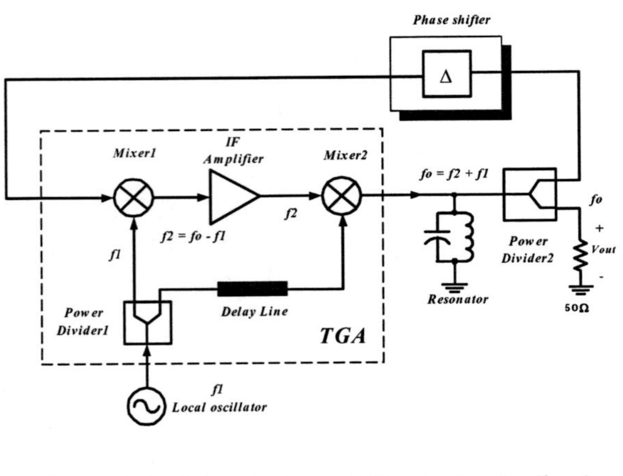

Figure 6 Additional phase shifter in the feedback path.

5. EXPERIMENTS

Figure 7 shows a 670 MHz voltage-controlled TGO. The resonator comprises a back-to-back connection of BB833 varactor diodes and a 22nH Coilcraft air-core inductor. The loaded Q of the voltage controlled resonator is around 15.

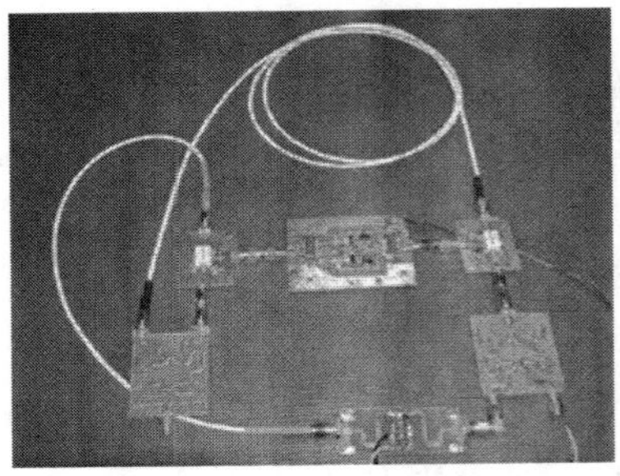

Figure7 670 MHz voltage-controlled TGO.

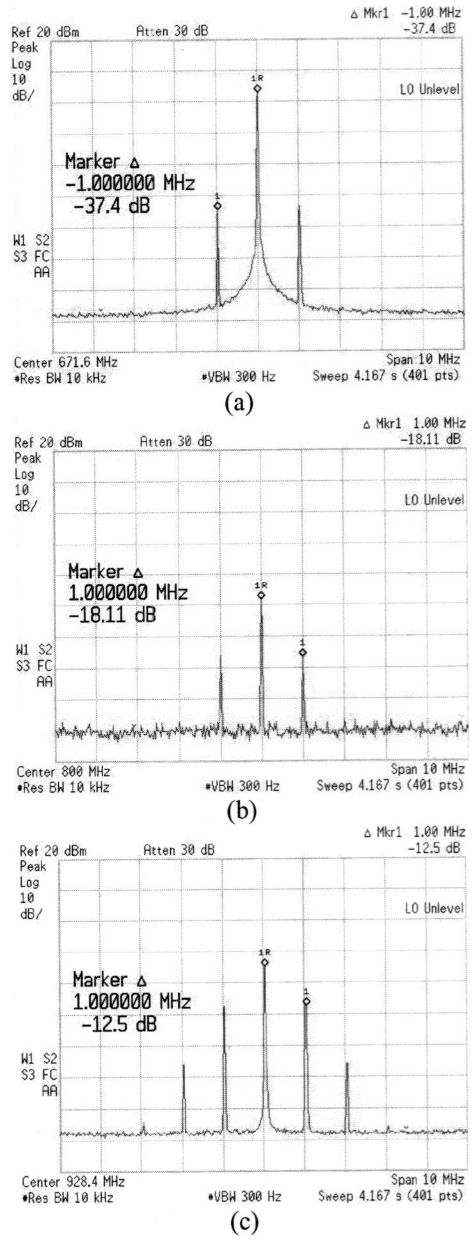

Figure 8 (a) TGO output signal, (b) LO leakage and (c) residual upper-sideband signal.

Phase shifter is included in the feedback path to make circuit oscillate.

The oscillator is driven by 20dBm 800 MHz FM signal. LO carrier has 1 MHz modulating signal with a modulation index of 0.25. Figure 8 (a) shows the experimental results for sideband noise reduction of the LSB output signal when fm = 1MHz, $\beta = 0.25$, and delay mismatch is set to 16 ns. The carrier to sideband is 37.4 dB, as can be seen, the modulating signal is now suppressed by 20dB. Using equations (2) and (3), the calculated carrier to sideband ratio is 39.95 dB. We can see from Figure 8(c) that upper sideband signal has modulation index doubled [5].

6. CONCLUSIONS

The LO sideband noise suppression in the transposed gain oscillator depends on IF-LO path delay mismatch. Phase response of the TGA is a strong function of LO path's delay. The optimum delay matching TGO may not oscillate and phase shifter in the feedback path is required

ACKNOWLEDGMENTS

The authors wish to acknowledge Mahanakorn University of Technology for financial support.
The authors would also like to thank WJ communications, San Jose, CA, for SME1400B-17 high-level double balanced diode mixers.

REFERENCES

[1] Jing-Jong Pan, M. P. Arnold, "High Q Multi-Mode Resonator Controlled Source," U.S. Patent 4,245,193, January 1981.

[2] M.M Driscoll, R.W. Weinert, "Spectral Performance of Sapphire Dielectric Resonator-Controlled Oscillators Operating in the 80K to 275K Temperature Range," *Proc. Frequency Control Symposium*, 1995, pp 404- 412.

[3] J.K.A. Everard, M.A. Page-Jones, "Transposed Gain Microwave Oscillators with Low Residual Flicker Noise," *Proc. Frequency Control Symposium*, 1995, pp 374 - 378.

[4] W.P. Robin, *Phase Noise in Signal Sources*, Exeter UK: Peter Peregrinus, 1998.

[5] S. Bunnjaweht, M. J. Underhill and I. D. Robertson, "Sideband Noise Reduction in Transposed Gain Oscillators," *IEEE Proceeding ISCAS2003*, pp713-716, May. 2003.

[6] A. Hajimiri, T.H. Lee, "A General Theory of Phase Noise in Electrical Oscillator," *IEEE J. Solid State Circuits*, vol. SC-33, pp197-194, Feb. 1998.

[7] A. Hajimiri, T.H. Lee, *The Design of Low Noise Oscillator*, Boston, MA: Kluwer Academic Publisher, 2000.

[8] H. B. A van der Ziel, K. Amberiadis," Oscillator-with Odd-Symmetrical Characteristics Eliminates Low-Frequency Noise Sidebands," *IEEE Trans. on Cir. and Syst.*, vol. CAS-31, no. 9, Sept. 1984.

PHASE NOISE RESONATOR MEASUREMENT BENCH USING DDS AS EXCITATION SOURCE

F. STHAL*, D. VERNIER†, D. GILLET†, N. TRENQUE*, S. GALLIOU*

* Département Laboratoire de Chronométrie, Electronique et Piézoélectricité (LCEP)
† Département Laboratoire de Physique et Métrologie des Oscillateurs (LPMO),
FEMTO-ST, UMR CNRS 6174
ENSMM, 26 Chemin de l'Épitaphe, 25030 Besançon cedex – France.

Keywords: Direct Digital Synthesis, phase noise, resonator.

Abstract

Accuracy of resonator phase noise measurement systems in a passive method depends on the stability of the excitation source. Good accuracy over a quite large frequency range is needed because of the various resonant frequencies of the resonators studied. In this paper, Direct Digital Synthesis solution is investigated as excitation source of measurement bench. Various stability measurements on this source are presented. Phase noise floor of the passive measurement system is given as well as results of pairs of quartz crystal resonators. Finally, limitations presented by this synthesizer applied to phase noise measurement systems, are discussed.

1 Introduction

Direct Digital Synthesizers (DDSs) have been under investigation for many years [1-12]. A lot of works talk about this method to obtain a sine wave with a good spectral purity. The phase noise of this kind of source has been also studied [13-18]. Recent off-the-shelf devices with higher speed allows now an easy implementation [19-20].
Usually, resonators dedicated to ultra stable oscillators have a resonant frequency lower than the nominal value of the oscillator frequency. As a consequence, excitation source (ES) of the passive measurement bench has to be tuned over a frequency range larger than the VCXO's. Moreover, the bench frequency adjustment, performed by classical capacitor tuning, is commonly too much limited. Since we also need some good accuracy over a quite large frequency range, we decided to investigate a DDS solution for this purpose.
A dedicated printed circuit board was developed using an off-the-shelf DDS, driven by a 100 MHz source, and programmed via a specific parallel link from a personal computer.
Stability measurements on this source are presented in the paper. The phase noise floor of the passive measurement system of resonator phase noise is given as well as results of pairs of 10 MHz and 5 MHz quartz crystal resonators. Finally, limitations of the synthesizer applied to phase noise measurement systems, are discussed.

2 Direct Digital Synthesizer

2.1 Specifications of the device

For Direct Digital Synthesis, an AD9852 commercial model from *Analog Devices* company is used. It provides a 48 bits phase accumulator, a 14 bits output converter and a working frequency that can be as higher as 300 MHz. This model has been designed for numerical telecommunication applications. Several modes can be used, the fundamental one called single-tone is the only retained for our application. It generates a sinusoidal signal by scanning a memorized wave-table and using a digital-analog conversion.
The frequency given by the DDS is :

$$F_{out} = (p/2^{48})\cdot F_{clock} \qquad (1)$$

where p is the step of the phase accumulator defined as a 48 bits unsigned integer. As 2^{48} is closed to $10^{14.5}$, the frequency resolution at 10 MHz is closed to one μHertz [21].
We developed a printed circuit board with a driving module based on a Electrical Programmable Logic Device (EPLD) that manages the parallel port from a personal computer. It allows to program the DDS registers (Fig. 1).

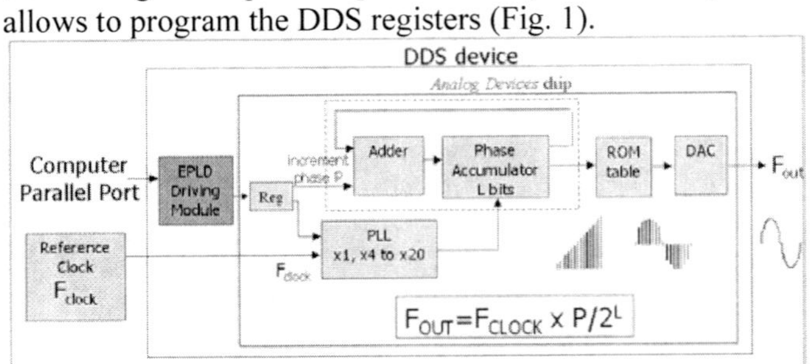

Fig. 1. Schematic of the DDS device.

In order to sink heating due to the electrical consumption of the DDS, a copper thermal drain bridge was mounted in the case. The 5 or 10 MHz output signal power is only –5 dBm. So, an external amplifier stage has to be placed at the DDS output after the external anti-aliasing filter. Then, the rejection of this low-pass-amplifier stage is greater than 40 dBc on the harmonics of the 10 MHz Intermediary Frequency (IF).

Front panel Rear panel

Fig. 2. DDS included AD9852, EPLD and Parallel Port Connector.

2.2 Characterization of the DDS

2.2.1 Spurious modes

Output of the DDS device is measured using a network analyzer to obtain the discrete line spurs normally present when the signal is digitalized (Fig. 3a). These spurs due to the clock pulses must be suppressed using a low-pass filter (Fig 3b). The DDS reference frequency 100 MHz, provided by a Hydrogen-maser, and the output signal frequency is 10 MHz. The filter attenuation is better than 50 dB.

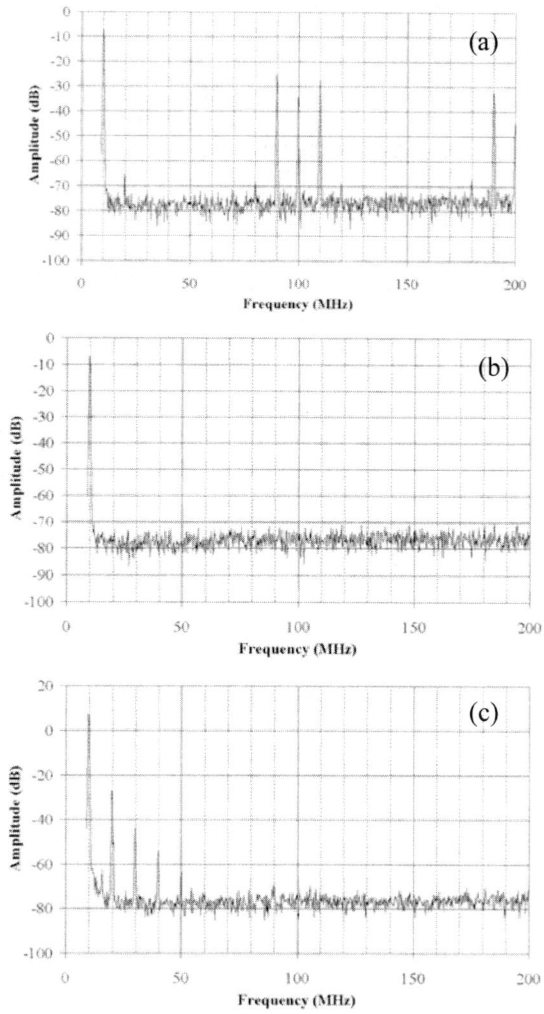

Fig. 3. Spurious modes far from the output frequency of the DDS (a) DDS output, (b) DDS output + low pass-filter, (c) DDS output + low pass filter + 15 dB external amplifier.

Because of the low level of the DDS output, 15 dB or 20 dB external amplifier are used to obtain a sufficient level to drive the measurement bench of resonator phase noise which must be about 4 dBm. Fig. 3c shows the spectrum of the output sine wave. Attenuation between the fundamental and the second harmonic allows us to consider output signal as a sine wave with an acceptable purity.

2.2.2 Phase noise of the source

Signal analysis closed to the carrier frequency is performed using a classical measurement bench of oscillator phase noise. Output phase noise of the DDS device are compared with different oscillator ones.

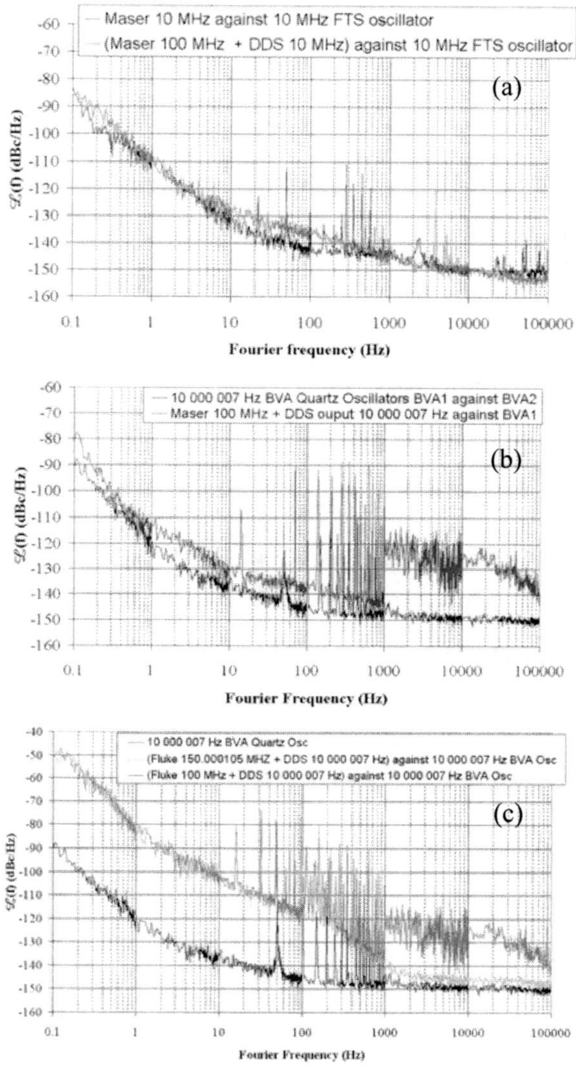

Fig. 4. Phase noise measurement of the DDS output.
(a) H-maser 100 MHz input, output 10 MHz.
(b) H-maser 100 MHz input, output 10 MHz + 7 Hz.
(c) Fluke 6160B 100 MHz input, output 10 MHz + 7 Hz and Fluke 6160B 150 000 105 Hz input, output 10 MHz + 7 Hz.

Fig. 4a shows the Fourier spectrum of the DDS output tuned at 10 MHz. The DDS is driven by a 100 MHz H-maser and the internal phase lock loop (PLL) is not used. The phase noise of the DDS output is measured against a quartz crystal oscillator (FTS oscillator) which has been compared directly to the 10 MHz H-maser signal. The DDS doesn't increased the phase noise level of the H-maser except in the 10 to 1000 Hz frequency range. Because the ratio between the DDS input frequency and DDS output frequency is here an integer, the comb of spurious is limited.

Fig 4b shows the phase noise of the DDS output when the ratio between the input clock and the output frequency is not an integer. The DDS output frequency is set at 10 MHz + 7 Hz. The resulting phase noise is obtained against a BVA oscillator with a resonant frequency upper 10 MHz (10 MHz + 7 Hz). In this case, the comb of spurious signals is really heavy. For Fourier frequencies from 10 to 1000 Hz, the measurement bench resolution allows to distinguish all the spurious frequencies. Above 1000 Hz, spurious frequencies cannot be distinguished and we observe a continuum spectrum due to the integration of the discrete spurs. The comb of the spurious signals can be suppressed as shown in Fig. 4c using an integer ratio between the sampled clock and the output frequency (Fluke 6160B input clock at 100 MHz compared with 150 000 105 Hz).

Because the output transitions of the DDS can only occur coincidentally with clock pulses, generally zero crossings of its output do not coincide with theoretical zero crossings of the ideal waveform to be approximated.

Considering that, a simple qualitative approach gives:

$$\frac{F_{clock}}{F_{out}} = \frac{Q}{P} = \frac{T_{out}}{T_{clock}}$$

where F_{clock} is the sampled frequency (frequency of the DDS input clock), F_{out} is the frequency of the output signal. Q and P are the smallest two integer numbers the ratio of which is equal to the ratio F_{clock}/F_{out}.

If k^- is the integer part of Q/P and $k^+ = k^- + 1$, we can write:

$$N^+ \cdot k^+ + N^- \cdot k^- = Q$$

$$N^+ + N^- = P$$

Taking into account the 1 μHz DDS resolution, P and Q used in decimal base is sufficient and simplify the calculus. Thus in Fig. 5, with a division ratio of

$$100\,000\,000/10\,000\,007 = 9.999993$$

between the input clock frequency and the output frequency, only an number N^- of $k^- = 9$ periods of input clock and a number N^+ of $k^+ = 10$ periods of input clock pulses can be associated together to generate the "true" basic periodicity of the signal which is equal to $Q \cdot T_{clock} = 1$ s.

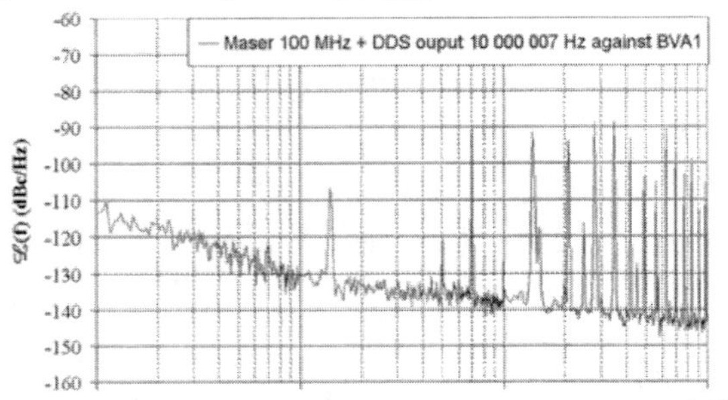

Fig. 5. Phase noise measurement of the DDS output, Fig 4b zoom between 1 to 1000 Hz . DDS input frequency 100 MHz with Maser, DDS output frequency 10 MHz + 7 Hz.

In this case, one find $N^- = 70$ and $N^+ = 9\,999\,937$ during the time of 1 second. As a consequence, a modulation of the output signal appears at 70 Hz and its harmonics in the close to the carrier spectrum (Fig. 5). Moreover, due to the integer number of clock periods, an intermodulation between the output signal and the frequency generated by the integer number of clock periods ($F_{clock}/10$) is provided at 7 Hz and its harmonics.

At 5 MHz, this phenomenon is observed again when the output frequency of the DDS is moved around 5 MHz (Fig. 6).

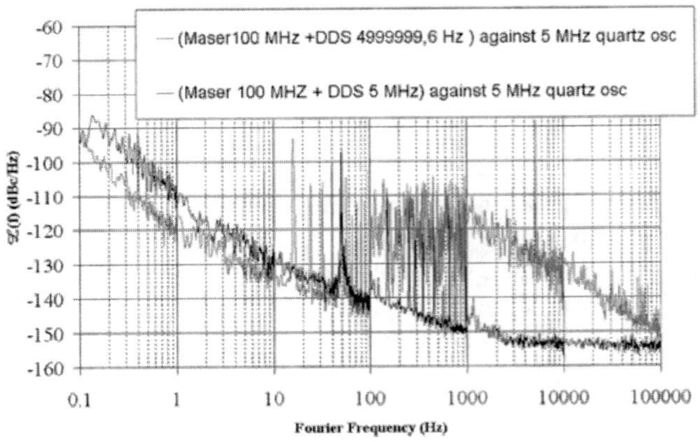

Fig. 6. Phase noise measurement of the DDS output.

In this case, modulation of the output signal is measured at 8 Hz and intermodulation occurred at 0.4 Hz.

3 Measurement bench of resonator phase noise

3.1 Measurement system principle

Phase noise of both couples of quartz crystal resonators is measured in a commercial version of carrier suppression bench [22] (Fig. 7). The carrier signal of the ES is split into equal parts to drive both devices under test (DUT). DUTs are resistors to measure the noise floor of the system or crystal resonators.

A carrier suppression of 10 to 60 dB can be reached but a sufficient level at the input of the phase noise detector is necessary to maintain a low noise floor (4 dBm is the minimum level). The struggle in using this system is to get enough carrier suppression to get a good measurement of resonator noise without saturating the output amplifiers.

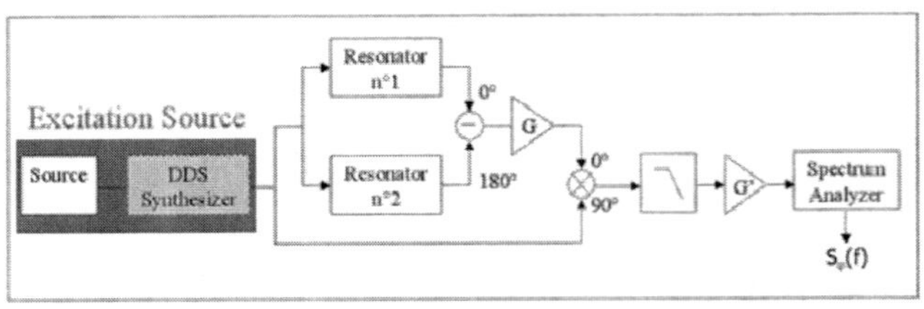

Fig. 7. Resonator measurement system.

Dissipated powers into resonators are adjusted according to their working powers in oscillator and each resonator is separately temperature-controlled in order to work at its own turn-over temperature.

Each arm of the bench is tuned at the ES frequency. A quartz crystal oscillator with low phase noise is usually used to measure resonators but the bench frequency adjustment, performed by classical capacitor tuning, is limited. In some cases, VCXOs and tuning capacitors don't allow the measurement of resonator which have a resonant frequency outside the range of the bench. Thus, 10 MHz and 5 MHz resonators have been tested using H-maser and DDS as ES.

3.2 Experimental results

3.2.1 Noise floor of the bench

Noise floor of the measurement bench of resonator phase noise is given at 10 MHz (Fig. 8). This result is obtained using resistors equal to 91 Ω instead of resonators. Comb of spurious signals appears when ratio between the clock frequency and the DDS output frequency is not an integer.

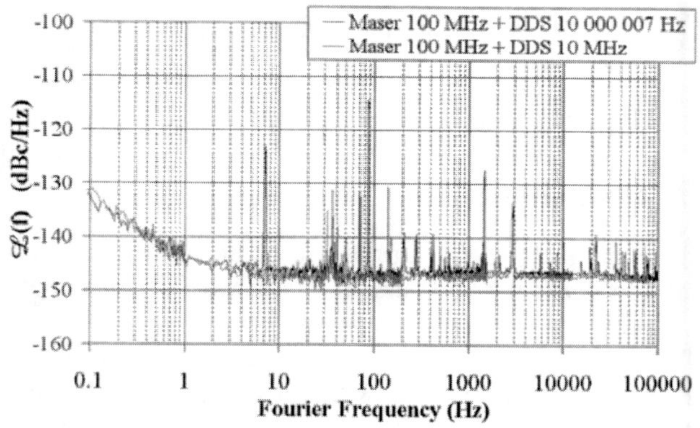

Fig. 8. Noise floor of the measurement system at 10 MHz.

As expected, peaks appear at 0.4 Hz, 0.8 Hz and 8 Hz for an output frequency equal to 4 999 999.6 Hz (Fig. 9). This kind of spurious, which will be in the resonator bandwidth, could mask the resonator noise. The DDS device doesn't increase significantly the floor of the measurement bench.

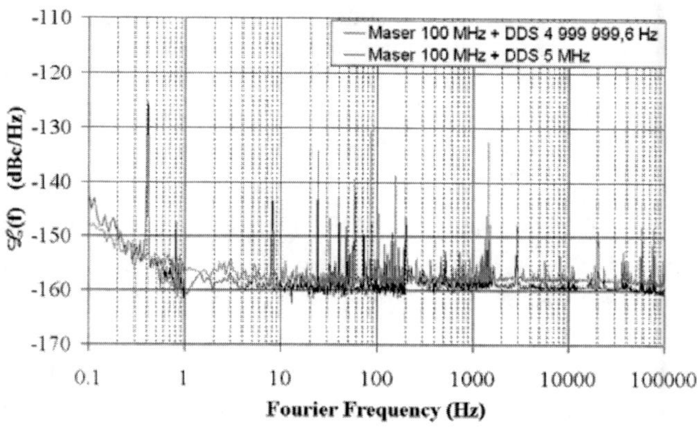

Fig. 9. Noise floor of the measurement system around 5 MHz.

3.2.2 Measures of resonator noise

Typically, crystal frequencies are 50 Hz below 10 MHz and these resonators cannot be measured using a 10 MHz VCXO's source. These resonators exhibit typical inherent quality factors of about $Q \approx 1.2 \cdot 10^6$ and motional resistors $R \approx 85\ \Omega$.

The frequency of the ES is set at 9 999 960 Hz. Fig. 10 shows the phase noise of this pair of resonators measured using three different sources: an Agilent 33250A which is a commercial DDS generator, a high purity Fluke 6160B synthesizer locked on a Cesium clock and our DDS device driven by a 100 MHz H-maser.
The low-pass filter effect of the quartz crystal is observed. Number of spurs is limited when the DDS device is used.

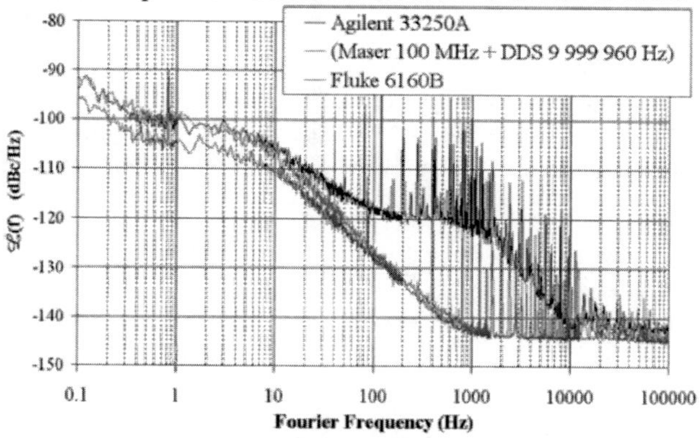

Fig. 10. Phase noise of 9 999 9950 Hz quartz crystal resonators.

When the phase noise of the ES is too high, the partial carrier suppression, using in the bench, leads to a bad noise source rejection at the output of the measurement system (see the "Agilent" curve in Fig. 10).
Loaded Qs of quartz crystals are found using the Leeson's frequencies (Fig. 11).

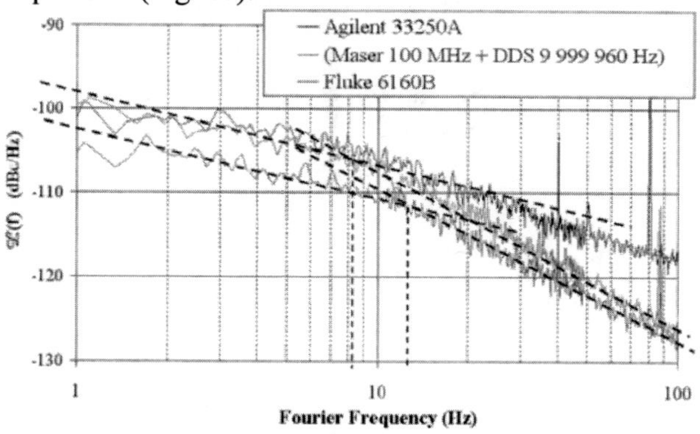

Fig. 11. Leeson's frequency measurements.

It allows to find the standard deviation of the resonators presented in table 1. Because of the 1/f frequency noise of the resonators, the measured phase noise depends on the loaded Q. Inside the resonator bandwidth, the lower the loaded Q is, the smaller ℒ(f) is. $\sigma_y(1\ s)$ of the resonators is about $1.3\ 10^{-11}$.

Source	Fluke 6160B	Maser + DDS
$\mathcal{L}_{mes}(1\ Hz)$ (dBc/Hz)	-105	-100
$S_\phi(1Hz)$ (dBrad²/Hz)	-102	-97
f_L (Hz)	13.4	8.2
$Q_L = f_0/(2*f_L)$ (10^6)	0.4	0.6
$\sigma_y(1s) = \sqrt{2 \cdot \ln 2 \cdot S_\phi(1 Hz)} / (2 \cdot Q_L)$	$1.25 \cdot 10^{-11}$	$1.36 \cdot 10^{-11}$

Table 1. Phase noise of the 10 MHz resonator pair.

Similar phase noise measurements have been achieved on 5 MHz BVA resonators. The resonant frequencies of both resonators are 4 999 994.7 and 4 999 992.5 Hz, respectively. Inherent Q factors are about $2.4 \cdot 10^6$ and the motional resistances are about 60 Ω. Fig. 12 shows the measures of this couple of resonators. Quartz crystals have been measured with a 5 MHz ultra-stable quartz crystals oscillator and with the DDS device. In this last measurement, the frequency of the ES is set at 4 999 999.6 Hz and the input clock is H-maser 100 MHz. Because of the non integer ratio between F_{clock} and F_{out}, the comb of spurs appears in the spectrum even if the phase noise of the source is cancelled by the principle of the measurement bench (0.4 Hz and 8 Hz fundamental frequencies).

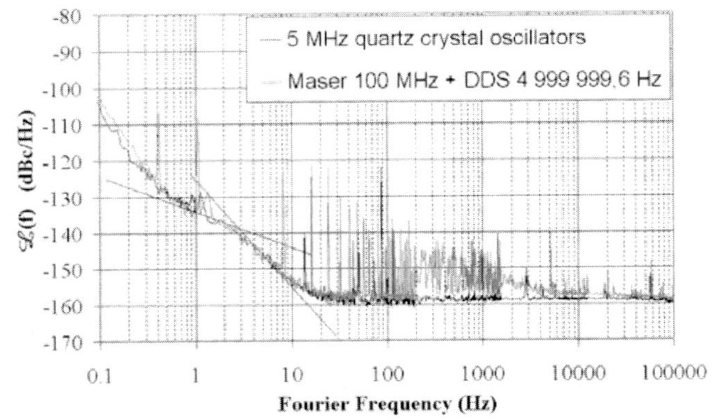

Fig. 12. Resonator measurement system.

$\sigma_y(1\ s)$ is shown in Table 2. Results obtained with the DDS device is well correlated with the results given by a classical oscillator as an excitation source.

Source	Oscillator	Maser + DDS
$\mathcal{L}_{mes}(1\ Hz)$ (dBc/Hz)	-134	-134
S_ϕ (dBrad²/Hz)	-131	-131
f_L	2.8	2.8
$Q_L = f_0/(2 * f_L)$ (10^6)	0.89	0.89
$\sigma_y(1s) = \sqrt{2 \cdot \ln 2 \cdot S_\phi(1 Hz)} / (2 \cdot Q_L)$	$1.86 \cdot 10^{-13}$	$1.86 \cdot 10^{-13}$

Table 2. Phase noise characteristics of 5 MHz resonator pair.

The short-term stability of this 5 MHz crystal pair is about $1.8 \cdot 10^{-13}$ and can be considered as ultra-stable resonators.

4 PLL Limitation

In this work, we have seen that H-Maser 100 MHz as input clock of the DDS device can be used to measure the phase noise of ultra-stable resonators. One advantage of the *Analog Devices* DDS is that a PLL is integrated, allowing an input clock with lower frequency. This one could be a 10 MHz ultra-stable oscillator. Fig. 13 shows test results of this solution. The 10 MHz H-maser is used as the DDS input clock and the PLL function is activated with two ratios (x20 and x10). It seems that the PLL function increases considerably the noise floor of the ES. This result is in accordance with manufacturer results [20].

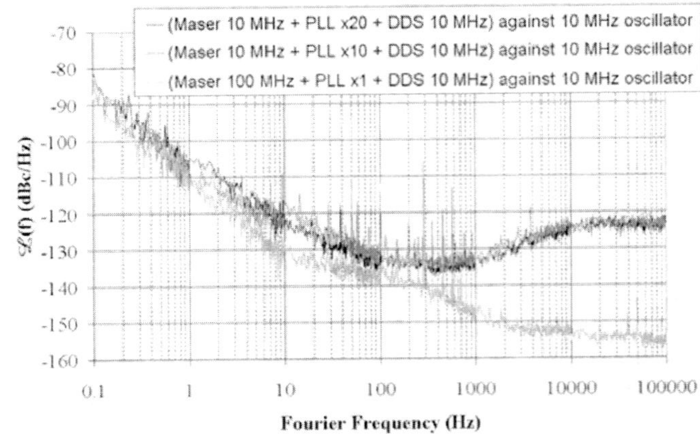

Fig. 13. PLL limitation of the DDS device.

Using this kind of source leads to have a source like the Agilent synthesizer which is unusable in our case.

5 Conclusion

The measurements of ultra-stable resonators using a DDS device as an excitation source in a passive measurement bench of phase noise have been demonstrated.

The DDS system generates a lot of spurious signals in the case of a non integer ratio between the input frequency and the DDS output frequency. These spurs can mask the phase noise of the resonator. In all cases, it is necessary to use DDS carefully and trying to adjust the output frequency according to the constant frequency of the DDS reference input, resonator frequencies and tuning capacitors of the bench.

The DDS device must be driven with a low noise input source and unfortunately the PLL function of this kind of devices seems to be unusable.

References

[1] C.E. Weahtley III, D.E. Phillips, "Spurious suppression in direct digital synthesizers", Proc. 35th IEEE Ann. Freq. Cont. Symp., Philadelphia, Pennsylvania, 27-29 May, pp. 428-435, (1981).

[2] D. P. O'Rourke, "SOS frequency synthesizers development", Proc. 40th IEEE Ann. Freq. Cont. Symp., Philadelphia, Pennsylvania, 28-30 May, pp. 370-372, (1986).

[3] H. T. Nicholas III and H. Samueli, "An analysis of the output spectrum of direct digital frequency Synthesizers in the presence of phase-accumulation truncation", Proc. 41st IEEE Ann. Freq. Cont. Symp., Philadelphia, Pennsylvania, 27-29 May, pp. 495-502, (1987).

[4] V. S. Reinhardt, "Spur reduction techniques in direct digital synthesizers", Proc. 47th IEEE Ann. Freq. Cont. Symp., Salt Lake City, Utah, 2-4 June, pp. 230-241, (1993).

[5] L. J. Kushner and M. T. Ainsworth, "A spurious reduction technique for high-speed direct digital synthesizers", Proc. 50th IEEE Ann. Freq. Cont. Symp., Honolulu, Hawai, 5-7 June, pp. 920-927, (1996).

[6] K. A. Essenwanger and V. S. Reinhardt, "Sine output DDSs a survey of the state of the art", Proc. 50th IEEE Ann. Freq. Cont. Symp., Honolulu, Hawai, 5-7 June, pp. 370-378, (1996).

[7] J. Vankka, "Spur reduction techniques in sine output direct digital synthesis", Proc. 50th IEEE Ann. Freq. Cont. Symp., Honolulu, Hawai, 5-7 June, pp. 951-959, (1996).

[8] J. Vankka, "A direct digital synthesizer with a tunable error feedback structure", IEEE Trans. on Communications., vol 45, no. 4, April, pp. 416-420, (1997).

[9] V. F. Kroupa, V. Cizek and H. Svandova, "Spurious signals in direct digital frequency synthesizers due to phase truncation", IEEE Trans. on Ultrasonics, Ferroelectrics and Freq. Cont., vol 47, no. 5, Sept., pp. 1166-1171, (2000).

[10] M.A. Taslakov, "Direct digital synthesizer with improved spectrum at low frequencies", Proc. 54th IEEE Ann. Freq. Cont. Symp., Kansas City, Missouri, 7-9 June, pp. 280-283, (2000).

[11] R. Barillet, F. Hamouda, D. Venot and C. Audoin, "Limitation of the frequency stability by local oscillator phase noise: new investigations and natural improvements", IEEE Trans. on Ultrasonics, Ferroelectrics and Freq. Cont., vol 47, no. 5, Sept., pp. 1152-1158, (2000).

[12] J.M.P. Langlois and D. Al-khalili, "A new approach to design of low power direct digital frequency synthesizers", Proc. 56th IEEE Ann. Freq. Cont. Symp., New Orleans, Louisiana, 29-31 May, pp. 654-661, (2002).

[13] E. M. Mattison and L. M. Coyle, "Phase noise in direct digital synthesizers", Proc. 42th IEEE Ann. Freq. Cont. Symp., Baltimore, Maryland, 1-3 June, pp. 352-356, (1988).

[14] V. N. Kuleshov, H. Y. Liu and B. E. Leshukov, "1/f fluctuations sources in direct digital frequency synthesizers and their contribution to the output oscillations power spectral density", Proc. 49th IEEE Ann. Freq. Cont. Symp., San Francisco, California, 31 May - 2 June, pp. 934-941, (1995).

[15] V. N. Kuleshov, H. Y. Liu, "Fundamental noise in direct digital frequency synthesizers", Proc. 49th IEEE Ann. Freq. Cont. Symp., San Francisco, California, 31 May - 2 June, pp. 288-293, (1995).

[16] V. F. Kroupa, "Close to the carrier", Proc. 50th IEEE Ann. Freq. Cont. Symp., Honolulu, Hawai, 5-7 June, pp. 934-941, (1996).

[17] V. F. Kroupa, "Phase and amplitude disturbances in direct digital frequency synthesizers", IEEE Trans. on Ultrasonics, Ferroelectrics and Freq. Cont., vol 46, no. 3, May, pp. 481-486, (1999).

[18] A. SenGupta and F.L. Walls, "Effect of aliasing on spurs and PM noise", Proc. 54th IEEE Ann. Freq. Cont. Symp., Kansas City, Missouri, 7-9 June, pp. 541-548, (2000).

[19] Analog devices, "A technical Tutorial on Digital Signal Synthesis", Analog Devices, Inc., (1999)

[20] Analog devices, "CMOS 300 MSPS, Complete-DDS, AD9852", Datasheets, Analog Devices, Inc., (2002).

[21] F. Lardet-Vieudrin, P. Salzenstein, D. Vernier, D. Gillet, M. Chaubet and V. Giordano. "Design and realization of a 100 MHz Synthesis chain from an X-band reference signal", Proc. IEEE Int. Freq. Cont. Symp. jointly with the 17th EFTF, Tampa, Florida, 4-8 May, pp. 560-564, (2003).

[22] F. Sthal, M. Mourey, F. Marionnet, W.F. Walls, "Phase Noise Measurements of 10 MHZ BVA Quartz Crystal Resonators", IEEE Trans. on Ultrasonics, Ferroelectrics and Frequency Control, vol 47, no. 2, march, pp. 361-365, (2000).

MEASUREMENT OF SHORT-TERM STABILITY OF ULTRA STABLE OSCILLATORS

P.Y. Bourgeois[1], **Y.Kersalé**[1], **N.Bazin**[1],
M.Chaubet[2], and **V.Giordano**[1]
[1]Institut FEMTO-ST, UMR 6174, département L.P.M.O., Besançon, France
[2]Centre National des Etudes Spatiales, Toulouse, France
Email: pierre-yves.bourgeois@lpmo.edu

Keywords: Sapphire Resonator, Whispering Gallery modes, short term stability

Abstract

We present in this paper the progress report of the building of two cryogenic sapphire oscillators (CSO) at LPMO. A description of our original resonator configuration is discussed. The measurement of a short-term frequency instability better than $2 \cdot 10^{-14}$ for short integration times is presented. We also demonstrated an exceptional long term frequency stability of better than $5 \cdot 10^{-14}$ over one day. We finally report the usefullness of CSO in the characterization of state of the art ultra stable quartz cristal oscillators in a simple way of measurement.

1 Introduction

At the LPMO, we have recently undertaken the construction of two cryogenic sapphire oscillators with the support of the french Space and Metrology agencies, *i.e.* the CNES and the BNM. Physical properties of sapphire single crystal, which exhibits ultra low dielectric losses, enable to achieve very high Q-factor at liquid helium temperature. As a consequence, the sapphire resonator constitutes an excellent terrestrial frequency reference for applications such as Lorentz invariance tests (ESR) [1], time and frequency metrology, or the qualification of instruments. In previous works [2], our cryogenic sapphire oscillator (CSO) presented a less than $2 \cdot 10^{-14}$ mean term frequency instability with long term presenting a random walk $\tau^{1/2}$ process corresponding to a frequency instability of the order of $2 \cdot 10^{-12}$ per day integration time (drift not removed in calculation). The short-term was limited by the measurement system noise floor constituted by a microwave synthesizer locked on a Hydrogen maser. In this paper, we describe the implementation of our two CSO, especially the cryogenic resonator and oscillator structure, and the coupling adjustements. Eventually the short term and long term frequency stability in terms of fractional frequency instability are commented. Finally we demonstrate the usefullness of such a frequency reference to qualify other types of ultra stable oscillators (USO).

2 Resonator Structure

The resonator that we used is a $50mm$ diameter $20mm$ height disk machined high purity sapphire single crystal. These monocristalline dielectric resonators (DR) exhibit very high Q-factors as high as 10^9 (*i.e.* low dielectric losses) especially at liquid helium temperature [3]. The main disadvantage of this kind of DR is the frequency sensitivity to the temperature. But providing some paramagnetic impurities [4] due to the growing process, we can take the advantage of a turnover temperature just over the boiling point of the liquid helium, *i.e.* around $6K$. We have choosen to operate in the Whispering Gallery (WG) mode configuration. WG modes are characterized by a high level of confinement of the electromagnetic fields in the sapphire rod near the curved air-dielectric interface and by a Q-factor essentially limited by the sapphire losses. A $d = 5mm$ diameter hole has been drilled along the rod axis to enable a rigid mechanical mounting. Sapphire resonator is usually enclosed in a cylindrical metallic cavity made in copper or niobium [5] [6]. This cavity prevents from radiation losses, ensures thermal shielding and enables a stable mechanical mounting of the resonator and of its coupling structure. In our preliminary experiments, most of the WG modes have been identified with such a closed cavity. Nevertheless most of them appeared perturbed by low-Q spurious resonances corresponding to the empty cavity modes. After few trials we rapidly realized that it would be difficult to simultaneously achieve an optimal resonator coupling and a low densisty of spurious modes. It is well known that these spurious modes can induce Q-factor degradation and thermal sensitivity enhancement of the main resonance [7] [8]. Spurious modes interaction with the main resonances should be reduced. Several methods for suppressing undesired modes have been already demonstrated. The deposition of thin metallic lines on the sapphire disk has been proposed by Di Monaco *et al* [9] [10]. This technique is efficient for room temperature resonators but induces detrimental extra losses at low temperature and so is not suitable for our purpose. Tobar *et al.* proposed a modified cavity where slots have been machined to make the cavity radiating for the spurious modes. The slots position depends on the modes we want to excite and then a particular cavity is not suitable for the entire WG mode family. In a preceeding work devoted to the measurement of the dielectric properties of langasite crystal [11], we demonstrated the efficiency of an open cavity structure to excite whispering gallery modes in good conditions. We then decided to test the sapphire resonator in

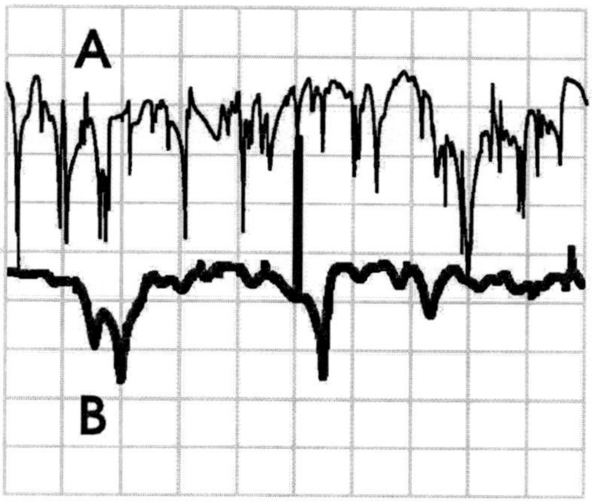

Fig. 1. Resonator Transmission spectra around the $WGH_{17,0,0}$ at 4.2K. A)in a closed cavity. B) in an open cavity. Frequency span : 1 GHz. Center frenquency: 12.17 GHz. Vertical scale 10dB/div. Vertical reference value (center of the screen): -50dB.

this configuration. In this structure, the cylindrical wall of the cavity was removed and replaced by microwave absorber. Spurious modes that are not well confined in the dielectric rod are greatly affected by the absorber layer. Conversly, high order WG modes are not drastically altered. Figure 1 demonstrates the advantage of the open cavity compared to the closed one. It represents the magnitude of the resonator transmission coefficient recorded at 4.2K around the quasi-TM $WGH_{17,0,0}$ mode at 12.17 GHz for the two structures. In the closed cavity (fig.1A), low–Q resonances are numerous yielding to an apparent transmission background of -30dB completely masking the main resonance. Although the later can be observed by reducing the frequency span, its shape is altered by nearby spurious resonances. In the open cavity (fig.1B), spurious resonances are completly eliminated and the main resonance clearly appears surimposed on a –50dB background.

Despite its simplicity the open cavity enables to achieve high–Q factors at low temperature. The experimental Q–factor of the quasi–Transverse–Magnetic (quasi–TM or $WGH_{m,0,0}$) modes measured at 4.2K are given in figure 2. The Q–factor regularly increases with the azimuthal number m. When m is higher than 15, the Q-factor no longer increases and for $m = 19$ it reaches an asymptotical value of the order of 5×10^8.

To explain this behavior, we simulated the effect of the absorbing layer thanks to a Mode Matching Model similar to those described in [12]. Frequencies and electromagnetic fields have been computed for a closed cavity with the sapphire rod inside. The absorbing layer, assumed to cover the inner cylindrical wall, is treated as a perturbation. The

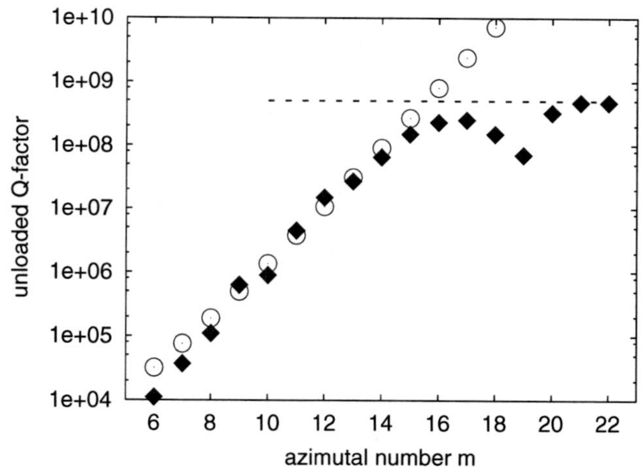

Fig. 2. Unloaded Q–factors vs azimuthal mode number m at 4.2K. ♦: experimental Q–factors. ○: Q_a resulting from the absorber layer.

resulting Q–factor limitation Q_a is then calculated as:

$$Q_a = \frac{W_{tot}}{\int_{V_a} (\varepsilon'' EE^* + \mu'' HH^*)\, dv} \qquad (1)$$

where ε'' (μ'') is the imaginary part of the absorber relative permittivity (permeability). We assumed $\varepsilon'' = 6.47$ and $\mu'' = 2$, *i.e.* the values provided by the manufacturer. Moreover ε'' and μ'' are assumed constant with the temperature. V_a is the volume of the absorber. E and H are the unperturbed electric and magnetic fields and W_{tot} the total energy stored in the structure.

In figure 2, the computed values Q_a for the $WGH_{m,0,0}$ modes are compared to the experimental Q–factors. As expected for low order modes the experimental Q–factor is effectively limited by the losses induced by the microwave absorber. The Q–factors for $m = 18$ and $m = 19$ modes do not follow the expected values. We do not know yet the exact cause of this discrepancy. It could be due to some extra losses induced by a non–perfect coupling or by some paramagnetic resonance of included impurities. The observed asymptotical value, *i.e.* 5×10^8 is lower than some other published results observed with high-quality sapphire crystal [5].

The high degree of confinement of the electromagnetic energy very close to the air-dielectric interface enables also a simple screw-mounting structure in the centre of the dielectric resonator. For this experiment we have choosen to operate $WGH_{16,0,0}$ at 11.56 GHz for both resonators. The 2 resonators are separated by a few $104kHz$ that reveals to be sufficient for the good operation of the electronic counter. The two cryogenic insert structures are identical and schematized in figure 3. The sapphire crystal is maintained with a nut on a OFHC copper cold-finger thermally connected to the top flange of the stainless steel can. The cold finger also supports a germanium thermometer and a heater for thermal control. The axial dimension of the resonant structure is limited by two copper lids placed at $20mm$ from the sapphire flat surfaces. The upper lid is fixed to the cold finger. The bottom one is maintained by four $10mm$ diameter copper posts. The internal

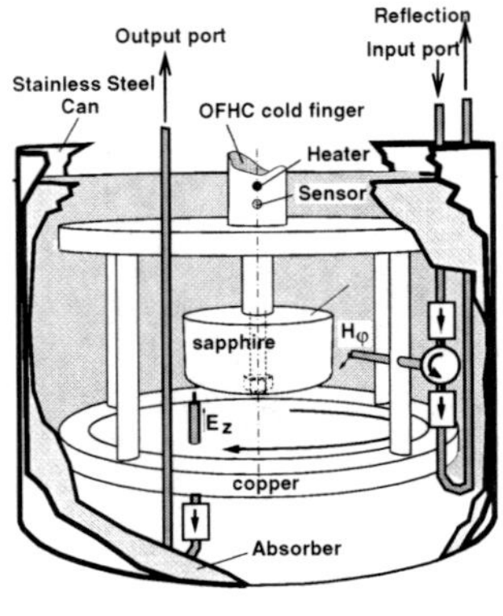

Fig. 3. Schematic of the resonator in the cryogenic can.

walls of the stainless steel can are covered with $2mm$ thick microwave absorber sheets. Its external wall is in contact with the helium bath. For the first resonant structure, a radial H_φ field is excited by a small magnetic loop in the resonator equatorial plan and constitutes the input coupling port with $\beta_1 \approx 0.2$. This value, which is normally set close to the unity for the Pound operation is relatively small. This induces the sensitivity of the Pound discriminator may be degraded by a factor of 3 which may have a direct impact on the short term frequency stability. A second E_z field is detected by an electric probe with an output coupling port $\beta_2 \approx 6\cdot 10^{-4}$. The $4.2K$ loaded Q-factor was measured around $160\cdot 10^6$ and the insertion losses at about $-35dB$. The second structure has a quite similar configuration with $\beta_1 \approx 0.45$, $\beta_2 \approx 2\cdot 10^{-4}$, $Q = 160\cdot 10^6$ and $I.L. = -33dB$. We can notice elsewhere that the given values of the insertion losses concern all the cryogenic resonator strucure including probes, the resonator, isolators, one circulator and two $1.5m$ long semi-rigid microwave copper cables. The choice of this kind of cables instead of stainless steel microwave cryogenic cables limits the signal losses without degrading drastically the thermal insulation of the helium bath. As a matter of fact, the main source of evaporation comes from the $200mm$ diameter neck of our dewar. Multiple thermal shields are placed between the $300K$ top flange and the cryogenic insert to limit the heating power. Actually the liquid helium evaporation rate is $9.6l/day$. With this system, we can operate the oscillator during one week without refilling. Each cryogenic insert is placed in the bottom of a large $100l$ liquid helium dewar isolated from the low-frequency vibrations. (the ensemble rests on marble stones mounted on springs)

3 Oscillator implementation

For the measurement of the frequency short term instabilities, the 2 CSO should have an equivalent structure. Nevertheless we decided to combine both short term stability evaluation and oscillator improvement. So our 2 oscillators structures present some differences. The first one has got full adjustements like a pound servo, a power servo and several thermal controls. The second one has just got a pound servo loop, this last CSO has only been recently implemented. We encountered also some difficulties to get 2 cryostats cold at the same time because of the complexity of provisioning liquid helium. Anyway, the comparison between the 2 CSO will give us an under-estimated value of the short-term stability. The basic oscillator sustaining loop is mainly composed of 2 commercial $GaAs$-type microwave amplifiers to satisfy the oscillating condition, and a mechanical phase shifter to realize the phase condition. A $11.565GHz$ centered $200MHz$ bandwidth selective filter is employed to be sure of the oscillation on the interesting $WGH_{16,0,0}$ mode. A fraction of energy is extracted to become a treatable signal. As the electric length greatly varies with high temperature gradients along the cables, the phase rotations have to be corrected to be equal to $2m\pi$. This is done by implementing a Pound servo loop to lock the oscillating signal on the sapphire resonance. Using the heterodyne technique, the microwave is low frequency $52kHz$ modulated ; a fraction of the incident power is then extracted by reflexion and is converted into a low-frequency modulated signal incoming a numeric phase locking. The error signal is integrated and finally sent back to a voltage controlled phase shifter (VCPS). We can notice elsewhere that we used 2 vcps, the first one realizing the modulation index, the second one correcting the phase fluctuations.

4 Turnover temperature determination

A high quality sapphire resonator generally presents a turnover temperature T_0 around $6K$. Indeed paramagnetic impurities resulting from the growth process are always present in small amounts in the crystal. These impurities yield to null temperature coefficient of frequency of the resonator at a given temperature which depends on the amount of impurities and to the exact electromagnetic field configuration inside the crystal. To reach a high frequency stability, the resonator has to be

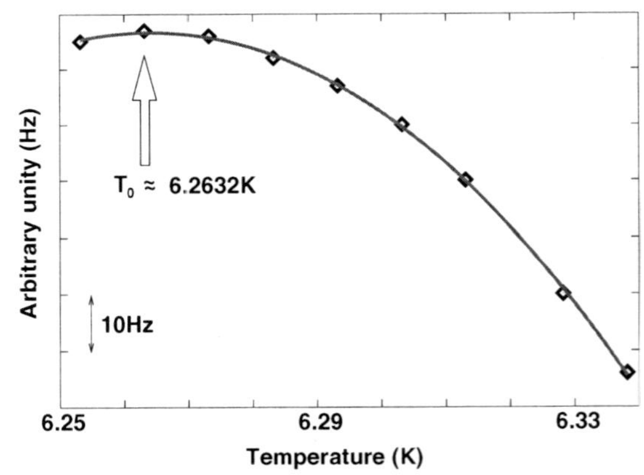

Fig. 4. Example of measurement and fit of the turnover T_0 for the first structure

stabilized very near its turnover temperature. But considering the case of a frequency drift of the order of $1.10^{-13}/day$, the estimation of the turning point can be done relatively quickly. We realized series of $10mK$ steps around the pre-evaluated turnover value during the cool down. Then the data have been approximated with a second-order polynomial law. We get for the first structure $T_0 = 6.2632K$ with a curvature of the frequency-vs-temperature characteristic around T_0, *i.e.* $\frac{1}{f}\frac{\partial^2 f}{\partial T^2}$ of the order of $A = 1.93 \cdot 10^{-9} K^{-2} K$ (see figure 4), and $T_0 = 5.7983K$ with an equivalent curvature for the second structure. At this point we can consider that if a frequency instability of $\frac{\partial f}{f} = 1 \cdot 10^{-14}$ has to be obtained assuming a thermal regulation (with a commercial PID facility) of $\pm 1mK$, the turning point must be determined with an accuracy better than $5mK$. Taking into account all the uncertainties of our fit(T_0 is given with an accuracy of $\pm 7 \cdot 10^{-4} K$), we thus can reach in theory $1 \cdot 10^{-15}$ frequency stability.

5 Stability Measurements

In a previous experiment [2], we have compared a cryogenic sapphire oscillator to a microwave synthesizer locked on a Hydrogen maser. The short term stability was limited by the measurement system noise floor, *i.e.* the hydrogen maser (HM) placed in a nearby laboratory $1km$ apart. The $10MHz$ signal of the HM reference is indeed transfered by an optical link. Then the CSO presented an estimated flicker noise at around $2 \cdot 10^{-14}$ until $100s$. The long term frequency stability was also limited by a random walk process which leads to a frequency instability of the order of $2 \cdot 10^{-12}/day$. (see figure 5)

5.1 Short-term frequency stability

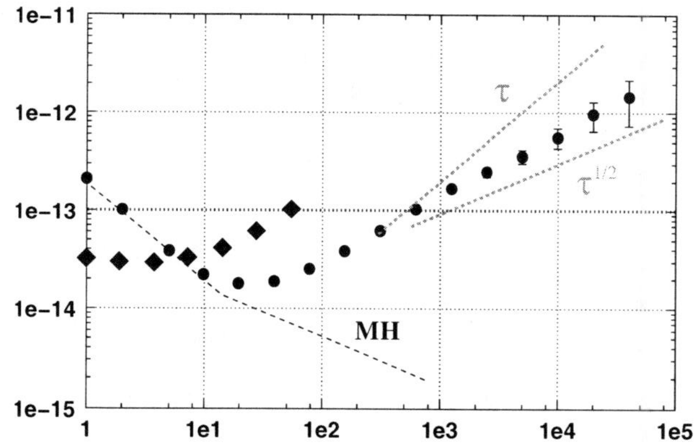

Fig. 5. Previous result (•) and measurement of short term stability (♦) which confi rms the noise fbor value. The dashed line represents the instrinsic HM frequency stability

Figure 5 shows the previous result and the comparison of the couple of CSO to have an estimation of the short term frequency instability. This result confirms the presence of a flicker noise floor which seems to be better than $3 \cdot 10^{-14}$ over $1s$ integration time. As the second oscillator is not finalised, it presents large instabilities in mean term which turns also to increase the short-term instability. Moreover as we had some difficulties to maintain at the same time the two cryogenic sapphire oscillators running well at low temperatures, it results that resonator 2 was not totally in thermal equilibrium. This could explain also its large frequency drift.

5.2 Long-term frequency stability

As the second oscillator presented large frequency fluctuations, we concentrated our efforts to the first one. The oscillator signal is mixed with the signal of an Agilent HP8254A microwave synthesizer referenced to the hydrogen maser. The $46kHz$ beat note is sent to a high resolution frequency counter (HP53132A). The fractional frequency Allan deviation is then computed. Figure 6 shows the results of the first oscillating structure. At short integration times $\tau \leqslant 100s$ the measurement

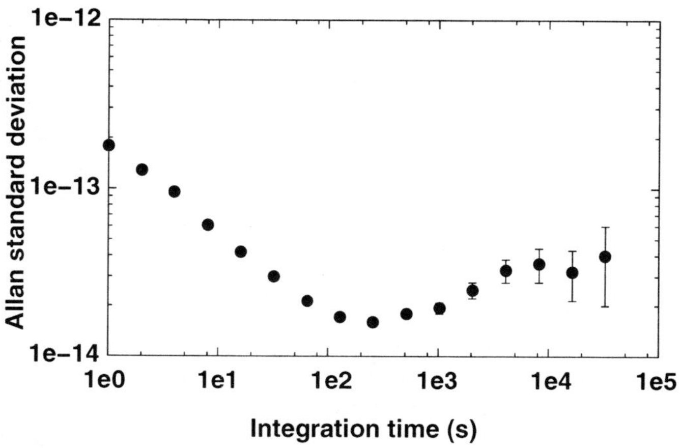

Fig. 6. Allan standard deviation of optimized CSO

is limited by the intrinsic HM instabilities and by the phase fluctuations along the optical link. A plateau at $1.5 \cdot 10^{-14}$ is observed around $200s$ and then the frequency instability increases slowly with τ to reach $4 \cdot 10^{-14}$ over $40000s$. No frequency drift is clearly identified in this diagram although Allan variance was computed without removing the frequency drift. The same conclusion can be drawn by observing directly

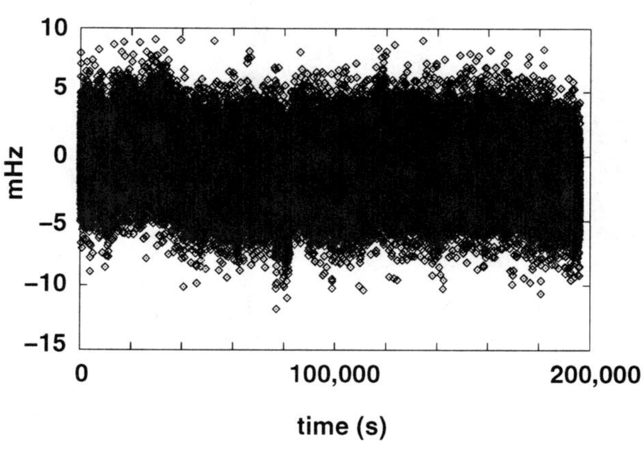

Fig. 7. Beat note frequency variations around the mean value equal to $46,374.779kHz$

the frequency data recorded during $3days$ measurement (see figure 7). Each point corresponds to the averaging of the beat note frequency over $1s$ gate time. The relatively large scattering (about $\pm 5mHz$) is due to the HM short term instability. Some low frequency variations have been also observed. This can probably be due to some oscillator residual sensitivity to the room temperature fluctuations. To give an under-estimated value of our oscillator drift we fitted the frequency data with a linear function and found a slope lower than $6 \cdot 10^{-14}/day$.

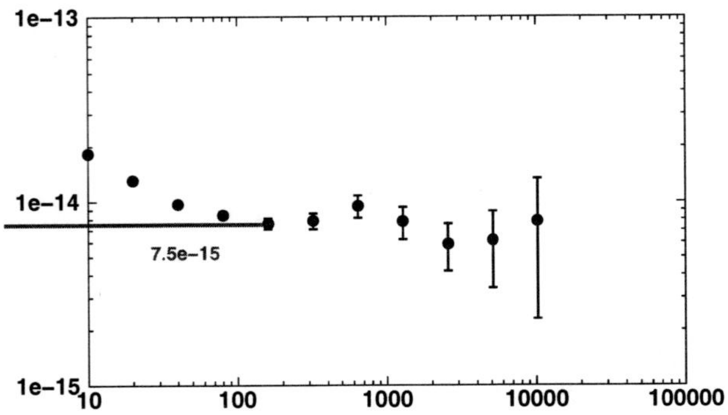

Fig. 8. Allan standard deviation of optimized CSO-second experiment (one day integration).

We re-run the experiment to verify the reproductibility by implementing also some other temperature controls for the room-temperature sustaining loop. We have choosen to operate a $WGH_{15,0,0}$ mode with a loaded Q–factor around $450 \cdot 10^6$ with $\beta_1 \approx 0.6$ and $\beta_2 \approx 7 \cdot 10^{-4}$. Results are presented in figure 8 with one day integration time and no drift removed. We subsequantly improved the short term stability with a flicker noise floor felt downto $7.5 \cdot 10^{-15}$. The long term frequency stability also presents the same characteristics as the previous experiment.

6 Application

The knowledge of short term stability of other types of ultra stable oscillators like Quartz oscillators can be usefull for their improvements. We proposed to characterized in a simple way the frequency stability of a state of the art $5MHz$ ultra-stable quartz crystal oscillator. The quartz oscillator was frequency

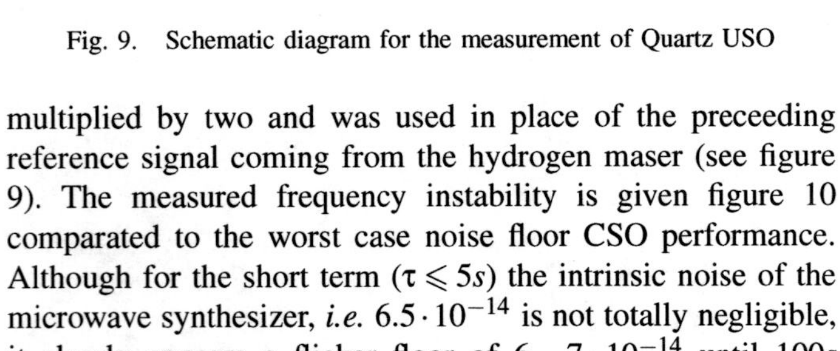

Fig. 9. Schematic diagram for the measurement of Quartz USO

multiplied by two and was used in place of the preceeding reference signal coming from the hydrogen maser (see figure 9). The measured frequency instability is given figure 10 comparated to the worst case noise floor CSO performance. Although for the short term ($\tau \leqslant 5s$) the intrinsic noise of the microwave synthesizer, *i.e.* $6.5 \cdot 10^{-14}$ is not totally negligible, it clearly appears a flicker floor of $6 - 7 \cdot 10^{-14}$ until $100s$ which can be attributed to the quartz crystal oscillator.

7 Conclusion

We presented an under-estimated value of the short-term frequency stability of our open cavity cryogenic sapphire

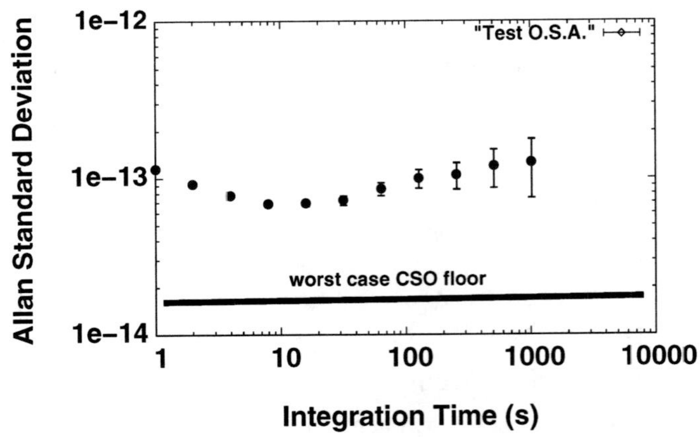

Fig. 10. Allan standard deviation of Quartz USO. The worst case CSO noise floor is also represented.

oscillator at better than $2 \cdot 10^{-14}$. Despite no AM servo was used, the oscillator presents an exceptionnal long term frequency stability without any clearly identified frequency drift. Indeed, a frequency drift is generally observed for long integration times limiting the frequency stability at typically a few $1 \cdot 10^{-13}$ over one day [13] [14] [15]. The reason of this performance is not yet understood but could be reliated to the open cavity configuration and the abscence of spurious modes nearby the main resonance which could induce some frequency pulling. Actually as it has been already pointed out [8], the coupling between cavity-like modes and the main resonance can induce a degradation of the intrinsic resonator frequency stability.

References

[1] P.Wolf, S.Bize, A. Clairon, A. N. Luiten, G. Santarelli, and M.E.Tobar, "Test of lorentz invariance using a microwave resonator," *Physical Review Letters*, vol. 90, pp. 060402–1–4, Feb. 14 2003.

[2] P.Y.Bourgeois, Y.Kersalé, N.Bazin, J.G.Hartnett, M.Chaubet, and V.Giordano, "Progress in the building of sapphire-helium clock at lpmo," *Proceedings of the Joint meeting 17th E.F.T.F and 2003 IEEE IFCS, Tampa, Fl*, pp. 355–359, May 4-8 2003.

[3] M.E.Tobar, A.Giles, and S.E.J.Searls, "High-Q thermo–electric stabilized sapphire microwave resonators for low noise applications," *IEEE Transactions on Ultrasononics, Ferroelectrics and Frequency Control*, vol. 41, no. 3, pp. 391–394, 1994.

[4] A.N.Luiten, A.G.Mann, and D.G.Blair, "Paramagnetic susceptibility and permittivity measurements at microwave frequencies in cryogenic sapphire resonators," *J. Phys. D. Appl. Phys.*, vol. 29, pp. 2082–2090, Jan. 1996.

[5] A.N.Luiten, A.G.Mann, and D.G.Blair, "Improved sapphire dielectric resonators for ultrastable oscillators," *Proc. 47th Annual Frequency Control Symposium, Salt Lake City, UT*, pp. 757–762, May 2-4 1993.

[6] K.Watabe, Y.Koga, S.Ohshima, T.Ikegami, and G.J.Hartnett, "Cryogenic whispering gallery sapphire oscillator using 4K pulse-tube cryocooler," *Proc. of the 2003 IEEE International Freqeuncy Control Symposium, Tampa, Fl.*, pp. 388–390, May 4-8 2003.

[7] M.E.Tobar and D.G.Blair, "A generalized equivalent circuit applied to a tunable sapphire–loaded superconducting cavity," *IEEE Transactions on Microwave Theory and Techniques*, vol. 39, no. 9, pp. 1582–1594, 1991.

[8] M.E.Tobar, "Effects of spurious modes in resonant cavities," *J. Phys. D. Appl. Phys.*, vol. 26, pp. 2022–2027, 1993.

[9] O.DiMonaco, W.Daniau, I.Lajoie, Y.Gruson, M.Chaubet, and V.Giordano, "Modal selection for a whispering gallery mode resonator," *Electronics Letters*, vol. 32, pp. 669–670, Mar. 28 1996.

[10] O.DiMonaco, Y.Kersalé, and V.Giordano, "Resonance degeneration and spurious mode suppression in a cryogenic whispering gallery mode sapphire resonator," *IEEE Microwave and Guided Wave Letters*, vol. 10, pp. 368–370, Sept. 2000.

[11] V.Giordano, G.J.Harnett, J.Krupka, Y.Kersalé, P.Y.Bourgeois, and M.E.Tobar, "Whispering gallery mode technique applied to the measurement of the dielectric properties of langasite between 4K and 300K," *IEEE Transactions on Ultrasononics, Ferroelectrics and Frequency Control*, vol. to be published, no. TUFFC-00214-2003.R1, 2004.

[12] E.Ivanov, D.Blair, and V.Kalinichev, "Approximate approach to the design of shielded dielectric disk resonators with whispering gallery modes," *IEEE Transactions on Microwave Theory and Techniques*, vol. 41, pp. 632–638, Apr. 1993.

[13] A.N.Luiten, A.G.Mann, N.J.McDonald, and D.G.Blair, "Latest results of the uwa cryogenic sapphire oscillator," *Proc. of the 49th Annual Frequency Control Symposium, San Fransisco, CA, USA*, pp. 433–437, May 31 - June 2 1995.

[14] S.Chang and A.G.Mann, "Mechanical stress caused frequency drift in cryogenic sapphire resonators," *Proc. of the 2001 IEEE International Frequency Control Symposium*, pp. 710–714, 2001.

[15] G.J.Dick, R.T.Wang, and R.L.Tjoelker, "Cryo-cooled sapphire oscillator with ultra-high stability," *Proc. of the 1998 International Frequency Control Symposium, Passadena, CA, USA*, pp. 528–533, May 27-29 1998.

PREPARATION AND REVIEW OF EUROMET CMC FILES FOR TIME AND FREQUENCY

R. Lapuh[1], F. Cordara[2], and G. de Jong[3]

[1]SIQ, Tržaška 2, Ljubljana, Slovenia, rado.lapuh@siq.si
[2]IEN, Strada delle Cacce 91, Torino, Italy, cordara@ien.it
[3]NMi-VSL, P.O. Box 654, Delft, The Netherlands, gdejong@nmi.nl

Keywords: calibration, time, frequency, uncertainty, quality system

Abstract

In this paper, a process of preparing and reviewing calibration and measurement capabilities of European time and frequency laboratories is described. Detailed review procedures are described. All major events are given which determined both the timing and content of this international project.

1 Introduction

At a Comité International des Poids et Mesures (CIPM) meeting held in Paris on 14 October 1999, the directors of the national metrology institutes (NMIs) of thirty-eight Member States of the Metre Convention and representatives of two international organizations signed a Mutual Recognition Arrangement (MRA) for national measurement standards and for calibration and measurement certificates issued by NMIs [1]. The MRA was signed to establish the degree of equivalence of national measurement standards, to provide for mutual recognition of calibration and measurement certificates issued by NMIs and to provide governments and other parties with a secure technical foundation for wider commercial agreements.

2 Calibration and measurement capabilities

Implementation of the MRA and thus recognition of national standards and certificates comprises 3 interconnected processes taking place at the NMIs, RMOs and at global level:

- Key Comparisons (KC's) (currently 408) of national standards serve to test capabilities of NMIs (to determine their degree of equivalence) in various metrological areas
- NMI Calibration and Measurement Capabilities (CMC) are thoroughly reviewed by RMO experts including the coverage by a Quality System based on ISO 17025 standard.
- NMI Quality Systems need to be in place and successfully reviewed by RMO's.

Outcome of this process are statements of CMCs of each NMI quoted in the BIPM key comparison and calibration database (KCDB), which is maintained by BIPM and is available to the public on the BIPM website (www.bipm.fr).

Key comparisons and CMCs are grouped under the following fields:

- Acoustics, ultrasound and vibrations,
- Electricity and magnetism,
- Flow,
- Length,
- Mass,
- Metrology in chemistry,
- Photometry and radiometry,
- Ionizing radiation,
- Thermometry and
- Time and frequency.

3 CMCs in the field of time and frequency

In the field of time and frequency, no CMC tables were prepared before year 2001 because there was no key

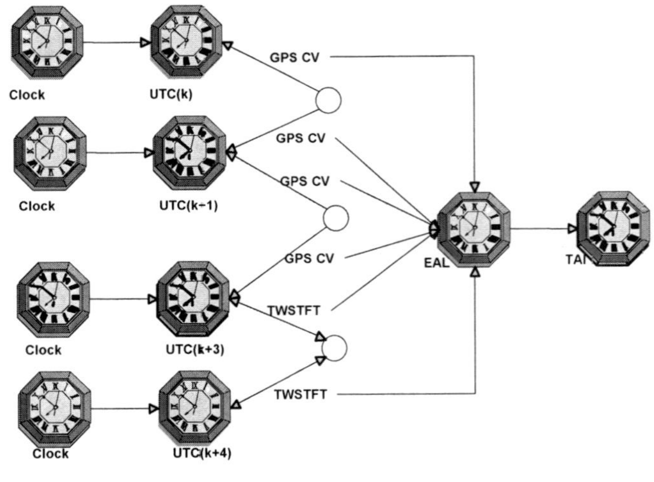

Figure 1. From the results of the measurements UTC(NMI)-Clock and GPS Common View data and TWSTFT data, the BIPM computes the international atomic time scale TAI .

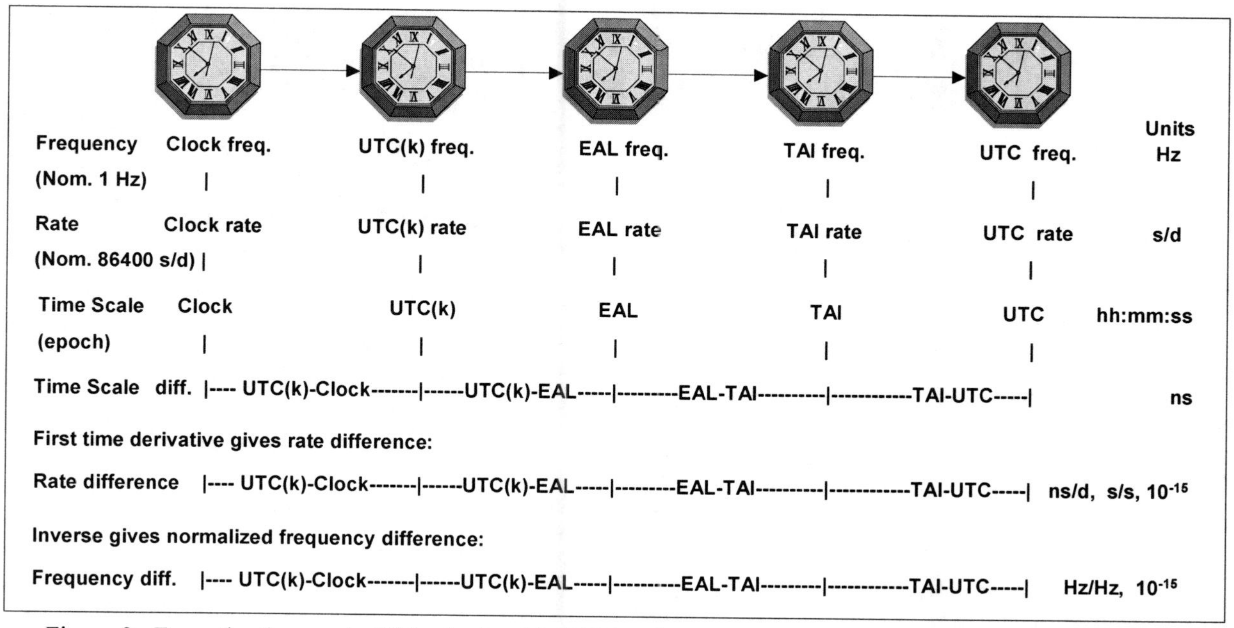

Figure 2. From the time scale TAI calculated by BIPM, the Key Reference Value UTC is derived and also UTC –UTC(NMI) which is the result of the Key Comparison for time, CCTF-K2001.UTC (for 2001)

comparison (KC) defined in this field. However, at the 14th Consultative Committee for Time and Frequency (CCTF) meeting (April 1999) the Working Group (WG) on the consequences of the global MRA was created to examine and report on this matter to the next CCTF. A first report of this WGMRA was presented at the 15th CCTF (June 2001). This report and the consequent proposals were discussed at the CCTF meeting as reported under section 11 of the Report of the 15th Meeting of CCTF. It was decided that the Key Comparison for time and frequency is the outcome of the computation of UTC – UTC(k) , its designation is CCTF-K2001.UTC (for the year 2001) and the Key Comparison Reference Value is UTC as computed by the BIPM Time Section.

EUROMET and APMP RMOs have then prepared CMC tables for their NMIs. However, the format of the CMC tables was not yet harmonized by the WGMRA so this process, already in the stage of inter-RMO review, was stopped in 2002.

For that reason, WGMRA met during the PTTI Meeting in December 2002. On this occasion the final list of the three main t&f quantities (time scale difference, frequency and time interval) was discussed and adopted. It was published as WGMRA Guideline 1, Rev. 20021209 [2]. Two more Guidelines, related to the first, were also adopted then, namely:

- WGMRA Guideline 2 (Rev. 20021205) clarifies the estimation of the uncertainty to be taken for the Best Measurement Capability (BMC);
- WGMRA Guideline 3 (Rev. 20021210) clarifies how to extrapolate the uncertainty from the KC results for averaging times shorter than the 5 days interval of the BIPM circular T data.

4 Structure and contents of the t&f CMC files

CMC files have the following main titles with corresponding subtitles:

- Calibration or Measurement Service
  - Quantity
  - Instrument or Artifact
  - Instrument Type or Method
- Measurand Level or Range
  - Minimum value
  - Maximum value
  - Units
- Measurement Conditions / Independent Variable
  - Parameter
  - Specifications
- Expanded Uncertainty
  - Value
  - Units
  - Coverage Factor
  - Level of Confidence
  - Is the expanded uncertainty a relative one?
- Reference Standard used in calibration
  - Standard
  - Source of Traceability
- List of Comparisons supporting this measurement / calibration service
- Euromet Time and Frequency Services Administration
  - NMI Service Identification
  - Service Category
  - NMI
  - Internal Comment
- Comment for the user

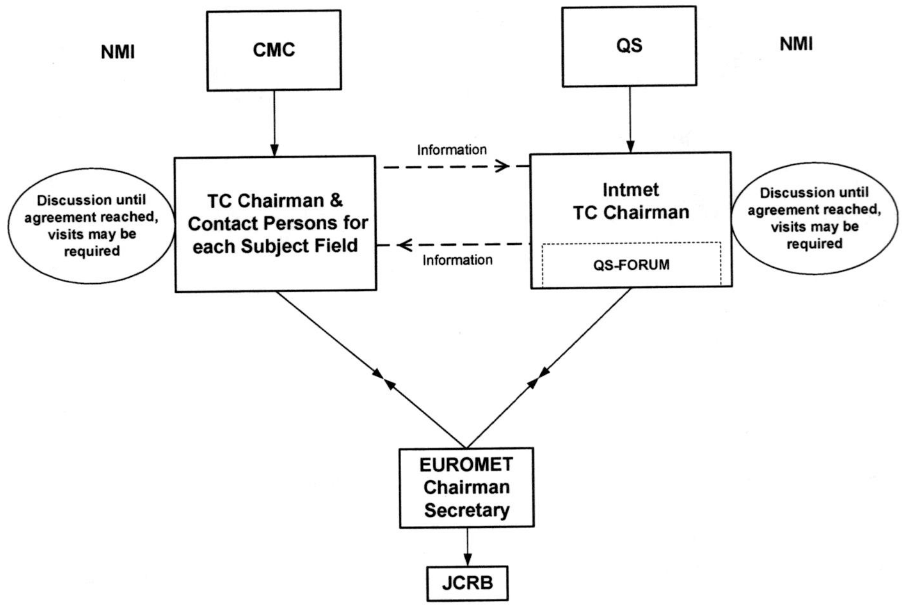

Figure 3. Overview scheme of CMC and QS reviews

The WGMRA Guideline 1 is specifying only Quantity and Instrument or Artifact under the Calibration and Measurement Service as well as Service Category under the Time and Frequency Services Administration. These entries should strictly follow the WGMRA Guideline 1.

5 Work of EUROMET Technical Committee (TC)

EUROMET has prepared a detailed procedure for reviewing CMC files in the EUROMET Directory [3]. Overview scheme is shown on Figure 3 for both CMC and QS reviews. The same procedures apply for self-declared and third party assessed quality systems covering the NMI services providing the declared CMCs.
Work of the EUROMET TC was divided into three logical steps: preparation of CMC files, regional review of those files and inter-regional review of other RMOs CMC files. Although the work started in year 2000, not much progress was done due to nonexistent KC for time and frequency field. In the followings only the work carried out after the adoption of WGMRA Guideline 1 is presented.

5.1 Preparation of CMC files

In order to prepare CMC files, some more detailed rules were established prior starting the work. EUROMET t&f TC took WGMRA Guideline 1 as a reference for preparing CMC files. Based on this document, a more detailed list for each classification was prepared in which all the details, starting from those reported in chapter 4, and necessary to the preparation of the CMC files have been defined.

5.1.1 Time scale difference

The following service categories were defined for time scale quantity:

Calibration or Measurement Service		Administration
Quantity	*Instrument or Artifact*	*Service Category*
Time scale difference	Local clock vs. UTC (NMI)	1.1.1
Time scale difference	Local clock vs. UTC	1.1.2
Time scale difference	Remote clock vs. UTC (NMI)	1.2.1
Time scale difference	Remote clock vs. UTC	1.2.2

Table 1. Time scale difference calibration services

Measurement methods used are:

- for Local clock vs. UTC(NMI) or UTC:
  - Time interval measurement
- for Remote clock vs. UTC(NMI) or UTC:
  - GPS common view
  - Two-way satellite time and frequency transfer (TWSTFT)
  - standard signal emission
  - telephone time signal, etc.

Measurand level or range is typically specified from –1 s to +1 s. Parameter under measurement conditions is typically prediction time for results related to UTC and averaging time for remote calibrations. Expanded uncertainty is specified as an absolute uncertainty in ns, with coverage factor of 2, which gives 95% level of confidence. The reference is typically a ceasium beam clock and list of comparisons supporting this measurement calibration service is CCTF-K2001.UTC. In the comment for the user, it is noted that the result of a calibration is fractional part of modulo 1 second difference, but additionally an integer part can be provided also.

5.1.2 Frequency

For frequency calibrations, the following services have been defined:

Calibration or Measurement Service		**Administration**
Quantity	*Instrument or Artifact*	*Service Category*
Frequency	Local frequency standard	2.1.1
Frequency	Remote frequency standard	2.1.2
Frequency	General frequency source	2.2.1
Frequency	Frequency counter	2.3.1
Frequency	Frequency meter	2.3.2

Table 2. Frequency calibration services

Measurement methods used are:

- for Local frequency standard:
  - phase / time measurement
  - frequency difference multiplication
- for Local generic frequency source:
  - direct frequency measurement using electronic counter
- for Remote frequency standard:
  - GPS common view
  - TWSTFT
  - standard signal emission
  - telephone time signal, etc.

Measurand level or range specifies the applicable frequency range of the service. Measurement conditions are specified by averaging time corresponding to the gate time of the frequency counter used. Uncertainty is typically specified in relative form and is given in Hz/Hz. Other parameters, except comments for the user, are identical as for time scale difference.

5.1.3 Time interval

For time interval calibrations, the services envisaged are found in table 3.
Measurement methods used are:

- passive and active sources / meters:
  - Time interval measurement with electronic counter
- for chronometers:
  - stopwatch calibrator

Measurand level or range specifies minimum and maximum time interval applicable for the service. Under measurement conditions / independent variable a pulse amplitude and signal rise or fall time are specified. Uncertainty is specified as absolute uncertainty in ns with 95% level of confidence. Other parameters are equivalent to those specified for frequency service.

Calibration or Measurement Service		**Administration**
Quantity	*Instrument or Artifact*	*Service Category*
Time interval	Period source	3.1.1
Time interval	Rise/fall time source	3.2.1
Time interval	Pulse width source	3.2.2
Time interval	Time difference source	3.2.3
Time interval	Delay source	3.2.4
Time interval	Period meter	3.3.1
Time interval	Rise/fall time meter	3.4.1
Time interval	Pulse width meter	3.4.2
Time interval	Time difference meter	3.4.3
Time interval	Delay meter	3.4.4

Table 3. Time interval calibration services

5.2 Regional review of CMC files

When the new classification was defined in April 2003 Contact Person Meeting in Ljubljana, a new Analysis Working Group (AWG) was elected from its members. It consists of six time and frequency experts from leading European time and frequency laboratories. A first task of AWG was the review of the new EUROMET CMCs files. A second task will be inter-regional review of other RMOs CMCs files.
When CMC files had been prepared and collected by TC Chairman, they were distributed among members of AWG to be reviewed. Additionally, with CMCs files an Appendix1 document was prepared by NMIs for each quantity included in the CMCs file. In this document, the laboratory had to specify details regarding the applied quality system, the publications relevant to the given CMCs, peer review or expert visit and other relevant information, and confirming laboratory abilities for specified CMCs. The review process consists of analyzing both technical abilities and quality system (Figure 1). Quality system verification within EUROMET is covered by QS Forum and therefore most of the review activities was focused on specifications given in CMC files. A direct communication between AWG member and the laboratory which CMCs were under the review process was established via e-mail. 16 CMCs files were thoroughly reviewed in three months. These CMCs files were then sent at the end of December 2003 to EUROMET Chairman, who started the process of inter-regional review.

5.3 Inter-regional review of CMCs files

The inter-regional review of EUROMET time and frequency CMCs files started in the beginning of 2004. Questions raised from other RMOs are answered directly by laboratories who prepared CMCs. When necessary, CMC files are corrected and updated. Communication is again established through e-mail and the TC Chairman is coordinating activities within EUROMET.

6 Current status of EUROMET t&f CMCs

Currently, there are 16 CMCs files for time and frequency covered by the EUROMET review process. These CMC files are now under the inter-regional review. Within these 16 CMCs files, there are 60 entries for time scale difference, 127 entries for frequency and 55 entries for time interval. Altogether, there are 242 CMCs entries for time and frequency in EUROMET CMCs files.

7 Future developments

One of the MRA goals is certainly a working calibration database (KCDB) in which all CMCs tables would be listed. Since no time and frequency CMCs tables are currently present in the KCDB, it is foreseen to conclude this process for EUROMET time and frequency CMCs tables, as well as for all other RMOs. The main workload is therefore expected in inter-RMOs review of CMCs tables. At the same time, the maintenance of CMC tables, when included in KCDB, will certainly present additional challenge.

Acknowledgements

Authors would like to thank all the EUROMET time and frequency contact persons as well as EUROMET Chairman and EUROMET Secretary for extraordinary cooperation in the process of preparing and reviewing CMC files.

References

[1] Comité International des Poids et Mesures. "Mutual Recognition Arrangement", Paris, 14 October 1999

[2] CCTF. "WGMRA Guideline 1 (Rev. 20021209), The Service Category classification scheme for T&F CMC entries", WGMRA, December 2002

[3] "EUROMET Guide Nr. 8", National Metrology Laboratory, Enterprise Ireland Campus, 31 May 2002

[4] Draft "EUROMET Time and Frequency Supplement Guide to the JCRB Instructions for Appendix C of MRA", 2001

RECENT DEVELOPMENTS IN TIME & FREQUENCY DISSEMINATION SYSTEMS

D. GIUNTA (1), G. BUSCA (2), A. DELLA TORRE (3), M. VITTA (3), V. DE PERINI (3), R. GRIMOLDI (3), X. STEHLIN (4), P. ROCHAT (4)

(1) ESA-ESTEC, P.O. Box 299, 2200AG Noordwijk, (NL), email: domenico.giunta@esa.int;
(2) KYTIME Sàrl, Ecluse 11, 2022 BEVAIX, (CH), email: tec.kytime@net2000.ch;
(3)Carlo Gavazzi Space S.p.A., v. Gallarate, 150 - 20151 Milano (I), email:adellatorre@cgspace.it; (4) TEMEX Neuchatel Time, 2000 Neuchatel (CH), email:stehlin@temex.ch

Keywords: Time & Frequency Dissemination, TDMA, PLL, Atomic Clock, Field Trial, On board clock module.

Abstract

The paper addresses the activities performed on a novel Time & Frequency Dissemination (T&F D) System based on ESA Patent 407, including its most recent developments.
The method allows to transfer the precision and the stability features of a state-of-the-art ground clock to a less performant clock module on board a satellite. By implementing the patent methodology, the on board clock module becomes the "master Time & Frequency dissemination source" for all the ground terminals located in the footprint of the satellite hosting the on board clock module.
The paper, after having addressed the basic principles of the method, provides a summary description of the Time & Frequency Dissemination System, giving an overview of the applications that can benefit of it and comparing its performances against the ones of the main competing systems.
The actions for the definition of the system and for assessing the related performances are described as well.
The final part of the paper is dedicated to the description of the most recent activity, consisting in the realisation of the experimental setup and in the performance of a Field Trial campaign for the assessment of the actual system performances. The paper is concluded by highlighting future planned activities following the Field Trial Campaign.

1. Innovative Principles for Time & Frequency Distribution

To better exploit bandwidth spectrum and to ease interconnection between different operators, telecom operators have always looked for an "independent" infrastructure able to distribute in real time an accurate and stable clock reference. In addition to telecom operators a wide range of disciplines would benefit from the availability of such an infrastructure.

This is not the case as of today; available solutions are mostly spin-off of positioning systems as GPS[1], they are usable only close to ground and offer limited guarantees in terms of accuracy, stability and reliability for many applications. Even if the planned Galileo Navigation System, as a civil navigation system, is expected to improve the situation the Galileo Time distribution function , also in this case, will not be the primary system function
The ESA Patent 407 proposes a novel approach and offers a solution to this need.
The T&F D system concepts are based on the innovative principle to distribute a very accurate ground-based time reference by means of a satellite network. This is done by synchronizing a low performance on-board oscillator (slave clock) with a very accurate Ground Time Reference (master clock) and then distributing the space clock time to ground users. Fig. 1.1 shows the overall Time & Frequency Distribution system architecture.

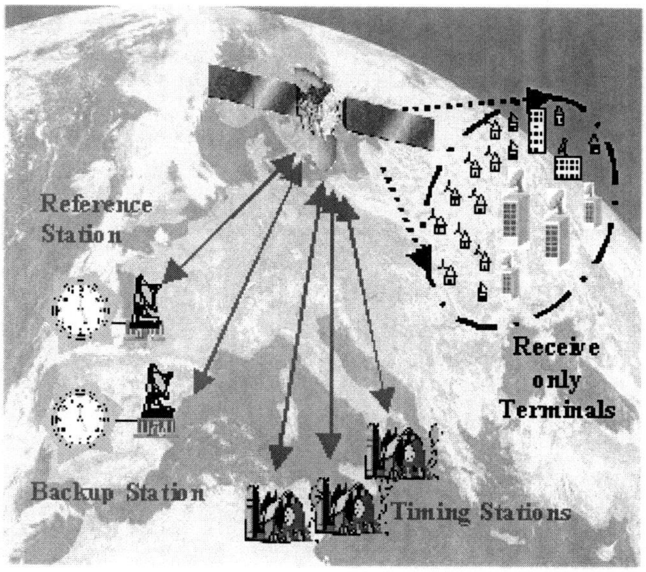

Fig. 1.1 Time & Frequency Distribution System Architecture

[1] Despite the diffusion of GPS-based receivers, it is worth reminding that GPS is a military system and not under European control.

The system concept can be broadly summarized in the following points:

1. A state-of-the-art atomic clock located on ground in the Master Clock Station (MCS) is used as time reference.
2. A Backup Station is provided for availability reasons in order to take the role of MCS in case of its failure.
3. A less accurate clock is embarked on a satellite. The on board clock is synchronised to the ground-based time reference, using a special PLL. By implementing the Patent concepts the on board clock inherits the accuracy and stability features of the ground one and can be used as reference for the following time and frequency dissemination process.
4. The radio link connecting the ground stations and the space-based time reference is operated using existing communication techniques; in particular a TDMA access scheme is used for the uplink while TDM is used for the downlink.
5. After Satellite Clock synchronization the time/frequency reference is provided to the following categories of users:
   - "Clock Stations" and "Timing Stations"; these are transmitting/receiving stations, similar to the Master Control Station and using the same synchronisation principle. Such stations are required by users with more demanding performance requirements.
   - One-way user stations; these are receiving only stations using the timing and data information from the satellite signal to lock their clocks. Such stations are required by users with less performance requirements.

The above mentioned mechanism implies that Doppler compensation techniques are properly applied at each transmit and/or receive station in order to compensate the effects of Satellite / User Stations relative movement.

As it can be seen from the above system description, it is not necessary to set-up a complete new-from-scratch system for the implementation of the Patent concept, but this system can be seen as a "plug-in" of existing infrastructures, as the system will require a small on-board payload and traditional TDMA ground equipment (with some customisations and ad-hoc interfaces required by the application).

Being the system part of an existing telecommunications network there is no need to apply for specific frequency to Frequency Regulatory Boards.

The required mass for overall on-board clock module payload is estimated to be approximately 5kg. The on-board clock module would be a piggy-back in a standard telecommunications satellite, and it will share IF to IF interfaces of "host" satellite. The on-board clock module will consist of the following main components: (a) a clock recovery module; (b) a phase detector; (c) a loop filter (d); a DAC for controlling the local clock; (e) a local clock.

The local on board clock would be based on a "Precision quartz oscillator " or a Rubidium clock, so there is no more need to embark high-accuracy (and then heavy and difficult to control) clocks.

It could be envisaged to have one ground clock placed under the footprints of two satellites. This will allow synchronizing clocks in one geographical area with clocks in another geographical area. This leads that with 3 GEO satellites (opportunely located) it will become possible to realise a real-time universal time.

2. Time & Frequency Distribution System Main Characteristics

The system configuration to implement this T&F distribution method consists of a ground segment and of a space segment.

Ground segment

One (or more) ground MCS(s), equipped with state of the art atomic clock, are in charge of :

- Estimating the time/frequency error at frame, symbol and sub-symbol level (delay/shift due to all propagation effects from ground to satellite) by using received loop-back timing & data from Satellite On Board Clock Module
- Transmitting a TDMA signal corrected at frame/symbol level and providing sub-symbol compensation info into TDMA burst data for further correction by Satellite On Board Clock Module

A backup of the MCS is foreseen: this one operates normally as a clock/timing station, in order to shorten the switchover phase. When the satellite looses the contact with the MCS it commands the backup station to behave as a reference. It then stops adjusting its local clock and the satellite in turn takes the backup station clock samples to synchronize its own clock.

The Two-way Clock /Timing User stations exploit the same two way method to discipline their local clocks with the Satellite clock, which is already synchronized with the MCS reference clock.

The One-way User Stations use the system in one-way mode, taking advantage of the precise reference coming from the satellite, but with lower performance with respect to two-way users. In this case the downlink delay is estimated and corrected using standard techniques.

Space segment

A GEO/MEO/LEO Satellite with one On Board Clock Module is in charge of:

- Recovering the clock from the received TDMA burst signal (data clock) and performing a further sub-symbol fine adjustment by exploiting correction data computed and transmitted by MCS
- Measuring the Time of Arrival shift wrt nominal position and transmitting it to the Ground Clock station via TDM(A) for exploitation in time/frequency error estimation.

The synchronization technique

As above mentioned, the synchronization technique relies on the measurement of the 2-way delay for estimating the uplink delay and compensating for all the asymmetries by means of either estimations or measurements in order to synchronize the Satellite Clock with the Ground Clock.

The MCS establishes a communication link with the satellite using the TDMA access. The satellite TDMA payload is regenerative, i.e. the TDMA frame is built on the satellite and

retransmitted to ground in TDM mode.
The ground station is able to adjust the transmission time to the correct position by looking at the return signal.
Considering Fig. 2.2, measurements on board of the pseudodelay (PDS) and measurements on ground of the pseudodelay (PDG) are taken each time frame.

$$PDS-PDG = AS+2dT \quad (1)$$
$$PDG+PDS = \text{2-way delay} = 2\Delta+AS \quad (2)$$

where AS is the link asymmetry and dT is the clock offset between the satellite time Ts and the ground station time Tg.

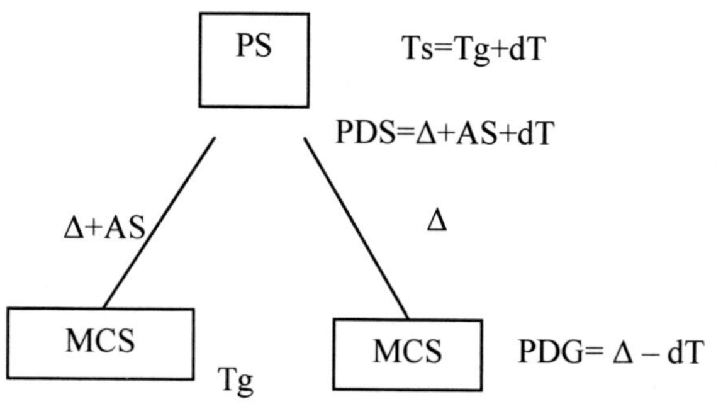

Fig. 2.2: The Synchronization Technique

The link asymmetry (AS) has contributions from the Geometry, Ionosphere and Electronic Delays.
The geometrical AS is estimated from the time evolution of the 2-way delay measurements performed at the frame rate. Other AS are estimated with link asymmetry models.
The clock offset dT is derived from Eq.(1).
The MCS sends time and data to the payload synchronized with its local clock, used as a reference for the system.
The uplink delay is compensated at different levels.
A gross delay adjustment is performed on ground by:

- Changing the frame number in the sent data packets (at frame period rate)
- Adjusting the phase of the sent packets (at symbol level)

A fine delay adjustment (under symbol level) is performed on board by adjusting the phase of the payload clock by using an information inserted in a burst field by MCS.
The clock offset is used for locking the S-Clock to the G-Clock via an on board PLL.
This circuit is fed with windowed ground clock samples recovered by the TDMA demodulator: in this way the phase and frequency differences between the satellite clock and the ground clock are compensated.
At this point the satellite clock has been synchronized with the reference clock and the synchronization of user terminals can start.

3. Applications Overview

Basically, all the applications requiring a Time & Frequency dissemination service providing high reliability and availability features can benefit of the proposed system.

Among others, the following main applications are in particular identified:

- Synchronisation of a Satellite Clock to a Master Ground Clock.
- Synchronisation of communication satellites using the technique of the on board switching.
- Independent measurements of the Galileo Satellite Clocks synchronisation, particularly during the Test Bed and IOV phases.
- Synchronisation of Clocks on board of Deep Space Probes for scientific missions.
- Timing for a Martian Navigation system.
- Time Distribution System from Space: High Accuracy (2-way); Lower Accuracy .
- Simultaneous Accurate Time comparison between an ensemble of ground stations for metrology applications.
- Digital television
- Medical applications
- Satellite and inter-satellite ranging applications

Furthermore, the T&FD system can be used as a backup for existing GPS-based applications in order to increase reliability and provide the service whenever the GPS is unavailable due to political/military reasons.

4. Comparison of T&F D System Performances vs. the ones of the main competing systems

In Table 4.1 the performances achievable by the main existing time and frequency dissemination systems, like TWSTFT and the Navigation Satellite System, like GPS, are compared with the expected performances of the T&F Dissemination system subject of the paper.
T&FD system should associate both performance and cost interests with respect to competitive solutions like GPS for mass market applications and TWSTFT for high precision clock comparisons.

	TFD	**TWSTFT**	**Navigation Sat. Sys.**
Availability	Continuous	On demand, subject to the link availability	Continuous
Performance	Real time	Presently it requires post-processing	Real time / Post processing
Users	Several stations	Two stations	Several stations
Coverage	Satellite footprint	Intercontinental	Global
Accuracy	0.1 ns (Goal)	1 ns	30 ns/ 1ns
Stability over 1 day	10^{-15}	10^{-14}	10^{-13}/ 10^{-14}

Table 4.1:T&FD system and main competitors characteristics

At the current time TWSTFT is the preferred method for time and frequency comparisons between remote sites for metrological applications, however, TWSTFT is not widely used because of the leasing cost and the limited availability of satellite communication channels. In addition, the time comparison can be performed only between two stations.
The proposed method offers, with respect to techniques already available, comparable or even better performance along with the capability of supporting several ground-stations simultaneously. This capability constitutes a significant improvement with respect to the traditional two-way systems, which support the synchronization of two stations only.
As anticipated in the introductory sections, this new feature is obtained by the adoption of already existing communication techniques and with a little payload embarked on a LEO\GEO telecommunication satellite. Furthermore, the TFD network could be uninterruptedly available in real time without the necessity of post-processing.
Moreover, with the launch of three T&FD payloads on suitably positioned geostationary satellites, the system may broadcast a global real time reference.

5. ESA Actions on T&F Dissemination Systems based on ESA Patent 407 and Main Objectives of the Field Trial Campaign

After having been granted with the patent related to this new method, ESA has funded some studies to exploit its performances and feasibility.
Via different Agency programmes, ESA has already funded the following studies:

- *Prototype development o fan On-board module (1999)*
- *Two Parallel System Studies (2000)*
- *Field Trial Campaign (2003)*

The last one of these studies, (which is object of this paper), is related to a field trial campaign: the space segment in this case is replaced by a double-hop satellite access.
It means using on ground a representative prototype of an on-board clock module.
This will allow to proof the concept with an easy deployment and also to learn lesson before embarking the "real space segment" on satellite. (The study is funded under the "ARTES-5 [2]" programme of the Agency).
At the end of this campaign it is expected not only to validate the method and the algorithms needed for its implementation, but also to use the field trial campaign configuration as a preliminary distribution system.
Once (or in parallel with) the preliminary distribution has started, it might be possible to look for a "host satellite" embarking the On-board Clock module.
The Field Trial Project team is leaded by Carlo Gavazzi Space and is composed by TEMEX Time, KYTIME and ALCATEL Space.

[2] To find more on ESA telecom programmes, please visit the website: telecom.esa.int

Carlo Gavazzi Space (CGS), as Prime Contractor, is responsible for project management and technical co-ordination, system definition, definition of design, development and AIT plan, system integration and testing, management of the links with the satellite operator and field trial campaign results evaluation. Moreover, CGS is developing the TDMA Terminal and the Station Controller which are the core of the MCS, Satellite Payload Simulator and Auxiliary/Timing stations design.
TEMEX Time is responsible of the specification and setup of the instruments to be exploited for the test campaign, including Atomic Clocks (RAFS, LPFRS), Measurement Subsystem (Picotime) and Micro Phase Stepper.
KYTIME supports CGS in the implementation of the concepts defined in the ESA Patent 407, in the theoretical analyses, in the definition of the measurement methods and in the interpretation of the test results.
ALCATEL Space supports CGS on system definition aspects.

6. Main differences between Field Trial Configuration and Final Target System configuration

The Field Trial envisaged topology is based on satellite double hopping access, allowing to use one On Board Clock (OBCK) module placed on ground resembling as much as possible the future flight OBCK module.
The following figure 6.3 shows how the different up and down links are exploited in the field trial configuration in order to synchronise at first the on board clock module with the ground reference and then to disseminate time / frequency signals to the secondary stations.

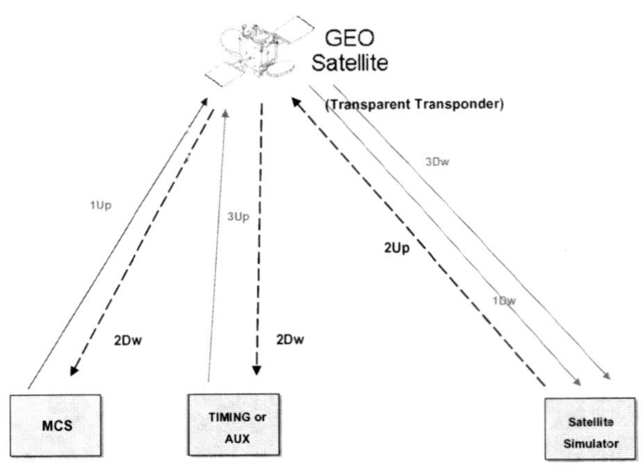

Fig. 6.3 Field Trial configuration links exploitation

It can be seen that respect to the target system the payload is now hosted on ground. The MCS-Satellite uplink has been replaced with the sum of MCS-Satellite and Satellite-Satellite Simulator (SS) links (1Up+ 1Dw), while the Satellite- MCS downlink is replaced with the sum SS-Satellite and Satellite-MCS links (2Up+2Dw).

For the Timing/Aux station the up and downlinks have been replaced with (3Up+3Dw) and (2Up+2Dw) links respectively.
The Field Trial System is intended to provide the user community with the confidence that the Target System will offer the performances that have been analytically envisaged in the system studies.
It can be also exploited to detect practical problems that have not been identified during theoretical studies and to propose solutions to such problems before committing to the final configuration.
However, it is clear that the layout of the Field Trial System is different with respect to that of the Target System.
The following table reports the main differences between the two configurations.

Parameter	Target System (TS)	Field Trial (FT)	Notes
Doppler	Asymmetry induced by the different frequencies	Symmetry induced by the double hop.	The FT should have better performances than the TS.
Ionosph. Delay	Asymmetry	Compensated by the double hop	The FT should have better performances than the TS.
Location	The stations are in different locations	All the stations can be located in the same site	FT could allow an easier evaluation of the achieved synchronization level
Relativistic Effect	Being the Clock on board, the relativistic effect shall be compensated.	Clocks on ground (and possibly in the same location) => no need of relativistic effect compensation	The FT should have better performances than the TS.
Noise	Lower, because the single hop	Higher because the double hop	The TS should have better performances than the FT.
Down-Link	Continuous clock on the downlink	Because of the TDMA access, the clock is windowed also in the "down-link"	The TS should have better performances than the FT.

Table 6.2- Comparison between Target System and Field Trial

7. Theoretical Field Trial System Performances Evaluation

In order to assess the system performances it has to be computed the time error of the 10 MHz Payload Simulator locked oscillator with respect to the reference oscillator in the MCS.
This error is due to the overall noise introduced by elements of the system.
The following main items contribute to the overall performances:

- Delay and Doppler compensation system
- Demodulator phase recovery accuracy with noisy signal. The demodulator takes into account the effect of the link noise.
- Phase noise of the slaved clock

Moreover, the up-link transmission rate determines the achievable overall system time stability (see Table 7.3).
The Field Trial Performance Evaluation Model is shown in Fig 7.4 where:
MOD= Modulator part, DEM= Demodulator part
PROC: Processing part, SPL: Splitter
DOWN CONV: Downconverter, UP CONV=Upconverter
PLL= Phase Lock Loop
CNTR=Counter

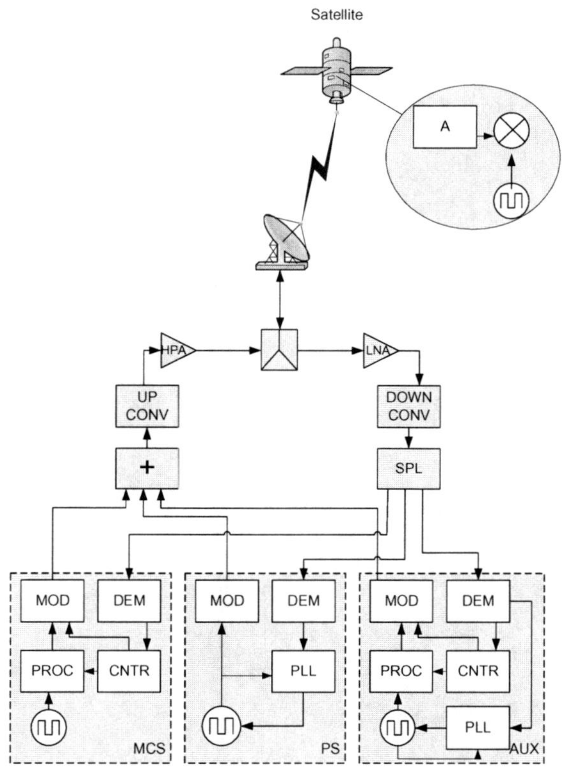

Fig. 7.4 TFD Field trial system performances evaluation model

The main entities involved in the loop are:

- Master Control Station (MCS)
- Satellite (in the Field Trial – FT - a transparent transponder)
- Payload Simulator (in the FT it is located on ground)
- Secondary Station (SS or AUX)
  This station shall perform two functions in relation to the actual test phase:
  - backup station for the MCS, in order to verify the hand-over procedure
  - clock/timing station, in order to verify the user stations achievable performances.

Table 7.3 gives an indication of the performances (Reference Clock vs. On board Clock Relative Frequency Error ε over 1 day averaging time) that can be achieved by the Target System (TS) and by the Field Trial configuration (FT) by the application of this approach in relation to the up-link access rate and the timing recovery error.
The method is frequency independent, i.e. it is limited only by the state of the art technology for phase tracking of both on board module and ground station module. This means that the results shown in Table 7.3 may evolve when a most performant technology allows a smaller peak-to-peak phase jitter for on board tracking (e.g. faster comparators, more accurate equalization of delay paths within the clock module, higher clock frequency).

Timing error	**Symbol rate MHz**	**Relative Frequency Error over 1 day**	
		TS	**FT**
0.1°	2	1.56E-15	2.86E-15
1°	2	1.56E-14	2.78E-14
0.1°	4	7.85E-16	1.54E-15
1°	4	7.85E-15	1.39E-14
0.1°	8	3.96E-16	9.72E-16
1°	8	3.96E-15	6.99E-15

Table 7.3 Preliminary performances estimation as function of symbol rate and of phase error in timing recovery

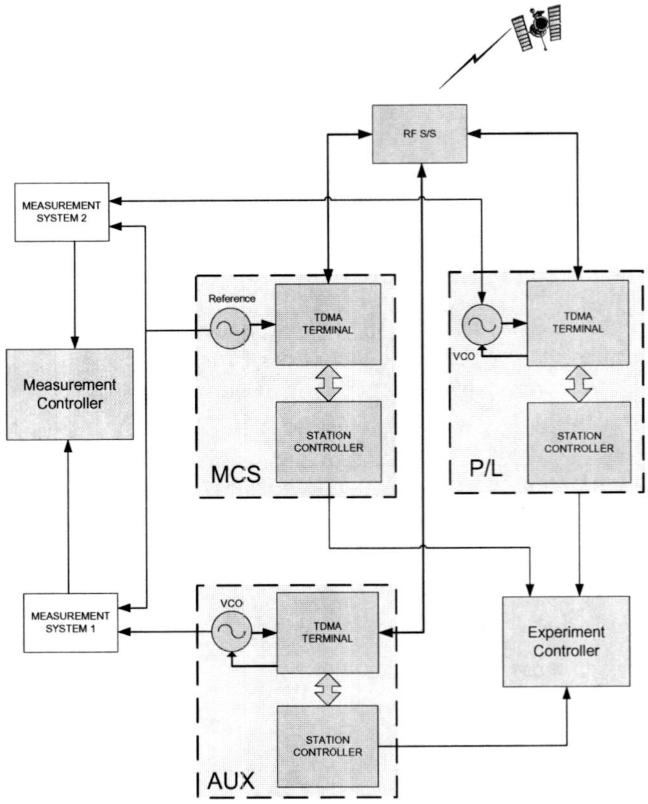

Figure 8.1 Field Trial System Block Diagram

8. T&FD Field Trial System Description

The Time & Frequency Dissemination Field Trial System configuration is physically composed of the following main parts:

1. A Master Control Station (MCS)
2. A Payload Simulator (P/L)
3. A Secondary Station, that can act as an Auxiliary Station as well (AUX).
4. A Ground Station (RF part, including antenna) for accessing the Satellite exploited during the Field Trial campaign
5. A Satellite (ASTRA 1-A) providing a transparent transponder.

All the stations (MCS, PL, Aux) are located on ground in the same premises, in order to allow an accurate phase measurement between:

a. The MCS reference clock and the P/L clock
b. The MCS reference clock and the AUX clock

All the measurements are performed by means of two dedicated Measurement Systems, based on the Picotime Unit. The overall system is depicted in figure 8.1.

The Experiment Controller (EC) provides the man machine interface: it shows to the experiment conductor the overall system behaviour including the contents of the three stations internal registers and the results of the Doppler compensation algorithm. The data can be displayed as numbers or plotted as XY graphs. All the acquired data are stored in log files, which can be retrieved also by remote data communication links.
The EC permits also the sending of commands to the stations for debugging purposes, e.g. it can force each station in an arbitrary state.
The EC is implemented using a personal computer in a Windows environment, running a program written in ANSI C and exploiting off-the-shelf tools for the graphical part.
All the three stations are composed of a Station Controller and of a TDMA terminal.
The Station Controller (SC) is in charge of implementing the TDMA burst assembly, the TDMA slot acquisition strategy and the Doppler compensation algorithm. This module manages also the communication with the experiment controller. The SC is physically implemented in an Analog Devices TS101S Tiger Shark DSP based board running a software developed in ANSI C. The board is linked to the TDMA terminal by means of two DSP high speed serial link ports.
The TDMA Terminal is implemented on two boards, one including the digital parts and the other hosting the filters and the analog circuits.
The Digital Board hosts a Xilinx Virtex II 4MGates FPGA that contains the demodulator circuit and the modulator

control interface. The Modulator is completed by a commercial DDS IC from Analog Devices.
The P/L Simulator and the AUX stations are equipped with a DAC, necessary to generate the control signal for the local VCO.
The MCS is timed using an high-performance Rubidium Atomic Frequency Source (RAFS). The P/L and the Aux stations receive their reference from Low Performance Rubidium Frequency Source. Those devices are controlled using a built in analog input.
The measurement system is composed of two Picotime devices that permit to measure the Allan variance of a clock under test with respect to a reference. Both the Picotime Units use the RAFS as a reference: they measure the performances of the P/L and the AUX clock respectively.
The system is completed by a Power Supply unit and a power Distribution unit.
All the equipments are hosted in a standard 19" rack in order to ease the transportation and installation.

9. Field Trial Program Schedule and Status

The Field Trial Program was started on the 2nd quarter 2003 and is planned to be concluded within June 2004.
During the first program phase a preliminary theoretical analysis was performed for assessing the expected system performances prior to committing to a final system configuration. Then, the System and Subsystems Specifications were established, being the starting point for the setup of the Time & Frequency Dissemination system to be validated during the test campaign.
Based on the established specifications, the detailed design and MAIT of the items to be specifically designed for the Field Trial Campaign were performed (TDMA Terminals, Station Controller, Experiment Controller, Micro Phase Stepper).
In parallel, the off-the-shelf items complementing the custom design items were procured (Measurement SS, Precision Clock Sources).
Currently, all the custom design units have been successfully setup and the system integration is in progress at Carlo Gavazzi Space premises. At the beginning of April 2004 a first station has been successfully validated at SES ASTRA premises, in Betzdorf, while the Field Trial Test Campaign exploiting ASTRA 1-A Satellite is planned to start in the middle of May 2004.
The Final Report of the program will be available within June 2004.

10. Future Plans

After the conclusion of the Field Trial Campaign the short – medium term plans foresee the upgrading of the Field Trial Configuration for providing of a pre-operational service.
This upgraded configuration will take benefit of the lessons learnt by performing the Field Trial test campaign.
Apart from the re-engineering of the Field Trail configuration, this upgraded system would not require a long time period to be setup as it will not be required to fly the clock module on board a satellite but a transparent transponder will be exploited as for the Field Trial campaign.
At medium – long term it can be foreseen the implementation of a pre-operational mission by embarking the Time & Frequency Dissemination Payload hosting the on board clock module onto a Satellite.

Acknowledgements

The authors will thank ESA for the support given to the Time and Frequency Dissemination System initial studies and the definition work made so far and the Italian, Swiss and French Delegations that with their support made possible the implementation of the Field Trial campaign in the frame of ARTES V program.

References

[1] G. Garofalo , G. Busca, "Novel High Performance Satellite Time & Frequency Distribution and Synchronization System"; Proc of 14th EFTF, 14-16 Mar 2000, Torino (I)

[1] D. Giunta, P. Erhard, A. Della Torre, R. Grimoldi and M. Vitta, D. Levavasseur, O. de la Burgade "Signal Processing Techniques for Time and Frequency References Distribution by Satellite"; DSP 2001 Conf. Proc., Sesimbra (Portugal)

[3] D. Giunta, G. Busca, M. Vitta, V. De Perini, R. Grimoldi, X. Stehlin "Time Frequency And Distribution"; SPSC 2003 Conf. Proc., Catania (I)

[4] C.Mosquera, S.Scalise, G.Taricco, G.Garofalo, D.Giunta; "Time-Transfer Performance in Burst-Mode Communication Systems"; IEEE JSAC, Vol. 19, No. 12, December 2001

[5] G. Busca - "Advanced Clock Generation and Distribution for Science Missions" - ESA Contract no 12463/97/NL/MV Final Report

[6] Carlo Gavazzi Space - "New Methods and Systems for Time & Frequency Distribution via Satellite" - ESA Contract no 14427/00/NL/PB Final Report

[7] P. Erhard - "New Methods and Systems for Time & Frequency Distribution via Satellite" - ESA Contract no 14428/00/NL/PB Final Report

FREQUENCY STANDARD TRANSFER OVER A 43-KM OPTICAL FIBER

C. DAUSSY, O. LOPEZ, and C. CHARDONNET, Laboratoire de Physique des Lasers, UMR 7538 C.N.R.S.,Université Paris 13, 99 av. J.-B. Clément, 93430 Villetaneuse, France
F. NARBONNEAU, M. LOURS, and G. SANTARELLI, BNM-SYRTE, UMR 8630 C.N.R.S., Observatoire de Paris, 61 av. de l'Observatoire, 75014 Paris, France
e-mail : daussy@lpl.univ-paris13.fr

Applications including : coherently arrayed antennas, radio and radar astronomy, very long baseline interferometry and geodynamic measurement require an ultra stable frequency and distribution system. Phase stable and low noise optical distribution systems would be very useful in the metrological domain. Indeed the opportunity to compare clocks when laboratories are in a range of 100 km or more would greatly help the evaluation of their accuracy and stability performances [1]. Since the recent development of absolute frequency measurement with femtosecond laser comb generators, it is now possible to compare directly a frequency standard in the microwave domain to a frequency standard in the IR or visible domain. For these measurements an absolute frequency reference is needed, either a Cs- clock or a Rb- clock, the frequency of which is compared to TAI (Temps Atomique International). Here we present an ultra-stable fiber optic frequency link used to transfer such reference frequency signal through a standard telecommunication optical fiber between laboratories at some 20-km distance. The characterization of new ultra-stable frequency standards under progress at LPL [2], such as the CO_2/SF_6 two-photon frequency standard in the 30-THz spectral region or Ar/I_2 requires such ultra stable frequency distribution system that surpasses the phase stability of all systems previously developed.

The optical fiber is 43 km long and propagates a carrier at 1.55 μm from SYRTE (Observatoire de Paris) to LPL(Villetaneuse). The carrier is amplitude modulated by a 100-MHz signal, itself referenced to a cryogenic sapphire oscillator and Cs atomic fountain (accuracy of 10^{-15}) at SYRTE. This frequency is used at LPL to phase-lock a low phase noise quartz oscillator at 100-MHz. This signal which reproduces the frequency stability of the 100-MHz received signal for integration time longer than 1 s is then used to phase-lock an oscillator at 1-GHz (SAW) for direct femtosecond repetition rate measurements. The quartz oscillator at 100-MHz is also used at LPL to phase-lock a low phase noise quartz oscillator at 5 MHz which second harmonic is distributed through the laboratory and is used as an external reference clock for all the RF synthesisers and the counter involved in the frequency measurements. This new measurement method gives us the possibility to reach a relative accuracy much better than can be obtained with a GPS receiver or a Rb clock for averaging time as short as a few s.

The fiber transfer is sensitive to both the mechanical stress and the temperature variations, we have thus developed an actively stabilized frequency distribution system to suppress the phase changes. Fluctuations in phase are monitored continuously by comparing the source beam with the return beam after a round trip. A complete characterisation of the residual phase noise introduced by the optical link has been performed.

[1] J. Ye et al: J. Opt. Soc. Am. B 20, 1459 (2003).
[2] A. Amy-Klein et al: Appl. Phys. B, to be published (2003).

TIMEKEEPING IN FUTURE NASA MISSIONS

Al Gifford,[1] Samuel R. Stein,[2] and Robert A. Nelson[3]

[1]National Aeronautics and Space Administration , Code M-3, 300 E Street SW, Washington, DC 20546 USA
[2]Timing Solutions Corporation, 4775 Walnut Street, Suite 1B, Boulder, CO 80301 USA
[3]Satellite Engineering Research Corporation, 7701 Woodmont Avenue, Suite 208, Bethesda, MD 20814 USA

Keywords: Time, clocks, relativity, space navigation

Abstract

The GPS has provided a model for precise time dissemination in the environment of the Earth. In the future, principles analogous to those now used in the GPS will need to be applied to space missions beyond the Earth. This paper presents some of the theoretical and technical issues that will be important for time synchronization among atomic clocks on Earth, on the surface of Mars, and on satellites in orbit about Mars. These issues include relativistic time transformations, environmental effects, and hardware factors. Potential applications are solar system Very Long Baseline Interferometry (VLBI), navigation, and refined tests of general relativity. The paper also considers issues regarding the formation of a distributed time scale using clocks on Earth and Mars, with an eventual extension of the time scale throughout the solar system.

1 Introduction

Space exploration is entering the next generation with a new initiative for human exploration of the Moon and Mars. Alternative architectures for communication, navigation, and time dissemination are being investigated. The Moon will be the site of future astronomical observatories and will provide a near-Earth platform for testing technologies and procedures that will eventually be used for Mars.

Among the many technical considerations is the need for time synchronization and dissemination. Time is central to communication and navigation. It is also fundamental to scientific applications such as Very Long Baseline Interferometry (VLBI) that involve the assimilation of data from widely separated sources.

The purpose of this paper is to discuss some notional concepts for solar system time and navigation under consideration by the National Aeronautics and Space Administration (NASA). In particular, we wish to call attention to the need for relativistic transformations for precise timekeeping. The Global Positioning System (GPS) provides a model for navigation and the appropriate principles for relativistic time transfer. We shall also discuss the use of communication links for time synchronization, the need to consider environmental and hardware factors, and the requirements for the formation of a distributed time scale throughout the solar system.

2 Relativity

In the theory of relativity, there are two kinds of time: proper time and coordinate time. Proper time is the reading of a clock in its own rest frame. It is different for clocks in different states of motion and in different gravitational potentials. Coordinate time is the time coordinate in the arbitrary space-time coordinate system chosen on the basis of symmetry or convenience. It is a global coordinate and has the same value everywhere for a given event.

Three effects contribute to the net relativistic effect on a transported clock [1, 2]. These are:

(1) *Time dilation (velocity).* This effect makes a moving clock run slow relative to a stationary clock and is independent of direction.

(2) *Redshift (gravitational potential).* This effect makes a clock in a weaker potential run fast relative to a clock in a stronger potential. In the frame of reference of the Earth it is a function of altitude. (The term "redshift" is generic, regardless of whether the clock rate is faster or slower.)

(3) *Sagnac effect.* This effect is intrinsic to a rotating frame of reference. It makes a transported clock run fast or slow depending on the direction of the path taken. It is zero for a closed path that retraces itself.

For each space-time coordinate system there is a relativistic transformation between proper time τ and coordinate time t. In the Earth-Centered Inertial (ECI) frame of reference, the elapsed coordinate time for a transported clock along a path between points A and B to order v^2 / c^2 is

$$\Delta t = \int_A^B \left(1 + \frac{1}{2}\frac{1}{c^2} v^2 + \frac{1}{c^2} U \right) d\tau$$

where v is the clock velocity, U is the gravitational potential, and c is the speed of light. The corrections under the integral represent the time dilation and redshift effects, respectively. In a rotating Earth-Centered Earth-Fixed (ECEF) frame of reference, the elapsed coordinate time to the same order is

$$\Delta t' = \int_A^B \left(1 + \frac{1}{2}\frac{1}{c^2} v'^2 - \frac{1}{c^2} g\, h \right) d\tau + \frac{\omega}{c^2} \int_A^B R^2 \cos^2 \phi \; d\lambda$$

where v' is the velocity in the rotating frame, g is the local acceleration of gravity, h is the height above the geoid, ω is the angular velocity of the Earth, R is the distance from the center of the Earth, ϕ is the latitude, and λ is the longitude. The corrections under the first integral represent the time dilation and redshift effects in the rotating frame. The second integral represents the Sagnac effect. It can be expressed as $2\,\omega\,A\,/\,c^2$, where A is the equatorial projection of the area bounded by the path

3 The Global Positioning System

The Global Positioning System (GPS) has served as a laboratory for relativity and has provided a model for theoretical algorithms. Each satellite carries a suite of caesium or rubidium atomic clocks. The reference frequency is precisely 10.23 MHz. The GPS satellites transmit signals at two carrier frequencies in L-band: the L1 component with a center frequency of 1575.42 MHz and the L2 component with a center frequency of 1227.60 MHz. Additional signals and a third carrier known as L5, to be centered at 1176.45 MHz, will be implemented as part of GPS modernization. The basic measurement is the alignment of a pseudorandom noise code from the satellite and a replica code generated in the receiver to determine the signal time of flight. The navigation solution may be characterized as the process of triangulation using range measurements from four or more satellites, corrected for satellite clock error with respect to GPS Time, ionospheric delay, tropospheric delay, and relativity [3 – 5].

The nominal constellation consists of 24 satellites in circular orbits, distributed into 6 planes inclined at 55° with 4 satellites per plane. The period of revolution is 12 sidereal hours (11 h 58 min), which corresponds to an orbital radius of 26 560 km and an altitude of 20 184 km. The satellite velocity is 3.874 km/s.

On account of time dilation, the satellite clocks run slow by 7.1 μs per day compared to a clock on the geoid. In addition, on account of the difference in gravitational potential, they run fast by 45 μs per day. The net secular effect is fast by 38 μs per day. This is an enormous effect for clocks that maintain a precision on the order of 10 ns over several hours. Thus prior to launch, the clocks are given an offset of 4.465 parts in 10^{10}, so that on average the satellite clocks run at the same rate as a clock on the geoid. The resulting frequency is 10 229 999.995 4326 Hz

The orbits inevitably have a small eccentricity, which causes the velocity and altitude to vary slightly with time. Thus there is a residual periodic time dilation and redshift effect. For an eccentricity of 0.02, the amplitude is 46 ns. This effect is corrected in the user's receiver.

Relativistic rate changes have also been observed during stationkeeping maneuvers. Recently, corrections have been applied to model these rate changes to improve the response of the Kalman filter used in the orbit determination algorithm.

In addition, there is the Sagnac effect in the Earth's rotating frame of reference. For a receiver at rest on the surface of the geoid, the maximum Sagnac effect is 133 ns, which occurs when the receiver is on the equator and the satellite is on the horizon. This correction is applied in the receiver. From the point of view of an inertial observer, the correction is due to the motion of the receiver with a velocity of 465 m/s relative to an ECI frame of reference during the 86 ms signal propagation time. If the receiver is moving relative to the geoid, there must also be a range rate correction.

These relativistic corrections comprise all those currently incorporated in the GPS and are sufficient for measurement of time down to the nanosecond level. For sub-nanosecond time measurement, the next most important factor is the contribution to the gravitational redshift due to the Earth's J_2 oblateness perturbation in the potential U. For a GPS orbit, there is a periodic term that has an amplitude of 24 ps.

The gravitational potentials of the Moon and Sun do not directly affect a clock on a satellite in orbit about the Earth because the Earth constitutes a freely falling frame of reference in those potentials. This is a manifestation of the Principle of Equivalence, which states that a freely falling frame of reference is locally equivalent to an inertial frame of reference. Thus the effects of the lunar and solar potentials appear instead as tidal effects. For a GPS clock, the amplitudes of the periodic tidal terms for the Moon and Sun are approximately 1 ps and 0.5 ps, respectively.

The gravitational potential also modifies the speed of propagation of the electromagnetic signal itself. The gravitational delay with respect to a clock on the geoid is 4.7 ps for a GPS satellite at the zenith. Although the effect is negligible at the current level of precision in the GPS, it is significant for solar system measurements. The round trip gravitational delays for ranging to Mercury and Venus at superior conjunction are approximately 240 μs and 180 μs, respectively.

Frequency measurements are affected by both classical and relativistic Doppler effects. For a GPS satellite with an orbital eccentricity of 0.02, the L1 carrier frequency undergoes a modulation due to relativity with an amplitude of 0.011 Hz.

In the future, GPS satellites may have intersatellite links. The analysis of signals used in cross-link ranging, time transfer among satellites and ground stations, and interoperability across satellite constellations involves three steps: (1) a relativistic transformation from the proper time reading of the clock at the transmitter to the coordinate time of transmission in the adopted coordinate system (e.g., ECI or ECEF); (2) calculation of the coordinate time of signal propagation, including both relativistic and nonrelativistic effects; and (3) a relativistic transformation from the coordinate time of reception in the adopted coordinate system to the proper time reading of the clock at the receiver.

Relativistic effects on satellite clocks and signal propagation for some representative satellites are summarized in Table 1.

Table 1: Relativistic Effects on Satellite Clocks and Signal Propagation

Satellite orbital properties									
Satellite		**ISS**	**TOPEX**	**Glonass**	**GPS**	**Galileo**	**Molniya**	**GEO**	**Tundra**
Semimajor axis	km	6766	7715	25510	26562	29994	26562	42164	42164
Eccentricity		0.01	0.001	0.02	0.02	0.02	0.722	0.01	0.2684
Inclination	deg	51.6	66.0	64.8	55.0	56.0	63.4	0.1	63.4
Argument of perigee	deg	0	0	0	0	0	250	0	270
Apogee altitude	km	456	1345	19642	20715	24216	39362	36208	47103
Perigee altitude	km	320	1329	18622	19652	23016	1006	35364	24469
Ascending node altitude	km	320	1329	18622	19652	23016	10507	35364	32748
Period of revolution	s	5539	6744	40549	43082	51697	43083	86164	86164
Mean motion	rev/d	15.6	12.8	2.1	2.0	1.7	2.0	1.0	1.0
Mean velocity	km/s	7.675	7.188	3.953	3.874	3.645	3.874	3.075	3.075
Clocks									
Secular time dilation	μs/d	− 28.2	− 24.7	− 7.4	− 7.1	− 6.3	− 7.1	− 4.4	− 4.4
Secular redshift	μs/d	3.5	10.4	45.1	45.7	47.3	45.7	51.0	51.0
Net secular effect	μs/d	− 24.7	− 14.3	37.7	38.6	41.1	38.6	46.6	46.6
Periodic eccentricity effect	ns	12	1	45	46	49	1653	29	774
Secular oblateness redshift	ns/d	2.1	− 4.6	− 0.1	0.0	0.0	− 0.1	0.1	0.0
Periodic oblateness redshift	ps	171	191	31	24	21	30	0	14
Periodic tidal effect of Moon	ps	0.0	0.0	1.0	1.2	1.8	1.2	6.1	6.1
Periodic tidal effect of Sun	ps	0.0	0.0	0.5	0.5	0.8	0.5	2.7	2.7
Signal propagation									
Maximum Sagnac effect	ns	13	23	131	136	155	234	218	275
Gravitational delay along radius	ps	0.8	2.5	− 3.5	− 4.7	− 9.1	− 4.7	− 27.3	− 27.3
Periodic fractional Doppler shift	10^{-12}	13.1	1.1	7.0	6.7	5.9	241.1	2.1	56.5

4 TWTT flight tests

An active data communications channel can be utilized as a vehicle for Two-Way Time Transfer (TWTT) using time-based communications technology [6]. This technique provides a means of precise time transfer in the background of an active data communications channel. The two ends of a communications link can thus be precisely synchronized without fielding an independent timing system.

In November, 2002 a series of six flight tests were conducted by NASA, AFRL, Timing Solutions Corporation, and Zeta Associates to evaluate the TWTT technique between a ground station and an airborne platform. Data were collected using a high quality atomic clock and a GPS receiver carried onboard a C-135 aircraft flown from Wright-Patterson Air Force Base. The measurements used both a direct line-of-sight radio frequency (RF) link as well as a geostationary satellite communications relay. The flight tests provided an opportunity to monitor clock behaviour throughout the entire flight by virtue of using the satellite time transfer method from the mobile platform.

As in time transfer involving satellite clocks, relativistic corrections are necessary. Although it is well known that special and general relativity cause satellite clocks to run fast or slow compared to clocks at rest on the surface of the Earth,

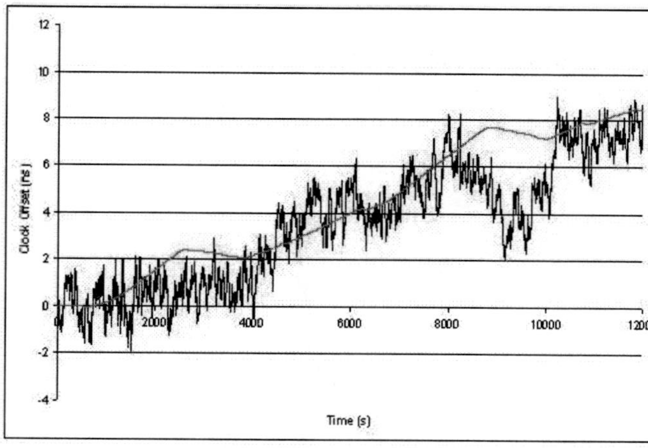

Figure 1. Comparison of measured and predicted net relativistic correction (Flight Clock – Reference Clock). The vertical scale is 2 ns and the horizontal scale is 2000 s. The theoretical relativity model is shown as the curve superimposed on the measured data.

it may not be as generally appreciated that terrestrial motion associated with, for example, an aircraft can also result in measurable relativistic effects. The relativity model was illustrated by the experiments to demonstrate TWTT in the background of an active communications channel [7, 8].

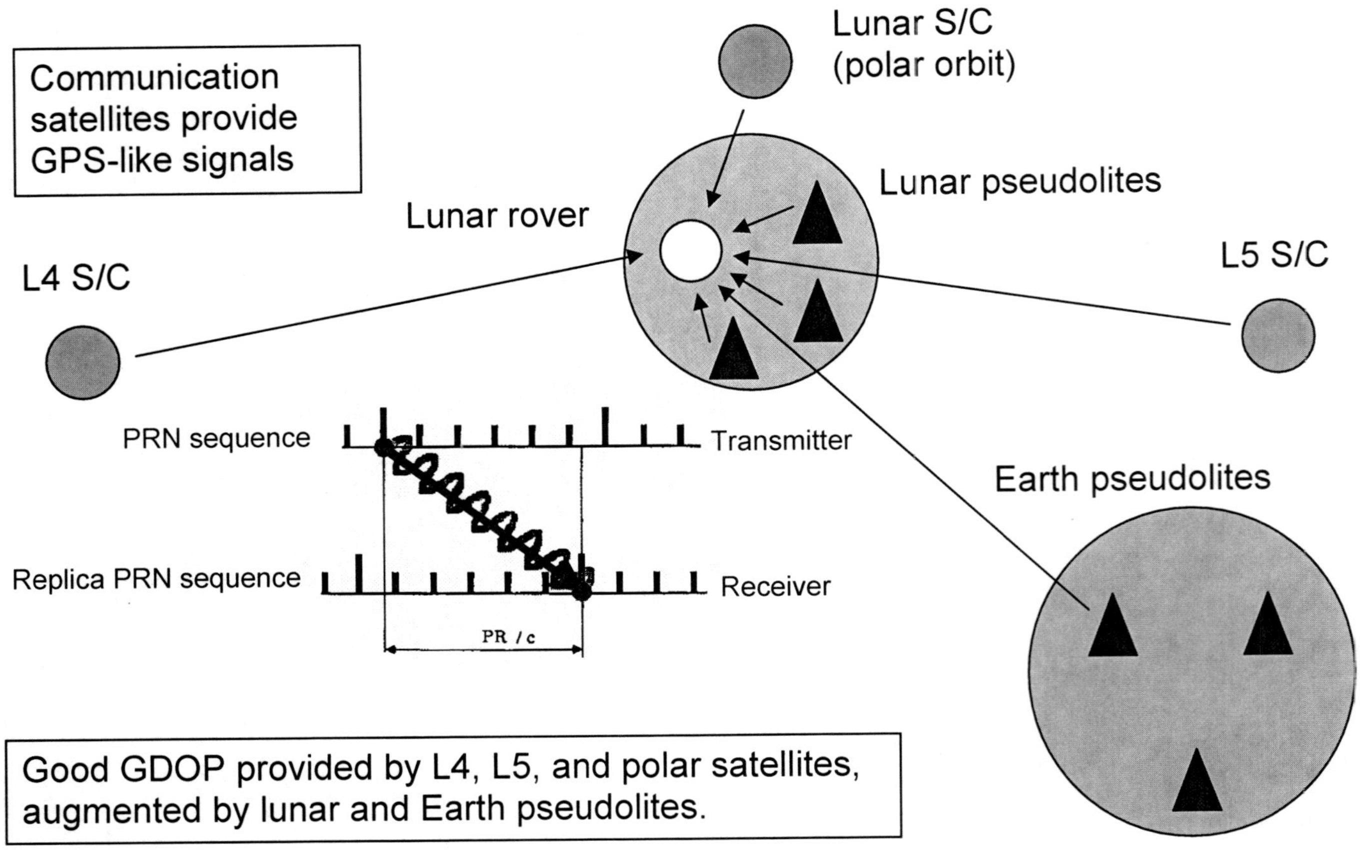

Figure 2. Notional space navigation architecture.

The relativistic corrections for time dilation, redshift, and the Sagnac effect in the rotating ECEF reference frame were integrated numerically over the flight path as given by the formulas (Flight Clock – Reference Clock)

$$\Delta T_v = -\frac{1}{2c^2}\sum_{i=1}^{N} v_i^2\,\Delta t_i$$

$$\Delta T_g = \frac{g}{c^2}\sum_{i=1}^{N}(h_i - h_0)\,\Delta t_i$$

$$\Delta T_S = -\frac{\omega}{c^2}\sum_{i=1}^{N} R_i^2 \cos^2\phi_i\,\Delta\lambda_i$$

Data from the first flight, representative of the results in general, are shown in *Figure 1*. The aircraft flew for nearly three hours at a cruising altitude of 10 700 m (35 000 ft). The average speed was 790 km/h (425 knots). The calculated total relativistic corrections were – 3.03 ns due to time dilation, + 11.46 ns due to redshift, and – 0.01 ns due to the Sagnac effect around the closed path. The net effect was thus + 8.42 ns. The measured test results, after calibrating the drift between the two clocks using data collected before and after the flight was + 8.3 ns. The anomalous dip of about 3 ns near the end of flight is not inconsistent with typical caesium clock behaviour. Similar artifacts of this magnitude were observed during comparisons between the clocks in the laboratory.

5 Notional space navigation architecture

Besides communication, the system architecture must include a method for navigation. One alternative architecture for space navigation is depicted in *Figure 2*. This notional option is based on proven GPS technology using triangulation by means of pseudorandom noise codes. The signals may be provided by communications satellites at the Lagrange points of the Earth-Moon system and in orbit about the Moon.

The Lagrange points are points of equilibrium where the combined gravitational forces of the Earth and Moon balance the centrifugal force in the rotating frame of reference in which the Earth and Moon are at rest with origin at their center of mass. As illustrated in *Figure 3*, there are five Lagrange points, whose distances from the Earth and the Moon are given in Table 2. The points L1, L2, and L3 are unstable points of equilibrium. L1 and L2 are at roughly 60 000 km on either side of the Moon. They provide good coverage of the Moon but require long-term stationkeeping. It is possible to place the satellites in "halo" orbits about these points to enhance the lunar visibility. The points L4 and L5 are stable points of equilibrium. Each of these points is at the vertex of an equilateral triangle with the Earth and Moon at the other two vertices. Therefore, they are equidistant from both the Earth and the Moon. They have the advantage that together they see two thirds of the Moon's far side.

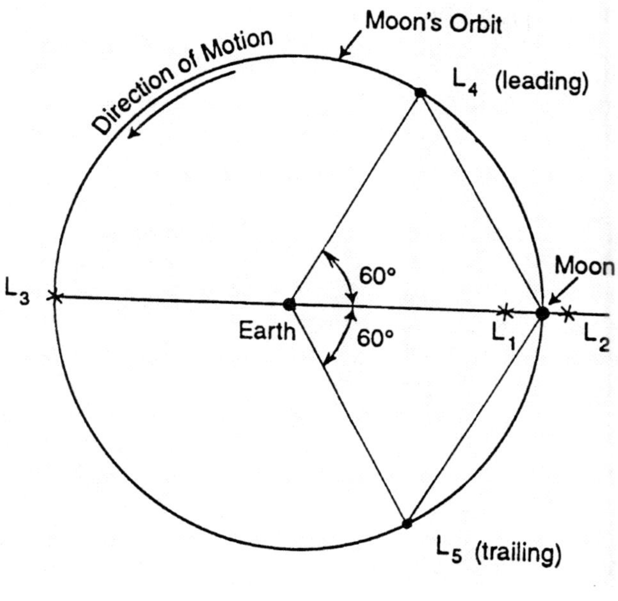

Figure 3. The Lagrange points of the Earth-Moon system.

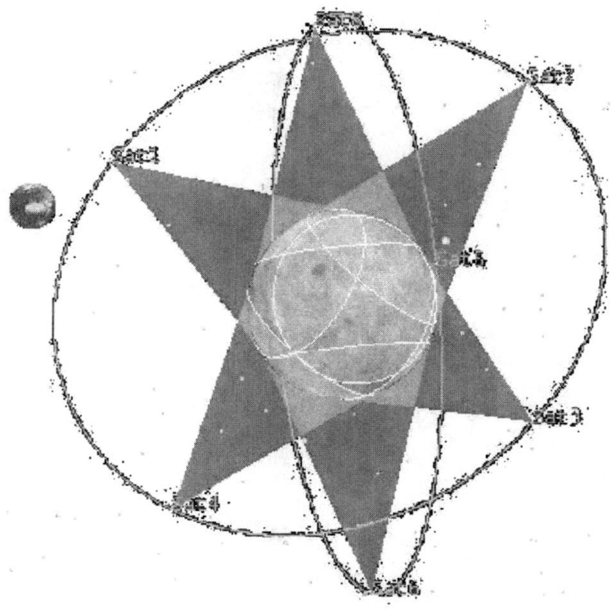

Figure 4. Polar constellation about the Moon.

Table 2: Positions of the Lagrange points

Point	*Distance from the Earth* Orbit radius	km	*Distance from the Moon* Orbit radius	km
L1	0.849 066	326 385	0.150 934	58 020
L2	1.167 833	448 921	0.167 833	64 516
L3	0.992 912	381 680	1.992 912	766 085
L4	1.000 000	384 405	1.000 000	384 405
L5	1.000 000	384 405	1.000 000	384 405

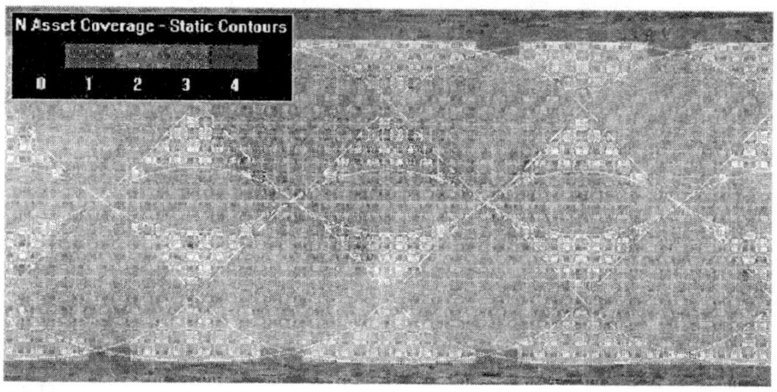

Figure 5. Level of coverage provided by polar constellation.

Satellites may also be deployed into constellations about the Moon. Of particular interest is the lunar South Pole, where there is some evidence that there may be water ice in deep craters. Thus polar orbits may be desirable.

One example of a polar constellation is shown in *Figure 4*. In this constellation there are 8 satellites in 2 orbital planes with 4 satellites per plane. The orbital radius is 3 lunar radii, or 5214 km. A preliminary analysis of the orbital perturbations due to the Earth and Sun indicates that the orbit is stable. At this radius, the constellation geometry ensures that the angle of elevation at edge of coverage is somewhat greater than 10°, a minimum requirement for satellite visibility. The level of coverage is shown in *Figure 5*. At the poles there is continuous coverage by four satellites. Over most areas in mid-latitudes, continuous coverage is provided by two satellites.

Pseudolites may be used on the lunar surface and on the Earth to supplement satellite coverage for navigation using GPS-like signals. In some cases it may not be necessary to require four simultaneous signals, as in the case where a precise time reference is independently available or when a three-dimensional position determination is not necessary.

An example of a satellite constellation about Mars is shown in *Figure 6*. This is a Walker 12/3/1 "delta" pattern [9], consisting of 12 satellites in 3 planes with a relative satellite phasing between neighboring planes of 30°. The optimum inclination is 50.7°. For a minimum angle of elevation at edge of coverage of 20°, the orbital radius is 2.5 Mars radii or 8317 km. The period of revolution is 6.4 hours. The level of coverage is shown in *Figure 7*. At least two satellites are simultaneously visible over most of the planet.

Another possible satellite orbit is the areostationary orbit, the Mars equivalent of a geostationary orbit. The period of rotation of Mars is 24 h 37 min. Thus by Kepler's third law the orbital radius is 6.0 Mars radii or 20 427 km. The altitude is 17 030 km. By comparison, for a geostationary orbit the radius is 42 164 km and the altitude is 35 786 km. As the required antenna size is proportional to orbital altitude, an

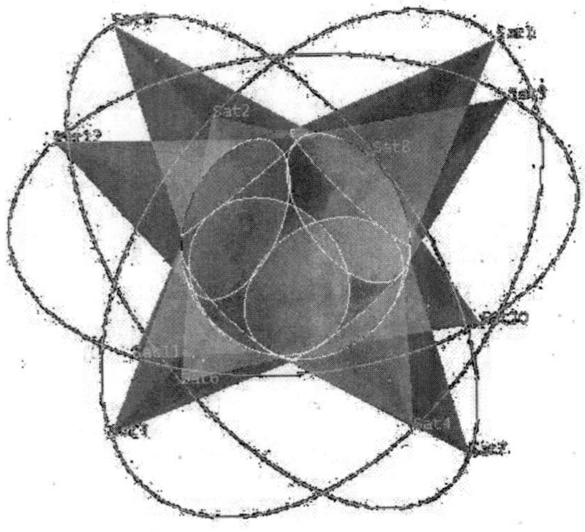

Figure 6. Walker 12/3/1 constellation about Mars.

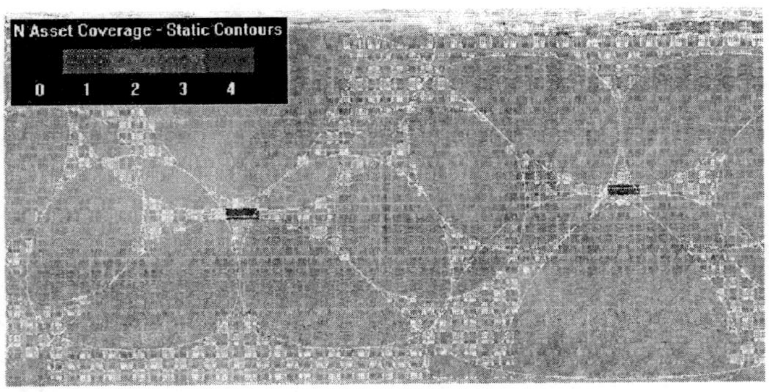

Figure 7. Level of coverage provided by a 12/3/1 Walker constellation.

antenna on Mars need have only half the diameter as an antenna on the Earth. Satellites in areostationary orbits could supplement those in an inclined orbit constellation.

6 Relativistic corrections to solar system time

Synchronization of clocks throughout the solar system requires relativistic transformations between proper time and coordinate time in a common coordinate system. In the coordinates of the solar system barycenter the difference between barycentric coordinate time (TCB) and terrestrial time (TT), as maintained by atomic clocks on the surface of the Earth, is [10 – 12]

$$\mathrm{TCB}-\mathrm{TT} = \frac{1}{c^2}\int\left(\frac{v_E^{\;2}}{2}+U_{ext}(\mathbf{r}_E)\right)dt + L_G\,\Delta D + \frac{1}{c^2}\mathbf{v}_E\cdot(\mathbf{r}-\mathbf{r}_E)$$

where $\mathbf{v}_E$ is the velocity of the Earth's center and $U_{ext}(\mathbf{r}_E)$ is the net gravitational potential of all the bodies in the solar system other than the Earth evaluated at the Earth's center. The potential includes the contributions of the Sun, the planets, and the principal minor planets. In practice, the integral is computed numerically using the relativistic ephemerides of the planetary orbits. The second term is the time difference between a clock at the geocenter and a clock on the geoid, where L_G is approximately 60.2 μs per day and ΔD is the time interval in days. The third term depends on the clock's position $\mathbf{r}$ relative to the position of the center of the Earth $\mathbf{r}_E$. It corresponds to the special relativistic clock synchronization correction in the moving geocentric frame when observed from the barycentric frame. For a clock on the geoid it varies diurnally with an amplitude of 2.1 μs. Similarly, for a clock on Mars there is a relativistic transformation of the form

$$\mathrm{TCB}-\mathrm{MT} = \frac{1}{c^2}\int\left(\frac{v_M^{\;2}}{2}+U_{ext}(\mathbf{r}_M)\right)dt + L_M\Delta D + \frac{1}{c^2}\mathbf{v}_M\cdot(\mathbf{r}-\mathbf{r}_M)$$

Therefore, to synchronize a clock on Mars with a clock on the Earth, one must perform a transformation from the surface of Mars to the solar system barycenter and thence from the solar system barycenter to the surface of the Earth.

The signal propagation time must also be taken into account, including the gravitational propagation delay. The orbital semimajor axis of Mars is 1.524 AU or 2.280×10^8 km. Thus the maximum light time is 21.0 minutes and the minimum light time is 4.4 minutes.

The theoretical details have been evaluated to a level of precision on the order of femtoseconds [13, 14]. Therefore, the required theoretical physics is known.

7 Requirements for a distributed time scale

A clock is a device capable of counting periods of a repeatable phenomenon whose motion or change of state is observable and obeys a definite law. Caesium or rubidium clocks of the type carried onboard GPS satellites have a precision on the order of a few parts in 10^{14}, or a few nanoseconds over a day. Hydrogen maser clocks in the laboratory have a precision approaching parts in 10^{16}. Precision clocks have two types of imperfections: variations in period due to noise and those due to environmental influences.

Time scale technology is used to create a robust scale of time despite inevitable clock failures and power interruptions. The times of multiple clocks are combined after first removing all modelable deterministic effects. For co-located clocks these include frequency differences and frequency aging (drift) differences. For clocks distributed within the solar system, it is also necessary to transform from proper time to coordinate time in a common frame of reference (barycentric coordinate time).

The time scale algorithm apportions residual noise among the member clocks. The delay in receiving the measurements affects the computation. As the theoretical algorithms have been calculated to high precision, the performance of the time scale is not limited by the precision of the relativistic transformations. Rather, it is limited by the customary hardware and environmental factors well known from the formation of distributed time scales on the Earth [15].

8 Applications

Solar system navigation will require technologies and procedures for the synchronization of clocks and the dissemination of time. An important potential application for clocks on Mars is the possibility of Very Long Baseline Interferometry (VLBI). Any form of data collection where local time-tagging is involved, such as in the gathering of imagery from widely separated antennas, will require precise time synchronization. In addition, there are purely scientific applications, such as refined tests of general relativity and alternative metric theories of gravitation.

9 Summary

The future exploration of space will depend on the development of new architectures for communications, navigation, and time synchronization. This paper has identified some of the principles required for time measurement and has discussed a notional architecture for space navigation.

The GPS provides a model for timekeeping and time dissemination. In parallel to advances in clock technology, there must be a corresponding development of theoretical methods of time dissemination within the adopted space-time coordinate system. These methods involve considerations of hardware biases, environmental effects, propagation delays, orbital perturbations, and relativity physics. A common reference system with appropriate relativistic transformations is required.

The GPS timekeeping paradigm can be extended to support NASA space exploration objectives. This is one alternative among many currently under investigation.

References

[1] R. A. Nelson, *Handbook on Relativistic Time Transfer* (May, 2003).

[2] R. A. Nelson, "Relativistic Effects," to be published in *ITU Handbook on Satellite Time and Frequency Transfer and Dissemination* (International Telecommunication Union, Geneva, 2004).

[3] B. W. Parkinson and J. J. Spilker, Jr. (editors), *Global Positioning System: Theory and Applications* (American Institute of Aeronautics and Astronautics, Washington, DC, 1996).

[4] E. D. Kaplan (editor), *Understanding GPS: Principles and Applications* (Artech House, Boston, 1996).

[5] P. Enge and P. Misra (editors), "Special Issue on Global Positioning System," *Proc. IEEE* **87** (January, 1999).

[6] T. Celano, J. Warriner, S. Francis, A. Gifford, P. Howe, and R. Beckman, "Two-Way Time Transfer to Airborne Platforms Using Commercial Satellite Modems," *Proc. 34th Annual Precise Time and Time Interval (PTTI) Meeting*, December 3 – 5, 2002 (U. S. Naval Observatory, Washington, DC, 2003), pp. 353 – 366.

[7] S. Francis, B. Ramsey, S. Stein, J. Leitner, M. Moreau, R. Burns, R. A. Nelson, T. R. Bartholomew, and A. Gifford, "Timekeeping and Time Dissemination in a Distributed Clock Ensemble," *Proc. 34th Annual Precise Time and Time Interval (PTTI) Meeting*, December 3 – 5, 2002 (U. S. Naval Observatory, Washington, DC, 2003), pp. 201 – 214.

[8] S. Francis, S. Stein, T. Celano, J. Warriner, J. Leitner, M. Moreau, R. Burns, R. A. Nelson, and A. Gifford, "Analytical Tools for Clocks in Space," *Proc. 2003 International Frequency Control Symposium and PDA Exhibition Jointly with the 17th European Frequency and Time Forum*, May 4 – 8, 2003, pp. 312 – 316.

[9] J. G. Walker, "Continuous Whole Earth Coverage By Circular Orbit Satellites," *Proc. International Conference on Satellite Systems for Mobile Communications and Surveillance*, March 1973 (Institution of Electrical Engineers, London), pp. 35 – 38.

[10] J. B. Thomas, "Reformulation of the Relativistic Conversion Between Coordinate Time and Atomic Time," *Ap. J.* **80**, pp. 405 – 411 (1975)

[11] T. D. Moyer, "Transformation from Proper Time on Earth to Coordinate Time in Solar System Barycentric Space-Time Frame of Reference," *Celestial Mech.* **23**, pp. 33 – 68 (1981).

[12] E. M. Standish, "Time Scales in the JPL and CfA Ephemerides," *Astron. Astrophys.* **336**, pp. 381 – 384 (1998).

[13] T. Fukushima, "Time Ephemeris," *Astron. Astrophys.* **204**, pp. 895 – 906 (1995).

[14] A. W. Irwin and T. Fukushima, "A Numerical Time Ephemeris of the Earth," *Astron. Astrophys.* **348**, pp. 642 – 652 (1999).

[15] C. Audoin and B. Guinot, *The Measurement of Time: Time, Frequency and the Atomic Clock* (Cambridge University Press, New York, 2001).

TIME AND FREQUENCY MEASUREMENTS UNIFICATION IN RUSSIA A KEY ROLE OF NATIONAL SERVICE FOR TIME AND FREQUENCY

V.Krutikov*, V.Kostromin†, N.Koshelyaevsky†

* Gosstandard of Russia, Russia, Moscow, 119991
† Institute of Metrology for Time and Space FGUP «VNIIFTRI», Russia, Mendeleevo, 141570,
nkoshelyaevsky@imvp.aspnet.ru

Keywords: Time scale, Time Transfer

1 Introduction

In response to challenges of new technologies just after World War II the National Service for Time and Frequency (NSTF) have been established in 1947 in Russia. The NSTF joined all available human, scientific and instrumentation resources of the post-war Russia. Today NSTF consolidates efforts of thirteen governmental departments and ministries in time and frequency measurements unification through the whole territory of Russian Federation.
Legal, technical and metrology aspects of the NSTF current activity as well as closest future perspectives will be presented in this paper.

2 The legal regulation

According to the Constitution of the Russian Federation, Chapter 3, Federal Structure, Article 71- " The jurisdiction of the Russian Federation includes: ...p) ...norms, measurement standards, metric system and chronometry accounting..." [1]. To realise this statement and for measurements unification inside country the law of the Russian Federation on Measurement Unification was adopted in 1993 [2]. The article 10 "The National Metrology Service and other National Services for ensuring the measurement unification" of it says that " The National Service for Time and Frequency and the Earth Orientation Parameters Determination realises interregional and interdepartmental activity co-ordination for ensuring time and frequency measurements and the Earth Orientation Parameter unification". Further more in spring 2001 new Statute of the National Service for Time, Frequency and EOP determination was adopted by Decree No 225 of the Government of the Russian Federation [3]. It says that:

"2. The National Service for Time realises scientific and metrology activity for uninterruptable reproduction and keeping of the Russian Federation National time scale and standard frequencies... "

Apart from specifying the main designation of the NSTF the Statute determined the operational authority of the Service – Main Metrology Centre of the NSTF:

"5. The Main Metrology Centre of the National Service for Time realises scientific, methodology and operational activity supervising of the National Service for Time."

And perhaps the most important article of the Statute is that:

"10. An information of the National Service for Time regarding time, frequency and the Earth Orientation Parameters is obligatory in the Russian Federation."

3 National Service for Time and Frequency

The structure of the National Service for Time and Frequency is presented on Fig1.
As it was mentioned previously it consists of thirteen governmental departments and ministries. The moderator of the community is so called inter-department commission for time and frequency. The authorities of the Commission is defined as [4]:
"The Inter-department commission for time and frequency is co-ordination body in the area of developing and utilisation the state means and systems for reproduction and keeping of time scale, standard frequencies and EOP determination ... "
Every partners are responsible for appointed field of activity, but only two of them: Department of Defence (DoD) and State Committee of Standards (Gosstandard) have their own operational centres. The operational centre of the DoD - ACCS - is responsible for time keeping and dissemination for the needs of defence, including support of such navigation system as GLONASS, Loran and Omega like stations. It works in close connection with the Main Metrology Centre of the NSTF.
The main scope of the operational responsibility for time and frequency measurement unification in civil segment, including maintenance of the state standard of time and frequency, National time scale generation and time signals and standard frequency dissemination, supervision of the state secondary time laboratories belonging to the Main Metrology Centre. The authorities of the Main Metrology Centre realises Institute of Metrology for Time and Space FGUP "VNIIFTRI".

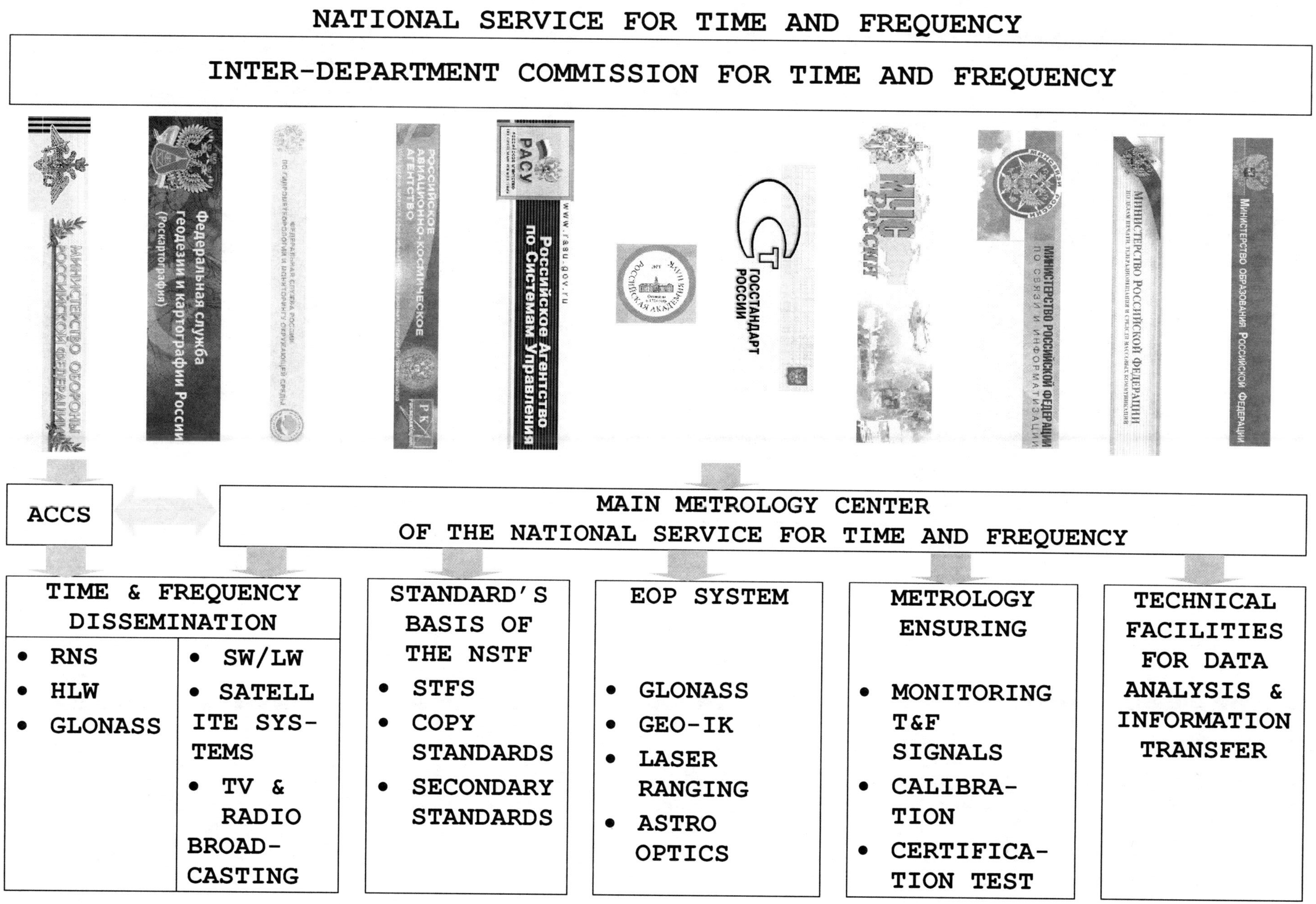
NATIONAL SERVICE FOR TIME AND FREQUENCY
INTER-DEPARTMENT COMMISSION FOR TIME AND FREQUENCY
МИНИСТЕРСТВО ОБОРОНЫ РОССИЙСКОЙ ФЕДЕРАЦИИ
Федеральная служба геодезии и картографии России (Роскартография)
РОССИЙСКОЕ АВИАЦИОННО-КОСМИЧЕСКОЕ АГЕНТСТВО
РАСУ
www.rasu.gov.ru
Российское Агентство по Системам Управления
ГОССТАНДАРТ РОССИИ
МЧС РОССИИ
МИНИСТЕРСТВО РОССИЙСКОЙ ФЕДЕРАЦИИ ПО СВЯЗИ И ИНФОРМАТИЗАЦИИ
МИНИСТЕРСТВО РОССИЙСКОЙ ФЕДЕРАЦИИ
Министерство образования Российской Федерации
ACCS
MAIN METROLOGY CENTER
OF THE NATIONAL SERVICE FOR TIME AND FREQUENCY
TIME & FREQUENCY DISSEMINATION
• RNS
• HLW
• GLONASS
• SW/LW
• SATELLITE SYSTEMS
• TV & RADIO BROADCASTING
STANDARD'S BASIS OF THE NSTF
• STFS
• COPY STANDARDS
• SECONDARY STANDARDS
EOP SYSTEM
• GLONASS
• GEO-IK
• LASER RANGING
• ASTRO OPTICS
METROLOGY ENSURING
• MONITORING T&F SIGNALS
• CALIBRATION
• CERTIFICATION TEST
TECHNICAL FACILITIES FOR DATA ANALYSIS & INFORMATION TRANSFER

Fig.1

4 The Main Metrology Centre of National Service for Time and Frequency

The main operational technical task of the Institute of Metrology for Time and Space (IMVP) is to maintain the State Time and Frequency Standard (STFS) and generate the National time scale UTC(SU) of Russian Federation. According to the Traceability Chart [5], fig.2, STFS plays constitutive role not

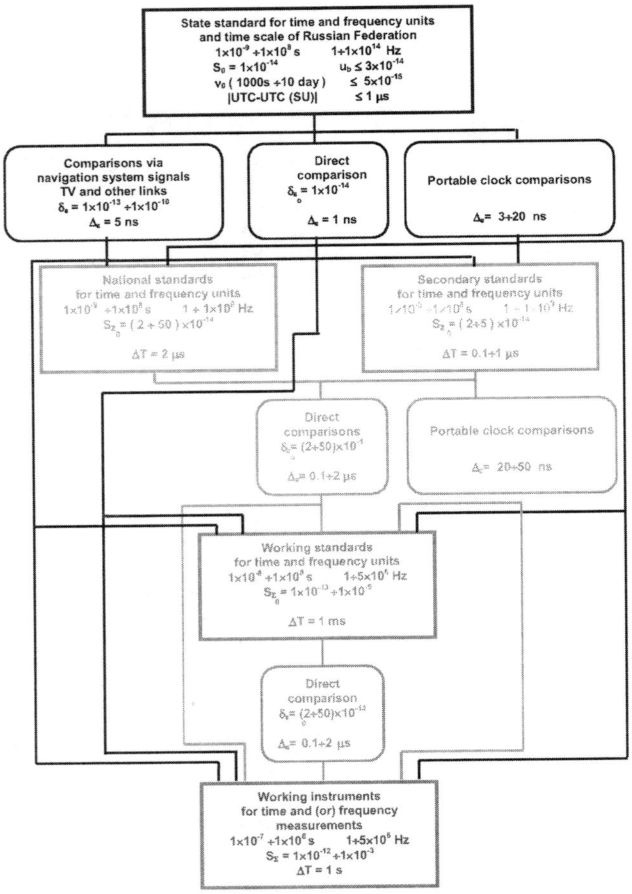

Fig.2

only within Russian Federation but according to the Intergovernmental Agreement on Time and Frequency for the whole Commonwealth of the Independent Countries [6]. The most striking feature of the Traceability Chart is that the highest level of accuracy is accessible in a direct way by any measurement tool without any intermediaries by means of time and frequency transfer via signals of the GNSS. Keeping in mind the constitutive role of the STFS let's look a little inside it.

5 The State Time and Frequency Standard of Russian Federation

The most principal parts of it are:

- the primary laboratory caesium standard
- the ensemble of continuously operating clocks
- the clock intercomparison system
- the remote time and frequency comparison system
- the infrastructure instrumentaions

The primary laboratory caesium standard

The caesium standard – CS 102, fig.3, is based on old principles of thermal atomic beam and magnetic state selection.

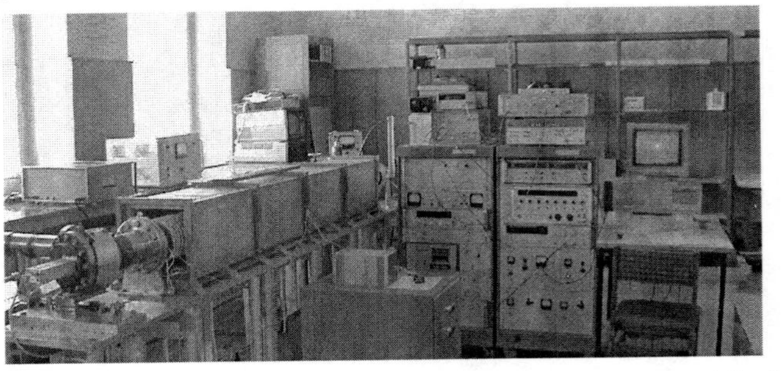

Fig.3

The width of central Ramsey fringe is about 100 Hz and the uncertainty u_b (accuracy) of the instrument 3×10^{-14}. Stability level of the instrument relative to TAI exceeds 1×10^{-14} for sample time 100 days and more, fig.4. The existence of the

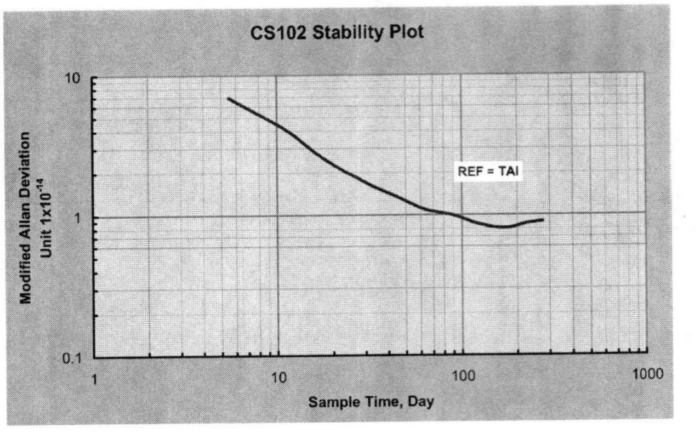

Fig.4

primary caesium standard enables State Standard reproduce the unit of time in a full accordance to it's SI definition. Despite inferior accuracy level frequency difference relative to TAI for last few years does not exceed 1×10^{-14}.

The ensemble of continuously operating clocks

The ensemble of continuously operating H-masers, the whole amount of 10 instruments, fig. 5, is a source for time unit

Fig. 5

keeping and scale generation. All instruments are employed cavity autotuning system. The flicker floor level of the best time-keeping instruments about $1\div2\times10^{-15}$, fig. 6, and is reached for sample time of few days. Each H-maser reveals detectable frequency drift. The value of it is quite individual, typically about $\sim 2 \div 3 \times 10^{-16}$/day, depending on particular in-

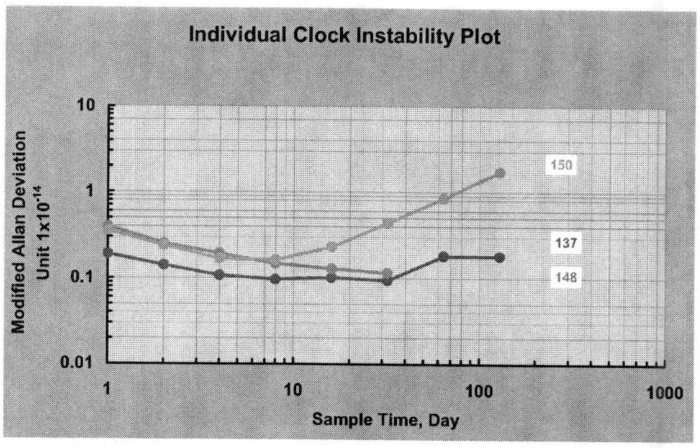

Fig.6

strument, but the sign of drift is usually the same. Because drift is more or less stable and predictable there is no problem to take it into account for time scale calculations.

The clock intercomparison system

The clock intercomparison system processes two signal streams: i. clock signal, i.e. 1 ppS output signal of H-masers and ii. 5/100 MHz signals. Time resolution of the i. system equals to 100 pS, i.e. equivalent frequency resolution about 1×10^{-15} for one day basis. On other hand it enables the cycle skipping detection of the crystal oscillator locked to quantum transition in H-maser. Frequency resolution of 5/100 MHz signal measurements exceeds 1×10^{-15} for sample time 1 hour.

The remote time and frequency comparison system

The time link ties UTC(SU) to UTC is based on GPS single cannel receiver TTR 6 type manufactured by Allan Osborne Associate. Based on BIPM Circular T UTC – UTC(SU) data the estimated time link uncertainty type u_a is about 3 ns, and corresponding frequency uncertainty about 1×10^{-14} for sample time 5 days, fig.7. These values are in conformity with re

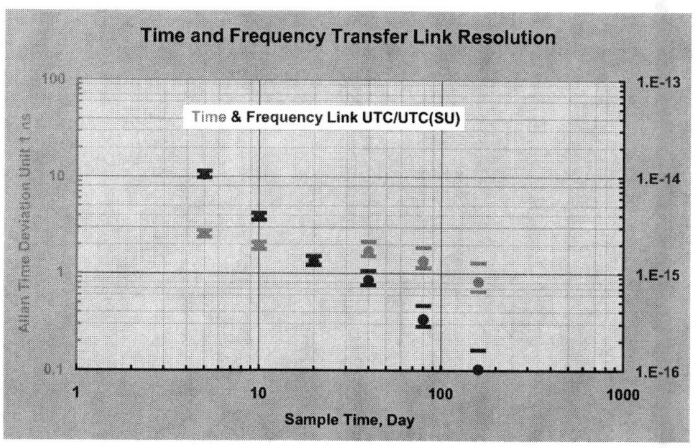

Fig. 7

cent BIPM publication [7]. Along with it GLONASS common view links connect UTC(SU) time scale with scales of secondary laboratories in east part of Russia. Up to now within Moscow region TV common view links are still in use. Its uncertainty u_a is ≤ 3 ns for 1 day sample time.

The National time scale UTC(SU)

The National time scale UTC(SU) is based on the ensemble of above mentioned H-masers, time intercomparison system and simple algorithm with binary weighting. The frequency drift of H-masers is taken into account. The performances of UTC(SU) demonstrated relative to UTC on Fig. 8 and 9 and

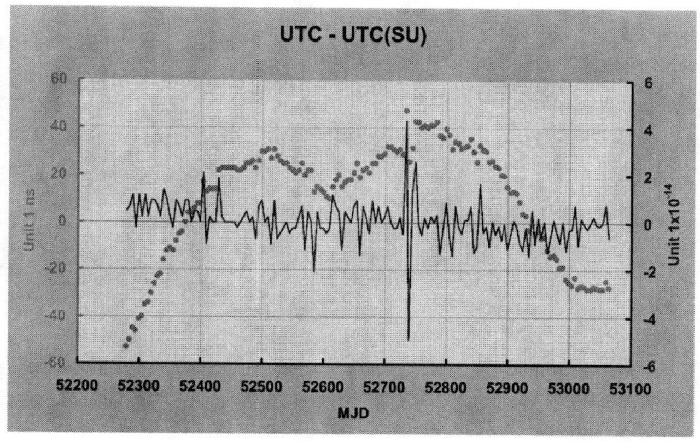

Fig.8

are based on official BIPM publications. During at least two last years |UTC-UTC(SU)| does not exceed about few tens nanoseconds and UTC(SU) time unit difference relative to TAI less than 1×10^{-14} . The obvious spikes at about 5270 MJD is due to GPS receiver failure. On other hand during last

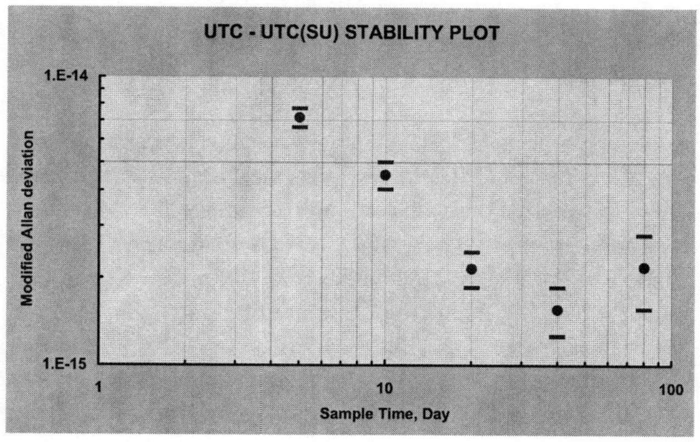

Fig.9

two years there is obvious residual frequency drift about $\sim 3\times10^{-17}$/day. We would like to stress that during last few year there were no steering corrections introduces to the UTC(SU).

6 The Secondary Time and Frequency Laboratories

Due to enormous Russian territory and lack of accurate enough and reliable time links at the end of sixties few secondary time laboratories (SL) were arranged in Gosstandard's establishments in eastern part of country from the West Siberia – Novosibirsk (Nm) through the East Siberia – Irkutsk (Im) up to the Far East – Khabarovsk(Khm) and Kamchatka peninsula – Petropavlovsk-Kamchatsky (P-K), fig. 10. All these laboratories are under jurisdiction of Gosstandard and are supervised by IMVP as Main Metrology Center.

The most equipped laboratories locate in Novosibirsk, Irkutsk and Khabarovsk. Their instrumentation looks quite similar to each other and consists of ensemble of H-masers, time and frequency intercomparison system and time link instruments. At present time the main time links tied timescales of the secondary laboratories to UTC(SU) are based on GLONASS signals. On other hand within EOP activity each SL is supplied with GPS/GLONASS carrier phase receiver. Each laboratory maintain atomic (autonomous) and coordinated time scales.

Main performances, such as coordinated time difference and atomic time scale stability plots of these laboratories are depicted and fig. 11 and 12.Apart from time scale keeping duties these laboratories are strongly involved in time signal monitoring and dissemination process. For example, laboratory in Irkutsk supports emitting station RTZ, laboratories in Khabarovsk and Petropavlovsk-Kamchatsky work in close co-operation with ACCS and support Sakhalin Loran like chain.

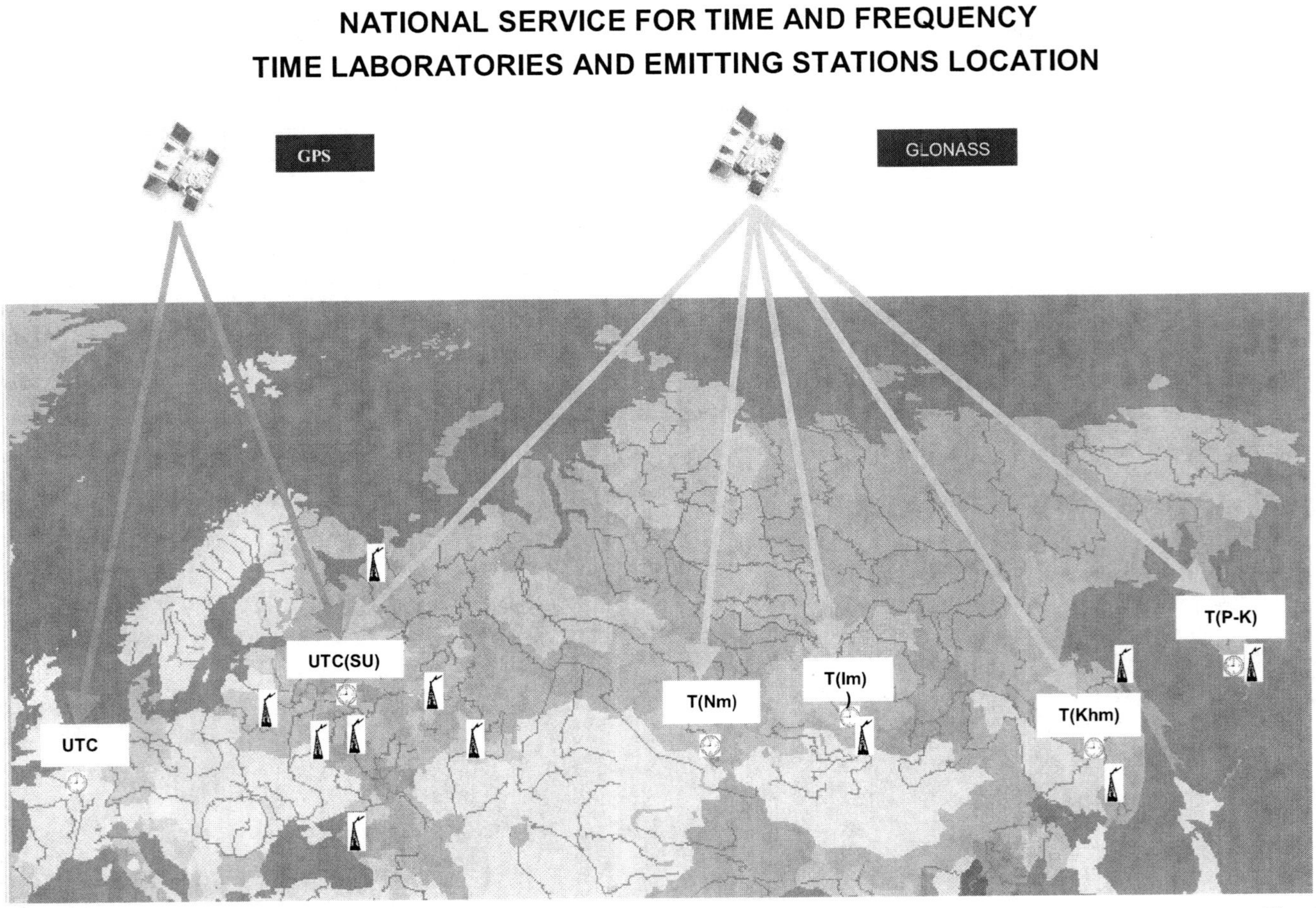

Fig. 10

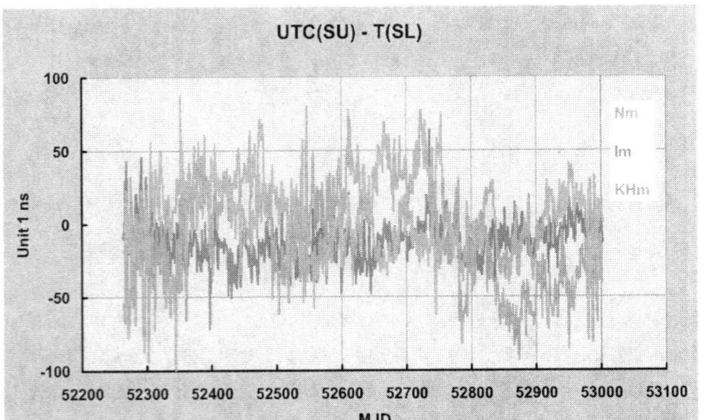

Fig. 11

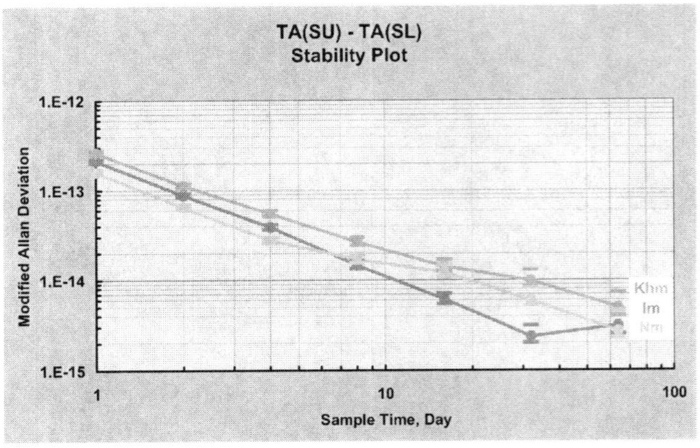

Fig. 12

7 Dissemination of the Standard Frequencies and Time Signals

Along with generating and maintaining the National time scale UTC(SU) the other important task of NSTF faces to enormous users is to disseminate legal time code and standard frequencies.

First of all it worth to remind that "The time code and standard frequencies in Russian Federation and Commonwealth of Independent States broadcasted by TV, radio, on land and space navigation systems and other communications means are emitted referring to the UTC(SU) time scale" [5].

To realize this statement two NSTF's laboratories under jurisdiction of Gosstandard: IMVP and Irkutsk control specialized LW and SW stations: RBU and RWM in vicinity of Moscow and RTZ and RID in vicinity of Irkutsk. The typical uncertainty time level about 10 μS and between $2\text{-}10 \times 10^{-12}$ for standard frequencies.

Apart from it IMVP controls TV transmission from Ostankino station – the main TV station in Moscow. The time code and standard frequencies are embedded in 6 line of uneven field of the TV signal format. The time uncertainty ≤ 0.5 μS and $\leq 2\times10^{-12}$ for standard frequencies.

Lions share of other stations emitted time code and standard frequencies at secondary basis are under DoD jurisdiction, including on land navigation stations, and are controlled by ACCS, which time scale in turn according to Traceability Chart is tied within predetermined limits to UTC(SU).

At present the most widespread source of time and standard frequency information moved from on land means to GLONASS. For endless territory of Russia this is the most convenient way of accurate time and frequency dissemination. Today UTC(SU) is disseminated by the GLONASS through the whole world with uncertainty few hundred ns. We hope that in close future the uncertainty level will be ten times improved.

8 The closest perspectives of the National Service for Time and Frequency

The above mentioned configuration and performances of the NSTF's technical means reflect requirements of the middle of nineties years of last century. Today user's demands are considerably higher – not only regarding accuracy, but first of all regarding accessibility also.

To meet the growing requirements of potential users NSTF supposes in close future realise the following improvements.

First off all a new primary caesium standard will be realised. Today we have an experimental prototype, fig. 13, and the most sophisticated parts of the instrument are successfully tested. Successful experiments were carried out to cool and to launch atoms. Near 1 μK temperature was achieved. Atoms were launched to a working height and Ramsey oscillations were observed. Ramsey pick width was near 1Hz and Rabi pedestal is near 60 Hz. A special room is prepared for installation and assembling the operational instrument in close vicinity of the ensemble of H-masers.

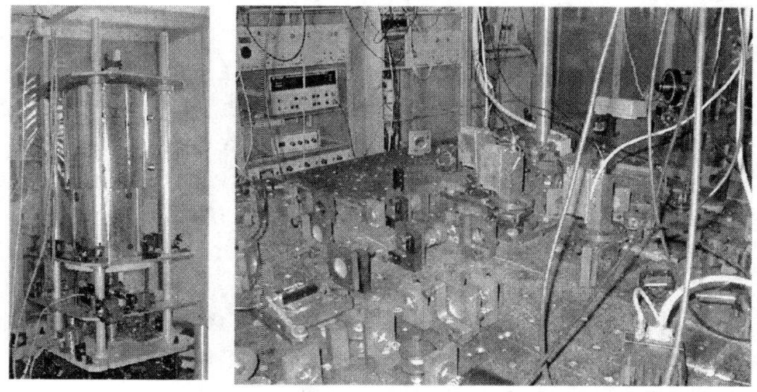

Fig.13

The next one, strongly tied to previous problem, is modification and improvement performances of the operational time and frequency keeping H-maser ensemble. The goals here are on one hand to improve stability level of the time keeping instrument up to the level $\sim 1\times10^{-15}$ for one day and $\sim 3\times10^{-16}$ for 10 days, and, on other hand, improve time links to TAI frequency resolution to the level $\sim 1\times10^{-15}$/day.

These both goals seem to be quite feasible. Even now the most reliable H-masers stability reaches level of $2\text{-}3\times10^{-15}$ for 1 day sample and $\sim 1\times10^{-15}$ for 10 days. Keeping in mind that instrument temperature sensitivity is $\sim 1\times10^{-14}$/K and temperature variations may be about or even exceed 1 K per day simple individual H-maser temperature stabilisation may results in expected goal.

Concerning time link we also have optimistic perspectives. Today in our disposal are few GLONASS/GPS, fig. 14, mul

Fig.14

tichannel sensors, some of them operates carrier phase and two frequencies. Because primary goal is to get frequency transfer very accurate time calibration is not necessary at least at the beginning and key question will concern developing a software which meets requirements of the CGGTTS for existing sensors and time measuring instruments and, once more, temperature stabilisation of the most critical parts of the instrument – antenna and cables.

As a part of clock keeping ensemble improvement we are going to improve time and frequency intercomparison and data collection system. The most striking features of future system, strictly speaking it have been just delivered, is simultaneity and high resolution of frequency comparisons of all clocks in the ensemble with sample time up to 1 second. Time signals will compared across each other in series as previously. This new system will produce enormous data flow. Our closest task here is to develop proper software to manipulate and process this data flow and to test the whole measuring system. We hope the intercomparison system will help considerably in more detailed evaluation of H-maser behaviour.

As far as multichannel GNSS time transfer techniques will be introduced into secondary time laboratories operational activity we hope to involve into the time keeping process more and more number of the remote clocks. As a first step it will be Moscow region time laboratories clocks (expected number few tens), then more distant laboratories.

There are few novelties in future changes of time algorithm. Nowadays we consider idea to break off rigid link between TA(SU) and UTC(SU). Up to now the time unit in both time scales is the same. We consider to generate UTC(SU) referring to TA(SU) but introduce a steering corrections into UTC(SU). The estimations based on few last years time difference UTC – UTC(SU) demonstrated that existing UTC(SU) stability and GPS common-view single channel time link to UTC enable us to achieve $\sigma_x(\tau) \leq 10$ ns for UTC - $UTC(SU)_{steering}$ $\tau \geq 5$ days.

The expanding clocks number inevitably leads us to the changes in operational time algorithm – at that step it looks feasible to introduce statistical weighting for clock instead of existing binary.

The above mentioned future developments deals first of all with time and frequency standard itself. The user level is not forgotten also.

Regarding time dissemination facilities within federal program on GLONASS modification we do all our best to improve its time dissemination performances for anonymous users.

On other hand keeping in mind the explosive growth of Internet users (today more than 10 000 000) and network technologies in Russia a time servers will be run in closest future in main NSTF facilities.

9 Conclusions

For more than 50 years history the National Service for Time and Frequency have proved ability to solve the most sophisticated problems in the field of time metrology and technology. It survived during revolutionary changes in economy and politics of Russia during the end of nineties.

Today apart from traditional duties NSTF is strongly involved in different federal programs required time and frequency support. These are further GLONASS generation – first of all as monitoring stations for ephemeris/time data source for GLONASS control segment.

Apart from it the above mentioned time laboratories are included into the list of federal reference stations for GLONASS/GPS differential navigation.

So each of us, fig. 15, have optimistic views on future and

Fig.15

are sure that National Service for Time and Frequency is a right partner in Russia in any project required accurate and precise timing and standard frequencies.

References

[1] The Constitution of the Russian Federation, Chapter 3 The Federal Structure, Article 71, (was Adopted at National Voting on December 12, 1993)

[2] The law of the Russian Federation on Measurement Unification, President of the Russian Federation Yeltzin, 27 April, 1993.

[3] On approval of the Statute of the National Service for Time and Frequency, The Government of the Russian Federation, Decree No. 225, 23 March 2001, Moscow.

[4] On approval of the Statute of the Inter-department commission for time and standard frequencies, The Government of the Russian Federation, Decree No. 125, 19 February 2001, Moscow.

[5] State standard 8.129-99

[6] Agreement on Cooperation on Measurements Unification Ensuring for Time and Frequency, Bishkek, October 9, 1992.

[7] Circular T 194 ISSN 1143 - 1393, 2004 March 11, Bureau International des Poids et Mesures, Organisation Intergovernementale de la Convention du Metre, Pavillon de Breteuil F- 92312 Sevres CEDEX, tai@ bipm. org

STATISTICAL ESTIMATES OF PHASE ERRORS OF PASSIVE WIRELESS SAW SENSORS EMPLOYING DIFFERENTIAL MEASUREMENT

YURIY SHMALIY and OSCAR IBARRA-MANZANO,
Guanajuato University, FIMEE, Salamanca, 36730, Mexico, shmaliy@salamanca.ugto.mx

Passive wireless surface acoustic wave (SAW) sensors are used to measure temperature, pressure and torque, identify the railway vehicle at high speed, *etc.* with a resolution of about 1%. Most frequently, the information bearer in such sensors is a time delay of the SAW estimated at the receiver. The basic principle utilized in such a technique combines advantages of the precise piezoelectric sensors, high SAW sensitivity to the environment, passive (without a power supplied) operation, and wireless communication between the sensor element and the interrogator.

In this report, we address a statistical analysis of the estimate errors of the time delay (phase difference) between two reflectors of the passive wireless surface acoustic wave (SAW) sensor. The estimation is provided in a sense of the maximum likelihood function at coherent receiver in presence of Gaussian noise. Assuming the Gauss shape interrogating radio frequency (RF) pulse, we bring an important proof that the first time derivatives of its amplitude and phase do not affect the sensor phase (and phase difference) at the receiver. Rigorous and approximate relations for the mean error and mean square error (MSE) along with the probability for the estimate error to exceed a threshold are derived and studied in detail. The plots to evaluate the errors in a wide range of signal-to-noise ratios (SNRs) and thresholds are given. Practical findings for designers of the SAW sensor systems exploiting differential measurement are also pointed out.

Finite Element Analysis for a New Design of NLC-cut Quartz Crystal Resonant Temperature Sensor

B. Dulmet*, L. Spassov[†], Ts. Angelov [†]

*Ecole Nationale Superieure de Mecanique et des Microtechniques, FEMTO-ST Dept. LCEP, France
bernard.dulmet@ens2m.fr, Fax 33 381 88 57 14

†Institute of Solid State Physics, Acoustoelectronics Laboratory, Sofia, Bulgaria

Keywords: Design of resonant sensors, dispersion curves, temperature sensors.

Abstract

In the process of designing resonant temperature sensors, getting minimal dimensions is required to decrease calorific capacity and response time. In this view, we investigated new shapes of sensing elements essentially obtained by modifying an initial resonator's design. This design consists in plane parallel circular disk plates in NLC–cut for energy trapping by electrodes mass loading. Then, the disks are cut with two flats parallel to X crystallographic direction, thereby leading to designs intermediate between the common circular plates and so–called strip–resonator design. A combination of FEA simulations and of semi-analytical analyses were used to master the influence of width–to–thickness ratio onto the spectral characteristics of sensors, including the control of activity-dips occurrence. Q factors near 75000 and motional resistances near 9 Ω were obtained with sensors 4 mm–wide, operating on fundamental B–mode at 29.3 MHz, and exhibiting a temperature sensitivity close to 1000 Hz/K.

1 Introduction

Examples of studied designs of thermo–sensitive quartz resonator (TSQR) used as temperature sensor are shown on Fig. 1. The pursued goal was to significantly reduce at least one dimension of TSQR from a previously–established design [1] consisting of circular disks of diameter ∅=5mm, thickness t = 2h ≈ 82 μm, in NLC-cut (φ=0, θ=-31°30'). For this singly–rotated cut, the pure thickness–shear (TS), piezo-electrically excitable (by vertical field) modes are so–called B modes (fast shear). Electrode mass–loading provides with energy–trapping and good electrical characteristics (R1=10Ω for instance) with silver electrodes of diameter 2.0 < ∅ < 2.5mm, while keeping the electrode thickness close to *160 nm.* Decreasing dimensions by means of two flats perpendicular to $x_1 \equiv X$ (crystallographic) axis was worked out as a technically affordable solution to aim at a smaller enclosure. Even gaining only *1 mm* in width was found useful from a practical point of view. In all what follows, we use local axes for the NLC–cut, with $x1 \equiv X$ crystallographic, $x2$ the plate normal, and $x3$ the second in–plane axis. w is the width along $x1$.

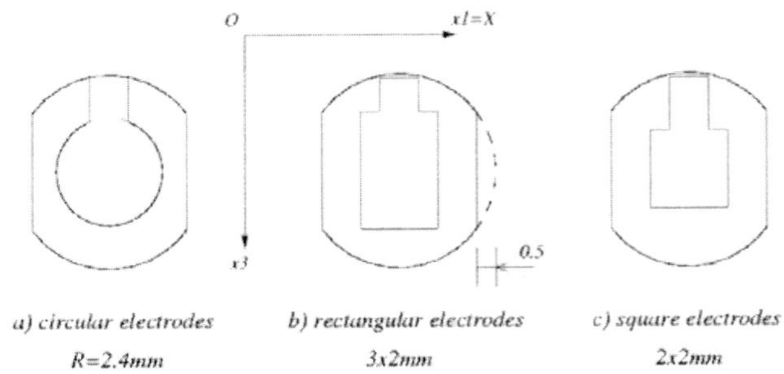

Fig. 1 TSQrs with reduced dimensions.

FEA (finite element analysis) was intended to systematically support the dimensioning of TSQRs from the double point of view of spectral/electrical parameters and of smoothness of sensor function (frequency–temperature characteristic). First FEA simulations quickly led to typical mode patterns predictions shown on Figs. 2 (1,0,0) mode, and 3 (1,2,0) mode.

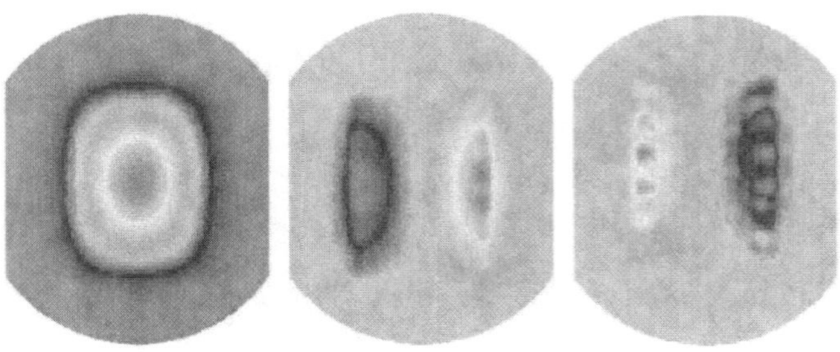

Fig. 2 Predicted distribution of u_1, u_2, u_3 components (from left to right) of main mode (B,1,0,0) on top surface of TSQR with *3x2mm* electrodes, and $t_e \approx 70$nm.

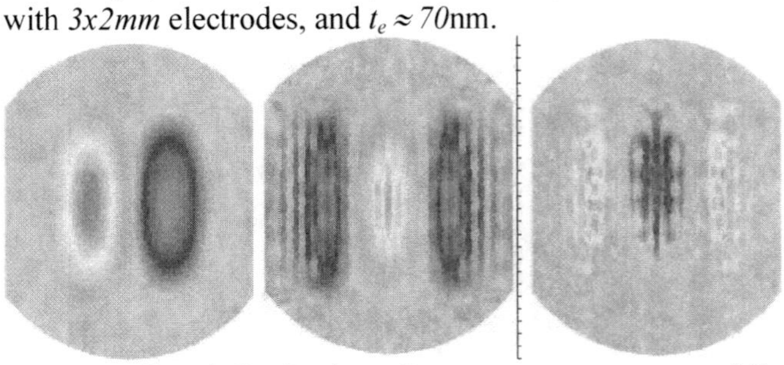

Fig. 3 Predicted distribution of u_1, u_2, u_3 components of first antisymmetric anharmonic (B,1,1,0) on top surface of TSQR with *3x2mm* electrodes, and $t_e \approx 70nm$.

The moderate thickness of electrode coating used for those simulations did not seem to compromise a fairly sufficient trapping, according to the smooth distributions predicted for the essential component u_1. On Figs. 2–3, each component pattern was separately normalised so that the full scale of colouring is used in each case. Nevertheless, we obtain $|u_1| >> (|u_2|,|u_3|)$ in any case. A satisfactory theoretical interpretation of all those patterns requires some serious thinking. At this point, we can formulate two remarks.

a) The widely recognised model from Tiersten *et al* [2,3] does not allow for a direct closed form prediction of so–called minor components of vibration, *i.e.* the components within a plane orthogonal to the polarisation vector obtained as solution of Christoffel's problem for the family of modes under study. Tiersten's analysis implies some symmetry conditions imposed onto the minor components, but two previous studies independently performed at LCEP [4,5] have shown that those expected symmetries may not be experimentally verified for a simple reason : Tiersten's model is rigorous but it is *asymptotic*, and in practice, the difference between the *actual* solution and the *asymptotic* one may be noticeable, depending on the strength of trapping, in other words on the relative importance of transverse variation of vibration *w.r.t.* its thickness variation.

b) Figs. 3*b* and 3*c* contain an implicit questioning : They indicate an oscillatory behaviour of the minor components of mode pattern in the x_1 direction. Unfortunately, Tiersten's model does not provide with any handle to account for separate trapping behaviour of major and minor components of the vibration. In the present case, FEA predicted mode patterns seem to involve a compound vibration based upon several branches of dispersion curves, whereas Tiersten's analytical expansions can take into account only one branch. Nevertheless, since the FEA mesh — in that case, it was of moderate size — was not performed purposely to account for such a refined behaviour, this could have been a mesh–induced artefact as well.

In this situation, it became obvious that it was more than tricky to going on pursuing FEA simulations without a better established theoretical analysis of the whole class of here–studied configurations.

2 Two-dimensional analytical dispersion curves for NLC-cut

In simply–rotated cuts, it is very well known that pure thickness shear propagating along the thickness cannot satisfy stress–free boundary conditions on X–oriented edges. Thus, an immediate goal for the modelling of shallow–trapped resonators consists in determining which harmonics of guided waves along X-axis can complement the thickness–shear branch and constitute a compound standing wave system satisfying the boundary conditions on X–oriented edges. Knowing involved harmonics of guided waves is greatly useful to properly design the mesh for FEA simulations. This comes out from the fact that shape functions used in FEA should be able to accurately simulate the expected oscillatory behaviour of solutions along the width of plate. By the time of the conference, we have set up a simple model which does not take piezoelectric effect into account. Of course, the model is liable to further improvements. If we restrict our interest to modes with *u1* antisymmetric *w.r.t.* x_2 and symmetric *w.r.t.* x_1, the symmetries of material constants in singly–rotated cut allow us to consider guided waves with the following form :

$$\begin{aligned} u_1 &= \cos(\xi x_1) \sum_{n=1}^{3} C_n \beta_1^n \sin(\eta_n x_2) \\ u_2 &= \sin(\xi x_1) \sum_{n=1}^{3} C_n \beta_2^n \cos(\eta_n x_2) \\ u_3 &= \sin(\xi x_1) \sum_{n=1}^{3} C_n \beta_3^n \cos(\eta_n x_2) \end{aligned} \tag{1}$$

Substituting such solutions into the PDEs of linear elasticity turn the latter ones into a secular and homogeneous algebraic system. This system algebraically determines the sets $\{ \eta_n ; \beta_i^n \}$ as function of the given angular frequency ω and of the x1–component of the wave vector ξ, equivalently called "lateral wavenumber" :

$$\left.\begin{aligned} \frac{\partial T_1^n}{\partial x_1} + \frac{\partial T_6^n}{\partial x_2} + \rho\omega^2 u_1 = 0 \\ \frac{\partial T_6^n}{\partial x_1} + \frac{\partial T_2^n}{\partial x_2} + \rho\omega^2 u_2 = 0 \\ \frac{\partial T_5^n}{\partial x_1} + \frac{\partial T_4^n}{\partial x_2} + \rho\omega^2 u_3 = 0 \end{aligned}\right\} \Rightarrow \{ \eta_n ; \beta_i^n \} = f(\xi;\omega) \tag{2}$$

Due to the order of algebraic system, there exist three mutually independent sets $\{ \eta_n ; \beta_i^n \}$ (*n=1...3*) for each given pair $\{ \xi ; \omega \}$. Each set defines a so–called *partial wave*. As shown in Eq. (1), the most general possible solution of PDEs consists of a linear combination of the three partial waves, each of them separately satisfying PDEs. The respective weights of partial waves – the coefficients of the linear combination – are obtained from the boundary conditions on top and bottom surfaces of the plate :

$$\left.\begin{aligned} T_6\big|_{x_2=\pm h} = 0 \\ T_2\big|_{x_2=\pm h} = 0 \\ T_4\big|_{x_2=\pm h} = 0 \end{aligned}\right\} \Leftrightarrow \left\{\begin{aligned} \textstyle\sum_{n=1}^{3} C_n L_1^n(\xi,\omega) \cos\eta_n h = 0 \\ \textstyle\sum_{n=1}^{3} C_n L_2^n(\xi,\omega) \sin\eta_n h = 0 \\ \textstyle\sum_{n=1}^{3} C_n L_3^n(\xi,\omega) \sin\eta_n h = 0 \end{aligned}\right.$$

$$\Downarrow$$

$$(C_n ; \omega) = f(\xi) \tag{3}$$

The angular frequency ω is determined as a root of the determinant of the homogeneous linear system of boundary conditions at a given lateral wavenumber ξ. The combination's coefficients C_n are just the non trivial solutions of the linear system, conditioned by the cancellation of its determinant. In this very classical way, we obtained the set dispersion curves for NLC–cut represented on Fig. 4.

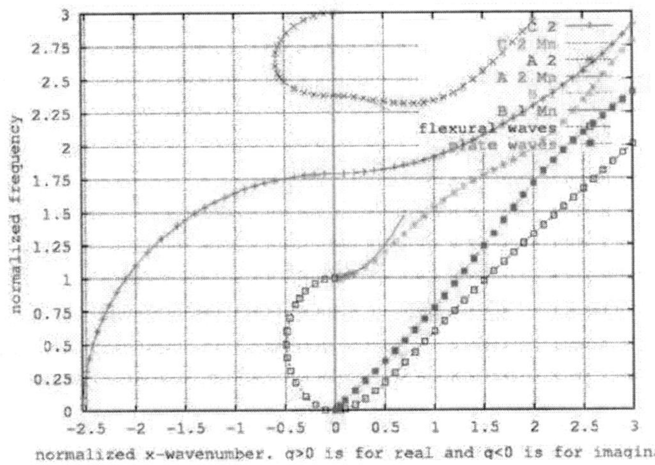

Fig. 4 Dispersion curves for x_2–antisymmetric, x_1–symmetric guided waves in NLC–cut.

On those curves, the normalized frequency is the dimensionless quantity defined by

$$\Omega = \frac{\omega}{\omega_B^F} \qquad (4)$$

where ω_B^F denotes the fundamental resonant frequency of pure–shear (in NLC–cut, this is B mode) in pure thickness regime (ξ=0). Similarly, we introduced the normalised wavenumber γ defined as the ratio of lateral wavenumber ξ to the vertical wavenumber for pure shear (B), pure thickness and fundamental mode :

$$\gamma = \frac{\xi}{\eta_B^F} \qquad (5)$$

Of course, one obtains $\eta_B^F = \frac{\pi}{2h}$ for the fundamental of pure TS mode in an infinite plate with free surfaces

In Fig. 4, the left region of horizontal axis is for γ purely imaginary, while the right region is for γ purely real. In the regions near γ=0, we have also indicated the second order asymptotic expansion of dispersion curves resulting from the calculation of Tiersten's dispersion constant *Mn* for *B1*, *A2*, *C2* pure thickness modes. In NLC cut, the actual curves remains closer to the parabolic asymptote for *C2* than for *B1* and *A2*.

3 Variational treatment of edge BCs

Each curve appearing on Fig. 4 represents by definition a separate branch of dispersion curves. The whole set encompasses all kinds of guided waves in the (x_1,x_2) cross–sectional plane of the cut. Thus, combinations of such branches constitute the natural but complex structure of vibrations in the most general case. Only in fortunately simple cases can the vibration be described by a single branch only. To approximately satisfy edge boundary conditions, a natural approach consists in building a compound solution by superposing a maximum amount of branches of the dispersion curves. Then, to determine an optimal superposition, we can use Hamilton's principle in its well–known form previously proposed by H. F. Tiersten [6]:

$$\int_V \left(T_{ij,i} + \rho\omega^2 u_j\right)\delta u_j + \int_S \left(t_j - n_i T_{ij}\right)\delta u_j = 0 \qquad (6)$$

Because each branch is itself a combination of partial waves which do satisfy the volume PDEs as well as the boundary conditions on top and bottom surfaces, the sought compound solution will identically cancel the volume integral and the integrals on surfaces at $x_2 = \pm h$, so that what is left of (6) is only :

$$\int_{-h}^{+h} \left(T_1\, \delta u_1 + T_6\, \delta u_2 + T_5\, \delta u_3\right)_{x_1 = \pm\frac{w}{2}} dx_2 = 0 \qquad (7)$$

i.e. an integral performed on the edge surfaces. In case of plates with significant – if not total, though – trapping by mass loading, we will expect resonant modes with oscillatory behaviour along x1. In addition, both ratios of the plate and the electrode widths to thickness are rather large, and we will expect $0 < \gamma << 1$ for the mainly TS branch, which connects vertical axis at the point Ω=1. This yields Ω slightly larger than 1, and because the frequency should be everywhere identical in a linear analysis, we may only use the two branches connecting origin in the real side of diagram. For the symmetries at hand, this corresponds to the branches of plate waves (upper branch) and flexure (lower branch). Thus, we can properly update our notations to rewrite the compound vibration as :

$$\left.\begin{aligned}
u_i &= \sum_{m=1}^{3} K_m\, u_i^m \\
u_1^m &= \sum_{j=1}^{3} \left(C_{mj}\, \beta_1^{mj} \sin \eta_{mj} x_2\right) \cos \xi_m x_1 \\
u_2^m &= \sum_{j=1}^{3} \left(C_{mj}\, \beta_2^{mj} \cos \eta_{mj} x_2\right) \cos \xi_m x_1 \\
u_3^m &= \sum_{j=1}^{3} \left(C_{mj}\, \beta_3^{mj} \cos \eta_{mj} x_2\right) \cos \xi_m x_1
\end{aligned}\right\} \qquad (8)$$

where m is a symbolic index to denote the branch. Then, the relations between the variations δu_i and the coefficients of combination become quite obvious :

$$\left.\begin{aligned} \delta u_1 &= u_1^1\,\delta K_1 + u_1^2\,\delta K_1 + u_1^1\,\delta K_1 \\ \delta u_2 &= u_2^1\,\delta K_1 + u_2^2\,\delta K_1 + u_2^1\,\delta K_1 \\ \delta u_3 &= u_3^1\,\delta K_1 + u_3^2\,\delta K_1 + u_3^1\,\delta K_1 \end{aligned}\right\} \quad (9)$$

and the variational equation (7), which must be satisfied whatever is the arbitrary δu_i , readily turns into an algebraic system :

$$\begin{aligned} &A_{im} K_m = 0 \;\Rightarrow\; \left|A_{im}\right| = 0 \\ &A_{im} = \int_{-h}^{+h} \left(u_1^i\, T_1^m + u_2^i\, T_6^m + u_3^i\, T_5^m \right) dx_2 \end{aligned} \quad (10)$$

For instance, we obtain the following detailed form regarding terms factoring δu_1 inside the variational equation :

$$\begin{aligned} &\int_{-h}^{+h} u_1^i\, T_1^m\, dx_2 = -\left(M_{ij}^1 P_{jn}^{im} N_{nm}^1 \right) \sin \xi_m l \cos \xi_i l \\ &M_{ij}^1 = C_{ij}\, \beta_1^{ij} \quad P_{jn}^{im} = \int_{-h}^{+h} \left(\sin \eta_{ij} x_2 \sin \eta_{mn} x_2 \right) dx_2 \\ &N_{nm}^1 = C_{mn} \left[c_{11}\, \gamma_m\, \beta_1^{mn} + \left(c_{12}\, \beta_2^{mn} + c_{14}\, \beta_3^m \right) \eta_{mn} \right] \end{aligned} \quad (11)$$

where $l = w/2$. Then, the resonant frequencies are obtained by numerical determination of the roots of the determinant of the K_ms. After that, the compound–wave structure is fully determined once the K_ms are known. In our work plan, we did not intend to complete a three dimensional model of the resonators at hand, but simply to determine how many brick elements have to be used in the width direction to grant the accuracy of FEA *w.r.t.* transversal mode couplings which are induced by the boundary conditions on x_1–oriented in case of shallow trapping. In here–considered kinds of resonators, when the distance between flats is near 4mm, we got by this semi–analytical model the u_1 amplitude distributions along x_1 (alias crystallographic X in that case) represented on Figs. 5–6. The first one is the "main mode" amplitude distribution in absence of trapping, and the second one represents the amplitude distribution of the first "symmetric anharmonic" along x_1.

Now, we must consider that the FEA program developed at LCEP for the modal analysis of piezoelectric resonators allows cubic interpolations in any direction. Previous convergence studies have shown that accuracy and stability of results begin to show up when one uses at least one element layer (4–nodes each direction) for half a pseudo wavelength of the vibration. Also, for singly–rotated cuts such as NLC, we can pursue FEA for the X–half of the resonator, separately looking for symmetric and antisymmetric modes (still along X, indeed).

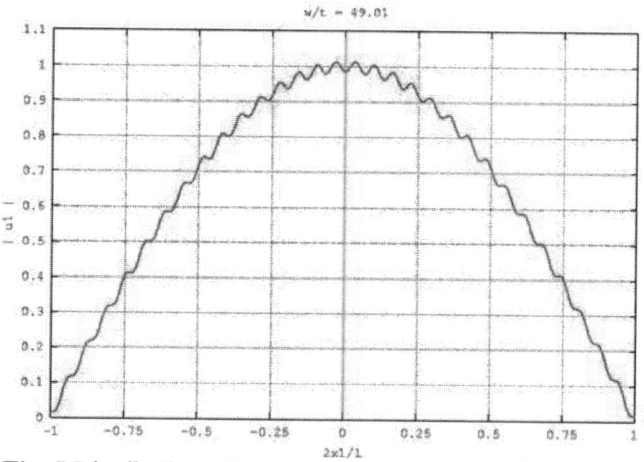

Fig. 5 Distribution of main coupled–B mode in fundamental NLC cut, w/*t=49.01. l=w/2.*

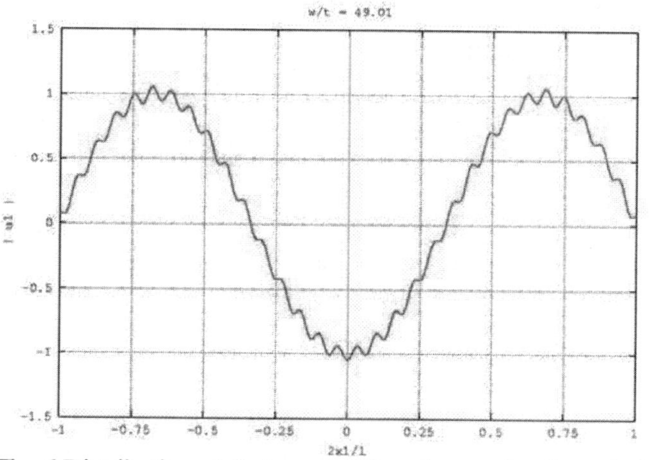

Fig. 6 Distribution of first symmetric anharmonic of coupled–B mode in fundamental NLC cut, w/*t=49.01. l=w/2.*

Since the dispersion analysis predicted 30 pseudo–wavelengths from edge to edge, as can be seen on Figs. 5–6, we chose as optimal FEA model a mesh design with 32 bricks along the half–width of resonator. For *70nm*–thick electrodes conforming to the abstract design of Fig. 7a, FEA produced the main mode pattern shown on Fig. 8. The number of predicted oscillations along x1–axis is in fairly good agreement with the prediction from above–presented model based on dispersion curves. Since, in this design, electrodes cover the whole width of resonator, there is no trapping along that direction and the oscillatory behaviour of the compound vibration has to be *maximum*. Now, if we try a design with electrodes not covering the whole width, it can only provide with a limited energy–trapping effect in that direction. Keeping the same design as before, but with an electrode coating interrupted 0.5 mm before the x1–oriented edges (design of Fig.7b), we obtained the results of Fig. 9.

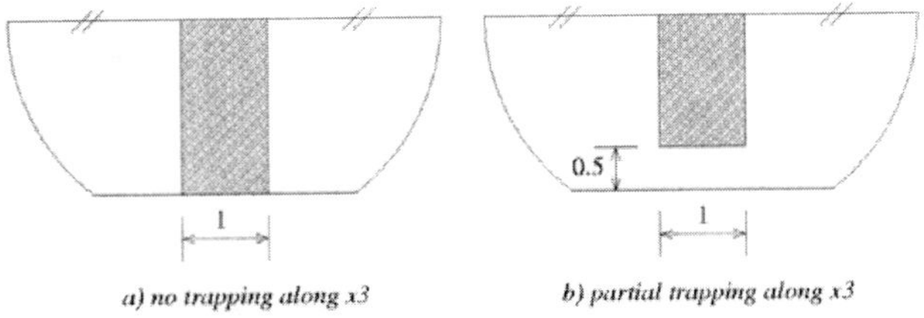

Fig. 7 Test designs for FEA

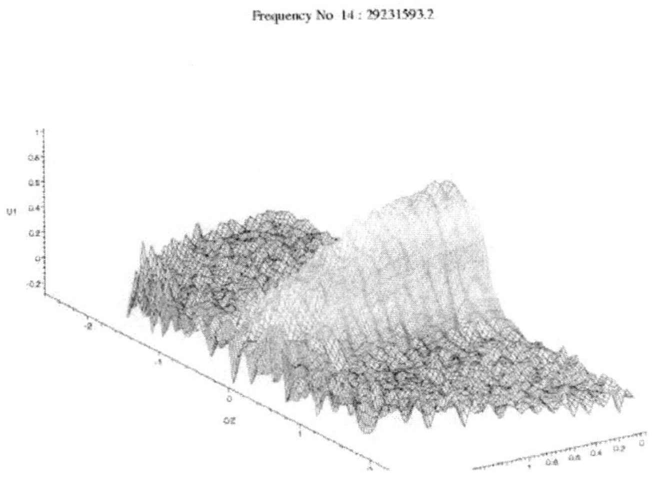

Fig. 8 FEA–predicted mode pattern of main mode, design a).

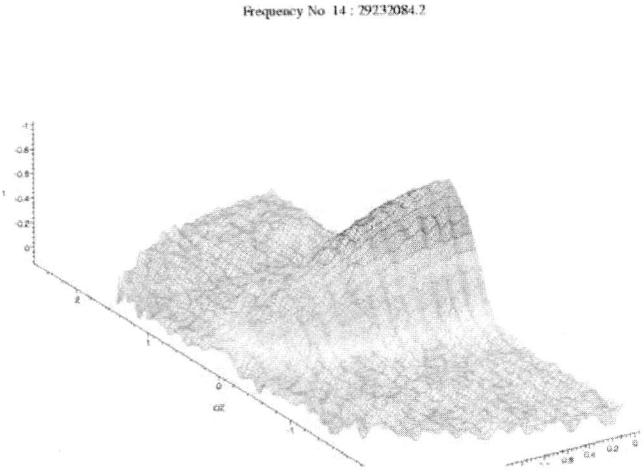

Fig. 9 Predicted pattern of main mode in case of partial energy–trapping along x1, design b).

For the considered semi–circular design, it was experimentally found that an optimal trapping in all directions is obtained with *2x3mm* or *2x2mm* electrodes having a thickness of *160nm* (silver electrodes). Nevertheless, even with only *70nm*–thick electrodes, the TS–component trapping along *x3* provided by *1–mm* long electrodes should also be excellent. A simple look at FEA predicted Figs. 8–9 shows that that if the mass loading effect is efficient to trap the TS component of vibration along x3, it does not trap at all the "oscillatory" part of vibration which actually represents the contribution of other branches of the dispersion curves in the 2D semi–analytical model. The mechanism at hand can be qualitatively described as follows:

If the TS–component of the actually compound vibration is insufficiently trapped along x1, then the contributions of other branches becomes higher due to edge boundaries coupling and induces a marked oscillatory behaviour of the compound vibration. In turn, those contributions are not trappable by electrode mass loading along the second in–plane axis (x3).

This would claim for the use of over–trapping and additional reshaping techniques in further studies aimed to optimising the design of so–called strip resonators (narrower than here–presented intermediate design).

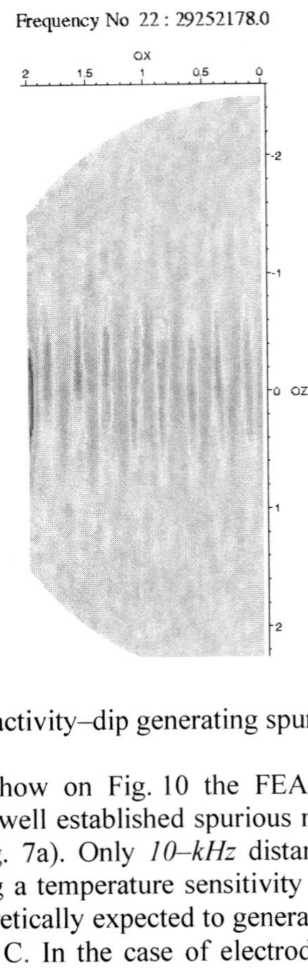

Fig. 10 An activity–dip generating spurious mode.

Eventually, we show on Fig. 10 the FEA–predicted mode pattern of a very well established spurious mode, found with the design of Fig. 7a). Only *10–kHz* distant from the main mode, and having a temperature sensitivity of opposite sign, this mode is theoretically expected to generate an activity–dip at a few degrees C. In the case of electrodes not extending along the whole width of resonator (Fig. 7b), this spurious mode moves *34kHz* up and its FEA–predicted pattern becomes extremely noisy, indicating a much less physical significance. The kind of spurious mode shown on Fig. 10 is recurrent in further systematic studies now under way for actual rectangular strip resonators (*w/t* near 24.5) and shortly outlined in the next section.

4 Obtained results on here–studied design and further research trends

Q factors near 75000 and motional resistances near 9 Ω were obtained with the presented 4 mm–wide design, operating on fundamental B–mode at 29.3 MHz, and exhibiting a temperature sensitivity close to 1000 Hz/K. Thus the reduction of dimension resulting from the realisation of x1–normal edges did induce a performance decrease *w.r.t* essentially circular TSQRs with ∅ 5mm since the latter design allows Q factors higher than 100 000. Nevertheless, a further drastic reduction of dimension was recently undertaken, conforming to the design outlined on Fig. 11. Then, the actual width was sought for in the vicinity of *2mm*, and one falls in the regular case study of so–called strip resonators. Technical realisation becomes more difficult, but FEA becomes easier, due to the smaller width–to–thickness ratio. For instance, RAM requirements for the simulations shown on Figs. 8–10 were close to 2 Gigabyte and it took typically 1.5 day to get a bunch of 40 mode patterns at a given temperature, whereas lowering the *w/t* ratio down to the 24–25 reduced the RAM requirements below 1 Gigabyte, keeping the time needed to simulate one temperature point close to 6–10 hours. As an example, we show on Figs. 12–13 semi–analytically–computed and FEA–computed dispersion curves for strip design. Whereas the curves of Fig.12 are obtained in the purely two–dimensionnal model from Mindlin's and Gazis [9], the curves of Fig. 13 take into account the tridimensional character of vibration.

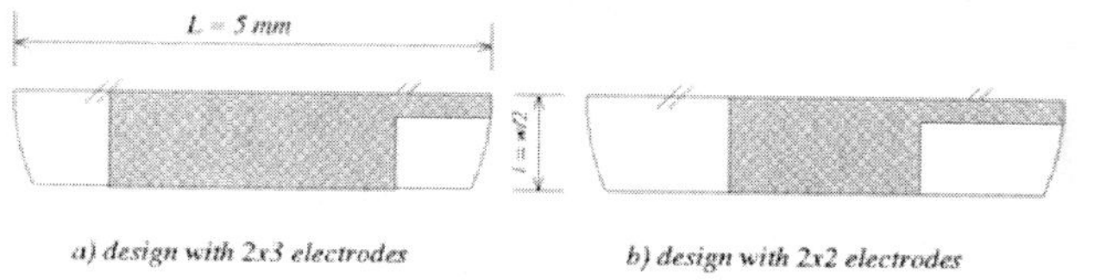

Fig. 11 Recently investigated pure strip resonator designs

Computing the spectrum at several temperatures allowed to determine regions of *w/t* for which thermally–induced mode crossings are expected. For instance, on Figs. 12–13, the regions of dispersion curves where the compound vibration existing in the strip resonator correspond to the "plateau", i.e. the vicinity of inflexion point, where the slope of the branch is minimal. The crossing curve near that point on Fig. 12 (2D analysis) is for modes of different symmetry and can left over. Conversely, on Fig.13, the curve labelled (F,n,0) is of a different nature and, near its crossing point with the curve labelled n(B,1,0), both involved modes interact, creating an activity-dip. This is due to the fact that both kinds of curves have not the same predicted thermal sensitivities, as shown on Fig. 14. The spurious mode involved in this effect is of the same nature as the one depicted on Fig. 10. A more detailed analysis in this class of more miniaturised resonators, including identification by X-Ray topography is under way and will be the subject of a paper proposed at IEEE sensor conference, to be held in Vienna next fall.

5 Conclusion

The main result of here–presented analysis may be the theoretical one : all components of the vibrations may not be similarly trapped in some configurations, so that an apparently trapped vibration may include a small non–trapped part. This was experimentally observed by scientists like Detaint *et al* in EFTF reports delivered during the nineties. Furthermore, this cannot be explained without calling for several branches of dispersion curves, and not a single one. When this happens, overloading the plate may not be sufficient to avoid acoustic losses at holding points of resonators. This certainly calls for very refined further analyses and changes in design to remedy this situation. To have FEA properly simulate such refinements, one has to get at least a rough idea of the number of brick elements required in a given direction. A relatively simple semi–analytical model such as here–presented one is of great help in reducing the required time to establish sufficient convergence of FEA model.

Acknowledgements

Here–presented work is part of QxSens team work. QxSens is an European Union RDTD Project of the 5-th framework Project, part of GROWTH Project. Project Number is GRD1-2001-41816, contract number is G6RD-CT-2002-00648, and QxSens is the acronym standing for the project's full title : Multi-channel measurement and control system based on resonant piezoelectric crystal sensors.

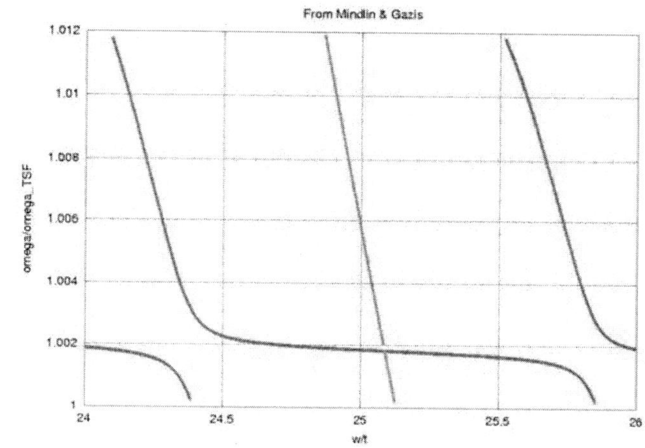

Fig. 12 Dispersion curves for true strip design from implementing Mindlin and Gazis [9] method.

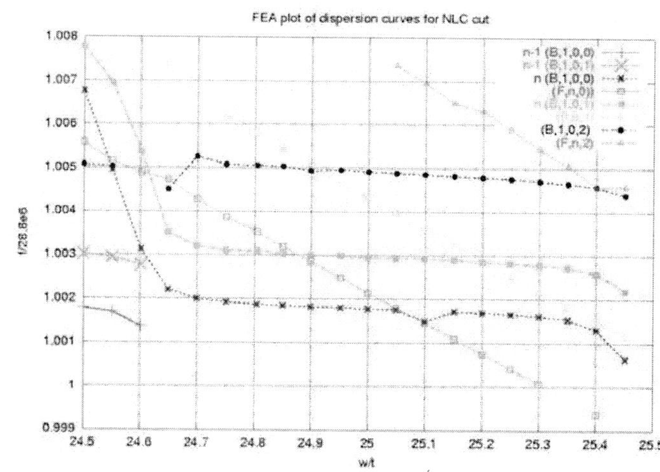

Fig. 13 FEA–generated dispersion curves for true strip–resonator design

References

[1] R. Bourquin, B. Dulmet, R. Velcheva, L. Spassov, *Study of NLC–cut resonators by X–ray Topography*, Proc. Eur. Freq. and Time Forum, St. Petersburg, (2002).

[2] H. F. Tiersten, *Analysis of Trapped Energy Resonators Operating in Overtones of Coupled Thickness Shear and Thickness Twist*, J. Acoust. Soc. Am., 59, P. 879 sq., (1976).

[3] D. Stevens, H. F. Tiersten, *An Analysis of Doubly Rotated Quartz Resonators utilizing Essentially Thickness Modes with Transverse Variations,* J. Ac. Soc. Am., 79, n° 9, pp. 1811-1826, (1986).

[4] R. Bourquin, D. Nassour, *Etude tridimensionnelle d'un résonateur à quartz de coupe AT de forme plan-convexe,* C. R. Acad. Sci., Paris, Serie II, n° 12, pp. 517-520, (1984).

[5] B. Dulmet, *A general model for quasi–thickness vibrations in contoured quartz resonators*, Proc. 6-th Eur. Freq. and Time Forum, pp. 164–172, (1991).

[6] H. F. Tiersten, "Linear Piezoelectric Plates Vibrations", Plenum Press Ed., New York, (1969).

[7] R.D. Mindlin, *Thickness-Shear and Flexural Vibrations of Crystal Plates,* J. A. P., pp. 316-323, 1951.

[8] R.D. Mindlin, *High Frequency Vibrations of Crystal Plates*, Quarterly of Applied Mathematics, April 1961, N°1, pp. 51-61.

[9] R.D. Mindlin, D.C Gazis, *Strong resonances of rectangular AT-cut quartz plates*, Proc. 4-th U.S. Nat. Congress of Appl. Mech., pp. 305-310, 1962.

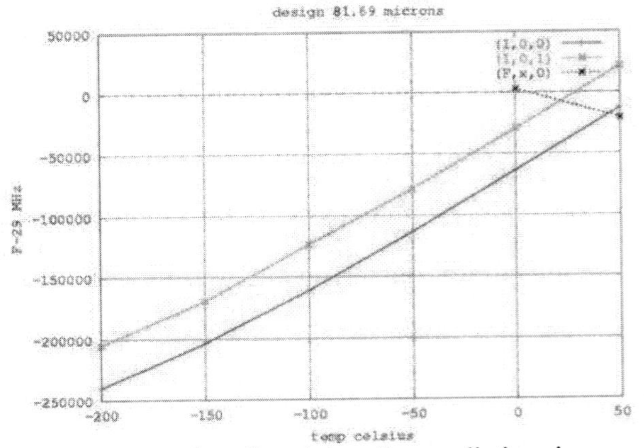

Fig. 14 Example of activity–dip prediction in pure strip design (w/t ≈ 24.6)

INVESTIGATIONS OF ELECTRICAL PARAMETERS AND MODE PATTERNS OF NLC- CUT QUARTZ RESONATOR AT LOW TEMPERATURE

L. SPASSOV*, R. VELCHEVA*, L. VERGOV*, R.BOURQUIN**, A.IVAN**
*Institute of Solid State Physics, Bulgarian Academy of Sciences
Tzarigradsko Chaussee 72, Sofia 7184, Bulgaria
e-mail: lspassov@issp.bas.bg
** Ecole Nationale Supérieure de Mécanique et des Microtechniques,
26 Chemin de l'Epitaphe, F25000,Besançon, France
e-mail: rbourquin@ens2m.fr

Keywords: Miniature quartz resonator, thermal sensor.

Abstract

The scope of this paper is to investigate the motional parameters and the frequency spectrum of small size resonators designed to be used as thermal sensors on a wide range of temperature from ambient down to cryogenic temperatures. Different electrodes configurations are used to trap the vibration .

1 Introduction

NLC Cut [1] has been introduced for the realisation of thermal resonant sensors for use at temperatures as low as the one of liquid helium. The interest of this cut comes from its monotonous relationship between temperature and frequency, while its singly-rotated nature simplifies the technology to produce miniature sensing resonators. In order to obtain precise sensing functions, it is necessary to ensure that the motional parameters of resonators do not exhibit uncontrolled variations in the temperature range of measurement, so that the frequency variation of the oscillator associated to the sensor be not altered by any change of impedance.

To reduce the size of the sensor and its response time, the sensor is made from a resonator with an altogether circular shape, where two parallel flats cut the disk perpendicular to the X crystallographic direction. Then the trapping of the vibration in the X direction, obtained by mass loading effect, is critical due to the limited width. Different shapes of electrodes can be used, but in this kind of resonator, the tabs used for the electrical connection of electrodes can play an important role.

Variations of motional resistance and other elements of the electrical equivalent circuit can arise from modes coupling, which further cause some shifts of frequency. At low temperature, variations of motional resistance can also arise from acoustic losses due to relaxation process associated to impurities into the quartz material [2,3]. Predictions of the motional parameters can be made by analytical or numerical methods [4].

2 NLC Cut resonator design

The NLC cut [1] is a singly rotated cut ($\Phi = 0$, $\Theta = -31.5°$) operating on B mode, first overtone. The figure (1) depicts the dimension and shape of thermal sensors. Three electrode shapes have been tested.

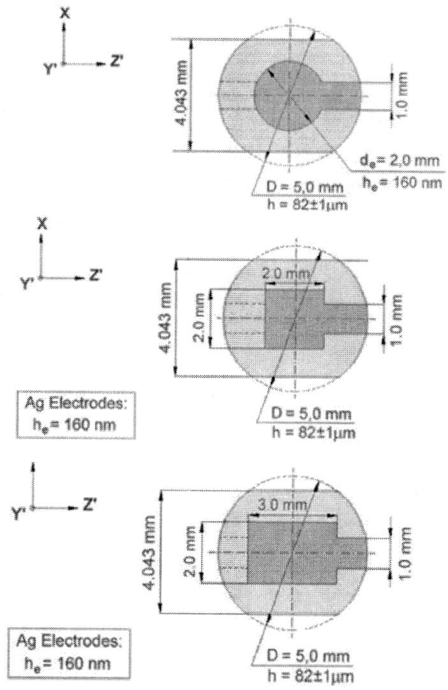

Figure 1: Resonators configurations

The thickness of plates is 82 μm and the diameter 5 mm. The width is limited by two flats (semi strip resonator). Electrodes are made of silver (thickness 160nm). The thickness corresponds to a nominal frequency of 29 MHz and a thermal sensitivity of +1kHz / °C. Using the Tiersten's model of trapped energy resonators [5], the wave numbers $\bar{\xi}$ and $\bar{\upsilon}$ in the plane of the resonator are obtained by solving the transcendental equations:

$$\bar{\xi}\tan\bar{\xi}l = \left[k\,M_1/\Delta_1-\bar{\xi}^2\right]^2$$
$$\bar{\upsilon}\tan\bar{\upsilon}l = \left[k\,P_1/\Delta_1-\bar{\upsilon}^2\right]^2 \qquad (1)$$

with

$$k = \pi\sqrt{c_{66}/\rho}\,/\,h$$
$$\Delta_1 = \frac{\pi}{2h}\sqrt{\frac{c_{66}}{\rho}}\left[\frac{4k_{26}^2}{\pi^2}+R\right]$$

where k_{26} is the coupling factor and R the ratio between the total mass of electrodes and the total mass of the resonator. The mechanical displacement is:

$$u = \sin\left(\frac{\pi}{2h}y\right)\cos(\bar{\upsilon}x)\cos(\bar{\xi}z) \qquad x=-l...l,\quad z=-b...b$$
$$u = \sin\left(\frac{\pi}{2h}y\right)e^{-\upsilon x}\cos(\bar{\xi}z) \qquad |x|\geq l,\quad z=-b...b \qquad (2)$$
$$u = \sin\left(\frac{\pi}{2h}y\right)\cos(\bar{\upsilon}x)e^{-\xi z} \qquad |z|\geq b,\quad x=-l...l$$

The values of ξ et υ are such that the continuity of the derivatives of u with respect of x and z is fulfilled at the contour of electrodes. For the NLC cut the dispersion constants are M_1= 118 GPa, P_1 = 31.8 Gpa.

	Square electrodes	Rectangular electrodes
Wave number $(\bar{\xi})$ along X	1327 m^{-1}	1327 m^{-1}
Wave number $(\bar{\upsilon})$ along Z'	1433 m^{-1}	984 m^{-1}
Normalized amplitude along X'	1.10^{-3}	1.10^{-3}
Normalized amplitude along Z'	5.10^{-5}	$2.5.10^{-3}$

Table 1: Energy trapping of 100 (usual) mode

From the wave numbers in the electroded region of the resonator, we can compute the amplitude of vibration on the border of the resonator (assuming the value one at the centre). direction. We must note that the results shown on table (1) don't take into account the mass loading effect of the electric tabs in the Z' direction.

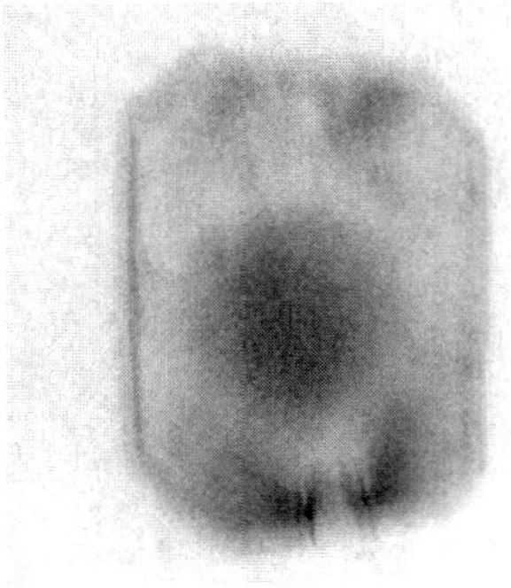

f = 29.197 MHz

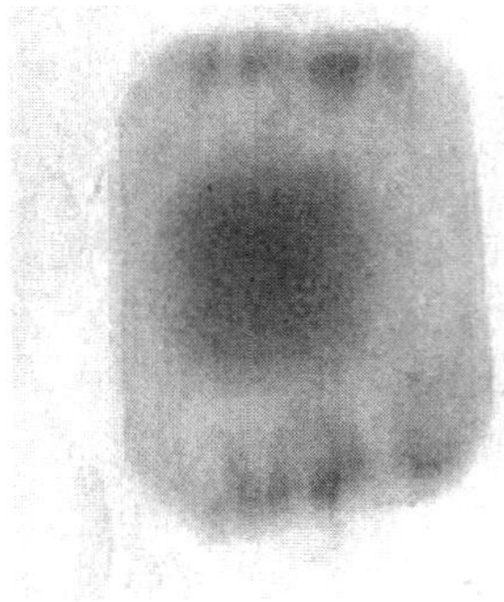

f = 29.002 MHz

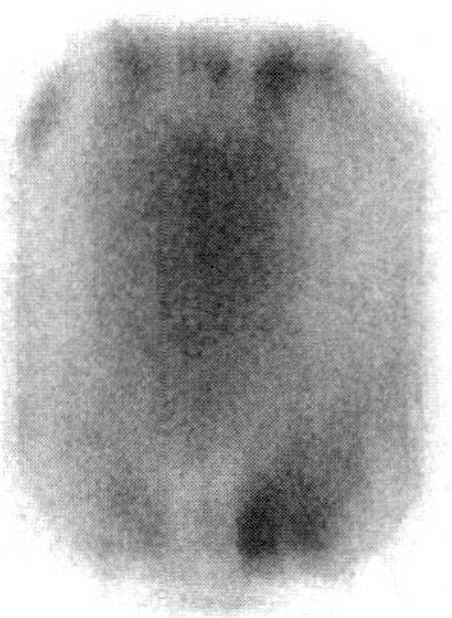

f = 29.147 MHz

Figure 2: X-Ray topographies of (100) mode shapes (Z' axis is vertical)

2.1 Modes patterns

Figure (2) depicts main mode patterns for resonators having circular, square and rectangular electrodes, obtained by X-Ray topography. Mode shapes are in accordance with the model of trapped energy resonators, which indicates that there is no significant coupling with other modes like face shear or flexure modes. Topographies are similar at room and liquid nitrogen temperatures.

2.2 Frequency spectrums

Since the resonators must be used on a wide range of temperature, it is important that no parasitic mode appears in the vicinity of the metrological mode. Figures (3-5) report the

frequency spectrum of the three kinds of sensors, recorded at room and liquid nitrogen temperature.
Curves on the graphs represent the admittance (solid curve) in logarithmic scale and the phase (dotted line) of the resonators. As expected, the resonator with circular electrodes has less spurious modes than the ones with square and rectangular electrodes. The main change between room and low temperatures, apart the translation towards the low frequencies, is the slight increase of admittance in relation with the Q factor increase.

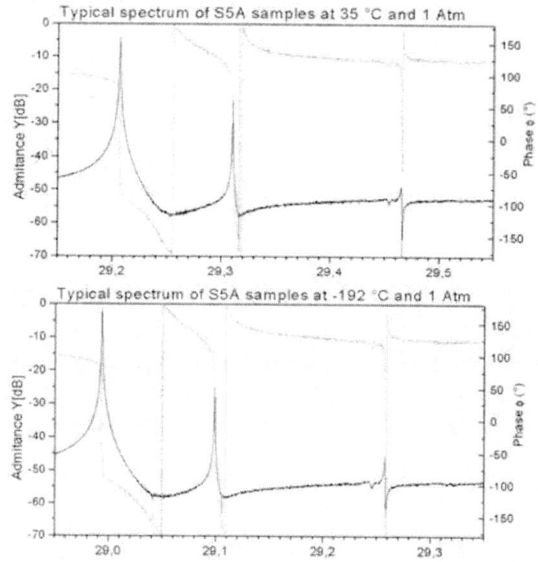

Figure 3 : Frequency spectrum of resonator with circular electrodes

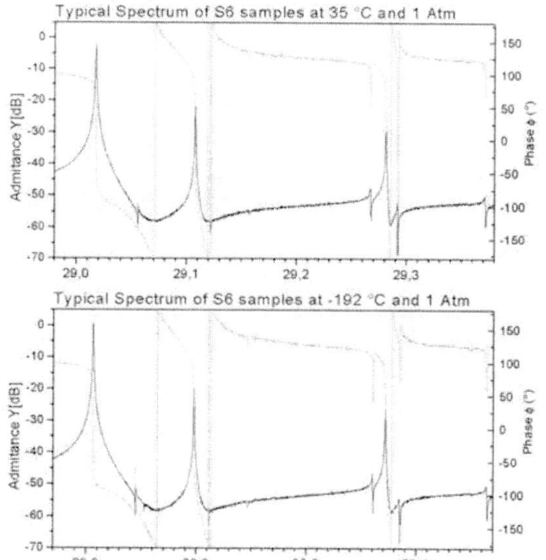

Figure 4 : Frequency spectrum of resonator with square electrodes

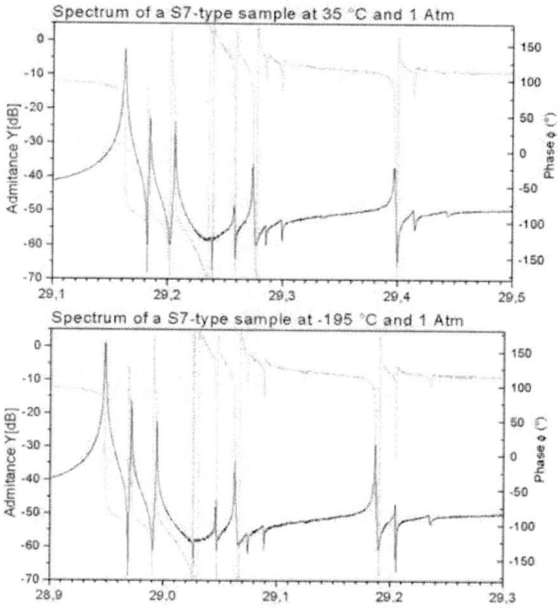

Figure 5 : Frequency spectrum of resonator with rectangular electrodes

3 Thermal behaviour

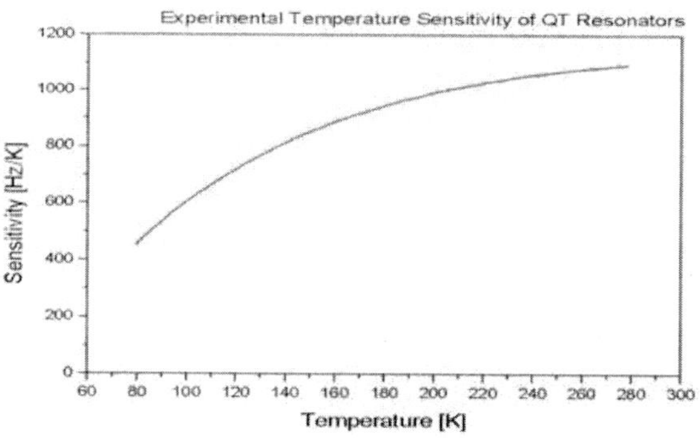

Figure 6: Thermal sensitivity

The thermal sensitivity variation of the sensors with the temperature is shown on the figure (6). There is a continuous decrease of sensitivity at low temperatures down to 2 Hz/°C at liquid helium temperature [6]. Some specimens exhibit activity dips on their frequency-temperature curve. A typical defect is reported on figure (7). Activity dips can occur by coupling of the main mode with other having the same resonant frequency in the vicinity of a given temperature. Some possible modes responsible for these defects can be high anharmonic modes of fundamental C mode or flexure, even if these mode are not electrically driven, the coupling occurring by mechanical effect [7]. The prediction of possible coupling is difficult for several reasons. First one, the lack of precision of the analytical models of high rank anharmonic modes. Second one, the imprecision of thermal sensitivity of coefficients of quartz on a wide range of temperature, especially at cryogenic temperature and, third, the influence of possible static stress induced in the resonator by its mounting, which, by non linear effect can change the frequency of a mode. This last point is illustrated by the figure (8), obtained by using the Source Image Distortion (SID) X ray topography method [8]. The distortion of the line corresponds to a local rotation of the plate of the resonator around the local axis parallel to the lines.

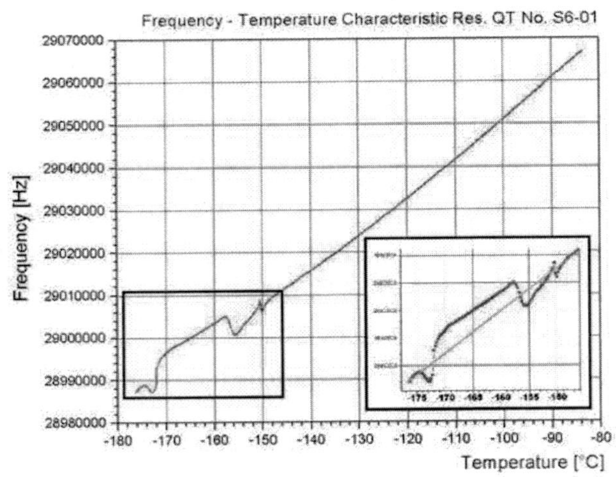

Figure 7: Frequency-temperature characteristic of a device exhibiting activity dips

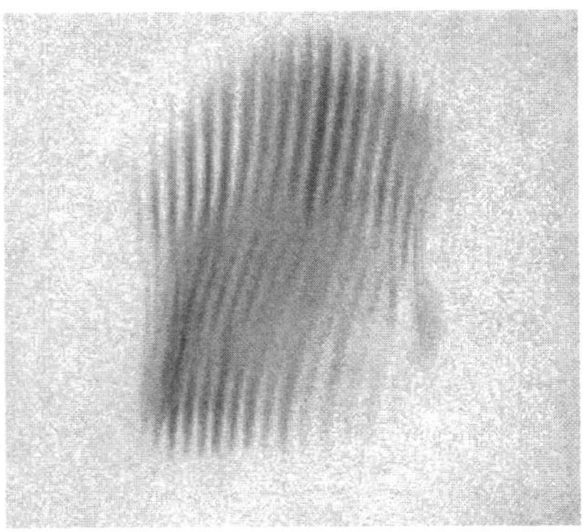
Figure 8 : SID X-Ray topography showing static strain (Z' direction is vertical)

4 Electrical parameters

On resonators free from activity dips, the motional resistance exhibits a small decrease at temperature below −100°C while the Q factor increases. The table (2) reports typical results for good sensors. Measurements at room temperature have been made by using both a Pi network or a HP 4396A network analyser. At low temperature , only the Pi network was used when the device under test was put in the cryogenic chamber. In both cases, the resonator was placed in helium gas at atmospheric pressure.

The motional inductance L remains constant. This is an indication that the mode shape doesn't change significantly when the temperature varies.

Electrode Type	T = + 35°C		
	Q	R (Ω)	L (mH)
Circle	91600	6.4	3.18
Square	85200	5.32	2.485
Rectangular	61600	5.05	1.54
	T = - 195°C		
	Q	R (Ω)	L (mH)
Circle	108000	5.35	3.17
Square	99700	4.55	2.50
Rectangular	68900	4.22	1.60

Table 2 : Electrical parameters

The motional resistance can be computed by using values of the wave numbers indicated above and the experimental values of the Q factors. With the following formula easily obtained from ref. [9]:

$$R = \frac{1}{Q} \frac{\int_V c_{66} S_6^2 dV}{\left[\int_V e_{25} E_2 S_6 dV\right]^2} \frac{1}{\omega} \tag{3}$$

where E_2 is the electrical field for a unity driving potential. By using (2) , we obtain

$$R = \pi^2 c_{66} h [\cos(\overline{\nu} b)\sin(\overline{\nu} b)\cos(\overline{\xi} l)\sin(\overline{\xi} l) + \sin(\overline{\nu} b)\cos(\overline{\nu} b)\overline{\xi} l + \sin(\overline{\xi} l)\cos(\overline{\xi} l)\overline{\nu} b + \overline{\nu} b \overline{\xi} l] \overline{\xi} \overline{\upsilon} \,/\, 64 Q e_{26}^2 \sin(\overline{\xi} l)^2 \sin(\overline{\nu} b)^2 \omega \tag{4}$$

where the integration over the volume V of the resonator is restricted to the part located under the electrodes since in well trapped energy resonator the contribution of the other part of the volume is negligible. The motional inductance is derived from

$$L = \frac{Q R}{\omega}$$

The computed values are:

Square electrodes : R = 5.39 Ω, L = 2.51 mH
Rectangular electrodes: R = 5.11 Ω, L = 1.60 mH

The variation of motional resistance is illustrated on figure (9), which give the output voltage of a Pi network containing a sensor (the voltage is proportional to 1/R).

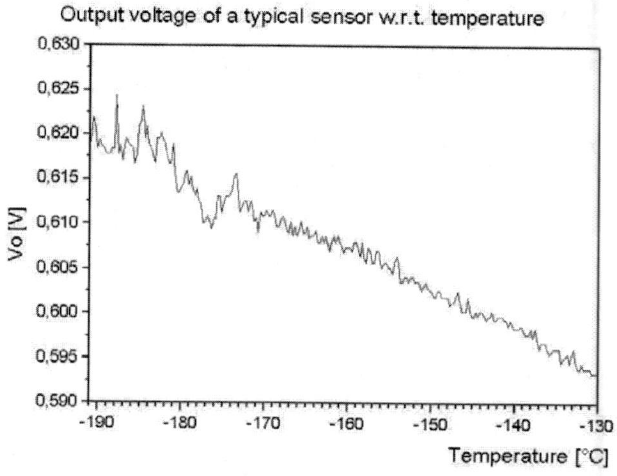

Figure 9 : Experimental effect of motional resistance variation at low temperatures

5 Conclusion

The NLC cut allows us to find designs that are suitable to make thermal quartz sensor for low temperature with a good monotonicity of the frequency vs. temperature characteristic. Designs are intermediate between classical (circular) resonators and strip resonators.

Acknowledgements

This work was supported by the European Community in the framework of the QxSens Contract G6RD-CT-2002-00648.

References

[1] L.Spassov, E.Yossiffov, V.Georgiev, L.Vergov, "A rotated Y-cut quartz resonator with a linear temperature-frequency characteristic", Sensors & Actuators, A58, pp.185-189,(1997)

[2] J.M. Stevels, J. Volger, "Further experimental investigations on the dielectric losses of quartz crytals in relation to their imperfections", *Philips Res. Rept.*, **17**, pp. 283-314, (1962).

[3] D.B.Fraser, "Impurities anaelasticity in crystalline quartz", Physical Acoustics, Volume V, pp.59-110, Academic Press, New York, (1968)

[4] B.Dulmet, L.Spassov, Ts.Angelov , "Finite element analysis for a new Design of NLC-cut Quartz Crystal Resonant Temperature Sensor ", this proceeding.

[5] H.F.Tiersten, "Analysis of trapped-energy resonators operating in overtones of coupled thickness shear and thickness twist", J. Acoust. Soc. Am., pp. 879-888, (1976)

[6] R.Velcheva, " Investigations of temperature sensors based on thermosensitive quartz resonators" Ph. D. Thesis, Sofia University, (2003)

[7] R.Bourquin, B.Dulmet, G.Genestier," Mechanical couplings involving discontinuities of the frequency temperature curves of contoured resonators ", Proc. 39th Annual Frequency Control Symposium, Philadelphia, U.S.A., pp. 405-414, (1985)

[8] W.J. Spencer, " Observation of resonant vibrations and defects ", Physical Acoustics, Volume V, pp.111-161, Academic Press, New York, (1968)

[9] J.A. Lewis, " The effect of riving electrode shape on the electrical properties if piezoelectric crystals", Bell Syst. Tech. J., pp.1259-1580, (1961)

MICROBALANCE SENSOR APPLICATIONS USING THE PIEZOELECTRIC CRYSTAL MATERIAL $GaPO_4$

F. Krispel*, C. Reiter*, R. Schauperl*, I. Mannelli[+] and W. Wallnöfer*

* Piezocryst Gesellschaft für piezoelektrische Technologien mbH, Hans-List-Platz 1, A-8020 Graz, Austria
[+] Dip. di Chimica - Università di Firenze, Polo Scientifico, Via della Lastruccia 3, 50019 Sesto Fiorentino (Firenze), Italy

email: ferdinand.krispel@piezocryst.com
Fax: ++43(0)316 787-591

Keywords: $GaPO_4$, micro balance, high temperature, thermo gravimetric, affinity sensor .

Abstract

Gallium phosphate, $GaPO_4$, is a piezoelectric crystal material homeotypic to quartz which features outstanding thermal stability up to 970°C, higher coupling than quartz and good temperature compensation for resonant applications. We report results in the research and development for high-temperature micro balance including thermo gravimetric analysis and for affinity sensor applications in aqueous liquids, such as biosensors.
A sensor head for high-temperature micro balance applications with an integrated heater was developed. The clamping and heating device for the $GaPO_4$ resonator was designed for mechanical stability, reliable contacting and homogeneous temperature distribution. Cycles up to 600°C allow separation of volatile and fixed parts of layers deposited on the resonator surface and pyrolytic cleaning from carbonaceous residues.
Another interesting field is the application of $GaPO_4$ for affinity measurements in liquids, e.g. for biosensors. An optimized crystal cut was developed and tested which allows temperature compensation in contact with an aqueous solution. A flow cell with a design for better reproduceability was constructed.

1 Introduction

Gallium phosphate ($GaPO_4$, also gallium orthophosphate) is a piezoelectric crystal material homeotypic to quartz. Some material constants are similar to quartz thus $GaPO_4$ can be used for all quartz applications without changes in the arrangement of the devices. But the well known $\alpha-\beta$ phase transition, which changes the structure in SiO_2, is absent in $GaPO_4$, thus the novel crystal can be used up to 970°C where another phase transition occurs and the low quartz structure is changed into a cristobalite like one. The most physical properties of the material are very stable up to at least 900°C (Fig. 1).
Other advantages compared to quartz are the twice piezoelectric effect and thus the doubled electric mechanical coupling coefficient which can be used to produce e. g. VCXO's with pulling ranges up to 1500 ppm [1].

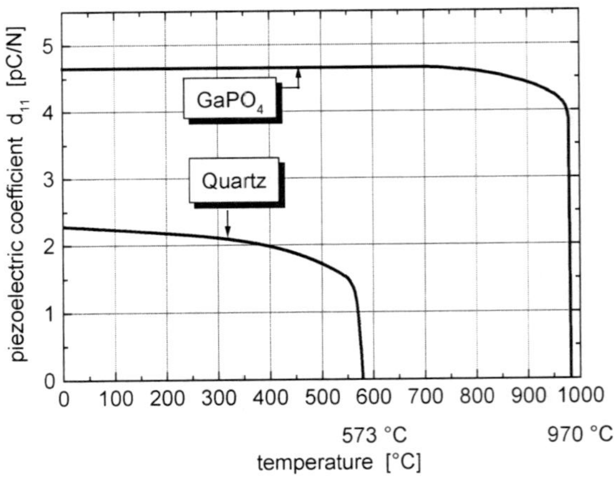

Fig. 1: Piezoelectric constant of quartz and $GaPO_4$

Because of the lower elastic constants the frequency constant is one quarter lower and the thickness of resonators is also one quarter smaller compared with quartz resonators of a similar frequency. The Q factor is at least comparable to quartz but studies are in progress which show a potentially increase of the frequency quality product against quartz. The dynamical resistance R_1, which is used in the simple Butterworth van Dyke electrical equivalent circuit for a crystal resonator (Fig. 2), is also lower for a given size and frequency.
This leads e. g. to well working 10 MHz/3rd harmonics resonators in HC-52/U crystal holders [2]. The drive level dependence (DLD) is in high coupling $GaPO_4$ resonators (temperature compensated cut for ambient temperature) only one hundredth of comparable quartz resonators.

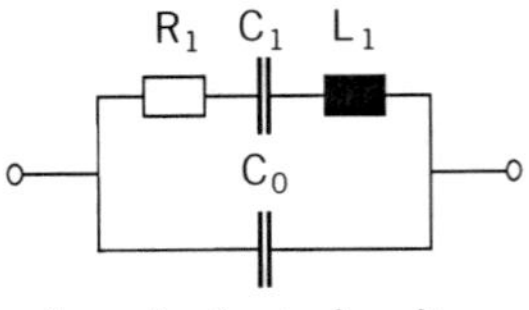

Fig. 2: Simple crystal equivalent circuit

These outstanding properties make $GaPO_4$ very suitable for micro balance applications like affinity sensors for biological investigations or micro balances for very sensitive particulate measuring systems. But the most important property of $GaPO_4$ is the usability up to at least 900°C. New applications can be developed as e. g. high temperature micro balances or particulate measurement systems with the possibility of high temperature cleaning.

A necessary requirement for micro balances is the temperature stability. If the temperature frequency behaviour of the resonator is flat at the used temperature (compensated cut), it is possible to make very sensitive measurements. For each temperature up to 500°C exists a compensated cut with the crystal $GaPO_4$. At higher temperatures the 2nd order parabolic behaviour changes to a 3rd order one (Fig.3).

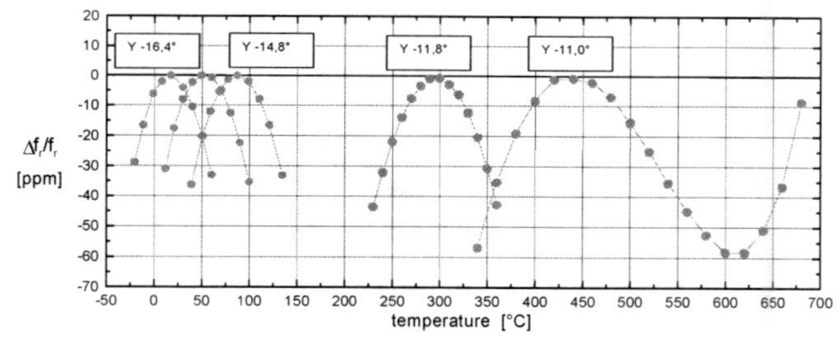

Fig. 3: Temperature compensated cuts with $GaPO_4$

Another necessary material property is the electrical insulation. Most materials, for example some kinds of ceramics, have a high insulation, but for higher temperatures (>500°C) the insulation drops down and thus the quality factor of the resonator. The resistivity of $GaPO_4$ is shown in Fig. 4. At 900°C the resisitivity is still higher than 1 MΩ.

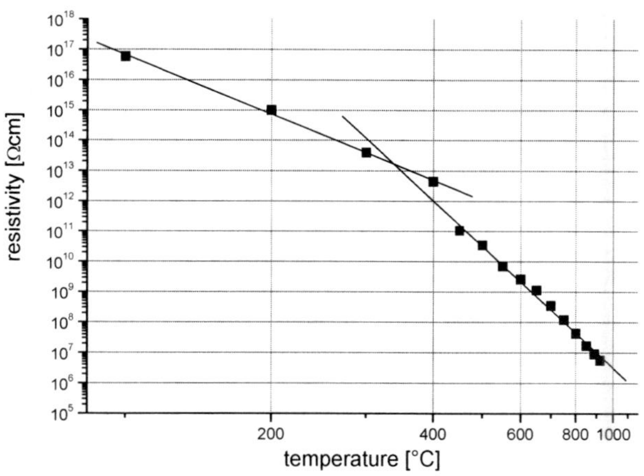

Fig. 4: Resistivity of $GaPO_4$ depending on temperature

2 Samples

The resonators are prepared out of $GaPO_4$ single crystals which are grown at elevated temperatures in a phosphoric acid solution. These crystals were cut with a diamond saw, lapped and cylindrical grained. Diameters of 7.4 mm and 14 mm were chosen. Electrodes, made of gold, were sputtered at the resonators with a 20 percent adhesion layer of NiCr.

The cut angle was Y-17.4° which leads to a turnover temperature of 25°C. This cut is especially designed for the use in liquids for temperature compensated measurements. In Fig. 5 two resonators are compared, the first made of $GaPO_4$ the other one made of quartz. Measurements have been performed in a temperature controlled liquid cell, but the control is switched off to allow a thermal drift from 4°C to 21°C.

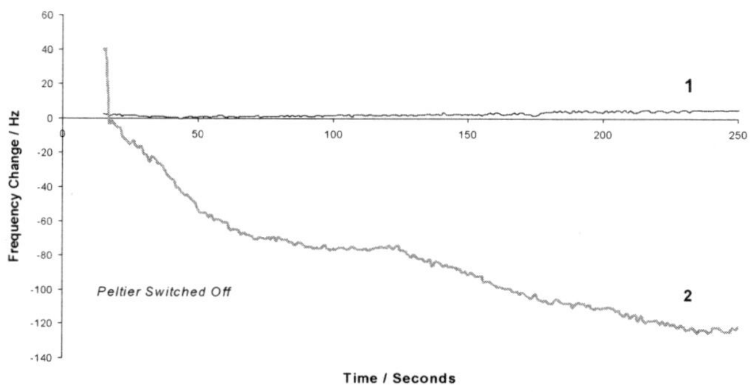

Fig. 5: Temperature stability of $GaPO_4$- (1) and quartz-resonators (2) in liquids [3]

Another cut angle was chosen for the particulate monitoring system with a turnover point at 300 °C. The thickness of the resonators was 0.200mm to get a resonant frequency of 6 MHz.

3 Principle of micro balances with $GaPO_4$

A micro balance based on $GaPO_4$ consist of the same elements like another micro balance. In principle the frequency and the damping of the resonator are changed because of the loading (Fig. 6).

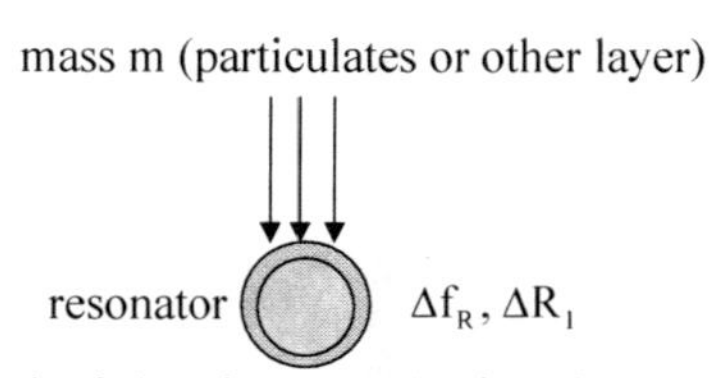

Fig. 6: Principle of a crystal micro balance

This loading is in the simplest case a rigid layer, which leads to a shift in frequency (1) because of the increased mass [4].

$$\Delta f_R = -\text{const.}\, m = -\frac{2 f_R^2}{\sqrt{\rho c_{eff}^D}} m \qquad (1)$$

Where ρ and c_{eff}^D are the density and the elastic constant of $GaPO_4$. Improved models are dealing with the damping and the acoustic impedance, but no model can describe the behaviour in a correct manner.

In this work the micro balance is used to make an affinity sensor with $GaPO_4$ and to demonstrate the possibility of particulate measurements with thermo gravimetric analysis in principle.

Particulates are measured with the increase of mass but the affinity sensor uses another principle. Some molecules are bonded to the surface of the coated resonator (in most cases gold electrodes) and thus influence frequency and damping. The sensitivity of this method should depend on the coupling coefficient. Thus, materials with higher coupling than quartz should be more sensitive.

$GaPO_4$ resonators (Y-17.4°) are made with gold electrodes and a resonant frequency of 6 MHz. The coupling is 13%, the damping R_1 in air 8 Ω and in the cell, but still in air, 80 Ω (clamped). The Q factor drops from 15000 down to 1500 in the cell.

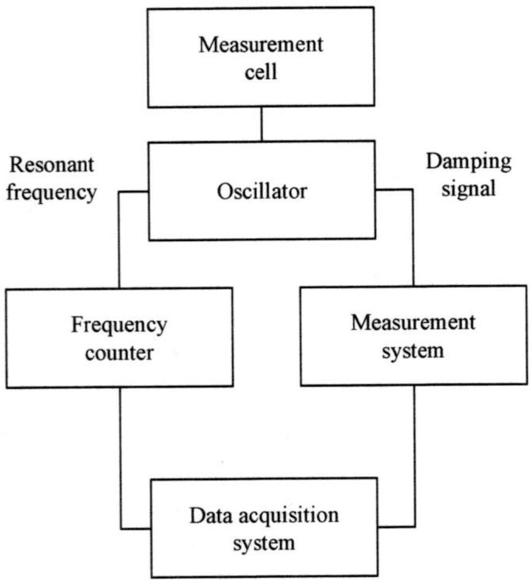

Fig. 7: Measurement system

The measurement system consists of the cell, an oscillator and the data acquisition system where frequency and damping are recorded. The oscillator is home made with a very high amplification, a cable compensation and a possibility to control the resonator excitation (power or current).

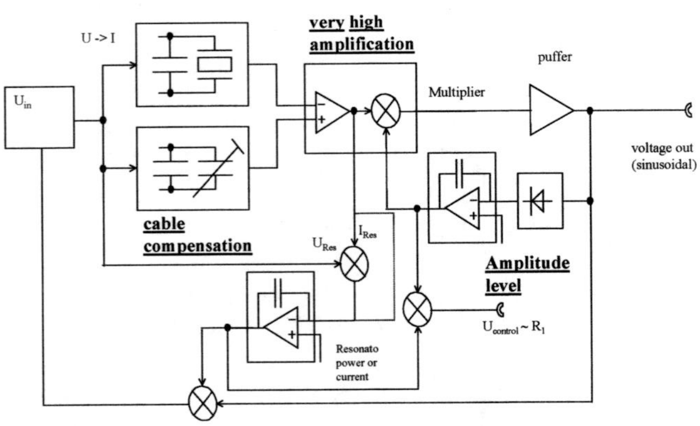

Fig. 8: Oscillator schematics

Very high damping of the resonator can be compensated with the high gain of the oscillator. The feed back loop is controlled by a voltage which delivers the damping signal.

4 Affinity sensor with $GaPO_4$

For the first use of $GaPO_4$ as an affinity sensor a new liquid cell was constructed. Most important view have been given to a simple manageability of the cell and to change resonators very easy (Fig. 9).

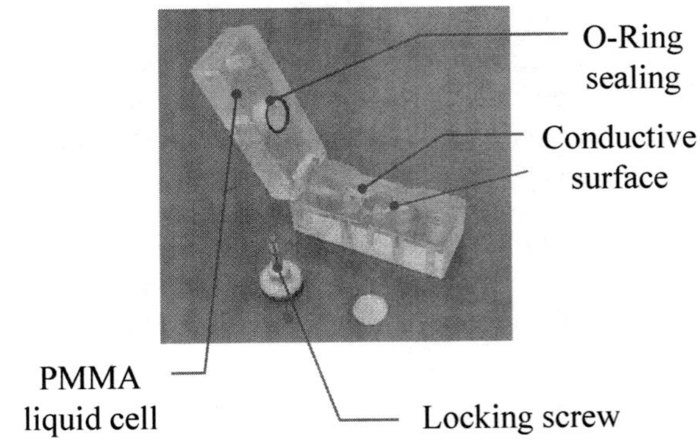

Fig. 9: Improved liquid cell

The cell can be closed with a locking screw, an O-ring holds the resonator, which is lying on a conductive surface. The probe volume is very small (0.2 mm^3) because of the very expensive chemicals which should be investigated. Electrical conducts (both electrode contacts at one side of the resonator) are made by built in sockets at the bottom of the cell. Especially designed wrapped around electrodes are used (Fig.10).

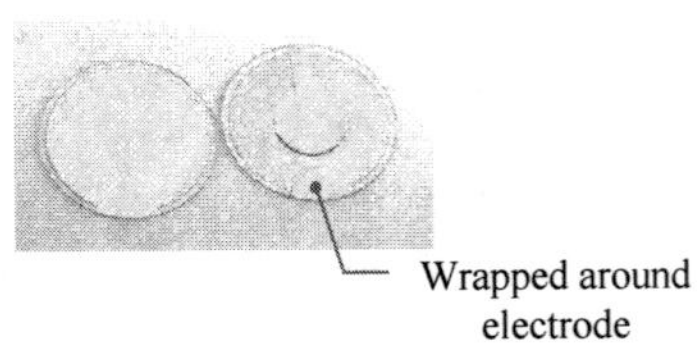

Fig. 10: $GaPO_4$ resonators with electrodes

First measurements have been done with an anti-rabbit genom and also first result can be shown. For this kind of measurements the surface must be very clean. Then an immobilisation procedure is used to make the gold surface sensitive for the molecules which should be investigated. This immobilisation procedure was done with Rabbit-lgG. After removing the not bonded molecules the Antirabbit-lgG (1:500) is filled into the cell. Molecules which are not bonded to the immobilised layer are removed and the shift in frequency is measured (Fig. 11).

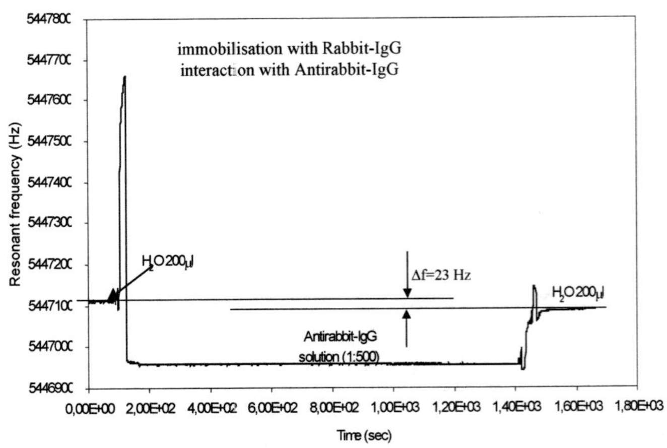

Fig. 11: First result with $GaPO_4$ as affinity sensor

The change in resonant frequency was with 23 Hz clearly more than the frequency instabilities. Of course this result must be checked for linearity with different solutions. But in principle the system is working with $GaPO_4$. Further measurements will be done to compare the sensitivity of $GaPO_4$ and quartz. Further work is planned also to improve the stability of the system for an increased sensitivity.

5 Particulate measurement principle

Another very promising application with micro balances based on $GaPO_4$ is a system to measure particulates (at room or increased temperatures) and to make a thermo gravimetric analysis of their chemical composition. This is possible because with $GaPO_4$ temperatures up to 900°C are allowed. During the heating process an auto cleaning is done and the system can be used again without changing the resonator.
The measurement principle is shown in Fig. 12.

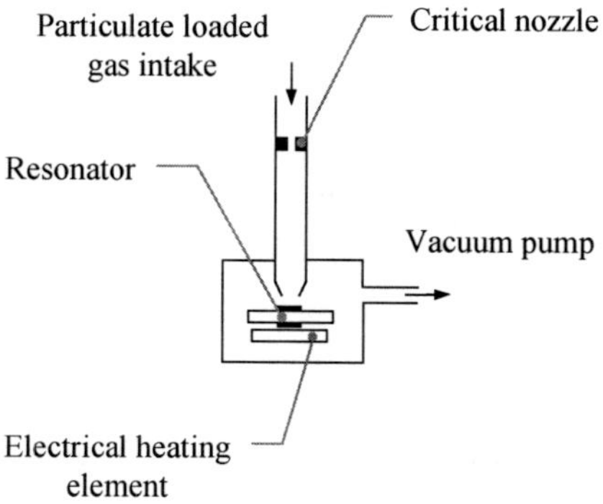

Fig. 12: Principle of particulate measurements with a micro balance

The gas flow through the measurement chamber is generated by a vacuum pump. The critical nozzle stabilises the flow rate and generates a low pressure inside the chamber. This is necessary to precipitate the particulates at the electrode of the resonator. The heating element below the resonator is used to heat up the particulates for the thermo gravimetric analysis.

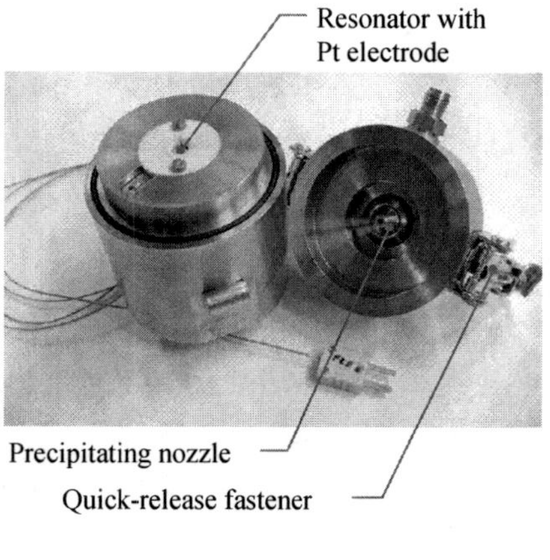

Fig. 13: Sensor head and precipitator

The sensor head is shown in Fig. 13. The resonator is coated with platinum because of the high stability of the electrode at high temperatures. The frequency temperature behaviour of the resonator depends on application. To get the information of the particulate concentration with high accuracy the temperature compensation should be at the sampling temperature. Temperature compensation between room temperature and about 500°C is possible (Fig. 3).
First measurements will be done with the measurement system shown in Fig. 7 in next months but sensor heat tests are performed up to 650°C (Fig. 14).

Fig. 14: Glowing resonator at 650°C

6 Conclusions

$GaPO_4$ is a very promising material for micro balance application. Advantages are expected especially in high temperature applications or in applications where a high coupling coefficient is necessary.
A first demonstration was shown with $GaPO_4$ as affinity sensor and another application is in progress where particulates can be analysed by the use of thermo gravimetry.

Acknowledgements

We gratefully acknowledge EU funding of the GAPO-GROWTH project under contract No. G5RD-CT-2002-00709 (GROWTH programme).

References

[1] for more details see: www.axtal.com

[2] F. Krispel, C. Reiter, J. Neubig, F. Lenzenhuber, P. W. Krempl, P. M. Worsch and W. Wallnöfer, "Properties and Applications of Singly Rotated $GaPO_4$ Resonators", Proc. IEEE FCS and PDA Exh. jouintly with 17th EFTF (Tampa, 2003), 668 – 673; and www.gapo4.com

[3] F. Paul, "Circumnavigating the 'big S': practical applications of acoustic sensor technology for the Pharmaceutical Industry ", QCM2002, Brighton (July, 2002)

[4] G. Sauerbrey, "Verwendung von Schwingungen zur Wägung dünner Schichten und zur Mikrowägung", Z. Phys. 155 (1959), 206 - 222

BAW MICROWAVE TEMPERATURE SENSOR

S.G.ALEKSEEV, G.D.MANSFELD, I.M.KOTELANSKII, V.S.VERETIN
Institute of Radioengineering and Electronics, RAS
11, Mokhovaya st., 125009, Moscow, Russia
mans@mail.cplire.ru

This work is aimed at the study of a potential possibilities of microwave Bulk Acoustic Wave resonators for temperature measurements. The resonators under study had non-standard high overtone BAW resonator configuration: a thin piezoelectric film ZnO with electrodes was deposited onto the surface of the substrate made of the (100) oriented YAG plate with the thickness 480… 980 μm. The new technical solution was that resonator was acoustically isolated and protected from the ambient mechanical influence. For this purpose the bottom side of the plate and the top electrode of the resonator structure were isolated from ambient by a nine-layer Bragg Ti-W quarter-wavelength thin film reflectors. Due to the isolation it can become no sensitive even to liquids getting on the surfaces of the resonator. Total thickness of all deposited layers was less than 12μm. The resonator aperture was less than 600μm.

The structure was excited on third BAW harmonic of ZnO film near the frequency 430MHz. With the change of the temperature the frequency of resonator was monotonously changed.

In case of BAW resonators two types of taking out of the data were possible. The first one was based on the direct measurement of the resonant frequency change under external influence. In this case the data obtained can be analyzed with a network analyzer. In a narrow frequency (and hence measurand) interval it is possible to interrogate the sensor on a single frequency within the bandwidth frequency interval of the resonator. The parameter under measurement is the phase of the reflected electromagnetic signal. In this case the resolution of the measurand can be increased at least two orders of magnitude.

It is evident that the temperature sensitivity of the sensors based on HBAR structures is governed by the temperature sensitivity of the substrate due to the big ratio of the thickness of the substrate to that of all the other layers. The results of measurements illustrated the possibility to achieve a good temperature resolution. The quantity characterizing the sensor sensitivity the ratio of frequency change for one measurand unit (in our case 1K) to the bandwidth of the resonator $S_Q(1K)$ was introduced (E. Benes et al.) For our temperature sensor based on HBAR the factor was found to be $S_Q(1K)$ =5. It is much better than that for typical SAW sensors (S_Q=0.3). It is owing to the facts that for HBAR's have much higher value of the quality factor. The phase method of measurement increases the temperature sensitivity at least two order of magnitude.

The theory based on matrix multiplication formalism was developed and was used for calculations of temperature behavior of the modulus and phase of input electric impedance of the temperature sensor. It was in a good agreement with the experimental data.
BAW temperature sensors are more sensitive and miniature than known SAW-sensors; they are compatible with small-size microwave aerials for wireless distance probing.

DRIFT AND UMP CONTROL IN MULTIPLE CR STAL OSCILLATOR S STEMS

Hedley Rokos

Adaptalog Consultancy
Hole Farm, Bush Lane, Little Sampford, CB10 2R , UK
a l he ley a ap alog co

Abstract

Techniques are proposed that allow systems that use multiple oscillators to correct the effects of occasional frequency jumps, or of changes in drift rates in any of the oscillators. Existing methods of drift prediction may readily be adapted to suppress drift changes in individual oscillators. In addition, previously unreported jump detection schemes are shown to be capable of suppressing large jumps, and with very low levels of additional noise. By way of example, a system that detects jumps with single Allan-variance type measurements will limit individual jump amplitude at the output to 4· $_{out}$/ (N-1). For Flicker-FM noise, the random walk noise that this scheme generates would reach the Allan variance after about 80 000·τ. More complex schemes, but that still need at most 16 samples to be stored, would reduce the jump height by a factor 2, while extending the time by several orders of magnitude. This allows systems using three or more oscillators to operate in the presence of these jumps and with negligible increase in total system error. These techniques have the potential to avoid the need for costly or bulky perturbation-free oscillators in some intermediate applications.

1 Introduction

Local frequency references can use multiple sources to provide better performance than would be available from any of the individual sources. Generally, such references use each type of source to control the output over the time range for which it provides the best characteristics. In addition, multiple sources with similar characteristics may be averaged to provide better stability than would be available from any individual. Further, systematic drift may be measured against a remote standard when signal is available, and its effects compensated. Up to the present, at least one of the local sources has been required to provide wholly predictable aging if long-term performance is to be maintained,. Unfortunately, this generally implies a requirement for atomic sources, as more compact, economic, and robust frequency sources (such as quartz crystal oscillators) can suffer from occasional frequency jumps and sudden changes in aging performance. Clearly, there could be significant benefits if we can learn to live with these effects.

Where several (N) oscillators are available, it is possible to characterise each of them with respect to the average of the remainder. If the oscillators are similar, the measurement variance on these measurements will be $N^2/(N-1)$ times the measurement variance that would be achievable for the ensemble - assuming an ideal reference was available for this measurement. Suppose that the measurements, equivalent to those for the ensemble, indicate that it is necessary to adjust one of the oscillators. The resultant error at the output of the ensemble would be just 1/(N-1) times the ultimate accuracy for the ensemble. If this is a significant degradation, it could be corrected by adding just one oscillator to the ensemble. The bigger issue is probably that, if errors of moderate size are to be detected and corrected, unnecessary adjustments will inevitably be made, resulting in long-term performance degradation due to random walk.

In this paper, we consider general techniques for minimising the random walk. These may be applied in conjunction with minor adaptations to previously published work on the detection of changes in the aging rate [3,4]. In addition, as there is little in the literature on the measurement of frequency jumps, initial work is presented that scopes the available precision and relative data storage costs. We also consider how some practical issues that could degrade system performance can be avoided.

2 Origins, averaging, and practical issues

2.1 Errors in individual sources

In terms of the behaviour of the errors, effects may be generally divided into four groups [1,5] stationary stochastic noise, drift (or aging), environmental effects, and instantaneous steps in the output frequency (or frequency jumps).

Stationary stochastic noise exhibits different behaviour at different timescales [1]. This noise represents one of the limits in the accuracy with which frequency jumps can be measured or corrected. As frequency jumps are best measured over periods where the noise characteristic approximates flicker FM, analysis will mostly concentrate on behaviour in this region.

For a precision frequency source, the rate of systematic frequency drift generally reduces with elapsed time. As there may be many mechanisms involved, each with different time evolutions, prediction of drift requires the analysis of relatively large amounts of data. This has been extensively studied elsewhere [3,4], and we assume with minimal discussion that drift can be adequately modelled.

For the purposes of this paper, we mostly assume that environmental effects appear as contributions to the other errors in so far as this applies, they need not be considered separately. Suffice it to observe that drift due to environmental effects tends to display a mix of cyclic and random behaviour that may locally resemble aging with the notable exception that it does not decay with time. We note also that multiple sources may be required for environmental reasons, such as offsetting the sensitivity to acceleration.

Significant sudden changes are also observed in some frequency sources - notably quartz crystal resonators. The origins of these changes, known as frequency jumps range from crystalline defects (including colour centres) in the quartz, through discontinuous annealing in the support structure or electrodes, to the impact of high-energy ionising radiation. [There are many other potential causes, including stiction if the nominally free end of a cantilevered blank rests against a shelf]. Larger jumps can also trigger an increase in medium-term drift (or aging). Historically, the size of these jumps has been a selection criterion for precision quartz oscillators, sometimes resulting in a significant proportion of reject devices.

2.1 Output averaging, and observational notes

We now consider averaging the outputs from an array of N similar oscillators at frequency f. For a uniform scheme, the Allan variance at the output is

$$ {}^{2}_{out} = {}^{2}_{osc} / N \qquad (1)$$

where ${}^{2}_{osc}$ is the Allan variance[1] of each source
${}^{2}_{out}$ is the Allan variance[1] seen at the output,
and N is the number of oscillators in the ensemble.

Theoretically, this may be achieved in a number of ways, including mixing the outputs from a number of free-running oscillators to provide output at frequency N.f. In practice, if the oscillator frequencies are close to each other, signal coupled from one oscillator into another will shift its frequency, an effect known as injection pulling. As levels of injection pulling are typically not matched, the effects of the oscillators at the output will be unequal, degrading the noise. This may be overcome in two ways by ensuring that the oscillator frequencies are sufficiently far apart for injection pulling to be minimal, or by introducing a sufficiently strong source of coupling that the overall interaction becomes well matched. Either of these approaches is equally effective, and both are compatible with additional processing. In this section, we consider some of the properties of strongly coupled arrangements. Consider first locking[2,3] the phase of every oscillator to the average of the outputs. If the loop bandwidths are sufficient and known, the control voltage[4] can be used as the sole variable for detection and correction.

It is now worth considering some practical detail of making corrections in PLL implementations. When using tightly coupled PLLs, intrinsic frequency differences are seen as changes in the tuning voltage[5]. As expression in terms of these Voltages would simply require continuous re-normalisation, it remains preferable to cast all equations in terms of frequency errors. This clearly neglects the effects of tuning non-linearity. We can see that the sizes of observations are assessed against the same measure as their correction in this respect, tuning non-linearity is self-cancelling. If noise characteristics are pre-measured and used as system constants, uncorrected non-linearity will cause errors in its application these may be avoided by measuring individual variances on an ongoing basis. Unless the non-linearity is very large, however, or an individual frequency jump explores the non-linear range, the residual effect is merely unequal contributions of the oscillator variances to the final output.

Another feature of the PLL implementation is that there are two convenient measures for the relative frequency of an individual oscillator with respect to the global average, or as an absolute potential. As the average should be stationary, both attenuate the signal by the same amount

$$\text{Amplitude}_{measured} = (N-1)\ \text{Amplitude}_{differential} / N \qquad (2)$$

As there would appear to be no specific advantage in measurement w.r.t. the average, it will generally be convenient to exploit absolute measurements to allow the system to counteract the drift that will inevitably be induced by small mismatches in the gains of the PLLs.

---

[1] (at corresponding measurement times)

[2] If the noise floor is to be optimised, this requires that the individual loops lock the oscillators approximately in phase with the averaged output.

[3] In principle, this arrangement allows the contributions of different oscillators to be matched to their performance. In practice, it is relatively simple to optimise both the noise floor (by weighted addition of oscillator outputs), and one other noise contributor (relative contribution proportional to amplitude/PLL-gain).

[4] This can require ADC and DAC resolutions that exceed 20 bits. However, if the converters are reasonably fast, simple self-correcting schemes can be used so that only local linearity and monotonicity is required. This allows low-cost converters to be used.

[5] i.e. f V pulling-gain

3 Measurement impact on correction

3.1 Benefit and side effects of non linearity

The intrinsic reason that these correction schemes can be viable is that the accuracy limit for sensing an individual jump is based on the measurement noise for a single oscillator, or

$$^{2}_{eff} = N \; ^{2}_{osc}/(N-1) \qquad (3)$$

and the averaging gives an attenuation of $1/N^2$, giving an effective correction noise limit at the combined output of

$$^{2}_{effout} = \; ^{2}_{osc}/(N-1)/N \qquad (4)$$

This is a factor of (N-1) lower than the aggregate noise of equation (1), which may be seen as the measurement noise when the aggregate is compared against a perfect reference. Thus, for equivalent measurement schemes, the error on a correction would be equivalent to reducing the number of oscillators by one and only in regions where the performance is impacted by the correction.

Unfortunately, this is by no means the whole story. First, if we were to correct all apparent changes, we would simply be aligning the frequencies of all the oscillators on a continuous basis which is equivalent to adding a frequency-locking component to a PLL. In this case, we would get no correction of the observed effect. Clearly, the benefit applies only to individual corrections, and this requires a decision to be made. The decision process itself carries noise. For example, if a threshold is set, above which effects are corrected, any change less than the threshold will remain uncorrected. In addition, there will be spurious corrections triggered by the background noise. Although the individual events will be small, at least a part of their effects will accumulate as Random Walk FM. The size of the individual steps may be limited by correcting only the part of the error that exceeds the threshold [6]. In this case, the Random Walk FM error will grow as

$$^{2}(time) = k \quad time/\tau \; Var_{truncation} \qquad (5)$$

where k is a constant that depends on the time evolution of the noise[7], and $Var_{truncation}$ (the noise variance due to truncation) is

$$2 \;\; _{Ktr} \; (0.399\,(x - K_{tr})^2 \exp(-x^2/2))dx \qquad (6)$$

Truncation level - (K_{tr})	3	3.5	4	4.5
Variance/ $^{2}(\tau)$	4.1E-4	5.6E-5	6.2E-6	5.5E-7

Table 1: Variance of Truncation Error ($Var_{truncation}$) for Normal Distribution (Equation 6)

[6] If the system works to a single threshold, there would appear to be no disadvantage to this arrangement. It is possible that a system whose correction is a more complex function of the sensed level would give better results.

[7] If Allan variance measurements were to be made many times per measurement period, k would be around 4.

By way of example, we consider the effect of forcing truncation on Allan variance, with measurements spaced by much less than τ this may be regarded as a relatively crude jump correction algorithm). The output noise is

$$^{2}_{output}(time)$$
$$= k \;\; ^{2}_{out} \; time/\tau \; Var_{truncation} \; N/(N-1) \qquad (7)$$

Observe that the relative noise becomes virtually independent of the number of oscillators [N/(N-1) varies only from 1.5 (for the minimum possible case of three sources) down to 1]. Observe also that for the important case of flicker FM noise (Allan variance exponent =0), the relative phase error at time t is half the frequency error given in equation (7). Truncating at $_{meas}$, the expected phase error corresponds to the Allan variance after about 80000 $_{meas}$.

Also of interest is the size of residue of truncated Allan variances:

$$\text{Jump Residue} = K_{tr} \;\; _{out} / \; (N-1) \qquad (8)$$

As a measure of jump correction, these results are purely indicative, as they would apply to the purely theoretical situation of jumps being sensed continuously. There is also the issue of handling multiple close-spaced peaks.

3.2 Avoiding multiple detection

Clearly, any sensing algorithm will need, so far as possible, to discriminate between actual and (noise-induced) spurious events. Thus far, the paper has examined only the magnitude of noise, and the consequent probability that it will be detected as a frequency jump. Two other potential sources of spurious detection have been identified multiple detection of the same event, and large jumps on one source being simultaneously interpreted as apparent smaller ones on other sources. Recognising that umpteen suitable algorithms can be found, a single example (for each class of error) is described here.

The simplest method for eliminating real changes on one oscillator causing spurious ones on another is to accept only the largest change at any given time. Given an adequately high detection threshold, this will often be perfectly satisfactory, as we only correct effects that are intrinsically rare it is also the only possible strategy if only three oscillators are used. Sometimes, however, it may be desirable to detect more than one change at a particular time, for example if an external event can trigger changes that require correction. Here, starting with the largest change, the excess of each change may be removed from the average and from any changes previously truncated. This would continue until either there are no more candidate jumps of sufficient size, or the residual number of oscillators falls below a predefined threshold.

It is desirable to avoid both multiple detection of peaks and the screening of the true peak by a subsidiary side peak. The method chosen will also determine both the processing load and the accumulated phase error at the point of detection. The detail will inevitably depend on the type of error being detected, and on the method. The principle, however, is simply to find the first peak, and then to accept the highest peak that occurs within a defined time thereafter. In some cases, it may be beneficial for this time to depend on the history of the individual group. Generally the error would be corrected and the correction refined throughout the period. A specific case for jump correction is considered later.

3.3 Further limitation on aging change measurement

In principle, measurement of aging over long periods provides significant accuracy advantages. Clearly, this is only possible while the aging is well behaved. This span of data will not be available for measuring frequency changes. The effective variance on measurement of aging changes will therefore be significantly greater than the variance on the original (presumed continuous) aging measurements. This paper contains no explicit work on aging changes per se. However, the scale of the residual error will be similar to that observed with frequency jumps albeit the measurement timescale and weighting requirements will be very different.

4 Measurement impact on correction

4.1 Basic accuracy of simple measurement schemes

In section 3.1, we considered the (purely conceptual) performance of a frequency jump detection scheme that used differences between frequency averages over equal, contiguous measurement periods, with measurements made continuously. In this section, we consider the impact of taking finite numbers of samples in any period.

Frequency jumps may be detected and measured using contiguous sampling, or with gaps (or dead time) between the samples. There are significant differences between the behaviour of the two schemes. For arrangements that use contiguous measurements, the error is essentially a random gain reduction caused by the jump occurring during one of the measurement intervals. This adds to the measurement error of the frequency jump, but does not increase the random walk contribution in the absence of such jumps. The variance of the error contribution, due to this asymmetry, for jump amplitude J and time between successive measurements τ is

$$J^2/3 \qquad (9)$$

For measurement with dead time, the noise increases according to Barnes bias function $B_2(r, \ )$ [2], giving

$$^2{}_{output}(time) =$$

$$k \ {}^2{}_{out} \ B_2(r, \ ) \ time/\tau \ Var_{truncation} \ N/(N-1) \qquad (10)$$

Clearly, the number of frequency samples that need to be stored will influence the choice of scheme, particularly if the number of sources to be monitored is large. The storage requirements for uniform measurements are:

$$= 2/ \quad \text{(for zero dead time)} \qquad (11)$$

$$\text{or} \quad = 1+2/(r-1) \quad \text{[for 1/(r-1) an integer]} \qquad (12)$$

In order to ease comparison between the two schemes, values for the bias function and for the detection error bias due to asynchrony are given in table 2, for some typical values of r and .

4.2 Controlling multiple peaks

The method described here gives a relatively low (although not the lowest possible) phase error. The record of frequency is built up piecemeal by repeatedly adding fractions of the detected frequency[8] to the stored value. Following each measurement, the accumulated phase error is calculated w.r.t. the frequency measurement over a full measurement interval. When it first exceeds $K_{tr}\ \tau$, the frequency is adjusted sufficiently to reduce the frequency to a point where it would have reached exactly the excess phase threshold, and the adjustment is recorded. If, within a period τ, the phase again exceeds the phase limit, the adjustment is updated (also the record) so that the total excess phase is limited to $K_{tr}\ \tau$. Once no adjustments have been made for a further period τ, the record can be deleted.

4.3 Memory and noise reduction

Thus far, only uniform[9] schemes have been considered. Significant benefit can be obtained, however, by using non-uniform measurement arrangements.

Consider first the impact of using a single decaying store (equivalent to a discrete RC network) for measurements before the jump. The decaying store is updated once per inter-measurement time τ as

$$F_{stored}(T + \ \tau) - F_{stored}(T) =$$

$$(F_{meas}(\text{over} \ \ \tau) - F_{stored}(T)) \quad \tau/(\text{Integration time}) \qquad (13)$$

This has a number of advantages:

a) If we simply maintain the effective integration time, it reduces the required memory use per frequency source by 1/ [no dead time] or by 1/(r-1) [with dead time].

b) For noise types where the generation of random walk truncation noise is an issue (< 1), the decaying store carries less noise than a uniform average with the same integration time. Table 3 shows the idealised[10] bias function for some noise types with the decay symmetrical about the detection time. Note though, that for Flicker FM the practical improvement will be less than would be expected from these values, because of the implicit asymmetry between the uniform and decaying characteristics. Simulations with = 0.1 and 0.2 gave values of 0.876 and 0.879 respectively.

---

[8] Possibly stored as the difference from a running value to minimise memory usage

[9] i.e. schemes where the measurement intervals before and after the detection times are equal, and where weighting within the measurement intervals is uniform

[10] i.e. with no dead time and continuous sampling

Noise Type	White FM	Flicker FM	Random Walk FM
Allan Exponent	-1	0	1
Continuous B_D =	0.5	0.707	1
= 0.1 B_D	0.525	0.734	1
= 0.2 B_D	0.554	0.736	1

Table3 Variance bias multiplier B_D for symmetrical decay averaging (relative to Allan Variance)

c) It allows the integration time before the sample to be increased without storage penalty. This reduces the effective value of or r without delaying final measurement of a jump, though some of that benefit will be lost due to increased asymmetry. As an initial estimate, the impact of asymmetry has been calculated for uniform weighting in the measurement intervals. Values for various noise types are given in table 4.

The main disadvantage of using decaying stores appears to be that this complicates the algorithms[11] for discriminating between multiple peaks.

For 0, further improvement is available by instituting additional measurement following a detected jump[12]. Details of specific cases are given in sections 4.3 and 4.4

4.3 Optimising for White FM noise (μ 1), and for Random Walk FM noise (μ 1)

For white FM noise, measurement noise is minimised with uniform weighting through the longest practical measurement period. If storage is limited, a good strategy would be to combine uniform sensing after the detection region with a reference decaying store that has the longest practical integration time[13,14]. Once it has been determined that there has been a frequency jump, measurement may continue for an extended period to refine the data. The refined data may then be used to re-correct the post-jump frequency as described in section 3.2. However, the system needs to continue checking for new jumps, so additional measurement storage will be needed as this is used only for sources where a jump has been detected, the overhead will be quite small.

For Random Walk FM noise, the most accurate measurements are made at the shortest measurement times. As the error due to miscorrection grows in the same way as the background, this would not appear to be a problem.

4.4 Optimising for Flicker FM noise (μ)

The optimum integration weighting for jump detection with Flicker FM noise[15] is 1/(time difference from centre). For uniform weighting from 0 to t_1, and continuous 1/time distribution from t_1 to t_2

$$B_H \quad \ln(2)/\ln(1+(t_2/t_1)^{0.25}) \tag{14}$$

Simulation results for various time ranges using weighted averages of standard Allan variances are given in table 5. Clearly, the correlation between measurements over different periods means that quite sparse data can give near-optimum results. This means that the pre-jump reference can readily be made up from the sum of decaying stores with well-spaced integration times . Given the results with standard Allan samples, and the wider spectrum of the decaying store, a spacing factor around 16 might be expected to allow a good compromise between storage requirements, response time, and ultimate precision[16]. Post-jump weighting can be optimised empirically for the granularity of the stores. If we were to use three pre-jump decaying stores and eight sliding post-jump stores, I anticipate reducing the variance at the jump-decision point by a factor of two (relative to uniform sampling). Post-decision processing, using three extra pre-jump stores and a shared additional store[17], should allow measurement variance to be reduced by a further factor of two. In principle, these improvements would allow reduction of the absolute truncation level at the same time as reducing the frequency at which spurious corrections are allowed to remain. Unfortunately, there are no current plans to develop optimum strategies for such schemes.

5 Conclusions

Schemes have been proposed that will allow frequency jumps and changes in aging to be detected and corrected when using ensembles of similar oscillators, even during the absence of a more stable reference. Some of the detail has been filled in for detecting and correcting frequency jumps. In addition, the work of previous authors on aging prediction may be adapted to suppress aging changes.

This work holds out the promise that arrays of inexpensive compact crystal oscillators may suffice where formerly a supplementary atomic reference would have been needed.

References

[1] D. W. Allan, Time and frequency characterization, estimation, and prediction of precision clocks and oscillators IEEE Trans. UFFC-34 pp647-654 (1987)

---

11 Mainly by requiring the decay value to be reset. It is probably best to reset it to the post-jump measurement in the uniform store before resuming the decay algorithm.

12 Clearly, this requires additional storage, but only for the jumping oscillators.

13 The variance of this jump measurement being simply the sum of the variances before and after the jump

14 Although not optimal for other values of , extending measurement time can provide arbitrarily small errors for any <0.

15 ignoring end corrections

16 The variance can be proportional to $1/(N_{prejump})^{(2-}$, 0 as separation

17 Unlike the sliding stores, data weighting is time-dependent as it is loaded into this store.

[2] J. A. Barnes, Tables of bias functions B1 and B2 for variances based on finite samples of processes with power law densities NBS Tech. Note 375 (1969)

[3] S. R. Stein, Kalman filter analysis of precision clock with real-time parameter estimation Proc 43 Annual Symposium on Frequency Control, pp232-236 (1989)

[4] W. Su, R. L. Filler, A new approach to clock modeling and Kalman filter time and frequency prediction Proc. 1993 IEEE International Frequency Control Symposium, pp. 331-334 (1993)

[5] J. R. Vig, Introduction to quartz frequency standards Proc 23rd Annual PTTI applications and planning meeting (1991)

Measurements without dead-time							
Storage requirements (frequency samples)		21	9	7	5	3	2
(inter-measurement time) / τ	0	0.095	0.182	0.222	0.286	0.4	1
Threshold scaling =	0	0.003	0.016	0.027	0.053	0.148	0.333
Asynchrony bias @ 4 detection threshold	1	1.048	1.263	1.435	1.853	3.370	6.333
Asynchrony bias @ 3 detection threshold	1	1.027	1.148	1.245	1.48	2.333	4.000
Asynchrony bias @ 2 detection threshold	1	1.012	1.066	1.108	1.213	1.593	2.333
Measurements with dead-time							
r	1	1.1	1.25	1.333	1.5	2	N/A
$B_2(r,-1)$ [White FM] [2]	1	1	1	1	1	1	
$B_2(r,0)$ [Flicker FM] (simulation)	1	1.090	1.200	1.233	1.348	1.568	
$B_2(r,1)$ [Random walk FM] (1.5r - 0.5) [2]	1	1.150	1.375	1.500	1.750	2.500	

Table 2: Variance bias for uniform measurements with and without dead times.
Note: The table is arranged such that schemes requiring the same amount of data storage are vertically aligned

Time Ratio [t1/t2] [(t1-t2)/2/τ_{ave}]	1 0	1.222 0.1	1.414 0.172	1.5 0.2	1.857 0.3	2 0.333	2.333 0.4	3 0.5
White or Flicker PM (contiguous) B_H	1	1.051	1.155	1.215	1.534	1.688	2.098	3.111
White/Flicker PM (separated) B_H	1	1.031	1.093	1.128	1.316	1.406	1.644	2.222
White FM B_H	1	1.010	1.030	1.042	1.099	1.125	1.190	1.333
Flicker FM: for < 0.8, B_H	1	1.003	1.009	1.011	1.028	1.033	1.051	1.084
Random walk FM B_H	1	1	1	1	1	1	1	1

Table 4: Effect on variance of asymmetric measurement of frequency change
Bias function B_H for some Allan variance exponents
Note: The values for Flicker FM are taken from a fit to simulated values

Range = N_{max}/N_{min} (3)	1	2	4	8	16	32	64	128	256
spacing = 2:1 equal weights	-	0.890	0.794	0.712	0.644	0.585	0.531	0.484	0.447
spacing = 4:1 equal weights	-	-	0.759	-	0.689	-	0.502	-	0.427
spacing = 8:1 equal weights	-	-	-	0.649	-	-	0.491	-	-
spacing = 16:1 equal weights	-	-	-	-	0.597	-	-	-	0.425
ends only	-					0.555	0.532	0.518	0.510
2:1 optimum weight (2)	-		N. A.	0.648	0.585	N. A.	N. A.	N. A.	N. A.
approximate formula $\ln(2)/\ln(1+\text{Range}^{0.25})$	1	0.885	0.786	0.703	0.631	0.569	0.516	0.470	0.431

Table 5: Reduction in variance of Flicker-FM frequency-jump estimates using samples spaced 2^N.
Notes: 1. Values are from limited simulations, so are indicative only
2. Optimum weightings were determined empirically
3. White FM and Flicker FM give minimum variance with a single sample. For White FM this should be at the longest available time, and for Flicker FM it should be at the shortest [so the limit on accuracy is determined by measurement at intermediate sample times]

POWER LAW NOISE IDENTIFICATION USING THE LAG 1 AUTOCORRELATION

W.J. Riley* and C.A. Greenhall†

* Symmetricom, Inc., 34 Tozer Road, Beverly, MA 01915 USA, wriley@symmetricom.com
† Jet Propulsion Laboratory, 4800 Oak Grove Drive, Pasadena, CA 91109 USA, Charles.Greenhall@jpl.nasa.gov

Keywords: Power Law Noise, Autocorrelation, Frequency Stability Analysis.

Abstract

This paper describes a new method for power law noise identification, based on the lag 1 autocorrelation function, that can determine the dominant noise type for all common noise processes, from phase or frequency data, for all averaging factors, in a consistent and analytic manner.

1 Background

It is often necessary to identify the dominant power law noise process (WPM, FPM, WFM, FFM, RWFM, FWFM or RRFM) of the spectral density of the fractional frequency fluctuations, $S_y(f) = h_\alpha f^{\alpha}$ ($\alpha = 2$ to -4), to perform a frequency stability analysis. For example, knowledge of the noise type is necessary to determine the equivalent number of chi-squared degrees of freedom (edf) for setting confidence intervals and error bars, and it is essential to know the dominant noise type to correct for bias in the newer Total and Thêo1 variances. While the noise type may be known a priori or estimated manually, it is desirable to have an analytic method for power law noise identification that can be used automatically as part of a stability analysis algorithm.

2 Prior Art

There is little literature on the subject of power-law noise identification. The most common method for power law noise identification is simply to observe the slope of a log-log plot of the Allan or modified Allan deviation versus averaging time, either manually or by fitting a line to it. This obviously requires at least two stability points. During a stability calculation, it is desirable (or necessary) to automatically identify the power law noise type at each point, particularly if bias corrections and/or error bars must be applied. Previous methods for power law noise identification [1], based on the Barnes B_1 and R(n) bias ratios [2], have been ad hoc, have not used a consistent methodology for all noise types, and have not handled all cases (e.g. resolving white and flicker PM at unity averaging factor).

3 The Autocorrelation Function

The autocorrelation function (ACF) is a fundamental way to describe a time series by multiplying it by a delayed version of itself, thereby showing the degree by which its value at one time is similar to its value at a certain later time. More specifically, the autocorrelation at lag k is defined as

$$\rho_k = \frac{E[(z_t - \mu)(z_{t+k} - \mu)]}{\sigma_z^{\,2}} \tag{1}$$

where z_t is the time series, μ is its mean value, σ_z^2 is its variance, and E denotes the expected value. The autocorrelation is usually estimated by the expression

$$r_k = \frac{\dfrac{1}{N}\sum_{t=1}^{N-k}(z_t - \bar{z})(z_{t+k} - \bar{z})}{\dfrac{1}{N}\sum_{t=1}^{N}(z_t - \bar{z})^2} \tag{2}$$

where $\bar{z}$ is the mean value of the time series and N is the number of data points [3].

4 The Lag 1 Autocorrelation

The lag 1 autocorrelation is simply the value of r_1 as given by the expression above. For frequency data, the lag 1 autocorrelation is able to easily identify white and flicker PM noise, and white (uncorrelated) FM noise, for which the expected values are $-1/2$, $-1/3$ and zero respectively. The more divergent noises have positive r_1 values that depend on the number of samples, and tend to be larger (approaching 1). For those more divergent noises, the data are differenced until they become stationary, and the same criteria as for WPM, FPM and WFM are then used, corrected for the differencing. The results can be rounded to determine the dominant noise type or used directly to estimate the noise mixture.

5 Noise Identification Using r_1

An effective method for identifying power law noises using the lag 1 autocorrelation is based on the properties of discrete-time fractionally integrated noises having spectral densities of the form $(2 \sin \pi f)^{-2\delta}$. For $\delta < ½$, the process is stationary and has a lag 1 autocorrelation equal to $\rho_1 = \delta / (1-\delta)$ [4], and the noise type can therefore be estimated from $\delta = r_1 / (1+r_1)$. For frequency data, white PM noise has $\rho_1 = -1/2$, flicker PM noise has $\rho_1 = -1/3$, and white FM noise has $\rho_1 = 0$. For the more divergent noises, first differences of the data are taken until a stationary process is obtained as determined by the criterion $\delta < 0.25$. The noise identification method therefore uses p = -round (2δ) –2d, where round (2δ) is 2δ rounded to the nearest integer and d is the number of times that the data is differenced to bring δ down to < 0.25. If z is a τ-average of frequency data y(t), then $\alpha = p$; if z is a τ-sample of phase data x(t), then $\alpha = p + 2$, where α is the usual power law exponent f^{α}, thereby determining the noise type at that averaging time. The properties of this power law noise identification method are summarized in Table 2. It has excellent discrimination for all common power law noises for both phase and frequency data, including difficult cases with mixed noises.

6 Noise ID Algorithm

The basic lag 1 autocorrelation power law noise identification algorithm is quite simple. The inputs are a vector $z_1, \ldots, z_N$ of phase or frequency data, the minimum order of differencing dmin (default = 0), and the maximum order of differencing dmax. The output is p, an estimate of the α of the dominant power law noise type, and (optionally) the value of d.

Done = False, $d = 0$

While Not Done

$$\bar{z} = \frac{1}{N}\sum_{i=1}^{N} z_i$$

$$r_1 = \frac{\sum_{i=1}^{N-1}(z_i - \bar{z})(z_{i+1} - \bar{z})}{\sum_{i=1}^{N}(z_i - \bar{z})^2}$$

$$\delta = \frac{r_1}{1+r_1}$$

If d >= dmin And (δ < 0.25 Or d >= dmax)

$$p = -2(\delta + d)$$

Done = True

Else

$$z_1 = z_2 - z_1, \ldots, z_{N-1} = z_N - z_{N-1}$$

$$N = N - 1$$

$$d = d + 1$$

End If

End While

Note: May round p to nearest integer

The input data should be for the particular averaging time, τ, of interest, and it may therefore be necessary to decimate the phase data or average the frequency by the appropriate averaging factor before applying the noise identification algorithm. The dmax parameter should be set to 2 or 3 for an Allan or Hadamard (2 or 3-sample) variance analysis respectively. The alpha result is equal to p+2 or p for phase or frequency data respectively, and may be rounded to an integer (although the fractional part is useful for estimated mixed noises). The algorithm is fast, requiring only the calculation of one autocorrelation value and 1st differences for several times. It is independent of any particular variance.

7 Results

The lag 1 autocorrelation method yields good results, consistently identifying pure power noise for α = 2 to –4 for sample sizes of about 30 or more, and generally identifying the dominant type of mixed noises when it is at least 10% larger than the others. For a mixture of adjacent noises, the fractional result provides an indication of their ratio. For those reasons, and because it can handle all averaging factors, the new lag 1 autocorrelation method has replaced the $B_1/R(n)$ bias ratio method in Version 1.42a and higher of the Stable32 program [5].

8 Examples

Examples of the lag 1 autocorrelation method for power law noise identification are shown in Figures 1 and 2 for 100 sets of 1024 points of pure and mixed simulated noise, as generated by the Kasdin-Walter method [6], with approximately equal Allan variances (τ=1) summed for the mixed noises. Figure 4 shows a composite plot of overlapping Allan deviation and lag 1 ACF noise type for a pair of SAO VLG11B hydrogen masers. The noise type varies from white/flicker PM at short averaging times to more divergent random walk/flicker walk FM at longer averaging times. Figure 5 shows a composite plot of overlapping Allan deviation and Lag 1 ACF noise type for a Symmetricom Cs-III high performance laboratory cesium frequency standard vs. a Symmetricom MHM 2010 hydrogen maser. As is typical for such devices, it displays white FM noise out to an averaging time of several days before reaching a "flicker floor". Figure 6 shows a composite plot of overlapping Allan deviation and Lag 1 ACF noise type for a Symmetricom Model 8130 militarized rubidium frequency standard, again vs. a Symmetricom MHM 2010 hydrogen maser. It displays white FM noise before reaching a region of flicker and random walk FM noise at an averaging time of about 10^4 seconds.

9 Limitations

Before analysis, the data should be preprocessed to remove outliers, discontinuities, and deterministic components. Simulations using 100 sets of white FM noise of various sample sizes have shown that acceptable results can be obtained from the lag 1 autocorrelation noise identification method for N ≥ 32, where N is the number of data points, as shown in the Table 1. The table shows the percentage of estimates that differ from the expected α value of 0 by more than a half-integer noise type. This percentage includes whatever error is due to the simulated noise data itself.

N	%	α Range
32	16	-0.4 to +1.4
64	6	-0.4 to +0.7
128	1	-0.3 to +0.5
256	0	-0.2 to +0.3
512	0	-0.2 to +0.2
1024	0	-0.1 to +0.2

Table 1: Percent Incorrect Power Law Noise Estimates vs. Sample Size

It would be interesting to devise an improved lag 1 autocorrelation noise identification method (perhaps based on reflected data like the Total variance) that could work down to smaller sample sizes. The algorithm tends to produce jumps in the estimated alpha for mixed noises when the differencing factor, d, changes (although the alpha value when rounded to an integer is still consistent). As shown in Figures 3, this can be avoided by using the same d for the entire range of averaging times, at the expense of higher variability when a lower d would have been sufficient. Further research is encouraged to develop even better ways to identify power law noise.

10 Conclusion

This paper has described a method for power law noise identification based on the lag 1 autocorrelation function. It is a fast and effective way to support the setting of confidence intervals and to apply bias corrections during a frequency stability analysis.

References

1. D. Howe, R. Beard, C. Greenhall, F. Vernotte and B. Riley, "A Total Estimator of the Hadamard Function Used by GPS Operations", *Proc. 32nd Precise Time and Time Interval (PTTI) Systems and Applications Meeting*, November 2000, pp. 255-268.
2. J.A. Barnes, "Tables of Bias Functions, B1 and B2, for Variances Based on Finite Samples of Processes with Power Law Spectral Densities", *NBS Technical Note 375*, Jan. 1969.
3. G. Box and G. Jenkins, *Time Series Analysis, Forecasting and Control*, Chapter 2, Holden-Day, Oakland, California, 1976, ISBN 0-8162-1104-3.
4. P. Brockwell and R. Davis, Time Series: Theory and Methods, 2nd Edition, Eq. (13.2.9), Springer-Verlag, New York, 1991.
5. Stable32, Software for Frequency Stability Analysis, Hamilton Technical Services, 195 Woodbury Street, South Hamilton, MA 01982 USA, http://www.wriley.com.
6. N.J. Kasdin and T. Walter, "Discrete Simulation of Power Law Noise", *Proc. 1992 IEEE Frequency Control Symposium*, pp. 274-283, May 1992.

Noise Type	α	Phase Data* x(t)	d=0 ACF of Phase Data	Lag 1 Autocorrelation, r_1† d=0 x(t)	d=0 y(t)	d=1 x(t)	d=1 y(t)	d=2 x(t)	d=2 y(t)
W PM	2	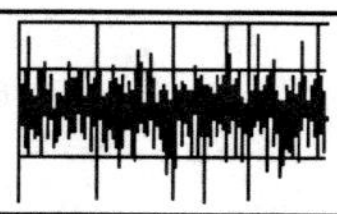	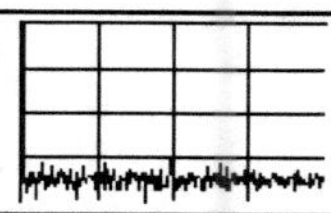	0	-1/2	-1/2	-2/3	-2/3	-3/4
F PM	1	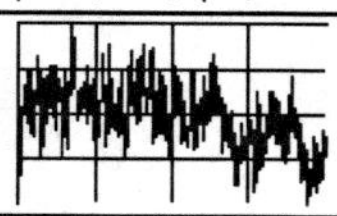	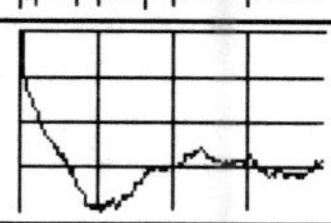	≈0.7	-1/3	-1/3	-3/5	-3/5	-5/7
W FM	0	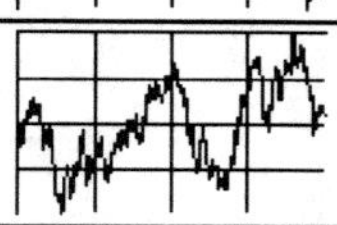	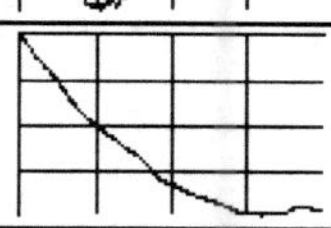	≈1	0	0	-1/2	-1/2	-2/3
F FM	-1	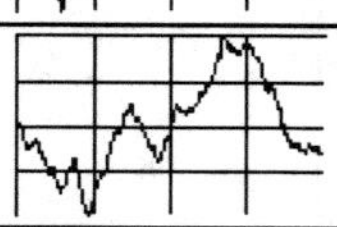	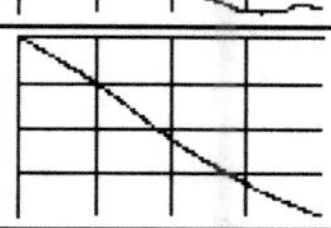	≈1	≈0.7	≈0.7	-1/3	-1/3	-3/5
RW FM	-2	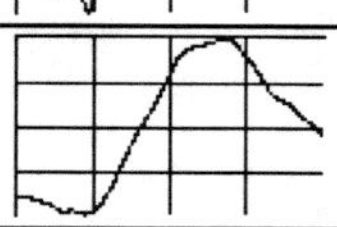	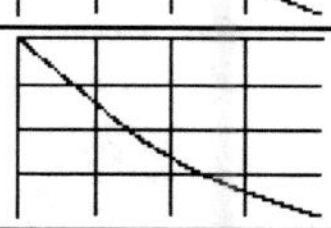	≈1	≈1	≈1	0	0	-1/2

* The differencing operation changes the appearance of the phase data to that shown 2 rows higher.
† Shaded values are those used for noise ID for the particular noise and data type

Table 2: Lag 1 Autocorrelation for Various Power Law Noises and Differences

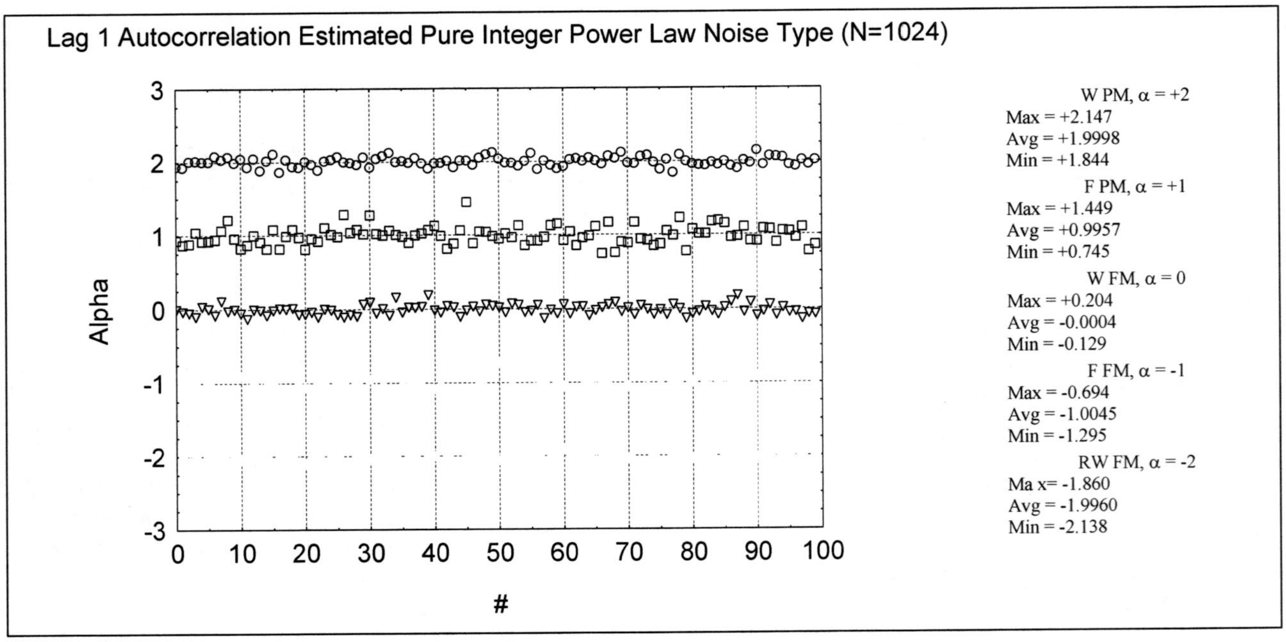

Figure 1: Examples of Lag 1 Autocorrelation Noise ID for Pure Integer Simulated Power Law Noises

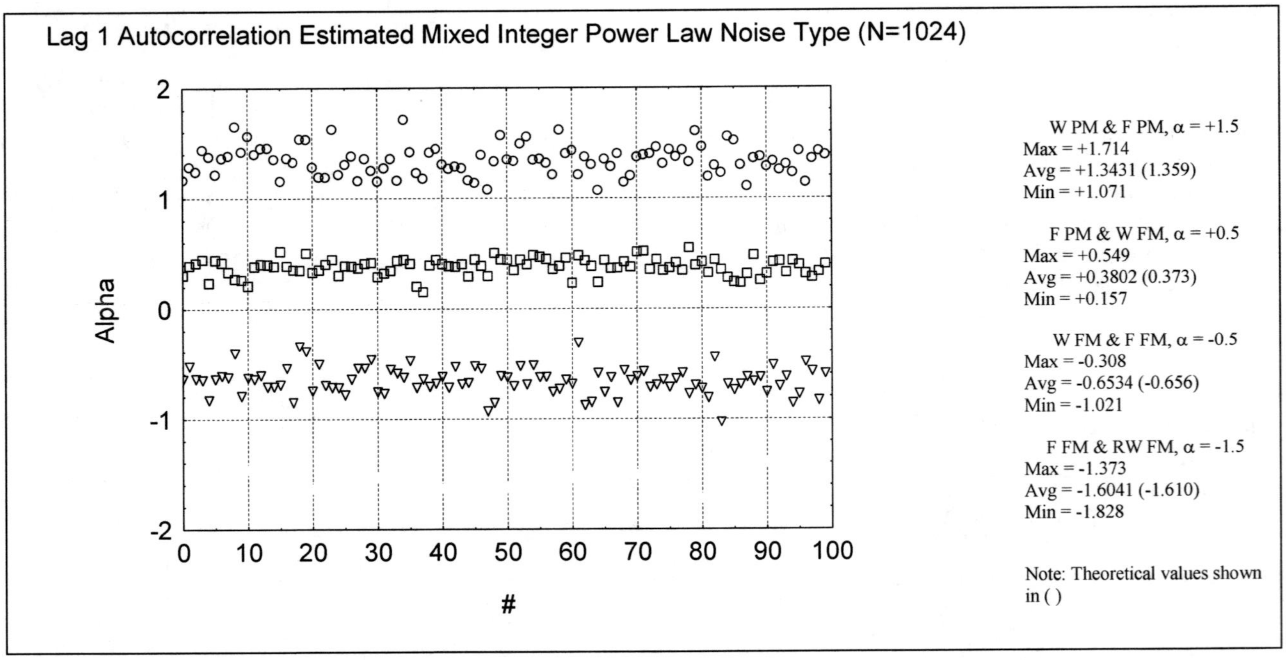

Figure 2: Examples of Lag 1 Autocorrelation Noise ID for 50% Mixture of Adjacent Integer Power Law Noises

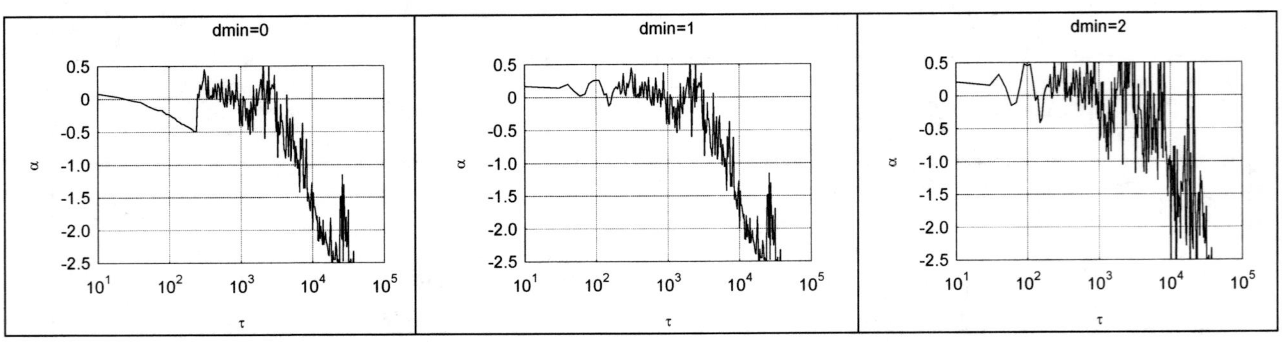

Figure 3: Noise Analysis of the Rubidium Frequency Standard of Figure 6 for dmin = 0, 1 and 2

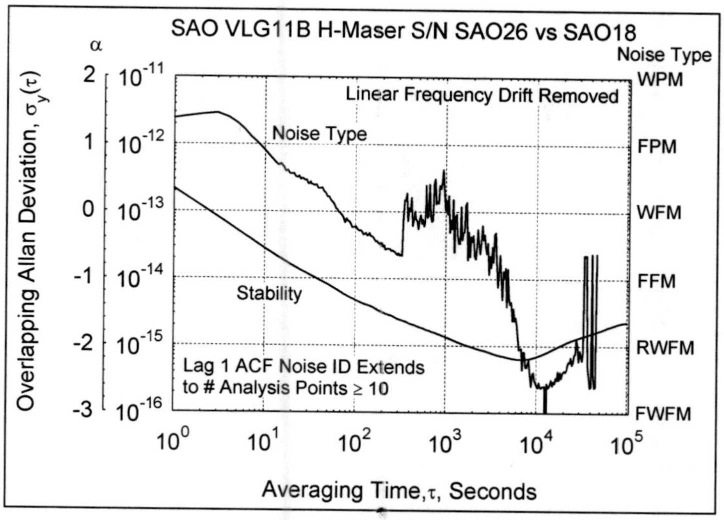

Figure 4: Frequency Stability and Noise Analysis of Two Hydrogen Masers

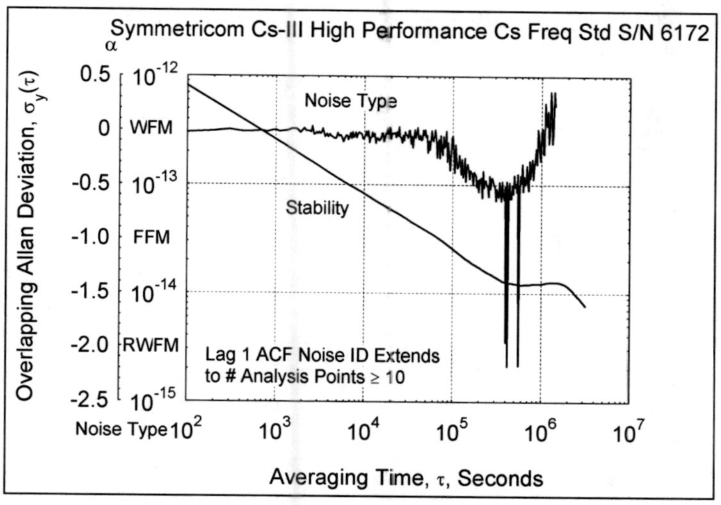

Figure 5: Frequency Stability and Noise Analysis of a Cesium Frequency Standard

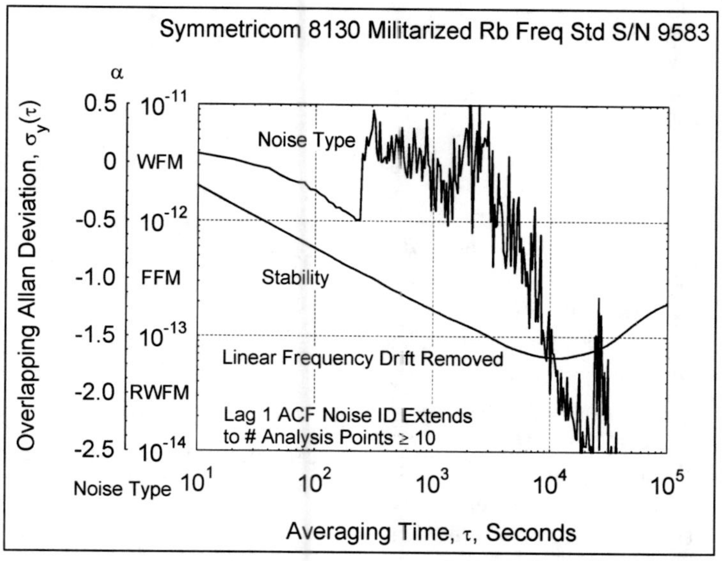

Figure 6: Frequency Stability and Noise Analysis of a Rubidium Frequency Standard

$\widehat{\text{THEO1}}$: CHARACTERIZATION OF VERY LONG-TERM FREQUENCY STABILITY*

D.A. Howe, T.N. Tasset

National Institute of Standards and Technology, 325 Broadway, Boulder, CO 80305 USA

Keywords: Allan Variance, Frequency Stability, $\widehat{\text{Theo1}}$, Total Variance, Oscillator Stability

Abstract

$\widehat{\text{Theo1}}$ is the first new species of variance that addresses a particularly difficult measurement problem, namely, obtaining reliable estimation of frequency stability for sample periods that are long compared to the length of a data run. $\widehat{\text{Theo1}}$ has statistical properties that are like the Allan variance (Avar), but $\widehat{\text{Theo1}}$ also has two advantages over other estimators of frequency stability: (1) it can evaluate frequency stability at a sample period (τ) of 3/4 the length of a data run, and (2) it presently attains the highest equivalent degrees of freedom (edf) of any estimator of frequency stability including Total-var and overlapping-Avar. $\widehat{\text{Theo1}}$ is unbiased relative to Avar for WHFM noise. $\widehat{\text{Theo1}}$ is biased slightly low with FLFM and RWFM, and we present a formula for a hybrid statistic (TheoH) made up of a combination of $\widehat{\text{Theo1}}$ and Avar in which bias is automatically removed. We explain the sampling function used in $\widehat{\text{Theo1}}$ and show that its frequency response is nearly ideal for extracting power-law noise processes of the types encountered with precision oscillators and clocks. We present results which, for a given data run, show how $\widehat{\text{Theo1}}$ anticipates the levels of frequency stability that are determined by Avar when given a longer data run from the same set of clocks.

1 Introduction

The primary strength of the Allan variance is its half-octave frequency response for a fixed τ_s, where τ_s is a "stride" described in Section 3. Avar's peak response is at reciprocal period of $f_p = \frac{1}{2\tau_s}$. A weakness of the Allan variance is that it cannot characterize frequency stability over an interval τ greater that one-half the length of the data run. For example, suppose we measure the time error between two clocks or oscillators, say, every couple of hours for one month. The maximum-overlap Allan variance estimator of frequency stability cannot report frequency stability for intervals longer than half the month, or two weeks [9, 15, 16]. By definition, a zero-dead-time average frequency difference for averaging interval τ cannot possibly extend beyond 50 % of the length of the data run T, that is, beyond $\tau = \frac{T}{2}$. Furthermore, this estimate is often too low. This is because the chi-square distribution function associated with an estimate composed of only one sample at $\tau = \frac{T}{2}$ (representing one degree of freedom) is so negatively skewed that it is twice as likely to be lower than above the FM noise level's true value [10, 23]. In addition, if a sample estimate of frequency drift is removed, Avar is likely to respond with levels too low at longest-term compared to the expected or true underlying characteristic level [10]. The overlapping estimator for the Allan variance has sufficiently good confidence at short- and medium-term τ averaging intervals but, to be conservative in light of the reasons just stated, it is not recommended for τ beyond 10 % of a data run T [15]. In the one-month example above, this amounts to only a three-day τ-average. In this situation, the best estimator of the Allan variance, which is the Total variance, or Totvar [7, 10, 15], is recommended. Use of the Total approach yields improved confidence between 10 % and 50 % of a data run, or up to two weeks in a one-month data run. At this writing, analysts in our field are confident of Totvar's properties. Easy-to-use 32-bit Windows software is commercially distributed that implements Totvar on large data sets, computes its confidence intervals, and automatically adjusts for bias [22].

It would seem preposterous to report a reliable estimate of frequency stability at a τ of <u>three weeks</u>, given a one-month data run, again considering the reasons stated, not to mention that this is theoretically impossible with the Allan variance! In this paper, we discuss $\widehat{\text{Theo1}}$, a special-purpose statistic that evaluates very-long-term frequency stability at τ between $\frac{T}{2}$ and T, is less susceptible to drift removal, and has a more symmetric distribution function than that of chi-square [6]. At this writing, the statistic has the highest confidence in estimating long-term frequency stability.

2 Sampling Function

Based on the experience gained from Totvar [5, 10, 12] we can manipulate frequency response while maintaining desirable statistical properties. The development of $\widehat{\text{Theo1}}$ involved the following issues: First, it is common practice to measure samples of the time-error function $x(t)$ between two oscillators and then derive frequency stability. For example, Avar is usually calculated as a normalized second difference of time-error measurements $\{x_n\}$. Measuring in this way assures Avar's statistical requirement for zero dead-time between average frequency differences [2]. Second, we desire a frequency response that efficiently extracts levels of common nonstationary FM power-law noise types [1, 14, 18, 19, 21, 24] while retaining simple, distinct straight-line mapping (on log-log plots) to $S_y(f)$, which is the recommended characterization of frequency stability [3]. Finally, we want to maximize equivalent degrees of freedom (edf) for a data run while minimizing bias relative to the conventional Allan variance. We can accomplish these goals by using most, and preferably all, of the available $\{x_n\}$ data, including small sampling interval $\tau_0 \ll T$. Starting with a sequence of time-error samples $\{x_n : n = 1, \ldots, N_x\}$ with a sampling period between adjacent observations given by τ_0, $\widehat{\text{Theo1}}$ averages every permissible squared second-difference of time errors in a given span or stride $\tau_s = 0.75 m\tau_0$ as shown in Figure 1. It is defined in terms of $\{x_n\}$ data by

$$\widehat{\text{Theo1}}(m, \tau_0, N_x) = \frac{1}{0.75(N_x - m)(m\tau_0)^2} \sum_{i=1}^{N_x - m}$$

$$\sum_{\delta=0}^{\frac{m}{2}-1} \frac{1}{(\frac{m}{2} - \delta)} \left[(x_i - x_{i-\delta+\frac{m}{2}}) + (x_{i+m} - x_{i+\delta+\frac{m}{2}})\right]^2, \quad (1)$$

for m even, $10 \leq m \leq N_x - 1$. The sampling functions of $\widehat{\text{Theo1}}$ are easier to understand intuitively in terms of fractional frequency measurements $\{y_n\}$ as shown in Figure 1, where $\{y_n\}$ is defined in terms of $\{x_n\}$ as

$$y_n(\tau) = \frac{x_n - x_{n-1}}{\tau}. \quad (2)$$

Figure 1 shows $\widehat{\text{Theo1}}$'s sampling of fractional-frequency

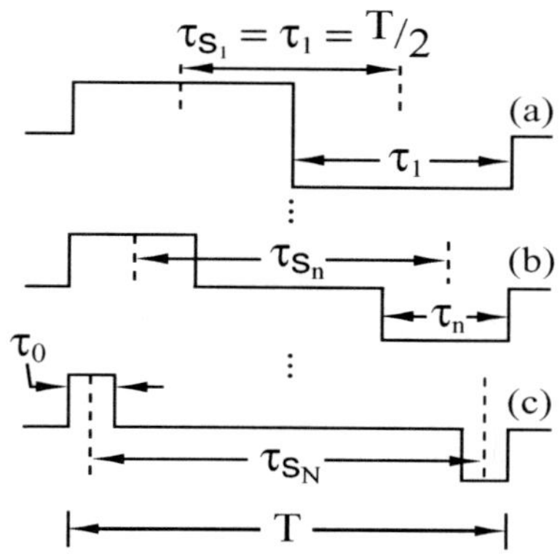

Figure 1: Sampling using $\widehat{\text{Theo1}}$ of fractional-frequency measurements $\{y_n\}$ at varying *stride* values, $\tau_{s1,s2,etc.}$.

measurements $\{y_n\}$. $\widehat{\text{Theo1}}$ computes frequency differences in interval T at varying *stride* $\tau_{s1,s2,etc.}$ and corresponding averaging time $\tau_{1,2,etc.}$ given by the inner summation in Equation (1). The summation's first term ($\delta = 0$) is the sampling in (a) which is that of the classical Allan variance. In this case, stride τ_{s1} equals averaging time τ_1, and both equal $\frac{T}{2}$. For $1 < \delta \leq \frac{m}{2}$, intermediate sampling functions are illustrated by (b) in which $\tau_{s(\cdot)} > \frac{T}{2}$. The summation's last sampling function is (c) in which $\tau_{s(N)} = T - \tau_0$. Therefore, the effective τ-value of the individual frequency differences averaged in $\widehat{\text{Theo1}}$ is between $\frac{T}{2}$ and $T - \tau_0$.

3 Bias

3.1 The Bias Function

Because of the novel data sampling of $\widehat{\text{Theo1}}$ (see Figure 1), there is an inevitable bias with respect to the Allan variance or "Avar". Bias in this instance is the ratio of the expected value of Avar to $\widehat{\text{Theo1}}$. $\widehat{\text{Theo1}}$ is formulated in order to be unbiased with respect to Avar in the case of the white FM (WHFM) noise type. A single bias value was previously reported for each of the five noise types [13]. However, it has since been found that the bias has a slight dependence on τ. The dependence can be described by the function

$$\text{bias}(\tau) = a + \frac{b}{\tau^c}, \quad (3)$$

where a, b, and c are constants. These constants are summarized for each of the five noise types in Table 1. The bias functions were fits to results of Monte Carlo simulations. Thousands of data sets of different lengths (up to 10^5) were used for computer-generated realizations of the five noise types listed in Table 1.

Table 1: Constant values for the bias functions of $\widehat{\text{Theo1}}$ defined in Equation (3).

Noise	a	b	c
WHPM	0.09	0.74	0.40
FLPM	0.14	0.82	0.30
WHFM	1	0	0
FLFM	1.87	-1.05	0.79
RWFM	2.70	-1.53	0.85

3.2 TheoBR and TheoH

A strategy we might follow for using $\widehat{\text{Theo1}}$ as an estimate for Avar at long values of τ is to calculate $\widehat{\text{Theo1}}$ and correct for the bias using the above table. However, this method assumes we know the noise type at a particular value of τ, which leads us to a difficulty at $\tau > \frac{T}{2}$. A useful method for determining noise type is the B_1 function [4]; however, this function is undefined beyond $\frac{T}{2}$ since Avar is undefined. Since B_1 does not exist for the longest τ values of $\widehat{\text{Theo1}}$, we cannot determine the noise type and hence correct for bias. We can estimate long-term noise types only by noting how much pre-whitening is required for a data run [20].

Another simpler strategy for using $\widehat{\text{Theo1}}$ as an estimator of frequency stability is to just remove a computed bias between $\widehat{\text{Theo1}}$ and Avar for a given data run. An unbiased version of $\widehat{\text{Theo1}}$, called TheoBR (for "Theo bias-removed"), can be written

$$\begin{aligned}&\text{TheoBR}(m, \tau_0, N_x)\\ &= \left[\frac{1}{n+1}\sum_{i=0}^{n}\frac{\text{Avar}(m=9+3i,\tau_0,N_x)}{\widehat{\text{Theo1}}(m=12+4i,\tau_0,N_x)}\right]\\ &\qquad\times \widehat{\text{Theo1}}(m,\tau_0,N_x), \quad (4)\end{aligned}$$

where $n = \lfloor \frac{0.1N_x}{3} - 3 \rfloor$ (where $\lfloor\cdot\rfloor$ denotes the floor function). In this equation, $\widehat{\text{Theo1}}$ is defined as in Equation (1), and Avar has its usual definitions as follows:

$$\begin{aligned}\text{Avar}\,(m,\tau_0,N_x) &= \frac{1}{2\,(m\tau_0)^2\,(N_x-2m)}\\ &\times\sum_{n=m+1}^{N_x-m}\left(x_{n+m}-2x_n+x_{n-m}\right)^2, \quad (5)\end{aligned}$$

for $\tau = m\tau_0$.

In order to get the most complete information over the entire data range, we define a hybrid frequency stability estimator called TheoH as a composite of Avar (m,τ_0,N_x) and TheoBR(m,τ_0,N_x), namely,

$$\begin{aligned}&\text{TheoH}(m,\tau_0,N_x)\\ &= \begin{cases}\text{Avar}(m,\tau_0,N_x), & \text{for } 1\le m<\frac{k}{\tau_0},\\ \text{TheoBR}(m,\tau_0,N_x), & \text{for } \frac{k}{0.75\tau_0}\le m\le N_x-1, \quad (6)\\ & m \text{ even},\end{cases}\end{aligned}$$

where k is the largest $\tau \le 10\%T$ where Avar (m,τ_0,N_x) has sufficient confidence. From Equation (4), TheoBR(m,k,N_x) = Avar (m,k,N_x); thus TheoH can be plotted vs. τ as one function with no discontinuity using (6); however, it must be noted that within the definition, Avar and TheoBR have different dependence on τ of $\tau = m\tau_0$ and $\tau = 0.75m\tau_0$ respectively. We note that deviation or square root of Equations (4)-(6) will be reported.

4 Properties of $\widehat{\text{Theo1}}$

4.1 Criteria for $\tau_s = 0.75m\tau_0$

Response of a statistic is the Fourier transform of its sampling sequence that, in some cases, can be nearly impossible to interpret in the time domain but easier to understand in the frequency domain [11]. Recall that it is desirable to maintain Avar's half-octave frequency response with peak at a reciprocal period of $f_p = \frac{1}{2\tau}$. The dashed line in Figure 2 shows the response of a constant-Q, half-octave pass-band filter considered to be ideal for extracting typical power-law noise levels [1, 14, 18, 19, 21, 24].

Frequency-response functions associated with $\widehat{\text{Theo1}}$ and Avar are shown in Figure 2. Prior to Equation (1), we obtained a high-edf, low-bias prototype variance, whose frequency response peak was shifted above $f_p = \frac{1}{2\tau_s}$. We found that if $\tau_s = 0.75m\tau_0$ and the amplitude of the response is adjusted by 0.75 (in the denominator of the amplitude coefficient of definition in Equation (1)), then the frequency response could be shifted to be precisely $f_p = \frac{1}{2\tau_s}$.

4.2 Response to Data Periodicity

Avar has deep nulls in its response to periodic or cyclical variations in $\{x_n\}$ at frequencies $f = \frac{\text{int}}{\tau}$, int = 1, 2, 3, ... , whereas $\widehat{\text{Theo1}}$ does not (see Figure 2). This means that the response of $\widehat{\text{Theo1}}$ to a periodic term in the data with frequency near $f = \frac{\text{int}}{\tau}$ is going to be more accurate than if Avar is used. In the end, $\widehat{\text{Theo1}}$'s frequency response is closer to the response of the ideal pass-band filter that Avar attempts to approximate. This closer approximation explains why $\widehat{\text{Theo1}}$ is so efficient in extracting power-law noise levels and types.

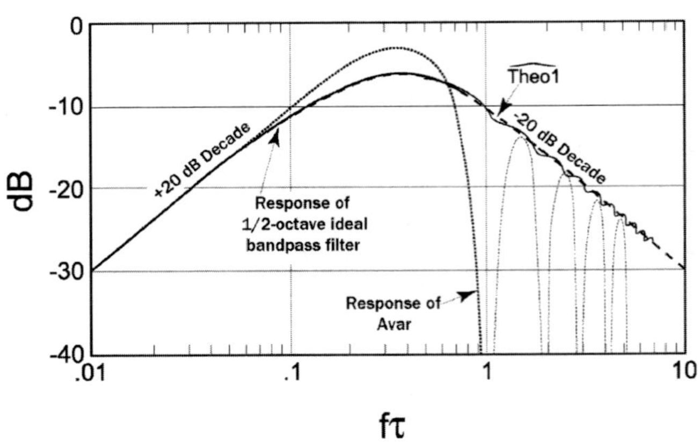

Figure 2: A comparison of frequency responses of $\widehat{\text{Theo1}}$, Avar, and a passband variance consisting of a simple cascade of a single-pole high-pass followed by a low-pass filter with identical break points at $RC = \tau/2$ [24].

5 Extraction of Noises and Other Errors

If the plot of frequency stability is not a straight line over a particular range (for example, it has nulls, "structure" or oscillations), then it is probably not a power-law process and we must try to estimate the spectral density $S_y(f)$ using another method such as a DFT or FFT in addition to using the $\widehat{\text{Theo1}}$-deviation plot. In general, any number of random noise components may be present in the data, depending on the type of test and reference oscillators being compared and the environment in which the data are obtained.

After using a method to extract the level of contribution for each component, relationships of $\widehat{\text{Theo1}}$-deviation slopes, various noise sources, and the corresponding noise level as previously discussed above are summarized in Table 2. Assuming that the computed $\widehat{\text{Theo1}}$-deviation is expressed in *fractional frequency fluctuation* and that τ is in seconds, then the expressions can be used for 1σ measures of level.

Table 2: Mapping of $\widehat{\text{Theo1}}$-deviation level and slope on log-log plot to the square root of noise spectrum, $\sqrt{S_y(f)}$, or rms fractional frequency fluctuations, $\frac{\Delta\nu}{\nu_0}$ of an oscillator pair, in a 1 Hz BW evaluated at Fourier-frequency f. QPM is quantized phase modulation. QSFM(f_0) means quasi-sinusoidal frequency modulation at f_0.

Noise Type	$\widehat{\text{Theo1}}$ Level a_μ	$\widehat{\text{Theo1}}$ Slope $\frac{\mu}{2}$	$\sqrt{S_y(f)}$ or $\left(\frac{\Delta\nu}{\nu_0}\right)_{rms}$ vs. f
QPM*	a_{-2}	-1	$\frac{1}{2\sqrt{3f_h}}\tau a_{-2}$
WH/FL PM*	a_{-2}	-1	$\frac{\pi}{\sqrt{3f_h}}\tau a_{-2}$
WH FM	a_{-1}	$-\frac{1}{2}$	$\sqrt{2\tau} a_{-1}$
FL FM	a_0	0	$\frac{1}{\sqrt{ln2}} a_0$
RW FM	a_{+1}	$+\frac{1}{2}$	$\frac{1}{\pi}\sqrt{\frac{6}{\tau}} a_{+1}$
Drift	a_{+2}	$+1$	$\frac{2}{\tau} a_{+2}$
QSFM(f_0)	$\frac{\Delta\nu}{\nu_0}\left(\frac{sin^2\pi f_o\tau}{\pi f_o\tau}\right)$	$-1(avg)$	$2\frac{\Delta\nu}{\nu_0}(f_o)$

*requires f_h (high-freq. cutoff), $2\pi f_h\tau \gg 1$.

6 Examples

It is well known that interpreting the longest-term frequency stability from a $\sigma_y(\tau)$ plot can be problematic, even misleading, as frequency fluctuations become increasingly nonstationary [8, 17]. It is desirable to see how readily we can identify integer power-law noise types and levels at long-term extremes using log-log plots of $\widehat{\text{Theo1}}$-deviation on actual data as compared with the same data processed with the best (max-overlap) sample Allan deviation $\sigma_y(\tau)$. Accurate noise typing at long term remains essential for at least the following reasons:

1. predicting the evolution of time-error of a single clock,
2. calculating confidence intervals,
3. distinguishing and estimating frequency drift,
4. properly correcting for measurement-system deadtime, if any,
5. optimally combining clocks or frequency standards in an ensemble to form a time scale.

$\widehat{\text{Theo1}}(m, \tau_0, N_x)$ formulated in Equation (1) has been coded into NIST's time-scale computer for testing purposes. Various time-difference $\{x_n\}$ series of clock and time scale comparisons are readily available, data which have been recorded over many years at NIST, and a representative set of example plots are given that illustrate how $\widehat{\text{Theo1}}$-deviation characterizes frequency stability for averaging times up to the full length of a data run. 90 % confidence limits are set according to the straight-line slope projected off the preceeding adjacent power-of-2 τ-value. For slope $\mu \geq -0.5$, FLFM noise is assumed, otherwise WHFM is assumed. Finally, the confidence on the last point of $\widehat{\text{Theo1}}$-deviation plots assumes RWFM if a straight-line does not fit within the confidence limits of the preceeding two shorter-term power-of-2 τ-values, for example, if the last three points oscillate beyond a straight line that is able to fit inside their 90 % confidence limits.

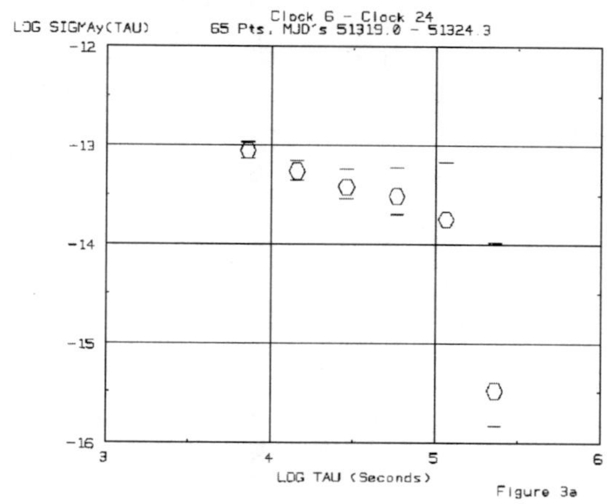

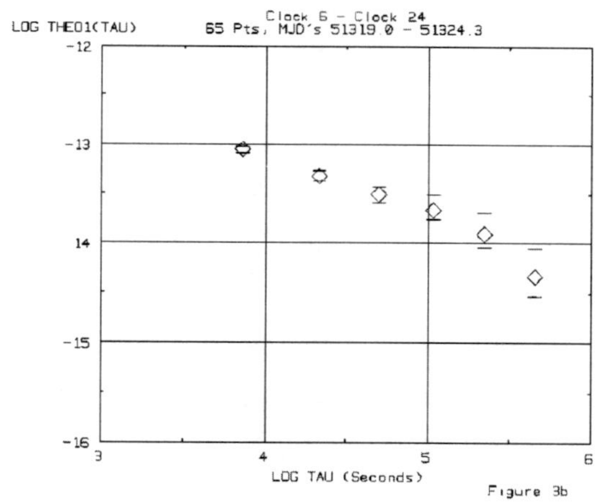

Figure 3: Two high-performance commercial Cs frequency standards in NIST's time scale taken over five days. The bottom plot, which estimates Theo1(τ)-deviation, can characterize fractional-frequency noise level and power-law noise type out to 3/4 of the length of the data run. $\widehat{\text{Theo1}}(\tau)$-deviation also has lower overall uncertainty compared to $\hat{\sigma}_y(\tau)$. Even though the last point drops, WHFM noise type is still supported within $\widehat{\text{Theo1}}(\tau)$'s uncertainties.

In addition to knowing the nature of the oscillators or time scales involved in a particular comparison, the narrower confidence limits of $\widehat{\mathrm{Theo1}}$-deviation are especially useful for singling out a likely underlying candidate power-law noise type. Table 2 allows us to relate the $\widehat{\mathrm{Theo1}}$-deviation amplitude to the amplitude of the spectral density of frequency noise modulation, namely $\sqrt{S_y(f)}$, arguably the most useful function involving frequency-standard noise modeling, oscillator synchronization or clock ensembling for example. Using $\widehat{\mathrm{Theo1}}$-deviation, we are able to confidently estimate spectral-density noise level and type at unprecedented low values of Fourier frequencies.

The first examples (figures 3 and 4) compare a $\sigma_y(\tau)$ plot and a $\widehat{\mathrm{Theo1}}$-deviation plot obtained from two commercial Cs standards for data runs of 5 days and 170 days. In both examples, the characteristic noise type is WHFM out to 3/4 of the 5-day and 170-day lengths of the data sets according to the $\widehat{\mathrm{Theo1}}$-deviation plot.

Figure 5 shows the evolution of frequency stability plots of two high-grade commercial Cs standards in the NIST time scale starting at the same time origin and advancing from one day to over two years of data. These are particularly interesting because the plots show $\widehat{\mathrm{Theo1}}$-deviation values beyond the last $\sigma_y(\tau)$ values and give some feel for whether $\widehat{\mathrm{Theo1}}$-deviation can anticipate longer-term $\sigma_y(\tau)$ values. To begin, figure 5-c shows that $\widehat{\mathrm{Theo1}}$-deviation indicates the onset of non-WHFM (nonstationary) long-term noise at $\tau \approx 85 - 150$, days whereas the $\sigma_y(\tau)$ plot does not to its last calculable value of $\tau = 85$ days. As the data run gets longer, figure 5-d shows that the $\sigma_y(\tau)$ indicates that flicker FM is occuring at this time. This would be reasonable as shown over the last four octaves of τ, out to its

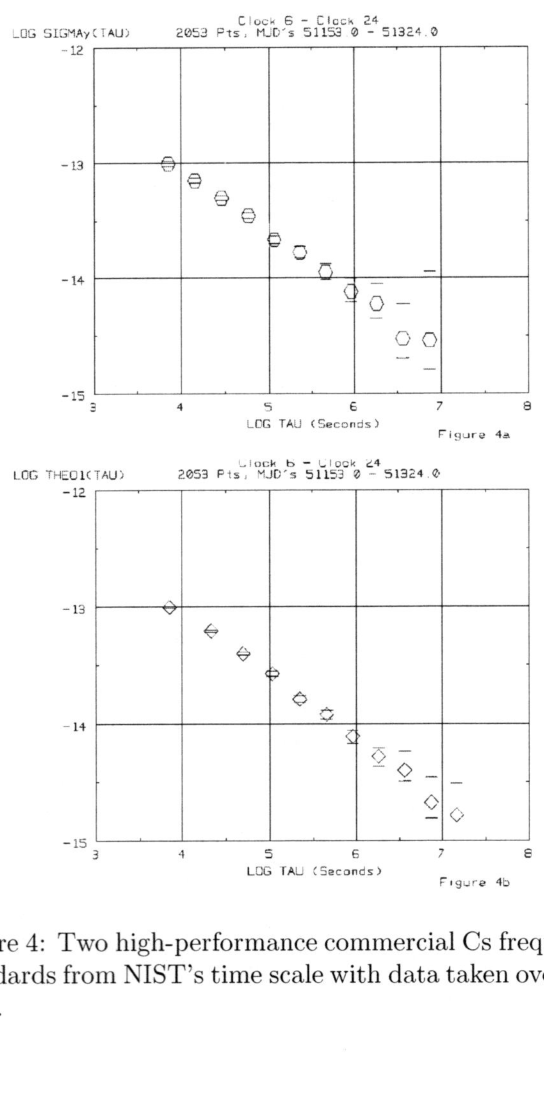

Figure 4: Two high-performance commercial Cs frequency standards from NIST's time scale with data taken over 170 days.

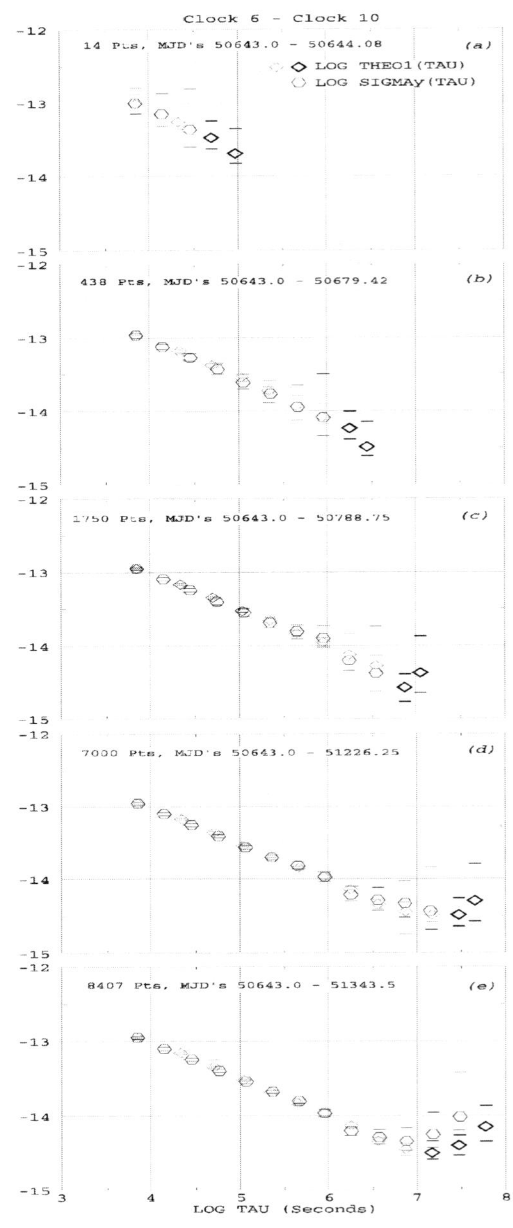

Figure 5: Frequency stability plots that start at the same time representing two high-performance commercial Cs standards in the NIST time scale for (a) one day, (b) 37 days, (c) 146 days, (d) 583 days, and (e) 701 days. The Allan deviation function is identified by the hexagons, and Theo1(τ)-deviation function is identified by the diamonds.

last $\sigma_y(\tau)$-value, corresponding to about 170 days, except that $\widehat{\text{Theo1}}$-deviation contradicts this hypothesis. In fact at the same time, $\widehat{\text{Theo1}}$-deviation supports a hypothesis of WHFM becoming RWFM, or possibly frequency drift, at $\tau \approx 170 - 450$ days since the slope is slightly steeper than RWFM $\propto \tau^{+\frac{1}{2}}$ but not quite as steep as Dr $\propto \tau^{+1}$. Full results using the remaining 120 days are shown in figure 5-e and indicate that $\widehat{\text{Theo1}}$-deviation continues to support long-term RWFM, and now, $\sigma_y(\tau)$ does as well.

7 Conclusion

$\widehat{\text{Theo1}}$ is effective to large τ-values, including 3/4 of the entire data run. This means that longest-term frequency stability can be obtained with only one-third more data-collection time. $\widehat{\text{Theo1}}$, like Avar and Totvar, is invariant to an overall shift in phase and frequency. $\widehat{\text{Theo1}}$, like Avar and Totvar, retains simple straight-line mapping (on log-log plots) to $S_y(f)$ for easily extracting the levels of the usual five FM power-law noise types by a linear-least-squares fit.

Acknowledgments

The authors thank Trudi Peppler of the National Institute of Standards and Technology for providing the clock data that is used in Figures (3)-(5).

References

[1] D. W. Allan, M. A. Weiss, and J. L. Jespersen, "A Frequency-domain View of Time-domain Characterizations of Clocks and Time and Frequency Distribution Systems," *Proc. 45th Freq. Cont. Symp.*, pp. 667–678, 1991.

[2] D.W. Allan, "Statistics of Atomic Frequency Standards," *Proc. IEEE* **54**, pp. 221–230, Feb. 1966.

[3] J. A. Barnes, A. R. Chi, L. S. Cutler, D. J. Healy, D. B. Leeson, T. E. McGunigal, J. A. Mullen, Jr., W. L. Smith, R. L. Sydnor, R. F. C. Vessot, G. M. R. Winkler, "Characterization of frequency stability," *IEEE Trans. Instrum. Meas.*, **IM-20**, pp. 105–120, 1971.

[4] J. A. Barnes, *Tables of Bias Functions, B1 and B2, for Variances Based on Finite Samples of Processes with Power Law Spectral Densities*, Natl. Inst. Stand. Technol. Technical Note NBS-375, 1969.

[5] C. A. Greenhall, D. A. Howe and D. B. Percival, "Total Variance, an Estimator of Long-Term Frequency Stability," *IEEE Trans. Ultrasonics, Ferroelectrics, and Freq. Control*, **UFFC-46**, no. 5, pp. 1183–1191, Sept, 1999.

[6] D. A. Howe and T. Peppler, "Characterization of Very Long-term Frequency Stability using $\widehat{\text{Theo1}}$," *Metrologia*, manuscript 1667, 2003, in process.

[7] D. A. Howe and T. Peppler, "Definitions of "Total" estimators of common time-domain variances," *2001 IEEE International Frequency Control Symposium*, pp. 127–132, June 2001.

[8] D. A. Howe and E. E. Hagn, "Limited Live-time Measurements of Frequency Stability," *Proc. 13th European Frequency and Time Forum* and *IEEE International Frequency Control Symposium*, pp. 1113–1116, April 1999.

[9] D. A. Howe, D. W. Allan, and J. A. Barnes, "Properties of Signal Sources and Measurement Methods," *Proc. 35th Freq. Cont. Symp.*, pp. 1–47 (1981; reprinted in [23]).

[10] D. A. Howe, "Total Deviation Approach to Long-Term Characterization of Frequency Stability," *IEEE Trans. Ultrasonics, Ferroelectrics, and Freq. Control*, UFFC-47 no. 5, pp. 1102–1110, Sept. 2000.

[11] D. A. Howe and C. A. Greenhall, "Total variance: a progress report on a new frequency stability characterization," *Proc. 29th Ann. Precise Time and Time Interval Meeting*, pp. 39–48, 1997.

[12] D. A. Howe, "Total Variance Explained," *Proc. 13th European Frequency and Time Forum* and *1999 IEEE International Frequency Control Symposium*, pp. 1093–1099, April 1999.

[13] D.A. Howe and T.K. Peppler, "Very Long-term Frequency Stability: Estimation using a Special-purpose Statistic", *Proc. 2003 Joint Mtg. IEEE Intl. Freq. Cont. Symp. and EFTF Conf.*, May 2003.

[14] D. A. Howe and D. B. Percival, "Wavelet Variance, Allan Variance, and Leakage," *IEEE Trans. Instrum. Meas.*, **IM-44**, pp. 94–97, 1995.

[15] "IEEE Std 1139-1999: Standard Definitions of Physical Quantities for Fundamental Frequency and Time Metrology—Random Instabilities," *IEEE-SA Standards Board*, J.R. Vig, chairperson, March, 1999.

[16] International Telecommunications Union, "Measures for Random Instabilities in Frequency and Time (Phase)", Recommendation ITU-R TF.538-3, 1994.

[17] J. Levine, "Time and Frequency Metrology," *Review of Scientific Instruments*, vol. 70, no. 6, June 1999.

[18] D. Percival and A. Walden, Spectral Analysis for Physical Applications, Cambridge University Press, New York, 1993.

[19] D. B. Percival, "Characterization of frequency stability: frequency-domain estimation of stability measures," *Proc. IEEE*, vol. 79, pp. 961–972, 1991.

[20] W.J. Riley and C.A. Greenhall, "Power Law Noise Identification Using the Lag 1 Autocorrelation," *18th European Frequency and Time Forum*, University of Surrey, Guildford, UK. 5 - 7 April 2004.

[21] J. Rutman, "Characterization of frequency stability: A transfer function approach and its application to measurements via filtering of phase noise," *IEEE Trans. Instrum. Meas.*, **IM-23**, pp. 40-48, Mar. 1974.

[22] Stable 32, Frequency Stability Analysis for Windows and NT, Hamilton Technical Services, 195 Woodbury Street, S. Hamilton, MA 01982, (978)-468-3703, http://www.wriley.com. The latest version contains $\widehat{\text{Theo}}1$. No endorsement is implied. Products are available from other manufacturers.

[23] D. B. Sullivan, D. W. Allan, D. A. Howe, and F. L. Walls (Editors), *Characterization of Clocks and Oscillators*, Natl. Inst. Stand. Technol. Technical Note 1337, 1990.

[24] R. G. Wiley, "A Direct Time-Domain Measure of Frequency Stability: The Modified Allan Variance," *IEEE Trans. Instrum. Meas.*, **IM-26**, pp. 38-41, Mar. 1977.

OPTIMIZATION OF DUAL-MIXER TIME-DIFFERENCE MULTIPLIER

L. Šojdr*, J. Čermák*, R. Barillet**

*Institute of Radio Engineering and Electronics, Czech Academy of Sciences, Chaberská 57, 182 51 Praha 8, Czech Republic
sojdr@ure.cas.cz, cermak@ure.cas.cz, fax: +420 84680222
**BNM-SYRTE, Observatoire de Paris, 61 avenue de l'Observatoire, 75014 – Paris, France
Roland.Barillet@obspm.fr, fax: + 33 (0)1 4325 5542

Keywords: Time and frequency metrology, frequency stability measurement

Abstract

The dual-mixer time-difference (DMTD) multiplication technique used in frequency stability measurement of precision frequency sources shows potential for achieving a frequency stability measurement floor as low as a few parts in 10^{15} at 1 s in terms of Allan deviation (ADEV) at 5 MHz at ambient temperature. The paper discusses the optimization of the DMTD multiplier using the frequency stability floor as a criterion for its efficiency. The instability sources of concern are all the noise sources as well as non linear effects and deterministic effects, such as phase noise in the common oscillator, noise in the mixers and amplifiers, inter-modulation products including the power-line pickup, electrical asymmetry (impairs the common-mode rejection of the phase and amplitude variations), phase dependence on environmental fluctuations, disturbing electromagnetic fields, and inappropriate construction (connectors, cabling, shielding, grounding). The optimization is demonstrated on the second version of DMTD multiplier designed at the Institute of Radio Engineering and Electronics (IREE). The IREE system has the "classical" DMTD structure, i.e. with a time-interval counter measuring the time difference between the zero crossings of the two beat-note signals. The modular construction allows modifications of the DMTD critical elements. The common oscillator as well as the time-interval counter are located outside the mixer-amplifier unit. The system can be operated at 5 and 10 MHz with the beat frequencies of 5 and 10 Hz, respectively, giving the multiplication factor of 10^6. The frequency stability measurement floor achieved through the optimization shows ADEV(τ = 1 s) = 6.9×10^{-15} at 5 MHz. At the basic sampling interval, τ_0 = 0.2s, the time deviation TDEV is 3.5 fs with the TDEV floor of 2 fs. This outstanding performance makes it possible to measure the frequency stability of future 5MHz quartz crystal oscillators of a few parts in 10^{-14} at 1s.

1 Introduction

Precise time and frequency applications require ultra-sensitive frequency/phase stability measurement of low-noise frequency sources, frequency conversion systems, time/frequency transfer elements etc. Since the measurement of short-term frequency/phase stability of the above systems is an integral part of time and frequency metrology, appropriate attention should be paid to it by metrology laboratories.

The IREE has recently decided to establish a special laboratory, within its Department of Standard Time and Frequency, that would allow high-precision short-term frequency/phase stability measurements in both the time and the frequency domains. The laboratory has been equipped with two ultra-stable 5 MHz Oscilloquartz 8600-BC5GE BVA oscillators each with ADEV(τ = 1 s) ≈ 8×10^{-14} and with a frequency-domain measurement system based on the FSS 1000E Femtosecond Systems assembly which makes use of the phase detection method and shows the background noise at 5 MHz: L(f)= -145dBc/Hz at f=1Hz and L(f)= -177dBc/Hz at f=100kHz [11].

Concerning the time domain measurement, it was decided to develop an ultra-low noise system based on the DMTD multiplier of IREE's own design. The first version of the multiplier achieved the stability floor of ADEV(τ = 1 s) = 4.2×10^{-14} at 5 MHz with 100 Hz low-pass cutoff, and it showed successful in comparison with similar time or frequency difference multiplication systems that have been tested at IREE [6].

The relative success of the first version was motivation to continue the work on the IREE DMTD system, which has resulted in its second version whose design and performance are discussed in this paper.

2 DMTD multiplication

2.1 Principle

The DMTD multiplication for the time-domain frequency-stability measurement has been known and used for years [1-8]. The principle is apparent from the block diagram in Fig. 1. Assume that the phases of the two sine-wave signals from oscillators 1 and 2 evolve as

$$\Phi_1(t) = 2\pi\nu t, \qquad (1)$$
$$\Phi_2(t) = 2\pi\nu\, t + \varphi(t), \qquad (2)$$

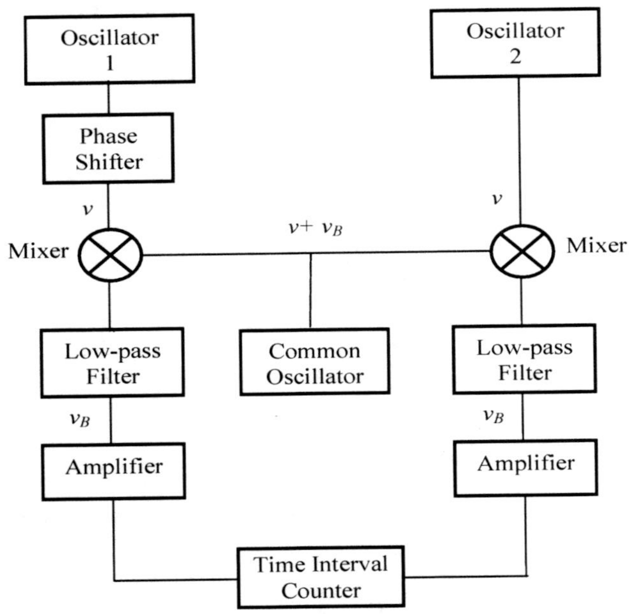

Fig.1: Block diagram of the DMTD multiplication

where ν is the frequency and $\varphi(t)$ is a time-dependent phase difference. Assume further that the signals can be made quasi-synchronous during the measurement, i.e. $|\varphi(t)| << 1$. The two signals are mixed with the sine-wave signal from the common oscillator, whose phase evolves as

$$\Phi_C(t) = 2\pi(\nu + \nu_B)t + \varphi_C(t), \qquad (3)$$

to provide two beat-note signals at frequency ν_B with the phases

$$\Phi_{B1}(t) = 2\pi\, \nu_B t + \varphi_C(t), \qquad (4)$$

$$\Phi_{B2}(t) = 2\pi\, \nu_B t + \varphi_C(t) - \varphi(t). \qquad (5)$$

The beat time difference between the zero crossings at instants t_0 and $t_0 + x_B(t_0)$ can be written in terms of phase differences as

$$x_B(t_0) = [\varphi(t_0) + \varphi_C(t_0 + x_B(t_0)) - \varphi_C(t_0)]/(2\pi\nu_B). \qquad (6)$$

Assuming that $\varphi_C(t_0 + x_B(t_0)) \approx \varphi_C(t_0)$, which obviously holds better for small x_B and stable common oscillator, we obtain

$$x_B(t_0) \approx \varphi(t_0)/(2\pi\nu_B). \qquad (7)$$

Since the "input" time difference at frequency ν is

$$x(t_0) = \varphi(t_0)/(2\pi\nu) \qquad (8)$$

we have

$$x_B(t_0) \approx m\, x(t_0) \qquad (9)$$

where $m = \nu/\nu_B$ is the multiplication factor.

2.2 IREE Design

The IREE DMTD design is in the main based on the "classical" architecture depicted in Fig.1. The block diagram of one channel is shown in Fig. 2.

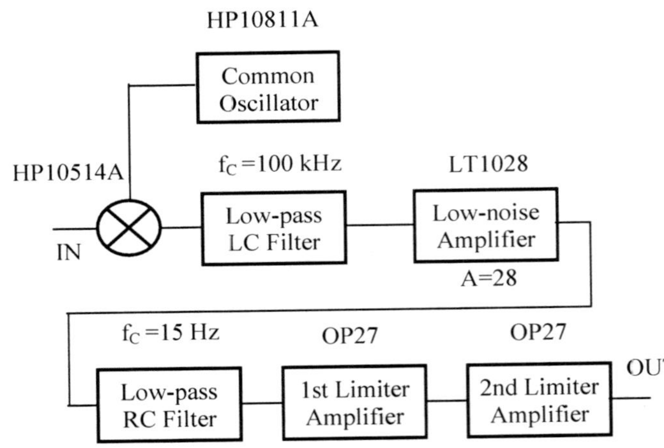

Fig.2: Block diagram of the IREE multiplier (one channel)

The common frequency source is a HP10811A quartz oscillator that provides 10 MHz at +7 dBm power. Part of the oscillator block is a built-in modulo-2 divider that provides 5 MHz at +11 dBm level. The oscillator frequency is offset by +10 Hz which gives $\nu_B = 5$ Hz at $\nu = 5$ MHz and 10 Hz at 10 MHz. Thus in both cases the multiplication factor is 10^6. The coefficient h_2 characterizing the white phase noise modulation in the frequency domain is 4×10^{-29} calculated from the L(f) floor of -153 dBc/Hz at 5MHz measured with the Femtosecond Systems equipment.

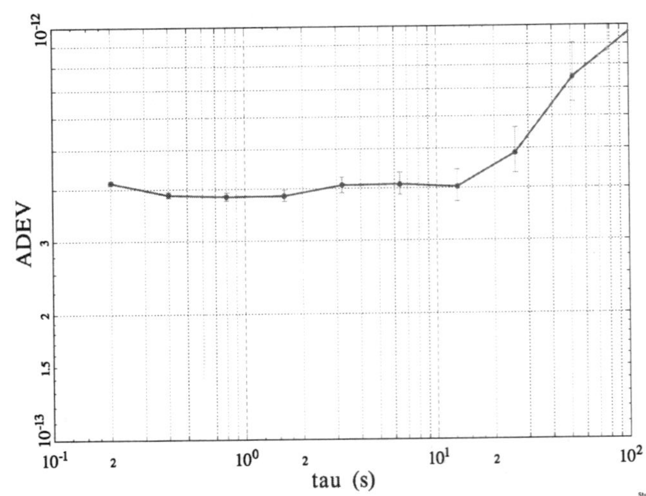

Fig.3: Frequency stability of the common oscillator.

The frequency stability of the HP10811A at 5 MHz is shown in Fig. 3. In both cases the reference was the 8600-BC5GE BVA oscillator whose contribution can be neglected.

The mixer is a double-balanced HP10514A. To our knowledge this model exhibits the lowest flicker phase modulation out of the mixers available at the market, most likely because it is assembled from separate diodes. The mixer output has a capacitive loading (22 nF) that increases the zero-crossing slope [2]. The low-pass filter following the mixer is a two-stage LC structure with the cutoff frequency around 100 kHz. It is used to efficiently attenuate the upper-band products. A single-pole RC filter (not shown in Fig. 2) with a cut-off frequency of 2.2 kHz is added right after the LC to compensate for imperfection of the L,C components.

The low-noise (LN) amplifier is a LT1028 connected in non-inverting mode (voltage gain A=28). The LT1028 amplifier exhibits a voltage noise of less than 1nV/√Hz, the 1/f corner frequency around 3 Hz, and the gain-bandwidth product (GBW) of 75 MHz. It is followed by a RC single-pole low-pass filter with a cut-off frequency of 15 Hz. In the version 2 the cut-off frequency can be switched to 140 Hz for experimental reasons.

The limiter amplifiers are the OP27 (Analog Devices). The 1st OP27 limiter amplifier is connected in non-inverting mode while the 2nd is in inverting mode. The OP27 exhibits the voltage noise of 3nV/√Hz, the 1/f corner frequency of 3Hz and GBW= 8 MHz . The DC voltage gains are set to 130 and 100, respectively. The 1st limiter amplifier contains a single-pole RC filter in the feed-back with a bandwidth of 160 Hz in the zero-crossing area. The 2nd limiter amplifier provides a trapezoidal waveform with a slope of ± 7×10^5V/s at the zero crossings.

3 Sources of instability

3.1 Noise from common oscillator

A comprehensive analysis of the rejection of the phase variations $\varphi_C(t)$ from the common oscillator has been made in [9]. It has been shown that the residual Allan deviation of the relative frequency deviation

$$y_C(t) = \frac{1}{2\pi(\nu + \nu_B)} \dot{\varphi}_C(t) \qquad (10)$$

takes the form

$$ADEV_{CR}(T_B, x_B) = [\int_0^\infty S_{yC}(f)\, G_C(f,T_B)\, G_R(f,x_B)\, df]^{1/2} \qquad (11)$$

where S_{yC} is the power spectral density of $y_C(t)$ and

$$G_C(f,T_B) = \frac{2\sin^4 \pi T_B f}{(\pi T_B f)^2} \qquad (12)$$

is the well-known power transfer function of the equivalent "filter" of $y_C(t)$ which corresponds to the algorithm for computing the Allan deviation. The rejection is achieved through the multiplication by the transfer function

$$G_R(f, x_B) = 2\,(1 - \cos 2\pi x_B f). \qquad (13)$$

Apparently, for x_B approaching zero also $ADEV_{CR}$ approaches zero.

Solving the integral (11) for the case of white phase modulation where $S_{yC}(f)=h_2f^2$, and for an ideal low-pass filter with the cutoff frequency f_C, and considering $x_B \ll T_B$, we obtain

$$ADEV_{CR} = [\frac{3h_2 f_C}{2(\pi T_B)^2}(1 - \frac{\sin 2\pi x_B f_C}{2\pi x_B f_C})]^{1/2}. \qquad (14)$$

Again for $x_B \rightarrow 0$ also $ADEV_{CR} \rightarrow 0$.

The theoretical values of $ADEV_{CR}$ calculated from equation (14) as a function of x_B for various f_c are shown in Fig. 4. The parameters h_2 and T_B are those used in the IREE DMTD system, i.e. $h_2 = 4\times10^{-29}$ (HP10811A oscillator) and T_B = 0.2 Hz. It should be noted that for large f_C and x_B, the value of $ADEV_{CR}$ becomes independent of x_B while the condition $x_B \ll T_B$ still holds.

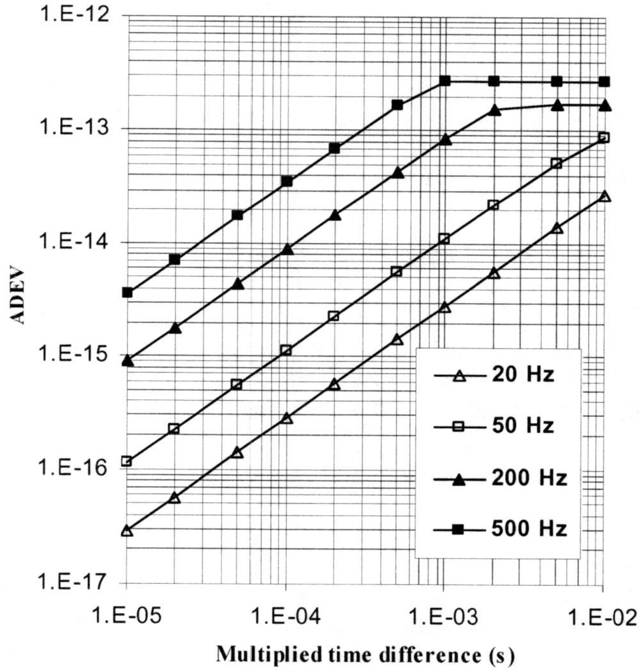

Fig.4: Theoretical plots of $ADEV_{CR}$ (x_B,f_C) .

For low and very low Fourier frequencies, where flicker frequency noise and random walk of frequency noise dominate the fluctuations, a limited expansion of the right hand side in (11) shows that the corresponding contributions in ADEV remain extremely low and do not affect the noise floor of the DMTD system.

Due to the system symmetry, the noise impact of the test oscillator is described by the same formulas as that of the common oscillator. Since the Oscilloquartz 8600-BC5GE used as the test oscillator has $h_2 = 1.3\times10^{-29}$ (about three times less than the common oscillator), its noise contribution can be neglected.

3.2 Noise from time interval counter

The multiplied time difference x_B is eventually measured with a time interval counter whose noise impact (primarily the trigger noise) on the DMTD multiplication must be evaluated. Experimental evaluation can be made by applying one of the beat signals to both start and stop counter inputs. A suitable delay in the stop input can be ensured by a small voltage offset in the triggering level.

The result in terms of TDEV(τ) obtained for the 7×10^5V/s slope and $T_B = \tau_0 = 0.2$ s with the SR620 counter used in the IREE DMTD system is shown in Fig. 5.

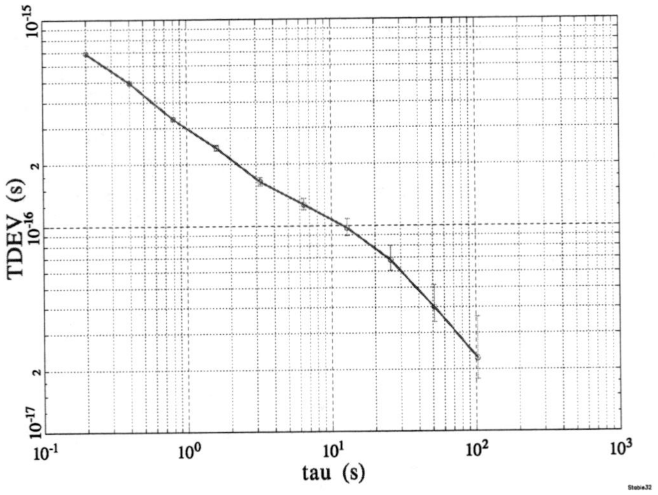

Fig.5: Time stability of the SR620 counter.

It is apparent from the TDEV(τ) slope that in the short run the counter shows mainly the white phase modulation. The corresponding frequency stability is ADEV(τ) = $1.2 \times 10^{-15} \tau^{-1}$. Thus the noise contribution of the counter is negligible.

3.3 Noise in mixers and amplifiers

In the configuration shown in Fig. 2, the overall noise includes two main components: the noise in the first elements of the channel with a narrow bandwidth, and the noise in the limiter amplifiers with a wide bandwidth. The first case concerns the proper noise of the mixers (specially the flicker phase modulation), the noise of the lossy elements used to filter the high frequency components at the output of the mixer, the noise in the resistors around the LN amplifier, and the proper (voltage) noise of the LN amplifier.

The wideband noise contribution is provided by the elements located after the low pass filter following the low-noise amplifier. Generally speaking, the noise contribution of all the elements located between the mixer and the counter is reduced by the gain K_d (in V/rad) of the mixer, which must be as high as possible.

The 1st limiter amplifier increases the zero-crossing slope and therefore it requires a higher bandwidth (in the linear zone). Experimental optimization of the bandwidth (B) can be made at this stage: narrowing the bandwidth without affecting the amplitude of the harmonics that allow a steep slope makes the time noise decrease. However, excessive narrowing causes a less steep slope and an overall increase in time noise. The same applies to the 2nd limiter amplifier.

If the slope of the output signal of the 2nd limiter amplifier reaches the amplifier's slew rate (2.8×10^6 V/s in OP27), an additional non-linear effect may take place which can reduce the system background noise but also the fluctuations of interest that are to be measured. Currently, the slope at the output of the last op amp is 7×10^5V/s and this effect need not be considered.

3.4 Interference

Interference signals belong to important degradation factors of the multiplier performance. At this extremely-low noise level one should expect to encounter a variety of mechanisms for interference many of which are difficult to identify and quantify. Apparently the more perturbing is the interference that takes place only in one channel and thus cannot be rejected by the symmetry of the setup. Difficult to identify is the interference involving unstable sources since their effect is not stationary.

One of the known mechanisms is intermodulation in the mixers due to their imperfect (i.e. not quadratic) mixing characteristics. Intermodulation may also take place in the amplifiers. Obviously, of concern are all the intermodulation products

$$\nu_S = r_1 \nu_1 + r_2 \nu_2 + \dots, \quad r_i = 0, \pm 1, \pm 2, \dots \qquad (15)$$

that fall in the system bandwidth. In fact their first contribution improves the performance since they increase the gain of the mixer. They do not directly provide spurious fluctuations, but they may provide an unexpected beat note with a disturbing signal. The frequencies ν_1, ν_2, ... may come not only from the oscillators involved in the DMTD system, but also from other frequency sources in the laboratory including the power-line frequency.

Fig. 6 shows an example of interference with a stable source we have experienced in one of our measurements. It turned out that the source of interference was another 5 MHz BVA oscillator which happened to be offset by 3.8×10^{-9} or 19mHz, and which was located close to the test BVA oscillator. The interfering oscillator had an unterminated coaxial cable of about 1 m length connected to its output and the interference disappeared only after the cable had been removed.

Interference may cause an undesired phase lock by an injection locking process between the measured oscillators. Fortunately these undesired phase locks can be easily detected. The more dangerous are near-false phase locks where the phase is only disturbed.

The use of battery supplies for all DMTD blocks (test oscillator, common oscillator and the mixer-amplifier unit) is a natural remedy against power-line pickup but there remains always a portion of power-line pickup that comes from the radiated low frequency magnetic fields, that can be shielded only by magnetic packages.

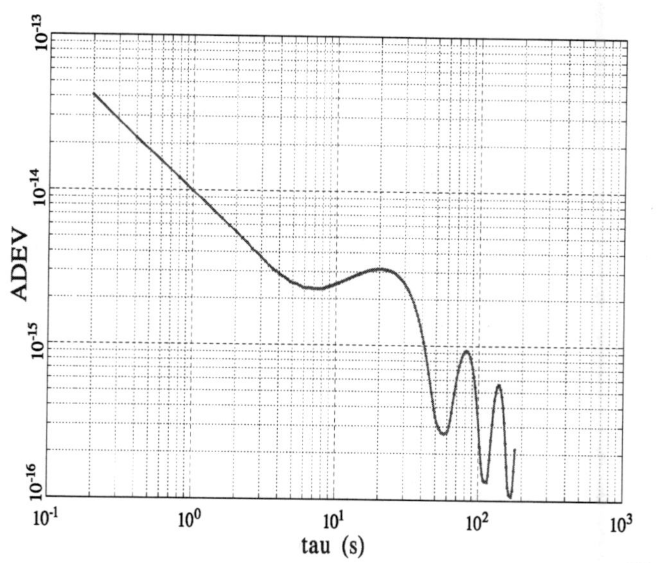

Fig.6: Interference with a nearby oscillator.

3.5 Environment

It is apparent that such a sensitive electronic system is affected by the environment. One would expect that vibrations and electromagnetic disturbances will have an impact on short-term stability while changes in temperature will rather affect the long-term performance. The problem is that the short-term manifestation of the environment can hardly be discerned from other short-term variations and the long-term effects can be masked by large-period spurious signals. Also the question arises about what kind of fluctuations we obtain when translating the temperature variations into DMTD.

In Fig. 7 we can see the background frequency stability (ADEV) and the corresponding time stability (TDEV) of the IREE DMTD system calculated from the longest available measurement. The right-hand slopes are roughly similar to the effect of a random-walk frequency noise. With possible deterministic effects and with the limited data length, however, we can just speculate about the cause of the long-term impairment: most likely it is not a random-walk flicker frequency modulation but rather a manifestation of some spurious signal with a period larger than 2000 s (compare the character of the ADEV plot with that in Fig. 6), or a spurious thermal effect not identical in the two channels.

3.6 Construction

Shielding, grounding, connectors, cables, printed circuit design etc. all may have an impact on the multiplier noise performance. Here in particular the result depends on the designer experience and know-how. Because of the optimization, the version 2 is constructed so as to allow an access to critical elements of the system. This, on the other hand, remains a weak point of version 2 since we cannot yet use semi-rigid cables and, consequently, neither SMA connectors (we use common BNCs). In a number of measurements a low performance has been observed because of mechanical effects.

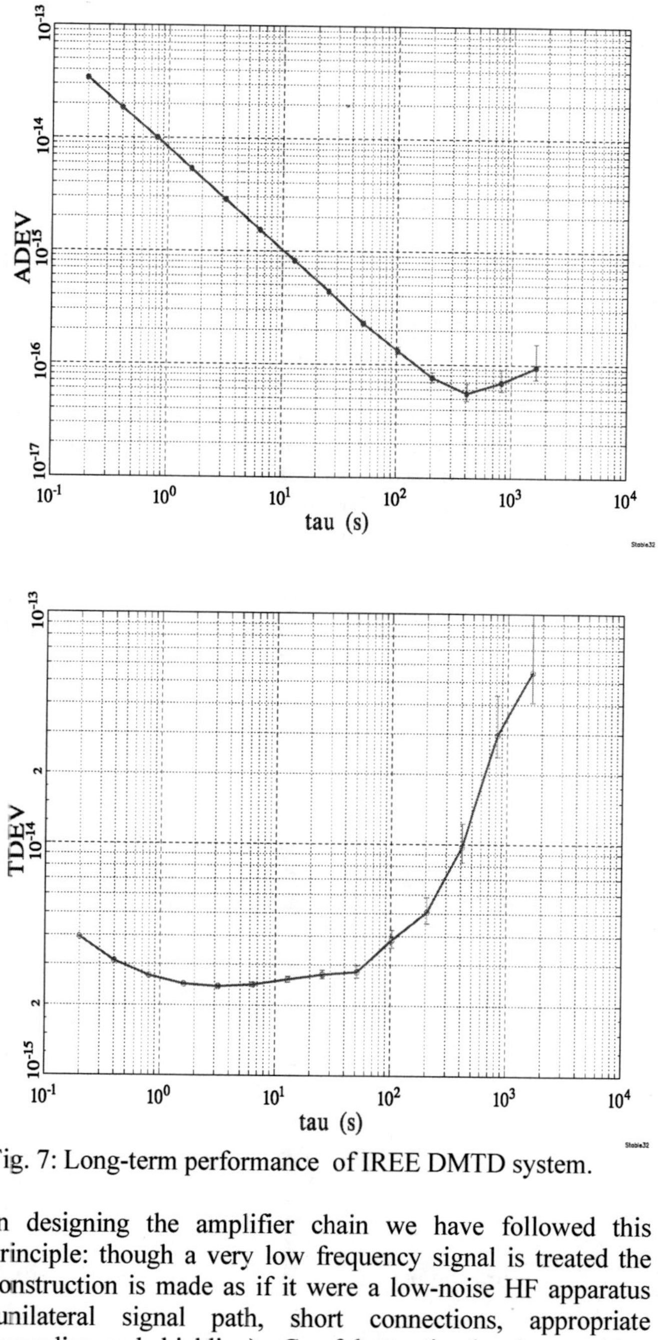

Fig. 7: Long-term performance of IREE DMTD system.

In designing the amplifier chain we have followed this principle: though a very low frequency signal is treated the construction is made as if it were a low-noise HF apparatus (unilateral signal path, short connections, appropriate grounding and shielding). Careful attention has been paid to avoiding parasitic feedbacks and the leakage of RF signals from the mixer. Also, the two DMTD channels must be prevented from cross-talks (separate mounting, separate power supply).

4 Optimization

4.1 Methodology

The term "optimization" reflects a dual approach: the conceptual approach which is not based on the classical overall noise figure since the optimum way to operate the DMTD multiplier involves non-linearity, but on considering several noise processes (with various bandwidths) and components whose overall impact must be understood and minimized [2]. A similar problem exists in heterodyne receivers, where the noise performance does not depend only on the noise figure of the first amplifier.

The engineering approach complements the conceptual approach and is based on gradual adjustment of the system in order to practically achieve the best noise performance. This is assessed by the magnitude (and character) of the fluctuations in the multiplied time difference $x_B(t)$ in the time domain by using the standard statistical measures: Allan deviation (ADEV) for frequency stability and time deviation (TDEV) for time stability. The criterion for whether the optimization step was good or bad was the "standard" value of ADEV at 1 s. By employing robust data the uncertainties obtained are very small. For calculations and simulations we have made use of Stable32 [10].

Throughout the optimization the test signal was generated from the same output of the 8600-BC5GE (sn 315) oscillator via a T-connector to both multiplier inputs.

4.2 Results

All the measurements mentioned below have been made at ν = 5 MHz , T_B = 0.2 s (i.e. m = 10^6) and f_C = 15 Hz (- 3dB cut-off frequency of the dominant single-pole low-pass RC filter after the LN amplifier – see Fig. 2). The value of T_B is about the minimum obtainable due to the limited tuning range of the common oscillator.

In what follows only reproducible steps that lead to smaller ADEV(1s) are reported. The steps intended to improve the performance but which have a negative impact (e.g. implementation of separation amplifiers in the common signal) are not mentioned. The value of the TDEV floor is about an average within the flicker-phase interval. Taking the minimum TDEV would give a misleading picture because of the wavy character of the flicker floor (see Fig. 8)

- The starting point was the first version described in [6] with

  ADEV(1s)= $4.2\text{x}10^{-14}$.

- In the first optimization step a number of changes have been made in the amplifying channel: different regime of LT1028 (linear, no band limitation, RC low-pass filter after it), decrease of low-pass cut-off frequency (from ≈ 100 Hz to 15 Hz), simplification of the output architecture (two OP27 used), adn separation of power supplies for the two channels. The result was:

  ADEV(1s)= $2.6\text{x}10^{-14}$, TDEV floor ≈ 8 fs

- Test signals fed through a simple T connector instead of (-3 dB) power splitter:

  ADEV= $2.2\text{x}10^{-14}$, TDEV floor ≈ 7 fs

- Capacitive loading (C=22 nf) used at mixer outputs:

  ADEV= $1.9\text{x}10^{-14}$, TDEV floor ≈ 5 fs

- Replacement of the SRA1 integrated mixers with separate-diode HP10514A mixers:

  ADEV= $1.1\text{x}10^{-14}$, TDEV floor ≈ 3 fs

- Optimization of the bandwidth of the 1st limiter amplifier (f_C=160 Hz):

  ADEV= $8.5\text{x}10^{-15}$, TDEV floor ≈ 2.5 fs

- Replacement of the 10 MHz Milliren MTI 250-0502A common oscillator with the 10 MHz HP10811 oscillator. The improvement, however, was not because of better oscillator noise performance (which is about the same) but because of the higher power level at 5 MHz obtained from the modulo-2 dividers, which gives a larger gain Kd of the mixers. Namely, the low-noise regenerative divider (RD) connected to the Milliren oscillator provides +7 dBm (which shows not sufficient) while the built-in digital divider (DD) in the HP10811 oscillator block provides +11 dBm. Thus since RD is by 7 dB less noisier than DD, a good solution would be to low-noise amplify the signal from RD. This could not be done in this version because of its construction. The result of the replacement change was:

  ADEV= $7.5\text{x}10^{-15}$, TDEV floor ≈ 2 fs

- Mechanical revision (fixing the connectors, cables, covers):

  ADEV= $6.9\text{x}10^{-15}$, TDEV floor ≈ 2 fs.

The frequency and time instability corresponding to the best result is shown in Fig. 8.

We have to realize that we are treating a very sensitive system which makes it possible to measure very tiny effects including those (unwanted) from environment. Therefore the performance is always slightly environment-dependent. Therofore it should be pointed out that the above results have been obtained in very favorable environmental conditions that the IREE's special metrology room for frequency stability measurement offers (located under ground, partly Faraday cage, no electronics active besides the measurement system).

At the same time, it should be noted that no additional precautions were taken : the measurements were performed during the working hours and common AC-DC power sources were used.

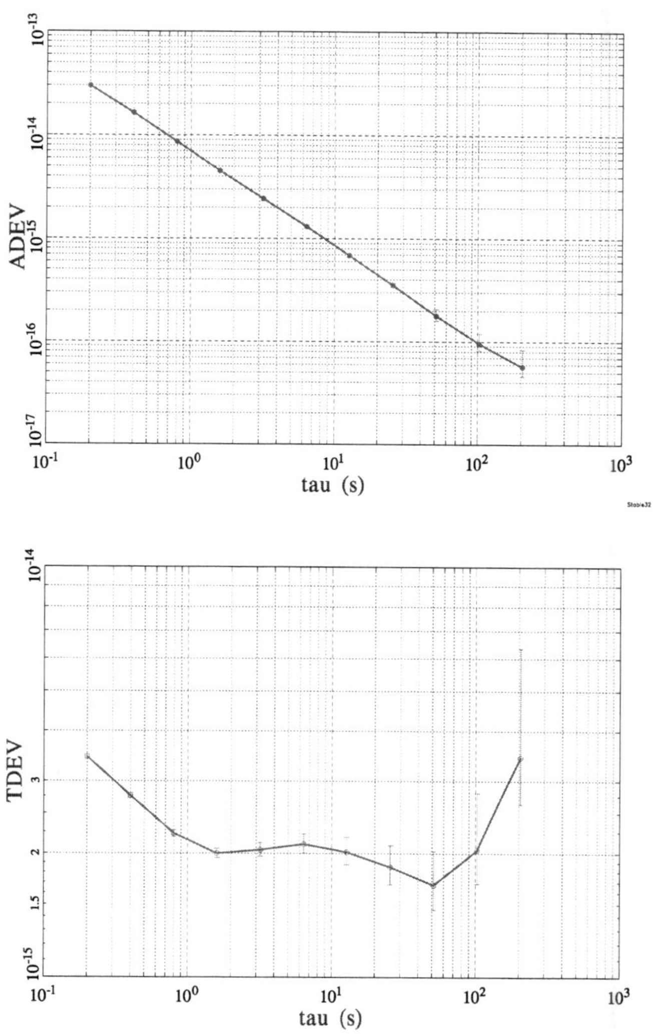

Fig. 8: Best measurement capability of IREE DMTD system.

5 Conclusions

The results have confirmed the excellent potential of the DMTD multiplication. The optimization has shown that an outstanding noise performance can be obtained from a relatively simple apparatus.

Concerning further investigation, it is of interest to determine the instability limits of this DMTD system, both long-term and short-term. It is naturally appealing to make use of the extreme sensitivity achievable with the DMTD multiplication also in other types of measurement (i.e. not only for frequency stability). This concerns primarily its possible use in accurate measurement of small time delays in electronic elements (connectors, cables etc.). A challenge would be to employ the DMTD system in the cases where the measured magnitude (temperature, vibrations) can be translated via an appropriate sensor into the change in time difference between the zero crossings of the input signals.

Acknowledgements

The Czech authors wish to thank Dr. Giorgio Brida of Istituto Elettrotecnico Nazionale "Galileo Ferraris", Turin, Italy for his valuable advice in the initial period of investigation and also Dr. Otokar Buzek of IREE for his comments and recommendations.

The work at IREE has been supported by the Czech Grant Agency (Project #101/02/0672). The collaboration between IREE and BNM-SYRTE takes place under the EUROMET Project #651.

References

[1] D.W. Allan, H. Daams. "Picosecond time difference measurement system", *Proc. 29th Annu. Symp. Frequency Contr.*, Atlantic City, USA, pp. 404-411, (1975).

[2] R. Barillet. "Comparateur de phase ultra faible bruit pour les futurs étalons de frequence", *Proc 3th European Time and Frequency Forum*, Besancon, France, pp. 249-254, (1989).

[3] G. Brida. "High resolution frequency stability measurement", *Rev. Sci. Instrum.*, **volume** 73, pp. 2171-2174, (2002).

[4] L.S. Cutler, C.L. Searle. "Some aspects of the theory and measurement of frequency fluctuations in frequency standards. *Proc. IEEE*, **volume** 54, pp. 136-154, (1966).

[5] C.A. Greenhall. "Common-source phase error of a dual-mixer stability analyzer", *TMO Progress Report 42-143*, Jet Propulsion Laboratory, Pasadena, XA, (2000).

[6] L. Sojdr, J. Cermak, G. Brida. "Comparison of high-precision frequency-stability measurement systems. *Proc. IEEE International Frequency Control Symposium and 17th European Frequency and Time Forum*, Tampa, USA, pp. 317-325, (2003).

[7] S.R. Stein. "Frequency and time – their measurement and characterization", in *Precision Frequency Control*, **volume** II, E.A. Gerber and A. Ballato, Eds. New York: Academic Press, pp.229-231, (1985)

[8] S. Stein, D. Glaze, J. Levine, J. Gray, D. Hilliard, D. Howe, L.A. Erb. "Automated high-accuracy phase measurement system", *IEEE Trans. Instrum. Meas.* **volume** IM-32, pp. 227-231, (1983)

[9] L. Sze-Ming, "Influence of noise of common oscillator in dual-mixer time-difference measurement system", *IEEE Trans. on Instr. and Meas.*, **volume** IM-35, pp. 648-651 (1986).

[10] "Stable32 version 1.35: frequency stability analysis," *Hamilton Technical Services*, S. Hamilton, MA 01982 USA, (2002).

[11] "FSS1000E Phase noise detector", Operation Manual Rev 3.2, Femtosecond Systems, Inc., p.4-6 (1999)

EXTRA HIGH PRECISION DIGITAL PHASE RECORDER

G. Kramer*, W. Klische[†]

*retired from PTB, [†]K+K Messtechnik GmbH, Braunschweig, Germany

Keywords: Phase comparison, frequency counter, frequency stability

Abstract

The paper describes a new type of instrument for measuring phase and frequency, that was presented for the first time in 2001 [2]. After recalling in the introduction the principles of operation, the paper explains in some detail the component elements and the latest enhanced versions of the instrument, which now allow the comparison of clock signals with sub-picosecond precision.

1 Introduction

The 'Multi-Channel Synchronous Digital Phase Recorder' (instrument name: FXM), that we have developed and are presenting here, is still a novel type of instrument. So a description will first have to explain the general principle of operation.

Basically, FXM is an electronic frequency meter that determines the frequency of signals by counting the periods occurring during a certain length of time, the 'gate' interval. 'Multi-channel' indicates that a number of such frequency counters work in parallel, 'synchronous' meaning that the gates of these frequency counters are all operated at exactly the same times. The readings of such synchronized counters will then represent frequency averages taken over one and the same interval of time.

The reason why we call it a phase recorder, reflects the fact that we are not really using gates, that would pass the input pulses during the gate interval only. Instead we count all periods, and we use strobe pulses to periodically transfer the contents of the counters to a computer, without interrupting the counting process and without resetting the counters.

This reading without stopping the counters makes the process equivalent to recording a (multi-dimensional) phase. If we were reading the digital state of the period counters only, the phase resolution would be 1 period. Such resolution of one period had been sufficient for recording the optical beat frequencies in the optical frequency measurement chain, for which the system was originally developed at PTB [1].

The read-out mechanism, however, now includes electronic circuitry for analog interpolation, which resolves the phase of the measured periodic signal to small fractions of a period. In the original FXM the resolution is 0.001 period, which for a 10 MHz signal corresponds to a time resolution of 0.1 ns.

FXM reads the phases of the periodic signals at the instant of particular, common, trigger events. We are using here a possibly unusual, but quite universal definition of the quantity 'phase':

Phase is the time-integral of frequency.

For constant frequency, phase linearly increases with time. Constant phase can be constant phase difference only.

If the signal is a sine wave: $u = \hat{u} \cdot \sin(\omega \cdot t + \varphi)$, for example, then the phase is the argument of the sine function. With sine-waves the radian may be the natural unit. But we normally do not use sine-waves. With pulse trains or square-waves the factor 2π would look artificial, and we prefer the period (cycle, repetition) as the unit of phase. Frequency accordingly is measured in Hertz rather than in radians per second.

The Principle of Synchronous Phase Readings

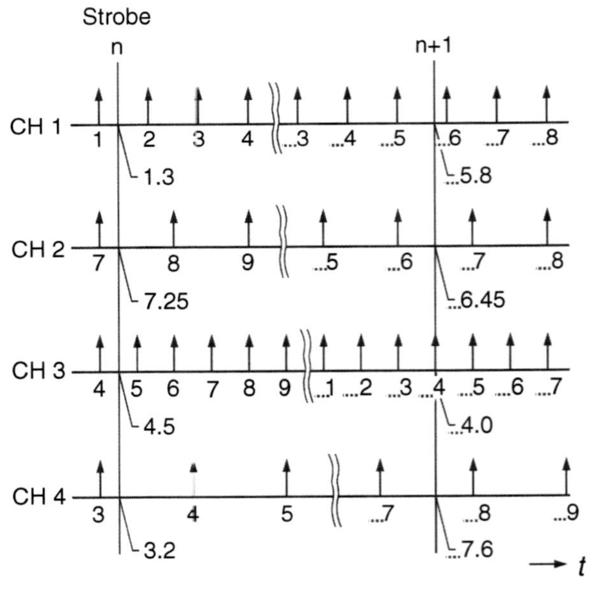

Figure 1: Four-dimensional strobed phase readings

In figure 1 we show what we mean by 'Synchronous phase readings':

The 4 horizontal lines represent the 4 signals at the 4 channels of the basic instrument. These signals all have different frequencies, normally in the range between 5 MHz and 50 MHz (FXM, FX80). The figure does not reproduce the wave forms like displayed on an oscilloscope. The signals can be sine-waves, pulses, or any other periodic function of time. The arrows mark 'events', i.e. the periodic event of a well defined phase, like a zero-crossing, or, even better, a logic level transition. We may imagine the events to be numbered.

The long vertical lines now stand for the strobe events. FXM determines the phases that the signals have reached at exactly the times of these read commands (strobes, trigger events). In a way, the strobes act like the gate of a normal frequency counter. Like in a normal counter they are derived from a reference frequency, and their rate can be selected between 1 Hz and 1 kHz .

Each point, where the lines cross, then defines a phase value. The apparatus converts these values to digital numbers and reports them to the computer. The numbers printed near the crossing points should illustrate what we mean, and how we define the quantity 'phase'.

The unit of phase is unity, one period or one cycle. And these periods are counted as integer numbers. With (quasi-) periodic signals the fractions are also easily and clearly defined by interpolation. As shown in **Figure 1**: between events #4 and #5 the phase is 4.5 for example.

We can say that the new phase recorder measures a coherent matrix (2-dim array) of phase values, which can be processed and evaluated, online or offline, in a great variety of ways. If, for example, the report rate is set to 1 Hz, then the difference between consecutive readings is the frequency in Hertz.

With a normal frequency meter, the user has to set the gate time first, in the set-up of the instrument. If later he finds that he would need longer term averages, higher resolution, or some other kind of analysis, then the measurement must be repeated. Not so if he has a coherent phase record in a computer file. (It does not actually matter, whether phase or coherent frequency values are stored in the data file. Coherence is preserved as long as the counters have no dead times and as long as the computer does not introduce rounding errors.)

The fact, on the other hand, that the phase values grow beyond the 32 bit capacity of the hardware counters, does also not pose a real problem. As long as the counters are read and their contents transferred to the computer at least once per second, the driver program can handle the overflows and without problem can hold and process integer numbers of virtually infinite magnitude.

The difference between phase values taken at one particular strobe, but at two different channels, is the relative phase of the two signals (or the phase of a virtual beat frequency).

Channel-to-channel coherence allows precise determination of phase difference (also of frequency ratio). For recording more than 4 channels several boards can be concatenated for recording an, in principle, unlimited number of channels, inter-channel bias is smallest and most stable though, if the channels are from the same set of four channels situated on a single PCB and in a single FPGA.

2 Elements of the Phase Recorder

2.1 Strobed Counter Read-Out

Figure 2 is the block diagram of a single FXM channel. **fx** is the channel input, the unknown frequency. The main novel features are in the "counter control" block

Beat frequency counter (One channel FXM)

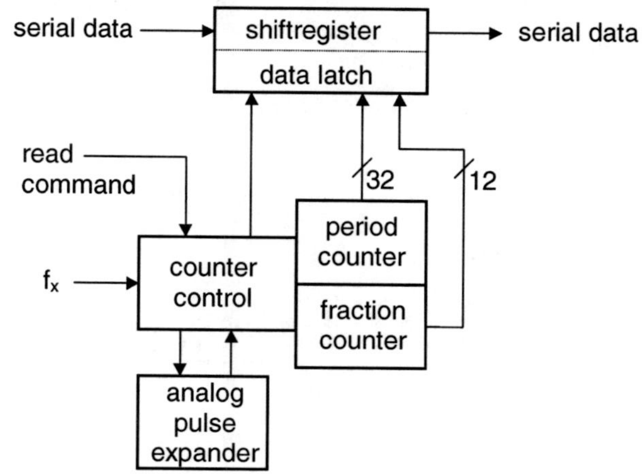

Figure 2: Function blocks of the phase recorder.

In a normal counter there would be a 'gate', which would open for a certain length of time to pass the fx pulses. Here these pulses are counted/accumulated continuously in the 32bit period counter. On a read command, the contents of this counter is latched in a register.

It is a well known problem that in a normal pulse counter (a so-called ripple counter), the states of the flip-flop registers do not at every point of time represent the number of clock pulses that have been applied at the counter input. Particularly in the moment when the counter overflows and the flip-flops have to change from all ones to all zeros, this switching process 'ripples' down the stages like a domino effect. If there are many stages, this 'ripple' time may easily fill a full period of the counted frequency. Only in a synchronous counter does there exist a phase, when the states of the flip-flops reliably represent the input pulse count.

A straight-forward synchronous counter is no solution however. It would require gates with a total of 500 inputs and still slow down the counting. Instead of using 'carry-look-ahead' or a similar approach, our counter control here solves the problem by synchronizing the whole reading procedure to the fx signal. On a read command, the control circuit simply cuts out an exact number of 16 pulses.

This is time enough for all flip-flops to settle (even in the worst case). The data is then latched, and the counter is advanced by 16, simply by adding one pulse into the appropriate counter stage.

The function of the 'fractions counter' and the pulse expander will be explained further down with reference to figure 3. (In the new FX80 the fractions counter has 15 bits, while 16 bits suffice for the period counter.)

We recall: fx is the signal individual to the channel, the read command is common to all channels, and the shift registers are daisy-chained to form one long record of the data of all channels. This record contains start bits and stop bits, so the data can be read directly at the COM port of a PC. The transmission time sets a limit to the measurement rate.

2.2 Analog Phase Interpolation

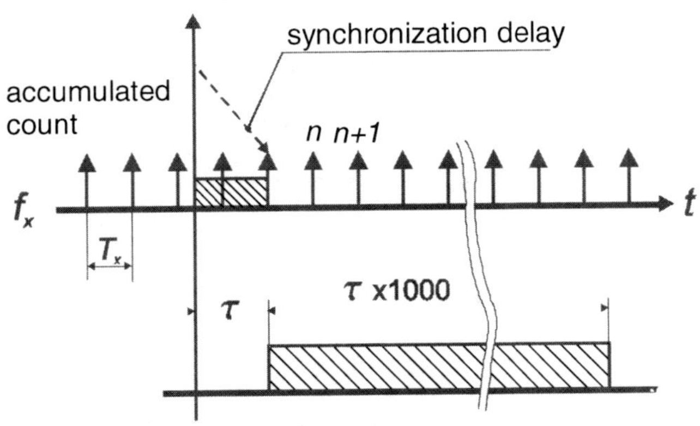

Figure 3: The sync. delay varies from $1 < \tau < 2$ fx-periods.

Figure 3 shows the principle of the phase interpolation. Where, in a standard resolution system without interpolation, the controller could only distinguish, whether an fx event occurred before or after the strobe, we now measure with high resolution, how long after the strobe the next fx pulse arrives. The synchronization delay τ begins with the strobe event, and it ends with one of the next fx events.
The analog circuitry determines this short time interval by charging a capacitor at constant current. The charge then is a linear function of the duration τ. Effectively the module works as a pulse expander (a pulse length multiplier), which stretches the pulse by an exact factor of 1000 (or 1024). The length of the expanded pulse is then measured in terms of the fx period, thereby resolving the phase to exactly one thousandth of a period . (FXM only)
Input frequencies can be between 5 and 50 MHz with a slight loss in phase accuracy at the higher end of the frequency range. For still higher frequencies a prescaler may be used. Lower frequencies require fitting of larger value capacitors.

3 New elements of models FX80 and FXE

3.1 Onboard microprocessor.

In the original design of the multi-channel instrument, as used in the optical frequency synthesis chain [1], the shift registers containing the counter readings of all channels are concatenated so as to form a long serial (RS-232) telegram, that can be read directly by the COM-Ports of a PC. This serial data transfer, other than GPIB (IEEE—488), allows to join together (in real-time) experiments going on in different laboratories, even in separate buildings. This worked reliably with DOS-computers. Windows PCs however, particularly when connected to a LAN, are not suited for real-time applications. We have therefore added to the system a dedicated micro processor (MC68306) . It receives the counter data at a much higher rate (4 MBd) and after pre-processing, reports phase values to the PC via RS232 or USB under a hand-shake protected protocol. In the PC a special DLL receives the reports and passes them on for the high-level language application program. The new versions (FX80, FXE) now all take samples at a constant rate of 1000 per second in up to 16 channels. The processing in the μP includes data reduction by averaging, and all other operations that must be carried out in real-time. The μP program is stored in a Flash ROM and can be downloaded from the PC. So special application code and updates can be sent through the internet.

3.2 FXE Extended frequency range

FXM and FX80 both determine fractions of periods, i. e. the fractions part of the phase value, by expanding the synchronization delay (see Fig. 3) with the help of an analog time magnifier. With a given capacitor value these magnifiers work fine for frequencies down to 4 MHz. For lower frequencies other capacitors would have to be fitted. Also, for low frequencies the duration of the expanded pulse would stretch the conversion time and finally would get into conflict with the measurement rate. In FXE, which is designed for measuring frequencies from 50 MHz to as low as 50 kHz, the synchronization delay is digitized in a two-step process. In a first step the number of full 10 MHz reference periods that fit in the interval, is determined (and counted in a 3rd counter). Then, in the 2nd step, the remainder is again measured by the analog time magnifier. FXE and FX80 both keep constant time resolution across their frequency range, phase resolution varies accordingly.
Extension of the frequency range at the high frequency end can always be accomplished by using a prescaler.

4 Application

Figure 6 shows the setup for the measurement of propagation delays.
A synthesizer serves as the signal generator for both the test frequency and for the reference frequency. For test purposes only, i.e. for detecting possible dependencies of the result on the reference phase, a computer controlled 10 MHz phase shifter is inserted in the reference line.
A distribution buffer generates precisely synchronous square-wave signals with 50 Ω source impedance.
The Device Under Test DUT can be a length of cable, a buffer amplifier, a variable delay generator, or another phase shifter.

Measurement of Propagation Delay

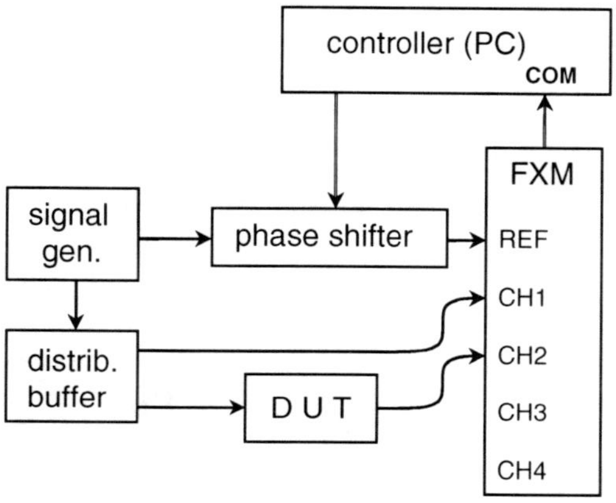

Figure 4: Delay measurement in differential mode

5 Precision of the phase measurement

Phase Interpolation

For enhancing the resolution, Time Interval Counters (TIC) as well as universal counters that determine frequency by calculating it as the reciprocal of the measured duration of a number of signal periods, often use some kind of interpolation. At the present state of art, resolution has to be traded against speed, power consumption and complexity.
On the high speed side are some purely digital Time-to-Digital Converters (TDC) that use delay chains and logic elements only. They employ similar encoders as flash A/D converters that produce the digital output word in one machine cycle. Typical LSB resolution is 1 ns.
Not quite so fast but with the potential for much higher resolution are analog methods.

5.1 Analog Interpolation

Analog measurement of short time intervals Tau has been in use for many years. [3]
One common method has been, to charge for the time Tau a capacitor at constant current and then to determine the charge by measuring the voltage across the capacitor by means of a DAC.
The method can be quite fast and precise. The calibration factor, however, is based on such disparate quantities as the value of the capacitance , the charging current, and the reference voltage for the D/A Converter.
We are using a dual-ramp-technique, instead. Like in the digital voltmeter, the main advantage of this technique is that the capacitance does not have to be precisely known nor particularly stable. It simply has to be free of hysteresis. Our circuit acts as a pulse stretcher or time (interval) magnifier that produces a pulse 1000 (1024) times the duration of the input pulse.
Here the magnification factor is embodied in the ratio of two resistances, which can be very stable and precisely adjusted. The length of the stretched pulse is then digitized by using it as the gate for counting an 80 MHz clock.
In our older FXM model, this counter counted the channel frequency Fx, thereby automatically performing the multiplication which converts time to phase. In our new FXE and FX80 models this multiplication is carried out in the onboard microcomputer. The time resolution (LSB, digitization step) is now a fixed 12.2 ps, irrespective of the frequency being measured.

5.2 Noise

With all input signals and conditions kept fixed and stable, we still observe a distribution of the digital output values (Fig.5), which is indicative of noise processes additional to the quantization noise (=3.7 ps rms). We have not fully analyzed these noise processes yet, but considering simply the fact that the propagation time through a chain of HC gates is noise modulated by a noisy Vcc supply, we are not surprised that random phase noise originating in the circuitry has the same order of magnitude as the digitization noise.

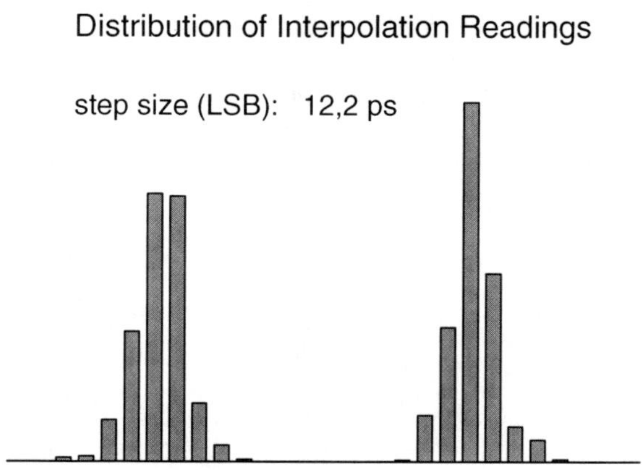

Figure 5: Histograms show the phase noise

In the TI 1991 data book for the 74HCxx logic gates is described a simple ring oscillator VCO, which is tuned by varying the Vcc supply voltage only. From the tuning sensitivity (5.8 MHz/V) it is possible to calculate the round-trip delay sensitivity to be -7 ns/V. The usual (7805) voltage regulator noise on the other hand amounts to 200 μV, which would explain a phase noise level of 1.5 ps rms already.

5.3 Linearity

Like in a typical metrological problem, no superior 'master' phase standard is available for the test and calibration adjustment of the magnifiers.
We however found that a few meters of RG58 cable form a propagation delay that is stable enough for the verification of linearity. This cable delay can be recorded by making use of the multi-channel capability of our phase recorder. By feeding two channels from the same standard frequency source but through different lengths of cable (Fig 4), we record the phase difference, which reflects the delay difference (27 ns) of the two cables. This difference should come out constant and independent of the relative phase of the reference signal with respect to the test signal. The residual dependency thus reveals either misadjustment or a systematic distortion/non-linearity in the magnifiers.

The figure (Fig 6) shows the residual imperfections. The quantization noise is barely noticeable in this picture, because it is reduced to insignificant level by averaging 128 samples for each point.
1.) The sharp steps in ***trace a*** reveal a residual discrepancy between the real, incorporated magnifications (one for each channel) and the factors used in the evaluation program.

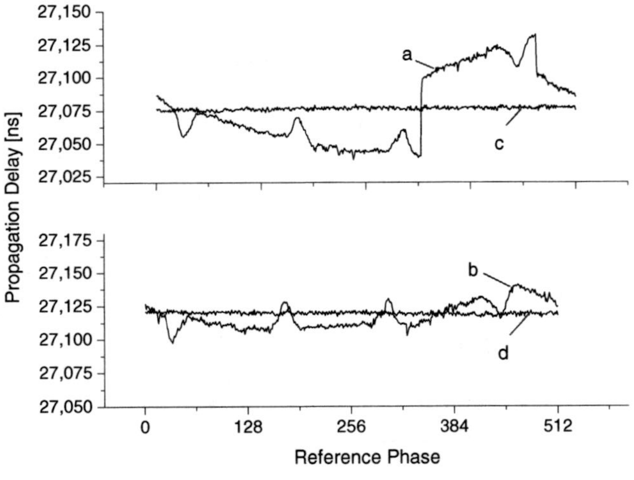

The steps disappear with proper adjustment.(***trace b***)
Figure 6: Residual non-linearity of the interpolation

2.) A systematic non-linearity of the magnifier transfer function, which certainly exists and can be estimated from transistor properties, should, in the lowest order, appear as a triangular curve (steps in the 1st derivative). Such triangular feature is visible but not really salient.

3.) Instead, what we see is 4 little bumps, ~ 25 ps high and ~4 ns wide, that must be attributed to internal cross-talk of the sharp logic level transitions (~1 ns rise time) that carry the timing information in the circuitry. Because we are displaying the difference of two phase measurements, two of the bumps appear as holes.

Quite a number of cross-talk mechanisms and paths can be imagined. Most of them have already been identified and minimized in the circuit design. The interference responsible for the 'bumps', appears to be happening inside the synchronizer module, where signals cannot be isolated. The synchronizer, a two stage shift register, acts as a time comparator and decider that determines, which of two events (level transitions) comes first or second. So it is a kind of a mixer, a nonlinear multiplier, that needs both factors, at the clock and the data input. The residual imperfections of the synchronizers thus appear to set a limit to the precision possible with the generally available logic circuits.

We found a strategy however to completely suppress these and other systematic imperfections of the interpolation. Before explaining it in the last paragraph, I have to describe the phase shifter module that is needed for its realization.

6 The digitally controlled phase shifter.

The test plot (Fig 6) showing the small systematic linearity deviations of the analog interpolators could have been generated by simply letting the phase of the reference input drift slowly with respect to the signals being measured. Instead we use for this purpose (set-up: Fig.4) another instrument module that we have developed: a digitally controlled continuous phase shifter.
This (patented) phase shifter for clock signals [4] generates a clock signal precisely phase-shifted with respect to its input. The phase shift is set by a digital number, a 12 bit word, and can thus be controlled by a micro-computer. The digital phase shifter can be considered to be the actor complement of the digital phase recorder as the sensor. It is continuous in the sense that it can accumulate an infinite number of cycles. In what some people call the natural binary code for angles, the most significant bit has a value of one semi-circle or half a cycle.
If the phase angle is increased in small steps, the digital code simply rolls over at every completion of a period. The resolution of our 10 MHz phase shifters is 12 bit, corresponding to a step size of 24.4 ps. (Such phase shifters can be built for any frequency).

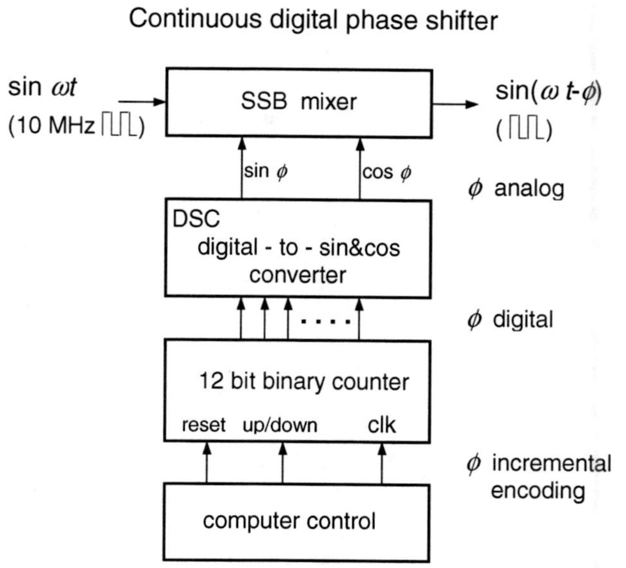

$$\sin(\omega t+\Phi) = \sin(\omega t)\cos(\Phi) + \cos(\omega t)\sin(\Phi)$$

Figure 7: Block diagram of the digital phase shifter.

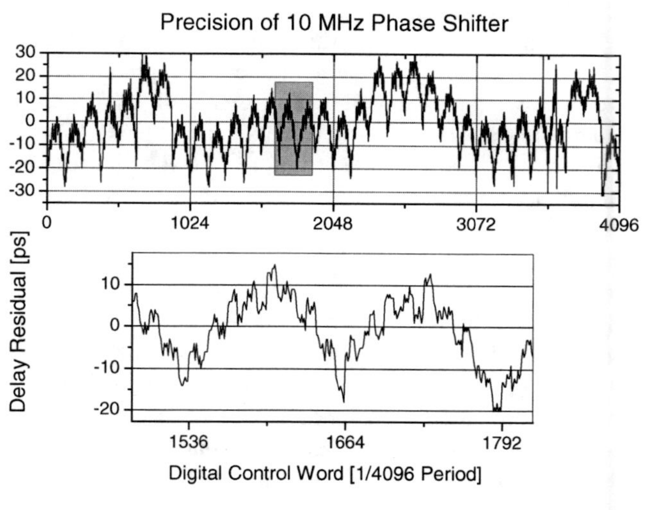

Figure 8: Precision of the 10 MHz digital phase shifter

With the help of FX80, actually employing another phase shifter as described in the next paragraph, we now can measure the phase shift very precisely and compare it with its nominal value. Fig.8 shows the residual non-linearity error: The high precision measurement possible only now with FX80, clearly reveals the origins of the various imperfections: The periodic structure with 32 periods per cycle
and 30 ps p-to-p is easily identified as being caused by the approximation method used in generating the sine and cosine functions. Similar identification is possible of other imperfections, even that of individual resistor values, switches and transistors.
When properly trimmed (by setting four potentiometers), the error of this digital-to-phase converter stays within a band of about 60 ps p-to-p, which is adequate for many applications, although this is still more than double the step size, i.e. the value of one LSB.

7 The method of systematically spreading the sampling phase

Even though the precision of the phase shifter does not reach that of the FX80 phase recorder, it can, – and we find it rather noteworthy that it can – as a companion module enhance the precision, i.e. reduce the systematic errors of the phase recorder by a factor of more than a hundred.
The reason, why this is possible, lies in the periodic nature of the quantity. Phase angles like geometrical angles, are in a sense self-calibrating. No standard is needed for calibrating 360° or a full period. The only possible systematic errors are deviations from linearity in subdividing the period.
There exists a close analogy of our method with the century-old strategy of surveyors to get around the imperfections of the available theodolites, a method that enabled them to measure angles for triangulation with an accuracy far better than the basic accuracy of their instruments. The method is based on averaging the results of repeated measurements of the same angle. By carrying out the measurements at different positions of the reference circle, the effect of systematic imperfections of the reference circle is reduced, in some cases completely neutralized. The prescription for efficient suppression is, that after each set of observations (2 bearings form 1 angle) you should advance the reference circle by a precise subdivision of 360°, let us say by 10°. Then, after 36 measurements, the averaging procedure not only reduces the statistical error of the observation (by a factor of 6), but also suppresses systematic errors of the instrument.
The degree of suppression depends on the character of the imperfection, which can be described by its Fourier transform. (The lowest Fourier coefficient, f. ex., describes a possible offset of the axis from the center of the reference circle.)
In our phase/time recorder, the equivalent of rotating the pitch circle is the shifting of the reference phase. By means of what we call the 'Scrambler' module, the μP shifts back and forth the timing of the strobe pulses. It then calculates the mean of e.g. 128 phase samples and reports it to the PC. The traces **c** and **d** in Fig. 6 show that the scrambling completely flattens the steps and bumps.
The scrambling program is designed in such a way ('back and forth'), that for consecutive reports the centers of gravity of the samples are spaced by the exact number of milliseconds. So it is still possible to calculate the frequency from consecutive reports of a single channel. The

sampling phases should, on the other hand, be evenly spread over a full period of the frequency being measured. Quite a different question is the order, in which the samples are taken. Here a pseudo-random sequence would be the optimum in minimizing aliasing products with possible periodic phase modulation of the signal being measured. The scrambling program is still being further refined and developed.

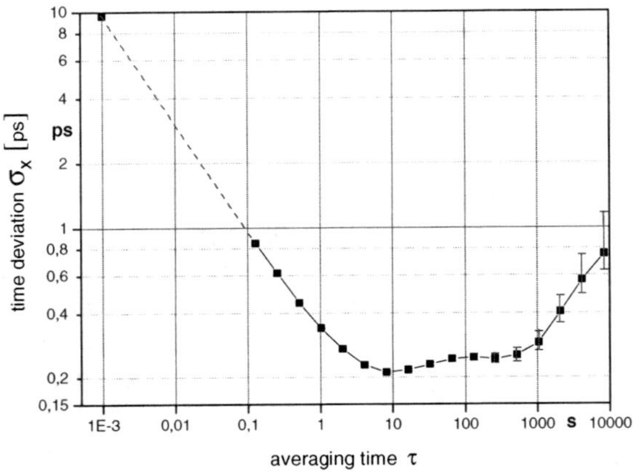

Figure 9: Measurement error of phase difference at 10 MHz

For uncorrelated fluctuations, the standard deviation of the mean decreases as the (inverse) square root of the number of samples. In our phase recorder this is the case for fluctuations caused by white phase noise of the signals. If the quantization noise is about as high the random fluctuations, it will decrease as in the same way. Fig. 5 is a histogram showing that in FX80 the resolution and the internally generated noise level are of the same order of magnitude.

FX80 (as well as FXE) always takes 1000 phase samples per second (in each of up to 16 channels). Only the mean values of a selectable (in powers of 2) number of samples are normally reported to the PC. At a report interval of 128 ms the 'scrambler' works nicely already, and the standard deviation due to internal noise of the instrument is reduced by a factor 11.

The noise level we are actually observing is shown in the figure (Fig 9) . For short averaging times this Allan plot of a phase difference shows the expected slope for white phase noise. The dashed line is a theoretical extrapolation. At averaging times of several seconds, flicker of phase takes over with a level of 0.2 ps, until at 1000 s drifts are beginning to limit the stability. These drifts are caused by a residual temperature sensitivity of the magnifier modules, or rather a residual imbalance of this sensitivity between the channels.
The prime consideration in the design of the pulse magnifiers had been to make the magnification factor as stable as possible, i.e. to make it insensitive to changes in temperature or supply voltage, thus avoiding the necessity for frequent re-calibration. With the 'scrambler' method, all linearity deviations are reduced (by a factor of 128). But not reduced is the temperature sensitivity of the additive (delay) term:
In phase measurements of an individual channel with respect to the reference we have observed a temperature sensitivity as high as ~100 ps/K. Instead of trying to reduce this sensitivity by compensation or a thermostat, the idea was to use for precision phase measurements the differential mode, where the temperature sensitivity of one channel and its magnifier is compensated by the other one. For this reason the magnifiers have been fabricated as equal as possible and are mounted under a common thermal shield. Special measures are taken for minimizing variable thermal gradients (convection) inside the shield, so that in differential mode we now observe a temperature sensitivity as low as 1 ...2 ps/K, which corresponds to the sensitivity of the propagation delay through a single HC gate or through 1 meter of RG58 coaxial cable. (Actually the phase delay recorded for the stability measurement was that of a cable 10 m long).
The rise of $\sigma_x(\tau)$ at $\tau > 1000$ s thus does not indicate an aging divergence. As can be seen from a simple time plot of the data, it reflects the warm-up time and the diurnal cycle of the economy heating system, programmed for lowering the room temperature at night time. The most temperature sensitive part was found to be the DUT, the cable delay we had been recording for the stability test. In differential mode the phase recorder shows a sensitivity of ±1 ps/K only.

Another common noise/interference phenomenon that a phase recorder has to deal with, is power line related phase modulation of the signals, which it should either record truthfully or reject completely. An effective filter for simply suppressing 50 Hz modulation would be to record phase averages taken over exact time intervals of 20 ms. Because of the sampling process this can be done approximately only. (Harmonics of the 50 Hz will form aliasing products). In any case, good suppression will require averaging intervals that are not in the powers of two series of milliseconds, that we had chosen for the simplicity of the averaging arithmetic. The latest software versions now offer selection of the averaging interval from a 1 - 2 - 5 - 10 ... series. An averaging interval of 100 ms will suppress both 50 Hz and 60 Hz interference (60 Hz in North America).

8 Conclusion & outlook

Apart from its multi-channel capabilities, the new, high precision synchronous phase recorder also appears to represent the state of the art as a universal phase/frequency meter, interfacing directly with a Windows PC.

The main direction for further development would be to make possible higher sampling rates for recording phase modulated signals.

Acknowledgements

This work was partly funded by the European Union [5]. The authors acknowledge that PTB permitted the work to continue after the retirement of one of us. G.K. also particularly acknowledges the contribution of B. Lipphardt, who had built and operated the prototype counters in the conventional optical frequency synthesis chain [1] and is now testing FXE with the new femtosecond comb.

References

[1] G. Kramer, B. Lipphardt, U. Huebner. "Phase Coherence in Optical Frequency Measurement and Synthesis", *Proc. 1999 Joint Meeting EFTF - IEEE IFCS*, pp. 651-654

[2] G. Kramer, W. Klische. "Multi-Channel Synchronous Digital Phase Recorder", *Proc. 2001 IEEE IFCS*, pp. 144-151

[3] G. Kramer: „Ein hochauflösender elektronischer Zeitmesser", *NTZ* **9**, p. 433, (1970)
(A modern version of this instrument is now available at K+K Messtechnik GmbH, D-38116 Braunschweig e-mail: kplusk@t-online.de)

[4] German Patent P 38 34 834.9-09

[5] Contract SMT4-CT98-5521,
Coordinator: Clive Green, Quartzlock, UK

EXTENDING THE RANGE AND UTILITY OF FEMTOSECOND COMBS

S. N. Lea, H. S. Margolis, H. A. Hodgkins*, G. Huang and P. Gill

National Physical Laboratory
Queens Road, Teddington, Middlesex TW11 0LW
United Kingdom
stephen.lea@npl.co.uk
fax: +44 20 8943 2945

Keywords: femtosecond comb; mode-locked lasers; optical frequency metrology; GPS receiver.

Abstract

This paper reports on progress at NPL in extending the range and utility of optical frequency combs. To enable measurements to be made at locations other than our own laboratory, a transportable comb system is being built whilst to extend the spectral range to cover the telecommunications bands we are also building a comb based on a Cr:forsterite laser.

1 Introduction

Optical frequency combs based on mode-locked femtosecond lasers have, over the last five years, revolutionised the field of optical frequency metrology. At NPL, a femtosecond comb based on a mode-locked Ti:sapphire laser has been used to make absolute frequency measurements of a variety of optical frequency standards, including lasers stabilised to ultra-narrow transitions in single cold trapped ions of $^{88}Sr^+$ at 674 nm [1] and $^{171}Yb^+$ at 467 nm [2] and to transitions in I_2 at 633 nm [3] and $^{13}C_2H_2$ at 1542 nm [4]. The spectral range covered by the Ti:sapphire-based comb does not extend beyond about 1.2 μm. Consequently, a short frequency chain was required in order to measure the acetylene-stabilised standard at its second harmonic, 771 nm. The additional complexity of this measurement scheme illustrates the desirability of extending the range of optical frequency combs to cover the mid-infrared spectral region, in particular the telecommunications bands. Development of a mid-infrared optical frequency comb based on a mode-locked Cr:forsterite laser is described in section 2.

Measurements have also been made of an I_2-stabilised laser reference at 656 nm for spectroscopy of He^+ at the University of Sussex, UK [5], and of a commercial side-of-peak-stabilised HeNe laser at 594 nm. These two measurements are typical of numerous applications where a lower level of accuracy is adequate and it would be more convenient to make measurements *in situ* rather than to transport the stabilised laser system to the standards laboratory. To address this requirement, a transportable, GPS-referenced femtosecond comb is being developed and is described in section 3.

2 Mid-infrared Cr:forsterite-based comb

The spectral coverage of combs based on Ti:sapphire lasers does not extend beyond about 1.2 μm. Various laser systems can potentially be used to extend this range to at least 2 μm, providing complete coverage of the telecommunications bands. Compact mode-locked Cr:YAG laser systems with several-GHz pulse repetition rate at 1.54 μm have been demonstrated [6]. Whilst these systems are conveniently centred on the telecommunications C-band, the spectral bandwidth is limited. Mode-locked fibre lasers offer the prospect of compact all-fibre devices: an octave-spanning comb using an erbium fibre laser and highly non-linear fibre has been reported at NIST [7]. For absolute frequency metrology applications, the challenge is to develop higher repetition-rate devices. Mode-locked Cr:forsterite lasers [8] offer a promising combination of higher repetition rate and short, high-energy pulses. Pulse lengths as short as 14 fs, centred near the 1.3 μm zero-dispersion wavelength of the laser crystal, have been demonstrated at MIT [9]. An octave-spanning comb based on a ring laser with 420 MHz repetition rate has been reported at NIST [10].

We have constructed a mode-locked Cr:forsterite laser. The laser is a near-symmetrical linear-cavity design, with dispersion compensation provided by chirped mirrors and a pair of intra-cavity prisms (fig. 1). The 10 mm-long Brewster-cut Cr:forsterite crystal is cooled to –10°C in a nitrogen-purged enclosure to enhance the gain. The chirped mirrors each have GVD of $-70\ fs^2$ over 100 nm bandwidth. The pair of intra-cavity isosceles Brewster prisms are made of a high-index glass, PBH71. This material has its zero dispersion wavelength around 2 μm. The apex-to-apex separation is about 0.1 m, providing compensation for the third-order dispersion of the Cr:forsterite crystal [9]. The cavity length is adjusted to give a pulse repetition rate of about 88 MHz.

Pumped by 7 W of single frequency radiation at 1053 nm from a Nd:YLF laser, this laser generates 800 mW of cw radiation at 1260 nm when an output coupling mirror with

*Also at: School of Engineering and Physical Sciences, David Brewster Building, Heriot-Watt University, Riccarton, Edinburgh EH14 4AS, United Kingdom.

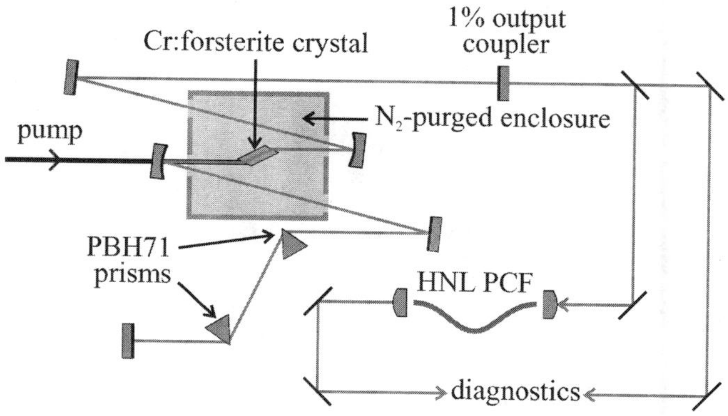

Fig. 1. Schematic of the mode-locked Cr:forsterite laser constructed at NPL. With the exception of the output coupler, all the cavity mirrors have GVD of -70 fs^2. HNL PCF: highly non-linear photonic crystal fibre.

reflectivity $R = 95\%$ is used. An output coupler with $R = 99\%$ is used to achieve the higher intra-cavity power required for mode-locking. Mode-locked pulses with up to 40 nm spectral bandwidth at 1280 nm are obtained at a repetition rate of 88 MHz (fig. 2). The corresponding Fourier-transform-limited pulse duration is 60 fs FWHM. The average mode-locked power is 100 mW and the pulse energy is about 1.2 nJ. Further work is required to suppress cw components in the output spectrum.

To perform absolute frequency measurements throughout the telecommunications bands, it is desirable to expand the comb width to cover an octave in optical frequency. The self-referencing technique to measure and stabilise the carrier offset frequency can then be used [11]. Several highly non-linear photonic crystal fibres have been tested with this laser. Fibres providing octave broadening for the Ti:sapphire laser have zero dispersion in the region of 720 – 760 nm do not provide spectral broadening at the Cr:forsterite wavelength. Samples of fibre designed to have zero dispersion near 1.55 μm, fabricated at the Optoelectronics Research Centre, Southampton, UK, have been tested. The light grey curves in fig. 2 show the broadening obtained with 20 cm lengths of two samples. A rather smooth spectrum is obtained with one sample but the other sample yields a strongly spectrally-fluctuating intensity profile, suggesting that self phase modulation is a dominant process in the fibre.

It is characteristic of highly-non-linear photonic crystal fibres to have a second dispersion zero, with the opposite sign of second order dispersion, at longer wavelength. This second zero-dispersion is usually at a wavelength of several μm; however a fibre manufactured by Crystal Fibre, type NL-PM-750, has its second dispersion zero at 1260 ± 20 nm, close to the Cr:forsterite wavelength [12]. Using a 20 cm length of this fibre, a spectrum covering 300 nm is generated (dark grey curve in fig. 2). Further work is required to optimise this spectrum: the offset of the centre wavelength of the laser pulses from the zero dispersion wavelength is known to be a critical parameter in optimising the spectral broadening in Ti:sapphire-based combs.

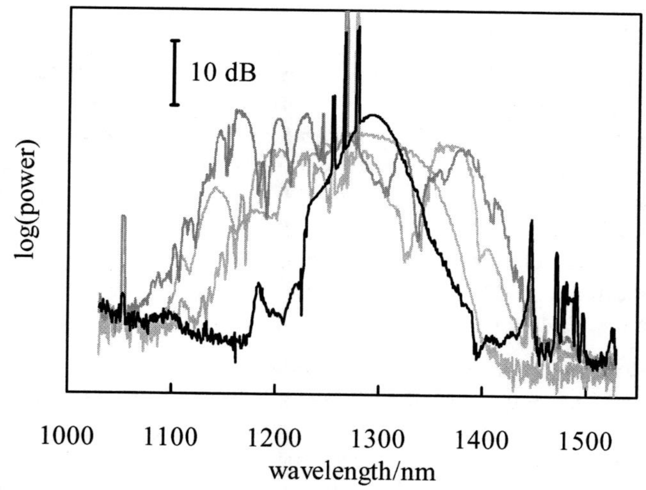

Fig. 2. Wavelength spectrum of pulses from the Cr:forsterite laser (solid black line) and of pulses spectrally broadened in highly-nonlinear photonic crystal fibre with zero dispersion at 1.55 μm (light grey lines) and 1.26 μm (dark grey line).

3 Transportable Cr:LiSAF-based comb

The Ti:sapphire-based comb is phase-locked to the 10 MHz reference of a hydrogen maser, providing traceability to the SI second, either through the timescale UTC(NPL) or by comparison with NPL's caesium fountain primary standard. In addition, phase-locking to the maser provides good short-term frequency stability of the comb modes. Whilst this set-up enables measurements of the highest accuracy to be made, the infrastructure of optical and microwave standards and comb has to be co-located on a single site. There are a number of metrological applications which can benefit from the flexibility of femtosecond comb technology but which do not require the highest level of absolute accuracy. To meet this need we are developing a transportable comb system, based on a diode-pumped mode-locked Cr:LiSAF laser, which will derive its frequency traceability from the GPS.

3.1 Mode-locked Cr:LiSAF laser

The Ti:sapphire mode-locked laser requires a high power green solid-state pump laser, making it an unsuitable basis for a readily-transportable femtosecond comb. Cr^{3+}-doped colquiriites such as $LiSrAlF_6$ (LiSAF) have similar laser characteristics to Ti:sapphire but have peak absorption in the red, around 670 nm, and hence can be directly diode-pumped. Kerr-lens mode-locking of a diode-pumped Cr:LiSAF laser was first reported at the University of Strathclyde [13]. Compact, low threshold systems have been demonstrated at the University of St Andrews [14]. Generation of an octave-spanning comb has been demonstrated by the Garching group [15]. We are constructing a laser which will initially be a linear cavity design with dispersion compensation provided by both an intra-cavity prism pair and mirrors with group velocity dispersion (GVD) of -50 fs^2. For a compact, transportable system, it would clearly be desirable to progress to a higher repetition rate design relying solely on chirped mirrors for dispersion compensation.

3.2 Referencing to GPS

The transportable comb has to be referenced to a disseminated timing signal such as that provided by the GPS. Issues arising in the design of a suitable disciplined local oscillator (LO) include the short-term stability of the LO, the steering algorithm and the long-term frequency stability of the undisciplined LO. The latter is important since the time taken for the LO to pull in after setting up the comb at a new location could be the limiting factor in the time taken to make a measurement.

We have assessed the performance of our 88 MHz repetition rate Ti:sapphire-based femtosecond comb when referenced to the 10 MHz output of two GPS-disciplined LO configurations. Measurements were made of the absolute optical frequency of the laser system used to interrogate the 674 nm quadrupole transition of the NPL Sr^+ ion trap standard [1]. This system consists of an extended cavity diode laser stabilised to a mode of a high finesse reference cavity constructed from ultra low expansion (ULE) material and exhibits very low drift rate, typically below 0.3 Hz/s.

Measurements referenced to a GPS-disciplined LO were compared to measurements referenced to the maser; in fig. 3 the fractional frequency offset of the GPS-referenced measurements from the maser-referenced measurements is plotted. The maser-referenced data is visible as periods of low scatter at the start and end of the data in fig. 3(a) and for three periods in fig. 3(b). The data is corrected for the linear frequency drift of the ULE cavity, determined by interpolation from the maser-referenced data. The 10 MHz output of the GPS-disciplined local oscillator was simultaneously phase-compared to the maser 10 MHz signal: the fractional frequency offset of the GPS-disciplined signal from the maser is also plotted in fig. 3. Longer periods of phase comparison at 10 MHz were used to determine the Allan deviation of the GPS-disciplined signals shown in fig. 4.

Initially, a GPS-disciplined 10 MHz frequency source incorporating a standard-performance quartz crystal oscillator (Rapco 2804A [16]) was tested. In this unit, the frequency of the quartz oscillator is corrected to the GPS 1 pps signal every 512 s. This step correction is apparent in the Allan deviation at 10 MHz (dashed line in fig. 4) and the measurement data (fig. 3(a)) The result of typical 1000 s measurement run (consisting of 100 ten-second measurements) may deviate from the maser-referenced value by several 10s of kHz.

Superior medium-term stability can be obtained using a rubidium LO. The Allan deviation at 10 MHz of a Rapco 2804AR GPS-disciplined oscillator is shown as the chain-dotted line in fig. 4. However, the rubidium oscillator has relatively poor stability at short (<1 s) timescales. To achieve adequate short-term stability of the comb beat frequencies for accurate locking and frequency counting, a low-phase-noise quartz oscillator (Rapco 803M) is locked to the rubidium oscillator. The Allan deviation of this system is given by the solid line in fig. 4 and typical measurement data in fig. 3(b). The performance of this system is an order-of-magnitude better than that of the standard quartz unit: the measurement

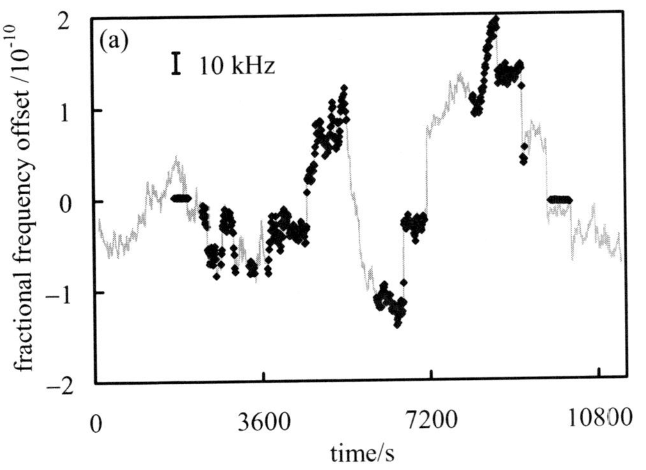

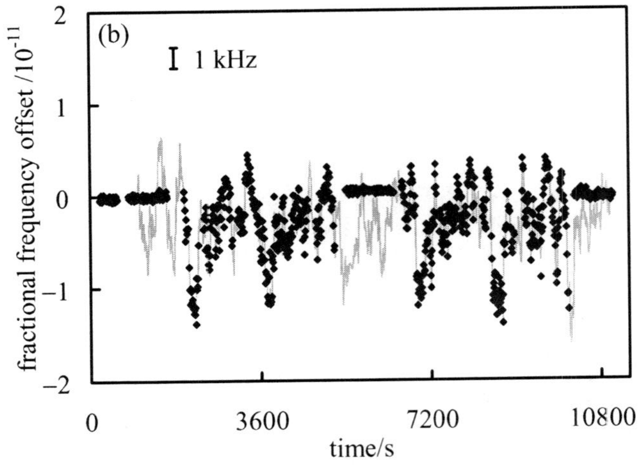

Fig. 3. GPS-referenced frequency measurements of a 674-nm cavity-stabilised laser using a femtosecond comb, normalised to maser-referenced measurements (solid points), compared to the fractional frequency offset of the GPS-derived 10 MHz reference frequency from the maser 10 MHz output (grey line). (a): standard quartz LO. (b): low-phase-noise quartz LO phase-locked to Rb LO. Note × 10 change of scale.

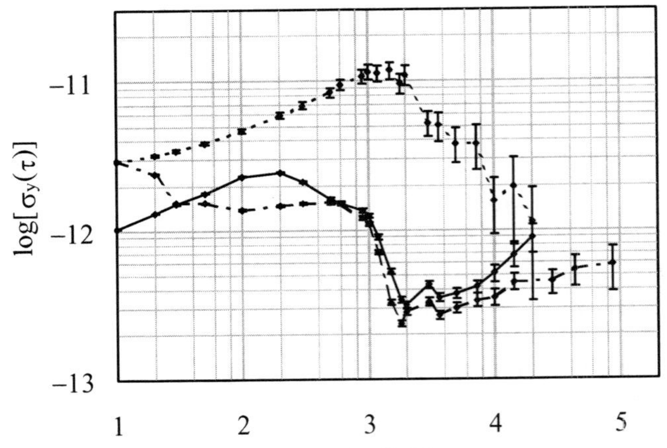

Fig. 4. Allan deviation of GPS-disciplined LOs at 10 MHz: Rapco 2804A standard quartz (dotted line); Rapco 2804AR Rb LO (chain-dashed line); low-phase-noise quartz phase-locked to Rb LO (solid line).

statistics and offset from the maser-referenced measurement reach the few kHz level for a typical 1000 s measurement. The 10 s optical frequency measurement data points closely follow the 10 MHz phase comparison with the maser, showing that the stability of the measurement is limited purely by the GPS receiver.

The level of accuracy obtained is comparable with that obtained using a transportable 633 nm iodine-stabilised helium-neon laser standard [3] but the GPS-referenced femtosecond comb will enable this level of accuracy to be obtained at arbitrary optical frequency.

4 Summary

Femtosecond combs based on mode-locked Ti:sapphire lasers are an established part of the infrastructure for the realisation of optical frequency standards in national metrology institutes such as NPL. To extend the spectral range of measurement to span the infrared telecommunications bands, we are developing a femtosecond comb based on a mode-locked Cr:forsterite laser. This laser produces 1.2 nJ pulses of 40 nm FWHM at 88 MHz repetition rate, which have been broadened in highly-nonlinear photonic crystal fibre to span a 300 nm bandwidth. Further development of this system is required to suppress cw oscillation in the laser and to achieve an octave-spanning comb. To make femtosecond comb measurement technology more widely available to users in industry and academia, a transportable comb system is being constructed. We have demonstrated that, with an appropriate choice of receiver and local oscillator, a GPS-referenced comb can be used to make optical frequency measurements with a few kHz uncertainty.

Acknowledgements

We acknowledge useful discussions on Cr:forsterite lasers with J. R. Taylor, P. E. G. Baird, and S. A. Diddams, on Cr:LiSAF lasers with W. Sibbett and B. Agate, and on GPS-disciplined oscillators with J. A. Davis, P. Baker and A. Turner. We thank D. J. Richardson for supplying samples of 1.5 μm photonic crystal fibre. This work is supported by the UK NMS 2002–2005 Length Programme and 2001–2004 Photonics Programme. HAH acknowledges the support of EPSRC and of his Eng. D. supervisor at Heriot-Watt University, D. T. Reid.

References

[1] H. S. Margolis, G. Huang, G. P. Barwood, S. N. Lea, H. A. Klein, W. R. C. Rowley, P. Gill and R. S. Windeler, "Absolute frequency measurement of the 674 nm $^{88}Sr^+$ clock transition using a femtosecond optical frequency comb", *Phys. Rev. A*, **67**, 032501 (2003).

[2] P. J. Blythe, S. A. Webster, H. S. Margolis, S. N. Lea, G. Huang, S.-K. Choi, W. R. C. Rowley, P. Gill and R. S. Windeler, "Subkilohertz absolute-frequency measurement of the 467-nm electric octupole transition in $^{171}Yb^+$", *Phys. Rev. A*, **67**, 020501 (2003).

[3] S. N. Lea, W. R. C. Rowley, H. S. Margolis, G. P. Barwood, G. Huang, P. Gill, J.-M. Chartier and R. S. Windeler, "Absolute frequency measurements of 633 nm iodine-stabilised helium-neon lasers", *Metrologia*, **40**, 84–88 (2003).

[4] C. S. Edwards, H. S. Margolis, G. P. Barwood, S. N. Lea, P. Gill, G. Huang and W. R. C. Rowley, "Absolute frequency measurement of an acetylene standard by use of a combined frequency chain and femtosecond comb", *Opt. Lett.*, **29**, 556–558 (2004).

[5] S. A. Burrows, S. Guérandel, E. A. Hinds, F. Lison and M. G. Boshier, in "The Hydrogen Atom: Precision Physics of Simple Atomic Systems", eds. S. Karschenboim, F. S. Pavone, F. Bassoni, M. Inguscio and T. W. Hänsch (Berlin, Springer) 303–313 (2001).

[6] T. Tomaru, "Two-element-cavity femtosecond Cr^{4+}:YAG laser operating at a 2.6-GHz repetition rate", *Opt. Lett.*, **26**, 1439–1441 (2001).

[7] B. R. Washburn, S. A. Diddams, N. R. Newbury, J. W. Nicholson, M. F. Yan and C. G. Jørgersen, "Phase-locked, erbium-fibre-laser-based frequency comb in the near infrared", *Opt. Lett.*, **29**, 250–252 (2004).

[8] Y. P. Tong, P. M. W. French, J. R. Taylor and J. O. Fujimoto, "All-solid-state femtosecond sources in the near infrared", *Opt. Commun.*, **136**, 235–238 (1997).

[9] C. Chudoba, J. G. Fujimoto, E. P. Ippen, H. A. Haus, U. Morgner, F. X. Kärtner, V. Scheuer, G. Angelow and T. Tschudi, "All-solid-state Cr:forsterite laser generating 14-fs pulses at 1.3 μm", *Opt. Lett.*, **26**, 292–294 (2001).

[10] I. Thomann, A. Bartels, K. L. Korwin, N. R. Newbury, L. Hollberg, S. A. Diddams, J. W. Nicholson and M. F. Yan, "420-MHz Cr:forsterite femtosecond ring laser and continuum generation in the 1–2 μm range", *Opt. Lett.*, **28**, 1368–1370 (2003).

[11] D. J. Jones, S. A. Diddams, J. K. Ranka, A. Stentz, R. S. Windeler, J. L. Hall and S. T. Cundiff, "Carrier-envelope phase control of femtosecond mode-locked lasers and direct optical frequency synthesis", *Science* **288**, 635–639 (2000); R. Holzwarth, Th. Udem, T. W. Hänsch, J. C. Knight, W. J. Wadsworth and P. St. J. Russell, "Optical frequency synthesizer for precision spectroscopy", *Phys. Rev. Lett.*, **85**, 2264–2267 (2000).

[12] Datasheet for Nonlinear Photonic Crystal Fiber – NL-PM-750, Crystal Fibre A/S, Blokken 84, DK-3460 Birkerød, Denmark.

[13] M. J. P. Dymott and A. I. Ferguson, "Self-mode-locked diode-pumped Cr:LiSAF laser", *Opt. Lett.*, **19**, 1988–1990 (1994).

[14] B. Agate, B. Stormont, A. J. Kemp, C. T. A. Brown, U. Keller and W. Sibbett, "Simplified cavity designs for efficient and compact femtosecond Cr:LiSAF lasers", *Opt. Commun.*, **205**, 207–213 (2002).

[15] R. Holzwarth, M. Zimmermann, Th. Udem, T. W. Hänsch, P. Russbüldt, K. Gäbel, R. Poprawe, J. C. Knight, W. J. Wadsworth and P. St. J. Russell, "White-light frequency comb generation with a diode-pumped Cr:LiSAF laser", *Opt. Lett.*, **26**, 1376–1379 (2001).

[16] Rapco Electronics Ltd, Joule Road, Basingstoke, Hants RG21 6XF U.K.

PRECISION FREQUENCY MEASUREMENTS WITH ENTANGLED IONS

CHRISTIAN ROOS, MICHAEL CHWALLA, MARK RIEBE, GAVIN LANCASTER, CHRISTOPH BECHER, HARTMUT HÄFFNER, WOLFGANG HÄNSEL, FERDINAND SCHMIDT-KALER, RAINER BLATT,
Universität Innsbruck, Institut für Experimentalphysik, Technikerstr. 25, 6020 Innsbruck,
Austria
Christian.Roos@uibk.ac.at

The use of entangled states can provide an increased sensitivity in quantum-limited spectroscopic measurements, leading to an uncertainty that is inversely proportional to the number of particles instead of the usual square-law dependence [1].

In this contribution, we show that spectroscopy with maximally entangled states of atoms also offers significant advantages over experiments done with single atoms. As a first example, we demonstrate that entanglement can be used to effectively eliminate first-order Zeeman shifts in spectroscopy with $^{40}Ca^+$ even though there are no m=0 -> m=0 transitions. Secondly, we discuss how maximally correlated states of two $^{40}Ca^+$ could be used for measuring tiny frequency shifts of the S1/2 – D5/2 transition arising from second-order Zeeman shifts and electric quadrupole shifts due to the trapping potential.

[1] See for example: J.J. Bollinger et al., Phys. Rev. A **54**, R4649 (1996)

ABSOLUTE FREQUENCY MEASUREMENT AROUND 30 THZ WITH A FEMTOSECOND LASER COMB

A. AMY-KLEIN, C. GRAIN, R.J. BUTCHER[1], A. SHELKOVNIKOV[2], C. DAUSSY, O. LOPEZ, M. GUINET, A. GONCHAROV[3], C. CHARDONNET

Laboratoire de Physique des Lasers, C.N.R.S., Université Paris 13, 99, av. J.-B. Clément, 93430 Villetaneuse, France.
[1] Permanent address : The Cavendish Laboratory, Cambridge, UK.
[2] Permanent address : Lebedev Physical Institute, Moscow, Russia.
[3] Permanent address : Institute of Laser Physics, Novosibirsk

Keywords: Optical frequency standard, SF_6, 30 THz.

Abstract

Recebt advances concerning two-photon Ramsey fringes in SF_6 at 30 THz obtained with a 1m separation between the two interaction zones are presented. The experiment is referenced to a primary standard, the Hydrogen maser / Cs fountain located at BNM-SYRTE, Observatoire de Paris, via a femtosecond laser frequency comb generator. This results in an absolute frequency measurement of the central fringe to ± 1.4 Hz which corresponds to 5×10^{-14} in relative frequency.

1 Introduction

This paper presents recent advances of a two-photon Ramsey fringe experiment on a supersonic beam of SF_6, which are essentially due to two experimental developments [13]. First, the distance between the interaction zones has been increased to 1m, so that the fringe periodicity is now 200 Hz. With a detection by stimulated emission on the upper one-photon transition a signal-to-noise ratio (SNR) of 20 in a bandwidth of 1 Hz was obtained. Second the entire system can now be directly related to the frequency comb of a femtosecond laser, itself referenced to a primary standard, H-maser or Cs fountain, of the BNM-SYRTE, at Observatoire de Paris.

2 The two-photon Ramsey fringe experiment

The two-photon Ramsey fringe experiment employing a supersonic beam of SF_6 has been developed over number of years [14]. The set-up is shown in Fig. 1. In brief, the molecular beam interacts successively with two standing waves, which are tuned to a two-photon resonance. The excitation probabilities for the two zones of interaction interfere so that the population of the upper level oscillates with a periodicity P which depends on the transit time between the two zones : $P = u/_{2D}$, where u is the mean velocity of the beam and D the distance between zones. In order to obtain the central fringe in exact coincidence with the two-photon resonance the relative phase of the two standing waves must cancel [3]. This is ensured by using a single

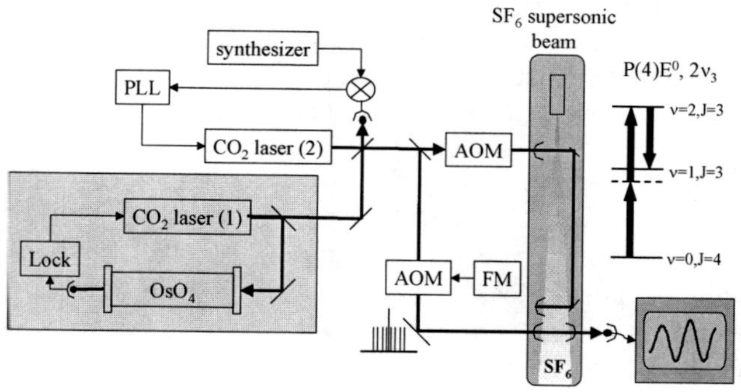

Figure 1 : Schematic of the two-photon Ramsey fringe experiment

folded Fabry-Perot cavity, comprising four mirrors in a U configuration. The two-photon transition is the P(4)E^0 line of the $2\nu_3$ band of SF_6, excited by the P(16) CO_2 laser line (see insert in Fig. 1). This is particularly favourable because of its low J value and because of the enhancement due to the near-resonant intermediate transition, P(4) ν_3. Fringes in the two-photon absorption are detected by stimulated emission in a separate Fabry-Perot cavity. For this purpose the laser beam used for detection is strongly modulated in order to sample the whole Doppler profile. This technique is crucial in obtaining a good SNR [7].

Two CO_2 lasers are employed on the 10μm P(16) transition. The first is locked to the P(55) line in OsO_4 while the second is phase locked to the first with a radio frequency offset. The spectral purity for the whole experiment is thus determined by the first laser, while the second confers tunability under computer control. An electro-optic and three acousto-optic modulators are incorporated to shift the laser frequencies to the exact points required and to introduce various modulations.

The current experiment incorporates a new folded cavity with a separation of 1 m between interaction zones. Optimum operation was obtained by frequency modulating Laser 2 at 115 Hz with an index of 0.43, 12 mW in the U cavity, corresponding to a $\pi/2$ pulse, and 75 μW in the detection cavity.

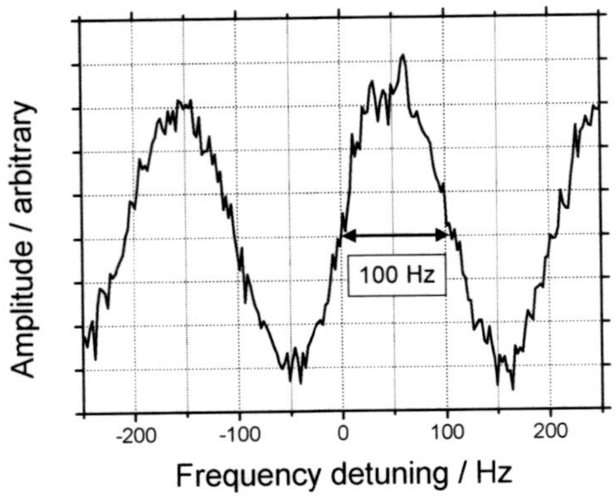

Figure 2 : Fringes at 200 Hz obtained on the central hyperfine component. Experimental conditions: SF_6 input pressure $5x10^5$ Pa, 12 mW inside U cavity, FM modulation at 115 Hz index 0.43, 75 μW inside the detection cavity, 1s per point.

Figure 2 shows a recording of the central fringes of the main hyperfine component, using a beam of pure SF_6, generated from a reservoir pressure of 5×10^5 Pa, 10 scans of 200 data points, lock-in time constant 100 ms. The fringe periodicity is 200 Hz and the SNR is 20. The recording is thus obtained in less than 4 minutes, using 1 second of averaging per point.

3 Absolute frequency measurements

The principle of frequency measurements using a femtosecond laser comb has been widely described[4, 8, 11], and the specific technique used in this experiment was initially employed in the measurement of an OsO_4 frequency, as recently published in detail [1].

3.1 The optical link to the primary standard

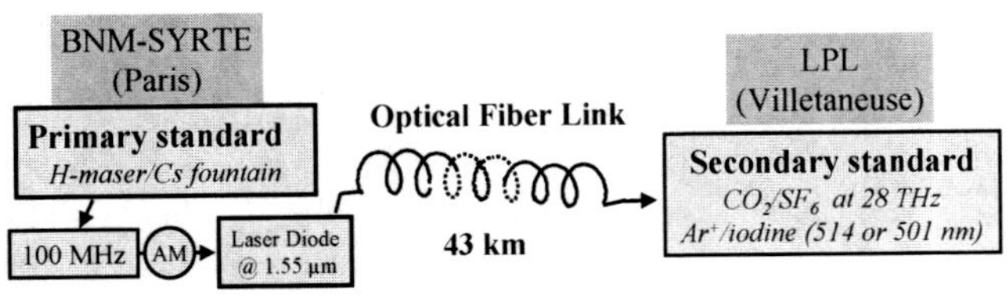

Figure 3 : optical fiber link between LPL and BNM-SYRTE

The LPL is located in the surroundings of Paris, whereas in the centre of Paris the BNM-SYRTE develops cold atoms frequency standards in the microwave domain. Their Cs atomic fountain has a demonstrated accuracy of 10^{-15}, and a frequency stability of $5\times10^{-14}\tau^{-1/2}$ [12]. They have developed an internal reference signal at 100 MHz, which is distributed around their whole laboratory for various metrological experiments. Its stability is given by an H-Maser or a cryogenic oscillator, Allan deviations in Fig. 4, and the accuracy is controlled either with the Cs fountains or with a comparison to the local time scale and with GPS. Typical accuracy of the 100-MHz signal is below 10^{-14}.

It was very attractive to take advantage of this excellent reference for our measurements. Two optical links has been developed to transmit the 100-MHz signal from SYRTE to LPL (Fig. 3). Interconnecting many different sections of existing standard single mode optical fibres results in two close optical links of 43 km, which is much larger that the physical distance between the two laboratories of 13 km.

The complete optical link includes a distributed feedback (DFB) semiconductor laser, emitting at 1.55 μm, and a p-i-n photodiode. The current of the laser is modulated at 100 MHz directly controlled by the H-Maser. The amplitude modulation is then converted back into the electrical domain by the photodetector at the output of the link.

This optical link must not degrade the spectral quality of the metrological signal, while the background noise could impact on the 100-MHz stability along the 43 km propagation. The metrological features of this link were measured by a complete round trip (86-km) of the 100 MHz signal. The phase difference between the input and return signal was detected and its variation was analyzed in terms of frequency fluctuations. The noise of both fibers are strongly correlated : the difference of input and return fiber phase fluctuations is an order of magnitude below the one-way phase fluctuations. The Allan deviation for one path is reported in fig. 4. It exhibits lower deviation than the H-maser. This demonstrates that the optical link does not degrade the performance characteristics of the H-Maser reference frequency for measurement times below 10000 s. However the free-running optical link does not permit to transfer the stability of the cryogenic sapphire oscillator [15], or the Cs fountain. Active control of the phase fluctuation is under progress [10].

This signal is then divided down at LPL and used to phase lock the harmonic of a 5 MHz quartz oscillator, which thus provides the 10 MHz reference for the synthesizers and counter. It is also used to phase-lock a 1 GHz oscillator for the measurement of the fs laser repetition rate. The relative

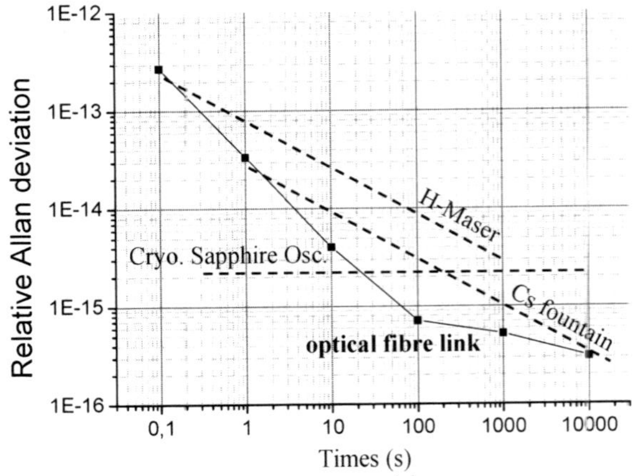

Figure 4 : Relative Allan deviation for the optical link, the H-maser, the cryogenic sapphire oscillator and the Cs fountain.

Allan deviation of these systems is less than 10^{-13} for 1s

3.2 The fs comb

The emission spectrum of a mode-locked femtosecond laser is composed of modes which are spaced by the laser repetition rate f_r (around 1 GHz in our case). It has been demonstrated that this space f_r is constant from one side to the other side of the spectrum at a level better than 3×10^{-17} [11]. This fundamental property permits us to use the femtosecond comb as a "frequency ruler".

Here no direct beatnote can be observed between the CO_2 laser emitting at 10 μm, that is 28 THz, and the frequency comb centered at 800 nm. However, we take advantage of the fact that the comb frequency width is larger than the CO_2 frequency. Then the CO_2 frequency can be compared to the frequency difference (n-m)f_r between two modes, where n and m are integers. This method avoids any measurement of the comb offset, which simplifies the experimental arrangement. The CO_2 laser frequency must just be transferred to the spectral range of the comb.

The measurement setup is depicted on Fig. 5. The laser diode at 852 nm is phase locked to a fs mode. The sum of the diode and CO_2 frequencies is generated in a crystal of $AgGaS_2$ and a second laser diode, at 788 nm, is phase locked to the sum. Finally a second fs mode is phase locked to the diode at 788 nm by feeding back to the fs cavity length. With the notation of Figure 5, the frequencies are related as: $f(CO_2) = qf_r \pm \delta_1 \pm \delta_2$.

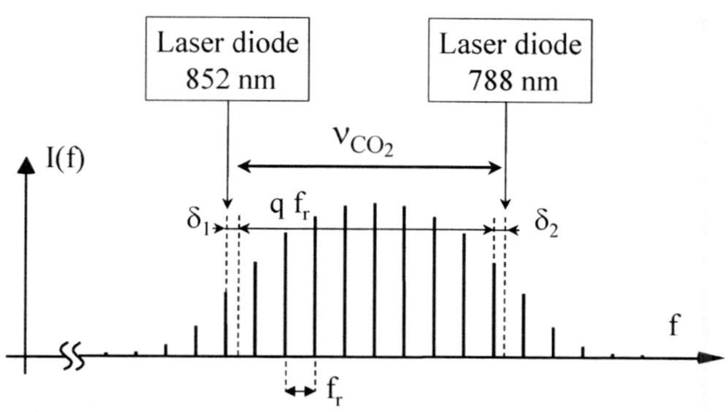

Figure 5 : Simplified schematic of lasers and frequencies involved in locking of the fs system to a CO_2 laser frequency. Frequency difference between two laser diodes is locked to the CO_2 frequency.

The primary datum measured is f_r. The integer *q* (approx. 28400) and the signs are established unambiguously because the CO_2 frequency is already well known.

A counter with 1 s gate is used to measure f_r. A computer then calculates the mean frequency and the Allan deviation from a long series of 1 s gate measurements.

3.3 A molecular clock

To measure the central fringe and to characterize the potential performance of the apparatus as a frequency standard, the central SF_6 fringe was employed as a discriminant. The OsO_4 continued to act as a fast discriminant to narrow the laser line while the absolute frequency was controlled by the SF_6 fringe. That way a molecular SF_6 clock was demonstrated. Logging of f_r then indirectly measured the central SF_6 fringe [13].

4 Metrological performance

4.1 A new local oscillator at 30 THz

In obtaining the highest resolution the most critical component is the first laser locked to OsO_4. To obtain fringes of periodicity 200 Hz at good contrast a stability of better than 20 Hz is required, which is difficult to obtain starting from a reference of FWHM 20kHz. The most significant requirement is spectral purity over times of order 1s, which can certainly be obtained using OsO_4 as the discriminant [2]. Stability, however, is also required over longer times, 100s and more for data collection and averaging. For that purpose, an additional active long-term stabilisation has been implemented by using the femtosecond comb.

The CO_2 reference laser, (1) in Figure 1, was phase locked by comparing the repetition frequency f_r with a synthesized frequency and using the error signal in a slow feedback loop, time constant approximately 10s, to the CO_2 laser (1). This provided the absolute frequency scale while the OsO_4 acted as a fast discriminant to narrow the laser line. The fringe spectrum could then be scanned by computer control of the frequency of synthesizer in the phase lock connecting the two CO_2 lasers.

4.2 Measurement of the central fringe

The central fringe was measured by alternating 5 up- and 5 down-scans of 500 Hz over the central fringe. 200 data points were recorded with a lock-in time constant of 0.1s, thus giving an averaging of 1s in a measurement lasting 200s. The individual measurements comprised 31 groups, each group consisting of successive measurements without any changes of experimental conditions. For maximum fringe signal, which corresponds to a pulse of $\pi/2$ in the absorption, the absolute frequency is:

$\nu(SF_6$, P(4) E^0, central fringe)=28 412 764 347 322.9 (14) Hz.

With this standard deviation of 1.4 Hz, that is 5×10^{-14} in relative value, the measurement technology is being pushed close to its limits, both in frequency measurements and in diagnostics. The limit is not yet in the performance of the Ramsey fringe experiment. The relative precision in the measurement of the 1GHz frequency is required to be better than the fringe precision; but this was clearly not the case for the first measurements. In addition the phase noise on the fibre link to BNM-SYRTE was not monitored during the measurements. It could result in significant frequency shifts, although these are below 1 Hz under our experimental conditions. Further the fs system is being used at the level of 1 Hz, a level of precision where satisfactory operation is non-trivial. There thus clearly remains a range of modifications and careful work to realize the ultimate performance.

4.3 Light shift

Light shift on the 2-photon transition was measured by direct measurement of the frequency and amplitude of the central fringe as a function of intensity in the cavity in U, as displayed in Figure 6. The plot of fringe amplitude against intensity then locates the $\pi/2$ point which agrees, well within its error, with the value calculated for the operational conditions. The shift in frequency from zero power to $\pi/2$ is $+0.75\pm0.5$ Hz where the accuracy is limited by the location of the $\pi/2$ point. The curve is almost flat around the $\pi/2$ point, having a slope of +0.06 Hz/mW. Indeed this is particularly significant for a potential standard. The current measurements are smaller than those earlier reported [14] for a He-seeded beam, which is undoubtedly due to the resolution. Most importantly the saturation power varies with hyperfine component. Thus, when the light power is increased in a situation of incomplete resolution the centre of gravity of the fringe pattern will shift. A simple calculation predicts a global shift of +5Hz for the fringes. However this must be modified to apply directly to individual hyperfine components and we are continuing to work on this question.

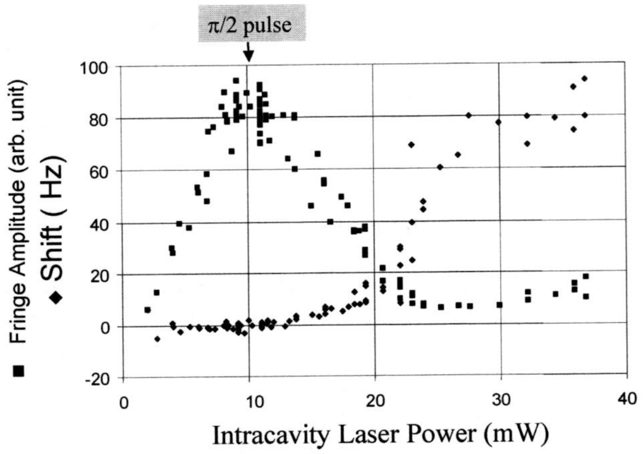

Figure 6 : amplitude and shift of central fringe as a function of laser power in the U Cavity.

5 Test for a time variation of the fundamental constants

With the resolution and reproducibility now obtained very sensitive tests of fundamental physics are a realistic goal. Time variation of fundamental constants presents an intriguing possibility. The current Ramsey fringe experiment measures the ratio of a vibrational frequency in the SF_6 molecule to a hyperfine frequency in an atom, since the Cs atomic fountain gives the ultimate reference. This ratio is mainly sensitive to the fine structure constant and to the ratio of proton to electron masses.

Following the line of argument of Flambaum [6], the variation of the frequency ratio can be expressed with the variation of α, the fundamental constant of the electromagnetic interaction, and with the quark mass m_q and the parameter Λ_{QCD}, which characterize the weak and strong interaction respectively :

$$\frac{\delta(\nu(Cs)/\nu(SF_6)))}{\nu(Cs)/\nu(SF_6))} = 2.83\frac{\delta\alpha}{\alpha} + 0.61\frac{\delta(m_q/\Lambda_{QCD})}{m_q/\Lambda_{QCD}}.$$

Moreover, modern theory of unification imposes constraints on the relative variation of this different couplings. Then the variation of the frequency ratio can be linked only to the variation of α, with an amplification factor of 25 [5, 9] : $\frac{\delta(\nu(Cs)/\nu(SF_6))}{\nu(Cs)/\nu(SF_6)} \approx 25\frac{\delta\alpha}{\alpha}$.

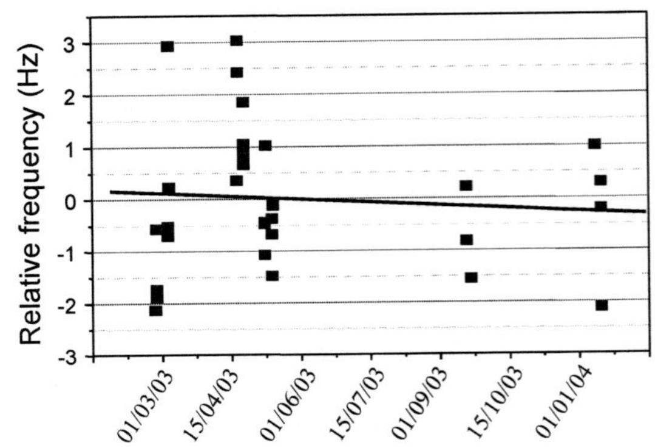

Figure 7 : Absolute frequency measurement of the P(4)E transition of the $2\nu_3$ band of SF_6 along 10 months

Figure 7 displays the SF_6 frequency measurements versus time over a period of 10 months. The relative time variation is below 10^{-13} per year which should result in an upper limit for the relative time variation of α of 4×10^{-15}.

6 Conclusion

The 2-photon Ramsey fringe experiment on SF_6 currently presents the best frequency standard available at 30 THz or 10 µm. The central fringe reproduces to 1.4 Hz. This situation results from two significant advances: the use of a 1 m zone separation giving a basic fringe periodicity of 200 Hz, and the implementation of a fs comb locking and measuring system related directly to the H maser standard.

With the reproducibility now obtained, limits has been placed on the time variations of the fine structure constant α. This result will certainly be refined in the near future. The reference to a better primary standard, the active control of the optical link to the primary standard, and the precise control of the SF_6 experiment parameters may pushed this limit an order of magnitude below.

References

[1] A. Amy-Klein *et al.*, "Absolute frequency measurement in the 28 THz spectral region with a femtosecond laser

comb and a long-distance optical link to a primary standard", *Appl. Phys. B* , **78**, pp. 25-30, (2004).
[2] V. Bernard *et al.*, "CO_2 laser stabilization to 0.1-Hz level using external electrooptic modulation", *IEEE J. of Quant. Electron.* , **QE-33**, pp. 1282-1287, (1997).
[3] Ch.J. Bordé, "Sur les franges de Ramsey en spectroscopie sans élargissement Doppler", *C. R. Séances Acad. Sc. Paris* , **B 284**, pp. 101-104, (1977).
[4] S. T. Cundiff, J. Ye, and J. L. Hall, "Optical frequency synthesis based on mode-locked lasers", *Review of Scientific Instruments* , **72**, pp. 3749-3771, (2001).
[5] V.V. Flambaum, private communication.
[6] V.V. Flambaum, "Limits on temporal variation of fine structure constant, quark masses and strong interaction", *Proceedings of the XVI° International Conference on Laser Spectroscopy* , pp. 49-57, (2004).
[7] C. Grain *et al.*, "High-sensitivity detection of two-photon Ramsey fringes at 30 THz by frequency-comb assisted stimulated emission", *IEEE J. of Quant. Electron.* , **38**, pp. 1406-1411, (2002).
[8] D.J. Jones *et al.*, "Carrier-envelope Phase Control of Femtosecond Mode-Locked Lasers and Direct Optical Frequency Synthesis", *Science*, **288**, pp. 635-639, (2000).
[9] P. Langacker, G. Segré, and M.J. Strassler, "Implications of gauge unification for time variation of the fine structure constant", *Physics Letters B* , **528**, pp. 121-128, (2002).
[10] F. Narbonneau et al, *Proceedings of EFTF04*, (2004).
[11] J. Reichert *et al.*, "Measuring the frequency of light with mode-locked lasers", *Optics Comm.*, **172**, pp. 59-68, (1999).
[12] G. Santarelli *et al.*, "Quantum Projection Noise in an Atomic Fountain : a High Stability Cesium Frequency Standard", *Phys. Rev. Lett.*, **82**, pp. 4619-4622, (1999).
[13] A. Shelkovnikov *et al.*, "Two-photon Ramsey fringes at 30 THz referenced to an H Maser / Cs fountain via an optical frequency comb at the 1 Hz level", *to be published in IEEE J. of Quantum Electronics* , (2004).
[14] A. Shelkovnikov *et al.*, "500-Hz two-photon Ramsey fringes with a SF_6 beam : towards a new frequency standard in the 30-THz spectral region", *Appl. Phys. B* , **73**, pp. 93-98, (2001).
[15] P. Wolf *et al.*, "Tests of Lorentz invariance using a microwave resonator", *Phys. Rev. Lett.*, **6**, pp. 9006, (2003).

HIGH STABILITY Nd:YAG LASER

S. A. WEBSTER, M. OXBORROW, P. GILL, National Physical Laboratory
Queen's Road, Teddington, Middlesex, TW11 0LW, UK
stephen.webster@npl.co.uk

We report on a high stability optical frequency source based on a Nd:YAG laser. A stable local oscillator is an essential component of an optical frequency standard and the techniques acquired through this research will inform the development of wavelength specific standards referenced to atomic absorptions. Light sources with a high degree of coherence are also of interest to the space-science community, namely in applications such as navigation, deep-space communication and gravitational-wave detection.

Light from a non-planar ring oscillator Nd:YAG laser is independently stabilised to two high finesse Fabry-Perot etalons constructed from ultra-low-expansivity glass. These are situated on separate vibration isolated platforms within an acoustic and vibration isolated chamber. A heterodyne beat between light separately stabilised to the two etalons has a relative fractional stability of 2×10^{-15} at 1s. The performance of active and passive vibration isolation platforms are compared and the limitations posed by servo control accuracy, vibrational instability and long-term drift will be discussed.

FIRST OBSERVATION OF THE $4d^{10}\,5s\;^2S_{1/2} \rightarrow 4d^9\,5s^2\;^2D_{3/2}$ TRANSITION BY DOPPLER-FREE TWO-PHOTON SPECTROSCOPY IN A SILVER ATOMIC BEAM

T. Badr*, Y. Louyer*, M.D. Plimmer*, P. Juncar*, M.E. Himbert*, G.D. Rovera†, J.D. Silver ††

* BNM-INM, CNAM, 292 rue Saint-Martin, F75003 Paris, France
† BNM-SYRTE, 61 avenue de l'Observatoire, F75014 Paris, France
†† Clarendon Laboratory, Parks Road, Oxford OX1 3PU, United Kingdom

Keywords: Frequency measurement, isotope shift, hyperfine structure, laser spectroscopy, silver atom.

Abstract

We have measured the frequencies of the hyperfine components of the transition $4d^{10}\,5s\;^2S_{1/2} \rightarrow 4d^9\,5s^2\;^2D_{3/2}$ in ^{107}Ag and ^{109}Ag, by Doppler-free two-photon laser spectroscopy of a thermal atomic beam and heterodyne calibration with respect to a molecular iodine line near 520THz. We deduce the value 1 040 706 327 (3) MHz for the centre of gravity of a mixture of natural abundance. For the isotope shift, we obtain $\nu(^{109}\text{Ag})-\nu(^{107}\text{Ag})=+599.6$ (2) MHz. We find the magnetic hyperfine splitting constants of the excited state to be $A(^{107}\text{Ag}(^2D_{3/2}))=-315.9$ (1) MHz and $A(^{109}\text{Ag}(^2D_{3/2}))=-363.3$ (1) MHz, an order of magnitude improvement over previous authors.

1 Introduction

The silver atom is an interesting candidate for an optical frequency standard based on narrow two-photon transitions from the $4d^{10}\,5s\;^2S_{1/2}$ ground state (*cf.* figure 1). Excitation of the $4d^9\,5s^2\;^2D_{3/2}$ state, whose natural width is about 4 kHz, requires two photons near 576 nm. Detection of this transition via cascade fluorescence is straightforward. The narrower $4d^9\,5s^2\;^2D_{5/2}$ level (natural width less than 1 Hz) can be excited using two photons near 661 nm and was proposed as a frequency standard many years ago [1]. This transition is harder to detect because the upper level is metastable. The group of H. Walther (Max Planck Institut für Quantenoptik, Garching) has cooled and trapped a sample of about 10^6 silver atoms [8]. In our laboratory, we use a thermal atomic beam with a first aim of observing both two-photon transitions and measuring their frequencies.

2 Current experiment

In this article, we describe a measurement of the frequency, isotope shift and hyperfine structure of the $4d^{10}\,5s\;^2S_{1/2} \rightarrow 4d^9\,5s^2\;^2D_{3/2}$ transition using Doppler-free two-photon laser spectroscopy. Our aim was to perform a pilot experiment to test the performance of our laser system, power build-up cavity, light collection optics and atomic beam apparatus for a narrow transition without the need to detect metastable atoms. For a given homogeneous linewidth, the excitation probability per unit laser power squared is expected to be similar for double quantum excitation of the $4d^9\,5s^2\;^2D_{3/2}$ and $4d^9\,5s^2\;^2D_{5/2}$ states. In other words, successful excitation of the former transition would be a stepping stone to an observation of the latter.

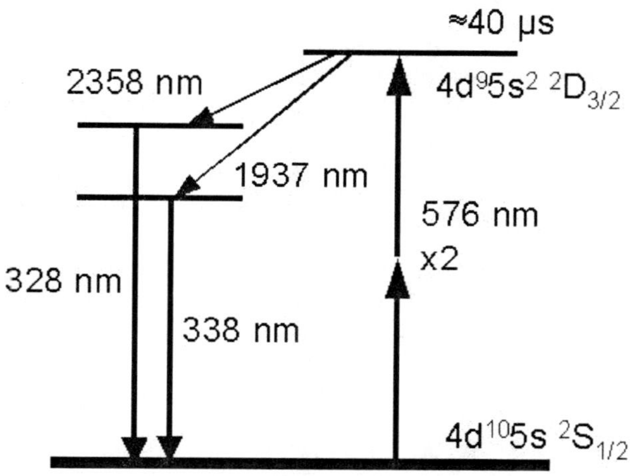

Figure 1 : partial energy diagram of silver.

2.1 Prior estimation of the transition frequency

Measurements by Nave [6] of the lines at 1937 and 2358 nm, combined with those of Pickering [7] at 328 and 338 nm provided two independent, accurate determinations of the transition frequency.

2.2 Apparatus

The apparatus is shown in figure 2. The atomic beam unit is similar to that described in [5]. A tantalum oven filled with silver powder is placed in a vacuum chamber and heated up to 1200°C to produce a thermal beam.
The laser used to probe the silver transition is a continuous ring dye laser. A microwave sideband method [3] is used to lock the output frequency of the laser to the centre of a fringe of a hemiconfocal Fabry Perot cavity (finesse 600 at 576 nm).

The error signal gave an estimate of the frequency jitter with respect to the cavity of below 10 kHz for a 1 s integration time. We scan the probe laser frequency by varying the length of the FP cavity.

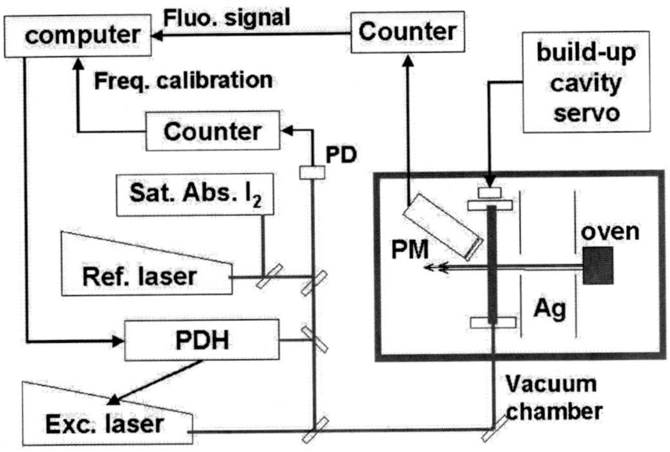

Figure 2: Experimental setup. PDH: Pound-Drever-Hall stabilisation system. PM: photomultiplier. PD: photodiode. Exc laser: excitation laser. Ref laser: reference laser. Sat abs I_2:saturated absorption spectroscopy in molecular iodine

The reference laser is a second continuous ring laser. We lock the frequency of this laser to that of the hyperfine component a_3 of the line R111(18-1) in $^{127}I_2$. This component is observed using saturated absorption spectroscopy in an iodine cell of length 50 cm. The beat signal between the probe laser and the reference laser provides direct calibration of the frequency axis.
To increase the two-photon transition probability a FP cavity (finesse ~300) is placed in the vacuum chamber around the interaction region. The length of the cavity is servo-locked to maintain resonance with the laser. In this way we could obtain intra-cavity powers of over 12 W for a dye laser output power of 300 mW.
We detected the fluorescence at 328 and 338 nm resulting from the cascade decay of the 6p $^2P_{3/2}$ level using a photomultiplier in pulse counting mode linked to a counter. A typical background of 50 000 counts per second arose from unfiltered oven light. The noise on this background, a few hundred counts per second, is negligible in comparison with the signals observed.

2.3 Experimental procedure and data analysis

We have scanned the probe laser frequency over a range of about 40 MHz around each of the three components for each of the two stable isotopes ^{107}Ag and ^{109}Ag. For instance, the ^{107}Ag(F=0→F'=2) component is shown on figure 3. Signal to noise ratio was about 100. Although the theoretical lineshape arising from velocity distribution in the atomic beam, radial distribution of energy in the probe laser beam, is somewhat complicated, to analyse the data, we fitted gaussian profiles to the individual components. . Typical single-photon linewidths were then 500-700 kHz, mainly limited by transit-time broadening and Zeeman splitting. Each of the six components has been scanned and fitted five times. Given the excellent signal to noise ratio, we could determine the line centre of each component accurately, with a standard deviation over the five scans of less than 50 kHz.

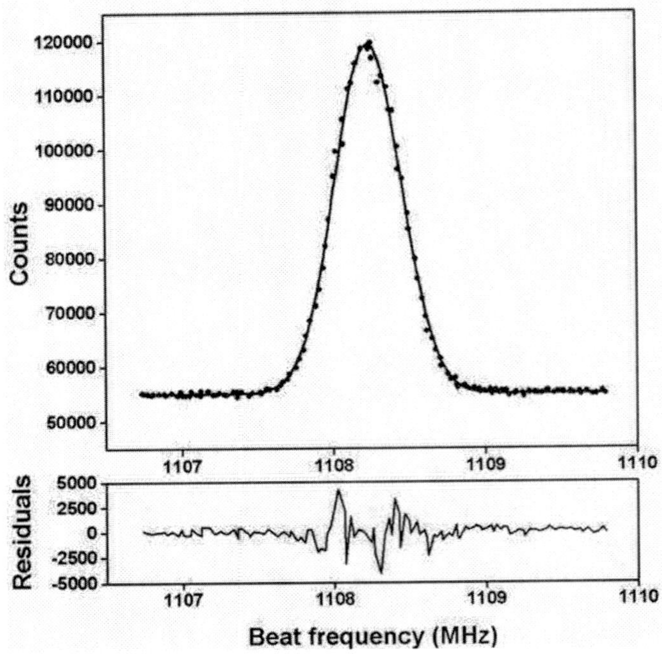

Figure 3: Upper graph : hyperfine component F=0→F'=2 in ^{107}Ag. Dots are experimental data, solid line is a fit by a gaussian profile. Lower graph: residuals of the fit.

2.4 Systematic effects

In the following section we discuss the various perturbations affecting the values of the transition frequencies.
The accuracy of the frequency of the $^{127}I_2$ reference line, deduced using the IodineSpec programme from TOPTICA is considered by its developers to be 1.5 MHz. A further consideration is the lock point of the reference laser. Drifts of the cell temperature (about 500 Hz / °C) have a negligible effect upon the frequency compared with the uncertainty of the iodine line itself.
The first-order Doppler shift is absent because we use counter propagating laser beams and excitation occurs at the beam waist. The second-order Doppler shift, about 2.5 kHz for our thermal beam (most probable velocity 660 ms^{-1}) is negligible. The D.C. Stark shift, about 1 kHz, is negligible too. In fact, only two atomic perturbations have any importance at our level of accuracy: the light shift due to the excitation laser and the Zeeman shift.
For an optical power in the build-up cavity of 12 W concentrated in a waist of w0=320 μm, the light shift is estimated to be about -15 kHz, which is comparable with curve fitting uncertainties.
We neither shielded nor compensated the ambient magnetic field, measured to be 800 (50) mG using a 3-axis Hall effect probe. As expected, the Zeeman effect leads to a splitting of the components of the (F=1→F'=1) and (F=1→F'=2) transitions. Since we use linearly polarised light, this splitting

is symmetric and leads to a broadening but no overall shift of the line centre to first order. In addition, because there is an angle of 20° between the vertical electric vector of the laser and the magnetic field, the component ($F=0 \rightarrow F'=2$) is also broadened.

3 Measurements and uncertainties

We measured individually the frequencies of the three hyperfine components for the stable isotopes ^{107}Ag and ^{109}Ag, with respect to the component a_3 of the *R111(18-1)* line in iodine. Uncertainty upon these frequency is 80 kHz due to statistical fluctuations of the determination of the line center, and possible asymetry of the iodine reference line. From their separations we calculated the hyperfine splittings of the ground and excited states and hence the magnetic dipole hyperfine constants A. The difference of the centres of gravity for each isotope gave the isotope shift. Finally, to compare our results with previously published values, we calculated the centre of gravity of the spectrum of a naturally abundant sample containing 51.84 % ^{107}Ag and 48.16 % ^{109}Ag. Our measurements are summarized in table 1 and table 2.

Level	*Isotope*	*This work*	*Others*
$4d^{10}\ 5s\ ^2S_{1/2}$	107	-1712.54 (20)	-1712.512111 (18)
	109	-1976.85 (20)	-1976.932075 (17)
$4d^9\ 5s^2\ ^2D_{3/2}$	107	-315.9 (1)	-315.7 (1.6)
	109	-363.3 (1)	-363.1 (1.8)

Table 1: Hyperfine coupling constants (MHz) for the levels $4d^{10}\ 5s\ ^2S_{1/2}$ and $4d^9\ 5s^2\ ^2D_{3/2}$ in ^{107}Ag and ^{109}Ag.

Our values of the hyperfine constants of the ground state (see table 1) lie within 100 kHz of the far more accurate radio-frequency measurements of Dahmen and Penselin [2]. The discrepancy is somewhat larger than our fitting errors and is most probably due to a drift in the lock point of the reference laser frequency and differential light shifts. Our values of the hyperfine constants for the excited state $4d^9\ 5s^2\ ^2D_{3/2}$ are more than an order of magnitude more accurate than those of [4] with which they agree. Our measurements enable the first direct determination of the isotope shift of the transition. The value we find 599.6 (0.2) MHz agrees remarkably well with that we estimated from the sum of isotope shifts of the lines at 328 nm [8] and 2358 nm [6], namely 599.5 (16.0) MHz.
of the lines at 328 and 338 nm [7] and 1937 and 2358 nm [6] whose standard uncertainties are given as 10 to 30 MHz.

	This work	*Others*
Freq. ^{107}Ag	1 040 706 038.0 (3.0)	1 040 706 159 (32)
Freq. ^{109}Ag	1 040 706 637.6 (3.0)	1 040 706 758 (32)
Freq. of both	1 040 706 326.8 (3.0)	1 040 706 458 (32)
Isotope shift	+599.64 (14)	-

Table 2: Frequency (MHz) and isotope shift (MHz) of the $4d^{10}\ 5s\ ^2S_{1/2} \rightarrow 4d^9\ 5s^2\ ^2D_{3/2}$ transition.

By contrast, as regards the optical frequency of the transition (1 040 706 327 (3) MHz) there is no close agreement with other measured values deduced from Fourier transform spectroscopy hollow cathode discharge emission lines (see table 2). Our uncertainty of 3.0 MHz is dominated by that of the molecular iodine reference line. The disagreement with the value of [7] (1 040 708 682 (210) MHz a difference of 2 GHz or 10 standard deviations) confirms their suspicion of problems due to line blending. Finally, our values lie about 120 MHz below those deduced from recent measurements of the frequencies.

4 Conclusion

We have performed measurements of the frequency, isotope shift and hyperfine coupling constants of the $4d^{10}\ 5s\ ^2S_{1/2} \rightarrow 4d^9\ 5s^2\ ^2D_{3/2}$ transition in atomic silver. This work can be regarded as an encouraging step towards a study of the very narrow $4d^{10}\ 5s\ ^2S_{1/2} \rightarrow 4d^9\ 5s^2\ ^2D_{5/2}$ transition. Our measurement of the frequency, obtained using laser heterodyne techniques lies about 120 MHz below values deduced from measurements of different transitions by other authors using Fourier transform spectroscopy. We have performed the first direct measurement of the isotope shift of this transition. Our value is compatible with those estimated by comparison with other transitions in silver. Our values for the hyperfine splitting constants of the $4d^9\ 5s^2\ ^2D_{3/2}$ level are 15 times more accurate than those of previous results.

References

[1] P.L. Bender *et al*, "Candidates for two-photon optical frequency standards", Bull. Am. Phys. Soc., 21:599, 1976.

[2] H. Dahmen and S. Penselin, "Measurement of nuclear magnetic dipole moment of ^{197}Au and hyperfine structure measurements in the ground states of ^{197}Au, ^{107}Ag, ^{109}Ag and ^{39}K", Z. Phys. 200:456-466, 1967.

[3] R.W.P. Drever *et al*, "Laser phase and frequency stabilization using an optical resonator", Appl. Phys. B, 31:97-105, 1983.

[4] W. Fischer *et al*, Die Hyperfeinstruktur des Übergangs Ag I", Z. Phys., 216:136-141, 1968.

[5] S. Guérandel *et al*, "Frequency measurement, isotope shift and hyperfine structure of the $4d^9\ 5s^2\ ^2D_{5/2} \rightarrow 4d^{10}\ 6p\ ^2P_{3/2}$ transition in atomic silver", Eur. Phys. J. D, 10:33-38, 2000.

[6] G. Nave, 2003, private communication.

[7] J.C. Pickering and V. Zilio, "New accurate data for the spectrum of neutral silver", Eur. Phys. J. D, 13:181-185, 2001.

[8] G. Uhlenberg, J. Dirscherl and H. Walther. "Magneto-optical trapping of silver atoms", Phys. Rev. A, 62(6):063404-1-063404-4, 2000.

A SIMULTANEOUS CALIBRATION OF THE IEN/PTB TIME LINK BY GPS CV AND TWSTFT PORTABLE EQUIPMENT

W. Lewandowski[1], F. Cordara[2], L. Lorini[2], V. Pettiti[2], A. Bauch[3], D. Piester[3], O. Koudelka[4]

[1] Bureau International des Poids et Mesures, Sèvres, wlewandowski@bipm.org
[2] Istituto Elettrotecnico Nazionale Galileo Ferraris, Italy, cordara@ient.it
[3] Physikalisch-Technische Bundesanstalt, Germany, andreas.bauch@ptb.de
[4] Technical University Graz, Austria, koudelka@inw.tu-graz.ac.at

Keywords: clock comparison, equipment calibration, GPS, TWSTFT.

Abstract

The differential calibrations of GPS Common-View time-receiving equipment have been made for over twenty years. Until recently, TWSTFT calibrations have been rare due to a lack of suitable equipment. Now, the Technical University of Graz is providing such a calibration service, and this paper describes a simultaneous differential calibration of the IEN/PTB time link by GPS CV and TWSTFT. The GPS calibration, although of higher uncertainty that the TWSTFT one, provides an independent check of TWSTFT calibration technique.

1 Introduction

Two-Way Satellite Time and Frequency Transfer (TWSTFT) has been used for almost five years for the computation of International Atomic Time (TAI). At present, there are twelve TWSTFT time links contributing to TAI and several others are being prepared. Most of these links have been calibrated by GPS C/A-code Common-View (CV) with a standard uncertainty of about 5 ns. However, TWSTFT links can be calibrated with an uncertainty of 1 ns or better if portable TWSTFT equipment is used. Until recently, such calibrations have been rare due to a lack of suitable equipment. The first TWSTFT calibration was performed in 1991 between the Technical University of Graz (TUG) and the Observatoire de la Côte d'Azure (OCA) [1]. Now, the TUG is providing a TWSTFT calibration service available on request. Also the United States Naval Observatory (USNO) is involved in a number of TWSTFT calibrations in the United States and between the USNO and the Physikalisch-Technische Bundesanstalt (PTB): the calibration of June 2002 is described in [2] and that of January 2003 in [3].

This paper describes a simultaneous differential calibration of GPS CV and TWSTFT equipment located at the Istituto Elettrotecnico Nazionale Galileo Ferraris (IEN), Turin, Italy, and the PTB, Braunschweig, Germany. The two laboratories are separated by about 835 km. The BIPM portable GPS receiver and the TUG portable TWSTFT station were used for this exercise. The GPS calibration was performed between 28 May - 7 June 2003, and the TWSTFT calibration between 30 May - 6 June 2003. The TWSTFT calibration is described in detail in [4, 5].

2 Uncertainties of TAI time links

There are three major elements in the construction of International Atomic Time (TAI) and ultimately Coordinated Universal Time (UTC): clocks, some means of comparing remote clocks (time transfer), and a time-scale algorithm. The uncertainties of time transfer can affect the stability of TAI and the accuracy of the access to UTC of the contributing laboratories. The uncertainties of Type A (uA) and Type B (uB) of TAI time links are the subject of several studies [6], and have been published in the BIPM Circular T since February 2004.

Table 1 provides time link uncertainties published in the BIPM Circular T 194 for IEN/PTB and USNO/PTB time links [7]. The IEN reference clock is a commercial cesium atomic clock, the PTB clock is a laboratory cesium clock, and the USNO clock is a hydrogen maser. The type of clock might have some impact on the determination of statistical uncertainty uA. uB, the systematic part of the uncertainty, is driven mainly by the quality of calibration. The IEN/PTB TWSTFT link was calibrated by GPS C/A-code common-view (GPS CV) at the beginning of 2002. This means that the TWSTFT and GPS equipment were referenced to the same clocks at each laboratory, and the TWSTFT equipment delay was determined so that both time comparison results agreed with each other. The

USNO/PTB TWSTFT link was initially calibrated by a transportable TWSTFT station, but this calibration was lost by the change of Intelsat satellite in September 2003, and some changes in the PTB TWSTFT hardware [3]. This calibration could be partially recovered by bridging the USNO/PTB TWSTFT link between the two Intelsat satellites using a so-called GPS P3 link [3, 8]. The calibration dates in Table 1 indicate the most recent calibration results.

2 Set-up of differential calibration

For GPS calibration, the portable equipment comprised the BIPM GPS CV receiver, its antenna, and a calibrated antenna cable. The laboratories visited supplied: (a) a 10 MHz reference signal and (b) a series of 1 s pulses from the local reference, UTC(k), via a cable of known delay. In each laboratory, the portable GPS CV receiver was connected to the same clock as the local GPS CV receiver and the antenna of the portable receiver was placed close to the local antenna. The differential coordinates of the antenna phase centres were known at each site with standard uncertainties (1σ) of a few centimetres.

The processing of the comparison data obtained in laboratory k consisted of computing, for each track i, the time differences:

$$dt_{k,i} = [UTC(k) - GPS\ time]_{\mathrm{BIPM},i} - [UTC(\mathrm{k}) - GPS\ time]_{k,i}\,.$$

Then from sets of above data recorded at two visited laboratories a differential time correction to be added to [*UTC*(IEN) – *UTC*(PTB)] was computed. More extensive description of this calibration is provided in [9].

A detailed description of the TWSTFT equipment and the calibration procedure conducted by the TUG are provided in [4, 5].

Results of TWSTFT and GPS CV differential calibrations were referenced to the same clock at each laboratory.

3 Results

Final results of the TWSTFT and GPS CV simultaneous differential calibration exercise described in this paper are given in Table 2. In order to obtain accurate time comparisons between the IEN and PTB, respective corrections should be added to TWSTFT and GPS CV time transfer results between these two laboratories.

Concerning GPS CV differential calibrations between the IEN and PTB, several exercises have been conducted since 1986. Results of these are given in Table 3. We observe a discrepancy in two cases exceeding the uncertainty of calibration. This might be due to delay changes of GPS equipment as well as to some unnoticed delay changes during laboratory hardware updates. Table 3 underlines the need for frequent hardware checks by means of calibrations.

There are a few repeated TWSTFT differential calibrations, but a consistency of about 1 ns has already been seen at the USNO [2].

4 Comparison of TWSTFT and GPS CV

The ultimate goal of this paper is to compare TWSTFT and GPS CV time transfer between the IEN and PTB after correcting the results of each of these techniques by the calibration corrections given in Table 2. Tables 4 to 6 show application of these corrections from the beginning of June 2003 (period of calibration) to the end of July 2003. The differences between TWSTFT and GPS CV are affected only by the noise of the two techniques, but no longer from the instabilities of the time scales involved. The noise of compared clocks was cancelled in consecutive differences. In Table 5 and 6 we document the excellent agreement between the two techniques after application of respective calibration corrections: the mean difference for the period of two months following the calibration is 0.8 ns with a rms deviation of 1.7 ns. This is a result far below the involved uncertainties of TWSTFT calibration (1 ns) and GPS CV calibration (3 ns).

Next, the comparison of the two techniques was extended to a period of nine months (4 June 2003 – 29 February 2004). This is illustrated by Figure 1 and 2. We observe a departure of a few nanoseconds with an rms of 2.6 ns. This slight deterioration of the difference between the two techniques during a longer period of time might be explained by some delay changes in the GPS CV hardware. This kind of behaviour of GPS CV hardware is frequently observed as described in [6, 10, 11] and shown by Table 3. Usually the TWSTFT technique exhibits much lower noise. It should be also noted that in September 2003 due to the satellite change used by TWSTFT, some TWSTFT delays (as Sagnac delay) needed to be readjusted. These delays were revaluated with uncertainty of 1 ns [5] and should not be the cause of the slight 'departure' between the two techniques observed in Figures 1 and 2.

It should be noted that the independent TWSTFT calibration reported in this paper differs by 1 ns with the calibration of TWSTFT by GPS CV used in Circular T of February 2004 [5, 10]. It is an excellent agreement and worth noting. However, the GPS CV calibration used by Circular T at the beginning of 2002 was some years 'old', and regarding data of Table 3 and Figures 1 and 2, we should consider this excellent agreement more as a fortunate coincidence than proof of long-term longevity of GPS CV calibration. The excellent agreement of simultaneous TWSTFT and GPS CV calibrations reported in this paper is more realistic, but again we should be aware that this is for the period of comparison and may not remain the same in the future.

5 Conclusion

A simultaneous differential calibration of the IEN/PTB time link by GPS CV C/A-code and TWSTFT, reported in this paper, differ by about 0.8 ns. The difference between the two calibration techniques is well below the involved uncertainties: 3 ns for GPS CV C/A-code and 1 ns for TWSTFT. This is an excellent check of the performance of the two methods of calibration.

New TWSTFT calibrations are expected in summer 2004. It would be of interest to conduct at the same time GPS calibrations. Also repeated calibrations of the same link are necessary to monitor possible hardware or other delay changes ('aging' of calibration), as illustrated by Table 3.

The results of the IEN/PTB TWSTFT calibration were applied in March 2004 to the IEN/PTB TWSTFT link time contributing to TAI. Consequently, Section 6 of BIPM Circular T (see Table 1) will be updated regarding the stated Type B uncertainty.

References

[1] D. Kirchner, H. Ressler, P. Grudler, F. Baumont, C. Veillet, W. Lewandowski, W. Hanson, W.J. Klepczynski, P. Uhrich, "Comparison of GPS Common-View and Two-Way Satellite Time Transfer Over a Baseline of 800 km", *Metrologia*, **30,** pp. 183-192, (1993).

[2] D. Matsakis, "Time and Frequency Activities at the U.S. Naval Observatory", Proceedings of the 34th PTTI, pp. 437-456, (2003). (Note that the Proceedings appear in the calendar year after the Conference).

[3] D. Piester, A. Bauch, J. Becker, and T. Polewka, A. McKinley and D. Matsakis, "Time Transfer Between USNO and PTB: Operation and Calibration Results", to be pulished in the *Proceedings of the 35th PTTI*, (2004).

[4] D. Piester, A. Bauch, J. Becker, and T. Polewka, "An Update on PTB's Activities in Time and Frequency", to be pulished in the *Proceedings of the 35th PTTI*, (2004).

[5] F. Cordara, L. Lorini, V. Pettiti, P. Tavella, D. Piester, J. Becker, T. Polewka, G. de Jong, O. Koudelka, H. Ressler, B. Blanzano, C. Karel, "Calibration of the IEN-PTB TWSTFT link with a portable reference station", in these *Proceedings*, (2004).

[6] J. Azoubib, W. Lewandowski, "Uncertainties of time links used for TAI", *34th PTTI*, Reston, Virginia, pp. 413-424, (2003).

[7] BIPM Circular T 194, available at ftp://ftp2.Bipm.fr/pub/tai/publication/cirt.194, (March 2004).

[8] P. Defraigne, G. Petit, "Time transfer to TAI using geodetic receivers", *Metrologia*, **40**, pp. 184-188, (2003).

[9] W. Lewandowski, L. Tisserand, 'Determination of the differential time corrections for GPS time equipment located at the OP, NPL, IEN, PTB, and VSL", *Rapport BIPM-2004/05*, (2004).

[10] J. Azoubib, W. Lewandowski, *23rd BIPM TWSTFT Report*, available at http://www.bipm.fr:/en/scientific/tai/time_ftp.html, (2003).

[11] W. Lewandowski, R. Tourde, "Sensitivity to the external temperature of some GPS time receivers", *Proc. 22nd PTTI*, pp. 307-316, (1991).

Link	Type	*u*A/ns	*u*B/ns	Calibration Type	Calibration Dates
IEN /PTB	TWSTFT	1.0	5.0	LC (GPS SC)	2002 Feb
USNO/PTB	TWSTFT	0.5	3.0	BC(GPS P3)	2003 Sep

Table 1. Some time link uncertainties of Type A (*u*A) and Type B (*u*B) published in the BIPM Circular T.

Method	*d*/ns	*u*(*d*)/ns
TWSTFT	–1.0	1.0
GPS C/A-code	+3.2	3.0

Table 2. Results of TWSTFT and GPS C/A-code differential calibrations of the IEN/PTB time link. Differential time correction *d* to be added to [*UTC(IEN) – UTC(PTB)*], and its estimated uncertainty *u*(*d*) for the period of comparison (1σ).

Date	*d*/ns	*u*(*d*)/ns
Oct 1986	–9.4*	3.0
July 1997	–1.0*	3.0
Dec 1997	–1.0*	3.0
Mar 1998	+3.0*	3.0
Mar 2002	+7.0	3.0
Jun 2003	+3.2	3.0

Table 3. Some past GPS calibrations between IEN and PTB. *d* is the differential time correction to be added to [*UTC*(IEN) – *UTC*(PTB)], and *u*(*d*) is the estimated uncertainty for the period of comparison.

Note*: In July 1999 the IEN GPS receiver delay was corrected by +18 ns

Date 2003 (MJD)	[*UTC(IEN) – UTC(PTB)*] /ns		(TWSTFT–GPS) /ns
	TWSTFT	GPS	
4 June (52794)	–30.1	–35.4	5.3
9 June (52799)	–38.6	–43.8	5.2
14 June (52804)	–47.1	–49.9	2.8
19 June (52809)	–52.6	–59.4	6.8
24 June (52814)	–49.5	–53.2	3.7
29 June (52819)	–49.0	–54.5	5.5
4 July (52824)	–53.6	–58.1	4.5
9 July (52829)	–48.4	–55.3	6.9
14 July (52834)	–55.2	–60.3	5.1
19 July (52839)	–61.9	–70.1	8.2
24 July (52844)	–54.0	–57.1	3.1
29 July (52849)	–56.7	–59.8	3.1

Table 4. IEN/PTB TWSTFT and GPS C/A-code links before applying calibration corrections.

Date 2003 (MJD)	[*UTC(IEN) – UTC(PTB)*] /ns		(TWSTFT–GPS)/ns
	TWSTFT calibrated	GPS calibrated	
4 June (52794)	–31.1	–32.2	1.1
9 June (52799)	–39.6	–40.6	1.0
14 June (52804)	–48.1	–46.7	–1.4
19 June (52809)	–53.6	–56.2	2.6
24 June (52814)	–50.5	–50.0	–0.5
29 June (52819)	–50.0	–51.3	1.3
4 July (52824)	–54.6	–54.9	0.3
9 July (52829)	–49.4	–52.1	2.7
14 July (52834)	–56.2	–57.1	0.9
19 July (52839)	–62.9	–66.9	4.0
24 July (52844)	–55.0	–53.9	–1.1
29 July (52849)	–57.7	–56.6	–1.1

Table 5. IEN/PTB TWSTFT and GPS CV links after adding calibration corrections: –1.0 ns to TWSTFT link and +3.2 ns to GPS link.

IEN/PTB Links	(TWSTFT – GPS) /ns	
	Mean	rms
Before calibration	5.0	1.7
After calibration	0.8	1.7

Table 6. Differences between TWSTFT and GPS CV before and after applying calibration corrections to IEN/PTB links for the period 4 June – 29 July 2003.

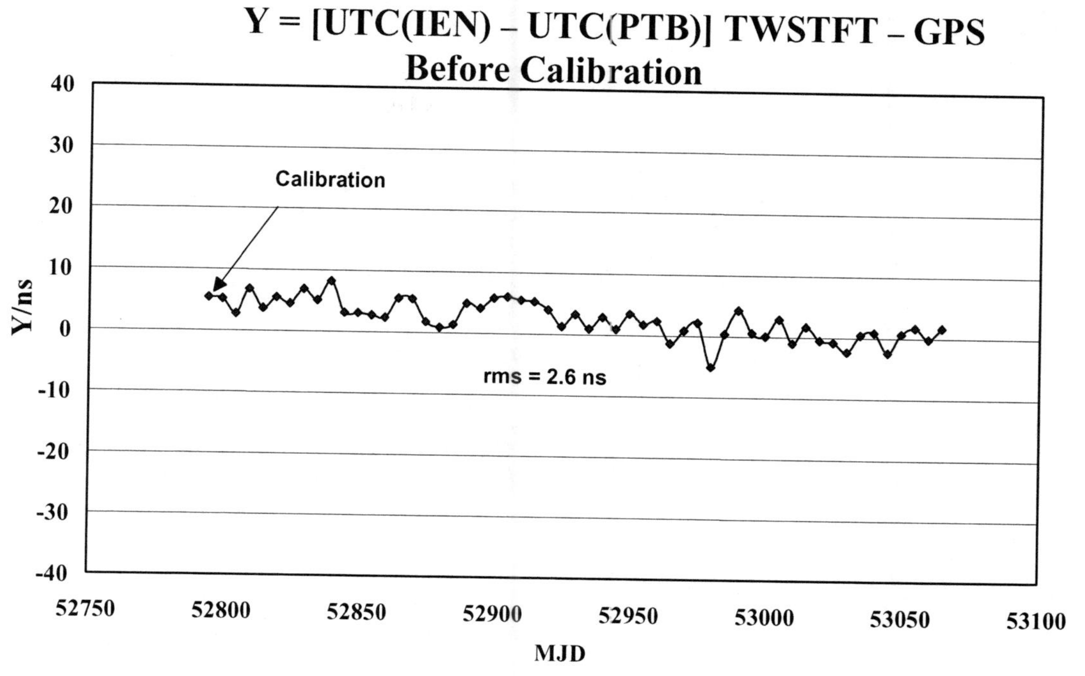

Figure 1. Differences between TWSTFT and GPS CV before and after applying calibration corrections to IEN/PTB links for the period 4 June 2003 – 29 February 2004.

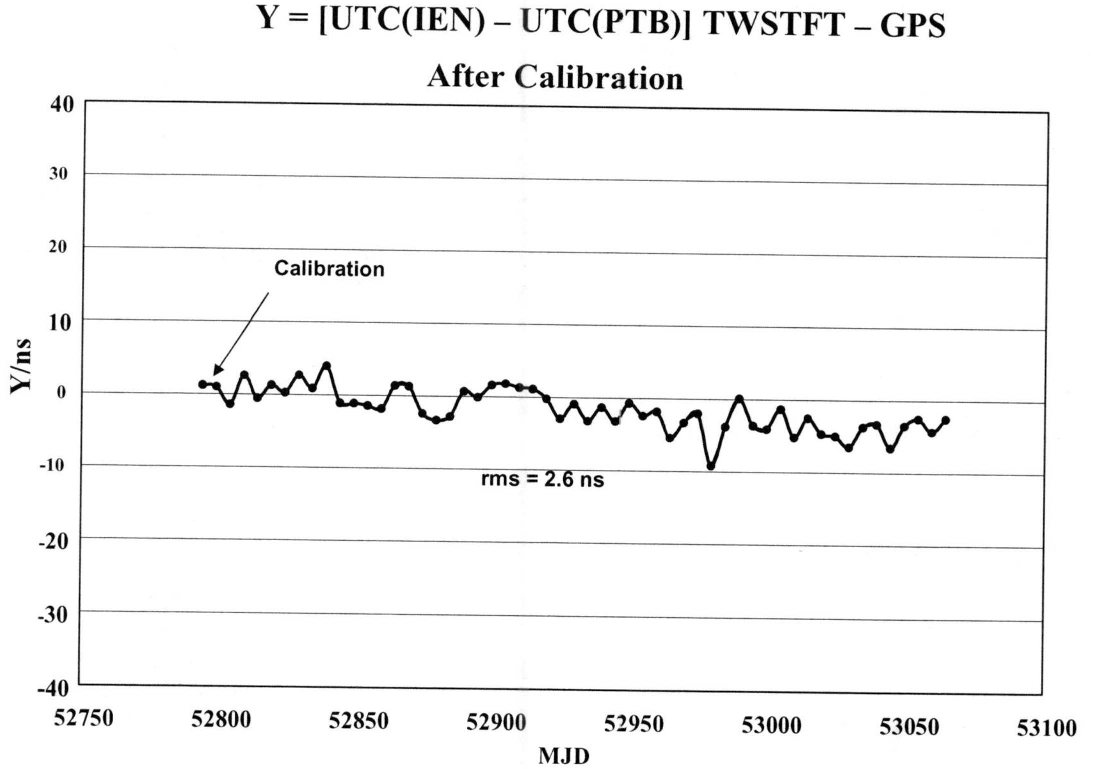

Figure 2. Differences between TWSTFT and GPS CV after applying calibration corrections to IEN/PTB links for the period 4 June 2003 – 29 February 2004.

FREQUENCY TRANSFER USING PRECISE POINT POSITIONING*

M.Weiss[1], P. Fenton[2], E. Powers[3], K. Senior[4]

[1]National Institute of Standards and Technology, Boulder CO, USA, ++1-303-497-6461 (Fax), mweiss@boulder.nist.gov
[2]Novatel Inc., Calgary, Alberta, CANADA, Patrick.Fenton@novatel.ca
[3]U.S. Naval Observatory, Washington, DC, USA, powers.edward@usno.navy.mil
[4] Naval Research Laboratory, Washington, DC, USA, Ken.Senior@nrl.navy.mil

Keywords: carrier phase, frequency transfer, global positioning system, precise point positioning, time transfer

Abstract

Using Precise Point Positioning (PPP) we show the potential for comparing frequency standards with time stabilities of under 300 ps in flicker phase noise for time periods of 1 to 10 d. This leads to comparisons at 1 part in 10^{15} or better at 10 d. We compare short baseline PPP results with 2 hour data from the NIST Measurements System. We compare results from the NIST – USNO baseline with hourly Two-Way Satellite Time and Frequency Transfer results.

1 Introduction

The focus of this work is on aiding the comparison of state-of-the-art frequency standards. Current standards require comparisons at 1 part in 10^{15} or better. Frequency comparisons of such standards are needed across intercontinental baselines, such as between the National Institute of Standards and Technology (NIST) in Boulder, Colorado, USA and Europe. Currently, the best methods for comparison are Two-Way Satellite Time and Frequency Transfer (TWSTFT) and GPS carrier-phase frequency transfer using a geodetic software. Such software packages have been designed for the best possible global positioning using GPS for applications such as tracking movement of continental plates. A comparison between the standards at NIST and PTB showed transfer noise was negligible after 15 d [8].

Another method was reported by Fenton et. al. [2] using the Wide Area Augmentation System (WAAS) satellite and high-gain dishes. They performed an experiment between the National Research Council (NRC) of Canada and the US Naval Observatory (USNO) in Washington DC, USA, using the code-carrier difference to estimate changes in the ionospheric delay. The Allan deviation of the time transfer using this method was less than 10^{-15} after an averaging time of 8 h.

Whereas previous studies of GPS carrier-phase frequency transfer have generally been done using a network of stations [3,5,6], in this case we compute for each single location separately. We accept the International GPS Service (IGS) post-processed estimates of satellite ephemerides and clocks [4,12], hold the receiver coordinates fixed, and estimate the local clock against the IGS time scale. We do this for two receivers, then difference the results to cancel the IGS time, obtaining the comparison of the clocks driving the receivers. We evaluate these results by comparing with independent measurement.

The receivers used in this experiment were all of the same manufacture and model, though with somewhat different firmware. The same company made the antennas that made the receivers. The antenna models were different, but of similar design. The antennas were not temperature-controlled at all. The receiver temperatures were controlled only in that they were inside laboratories.

We report data using two baselines: a short-baseline comparison at NIST, and comparisons between NIST and at USNO when hourly TWSTFT measurements were also available.

2 Precise Point Positioning

In Precise Point Positioning (PPP) we are able to estimate the clock driving a single receiver against the IGS Time Scale, typically with deviations across satellites at a common time of under 2 cm, about 67 ps. To do this we minimize the variables that we must estimate by fixing previously estimated values for antenna phase center position, satellite orbits, and satellite clocks. The software must estimate or obtain the usual parameters used for geodetic positioning: satellite antenna phase center, phase wind up, Earth tides, and C/A to P code offsets. Errors in these estimates are small compared to the instability of the results. The need to estimate ionospheric delay is eliminated by using the ionosphere-free combination of the phases at the two L-band frequencies, P1 and P2. The main sources of instability can be separated into hardware and software. Hardware instabilities in delay due to temperature dependence, multi-path interference, and impedance mismatch produce instabilities in the clock measurements. Likely candidates for instabilities from software include tropospheric delay estimates and effects due to the software design. In particular, the use of a continuous filter spanning many days is a significant departure from the more common use of independent 1-day arcs. The comparison of these two modes merits further study.

We report here two different software systems run in PPP mode. We estimated frequency-transfer results over all intervals discussed in this paper using a commercial software. This software was developed based on code from the Canadian Geodetic Survey Division, Geomatics Canada, Natural Resources Canada. The GIPSY software [11], developed by the Jet Propulsion Laboratory (JPL), was used on a 40 d run of the short-baseline experiment.

3 Short-Baseline Experiment

The short-baseline experiment at NIST involved a receiver driven by an H-maser and another driven by an OCXO in two comparisons, the first lasting 40 d and the second 254 d. Unfortunately, the OCXO receiver had no system in place for measuring or controlling phase changes upon reset. The H-maser receiver, however, had a 1 pps measurement system. The two receivers, of the same manufacture and model, though slightly different firmware versions, were located in separate rooms and different floors of the NIST laboratory in Boulder. Because of the resets in the OCXO-locked receiver, we analysed the data in segments. The two clocks were compared in the NIST measurement system every two hours with an accuracy better than 50 ps.

3.1 A 40 Day Short-Baseline Experiment at NIST

As we see in Figure 1, the difference between the PPP results and the measurement system were peak-to-peak about 1 ns. The Time Deviation (TDEV) is a statistic useful for characterizing the time stability of time and frequency comparison systems [1]. Looking at TDEV of the longest segment in Figure 2, we see that the noise type was consistent with a flicker phase modulation (PM) model at 100 ps or less [9].

3.2 A 48-day interval without breaks

This interval was part of a 205 d interval of study. Again, the receiver locked to the OCXO had numerous resets. We were able to obtain a 48 d interval without breaks to compare the commercial PPP results to the NIST measurement system. The difference between these two data sets is shown below in Figure 5. The TDEV of these data is shown in Figure 6. The very short term here shows a much higher noise level. But from 1 d on, the data are consistent with a flicker PM model under 300 ps.

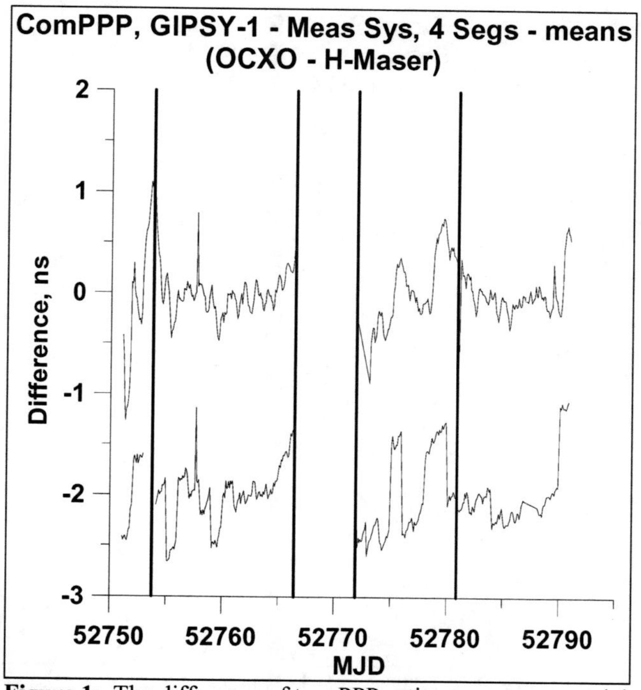

Figure 1: The difference of two PPP estimates, a commercial software and GIPSY, minus data from the local NIST measurement system 2-hour points. The GIPSY results were offset by –2 ns. The data were analyzed in four segments. The vertical lines mark the segments.

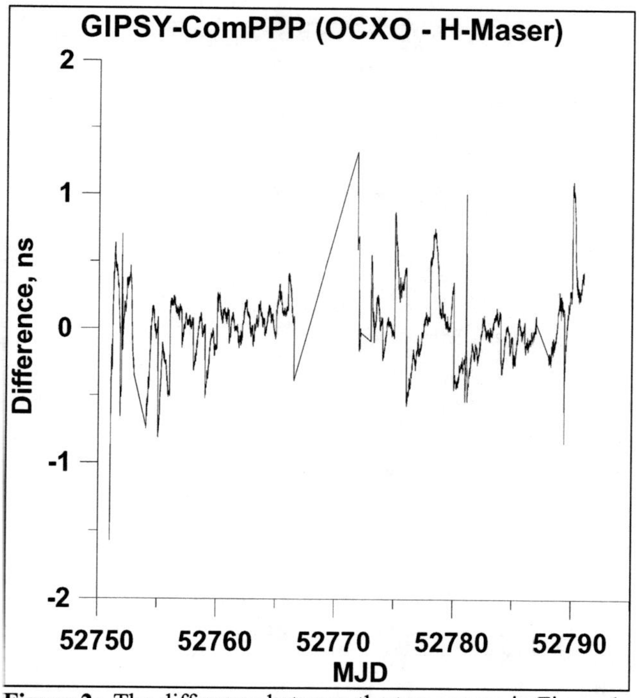

Figure 2: The difference between the two curves in Figure 1. The upper curve, the commercial PPP data, was subtracted from the lower curve, the GIPSY data.

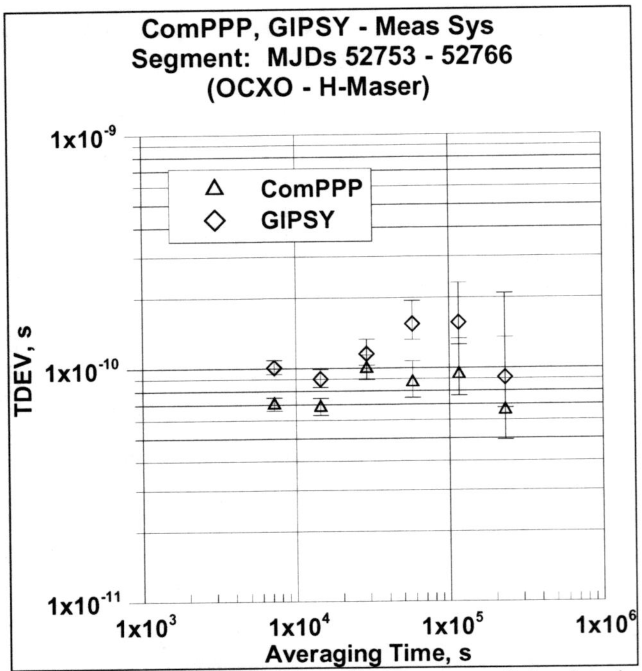

Figure 3: TDEV of the longest segment of the data in Figure 1.

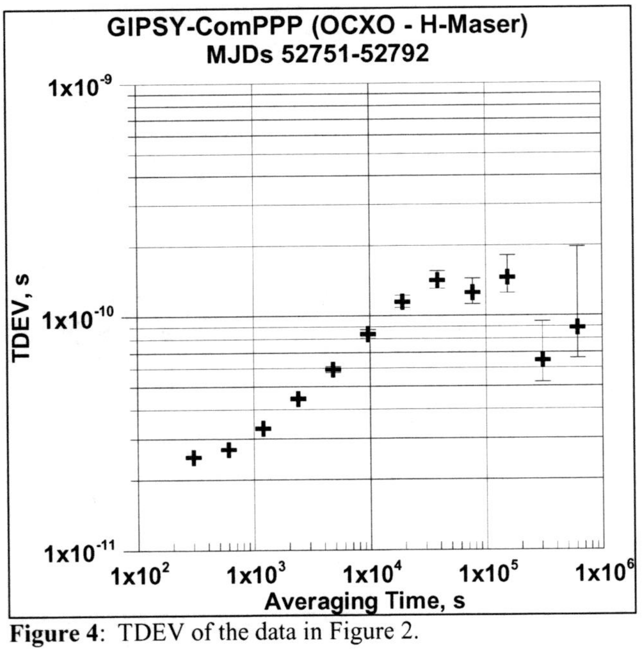

Figure 4: TDEV of the data in Figure 2.

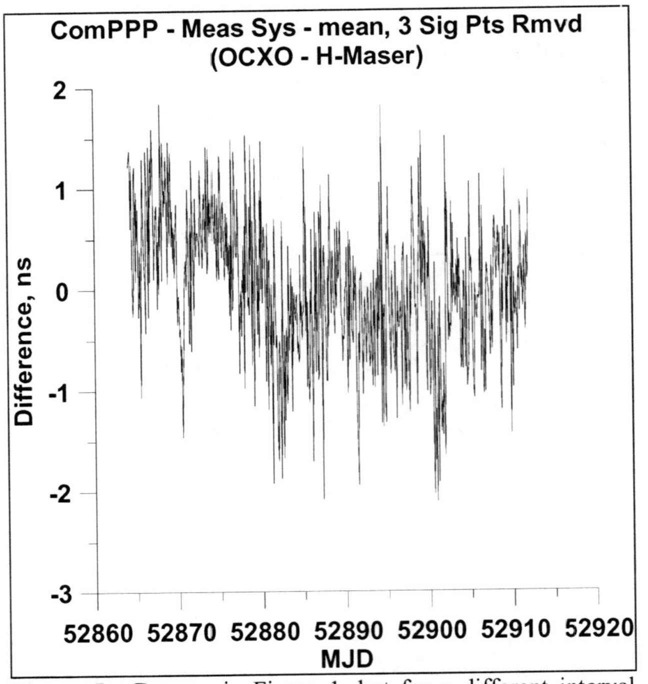

Figure 5: Data as in Figure 1, but for a different interval without resets. A commercial software's clock difference estimates, minus data from the local NIST Measurement System's 2 hour points. The clocks estimated were an OCXO minus an H-Maser. Nine points have been removed that exceeded three standard deviations from the data mean.

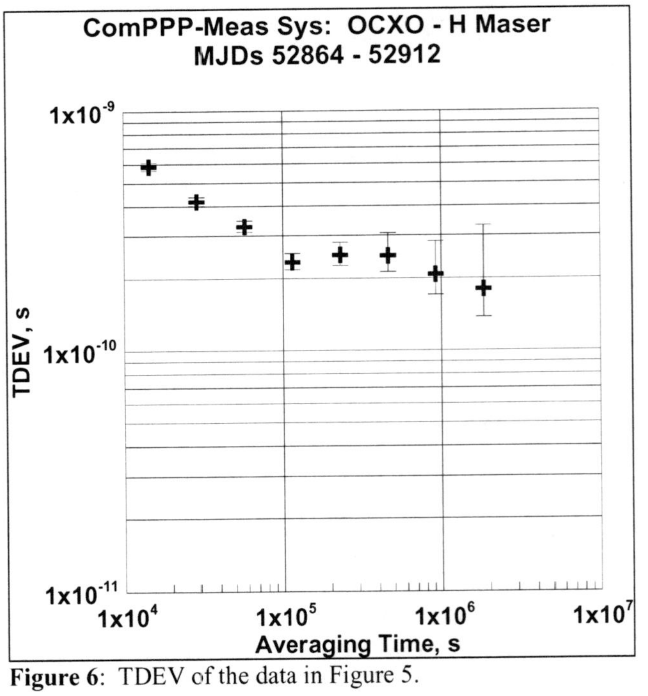

Figure 6: TDEV of the data in Figure 5.

4 NIST-USNO Experiment

We compared H-masers between NIST and USNO using the commercial PPP software in three intervals when there were hourly two-way time transfer data. The signals from these H-masers were steered to time and frequency of each lab's UTC. We computed the difference between the two transfer techniques: two-way minus PPP.

There were two relatively short intervals: MJDs 52556 – 52566 (October 9 – 19, 2002), and MJDs 52642-52657 (January 3 – 16, 2003). Finally, we look at the 252 d interval from June 2003 – January 2004.

4.1 Short interval experiments comparing UTC(USNO) and UTC(NIST)

In the October 2002 experiment, we can see significant intervals with missing two-way data in Figure 7, making it difficult to draw conclusions. For example, we have not computed any variances.

In the second experiment, there were again periods of missing two-way data. However, we were able to find a 13 d interval without gaps. Figure 8 shows the two-way minus PPP data for the entire interval. Figure 9 shows TDEV for the interval without breaks. The noise type of the difference between the commercial PPP and TWSTFT systems was consistent with a flicker PM model at 300 ps or less.

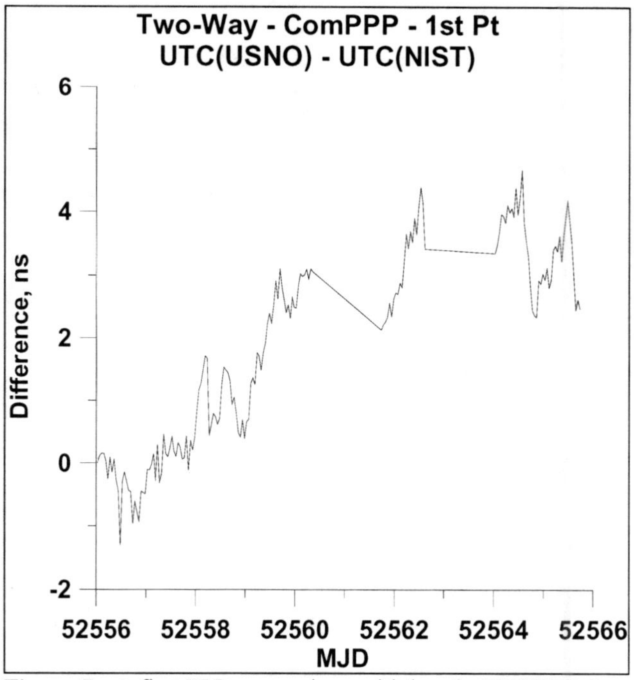

Figure 7: A first PPP comparison with hourly TWSTFT data between UTC(USNO) and UTC(NIST).

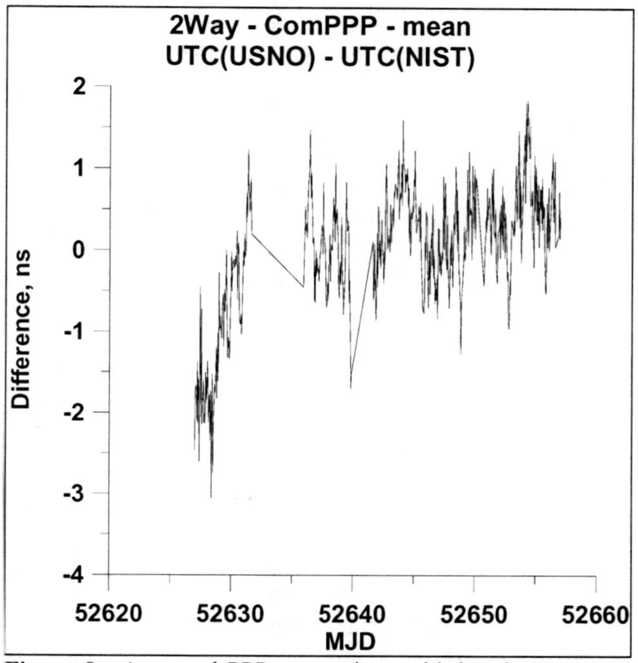

Figure 8: A second PPP comparison with hourly TWSTFT data between UTC(USNO) and UTC(NIST). There is a 13 d interval without gaps from MJD 52642 to 52657.

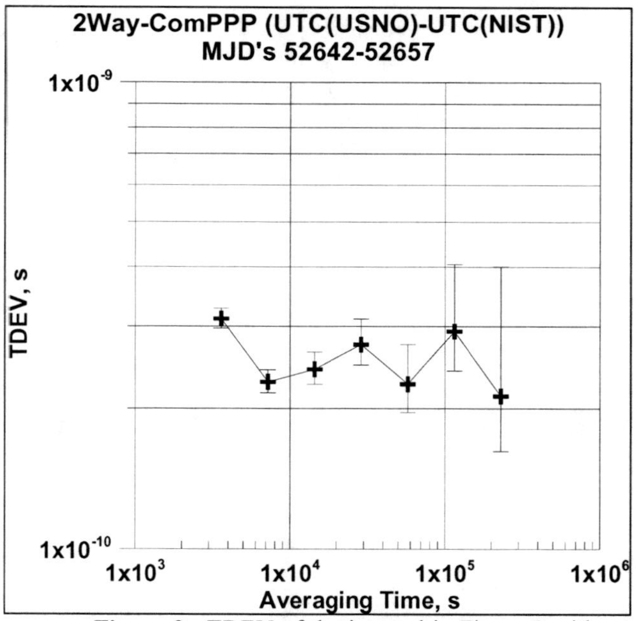

Figure 9: TDEV of the interval in Figure 8 without gaps in the data. The noise type is consistent with a flicker PM model at 300 ps or less.

4.2 A 205 day comparison between PPP and TWSTFT

We were able to obtain both commercial PPP results without uncorrectable resets and hourly TWSTFT data with no significant gaps from MJD 52802 to 53007, June 12, 2003 – January 3, 2004. The comparison, in Figure 10 below, shows a drift in the offset between the two systems, particularly in the months of November and December when the weather is colder. The slope during this latter interval is about 35 ps/d or $4\bullet10^{-16}$. This may indicate a temperature dependence on the delay through one or both of the systems.

We do not know whether the cause is in the GPS receivers or the TWSTFT hardware. We looked at the code common-view time transfer to see if it could offer a third vote. That method is much noisier, and varied between the results of the PPP and the TWSTFT data, yielding no preference for agreement with the bias of either PPP or TWSTFT.

We compute the TDEV of the data in Figure 10 to characterize the transfer systems. The commercial PPP results are best in short-term out to 1 d. The TWSTFT results are somewhat better after 1 d to where the clock noise dominates. The differential transfer after 1 d is again consistent with a flicker PM model of under 300 ps. This is shown in Figure 11, along with the TDEV values of the transfer data. Figure 12 is the modified Allan deviation (MDEV) of these data [9]. This allows us to see the capabilities for frequency transfer in a second-difference variance. It appears that frequency transfer at 1 part in 10^{15} is possible at 10 d of averaging.

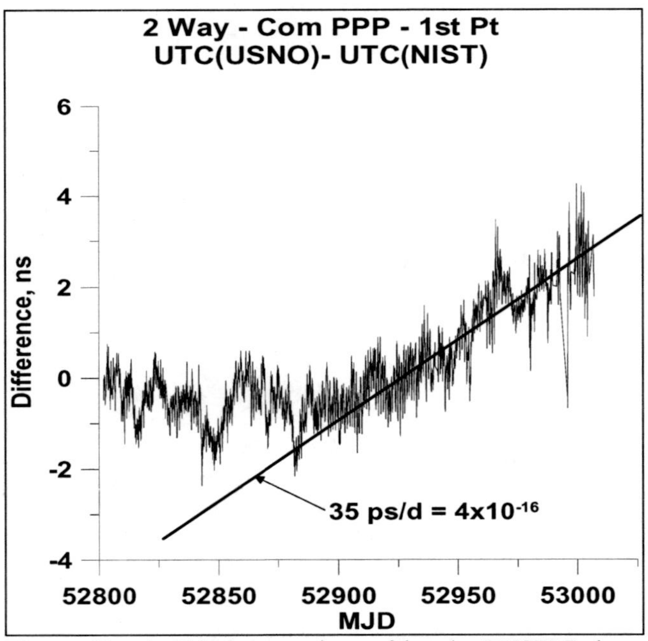

Figure 10: A 205 d comparison of hourly TWSTFT data with the commercial PPP results, between UTC(NIST) and UTC(USNO). We see a change in delay between the two systems over the colder months of November and December of 35 ps/d or $4\bullet10^{-16}$.

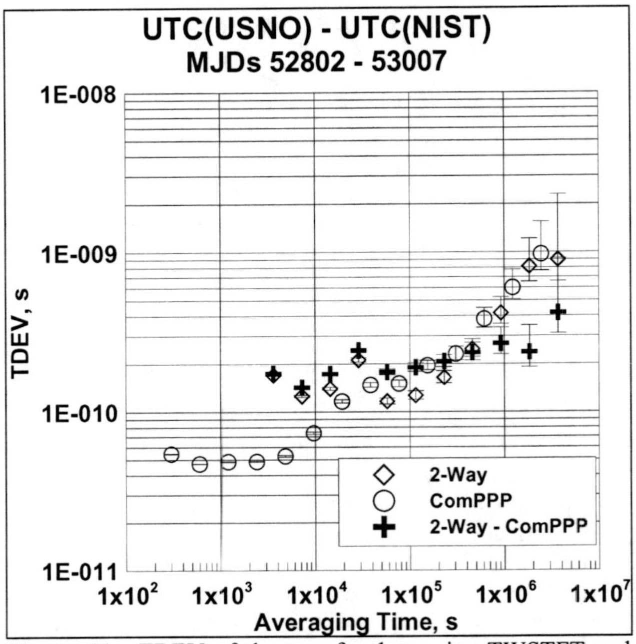

Figure 11: TDEV of the transfer data using TWSTFT and PPP, and of the difference. The commercial PPP results are best in short-term out to 1 d. The TWSTFT results are somewhat better after 1 d. The differential transfer is again consistent with a flicker PM model of under 300 ps.

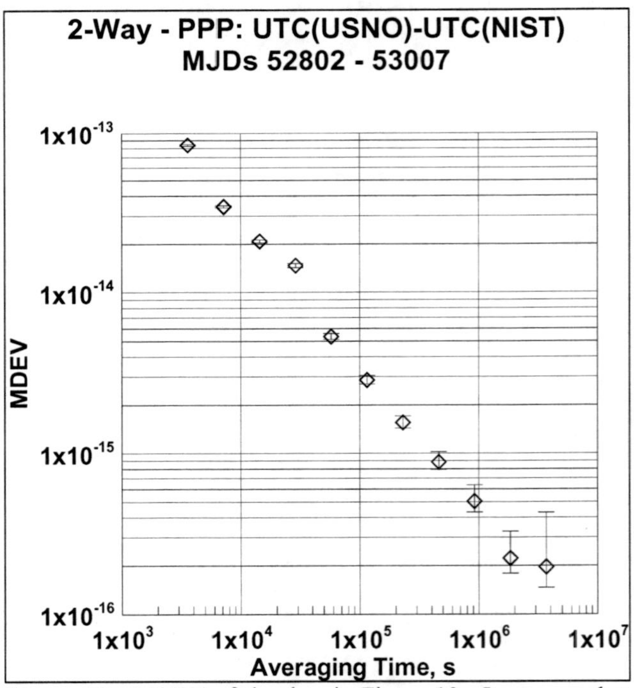

Figure 12: MDEV of the data in Figure 10. It appears that frequency transfer of 1 part in 10^{15} is possible after 10 d of averaging.

4.3 A frequency transfer statistic

Parker, Howe and Weiss discussed the use of a statistic that may be more appropriate for characterizing frequency transfer than MDEV and TDEV which are second-difference variances [8]. A more direct way to estimate frequency transfer uncertainty is using a first difference of averaged time data. This frequency transfer statistic averages the frequency transfer squared for different transfer intervals. T is the total time for the transfer experiment. We average measured time differences over equal intervals, A, at the beginning and end of the test time T. This gives us a frequency average over a running time of τ = T – A. The definition is

$$\sigma^2_{tt}(T,\tau) = \frac{\left\langle (\bar{x}_{t+\tau} - \bar{x}_t)^2 \right\rangle}{\tau^2}. \qquad (1)$$

This is just the mean-squared fractional frequency of the time transfer data set. The definition is illustrated in Figure 13, where A is the averaging time for each of the end-points,

$$A = T - \tau.$$

Figure 14 gives this statistic for the data in Figure 10. This is a combined uncertainty due to frequency transfer error both in TWSTFT and PPP. From Figure 14 we see that frequency transfer at 1 part in 10^{15} is possible after 10 d.

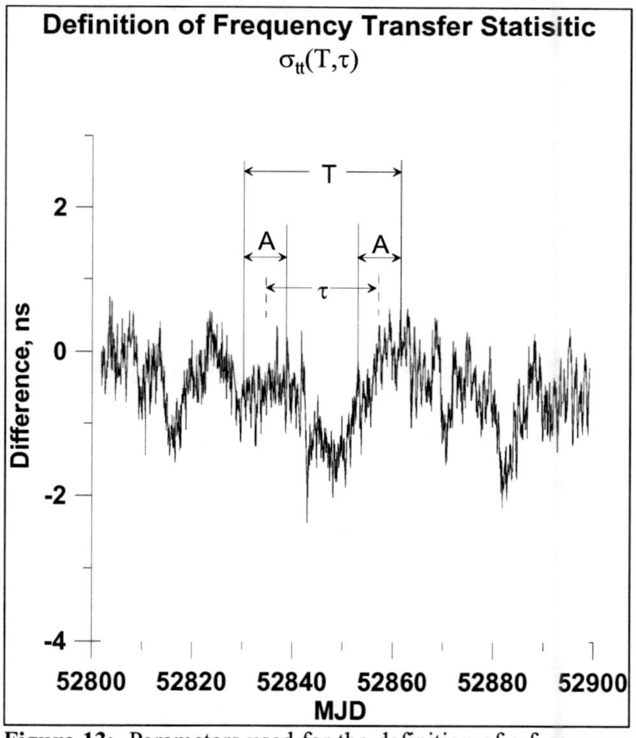

Figure 13: Parameters used for the definition of a frequency transfer statistic, $\sigma_{tt}(T,\tau)$. A is the averaging time for each of the end-points, $A = T - \tau$.

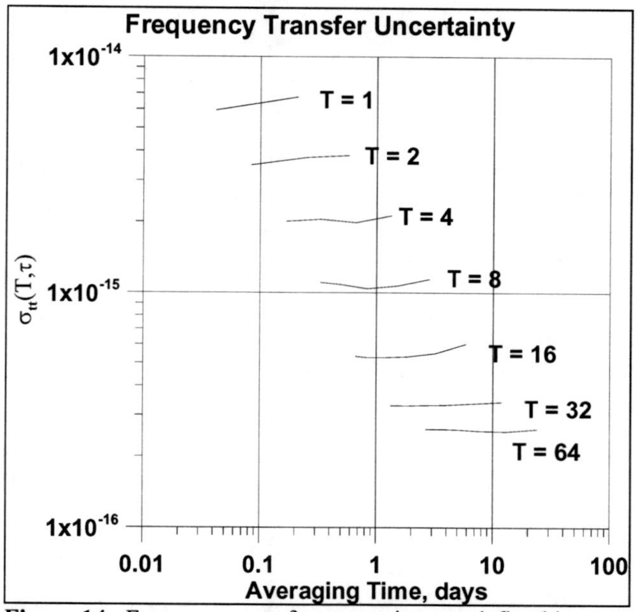

Figure 14: Frequency transfer uncertainty, as defined in eqution (1), for the data of Figure 10.

Acknowledgements

We thank Dr. T. Parker of NIST and Dr. D. Matsakis of USNO for allowing us to use the hourly two-way data for this publication. Thanks also to Dr. Parker for ideas and help in the development of this work. Thanks to Dr. J. Levine and V. Zhang for assistance in reading the paper.

References

[1] ANSI (Telecommunications) Standard T1.101-1994, available from American National Standards Institute, 11 West 42nd Street, New York, New York 10036, or on the Internet at http://web.ansi.org/.

[2] P. Fenton, E. Powers, B Klepczynski, W. Lewandowski, R. Douglass, "Capabilities of the WAAS and EGNOS using high gain antenna for time distribution and frequency transfer," *Proc. Institute of Navigation (ION) Meeting, GPS 2001* (2001).

[3] C. Hackman, J. Levine, "New frequency comparisons using GPS carrier-phase time transfer", *Proc. 2003 Joint Meeting IEEE IFCS-EFTF*, pp. 258-265, (2003).

[4] International GPS Service web site, http://igscb.jpl.nasa.gov/

[5] W. Lewandowski, J. Azoubib, "Time transfer and TAI," *Proc. 2000 IEEE International Frequency Control Symposium*, pp 586-597 (2000).

[6] Nelson, L.M., J. Levine, K.M. Larson, P. Hetzel, "Comparing primary frequency standards at NIST and PTB, *Proc. IEEE IFCS 2000*, pp. 622-628 (2000).

[7] T.E. Parker, P. Hetzel, S. Jefferts, L. Nelson, A. Bauch, J. Levine, "First comparison of remote cesium fountains," *Proc. 15th European Forum on Time and Frequency,* 2001, pp. 57-61.

[8] T.E. Parker, D.A. Howe, M.A. Weiss, "Accurate frequency comparisons at the 10^{-15} level, *Proc. 1998 IEEE International Frequency Control Symposium*, pp. 265-272 (1998).

[9] D.B. Sullivan, D.W. Allan, D.A. Howe, F.L. Walls, ed., *Characterization of Clocks and Oscillators*, NIST Tech Note 1337, U.S. Government Printing Office, Washington D.C., USA, (1990).

[10] C. Thomas, "The accuracy of International Atomic Time TAI," *Proc. 11th European Forum on Time and Frequency,* 1997, pp. 283-289.

[11] F.H. Webb and J.F. Zumberge, ed., *An Introduction to GIPSY/OASIS-II: Precision Software for the Analysis of Data from the Global Positioning System*, JPL D-11088 (internal publication), Jet Propulsion Laboratory, Pasadena, CA, USA (July 1997).

[12] P. Wolf, G. Petit, "Use of IGS products in TAI", *Proc.31st Precise Time and Time Interval (PTTI) Meeting*, pp. 419-430 (1999).

EVALUATION OF TTS-3, A NEW MULTI-SYSTEM TIME RECEIVER

J. NAWROCKI, P. NOGAŚ, Astrogeodynamical Observatory, Polish Academy of Sciences, Borowiec, 62-035 Kórnik, Poland.
nawrocki@cbk.poznan.pl

W. LEWANDOWSKI, Bureau International des Poids et Mesures, 92310 Sèvres, France
wlewandowski@bipm.org

There are now four types of navigation satellite available for time and frequency transfer: GPS, GLONASS, WAAS and EGNOS. This paper describes briefly a new time-transfer receiver TTS-3, based on a Javad Legacy plate, which allows observations of these satellites simultaneously in multi-channel, multi-frequency mode. The following codes are used: C/A-code for GPS, WAAS, EGNOS and GLONASS, P-code for GLONASS, and reconstructed P-code for GPS. The receiver hardware, the treatment of the observations, and the output data fulfil the recommendations of the CCTF Group on GNSS Time Transfer Standards (CGGTTS). The paper reports an on-site evaluation of the TTS-3 stability, and first tests of multi-system time-transfers.

A CONTRIBUTION FOR THE IMPROVEMENT OF GPS CV TIME TRANSFER TECHNIQUE

J Palacio, J Galindo & J A Lima

Real Instituto y Observatorio de la Armada
Cecilio Pujazón s/n
11.100 San Fernando, Spain
Phone: + 34 956 599 286
Fax: +34 956 599 366
jpalacio@roa.es, jgalindo@roa.es, jalima@roa.es

Keywords: Time transfer, GPS Common view technique, calibration.

Abstract

Most of the GPS links between timing laboratories that contribute to TAI are performed by single channel GPS timing receivers. These are based in the original NBS and JPL designs, and can perform and store 48 tracks per day.
We have increased the numbers of tracks per day using an external computer so reaching the maximum of 91 tracks. We are now able to follow the GPS schedule issued by BIPM and fill up all the gaps between tracks selecting additional satellites.
In this paper we are exposing the studies done after an experience using a common clock link at our site and the results on the remote calibration of a time standard. It also contains a study of the impact of the increasing of the data per day in the uncertainty of the comparison using GPS CV technique. Finally, a comparison with a multi-channel GPS receiver is also detailed.

1 Introduction

Most of the Timing Laboratories contributing to TAI, uses for the time link a single channel timing GPS receiver. These receivers were developed from the original NBS type receiver. Most of the performances limitations of the receiver are due to the technology at hand at the development epoch, mainly the amount of memory allocation that the microprocessor address to store data.
For this reason these receiver can store information to perform 48 tracks per day, and can store data from the last 150 tracks.
Using the serial communication port to upload tracking parameters, one can increase the total number of track per day the receiver can perform.

2 Operational procedure.

BIPM Time Section distribute to Timing laboratories the Common View GPS tracking schedule that receivers must follow at a determined day. The following steps must be followed to implement the procedure:

1. Obtain an almanac of the GPS constellation valid for the first day of measurements.
2. Produce a list of the visible satellites for your location, at the mid time of the every track period. For this propose we have found a very appropriate tool from Trimble at: http://www.trimble.com/planningsoftware.html
3. Using this list, circle the elevation and azimuth of the selected satellites for CV contained in the BIPM schedule.
4. When finished, you will be in position to disregard those satellites crossing the 20 degrees of elevation during the track time, selecting other at better elevation. We use class 'EA' to designate all these 'alternative' satellite.
5. Fill the gaps of CV tracking time with the satellites that better fit the time transfer need for your locations or the region you are going to work in. Use class 'EE' to designate these 'extra' satellites and also the first 'EA' class of the list that must be changed to 'EE' class for the reason we will exposed later.
6. There is a gap of 12 minutes time that can also be filled with a 09 minutes effective (11 nominal) length track under your consideration.
7. Build a new list including in chronological order the satellites form the original BIPM schedule, the class 'EA' and the class 'EE' satellites and store it into a file. It is convenient that the first line of this file be the MJD of the schedule, and that the starting time be adjusted by the 2 minutes the receiver takes to make the counter calibration, and locking procedure.
8. One of the memory registers that contains the data of the CV tracks (TTR-6: that corresponding to the track we changed from class 'EE' to 'EA'; NBS: register 01 that correspond to memory allocation 7700), will be used to dynamically upload the data of the 'EE' track in the way we will exposed following. Let us consider that we are using registers 07 or 01.
9. Before the beginning of the implementing time, upload to the receiver the tracking schedule, filtering the 'EE' tracks that must not be uploaded. It must

also be considered that register determined at point 8 (07 –TTR-6- or 01 –NBS-) must not be used to store data, and must be skipped when uploading data.

10. Immediately after the scheduled be uploaded, upload the data corresponding to the first 'EE' class track to the reserved register. Choose one of the following commands depending on the type of your receiver is:
    a. TTR-6: T07,EE,SS,HHMM,LL
    b. NBS : EESSHHMMLL057700FLCMD
11. The computer will continue waiting until HH(MM+2) to upload the data corresponding to the next 'EE' class track in the table.
12. Repeat from step 10 for every 'EE' track time.
13. Update the tacking table before the beginning of the next day.
14. As the receiver will store data of the last 150 tracks when regular working and of the last 12 tracks after a power down or fail, the data must be downloaded from it at least every three hours. Attention must be paid to do not try to access the communication port by two different processes.

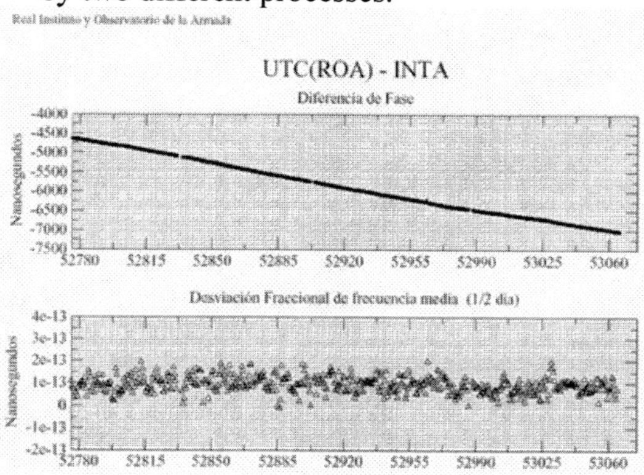

Figure 1 Graph of the phase difference and fractional frequency deviation (1/2 day average) of a remote caesium clock, located at INTA (Madrid).

3 Results

To get the best results from a receiver following this extended schedule method, obviously must be compared to other receiver performing in the same way. This has been a difficult task for us to get cooperation from other laboratories to implement the method. Anyway we have exploited the method doing remote clock calibrations of few accredited T&F laboratories using our own GPS receiver at both sides.

Figure 1 shows the result of a calibration of a caesium clock located at Madrid. The obtained uncertainty of the calculated average for a ½ day period of time is lower that when considering only 'BIPM tracks', obviously due to increase of the number of comparisons per day and to the quality of the measurements using high satellites..

Since February 2004 a EUROMET Project was launched, lead by ROA and VSL, to circulate a single channel GPS receiver controlled by an external computer to perform 'extended' tracks. In Figure 2 the total number of tracks per day done by the travelling receiver while it was at ROA is plotted, compared to the number of BIPM regular tracks. As ROA have two receivers working this procedure, we can compare the results of the common clock at ROA and at visited laboratory. Table 1 exposes a resume of the measurements done during three days. Although the differences of the calculated sigma values are not directly related to the number of tracks, but to the presence of other influences in the measurements, an idea of the performances of the method is visible looking at the values at ROA.

MJD	ROA			VSL		
	#track	Mean	Sigma	#track	Mean	sigma
53065	83	2.22	1.416	43	-23.12	2.998
53066	83	2.16	1.707	43	-23.65	2.843
53067	84	2.21	1.758	43	-22.92	2.978

Table 1.- Results of the common clock differences of two receivers collocated at ROA an other two at VSL. One of the later is the travelling EUROMET Project 529 GPS receiver.

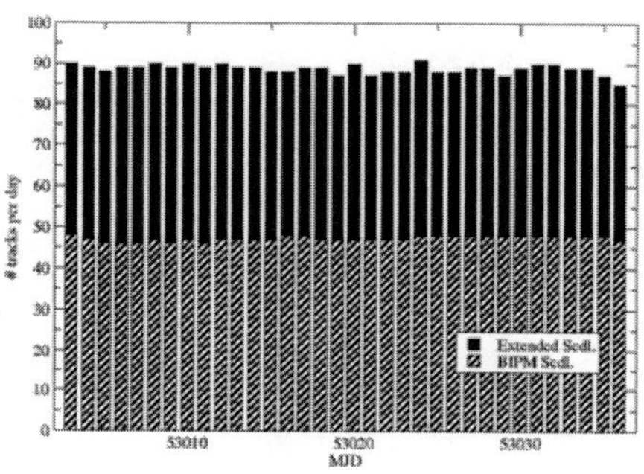

Figure 2 Number of tracks per day a single channel GPS receiver performed and stored at ROA from MJD 53003-53037.

The complete series of common clock measurements are shown in Figure 3.

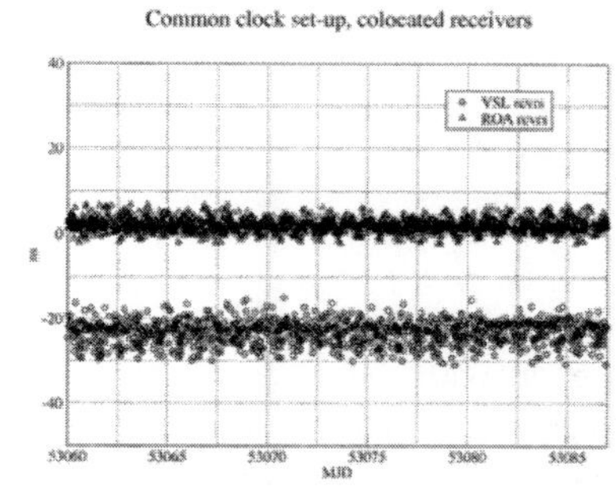

Figure 3 Raw measurements of common clock setups at ROA and at VSL. Triangles are ROA's data and circle VSL's data.

The stability of these sets of data have been studied, after detecting and filling the gaps; the results are shows in Figures 4 and 5. It is shown that the 'extended' schedule offers better stability in both frequency and time domain. For small averaging times, the original BIPM schedule seems to be

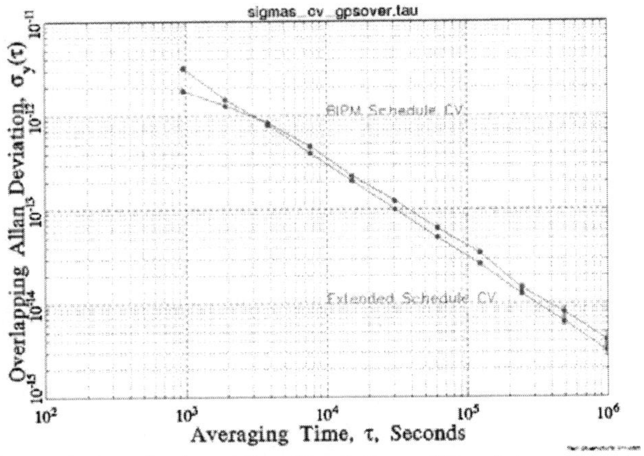

Figure 4. Overlaping Allan Deviation of the data series shown in Figure 3.

better, but for these σ values the interpolation points introduced in the series are dominant (we used 2202 points for ROA study and 854 point for VSL)so this part of the curves are not relevant.

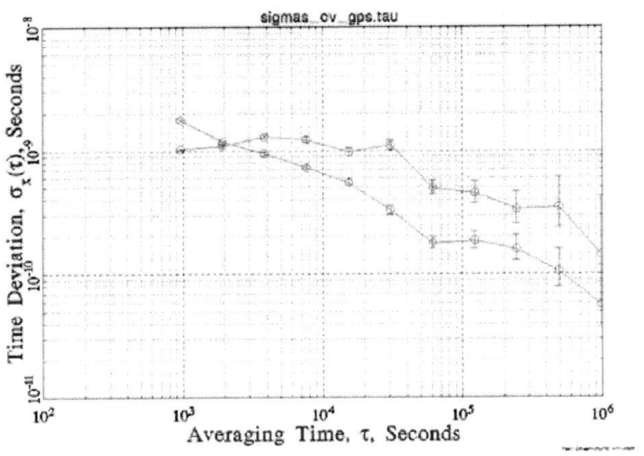

Figure 5.- Time deviation of the series of data shown in Figure 3.

We have also compared a single channel GPS receiver to a multi-channel GPS receiver. The time transfer with laboratories equipped with these devices have increased the number of tracks per day. Using these data we can now detect and measure some diurnal effects, taking advantage from the number of high elevation satellites we are now using, after having removed the original low elevation tracks and selecting appropriated one for filling the gaps.

During these comparisons with multi-channel receiver we have detected some strange measurements that do not agree to a common sense and need to be investigated. In figure 6 we are showing the DSG value [1], of one of these multi-channel receiver compare to those of a single-channel. Multi-channel receiver concentrate the DSG values around two different man points: 1.4 and 5 nanoseconds.

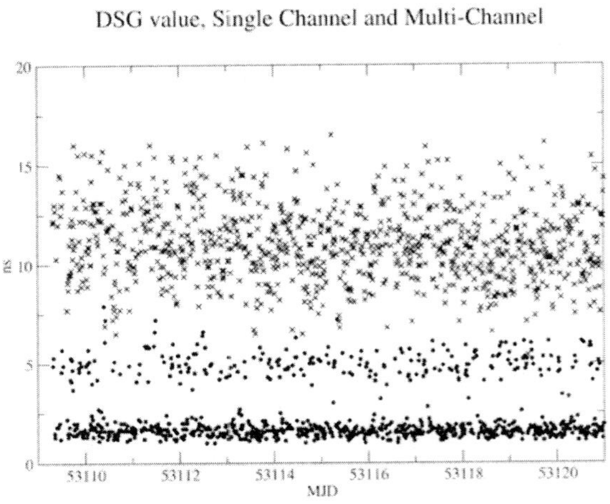

Figure 6.- Plot of the DSG column of a single channel (x) and multi-channel (dots). Last one concentrate its values around two different points.

3. Conclusions

The performances of a single channel receiver can be extended using an external computer to overcome the lack of memory to store data for 91 tracks per day.

The procedure offers advantages to decrease the uncertainty of the calibration of remote clocks.

This procedure is now being used and will be used for next calibration trips around EUROMET under Project 529.

A receiver operated in the way we are exposing can be used as reference when comparing to multi-channel receiver to characterise delays, performances, computation method, etc.

A complete Power Point description of the procedure is available at [2].

Acknowledgements

The authors want to express their gratitude to Dr. Mark Weiss, from NIST, for providing us the memory map and the command set of NBS receivers. Also want to thank to Dr. Gerrit de Jong for discussions

References

[1] D. Allan and C. Thomas, "Technical Directives for Standardization of GPS Time Receiver Software", *Metrología*, **31**, pp. 69-79, (1994)

[2]ftp://ftp.roa.es/pub/hora/papers/EFTF04_ROA_archivos/frame.htm